D1356451

BAYESIAN STATISTICS 4

BAYESIAN STATISTICS 4

Proceedings of the Fourth Valencia International Meeting

Dedicated to the memory of Morris H. DeGroot, 1931–1989

April 15–20, 1991

Edited by

J. M. Bernardo
J. O. Berger
A. P. Dawid
and
A. F. M. Smith

CLARENDON PRESS · OXFORD
1992

Oxford University Press, Walton Street, Oxford OX2 6DP
Oxford New York Toronto
Delhi Bombay Calcutta Madras Karachi
Petaling Jaya Singapore Hong Kong Tokyo
Nairobi Dar es Salaam Cape Town
Melbourne Auckland
and associated companies in
Berlin Ibadan

Oxford is a trade mark of Oxford University Press

Published in the United States
by Oxford University Press, New York

A catalogue record for this book is available from the British Library

Library of Congress Cataloging in Publication Data
Bayesian statistics 4: proceedings of the Fourth Valencia International Meeting, April 15–20, 1991 / edited by J. M. Bernardo . . . [et al.].
'Dedicated to the memory of Morris H. DeGroot, 1931–1989.'
'Fourth Valencia International Meeting on Bayesian Statistics . . . held at the Hotel Papa Luna, Peñíscola . . . 15th to 20th April 1991'—Pref.
1. Bayesian statistical decision theory—Congresses. I. Bernardo, José M. II. DeGroot, Morris H., 1931– . III. Title: Bayesian statistics four.
QA279.5.B394 1992 519.5'42—dc20 92–16010
ISBN 0–19–852266–5

Author supplied CRC
Printed in Great Britain by
Biddles Ltd, Guildford & King's Lynn

OPENING ADDRESS BY THE GOVERNOR OF THE STATE OF VALENCIA

Monday, April 15, 1991, Gothic Room, Peñíscola Castle

Ladies and Gentlemen,

It is my pleasure to welcome you again to Valencia. The Government of the State of Valencia is proud to sponsor, every four years, the International Meetings on Bayesian Statistics, and is very conscious that Valencia has become internationally known as the accepted meeting point for Bayesian specialists from all over the world.

As an economist, I am able to appreciate the importance of a coherent theory for the analysis of the decision problems which we continually have to face. The mere statement of a problem within the structure suggested by decision theory often yields powerful advances towards its solution. Indeed, a disciplined attitude, where all alternatives are systematically considered, without a priori exclusions, where all non-controllable factors which may influence the final outcome are studied, and, above all, an attitude which requires a comparative evaluation of all possible consequences, creates the appropriate conditions for rigorous decision-making.

As a politician, I am especially interested in the use of statistical methods to learn scientifically about the people's opinions on the different questions which successively arise. Recently, we have had in this country a debate on the possible differences between public opinion and published opinion; obviously, for a Government genuinely interested in responding to the needs of its citizens, it is fundamental to know their true preferences among the alternatives which are available at a given moment. In a very precise sense, the use of modern statistical methods makes it possible to deepen the democratic system –in the sense suggested by Eric Fromm– since it allows a continuous expression of the people's will.

Bayesian statistical methods are especially appropriate for analyzing and describing the information which is available when a political decision has to be made. Indeed, the probability concept, which we need to describe the uncertain elements of the problem, must be very general; obviously, in social matters there are no repetitions or relative frequencies to determine probability, since the relevant events are typically unique and not repeatable. Moreover, in political language, the concept of probability naturally refers to a reasonable degree of belief, given the available information; I do not have to explain to you that this is only possible within the Bayesian framework.

I am firmly convinced that, in the citizens' interests, public administrations should make a special effort to use –at all levels– the most advanced technology available; it is necessary to destroy the image of outdated, obsolete administration. Bayesian statistical methods are the scientific avant-garde in their field; consequently, it would be in the interest of public administrations to incorporate them into their tool kit. I am sure that, in this meeting, you are going to provide additional reasons for so doing.

I hope that your extensive scientific programme will not prevent you from appreciating the environment you are in. In the Mediterranean, we firmly believe in the importance of the quality of life, and let me assure you that this land has a lot to offer to those who give it their attention.

I wish you all a productive conference, and a very pleasant stay among us.

I declare open the Fourth Valencia International Meeting on Bayesian Statistics.

Joan Lerma

Governor of the State of Valencia

PREFACE

The Fourth Valencia International Meeting on Bayesian Statistics was held at the Hotel Papa Luna, Peñíscola, 140 km. north of Valencia, Spain, from 15th to 20th April 1991. It followed a similar pattern to the three previous meetings[1], though with a change in the organizing committee of J. M. Bernardo, M. H. DeGroot, D. V. Lindley and A. F. M. Smith. The whole statistical community was saddened by the death of Morrie DeGroot in 1989[2]. He had played a key role in previous meetings, both academically and socially. This fourth meeting was dedicated to his memory and the second session contained two papers of which he was the co-author. **Bayarri** and **DeGroot** discuss weighted and selection models; **Goel, Gulati** and **DeGroot** consider optimal stopping rules for a team. **Lindley** retired from the committee but gave an address, included here, as President of the meeting. DeGroot and Lindley were replaced on the committee by J. O. Berger and A. P. Dawid.

The committee invited 29 leading experts in the field to give papers, each of which was followed by a discussion. In addition, there were 4 poster sessions featuring over 150 contributed papers. These Proceedings contain all the invited papers, with associated discussions. The contributed papers were subjected to a refereeing process. Owing to the high quality of the poster sessions, this process had to be rather severe, and space only allows for the publication of 33 of them. The Editors believe that the Proceedings provide a definitive, up-to-date overview of current concerns and activity in Bayesian Statistics, encompassing a wide range of theoretical and applied research.

Two papers are concerned with the foundations of the subject. **Royall** discusses the role of evidence and likelihood; **Gilio** explores the basic idea of coherence.

The emphasis in the Proceedings is much more on developing operational methods and, in particular, on producing usable techniques. These often involve considerable computing power, and numerical work is emphasized in many papers.

The special technique of the Gibbs sampler proved very popular: **Racine-Poon** uses it, and sample-resample techniques, to explore the posterior distribution; **Geweke** considers it, and data augmentation, in the evaluation of posterior moments; **Gelman** and **Rubin** draw attention to some implementation difficulties; **Gilks** relates it to adaptive, rejection sampling; **Raftery** and **Lewis** advise on the number of iterations needed; **Roberts** explores theoretically the rapidity of convergence; **Thomas, Spiegelhalter** and **Gilks** present a prototype computer program.

Other papers on numerical ideas include one in which **O'Hagan** treats numerical analysis problems as Bayesian inference problems; numerical integration is used in Bayesian analyses by **Dellaportas** and **Wright**; **Meng** and **Rubin** extend the EM algorithm so that the dispersion matrix can be more easily evaluated; **Wooff** describes a software environment for subjective linear analyses.

Statisticians outside the Bayesian school often criticize the use of a prior which, they claim, is typically unknown. Several attempts to surmount this difficulty have been proposed. Two papers here use the method of reference priors: **Ghosh** and **Mukerjee** relate them to entropy and other concepts; **Berger** and **Bernardo** review and extend the basic ideas. **Consonni**

1 The Proceedings of these have been published: the first by the University Press, Valencia (1980); the second by that Press in conjunction with North-Holland, Amsterdam (1985); the third by the Clarendon Press, Oxford (1988). The editors in each case were the members of the organizing committee.

2 A biography appeared in *Statist. Sci.* **6**, 3–14 (1991).

and **Veronese** discuss the difficulties with the integrated likelihood ratio when the prior is improper.

One way of handling the problem posed by the prior is to explore the effect it has on the analysis; in particular, the robustness of the procedure to changes in the prior. **Wasserman** provides a review and some new ideas; **Moreno** and **Pericchi** consider the resulting range of posteriors; **Sivaganesan** examines the phenomenon in Empirical Bayes situations.

A Bayesian inference is a posterior distribution and it is not surprising that several papers focus on its interpretation, especially in the multi-parameter case. **Kass** and **Slate** define diagnostics that indicate departure from normality; **Hills** and **Smith** invoke reparameterization to help appreciate the distribution; asymptotic normality is the topic of the paper by **Sweeting**.

Besides the prior, another source of robustness, or lack of it, resides in the likelihood, itself obtained from a model. Model diagnostics occupy the attentions of **Gelfand, Dey** and **Chang**, both in respect of adequacy and selection; **Peña** and **Tiao** introduce two new diagnostic tools for outliers; the paper of **Gómez-Villegas** and **Maín** is concerned with the effects of the tails of a distribution; **Gutiérrez-Peña** evaluates the Kullback-Leibler divergence for exponential families. The influence of individual observations on the inference is investigated by **Carlin** and **Polson** using an expected-utility approach; by **Girón, Martínez** and **Morcillo** in linear regression, using the Kalman filter; and by **Pettit** for the transformation parameter.

One view of inference is that it fundamentally concerns the nature of future data. Three papers discuss prediction: **Brown** and **Mäkeläinen** consider the case of a large number of explanatory variables; **Zidek** and **Weerahandi** extend a method for predicting a sample path; **Dawid**, in a wide-ranging paper, relates prequential analysis (which is allied to forecasting and prediction) to coding theory, scoring rules and stochastic complexity.

With the increasing confidence in the paradigm that Bayesians have gained by developing successful methods and applications, there has been less attention paid to the relationship with classical methods. **Hodges** casts a favourable eye on significance tests; **Schervish** looks at the Bayesian replacement of F-tests in the linear model.

The (generalized) linear model remains popular. Estimation is discussed by **Angers**, and by **Strawderman**. **Stephens** and **Dellaportas** introduce the complexity of measurement errors in covariates; **A. J.** and **C. A. van der Merwe** explore the effect of strong prior knowledge about a subset of parameters; **Pérez** and **Pericchi** employ a hierarchical, linear model in a multi-stage survey; **Morris** and **Normand** use hierarchical models for meta-analysis, a topic that, with its coherence requirements between the separate analyses, seems natural for the Bayesian approach. Mixture models occupy the attentions of **Florens, Mouchart** and **Rolin**, who investigate estimability and identification; and **West**, who ranges widely over many problems concerned with modelling.

Experimental design is a problem which seems particularly suited to the Bayesian framework, for what does one have before design but the prior? **Verdinelli** produces old and new design criteria using a utility function; **Farrow** and **Goldstein** explore the balance between costs and benefits; whereas **Parmigiani** and **Polson** express the trade-off between cost and information in a random walk design.

The Bayesian approach to expert systems is exemplified by the probabilistic approach to learning given by **Spiegelhalter** and **Cowell**; and **Cowell**, separately, describes a companion computer program, BAIES (Bayesian Analysis In Expert Systems); **French, Cooke** and **Voght** are concerned with the combination of the views of several experts.

Stochastic processes and sequential methods provide the stimulus for four papers. **Irony, Pereira** and **Barlow** discuss additive and multiplicative models for quality assurance; **Jones** provides optimality results for the two-armed bandit; **Queen** and **Smith** use dynamic, graphical models to forecast time series; **Spizzichino** considers the phenomenon of wear-out in survival data.

Everyone who has had experience in editing a journal knows how difficult it is to obtain good papers on applications. Evidence of the success of Bayesian ideas in relating to the real world is provided by the number of wide-ranging papers on practical problems. **Wolpert** and **Warren-Hicks** combine laboratory and field data for brook trout; **Bernardo** and **Girón** discuss election-night forecasting; **Berry, Wolff** and **Sack** raise the important issue of when a sequential trial should terminate, referring, in particular, to an influenza trial with Navaho Indians; **Singpurwalla** and **Wilson** are occupied with the nature of warranties and the reserves needed to cover them; **Kadane** and **Schum** consider a legal case with disputed opinions; **Mendel** looks at the modelling of lifetimes; **Quintana** discusses currency contract portfolios; and **Sølna** introduces prior knowledge into the study of rock deformation.

One session of the meeting, not included in this volume, was an open discussion of proposals to form a Bayesian Society and publish a Bayesian Journal. The former was generally agreed to be a sound idea. After debate, and a vote, the idea of a journal was opposed by a large majority. The voting on a show of hands, was about 150 against, 40 in favour and 40 abstaining.

About 300 people attended the Conference and there was discussion about the effect of this encouraging growth upon the nature of future Valencia meetings. The ease of communication provided by all participants being gathered together for all functions, social and professional, has to be balanced against the increasing interest in Bayesian methods (and, hence, conference attendance) shown by the statistical profession and users of statistical ideas.

As usual, after the final conference dinner the Bayesian versus non-Bayesian debate found musical expression. On this occasion to the music of John Lennon's "Imagine", with new lyrics by **Carlin** and **McCulloch**, which began:

"Imagine you're a Bayesian –
    It's easy if you try,
You just adopt a prior,
    And the data updates π.
Statistics is so simple
    With subjective probability . . .

Now imagine you're a frequentist,
    Worrying about what might have been,
Spending your whole lifetime
    Analyzing data you've never seen.
. . .

We are most grateful for the financial and material support provided by: the Government of Spain; the Government of the State of Valencia; the Council of Peñíscola; the University of Valencia; the United States National Science Foundation; the United States Office of Naval Research; the British Council. In particular, we are delighted to have, once again, the personal message of encouragement from the Governor of the State of Valencia, reprinted in this volume in translation from the original Spanish.

We are also most grateful to Macu Gisbert and María Dolores Tortajada for administrative and technical assistance, in relation both to the organization of the meeting and in the preparation of these Proceedings.

The *Fifth Valencia International Meeting on Bayesian Statistics* is planned to take place in the State of Valencia, in early summer 1994.

J. O. Berger
J. M. Bernardo
A. P. Dawid
D. V. Lindley
A. F. M. Smith

CONTENTS

I. INVITED PAPERS (with discussion)

II. CONTRIBUTED PAPERS

INVITED PAPERS

BAYESIAN STATISTICS 4, pp. 1–15
J. M. Bernardo, J. O. Berger, A. P. Dawid and A. F. M. Smith, (Eds.)

Is our View of Bayesian Statistics too Narrow?*

DENNIS V. LINDLEY
Somerset, UK

SUMMARY

The paper considers the scope of the Bayesian paradigm and whether our present use of it is too restricted. After a brief account of the paradigm, the role of probability, and in particular its use by frequentists, is discussed; followed by a similar consideration of utility. It is suggested that the paradigm provides a reasonable account of scientific method. The role of a statistician in providing advice is discussed. The measurements of probability and utility are considered. Suggestions are made as to how the method can be used outside science, and the roles of coherence and comparison emphasized.

Keywords: PROBABILITY; UTILITY; MAXIMIZATION OF EXPECTED UTILITY; SCIENTIFIC METHOD; ROLE OF A STATISTICIAN; ASSESSMENT; COHERENCE; COMPARISON.

1. INTRODUCTION

The purpose of this paper is to explore the idea that we have failed to take a wide enough view of Bayesian statistics; that the potentiality of maximization of expected utility has not been adequately exploited. The title is in the form of a question because the attitude adopted is one in which the idea of unrealistic restrictions is explored, rather than asserted as a truth.

Classical statistics is deliberately restricted in at least two respects. First, the ideas are almost always confined to repeatable experiments or surveys, where the concept of probability is based on frequencies or populations. Second, the statistician's role is seen as that of a commentator, not as a decision-maker. This attitude is clearly seen in the ideas behind the founding of statistical societies in both Britain and the United States, where the statistician merely gathered and presented data neutrally, taking no sides.

Bayesians should query both restrictions: the first because probability is a measure of belief and only in the restricted class where the beliefs include exchangeability is the frequency concept relevant; the second because decision-making plays such a central role in the paradigm. I suggest that Bayesians have accepted the two constraints needlessly. But there may be more to it than that; even within the frequentist commentary, classical ideas may have been too easily accepted and the power of a personalistic analysis not fully appreciated.

Before exploring these ideas in detail, let me make two personal remarks. I hope that nobody will feel that they are being attacked or criticized in what follows. If anyone can be accused of having taken too narrow a view, it should be myself. For many years my attitude was that Bayesian statistics would provide the logic required to underpin the majority of statistical techniques that had been successful in helping us to appreciate data. It took a long time for me to realize that, far from supporting them, the coherent view showed them

* Presidential address, delivered in the Gothic Room of Peñíscola Castle on Monday, April 15, 1991.

to be unsound or, at best, reasonable approximations to a proper solution. Even with this appreciation, there has been a tendency to work within categories more suitable to the narrow, classical view, rather than see practical problems in the full light of probability as belief and utility as a measure of worth.

A second personal note is that I am taking advantage of my position at this conference to indulge in speculation. There are no new results here; no wondrous formulae with sophisticated notation; no theorems, not even a lemma. Instead, an attempt is made to look at maximization of expected utility, to see what has been done with it and especially to appreciate what has *not* been done and perhaps needs doing.

2. THE BAYESIAN PARADIGM

It is sensible to begin with a statement of the Bayesian position so that it is clear what is under discussion. That position has three features.

(a) All quantities whose value is not known to you ought to have their uncertainties described by numbers that obey the rules of the probability calculus. Thus the correct measure of uncertainty of a hypothesis is the probability of that hypothesis, not a significance level. The uncertainty of a defendant's guilt should similarly be a probability of guilt.

(b) All consequences should have their worth to you described by numbers on a utility scale; a scale that itself is based on probability. The relationship between utility and probability is essential for the third feature.

(c) Decisions, or acts, should be based on maximization of expected utility; the expectation being in accord with the probabilities described in (a).

All these features are appropriate for a single decision-maker, conveniently called 'you', in a position of uncertainty. The use of the personal pronoun does not imply an individual as decision- maker. It may be a board, a court or even a government. The probabilities and utilities are those perceived by that decision- maker, 'you'. The statistician's role is to advise you on your appreciation of the situation and to help assess your probabilities and utilities, which are truly yours and not the statistician's.

These features are not compelling when several decision- makers are involved, especially in an adversarial situation, though even there the Bayesian approach can shed some illumination. This paper will not consider such legitimate limitations to the personalistic view but will rather argue that the three features of the paradigm have wider relevance than has perhaps been understood. The basic feature is the first, probability, for it underlies the others, so we begin with it.

3. CLASSICAL IDEAS: PROBABILITY

The standard, mathematical model for most of statistics is a sample space of elements x and a set of probability measures over the space indexed by a parameter θ, leading ordinarily to a density $p(x|\theta)$ of x for each θ . Bayesians have adopted this model, which seems entirely appropriate within the paradigm, though the interpretation of probability may change. For many purposes the underpinning sample space is irrelevant and only the likelihood $p(x|.)$ as a function of θ for fixed x matters. In dispensing with the space, the Bayesian has added, in accordance with feature (a) above, a probability distribution $p(\theta)$ for θ.

Upon this model the classical statistician has heaped other ideas, such as hypothesis-testing, point- and interval-estimation. Sometimes these are treated entirely within the probabilistic frame (as with Fisher's approach to significance tests). Sometimes they include decisions and a loss structure, that roughly corresponds to the Bayesian's utility, so including features (b) and (c) above. (The Neyman-Pearson view of hypothesis-testing does just this.)

Indeed , the classical statistician often adopts the terminology of Bayes estimates. All these ideas seem less relevant to the Bayesian than the basic model itself. The difference between hypothesis-testing and estimation surely lies in the nature of the distribution of θ. Thus, in the former there is a concentration of probability on the null value whereas in the latter there is a smoother distribution for the parameter. The Bayesian equivalent of the Fisherian view, both for hypothesis-testing and estimation, is surely an argument entirely within the probability calculus in which $p(\theta)$ is carefully assessed and then combined with the likelihood by Bayes formula to produce $p(\theta|x)$ or relevant margins thereof. This is the way in which the modern Bayesian has proceeded and it seems unobjectionable except perhaps in regard to the choice of $p(\theta)$.

There are two aspects to this: the assessment of any probability, a topic discussed in Section 7; and the particular form to be adopted here. Bayesians have tended to avoid thinking about $p(\theta)$ directly and to have concentrated their efforts on producing distributions that either reflect ignorance of θ or form a reference distribution. This seems to me to be unnecessarily narrow. Modern work has substantially reduced my resistance to such procedures and I see that 'reference priors' have an important role to play, but I feel that we should widen our thinking and consider the value of $p(\theta)$ in the specific problem. This seems particularly important in problems with many parameters. A reference prior does not justify shrinkage. In any case, why should $p(\theta)$ depend on the likelihood or sample space, as reference priors do? The parameter often has a reality of its own that is independent of the experiment designed to learn about it and that fact should be observed. Essentially what is needed is a joint distribution for x and θ.

(As an aside, let us notice another model that depends not on x and θ, but on x, past data, and y, future data. Here $p(y|x)$ is required. This is an example where the Bayesian has profitably extended the classical view and argued more realistically in terms of data instead of the less directly-observable parameter. See, for example, Dawid (1991). Nevertheless the Bayesian analysis often introduces a parameter as a means of simplifying the analysis.)

A possible criticism of the way the Bayesian has used the classical model lies in the uncritical acceptance of the likelihood. According to de Finetti's view there is no distinction between $p(x|\theta)$ and $p(\theta)$; they both describe your beliefs, one about the data, were the value of the parameter known to you, the other about the parameter. Whilst we have thought, perhaps inadequately, about $p(\theta)$, we have tended to accept the classical view of the likelihood as known. It can happen that there is available data to support strongly a belief about x, given θ, as with a uniformity field trial suggesting normality. But typically this is not so. I know of little evidence to suggest normality for laboratory errors. What information there is, is more compatible with the longer tails of a t-distribution.

Another point, that has been recognized, for example in Bayarri *et al.* (1988), is that it is often not clear what is in the likelihood and what in the prior. My favourite example is the distinction between models I and II analyses of variance. One Bayesian model can cover both cases. In model I the distribution of the means is part of the prior; in model II, it is contained in the likelihood. The lesson to be learnt from these considerations is that what is required is an honest assessment of the joint distribution $p(x, \theta)$, or $p(x, y)$ in the pure-data case.

The classical statistician's limitation to frequency ideas means that it is only possible for the frequency component to be quantified. For example, in quoting a standard error, the only variation contemplated is that ascribable to sampling error. As a result, the error may be much too low. In the determination of physical constants, the frequency error quoted has consistently been too small, modern values of the constant falling outside the intervals quoted even 50 years ago. The reason is uncritical acceptance of a model; in particular, the failure to recognize the existence of bias. The simplest form erroneously assumes $x \sim N(\theta, \sigma^2)$ where θ is the constant: whereas more realistically $x \sim N(\theta + \beta, \sigma^2)$ with β a bias about which there is little or no data and which is not amenable to a frequency analysis.

4. CLASSICAL IDEAS: UTILITY

The model with $p(x, \theta)$ seems adequate for many purposes and the approach may be described as inference. But for some purposes it is natural to introduce decisions and utilities in partial imitation of the Neyman-Pearson scheme. However a Bayesian can roam much wider than the usual zero-one, or quadratic losses in that scheme; and he can introduce with profit decision situations that go beyond hypothesis-testing and estimation. For example, the hybrid in which a hypothesis is tested, and if rejected, the alternative value is estimated, has not been adequately treated. As with inference, what is needed is a thoughtful structuring of the practical reality. There $p(x, \theta)$ was enough. Here the space of decisions d has to be considered together with the utility $u(d, x, \theta)$.

An important example of a hybrid is that series of techniques usually termed 'analysis of variance'. It is interesting that although Bayesians have generally recognized the inadequacy of least squares, especially in high dimensions, they have not seriously tackled Fisher's important extension of least squares into the analysis of variance. This technique is surely an important practical tool yet there appears to be no Bayesian substitute. If the least-squares estimates are inadmissible then the usual sums of squares involving them are presumably not the appropriate ones. In any case, what replaces the F-tests and what estimate of main effect is reasonable if an interaction involving it is not significant? Here are elements of hypothesis-testing and estimation that may not have been adequately explored.

Caution is necessary. Much of the decision apparatus that has been added seems to me to be spurious and misleading. For example, point estimation has been studied as a decision problem with usually quadratic loss and much effort has been devoted to finding the 'best' estimator. For a Bayesian this is irrelevant for inferential purposes. All that is required is the conditional distribution of θ, given x. This is a complete description of your uncertainty of θ after the data x have been seen. There is no need to optimize, no need for standard errors. $p(\theta|x)$ says it all. I have elsewhere, Lindley (1990), shown how unsatisfactory point estimation can sometimes be. There are genuine decision problems that imitate point estimation, as when a dial has to be set, but there the decision is real and not artificially imposed by the statistician.

My conclusion is that although Bayesians have been somewhat constrained by classical concepts of the problems, the approach has not seriously suffered as a result. Where they may have been restricted is in their concentration on the limited class of problems studied in sampling-theory statistics. To this topic we now turn.

5. APPLICATIONS TO SCIENCE

Originally statistics dealt with the collection of data. Then the emphasis passed to the analysis of data, and especially that of experimental data. The word 'experiment' is here used broadly to cover any procedure to collect data under controlled conditions. Laboratory experiments, agricultural field trials, clinical trials and sample surveys provide examples. A basic idea nearly always present here is that of repetition. First, in the narrow sense of repetition within a single experiment; a circumstance familiar to a statistician in the concept of a random sample. Second, in the wider sense that another scientist, repeating this or a similar experiment, should reach broadly the same results. Nearly all of Bayesian statistics is confined to such 'repetitive' cases, although sometimes the methods there developed are applied perhaps without enough care to observational studies where no experimental design is present. There are branches of science in which the possibilities for planned experiments are severely limited, the bulk of the available data being outside the scientists' jurisdiction. Palaeontology is an example. Experimental scientists are often scornful of their colleagues in these fields. A physicist has compared palaeontology with stamp-collecting. To me, this appears grossly unfair, for there is surely something 'scientific' in the study of fossils that is lacking in the attitude of, say, religious fundamentalists to these creatures.

My opinion is that the Bayesian paradigm is the correct explanation of the scientific method; that it applies both to experimental science and to observational studies in disciplines like palaeontology or sociology. Kadane and Hastorf (1988) provide an example. A worker in these subjects has beliefs which he changes in the light of his data. He formulates hypotheses and tests them as far as he is able by looking for additional evidence. I suggest that our natural concern with experimental science, with its repetition and frequency ideas, may have caused us to forget the scientific work that does not have many facilities for repetition. I am hopeful that the Bayesian paradigm is capable of providing a reasonably complete account of scientific procedures; in particular by demonstrating how scientists of differing views can be brought into agreement by adequate data.

There is one field, not ordinarily thought of as scientific, with observational data where the Bayesian paradigm has had some impact. This is law, or, more narrowly, the trial aspect of the law. Here the uncertain quantity is the defendant's guilt and the data are the witnesses' statements. Most progress has been made in the case where the witness is a forensic scientist but there are possibilities, not yet much explored, of quantification of other, if not all, evidence. Perhaps we need a person of the stature of Fisher to perform a revolution in the criminal trial comparable to what he did for the agricultural trial.

6. THE STATISTICIAN'S BRIEF

The discussion so far has concentrated mainly on probability; feature (a) of the trichotomy in Section 2. Although classical, mathematical statistics, as we have seen, contains a lot of decision terminology, practice has largely been confined to the analysis of data, with some exceptions such as quality control. This is in accord with the concept of neutrality that a statistician is supposed to adhere to; presenting data but not advocating one action above another. There are perhaps two reasons for this. The first is historical. The second is that classical statistics just does not match with decision-making. Of what use is a significance level in initiating action? The two kinds of error provide only two snapshots from a holistic picture. The situation is quite different in the personalistic framework where the posterior distribution is the sole contribution of the data to action, only the utility needing to be added.

Here then is a place where the Bayesian has been too narrow. Usually the historical path followed by the sampling-theorist has been trodden without adequate exploration being made of the actions that might follow. It is my view that statisticians have, in the principle of maximization of expected utility, a tool of wide applicability, in which they are experts. Their clients could be advised about utilities in much the same way that they have been advised about uncertainty. Public relations are important here. We should not present ourselves as decision-makers but merely as advisors on uncertainty and worth and the computational problems of their combination in expected utility. Here is an example.

There has recently been a lot of valuable analyses of the data on AIDS involving predictions of likely trends. Here the statistician has played an important role. However, governments, the medical profession and insurers, amongst others, have to make decisions,

decisions which will affect future outcomes and hence predictions. Classical statisticians are handicapped here because they are confined to frequency concepts and their errors of prediction include only that type of uncertainty. Their standard errors may be too low (Section 3). It is hard for them to predict what will happen if a government took an action. Bayesians can include such uncertainty, can discuss quantitatively the merits of possible outcomes and provide a more coherent picture of the circumstance; all the time as consultants. I have discussed this elsewhere, Lindley (1984).

7. MEASUREMENT

If the question mark of my title serves to indicate suggestion as opposed to fact, there is one aspect of Bayesian statistics where it can be omitted and where current work is surely inadequate. Bayesian statistics is based on the measurement of uncertainty and of worth. Yet the literature on measurement is minimal. We are like geometers who study space and never measure. Where are the surveyors amongst us? A lot of work has been put into the measurement of distances. Where is the comparable work in probability?

The most commonly referenced work on probability assessment is that of Tversky and his colleagues, Kahneman *et al.* (1982), which broadly speaking shows that people make a mess of it, whilst at the same time demonstrating reasons for their folly. Maybe people cannot assess probabilities except in limited circumstances (as when a frequency basis is present) and the broad Bayesian paradigm must remain unimplementable. The alternative view is that people cannot perform the task without training and tools, any more than they can measure distance without them. Surely our task must be to provide the tools and the training in their use. The tools are lacking at the moment and little effort is put into providing them. Is it better, in the case of a single quantity, to assess fractiles or to provide moments? Or is there a better procedure? Another difficulty is that of assessing the tails of a distribution. We know a lot about the effects of heavy tails, for example O'Hagan (1988), which might be used for assessment, but exactly how is unclear. Similar difficulties apply to utility evaluations. The situation becomes even more obscure when several quantities are involved and considerations are necessarily multivariate.

It is likely that probability assessment is not a task that can be undertaken by statisticians working alone. Involving human beings, it may be essential to have psychologists working alongside of us, the better to appreciate the interaction going on between the human brain and external reality. There is some important, original work on utility assessment, Keeney and Raiffa (1976), especially in the multivariate context. The application of these ideas to situations considered by statisticians appears to be minimal.

There is a basic idea that I feel has to be used and which has been under-utilized. That is coherence. This is the fundamental concept upon which the Bayesian paradigm is based. The axioms are essentially rules for fitting judgements sensibly together. It follows that a reasonable expectation is that it would also prove a useful tool in practice. If so, we should not expect to assess a probability in isolation, nor even a probability distribution. Rather several probabilities should be jointly assessed and tests made for coherence using the rules of the probability calculus. Dawid (1988) provides some good illustrations.

There is a major puzzle about the assessments. Whilst there is literature on the separate assessment of probability and utility, few people seem to have recognized that, without special assumptions, their separation is not possible. The reason is easy to see. We can only judge people by their actions. Even the statement *"p is my personal probability"* is a decision to give the value p. But in a choice of action using expected utility, only the product $u(d,\theta)p(\theta)$ of $u(d,\theta)$, the utility of decision d when θ is the state of nature, and the probability $p(\theta)$ for

θ appear, so that their separation is not possible. For example, the methods of Brier and de Finetti using a quadratic scoring rule tacitly assume that the subject is using that loss. For all one knows a monotone function of the total score might be used as utility, so violating the probability assessment. The difficulty expresses itself practically, for it is often true that your perception of the uncertainty of an event is influenced by the consequences of that event being true or false. Yet, at a practical level, the separation into inference, involving only probability, and decision-analysis, incorporating utility as well, seems highly useful. If these two ideas cannot be separated, is the division sensible?

8. APPLICATIONS OUTSIDE SCIENCE

In this section I stick my neck out. I have been accused of seeing Bayes everywhere. This is true. Whilst I do not think personalistic thinking is a panacea for everything, I do feel it has contributions to make to much of human activity. In consideration of some topic, it often seems useful to look at it from a Bayesian perspective. There are two features of that view that are especially valuable: coherence and comparison.

Coherence is the basis of our attitude. We are Bayesians because we do not wish to make a series of statements that would lose us money for sure. A single probability can have any value between 0 and 1; a set of probabilities cannot, they must obey the rules. It is always worth having a look at a situation through the eyes of coherence. As I write these words, a war is being waged in the Gulf. Neither of the participants is exhibiting much coherence. The United States is invoking the United Nations, when over the past 25 years it has been the main obstruction to that body's work, having used its veto much more frequently than any other nation and has conspicuously failed to pay its dues. It has ignored the International Court of Justice and has suppressed democracy in Central America whilst purportedly acting in its name. Equally the Iraqis have criticized the coalition for killing civilians when they have been responsible for massive loss of life amongst Kurds and Kuwaitis.

Examples abound of man's apparent violation of coherence. We allow thousands to be killed on the roads yet spend great sums of money to save a baby born defective or rescue a lone yachtsman whose folly leads him into difficulties. In Britain we cannot afford decent hospital care yet we can spend enormous sums killing Iraqis.

Some of these actions may not be incoherent. The behaviour may be based on a utility function different from that implicitly declared. Thus, the apparently unsatisfactory reasoning of the United States cited above may not be based on uniting with other countries and considerations of democracy but involve capitalist considerations of selfish, financial avarice. The Gulf war is not about democracy but about the control of oil. Similarly the Iraqi leadership may have high utility for some lives and low for others. Whatever the explanation, incoherence or deceived utilities, the Bayesian approach can help us to understand and criticize. It can, of course, by considerations of probability and utility, help us positively to formulate procedures.

A lesson that the personalistic attitude reveals is that all actions are based on comparisons. There are no absolutes in the world of Bayes. This can be seen in the narrow world of significance tests. Fisher essentially considers only behaviour on the null hypothesis. Bayes compares the null with the alternative in a likelihood ratio or by odds. We cannot but live with a hypothesis until there is an alternative to substitute. But the concept is wider. All action within the Bayesian view is based on maximization of expected utility where the maximum value is irrelevant, all that matters is that it is the largest in comparison with everything else. We do something not because it is 'good' but because it is 'better' than anything else we can think of. The religious idea of an absolute 'good' appears foreign to a utility approach. I am even suspicious of 'goals', rather we should compare the attainables.

As with coherence, there are many applications of the comparison technique. The law should not concentrate on innocence until proved guilty or worry overmuch about convicting the innocent, but should recognize that someone committed the crime and that there is folly in letting the guilty go free. Lawyers might also recognize the comparison between their own fees and social costs. A case in Britain, lasting five years, resulted in $25K fines against the police and one million in legal fees. Is it worth the expenditure of so much money to redress a few damaged vehicles and heads when the police went beserk? Perhaps the clearest cases occur in politics. One may vote for a political party, not because its views closely match your own but because they do so more nearly than those of the opposing parties. An action involving environmental damage can be supported if its denial would deprive people of decent housing.

Many of you will feel that these topics do not belong to a conference on Bayesian statistics. I disagree. The twin concepts of probability and utility provide useful tools for thinking about all (or most) aspects of the world. We know and understand them and should recognize their value beyond the limited world of likelihoods from random samples and priors.

REFERENCES

Bayarri, M. J., DeGroot, M. H. and Kadane, J. B. (1988). What is the likelihood function? *Statistical Decision Theory and Related Topics IV* **1**, 3–27. (S. S. Gupta and J. O. Berger, eds.) New York: Springer.

Dawid, A. P. (1988). The infinite regress and its conjugate analysis. *Bayesian Statistics 3* (J. M. Bernardo, M. H. DeGroot, D. V. Lindley and A. F. M. Smith, eds.), Oxford: University Press, 95–110.

Dawid, A. P. (1991). Fisherian inference in likelihood and prequential frames of reference. *J. Roy. Statist. Soc. B* **53**, 79–109, (with discussion).

Kadane, J. B. and Hastorf, C. A. (1988). Bayesian paleoethnobotany. *Bayesian Statistics 3* (J. M. Bernardo, M. H. DeGroot, D. V. Lindley and A. F. M. Smith, eds.), Oxford: University Press, 243–260.

Kahneman, D., Slovic, P. and Tversky, A. (1982). *Judgment under Uncertainty: Heuristics and Biases*. Cambridge: University Press.

Keeney, R. L. and Raiffa, H. (1976). *Decisions with Multiple Objectives: Preferences and Value Tradeoffs*. New York: Wiley.

Lindley, D. V. (1984). The next 50 years. *J. Roy. Statist. Soc. A* **147**, 359–367.

Lindley, D. V. (1990). The present position in Bayesian statistics. *Statist. Sci.* **5**, 44–89, (with discussion).

O'Hagan, A. (1988). Modelling with heavy tails. *Bayesian Statistics 3* (J. M. Bernardo, M. H. DeGroot, D. V. Lindley and A. F. M. Smith, eds.), Oxford: University Press, 345–359.

DISCUSSION

M. GOLDSTEIN (*University of Durham, UK*)

I would like to thank Dennis Lindley for choosing such a pertinent topic for his presidential address and confronting so honestly the crucial issues that his title raises. I would like to concentrate most of my comments on one issue that I consider to be fundamental.

Bayesian statistics was developed as a way of using beliefs to help us to make better analyses of data. (There were, of course, other conceptual inputs, from philosophical and decision-theoretic sources, but my impression is that the distinctive flavour of Bayesian statistics has been most strongly influenced by the work of those, such as Savage and Lindley, who introduced subjective elements to make better sense of actual statistical practice.)

Why is a data analysis of interest? Because it can influence someone's beliefs! Our beliefs and how they change are what is fundamental – data analysis is simply a means, not an end in itself. Therefore, isn't it natural to move from using beliefs to help us to analyse data to using data to help us analyse our beliefs?

Have we reached this stage? I don't think so. Look, for example, at the invited papers at this conference. I think that you will be hard pressed to find any beliefs believed by anybody about anything. Even those beliefs which can be attributed to individuals are rarely the focus of the enquiry –somehow they happened off-stage, specified before the work of the paper was started. This is not intended as a criticism of the papers –it is just a comment as to what Bayesian statistics is about, as reflected in how we choose to write about it. Dennis Lindley refers to the inadequacy of current work on the measurement of uncertainty. To me, this reflects our basic confusion as to whether we are primarily statisticians, who analyse data, or 'belief analysts' (for want of a better term) who analyse beliefs.

Why do we accept such a restricted role? Mainly, I think, because many of us live in a comfortable statistics niche, where we know what is expected of us, and where we can perform rather better than those other guys who don't use prior beliefs at all. However, when we try to elicit genuine beliefs about serious issues (for example, just what is going to happen to the ozone layer?) then we find the process to be hard, painful and, mostly, we don't know how to do it.

Can we do better? I was attracted to the following quote in the paper:

> We do something not because it is 'good' but because it is better than anything else we can think of. ... I am even suspicious of 'goals', rather we should compare the attainables.

What is attainable in most cases is not the elicitation of very complicated beliefs, but elicitation of limited aspects of such beliefs. Therefore the better methods that we need to develop must focus on ways to use data to modify such limited aspects of our beliefs. Within such a framework, Bayes theorem would simply stand as a special case of much more general methods of belief revision, and perhaps the 'narrow' view of Bayesian statistics could be overcome.

Finally, I agree with the general comments in this paper on the central role of values and utilities. However, in practice, values are tricky things. For example, there is one sentence in this paper which gave me a pause:

> The law should not concentrate on innocence until proved guilty or worry overmuch about convicting the innocent, but should recognize that someone committed the crime and that there is folly in letting the guilty go free.

This is a somewhat emotive issue in the UK at the moment, with the cases of the Birmingham six and the Guildford four on our minds. The police seemed to follow the above quoted advice in those cases, and certainly didn't worry overmuch about convicting the innocent. Somehow, however, this hardly seems a triumph of rationality. Now, I know that Lindley did not intend his remarks as offering a justification for evil. However, such examples do illustrate how careful we need to be. Indeed, if beliefs are hard to specify for substantive problems, then we haven't begun to get close to the issues of specifying values. Maybe our view of Bayesian statistics is too narrow because, like every other serious field of human endeavour, we need to build in some notion of morality.

J. O. BERGER (*Purdue University, USA*)

Surely any Bayesian will agree with virtually all the insightful observations in this paper. Also, because of space constraints, many statements in the paper are understandably not fully argued. Probably the most useful comments for a discussion, therefore, are simply to note which statements in the paper are still open to debate among Bayesians and why.

In Section 4, Lindley states that concepts such as point estimation and standard errors are superfluous, since one has the entire posterior distribution available. I would argue otherwise, for two major reasons. The first is that, in complicated high dimensional problems, the posterior is typically uninterpretable; even consideration of all marginals or bivariate marginals of the posteriors may be visually overwhelming or computationally infeasible. Simple numbers, such as point estimates, standard errors and correlations, are then needed to summarize the posterior. Even in one-dimensional problems reporting the posterior is often rather gratuitous, since typical consumers of the information from a statistical study are not able to process information in the form of a density.

My typical consulting experience in this regard is that I spend a considerable amount of time trying to teach the users how to interpret the posterior density, and urging them to use it, but when I return later to check on their implementation of the methodology they have reverted to point estimates and standard errors because "nobody could really understand these graphs." Proper interpretation of densities is a rather high-level skill.

I have some uneasiness with the argument that our main difficulty is the lack of a theory of measurement of uncertainty. It is true that development of such has received scant attention, while comparatively enormous effort has been spent on the much less useful enterprise of modelling how people, without training, actually measure uncertainty. My feeling, however, is that the problem is too immense to be amenable to a frontal attack, in the sense that we can ever learn how to "measure" high-dimensional uncertainty. I am more attracted to the "solution" of finding ways of determining which features of the uncertainty *need to be measured*, the point being that since we will only be able to make a finite number of measurements of uncertainty it is important to make sure we measure the important things; and these cannot typically be determined a priori, but require interaction with the data.

While appreciating the artificiality of the axioms that force separation of probability and utility, it seems to me to be a practical necessity to derive them separately. I believe that it is possible to train individuals to think separately about probabilities and utilities, but impossible to train people to adequately think about their joint product. Separation should thus be considered to be a "measurement" or "computational" tool, not a logical consequence of coherence.

The view that coherence can be widely applied in life and society is highly attractive and important to promote. However, making applications outside of science does require great care, since then it is especially important to "... not present ourselves as decision-makers but merely as advisors on uncertainty and worth and the computational problems of their combination in maximization of expected utility."

At least some of the examples in Section 7 fail to adhere to the spirit of this guideline. For instance, the discussion about the role of the United States in the Gulf War does not seem to be oriented *only* towards helping the United States determine if it is acting coherently, since the discussion includes personal opinions about the motives of the United States. The United States gave many reasons for initiation of the war, including those mentioned by Lindley, and the job of a "coherency advisor" would be to see if these reasons could be molded into a utility function for which expected utility could be utilized. The coherency advisor is certainly allowed to question the client concerning this utility by considering past actions of the client, but the coherency advisor must refrain from personally choosing the utility function, as is implicit in the statement "The Gulf War is not about democracy but about the control of oil." Statistics has suffered because of its casual and unsound use by politicians to make political points; it would be a disservice if, through examples like this, we encourage similar and casual unsound use of coherency.

S. FRENCH (*University of Leeds, UK*)

There are two sides to the Bayesian coin equal in importance: probability and utility.

Looking at our conferences, however, one cannot but subscribe to the view that equal in importance they may be, but probability is more fundamental than utility. This is clear in Professor Lindley's remark that the utility scale is itself "based upon probability" (Section 2, remark b). Reading many of the axiomatic derivations of the Bayesian paradigm it is easy to fall prey to such a view. The utility of an outcome is found to be the probability associated with a reference outcome in a hypothetical gamble. But such a view does not stand up to further inspection.

Utility and probability are numeric representations of different cognitive judgements. The first represents preferences, the second beliefs. Utility is unique up to positive affine transformations; probability is a unique measurement scale with range $[0, 1]$. So we should not be surprised that their scales differ fundamentally. Certainly they are compatible scales in the sense that forming expected utilities provides a rational ordering of the alternatives in a problem. But to argue that this means they are fundamentally the same is to argue that mass and velocity are the same because they can be combined to form momentum.

Many axiomatisations, which lead to the use of expected utilities, begin by assuming the decision maker has a weak preference ordering over the set of outcomes and a weak belief ordering over the set of states. Note that each outcome and each state are assumed to be holistic entities. In any decision model we build, however, outcomes are represented as multi-dimensional vectors of attribute levels and the state as multi-dimensional vectors of parameters. Ask any decision maker to produce a weak ordering over either of those and he or she will look dumbfounded at the complexity of judgements required. So we help them by introducing further structure. We build multi-attribute value models to provide weak preference ordering over the outcomes and multi-parameter belief models to provide the weak belief ordering over the states. The 'trick' of equating utility with a probability in a hypothetical gamble is then the mechanism whereby the two weak orderings are related and the expected utility hypothesis justified.

Looking more closely at the two weak orderings it is clear that they are fundamentally different. The axioms and independence conditions that underpin multi-attribute value models have a quite different structure to the axioms and probabilistic independence or exchangeability conditions that underpin multi-parameter belief models. Compare, for instance, the seminal books of Keeney and Raiffa (1976) and de Finetti (1973, 1974).

My reason for making this point is not esoteric, it is practical. We do have the tools to comment on the problems of our time: justice versus cost in the legal system; quality of life versus cost in health care systems; the Gulf war; etc. But those comments will in many instances be based on considerations of equity, social responsibility, morality, etc. These we need to explore through the preferential independence conditions in the multi-attribute preference model not simply through our probability modelling.

TRAN VAN HOA (*Wollongong University, Australia*)

As one of Professor Lindley's admirers since our first meeting at the Center for Operations Research and Econometrics at the Catholic University of Leuven in 1972, I now admire him more, for his stamina, dedication, and intellectual weight to popularize and advance Bayesian statistics as a respectable branch of science (he often says Bayesian statistics is not a branch of statistics). However, on the question of whether the view of many Bayesian statisticians is too narrow in these days, I have to agree with him in the affirmative, but from another perspective.

First, in the age of information technology, the idea of collecting and using more information for better decision analysis should not be rejected by any rational person. Using more information is obviously better than using less information (even if the extra information is wrong). In this context, Bayesian statistics has an unquestionable proper role. The direction of Bayesian statistics therefore should not lie in its further epistemological justification but in resolving the still unresolved (and popular with detractors) problem of what is the best prior, what is the best prior for the problem at hand, or what is the best class of priors that can provide consistency and dominance in average posterior analysis. On the second category of priors mentioned above, one promising line of research is our current work on the evaluation of priors based on the best *ex post* forecasting criterion for an observed finite sample dataset.

Secondly, in doing posterior analysis, it would be much more convincing to the statisticians and the users of statistical tools alike if the results have some empirical real-life content. After all, statistics is the art of collecting and analyzing observed data. In this context, the future direction of Bayesian statistics should give applications a more prominent role. Work in this direction has been undertaken by us in the area of economics with highly significant results. Our work includes Bayesian or more properly empirical-Bayes studies of the measurement of economic welfare, the business cycles, unemployment, the rate of interest, and energy consumption in developing countries (Tran Van Hoa, 1990, 1991) to cite a few.

Finally, my comment is on the use of data analysis to do belief analysis. While the idea is fine in theory for Bayesian statisticians, it cannot be taken seriously in many applications in real life. One obvious example is in economic modelling and predictions in which the dataset is inherently small. Clearly, using a sample size of ten or even twenty data points to adequately model and predict multi-sector economy-wide behaviour or to question time-tested fundamental economic postulates is not right in any context.

D. WOOFF (*University of Durham, UK*)

It seems to me that Professor Lindley's title serves as both question and answer, for what matters is not especially *our* view, but the views of those who would use the fruits of our efforts. In the former sense, this paper serves well in drawing to our attention several aspects of the Bayesian paradigm which deserve rather more than the occasional airing, in particular the thorny, but central problem of how we make good prior probability assessments; something of a millstone around our collective Bayesian necks. It is in a wider sense that we are in danger of pushing ourselves into ever narrower channels, and thus of losing our professional *raison d'être*, by being subjected to a constant pressure which acts to divert our ideas and methodologies away from general statistical practice into a Bayesian ghetto. How does this pressure arise?

Firstly, many of us are academics, subject to the market forces operating on ivory towers: we must produce streams of papers to survive - quantity is essential, quality a welcome bonus. Where we suffer in comparison with frequentist statisticians is that we produce hard but meaningful analyses rather than easy, but arguably worthless analyses. In short, the easiest way for a Bayesian to publish is to publish theory, rather than to go to the trouble of performing and reporting Bayesian applications. In this way we dig our own graves: we simply cannot convince users of statistical methodologies of the efficacy of the Bayesian way without adding meat to our theoretical bones in the way of large numbers of successful applications.

The tilting of the balance toward theory rather than application is compounded by a belief that to do the former is somehow smarter. We must avoid the folly of the theoretician sneering at the practitioner: we should instead codemn the minimal stature of the theoretician divorced from reality. (In Britain we have a proverb which avers "those who can, do; those who can't, teach.")

Such self-induced narrowing is also self-perpetuating because of the way in which ideas are promulgated: mainly via publication and disciple. Research supervisors tend to make graduate students in their own image, so we shouldn't be surprised that they tend to work in the same fairly narrow areas of specialisation, even unto their acquiring their own disciples, into whom similar wisdom is passed. In this regard I am a fourth-generation descendant of Professor Lindley, although he may argue that I have acquired some bad blood along the way! At the same time, to be Bayesian simply because one's thesis supervisor is a Bayesian is unsatisfactory for obvious reasons.

I believe that we do ourselves no favours by concentrating upon such things as reference priors (as a methodology), *particularly* when they are claimed to be something "objective" which might attract frequentists. If the Bayesian way is synonymous with the holding of beliefs which we might change when we see pertinent data, how can reference priors be Bayesian? We might, just as sensibly, consider the idea of "reference likelihood" in combination with genuine beliefs. No, to tempt a frequentist with reference priors is to offer sorcery as an alternative to witchcraft. Our task should not be to kowtow to the demands of (uneducated) users in providing pseudo-objective procedures, but to persuade them that our philosophy and practice is sound; to do otherwise is to fall into the trap of permitting our methods to be seen as just one of many valid tools. Herein lies the true Bayesian ghetto.

Bayesian statistics is hard! It is no wonder that we find so few adherents to our cause: not only is the thinking hard, but the analysis thereafter may prove impossible without recourse to sophisticated algorithms and computers. This leads to a further narrowing in two ways: firstly our methods are increasingly being seen as too tricky and involved; and secondly our research may be top-heavy in numerical methods, and light in tackling actual statistical problems. I do not know whether it is realistic to demand that a statistical approach be both simple to understand and simple to apply, but my instinct is to reject complexity: must the path between theoretical soundness and pragmatism be so narrow?

Lastly we come to the question of the statistician's neutrality. The introduction to the paper hinted that neutrality should be regarded as a questionable, perhaps limiting virtue. However, the advice contained in a later section, *The Statistician's Brief*, is unclear: it seems to suggest that we should abandon our traditional role of passive independent commentator in favour of an active decision-making role, something at which we are supposedly expert. I am not sure whether we are being advised to "go home and prepare for government" in the way of British Liberals of recent years, or whether we are being requested to ensure a more rigorous application of utility theory to decision making. Whichever is the case, the proviso "Public relations are important here" makes me extremely nervous.

REPLY TO THE DISCUSSION

It is reported that, after hearing someone else making a brilliant remark, Oscar Wilde murmured "I wish I had said that"; to which Whistler responded "You will, Oscar, you will." The brilliant remark at the conference was contained in Michael Goldstein's comment that we should be belief analysts. I propose to emulate Wilde and repeat the idea for it is a strikingly correct way of regarding our subject. Bayesian statistics is a systematic way of handling beliefs. Since beliefs can be sensibly held about most things, the scope for the personalistic method is enormously wide. If this were generally recognized in the Bayesian community, we should find ourselves dealing with many topics that are scarcely handled today. Goldstein is also right to emphasize that we can only elicit limited aspects of beliefs. Berger appears to share this view when he mentions the conceptual difficulties with multivariate distributions that require us to summarize their main features. But I feel that he is wrong when he connects these difficulties with methods developed by frequentists for point estimation. The posterior distribution stands as a complete inference and the genuine problems in its appreciation are not assisted in their resolution by optimality properties, like BLUE, of point estimates. While 'belief analysts' is a brilliant phrase, it only encapsulates part of our activities. Our beliefs, unlike those of frequentists, can be used for decision-making through combination with utilities. My view, in answer to Wooff, is that advice to our clients should embrace both opinion and action. That just as we advise on the influence of data on opinions, so we should explain the consequences for action. Our modelling for the data could be paralleled with that for decision. Could we not be "passive, independent commentators" over utility as we claim to be over probability?

Perhaps this point of disagreement arises because my original wording was lacking in precision and the responsibility for the misinterpretation lies with me. This is surely true of Goldstein's comment about the Guildford four, and Berger's remark about the Gulf war. In the latter case, I ought to have said that a utility function based mainly on oil might lead to coherence, rather than expressing it as my opinion (which it is).

Wooff is correct when he points out the effect that the pressure to publish has on the content of journals. In today's climate, it is better, and easier, to publish six papers each dealing with one minor, technical matter, than to think deeply and only have a single, substantial contribution. Deans can count but not judge quality. I have no suggestion as to how this trap can be avoided except to make the standard remark of the elderly that we did not have that trouble when we were young. Wooff is also right when he remarks that graduate students think like their supervisors. I have often thought that graduate students are too subservient and that there is not enough dissent in academia.

French and I have jousted before over the relationship between probability and utility. We completely agree that the conditions underpinning multi-attribute models have a different structure from those for multiparameter, belief models; and that the elicitation procedures are of practical necessity different. Where we perhaps disagree is over the relationships between the concepts of value and belief; utility and probability. There are many possible measures of value. A key idea about the particular measure we call utility is that it is dependent on probability. This dependence is essential if maximization of expected utility is to be our criterion for decision-making because it provides the link between belief and action mentioned earlier. Similarly, I do not see any way of determining a subject's probabilities except with an implicit assumption of what their utility structure is. Of course, as French says, the two weak orderings, of states and outcomes, are fundamentally different. Nevertheless, they are connected, and the connection is vital. The unanswered question is whether they can be, or need to be, separated. Bayesians, as this conference confirms, have concentrated on probability rather than utility. This may be unwise.

The Bayesian paradigm refers to a single decision-maker, where all information is expected to be of value, as Hoa reminds us. But is this true of society, a collection of decision-makers? Even in democracies, there is reluctance to provide information for fear that it will be used against one. In the theory of games it is known that a player can be at a disadvantage if information that he has passes to the opponent. Whilst agreeing with Hoa over the real-life content of our research, I cannot support his search for a 'best prior', except in the sense that what is best is what is true. My disagreement with Berger on the point he makes about measuring uncertainty is more fundamental. It is not a problem only in high dimensions. We do not have satisfactory procedures for the measurement of probability for a single event. How can scientists express in quantitative form their opinions about the effect of ozone depletion on the growth of plants? The problem appears to be immense, but surely that is a reason for devoting a lot of research to it rather than denying the reality of the knowledge we undoubtedly do possess and searching for non-informative priors. Many measurement problems have been solved despite their enormous difficulties. For example, the 'distances' between species were unsatisfactorily treated by taxonomists but may be solved by genetic methods. The difficulty with studying particular features is that what may be important for one application may be irrelevant for another. Berger's idea, that the separation of probability and utility is a computational tool, is excellent.

My experience with probability densities in one or two dimensions is the opposite of Berger's. The reason may be cultural, since Europeans are familiar with contour maps, whilst Americans are not. I find people can appreciate the implied areas under the curves and see which values are most likely, which unusual. In fact, they find it easier to understand densities than means or standard deviations. The integrals in the latter are hard to appreciate. It is not easy to see the difference in standard deviations between a normal and a t-distribution of the same interquartile range. Because a variance is mathematically easier than other measures of spread does not entitle it to an exclusive position.

In an earlier discussion of a paper of mine, Lindley (1990), Berger said that, in comparison with frequentist statistics, Bayesian statistics "is much easier to understand and yields sensible answers with less effort". Wooff here says "Bayesian statistics is hard". Conceptually, the Bayesian view is simpler but its operation is surely harder because it deals with reality and not with mathematical fiction. If the former is neglected in comparison with the latter, it may appear simpler, but this can only lead to a distorted and unsatisfactory science. Mathematics is a marvellous handmaiden but a disastrous dictator.

The paper was, of course, written before the conference in Peñíscola: these remarks are being written afterwards. Attendance there leads me to think that the question mark of the title can be removed: that we are too narrow. The standard of the papers was high. It is the balance that I find worrying. Too many of the papers dealt with narrow, technical issues: too few addressed practical problems with real beliefs to be understood. Hardly any talked of decision and utility. It is not that the technical papers were irrelevant or wrong; far from it. We need them and the understanding they undoubtedly attain. But the sparkle of symbols should not make us forget the reality that hopefully lies behind them. Do not let Spanish conferences be known as the home of theoretical ideas, leaving practicality for Pittsburgh later this year or Nottingham next.

ADDITIONAL REFERENCES IN THE DISCUSSION

De Finetti, B. (1973). *Theory of Probability* **1**. New York: Wiley.

De Finetti, B. (1974). *Theory of Probability* **2**. New York: Wiley.

Tran Van Hoa (1990). Modelling output growth: a new approach. *Economic Letters*, (to appear).

Tran Van Hoa (1991). Energy consumption in Thailand: estimated structure and improved forecasts for enviromental impact studies. *Tech. Rep.* Thammasat University, Bangkok, Thailand.

BAYESIAN STATISTICS 4, pp. 17–33
J. M. Bernardo, J. O. Berger, A. P. Dawid and A. F. M. Smith, (Eds.)

A "BAD" View of Weighted Distributions and Selection Models*

M. J. BAYARRI and M. H. DeGROOT
Universidad de Valencia, Spain and *Carnegie Mellon University, USA*

SUMMARY

Weighted distributions and selection models arise when a random sample from the entire population of interest cannot be obtained or it is not desired. They model situations in which the probability or density of a potential observation gets distorted so that it is multiplied by some (non–negative) weight function. Selection models are the important particular case in which the weight function is the indicator function of some selection set. Bayarri and DeGroot (BAD) have studied, in a number of papers, many statistical issues that arise with these models. This paper will be a brief, unifying review of their work in the subject.

Keywords: COMPARISON OF EXPERIMENTS; CONJUGATE PRIOR DISTRIBUTIONS; FILE-DRAWER PROBLEM; SELECTION BIAS; SIZE-BIASED SAMPLING.

1. INTRODUCTION

Over the years, friends and colleagues have often asked me when and how Morris DeGroot and myself first became interested in weighted distributions and selection models. Our interest actually began on the very first day of my one–year visit to Carnegie Mellon University during the academic year 1985/86. Morrie was showing me around CMU, introducing me to the faculty and explaining the system, and all the while was chatting in an informal way about various stimulating statistical problems of the kind that Morrie always produced. One such problem was the question: "Do you think that, given the choice, we should go for a sample that has been obtained in some selected subset of the sample space, maybe some set in which observations are very difficult to obtain (as in the tails), or should we always stick to the usual random sample over the whole population?" I had no clue at the time as to the answer, but this was the starting point of a long and productive (and for me extremely interesting) study of the subject. It was, indeed, only a few months later that Morrie gave our first joint talk at a meeting on weighted distributions. The name BAD (Bayarri and DeGroot) was given to us by Nozer Singpurwalla in his discussion of our paper presented at the third Valencia Meeting, and it was a name we always enjoyed.

In our work on selection models and weighted distributions, we treated a variety of different statistical issues. Our first incursions in the subject appeared in BAD (1987a, 1988), where we formulated the problem, gave several examples, and discussed inferences in various selection models.

We soon became interested in the question of whether a selection sample (that is, a random sample obtained in some subset of the sample space) could be more informative than

* Research supported in part by National Science Foundation Grant No. DMS 88–02535 and Spanish Ministry of Education and Science, DGICYT Grant BE91-038.

a random sample over the whole population. Our first results pertained to the comparison of the information in a random sample from the whole population with the information in a selection sample, according to the criteria of Fisher information and pairwise sufficiency. These results appeared in BAD (1987b). We were working on this subject while Morrie was visiting Ohio State University, and Prem Goel also became interested in comparing selection experiments with unrestricted experiments. Eventually, a by–product of our study of Fisher information in selection models became a simple and interesting new property of discrete distributions, appearing in Bayarri, DeGroot, and Goel (1989).

In the meantime, we encountered various striking statistical features that appear when analyzing selection models; also, our favorite example (that of properly analyzing statistically significant results that are published in the scientific literature) was studied and discussed from a variety of different points of view. (I collected part of these results in BAD, 1991a.) This example also interested us in the issue of whether we should worry about observations that we do not observe, in apparent contradiction with the likelihood principle (at the time we were also pursuing some work on the likelihood function and likelihood principle). This led us to explicitly consider the selection mechanisms that could have produced the actual selection sample. All these topics are considered in BAD, 1990.

In the last paper that we jointly wrote on the subject (BAD, 1989), we returned to the original question, namely, whether weighted samples can be more informative than random samples. In that paper we moved from selection models to more general weighted distributions, and we abandoned Fisher information in favor of sufficiency of experiments. We wrote that paper while Morrie was fighting against cancer. Through all those years, our work benefited from comments and conversations with many friends, more than can be listed here; I would nevertheless like to especially acknowledge James Berger, Prem Goel, Joel Greenhouse and Satish Iyengar, with all of whom we spent many hours in fruitful and interesting discussions on the subject and whose remarks directly influenced, improved or motivated some of our results.

The need for writing a review paper was apparent, but we never got it written. This paper hopefully fills that gap of reviewing BAD's work. The orientation here is to stress the ideas and inferential aspects more than the specific results. Thus, derivations of posterior distributions, proofs, etc., are not given, and technical details have been kept to a minimum. Due to severe space restrictions, few non–BAD references can be given. The interested reader can find them in BAD's original papers and in the technical report BAD (1991b) of which this paper is a reduced version.

The paper has 5 sections the first of which is this introduction. In Section 2 we present the BAD approach to weighted distributions and selection models, give examples, and motivate their use. In Section 3 some interesting statistical issues that arise when carrying out a Bayesian analysis of selection models are discussed. Section 4 is devoted to the analysis of published significant results and Section 5 to the comparison of the information in weighted samples with the information in a random sample.

2. WEIGHTED DISTRIBUTIONS AND SELECTION MODELS

Suppose that the random variable (or random vector) X is distributed over some population of interest according to the (generalized) density $g(x|\theta)$, and that it is desired to make inferences about θ. The usual statistical analysis assumes that a random sample $X_1, \ldots, X_n$ from $g(x|\theta)$ can be observed. There are many situations, however, in which a random sample might be too difficult or too expensive or impossible to obtain and statistical models have to then be developed to incorporate the non–randomness or bias in the observations. (We will see later that, in some situations, even if a random sample can be obtained, the experimenter may decide not to do so, since a carefully biased sample may be more informative about θ than a random sample.) *Weighted distributions* arise when the probability (or density) of the potential observation x gets distorted so that it is multiplied by some (non–negative) weight function $w(x)$, which in turn may involve some unknown parameters. Thus, the observed data is a random sample from the following weighted version of $g(x|\theta)$:

$$f(x|\theta) = \frac{w(x)\ g(x|\theta)}{E_\theta[w(X)]}, \tag{2.1}$$

where the expectation in the denominator is just a normalizing constant so that $f(x|\theta)$ integrates to 1. Distributions with densities of the form (2.1) are called *weighted distributions* by Rao (1965) who first unified the available results, although their use can be traced to Fisher (1934). Good surveys on the topic are Patil (1984) and Rao (1985).

As an example, suppose that the number of talks X that participants in a given statistical meeting would attend has distribution $g(x|\theta)$ and that, while at the meeting, we are interested in inferring about θ. In statistical meetings a list of participants is usually available, so that an obvious way to get a random sample $X_1, X_2, \ldots, X_n$ from $g(\cdot|\theta)$ would be to select n participants at random from that list and ask them how many talks they are attending. This procedure can, however, be inconvenient, cumbersome and time consuming, since the particular participants thus selected have to be located and interviewed. An easier and simpler way to get observations X_i would be to select one talk (or several talks) at random and then interview n people among those attending the talk. Notice, however, that in this case the more talks a participant attends (the larger the value of X) the more likely it is for him or her to be present in the sample. In other words, each observation is *size–biased* in the sense that its density is given by (2.1) where $w(x)$ is an increasing function of x. Under certain equilibrium conditions, it is standard to assume that $w(x) = x$; this particular case is sometimes referred to in the literature as *length–biased* sampling. A more general formulation would be to take $w(x) = x^\tau$ and to treat τ as an unknown parameter.

A particular class of weighted distributions to which BAD gave special emphasis are the the so–called *selection models* (or truncation models) in which observations are obtained only from some "selected" portions of the population of interest. For example, suppose that the distribution of a certain vector X of characteristics in a given population is represented by the density $g(x|\theta)$ and that individuals for whom the value of X lies in some set S manifest a certain disease. Assume that we are interested in the distribution of X in the whole population but that X is only measured for individuals who manifest the disease because the observation of X is expensive, painful, or even dangerous (involving perhaps some sort of surgery).

In this example, inferences about θ and S have to be based on data obtained from people who manifest the disease because X will be observed for these persons in the course of their treatment. The density of x for such a person is given by (2.1) where the weight function $w(x)$ is of the form:

$$w(x) = \begin{cases} 1, & \text{if } x \in S, \\ 0, & \text{otherwise.} \end{cases} \tag{2.2}$$

When selection occurs, data consist of observations of a random sample $X_1, \ldots, X_n$ of the selection model $f(x|\theta, S)$ which, from (2.1) and (2.2) is given by

$$f(x|\theta, S) = \frac{g(x|\theta)}{\Pr(X \in S|\theta)} \qquad \text{for } x \in S, \tag{2.3}$$

and $f(x|\theta, S) = 0$ otherwise. A random sample from (2.3) will be called a *selection sample* and the set S, the *selection set.* The name "selection models" in this context is due to Fraser (1952, 1966) although the term "selection" was used in a more general setting by Tukey (1949). In the rest of the paper we will restrict ourselves to problems in which $g(x|\theta)$ is a univariate distribution.

Simple, yet common, selection models arise when selection occurs in one of the tails, say the upper tail, of the distribution. In this case, the selection model (2.3) is given by:

$$f(x|\theta) = \frac{g(x|\theta)}{1 - G(\tau|\theta)} \qquad \text{for } x \geq \tau, \tag{2.4}$$

and $f(x|\theta) = 0$ otherwise, where G is the c.d.f. corresponding to g. Selection models of this type arise naturally when sampling from historical records. In such problems, observations from the entire historical population are usually not available and inferences have to be based on observations that were recorded just for some selected group. For example, estimation of the distribution of the height of the population might have to be based on the historical observations of the heights of members of the army, for which there was a known minimum required height τ.

Other practical problems in which τ is known arise in the analysis of data reported to reinsurance companies, which only receive claims that exceed τ, with the value of τ being fixed by agreement with the original insurance company. The reinsurance company, however, is often interested in making inferences about the entire distribution of claims. Still another example is provided by situations in which measurements are made with instruments of a given sensitivity so that some items cannot be recorded; for example, in astronomy, only sufficiently bright objects can be observed. Sampling from a truncated binomial or Poisson distribution in which the zero class is missing ($\tau = 1$) is another selection model widely treated in the literature. BAD's favorite example of selection models with known τ occurs when only experimental results that are found to be "statistically significant" are published in the scientific literature; we will turn back to this example in Section 4.

3. INADEQUACIES OF "ROUTINE" PRIOR DISTRIBUTIONS

In this section we will show that, when dealing with selection models, the naïve use of "routine" types of prior distributions can be inadequate. By "routine" priors we mean the ones that are usually used in Bayesian work, namely improper priors and conjugate priors.

Selection models provide very simple examples where improper priors typically lead to improper posteriors, no matter which observations or how many observations have been obtained. Consider the case in which selection occurs in the upper tail of a completely specified $g(x)$, so that the selection model is given by (2.4) for known θ and unknown τ. The natural choice for a conjugate prior density for τ is

$$\xi_1(\tau) \propto [1 - G(\tau)]^a \qquad \text{for } \tau \leq b, \tag{3.1}$$

where the hyperparameters a and b are constants to be specified. A distribution of this form may be appropriate in problems in which the support of $g(\cdot)$ is bounded on the left, so that the possible values of τ are also bounded on the left. Otherwise, the density $\xi_1(\tau)$ will be improper for every real number a. Although Bayesians are used to improper prior distributions, this particular one is completely useless, since the posterior distribution will also be improper for all values of n and all values of $x_1, x_2, \ldots, x_n$. Thus, if it is desired to carry out an analysis under a conjugate distribution, a different form for the conjugate prior has to be selected.

There are, indeed, many different families of prior distributions that are closed under sampling. For example, any density of the form

$$\xi(\tau) \propto h(\tau)[1 - G(\tau)]^a \qquad \text{for } \tau \leq b \tag{3.2}$$

will have this property, where h could be *any* function such that the right–hand–side of (3.2) is integrable over τ. In particular, h could be taken to be of the form

$$h(\tau) = \begin{cases} 0 & \text{for } \tau < c \\ 1 & \text{for } \tau \geq c \end{cases} \tag{3.3}$$

so as to specify a lower bound c for the possible values of τ. However, a more suitable choice in this case (see BAD, 1987a) is to simply take h to be g. The resulting prior density:

$$\xi_2(\tau) \propto g(\tau)[1 - G(\tau)]^a \qquad \text{for } \tau \leq b \tag{3.4}$$

is proper for all values of a and the Bayesian analysis is straightforward.

It should be emphasized that it is not the particular form of (3.1) which makes the posterior improper. As a matter of fact, *any* improper density $\xi_u(\tau)$ for which the area under the lower tail is infinite will result in posterior distributions that are improper for all n and all $\boldsymbol{x} = (x_1, x_2, \ldots, x_n)$. Moreover, if $\xi(\tau)$ is a proper prior for which exactly the first k moments exist ($k = 0, 1, \ldots$) and for which higher moments do not exist because the integral $\int_{-\infty}^{c} |\tau|^{k+1} \xi(\tau) d\tau$ is infinite for all c, then the corresponding posterior distribution $\xi(\tau|\boldsymbol{x})$ will also have exactly k moments for all n and all $\boldsymbol{x}$ (see BAD, 1990).

In the example just presented, one might say that data have little influence on the posterior distribution, in the sense that they cannot make proper an improper prior. Selection models also provide examples of the opposite situation in which, under conjugate analysis, the posterior distribution is not *at all* influenced by the hyperparameters of the prior after just a finite number of observations are obtained (see BAD, 1990).

Still another way in which conjugate prior distributions can work poorly with selection models is by exhibiting an inappropriate dependence on the experiment to be performed. Assume, for example, that the underlying density $g(x|\theta)$ is a member of an exponential family so that

$$g(x|\theta) = r(x)s(\theta) \exp\{u(x)v(\theta)\}, \tag{3.5}$$

and that the value x is observed only if $x \geq \tau$, where τ is a known value. In this case, the density $f(x|\theta)$ is given by (2.4), and it follows that a conjugate prior density for θ based directly on the form of the likelihood function is

$$\xi(\theta) \propto \frac{[s(\theta)]^a \exp\{bv(\theta)\}}{[1 - G(\tau|\theta)]^c}, \tag{3.6}$$

where the hyperparameters a, b and c are specified constants. Although the form of any conjugate prior distribution is always related to the experiment to be performed, the form in (3.6) seems especially unsuitable because of its explicit dependence on the value of τ. It is true that, in some cases, τ and θ are related, as in the example of sampling from historical records where τ was the minimum required height to join the army; obviously τ should be assumed to be related to the mean height of the population. In most problems, however, the *fixed* value of τ conveys no information about θ: for example, it would usually be unsuitable to assume, in the astronomy example, that the mean number of stars in regions of interest is related to the sensitivity of the observing instrument.

It seems more appropriate in these problems to use a prior distribution that is conjugate with respect to the unrestricted model, which is equivalent to using the value $c = 0$ in (3.6). With this choice we have eliminated the dependence of the prior distribution on τ without unduly complicating the statistical analysis.

4. SELECTION MODELS FOR SIGNIFICANT RESULTS

As mentioned in Section 2, an important example of selection models with known value of τ occurs when only experimental results that are found to be "statistically significant" are published in the scientific literature. This situation is very common both because the editors of some journals encourage the publication of articles in which statistical significance has been obtained, and because many experimenters themselves regard their results as being useless unless the results are statistically significant and will not even submit them for publication.

The unfortunate effects that these policies can have on scientific learning have been discussed by several authors and some references are given in BAD (1987a, 1991a). The careful modeling of publication bias should be especially important when a meta–analysis (combining the results from different experiments) is carried out, since meta–analyses are usually based on published experiments. Unfortunately, this is seldom the case. An interesting model for meta–analysis can be found in Iyengar and Greenhouse (1988).

To present our discussion of this problem, we will use a very simple (unrealistically simple) example so as to highlight the main points we would like to stress. Real situations are seldom so clear–cut and the effects of publication bias therein might be less dramatic than those we will see here. Our framework will be that of one–sided hypothesis testing for the mean of a normal distribution with known variance, and we will restrict ourselves to the analysis of a single published significant result.

Assume that independent experiments are carried out by the same or different experimenters around the world. In each of them a random sample $y_1, y_2, \ldots, y_m$ of size m is taken from a normal distribution with unknown mean μ and known variance σ^2 and the uniformly most powerful test, at some level α, is used for testing

$$H_0\colon \mu \leq 0 \text{ versus } H_1\colon \mu > 0. \tag{4.1}$$

In this case, the distribution $g(t|\theta)$ of the test statistic $T = \sqrt{m}\ \overline{Y}_m/\sigma$ is normal with mean $\theta = \sqrt{m}\ \mu/\sigma$ and variance 1, where $\overline{Y}_m$ represents, as usual, the sample mean in a given experiment. Assume that the results of one such experiment appear published in some scientific journal rejecting the null hypothesis H_0 and declaring the data significant because they yield a small p–value. Assume also that only experimental results that are found "statistically significant" get published.

As readers of the journal, we will not learn the results of this experiment unless they lead to the rejection of H_0; that is, unless $T \geq \Phi^{-1}(1-\alpha)$. In this situation, the density of the T that we will actually get to observe is not $g(t|\theta)$, but is given by the selection model:

$$f(t|\theta) = \frac{\phi(t-\theta)}{1-\Phi(\tau-\theta)} \qquad \text{for } t \geq \tau, \tag{4.2}$$

and $f(t|\theta) = 0$ for $t < \tau$, where ϕ and Φ are the p.d.f. and c.d.f. respectively, of the standard normal distribution, and $\tau = \Phi^{-1}(1-\alpha)$.

To see how a "significant" value of T when properly analyzed can support the null hypothesis H_0, it is enlightening to calculate the maximum likelihood estimate for θ. The values of $\hat{\theta}$ for $\alpha = 0.01$, 0.05, and 0.10 and various observed values of t are given in Table 1. Of course, the most widely used criterion for statistical significance is $\alpha = 0.05$. In Table 1 we also give the p–values corresponding to the observed values of t as they would be reported in the literature, that is, calculated from the standard normal distribution. Since only "significant" values of T can be observed, all the p–values must be less than α.

	$\alpha = 0.01$		$\alpha = 0.05$		$\alpha = 0.10$	
$\hat{\theta}$	t	p	t	p	t	p
3	3.424	0.0003	3.175	0.0008	3.095	0.0010
2	3.017	0.0013	2.586	0.0049	2.403	0.0081
1	2.792	0.0026	2.249	0.0124	1.985	0.0236
0	2.665	0.0038	2.063	0.0195	1.755	0.0397
−1	2.588	0.0048	1.955	0.0253	1.625	0.0526
−2	2.538	0.0056	1.888	0.0305	1.546	0.0611
−3	2.503	0.0062	1.844	0.0326	1.496	0.0675
−5	2.453	0.0071	1.789	0.0368	1.434	0.0759

Table 1. *Maximum likelihood estimates of θ for various observed values t and p.*

Since $\hat{\mu} = \sigma\hat{\theta}/m^{1/2}$, it follows from (4.1) that negative values of $\hat{\theta}$ support the null hypothesis H_0. The basic conclusion to be drawn from the discussion in this section is that even observed values of t that appear to be highly significant can yield maximum likelihood estimates that are very negative and give strong support to H_0. For example, it can be seen from Table 1 that when $\alpha = 0.05$, a p–value as small as 0.033 yields $\hat{\theta} = -3$, an estimate that is at least 3 standard deviations away from the parameter values in H_1. Similar behavior is found for $\alpha = 0.01$ and $\alpha = 0.10$. In fact, as the p–value approaches α in this selection model, $\hat{\theta} \to -\infty$.

The analysis just presented is appropriate when we assume that the experiment was performed repeatedly (not necessarily by the same experimenter) until one significant report was obtained and published. Under different conditions, the analysis should proceed differently. For example, suppose at the other extreme that we know beforehand that there is just a single experimenter who performs the experiment just once. In this case, when we read his or her published report of a significant result, we know that this is the actual outcome of the only experiment that was performed and therefore it should be accepted at face value and analyzed using the density $g(t|\theta)$.

Intermediate conclusions between these two extremes can be obtained by taking into consideration the number N of performed experiments. Problems of this type, dubbed the "file drawer problem" by Rosenthal (1979), have received considerable attention in the literature. Of course, this is only one example of a more general phenomenon that can be present when analyzing selection samples, namely that the way the selection sample was actually produced, that is, the selection mechanism generating the observed sample, can have a decisive influence in the statistical analysis of the data. In BAD (1990), selection mechanisms are considered and discussed in the general scenario of selection models, as well as conditions under which the selection mechanism can be ignored in the analysis of data, either because it does not provide additional information about θ, or because, even if it does, the particular form of the prior distribution makes it ignorable when making inferences about θ. We will content ourselves here with one particular example.

Suppose that we observe n published results $\boldsymbol{t} = (t_1, \ldots, t_n)$, all of them significant, and that we are uncertain about the total number N of experiments that have been performed. Included in the $N - n$ unpublished results there might be both significant and non–significant outcomes. Under the Bayesian approach to this problem, a joint generalized density $p(\boldsymbol{t}, n, N, \theta)$ is specified, which can be factored in the following convenient way:

$$p(\boldsymbol{t}, n, N, \theta) = p(\boldsymbol{t}|n, N, \theta)p(n|N, \theta)p(N|\theta)p(\theta). \tag{4.3}$$

In (4.3), we are using the symbol p to generically denote a density without any implication that it is the same one for all variables. We assume that τ is fixed and known throughout the analysis, and hence it is omitted from the notation.

Next, we make the basic assumption that, for the observed value of n, the observed significant results $t_1, \ldots, t_n$ form a random sample from the selection model given by (4.2). Therefore, we have $p(t|n, N, \theta)$, and

$$p(t|n,\theta) = \prod_{i=1}^{n} f(t_i|\theta) = \frac{\prod_{i=1}^{n} \phi(t_i - \theta)}{[1 - \Phi(\tau - \theta)]^n} \quad \text{for } t_i > \tau \quad (i = 1, \ldots, n). \tag{4.4}$$

The two situations that we previously discussed are special cases of the model (4.3) under the assumption (4.4): In the first situation, it is known beforehand that n will have the fixed value $n = 1$. Thus, (4.3) becomes

$$f(t_1|\theta)p(N|\theta)p(\theta), \tag{4.5}$$

where $f(t_1|\theta)$ is given by (4.2), and the joint density of t_1 and θ reduces to

$$p(t_1,\theta) = f(t_1|\theta) \left[\sum_{N=1}^{\infty} p(N|\theta)\right] p(\theta) = f(t_1|\theta)p(\theta). \tag{4.6}$$

It follows from (4.6) that analysis of this published significant result is based on the selection model $f(t_1|\theta)$, as discussed previously.

In the second situation, it is known beforehand that $N = 1$. In this case, the probability of obtaining $n = 1$ is

$$p(n = 1|N = 1, \theta) = 1 - \Phi(\tau - \theta). \tag{4.7}$$

Hence, from (4.7) and the fact that $p(N = 1|\theta) = 1$, the joint density (4.3) becomes

$$f(t_1|\theta)[1 - \Phi(\tau - \theta)]p(\theta) = g(t_1|\theta)p(\theta) = \phi(t_1 - \theta)p(\theta), \tag{4.8}$$

where $g(t_1|\theta)$ is the original, unrestricted density of T. It follows from (4.8) that the observed significant result t_1 should be analyzed at face value, as common sense suggested.

In the file–drawer problem, none of the two situations above is usually assumed. Instead, it is assumed that there is an unknown number N of performed experiments and subjective knowledge about N is required from the reader in order to properly analyze the significant result. From a classical point of view, a value of N that would reverse the conclusion (that is, that would make the result non–significant) is computed (called the *file–safe sample size*) and the reader is asked to compare that with his or her subjective opinion about N so as to conclude whether or not the significant data supports H_0. This approach can be very misleading (see, e.g., Iyengar and Greenhouse, 1988). From a Bayesian point of view, the reader is asked to specify his or her prior distribution for $N, p(N|\theta, \tau)$, which might depend on θ and τ. Inferences about θ will be based on the posterior distribution:

$$p(\theta|t, n = 1) \propto \phi(t - \theta)p(\theta) \sum_{N=1}^{\infty} N[\Phi(\tau - \theta)]^{N-1} p(N|\theta, \tau). \tag{4.9}$$

It can be seen from (4.9) that the analysis would simply be based on the selection model $f(t|\theta) = \phi(t - \theta)/[1 - \Phi(\tau - \theta)]$ if $p(N|\theta, \tau)$ is such that

$$\sum_{N=1}^{\infty} N[1 - \Phi(\tau - \theta)][\Phi(\tau - \theta)]^{N-1} p(N|\theta, \tau) \tag{4.10}$$

does not depend on θ, as can be shown to be the case when $p(N|\theta, \tau)$ is a Poisson distribution with mean $\lambda/\Phi(\tau - \theta)$, for any fixed λ, or when it is taken to be inversely proportional to N (which results in an improper prior). Otherwise the full analysis based on (4.9) must be carried out. For details, see BAD (1990, 1991a).

5. INFORMATION IN SELECTION MODELS AND WEIGHTED DISTRIBUTIONS

One constant in BAD work on selection models and weighted distributions has been comparison of the information about θ provided by a random sample from the underlying density $g(x|\theta)$ with the information in a selection sample or, more generally, in a random sample from a weighted version $f(x|\theta)$ of $g(x|\theta)$. We will see that a selection sample is *not* necessarily less informative about θ than an unrestricted random sample, and that in some statistical problems we are better off with a biased sample than with a random sample.

In order to compare both types of experiments it is useful to further elaborate the notation employed. We will continue to restrict ourselves to univariate problems. In this section, we will use X to denote a random variable whose distribution is the original, underlying $g(x|\theta)$ and will use Y to denote random variables whose distribution is characterized by some weighted version $f(y|\theta)$ of $g(x|\theta)$, as given in (2.1), where w will be some specific weight function. Also, $\mathcal{E}_X$ will denote the statistical experiment in which the observation X is to be obtained, with $\mathcal{E}_Y$ being similarly defined.

There are many different ways to measure and compare the information in statistical experiments. Probably the most common is to use Fisher information. We shall let $I_X(\theta)$ and $I_Y(\theta)$ denote Fisher information for the experiments $\mathcal{E}_X$ and $\mathcal{E}_Y$, respectively, under the standard regularity conditions. If $I_X(\theta) \geq I_Y(\theta)$ for all values of θ, then the unrestricted experiment $\mathcal{E}_X$ will always yield at least as much information about θ (in terms of Fisher information) as the weighted experiment $\mathcal{E}_Y$. In this case, we write $\mathcal{E}_X \succeq_F \mathcal{E}_Y$. The relation $\mathcal{E}_Y \succeq_F \mathcal{E}_X$ is defined analogously. Comparison of some weighted and unweighted distributions according to this criterion can be found in BAD (1987b) and Patil and Taillie (1987).

In spite of its widespread use in statistics, Fisher information does not have any clear–cut operational interpretation in statistical decision theory. More decision oriented although more restrictive, are the concepts of information based on the theory of comparison of statistical experiments, as developed originally by Blackwell (1951, 1953). The experiment $\mathcal{E}_X$ is said to be *sufficient* for the experiment $\mathcal{E}_Y$, denoted $\mathcal{E}_X \succeq \mathcal{E}_Y$, if there exists a stochastic transformation of X to a random variable $Z(X)$ such that, for each value of θ, the random variables $Z(X)$ and Y have identical distributions. The relation $\mathcal{E}_Y \succeq \mathcal{E}_X$ is defined in an analogous way.

Sufficiency of experiments is a very restrictive ordering, in the sense that it is relatively rare for one experiment to be regarded as more informative than another. Nevertheless, when this does occur, every decision maker would prefer the more informative experiment, regardless of his or her prior distribution and utility function, since the relation $\mathcal{E}_X \succeq \mathcal{E}_Y$ holds if and only if, for every decision problem involving θ and every prior distribution for θ, the expected Bayes risk from $\mathcal{E}_X$ is no greater than that from $\mathcal{E}_Y$.

A somewhat less restrictive partial order that still makes use of the basic concept of sufficiency is based in the notion of pairwise sufficiency. An experiment $\mathcal{E}_X$ is said to be *pairwise sufficient* for the experiment $\mathcal{E}_Y$, denoted $\mathcal{E}_X \succeq_2 \mathcal{E}_Y$, if for *every* pair of values θ_1, θ_2, $\mathcal{E}_X$ is sufficient for $\mathcal{E}_Y$ when the parameter space is restricted to contain just the two values θ_1 and θ_2. The relation $\mathcal{E}_Y \succeq_2 \mathcal{E}_X$ is defined in an analogous way.

Clearly, if $\mathcal{E}_X \succeq \mathcal{E}_Y$ then $\mathcal{E}_X \succeq_2 \mathcal{E}_Y$. Also, the relationship $\mathcal{E}_X \succeq \mathcal{E}_Y$ implies a similar ordering in terms of Fisher information; i.e., if $\mathcal{E}_X \succeq \mathcal{E}_Y$ then $\mathcal{E}_X \succeq_F \mathcal{E}_Y$. However, the converse relations do not necessarily hold. Moreover, since the Fisher information can be obtained from the Kullback–Liebler information by considering pairs of values of θ that are arbitrarily close to each other, it can be shown that, if $\mathcal{E}_X \succeq_2 \mathcal{E}_Y$, then $\mathcal{E}_X \succeq_F \mathcal{E}_Y$.

In general, we will be interested in the information in weighted and random samples, and not only in the information in single observations. Nevertheless, it will be sufficient for us to consider the simple experiments $\mathcal{E}_X$ and $\mathcal{E}_Y$ since it can be shown that, if any of the relations above between $\mathcal{E}_X$ and $\mathcal{E}_Y$ holds, then the same relation holds between the experiment in which a random sample $X_1, \ldots, X_n$ is obtained from the underlying density $g(x|\theta)$ and the experiment in which a random sample $Y_1, \ldots, Y_n$ is obtained from the weighted density $f(y|\theta)$.

We will next compare $\mathcal{E}_X$ and $\mathcal{E}_Y$ for various underlying models $g(x|\theta)$ and various weight functions and will show that the ordering of experiments can be very sensitive to the weight function used.

5.1. *Selection Models*

When comparing selection samples with random samples, several general results concerning Fisher information and pairwise sufficiency are available (see BAD, 1987b, 1989; Bayarri, DeGroot and Goel, 1989). Here we will restrict ourselves to examples of selection from normal populations.

We begin by considering an example in which the experiment based on a selection sample is pairwise sufficient for the analogous experiment based on an unrestricted random sample. Consider a population in which X has a normal distribution with a known mean, which we take to be 0, and an unknown precision θ. (The precision of a normal distribution is the reciprocal of its variance). For $i = 1, 2$, let $\mathcal{E}_i$ denote the experiment in which a single observation y_i is obtained by restricting X to the tails $X \geq \tau_i$ and $X \leq -\tau_i$, where $\tau_i > 0$ is given and fixed. (It should be noted that this experiment is equivalent to one in which X is restricted to the upper tail only.) It can be shown in this case that if $\tau_1 < \tau_2$ then $\mathcal{E}_2 \succeq_2 \mathcal{E}_1$, that is the experiment with a smaller selection set is pairwise sufficient for the experiment with a larger selection set. It follows that a selection sample from the upper tail of a normal distribution with known mean and unknown precision is pairwise sufficient for a selection sample of the same size from the upper tail with a smaller truncation point. Moreover, a selection sample from the upper tail is pairwise sufficient for a random sample from the entire distribution.

It is striking that the reverse relationship holds when the mean is unknown and the precision is known. Assume now that X has a normal distribution with unknown mean θ and known precision. For $i = 1, 2$, let $\mathcal{E}_i$ denote the experiment in which a single observation Y_i is obtained by restricting X to the upper tail $X \geq \tau_i$, where $-\infty < \tau_1 < \tau_2 < \infty$. Here it can be shown that $\mathcal{E}_1 \succeq_2 \mathcal{E}_2$. In particular, by letting $\tau_1 \to -\infty$, it follows that a random sample from the entire normal distribution is pairwise sufficient for a selection sample from the upper tail.

If X has a normal distribution with unknown mean θ and known precision, and the observation Y is obtained by restricting X to the two tails $X \leq \tau_1$ and $X \geq \tau_2$, then it can be shown that neither $\mathcal{E}_X \succeq_F \mathcal{E}_Y$ nor $\mathcal{E}_Y \succeq_F \mathcal{E}_X$, and therefore the stronger relations $\succeq_2$ and $\succeq$ cannot hold either.

5.2. *Other weight functions*

In this subsection we will present some results on sufficiency obtained when various weight functions are applied to standard statistical distributions. These will be used to demonstrate the wide variety of effects that weighting can have in the ordering of experiments.

Assume first that the distribution of X is exponential and that the distribution of Y is a weighted version of it. We will see that all the possibilities in the ordering of the experiments $\mathcal{E}_X$ and $\mathcal{E}_Y$ are attained for particular weight functions $w(x)$. Indeed, if we have exponential weight $w(x) = e^{-ax}(a > 0)$ then $\mathcal{E}_X \succeq \mathcal{E}_Y$, and a random sample is thus preferred to a "weighted" sample. On the other hand, for size–biased weights of the form $w(x) = x^a(a > 0)$ the relation is reversed and we have in fact $\mathcal{E}_Y \succeq \mathcal{E}_X$; notice that, if we are in such statistical situations, we should be very glad of having size–biased samples. Finally, we have equivalent experiments $\mathcal{E}_X \approx \mathcal{E}_Y$ when the weight function is the indicator function of any set of the form $x \geq a$, that is, when Y represents selection from the upper tail.

If the underlying distribution is a Poisson distribution, we can find a very simple family of weight functions that results in different orderings of the original and weighted experiments. Indeed, if the distribution of X is a Poisson distribution with parameter θ and Y is its weighted version with a weight function of the form $w(x) = a^x$ for some specific value of a, then if $a < 1$, a random sample is preferred since $\mathcal{E}_X \succeq \mathcal{E}_Y$, while the relation is reversed when $a > 1$, in which case we have $\mathcal{E}_Y \succeq \mathcal{E}_X$.

Our next example demonstrates that the effect of the *same* weight function on the information about different parameters of a multiparameter distribution can be dramatic. Assume that the distribution of the underlying X is Gamma with shape parameter α and scale parameter β and that the distribution of Y is a generalized size–biased version of it, that is, with $w(x) = x^a$ for some a. Then, if α is known (and β unknown), $\mathcal{E}_Y \succeq \mathcal{E}_X$, while if β is known (and α unknown), $\mathcal{E}_X \succeq \mathcal{E}_Y$.

We finally explore the consequences of using the widely used size–biased weight in the distributions belonging to the "Big 3" families of discrete distributions. Interestingly enough, the effect on each of them is different. Thus, assume that the distribution of X is one of the "Big 3": binomial, Poisson and negative binomial, and that Y is its size–biased version with $w(x) = x$. Then, if the distribution of X is Binomial, random samples are better, since $\mathcal{E}_X \succeq \mathcal{E}_Y$; if the distribution of X is Poisson, then both experiments are equivalent; if the distribution of X is negative binomial, then size–biased samples are better since $\mathcal{E}_Y \succeq \mathcal{E}_X$.

5.3. *The criminal example*

To finish the paper, consider another admittedly unrealistic (but cute) example that shows how what at first might seem to be a very bad sample might in fact provide more information about the parameter of interest than a random sample (that would anyway be impossible to obtain). The problem is to study the total population of criminals by sampling from just the criminals who are in jail. (We cannot, obviously, obtain a random sample from the entire population of criminals.)

We assume that each criminal commits crimes according to a Poisson process with his or her own rate, X, of crimes per year. We also assume that the density of X over the total population of criminals is exponential with parameter θ, that is $g(x|\theta) = \theta e^{-\theta x}$, and that we want to make inferences about θ.

Suppose that each time a crime is committed, there is a fixed probability q that the criminal is caught and sent to jail. We will sample from criminals who have been sent to jail during the year. Then, the distribution $f(x|\theta)$ for the crime rates, X, of criminals who are in jail will be a weighted version of the original $g(x|\theta)$, with the weight function $w(x)$ being simply the probability that a criminal who commits crimes at rate x is sent to jail during the year, that is:

$$\begin{aligned} w(x) &= \text{Pr(criminal is sent to jail during the year)} \\ &= 1 - \sum_{n=0}^{\infty} \text{Pr(commits } n \text{ crimes and does not get caught)} \\ &= 1 - \sum_{n=0}^{\infty} \frac{e^{-x} x^n}{n!} (1-q)^n = 1 - e^{-qx}. \end{aligned} \tag{5.1}$$

Therefore, the distribution of X over the population of criminals who are in jail is given by

$$f(y|\theta) \propto (1 - e^{-qy}) g(y|\theta) = \frac{\theta(\theta+q)}{q} (1 - e^{-qy}) e^{-\theta y}. \tag{5.2}$$

An easy calculation shows that

$$I_X(\theta) = \frac{1}{\theta^2}, \qquad I_Y(\theta) = \frac{1}{\theta^2} + \frac{1}{(\theta+q)^2}. \tag{5.3}$$

Since $I_Y(\theta) > I_X(\theta)$ for all θ, it follows rather surprisingly that, as far as Fisher information is concerned, it is *more* informative to sample from criminals who are in jail than to sample from the entire population of criminals.

Another interesting fact about the family of weights $\{w(y) = (1 - e^{-qy}),\ 0 < q < 1\}$ is worth mentioning. First notice from (5.3) that, as q decreases to 0, $I_Y(\theta)$ increases to $2/\theta^2$. Also, it can be seen from (5.2) that, as $q \to 0$, $f(y|\theta) \to \theta^2 y e^{-\theta y}$, but this is just the ordinary size–biased version ($w(x) = x$) of the exponential density $g(x|\theta)$ and we have seen that, in this case, $\mathcal{E}_Y$ is actually sufficient for $\mathcal{E}_X$. In other words, we have a family of weights for the exponential distribution that provides increasing Fisher information as $q \to 0$ and that in the limit leads to full sufficiency.

REFERENCES

Bayarri, M. J. and DeGroot, M. H. (1987a). Bayesian analysis of selection models. *The Statistician* **36**, 137–146.

Bayarri, M. J. and DeGroot, M. H. (1987b). Information in selection models. *Probability and Bayesian Statistics* (R. Viertl, ed.), New York: Plenum Press, 39–51.

Bayarri, M. J. and DeGroot, M. H. (1988). A Bayesian view of weighted distributions and selection models. *Accelerated Life Testing and Expert's Opinions in Reliability* (C. A. Clarotti and D. V. Lindley, eds.), Amsterdam: North-Holland, 70–82.

Bayarri, M. J. and DeGroot, M. H. (1989). Comparison of experiments with weighted distributions. *Statistical Data Analysis and Inference* (Y. Dodge, ed.), Amsterdam: North-Holland, 185–197.

Bayarri, M. J., and DeGroot, M. H. (1990). Selection models and selection mechanisms. *Bayesian and Likelihood Methods in Statistics and Econometrics* (S. Geisser, J. S. Hodges, S. J. Press and A. Zellner, eds.), Amsterdam: North-Holland, 211–227.

Bayarri, M. J., and DeGroot, M. H. (1991a). The analysis of published significant results. *Rassegna de Metodi Statistici ed Applicazioni* (W. Racugno, ed.), Bologna: Pitagora (in press).

Bayarri, M. J. and DeGroot, M. H. (1991b). A "BAD" view of weighted distributions and selection models. *Tech. Rep.* **91–12**, Department of Statistics, Purdue University.

Bayarri, M. J., DeGroot, M. H., and Goel, P. K. (1989). Truncation, information, and the coefficient of variation. *Contributions to Probability and Statistics* (L. J. Gleser, M. D. Perlman, S. J. Press and A. R. Sampson, eds.), New York: Springer, 412–428.

Blackwell, D. (1951). Comparison of experiments. *Proceedings of the Second Berkeley Symposium on Mathematical Statistics and Probability*, Berkeley, CA: University Press, 93–102.

Blackwell, D. (1953). Equivalent comparison of experiments. *Ann. Math. Statist.* **24**, 265–272.

Fisher, R. A. (1934). The effect of methods of ascertainment upon the estimation of frequencies. *Annals of Eugenics* **6**, 13–25.

Fraser, D. A. S. (1952). Sufficient statistics and selection depending on the parameters. *Ann. Math. Statist.* **23**, 417–425.

Fraser, D. A. S. (1966). Sufficiency for selection models. *Sankhya A* **28**, 329–334.

Iyengar, S. and Greenhouse, J. B. (1988). Selection models and the file drawer problem. *Statist. Sci.* **3**, 109–135.

Patil, G. P. (1984). Studies in statistical ecology involving weighted distributions. *Statistics: Applications and New Directions*, Calcutta: Indian Statistical Institute, 475–503.

Patil, G. P. and Taillie, C. (1987). Weighted distributions and the effects of weight functions on Fisher information. *Tech. Rep.* Department of Statistics, Pennsylvania State University.

Rao, C. R. (1965). On discrete distributions arising out of methods of ascertainment. *Classical and Contagious Discrete Distributions* (G. P. Patil, ed.), Calcutta: Statistical Publishing Society, 320–333.

Rao, C. R. (1985). Weighted distributions arising out of methods of ascertainment: What population does a sample represent? *A Celebration of Statistics: The ISI Centenary Volume* (A. G. Atkinson and S. E. Fienberg, eds.), New York: Springer, 543–569.

Rosenthal, R. (1979). The "file drawer problem" and tolerance for null results. *Psychological Bulletin* **86**, 638–641.

Tukey, J. W. (1949). Sufficiency, truncation and selection. *Annals of Mathematical Statistics* **20**, 309–311.

DISCUSSION

M. DELAMPADY (*University of British Columbia, Canada*)

It is a great honour to comment on this excellent review of the work on weighted distributions by Professors Bayarri and DeGroot (henceforth B&D). I will confine my comments to the work on selection models, an important special case of weighted distributions. B&D consider (i) comparison of selection models, and (ii) inference for selection models. In the following discussion we shall study how these two are related.

Comparison of selection models is done mainly using the concepts of sufficiency and pairwise sufficiency of experiments, and the Fisher information contained in experiments about the parameters of interest. It is claimed that "if an experiment $\mathcal{E}_X$ is sufficient for $\mathcal{E}_Y$, every decision maker would prefer the more informative $\mathcal{E}_X$, regardless of prior and utility." This probably is true most of the time, but if $\mathcal{E}_X$ is more expensive to conduct than $\mathcal{E}_Y$ then since this cost difference is part of the utility function also, the choice of the experiment is not clear. Otherwise sufficiency is a useful concept for comparing experiments. Since sufficiency defines a very strong ordering, B&D study pairwise sufficiency, a weaker but more useful ordering. In what follows we discuss an inference problem whee pairwise sufficiency exists but the choice of experiment is not clear.

Example: Suppose $X|\theta \sim N(0, \theta^{-1})$, so that $\theta > 0$ is the precision. Let Y denote X restricted to the set $\{x : x^2 > \tau^2\}$, where τ^2 is a known constant. B&D show that Y is pairwise sufficient for X. Consider the testing problem, $H_0 : \theta = 1$ (or $|\theta - 1| \leq \varepsilon$) versus $H_1 : \theta \neq 1$ (or $|\theta - 1| > \varepsilon$). Let us compare the inferences obtained from a random sample $X_1, \ldots, X_n$ with those obtained from $Y_1, \ldots, Y_n$.

In this case $g(x|\theta) = (2\pi)^{-1/2}\theta^{1/2}\exp(-\theta x^2/2)$, and $f(x|\theta) = g(x|\theta)/\ [1 - G_1(\tau^2\theta]$, $x^2 \geq \tau^2$, where G_1 denotes the χ^2_1 c.d.f. Consider a prior density $\pi(\theta)$ of the form $\pi_0 I(\theta = 1) + (1 - \pi_0)\pi_1(\theta)I(\theta \neq 1)$, where π_1 is a smooth density. Then the posterior probability of

H_o under the model g given the data $x_1, \ldots, x_n$ is

$$P_g^\pi(H_0|x_1,\ldots,x_n) = \frac{\pi_0 g(x_1,\ldots,x_n|\theta=1)}{\pi_0 g(x_1,\ldots,x_n|\theta=1) + (1-\pi_0)\int_{\theta\neq 1} g(x_1,\ldots,x_n|\theta)\pi_1(\theta)d\theta}$$

$$= \left[1 + \frac{(1-\pi_0)}{\pi_0} B_g^{\pi 1}(H_1 : H_0|x_1,\ldots,x_n)\right]^{-1},$$

where $B_g^{\pi 1}(H_1 : H_0|x_1,\ldots,x_n)$ is the Bayes factor of H_1 relative to H_0 with respect to the conditional prior density π_1. Similarly, the posterior probability of H_0 under the model f given the data $x_1,\ldots,x_n$ is

$$P_f^\pi(H_0|x_1,\ldots,x_n) = \left[1 + \frac{(1-\pi_0)}{\pi_0} B_f^{\pi 1}(H_1 : H_0|x_1,\ldots,x_n)\right]^{-1},$$

where $B_f^{\pi 1}(H_1 : H_0|x_1,\ldots,x_n)$ is the Bayes factor of H_1 relative to H_0 with respect to the conditional prior density π_1, and is given by

$$B_f^{\pi 1} = \frac{(1-G_1(\tau^2))^n \int_{\theta\neq 1}[g(x_1,\ldots,x_n|\theta)/(1-G_1(\tau^2\theta))^n]\pi_1(\theta)d\theta}{g(x_1,\ldots,x_n|\theta=1)}.$$

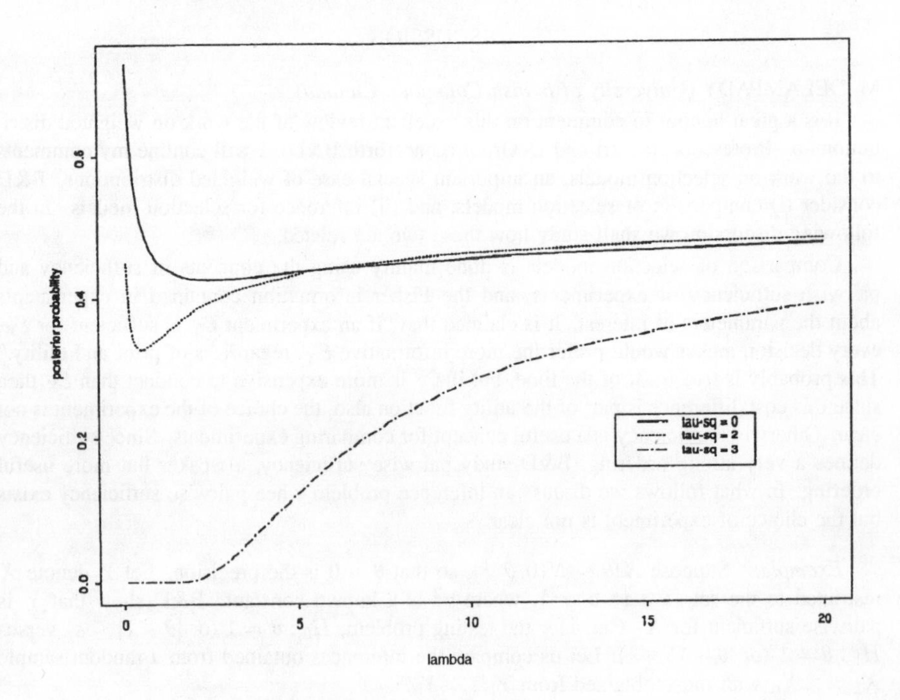

Figure 1. *Posterior Probability of the Null.*

Choice of π_1 and the robust Bayesian view of reporting evidence: As B&D indicate, priors conjugate to $f(\cdot|\theta)$ create serious problems, and are also not natural. Natural choice in this case is Gamma(α, λ) with density $\lambda^\alpha \Gamma(\alpha)^{-1} \exp(-\lambda\theta)\theta^{\alpha-1}$. Since under H_0, θ is 1, we shall consider all Gamma(α, λ) with $\alpha = \lambda$ so that $E^{\pi_1}(\theta) = 1$. Assume now that $x_1, \ldots, x_n$ is such that $s^2 = \sum_{i=1}^n x_i^2$ is some fixed quantile of χ_n^2. Then

$$B_g^{\pi_1} = \exp(s^2/2)\left(1 + \frac{s^2}{2\lambda}\right)^{-\lambda} (\lambda + s^2/2)^{-n/2}\Gamma(\lambda + n/2)/\Gamma(\lambda),$$

and

$$B_f^{\pi_1} = B_g^{\pi_1}(1 - G_1(\tau^2))^n E\left[(1 - G_1(\tau^2\theta))^{-n}\right],$$

where the above expectation is with respect to $\theta \sim \Gamma(\lambda + s^2/2, \lambda + n/2)$.

Consider a specific case where $n = 4$, $s^2 = 9.49$, and $\tau^2 = 0, 2, 3$. Then the behaviour of $B_g^{\pi_1}(\tau^2 = 0)$ and $B_f^{\pi_1}$, or equivalently those of P_g^π and P_f^π (with $\pi_0 = 0.5$) can be seen from Figure 1. For large λ (or $V^{\pi_1}(\theta) = 1/\lambda$ small), P_g^π and P_f^π are close (and close to 0.5), but when λ is small or the prior information is not very precise, the evidence for H_0 does depend heavily on τ^2 and the amount of prior information indicated by λ. In other words, if the choice is between g and f with $\tau^2 = 2$, then f should be chosen when the lowest possible λ is larger than 6, whereas g should be chosen if λ much smaller than 6 are possible. Further, for this value of s^2, one will not choose f with $\tau^2 = 3$ unless the prior information is extremely precise.

J. O. BERGER (*Purdue University, USA*)

The paper is an excellent survey, especially from the perspective of conceptual understanding. The surprising statistical features and relationships of selection models and weighted distributions are clearly explained. My comments refer to two of these features.

Undoubtedly because of my interest in improper priors, I was drawn to the interesting demonstration in Section 3 that improper priors cannot be used for the common selection model in which selection occurs in the upper tail $\{x > \tau\}$, where τ is unknown but the original density, $g(x)$, is known. Indeed, Bayarri states that the "data have little influence on the posterior distribution in the sense that they cannot make proper an improper prior."

This is all true, and is particularly fascinating in light of the fact that, for proper continuous priors, the data have much *more* influence on the posterior distribution than is common. Indeed, for a large sample $x = (x_1, \ldots, x_n)$, it can be shown that the posterior distribution of τ is approximately an exponential distribution, truncated at $x_{\min} = \min\{x_1, \ldots, x_n\}$, with αth quantile

$$x_{\min} + \frac{1}{n}(\log \alpha)(1 - G(x_{\min}))/g(x_{\min}).$$

The point here is that the posterior converges to a point mass at the rate of n^{-1}, rather than the common $n^{-1/2}$, indicating that the data are much more influential than is usual. Yet they cannot compensate for an improper prior!

A second startling feature of selection models that caught my attention was the drastic effect that selection for significant results has on the MLE, $\hat{\theta}$. Since selection for significant results typically arises in testing problems, a natural question is — what is the effect of selection for significant results on posterior probabilities of the hypotheses?

REPLY TO THE DISCUSSION

We thank both Professor Mohan Delampady and Professor James Berger for their kind words and nice discussions. Their comments highlight important aspects barely treated in this paper. We first address the two interesting concerns raised by Professor Delampady.

First, he claims that, even though an experiment is sufficient for another, the statistician might still prefer the latter if the former is very expensive. This is certainly true and similar concerns also exist when that experiment is more difficult or cumbersome to carry out, is unethical, etc. Due to lack of space we could not elaborate further concerning criteria for comparing information in statistical experiments, but sufficiency and pairwise sufficiency are, of course, meaningful only if the utility function to be used is the same no matter what experiment is carried out. Thus, although usual statistical decision problems such as estimation and tests of hypotheses can indeed be accommodated in this formulation, more complex situations in which the experiments have different costs, degrees of difficulty, etc., typically can not be accommodated, and sufficiency and pairwise sufficiency is then of little use. Of course, if the sufficient experiment is the cheaper of the two, it is to be preferred.

Second, Professor Delampady produces a nice example in which, even though pairwise sufficiency exists, which experiment is ultimately better depends on the prior distribution. Pairwise sufficiency is as powerful a criterion as full sufficiency in problems involving two values of the parameter (as in testing two simple hypotheses) or in monotone decision problems. Also, it indicates that the *expected* information, with respect to several criteria (see, e.g., Torgersen 1970, 1972, 1976, and Goel and DeGroot, 1979), is greater in the experiment which is pairwise sufficient than in the other; but, of course, if full sufficiency does not exist, a decision problem and a prior distribution can be found for which the pairwise sufficient experiment is inferior (has greater expected risk). These examples are typically *very* hard to produce. Professor Delampady has produced a nice example in which the *posterior* loss can be greater in the pairwise sufficient experiment. Unfortunately, it does not show anything about *expected* posterior loss and thus we still do not have an example in which the pairwise sufficient experiment is a priori expected to be the worse of the two experiments.

Professor Berger's very interesting comments provide new and fascinating insights into some features of selection models. In his first comment he points out to us that, when selection occurs on the upper tail of a completely specified distribution and if a proper (continuous) prior distribution is used, then the influence of the data on the posterior is greater than usual. This comment renders all the more interesting the fact that data can not make proper improper priors in this case.

His second comment made us further study how, in the selection model of Section 4, the same data can give rise to utterly different inferences when they are analyzed under the original, unrestricted model or under the selection model. The differences in the MLE's were shown in Table 1 and the fact that they differ so much is not a particular feature of the MLE's in these models. Similar patterns also occur with other inferences. For instance, it can immediately be seen from (4.1) and (4.2) that the p–value corresponding to an observed t in the selection model can be computed by simply dividing the p–value obtained under the original, unrestricted model by $1 - \Phi(\tau)$, which is the significance level α. Thus, in the common case of $\alpha = 0.05$, p–values under the selection model are obtained by multiplying the reported p–values by 20.

Professor Berger's question referred to what happens to posterior probabilities of the null hypothesis. It turns out that differences in posterior probabilities of H_0 are typically even more dramatic than differences in MLE's and p–values. Table 1b is an elaboration of Table 1 for the particular case of $\alpha = 0.05$ (so that $\tau = 1.645$). In this table, t is the published value of the test–statistic and it also corresponds to the MLE of θ when t is analyzed under the original model. $\hat{\theta}$ stands for the MLE of θ when t is analyzed under the selection model; p_0 refers to p–values as they would appear published, that is, calculated with respect to the original model; $p_s = 20p_0$ is the p–value under selection. Notice that p_0 would also be the posterior probability of H_0 (under a flat prior for θ) obtained when the original model is used; the column under $\Pr(H_0|t)$ displays the posterior probabilities of H_0 (with a flat prior for θ) under the selection model for the different values of t. Finally, $\Pr(H_0|t)/p_0$ gives the ratio between this posterior probability and the posterior probability of H_0 (p–value here) that would be obtained ignoring the selection effect. (Recall that the ratio between the corresponding p–values is 20 for all values of $t \geq \tau$).

t	$\hat{\theta}$	p_0	p_s	$\Pr(H_0\|t)$	$\Pr(H_0\|t)/p_0$
3.175	3	0.0008	0.0150	0.0326	25.68
2.586	2	0.0049	0.0971	0.2068	27.68
2.249	1	0.0123	0.2451	0.4716	29.19
2.063	0	0.0196	0.3911	0.6727	31.34
1.955	−1	0.0253	0.5958	0.7926	34.40
1.888	−2	0.0295	0.5903	0.8615	38.48
1.844	−3	0.0326	0.6518	0.9021	42.61
1.789	−5	0.0368	0.7361	0.9454	43.45

Table 1b. *Maximum likelihood estimates of θ, for $\alpha = 0.05$.*

It can be seen that, indeed, the most dramatic differences occur for posterior probabilities. This effect is due to the fact that the posterior distribution of θ is *very* flat to the left of its mode (here simply the MLE $\hat{\theta}$) while it goes relatively fast to 0 to the right of the mode. Thus, while the value of t that would make us indifferent between the two hypotheses in terms of the MLE ($\hat{\theta} = 0$) is $t = 2.063$, in order to have $\Pr(H_0|t) = 0.5$ we need a larger value of t, namely $t = 2.220$, for which the corresponding MLE is 0.88, and its published p–value would be 0.0132.

ADDITIONAL REFERENCES IN THE DISCUSSION

Goel, P. K. and DeGroot, M. H. (1979). Comparison of experiments and information measures. *Ann. Statist.* **7**, 1066–1077.

Torgersen, E. N. (1970). Comparison of experiments when the parameter space is finite. *Z. Wahrscheinlichkeitstheorie und Verw. Gebiete* **16**, 219–249.

Torgersen, E. N. (1972). Comparison of translation experiments. *Ann. Math. Statist.* **43**, 1383–1399.

Torgersen, E. N. (1976). Comparison of statistical experiments. *Scandinavian J. Statist.* **3**, 186–208.

BAYESIAN STATISTICS 4, pp. 35–60
J. M. Bernardo, J. O. Berger, A. P. Dawid and A. F. M. Smith, (Eds.)

On the Development of Reference Priors*

JAMES O. BERGER and JOSÉ M. BERNARDO
Purdue University, USA and *Generalitat Valenciana, Spain*

SUMMARY

The paper begins with a general, though idiosyncratic, discussion of noninformative priors. This provides the background for motivating the recent and ongoing elaborations of the reference prior method for developing noninformative priors, a method initiated in Bernardo (1979). Included in this description of the reference prior method is a new condition that has not previously appeared. Motivation for this new condition is found, in part, in the Fraser, Monette, Ng (1985) example. Extensive discussion of the motivation for reference priors and the specific steps in the algorithm are given, with reference to new examples where appropriate. Also, technical issues in implementing the algorithm are discussed.

Keywords: NONINFORMATIVE PRIORS; MULTIPARAMETER PROBLEMS; STEPWISE COMPUTATION; KULLBACK–LIEBLER DIVERGENCE.

1. INTRODUCTION

1.1. *Perspective on Noninformative Priors*

In some sense, Bayesian analysis is a distinct field only because of noninformative priors. This can certainly be argued from a historical perspective, noting that for virtually 200 years — from Bayes (1763) and Laplace (1774, 1812) through Jeffreys (1946, 1961) — Bayesian statistics was essentially based on noninformative priors. Even today, the overwhelming majority of applied Bayesian analyses use noninformative priors, at least in part. Indeed the only proper priors that are commonly used in practice are those in the early stages of hierarchical models, and these can virtually be thought of as part of the model. (Of course, thinking of such hierarchical distributions as priors rather than, say, random effects models is more natural and is inferentially superior.)

On a philosophical level, things are a bit murkier, but one can still argue for the centrality of noninformative priors. Basically, Bayesian analysis with proper priors is not clearly distinct from probability theory. Indeed, there have been a multitude of Bayesian analyses done throughout history that were viewed as simply being probability analyses. Bayesian analysis with noninformative priors typically does not fit within the usual probability calculus, however. Some Bayesians use foundational arguments to attempt to exclude noninformative priors from consideration, but this also is murky. While axiomatic perspectives typically do suggest that priors should be proper, sensible axiomatics do not rule out proper finitely additive distributions, which operationally can be equivalent to noninformative priors: cf.,

* Research supported by the National Science Foundation, Grants DMS8702620 and DMS8923071, and the Ministerio de Educación y Ciencia, Grant PB870607C0201. Professor José M. Bernardo in on leave of absence from the *Departamento de Estadística e I. O., Universidad de Valencia*, Spain.

Cifarelli and Regazzini (1987), Consonni and Veronese (1989), and Heath and Sudderth (1978).

Finally, even from a pragmatic viewpoint, it might pay to strongly associate Bayesian analysis with use of noninformative priors. How often do we hear "I'm not a Bayesian because statistical inference must be objective" or "I use Bayesian analysis if I actually have usable subjective information, but that is very rare." Statements such as these are, of course, contestable, but the rejoinders "Objectivity is a useless pursuit," and "It may be hard, but you always must try to quantify subjective information," are much less effective arguments than "If your statement were true, the best method of inference would nevertheless be Bayesian analysis with noninformative priors."

It is important, of course, to keep a balanced perspective. Thus today it is obviously to the advantage of Bayesians to claim as their own all true probability inference and to promote the use of subjective priors (especially for problems such as testing of precise hypotheses in which there are no remotely sensible objective answers). And it is important for noninformative prior Bayesians to acknowledge that they are treading on "improper" ground, upon which they do not have the automatic coherency protection provided by proper priors. The noninformative prior Bayesian can run afoul of the likelihood principle (see Berger and Wolpert, 1988, for discussion, but see Wasserman, 1991, for a contrary view), marginalization paradoxes (Dawid, Stone, and Zidek, 1973; but see Jaynes, 1980), strong inconsistency or incoherency (cf. Stone, 1971), and can even encounter the disaster of an improper posterior (see Ye and Berger, 1991, for an example.)

In recognition of these dangers, there are two types of safeguards that are typically pursued by noninformative prior Bayesians. The first, which is the subject of this paper, is the development of a method of generating noninformative priors that seems to avoid the potential problems. The second safeguard is to investigate robustness with respect to the prior, possibly by Bayesian sensitivity studies but more commonly by frequentist evaluation of the performance of the noninformative prior in repeated use. This last type of safeguard is obviously controversial and must be used and interpreted with caution, but it has historically been the most effective approach to discriminating among possible noninformative priors. (Note that the perspective of this second safeguard is that of studying a particular, or several, noninformative priors for a given model, and evaluating their sensibility or performance.)

1.2. *Perspective on Reference Priors*

Bernardo (1979) initiated the reference prior approach to development of noninformative priors, following in the tradition of Laplace and Jeffreys. This tradition is the pragmatic tradition that results are most important; the method should work. If examples are found in which the method fails, it should be modified or adjusted to correct the problem. Thus Laplace (1774, 1812) found that, for the problems he encountered, it worked exceptionally well to simply always choose the prior for θ to be the constant $\pi(\theta) = 1$ on the parameter space Θ. For very small sample sizes, however, it was observed that this led to a significant inconsistency, in that the answer could change markedly depending on the choice of parameterization. (A constant prior for one parameter will not typically transform into a constant prior for another).

This led Jeffreys (1946, 1961) to propose the now famous Jeffreys prior,

$$\pi(\theta) = \sqrt{\det(I(\theta))},$$

where $I(\theta)$ is the Fisher information (see (1.3.1)) and "det" stands for determinant. This method is invariant in the sense of yielding properly transformed priors under reparameterization, and has proved to be remarkably successful in one–dimensional problems. Jeffreys himself, however, noticed difficulties with the method when θ is multi–dimensional, and would then provide *ad hoc* modifications to the prior.

Bernardo (1979) sought to remove the need for *ad hoc* modifications by systematically dividing multi–dimensional $\theta = (\theta_1, \theta_2, \ldots, \theta_k)$ into the "parameters of interest" and the "nuisance parameters," developing the noninformative prior in corresponding stages. As with Jeffreys, this approach was based on information concepts, and indeed the approach yielded the Jeffreys prior in usual one–dimensional problems.

Over the subsequent years and scores of applications the reference prior method has been progressively defined and refined. The papers recording the evolutions in the method that are summarized here include Berger and Bernardo (1989, 1992a, 1992b), Berger, Bernardo, and Mendoza (1989), and Ye (1990). It is noteworthy that the primary impetus for refinement has come from examples, especially the continually–being–invented "counterexamples" to noninformative priors. This explains some of the apparent arbitrariness in the details of the current reference prior method; where different choices were possible, it was through extensive study of examples of application that the ambiguity was resolved. This ongoing process is reviewed in this paper, with several previously unpublished conditions and examples being highlighted.

The above should not be construed as an admission that the reference prior method is solely *ad hoc*. Far from it, the method is grounded in a very appealing heuristic which even today is the source of new insight. For instance, the condition (2.2.5) in Section 2.2 has only recently been added to our description of the reference prior method. This condition arose out of study of the delightful Fraser, Monette, Ng (1985) counterexample (discussed in Section 3.2), the resolution of which required us to return to the fundamental heuristic.

1.3. *Perspective on Methods for Deriving Noninformative Priors*

First, it is important to clarify that we are concerned here with *methods* of developing noninformative priors, not noninformative priors themselves. A method takes as input the statistical model (possibly including the design and / or stopping rule) and possibly the actual data, and produces as output a prior distribution. (Ultimately, of course, it is the posterior distribution which is desired; in some situations it might even be possible to directly develop a reference posterior.) Thus the Jeffreys "method" takes the sample density $f(x|\theta)$ for the data $X \in \mathcal{X}$, computes the Fisher information $I(\theta)$, i.e. the $(k \times k)$ matrix with elements

$$I_{ij}(\theta) = -E_\theta \left(\frac{\partial^2}{\partial\theta_i \partial\theta_j} \log f(X|\theta) \right), \tag{1.3.1}$$

with E_θ denoting expectation over X with θ given, and finally produces the prior density

$$\pi(\theta) = \sqrt{\det(I(\theta))}. \tag{1.3.2}$$

In comparing methods of producing noninformative priors, a variety of criteria are involved. The three most important criteria are simplicity, generality, and trustworthiness.

By far the simplest method is to follow Laplace and always choose $\pi(\theta) = 1$. In practice this is, indeed, often quite reasonable, since (as partly argued by Laplace) parameterizations are often chosen to reflect a vague notion of prior uniformity. This simple choice fails on enough problems of interest, however, that a more reliable general method is needed.

On the simplicity scale, the reference prior approach is at the opposite extreme. Indeed, computation of a reference prior is so complex that it typically requires the involvement of a research statistician. Of course, for each statistical model computation of the associated reference priors need be done only once, with the resulting reference priors (or perhaps posteriors) being made available in the literature.

In terms of generality, Laplace's method and the reference prior method are virtually universally applicable. The Jeffreys method is quite universal, but does require existence of $I(\theta)$ and, typically, additional regularity conditions such as asymptotic normality of the model. Other methods vary widely in terms of generality, some applying only in univariate problems, some requiring special group invariant or transformation structures, etc. Our goal has been the development of a universal method.

Trustworthiness of the method is a rather nebulous concept, essentially referring to how often the method yields a noninformative prior with undesirable properties. Undesirable properties include impropriety of the posterior (clearly the worst possibility), inconsistency or incoherency of resulting statistical procedures, lack of invariance to reparameterization, marginalization paradoxes, lack of reasonable coverage probabilities for resulting Bayesian credible sets, and unremovable singularities in the posterior. The best way to gauge the trustworthiness of a method is to try it on the large set of challenging "counterexamples" to noninformative priors that have been developed over the years. In this sense the reference prior method is very trustworthy; it does not yield a bad answer in any of the counterexamples.

Conspicuously absent in this discussion of methods for developing noninformative priors has been the notion of how to define "noninformative." Most methods begin with some attempt at measuring the amount of information in a prior or the amount of influence that the prior has on the answer. One could debate the sensibility or value of each such measure (and, of course, we are supporters of the measure underlying reference priors) but, on the whole, we feel that this is a somewhat tangential issue. No sensible absolute way to define "noninformative" is likely to ever be found, and often the most natural ways give the silliest answers (cf. Berger, Bernardo, and Mendoza, 1989).

Another aspect of this is the debate over the name "noninformative" versus, say, "reference." Many object to the former, feeling that it carries a false promise. Reference priors are sensibly named (see Bernardo, 1979) and less objectionable in this regard. Other names such as the "standard" or "default" prior have been proposed, the idea being that the profession should ultimately agree on a standard default prior for use with each particular model. Trying to change historical nomenclature is, however, generally a waste of time, so we have chosen to continue using "noninformative" to refer to the general area, and "reference" to refer specifically to reference priors.

No attempt is made here to survey the wide variety of methods for deriving a noninformative prior and to evaluate them by the above criteria. Those most similar to the reference prior approach, in the sense of explicitly considering parameters of interest and nuisance parameters separately, include Stein (1985), Tibshirani (1989), and Ghosh and Mukerjee (1992).

2. THE REFERENCE PRIOR METHOD

The reference prior method is presented here, in full detail. Unfortunately, it is notationally quite complex. The reader interested only in the ideas can skip to Section 3. For a description of the algorithm in the much simpler two-parameter case, see Berger and Bernardo (1989).

2.1. *Introduction and Notation*

In Section 2.2, the general reference prior method will be described. This method is typically very hard to implement. For the regular case, in which asymptotic normality of the model holds, a considerable simplification of the algorithm occurs. This simplification is given in Section 2.3, which is a review of Berger and Bernardo (1992a and 1992b).

We assume that the θ_i are separated into m groups of sizes $n_1, n_2, \ldots, n_m$, and that these groups are given by

$$\theta_{(1)} = (\theta_1, \ldots, \theta_{n_1}),\ \theta_{(2)} = (\theta_{n_1+1}, \ldots, \theta_{n_1+n_2}),$$
$$\ldots \theta_{(i)} = (\theta_{N_{i-1}+1}, \ldots, \theta_{N_i}), \ldots, \theta_{(m)} = (\theta_{N_{m-1}+1}, \ldots, \theta_k),$$

where $N_j = \sum_{i=1}^{j} n_i$. Also, define

$$\theta_{[j]} = (\theta_{(1)}, \ldots, \theta_{(j)}) = (\theta_1, \ldots, \theta_{N_j}),$$
$$\theta_{[\sim j]} = (\theta_{(j+1)}, \ldots, \theta_{(m)}) = (\theta_{N_j+1}, \ldots, \theta_k),$$

with the conventions that $\theta_{[\sim 0]} = \theta$ and $\theta_{[0]}$ is vacuous.

We will denote the Kullback–Liebler divergence between two densities g and h on Θ by

$$D(g, h) = \int_{\Theta} g(\theta) \log[g(\theta)/h(\theta)] d\theta. \tag{2.1.1}$$

Finally, let $Z_t = \{X_1, \ldots, X_t\}$ be the random variable that would arise from t conditionally independent replications of the original experiment, so that Z_t has density

$$p(z_t|\theta) = \prod_{i=1}^{t} f(x_i|\theta). \tag{2.1.2}$$

2.2. *The General Case*

The general reference prior method can be described in four steps. Justification and motivation will be given in Section 3.

Step 1. Choose a nested sequence $\{\Theta^\ell\}$ of compact subsets of Θ such that $\bigcup_{\ell=1}^{\infty} \Theta^\ell = \Theta$. (This step is unnecessary if the reference priors turn out to be proper.)

Step 2. Order the coordinates $(\theta_1, \ldots, \theta_k)$ and divide them into the m groups $\theta_{(1)}, \ldots, \theta_{(m)}$. Usually it is best to have $m = k$, and the order should typically be according to inferential importance; in particular, the first parameters should be the parameters of interest.

Step 3. For $j = m, m-1, \ldots, 1$, iteratively compute densities $\pi_j^\ell(\theta_{[\sim(j-1)]}|\theta_{[j-1]})$, using

$$\pi_j^\ell(\theta_{[\sim(j-1)]}|\theta_{[j-1]}) \propto \pi_{j+1}^\ell(\theta_{[\sim j]}|\theta_{[j]}) h_j^\ell(\theta_{(j)}|\theta_{[j-1]}), \tag{2.2.1}$$

where $\pi_{m+1}^\ell \equiv 1$ and h_j^ℓ is computed by the following two steps.

Step 3a: Define $p_t(\theta_{(j)}|\theta_{[j-1]})$ by

$$p_t(\theta_{(j)}|\theta_{[j-1]}) \propto \exp\left\{ \int p(z_t|\theta_{[j]}) \log p(\theta_{(j)}|z_t, \theta_{[j-1]}) dz_t \right\}, \tag{2.2.2}$$

where (using $p(\cdot)$ generically to represent the conditional density of the given variables)

$$\begin{aligned} p(z_t|\theta_{[j]}) &= \int p(z_t|\theta) \pi_{j+1}^\ell(\theta_{[\sim j]}|\theta_{[j]}) d\theta_{[\sim j]}, \\ p(\theta_{(j)}|z_t, \theta_{[j-1]}) &\propto p(z_t|\theta_{[j]}) p_t(\theta_{(j)}|\theta_{[j-1]}). \end{aligned} \tag{2.2.3}$$

Step 3b: Assuming the limit exists, define

$$h_j^\ell(\theta_{(j)}|\theta_{[j-1]}) = \lim_{t\to\infty} p_t(\theta_{(j)}|\theta_{[j-1]}). \tag{2.2.4}$$

Comment: In (2.2.2), p_t is only defined implicitly, since $p(\theta_{(j)}|z_t, \theta_{[j-1]})$ on the right hand side also depends on p_t (see (2.2.3)). In practice, it is thus usually very difficult to compute the p_t and find their limit. In the regular case discussed in the next section, however, this difficulty can be circumvented.

Step 4. Define a reference prior, $\pi(\theta)$, as any prior for which

$$E_\ell^X D(\pi_1^\ell(\theta|X), \pi(\theta|X)) \longrightarrow 0 \text{ as } \ell \to \infty, \tag{2.2.5}$$

where D is defined in 2.1.1 and E_ℓ^X is expectation with respect to

$$p^\ell(x) = \int_\Theta f(x|\theta)\pi_1^\ell(\theta)d\theta$$

(writing $\pi_1^\ell(\theta)$ for $\pi_1^\ell(\theta_{[\sim 0]}|\theta_{[0]})$). Typically one determines $\pi(\theta)$ by the simple relation

$$\pi(\theta) = \lim_{\ell\to\infty} \frac{\pi_1^\ell(\theta)}{\pi_1^\ell(\theta^*)}, \tag{2.2.6}$$

where θ^* is any fixed point in Θ with positive density for all π_1^ℓ, and then verifies that (2.2.5) is satisfied.

Comment: Note that (2.2.5) really defines a *reference posterior*; we convert to a reference prior mainly for pedagogical reasons.

2.3. *The Regular Case*

If the model is regular, in the sense that the replicated $p(z_t|\theta)$ is asymptotically normal, then Step 3 in Section 2.2 can be done in an explicit fashion. The following notation is needed, where $I(\theta)$ is the Fisher information matrix with elements given by (1.3.1) and $S(\theta) = (I(\theta))^{-1}$. Often, we will write just I and S for these matrices. Write S as

$$S = \begin{pmatrix} A_{11} & A_{21}^t & \dots & A_{m1}^t \\ A_{21} & A_{22} & \dots & A_{m2}^t \\ \vdots & \vdots & \ddots & \vdots \\ A_{m1} & A_{m2} & \dots & A_{mm} \end{pmatrix}$$

so that A_{ij} is $(n_i \times n_j)$, and define

$$S_j \equiv \text{ upper left } (N_j \times N_j) \text{ corner of } S, \text{ with } S_m \equiv S, \text{ and } H_j \equiv S_j^{-1}.$$

Then the matrices

$$h_j \equiv \text{ lower right } (n_j \times n_j) \text{ corner of } H_j, j = 1, \dots, m$$

will be of central importance. Note that $h_1 \equiv H_1 \equiv A_{11}^{-1}$ and, if S is a block diagonal matrix (i.e., $A_{ij} \equiv 0$ for all $i \neq j$), then $h_j \equiv A_{jj}^{-1}, j = 1, \dots, m$. Finally, if $\Theta^* \subset \Theta$, we will define

$$\Theta^*(\theta_{[j]}) = \{\theta_{(j+1)}: (\theta_{[j]}, \theta_{(j+1)}, \theta_{[\sim(j+1)]}) \in \Theta^* \text{ for some } \theta_{[\sim(j+1)]}\}. \tag{2.3.1}$$

We will use the common symbols $|A|$ = determinant of A, and $1_\Omega(y)$ equals 1 if $y \in \Omega$, 0 otherwise.

Step 3 from Section 2.2 can, in the regular case, by replaced by the following, which is essentially taken from Berger and Bernardo (1992b).

Step 3′: To start, define

$$\pi_m^l(\theta_{[\sim(m-1)]}|\theta_{[m-1]}) = \pi_m^l(\theta_{(m)}|\theta_{[m-1]}) = \frac{|h_m(\theta)|^{1/2} 1_{\Theta^l(\theta_{[m-1]})}(\theta_{(m)})}{\int_{\Theta^l(\theta_{[m-1]})} |h_m(\theta)|^{1/2} d\theta_{(m)}}. \tag{2.3.2}$$

For $j = m-1, m-2, \ldots, 1$, define

$$\pi_j^l(\theta_{[\sim(j-1)]}|\theta_{[j-1]}) = \frac{\pi_{j+1}^l(\theta_{[\sim j]}|\theta_{[j]}) \exp\{\frac{1}{2}E_j^l[(\log|h_j(\theta)|)|\theta_{[j]}\} 1_{\Theta^l(\theta_{[j-1]})}(\theta_{(j)})}{\int_{\Theta^l(\theta_{[j-1]})} \exp\{\frac{1}{2}E_j^l[(\log|h_j(\theta)|)|\theta_{[j]}\} d\theta_{(j)}} \tag{2.3.3}$$

where

$$E_j^l[g(\theta)|\theta_{[j]}] = \int_{\{\theta_{[\sim j]}:(\theta_{[j]},\theta_{[\sim j]})\in\Theta^l\}} g(\theta)\pi_{j+1}^l(\theta_{[\sim j]}|\theta_{[j]}) d\theta_{[\sim j]}. \tag{2.3.4}$$

The calculation of the m–group reference prior is greatly simplified under the condition

$$|h_j(\theta)| \text{ depends only on } \theta_{[j]}, \text{ for } j = 1, \ldots, m. \tag{2.3.5}$$

Then (see Lemma 2.2.1 in Berger and Bernardo, 1992b)

$$\pi^l(\theta) = \left(\prod_{i=1}^m \frac{|h_i(\theta)|^{1/2}}{\int_{\Theta^l(\theta_{[i-1]})} |h_i(\theta)|^{1/2} d\theta_{(i)}} \right) 1_{\Theta^l}(\theta). \tag{2.3.6}$$

3. MOTIVATION FOR THE REFERENCE PRIOR METHOD

3.1. *Information and Replication*

For simplicity, suppose there is a single parameter θ with a compact Θ (or that we are operating on the compact $\Theta^l \subset \Theta$). Suppose that it is desired to define a noninformative prior, $\pi(\theta)$, as that prior which "maximizes the amount of information about θ provided by the data, x." The most natural measure of the expected information about θ provided by X, when π is the prior distribution, is (Shannon, 1948; Lindley, 1956) $I^\theta = E^X D(\pi(\theta|X),\ \pi(\theta))$, where D is the Kullback–Liebler divergence defined in (2.1.1) and E^X stands for expectation with respect to the marginal density of X, $p(x) = \int_\Theta f(x|\theta)\pi(\theta)d\theta$.

Unfortunately, basing the analysis on I^θ is not very satisfactory, as is discussed in Berger, Bernardo, and Mendoza (1989). Indeed, it is shown therein that the $\pi(\theta)$ which maximizes I^θ (possibly with θ restricted to the compact Θ^ℓ) is typically a *discrete* distribution, even when Θ is, say, a connected subset of Euclidean space. Clearly such a $\pi(\theta)$ would be a very unappealing noninformative prior.

Bernardo (1979) considered a variant of this approach, defining

$$I_t^\theta = E^{Z_t} D(\pi(\theta|Z_t),\ \pi(\theta)), \tag{3.1.2}$$

where Z_t consists of t replicates of X as discussed in Section 2.1. The underlying idea is that, as $t \to \infty$, Z_t will typically provide perfect information about θ, in which case $I_\infty^\theta = \lim_{t\to\infty} I_t^\theta$ can be thought of as the missing information about θ when π describes the initial state of knowledge. Thus the π maximizing I_∞^θ could reasonably be called "least informative." Unfortunately, it is typically the case that I_∞^θ is infinite for almost all π, so that this approach also does not work. However, it suggests finding, for each t, the prior π_t which maximizes I_t^θ, and then passing to a limit in t. Using a variational argument it can be shown, under certain conditions, that π_t satisfies

$$\pi_t(\theta) \propto \exp\left\{\int p(z_t|\theta)\log\pi(\theta|z_t)dz_t\right\}. \tag{3.1.3}$$

This equation, reproduced in (2.2.2) for the multiparameter case, is the heart of the reference prior algorithm, and (2.2.4) defines the limit in t.

As observed in Section 2.2, (3.1.3) only defines π_t implicitly. However, as $t \to \infty$, both $p(z_t|\theta)$ and $\pi(\theta|z_t)$ will typically converge to their asymptotic distributions, and (3.1.3) will become an explicit equation. For instance, in the regular case of asymptotic normality of the posterior, it can be shown (cf. Bernardo, 1979, Berger and Bernardo, 1992a) that, for large t, $\pi_t(\theta)$ is approximately proportional to $\sqrt{I(\theta)}$, which is thus the reference prior.

For the case of two parameters, $\theta = (\theta_1, \theta_2)$, with $m = 2$ stages to be used in Section 2.3, the argument proceeds by first determining $\pi_2(\theta_2|\theta_1)$, the conditional reference prior for θ_2 assuming that θ_1 is given. This is done exactly as in the previous univariate argument, and results in the analogue of (2.3.2). The idea is then to use $\pi_2(\theta_2|\theta_1)$ to integrate θ_2 out of the model, leaving a marginal model $p^*(z_t|\theta_1)$, for which a reference prior $\pi(\theta_1)$ can (as $t \to \infty$) be found. The overall reference prior on Θ is then $\pi_1(\theta) \propto \pi_2(\theta_2|\theta_1)\pi(\theta_1)$, which is the analogue of (2.3.3); the expression for $\pi(\theta_1)$ in (2.3.3) follows from another asymptotic argument (cf., Bernardo, 1979, Berger and Bernardo, 1992a).

Extension to more than two groupings and multi–dimensional groupings is straightforward. The result is the algorithm described in Section 2.3.

3.2. *Compact Θ^ℓ and Condition (2.2.5)*

In Berger, Bernardo, and Mendoza (1989) it was shown that, for noncompact Θ, there typically exist priors for which I_t^θ in (3.1.2) is infinite, making useless any attempt to define "least informative prior" directly through I_t^θ. The most direct way to circumvent the problem is to operate on compact Θ^ℓ, passing to the limit as $\Theta^\ell \to \Theta$. The issue, then, is how to choose the Θ^ℓ. Usually the choice does not matter, but sometimes it does (cf., Berger and Bernardo, 1989 and 1992a). And even when the choice does matter, it seems to require quite pathological choices of Θ^ℓ to achieve different results.

Choosing the Θ^ℓ to be natural sets in the original parameterization has always worked well in our experience. Indeed, the way we think of the Θ^ℓ is that there is some large compact set on which we are really noninformative, but we are unable to specify the size of this set. We might, however, be able to specify a shape, Ω, for this set, and would then choose $\Theta^\ell = \ell\Omega \cap \Theta$, where $\ell\Omega$ consists of all points in Ω multiplied by ℓ.

Condition (2.2.5) is a new qualification that we have added to the reference prior method. The motivation for this condition is that the pointwise convergence in (2.2.6), that we had previously used in defining the method, does not necessarily imply convergence in an information sense, which is the basis of the reference prior method. Note that (2.2.5) is precisely convergence in the information measure defined by (3.1.1).

Because this is a new condition in the reference prior method, we present two examples, one in which the condition is satisfied and one in which it is not.

Example 1. Suppose $\mathcal{X} = \Theta = (-\infty, \infty)$ and X given θ is $\mathcal{N}(\theta, 1)$. Define $\Theta^\ell = [-\ell, \ell]$. It is easy to apply the reference prior method here, obtaining

$$\pi_1^\ell(\theta) = \frac{1}{2\ell} \text{ on } \Theta^\ell, \quad \pi_1^\ell(\theta|x) = \frac{f(x|\theta)}{[\Phi(\ell - x) - \Phi(-\ell - x)]} \text{ on } \Theta^\ell,$$
$$\pi(\theta) = 1, \quad \pi(\theta|x) = f(x|\theta),$$

and $p^\ell(x) = \int f(x|\theta)\pi_1^\ell(\theta)d\theta = [\Phi(\ell - x) - \Phi(-\ell - x)]/(2\ell)$, where Φ denotes the standard normal c.d.f. Thus

$$D(\pi_1^\ell(\theta|x),\ \pi(\theta|x)) = \int \pi_1^\ell(\theta|x) \log \frac{\pi_1^\ell(\theta|x)}{\pi(\theta|x)} d\theta = -\log([\Phi(\ell - x) - \Phi(-\ell - x)]),$$

and

$$\begin{aligned} E_\ell^X D(\pi_1^\ell(\theta|X),\ \pi(\theta|X)) &= \int p^\ell(x) D(\pi_1^\ell(\theta|x),\ \pi(\theta|x))dx \\ &= -\frac{1}{2\ell}\int_{-\infty}^{\infty} [\Phi(\ell - x) - \Phi(-\ell - x)] \log([\Phi(\ell - x) - \Phi(-\ell - x)])dx \\ &= -\int_1^{\infty} [\Phi(y\ell) - \Phi((y-2)\ell)] \log([\Phi(y\ell) - \Phi((y-2)\ell)])dy, \end{aligned}$$

the last step using symmetry and making the transformation $y = (\ell - x)/\ell$. Break this integral into $\int_1^3 + \int_3^\infty$. Since $-v \log v \le e^{-1}$ for $0 \le v \le 1$, the dominated convergence theorem can be applied to the first integral to show that it converges to 0 as $\ell \to \infty$. For the second integral, the inequality

$$1 - \tfrac{0.5}{v} e^{-\frac{1}{2}v^2} \le \Phi(v) \le 1 - \tfrac{0.3}{v} e^{-\frac{1}{2}v^2}$$

for large v can be used to prove convergence to 0 as $\ell \to \infty$. Hence Condition 2.2.5 is satisfied. ◁

Example 2. Fraser, Monette, and Ng (1985) considered a discrete problem with $\mathcal{X} = \Theta = \{1, 2, 3, \ldots\}$ and

$$f(x|\theta) = \tfrac{1}{3} \text{ for } x \in \{[\tfrac{\theta}{2}],\ 2\theta,\ 2\theta + 1\},$$

with $[v]$ denoting the integer part of v (and $[\frac{1}{2}]$ separately defined as 1). Note that, when x is observed, θ must lie in $\{[\frac{x}{2}],\ 2x,\ 2x + 1\}$, and that the likelihood function is constant over this set. It is immediate that, if one used the noninformative prior $\pi(\theta) = 1$, then

$$\pi(\theta|x) = \tfrac{1}{3} \text{ for } \theta \in \{[\tfrac{x}{2}],\ 2x,\ 2x + 1\}. \tag{3.2.1}$$

This is a very unsatisfactory answer, as discussed in Fraser, Monette, and Ng (1985). As a simple example of this inadequacy, consider the credible set $C(x) = \{2x,\ 2x + 1\}$, which according to (3.2.1) would have probability 2/3 of containing θ for each x. But it is easy to check that the frequentist coverage probability of $C(x)$, considered as a confidence set, is $P_\theta(C(X) \text{ contains } \theta) = \frac{1}{3}$ for all θ. This is an example of "strong inconsistency" (see Stone (1971) for other examples) and indicates a serious problem with the noninformative prior. For later discussion, it is interesting to note that the noninformative prior $\pi(\theta) = \theta^{-1}$ performs perfectly satisfactorily here, resulting in posterior probabilities and coverage probabilities that are in essential agreement. The prior of Rissanen (1983) is also fine here.

Now, to apply the reference prior method to this problem one must first choose compact subsets Θ^ℓ. Clearly any such sets will here be finite sets, and it can easily be shown that the $\pi_1^\ell(\theta)$ must be constant on finite sets. If now one attempted to pass to the limit in (2.2.6), the result would be the unsatisfactory $\pi(\theta) = 1$.

This turns out, however, to be a situation in which the limit from (2.2.6) violates (2.2.5). To see this take, for instance, the Θ^ℓ to be $\Theta^\ell = \{1, 2, \ldots, 2\ell\}$. As previously mentioned, $\pi_1^\ell(\theta)$ then becomes uniform on Θ^ℓ, so that (3.2.1) is modified to be

$$\pi_1^\ell(\theta|x) = \begin{cases} \frac{1}{3} \text{ for } \theta \in \{[\frac{x}{2}], 2x, 2x+1\} & \text{if } x < \ell \\ \frac{1}{2} \text{ for } \theta \in \{[\frac{x}{2}], 2x\} & \text{if } x = \ell \\ 1 \text{ for } \theta = [\frac{x}{2}] & \text{if } \ell < x \leq 4\ell + 1 \\ \text{nonexistent} & \text{if } 4\ell + 1 < x. \end{cases}$$

Also, it is easy to see that

$$p^\ell(x) = \sum_{\theta=1}^{\infty} f(x|\theta)\pi_1^\ell(\theta) = \begin{cases} 1/\ell & \text{if } x < \ell \\ 2/(3\ell) & \text{if } x = \ell \\ 1/(3\ell) & \text{if } \ell < x \leq 4\ell + 1 \\ 0 & \text{if } 4\ell + 1 < x. \end{cases}$$

An easy calculation then yields

$$\begin{aligned} E_\ell^X D(\pi_1^\ell(\theta|X),\, \pi(\theta|X)) &= \sum_{x=1}^{\infty} p^\ell(x) \sum_{\theta=1}^{\infty} \pi_1^\ell(\theta|x) \log[\pi_1^\ell(\theta|x)/\pi(\theta|x)] \\ &= \frac{2}{3\ell} \log\left(\frac{3}{2}\right) + \frac{(3\ell+1)}{3\ell} \log(3) \longrightarrow \log(3) \text{ as } \ell \longrightarrow \infty, \end{aligned} \tag{3.2.2}$$

so that (2.2.5) is violated.

At this point, all that can be concluded is that a reference prior, as we have defined it, does not exist. There is a fascinating hint, however, that our approach of approximating by compact sets and passing to a limit in "information divergence" may be too crude in this situation. The hint arises from consideration of priors $\pi(\theta) \propto \theta^{-\alpha}$. Repeating the computation done earlier for $\alpha = 0$ yields the interesting fact that the analogue of (3.2.2) does not converge to 0 for $\alpha < 1$ but does converge to 0 for $\alpha = 1$. This suggests that a more clever truncation or way of looking at the truncated problems would yield $\pi(\theta) \propto \theta^{-1}$ as the reference prior (which, as mentioned earlier, is perfectly satisfactory), but we have been unable to devise such a formulation. ◁

We have concentrated on condition (2.2.5) here because this is the first discussion of it in print. Our feeling, however, is that it would be highly unusual for $\pi(\theta)$, defined by (2.2.6), to lead to a violation of (2.2.5). Hence we hesitate to recommend routine verification of the condition, unless there is reason to suspect some pathology.

As one final comment, the need to use (2.2.5) rather than (2.2.6) to define a limit in ℓ suggests that an analogous condition might be needed to replace the pointwise limit in t in (2.2.4). As we have no examples of the necessity of such, however, we have stayed with the simple (2.2.4).

3.3. *Parameters of Interest and Stepwise Computation*

As mentioned in Section 1.2, the separation of θ into parameters of interest and nuisance parameters has been a cornerstone of the reference prior method. In the notation of Sections 2.1 and 2.2, θ would be divided into $m = 2$ groups, with $\theta_{(1)}$ being the parameters of interest and $\theta_{(2)}$ being the nuisance parameters. We begin the discussion of this with a historical example, that will subsequently be put to a new use.

Example 3. Neyman and Scott (1948) introduced an example that has since become a standard test for all new methods of inference. The model consists of $2n$ independent observations,

$$X_{ij} \sim \mathcal{N}(\mu_i, \sigma^2),\ i = 1, \ldots, n,\ j = 1, 2.$$

Reduction to sufficient statistics $X = (\overline{X}_1, \ldots, \overline{X}_n, S^2)$, where $\overline{X}_i = (X_{i1} + X_{i2})/2$ and $S^2 = \sum_{i=1}^{n}\sum_{j=1}^{2}(X_{ij} - \overline{X}_i)^2$, and use of the prior $\pi(\mu_1, \ldots, \mu_n, \sigma) = \sigma^{-\alpha}$ yields

$$\pi(\mu_1, \ldots, \mu_n, \sigma | x) \propto \frac{1}{\sigma^{(2n+\alpha)}} \exp\left\{ -\frac{1}{2\sigma^2}\left[s^2 + 2\sum_{i=1}^{n}(\overline{x}_i - \mu_i)^2 \right] \right\}, \tag{3.3.1}$$

for which the posterior mean of σ^2 is $E[\sigma^2|x] = s^2/(n + \alpha - 3)$.

The original interest in this example, from a noninformative prior perspective, is that the unmodified Jeffreys prior is $\pi(\mu_1, \ldots, \mu_n, \sigma) = \sqrt{\det I} \propto \sigma^{-(n+1)}$, leading to a posterior mean for σ^2 of $E[\sigma^2|x] = s^2/(2n - 2)$. This would be inconsistent as $n \to \infty$, since it can be shown that $S^2/n \to \sigma^2$ with probability one (frequentist) so that $S^2/(2n - 2) \to \sigma^2/2$.

Bernardo (1979) and Jeffreys (for related problems) overcame this difficulty by separately dealing with $\theta_{(1)} = \sigma$ and $\theta_{(2)} = (\mu_1, \ldots, \mu_n)$. To apply the reference prior algorithm to these two groups, compute $I(\theta)$ and write it as

$$I(\theta) = \begin{pmatrix} I_{11}(\theta) & 0 \\ 0 & I^*(\theta) \end{pmatrix}, \tag{3.3.2}$$

where $I_{11}(\theta) = 8n/\sigma^2$ and $I^*(\theta) = (2/\sigma^2)I_{(n-1)\times(n-1)}$. Computation yields $|h_1(\theta)| = 8n/\sigma^2$ and $|h_2(\theta)| = 2^n/\sigma^{2n}$, so that condition (2.3.5) is satisfied. Choosing $\Theta^\ell = (\ell^{-1}, \ell) \times (-\ell, \ell) \times \ldots \times (-\ell, \ell)$ (virtually any choice would give the same answer here), (2.3.6) can thus be used to yield, on Θ^ℓ,

$$\pi^\ell(\theta) = \frac{\sqrt{8n/\sigma^2}}{\int_{\ell^{-1}}^{\ell}\sqrt{8n/\sigma^2}d\sigma} \cdot \frac{\sqrt{2n/\sigma^{2n}}}{\int_{-\ell}^{\ell}\cdots\int_{-\ell}^{\ell}\sqrt{2n/\sigma^{2n}}d\mu_1\ldots d\mu_n} = k_\ell/\sigma, \tag{3.3.3}$$

where k_ℓ is a constant. Finally, applying (2.2.6) (verification of (2.2.5) is rather tedious here), yields $\pi(\theta) = 1/\sigma$.

This reference prior is perfectly satisfactory, yielding a posterior for which the posterior mean is the very sensible $s^2/(n - 2)$. Thus if σ^2 (or σ) is the parameter of interest with $(\mu_1, \ldots, \mu_n)$ being nuisance parameters, all is well with the reference prior algorithm.

Unfortunately, this simple method of grouping does not always work. Suppose, for instance, that $\theta_{(1)} = \mu_1$ and $\theta_{(2)} = (\mu_2, \ldots, \mu_n, \sigma)$, i.e., that μ_1 is the parameter of interest with the rest being nuisance parameters. Now, $I(\theta)$ is as in (3.3.2) but with $I_{11}(\theta) = 2/\sigma^2$ and $I^*(\theta) = \text{diag}\{\frac{2}{\sigma^2}, \ldots, \frac{2}{\sigma^2}, \frac{8n}{\sigma^2}\}$. Thus $h_1(\theta) = 2/\sigma^2$ and $h_2(\theta) = n2^{(n+2)}/\sigma^{2n}$. Define

$\Theta^\ell = (-\ell, \ell) \times \Theta^*$, where $\Theta^* = (-\ell, \ell) \times \ldots \times (-\ell, \ell) \times (\ell^{-1}, \ell)$. The start of the iteration for the reference prior yields (see (2.3.2))

$$\pi_2^\ell(\theta_{(2)}|\theta_{(1)}) = \frac{\sqrt{n2^{(n+2)}/\sigma^{2n}}}{\int_{\ell^{-1}}^{\ell}\int_{-\ell}^{\ell}\cdots\int_{-\ell}^{\ell}\sqrt{n2^{(n+2)}/\sigma^{2n}}d\mu_2\ldots d\mu_n d\sigma} 1_{\Theta^*}(\theta_{(2)}) = \frac{k_\ell}{\sigma^n} 1_{\Theta^*}(\theta_{(2)}).$$

Since $h_1(\theta)$ does not depend on $\theta_{(1)} = \mu_1$, it is easy to see that (2.3.3) becomes $\pi_1^\ell(\theta) = k_\ell \sigma^{-n} 1_{\Theta^\ell}(\theta)$. Passing to the limit in ℓ results in the reference prior $\pi(\theta) = 1/\sigma^n$.

For this prior, a standard Bayesian computation yields that the marginal posterior for μ_1 given x is a t–distribution with $(2n-1)$ degrees of freedom, median $\overline{x}_1$, and scale parameter $s^2/[2(2n-1)]$. Thus, for instance, a 95% HPD credible set for μ_1 is

$$C(\overline{x}_1, s) = \left(\overline{x}_1 - t_{(2n-1)}(.975)\frac{s}{\sqrt{2(2n-1)}},\ \overline{x}_1 + t_{(2n-1)}(.975)\frac{s}{\sqrt{2(2n-1)}}\right),$$

where $t_{(2n-1)}(.975)$ is the .975 quantile of a standard t with $(2n-1)$ degrees of freedom.

Now, from a frequentist perspective, it is easy to see that $(\overline{X}_1 - \mu_1)/(S/\sqrt{2n})$ has a standard t–distribution with n degrees of freedom. It follows that $C(\overline{X}_1, S)$ has frequentist coverage probability

$$P_\theta(C(\overline{X}_1, S) \text{ contains } \mu_1) = 2F_n\left(\sqrt{\frac{n}{(2n-1)}}t_{(2n-1)}(.975)\right) - 1,$$

where F_n is the standard t c.d.f. For large n, F_n is approximately the standard normal c.d.f. Φ, and $t_{(2n-1)}(.975) \cong 1.96$, so that $P_\theta(C(\overline{X}_1, S)$ contains $\mu_1) \cong 2\Phi\left((1.96)/\sqrt{2}\right) - 1 = 0.83$. This, again, is a strong inconsistency, indicating that the noninformative prior is highly inadequate. It is of interest to note that $\pi(\theta) = 1/\sigma$ would here result in perfect agreement between posterior probability and frequentist coverage. ◁

The above example clearly demonstrates that it is not sufficient to merely divide θ into parameters of interest and nuisance parameters. Once separation of θ into more groups is considered, the natural suggestion is to completely separate the coordinates into k groups of one element each.

Example 3 (continued). If one sets $m = k$, letting each coordinate of θ be a grouping for the reference prior algorithm, it can be checked that $\pi(\theta) = 1/\sigma$ is the resulting reference prior regardless of the ordering of the coordinates of θ. This one–at–a–time reference prior is thus excellent for this problem.

Example 4. In Ye (1990), the development of reference priors for problems in sequential analysis is considered. If N is the stopping time in a sequential problem with independent observations, the Fisher information matrix is $I(\theta) = (E_\theta N)I_1(\theta)$, where $I_1(\theta)$ is the Fisher information for a sample of size one. Then the Jeffreys prior becomes $\pi(\theta) = (E_\theta N)^{k/2}\sqrt{\det(I_1(\theta))}$, which can easily be terrible if k is large because of the presence of $(E_\theta N)^{k/2}$. Grouping and iterating the reference prior method will typically reduce the power of $k/2$, but does not necessarily cure the problem (see Ye, 1990, for examples). But if one uses the one–at–a–time reference prior, then under reasonable conditions (see Ye, 1990) the result is $\pi(\theta) = \sqrt{E_\theta N}\pi^*(\theta)$, where $\pi^*(\theta)$ is the one–at–a–time reference prior for the fixed sample size problem. This is a very reasonable prior. (Of course, use of this method of determining a prior violates the Stopping Rule Principle, but this appears to be one of the unavoidable penalties in use of noninformative priors.) ◁

Other arguments for use of the one–at–a–time reference prior can be found in Berger and Bernardo (1992a and 1992b). Bayarri (1981) gives an example where at least 3 groupings are necessary (and the one–at–a–time reference prior is fine). The bottom line is that we have not yet encountered an example in which the one–at–a–time reference prior is unappealing, and so our pragmatic recommendation is to use this reference prior unless there is a specific reason for using a certain grouping (see Berger and Bernardo, 1992b, for a possible example).

There remains the problem of how to order the parameters before applying the one–at–a–time reference prior algorithm. Currently, we recommend ordering the parameters according to "inferential importance," but beyond putting the "parameters of interest" first, this is too vague to be of much use. Using an average of the reference priors arising from the various acceptable orderings has some appeal, but seems a bit too *ad hoc*. In practice, we have typically computed all one–at–a–time reference priors (and, indeed, all possible reference priors). We have not yet encountered an example in which this could not be done. Having a variety of possible noninformative priors is actually rather useful, since it allows a sensitivity study to choice of the noninformative prior. For additional discussion of this issue see Ghosh and Mukerjee (1992) and the ensuing comments.

3.4. *Other Issues*

3.4.1. Technical Considerations

In computation of the reference prior in the regular case, the two most difficult steps would appear to be evaluation of the expectation E_j^ℓ in (2.3.3) and passing to the limit in (2.2.6). Fortuitously, the latter typically makes the former relatively easy. This is because the expectation in (2.3.3) is with respect to π_{j+1}, which typically is tending towards an improper prior as $\ell \to \infty$. When this happens, it will usually be the case that $E_j^\ell[(\log|h_j(\theta)|)|\theta_{[j]}]$ can be expanded in a Taylors series as

$$K_\ell + C_\ell\psi(\theta) + D_\ell(\theta),$$

where $K_\ell \to \infty$, $C_\ell \to C$, and $D_\ell(\theta) \to 0$ as $\ell \to \infty$. When inserted into (2.3.3), the K_ℓ term typically cancels in the numerator and denominator, and the $D_\ell(\theta)$ term is typically irrelevant (both because of the exponentiation of the E_j^ℓ term). Thus the contribution of the E_j^ℓ term to the final answer will be $\exp\{\frac{1}{2}C\psi(\theta)\}$. Many variants on this theme are possible. What is important is the recognition that (i) exact computation of the E_j^ℓ is typically not needed — computing the first few terms of a Taylors expansion (in ℓ) usually suffices; and (ii) since the expansion is then being exponentiated, all terms except those going to zero (in ℓ) are important.

3.4.2. Prediction and Hierarchical Models

Two classes of problems that are not covered by the reference prior methods so far discussed are hierarchical models and prediction problems. The difficulty with these problems is that there are unknowns (that are indeed even usually the unknowns of interest) that have specified distributions. For instance, if one wants to predict Y based on X when (Y, X) has density $f(y, x|\theta)$, the unknown of interest is Y, but its distribution is conditionally specified. One needs a noninformative prior for θ, not Y. Likewise, in a hierarchical model with, say, $\mu_1, \mu_2, \ldots, \mu_p$ being i.i.d. $\mathcal{N}(\xi, \tau^2)$, the $\{\mu_i\}$ may be the parameters of interest but a noninformative prior is needed only for the hyperparameters ξ and τ^2.

The obvious way to approach such problems is to integrate out the variables with conditionally known distributions (Y in the predictive problem and the $\{\mu_i\}$ in the hierarchical model), and find the reference prior for the remaining parameters based on this marginal model. The difficulty that arises is how to then identify parameters of interest and nuisance parameters to construct the ordering necessary for applying the reference prior method; the real parameters of interest were integrated out!

We currently deal with this difficulty by defining the parameter of interest in the reduced model to be the conditional mean of the original parameter of interest. Thus, in the prediction problem, $E[Y|\theta]$ (which will be either θ or some transformation thereof) will be the parameter of interest, and in the hierarchical model $E[\mu_i|\xi, \tau^2] = \xi$ will be defined to be the parameter of interest. This technique has worked well in the examples to which it has been applied, but further study is clearly needed.

3.4.3. Invariance

When $\pi(\theta) = \sqrt{\det I(\theta)}$ is the reference prior (typically recommended only for one–dimensional problems), one automatically has invariance with respect to one–to–one transformations of θ, in the sense that the reference prior for a different parameterization would be the correct transform of $\pi(\theta)$. For the iterative reference prior of Section 2.3, certain types of invariance also exist. For instance, in the case of two groupings, $\theta_{(1)}$ and $\theta_{(2)}$, the reference prior is invariant (in the above sense) with respect to choice of the "nuisance parameter" $\theta_{(2)}$, and is also invariant with respect to one–to–one transformations of $\theta_{(1)}$. The reference prior can depend dramatically, however, on which parameters are chosen to be $\theta_{(1)}$. Some results on invariance for more than two groupings are known, but the general issue is still under study.

REFERENCES

Bayarri, M. J. (1981). Inferencia Bayesiana sobre el coeficiente de correlación de una población normal bivariante. *Trab. Estadist.* **32**, 18–31.

Bayes, T. (1763). An essay towards solving a problem in the doctrine of chances. *Phil. Trans. Roy. Soc.* **53**, 370–418.

Berger, J. (1984). The robust Bayesian viewpoint. *Robustness of Bayesian Analysis* (J. Kadane, ed.), Amsterdam: North-Holland, 63–124.

Berger, J. and Bernardo, J .M. (1989). Estimating a product of means: Bayesian analysis with reference priors. *J. Amer. Statist. Assoc.* **84**, 200–207.

Berger, J. and Bernardo, J. M. (1992a). Ordered group reference priors with application to a multinomial problem. *Biometrika* **79**, (to appear).

Berger, J. and Bernardo, J. M. (1992b). Reference priors in a variance components problem. *Proceedings of the Indo–USA Workshop on Bayesian Analysis in Statistics and Econometrics* (P. Goel, ed.), New York: Springer, 323–340.

Berger, J., Bernardo, J. M. and Mendoza, M. (1989). On priors that maximize expected information. *Recent Developments in Statistics and Their Applications* (J. P. Klein and J. C. Lee, eds.), Seoul: Freedom Academy Publishing, 1–20.

Berger, J. and Wolpert, R. (1988). *The Likelihood Principle.* Hayward, CA: Institute of Mathematical Statistics.

Bernardo, J. M. (1979). Reference posterior distributions for Bayesian inference. *J. Roy. Statist. Soc. B* **41**, 113–147, (with discussion).

Cifarelli, D. M. and Regazzini, E. (1987). Priors for exponential families which maximize the association between past and future observations. *Probability and Bayesian Statistics* (R. Viertl, ed.), London: Plenum Press, 83–95.

Consonni, G. and Veronese, P. (1989). A note on coherent invariant distributions as non–informative priors for exponential and location–scale families. *Comm. Statist. Theory and Methods* **18**, 2883–2907.

Dawid, A. P., Stone, M. and Zidek, J. (1973). Marginalization paradoxes in Bayesian and structural inference. *J. Roy. Statist. Soc. B* **35**, 189–233, (with discussion).

Fraser, D. A. S., Monette, G., and Ng, K .W. (1985). Marginalization, likelihood, and structural models. *Multivariate Analysis VI* (P. R. Krishnaiah, ed.), Amsterdam: North-Holland, 209–217.

Ghosh, J. K. and Mukerjee, R. (1992). Non-informative priors. *Bayesian Statistics 4* (J. M. Bernardo, J. O. Berger, A. P. Dawid and A. F. M. Smith, eds.), Oxford: University Press, 195–210, (with discussion).

Heath, D. and Sudderth, W. (1978). On finitely additive priors, coherence, and extended admissibility. *Ann. Statist.* **6**, 333–345.

Jaynes, E. T. (1980). Marginalization and prior probabilities. *Bayesian Analysis in Econometrics and Statistics* (A. Zellner, ed.), Amsterdam: North-Holland, 43–78.

Jeffreys, H. (1946). An invariant form for the prior probability in estimation problems. *Proc. Roy. Soc. London A* **186**, 453–461.

Jeffreys, H. (1961). *Theory of Probability*. Oxford: University Press.

Laplace, P. (1774). Mémoire sur la probabilité des causes par les évenements. *Mem. Acad. R. Sci. Presentés par Divers Savans* **6**, 621–656. (Translated in *Statist. Sci.* **1**, 359–378).

Laplace, P. (1812). *Théorie Analytique des Probabilités*. Paris: Courcier.

Lindley, D. V. (1956). On a measure of the information provided by an experiment. *Ann. Math. Statist.* **27**, 986–1005.

Neyman, J. and Scott, B. (1948). Consistent estimates based on partially consistent observations. *Econometrica* **16**, 1–32.

Rissanen, J. (1983). A universal prior for integers and estimation by minimum description length. *Ann. Statist.* **11**, 416–431.

Shannon, C. E. (1948). A mathematical theory of communication. *Bell Systems Tech. J.* **27**, 379–423 and 623–656.

Stein, C. (1985). On the coverage probability of confidence sets based on a prior distribution. *Sequential Methods in Statistics*, Banach Centre Publications **16**, 485–514.

Stone, M. (1971). Strong inconsistency from uniform priors, with comments. *J. Amer. Statist. Assoc.* **58**, 480–486.

Tibshirani, R. (1989). Noninformative priors for one parameter of many. *Biometrika* **76**, 604–608.

Wasserman, L. (1991). An inferential interpretation of default priors. *Tech. Rep.* **516**, Carnegie-Mellon University.

Ye, K. Y. (1990). *Noninformative Priors in Bayesian Analysis*. Ph.D. Thesis, Department of Statistics, Purdue University.

Ye, K. Y. and Berger, J. O. (1991). Noninformative priors for inferences in exponential regression models. *Biometrika* **78**, 645–656.

DISCUSSION

R. McCULLOCH (*University of Chicago, USA*)

My discussion will touch on four points: i) the notion of a parameter of interest; ii) what I think the key idea is; iii) how a subjectivist might use the results; (iv) the role of compact sets.

The notion of a parameter of interest. From the outset the reference prior method has had the notion of a "parameter of interest" and the complementary notion "nuisance parameter" as central concepts. I did my graduate work at Minnesota under Seymour Geisser. If I ever uttered the phrase "paramer of interest" he would beat me up. Geisser smiles a lot and he makes jokes but he's a mean dude. If a parameter has no effect on predictions then by the principle of parsimony it would be eliminated from the model. If it does have an effect on predictions then it can hardly be called a nuisance parameter. Thus, the predictive viewpoint indicates that the need to identify parameters which are a "nuisance" severely restricts the applicability of the method.

I think that it is evident that much of statistical practice has been severely hurt by the nuisance parameter concept. Consider the simple $N(\mu, \sigma^2)$ model. In elementary statistics courses we teach methods for testing hypotheses about the parameter of interest μ without having to specify σ. But surely the conclusion that μ is close to 0 has quite different implications depending on whether σ is small or large.

What I think the key idea is. The reference prior method has become quite complicated. To me it ends up being something of a black box. The authors explicitly state that with all the asymptotic approximations and limits that the method should be viewed as heuristic and invite us to judge it by how well it works in various examples. The examples are interesting. By that I mean that they excite, in the reader, the psychological and emotional state we label with the word interest. In the product of means example I'm not sure I don't like the uniform prior best. In the Monette, Fraser, Ng example it works only in the sense that it warns us (rightly) that the resultant prior may not be quite right. I have never been happy about the Neyman Scott example because the number of parameters goes to infinity just as fast as the sample size.

And yet the method *does* generate interesting priors. If the method is applied to the entire parameter vector without breaking it up and iterating, the result is Jeffreys' prior. Jeffreys himself found fault with his prior is cases where there was more than one parameter and modified the result in *ad hoc* ways. By breaking up the parameter vector and more or less applying Jeffreys method iteratively the method produces, by a formal mechanism, the kind of results that Jeffreys himself seemed to prefer. This is the key idea in the method. Perhaps it is in one of the earlier papers that I didn't read, but I would like to know if the authors have any intuition for why this seems to work.

How a subjectivist might use the results. The first sentence of the paper is, "Bayesian analysis is a distinct field only because of noninformative priors". My reaction to this sentence was something of a personal epiphany. It revealed something of myself to me. I had thought that I was the kind of guy who was pretty much willing to try anything (an attitude that gets me into trouble in other walks of life). But upon reading the opening line my soul shuddered and then emitted a dark, twisted howl, of which, the only discernible syllable was "no".

Are these results of interest to a subjectivist? I think the answer is yes. Suppose I don't want to bother choosing my prior. Well, I will go ahead and see what I get from a "noninformative" prior. For example I might use the Laplace uniform prior, hopefully after giving some thought to the parametrization. Then, broadly, three things can happen to me. The simplest case is where my posterior distribution is quite tight. In which case I am tempted to conclude that the likelihood dominates the prior and if I bothered to elicit a subjective prior I would just get the same posterior anyway. The second case is where the posteriors is very diffuse. In this case I am tempted to conclude that there isn't much information in the data so that whatever prior I put in will be highly influential in that the posterior will be much like the prior so I might be better off getting more data than eliciting my prior. The third case is anything in between the other two. In this case we would like to check to see how influential the choice of prior is. Ideally we would like to know if the posterior is really any different from that which would be obtained from a carefully elicited prior. To gauge this (without eliciting a prior) we would compare the posterior based on the Laplace prior with that obtained from the reference prior. If the difference between these two posteriors is substantively important you probably can't get away without thinking about your prior.

Compact sets. Limits of compact sets play an important role in the reference prior method. It seems to me that this is one limit that, in our modern computational environment we could avoid taking. In the old days people wanted analytical results and given the set of mathematical tools it was actually more convenient to let the parameter space be infinite.

Now most of our work is done numerically so that, in effect, we are using a compact set. Also, it may be that the choice of a compact subset of the parameter space is something that could be done fairly easily based upon prior information even in high dimensional problems.

Well, all of the above seems like a lot of complaining and whining. We all use "noninformative" priors and this work is probably the most important current work in the area. I found the papers very um...er...ah...interesting. If the authors obtain impossible solutions it is because they are working on an impossible problem. It is comforting to see that Professor Bernardo is keeping the spirit of Don Quixote alive in Spain and I should not be surprised to see the aged Knight, some dark and stormy night, pursuing his quest yet, in the town of West Lafayette.

B. CLARKE (*Purdue University, USA*)

Introduction. Implicit in the work of Berger and Bernardo is a physical interpretation which merits direct examination. They note that in certain examples, the information-theoretic merging of two posteriors may depend on the sequence of compact sets supporting the prior which defines one them. This motivates the definition of reference priors given by expression (2.2.5). Although they have written that such dependence indicates the necessity for subjective input, it can also be given a physical interpretation, in terms of universal noiseless source coding.

In addition, the stepwise prior which appears in expression (2.2.5) can be given a physical interpretation in terms of the capacity of a certain information-theoretic channel. While expression (2.2.5) itself can also be interpreted physically in the context of channel coding, this seems somewhat artificial. Since channel coding and source coding are quite distinct, we raise the question of how to physically interpret the reference prior method.

As this may sound like a criticism, we also argue that the unsatisfactory results obtained in the Fraser-Monette-Ng example, and in the Neyman-Scott example are not a failure of the method, but instead reflect unreasonable expectations.

Some asymptotics will be used and we follow the notation of the paper. For instance, we use $Z_t = (X_1, \ldots, X_t)$, a vector of iid outcomes from $f(\cdot|\theta)$. Henceforth, we only note our occasional necessary departures.

Channel coding and source coding. First consider the function

$$K(t,l) = E_l^{Z_t} D\left(\pi_1^l(\cdot|Z_t), \pi(\cdot|Z_t)\right).$$

The criterion in (2.2.5) is that $\pi(\theta)$ satisfy

$$\lim_{l\to\infty} K(t,l) = 0,$$

so that π is a limit point of the sequence $< \pi_l > |_{l=1}^{\infty}$. The definition of $K(t,l)$ gives

$$K(t,l) = D(\pi_1^l, \pi) + \int \pi_1^l(\theta)\left[D\left(f(Z_t|\theta), p_l(Z_t)\right) - D(f(Z_t|\theta), p(Z_t))\right] d\theta. \qquad (1)$$

When Z_t is discrete, the integrand is essentially the change in redundancy due to using the Shannon code based on p_l, rather than the Shannon code based on p, when the true source is $f(\cdot|\theta)$. Integration over θ gives the Bayes redundancy, and the Shannon code based on a mixture of distributions with respect to a given prior is essentially the code achieving minimal Bayes redundancy, as defined by that prior. Consequently, if the integration over the second term of the integrand were with respect to π, we would say that the sequence of Bayes codes given by the sequence of mixtures p_l tends to the Bayes code for the entire family of $f(\cdot|\theta)$'s. However, both terms are integrated with respect to π_l which is intended to approximate the limit π.

The first term on the right in (1) represents the redundancy of coding with respect to π when the true prior is π_1^l. However, in the statistical context it is not clear what this means. Perhaps it is sensible to replace (2.2.5) with $D(\pi_1^l, \pi) \to 0$: If π is proper, regularity conditions already imply that for fixed $l, k(t, l) \to 0$ as t increases. The result might be finding rates at which l may be let to increase as a function of t.

Next we turn to a channel coding interpretation for the stepwise reference prior. A conditional density effectively defines a channel. The Shannon mutual information gives, typically, an achievable rate of transmission across the channel. The supremal value of that rate is called the channel capacity. For compact parameter spaces the reference prior is usually the source distribution which gives the channel capacity. In the two step case, there is a formula in Ghosh and Mukerjee (1992) which implies that π_1^l is the source distribution for the channel defined by $m(Z_t|\theta_1) = \int f(Z_t|\theta_1, \theta_2)\pi(\theta_2|\theta_1)d\theta_2$ which achieves the maximal rate of transmission, asymptotically in t.

This channel has the following interpretation. The message sent is θ_1. There are t receivers, and they pool their data to decode the message. The l defines the range of messages we are able to transmit. The effect of the mixing in $m(Z_t|\theta_1)$ amounts to saying that unbeknownst to the sender, once θ_1 is sent, an auxilliary message θ_2 is sent, with probability $\pi(d\theta_2|\theta_1)$. The decoding is affected in that the constant term in the expansion for the mutual information changes.

Contrasting the two interpretations, we note that adding and subtracting the integral $\int \pi(\theta)D(f(Z_t|\theta), p(Z_t))\, d\theta$ in (1) gives a difference of mutual informations which makes sense in terms of channel coding. However, the other terms are problematic. Also, the l defines an increasing sequence of parametric families in source coding, but a range of messages in channel coding. It is not clear what this means in a statistical context.

Comments on the examples. Finally, we pick a few knits. In the Neyman-Scott example, the two step reference prior approach is sensible when σ is the parameter of interest, but breaks down when μ_1 is the parameter of interest. This is not really surprising since the number of parameters is growing as a linear function of the data, so there is no hope to estimate all of them well. On a technical note, to control error terms in certain proofs, it is essential that the number of parameters grow slowly, if at all.

Regarding the Fraser-Monette-Ng example, it is important to note that the parameter space is discrete. The asymptotics in the discrete case are quite distinct from those in the continuous case: There is no dependence on t or $I(\theta)$. In the absence of nuisance parameters, the mutual information converges to the entropy of the prior. As a result, the reference prior is, asymptotically, the maximum entropy distribution. So, it is not surprising that anomalous results are obtained. Some constraint on the class over which the maximization occurs may be necessary.

In any event, the authors have made a valuable contribution, for which they are to be complimented.

M. GHOSH (*University of Florida, USA*)

The present article is yet another masterly contribution from Berger and Bernardo on the development of reference priors. These authors, over the last few years, have made several important extensions of the original work of Bernardo (1979), where the reference priors were first introduced. One of the major accomplishments of this ongoing research is a systematic development of reference priors in the presence of nuisance parameters, and the present article is yet another important step in that direction.

I will confine my discussion to the Neyman-Scott example, one of the major examples in this paper, a problem that has fascinated statisticians for more than four decades. I am particularly impressed by the simple reference prior $\pi(\mu_1, \ldots, \mu_n, \sigma) \propto \sigma^{-1}$ which leads to the consistent Bayes estimator $S^2/(n-2)$ of σ^2, consistency being achieved in a frequentist sense.

I now show that an alternative consistent estimator of σ^2 can be derived using a hierarchical Bayes approach, though the proof of consistency requires certain mild conditions on the μ_i's. The derivation proceeds as follows:

First note that $(\overline{X}_1, \ldots, \overline{X}_n, S^2)$ is minimal sufficient, where $\overline{X}_i = (X_{i1} + X_{i2})/2$, $i = 1, \cdots, n$, and $S^2 = \sum_{i=1}^n \sum_{j=1}^2 (X_{ij} - \overline{X}_i)^2$. Consider now the following hierarchical model:

I. Conditional on $\mu = (\mu_1, \ldots, \mu_n), \sigma^2, M = m$, and $\Lambda = \lambda, \overline{X}_1, \ldots, \overline{X}_n$ and S^2 are mutually independent with $\overline{X}_i \sim N(\mu_i, \frac{1}{2}\sigma^2), i = 1, \ldots, n$, and $S^2 \sim \sigma^2 \chi^2_n$.

II. Conditional on $M = m, \sigma^2$, and $\Lambda = \lambda$, μ_i's are iid $N(m, \lambda^{-1}\sigma^2)$.

III. Marginally M, σ^{-2} and $\Lambda\sigma^{-2}$ are mutually independent with $M \sim$ uniform $(-\infty, \infty)$, $\sigma^{-2} \sim$ Gamma $(0, \frac{1}{2}g_0)$, where $g_0 (\leq 0)$ is some specified number, and $\Lambda\sigma^{-2}$ is Gamma $(0, -1)$, where we use the notation Gamma(α, β) for a (possibly improper) distribution with pdf $f(y) \propto \exp(-\alpha y) y^{\beta - 1}$.

Based on the above hierarchical model, one obtains the following results:

(i) Conditional on $\overline{X}_i = \overline{x}_i (i = 1, \ldots, n), S^2 = s^2$, and $\Lambda = \lambda$,

$$\sigma^{-2} \sim \text{Gamma}\left(\frac{1}{2}\left(s^2 + \frac{2\lambda}{2+\lambda}\sum_{i=1}^{n}(\overline{x}_i - \overline{\overline{x}})^2\right), \frac{1}{2}(2n - 3 + g_0)\right),$$

where $\overline{\overline{x}} = n^{-1}\sum_{i=1}^n \overline{x}_i$;

(ii) conditional on $\overline{X}_i = \overline{x}_i (i = 1, \ldots, n)$, and $S^2 = s^2$, Λ has conditional pdf

$$f(\lambda|\overline{x}_1, \ldots, \overline{x}_n, s^2) \propto (\lambda/(2+\lambda))^{\frac{1}{2}(n-1)}\lambda^{-2}(s^2 + 2\lambda(2+\lambda)^{-1}\sum_{i=1}^{n}(\overline{x}_i - \overline{\overline{x}})^2)^{-\frac{1}{2}(2n-3+g_0)}.$$

It is convenient to reparametrize Λ into $U = \Lambda/(2+\Lambda)$ so that posterior pdf of U is

$$f(u|\overline{x}, \ldots, \overline{x}_n, s^2) \propto u^{\frac{1}{2}(n-5)}(1 + uF)^{-\frac{1}{2}(2n-3+g_0)} I_{[0<u<1]}. \tag{1}$$

Based on (i) and (ii), we obtain

$$E\left(\sigma^{-2}|\overline{x}_1, \ldots, \overline{x}_n, s^2\right) = s^2(2n - 5 + g_0)^{-1}\left[1 + E(UF|\overline{x}_1, \ldots, \overline{x}_n, s^2)\right] \tag{2}$$

where $F = 2\sum_{i=1}^n (\overline{x}_i - \overline{\overline{x}})^2/s^2$, a multiple of the usual F statistic. Using (1), it follows after some simplifications that

$$E(UF|\overline{x}_1, \ldots, \overline{x}_n, s^2) = \frac{\int_0^{\frac{F}{(1+F)}} v^{\frac{1}{2}(n-3)}(1-v)^{\frac{1}{2}(n-4+g_0)}dv}{\int_0^{\frac{F}{(1+F)}} v^{\frac{1}{2}(n-5)}(1-v)^{\frac{1}{2}(n-2+g_0)}dv} \tag{3}$$

Integrating by parts, it follows from (3) that

$$E(UF|\overline{x}_1,\ldots,\overline{x}_n,s^2) = (n-3)(n+g_0-2)^{-1} - \frac{2F^{\frac{1}{2}(n-3)}}{(n+g_0-2)(1+F)^{\frac{1}{2}(2n-5+g_0)}\int_0^{F/(1+F)} v^{\frac{1}{2}(n-5)}(1-v)^{\frac{1}{2}(n-2+g_0)}dv}. \tag{4}$$

Combining (2) and (4), one gets

$$E\left(\sigma^{-2}|\overline{x}_1,\ldots,\overline{x}_n,s^2\right) = s^2/(n+g_0-2) - \frac{s^2}{2n-5+g_0} \cdot \frac{2F^{\frac{1}{2}(n-3)}}{(n+g_0-2)(1+F)^{\frac{1}{2}(2n-5+g_0)}\int_0^{F/(1+F)} v^{\frac{1}{2}(n-5)}(1-v)^{\frac{1}{2}(n-2+g_0)}dv}. \tag{5}$$

As $n \to \infty$, the first term in the right hand side of (5) converges to σ^2 in probability for every fixed g_0. Using some heavy and tedious algebra, it can also be shown that the second term in the right hand side of (5) converges to zero in probability as $n \to \infty$ if $n^{-1}\sum_{i=1}^n(\mu_i - \overline{\mu})^2 \to A$ (some fixed positive number) as $n \to \infty$. Thus, the consistency of the Bayes estimate given in (5) holds under some mild conditions. It may be interesting to note that the first term in the right hand side of (5) equals the Berger-Bernardo reference prior estimate when $g_0 = 0$.

Barnard (1970) suggested that the Neyman-Scott problem could be resolved using an empirical Bayes approach. Barnard never did spell out how the empirical Bayes approach should be used, but it seems quite plausible that an empirical Bayes approach will also meet with success if one estimates m and λ rather than use a hyperprior on these parameters.

I wish to thank Professor J. K. Ghosh for posing a question which led to the development of (5).

M. GOLDSTEIN (*University of Durham, UK*)

The subject of reference priors seems to divide Bayesians between those who like and use them, and those who find them rather puzzling. As one of the latter group, I would like to pick up on the link asserted between reference priors and "objectivity", and the related claim that, in some general sense, reference priors "work".

Consider the Fraser-Monette-Ng example dealt with in the paper. The authors seem to imply that a reference-type prior distribution of form $\pi(\theta) \propto \theta^{-1}$ would be "perfectly satisfactory". Now, I can agree that such a prior distribution would not be inherently contradictory. However, what puzzles me is how such a prior could have some claim to objectivity. What is "objective" in placing four times higher probability on the smallest of the three allowable θ values, when we have seen x, than that placed on the other two values? Either this corresponds to a genuine prior judgement that small values are, a priori, more likely than large values, or it looks like an arbitrary fix. The idea that a scientist could "objectively" demonstrate that smaller values of θ were more likely than larger values, without making any subjective inputs, is rather weird, so maybe the authors might like to comment on what they view as the criteria for judging a successful reference prior for this problem.

R. E. KASS (*Carnegie Mellon University, USA*)

I have several comments on this interesting paper.

1. The authors' work is very much in the spirit of Jeffreys, who judged rules for determining prior distributions according to how they worked in specific examples. As a

matter of historical record, however, there is not much support for the claim that "Jeffreys himself ... noticed difficulties with the method, i.e., his general rule, taking the prior to be proportional to $\det[I(\theta)]^{1/2}$ when θ is multidimensional and would then provide ad hoc modifications to the prior". What Jeffreys did was (i) suggest that location parameters should be treated specially, (ii) note that simple alternative solutions exist in many problems (such as taking a uniform prior on the Binomial proportion), which do not agree with those produced by his general rule, and (iii) encourage further investigation of "invariance theory" in determining priors. The defects he noted in his general rule were present in one-parameter problems; application to problems of higher dimension *per se* did not seem to bother him.

2. As far as nomenclature is concerned, I think "noninformative prior" is sufficiently problematic that introducing an alternative is desirable. Since "reference prior" is often understood to refer to Bernardo's method, perhaps a better choice would be "conventional prior". This would clearly be true to Jeffreys's intent in his suggestion that such priors be determined "by international agreement ... as ... in the choice of units of measurement and many other standards of reference" (*Brit. J. Phil. Sci.*, 1955, p. 277).

3. Another matter of nomenclature involves the term "parameters of interest". Apart from location problems, it is by no means clear that parameters may be ordered according to "interest". Why not simply refer to ordered parameters, and let the arbitrariness in the choice remain obvious? (In Jeffreys's scenario, the specifics could be determined by the international committee that will table the results.)

4. In the Neyman-Scott example, it should be remarked that for any fixed n, Jeffreys's method was to take the prior proportional to $1/\sigma$. I do not see how his method is any more "ad hoc" than the authors'. Also, it would seem that a hierarchical prior would be of interest in this example; the authors treated the conjugate case in full detail in their previous work.

5. Hierarchical models present very important cases for any conventional prior methodology and I would hope to see further work in this direction. Computation, however, is likely to be a very serious difficulty. The information matrix is already hard to compute and the brief remarks made in Section 3.4.2 are not specific enough to offer much comfort: the matrix would have to be computed at a large number of values of θ and it is not clear how we would combine that computation with some Gaussian quadrature or simulation method for computing posterior quantities in a reasonably efficient scheme. (By the way, the problem when using asymptotic approximations is greatly reduced because the matrix need only be computed at a few points.) The authors note that their full iterative algorithm may not be feasible in analytically-intractable cases. What, then, are we to do in these commonly-encountered situations?

6. In answering such ultimately practical questions, it is perhaps inappropriate to separate prior selection from sensitivity analysis. In any real problem we will want to perform some kind of sensitivity analysis and we are then led to ask what the role will be for a conventional prior in such an alaysis. This seems to me to be an important outstanding problem. My own experience is dragging me toward subjective sensitivity analysis, but if one were to go to the trouble of performing a sensitivity analysis based on subjectively-determined priors, why would one need a conventional prior? It would seem that the best answer is that it would assist in scientific reporting. On the other hand, it may be possible to construct a method for assessing sensitivity that would be both useful and more convenient than one that requires detailed prior elicitation; perhaps conventional priors could play a role in this process.

G. KOOP (*Boston University, USA*) and
M. F. J. STEEL (*Tilburg University, The Netherlands*)

We congratulate Professors Berger and Bernardo for developing an elegant general approach to reference priors for independent experiments. Our comments are not so much a criticism of their approach as they are a query concerning an extension which we judge to be important. That is, among econometricians there has been a great deal of discussion lately on what constitute reasonable noninformative priors for non-independent experiments. One subject of controversy is the elicitation of noninformative priors for variants of the simple AR(1) model, $y_t = \rho y_{t-1} + \varepsilon_t$ (where the ε_t's are i.i.d. $N(0, \sigma^2)$ and $t = 1, \ldots, T$). Phillips (1991) develops Jeffreys prior for this model, which is often used to test for a unit root $(\rho = 1)$. It is our opinion that the development of noninformative priors for dynamic models such as the AR(1) model has great relevance indeed for practitioners of applied econometrics. In the following, we discuss the use of the reference prior in dynamic models and describe the problems that arise in this context.

In the simple AR(1) model, conditional on $y_0 = 0$, there are two parameters, ρ and σ^2. By treating either σ^2 or ρ as the nuisance parameter, the model satisfies (2.3.5) in Berger and Bernardo and hence (2.3.6) holds. The reference prior calculated using Berger and Bernardo's method is the same as the Jeffreys' prior, which possesses tails of $0(\rho^{T-2})$. Although it has all the advantages ascribed to it by Berger and Bernardo, the prior also has several disadvantages. First, the Jeffreys' prior depends on sample size, T, (i.e., is data based) and violates the Likelihood Principle. Second, this dependence on sample size occurs in such a way that the prior influences the posterior even as sample size gets large. That is, the likelihood does not dominate the posterior, even for large samples, which precludes "calibration" (i.e., two Bayesians can continue to disagree as information accrues). Third, econometricians are frequently interested in testing whether y_t is stationary against the hypothesis that it is non-stationary ($|\rho| < 1$ versus $|\rho| \geq 1$). The prior odds for the stationarity hypothesis against a hypothesis containing any finite interval of comparable length in the explosive region are virtually zero. Relative to what econometricians think is reasonable, the Jeffreys' prior places far too much weight on explosive alternatives. This is because the Jeffreys' procedure takes expectations over the sample space. In the AR(1) model, the sampling properties of explosive models dominate those from stationary models. Fourth, the Jeffreys' prior for the AR(1) model depends on the order in which data are observed. Sequential updating is precluded. For all of these reasons, many econometricians consider the Jeffreys' prior to be unreasonable and strongly criticize its indiscriminate use in dynamic models (see the discussion to Phillips (1991)).

On the basis of these objections, we contend that the reference prior approach described in Berger and Bernardo does not extend immediately to non-independent experiments. On reading the unit root literature in econometrics, we find that a great demand for noninformative priors appears to exist. Hence the development of reference priors that circumvent the above objections might just convince classical econometricians–a very challenging audience indeed–of the merits of Bayesian methods. Perhaps the authors could propose a convincing procedure for such models. Or should researchers just stick with the simple Laplace rule?

D. J. POIRIER (*University of Toronto, Canada*)

The authors readily acknowledge that pursuit of "reference" priors leads to priors that often have a variety of discomforting properties, two of which can include incoherency and violation of the Likelihood Principle. In addition the authors admit that reference priors are often not easy to derive, and are unlikely to achieve broad agreement because they depend on issues such as the parameters of interests. Given these latter pragmatic problems, coupled with the former distasteful theoretical violations, I think the reader deserves more elaboration on why this pursuit is worthwhile (other than the obvious reply that so many previous authors have gone before). If one is to violate basic principles, then at least the violator should outline the cases in which such violations may be palatable, and if pragmatic expediency is not the reason, what then is the reason?

L. WASSERMAN (*Carnegie Mellon University, USA*)

This is a very interesting paper. I have two comments; both are aimed at promoting more widespread use of the techniques in this paper. First, I think it may be possible to provide a rationale for the sample space dependence of reference priors. The argument goes like this. Let E be the experiment selected by the experimenter. Let I_E be the event that the experimenter preferred E to all other experiments and let π be the experimenter's prior. Suppose I try to guess π. Let J be my prior on the set $\mathcal{P}$ of all priors. It can be shown that, under suitable assumptions, my best guess at π conditional on I_E, is the Jeffreys' prior. In other words, $E_J(\pi|I_E) = \pi_E^*$ where π_E^* is Jeffreys' prior for experiment E. The details are in Wasserman (1991). Thinking of Jeffreys' prior as a guess at π conditional on I_E obviates the criticism that there is a violation of the likelihood principle. It might be possible to justify the stepwise prior in a similar way, by conditioning on the information that the experimenter has chosen a "parameter of interest".

My second comment is a minor point about terminology. As mentioned by the authors, the alernative "default prior" has been suggested in place of "non-informative prior" to refer to priors chosen by scientific convention. Kass (1989) uses this term too. I suggest we abandon the term "non-informative prior" and use "default prior" instead. The former is emotionally charged and, besides, we all agree that there is no such thing as a noninformative prior. Also, the term "reference prior" is ambiguous. Does it refer to (a) priors chosen by scientific convention, (b) priors chosen by the missing information argument or (c) priors chosen by a stepwise argument? To add to the confusion, Box and Tiao (1973) also use "reference prior". I suggest "default prior" for (a), "missing information prior" (MIP) for (b) and "stepwise prior" for (c).

REPLY TO DISCUSSION

We thank the discussants for their interesting comments and questions. Because several of the discussants raised certain common questions, we will respond by topic.

The Name. Kass suggests we replace the name "noninformative" prior with "conventional" prior, while Wasserman prefers "default" prior. Assuming "reference" is to be the name associated with the particular method we advocate for derivation of a noninformative prior, we would slightly prefer the name "conventional" to "default" for the general concept, simply because "default" sounds somewhat unscientific. Basically, however, it is so difficult to change a historical name that we do not advocate such a change. Perhaps when we finally have a true statistical convention to select our official noninformative priors, we should meet in Geneva and then we can call them the "Geneva Convention" priors.

Parameters of Interest. Kass and McCulloch express various concerns about the definition and meaning of "parameters of interest" and "nuisance parameters". We refer to these partly because the historical development of reference priors was heavily influenced by these notions, and partly because the concepts do still seem to provide some guidance in choosing the parameter ordering or the parameters (see Section 3.4.2) to which the reference prior algorithm should be applied. But we must admit that we are drifting away from these concepts; in particular we no longer recommend dividing the parameters exclusively into these two classes. And we are close to just recommending trying all parameters orderings, regardless of which parameters are of interest.

Dependent Data. The example discussed by Koop and Steel is fascinating, pointing out a serious potential difficulty in information-based methods of deriving noninformative priors. What seems to be happening is that the data can be made more and more informative by having the prior concentrate on larger and larger values of the parameter ρ. Therefore, as the sample size increases, the prior will shift to larger ρ to "increase the information provided by the data". We agree with Koop and Steel that the net effect of this does not seem to be good. Is there a solution within the reference prior theory? We will certainly think about it, but the answer might well be — No!

Why Do All This? This is a very good question, raised to different degrees by Kass, McCulloch, and Poirier. There are actually two distinct questions here. The first is: Has the reference prior method become so involved that we have lost the original motivation for noninformative priors — simplicity? The key to the answer is recognizing that noninformative priors are typically used in a "look-up" scenario, with the practitioner choosing a model and then searching the literature for the "correct" noninformative prior. It will be the job of the reference prior researchers to determine the reference priors for common models, and provide tables of such. The highly sophisticated practitioner who operates by inventing and studying many completely new models will probably find the reference prior algorithm too difficult to employ for each new model, but might well choose to derive it for the model ultimately selected.

The second important question here is: What is the alternative? Let us consider two possibilities:

(i) McCulloch considers use of a constant noninformative prior, lists three possible things that can happen, and suggests that the reference prior is at best useful only for checking if the answer is sensitive to the choice of a constant prior. There are at least two other possibilities that need to be considered, however. The first is that the posterior need not be proper for a constant prior, and impropriety of the posterior may not be easy to recognize in this age of analysis by computer. The exponential regression model referred to in the paper is an example. The other troubling possibility is that the posterior for a constant prior could be quite concentrated, but concentrated in the "wrong" place. The famous Stein example of estimating the squared norm of a multivariate normal mean is one such example; one of us recently even encountered a variant of this problem in a major consulting project. Reference priors are not guaranteed to avoid these difficulties, but their track record is certainly better.

(ii) Subjective Bayesian analysis, perhaps with sensitivity analysis as mentioned by Kass, is the obvious possible alternative. And as the noninformative prior theory grows in complexity, the difficulties in subjective Bayesian analysis start to seem less foreboding. We are not sure, however, if subjective Bayesian analysis is the cure to the difficulties in multi-dimensional problems. The point is simply that subjective elicitation is so difficult in even moderate dimensional problems (hierarchical prior and other structured scenarios excepted) that there is no guarantee that the subjective approach will even be superior to the noninformative prior approach. Only a small number of features of a multivariate prior are ever specified, and the simplifying assumptions that one typically must make (e.g., independence of the parameters) can be extremely influential without one being aware of it. It is nice to say that one will conduct a sensitivity study, but what is the chance, in a high dimensional space, of happening to encounter the truly influential features of the prior? Current research on Bayesian robustness may provide solutions to this concern, but the jury is still out. The surprising success of the reference prior method on the various known "difficult" multiparameter problems could be viewed as an indication that it somehow seeks and neutralizes potentially serious high-dimensional "confounding" of parameters, but at the moment this is sheer speculation. At the very least, it would be sound practice in a subjective Bayesian sensitivity study to include the reference prior (and probably the constant prior).

Violation of the Likelihood Principle and other Nasties. Several of the discussants express concern over the various foundational inconsistencies that can be encountered with common methods of developing noninformative priors. We have learned to live with these as one of the prices that must be paid. In this regard, we were extremely interested in the statement by Wasserman that, if one adopts a broad enough perspective, the information-based noninformative priors may not be in violation of these principles. His argument sounds plausible, and we await its fleshing out with considerable anticipation.

Koop and Steel discuss a number of unappealing properties of the Jeffreys (and probably reference) prior for the AR(1) model. Two of these properties fall in the category of general problems with noninformative priors, and are thus worth highlighting. First, the fact that the usual noninformative priors do not work well for testing is not too surprising, since noninformative priors rarely work well for testing. Indeed, they only work when there is symmetry between hypotheses or for eliminating nuisance parameters common to the hypotheses. Likewise, it is not uncommon for noninformative priors to be inconsistent with sequential updating, since they often will depend on the amount and even the nature of the data to be obtained. Of course, if the sequential sampling plan is completely known in advance, then one can obtain the noninformative prior for that sequential experiment —see Example 4. We are not saying that these are pleasing properties, but they do seem to be unavoidable.

The Neyman-Scott Example. Kass observes that Jeffreys method (not prior) for location-scale problems was to use $1/\sigma$ as the noninformative prior, and asks — why is this method (which gives the "right" answer for the Neyman-Scott example) more *ad hoc* than the reference prior method? In a sense, Jeffreys method here could be considered to be the stepwise reference prior method, since it is based on somehow attempting to separate the location and the scale parameters. Indeed, our recommended reference prior method could just be viewed as a general way to accomplish separation of parameters.

Kass also observes that a hierarchical model might be natural in this example. We agree, but it is important to note that choosing a hierarchical prior is a major subjective judgement, and is far from being noninformative.

Ghosh provides an interesting analysis showing that use of a hierarchical model here works well, in the sense of providing consistent estimators under weak assumptions. The key feature of his analysis to note is that he does not prove consistency only under the condition that the μ_i arise from the indicated hierarchical prior, but under much weaker assumptions. On a technical point, we suspect that his consistency condition can be weakened to

$$\limsup_{n\to\infty} \left[n^{-1} \sum_{i=1}^{n} (\mu_i - \overline{\mu})^2 \right] \leq k$$

which is of some interest because it states that the hierarchical prior works even if the μ_i are not arising as i.i.d. observations from a distribution with a finite variance.

Coding. The efforts by Clarke to explain, through notions of coding, the various motivations for the reference prior steps are very interesting. It is perhaps unfortunate that not everything seems to be completely explainable in this regard.

The suggestion that one might consider, instead of (2.2.5), the condition $D(\pi_1^\ell, \pi) \to 0$ is reasonable for proper priors, but for improper π this quantity typically converges to infinity. The suggestion of choosing ℓ to depend on the asymptotic repetition number is an interesting possibility, but in some sense there are already too many options in developing a reference prior; we would recommend adding more options only if the current structure proves to be inadequate.

Why Does the Stepwise Method Work?

McCulloch asks this question. We do not really know the answer. Examination of examples such as the Neyman-Scott example reveals the problem with considering all parameters jointly, but our insight is not much deeper than that. More generally, one of us has never been exactly sure why the entire reference prior method works, and has continually been very pleasantly surprised at its success. At the very least, one must agree with McCulloch's statement "and yet the method **does** generate interesting priors".

Miscellaneous Comments.

(i) The comments of Kass concerning the attitude of Jeffreys towards higher dimensional problems are interesting. To us, the key point is that Jeffreys was at least willing to modify his rule in higher dimensions.

(ii) The computational difficulties mentioned by Kass, especially in regards to determining reference priors for hierarchical models, are very real. Undoubtedly there are problems for which reference priors are not effectively computable, even numerically.

(iii) McCulloch suggests subjectively choosing a large compact set on which to operate, thereby avoiding the need to perform the limiting operation over compact sets. Actually, however, the limiting operation over compact sets is typically a simplifying operation as discussed in section 3.4.1; trying to do the exact computation of a reference prior over a fixed compact set would typically be much more difficult.

ADDITIONAL REFERENCES IN THE DISCUSSION

Barnard, G. A. (1970). Discussion of the paper by Kalbfleisch and Sprott. *J. Roy. Statist. Soc. B* **32**, 194–195.

Box, G. E. P. and Tiao, G. C. (1973). *Bayesian Inference in Statistical Analysis*. Reading, MA: Addison-Wesley.

Kass, R. E. (1989). The geometry of asymptotic inference. *Statist. Sci.* **4**, 188–234, (with discussion).

Phillips, P. C. B. (1991). To criticize the critics: an objective Bayesian analysis of stochastic trends. *Journal of Applied Econometrics*, (with discussion), (to appear).

BAYESIAN STATISTICS 4, pp. 61–77
J. M. Bernardo, J. O. Berger, A. P. Dawid and A. F. M. Smith, (Eds.)

Robust Sequential Prediction from Non-random Samples: the Election Night Forecasting Case*

JOSÉ M. BERNARDO and F. JAVIER GIRÓN
Generalitat Valençiana, Spain and *Universidad de Málaga, Spain*

SUMMARY

On Election Night, returns from polling stations occur in a highly non-random manner, thus posing special difficulties in forecasting the final result. Using a data base which contains the results of past elections for all polling stations, a robust hierarchical multivariate regression model is set up which uses the available returns as a training sample and the outcome of the campaign surveys as a prior. This model produces accurate predictions of the final results, even with only a fraction of the returns, and it is extremely robust against data transmission errors.

Keywords: HIERARCHICAL BAYESIAN REGRESSION; PREDICTIVE POSTERIOR DISTRIBUTIONS; ROBUST BAYESIAN METHODS.

1. THE PROBLEM

Consider a situation where, on election night, one is requested to produce a sequence of forecasts of the final result, based on incoming returns. Unfortunately, one cannot treat the available results at a given time as a random sample from all polling stations; indeed, returns from small rural communities typically come in early, with a vote distribution which is far removed from the overall vote distribution.

Naturally, one expects a certain geographical consistency among elections in the sense that areas with, say, a proportionally high socialist vote in the last election will still have a proportionally high socialist vote in the present election. Since the results of the past election are available for each polling station, each incoming result may be compared with the corresponding result in the past election in order to learn about the direction and magnitude of the swing for each party. Combining the results already known with a prediction of those yet to come, based on an estimation of the swings, one may hope to produce accurate forecasts of the final results.

Since the whole process is done in real time, with very limited checking possibilities, it is of paramount importance that the forecast procedure (i) should deal appropriately with missing data, since reports from some polling stations may be very delayed, and (ii) should be fairly robust against the influence of potentially misleading data, such as clerical mistakes in the actual typing of the incoming data, or in the identification of the corresponding polling station.

* This paper has been prepared with partial financial help from project number PB87-0607-C02-01/02 of the *Programa Sectorial de Promoción General del Conocimiento* granted by the *Ministerio de Educación y Ciencia*, Spain. Professor José M. Bernardo is on leave of absence from the *Departamento de Estadística e I.O., Universidad de Valencia*, Spain.

In this paper, we offer a possible answer to the problem described. Section 2 describes a solution in terms of a hierarchical linear model with heavy tailed error distributions. In Section 3, we develop the required theory as an extension of the normal hierarchical model; in Section 4, this theory is applied to the proposed model. Section 5 provides an example of the behaviour of the solution, using data from the last (1989) Spanish general election, where intentional "errors" have been planted in order to test the robustness of the procedure. Finally, Section 6 includes additional discussion and identifies areas for future research.

2. THE MODEL

In the Spanish electoral system, a certain number of parliamentary seats are assigned to each province, roughly proportional to its population, and those seats are allocated to the competing parties using a corrected proportional system known as the Jefferson-d'Hondt algorithm (see e.g., Bernardo, 1984, for details). Moreover, because of important regional differences deeply rooted in history, electoral data in a given region are only mildly relevant to a different region. Thus, a sensible strategy for the analysis of Spanish electoral data is to proceed province by province, leaving for a final step the combination of the different provincial predictions into a final overall forecast.

Let r_{ijkl} be the proportion of the valid vote which was obtained in the last election by party i in polling station j, of electoral district k, in county l of a given province. Here, $i = 1, \ldots, p$, where p is the number of studied parties, $j = 1, \ldots, n_{kl}$, where n_{kl} is the number of polling stations in district k of county l; $k = 1, \ldots, n_l$, where n_l is the number of electoral districts in county l, and $l = 1, \ldots, m$, where m is the number of counties (*municipios*) in the province. Thus, we will be dealing with a total of

$$N = \sum_{l=1}^{m} \sum_{k=1}^{n_l} n_{kl}$$

polling stations in the province, distributed over m counties. For convenience, let $\boldsymbol{r}$ generically denote the p-dimensional vector which contains the past results of a given polling station.

Similarly, let y_{ijkl} be the proportion of the valid vote which party i obtains in the present election in polling station j, of electoral district k, in county l of the province under study. As before, let $\boldsymbol{y}$ generically denote the p-dimensional vector which contains the incoming results of a given polling station.

At any given moment, only some of the $\boldsymbol{y}$'s, say $\boldsymbol{y}_1, \ldots, \boldsymbol{y}_n$, $0 \leq n \leq N$, will be known. An estimate of the final distribution of the vote $\boldsymbol{z} = \{z_1, \ldots, z_p\}$ will be given by

$$\hat{\boldsymbol{z}} = \sum_{i=1}^{n} \omega_i \boldsymbol{y}_i + \sum_{i=n+1}^{N} \omega_i \hat{\boldsymbol{y}}_i, \qquad \sum_{i=1}^{N} \omega_i = 1,$$

where the ω's are the relative weights of the polling stations, in terms of number of voters, and the $\hat{\boldsymbol{y}}_j$'s are estimates of the $N - n$ unobserved $\boldsymbol{y}$'s, to be obtained from the n observed results.

Within each electoral district, one may expect similar political behaviour, so that it seems plausible to assume that the observed swings should be exchangeable, i.e.,

$$\boldsymbol{y}_{jkl} - \boldsymbol{r}_{jkl} = \boldsymbol{\alpha}_{kl} + \boldsymbol{e}_{jkl}, \qquad j = 1, \ldots, n_{kl};$$

where the α's describe the average swings within each electoral district and where, for robustness, the e's should be assumed to be from a heavy tailed error distribution.

Moreover, electoral districts may safely be assumed to be exchangeable within each county, so that

$$\alpha_{kl} = \beta_l + u_{kl}, \qquad k = 1, \ldots, n_l,$$

where the β's describe the average swings within each county and where, again for robustness, the u's should be assumed to be from a heavy tailed error distribution.

Finally, county swings may be assumed to be exchangeable within the province, and thus

$$\beta_l = \gamma + v_l, \qquad l = 1, \ldots, m;$$

where γ describes the average expected swing within the province, which will be assumed to be known from the last campaign survey. Again, for robustness, the distribution of the v's should have heavy tails.

In Section 4, we shall make the specific calculations assuming that e, u and v have p-variate Cauchy distributions, centered at the origin and with known precision matrices P_α, P_β and P_γ which, in practice, are estimated from the swings recorded between the last two elections held. The model may however be easily extended to the far more general class of elliptical symmetric distributions.

From these assumptions, one may obtain the joint posterior distribution of the average swings of the electoral districts, i.e.,

$$p(\alpha_1, \ldots, \alpha_{n_m} \mid y_1, \ldots, y_n, r_1, \ldots, r_N)$$

and thus, one may compute the posterior predictive distribution

$$p(z \mid y_1, \ldots, y_n, r_1, \ldots, r_N)$$

of the final distribution of the vote,

$$z = \sum_{i=1}^{n} \omega_i y_i + \sum_{i=n+1}^{N} \omega_i(\alpha_i + r_i), \qquad \sum_{i=1}^{N} \omega_i = 1,$$

where, for each i, α_i is the swing which corresponds to the electoral district to which the polling station i belongs.

A final transformation, using the d'Hondt algorithm, $s = \text{Hondt}[z]$, which associates a partition

$$s = \{s_1, \ldots, s_p\}, \qquad s_1 + \cdots + s_p = S$$

among the p parties of the S seats allocated to the province as a function of the vote distribution z, may then be used to obtain a predictive posterior distribution

$$p(s \mid y_1, \ldots, y_n, r_1, \ldots, r_N) \tag{2.1}$$

over the possible distributions among the p parties of the S disputed seats.

The predictive distributions thus obtained from each province may finally be combined to obtain the desired final result, i.e., a predictive distribution over the possible Parliamentary seat configurations.

3. ROBUST HIERARCHICAL LINEAR MODELS

One of the most useful models in Bayesian practice is the Normal Hierarchical Linear Model (NHLM) developed by Lindley and Smith (1972) and Smith (1973). In their model the assumption of normality was essential for the derivation of the exact posterior distributions of the parameters of every hierarchy and the corresponding predictive likelihoods. Within this setup, all the distributions involved were normal and, accordingly, the computation of all parameters in these distributions was straightforward. However, the usefulness of the model was limited, to a great extent, by the assumption of independent normal errors in every stage of the hierarchy. In this section,

(i) We first generalize the NHLM model to a multivariate setting, to be denoted NMHLM, in a form which may be extended to more general error structures.

(ii) We then generalize that model to a Multivariate Hierarchical Linear Model (MHLM) with rather general error structures, in a form which retains the main features of the NMHLM.

(iii) Next, we show that the MHLM is weakly robust, in a sense to be made precise later, which, loosely speaking, means that the usual NMHLM estimates of the parameters in every stage are distribution independent for a large class of error structures.

(iv) We then develop the theory, and give exact distributional results, for error structures which may be written as scale mixtures of matrix-normal distributions.

(v) Finally, we give more precise results for the subclass of Student's matrix-variate t distributions.

These results generalize the standard multivariate linear model and also extend some previous work by Zellner (1976) for the usual linear regression model.

A k-stage general multivariate normal hierarchical linear model MNHLM, which generalizes the usual univariate model, is given by the following equations, each representing the conditional distribution of one hyperparameter given the next in the hierarchy. It is supposed that the last stage hyperparameter, Θ_k, is known.

$$\begin{aligned} \boldsymbol{Y} \mid \Theta_1 &\sim N(\boldsymbol{A}_1\Theta_1, \boldsymbol{C}_1 \otimes \Sigma) \\ \Theta_i \mid \Theta_{i+1} &\sim N(\boldsymbol{A}_{i+1}\Theta_{i+1}, \boldsymbol{C}_{i+1} \otimes \Sigma); \qquad i = 1, \dots k-1. \end{aligned} \tag{3.1}$$

In these equations $\boldsymbol{Y}$ is an $n \times p$ matrix which represents the observed data, the Θ_i's are the i-th stage hyperparameter matrices of dimensions $n_i \times p$ and the $\boldsymbol{A}_i$'s are design matrices of dimensions $n_{i-1} \times n_i$ (assuming that $n_0 = n$). The $\boldsymbol{C}_i$'s are positive definite matrices of dimensions $n_{i-1} \times n_{i-1}$ and, finally, Σ is a $p \times p$ positive definite matrix. The matrix of means for the conditional matrix-normal distribution at stage i is $\boldsymbol{A}_i\Theta_i$ and the corresponding covariance matrix is $\boldsymbol{C}_i \otimes \Sigma$, where $\otimes$ denotes the Kronecker product of matrices.

From this model, using standard properties of the matrix-normal distributions, one may derive the marginal distribution of the hyperparameter Θ_i, which is given by

$$\Theta_i \sim N(\boldsymbol{B}_{ik}\Theta_k, \boldsymbol{P}_i \otimes \Sigma), \qquad i = 1, \dots k-1,$$

where

$$\begin{aligned} \boldsymbol{B}_{ij} &= \boldsymbol{A}_{i+1} \cdots \boldsymbol{A}_j, \qquad i < j; \\ \boldsymbol{P}_i &= \boldsymbol{C}_{i+1} + \sum_{j=i+1}^{k-1} \boldsymbol{B}_{ij}\boldsymbol{C}_{j+1}\boldsymbol{B}'_{ij}. \end{aligned}$$

The predictive distribution of $\boldsymbol{Y}$ given Θ_i is

$$\boldsymbol{Y} \mid \Theta_i \sim N(\boldsymbol{A}_i^*\Theta_i, \boldsymbol{Q}_i \otimes \Sigma),$$

where

$$\boldsymbol{A}_i^* = \boldsymbol{A}_0\boldsymbol{A}_1\cdots\boldsymbol{A}_i \quad \text{with} \quad \boldsymbol{A}_0 = \boldsymbol{I};$$
$$\boldsymbol{Q}_i = \sum_{j=0}^{i-1} \boldsymbol{A}_j^*\boldsymbol{C}_{j+1}\boldsymbol{A}_i^{*\prime}.$$

From this, the posterior distribution of Θ_i given the data $\boldsymbol{Y}$, $\{\boldsymbol{A}_i\}$ and $\{\boldsymbol{C}_i\}$ is

$$\Theta_i \mid \boldsymbol{Y} \sim N(\boldsymbol{D}_i\boldsymbol{d}_i, \boldsymbol{D}_i \otimes \Sigma),$$

with

$$\boldsymbol{D}_i^{-1} = \boldsymbol{A}_i^{*\prime}\boldsymbol{Q}_i^{-1}\boldsymbol{A}_i^* + \boldsymbol{P}_i^{-1};$$
$$\boldsymbol{d}_i = \boldsymbol{A}_i^{*\prime}\boldsymbol{Q}_i^{-1}\boldsymbol{Y} + \boldsymbol{P}_i^{-1}\boldsymbol{B}_{ik}\Theta_k.$$

In order to prove the basic result of this section, the MNHLM (3.1) can be more usefully written in the form

$$\begin{aligned} \boldsymbol{Y} &= \boldsymbol{A}_1\Theta_1 + \boldsymbol{U}_1 \\ \Theta_i &= \boldsymbol{A}_{i+1}\Theta_{i+1} + \boldsymbol{U}_{i+1}; \qquad i = 1, \dots k-1, \end{aligned} \tag{3.2}$$

where the matrix of error terms $\boldsymbol{U}_i$ are assumed independent $N(\boldsymbol{O}, \boldsymbol{C}_i \otimes \Sigma)$ or, equivalently, that the matrix $\boldsymbol{U} = (\boldsymbol{U}_1, \dots, \boldsymbol{U}_k)$ is distributed as

$$\begin{pmatrix} \boldsymbol{U}_1 \\ \vdots \\ \boldsymbol{U}_k \end{pmatrix} \sim N\left[\begin{pmatrix} \boldsymbol{O} \\ \vdots \\ \boldsymbol{O} \end{pmatrix}; \begin{pmatrix} \boldsymbol{C}_1 & \dots & \boldsymbol{O} \\ \vdots & \ddots & \vdots \\ \boldsymbol{O} & \dots & \boldsymbol{C}_k \end{pmatrix} \otimes \Sigma\right]. \tag{3.3}$$

Predictive distributions for future data $\boldsymbol{Z}$ following the linear model

$$\boldsymbol{Z} = \boldsymbol{W}_1\Theta_1 + \boldsymbol{U}_W, \qquad \boldsymbol{U}_W \sim N(\boldsymbol{O}, \boldsymbol{C}_W \otimes \Sigma), \tag{3.4}$$

where $\boldsymbol{Z}$ is a $m \times p$ matrix and $\boldsymbol{U}_W$ is independent of the matrix $\boldsymbol{U}$, can now be easily derived. Indeed, from properties of the matrix-normal distributions it follows that

$$\boldsymbol{Z} \mid \boldsymbol{Y} \sim N(\boldsymbol{W}\boldsymbol{D}_1\boldsymbol{d}_1, (\boldsymbol{W}\boldsymbol{D}_i\boldsymbol{W}' + \boldsymbol{C}_W) \otimes \Sigma). \tag{3.5}$$

Suppose now that the error vector $\boldsymbol{U}$ is distributed according to the scale mixture

$$\boldsymbol{U} \sim \int N(0, \boldsymbol{C} \otimes \Lambda)\, dF(\Lambda), \tag{3.6}$$

where $\boldsymbol{C}$ represents the matrix whose diagonal elements are the matrices $\boldsymbol{C}_i$ and the remaining elements are zero matrices of the appropriate dimensions, i.e., the diagonal covariance matrix of equation (3.3), and $F(\Lambda)$ is any matrix-distribution with support in the class of positive definite $p \times p$ matrices. Clearly, the usual MNHLM (3.2) can be viewed as choosing a degenerate distribution at $\Lambda = \Sigma$ for F, while, for example, the hypothesis of $\boldsymbol{U}$ being distributed as a matrix-variate Student t distribution is equivalent to F being distributed as an inverted-Wishart distribution with appropriate parameters.

With this notation we can state the following theorem

Theorem 3.1 . *If the random matrix U is distributed according to (3.6), then*

i) the marginal distribution of Θ_i is

$$\Theta_i \sim \int N(B_{ik}\Theta_k, P_i \otimes \Lambda)\, dF(\Lambda) \qquad i = 1, \ldots k-1;$$

ii) the predictive distribution of Y given Θ_i is

$$Y \mid \Theta_i \sim \int N(A_i^*\Theta_i, Q_i \otimes \Lambda)\, dF(\Lambda \mid \Theta_i), \qquad i = 1, \ldots k-1;$$

where the posterior distribution of Λ given Θ_i, $F(\Lambda \mid \Theta_i)$, is given by

$$dF(\Lambda \mid \Theta_i) \propto |\Lambda|^{-n_i/2} \exp\Big\{ -\frac{1}{2} tr\Lambda^{-1}(\Theta_i - B_{ik}\Theta_k)' P_i^{-1}(\Theta_i - B_{ik}\Theta_k)\Big\} dF(\Lambda);$$

iii) the posterior distribution of Θ_i given the data Y is

$$\Theta_i \mid Y \sim \int N(D_i d_i, D_i \otimes \Lambda)\, dF(\Lambda \mid Y), \qquad i = 1, \ldots k-1;$$

where the posterior distribution of Λ given Y, $F(\Lambda \mid Y)$, is given by

$$dF(\Lambda \mid Y) \propto |\Lambda|^{-n/2} \exp\Big\{ -\frac{1}{2} tr\Lambda^{-1}(Y - A_k^*\Theta_k)' Q_k^{-1}(Y - A_k^*\Theta_k)\Big\} dF(\Lambda).$$

Proof. The main idea is, simply, to work conditionally on the scale hyperparameter Λ and, then, apply the results of the MNHLM stated above.

Conditionally on Λ, the error matrices U_i are independent and normally distributed as $U_i \sim N(O, C_i \otimes \Lambda)$; therefore, with the same notation as above, we have

$$\Theta_i \mid \Lambda \sim N(B_{ik}\Theta_k, P_i \otimes \Lambda),$$
$$Y \mid \Theta_i, \Lambda \sim N(A_i^*\Theta_i, Q_i \otimes \Lambda),$$

and

$$\Theta_i \mid Y, \Lambda \sim N(D_i d_i, D_i \otimes \Lambda); \qquad i = 1, \ldots, k.$$

Now, by Bayes theorem,

$$\frac{dF(\Lambda \mid \Theta_i)}{dF(\Lambda)} \propto g(\Theta_i \mid \Lambda), \qquad \frac{dF(\Lambda \mid Y)}{dF(\Lambda)} \propto h(Y \mid \Lambda),$$

where $g(\Theta_i \mid \Lambda)$ and $h(Y \mid \Lambda)$ represent the conditional densities of Θ_i given Λ and Y given Λ, which are $N(B_{ik}\Theta_k, P_i \otimes \Lambda)$ and $N(A_k^*\Theta_k, Q_k \otimes \Lambda)$, respectively.

From this, by integrating out the scale hyperparameter Λ with respect to the corresponding distribution, we obtain the stated results. $\triangleleft$

The theorem shows that all distributions involved are also scale mixtures of matrix-normal distributions. In particular, the most interesting distributions are the posteriors of the hyperparameters at every stage given the data, i.e., $\Theta_i \mid Y$. These distributions turn out to be just a scale mixture of matrix-normals. This implies that the usual modal estimator of the Θ_i's, i.e., the mode of the posterior distribution, which is also the matrix of means for those F's with finite first moments, is $D_i d_i$, whatever the prior distribution F of Λ. In this sense, these estimates are robust, that is, they do not depend on F. However, other parameters and characteristics of these distributions such as the H.P.D. regions for the hyperparameters in the hierarchy depend on the distribution F of Λ.

Note that from this theorem and formula (3.5) we can also compute the predictive distribution of future data $\boldsymbol{Z}$ generated by the model (3.4), which is also a scale mixture.

$$\boldsymbol{Z} \,|\, \boldsymbol{Y} \sim \int N(\boldsymbol{W}\boldsymbol{D}_1\boldsymbol{d}_1, (\boldsymbol{W}\boldsymbol{D}_1\boldsymbol{W}' + \boldsymbol{C}_W) \otimes \Lambda)\, dF(\Lambda \,|\, \boldsymbol{Y}). \tag{3.7}$$

More precise results can be derived for the special case in which the $\boldsymbol{U}$ matrix is distributed as a matrix-variate Student t. For the definition of the matrix-variate Student t, we follow the same notation as in Box and Tiao (1973, Chapter 8).

Theorem 3.2. *If $\boldsymbol{U} \sim t(\boldsymbol{O}, \boldsymbol{C}, \boldsymbol{S}; \nu)$ with dispersion matrix $\boldsymbol{C} \otimes \boldsymbol{S}$ and ν degrees of freedom, then*

(i) the posterior distribution of Θ_i given $\boldsymbol{Y}$ is

$$\Theta_i \,|\, \boldsymbol{Y} \sim t_{n_i p}(\boldsymbol{D}_i \boldsymbol{d}_i, \boldsymbol{D}_i, (\boldsymbol{S} + \boldsymbol{T}); \nu + n),$$

where the matrix $\boldsymbol{T} = (\boldsymbol{Y} - \boldsymbol{A}_k^* \Theta_k)' \boldsymbol{Q}_k^{-1} (\boldsymbol{Y} - \boldsymbol{A}_k^* \Theta_k)$;

(ii) the posterior distribution of Λ is an inverted-Wishart,

$$\Lambda \,|\, \boldsymbol{Y} \sim InW(\boldsymbol{S} + \boldsymbol{T}, \nu + n).$$

(iii) the predictive distribution of $\boldsymbol{Z} = \boldsymbol{W}_1 \Theta_1 + \boldsymbol{U}_W$ is

$$\boldsymbol{Z} \,|\, \boldsymbol{Y} \sim t_{mp}(\boldsymbol{W}\boldsymbol{D}_1\boldsymbol{d}_1, (\boldsymbol{W}\boldsymbol{D}_1\boldsymbol{W}' + \boldsymbol{C}_W), \boldsymbol{S} + \boldsymbol{T}; \nu + n).$$

Proof. The first result is a simple consequence of the fact that a matrix-variate Student t distribution is a scale mixture of matrix-variate normals. More precisely, if $\boldsymbol{U} \sim t(\boldsymbol{O}, \boldsymbol{C}, \boldsymbol{S}; \nu)$, then $\boldsymbol{U}$ is the mixture given by (3.6), with $F \sim InW(\boldsymbol{S}, \nu)$.

From this representation and Theorem 3.1. iii), we obtain that the inverted-Wishart family for Λ is a conjugate one. In fact,

$$\begin{aligned} \frac{dF(\Lambda \,|\, \boldsymbol{Y})}{d\Lambda} &\propto |\Lambda|^{-n/2} \exp\Big\{ -\frac{1}{2} \mathrm{tr} \Lambda^{-1} \boldsymbol{T} \Big\} \cdot |\Lambda|^{-(\nu/2+p)} \exp\Big\{ -\frac{1}{2} \mathrm{tr} \Lambda^{-1} \boldsymbol{S} \Big\} \\ &\propto |\Lambda|^{-((\nu+n)/2+p)} \exp\Big\{ -\frac{1}{2} \mathrm{tr} \Lambda^{-1} (\boldsymbol{T} + \boldsymbol{S}) \Big\}; \end{aligned}$$

and (ii) follows. Finally, substitution of (ii) into (3.7) establishes (iii). ◁

4. PREDICTIVE POSTERIOR DISTRIBUTIONS OF INTEREST

In this section we specialize the results just established to the particular case of the model described in Section 2. In order to derive the predictive distribution of the random quantity $\boldsymbol{z}$ let us introduce some useful notation. Let $\boldsymbol{Y}$ denote the full $N \times p$ matrix whose rows are the vectors $\boldsymbol{y}_i$ of observed and potentially observed results, as defined in Section 2. Partition this matrix into the already observed part $\boldsymbol{y}_1, \ldots, \boldsymbol{y}_n$, i.e., the $n \times p$ matrix $\boldsymbol{Y}_1$ and the unobserved part, the $(N - n) \times p$ matrix $\boldsymbol{Y}_2$ formed with the remaining $N - n$ rows of $\boldsymbol{Y}$. Let $\boldsymbol{R}$ denote the $N \times p$ matrix whose rows are the vectors $\boldsymbol{r}_i$ of past results and $\boldsymbol{R}_1$, $\boldsymbol{R}_2$ the corresponding partitions. By $\boldsymbol{X}$ we denote the matrix of swings, i.e., $\boldsymbol{X} = \boldsymbol{Y} - \boldsymbol{R}$ with $\boldsymbol{X}_1$,

$\boldsymbol{X}_2$ representing the corresponding partitions. Finally, let ω be the row vector of weights $(\omega_1, \ldots, \omega_N)$ and ω_1 and ω_2 the corresponding partition.

With this notation the model presented in Section 2, which in a sense is similar to a random effect model with missing data, can be written as a hierarchical model in three stages as follows

$$\begin{aligned} \boldsymbol{X}_1 &= \boldsymbol{A}_1\Theta_1 + \boldsymbol{U}_1, \\ \Theta_1 &= \boldsymbol{A}_2\Theta_2 + \boldsymbol{U}_2, \\ \Theta_2 &= \boldsymbol{A}_3\Theta_3 + \boldsymbol{U}_3; \end{aligned} \tag{4.1}$$

where $\boldsymbol{X}_1$ is a $n \times p$ matrix of known data, whose rows are of the form $\boldsymbol{y}_{jkl} - \boldsymbol{r}_{jkl}$ for those indexes corresponding to the observed data $\boldsymbol{y}_1, \ldots, \boldsymbol{y}_n$, Θ_1 is an $N \times p$ matrix whose rows are the p-dimensional vectors $\boldsymbol{\alpha}_{kl}$, Θ_2 is an $m \times p$ matrix whose rows are the p-dimensional vectors $\boldsymbol{\beta}_l$ and, finally, Θ_3 is the p-dimensional row vector $\boldsymbol{\gamma}$. The matrices $\boldsymbol{A}_i$ for $i = 1, 2, 3$ have special forms; in fact $\boldsymbol{A}_1$ is an $n \times N$ matrix whose rows are N-dimensional unit vectors, with the one in the place that matches the polling station in district k of county l from which the data arose. $\boldsymbol{A}_2$ is an $N \times m$ matrix whose rows are m-dimensional units vectors, as follows: the first n_1 rows are equal to the unit vector $\boldsymbol{e}_1$, the next n_2 rows are equal to the unit vector $\boldsymbol{e}_2$, and so on, so that the last n_m rows are equal to the unit vector $\boldsymbol{e}_m$. Finally, the $m \times 1$ matrix $\boldsymbol{A}_3$ is the m-dimensional column vector $(1, \ldots, 1)$.

The main objective is to obtain the predictive distribution of $\boldsymbol{z}$ given the observed data $\boldsymbol{y}_1, \ldots, \boldsymbol{y}_n$ and the results from the last election $\boldsymbol{r}_1, \ldots, \boldsymbol{r}_N$. From this, using the d'Hondt algorithm, it is easy to obtain the predictive distribution of the seats among the p parties.

The first step is to derive the posterior of the $\boldsymbol{\alpha}$'s or, equivalently, the posterior of Θ_1 given $\boldsymbol{Y}$ or, equivalently, $\boldsymbol{X}_1$.

From Theorem 3.2, for $k = 3$ we have

$$\begin{aligned} \boldsymbol{D}_1^{-1} &= \boldsymbol{A}_1'\boldsymbol{C}_1^{-1}\boldsymbol{A}_1 + (\boldsymbol{C}_2 + \boldsymbol{A}_2\boldsymbol{C}_3\boldsymbol{A}_2')^{-1} \\ \boldsymbol{d}_1 &= \boldsymbol{A}_1'\boldsymbol{C}_1^{-1}\boldsymbol{X}_1 + (\boldsymbol{C}_2 + \boldsymbol{A}_2\boldsymbol{C}_3\boldsymbol{A}_2')^{-1}\boldsymbol{A}_2\boldsymbol{A}_3\boldsymbol{\gamma}. \end{aligned}$$

The computation of $\boldsymbol{D}^{-1}$ involves the inversion of an $N \times N$ matrix. Using standard matrix identities, $\boldsymbol{D}^{-1}$ can also be written in the form

$$\boldsymbol{D}_1^{-1} = \boldsymbol{A}_1'\boldsymbol{C}_1^{-1}\boldsymbol{A}_1 + \boldsymbol{C}_2^{-1} - \boldsymbol{C}_2^{-1}\boldsymbol{A}_2(\boldsymbol{A}_2'\boldsymbol{C}_2^{-1}\boldsymbol{A}_2 + \boldsymbol{C}_3^{-1})^{-1}\boldsymbol{A}_2'\boldsymbol{C}_2^{-1}$$

which may be computationally more efficient when the matrix $\boldsymbol{C}_2$ is diagonal and m, as in our case, is much smaller than N.

Further simplification in the formulae and subsequent computations result from the hypothesis of exchangeability of the swings formulated in Section 2. This implies that the matrices $\boldsymbol{C}_i$ are of the form $k_i\boldsymbol{I}$, where k_i are positive constants and $\boldsymbol{I}$ are identity matrices of the appropiate dimensions.

Now, the predictive model for future observations is

$$\boldsymbol{X}_2 = \boldsymbol{Y}_2 - \boldsymbol{R}_2 = \boldsymbol{W}\Theta_1 + \boldsymbol{U}_W, \qquad \boldsymbol{U}_W \sim N(\boldsymbol{O}, \boldsymbol{C}_W \otimes \boldsymbol{S});$$

where $\boldsymbol{W}$ is the $(N - n) \times N$ matrix whose rows are N-dimensional unit vectors that have exactly the same meaning as those of matrix $\boldsymbol{A}_1$.

Then, using the results of the preceding section, the predictive ditribution of $\boldsymbol{Y}_2$ given the data $\boldsymbol{Y}_1$ and $\boldsymbol{R}$ is

$$\boldsymbol{Y}_2 \sim t_{(N-n)p}(\boldsymbol{R}_2 + \boldsymbol{W}\boldsymbol{D}_1\boldsymbol{d}_1, \boldsymbol{W}\boldsymbol{D}_1\boldsymbol{W}' + \boldsymbol{C}_W, \boldsymbol{S} + (\boldsymbol{Y}_1 - \boldsymbol{1}\boldsymbol{\gamma})'\boldsymbol{Q}_3^{-1}(\boldsymbol{Y}_1 - \boldsymbol{1}\boldsymbol{\gamma}); \nu + n)$$

due to the fact that the matrix $\boldsymbol{A}_3^* = 1$, where 1 is an n column vector with all entries equal to 1.

From this distribution, using properties of the matrix-variate Student t, the posterior of $\boldsymbol{z}$ which is a linear combination of $\boldsymbol{Y}_2$ is

$$\boldsymbol{z} \mid \boldsymbol{Y}_1, \boldsymbol{R} \sim t_{1p}(\omega_1 \boldsymbol{Y}_1 + \omega_2 \boldsymbol{R}_2 + \omega_2 \boldsymbol{W} \boldsymbol{D}_1 \boldsymbol{d}_1, \\ \omega_2(\boldsymbol{W}\boldsymbol{D}_1\boldsymbol{W}' + \boldsymbol{C}_W)\omega_2', \boldsymbol{S} + (\boldsymbol{Y}_1 - 1\boldsymbol{\gamma})'\boldsymbol{Q}_3^{-1}(\boldsymbol{Y}_1 - 1\boldsymbol{\gamma}); \nu + n).$$

This matrix-variate t is, in fact, a multivariate Student t distribution, so that, in the notation of Section 2,

$$p(\boldsymbol{z} \mid \boldsymbol{y}_1, \ldots; \boldsymbol{y}_n, \boldsymbol{r}_1, \ldots, \boldsymbol{r}_N) = \mathrm{St}_p(\boldsymbol{z} \mid \boldsymbol{m}_z, \boldsymbol{S}_z, \nu + n) \tag{4.2}$$

i.e., a p-dimensional Student t, with mean

$$\boldsymbol{m}_z = \omega_1 \boldsymbol{Y}_1 + \omega_2 \boldsymbol{R}_2 + \omega_2 \boldsymbol{W} \boldsymbol{D}_1 \boldsymbol{d}_1,$$

dispersion matrix,

$$\frac{\omega_2(\boldsymbol{W}\boldsymbol{D}_1\boldsymbol{W}' + \boldsymbol{C}_W)\omega_2'}{\nu + n}(\boldsymbol{S} + (\boldsymbol{Y}_1 - 1\boldsymbol{\gamma})'\boldsymbol{Q}_3^{-1}(\boldsymbol{Y}_1 - 1\boldsymbol{\gamma});$$

and $\nu + n$ degrees of freedom.

5. A CASE STUDY: THE 1989 SPANISH GENERAL ELECTION

The methodology described in Section 4 has been tested using the results, for the Province of Valencia, of the last two elections which have been held in Spain, namely the European Parliamentary Elections of June 1989, and the Spanish General Elections of October 1989.

The Province of Valencia has N = 1566 polling stations, distributed among m = 264 counties. The number n_l of electoral districts whithin each county varies between 1 and 19, and the number n_{kl} of polling stations within each electoral district varies between 1 and 57.

The outcome of the October General Election for the p = 5 parties with parliamentary representation in Valencia has been predicted, pretending that their returns are partially unknown, and using the June European Elections as the database. The parties considered were PSOE (socialist), PP (conservative), CDS (liberal), UV (conservative regionalist) and IU (communist).

	5%			20%			90%			**Final**
	Mean	Dev.	Error	Mean	Dev.	Error	Mean	Dev.	Error	
PSOE	40.08	0.46	−0.43	40.39	0.40	−0.13	40.50	0.16	−0.02	40.52
PP	23.72	0.49	−0.40	24.19	0.45	0.07	24.19	0.18	0.07	24.12
CDS	6.28	0.36	−0.20	6.33	0.33	−0.15	6.49	0.13	0.01	6.49
UV	11.88	0.50	0.44	11.62	0.46	0.17	11.42	0.17	−0.02	11.45
IU	10.05	0.40	0.03	9.93	0.37	−0.09	10.01	0.14	−0.02	10.02

Table 1. *Evolution of the percentages of valid votes.*

For several proportions of known returns (5%, 20% and 90% of the total number of votes), Table 1 shows the means and standard deviations of the marginal posterior distributions of the percentages of valid votes obtained by each of the five parties. The absolute error of the means with respect to the final result actually obtained are also quoted.

It is fairly impressive to observe that, with only 5% of the returns, the absolute errors of the posterior modes are all smaller than 0.5%, and that those errors drop to about 0.15% with just 20% of the returns, a proportion of the vote which is usually available about two hours after the polling stations close. With 90% of the returns, we are able to quote a "practically final" result without having to wait for the small proportion of returns which typically get delayed for one reason or another; indeed, the errors all drop below 0.1% and, on election night, vote percentages are never quoted to more than one decimal place.

In Table 2, we show the evolution, as the proportion of the returns grows, of the posterior probability distribution over the possible allocation of the S=16 disputed seats.

PSOE	PP	CDS	UV	IU	5%	20%	90%	**Final**
8	4	1	2	1	0.476	0.665	0.799	1.000
7	4	1	2	2	0.521	0.324	0.201	0.000
7	5	1	2	1	0.003	0.010	0.000	0.000

Table 2. *Evolution of the probability distribution over seat partitions.*

Interestingly, two seat distributions, namely {8, 4, 1, 2, 1} and {7, 4, 1, 2, 2}, have a relatively large probability from the very beginning. This gives advance warning of the fact that, because of the intrinsically discontinuous features of the d'Hondt algorithm, the last seat is going to be allocated by a few number of votes, to either the socialists or the communists. In fact, the socialists won that seat, but, had the communists obtained 1,667 more votes (they obtained 118,567) they would have won that seat.

Tables 1 and 2 are the product of a very realistic simulation. The numbers appear to be very stable even if the sampling mechanism in the simulation is heavily biased, as when the returns are introduced by city size. The next Valencia State Elections will be held on May 26th, 1991; that night, will be the *première* of this model in *real* time.

6. DISCUSSION

The multivariate normal model NMHLM developed in Section 3 is a natural extension of the usual NHLM; indeed, this is just the particular case which obtains when $p = 1$ and the matrix S is an scalar equal to 1. As defined in (3.1), our multivariate model imposes some restrictions on the structure of the global covariance matrix but, this is what makes possible the derivation of simple formulae for the posterior distributions of the parameters and for the predictive distributions of future observations, all of which are matrix-variate-normal. Moreover, within this setting it is also possible, as we have demonstrated, to extend the model to error structures generated by scale mixtures of matrix-variate-normals. Actually, this may be futher extended to the class of elliptically symmetric distributions, which contains the class of scale mixtures of matrix-variate-normals as a particular case; this will be reported elsewhere. Without the restrictions we have imposed on the covariance structure, further progress on the general model seems difficult.

One additional characteristic of this hierarchical model, that we have not developed in this paper but merits careful attention, is the possibility of sequential updating of the hyperparameters, in a Kalman-like fashion, when the observational errors are assumed to be conditionally independent given the scale matrix hyperparameter. The possibility of combining the flexibility of modelling the data according to a hierarchical model, with the computational advantages of the sequential characteristics of the Kalman filter deserves, we believe, some attention and further research.

As shown in our motivating example, the use of sophisticated Bayesian modelling in forecasting may provide qualitatively different answers, to the point of modifying the possible uses of the forecast.

REFERENCES

Bernardo, J. M. (1984). Monitoring the 1982 Spanish Socialist victory: a Bayesian analysis. *J. Amer. Statist. Assoc.* **79**, 510–515.

Box, G. E. P. and Tiao, G. C. (1973). *Bayesian Inference in Statistical Analysis.* Reading, MA: Addison-Wesley.

Lindley, D. V. and Smith, A. F. M. (1972). Bayes estimates for the linear model. *J. Roy. Statist. Soc. B* **34**, 1–41, (with discussion).

Smith, A. F. M. (1973). A general Bayesian linear model. *J. Roy. Statist. Soc. B* **35**, 67–75.

Zellner, A. (1976). Bayesian and non-Bayesian analysis of the regression model with multivariate Student-*t* error terms. *J. Amer. Statist. Assoc.* **71**, 400–405.

APPENDIX

Tables 3 and 4 below describe, with the notation used in Tables 1 and 2, what actually happened in the Province of Valencia on election night, May 26th, 1991, when $S = 37$ State Parliament seats were being contested.

	5%			20%			90%			**Final**
	Mean	Dev.	Error	Mean	Dev.	Error	Mean	Dev.	Error	
PSOE	41.5	3.6	−1.0	41.6	2.6	−0.9	42.4	2.2	−0.1	42.5
PP	23.5	3.1	0.0	23.4	2.8	−0.1	23.5	1.9	0.0	23.5
CDS	4.4	1.4	1.9	4.8	0.5	2.3	2.9	0.5	0.4	2.5
UV	14.4	2.3	−2.0	13.6	1.3	−2.8	16.0	2.0	−0.4	16.4
IU	9.2	2.0	0.9	9.4	2.2	1.1	8.6	1.9	0.3	8.3

Table 3. *Evolution of the percentages of valid votes.*

PSOE	PP	CDS	UV	IU	5%	20%	90%	**Final**
18	10	0	6	3	0.06	0.02	0.82	1.00
18	9	0	7	3	0.03	0.02	0.04	0.00
17	10	2	5	3	0.03	0.47	0.01	0.00
17	9	2	5	4	0.03	0.17	0.01	0.00
17	10	1	6	3	0.36	0.02	0.01	0.00
18	9	1	6	3	0.11	0.02	0.01	0.00

Table 4. *Evolution of the probability distribution over seat partitions.*

It is easily appreciated by comparison that both the standard deviations of the marginal posteriors, and the actual estimation errors, were far larger in real life than in the example. A general explanation lies in the fact that state elections have a far larger local component than national elections, so that variances within strata were far larger, specially with the regionalists (UV). Moreover, the liberals (CDS) performed very badly in this election (motivating the resignation from their leadership of former prime minister Adolfo Suarez); this poor performance was very inhomogeneous, however, thus adding to the inflated variances. Nevertheless, essentially accurate final predictions were made with 60% of the returns, and this was done over two hours before any other forecaster was able to produce a decent approximation to the final results.

DISCUSSION

L. R. PERICCHI (*Universidad Simón Bolívar, Venezuela*)

This paper addresses a problem that has captured statisticians' attention in the past. It is one of these public problems where the case for sophisticated statistical techniques, and moreover the case for the Bayesian approach, is put to the test: quick and accurate forecasts are demanded.

The proposal described here has some characteristics in common with previous approaches and some novel improvements. In general this article raises issues of modelling and robustness.

The problem is one on which there is substantial prior information from different sources, like past elections, surveys, etc. Also, exchangeability relationships in a hierarchy are natural. Furthermore, the objective is one of prediction in the form of a probability distribution of the possible configurations of the parliament. Thus, not surprisingly, this paper, as previous articles on the same subject, Brown and Payne (1975, 1984) and Bernardo (1984), have obtained shrinkage estimators, "borrowing strength", setting the problem as a Bayesian Hierarchical Linear model. Bernardo and Girón in the present article get closer to the Brown and Payne modelling than that of Bernardo (1984), since they resort to modelling directly the "swings" rather than modelling the log-odds of the multinomial probabilities. All this, coupled with the great amount of prior information, offers the possibility of very accurate predictions from the very begining of the exercise.

A limitation of the model, as has been pointed out by the authors, is the lack of sequential updating. The incoming data is highly structured —there is certainly a bias of order of declaration— producing a trend rather than a random ordering. This prompts the need for sequential updating in a dynamic model that may be in place just before the election, as the authors confirmed in their verbal reply to the discussion.

The second limitation is in our opinion of even greater importance and that is the lack of "strong" robustness (see below), protecting against unbounded influence of wrong information of counts and/or wrong classification of polling stations; i.e. gross errors or atypical data should not influence unduly the general prediction of the swings. The usual hierarchical normal model has been found extremely sensitive to gross errors, possibly producing large shrinkages in the wrong direction.

At this point a short general discussion is in order. The term 'Bayesian Robustness' covers a wide field within which it can have quite different meanings. The first meaning begins with the recognition of the inevitability of imprecision of probability specifications. Even this first approach admits two different interpretations (that have similarities but also important differences). One is the "*sensitivy analysis*" interpretation (Berger, 1990), which is widely known. The second is the *upper and lower probability* interpretation. The latter is a more radical departure from precise analysis, which rejects the usual axiomatic foundations and derives directly the lower probability from its own axioms for rational behaviour, (Walley, 1990). The second meaning of robustness is closer to the Huber-Hampel notion of assuming models (likelihoods and/or priors) that avoid unbounded influence of assumptions, but still work with a single probability model. The present paper uses this second meaning of robustness.

The authors address the need for robustness by replacing the normal errors throughout, by scale mixtures of normal errors. Scale mixtures of normal errors as outlier prone distributions have a long history in Bayesian analyses. They were, perhaps, first proposed as a Bayesian way of dealing with outliers by de Finetti (1961) and have been sucessfully used in static and dynamic linear regression, West (1981, 1984).

Let us note in passing that the class of scale mixture of normals has been considered as a class (in the first meaning of robustness mentioned above) by Moreno and Pericchi (1990). They consider an ε-contaminated model but the base prior π_0 is a scale mixture and the mixing distribution is only assumed to belong to a class H, i.e.

$$\Gamma_{\varepsilon,\pi_0}(H,Q) = \left\{ \pi(\theta) = (1-\varepsilon)\int \pi_0(\theta|r)h(dr) + \varepsilon q(\theta), q \in Q, h \in H \right\}$$

Examples of different classes of mixing distributions considered are

$$H_1 = \left\{ h(d_r) : \int_0^{r_i} h(d_r) = h_i, i = 1 \ldots n \right\}$$

$$H_2 = \left\{ h(d_r) : h(r) \text{ unimodal at } r_0 \text{ and } \int_0^{r_0} h(d_r) = h_0 \right\}$$

When π_0 is normal and $\varepsilon = 0$ then $\Gamma(H)$ is the class of scale mixtures of normal distributions with mixing distributions in H. The authors report sensible posterior ranges for probabilities of sets using H_1 and H_2.

Going back to the particular scale mixture of normals considered by Bernardo and Girón, they first conveniently write the usual Multivariate Normal Hierarchical model and by restricting to a common scale matrix (Σ in (3.3) or Λ in (3.6)), they are able to obtain an elegant expression of the posterior distributions (Theorem 3.1.). Furthermore in Theorem 3.2, by specializing to a particular combination of Student-t distributions, they are able to get closed form results. This would be surprising, were it not for Zellner's (1976) conjecture: "similar results (as those for regression) will be found with errors following a matrix Student-t". However, as with Zellner's results the authors get "weak" rather than "strong" robustness, in the sense that the posterior mean turns out to be linear in the observations (and therefore non-robust), although other characteristics of the distributions will be robust. However, "*strong*" robustness is what is required, and some *ad hoc* ways to protect against outlying data (like screening) may be required. Also, approximations on combination of models that yield "strong" robustness may be more useful than exact results. Having said that, we should bear in mind that compromises due to time pressure on election night, may have to be made given the insufficient development of the theory of scale mixtures of normals.

Finally, we remark that the elegant (even if too restricted) development of this paper opens wide possibilities for modelling. We should strive for more theoretical insight in the scale mixture of normals, to guide the assessment. For example O'Hagan's "Credence" theory is still quite incomplete. Moreover, scale mixture of normals offers a much wider choice than just the Student-t, that should be explored. So far Bernardo and Girón have shown us encouraging simulations. Let us wish them well on the actual election night.

A. P. DAWID (*University College London, UK*)

It seems worth emphasising that the "robustness" considered in this paper refers to the invariance of the results (formulae for means) in the face of varying Σ in (3.3) or (what is equivalent) the distribution F of (3.6). This distribution can be thought of either as part of the prior (Σ being a parameter) or, on using (3.6) in (3.2), as part of the model — although note that, in this latter case, the important independence (Markov) properties of the system (3.2) are lost. Relevant theory and formulae for both the general "left-spherical" case and the particular Student-t case may be found in Dawid (1977) — see also Dawid (1981, 1988).

At the presentation of this paper at the meeting, I understood the authors to suggest that the methods also exhibit robustness in the more common sense of insensitivity to extreme data values. One Bayesian approach to this involves modelling with heavy tailed prior and error distributions, as in Dawid (1973), O'Hagan (1979, 1988) —in particular, Student-t forms are often suitable. And indeed, as pointed out at the meeting, the model does allow the possibility of obtaining such distributions for all relevant quantitities. In order to avoid any ambiguity, therefore, it must be clearly realized that, even with this choice, this model does *not* possess robustness against outliers. The Bayesian outlier-robustness theory does not apply because, as mentioned above, after using (3.6) with $F \sim InW(S, \nu)$ the (U_i) are no longer independent. Independence is vital for the heavy-tails theory to work — zero correlation is simply not an acceptable alternative. In fact, since the predictive means under the model turn out to be linear in the data, it is obvious that the methods developed in this paper can *not* be outlier-robust.

S. E. FIENBERG (*York University, Canada*)

As Bernardo and Girón are aware, others have used hierarchical Bayesian models for election night predictions. As far as I am aware the earliest such prediction system was set up in the United States.

In the 1960s a group of statisticians working for the NBC televion network developed a computer-based statistical model for predicting the winner in the U.S. national elections for President (by state) and for individual state elections for Senator and Governor. In a presidential-election year, close to 100 predictions are made, otherwise only half that number are required. The statistical model used can be viewed as a primitive version of a Bayesian hierarchical linear model (with a fair bit of what I. J. Good would call ad hockery) and it predates the work of Lindley and Smith by several years. Primary contributors to the election prediction model development included D. Brillinger, J. Tukey, and D. Wallace. Since the actual model is still proprietary, the following description is somewhat general, and is based on my memory of the system as it operated in the 1970s.

In the 1960s an organization called the News Election Service (NES) was formed through a cooperative effort of the three national television networks and two wire services. NES collects data by precinct, from individual precincts and the 3000 county reporting centers and forwards them to the networks and wire services by county (for more details, see Link, 1989). All networks get the same data at the same time from NES.

For each state, at any point in time, there are data from four sources: (i) a prior estimate of the outcome, (ii) key precincts (chosen by their previous correlation with the actual outcome), (iii) county data, (iv) whole-state data (which are the numbers the networks "oficially" report). The NBC model works with estimates of the swings of the differences between % Republican vote and % Democratic vote (a more elaborate version is used for multiple candidates) *relative* to the difference from some previous election. In addition there is a related model for turnout ratios.

The four sources of data are combined to produce an estimate of $[\%R - \%D]/2$ with an estimated mean square error based on the sampling variance, historical information, and various bias parameters which can be varied depending on circumstances. A somewhat more elaborate structure is used to accomodate elections involving three or more major candidates. For each race the NBC model requires special settings for 78 different sets of parameters, for biases and variances, turnout adjustment factors, stratification of the state, etc. The model usually involves a geographic stratification of the state into four "substates" based on urban/suburban/rural structure and produces estimates by strata, which are then weighted by turnout to produce statewide estimates.

Even with such a computer-based model about a dozen statisticians are required to monitor the flow of data and the model performance. Special attention to the robustness of predictions relative to different historical bases for swings is an important factor, as is collateral information about where the early data are from (e.g., the city of Chicago vs. the Chicago suburbs vs. downstate Illinois).

Getting accurate early predictions is the name of the game in election night forecasting because NBC competes with the other networks on making forecasts. Borrowing strength in the Bayesian-model sense originally gave NBC an advantage over the raw data-based models employed by the other networks. For example, in 1976, NBC called 94 out of 95 races correctly (only the Presidential race in Oregon remained too close to determine) and made several calls of outcomes when the overall percentages favored the eventual loser. In the Texas Presidential race, another network called the Republican candidate as the winner early in the evening at a time when the NBC model was showing the Democratic candidate ahead (but with a large mean square error). Later this call was retracted and NBC was the first to call the Democrat the winner.

The 1980s brought a new phenomenon to U.S. election night predictions: the exit survey of voters (see Link, 1989). As a consequence, the television networks have been able to call most races long before the election polls have closed and before the precinct totals are available. All of the fancy bells and whistles of the kind of Bayesian prediction system designed by Bernardo and Girón or the earlier system designed by NBC have little use in such circumstances, unless the election race is extremely close.

REPLY TO THE DISCUSSION

We are grateful to Professor Pericchi for his valuable comments and for his wish that all worked well on election night. As described in the Appendix above, his wish was reasonably well achieved.

He also refers to the possibility of sequential updating, also mentioned in our final discussion. Assuming, as we do in sections 2 and 4, the hypothesis of exchangeability in the swings —which implies that the C_i matrices in the model are of the form $k_i I$— the derivation of recursive updating equations for the parameters of the posterior of Θ_1 given the data $y_1, \ldots, y_t$, for $t = 1, \ldots, n$, is straightforward. However, no *simple* recursive updating formulae seem to exist for the parameters of the predictive distribution (4.2), due to the complexity of the model (4.1) and to the fact that the order in which data from the polling stations arrive is unknown a priori and, hence, the matrix W used for prediction varies with n in a form which depends on the identity of the new data.

We agree with Pericchi that weak robustness, while being an interesting theoretical extension to the usual hierarchical normal model, may not be enough for detecting gross errors. As we prove in the paper, weak robustness of the posterior mean —which is linear in the observations— is obtained under the error specification given by (3.6), independently of $F(\Lambda)$.

To obtain strong robustness of the estimators, exchangeabilty should be abandoned in favour of independence. Thus, the first equation in model (4.1), should be replaced by

$$x_i = a_i'\Theta_1 + u_i, \qquad i = 1, \ldots, n,$$

where the a_i''s are the rows of matrix A_1, and the error matrix $U_i' = (u_1', \ldots, u_n')$ is such that the error vectors u_i are independent and identically distributed as scale mixtures of multivariate normals, i.e., $u_i \sim \int N(0, k_1\Lambda)\, dF(\Lambda)$.

Unfortunately, under these conditions, no closed form for the posterior is possible, except for the trivial case where $F(\cdot)$ is degenerate at some matrix, say, Σ. In fact, the posterior distribution of Θ_1 given the data is a very complex infinite mixture of matrix-normal distributions. Thus, in order to derive useful robust estimators, we have to resort to approximate methods. One possibility, which has been explored by Rojano (1991) in the context of dynamic linear models, is to update the parameters of the MHLM sequentially, considering one observation at a time, as pointed out above, thus obtaining a simple infinite mixture of matrix-normals, and then to approximate this mixture by a matrix-normal distribution, and proceed sequentially.

Professor Dawid refers again to the fact that the method described is not outlier-robust. Pragmatically, we protected ourselves from extreme outliers by screening out from the forecasting mechanism any values which were more than three standard deviations off under the appropriate predictive distribution, conditional on the information currently variable. Actually, we are developing a sequential robust updating procedure based on an approximate Kalman filter scheme adapted to the hierarchical model, that both detects and accomodates outliers on line.

We are grateful to Professor Fienberg for his detailed description of previous work on election forecasting. We should like however to make a couple of points on his final remarks:

(i) Predicting the winner in a two party race is *far* easier that predicting a parliamentary *seat distribution* among *several* parties.

(ii) In our experience, exit surveys show too much uncontrolled bias to be useful, at least if you have to forecast a seat distribution.

ADDITIONAL REFERENCES IN THE DISCUSSION

Berger, J. O. (1990). Robust Bayesian analysis: sensitivity to the prior. *J. Statist. Planning and Inference* **25**, 303–328.

Brown, P. J. and Payne, C. (1975). Election night forecasting. *J. Roy. Statist. Soc. A* **138**, 463–498.

Brown, P. J. and Payne, C. (1984). Forecasting the 1983 British General Election. *The Statistician* **33**, 217–228.

Dawid, A. P. (1973). Posterior expectations for large observations. *Biometrika* **60**, 664–667.

Dawid, A. P. (1977). Spherical matrix distributions and a multivariate model. *J. Roy. Statist. Soc. B* **39**, 254–261.

Dawid, A. P. (1981). Some matrix-variate distribution theory: notational considerations and a Bayesian application. *Biometrika* **68**, 265–274.

Dawid, A. P. (1988). The infinite regress and its conjugate analysis. *Bayesian Statistics 3* (J. M. Bernardo, M. H. DeGroot, D. V. Lindley and A. F. M. Smith, eds.), Oxford: University Press, 95–110, (with discussion).

de Finetti, B. (1961). The Bayesian approach to the rejection of outliers. *Proceedings 4th Berkeley Symp. Math. Prob. Statist.* **1**, Berkeley, CA: University Press, 199–210.

Link, R. F. (1989). Election night on television. *Statistics: A Guide to the Unknown* (J. M. Tanur *et al.* eds.), Pacific Grove, CA: Wadsworth & Brooks, 104–112.

Moreno, E. and Pericchi, L. R. (1990). An ε-contaminated hierarchical model. *Tech. Rep.* Universidad de Granada, Spain.

O'Hagan, A. (1979). On outlier rejection phenomena in Bayes inference, *J. Roy. Statist. Soc. B* **41**, 358–367.

O'Hagan, A. (1988). Modelling with heavy tails. *Bayesian Statistics 3* (J. M. Bernardo, M. H. DeGroot, D. V. Lindley and A. F. M. Smith, eds.), Oxford: University Press, 345–359, (with discussion).

O'Hagan, A. (1990). Outliers and credence for location parameter inference. *J. Amer. Statist. Assoc.* **85**, 172–176.

Rojano, J. C. (1991). *Métodos Bayesianos Aproximados para Mixturas de Distribuciones*. Ph.D. Thesis, University of Málaga, Spain.

Walley, P. (1990). *Statistical Reasoning with Imprecise Probabilities*. London: Chapman and Hall.

West, M. (1981). Robust sequential approximate Bayesian estimation. *J. Roy. Statist. Soc. B* **43**, 157–166.

West, M. (1984). Outlier models and prior distributions in Bayesian linear regressions. *J. Roy. Statist. Soc. B* **46**, 431–439.

BAYESIAN STATISTICS 4, pp. 79–96
J. M. Bernardo, J. O. Berger, A. P. Dawid and A. F. M. Smith, (Eds.)

Public Health Decision Making: a Sequential Vaccine Trial

DONALD A. BERRY*, MARK C. WOLFF** and DAVID SACK**
**Duke University, USA* and ***Johns Hopkins University, USA*

SUMMARY

A vaccine for *Hæmophilus influenzæ type B* is compared in a randomized trial with a placebo vaccine. The trial should end and the vaccine should be abandoned when it becomes clear that it is not sufficiently effective. And the trial should end when the available information makes it clear that the vaccine is sufficiently effective to be approved for general use by regulatory authorities and recommended for general use by medical authorities. We take a decision-theoretic point of view in which the objective is to minimize the expected number of cases of *H. influenzæ* among Navajo Indians during the next several years. We use the available historical information in determining the trial's design, and update this information as data accumulates from the trial. An important type of information to be assessed is that concerning actions the regulatory and medical authorities will take based on data from the trial. The data that accumulate during the trial includes the time each subject has spent at risk. The design of the trial is sequential; we find optimal designs by numerical calculations using backwards induction. We evaluate optimal designs in an actual trial and compare them with nonsequential designs and with the frequentist design that was actually used in the trial.

Keywords: BAYESIAN DECISION MAKING; SEQUENTIAL TRIALS; RANDOMIZED TRIAL; PREVENTIVE MEDICINE; BACKWARDS INDUCTION; VACCINES.

1. INTRODUCTION

Therapeutic clinical trials involve patients who have a particular disease or condition. The objective of treatment in such trials is to improve some aspect of the patient's circumstance. In prophylactic clinical trials, such as those involving vaccines, subjects do not have the disease in question, and the objective of treatment is to keep them disease-free. The incidence of some important diseases is less than one percent, and so success rates of a placebo vaccine may exceed 99%. Therefore, vaccine trials usually require large numbers of subjects to enable conclusive inferences concerning the vaccine's efficacy. Vaccine trials can be made more efficient by conducting them in populations that are highly susceptible to the disease in question. But even in such populations, determining whether a vaccine is efficacious may require many hundreds of subjects.

The value of a vaccine can be expressed as its protective efficacy (PE), which is the reduction in incidence of disease for vaccinated individuals:

$$\mathrm{PE} = 1 - \frac{\text{incidence in the vaccine group}}{\text{incidence in nonvaccinated group}}.$$

For example, if the rate of disease is 100 per 10,000 in a placebo group but only 30 per 10,000 in a comparable vaccinated group, the PE is 70%.

Protective efficacy is usually measured in placebo-controlled, randomized trials, and the incidence in the vaccine group is compared with the incidence in the placebo group. In less controlled studies such as case-controlled analysis of vaccination programs, outcome events (cases of disease) are observed and the rates of vaccination are ascertained in cases and matched controls. In these types of analysis the relative risk (RR) of vaccine status can then be estimated, and the PE approximates $1-$ RR. Such an estimate of PE is likely to be less accurate because it is subjected to several biases, the primary one being the rather intangible issue of the choice to become vaccinated. In a placebo-controlled trial, all of the participants in the analysis chose to be in the program. So placebo versus vaccine comparisons will be unbiased. But the results may not generalize to the entire population if trial participants may respond differently than nonparticipants.

When calculated from a randomized, placebo-controlled trial, PE provides an estimate of the biological power of the vaccine to protect the individual. However, this estimate may be different from the true effectiveness of the vaccine, which is its ability to prevent illness when it is used in a practical vaccine program. The true effectiveness, for example, may be lower than the PE if only a small proportion of the population agrees to become vaccinated because of perceived side-effects or if the program is poorly run. On the other hand, a vaccine's effectiveness may exceed the PE if the vaccine reduces transmission of the causative agent and thus induces "herd immunity."

When evaluating a vaccine for public use, several issues must be considered:

1. The disease which the vaccine is intended to prevent must have significant morbidity and/or mortality in the population to be vaccinated. The higher the rate of morbidity/mortality, the greater the urgency for vaccine development and implementation.
2. The vaccine must be compared with other alternative strategies for control of the disease. For example, typhoid fever has been largely controlled in the U.S. through improved sanitation, whereas, developing countries may need to utilize typhoid vaccine until sanitation improves. Another example of comparing alternative strategies is the use of vaccines for prevention as compared with antibiotics for treatment. If good antibiotics which would prevent serious sequelæ are available, one might reasonably choose not to vaccinate. However, if there is no effective treatment for the disease, and if the disease is frequently fatal (e.g., AIDS) a vaccine would be preferable.
3. Side-effects from vaccines must be minimal since healthy persons are being treated to prevent a disease they are unlikely to get. People will be unlikely to tolerate bothersome side-effects if they are frequent, or serious side-effects even if they are rare. Since healthy persons are receiving the vaccines, any symptoms occurring after the vaccine will usually be attributed to the vaccine. Patients and parents make subjective judgements of risk of illness with the vaccine versus without the vaccine. They include the possibility of side-effects in their comparison of risks.
4. Costs of vaccines are becoming increasingly important. From a patient's perspective, the costs of childhood vaccines has been increasing dramatically and may reach a level of resistance on the part of parents, especially since insurance companies generally do not pay for preventive care.
5. It may be appropriate to vaccinate only high risk subgroups. For example, pneumococcal vaccines would be of special importance to children with sickle cell disease.

Health planners also evaluate the cost-effectiveness of vaccines, especially when programs pay the costs of the vaccine program. Since so many persons must be vaccinated in a vaccine program, the cost of the program may be large when expressed as cost per case averted, even though the cost per subject vaccinated may be small. The cost per case averted must then be compared with the cost for treatment and care of cases should they not be prevented. For example, if an *H. influenzæ* vaccine program prevents a certain number of cases of disease at a certain cost per case averted, this cost should be compared with the cost of diagnosing and treating the cases when they occur. For *H. influenzæ* disease, which may be associated with long term neurologic complications, the cost to treat must include the long-term costs to society as well as the costs for acute medical care.

Costs and benefits of the vaccine programs may also take into account the desire of the community to control a particular problem. For example, in the mid 1950's, the polio vaccine was greeted with much enthusiasm, partly because there was widespread agreement of the need to "do something" about polio. The result was a general willingness to cooperate with the vaccine program. Other vaccines for less popular diseases might be greeted with less enthusiasm. A cost-utility analysis in which a vaccine's effectiveness is discounted to account for society's perception of the disease and the worth of the vaccine is beyond the scope of the current paper.

In Section 2 we describe *Hæmophilus influenzæ type B*, a vaccine that may prevent it, and a trial designed to evaluate the vaccine. Section 3 outlines our objectives. Section 4 compares the Bayesian and frequentist approaches to interim analyses of data from an experiment. Section 5 indicates the approach we suggest to public health decision making and vaccine studies in particular. Section 6 describes the technique of backwards induction and its application to the vaccine trial. Section 7 indicates the subjective assessments we use and their relation to the available historical data. Section 8 gives the results of and our analysis of an actual vaccine trial, and Section 9 contains some concluding remarks.

2. HÆMOPHILUS INFLUENZÆ TYPE B

H. influenzæ is a disease which usually affects infants and children. *H. influenzæ* is a bacterium which causes severe infections affecting the blood stream (septicemia), the central nervous system (meningitis), the bones and joints (osteomyelitis or septic arthritis), the upper airways (epiglottitis), and other sites. Collectively, these are known as "invasive *H. influenzæ* infections." If untreated or if treatment is delayed, the infection is often fatal. Among those treated for *H. influenzæ* meningitis, many develop neurologic complications, especially loss of hearing or a seizure disorder, and thus are impaired for life.

The illness occurs in all groups; however, groups with the highest incidence are certain tribes of American Indians, including the Navajo and the Apaches. Among these peoples the risk that a child will develop invasive *H.influenzæ* infection is about 1% at some time before reaching 3 years of age. This rate is from 10 to 50 times greater than that of other American populations. Thus, although a successful vaccine will be useful in the general U.S. population, it will be even more important to these tribes.

A new *H. influenzæ* vaccine (HIB-OMP, Merck, Sharpe and Dohme Research Laboratories) was recently tested for efficacy among Navajo infants living on the Navajo reservations in Arizona, New Mexico and Utah. It was found to be highly efficacious (see Section 8). The vaccine can be given to children beginning at 2 months of age and appears to protect from the time of first dose. Consisting of the capsular polysaccharide (the primary virulence antigen of the bacterium) conjugated to the outer membrane protein (to make it more immunogenic in young infants), the vaccine has very few side-effects when it is injected. The vaccine stimulates antibody to the bacterial capsule and this antibody neutralizes the bacteria's ability to invade the child's body, and thus protects the child from serious illness.

Thus the new vaccine fulfills many of the criteria for public health use. It is a vaccine for a serious and frequent problem which is not being controlled by alternative strategies. It can be given without significant side effects. The disease it controls is perceived by the population to be a serious health threat. Although cost-effectiveness studies have not been carried out in the Indian population, it is widely believed that the prevention of the cases and deaths will be worth the cost. Earlier studies with a vaccine believed to have a similar safety profile but less potential for preventing cases was shown to be cost-effective for the entire U. S. population.

Since the vaccine trial was carried out among the Navajo Nation, the vaccine has special relevance to American Indians who are at high risk. By demonstrating efficacy among the Indians, who are known to be poor responders to vaccines of this type, the vaccine also is now thought to be important for the general U.S. population who will also be receiving it. If the vaccine prevents illness among these Indian infants, it is reasoned, it will surely prevent disease among other groups who respond better and are less susceptible. Calculations of the cost per case averted will differ with other groups, but the fact of disease prevention should not change appreciably. Prior to the introduction of any HIB vaccine, HIB was responsible annually in the U. S. for 12,000 cases of bacterial meningitis and 7,500 cases of other systemic disease in children under five years old. Approximately 60% of cases occur in children under 18 months of age, the minimum age recommended for HIB vaccination at the outset of this trial. Therefore, younger infants must be vaccinated to provide maximum benefit.

3. OBJECTIVE

Even though the experimental vaccine may be used more generally, we will restrict consideration to the population of Navajo Indians in the U. S. who would participate in a vaccine program. The objective is to minimize the number of cases of *H. influenzæ* in this highly susceptible population over the next N months. This time horizon is unknown and so it has a probability distribution in the Bayesian approach. However, provided N is greater than the maximum length of the study with probability one, we can replace it in what follows with its expectation. For our numerical calculations we will consider a 20-year horizon and so take $N = 240$. There are approximately 450 Navajo babies born each month; we take the number of births per month to be fixed at this number during the N months in question.

Consider a randomized clinical trial that is being conducted in the Navajo population. This trial will last for at most T months, but it may stop sooner depending on the available data. We will assume $T = 60$. During the trial, two-month old babies are allocated in equal proportions to an active vaccine and a control (placebo vaccine). Approximately $2n = 210$ of the approximately 450 Navajo babies born each month are involved in the trial, and the other n_0 are not. We assume that $2n$ babies per month enter the trial until it is stopped–n of these are allocated to vaccine and n to control. If the vaccine has been approved for marketing, all $2n + n_0 = 450$ babies would be vaccinated. Between the time that the trial stops and approval, none of the $2n + n_0$ babies are vaccinated and so they would experience cases at the control or placebo rate.

The trial should be stopped and the vaccine's development abandoned should it become clear that the vaccine lacks efficacy; see Berry and Ho (1988) for a discussion of this problem. In the current paper we address the stopping problem that arises when the available data indicates that the vaccine is sufficiently efficacious that it should be recommended for general use.

If the trial is stopped before the available efficacy information is sufficiently convincing to the medical community then the vaccine will not be used, regardless of its actual efficacy. So stopping would be ill advised. On the other hand, if the trial is continued beyond the time that convincing data are available then general use will be delayed and unnecessary cases of disease will result. Continuing a trial once it becomes clear that the vaccine is beneficial has little counterbalancing benefit: although subjects receiving vaccine in the trial are protected from disease and so clearly benefit, control subjects do not. Moreover, administering placebo vaccine in such a circumstance raises serious ethical concerns.

For the purposes of this paper, "convincing the medical community" means convincing the U.S. Food and Drug Administration (FDA) and pediatric immunization advisory committees (Immunization Practices Advisory Committee of the U.S. Center for Disease Control and the Redbook committee of the American Academy of Pediatrics). The FDA is authorized by the U.S. Congress to regulate vaccines. The FDA allows the use of sufficiently promising experimental vaccines in closely monitored, controlled settings. But vaccines will not be widely used unless they are formally approved for marketing by the FDA and recommended by advisory panels. (We assume that any vaccine that is approved for marketing by the FDA and recommended by advisory panels will be sufficiently profitable to the manufacturer that it will in fact be marketed.) Before approving a vaccine, the FDA staff must be convinced of its safety and effectiveness. There are no fixed guidelines for this approval process, so any application to the FDA for marketing approval has stochastic elements. Not only is the matter of approval itself random, but the time consumed by the approval process varies, depending in part on the available safety and effectiveness information: an apparently more effective vaccine will usually be approved more quickly. Our approach to the stopping problem does not focus on whether the available data convinces us of the vaccine's efficacy, but on our perception of what will convince the FDA and the various advisory panels.

We assume that once application for marketing approval is submitted to the FDA, experimental use of the vaccine ceases, except that subjects still in clinical trials will continue to be followed according to the protocol. We also assume that a vaccine not approved for marketing will be abandoned. This is a simplifying assumption since in practice providing additional data at a later time could serve to reverse a negative FDA decision. Our model can be extended easily to include this more realistic possibility.

New vaccines arise rarely and their existence is a function of the success of the previously developed one. For example, polio vaccine is virtually unchanged since the early 1960's and only now is consideration given to developing better ones. On the other hand mediocre vaccines are often superseded much earlier. We will assume that if the current vaccine is approved, it will not be replaced during the horizon of N months. Allowing for the development of new vaccines depending on the efficacy of the current vaccine is a straightforward extension of our model.

If on the other hand the vaccine is not approved then we assume that a new vaccine may be developed during the horizon. Let T^* indicate the expected number of months required for such development; we take $T^* = 120$ (or 10 years). Including the possibility that no such vaccine is developed, we assume the average protective efficacy of a new vaccine is 50%.

4. EARLY STOPPING: BAYES VS. FREQUENTIST APPROACHES

It is undesirable and even unethical in medical settings to ignore accumulating data. Repeated analyses are problematical in a frequentist approach. Frequentist measures of evidence are conditional on hypotheses concerning particular levels of treatment effectiveness. For example, a significance level is a tail-area probability assuming the null hypothesis of no treatment effect. If there is no effect then repeated analyses will likely conclude that there is one. For a larger number of analyses, the actual error probability can be arbitrarily larger than the nominal significance level. Consequently, the standard frequentist approach is to adjust significance levels (for example, Lan and DeMets 1984), and frequentists recommend limiting the number of interim analyses (e.g., McPherson 1982). Such adjustments and restrictions on number of analyses limit the ability of investigators to react to available information, and may therefore themselves be unethical (Heitjan *et al.* 1991).

There are no such inhibitions in the Bayesian approach (Berger and Berry 1988; Berry 1985, 1987, 1988). In particular, Bayesian analyses are not affected by the stopping rule. Consequently, Bayesians can address stopping problems in a fully sequential manner.

Unfortunately, sequential Bayesian approaches to designing clinical trials have been seldom if ever used, at least not formally. Some of the reasons for this are spelled out in the discussion of Bather (1985) and Armitage (1985). One reason is that a decision-theoretic Bayesian approach requires an objective function, usually defined in terms of a "patient horizon": the set of patients who will be treated with one of the therapies under consideration. We too have assumed a horizon, with the slight variation that ours uses time rather than patients. But we explicitly consider the possibility that other vaccines will be used during the horizon.

Another objection to a decision-theoretic approach is that most are tailored to immediate response (a notable exception is Eick 1987), which is seldom appropriate. Our approach uses the partial information that is available on those subjects who are still in the trial but who are not yet cases.

5. STATEMENT OF THE PROBLEM

As indicated above, our objective is to decide when to stop a randomized trial in order to minimize the number of cases of *H. influenzæ* among the children of the Navajo nation over the next N months. Babies in the trial are vaccinated at 2 and 4 months of age. They are then followed until age 18 months. A case during the intervening $M = 16$ months is a failure of the vaccine. We assume that the subjects are at constant risk during this period, and let λ_1 stand for the failure rate per month for the active vaccine and λ_2 stand for the failure rate per month for placebo.

We will discretize the analysis by assuming each subject vaccinated in a month is at risk for the entire month. In month m, for $m \leq M$, the number of subjects at risk in each group, say r, is mn minus the number of previous cases in that group. In vaccine trials the numbers of cases are often negligible in comparison with nm; we take the number of subjects at risk to be $r = mn$ and so ignore the fact that cases are removed from the at-risk population. (Subtracting the number of cases and incorporating this into our analysis does not make the problem more difficult conceptually, but it does increase the size of the problem and hence the number of calculations required for its solution.) For $m > M$ the number of subjects entering the trial cancels the number leaving it and so r is constant at Mn. Therefore, $r = \min\{Mn, mn\}$.

Observations of cases of *H. M. C. influenzæ* within the treatment groups are assumed to be conditionally independent given λ_1 and λ_2. The λ_i's are themselves independent initially–and therefore also henceforth. The initial distribution of each λ_i is gamma (a_i, b_i):

$$f(\lambda_i|a_i, b_i) \propto \lambda_i^{(a_i-1)} \exp\{-b_i\lambda_i\}, \qquad \lambda_i > 0.$$

(See Section 7 for our assessment of the a's and b's from historical information.) The conjugate nature of the gamma family of distributions with respect to Poisson sampling means that λ_i continues to have a gamma distribution as observations are made. For example, if there are r subjects at risk in each of the treatment groups and c_i of them in group i become cases, the updated distributions of the λ_i are gamma $(a_i + c_i, b_i + r)$:

$$f(\lambda_i|a_i, b_i; c_i, r) \propto \lambda_i^{(a_i+c_i-1)} \exp\{-(b_i + r)\lambda_i\}, \qquad \lambda_i > 0.$$

The state space in this problem is five-dimensional: at any time, the state or pattern of information contains the current month and the effective numbers of cases and subjects on the two treatments. The initial state is $(a_1, b_1; a_2, b_2; 0)$; if during the first month there are c_i cases in the n subjects in treatment group i, the month number is increased by one and the state becomes $(a_1 + c_1, b_1 + n; a_2 + c_2, b_2 + n; 1)$. In the sequel we will let $(a_1, b_1; a_2, b_2; m)$ stand for a generic state of information–whether it be initial, intermediate, or final.

6. BACKWARDS INDUCTION EQUATIONS

Let $W(a_1, b_1; a_2, b_2; 0)$ stand for the expected number of cases of *H. influenzæ* in the Navajo population over the next N months when the current state is $(a_1, b_1; a_2, b_2; 0)$ and the stopping policy is optimal. This expectation is the minimum of two quantities, W_{STOP} and W_{CONT}:

$$W(a_1, b_1; a_2, b_2; 0) = \min\{W_{\mathrm{STOP}}(a_1, b_1; a_2, b_2; 0), W_{\mathrm{CONT}}(a_1, b_1; a_2, b_2; 0)\}. \quad (1)$$

We will address these two components in turn.

If the trial is stopped, the vaccine may or may not be approved for general use by the FDA, and if it is approved it may not be recommended for general use by pediatric advisory committees. Let $g(a_1, b_1; a_2, b_2; m)$ stand for the probability the vaccine will be eventually adopted when the current state is $(a_1, b_1; a_2, b_2; m)$–see Section 7. It follows that

$$\begin{aligned} W_{\mathrm{STOP}}(a_1, b_1; a_2, b_2; m) &= g(a_1, b_1; a_2, b_2; m) W_{\mathrm{APP}}(a_1, b_1; a_2, b_2; m) \\ &\quad + [1 - g(a_1, b_1; a_2, b_2; m)] W_{\mathrm{NOTAPP}}(a_1, b_1; a_2, b_2; m), \end{aligned} \quad (2)$$

where W_{APP} and W_{NOTAPP} are the expected numbers of cases assuming the vaccine is and is not adopted for general use.

Suppose the current state is $(a_1, b_1; a_2, b_2; m)$. The mean of the gamma (a_i, b_i) distribution is a_i/b_i. Were the possibility of other active vaccines becoming available during horizon $N-m$ ignored, the value of W_{NOTAPP} would be M (months per child) times a_2/b_2 times the number of babies born during in the next $N-m$ months. We *are* allowing the possibility that new vaccines are developed should the current one not be approved, and we are assuming that the rate of a new vaccine would be half that of placebo, on average:

$$\begin{aligned} &W_{\mathrm{NOTAPP}}(a_1, b_1; a_2, b_2; m) \\ &\quad = M(2n + n_O) \left\{ (T^* - m) \left[\frac{a_2}{b_2} \right] + (N - T^*) \left[\frac{a_2}{2b_2} \right] \right\} \quad \text{if } m < T^* \\ &\quad = M(2n + n_O)(N - m) \left[\frac{a_2}{b_2} \right] \quad \text{if } m \geq T^* \end{aligned}$$

We turn to the possibility of approval. There will be a delay averaging t months between the time the trial is stopped and the vaccine is approved. Children are not vaccinated during this period. Therefore

$$\begin{aligned} &W_{\mathrm{APP}}(a_1, b_1; a_2, b_2; m) \\ &= M(2n + n_O) \Big\{ [N - m - t(a_1, b_1; a_2, b_2; m)] \left[\tfrac{a_1}{b_1} \right] + t(a_1, b_1; a_2, b_2; m) \left[\tfrac{a_2}{b_2} \right] \Big\} \text{ if } N - m > t \\ &= M(2n + n_O)(N - m) \left[\tfrac{a_2}{b_2} \right] \quad \text{if } N - m \leq t. \end{aligned}$$

Here, $t(a_1, b_1; a_2, b_2; m)$ is the time from end of study until the vaccine is approved–see Section 7.

Suppose the current state is $(a_1, b_1; a_2, b_2; m)$ and the trial continues. Subsequent decisions (from month $m+1$ on) are made to minimize the expected number of future cases. So if the trial continues the expected number of future cases is

$$W_{\text{CONT}}(a_1, b_1; a_2, b_2; m) = \\ E\left[c_1 + c_2 + W(a_1 + c_1, b_1 + r; a_2 + c_2, b_2 + r; m+1)\right] + Mn_0\left[\frac{a_2}{b_2}\right], \tag{3}$$

where expectation E is over the predictive distribution of c_1 and c_2 and $r = \min\{Mn, mn\}$ is the number of subjects at risk in each group. The last term in (3) accounts for the babies who are born during the trial but are not randomized to one of the two groups; cases accumulate in this group at the rate of a_2/b_2 per month. We assume that these babies would be vaccinated if the vaccine were approved. Rewriting (3):

$$W_{\text{CONT}}(a_1, b_1; a_2, b_2; m) = \\ \sum_{c_1=0}^{\infty}\sum_{c_2=0}^{\infty} P(c_1, c_2)\left[c_1 + c_2 + W(a_1 + c_1, b_1 + r; a_2 + c_2, b_2 + r; m+1)\right] + Mn_0\left[\frac{a_2}{b_2}\right].$$

The predictive probability distributions of the numbers of cases c_1 and c_2 depend on r. When the current state is $(a_1, b_1; a_2, b_2; m)$, c_1 and c_2 are independent–that is, $P(c_1, c_2) = P_1(c_1)P_2(c_2)$–and the predictive distribution of each c_i is

$$P_i(c) = \frac{\Gamma(c + a_i)}{c!\Gamma(a_i)}\left[\frac{b_i}{b_i + r}\right]^{a_i}\left[\frac{r}{b_i + r}\right]^{c}, \quad c = 0, 1, 2, \ldots$$

Evaluating recursively we find that $P_i(0) = \left[\frac{b_i}{b_i+r}\right]^{a_i}$ and

$$P_i(c+1) = P_i(c)\frac{c + a_i}{c+1}\frac{r}{b_i + r}, \quad c = 0, 1, 2, \ldots$$

We indicated that the trial will not continue beyond month T. Hence, the backwards induction can start at that month, with the following boundary conditions:

$$W(a_1, b_1; a_2, b_2; T) = W_{\text{STOP}}(a_1, b_1; a_2, b_2; T). \tag{4}$$

Calculating using equations (1), (2), (3), and (4) determines all the $W(a_1, b_1; a_2, b_2; m)$, and also whether $W = W_{\text{STOP}}$ or W_{CONT} at each state. Taken together these calculations determine an optimal policy. Consider the initial state $(a_1, b_1; a_2, b_2; 0)$. Stopping is optimal initially (and so the trial should not be started at all) if

$$W(a_1, b_1; a_2, b_2; 0) = W_{\text{STOP}}(a_1, b_1; a_2, b_2; 0).$$

On the other hand, starting the trial is optimal if

$$W(a_1, b_1; a_2, b_2; 0) = W_{\text{CONT}}(a_1, b_1; a_2, b_2; 0).$$

(Both stopping and starting are optimal if $W_{\text{STOP}} = W_{\text{CONT}}$.) If the trial begins then the state reached at month one is $(a_1 + c_1, b_1 + r; a_2 + c_2, b_2 + r; 1)$, where $r = \min\{Mn, mn\} = n$

since $m = 1$. After observing c_1 and c_2, we note whether $W(a_1+c_1, b_1+r; a_2+c_2, b_2+r; 1)$ is $W_{\text{STOP}}(a_1+c_1, b_1+r; a_2+c_2, b_2+r; 1)$ or $W_{\text{CONT}}(a_1+c_1, b_1+r; a_2+c_2, b_2+r; 1)$, and act accordingly.

The example of Section 8 illustrates the calculations described here. The example uses the subjective assessments described in the next section.

7. SUBJECTIVE ASSESSMENTS

Subjective assessments are required for $g(a_1, b_1; a_2, b_2; m)$ and $t(a_1, b_1; a_2, b_2; m)$. These were based on the opinions of one of the authors (D.S.), who is an expert in vaccine trials. Historically, the interpretation of vaccine trial results has been based on frequentist methods. Recently, this has included an increased emphasis on the observed 95% confidence interval of efficacy. To assist in assessment, a computer program was written to graphically represent these parameters. The program assumed that the numbers of children in the vaccine and placebo groups were equal.

The probability of aproval $g(a_1, b_1; a_2, b_2; m)$ was deemed to depend only on the number of cases of disease actually observed during the trial in the true groups. In particular, it does not depend on the (common) number of subjects in the two groups or on the prior parameters. As usual denote the current parameters as a_1, b_1, a_2 and b_2, but for the purposes of this paragraph only, denote the initial values of a_1 and a_2 as a_{01} and a_{02}. Further, let the numbers of cases observed during the trial be $c_1 = a_1 - a_{01}$ in the vaccine group and $c_2 = a_2 - a_{02}$ in the placebo group. Given a group size and c_2, the program plotted the confidence bounds and P value as a function of c_1. In this way it was possible to assess the results of a large number of hypothetical trials. We fit the assessments of the probability of general use of the vaccine with the following function:

$$g(a_1, b_1; a_2, b_2; m) = \begin{cases} 0 & \text{if } c_2 < 8 \\ \max\left\{\frac{c_2 - 0.9c_1 - 7.5}{c_2 - 7}, 0\right\} & \text{if } c_2 \geq 8, \end{cases}$$

The assessed expected time consumed by the approval process depends on the numbers of cases only through g:

$$t(a_1, b_1; a_2, b_2; m) = \min\{12, 24 - 24g(a_1, b_1; a_2, b_2; m)\}.$$

The prior distributions of the λ's were assessed by another of the authors (M.C.W.) after considering experience with HIB-OMP and similar conjugate *H. influenzæ* vaccines prior to the trial. At the time the trial started, another *H. influenzæ* conjugate vaccine was licensed and in routine use for children 18 months and older. HIB-OMP had been tested in two-month old infants in immunogenicity studies in which serological responses were measured. There are two serological levels believed to be associated with vaccine protection. In one such study of Navajo and Apache children, 97% and 68% attained these levels. This suggested that the prior probability of negative efficacy in this younger age group should be low.

The distribution of the ratio of the λ's, or relative risk RR, is proportional to an F distribution. After examining a large number of possible models which satisfied this condition, one of the distributions shown in Figure 1 was selected. This was for $a_1 = 1, a_2 = 5$, and the ratio of the expected values of the λ's (that is, $(a_2/b_2)/(a_1/b_1)$) equal to 1.5. (The other two distributions shown in Figure 1 are for comparison, and one each will be considered in the sensitivity analysis of Figure 3.) Prior experience with the vaccine was approximately the equivalent of 200 infants observed for the 16-month period, so we chose $b_1 = 3200$. These considerations determined the other parameters as follows:

$$a_1 = 1, b_1 = 3200, a_2 = 5, b_2 = 10,700.$$

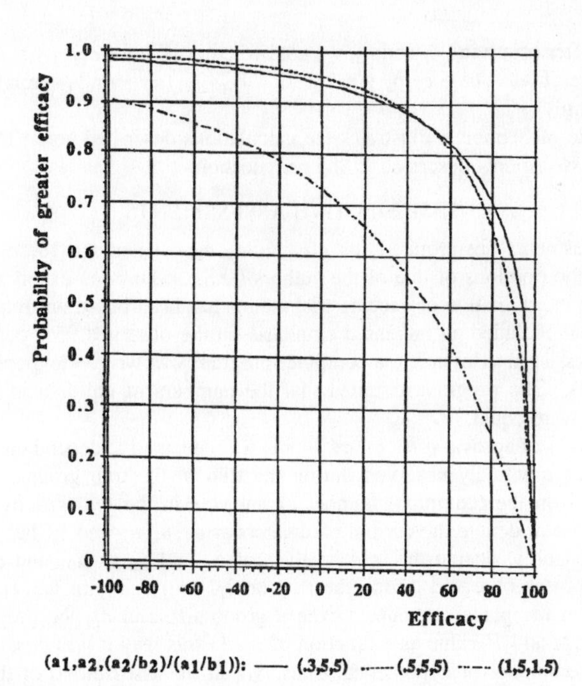

Figure 1. *Three prior distributions of efficacy of experimental vaccine as compared with placebo. One distribution in the third class listed (1, 5, 1.5) is used in Figures 2, 3, 4 and 5. One in each of the other two classes is used in a sensitivity analysis in Figure 3.*

These parameters are reasonably small and so will be dominated by the data after a moderately large period of time.

8. RESULTS

Table 1 shows the cumulative numbers of cases in the two treatment groups in a study discussed at various places in this paper. The study was conducted among Navajo Indians by the Department of International Health of The Johns Hopkins School of Hygiene and Public Health. (The data have been modified slightly to make them consistent with the assumption of uniform accrual of subjects.) The average accrual rate was $n = 105$ subjects per treatment group per month. We defined "month" as the time period during which 105 subjects were added to each treatment group. So the numbers of subjects at risk in each treatment group increased linearly until month 16 when this number remained constant at $16 \times 105 = 1680$.

The actual design of the study was to continue to a fixed total of 5000 subjects completing the trial. It was estimated that approximately 5600 subjects would have to be enrolled in the trial to reach this target and that approximately 40 months would be required to follow the subjects to completion. There was the possibility of stopping early based on confidence intervals for efficacy. The study actually accumulated about 5250 subjects; it was stopped when the vaccine/placebo comparison was "highly significant" and sufficient experience had accumulated in the oldest age group. As is evident from the fact that such earlier stopping did not occur even though the interim data were quite convincing, the criteria for early stopping were very stringent.

	Cumulative cases		Expected number of future cases	
Month	Vaccine(c_1)	Placebo (c_2)	W_{STOP}	W_{CONT}
0	0	0	605.6	**438.6**
1	0	0		
2	0	2	806.8	**504.9**
3	0	3		
4	0	4	953.6	**498.0**
5	0	4		
6	0	6	1042.4	**441.6**
7	0	6		
8	0	7	997.4	**366.1**
9	0	7		
10	0	7	864.0	**298.6**
11	0	7		
12	0	8	496.2	**245.9**
13	0	8		
14	0	10	238.1	**200.3**
15	1	10		
16	1	12	363.6	**312.7**
17	1	12		
18	1	13	297.1	**269.0**
19	1	15		
20	1	15	241.6	**233.5**
21	1	17		
22	1	18	**200.2**	201.3
23	1	20		
24	1	21	**171.8**	175.9
25	1	22	**153.5**	157.8

Table 1.

Table 1 also shows the results of the backwards induction described in Section 6, making the assumptions given in Sections 5, 6, and 7, and taking $N = 240$. Table 1 shows the expected numbers of future cases when stopping (W_{STOP}) and continuing (W_{CONT}). The smaller of the two is indicated *by putting it in boldface*. The optimal strategy indicates that stopping is first called for at month 22, only three months before the trial was actually stopped. (The calculations shown in Table 1 were carried out on a Macintosh SE personal computer. This computer has only about 20,000 memory locations, which inhibits handling a five-dimensional problem. Our program requires only two dimensions of computer memory.)

Figure 2 shows improvement of continuing as opposed to stopping, $W_{\text{STOP}} - W_{\text{CONT}}$, from Table 1. This figure also shows two other curves, one for each of two other values of horizon N: 120 and 360 months. The optimal strategy is not very sensitive to the choice of N, but the expected number of future cases is roughly proportional to N. Figure 3 addresses sensitivity of the optimal strategy to the prior distribution of the λ's by comparing the improvement from stopping for the prior used in Table 1 with two other distributions deemed to be possible priors.

The optimal sequential strategy should be compared with optimal fixed-sample strategies. This is done in Figure 4. The optimal length of this study is about 48 months. The original plan was for a study of about 24 months plus follow-up, and so was not too far from optimal

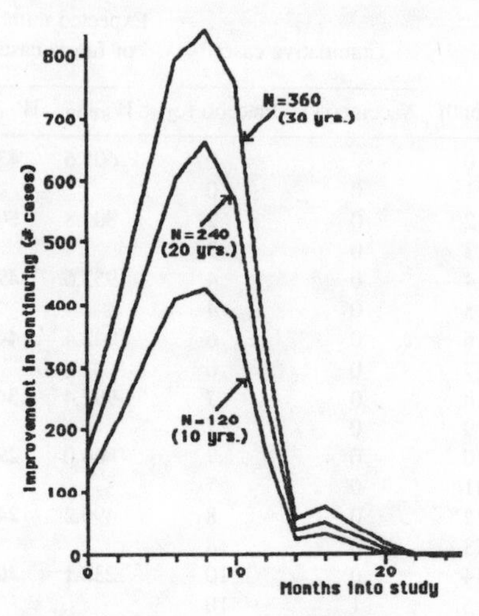

Figure 2. *Expected improvement in number of cases of H. influenzae of continuing the trial as opposed to stopping, for several values of horizon N; in each,* $a_1 = 1, b_1 = 3200, a_2 = 5, b_2 = 10,700$.

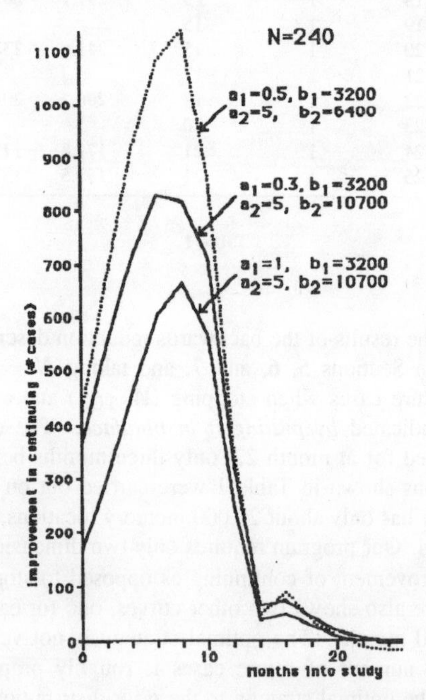

Figure 3. *Expected improvement in number of cases of H. influenzae of continuing the trial as opposed to stopping, for three priors, one each in the three classes shown in Figure 1. In each case,* $N = 240$.

(for this prior and N). But a fixed sample design is not really much better than no trial at all. The principal message of Figure 4 is that no fixed-length study compares favorably with a sequential study.

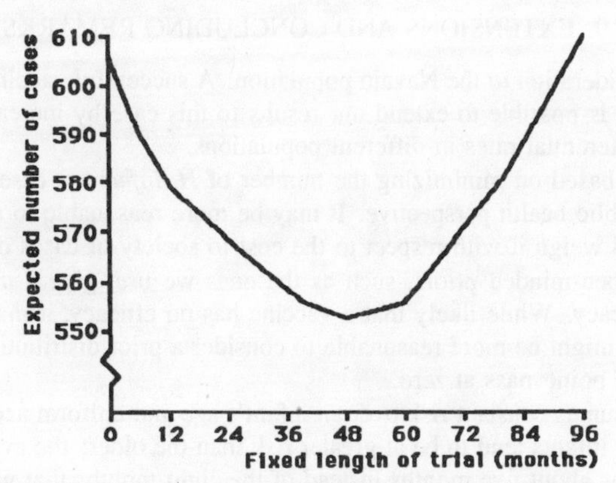

Figure 4. *Expected minimum of 553 cases above compares with 439 for a sequential trial and 606 for no trial. The prior distribution and the horizon are the same in both:* $a_1 = 1, b_1 = 3200, a_2 = 5, b_2 = 10,700$, *and* $N = 240$.

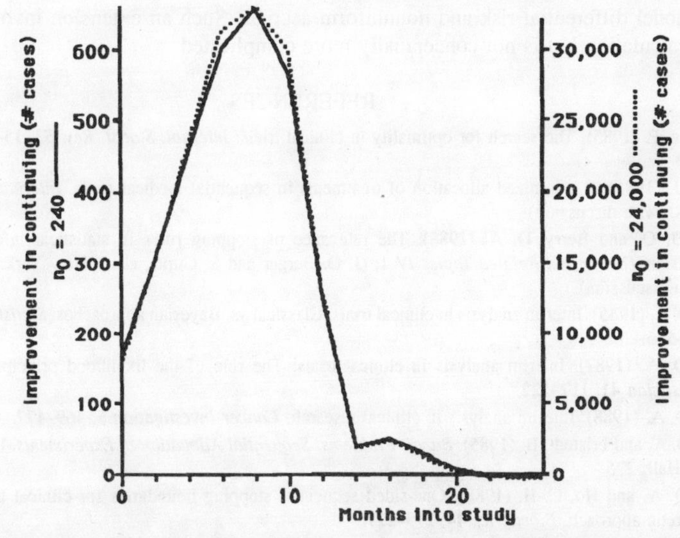

Figure 5. *Expected improvement in number of cases of H. influenzae of continuing the trial as opposed to stopping, for two values of* n_0*, the number of infants born per month who are not in the study. The prior distribution and the horizon are the same in both:* $a_1 = 1, b_1 = 3200, a_2 = 5, b_2 = 10,700$, *and* $N = 240$.

Finally, Figure 5 examines the sensitivity of the optimal stopping strategy as it depends on n_0, the number of infants born per month who are not in the study. The prior distribution and the horizon are the same in both: $a_1 = 1, b_1 = 3200, a_2 = 5, b_2 = 10,700$, and $N = 240$ months. If n_0 is increased by a factor of 100, the expected improvement in number of cases of *H. influenzae* of continuing the trial as opposed to stopping is increased approximately by a factor of 50. But the strategy stays about the same: Stopping should occur only one month sooner in the latter case.

9. EXTENSIONS AND CONCLUDING REMARKS

We restricted consideration to the Navajo population. A successful vaccine may well be used more generally. It is possible to extend our results to this case by increasing n_0 and N and accounting for differential rates in different populations.

Our model is based on minimizing the number of *H. influenzæ* cases. This may not be enough from a public health perspective. It may be more reasonable to account for the cost of the vaccine and weigh it with respect to the cost to society of cases of disease.

Reasonably open-minded priors, such as the ones we use, give a moderate probability with negative efficacy. While likely that a vaccine has no efficacy, it may be unlikely to be detrimental. So it might be more reasonable to consider a prior distribution that is a mixture of a gamma and a point mass at zero.

Our model assumes constant risk over an infant's age and uniform accrual. These are not accurate. Younger infants tend to be at greater risk than the older: the average time on study of the 23 cases was about five months instead of the eight months that would be expected if the subjects were at constant risk during the $M = 16$ months. Also, for logistical reasons the study was initiated in only two clinics and ultimately was conducted in ten. So the accrual rate increased during the course of the study. An obvious extension of the current approach is to model differential risk and nonuniform accrual. Such an extension involves somewhat more calculation but is not conceptually more complicated.

REFERENCES

Armitage, P. (1985). The search for optimality in clinical trials. *Internat. Statist. Rev.* **53**, 15–24, (with discussion).

Bather, J. (1985). Randomized allocation of treatments in sequential medical trials. *Internat. Statist. Rev.* **53**, 1–13, (with discussion).

Berger, J. O. and Berry, D. A. (1988). The relevance of stopping rules in statistical inference. *Statistical Decision Theory and Related Topics IV* **1**, (J. O. Berger and S. Gupta, eds.), New York: Springer, 29–72, (with discussion).

Berry, D. A. (1985). Interim analysis in clinical trials: Classical vs. Bayesian approaches. *Statistics in Medicine* **4**, 521–526.

Berry, D. A. (1987). Interim analysis in clinical trials: The role of the likelihood principle. *The American Statistician* **41**, 117–122.

Berry, D. A. (1988). Interim analysis in clinical research. *Cancer Investigation* **5**, 469–477.

Berry, D. A. and Fristedt, B. (1985). *Bandit Problems: Sequential Allocation of Experiments*. London: Chapman and Hall, 275.

Berry, D. A. and Ho, C.-H. (1988). One-sided sequential stopping boundaries for clinical trials: A decision-theoretic approach. *Biometrics* **44**, 219–227.

Eick, S. (1987). The two-armed bandit with delayed responses. *Ann. Statist.* **16**, 254–264.

Heitjan, D. F., Houts, P. S. and Harvey, H. A. (1991). A decision-theoretic evaluation of early stopping rules. (Unpublished *Tech. Rep.*).

Lan, K. K. G. and DeMets, D. L. (1984). Discrete sequential boundaries for clinical trials. *Biometrika* **70**, 659–663.

McPherson, K. (1982). On choosing the number of interim analyses in clinical trials. *Statistics in Medicine* **1**, 25–36.

DISCUSSION

S. E. FIENBERG (*York University, Canada*)

Throughout this meeting I kept on wishing that discussants would stop congratulating the authors and get on with the task of criticizing the paper. Thus it is with great diffidence that

I congratulate Don Berry and his co-authors for the excellent job they have done addressing a major practical problem in the domain of clinical trials. This is honest praise.

The paper takes a by-now standard class of Bayesian machinery and embeds it into the real world setting of restricted experimental populations, the FDA, and its drug-approval process, as well as related advisory committee activities, and it does so with great skill and care. This is one of the few papers at this meeting involving a *randomized* clinical trial, a trial in which Navajo babies are randomly allocated in *equal proportions* to vaccine and control groups. Berry *et al.* implicitly justify this on the grounds that the FDA requires such randomization. Given that the meeting is dedicated in the memory of my friend and former colleague, Morris DeGroot, I asked myself what he might have said about this feature. Morrie always argued that a Bayesian could do better than randomize by using his a priori knowledge. Others of us here would argue that randomization helps in other ways by uncoupling annoying covariates from the causal chain of inference (for technical details see Rubin (1978) and Fienberg and Tanur (1987)).

It is important to note, however, that the choice to employ randomization does not prevent the Bayesian from looking at the data along the way, and thus the authors now are allowed to ask the natural question of when is the optimal time to stop the trial in order to minimize the expected number of cases of influenza in the next several years among the Navajos. Other Bayesian features of the authors' approach include a careful specification of prior distributions, not only those capturing the "expert" views on the efficacy of the vaccine and the occurrence of influenza, but also on the expert's judgment on what it will take to convince the FDA to approve the vaccine for use. I will return to this feature in a moment.

The basic technical feature of the paper is its use of the method of backward induction, a tried and true approach in optimal stopping problems and one which Morrie DeGroot helped to popularize (see DeGroot, 1970). I should note that, as a discussant, I faced a related optimal stopping problem, or perhaps I should call it an optimal starting problem. Early on when I was asked to be a discussant I was told that I would receive the paper by the end of February. At that time I had not been told which paper I was to discuss and had certainly not received a paper. Several weeks later a copy of the preliminary program arrived, but no paper. Telephone calls to Santa Barbara and then Minneapolis subsequently revealed that the data and text were being accumulated but I wouldn't be allowed to examine them at an early stage. Finally, during the past two weeks, I have received four successive versions of the paper, each containing more details than the previous one.

My problem was: when is the optimal time to prepare a discussion?

I will not bore you with the technical details, but I spent several hours on the airplane to Spain solving the problem. The practical intervening factors that I was forced to consider included the possibility of sunshine in Peñíscola, the duration of each evening's disco party, and the probability of cancellation of the excursion of Morella. To make a long story short, I used backward induction beginning with Friday morning, and arrived at Tuesday afternoon as the optimal time to prepare my comments. (Nonetheless, I have, in good Bayesian fashion, updated my comments at least once.) Now I return to the details of the paper.

In a sense the analysis presented by Berry *et al.*, as compelling as I found it, comes in the form of a thought experiment. The actual clinical trial was carried out using a frequentist design, and the assessment of the priors was done retrospectively by the interrogation of the "experts", David Sack and Mark Wolff, who were asked to identify what they thought the prior distributions would have been had they made them at the beginning of the trial. Such retrospective judgments represent treacherous territory, and it doesn't help one much to be a Bayesian.

Recent experimental evidence from cognitive and social psychology suggests that retrospective judgments about past events of behavior are more likely to reflect current judgments about current events or behaviors other than those that might have been elicited in the past (e.g., see Pearson, Ross and Dawes, 1991).

Now when it comes to retrospection about prior judgments of past empirical situations, the problem may be more complicated—but still biased. Consider José Bernardo's prior judgment of the amount of wine participants in *Valencia 3* would drink per day. Unfortunately we ask him to consider what this prior was retrospectively, after *Valencia 4* occurs. Thus he knows that the consumption far exceeded his expectation. Then we might discover that he retrospectively made his prior for *Valencia 3* conform "too closely" with the *Valencia 3* data in order to put the blame on the participants for the mismatch between the prior judgment of wine consumption for *Valencia 4* and the actual data at *Valencia 4*.

This issue of retrospective assessment of priors turns out to matter quite a bit in the present context because of the prior quantity $g(a_1, b_1; a_2, b_2; 0)$, which represents the prior assessment that the FDA approves the vaccine. The stopping rules are reasonably robust to the choice of the priors on λ_1 and λ_2, but not necessarily to the choice of g, which to me seemed somewhat arbitrary. We need to ask if David Sack's judgment of g would have been different if the data had turned out differently and/or the FDA had not approved the vaccine.

Of course, the alternative to this kind of retrospective reanalysis of classical clinical trials is the advent of prospectively-conducted Bayesian trials. If Berry and his co-authors begin one almost immediately there might be just enough time to get the results ready for the next Valencia conference, four years from now.

D. ASHBY (*University of Liverpool, UK*)

As I have a joint appointment between a Statistics department with a strong interest in Bayesian statistics, and a Public Health department with a growing awareness of the role of public health workers in making decisions that affect the health of populations, I was delighted to see this paper. By explicit consideration of the health of a defined population the authors show how both the design and the decision to stop a vaccine trial could be better informed.

However, the challenge of Public Health is to make decisions which are rarely based on a single relevant randomised study, but on a collation of a variety of evidence. Even for a relatively narrow decision, such as "what health care should a health authority purchase for its residents for breast cancer, and from where?", evidence might include randomised studies, observational studies, routinely collected data and expert opinion. Relevant benefits could be measured in years of lives saved and quality of life. Costs include financial, both to the community and to the patient, the effects of some treatments, and the difficulties of attending a hospital some distance from home.

The debate also needs to consider screening for disease, and preventative strategies, which have both financial and other costs. It is widened further by considering the competing claims for finite health care resources, in the context of the limited impact that health professionals have in ameliorating ill-health partly determined by environment and social structure (Martin and McQueen, 1989).

The field of Public Health is developing rapidly, and the potential value of clear thought in the decision making process is enormous. I congratulate the authors on an interesting practical paper which shows how a Bayesian approach facilitates this, and I look forward to seeing more complex problems being considered in this framework.

YOU-GAN WANG (*University of Oxford, UK*)

In clinical trials sequential designs are important from the standpoints of both ethics and cost. Unfortunately there has been a gulf between theory and practice (Armitage, 1985; Simon, 1977.) The study presented by Professor Berry and his collaborators is therefore particularly welcome and most encouraging. Their model is appropriate; the analysis is convincing; and they are able to report a pleasingly positive interaction between clinicians and statisticians. The design is based on a decision-theoretic model and the purpose of this note is to suggest extending the model to include a discount factor. This can allow for the possibility of early termination of the study due to, for example, the strong side-effects on the patients. It can also incorporate the possibility of a new vaccine being developed during the trial. (Some other reasons are discussed in Wang, 1991.) If the occurrence times of such events are assumed to be independent and geometrically (exponentially) distributed, their overall influence may be expressed in terms of a single aggregate discount factor. For example, if the expected time until a new vaccine becomes available is 10 years, as in the paper, the corresponding discount factor should be $1 - 1/120$. If the possibility of termination of the trial due to side-effects discovered within 5 years is 5%, the corresponding discount factor is $(0.95)^{1/60} \approx 0.999$. The joint influence of both possibilities may be expressed by multiplying the two discount factors to give an overall discount factor of 0.991. The maximum time horizon can be truncated at some time, say, 20 years. The backwards induction equations take the same form as in §6.

REPLY TO THE DISCUSSION

We thank the discussants for their complimentary remarks. We have no real differences with any of the discussants, but we do have a few minor comments.

Response to Fienberg: Our probability assessments were indeed retrospective. We recognize the dangers involved in assessing probabilities after or during data collection (Berry 1988). In view of our sensitivity analyses this is not much of a problem for the distributions of the λ's, but it is an important concern as regards the function g. David Sack assessed this function in the context of a generic vaccine, but he indeed knew that the current vaccine had been approved by the FDA at the time of assessment.

We are currently planning vaccine trials prospectively and from the Bayesian point of view. In particular, we are assessing prior probabilities prospectively.

Response to Ashby: We too are interested in the "collation of a variety of evidence." However, vaccine decisions are indeed usually based on "a single relevant randomised study." In the case at hand the only other evidence, as described in our paper, was incorporated into our expert's prior opinion. There is always the question of whether this was done well, but our sensitivity analyses give some comfort that this does not present much of a problem.

Regarding "more complex problems," we are working on some! However, we point out that some (but not all) of Ashby's concerns apply to curative trials and not as much to preventative trials.

Response to Wang: We agree that discounting is appropriate in the context of our paper. Indeed, Chapter 3 of Berry and Fristedt (1985), which is dedicated to the theoretical and practical issues of discounting, makes that very point in the context of sequential experimentation. However, our choice not to discount was a conscious one. Had we been motivated by mathematical simplicity or esthetic appeal, we would have discounted. Instead, we want to convince health care workers of the appropriateness of our approach. And in our experience, however pleasing the notion of discounting is to us, it is a roadblock to less mathematical types. Most people are very comfortable thinking about reducing the number cases of disease over the next 10, 20 and 30 years, say.

ADDITIONAL REFERENCES IN THE DISCUSSION

Berry, D. A. (1988). Multiple comparisons, multiple tests, and data dredging: a Bayesian perspective. *Bayesian Statistics 3* (J. M. Bernardo, M. H. DeGroot, D. V. Lindley and A. F. M. Smith, eds.), Oxford: University Press, 79–94, (with discussion).

DeGroot, M. H. (1970). *Optimal Statistical Decisions*. New York: MacGraw Hill.

Fienberg, S. E. and Tanur, J. M. (1987). Experimental and sampling structures: Parallels diverging and meeting. *Internat. Statist. Rev.* **55**, 75–96.

Martin, C. J. and MacQueen, D. V. (1989). *Readings for a New Public Health*. Edinburgh: University Press.

Pearson, R., Ross, M. and Dawes, R. (1991). Personal recall and the limits of retrospective questions in surveys. *Questions About Survey Questions: Meaning, Memory, Expression, and Social Interactions in Surveys.* (J. M. Tamur, ed.) New York: Russell Sage, (to appear).

Rubin, D. B. (1978). Bayesian inference for causal effects: the role of randomization. *Ann. Statist.* **6**, 34–58.

Simon, R. (1977). Adaptive treatment assignment methods and clinical trials. *Biometrics* **33**, 743–749.

Wang, Y. G. (1991). Gittins indices and constrained allocation in clinical trials. *Biometrics* **78**, 101–111.

BAYESIAN STATISTICS 4, pp. 97–108
J. M. Bernardo, J. O. Berger, A. P. Dawid and A. F. M. Smith, (Eds.)

Regression, Sequenced Measurements and Coherent Calibration

P. J. BROWN and T. MÄKELÄINEN
University of Liverpool, UK and *University of Helsinki, Finland*

SUMMARY

We use the tractable Normal-Inverted-Wishart prior distribution to assign prior structure to $q+1$ variables. It is desired to predict one variable, the response, from the other q variables coherently with respect to varying q. Here we consider situations where these q variables are realistically a subsample from a continuous process. In this context we introduce the notion of *structural* coherence. Application is to spectroscopic analysis of a mixture where each datum is a spectral curve.

Keywords: STRUCTURAL COHERENCE; CONTINUOUS TIME ARMA MODEL; REPEATED MEASUREMENTS; NORMAL-INVERTED-WISHART; REFINED SUB-SAMPLED OBSERVATIONS; QR DECOMPOSITION; MATRIX-RIDGE REGRESSION; SPECTROSCOPIC DATA.

1. INTRODUCTION

We consider the situation of a researcher who wishes to predict a variable of special interest, Y_0, from a number, q, typically large, of further random variables, $Y_{(q)} = (Y_1, Y_2, \dots, Y_q)$. Here we think of these further variables as essentially some sub-sampled sequence of a continuous random process, so that there is an order to the variables and often the sub-sampling will be at equally spaced intervals of an underlying continuous variable. In practice $Y_{(q)}$ is quick and cheap to obtain, for example the absorbances at $q = 1168$ frequencies of infra-red light at 2 nanometer intervals. On the other hand Y_0, the concentration of a constituent, is expensive and slow to accurately determine by standard wet chemistry methods.

Training data for which both the interest variable and the further variables are observed on n specimens are available. For the purpose of this article we assume that this data may be viewed as obtained at random so that we are in the *random* or *natural* calibration case as opposed to the more problematic *controlled* calibration where the interest variable is strictly designed in the training data, see Brown (1982). In effect the calibration paradigm just reduces to standard regression prediction. One novelty for us is that nominally the number q of explanatory variables may be very much larger than the number of observations. The example described in Section 8 has $n = 12$ and $q = 43$, but q as large as 1000 is also typical. With this high degree of indeterminacy a well structured prior distribution appropriate to the sequenced nature of the data becomes essential. It is also natural to demand that this prior structure be coherent across sub-sequences of the q variables. If we observed only observations at 4 nanometer intervals instead of 2 nanometers so that $q/2$ (q even) alternate variables are utilised, then the prior probability assignment for this subset should (1) correspond to the marginalised prior distribution, (2) be structurally generated by the same prior considerations which led to the generation of the prior for the refined set of variables. Whereas (1) amounts to de Finetti's notion of coherence (see for example, de

Finetti, (1974)), (2) provides stronger requirements and might be termed *structural* coherence. It has similarities with the extendibilty notion in exchangeability. Readers may be reminded that such notions of coherence refer only to the prior distribution, the posterior distribution will certainly differ as a result of data at different resolutions.

The Normal-Inverted-Wishart distributional structure we shall adopt is that used in similar circumstances by Lindley (1978), Section 6, further reported by Dickey, Lindley and Press (1985). Dawid (1988) has shown that such a structure embodies an undesirable predictive determinism in the case of a countable number of variables. Whereas Dickey *et al.* (1985) adopt the rather special intraclass prior expected covariance, Mäkeläinen and Brown (1988) develop a class of prior expected covariance matrices which accept an ordering of the predictive variables. In the present article, the class of prior expected covariance is a rich class. It is as if it is sub-sampled from an underlying continuous parameter process. We have however avoided directly assigning a prior for a Gaussian process, rather remaining within the tractable if flawed Normal-Inverted-Wishart. This has an added bonus. We offer a more robust analysis by avoiding the strong assumption of a class of stationary Gaussian processes and relegating such structures to the hyperparameters of the inverted Wishart prior; see Chen (1979) for a similar attitude to imposing structural information.

The more usual approach to regression prediction works with the conditional distribution of the response given the predictive variables. By working less directly simply with the full joint distribution we are able to incorporate the two aspects of coherence mentioned above within the more complex assignment to implied regression coefficients. In fact the seemingly most natural autoregressive structure is ruled out. One class of priors we will work with, embodying autoregression of the predictive variables, allows computationally effective algorithms for feasible and speedy implementation even when implicitly one wishes to solve a system of regularised "normal" equations involving more than a thousand unknowns. A typical example might require predicting the amount of an ingredient of a detergent, where $q = 1168$ predictor variables are observed on training data of just $n = 12$ samples. This type of example has become important in the chemical and food industries and in process control. Existing techniques by chemometricians and statisticians involve, Principal Components Regression, Ridge Regression, Continuum Regression (Stone and Brooks, 1990), and Partial Least Squares, see Brown (1990) for a perspective. None of these approaches takes account of the contiguity of regressors, the focus of this paper. Our approach would seem to be applicable more generally to repeated measurements problems in medicine and social science, see Jones and Ackerson (1990) for a stationary Gaussian process model in this context.

2. THE MODEL

Following Dawid (1988) and Mäkeläinen and Brown (1988), we shall denote the data matrix, which is of order $n \times (q+1)$, by

$$X^q = (X_0, X_{(q)}). \tag{1}$$

Here X_0 is the vector containing the observations on the regressand and $X_{(q)}$ is the matrix of regressors. We shall consider random regressions and make the assumption that X^q is a sample from a $(q+1)$-variate normal distribution having mean zero. This latter assumption involves no loss of generality provided the mean vector is assumed known and we will relax this assumption when necessary later. In view of the supposition that regressions are in the centre of interest this amounts to making more extensive assumptions than are strictly necessary. As discussed in the introduction, they are necessary to enable us to specify prior assumptions coherently over submodels and the envisaged refined data.

The researcher also contemplates a future observation

$$Y^q = (Y_0, Y_{(q)}) \tag{2}$$

from the same distribution which is independent of (1) given the parameters of the distribution. Finally the covariance matrix of the distribution is taken to follow an inverse Wishart distribution with positive definite scale matrix M^q.

To facilitate Bayesian manipulations and to enable extension to countably infinite dimensional distributions, we use the notation introduced by Dawid (1981). In summary,

$$\begin{aligned} X^q \mid \Sigma &\sim \mathcal{N}_{n,q+1}(I_n, \Sigma), & (3)\\ \Sigma &\sim IW_{q+1}(\delta, M^q), \qquad \delta > 0, M^q > 0, & (4)\\ Y^q \mid \Sigma &\sim \mathcal{N}_{1,q+1}(1, \Sigma), & (5)\\ X^q \text{ and } Y^q &\text{ are independent, given } \Sigma. & (6) \end{aligned}$$

Here given Σ the rows are independent and within any row the covariance matrix is Σ. Precisely if a random matrix is $\mathcal{N}(A, B)$ then $a_{ii}B$ and $b_{jj}A$ are the covariance matrices of the ith row and jth column, respectively. The degrees of freedom δ above would be $\delta+(q+1)-1$ in the more standard notation, and thus avoids immediate notational difficulties when q is infinite. In some contexts it may be natural to consider an infinite sequence of regressors, when each of X^q, Y^q, and M^q are infinite with the understanding that for any finite subset appropriate conforming submatrices are chosen.

We utilise this model for prediction despite some well-known deficiencies in the range of beliefs which are expressible by means of the model. Equations (3)–(6) incorporate a rather limited parameterisation namely in terms of M^q the prior scale matrix of Σ and just one parameter δ for uncertainty. Furthermore, Dawid (1988) has elucidated a *determinism* of the prior in prediction in the infinite-dimensional case.

The future prediction problem envisaged is to predict Y_0 from $Y_{(q)}$ given the information about Σ as updated through observing X^q. We repeat the argument of Mäkeläinen and Brown (1987). The solution, a conditional distribution, is found by two consecutive conditionings in a matrix-variate T-distribution. To formulate the result needed suppose $Z \sim T_{r,s}(\delta; H, K)$. Here this matrix T distribution denotes the random matrix which given Σ is $\mathcal{N}(H, \Sigma)$ and Σ is $IW(\delta, K)$. Consider the partitions $Z = (Z_1, Z_2)$ and $K = [K_{ij}]$ with Z_i an $r \times s_i$ matrix and K_{ij} an $s_i \times s_j$ matrix $(i, j = 1, 2)$ so that $s_1 + s_2 = s$. Assuming that K_{11} is positive definite we have, see for example Dawid (1981),

$$Z_2 - Z_1 B'_{21} \mid Z_1 \sim T_{r,s_2}(\delta + s_1; H + Z_1 K_{11}^{-1} Z'_1, K_{22\cdot 1}), \tag{7}$$

where $B_{21} = K_{21}K_{11}^{-1}$ and $K_{22\cdot 1} = K_{22} - B_{21}K_{12}$. The formula indicates the linearity of the regressions in a t-distribution. Also note that $Z' \sim T_{s,r}(\delta; K, H)$.

By the assumptions (3)–(6) and by the definition of a T-distribution we have that

$$\begin{bmatrix} X^q \\ Y^q \end{bmatrix} \sim T_{n+1,q+1}(\delta; I_{n+1}, M^q).$$

By the transpose of (7),

$$Y^q \mid X^q \sim T_{1,q+1}(\delta + n; 1, N^q) \tag{8}$$

where

$$N^q = M^q + (X^q)'X^q. \tag{9}$$

Thus N^q is the predictive covariance matrix of Y^q. We shall write

$$M^q = \begin{bmatrix} m_{00} & m_{0(q)} \\ m_{(q)0} & M_{(q)(q)} \end{bmatrix}$$

and similarly for N^q. Using (7) in (8) we finally find that

$$Y_0 \mid Y_{(q)}, X^q \sim Y_{(q)}(b^n_{0(q)})' + T_{1,1}(\delta^*; a, c) \tag{10}$$

where

$$b^n_{0(q)} = n_{0(q)}(N_{(q)(q)})^{-1} \tag{11}$$

and

$$a = 1 + Y_{(q)}N^{-1}_{(q)(q)}Y'_{(q)}; \quad c = n_{00\cdot(q)}; \quad \delta^* = \delta + n + q.$$

In standard notation $T_{1,1}$ above is $\sqrt{(ac/\delta^*)}t_{\delta^*}$. Also notice that by the linearity of the regressions and since N^q is the covariance matrix of the predictive distribution of Y^q,

$$n_{00\cdot(q)} = n_{00} - b^n_{0(q)}n_{(q)0}$$

when scaled by appropriate degrees of freedom, is the residual variance in the regression of Y_0 on $Y_{(q)}$ in that distribution.

Equation (11) has a familiar form since by (9)

$$N_{(q)(q)} = M_{(q)(q)} + (X_{(q)(q)})'X_{(q)(q)}.$$

From standard Normal Bayesian regression analysis, see for example equation (7) of Lindley and Smith (1972), (11) is that estimate obtained from a prior covariance proportional to the *inverse* of $M_{(q)(q)}$, when the prior mean of the regression coefficients is (11) with M^q in place of N^q. We will have cause to utilise this later. In the case of an intraclass correlation structure of Lindley (1978) it explains the negative prior correlation between regression coefficients noted by him when ρ is non- negative. Naively it reflects a scale property: scaling X_j by a constant c scales the coefficient β_j of X_j in the regression of X_0 on $X_{(q)}$ by $1/c$.

3. REGRESSOR REFINEMENTS

We shall be interested in the submatrices of the $n \times q$ regressor matrix

$$X_{(q)} = (X_1, X_2, \ldots, X_q)$$

such as

$$X_{[q]} = (X_1, X_3, X_5, \ldots, X_q)$$

when q is odd which obtain on thinning of the regressors. Conceptually at the same time we are interested in possible refinements of the data, whereby had only $q/2$ regressors $X_{[q]}$ been observed we might wish to contemplate the prior distribution *had* q regressors $X_{(q)}$ been taken. More generally we might think of $[q]$ as any subset of (q), the integers $1, 2, \ldots, q$. In such contexts we shall generate coherence of regression specifications by the Normal–Inverted– Wishart prior of the previous Section with the prior covariance matrix that of a subsample from a random function. Details of the particular class of stationary random functions are discussed in the next section.

A special feature of T-distributions is that for any subset $[q]$,

$$Y_0 \mid Y_{[q]}, X^q \sim Y_0 \mid Y_{[q]}, X^{[q]},$$

where $X^{[q]} = (X_0, X_{[q]})$. That is, if, at a future time when Y_0 is to be predicted by means of $Y_{(q)}$, the researcher should decide to use only $Y_{[q]}$ there would be no advantage in having retained the regressors in $X_{[q]^c}$ where $[q]^c$ denotes the complementary even integers. This property, related to S-ancillarity is discussed by Mäkeläinen and Brown (1987), see also Brown (1982).

The posterior predictive distribution of Y_0 given the data $X^{[q]}$ and $Y_{[q]}$ is then analogous to (10) with $[q]$ replacing (q) and q replaced by $(q+1)/2$ or $q/2$ in the degrees of freedom depending on whether q is odd or even.

4. CONTIGUOUS PRIOR STRUCTURE

We will be concerned with the form of $M^q = (\delta - 2)E(\Sigma)$, where $E(\Sigma)$ is the prior expectation of the covariance of the $q+1$ variables. The prior covariances of the regressand with the regressors, $m_{0(q)}$ may often be such that all q covariances are judged equal. In reality it may be that certain regressors or regions of regressors are more predictively important than others and in principle such beliefs may be incorporated. With this proviso, we usually take the covariances equal and make the common value zero.

The $q \times q$ matrix $M_{(q)(q)}$ is our major concern and is generally assumed to correspond to the covariance kernel of a sub-sampled stationary Gaussian random function. This involves a rich class of possibilities, see for example Yaglom (1987). It ensures that the two coherence requirements of the Introduction are met. One subclass of these Gaussian random functions involves the kernel

$$\rho(\tau) = \exp(-\alpha \mid \tau \mid^{\kappa}), \quad \alpha > 0$$

where $0 < \kappa \leq 2$, with an AR(1) embedded in $\kappa = 1$. Note though that in discrete time negative correlation is allowable but precluded in continuous time. For discussion here our methods apply to more general autoregressive processes, but with the above $\kappa = 1$ the simplest and perhaps most important special case. Suppose for example one has equal spacing with $\kappa = 1$. The form of covariance is an AR(1),

$$M_{(q)(q)} = \{k/(1-\rho^2)\} \begin{pmatrix} 1 & \rho & \rho^2 & \cdots & \rho^{q-1} \\ \rho & 1 & \rho & \cdots & \rho^{q-2} \\ \vdots & \vdots & \vdots & & \vdots \\ \rho^{q-1} & \rho^{q-2} & \cdots & \rho & 1 \end{pmatrix}. \tag{12}$$

This has the same form as the prior covariance matrix $M_{[q][q]}$ for the $q/2$ thinned variables obtained by odd rows and columns of $M_{(q)(q)}$ except that the correlation parameter is now ρ^2, so that structural coherence is preserved. The same could not be said for example of a moving average structure for $M_{(q)(q)}$ when for a MA(1) the thinned covariance matrix would be proportional to an identity matrix and would not have the same structural form as for the q variables. Since the inverse of the covariance matrix of an MA(p) is the covariance matrix of an AR(p), by the remark at the end of Section 2, we preclude an autoregressive structure for a prior distribution describing the regression coefficients. The continuous time parameter version of an MA(1) is further precluded by stationarity of the random process. Doob (1953) Chapter 11, Section 10 in treating the continuous analogue of an autoregressive– moving average process, an ARMA(p, s), defines the process in terms of a rational spectral density,

$$f(\omega) \quad = \quad k\frac{\mid (\omega - a_1)\ldots(\omega - a_s) \mid^2}{\mid (\omega - c_1)\ldots(\omega - c_p) \mid^2}.$$

For stationarity it is necessary and sufficient that the spectral density be non-negative and *integrable*. Hence it is necessary that the autoregressive order p be greater than the moving average order, s. See also Yaglom (1987), Example 9, pp. 133–136. Except in the case of multiple roots of the spectral denominator, the covariance function is,

$$\rho(\tau) \quad = \quad \sum_1^p (C_j \cos(2\pi c_j' \tau) + D_j \sin(2\pi c_j' \tau)) \exp(-2\pi c_j'' \tau)$$

where C and D are constants not involving τ, and c_j' and c_j'' are respectively the real and imaginary parts of c_j.

Thus an ARMA(2,1) would be allowable as would pure autoregressive structures for $M_{(q)(q)}$ of second and higher order since they also allow structural coherence. For example with equal spacing for an AR(2) process, the autocorrelation function might be

$$A\rho_1^\tau + B\rho_2^\tau$$

where ρ_1 and ρ_2 are roots (assumed unequal) of an auxiliary quadratic and stationarity imposes conditions on allowable ρ_1 and ρ_2. Thus it is evident that the thinned process will have the same form of autocorrelation function.

For a simple non-stationary dynamic model applied directly to the regression coefficients in the calibration of the Olympic marathon see Smith and Corbett (1987); and, for a smoothness prior similar to theirs, see Polasek (1985).

5. AUGMENTATION AND COMPUTATION

The posterior mean of the regression coefficients is given by equation (11) and involves solving for the q-vector b the q equations given by

$$m_{(q)0} + X_{(q)}' X_0 \quad = \quad (X_{(q)}' X_{(q)} + M_{(q)(q)})\, b. \tag{13}$$

When q may be of order one thousand this is a formidable task as it stands. If M is proportional to the identity matrix, as in ridge regression, we may orthogonally transform the problem and solve $\min\{q, n\}$ equations. This represents a substantial saving as in our applications n is very much smaller than q. This route is not open with our more general structure for M. However an alternative route adapted from ridge regression in Marquardt (1970) not only enables us to speedily solve (13) but also to avoid setting up this "squared" equation in the first place. Suppose we can analytically specify a $q \times q$ matrix A, preferably upper triangular, such that $A'A = M_{(q)(q)}$. Now we augment the $n \times q$ data matrix $X_{(q)}$ with q rows A, forming an $(n+q) \times q$ matrix $X_{(q)}^*$. At the same time we augment the n-vector X_0 by a q-vector $A^{-1} m_{(q)0}$, forming an $(n+q) \times 1$ response vector X_0^*. Note that this latter augmentation is trivial if $m_{(q)0}$ is zero, and is easy if A and hence A^{-1} is triangular. Standard regression algorithms can now be applied as if $X_{(q)}^*$ is the design matrix and X_0^* the response vector. In particular one can apply the QR- decomposition and avoid "squaring" quantities, simply back-solving a set of triangular equations, see, for example, Thisted (1988). With the augmentation matrix A already triangular one can effect the QR decomposition by a series of Givens elementary planar rotations of order q^2 in number, instead of the order q^3 needed to solve equation (13) directly, with the added bonus of more numerical accuracy. Thisted rightly emphasises the instability of the Binomial Inverse theorem, see for example Press (1982), often used to good effect for analytical data augmentation.

Lemma 1. *We can analytically construct an upper triangular $q \times q$ matrix square root A for any AR(p) sequence.*

Proof. We first demonstrate this for an AR(1). Referring to (12), let $K = (1/k)M$. The requisite matrix square root of K is C where $C'C = K$ and

$$C = \begin{pmatrix} 1 & \rho & \rho^2 & \dots & \rho^{q-2} & \rho^{q-1}/\sqrt{(1-\rho^2)} \\ 0 & 1 & \rho & \dots & \rho^{q-3} & \rho^{q-2}/\sqrt{(1-\rho^2)} \\ 0 & 0 & 1 & \dots & \rho^{q-4} & \rho^{q-3}/\sqrt{(1-\rho^2)} \\ \vdots & \vdots & \vdots & \vdots & \vdots & \vdots \\ 0 & 0 & 0 & 0 & 0 & 1/\sqrt{(1-\rho^2)} \end{pmatrix}$$

This we derived by noting that the inverse of K is tridiagonal with $-\rho$ off diagonal and $(1+\rho^2)$ on diagonal apart from the (1,1) and (q,q) elements which are unity. The square root of this matrix is U where $U'U = K^{-1}$ and

$$U = \begin{pmatrix} 1 & -\rho & 0 & \dots & 0 \\ 0 & 1 & -\rho & 0 & \dots \\ \vdots & \vdots & \vdots & \vdots & \vdots \\ 0 & \dots & \dots & 1 & -\rho \\ 0 & \dots & \dots & 0 & \sqrt{(1-\rho^2)} \end{pmatrix}.$$

This structure is inherent in the autoregression

$$X_t = (1 - \rho L)^{-1} \epsilon_t$$

with L the lag operator and $\{\epsilon_t\}$ uncorrelated errors, augmented to q dimensions from the $q-1$ differences. The U matrix for an AR(2) will have first $q-2$ rows obtained from $1 - \phi_1 L - \phi_2 L^2$ and can be augmented by a further two rows so as to remain triangular and such that $U'U = K^{-1}$. The inverse of matrix U is then readily constructed and will be triangular. $\triangleleft$

Similar constructs to those above using Levinson's recursion may be found in Brockwell and Davis (1987), chapters 5 and 8. Other types of decomposition, but not including ours, are given by Bhansali (1990).

6. MEAN MODELLING

The models above all assume that the mean vector of Y^q and also for X^q is (i) known and (ii) constant from sample to sample. One may wish to relax these assumptions. When the mean vector is unknown it may be totally unstructured. Alternatively it may be assumed that $Y_{(q)}$ has a mean which is itself subsampled from some Gaussian process. If the mean is not constant from sample to sample then a hierarchical functional prior could be contemplated. This would be particularly apt in near infra-red spectroscopy with ground solid preparations where variation in particle size serves to bodily shift the functional form, see Fearn (1983) for the adverse effect of this on routinely applied Ridge regression, and the multivariate scatter corrections of Martens, Jensen and Geladi (1983).

However for this paper we will only explicitly consider a constant but unknown mean vector with an associated vague prior distribution, when posterior predictive distributions analogous to (10) may be produced with deviations measured from the training sample mean vector and degrees of freedom as before or reduced by one, depending on the prior limit adopted. A full natural conjugate approach may be extracted from Ando and Kaufman (1965).

7. HYPERPARAMETER ESTIMATION

The prior expected covariance matrix $M_{(q)(q)}$ as autoregressively structured in Section 4 has a number of unspecified hyperparameters. In the case of the AR(1) structure of (12), there are two parameters, k and ρ. We might envisage a further prior distribution on these hyperparameters. More pragmatically, following for example Lindley and Smith (1972), from the posterior Student distribution given in (12), we may form modal posterior estimates of k and ρ. Alternatively, as in the example described in the next Section, we choose these parameters by cross–validation. This is preferred as it is likely to be more robust to departures from the chosen simple Normal-Inverted-Wishart model.

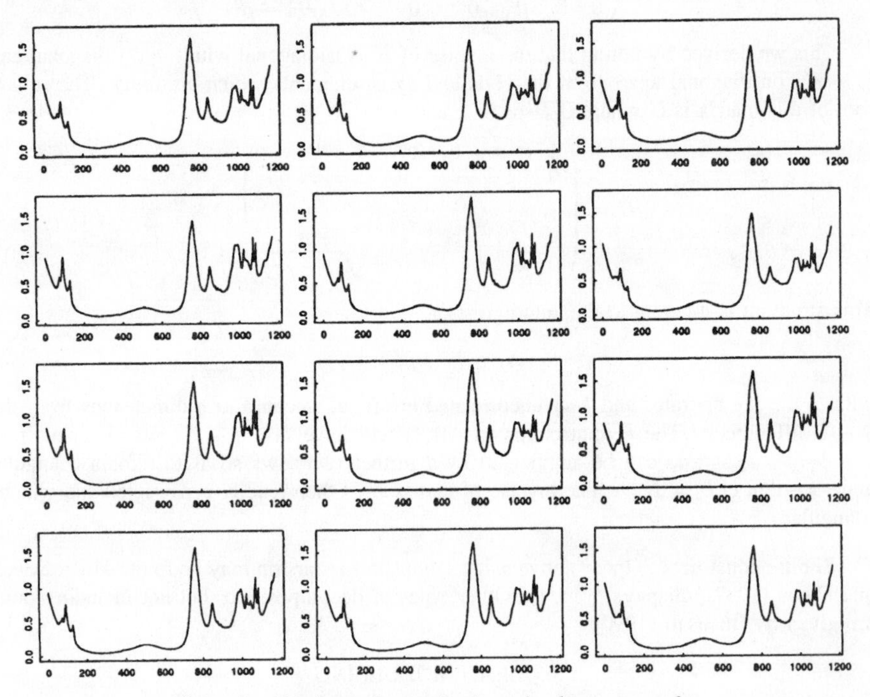

Figure 1. *Absorbance Spectra for the twelve detergent samples.*

8. DETERGENT EXAMPLE

The data set consists of seven bands of frequencies chosen from 1168 mid–infrared equally spaced frequencies, in the range 3100 to 750 cm^{-1}. The 43 frequencies were as chosen by Brown, Spiegelman and Denham (1992). Only twelve samples were available. The full data $X_{(q)}$ are plotted as 12 samples of absorptions at the 1168 wavelengths in Figure 1, and appears continuous to the resolution of printing. Figure 2 gives the mean and variance of the twelve absorptions plotted against frequency. The data, twelve detergent mixtures, contain 5 ingredients carefully designed. We however treat the data as randomly generated and focus on one ingredient. See Sundberg and Brown (1989) for a correspondence of estimation spaces for controlled and random calibration in underdetermined problems. The data X_0

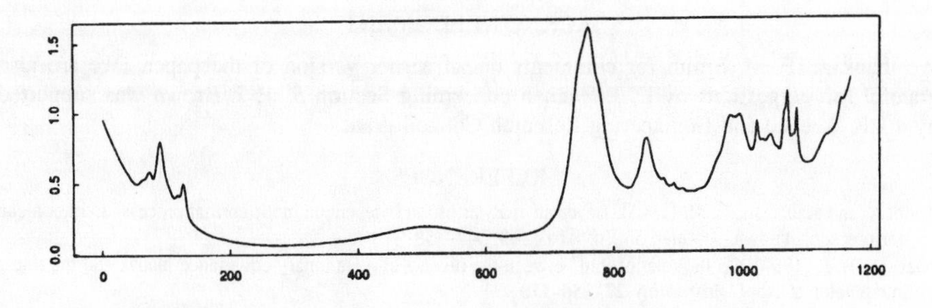

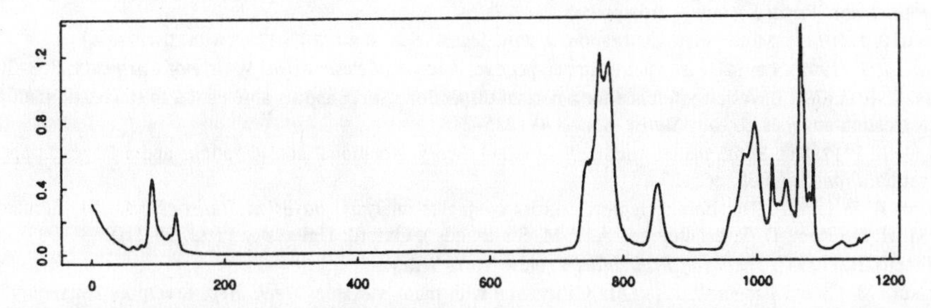

Figure 2. *Mean and Variance of twelve absorbance spectra.*

are (6.99, 13.10, 9.91, 13.09, 7.08, 6.89, 10.07, 10.08, 13.22, 7.91, 12.12, 9.49). The five-ingredient data are further analysed using splines and autoregressive error structure in Denham and Brown (1990).

Table 1 gives cross–validated, that is leave–one–out, sum of squares prediction errors for the 12 observations for a grid of values of k and ρ. The minimum value 0.371 at $k = 10^9$ and $\rho = 0.7$ gives a root mean square prediction error of $\sqrt{0.371/12} = 0.18$ and represents 99.5 percent of the variation explained. It demonstrates the impressive accuracy of the infra-red spectrometry. Statistically this is all the more reassuring given the 43 minus 12 degrees of indeterminacy in the implicit linear model. In this case our method of analysis has a 25 percent smaller prediction mean squared error than ridge regression, as embodied in the first row of Table 1 with $\rho = 0$. For prediction purposes distributionally we might also utilise the posterior predictive student form of (10).

ρ	k					
	5E5	2E6	8E6	35E6	14E7	E9
0.00	492	488	487	486	486	486
0.25	440	435	433	433	433	433
0.50	399	390	388	387	387	387
0.70	389	376	372	372	371	371
0.90	456	442	438	436	436	436

Table 1. *Prediction sum of squared errors × 1000 (where for example 8E6 is 8000000).*

ACKNOWLEDGMENT

We thank A. F. M. Smith for comments on an earlier version of the paper. We are also grateful for suggestions by E. J. Hannan concerning Section 5. P. J. Brown was supported by a UK Science and Engineering Research Council grant.

REFERENCES

Ando, A. and Kaufman, G. M. (1965). Bayesian analysis of the independent multinormal process– neither mean nor precision known. *J. Amer. Statist. Assoc.* **60**, 347–358.

Bhansali, R. J. (1990). On the relationship between the inverse of a stationary covariance matrix and the linear interpolator. *J. Appl. Probability* **27**, 156–170.

Brockwell, P. J. and Davis, R. A. (1987). *Time Series, Theory and Methods*. New York: Springer.

Brown, P. J., Spiegelman, C. H., and Denham, M. C. (1992). Chemometrics and spectral frequency selection. *Phil. Trans. Royal Society A*, (to appear).

Brown, P. J. (1982). Multivariate calibration. *J. Roy. Statist. Soc. B* **44**, 287–321, (with discussion).

Brown, P. J. (1990). Partial least squares in perspective. *Analytical Proc. Royal Society of Chemistry*, 303–306.

Chen, C.-F. (1979). Bayesian inference for a normal dispersion matrix and its application to stochastic multiple regression analysis. *J. Roy. Statist. Soc. B* **41**, 235–248.

Dawid, A. P. (1981). Some matrix-variate distribution theory, Notational considerations and a Bayesian application. *Biometrika* **68**, 265–274.

Dawid, A. P. (1988). The infinite regress and its conjugate analysis. *Bayesian Statistics 3* (J. M. Bernardo, M. H. DeGroot, D. V. Lindley and A. F. M. Smith, eds.), Oxford: University Press, 95–110.

de Finetti, B. (1974). *Theory of Probability* **1**. New York: Wiley.

Denham, M. C. and Brown, P. J. (1990). Calibration with many variables. *Tech. Rep.* Liverpool University.

Dickey, J. M., Lindley, D. V., and Press, S. J. (1985). Bayesian estimation of the dispersion matrix of a multivariate normal distribution. *Comm. Statist. Theory and Methods* **14**, 1019–1034.

Doob, J. L. (1953). *Stochastic Processes*. New York: Wiley.

Fearn, T. (1983). A misuse of ridge regression in the calibration of a near infrared reflectance instrument. *Appl. Statist.* **32**, 73–79.

Jones, R. H. and Ackerson, L. M. (1990). Serial correlation in unequally spaced longitudinal data. *Biometrika* **77**, 721–732.

Lindley, D. V. and Smith, A. F. M. (1972). Bayes estimates for the linear model. *J. Roy. Statist. Soc. B* **34**, 1–41, (with discussion).

Lindley, D. V. (1978). The Bayesian approach. *Scandinavian J. Statist.* **5**, 1–26, (with discussion).

Mäkeläinen, T. and Brown, P. J. (1987). Priors and choice of regressors. *Tech. Rep.* University of Helsinki.

Mäkeläinen, T. and Brown, P. J. (1988). Coherent priors for ordered regressions. *Bayesian Statistics 3* (J. M. Bernardo, M. H. DeGroot, D. V. Lindley and A. F. M. Smith, eds.), Oxford: University Press, 677–696.

Marquardt, D. W. (1970). Generalised inverses, ridge regression, biased linear estimation and nonlinear estimation. *Technometrics* **12**, 591–612.

Martens, H., Jensen, S. Å. and Geladi, P. (1983). Multivariate linearity transformation for near-infrared reflectance spectrometry. *Proc. Nordic Symposium on Applied Statistics*, Stravanger, Norway: Stokkand Forlag Publ., 205–234.

Polasek, W. (1985). Hierarchical models for seasonal time series. *Bayesian Statistics 2* (J. M. Bernardo, M. H. DeGroot, D. V. Lindley and A. F. M. Smith, eds.), Amsterdam: North-Holland, 723–732.

Press, S. J. (1982). *Applied Multivariate Analysis*. Melbourne, FL: Krieger.

Smith, R. L. and Corbett, M. (1987). Measuring marathon courses, an application of statistical calibration theory. *Appl. Statist.* **36**, 283–295.

Stone, M. and Brooks, R. J. (1990). Continuum regression, cross-validated sequentially constructed prediction embracing ordinary least squares, partial least squares and principal components regression. *J. Roy. Statist. Soc. B* **52**, 237–269.

Sundberg, R. and Brown, P. J. (1989). Multivariate calibration with more variables than observations. *Technometrics* **31**, 365–371.

Thisted, R. A. (1988). *Elements of Statistical Computing*. London: Chapman and Hall.

Yaglom, A. M. (1987). *Correlation Theory of Stationary and Related Random Functions* **1**. New York: Springer.

DISCUSSION

T. FEARN (*University College London, UK*)

First let me compliment the authors on a paper that brings some elegant mathematics to bear on an important practical problem. In particular it is refreshing to see so much careful attention to the computational details. Since it is my task to be critical, and since I find no fault with the theory, I shall concentrate on the example.

My reservations concern the appropriateness of the AR(1) form for $M_{(q)(q)}$ for this particular example, and indeed for spectroscopic examples in general. There are (at least) two sources of variability in the spectra of the samples: variation resulting from the different composition of the samples and variation due to what might loosely be described as measurement error. Here the former must dominate or the calibration would not be so successful. However the AR(1) variance matrix sems to me to be appropriate only to describe the latter. Consider the following argument.

The calibration samples, and presumably those to be predicted, are mixtures of $c = 5$ components: 4 chemicals and water. To a first approximation Beer's law implies that the spectrum of the mixture should be the appropriate linear combination of the spectra of the components, i.e.,

$$Y = \sum_{r=1}^{c} P_r z_r$$

where $Y(q \times 1)$ is the spectrum of the mixture, P_r is the proportion of substance r it contains, and $z_r(q \times 1)$ is the spectrum of the pure substance. Drawing Y from a population in which $\mathrm{var}(P_r) = v_r$ and $\mathrm{cov}(P_r, p_s) = c_{rs}$ would lead to a variance matrix for Y of the form

$$\mathrm{var}(Y) = \sum_{r=1}^{c} v_r z_r z_r^T + \sum_{r \neq s} c_{rs} z_r z_s^T$$

Deferring for the moment the question of whether we have the information required to quantify this expression, I suggest that, for this example at least and perhaps somewhat more generally, it would be a more appropriate form for $M_{(q)(q)}$, the prior expectation of the $q \times q$ submatrix of Σ that corresponds to the spectra.

This 'mixture' M, M^* say, has quite different (and arguably more appropriate) properties than the AR(1) form used:

(1) The correlations do not necessarily die away with distance. In the spectrum of each substance there will usually be several peaks, often widely separated, and there will be strong correlations between elements of Y not close to each other in the sequence.

(2) Variance is large where the mean is large, i.e., where there are peaks in the spectra. Compare Figure 2.

(3) The rank of M^* is $c - 1$ (noting that $\Sigma P_r = 1$) however large q becomes. Thus the inverse Wishart prior distribution for Σ has a singular scale matrix. This has philosophical attractions in that the determinism as $q \to \infty$ noted by Dawid (1988) would seem to be avoided (because the degrees of freedom no longer increase with q) but is hardly a technical asset. Pursuing the analysis with singular M would appear to lead in the direction of something resembling principal components regression (an approach widely and successfully used with spectroscopic data) but with the components defined *a priori* rather than extracted from $X^T X$. A more attractive alternative is to take not M^* but $M^* + \eta I$ as the scale matrix, possibly with η chosen by cross-validation, on the grounds that the Beer's law argument is only approximate and such drastic constraints on the

solution are not justified. Note that letting $\eta \to 0$ would amount to using the Moore-Penrose inverse for $N = M + X^T X$ in the expressions for predictive mean etc.

If I have understood what structural coherence is then the mixture choice for M is structurally coherent. A bonus is that it is rather easier to see how to proceed when the points at which the spectra are sampled are unequally spaced.

Do we have the information to quantify M^*? The answer in this particular case is probably "no but we could easily obtain it". All that would be needed for the z_r would be to take 5 more scans (of water and each of the 4 chemicals in water). Depending on the exact chemicals involved the relevant spectra may even be available in reference books. The P_r could be taken to be uniform, subject to $\Sigma P_r = 1$. It may be too late to acquire the relevant information for this example, thus saving me from having to try to explain more than 99.5% of the variation, but I hope to report elsewhere on another example in which the component spectra are available.

REPLY TO THE DISCUSSION

We welcome Dr. Fearn's thoughtful comments on the appropriate form of the prior scale matrix in the Normal-Inverted-Wishart prior. He suggests combining assumed known spectra of individual components linearly by amounts which are random. We have used the autoregressive moving average class of covariance structures in much the same spirit as Markov random field prior distributions in image analysis. We think of them as providing realistic *local* prior assumptions and implicitly trust that this is all that is needed to get good posterior results. Whilst it may not be worth pushing our small sample size detergent example further, there is some room for improved prediction on components other than the second component utilised in this paper. Some of the other four components do not have such dramatically good predictors, although they all show more than ninety percent variation explained.

Whilst there are aspects of Fearn's suggestion which are very appealing we might point to some difficulties. Underlying the linearity of Beer's law are two assumptions, (1) no interaction in the mixture and (2) complete dispersal of each component in the medium. The pure spectra are thus somewhat hypothetical, and subtraction of the spectrum of water from each component alone in water cannot strictly be justified by Beer's law. This renders the suggestion of a 'ridge constant' augmentation of the mixed spectra form of M even more necessary. We might here remark that the correctly alluded to avoidance of determinism disappears with this same augmentation and determinism returns.

Many frequency channels provide little information and although Fearn's suggestion will run into computational difficulties with 1168 channels, it should be quite feasible for the selected 43 channels in our analysed example. However such a selection was outside the Bayesian spirit of the paper, offering a computational short cut. In passing it may help the reader if we expand a little on the way this selection affected our implementation. The 43 selected channels chosen by the method of Brown, Spiegelman and Denham (1992), amounted to 7 separate bands of channels. Rather than thin out the AR(1) prior structure, we applied the structure separately to each of the bands and set the small between band correlations to zero.

Another difficulty with both the prior of the paper and Dr. Fearn's discussion is the single parameter δ for expressing uncertainty in the assumed prior scale matrix. This inherent weakness of the Normal-Inverted-Wishart prior may prove more critical when other than local properties are addressed. This may mitigate against Fearn's suggestion. It is however a suggestion within the framework of our paper that deserves further detailed investigation. Once again we thank the discussant for his perspicacious comments.

BAYESIAN STATISTICS 4, pp. 109–125
J. M. Bernardo, J. O. Berger, A. P. Dawid and A. F. M. Smith, (Eds.)

Prequential Analysis, Stochastic Complexity and Bayesian Inference

A. P. DAWID
University College London, UK

SUMMARY

Prequential Analysis addresses the empirical assessment of statistical models, and of their associated forecasting techniques, using techniques borrowed from the methodology of Probability Forecasting. In the theory of *Stochastic Complexity*, the empirical assessment of a model is based on the minimal length of a coded message needed to transmit the data. It turns out that this is essentially the same as prequential assessment based on the logarithmic scoring rule. These approaches are particularly well suited to model selection, where they provide further justification for the use of Bayes factors, or the asymptotically equivalent Jeffreys-Schwarz-BIC penalized log-likelihood criterion.

This paper surveys the current state of understanding of these new assessment methodologies, emphasizing connexions with Bayesian inference. In particular a general argument for the consistency of a Bayesian model selection procedure is given.

Keywords: PROBABILITY FORECASTING; OPTIMAL CODING; EMPIRICAL ASSESSMENT; CODE-LENGTH; LOGARITHMIC SCORE; LIKELIHOOD; CONSISTENT MODEL SELECTION.

1. INTRODUCTION

Recent years have seen the development of two new general approaches to problems of statistical inference: *prequential analysis* (Dawid (1984), Dawid (1991a), Dawid (1991b)) and *stochastic complexity* (Solomonoff (1978), Rissanen (1987), Rissanen (1989)). Although both approaches can be developed and applied in entirely non-Bayesian ways, they are in close sympathy with Bayesian thinking—prequential analysis in fact arose directly out of such thinking, whilst strong Bayesian connexions have become apparent during the development of stochastic complexity. But these two approaches also share a "Popperian" philosophical attitude more commonly associated with non-Bayesian theories of hypothesis testing: that any attempt at describing reality must be measured against empirical evidence, and be discarded if it proves inadequate. Such critical examination may be applied to subjective beliefs about the world no less than to statistical models of it.

I consider prequential analysis and stochastic complexity as conduits linking Bayesian and non-Bayesian approaches to inference, along which insights can flow in both directions. This paper presents a Bayesian overview of these theories and of the relationship between them.

2. PHILOSOPHICAL STANDPOINT

Prequential Analysis and Stochastic Complexity share the view that any theory or model (we use the terms interchangeably) is merely a human attempt to describe or explain reality; and that such models are to be assessed in terms of their success at this task. It is misguided, according to this view, to believe in Nature as "obeying" some theory or model (even an unknown one). Even if we can find a "completely successful" theory, this does not mean we have identified Nature's "true model"—some other, distinct theory might be just as successful.

Since, in this view, theories can only be distinguished by means of their predictions about observables, theories of seemingly different form which necessarily make identical predictions must be treated as identical. But it is also possible that two non-identical theories happen to make identical predictions in the particular circumstances of the empirical world, in which case we may call them *contingently identical*. We shall never, in this world, be able to distinguish between two contingently identical theories—but then again, we shall never need to (for any predictive purpose). Conversely, if two theories do ever make differing predictions, we should be able to distinguish them on the basis of empirical evidence, and thus prefer one over the other as explanations of the observed data. This property—that theories can be empirically distinguished when and only when it makes a difference that they should be—has been termed *Jeffreys's Law* by Dawid (1984). It is simultaneously trivial and profound.

My acceptance of the above philosophy in Probability and Statistics has been greatly influenced by de Finetti (de Finetti (1975)). In his view, it is meaningless to regard a coin as having an "intrinsic probability" of landing heads. Probability is a subjective (human) construction (theory), which can meaningfully be attached only to observable events, such as a specified sequence of outcomes. Even though, under exchangeability, the de Finetti representation theorem might be invoked to conjure up a "chance of landing heads" for the coin, this remains a mathematical artifact of the theory, and need not have any counterpart in the real world: while we may believe that the relative frequency of heads will tend to a limit, Nature is not obliged to humour us, and even if such a limit does exist, any particular coin toss could be embedded in different exchangeable sequences, with varying "chances" of heads (Dawid (1985b)). More generally, the (uniquely defined) subjective predictive model can be decomposable in several distinct ways into a "statistical model" and a prior distribution for its parameter (Dawid (1986b)), so that the concept of the "true model" must be a fiction.

De Finetti has also considered carefully the relationship between subjective opinions and the empirical world (see Dawid (1986a)). In particular, *proper scoring rules* provide the Bayesian with a means of comparing the success of different models and/or prior distributions in forecasting observables, while ideas of *calibration* and *refinement* may be used to measure the absolute success of a model.

3. THE PREQUENTIAL APPROACH TO PROBABILITY

In many contexts it is reasonable to regard Nature as producing, sequentially, an infinite data-string $\boldsymbol{x} = (x_1, x_2, \ldots)$. Of course, this is an idealization, since we shall never observe all of $\boldsymbol{x}$—some of the difficulties resulting from this have been discussed by Schervish (1985). But this will be our sole idealization—in particular, since we do not regard Nature as obeying any laws, probabilistic or otherwise, any consideration of "alternative data-strings which might have been produced" must be strictly theory-dependent. The relevant empirical evidence (past and future) is entirely contained in $\boldsymbol{x}$. In this view Nature, like History, is "just one darn thing after another".

Let now P be a joint distribution for a sequence of random variables $\boldsymbol{X} = (X_1, X_2, \ldots)$. We can enquire how successful P is as an explanation of the realized outcomes $x_1, x_2, \ldots$. It is natural to tackle this question sequentially. Let P_i be the conditional distribution, under P, of X_i, given $\boldsymbol{X}^{i-1} = \boldsymbol{x}^{i-1} \stackrel{\text{def}}{=} (x_1, \ldots, x_{i-1})$. Thus, after observing the first $i - 1$ elements of $\boldsymbol{x}$, the "theory" P outputs the *probability forecast* P_i for the next observable X_i. Considered as a rule for generating a distribution P_i of X_i for any values of i and $\boldsymbol{x}^{i-1}$,

P is a *Probability Forecasting System (PFS)*; and conversely any PFS determines a joint probability distribution P. We can now compare the realized forecasts (P_i) with the realized values (x_i), using probability assessment techniques such as scoring rules, calibration plots *etc.*, to obtain a global assessment of the success of P at explaining $\boldsymbol{x}$. This predictive, sequential methodology is termed *prequential assessment.* Note that since such assessments depend on P only through the actual sequence $\boldsymbol{P} = (P_1, P_2, \ldots)$ of probability forecasts it makes in the light of the empirical data $\boldsymbol{x}$, we do not need to know the full structure of P as a PFS—we make no use of conditional probabilities given histories which have not materialized.

3.1. *The Prequential Principle*

The last property, regarded as a desideratum for a method of assessing P against $\boldsymbol{x}$, is termed the *Prequential Principle*. We can consider various assessment methods, and investigate to what extent they respect it.

For example, suppose that, under P_i, X_i has mean μ_i and variance σ_i^2 (these values generally depending on $\boldsymbol{X}^{i-1}$). A simple and intuitively sensible statistic for testing the fit of P, based on $\boldsymbol{X}^n$, is

$$Y_n = \frac{\sum_{i=1}^{n}(X_i - \mu_i)}{(\sum_{i=1}^{n}\sigma_i^2)^{\frac{1}{2}}}.$$

Since the value y_n of Y_n, depends only on $\boldsymbol{P}$ and $\boldsymbol{x}$, calculation of y_n respects the Prequential Principle. However, any attempt to use this value to test the validity of P, by comparing y_n with the "null" distribution of Y_n under P, would appear to involve aspects of P over and above $\boldsymbol{P}$. Indeed, if we were only given the actual forecast sequence $\boldsymbol{P}$ we could not know, for instance, whether or not, under P, the (X_i) were being modelled as independent. If they were, then μ_i and σ_i^2 would not depend on $\boldsymbol{X}^{i-1}$, and we could (under mild conditions) take Y_n as asymptotically standard normal, thus yielding an approximate test of fit in which y_n was referred to the $\mathcal{N}(0,1)$ distribution. But what if P did not incorporate such independence? Seillier-Moiseiwitsch and Dawid (1992) have shown, using martingale arguments, that in this case also the limiting distribution of Y_n under P will (again under mild conditions) be $\mathcal{N}(0,1)$. Consequently, both the value and the asymptotic null reference distribution for Y_n may be calculated without any knowledge of P beyond its associated forecast sequence $\boldsymbol{P}$, and the test does therefore respect the Prequential Principle.

3.2. *Prequential Probability*

Results such as that above, which are numerous, suggest that it might be possible to define "prequential probabilities" for events defined only in terms of two sequences, one being a string of data $\boldsymbol{x}$, and the other an associated string of probability forecasts $\boldsymbol{P}$, no overall probability distribution being given. Such a programme, a genuine extension of the standard Kolmogorov framework, is currently being undertaken by Vovk (1990a, 1990b, 1991), who has given rigorous definitions of prequential probability and expectation, and has shown that the traditional limit theorems (laws of large numbers, central limit theorem, law of the iterated logarithm) can all be given valid interpretations within this new framework.

3.3. *Successful Explanations*

Kolmogorov and others (see Kolmogorov and Uspenskii (1987)), using ideas from algorithmic information theory, have addressed the idea that a specific infinite data-sequence x might be considered "random" with respect to a probability model P. In the light of Section 2, we might reinterpret this as stating that a posited model P is a "completely successful" explanation of the empirical data-sequence x.

There are various ways of explicating this. Dawid (1985a) formulates a criterion based on the concept of *computable calibration*. An alternative (stronger) requirement is that, for every alternative model Q, the observed likelihood ratio $(\mathrm{d}Q^n/\mathrm{d}P^n)(x^n)$, based on the first n terms of x, be bounded above as $n \to \infty$ (here both P and Q must be restricted to be suitably computable). This is reasonable if we interpret the likelihood ratio as measuring preference for Q over P, such preference becoming definitive when its value becomes infinite. Since $(\mathrm{d}Q^n/\mathrm{d}P^n)(\boldsymbol{X}^n)$ is an arbitrary (computable) non-negative martingale under P, this requirement may be termed the *martingale criterion* (Ville (1939)). Both these definitions of the success of P respect the Prequential Principle.

Both the above properties will be satisfied with probability 1 for sequences generated by P, thus respecting a compelling desideratum for any success criterion. In particular, if P and Q are *mutually absolutely continuous* over the space of infinite sequences, so that they agree as to which events have probability 1, then each will expect the other, as well as itself, to provide a fully successful explanation of the data to be observed. In this case we may regard P and Q as (essentially) equivalent

Under such a success criterion, two not necessarily equivalent probability models P and Q may both turn out to be completely successful explanations of the empirical data x. For consistency with Jeffreys's Law, we should expect that, in this case, P and Q yield essentially identical forecast sequences $\boldsymbol{P}$ and $\boldsymbol{Q}$ for the data-sequence x. In fact, this holds in an asymptotic sense: under the complete calibration criterion we can show $\rho(P_i, Q_i) \to 0$ (Dawid (1985a)), while under the martingale criterion we obtain the stronger result $\sum_{i=1}^{\infty} \rho(P_i, Q_i) < \infty$ (where ρ denotes Hellinger distance) (Vovk (1987)). In particular, when P and Q are equivalent, each assigns probability 1 to such essential identity of their forecasts (Blackwell and Dubins (1962), Shiryayev (1981)).

4. THE CODING APPROACH TO PROBABILITY

A seemingly quite different approach to assessing the success of a distribution P is based on the connexions between probability distributions and coding systems. We imagine a *sender*, who observes a (finite) data-string x, and wishes to transmit this, by means of a coded message, to a *receiver*. We suppose that the communication channel supports error-free serial transmission of sequences of binary digits, and that, in the case of real (x_i), the values may be rounded to some chosen precision, so that the set of possible messages is discrete: the actual process of discretization will prove to be asymptotically unimportant. We generically use the corresponding lower-case symbol (*e.g.* p) to denote the density or probability mass function of a distribution symbolized by an upper-case symbol (*e.g.* P).

Since data-transmission is expensive, the message should be as short as possible. More important, a short code has little redundancy, and so must capture any "pattern" which the data might possess. We therefore search for a *coding system* C, *i.e.* an injective map C from the space $\mathcal{X}$ of possible message strings into the space $\mathcal{B}^*$ of all binary strings, which will allow short encoding of x. This coding system must be agreed upon between sender and receiver before transmission can begin. After observation of x, the success of C is measured by the shortness of its code for x.

Clearly, the choice of C must depend on the nature and extent of the receiver's prior knowledge about $\boldsymbol{x}$: for example, if he already knows $\boldsymbol{x}$, a message of length 0 will suffice. More generally, the receiver will have certain expectations about $\boldsymbol{x}$, which could be expressed as a subjective probability distribution P for a random variable $\boldsymbol{X}$ over $\mathcal{X}$. Then we might choose C to minimize the expected code-length under P.

4.1. *Prefix Coding Systems*

Often it is desirable to transmit a string $(x_1, x_2, \ldots)$ by encoding each symbol separately and concatenating their codewords. If this is to admit unique decoding, then the coding system used for each symbol must be a *prefix coding system,* with the property that no code-word is the same as an initial segment of any other. This is also desirable when transmitting a single symbol, if the receiver is to know when transmission is over and he can proceed to decoding.

For a finite binary string $\boldsymbol{z}$, let $l(\boldsymbol{z})$ denote the length of $\boldsymbol{z}$. If Q denotes the model of Bernoulli trials with probability parameter $\frac{1}{2}$, then $Q(Z_i = z_i \text{ for } i = 1, \ldots, \; l(\boldsymbol{z})) = 2^{-l(\boldsymbol{z})}$. Given a prefix coding system C over $\mathcal{X}$, the associated *length function* L is defined by $L(y) \stackrel{\text{def}}{=} l(C(y)) \quad (y \in \mathcal{X})$. Since the events "an initial sequence of $(Z_1, Z_2, \ldots)$ forms a code-word for y" are disjoint as y ranges over $\mathcal{X}$, the sum of their Q-probabilities cannot exceed 1, so that we must have

$$\sum_{y \in \mathcal{X}} 2^{-L(y)} \leq 1. \tag{1}$$

If C is *complete* in the sense that every infinite binary string has an initial segment that is a code-word, then we shall have equality in (1). It is not difficult to show that, whenever (1) holds for an integer-valued function L, there exists a prefix coding system C (complete if we have equality in (1)) having L as its length function.

4.2. *Codes and Distributions*

Suppose that L is a non-negative but not necessarily integral function on $\mathcal{X}$, satisfying (1) with equality. Since the integer-valued function $L' \stackrel{\text{def}}{=} \lceil L \rceil$ satisfies (1), we can construct a prefix coding system whose length-function exceeds L by at most 1. Applying this argument to coding systems defined over the space $\mathcal{X}^n$ of sequences of n symbols from $\mathcal{X}$, we see that it is possible to encode long sequences with a per-symbol length function differing neglibly from that obtained using L. We shall therefore ignore the fact that L may be non-integral, and treat it as an achievable length function.

Let P be a distribution over $\mathcal{X}$. The function

$$L_P(y) \stackrel{\text{def}}{=} -\log p(y) \quad (y \in \mathcal{X}), \tag{2}$$

(where log denotes logarithm to base 2) satisfies (1) with equality, and hence may be considered as defining a complete prefix coding system. Conversely, given any such system C, with length function L, we can treat it as defining a distribution P_C over $\mathcal{X}$, where $p_C(y) = 2^{-L(y)}$.

The *entropy* of a distribution P is defined as

$$H(P) \stackrel{\text{def}}{=} E_P\left(L_P(X)\right) = -\sum_{y \in \mathcal{X}} p(y) \log p(y). \tag{3}$$

Proposition 1. *For any distribution P, and prefix coding system C with length function L,*

i)

$$E_P(L(X)) \geq H(P),$$

with equality if and only if $L \equiv L_P$;

(ii) for all $c > 0$

$$P(L(X) \leq L_P(X) - c) \leq 2^{-c}.$$

Proof. (i) The "information inequality",

$$\sum_{y \in \mathcal{X}} p(y) \log (p(y)/q(y)) \geq 0 \tag{4}$$

holds for all probability distributions P and Q over $\mathcal{X}$, with equality if and only if $Q = P$. Take $q(y) \stackrel{\text{def}}{=} k^{-1} 2^{-L(y)}$, with $k \stackrel{\text{def}}{=} \sum_{y \in \mathcal{X}} 2^{-L(y)} \leq 1$ by (1).

(ii) Whenever $L(y) \leq L_P(y) - c$, $p(y) \leq 2^{-c} 2^{-L(y)}$. Sum over all y satisfying this condition and again apply (1). ◁

Proposition 1 may be taken as establishing L_P as an optimal length function for a prefix coding system under the distribution P for the unknown message X. In particular, if we apply (ii) to the encoding of long sequences of symbols, then the per-symbol message length achieved by any prefix coding system C cannot improve on that given by L_P by more than a negligible amount, with arbitrarily high probability under P. These results suggest strongly that, when the receiver's state of uncertainty about the message to be transmitted is expressed by a distribution P, a coding system should be used with length function L_P.

A fruitful view of the above argument comes from turning it on its head. Suppose that sender and receiver agree to use a prefix coding system C, with length function L, to encode the message. We may suppose C to be complete, since otherwise we could shorten every code-word by using the coding system with length-function $L + \log \left\{ \sum_{y \in \mathcal{X}} 2^{-L(y)} \right\}$, which satisfies (1) with equality. The associated distribution P_C is then that distribution under which C is an optimal code. Thus the mere willingness to use C can be regarded as generating a distribution for that object. (Compare this argument with that of Savage (1971), which infers a subjective distribution from choices made in decision problems—*viz.* that for which these would be the optimal choices.) For example, Rissanen (1983) shows how one can construct a prefix coding system for an arbitrary non-negative integer n whose length-function is essentially $\log c + \log^*(n)$, where $\log^*(n) \stackrel{\text{def}}{=} \log n + \log \log n + \ldots$ (the sum including all positive iterates) and c is chosen to satisfy (1) with equality. The associated distribution P^* has $p^*(n) = (cn \log n \log \log n \ldots)^{-1}$, and can be shown to be optimal, in a certain sense, amongst all distributions having infinite entropy. Rissanen terms it a "universal prior distribution" for a completely unknown integer.

5. CONNECTIONS

For a particular message $\boldsymbol{x}$, we can compare coding systems in terms of the actual length of their codes for $\boldsymbol{x}$. If C is a complete prefix coding system, then this length is $-\log p_C(\boldsymbol{x})$. This is identical with the negative *logarithmic score* for assessing the success of the distribution P_C at explaining $\boldsymbol{x}$; it can also be interpreted as the observed negative *log-likelihood* of P_C, and as such the difference in code-lengths $C(\boldsymbol{x}) - C'(\boldsymbol{x})$ is just the observed *log likelihood-ratio* in favour of $P_{C'}$ as against P_C. We thus see that the criterion of code-length is identical with the use, in Probability Forecasting, of the logarithmic scoring function for assessing a distribution over $\mathcal{X}$ against the empirical data.

The individual terms of a string $\boldsymbol{x} = (x_1, x_2, \ldots)$ can be encoded and transmitted sequentially, using a prefix coding system for the i'th term which can be allowed to depend (in a way agreed between the parties before transmission) on the previously transmitted segment $\boldsymbol{x}^{i-1}$. Such a system for encoding strings of arbitrary length may be called a *prequential coding system.* When prior uncertainty is expressed in a joint distribution P for $\boldsymbol{X}$, we might try encoding x_i using a prefix code optimal for the conditional distribution, under P, of X_i given $\boldsymbol{X}^{i-1} = \boldsymbol{x}^{i-1}$. The length of the code for x_i is then $-\log p(x_i \mid x^{i-1})$, and so the overall code-length for $\boldsymbol{x}^n$ is $-\sum_{i=1}^{n} \log p(x_i \mid x^{i-1}) = -\log p(\boldsymbol{x}^n)$, showing that this prequentially optimal coding is in fact optimal for each n. Conversely, any complete prequential coding system determines a probability forecasting system, and hence a joint distribution for $\boldsymbol{X}$. The length of the code for each term may be interpreted as the contribution of that term to the overall negative logarithmic score, or negative log-likelihood, for the associated distribution. In this way we obtain a fully prequential interpretation of code-length.

For prequential coding systems, Proposition 1 (ii) can be strengthened as follows:

Proposition 2. *Let C be a prequential coding system with length function L_C, and P a distribution for $\boldsymbol{X}$. Then, with P-probability 1, $L_P(\boldsymbol{X}^n) - L_C(\boldsymbol{X}^n)$ is bounded above as $n \to \infty$.*

Proof. We may suppose C to be complete, since otherwise we could shorten the code-length term by term. Then (U_n), where $U_n = \exp\{L_P(\boldsymbol{X}^n) - L_C(\boldsymbol{X}^n)\} = p_C(\boldsymbol{X}^n)/p(\boldsymbol{X}^n)$, is a non-negative martingale under P, and hence is bounded above with P-probability 1. ◁

6. STATISTICAL MODELLING

Suppose Nature produces a data-string $\boldsymbol{x}$. In the prequential approach, we might search for a probability forecasting system for $\boldsymbol{X}$ which will be a good explanation of $\boldsymbol{x}$; from the coding viewpoint, we would like a coding system which yields a short code-length for $\boldsymbol{x}$.

In a sense, this problem is trivial: use the distribution which gives probability 1 to the actual data sequence—equivalent to a code of length 0. Of course this is only possible when the receiver knows the message before it is sent. However, the very existence of such a distribution or coding system does point up problems in the idea of minimizing code-length, or other probabilistic penalty function, when the possibilities are unrestricted.

We might therefore restrict the search to some parametric family $\mathcal{P} = \{P_\theta : \theta \in \Theta\}$ of joint distributions, or $\mathcal{C} = \{C_\theta : \theta \in \Theta\}$ of coding systems. However, there is still a problem. When the sender has observed $\boldsymbol{x}$, she will be able to discover the value of θ which yields the shortest code, or minimal penalty, for $\boldsymbol{x}$; but the receiver will not be in a position to apply this optimizing coding system without further information.

6.1. *Two-part Codes*

The first appproaches to this problem from a coding viewpoint (see Wallace and Freeman (1987), Rissanen (1985)) involved constructing a two-part code, having a *preamble*, encoding a value for θ using a prefix coding system C defined over some discrete subset $\boldsymbol{H}$ of Θ, and a *body*, which then encodes the data using the optimal code under P_θ. The sender should clearly choose that value of $\theta \in \boldsymbol{H}$ yielding the minimum total code length $l(C(\theta)) - \log p_\theta(\boldsymbol{x})$ for the data $\boldsymbol{x}$ at hand—the "minimum message length" estimator. The minimized total code-length can then be regarded as a measure of the success of the model (in conjunction with the subset $\boldsymbol{H}$ and coding system C) at explaining $\boldsymbol{x}$, yielding the "minimum description length" criterion.

The choice of $\boldsymbol{H}$ and C must be agreed in advance. They should be chosen in some *a priori* optimal way—say, so as to minimize the expected total code-length under some subjective distribution P^* for $\boldsymbol{X}$. The density of the discrete set $\boldsymbol{H}$ in Θ is particularly crucial: it is suboptimal to specify θ too accurately, since this will necessitate a long preamble. (In contrast, increasing the precision with which $\boldsymbol{x}$ is specified will add a quantity independent of θ to the length of the body, and so will not affect the choice of $\boldsymbol{H}$ and C.)

Suppose we have a regular statistical model for a sequence $\boldsymbol{X} = (X_1, X_2, \ldots)$, with k-dimensional parameter space, such that

(i) $\hat{\theta}_n \xrightarrow{L} \Pi$, a distribution with positive density $\pi(.)$, under P^*; and
(ii) with P^*-probability 1, $-\sup_\theta \log p_\theta(\boldsymbol{X}^n)$ is of order n,

where P^* is the subjective distribution for $\boldsymbol{X}$, and $\hat{\theta}_n$ is the maximum likelihood estimator of θ based on the length n initial segment $\boldsymbol{X}^n$ of $\boldsymbol{X}$. (These conditions will typically hold for $P^* = P_\Pi \stackrel{\text{def}}{=} \int P_\theta d\Pi(\theta)$, as would be appropriate for a Bayesian who fully believed in the model $\mathcal{P}$.) The optimal density of $\boldsymbol{H}$ in Θ then turns out to be of order $n^{k/2}$, the length of $C(\theta)$ is (constant) $- \log \pi(\theta) + \frac{1}{2} k \log n$ and the optimized total code-length for a data-string $\boldsymbol{x}^n$ is then asymptotically

$$-\sup_\theta \log p_\theta(\boldsymbol{x}^n) + (k/2) \log n, \tag{5}$$

which can thus be considered as expressing, approximately, the success of the model at explaining the data.

6.2. *Stochastic Complexity*

If the "theory" $\mathcal{P}$ is truly believed, and beliefs about θ are expressed by a prior distribution Π, then the predictive distribution of $\boldsymbol{X}$ is P_Π, and the optimal code-length for $\boldsymbol{x}^n$ is $L_\Pi(\boldsymbol{x}^n) \stackrel{\text{def}}{=} - \log p_\Pi(\boldsymbol{x}^n)$. If we are given any prefix coding system C for $\boldsymbol{X}^n$, with length-function L, then by Proposition 1 (ii), for any $\epsilon > 0$, $P_\Pi \left(L(\boldsymbol{X}^n) - L_\Pi(\boldsymbol{x}^n) \leq 2 \log \epsilon\right) \leq \epsilon^2$, whence, since $P_\Pi = \int_\Theta P_\theta d\Pi(\theta)$,

$$\Pi \left\{\theta : P_\theta \left(L(\boldsymbol{X}^n) - L_\Pi(\boldsymbol{x}^n) \leq 2 \log \epsilon\right) \leq \epsilon\right\} \geq 1 - \epsilon.$$

That is, for "most" θ, the probability is high under P_θ that the code-length obtained from C will not be significantly shorter than L_Π. This holds irrespective of the length n of the data-string.

Now suppose C is a prequential coding system. By Proposition 2, $P_\Pi(A) = 1$, where

$$A \stackrel{\text{def}}{=} \left(L_\Pi^n(\boldsymbol{X}^n) - L^n(\boldsymbol{X}^n) \quad \text{is bounded above as } n \to \infty\right).$$

Hence $\Pi \left\{\theta : P_\theta(A) = 1\right\} = 1$. In particular, if Π has an everywhere positive density, then

$$P_\theta \left(L_\Pi^n(\boldsymbol{X}^n) - L^n(\boldsymbol{X}^n) \quad \text{is bounded above as } n \to \infty\right) = 1 \tag{6}$$

for almost all $\theta \in \Theta$—this property is equivalent to the *prequential efficiency* of P_Π as defined by Dawid (1984). V'iugin and Vovk (1990) have shown a pointwise version of this result:

under weak computability requirements, for every θ random with respect to Π, the event A will hold for every data-string $\boldsymbol{x}$ random with respect to P_θ (here "random" means essentially the same as "successfully explained by", under the martingale criterion of Section 3.3). In this strong sense, P_Π thus defines an optimal prequential coding system when the model is supposed true. The corresponding optimal code-length $L_\Pi(\boldsymbol{x}^n) = -\log p_\Pi(\boldsymbol{x}^n)$ is called the *stochastic complexity* of the data $\boldsymbol{x}^n$ relative to the model $\mathcal{P}$ (this depends on Π, but the effect of changing Π is bounded as $n \to \infty$). It thus seems compelling to regard $p_\Pi(\boldsymbol{x}^n)$ as the "likelihood" of the model $\mathcal{P}$ on data $\boldsymbol{x}^n$, and to compare different models in terms of their likelihoods, or equivalently the stochastic complexities they yield for the data.

It may be shown (Jeffreys (1976), Schwarz (1976), Akaike (1978)) that, for a regular k-dimensional statistical model, under conditions (i) and (ii), with P^*-probability 1 the stochastic complexity is asymptotically equivalent to (5). This may be regarded as establishing the overall optimality of the two-part coding system (although it should be noted that this is not prequential). The negative of (5) can be regarded as a "penalized log-likelihood", the second term correcting for the over-optimism in the first engendered by using that value of θ which best fits the data observed. It is interesting to observe this correction arising automatically out of the coding approach.

Rissanen (1986a) showed that, in regular cases, for any prequential coding system C, for all $\epsilon > 0$, for almost all $\theta \in \Theta$

$$E_\theta \left(L^n(\boldsymbol{X}^n) - L_\theta^n(\boldsymbol{X}^n)\right) - (\frac{1}{2} - \epsilon)k \log n > 0 \tag{7}$$

for all but finitely many n. This can be regarded as an extension of (i) of Proposition 1. However, the use of the expectation in (7) violates the Prequential Principle. Our simple argument above, which respects that Principle, yields a stronger conclusion, extending Proposition 1 (ii): for almost all θ, the P_θ-probability is 1 that

$$L^n(\boldsymbol{X}^n) - L_\theta^n(\boldsymbol{X}^n) - (\frac{1}{2} - \epsilon)k \log n > 0$$

for all but finitely many n. Moreover, (6) is equally applicable to non-regular cases, where the negative stochastic complexity need not have asymptotic form (5).

6.3. *Prequential Coding*

For a model $\mathcal{P} = \{P_\theta\}$ and data-sequence $\boldsymbol{x}$, let $P_{\theta,i+1}$ be the conditional distribution of X_{i+1} given $\boldsymbol{x}^i$ under P_θ, and $\hat{\theta}_i$ an efficient estimate (*e.g.* the maximum likelihood estimate) of θ based on data $\boldsymbol{x}^i$. Dawid (1984) considered the PFS P defined by $P_{i+1} = P_{\hat{\theta}_i,i+1}$. The associated prequential coding system encodes x^n with total length

$$-\log p(x^n) = \sum_{i=0}^{n-1} -\log p_{\hat{\theta}_i}(x_{i+1} \mid x^i). \tag{8}$$

Under regularity conditions, (8) will (with P^*-probability 1) only differ from the stochastic complexity by a bounded quantity as $n \to \infty$—this is equivalent to the prequential efficiency of P for $\mathcal{P}$. Thus (8) can equally well be considered as the negative log-likelihood of the model $\mathcal{P}$. In general, for any prequentially efficient P, we can define the *prequential likelihood* of $\mathcal{P}$ on data $\boldsymbol{x}$ as $p(\boldsymbol{x})$, all such definitions being in asymptotic agreement up to a (data-dependent) finite positive scale-factor.

6.4. *Consistent Model Selection*

Suppose we wish to choose between a finite or countable collection $\{\mathcal{P}_j : j = 1, 2, \ldots\}$, of regular parametric statistical models for $\boldsymbol{X}$, with $\mathcal{P}_j = \{P_{j,\theta_j} : \theta_j \in \Theta_j\}$, Θ_j having finite dimension k_j. For each j, introduce a prior distribution Π_j with positive density $\pi_j(.)$ over Θ_j. Let

$$P_j = \int_{\Theta_j} P_{j,\theta_j} \pi_j(\theta_j) \mathrm{d}\theta_j$$

be the marginal distribution of $\boldsymbol{X}$ derived from $(\mathcal{P}_j, \pi_j)$, so that the stochastic complexity of $\boldsymbol{x}^n$ relative to $\mathcal{P}_j$ may be taken as $-\log p_j(\boldsymbol{x}^n)$.

We shall suppose the (P_j) to be mutually singular, since otherwise we could have no hope of distinguishing the various models even on the basis of infinite data. This property will generally be easy to verify when the models involve quite different distributions. Alternatively, suppose say $\mathcal{P}_1 \subset \mathcal{P}_2$, with Θ_1 a lower-dimensional submanifold of Θ_2. Suppose further that $\mathcal{P}_2$ admits consistent estimation, so that there exists a sequence of statistics $(T_n) = (T_n(\boldsymbol{X}_n))$ with $T_n \xrightarrow{\text{a.s.}} \theta_2$ under P_{2,θ_2}, all $\theta_2 \in \Theta_2$. Then the event "$\lim T_n \in \Theta_1$" has P_1-probability 1 but P_2-probability 0, so that P_1 and P_2 are mutually singular in this case.

Now take a sequence (α_j) with $\alpha_j > 0$, $\sum_i \alpha_j = 1$, and define $P_0 = \sum_j \alpha_j P_j$, the overall marginal distribution for $\boldsymbol{X}$ of a Bayesian who believes that, with probability α_j, the data are generated from a distribution in $\mathcal{P}_j$, and who, conditional on this, has prior distribution Π_j for its parameter. We can write $P_0 = \alpha_1 P_1 + (1 - \alpha_1) Q$, where $Q = \sum_{j>1} \{\alpha_j / (1 - \alpha_1)\} P_j$, and thus P_1 and Q are mutually singular. Hence with P_1-probability 1 the likelihood-ratio $q(\boldsymbol{X}^n)/p_1(\boldsymbol{X}^n)$ based on the first n observation tends to 0 as $n \to \infty$. From this we deduce that $(\alpha_j p_j(\boldsymbol{X}^n))/(\alpha_1 p_1(\boldsymbol{X}^n)) \to 0$, uniformly in i, with P_1-probability 1, and hence with P_{1,θ_1}-probability 1 for almost all $\theta_1 \in \Theta_1$. It follows that, with P_{1,θ_1}-probability 1 for almost all $\theta_1 \in \Theta_1$, the "adjusted prequential likelihood" $\alpha_j p_j(\boldsymbol{x}^n)$ will, for large enough n, be maximized at $j = 1$. Since any member of $\{\mathcal{P}_j\}$ may be taken instead of $\mathcal{P}_1$, we deduce that the model-selection method which proceeds by maximizing the adjusted prequential likelihood, or equivalently minimizing the "adjusted stochastic complexity" of $\boldsymbol{x}^n$, $-\log p_j(\boldsymbol{x}^n) - \log \alpha_j$, will be (almost everywhere) consistent.

Note that the above result will continue to hold if we replace P_j by any other prequentially efficient PFS for $\mathcal{P}_i$, for example that described at 6.3 above. In the case of a finite number of normal regression or autoregression models (where the correction term $-\log \alpha_j$ may be ignored) this yields the *predictive least squares* model selection method (Rissanen (1986b), Hermerly and Davis (1989)). The argument may similarly be used to show the consistency of the method of choosing j to maximize the penalized log-likelihood $\sup_{\theta_j \in \Theta_j} \log p_{j,\theta_j}(\boldsymbol{x}^n) - (k_j/2) \log n$, which approximates the negative stochastic complexity (Speed and Yu (1989)). When choosing from a countably infinite collection of models, the above argument continues to apply, so long as the correction term is not neglected. This provides a more straightforward consistent model-selection method than, say, that of Hemerly and Davis (1991).

7. FURTHER ISSUES

7.1. *Model Failure*

Arguments such as in Section 6.4 (amongst many others) are only relevant when the data-string $\boldsymbol{x}$ is well-explained by a member of one of the model-classes considered. In this case the use of the code-length criterion, with its connections to the logarithmic score, maximum likelihood estimation and efficiency, can be regarded as optimal. But it is also important to investigate the behaviour of such strategies when the model fails: for example, under the assumption that x is well-explained by some other distribution P^*, the "true model" (or the subjective distribution of a Bayesian, who will then believe, with probability 1, that this will hold). In such a case alternative strategies, say those based on other prequential assessment criteria, also deserve attention. Dawid (1991a) has investigated the "prequential consistency" of parameter estimates, based on a general loss-function, assuming a false model.

In model selection, we might want to apply predictive least squares, essentially a prequential quadratic score, to regression with non-normal errors: many of the results survive this extension. Alternatively, the "true model" might involve an infinite number of regressors (Breiman and Freedman (1983), Shibata (1980)). Further study of the general behaviour of model-based strategies when the model is false is desirable.

7.2. *Infinitely Many Parameters*

For a finitely parametrized regular model, any choice of prior distribution Π is essentially equivalent to any other—so long as it has *full support, i.e.* admits a positive density. Any two such priors will be mutually absolutely continuous on the parameter-space, implying the same for their associated predictive distributions over sequence-space, and allowing an essentially unique definition of stochastic complexity. However, when we turn to models having infinitely many parameters, there is no clear analogue of the concept of full support: any prior puts all its mass on some "thin" event. (For example, Dawid (1988) considers a normal regression with infinitely many potential predictor variables, and shows that any natural conjugate prior distribution assigns probability 1 to the event that the parameters have a certain "determinism" property which will, in many contexts, be unacceptable as a description of real beliefs.) Correspondingly, any two distinct priors—and likewise their associated predictive distributions—are generically mutually singular. We can still deduce that, given any prefix coding system C, the coding system based on the predictive density p_Π will be at least as good as C under P_θ for any $\theta \in \Phi$, where $\Phi \subseteq \Theta$ has Π-probability 1; but since Φ is "thin" in Θ, and dependent on Π, this result loses the force it had in the finite-parameter case. One possible approach is to set up a hierarchical structure for the prior, specifying the conditional distributions for θ given ω say, where ω is finite dimensional. Integrating out θ then yields a finitely parameterized model for $\boldsymbol{X}$, to which the earlier results apply. However, the stochastic complexity will still depend on the assumed "hyper-model" for θ. In full generality, specification and verification of appropriate optimality criteria for the case of infinitely many parameters present a deep challenge.

7.3. *Ad hoc Strategies*

When optimality theory fails us, or we have no model, we can still investigate the theoretical or empirical performance of intuitively sensible coding systems or prequential strategies. The coding approach is not restricted to data arriving in sequence, and can be applied to complex problems of data analysis where full probabilistic specification of both model and prior would be difficult or impossible: Patrick and Wallace (1982) describe an application to megalithic geometry. For sequence data, the prequential approach is more general, since its probabilistic assessment methods need not be based on the logarithmic scoring rule (Dawid (1991b)). Vovk (1990c) has investigated a form of "Bayesian mixture" over a countable set of strategies for forecasting, when faced with an arbitrary loss-function, and has shown that it will, in a certain sense, be almost as good as any in the original set, for any data. Work is needed further to define and characterize "good strategies" in the absence of a model.

REFERENCES

Akaike, H. (1978). A Bayesian analysis of the minimum AIC procedure. *Ann. Inst. Statist. Math.* **30**, 9–14.

Blackwell, D. and Dubins, L. E. (1962). Merging of opinions with increasing information. *Ann. Math. Statist.* **33**, 882–886.

Breiman, L. A. and Freedman, D. F. (1983). How many variables should be entered in a regression equation? *J. Amer. Statist. Assoc.* **78**, 131–136.

Dawid, A. P. (1982). Intersubjective statistical models. *Exchangeability in Probability and Statistics.* (G. Koch and F. Spizzichino. eds.), Amsterdam: North-Holland, 217–232.

Dawid, A. P. (1984). Present position and potential developments: some personal views. Statistical theory. The prequential approach. *J. Roy. Statist. Soc. A* **47**, 278–292, (with discussion).

Dawid, A. P. (1985a). Calibration-based empirical probability. *Ann. Statist.* **13**, 1251–1285, (with discussion).

Dawid, A. P. (1985b). Probability, symmetry and frequency. *Brit. J. Phil. Sci.* **36**, 107–128.

Dawid, A. P. (1986a). Probability forecasting. *Encyclopedia of Statistical Sciences* **7**, (S. Kotz, N. L. Johnson and C. B. Read eds.), New York: Wiley, 210–218.

Dawid, A. P. (1986b). A Bayesian view of statistical modelling. *Bayesian Inference and Decision Techniques*, (P. K. Goel and A. Zellner, eds.), Amsterdam: North-Holland, 391–404.

Dawid, A. P. (1988). The infinite regress and its conjugate analysis. *Bayesian Statistics 3* (J. M. Bernardo, M. H. DeGroot, D. V. Lindley and A. F. M. Smith, eds.), Oxford: University Press, 95-110, (with discussion).

Dawid, A. P. (1991a). Fisherian inference in likelihood and prequential frames of reference. *J. Roy. Statist. Soc. B* **53**, 79–109, (with discussion).

Dawid, A. P. (1991b). Prequential data analysis. *Issues and Controversies in Statistical Inference* (M. Ghosh and P. K. Pathak, eds.), (to appear).

de Finetti, B. (1975). *Theory of Probability*. New York: Wiley.

Hemerly, E. M. and Davis, M. H. A. (1989). Strong consistency of the PLS criterion for order determination of autoregressive processes. *Ann. Statist.* **17**, 941–946.

Hemerly, E. M. and Davis, M. H. A. (1991). Recursive order estimation of autoregressions without bounding the model set. *J. Roy. Statist. Soc. B* **53**, 201–210.

Jeffreys, H. (1936). Further significance tests. *Proc. Camb. Phil. Soc.* **32**, 416–445.

Kolmogorov, A. N. and Uspenskiĭ, V. A. (1987). Algorithms and randomness. *Theory Prob. Appl.* **32**, 389–412.

Patrick, J. D. and Wallace, C. S. (1982). Stone circle geometries: an information theory approach. *Archaeoastronomy in the Old World* (Heggie, D. C., ed.), Cambridge: University Press.

Rissanen, J. (1983). A universal prior for integers and estimation by minimum description length. *Ann. Statist.* **11**, 416–431.

Rissanen, J. (1985). Minimum description length principle. *Encyclopedia of Statistical Sciences* **5**, (S. Kotz, N. L. Johnson and C. B. Read, eds.), New York: Wiley, 523–527.

Rissanen, J. (1986a). Stochastic complexity and modeling. *Ann. Statist.* **14**, 1080–1100.

Rissanen, J. (1986b). A predictive least squares principle. *IMA J. of Math. Control and Information* **3**, 211–222.

Rissanen, J. (1987). Stochastic complexity. *J. Roy. Statist. Soc. B* **49**, 223–239 and 252–265, (with discussion).

Rissanen, J. (1989). *Stochastic Complexity in Statistical Inquiry*. Singapore: World Scientific Publishing.

Savage, L. J. (1971). Elicitation of personal probabilities and expectations. *J. Amer. Statist. Assoc.* **66**, 783–801.

Schervish, M. J. (1985). Discussion of "Calibration-based empirical probability", by A. P. Dawid. *Ann. Statist.* **13**, 1274–1282.

Schwarz, G. (1976). Estimating the dimension of a model. *Ann. Statist.* **6**, 461–464.

Seillier-Moiseiwitsch, F. and Dawid, A. P. (1992). On testing the validity of probability forecasts. *J. Amer. Statist. Assoc.* , (to appear).

Shibata, R. (1980). Asymptotically efficient selection of the order of the model for estimating the parameters of a linear process. *Ann. Statist.* **8**, 147–164.

Shiryayev, A. N. (1981). Martingales: recent developments, results and applications. *Int. Statist. Rev.* **49**, 199–233.

Solomonoff, R. J. (1978). Complexity-based induction systems: comparison and convergence theorems. *IEEE Trans. Information Theory* **IT-24**, 422–432.

Speed, T. P. and Yu, B. (1989). Stochastic complexity and model selection: normal regression. *Tech. Rep.* **207**, University of California, Berkeley.

Ville, J. (1939). *Etude Critique de la Notion de Collectif*. Paris: Gauthier-Villars.
V'iugin, V. V. and Vovk, V. G. (1990). On the efficiency of the Bayesian rule. (Unpublished *Tech. Rep.*).
Vovk, V. G. (1987). On a randomness criterion. *Soviet Math. Dokl.* **35**, 656–660.
Vovk, V. G. (1990a). Prequential probability theory. (Unpublished *Tech. Rep.*).
Vovk, V. G. (1990b). Prequential variants of the central limit theorem and the law of the iterated logarithm. (Unpublished *Tech. Rep.*).
Vovk, V. G. (1990c). Aggregating strategies. *Third Workshop on Computational Learning Theory*, Rochester, N.Y. (Unpublished *Tech. Rep.*).
Vovk, V. G. (1991). Finitary prequential probability: asymptotic results. (Unpublished *Tech. Rep.*).
Wallace, C. S. and Freeman, P. R. (1987). Estimation and inference by compact coding. *J. Roy. Statist. Soc. B* **49**, 240–265, (with discussion).

DISCUSSION

J. RISSANEN (*IBM Almaden Research Center, USA*)

This is an elegant survey of the two related ideas, prequential analysis and stochastic complexity, and their role in statistical inference, modeling, and the interpretation of probabilities. Of the several deep issues that deserve to be discussed extensively I shall touch here on just a few. I share Professor Dawid's philosophical view that there exist no unique 'true' models non intrinsic probabilities. Instead, any physical interpretation of the probability is to be done relative to a model. Although any model can legitimately be chosen, the selection should be done by criteria which admit a meaningful data dependent interpretation. I know of only two 'scoring rules' which satisfy this requirement, the code length and its special case the prediction error.

As the main point in my discussion I would like to try to clarify the somewhat subtle relationship between these ideas and model selection within the Bayesian framework. Consider an indexed family of classes $\{p(x|\theta_\alpha, \alpha)\}$ of parametric distributions as 'models' of the data x. The Bayesian principle amounts to picking the class with the maximum posterior probability $P(\alpha|x) = P(x|\alpha)\pi(\alpha)/P(x)$, for which we need, in addition to the prior $\pi(\alpha)$ for the classes, also the priors $\pi_\alpha(\theta_\alpha)$ with which to eliminate the parameters by integration

$$P(x|\alpha) = \int P(x|\theta_\alpha, \alpha)d\pi_\alpha(\theta_\alpha). \tag{1}$$

Although the principle has appeal, the difficulty in finding the two types of distributions to capture prior knowledge poses a severe restriction, which has led to a search for 'empirical' data dependent priors. However, since the same principle cannot be used for the task it is unclear what a 'good' prior should be, and instead the procedures employed are more or less *ad hoc*.

The principle of the minimum description length (MDL), again, declares in broad terms that the best model/class is the one which permits the shortest encoding of the observed *data* together with the model/class itself, when advantage is taken of the constraints in the data prescribed by the model/class. To make sure that the encoded data can also be decoded an agreement about the model classes as sort of *common* knowledge is required, which is shared by the imagined encoder and the decoder alike. This is in contrast with the *prior* knowledge about both the models and their classes in the Bayesian principle, which can even be of private subjective kind. In a widely applicable form the principle calls for evaluating first the code length $L(x|\alpha) = -\log P(x|\hat{\theta}_\alpha(x), \alpha) + L(\hat{\theta}_\alpha(x))$, where the second term is the code length for the optimally truncated ML parameter estimates, which at the same time serves as a formalization of the elusive idea of *model complexity*. To get this length, we may use a prior $\pi_\alpha(\theta_\alpha)$, which, however, may not provide the shortest encoding of $\hat{\theta}_\alpha(x)$, or we may encode the parameters by universal means, which then defines a data dependent 'prior'. In any event, the optimal 'prior' must reflect the complexity of the encoding task, and no integral of type (1) need be evaluated, which permits fitting even 'nonparametric' models. In the second and the final step the class $\hat{\alpha}(x)$ is sought which minimizes the code length $L(x) = L(x|\hat{\alpha}(x)) + L(\hat{\alpha}(x))$, where the second term denotes the code length needed to encode the optimal class.

As a final comment, I would like to point out that the prequential approach with whatever probability assessment is selected cannot be reduced to the posterior maximization principle. By contrast, the prequential probability is equivalent with a predictively calculated code length, and if the probability assessment is done by the MDL principle, the resulting model and the probability forecasting system have the basic asymptotic optimality properties required from a 'completely successful' explanation of the empirical data sequence, as so eloquently described by Professor Dawid. Moreover, experience shows that the results are generally very good, even non-asymptotically, although attention must be paid to an inherent start-up problem, which arises from poor initial predictions. Unless checked, these can grossly penalize a complex model, which at the later stages may turn out to provide superior predictions.

M. GOLDSTEIN (*University of Durham, UK*)

I am having a little difficulty understanding the prequential principle. Clearly, the comparison between what theory predicts and what actually happens should be an important component of any methodology. But, during the conference presentation of this paper, I got the impression (perhaps, wrongly) that the prequential principle claimed that this was the only comparison of any importance. Asymptotically, this may be true, but for actual real world observations surely it is false, because it omits our heuristic understanding as to why our various models are performing well or badly.

Perhaps the principle might be clarified by applying it to the following problem. We wish to forecast hours of sunshine in Spain. Our sampling method is to observe the weather each time we come to a Valencia meeting on Bayesian statistics. We have three forecasting methods. Method one always forecasts "very sunny". Method two follows the local weather forecast. Method three is more mysterious but so far has predicted sunshine with amazing accuracy.

On the prequential principle, method three is ahead. But suppose we happen to notice that the conditional forecast of method three for sunshine at *Valencia 6* given sunshine at *Valencia 5* (say) gives sunshine with probability 2. Should not this forecast, which has not yet occurred, and may never be observed, influence our assessment of this method?.

Go back to the comparison of methods one (always forecast sunny) and method two (use weather forecast). Overall, it may be that, on the current data, method one outperforms method two by a wide margin on all reasonable scoring rules. However, our sample of observations on *Valencia 4* was taken under somewhat different circumstances to the previous meetings (i.e., earlier in the year). Might this imply that good performance on rule two at *Valencia 4* should carry extra weight as compared to the preceding observations? If so, this extra weight derives from our heuristic understanding as to why rule two should outperform rule one under such circumstances. It is this general level of insight into the conceptual strengths and weaknesses of our forecasting methods which does not appear to be captured in prequential type principles.

D.V. LINDLEY (*Somerset, UK*)

This paper represents a substantial contribution to that method within the Bayesian paradigm that recognizes only observables and dispenses with parameters. It admirably succeeds in bringing together several strands of theory, yet always remains within the study of observables. I have two related questions. The coding theorem is rarely applicable because of noise in the system. The logarithmic scoring-rule has the defect of ignoring the topology of the space (it is local) and consequently does not allow for one's being nearly right. Can the ideas be generalized to incorporate these two features of noise and proximity, which themselves may be related?.

TRAN VAN HOA (*Wollongong University, Australia*)

As an econometrician who has done substantial analysis in economic modelling, I find the paper by Professor Dawid interesting and stimulating. It is interesting because it provides me (or other users of the statistical tools) with a good review of the foundation for the empirical assessment of alternative econometric models. It is stimulating because it gives impetus to statistical or econometric modellers to explore further methodologies or modifications to improve their model selection procedure in a context peculiar to their field of interest.

Professor Dawid's interpretation of the Popperian philosophical attitude associated with prequential analysis is akin to the idea of empiricism popularized by Nobel laureate Milton Friedman (formerly of Chicago University) in relation to economic hypothesis testing: any attempt at describing reality must be measured against empirical evidence. My concern is not with this line of argument but with the fact that, in practically all important economic modelling applications, empirical evidence is not what it should be. I can give a simple example of this. Suppose we are interested in measuring the standard of living of university academics in the United Kingdom. But the concept of "standard of living" is elusive or cannot be measured exactly. As a result, some approximation or even substitute is used. Since measurement errors are usually not known, how much is the result from prequential analysis different from what it should be when this phenomenon occurs?

When we deal with models fitted to data with unknown measurement errors, successful explanations or model failure may not tell the whole story. The problem is compounded when the models themselves are misspecified. In our area of economic research when only so much can be obtained from observed data or specificed models, we echo Professor DeGroot's motto of "practise, practise and practise" with "theorize, theorize and theorize more". The immediate relevance of this principle in economics is in the determination of the exchange rates in international finance and in the explanation of the business cycles.

REPLY TO THE DISCUSSION

I am very grateful to the discussants for raising a number of interesting points.

Rissanen effectively dismisses all scoring rules other than code length (= negative logarithmic score). I find this a most unimaginative position. As Lindley points out, the logarithmic criterion ignores the topology of the space—indeed it can be characterized as the unique proper scoring rule which does so. This locality property will often be undesirable, however. Thus suppose we are attempting to assess the prognosis of a sick person, and consider two possible forecast distributions:

	Full recovery	Partial recovery	Death
(i):	.7	.2	.1
(ii):	.1	.2	.7

If we observe the outcome "partial recovery", we might well regard distribution (i) as more successful than (ii), in spite of the fact that they both earn the same logarithmic score. This attitude could stem from the intuitive feeling that "full recovery" is closer to "partial recovery" than is "death", or could be formalized by developing a decision structure, with suitable loss function, for which the act optimal under distribution (i) would have smaller loss (when in fact the outcome is "partial recovery") than that optimal under (ii). The derivation of a proper scoring rule from such a decision structure is described by Dawid (1986a): it will typically be non-local. If we are concerned with using the data for decision-making, it would thus seem reasonable to use something other than the logarithmic score, or code-length, to assess performance. The prequential approach, unlike that based on coding, can handle this situation with ease (see for example Section 8 of Dawid (1991a)).

This said, there does appear to be something special about the logarithmic score in the particular case that we are convinced of the truth of a regular parametric model. For then choosing the parameter-value to maximize this score will deliver the maximum likelihood estimate, which, being fully efficient, will generally be superior (at least asymptotically) to that based on minimizing some other, loss-based, discrepancy—even when the purpose at hand is to plug in the chosen value and minimize expected loss. But this seemingly universal optimality is not robust to model failure, whereas the direct loss-based assessment typically will be.

Rissanen complains that the "Bayesian principle" (which I suspect few Bayesians would recognize as such) of maximizing posterior probability is inapplicable to the choice of a good prior distribution. But one of the purposes of my paper is to point out that—at any rate within the "small world" of a regular parametric model—any prior with full support is asymptotically as good as any other. The only possible criteria for choice of prior must therefore be subjective. And indeed Proposition 1 provides an argument for using, in the analysis, one's true subjective probabilities. When we move outside the small world, there arises the problem of choosing an optimal formal prior for use with a false model: this raises some new and important issues which deserve deeper attention, but the need to take prior opinion into account is in no way diminished.

The seemingly more objective approach of MDL relies on having a way to describe (encode) a non-random object such as a model-parameter or model-class descriptor. Although one can come up with clever and intuitively sensible coding systems, there remains something ultimately *ad hoc* about these, quite as arbitrary as the choice of a prior. What the coding approach does have to offer the Bayesian is a new approach to choosing a prior, based on constructing a code rather than direct subjective assessment. This allows application to complex and non-traditional problems, but its philosophical implications deserve further study.

I agree with Rissanen that the prequential approach does not reduce to posterior maximization, notwithstanding any impression to the contrary which may be left by Section 6.4 of the paper. Indeed the "compleat prequentialist" solution to the model selection problem does not attempt to maximize, but simply uses the full mixture distribution P_0 as a probability forecasting system. Since, for any j, P_j is absolutely continuous with respect to P_0, it follows as in Section 3.3 that the probability forecasts made by P_0 will be asymptotically indistinguishable from those made by P_j, with P_j-probability 1, and hence with P_{j,θ_j}-probability 1 for almost all θ_j. This simple argument does not even require the various $\{P_j\}$ to be mutually singular: if they are not, it is possible that no definitive decision as to the correct model can be made, but this will not matter since, by Jeffreys's Law, all remaining candidates will be making essentially identical forecasts.

Rissanen's last point concerns finite data-sequences and "start-up problems". It seems to me perfectly reasonable that a complex model should be heavily penalized in the initial stages, since the slow rate at which its parameters can be learned means that it may for a long time predict more poorly than a simpler "incorrect" model. In this case I would rather use the simple model until the data are sufficiently extensive as to demand more detailed description. In general, the complexity of the model used should increase with the amount of data available: prequential strategies for achieving this are considered by Dawid (1991b).

Lindley asks about the consequences of noise in the transmission system. My initial reaction to this point was that it was attempting to over-stretch the coding metaphor, but Tran Van Hoa's discussion suggests that there may be a real problem here, related to the possibility that the the receiver is not observing the variables he thinks he is. At one level, one might attempt to model all aspects of the system, including measurement errors and the relationships between variables and thier proxies, so as to arrive at a forecast distribution for whatever it is one is in fact observing: then the theories discussed in the paper become directly applicable. However, if the predictions are poor, it will be impossible to identify whether this is due to the poverty of the fundamental theory (about unobservables) or to that of the ancillary assumptions relating fundamental quantities with observables. It is just possible that the theory of communication across a noisy channel may be able to offer some insight into a better statistical treatment of this difficulty, but my feeling is that it is basic and irremovable.

Goldstein raises some sensible objections to the Prequential Principle—these echo some of the comments of Schervish (1985). But I think these objections are largely answered if one considers the Principle as applying at the level of "likelihood" rather than "posterior": that is, in so far as the *data* distinguish between different models, they will not separate those which have yielded identical forecasts. This is not, of course, to say that we cannot have any other reasons for choosing between them, whether based on their internal structure or on our prior beliefs. Asymptotically such additional considerations will be swamped by the data, and thus for extensive data the Prequential Principle can indeed be regarded as applying to our "posterior" assessments of models; but Goldstein's comments do point up the need for non-asymptotic circumspection.

BAYESIAN STATISTICS 4, pp. 127–145
J. M. Bernardo, J. O. Berger, A. P. Dawid and A. F. M. Smith, (Eds.)

Bayesian Analysis of Mixtures: Some Results on Exact Estimability and Identification

J.-P. FLORENS*, M. MOUCHART** and J.-M. ROLIN**
**Université de Toulouse, France* and ***Université Catholique de Louvain, Belgium.*

SUMMARY

Models frequently used for car insurance portfolios assume that the distribution of the expected number of accidents caused by a client is a mixture of Dirichlet processes. Given that the number of clients is typically rather large, insurance companies may find it relevant to inquire whether the posterior distributions (given the number of individual accidents) of the mixture, of the mixing parameter and the predictive distribution of a new observation, given the number of individual accidents, are consistently convergent. For most models of this rather general class, answering these questions requires unfathomable manipulations. In this paper, it is shown that answers are "easily" obtainable from general results, presented in the authors' monograph *Elements of Bayesian Statistics*. These answers rely on no specific assumption on the form of the distribution but only on simple conditions on latent processes. Some problems of approximations are also discussed.

Keywords: MIXTURES; EXACT ESTIMABILITY; IDENTIFICATION.

1. INTRODUCTION

Mixture models have proven to provide a useful structure in a wide range of empirical works. Analytical complexities of such models have however hindered a wider usage. Robustness of Bayesian models has also been the focus of much attention in recent years and again analytical complexities have hindered a more general reliance on such methods in applied work. These two features tend to substantiate the opinion that "there is nothing more practical than a good theory", meaning that a theory is "good" in as far as it does provide tools suitable to handle situations that would otherwise be untractable. It is therefore a primary concern to develop methods that not only address problems relevant in actual data analysis but also are built in an efficient way so that potential users are not deterred by irrelevant technicalities.

In this paper, we pay attention to problems traditionally considered as "difficult" in the Bayesian analysis of mixtures, namely problems of exact estimability and identification and problems related to the evaluation of posterior distributions. We suggest that some answers are "easily" obtainable from general results, presented in the authors' monograph *Elements of Bayesian Statistics* (to be referred to as EBS). These answers rely on no specific assumption on the form of the distribution but only on simple conditions on latent processes. Thus, the fringe benefit of that approach is to make operational a semi-parametric modelling (i.e., models entertaining both Euclidean and functional parameters) without relying on unduly complex analytical manipulations.

The kind of problems we have in mind can be briefly sketched as follows. Models frequently used for car insurance portfolios can be reformulated by assuming that the distribution of the expected number of accidents caused by a client is a mixture of Dirichlet processes. Given that the number of clients is typically rather large, insurance companies may find it relevant to inquire whether the posterior distributions (given the number of individual accidents) of the mixture, of the mixing parameter and the predictive distribution of a new observation, given the number of individual accidents, are consistently convergent. For most models of this rather general class, answering these questions requires unfathomable manipulations.

In the next section, this example, along with another one, is worked out with more details. In Section 3, a fairly general model, encompassing both examples of Section 2, is presented and general results on exact estimability are given using what we believe to be fairly simple arguments. From an asymptotic point of view, exact estimability is the natural Bayesian concept when investigating the relevant properties of almost sure convergence of posterior inferences. The main stream of thought, developed in EBS, consists of envisaging a Bayesian experiment as a unique probability (measure) on the product space "parameter × observations" and eventually deriving all results within the same space. In Section 4, some problems of approximations are discussed. The starting point is to remember a well-known issue in these kinds of models, namely that posterior distributions conditional on all available data are basically untractable. Next it is shown that conditioning on more information, containing also part of the latent variables, may suggest workable approximations retaining, at variance from other approximations, desirable asymptotic properties. The final section gathers some concluding remarks and opens avenues for future research efforts, in the direction of Monte Carlo simulations with due attention to the problem of guaranteeing consistent convergence, a property that has been lacking in some previous proposals.

2. EXAMPLES

2.1. *A First Example*

We consider X_i, the number of accidents caused during one year by the i-th client of an automobile insurance company and we suppose that X_i has a Poisson distribution with expectation θ_i. If we do not want to be too specific for the distribution of θ_i in the portfolio of this insurance company, we may suppose that the distribution of θ_i is a Dirichlet process. However such a process generates, with probability one, discrete distributions. One way to enrich the distribution of θ_i is to use a mixture of Dirichlet processes. So we may suppose for instance that θ_i is F distributed where F is a Dirichlet process the parameter of which is given by $\alpha\Gamma(\eta, b)$ where α is a positive real parameter, $\Gamma(\eta, b)$ is the Gamma distribution with shape parameter b and scale parameter η. This η is a positive random variable, the distribution of which is for instance $\Gamma(e, a)$. We end up with the following model.

$$\begin{aligned} &\eta \mid \alpha \sim \Gamma(e, a) && (2.1)(i) \\ &F \mid \eta, \alpha \sim \text{Di}[\alpha\Gamma(\eta, b)] && (ii) \\ &\theta_i \mid F, \eta, \alpha \sim F && (iii) \\ &X_i \mid \theta_i, F, \eta, \alpha \sim \text{Po}(\theta_i) && (iv) \end{aligned}$$

By setting:

$$G(B) = F(\eta^{-1}B), \qquad \nu_i = \eta\theta_i, \tag{2.2}$$

the above model is equivalent to the following one:

$$\begin{aligned} &\eta \mid \alpha \sim \Gamma(e, a) && (2.3)(i) \\ &G \mid \eta, \alpha \sim \text{Di}[\alpha\Gamma(1, b)] && (ii) \\ &\nu_i \mid G, \eta, \alpha \sim G && (iii) \\ &X_i \mid \nu_i, G, \eta, \alpha \sim \text{Po}(\eta^{-1}\nu_i). && (iv) \end{aligned}$$

Note that, integrating on η, shows that θ_i admits a density conditionally on G and α. Indeed

$$\begin{aligned} F(B) &= P[\theta_i \in B \mid G, \alpha] \\ &= E[G(\eta B) \mid G, \alpha] \\ &= \int_B p(t \mid G) dt \end{aligned}$$

where

$$p(t \mid G) = \int_0^\infty \frac{e^a}{\Gamma(a)} \frac{y^a}{t^{a+1}} e^{-(e/t)y} G(dy) \tag{2.5}$$

However the θ_i's are no longer independent conditionally on (G, α) but rather exchangeable.

2.2. *A Second Example*

Let X_i be the result of a student on an intellectual quotient test and θ_i be the actual intellectual quotient of this student. It is supposed that

$$X_i = \theta_i + \varepsilon_i \tag{2.6}$$

where

$$\varepsilon_i \sim N(0, \sigma^2).$$

Assuming a continuous nonparametric prior distribution for θ_i in the student population may be achieved by choosing θ_i to have a Dirichlet distribution with parameter $\alpha N(\eta, a^2)$ where η is distributed as a $N(m, v^2)$. We end up with the following model.

$$\begin{aligned} &\eta \mid \alpha, \sigma^2 \sim N(m, v^2) && (2.7)(i) \\ &F \mid \eta, \alpha, \sigma^2 \sim \text{Di}[\alpha N(\eta, a^2)] && (ii) \\ &\theta_i \mid F, \eta, \alpha, \sigma^2 \sim F && (iii) \\ &X_i \mid \theta_i, F, \eta, \alpha, \sigma^2 \sim N(\theta_i, \sigma^2) && (iv) \end{aligned}$$

Just as before, by setting

$$G(B) = F(B + \eta), \qquad \nu_i = \theta_i - \eta, \tag{2.8}$$

the above model is equivalent to the following one:

$$\begin{aligned} &\eta \mid \alpha, \sigma^2 \sim N(m, v^2) && (2.9)(i) \\ &G \mid \eta, \alpha, \sigma^2 \sim \text{Di}[\alpha N(0, a^2)] && (ii) \\ &\nu_i \mid G, \eta, \alpha, \sigma^2 \sim G && (iii) \\ &\varepsilon_i \mid G, \eta, \alpha, \sigma^2 \sim N(0, \sigma^2) && (iv) \\ &X_i = \eta + \nu_i + \varepsilon_i. && (v) \end{aligned}$$

Once again, conditionally on G and α, θ_i admits a density, since

$$\begin{aligned} F(B) &= P[\theta_i \in B \mid G, \alpha] \\ &= E[G(B - \eta) \mid G, \alpha] \\ &= \int_B p(x \mid G) dx \end{aligned} \tag{2.10}$$

where

$$p(x \mid G) = \int_{I\!R} \frac{1}{\sqrt{2\pi v^2}} e^{-(1/2v^2)(x-y-m)^2} G(dy) \tag{2.11}$$

If we reconsider (2.9)(v), the model can be reinterpreted as a normal model with heterogeneity in the population expressed by the presence of ν_i, a feature shared by many recent models in microeconometrics.

This paper addresses two kinds of problems. The first kind of problem in the above examples is to compute exact or approximate posterior distributions of F, η, λ knowing the data, i.e., $X_1^n = (X_1, X_2, \ldots, X_n)$. It is also of interest to compute, for a new potential observation (X_f, θ_f), the posterior predictive distribution of X_f. This is a rather difficult problem which will be examined in Section 4 and which has received several answers in different sampling models (see, e.g., Antoniak (1974), Bunke (1985), Lo (1984), Berry and Christensen (1979), Smith and Makov (1978), Simar (1982) and West (1990)).

A second kind of problem is to answer the question of consistency of these estimates and predictions obtained either by hard computations or by some appropriate approximations. These problems will be first treated in Section 3. We will show that very general results are provided using tools developed in EBS.

3. GENERAL MODEL. EXACT ESTIMABILITY

A general model encompassing examples 1 and 2 may be described as follows: $(\Omega, \mathcal{M}, P)$ is an abstract probability space and $(L, \mathcal{L})$, $(S, \mathcal{S})$, $(U, \mathcal{U})$, and $(V, \mathcal{V})$ are Standard Borel spaces, α is a positive random variable, γ is a random variable with values in $(L, \mathcal{L})$, η is a random variable with values in $(U, \mathcal{U})$, F is a random probability measure on $(S, \mathcal{S})$, θ_i, $1 \le i \le n$, are random variables with values in $(S, \mathcal{S})$ and X_i, $1 \le i \le n$, are random variables with values in $(V, \mathcal{V})$. Furthermore, m^η and $\ell^{\theta,\gamma}$ are families of transition probabilities on $(S, \mathcal{S})$ and on $(V, \mathcal{V})$ respectively. The model is specified as follows:

$$\eta \perp\!\!\!\perp \alpha \perp\!\!\!\perp \gamma \tag{3.1)(i}$$
$$F \mid \eta, \alpha, \gamma \sim \mathrm{Di}(\alpha m^\eta) \qquad (ii)$$
$$\perp\!\!\!\perp_{1 \le i \le n} \theta_i \mid F, \eta, \alpha, \gamma \qquad (iii)$$
$$\theta_i \mid F, \eta, \alpha, \gamma \sim F \qquad (iv)$$
$$\perp\!\!\!\perp_{1 \le i \le n} X_i \mid \theta_1^n, F, \eta, \alpha, \gamma \qquad (v)$$
$$X_i \perp\!\!\!\perp \theta_1^n \mid \theta_i, F, \eta, \alpha, \gamma \qquad (vi)$$
$$X_i \mid \theta_i, F, \eta, \alpha, \gamma \sim \ell^{\theta_i, \gamma}. \qquad (vii)$$

Conditions (v) and (vi) define X_1^n to be "$(F, \eta, \alpha, \gamma)$-*sifted*" by θ_1^n (see EBS Definition 7.6.11). Heuristically it means that, conditionally on $(F, \eta, \alpha, \gamma)$, the (θ_i)-process "sifts" the (X_i)-process in the sense that (X_i) behaves "nicely", i.e., independently and with each θ_i being "*allocated*" to each X_i (EBS, Definition 7.6.12). The sifting condition is a kind of minimal requirement for a non i.i.d. process to behave nicely from an inferential point of view. Investigating the properties of the present model provides an example of such a structure and an argument in favour of the practical relevance of this structure.

By Theorem 7.6.9 of EBS, assumptions (iii), (v), and (vi) are equivalent to the following assumption.

$$\perp\!\!\!\perp_{1 \le i \le n} (X_i, \theta_i) \mid F, \eta, \alpha, \gamma. \tag{3.2}$$

Now assumptions (iv) and (vii) imply

$$\theta_i \perp\!\!\!\perp (\eta, \alpha, \gamma) \mid F \tag{3.3)(i}$$
$$X_i \perp\!\!\!\perp (F, \eta, \alpha) \mid \theta_i, \gamma \tag{ii}$$

and assumption (ii) implies that

$$F \perp\!\!\!\perp \gamma \mid \eta, \alpha. \tag{3.4}$$

By Theorem 2.2.10 of EBS, (3.3) is equivalent to

$$(X_i, \theta_i) \perp\!\!\!\perp (\eta, \alpha) \mid F, \gamma \tag{3.5)(i}$$
$$\theta_i \perp\!\!\!\perp \gamma \mid F \tag{ii}$$
$$X_i \perp\!\!\!\perp F \mid \theta_i, \gamma \tag{iii}$$

and assumption (i) and (3.4) are equivalent to

$$\eta \perp\!\!\!\perp \alpha \tag{3.6)(i}$$
$$\gamma \perp\!\!\!\perp (F, \eta, \alpha). \tag{ii}$$

Another use of Theorem 2.2.10 and of Theorem 6.2.3 of EBS shows that (3.2) and (3.5)(i) are equivalent to

$$\perp\!\!\!\perp_{1 \leq i \leq n} (X_i, \theta_i) \; F, \gamma, \tag{3.7)(i}$$
$$(X_1^n, \theta_1^n) \perp\!\!\!\perp (\eta, \alpha) \mid F, \gamma. \tag{ii}$$

Therefore, the observations form a i.i.d. process conditionally on (F, γ) the distribution of which is given by

$$\begin{aligned} P[X_i \in B \mid F, \gamma] &= E[P[X_i \in B \mid \theta_i, F, \eta, \alpha, \gamma] \mid F, \gamma] \\ &= E[\ell^{\theta_i, \gamma}(B) \mid F, \gamma] \\ &= \int_S \ell^{\theta, \gamma}(B) F(d\theta). \end{aligned} \tag{3.8}$$

Therefore, if $\ell(x \mid \theta, \gamma)$ is a density of $\ell^{\theta,\gamma}$ with respect to a probability measure q on $(V, \mathcal{V})$, the observations have densities with respect to q given by

$$\ell(x \mid F, \gamma) = \int_S \ell(x \mid \theta, \gamma) F(d\theta), \tag{3.9}$$

i.e., the original densities mixed by a Dirichlet process.

Now, i.i.d. processes are known to have nice asymptotic properties. In order to correctly state these properties, we need some technical tools. For a random variable Y_i we denote by $\mathcal{Y}_i$ the σ-field generated by Y_i.

Definition 3.1. *The projection of $\mathcal{Y}_2$ on $\mathcal{Y}_1$, denoted by $\mathcal{Y}_1\mathcal{Y}_2$, is defined as the σ-field generated by $E[f(Y_2) \mid Y_1]$ for all f positive bounded measurable functions.*

This is the smallest sub-σ-field of $\mathcal{Y}_1$ which makes measurable all expectations of functions of Y_2 conditionally on Y_1 (see EBS section 4.3 for more details). In particular, if $\mathcal{Y}_2$ is the observation and $\mathcal{Y}_1$ the parameter, $\mathcal{Y}_1\mathcal{Y}_2$ represents the "identified" parameter.

The sampling concept of identification of Y_1 by Y_2 is the injectivity of the mapping that associates to each value of Y_1 the distribution of Y_2 conditionally on this value of Y_1. The Bayesian counterpart of this concept (see EBS Section 4.6.2) is provided by the following definition.

Definition 3.2. *The random variable Y_1 is identified by Y_2, denoted as $Y_1 < Y_2$, if $\mathcal{Y}_1\mathcal{Y}_2 = \mathcal{Y}_1$ and Y_1 is identified by Y_2 conditionally to a random variable Y_3, denoted as $Y_1 < Y_2 \mid Y_3$, if (Y_1, Y_3) is identified by (Y_2, Y_3), i.e., $(Y_1, Y_3) < (Y_2, Y_3)$.*

For a stochastic process $Y = \{Y_i : 1 \leq i < \infty\}$, the *tail-$\sigma$-field*, denoted by $\mathcal{Y}_T$, is defined as

$$\mathcal{Y}_T = \bigcap_n \bigvee_{m \geq n} \mathcal{Y}_m. \tag{3.10}$$

The tail-σ-field of Y is therefore the collection of Y-events the definition of which does not depend on the first k realisations, i.e., $\{Y_1, Y_2, \ldots, Y_k\}$, whatever k is.

For a sub-σ-field $\mathcal{N}$ of $\mathcal{M}$, we denote by $\overline{\mathcal{N}}$ the *completion* of $\mathcal{N}$ by the null-events of $\mathcal{M}$.

Now, Theorem 9.3.12 of EBS entails the following first result.

Proposition 3.3. *If $\mathcal{B} = \sigma(F, \gamma)$, i.e., is the σ-field generated by F and γ, then*

$$\overline{\mathcal{X}_T} = \overline{\mathcal{B}\mathcal{X}_1} \tag{3.11}$$

The idea of the proof is as follows. The independence implies by the Kolmogorov 0-1 law, that $\mathcal{X}_T \subset \overline{\mathcal{B}}$, i.e., any tail random variable is almost surely equal to a function of (F, γ). In such a case, we say that $\mathcal{X}_T$ is "*exactly estimating*". The stationarity implies by the ergodic theorem that $\mathcal{X}_T$ is *asymptotically sufficient* for (F, γ), i.e., $\mathcal{X}_1^\infty \perp\!\!\!\perp \mathcal{B} \mid \mathcal{X}_T$. Therefore the parameters identified by the X-process, i.e., $\mathcal{B}\mathcal{X}_1^\infty$, are the same as the parameters identified by $\mathcal{X}_T$, i.e., $\mathcal{B}\mathcal{X}_T$. But since $\mathcal{X}_T \subset \overline{\mathcal{B}}$, $\mathcal{B}\mathcal{X}_T = \overline{\mathcal{X}_T \cap \mathcal{B}}$. Finally for an i.i.d. process X, the parameters identified by X, i.e., $\mathcal{B}\mathcal{X}_1^\infty$, are the same as those identified by X_1, i.e., $\mathcal{B}\mathcal{X}_1$. We remark that (3.11) implies that $\mathcal{B}\mathcal{X}_1 \subset \overline{\mathcal{X}_1^\infty}$, i.e., the parameters identified by X_1 are almost surely equal to measurable functions of the X-process. This assures, by Doob's Martingale Convergence Theorem, that posterior expectations of identified parameters is an almost surely consistent sequence of estimators.

This sequence is therefore sampling consistent for almost all (for the prior) values of the parameter (see, for more details, EBS Section 7.4.2).

Proposition 3.3 entails the following straightforward corollary.

Corollary 3.4. *If F and γ are identified by X_1, they are consistently estimable in the following sense: For any random variable $g(F, \gamma)$,*

$$\lim_{n \to \infty} E[g(F, \gamma) \mid X_1^n] = g(F, \gamma) \quad \text{a.s.}P.$$

Identification of γ by X_1 is a classical problem. An easy case is the following: if $\gamma = E[f(X_1) \mid \theta_1, \gamma]$ for some measurable function f, then $\gamma = E[f(X_1) \mid F, \gamma]$ since $X_1 \perp\!\!\!\perp F \mid \theta_1, \gamma$ (relation (3.5)(iii)). This shows that $\sigma(\gamma) \subset \mathcal{B}\mathcal{X}_1$.

Identification of F by X_1 may be more difficult to check. This is the known problem of identification of mixtures (see, e.g., Chow and Teicher (1978)). It is shown in EBS (Theorem 5.5.20) that sampling identifiability of mixtures is equivalent to the completeness of the posterior distributions for any prior. In the Bayesian set-up, the next proposition provides a sufficient condition for the identifiability of F by X_1. We first need the following definition.

Definition 3.5. *Let Y_1, Y_2, Y_3 be three random variables. Then Y_1 is strongly-p-identified by Y_2 ($p > 0$) if the following condition holds:*

$$E[|f(Y_1)|^p] < \infty \quad \text{and} \quad E[f(Y_1) \mid Y_2] = 0 \quad \text{a.s.} \\ \text{imply} \quad f(Y_1) = 0 \quad \text{a.s.} \tag{3.12}$$

and Y_1 is strongly-p-identified by Y_2 conditionally on Y_3 if (Y_1, Y_3) is strongly-p-identified by (Y_2, Y_3).

With this definition, we may give the second result.

Proposition 3.6. *If $\sigma(\gamma) \subset \mathcal{B}\mathcal{X}_1$ and if θ_1 is strongly-2-identified by X_1 conditionally on γ, then (F, γ) is identified by X_1.*

Proof. First, note that F is identified by θ_1 since $F(B) = P(\theta_1 \in B \mid F)\ \forall\ B \in \mathcal{S}$. But $\theta_1 \perp\!\!\!\perp \gamma \mid F$ (Definition 3.5)(ii), implies by Corollary 4.3.6 of EBS, that (F, γ) is identified by (θ_1, γ). Definition (3.5)(iii) is equivalent to $(X_1, \gamma) \perp\!\!\!\perp (F, \gamma) \mid \theta_1, \gamma$. This along with θ_1 strongly-2-identified by X_1 conditionally on γ, imply by Theorem 5.4.14 of EBS, that the parameters identified by(X_1, γ) are the same as those identified by (θ_1, γ). Therefore (F, γ) is identified by (X_1, γ). This along with the identification of γ by X_1 implies that (F, γ) is identified by X_1. ◁

The advantage of Proposition 3.6 lies in the fact that the condition of strong identification does not involve the particularities of the distribution of F. Indeed, from (3.1)(ii) and (iv), we obtain (see Ferguson (1973)),

$$\theta_i \mid \eta, \alpha, \gamma \sim m^{\eta}. \tag{3.13}$$

But this along with (3.1)(i) implies

$$(\eta, \theta_i) \perp\!\!\!\perp \gamma \perp\!\!\!\perp \alpha. \tag{3.14}$$

Therefore, using (3.1)(vii), the distribution of θ_i conditionally on (X_i, γ) is easily obtained and standard techniques will be used to show the 2-completeness, i.e., the strong-2-identification, of this distribution.

The next step consists in showing the exact estimability of (η, α). This is based on the following lemma (the proof of which is given in the appendix) which incidentally tells us that the number of jumps greater than ε is of order $\ln \varepsilon$ for a Dirichlet process.

Lemma 3.7. *If $F \sim \mathrm{Di}(\rho)$, then, as r tends to infinity, $\frac{1}{\ln r} \sum_{x \in B} 1_{\{F(x) > \frac{1}{r}\}}$ converges in probability to $\rho(B)$.*

This leads to a third result.

Proposition 3.8. *If η is identified by θ_1, then α and η are measurable with respect to the σ-field generated by F and the null events and are therefore exactly estimable if F and γ are identified by X_1. Hence for any bounded random variable $h(\eta, \alpha)$*

$$\lim_{n \to \infty} E[h(\eta, \alpha) \mid X_1^n] = h(\eta, \alpha) \quad \text{a.s. } P.$$

Proof. By Theorems 7.4.7(ii) and 7.4.6(i) of EBS, there exist subsequences such that

$$\frac{1}{\ln r_k} \sum_{x \in S} 1_{\{F(x) > \frac{1}{r_k}\}} \longrightarrow \alpha \quad \text{a.s. } P.$$

$$\frac{1}{\ln r_{k'}} \sum_{x \in B} 1_{\{F(x) > \frac{1}{r_{k'}}\}} \longrightarrow \alpha m^{\eta}(B) \quad \text{a.s. } P.$$

Therefore α and $m^{\eta}(B) \quad \forall\, B \in \mathcal{S}$ are almost surely equal to measurable functions of F. But $m^{\eta}(B) = P(\theta_1 \in B \mid \eta)$ according to (3.13). Hence if η is identified by θ_1, η is almost surely equal to a measurable function of F. ◁

Finally, the posterior predictive distribution of X_f will also be consistent if (F, γ) is identified by X_1. Indeed, according to (3.7)(i),

$$P[X_f \in B \mid X_1^n] = E[P[X_f \in B \mid F, \gamma] \mid X_1^n]$$
$$= E\left[\int_S \ell^{\theta,\gamma}(B) F(d\theta) \mid X_1^n\right].$$

Therefore, by (3.8),

$$\lim_{n \to \infty} P[X_f \in B \mid X_1^n] = P[X_f \in B \mid F, \gamma] \quad \text{a.s.} \tag{3.16}$$

Similarly, if $\ell(x \mid \theta, \gamma)$ is a density of $\ell^{\theta,\gamma}$ with respect to a probability measure q on $(V, \mathcal{V})$, the posterior expectation of the density of X_f will be consistent. Indeed

$$E[\ell(x \mid F, \gamma) \mid X_1^n] = E\left[\int_S \ell(x \mid \theta, \gamma) F(d\theta) \mid X_1^n\right] \to \ell(x \mid F, \gamma) \quad \text{a.s. as} \quad n \to \infty. \tag{3.17}$$

Looking at the two introductory examples, we obtain, for the Poisson example,

$$P[X_1 = k \mid F] = \int_0^{\infty} \frac{\theta^k}{k!} e^{-\theta} F(d\theta),$$

and for the Normal example, when σ^2 is known,

$$\ell(x \mid F, \sigma^2) = \int_{-\infty}^{\infty} \frac{1}{\sqrt{2\pi\sigma^2}} e^{-(1/2\sigma^2)(x-\theta)^2} F(d\theta).$$

It is easily seen in these two cases that F is identified by X_1, since

$$E[(1-p)^{X_1} \mid F] = \int_0^{\infty} e^{-p\theta} F(d\theta)$$

$$E[e^{itX_1} \mid F, \sigma^2] = e^{-1/2(\sigma^2 t^2)} \int_{-\infty}^{+\infty} e^{it\theta} F(d\theta).$$

Clearly, in both examples, η is identified by θ_1.

4. ANALYTICAL ASPECTS

General properties of the Dirichlet process (Ferguson (1973)), applied to model (3.1), ensure the following results:

$$F \mid \theta_1^n, \eta, \alpha, \gamma \sim \text{Di}\left(\alpha m^\eta + \sum_{1\le i\le n} \varepsilon_{\theta_i}\right) \tag{4.1}$$

where ε_a denotes the measure giving unit mass at point a. Therefore

$$\theta_{n+1} \mid \theta_1^n, \eta, \alpha, \gamma \sim \frac{\alpha}{\alpha+n} m^\eta + \frac{1}{\alpha+n} \sum_{1\le i\le n} \varepsilon_{\theta_i}, \tag{4.2}$$

and from this we deduce that the distribution of θ_1^n conditionally on (η, α, γ) is given by

$$\begin{aligned} m^{\eta,\alpha}(d\theta_1, d\theta_2, \ldots, d\theta_n) &= \prod_{1\le i\le n}\left(\frac{\alpha}{\alpha+i-1} m^\eta(d\theta_i) + \frac{1}{\alpha+i-1}\sum_{1\le j\le i-1} \varepsilon_{\theta_j}(d\theta_i)\right) \\ &= \frac{\Gamma(\alpha)}{\Gamma(\alpha+n)} \prod_{1\le i\le n}\left(\alpha m^\eta(d\theta_i) + \sum_{1\le j\le i-1} \varepsilon_{\theta_j}(d\theta_i)\right). \end{aligned} \tag{4.3}$$

If $\ell(x \mid \theta, \gamma)$ is a density of $\ell^{\theta,\gamma}$ with respect to a probability measure q on $(V, \mathcal{V})$, the distribution of X_1^n conditionally on (η, α, γ) has the following density with respect to q^n,

$$\ell(x_1, x_2, \ldots, x_n \mid \eta, \alpha, \gamma) = \int_{S^n} \prod_{1\le i\le n} \ell(x_i \mid \theta_i, \gamma) m^{\eta,\alpha}(d\theta_1, d\theta_2, \ldots, d\theta_n). \tag{4.4}$$

It is then possible to obtain posterior distributions of (η, α, γ) given X_1^n and to obtain the distribution of θ_1^n conditionally on $(X_1^n, \eta, \alpha, \gamma)$. From this, one gets the distribution of F conditionally on $(X_1^n, \eta, \alpha, \gamma)$ as a mixture of Dirichlet processes and, finally, the distribution of F given X_1^n.

However, in practice, the distribution of θ_1^n conditionally on (η, α, γ), given in (4.3), is at most a sum of $n!$ partially degenerate measures and the amount of computations explodes quite rapidly as n increases. This is a standard issue when analyzing mixtures, see, e.g., Mouchart (1977), Simar (1982), Lo (1984; 1986); a very useful reference being Titterington, Smith and Makov (1985). Following Lo (1984) (see also Kuo (1980)), formula (4.4) can be simplified slightly. Let us define for $D \subset \{1, 2, \ldots, n\}$, $x_D = \{x_i : i \in D\}$ and the following densities

$$\ell^*(x_D \mid t, \gamma) = \prod_{i\in D} \ell(x_i \mid t, \gamma) \tag{4.5(i)}$$

$$\bar{\ell}(x_D \mid \eta, \gamma) = \int_S \ell^*(x_D \mid t, \gamma) m^\eta(dt) \tag{ii}$$

$$m(t \mid X_D, \eta, \gamma) = \frac{\ell^*(X_D \mid t, \gamma)}{\bar{\ell}(X_D \mid \eta, \gamma)}. \tag{iii}$$

For any partition of $\{1, 2, \ldots, n\}$, $P = \{D_\ell : 1 \le \ell \le k(P)\}$, let us also define

$$r(P) = \prod_{1\le \ell\le k(P)} (n_\ell - 1)! \tag{4.6}$$

where n_ℓ is the cardinality of D_ℓ, $1 \le \ell \le k(P)$, and the following densities

$$\ell_P(x_1^n \mid \eta, \gamma) = \prod_{1\le\ell\le k(P)} \overline{\ell}(x_{D_\ell} \mid \eta, \gamma). \tag{4.7}$$

Then (4.4) may be rewritten as

$$\ell(x_1^n \mid \eta, \alpha, \gamma) = \frac{\Gamma(\alpha)}{\Gamma(\alpha+n)} \sum_P \alpha^{k(P)} r(P) \ell_P(x_1^n \mid \eta, \gamma) \tag{4.8}$$

where the sum is over all partitions of $\{1, 2, \ldots, n\}$ into nonordered groups. Note that this is a convex combination of densities, the weights of which are given by

$$w_P = \frac{\Gamma(\alpha)}{\Gamma(\alpha+n)} \alpha^{k(P)} r(P). \tag{4.9}$$

For the inference on F conditionally on $(X_1^n, \eta, \alpha, \gamma)$ the following expression is obtained:

$$\begin{aligned} &E\left[\int_S f(\theta)F(d\theta) \mid X_1^n, \eta, \alpha, \gamma,\right] \\ &\quad = \frac{\alpha}{\alpha+n}\int_S f(\theta) m^\eta(d\theta) \\ &\qquad + \frac{n}{\alpha+n}\sum_P W_P \left\{ \sum_{1\le\ell\le k(P)} \frac{n_\ell}{n} \int_S f(t) m(t \mid X_{D_\ell}, \eta, \gamma) m^\eta(dt) \right\} \end{aligned} \tag{4.10}$$

where

$$W_P = w_P \, \frac{\ell_P(X_1^n \mid \eta, \gamma)}{\ell(X_1^n \mid \eta, \alpha, \gamma)} \tag{4.11}$$

are random weights.

Replacing in expression (4.10) $f(\theta)$ by $\ell(x \mid \theta, \gamma)$ will furnish Bayesian estimates of the densities of the X-process, i.e., $\ell(x \mid F, \gamma)$.

Expressions (4.8) and (4.10) show averages of all partitions. It suggests that some simplification may be obtained by conditioning on an appropriate random partition. This amounts to increasing the information given by the data by an unobservable supplementary observation but this seems to be a promising approximation.

So let us define the following random variables

$$\mu_1 = \theta_1 \tag{4.12}$$

and for $k > 1$,

$$\begin{aligned} \mathrm{I}_k &= \inf\{i : \theta_i \ne \mu_\ell : \forall\, 1 \le \ell \le k-1\} \\ \mu_k &= \theta_{\mathrm{I}_k}. \end{aligned} \tag{4.13}$$

Thus for $\{\theta_i : 1 \le i \le n\}$, $\{\mu_k : 1 \le k \le K_n\}$, represent the distinct values by order of appearance, and

$$K_n = \sup\{\ell : \mathrm{I}_\ell \le n\}. \tag{4.14}$$

Let us also define the following random variables

$$C_i = k \quad \Leftrightarrow \quad \theta_i = \mu_k. \tag{4.15}$$

Thus $\{C_i : 1 \leq i \leq n\}$ totally describes the equality structure of $\{\theta_i : 1 \leq i \leq n\}$. Note that

$$K_n = \sup_{1 \leq i \leq n} C_i. \tag{4.16}$$

Therefore, the σ-field generated by θ_1^n is the same as the σ-field generated by $(C_1^n, \mu_1^{K_n})$. By defining

$$\begin{aligned} D_{n,k} &= \{i \leq n : C_i = k\} \\ \nu_{n,k} &= \sum_{1 \leq i \leq n} 1_{\{C_i = k\}} \end{aligned} \tag{4.17}$$

we see that C_1^n generates a random nonordered partition $Q_n = \{D_{n,k} : 1 \leq k \leq K_n\}$ of $\{1, 2, \ldots, n\}$. For the rest of this section, we will make the following assumption.

(A) *m^η is diffuse.*

Under this assumption, nice relations are obtainable (see Ferguson (1973), West (1990)):

$$\begin{aligned} P[C_{n+1} = k \mid \theta_1^n, \alpha, \eta, \gamma] \\ &= \frac{\nu_{n,k}}{\alpha + n} \qquad \text{if} \quad k \leq K_n \\ &= \frac{\alpha}{\alpha + n} \qquad \text{if} \quad k = K_n + 1 \end{aligned} \tag{4.18}$$

From this, it can be deduced that

$$C_1^n \perp\!\!\!\perp (\eta, \gamma) \mid \alpha \tag{4.19}$$

and that

$$P[Q_n = Q_0 \mid \alpha] = w_{Q_0} \tag{4.20}$$

where w_{Q_0} is given by (4.9).

On the other hand

$$\begin{aligned} &\perp\!\!\!\perp_{1 \leq k \leq K_n} \mu_k \mid C_1^n, \alpha, \eta, \gamma \\ &\mu_k \mid C_1^m, \alpha, \eta, \gamma \sim m^\eta. \end{aligned} \tag{4.21}$$

This implies that

$$\mu_1^{K_n} \perp\!\!\!\perp (\alpha, \gamma, C_1^n) \mid K_n, \eta. \tag{4.22}$$

By Theorem 2.2.10 of EBS, (4.19), (4.22) and 3.1(i) are equivalent to

$$\eta \perp\!\!\!\perp K_n \perp\!\!\!\perp \gamma \tag{4.23)(i}$$

$$(\alpha, C_1^n) \perp\!\!\!\perp (\eta, \mu_1^{K_n}) \perp\!\!\!\perp \gamma \mid K_n. \tag{ii}$$

Expression (4.23)(ii) implies in particular that, for inference on α, C_1^n is sufficient and that for inference on η, $(\mu_1^{K_n}, K_n)$ is sufficient.

Now looking at the data conditionally on C_1^n, we define

$$X_{n,k} = \{X_i : i \in D_{n,k}\}. \tag{4.24}$$

Clearly

$$X_i \mid \theta_1^n, \eta, \alpha, \gamma \sim \ell^{\mu_k, \gamma}, \qquad i \in D_{n,k}, \tag{4.25}$$

and the densities of $X_{n,k}$ conditionally on $(\theta_1^n, \eta, \alpha, \gamma)$ are given by $\ell^*(x_{n,k} \mid \mu_k, \gamma)$ defined in (4.5)(i). Now

$$\perp\!\!\!\perp_{1\le k\le K_n} X_{n,k} \mid \theta_1^n, F, \eta, \alpha, \gamma, \tag{4.26)(i}$$

$$X_{n,k} \perp\!\!\!\perp (\theta_1^n, F, \eta, \alpha) \mid C_1^n, \mu_k, \gamma. \qquad (ii)$$

This, along with (5.20), by Theorem 7.6.9 of EBS, are equivalent to

$$\perp\!\!\!\perp_{1\le k\le K_n} (X_{n,k}, \mu_k) \mid C_1^n, \alpha, \eta, \gamma, \tag{4.27)(i}$$

$$(X_{n,k}, \mu_k) \perp\!\!\!\perp \alpha \mid C_1^n, \eta, \gamma. \qquad (ii)$$

Hence the densities of $X_{n,k}$ conditionally on $(C_1^n, \eta, \alpha, \gamma)$ are given by $\overline{\ell}(x_{n,k} \mid \eta, \gamma)$ defined in (4.5)(ii) and those of X_1^n conditionally on $(C_1^n, \eta, \alpha, \gamma)$ are given by $\ell_{Q_n}(x_1^n \mid \eta, \gamma)$ defined in (4.7). A posteriori, we obtain the distribution of (η, γ) conditionally on (X_1^n, C_1^n) and

$$\perp\!\!\!\perp_{1\le k\le K_n} \mu_k \mid C_1^n, X_1^n, \alpha, \eta, \gamma. \tag{4.28}$$

Moreover the densities of μ_k conditionally on $(X_1^n, C_1^n, \eta, \alpha, \gamma)$ are given by $m(t \mid X_{n,k}, \eta, \gamma)$ defined in (4.5)(iii).

Finally, since $\sum_{1\le i\le n} \varepsilon_{\theta_i} = \sum_{1\le k\le K_n} \nu_{n,k}\, \varepsilon_{\mu_k}$, it is easy to obtain the distribution of F conditionally on $(C_1^n, X_1^n, \eta, \alpha, \gamma)$ as a mixture of Dirichlet process. In particular, we obtain

$$\begin{aligned} E&\left[\int_S f(\theta)F(d\theta) \mid C_1^n, X_1^n, \eta, \alpha, \gamma\right] \\ &= \frac{\alpha}{\alpha+n} \int_S f(\theta) m^\eta(d\theta) + \\ &\sum_{1\le \ell\le K_n} \frac{\nu_{n,k}}{\alpha+n} \int_S f(\theta) m(\theta \mid X_{n,k}, \eta, \gamma) m^\eta(d\theta). \end{aligned} \tag{4.29}$$

We deduce from this that

$$P(Q_n = Q_0 \mid X_1^n, \eta, \alpha, \gamma) = W_{Q_0}, \tag{4.30}$$

where W_{Q_0} is defined in 4.11.

Replacing in expression (4.29) $f(\theta)$ by $\ell(x \mid \theta, \gamma)$ will furnish Bayesian estimates of the densities of the X-process, i.e., $\ell(x \mid F, \gamma)$.

Since we have conditioned the parameters on a bigger filtration of σ-fields, all the consistency results obtained in Section 3 still apply under the same conditions. For instance, if F is identified by X_1 conditionally on γ, then expression (4.29) produced a.s. consistent estimations of $\int_S f(\theta)F(d\theta)$.

Going back to the first example, usual computations will give the following results.

$$E\left[\int_0^\infty f(\theta)F(d\theta) \mid C_1^n, X_1^n, \eta, \alpha\right] = \int_0^\infty f(\theta) m(\theta \mid C_1^n, X_1^n, \eta, \alpha) d\theta \tag{4.31}$$

where $m(\theta \mid C_1^n, X_1^n, \eta, \alpha)$ is the density of the following distribution:

$$\frac{\alpha}{\alpha+n}\,\Gamma(\eta, b) + \frac{n}{\alpha+n} \sum_{1\le k\le K_n} \frac{\nu_{n,k}}{n}\,\Gamma(\eta + \nu_{n,k}, b + \widetilde{X}_{n,k}) \tag{4.32}$$

where

$$\widetilde{X}_{n,k} = \sum_{i \in D_{n,k}} X_i. \tag{4.33}$$

The posterior predictive distribution of X_f is given by

$$X_f \mid C_1^m, X_1^n, \eta, \alpha \tag{4.34}$$
$$\sim \frac{\alpha}{\alpha+n} G\left(b, \frac{\eta}{1+\eta}\right) + \frac{n}{\alpha+n} \sum_{1 \le k \le K_n} \frac{\nu_{n,k}}{n}\, G\left(b + \widetilde{X}_{n,k}, \frac{\eta + \nu_{n,k}}{1+\eta+\nu_{n,k}}\right),$$

where $G(a,p)$ is the geometric distribution for $a \in \mathbb{R}_0^+$ and $p \in (0,1)$, i.e., $X \sim G(a,p)$ if

$$P[X = x] = C_{a+x-1}^{x}\, p^a (1-p)^x.$$

Now, as $n \to \infty$, $K_n \to \infty$ a.s. and $\nu_{n,k} \to \infty$ a.s. $\forall\ k$. Therefore if

$$\frac{1}{\nu_{n,k}} \widetilde{X}_{n,k} \longrightarrow R_k \quad \text{a.s.} \tag{4.35}$$
$$\Gamma(\eta + \nu_{n,k}, b + \widetilde{X}_{n,k}) \longrightarrow \varepsilon_{R_k} \quad \text{weakly and}$$
$$G\left(b + \widetilde{X}_{n,k}, \frac{\eta + \nu_{n,k}}{1+\eta+\nu_{n,k}}\right) \longrightarrow \mathrm{Po}(R_k) \quad \text{weakly.}$$

This shows that (4.34) is some kind of kernel approximation of the distribution of X_f.

At last, it may be shown that $\eta \mid C_1^n, X_1^n, \alpha$ has a density proportional to

$$\frac{\eta^{a+bK_n-1}\, e^{-e\eta}}{\prod_{1 \le k \le K_n} (\eta + \nu_{n,k})^{b+\widetilde{X}_{n,k}}} \tag{4.36}$$

which, for large n, may be approximated by

$$\Gamma\left(e + \sum_{1 \le k \le K_n} \frac{\widetilde{X}_{n,k}}{\nu_{n,k}}, a + bK_n\right). \tag{4.37}$$

5. FINAL REMARK

In view of approximate expression (4.10), L. Kuo (1986) proposed using a Monte Carlo method by selecting random partitions according to the distribution defined in (4.9). But looking at expression (4.29), we realize that the process X is i.i.d. conditionally on F, γ and that the distribution of C_1^m depends only on α. Therefore, we propose the following simulation:

(i) Choose α at random according to some distribution on the positive axis.
(ii) Given C_1^k, choose C_{k+1} at random according to the distribution given in (4.18), $1 \le k \le n-1$.
(iii) Then, group the observations $X_{n,k}$, $1 \le k \le K_n$, according to (4.17) and (4.24).
(iv) Compute the estimate given in expression (4.29).

We do not know whether such a procedure gives a consistent estimation of the posterior predictive distribution of X_f but, at least, K_n and $\nu_{n,k}$ have the proper asymptotic behaviour.

Other choices of C_1^n have been proposed in the literature, corresponding to extreme cases. The first one corresponds to $\alpha = 0$. In this situation, $C_1 = 1$ a.s. $\forall\, 1 \le i \le n$ by (4.18). Therefore, $K_n = 1$ a.s. and $\nu_{n,1} = n$ a.s. Such a configuration has an asymptotic probability equal to zero since

$$P[K_n = 1 \mid \alpha] = \frac{\Gamma(\alpha)}{\Gamma(\alpha+n)}\alpha \cdot (n-1)! \tag{5.1}$$

The second one corresponds to $\alpha = \infty$. In this situation, $C_i = i$ a.s. $\forall\, 1 \le i \le n$ by (4.18). Therefore, $K_n = n$ a.s. and $\nu_{n,k} = 1$ a.s. $\forall\, 1 \le k \le n$. Such a configuration has also an asymptotic probability equal to zero since

$$P[K_n = 1 \mid \alpha] = \frac{\Gamma(\alpha)}{\Gamma(\alpha+n)}\alpha^n. \tag{5.2}$$

Such procedures are however not consistent. Indeed, for example 1, it may be seen from (4.34) that the posterior predictive distribution of X_f converges weakly to $\mathrm{Po}[E(X_1 \mid F)]$ in the first case and to

$$E[G(b + X_1, \frac{\eta+1}{\eta+2}) \mid F]$$

in the second case.

Consistency requires at least that K_n and $\nu_{n,k}$ tends to infinity to obtain a kernel behaviour of the estimator.

Appendix: A Path Property of Purely Random Measures

Let A be a random measure on a Standard Borel Space which is purely random, i.e., $\forall\ \{B_i : 1 \le i \le k\}$ measurable partition of S,

$$\perp\!\!\!\perp_{1 \le i \le k} A(B_i). \tag{A.1}$$

If $\alpha(B) = E[A(B)]$ is a diffuse σ-finite measure and if A has no continuous deterministic part, what is achieved if

$$\lim_{p\to\infty} \frac{1}{p} \ln E[e^{-pA(S)}] = 0,$$

then the Levy measure of this random measure (see, e.g., Karr (1986) or Daley and Vere-Jones (1988)) has the following form

$$\nu(dx, du) = \frac{1}{u}\rho(x, du)\alpha(dx), \tag{A.2}$$

where $\rho(x, \cdot)$ is a transition probability on $(\mathbb{R}_0^+, \mathbb{B}_0^+)$ so that

$$-\ln E[e^{-pA(B)}] = \int_B \alpha(dx) \int_0^\infty \frac{1-e^{-pu}}{u}\rho(x, du). \tag{A.3}$$

Lemma A.1. *If* $\int_0^\infty \frac{1}{u}\rho(x, du) = \infty$ *and if*

$$\lim_{\varepsilon \downarrow 0} g(x, \varepsilon) \int_\varepsilon^\infty \frac{1}{u}\rho(x, du) = 1 \quad \forall\, x \in S,$$

then as $\varepsilon \downarrow 0$, $\sum_{x\in B} g(x, \varepsilon) 1_{\{A(x)>\varepsilon\}}$ *converges to* $\alpha(B)$ *in probability* $\forall B \in \mathcal{S}$ *such that* $\alpha(B) < \infty$.

Proof. If $M_\varepsilon(B) = \sum_{x\in B} g(x,\varepsilon)1_{\{A(x)>\varepsilon\}}$,

$$E\left[e^{-pM_\varepsilon(B)}\right] = \exp - \int_B \alpha(dx)\{1-e^{-pg(x,\varepsilon)}\}\left\{\int_\varepsilon^\infty \frac{1}{u}\rho(x,du)\right\}$$
$$\to \quad e^{-p\alpha(B)} \quad \text{as} \quad \varepsilon \downarrow 0. \quad \triangleleft$$

Corollary A.2. *If A is Gamma measure, i.e., $A(B) \sim \Gamma(1,\alpha(B))$, then, as $\varepsilon \downarrow 0$, $\frac{1}{\ln \frac{1}{\varepsilon}} \sum_{x\in B} 1_{\{A(x)>a\varepsilon\}} \to \alpha(B)$ in probability $\forall\ B \in S$ such that $\alpha(B) < \infty$, $\forall\ a > 0$.*

Proof. Indeed, for a Gamma measure, $\rho(x,du) = e^{-u}du$ and

$$\lim_{\varepsilon\downarrow 0} \frac{1}{\ln\frac{1}{\varepsilon}} \int_{a\varepsilon}^\infty \frac{e^{-u}}{u} du = 1. \quad \triangleleft$$

Now, we know that if $F \sim \mathrm{Di}(\alpha)$, then F may be represented as

$$F(B) = \frac{A(B)}{A(S)}, \tag{A.4}$$

where A is a Gamma measure with parameter α and moreover, $A(S) \perp\!\!\!\perp F$. Since $F[x] > \varepsilon$ if and only if $A(x) > \varepsilon\ A(S)$, the same result as in Corollary A.2.3 may be shown to be true for a Dirichlet process, i.e.,

Corollary A.3. *If $F \sim \mathrm{Di}(\alpha)$, then as $r \uparrow \infty$,*

$$\frac{1}{\ln r} \sum_{x\in B} 1_{\{F(x)>\frac{1}{r}\}} \longrightarrow \alpha(B)$$

in probability.

Proof. Let $N_r(B) = \frac{1}{\ln r} \sum_{x\in B} 1_{\{F(x)>\frac{1}{r}\}}$ and $M_{\frac{a}{r}}(B) = \sum_{x\in B} 1_{\{A(x)>\frac{a}{r}\}}$. Then

$$\begin{aligned} &E[e^{-pN_r(B)}]P[a < A(S) < b] \\ &\quad = E\left[e^{-(p/\ln r)M(A(S)/r)}1_{\{a<A(S)<b\}}\right] \\ &\quad \le E\left[e^{-(p/\ln r)M(b/r)}1_{\{a<A(S)<b\}}\right] \\ &\quad \ge E\left[e^{-(p/\ln r)M(a/r)}1_{\{a<A(S)<b\}}\right] \end{aligned}$$

But as $r \to \infty$, these two last terms tend to $e^{-p\alpha(B)}P[a < A(S) < b]$. Therefore, as $r \to \infty$,

$$E[e^{-pN_r(B)}] \quad \longrightarrow \quad e^{-p\alpha(B)}. \quad \triangleleft$$

REFERENCES

Antoniak, C. E. (1974). Mixtures of Dirichlet process with applications to Bayesian nonparametric problems. *Ann. Statist.* **2**, 1152–1174.

Berry, D. and Christensen, R. (1979). Empirical Bayes estimation of a binomial parameter via mixtures of Dirichlet processes. *Ann. Statist.* **7**, 558–568.

Bunke, O. (1985). Bayesian estimators in semiparametric problems. *Tech. Rep.* **102**, Humboldt-Universität, Berlin.

Chow, Y. S. and Teicher, H. (1978). *Probability Theory*. New York: Springer.

Daley, D. J. and Vere-Jones, D. (1988). *An Introduction to the Theory of Point Processes*. New York: Springer.

Ferguson, T. J. (1973). A Bayesian analysis of some nonparametric problems. *Ann. Statist.* **1**, 209–230.

Ferguson, T. S. (1983). Bayesian density estimation by mixture of Normal distributions. *Recent Advances in Statistics*. New York: Academic Press, 278–302.

Florens, J.-P., Mouchart, M. and Rolin, J. M. (1990). *Elements of Bayesian Statistics*. New York: Marcel Dekker.

Karr, A. F. (1986). *Point Processes and their Statistical Inference*. New York: Marcel Dekker.

Kuo, L. (1986). Computations of mixtures of Dirichlet processes. *SIAM Journal of Sciences and Statistical Computation* **7**, 60–71.

Lo, A. Y. (1984). On a class of Bayesian nonparametric estimates: I. Density estimates. *Ann. Statist.* **12**, 351–357.

Lo, A. Y. (1986). Bayes method for mixture models. *International Seminar on Bayesian Analysis*, Stresa, Italy.

Mouchart, M. (1977). A regression model with an explanatory variable which is both binary and subject to errors. *Latent Variables in Socio-Economic Models* (D. J. Aigner and A. S. Goldberger, eds.) Amsterdam: North-Holland, 49–66.

Simar, L. (1982). Some results on the Bayesian estimation of mixtures. *Tech. Rep.* **8306**, CORE.

Smith, A. F. M. and Makov, U. E. (1978). A quasi-Bayes sequential procedure for mixtures. *J. Roy. Statist. Soc. B* **40**, 106–112.

Titterington, D. M., Smith, A. F. M. and Makov, U. E. (1985). *Statistical Analysis of Finite Mixture Distributions*. New York: Wiley.

West, M. (1990). Bayesian kernel density estimation. *Tech. Rep.* **90–A02**, Duke University.

DISCUSSION

M. D. ESCOBAR* (*Carnegie Mellon University, USA*)

1. Introduction. I would like to restrict my discussion to (a) calculating values from a mixture of Dirichlet process priors, and (b) discussing the importance of the parameter α when we are interested in the number of mixtures.

2. Calculating Values. In general, if we wish to calculate the posterior expectation of some function of θ_1^n, say $\psi(\theta_1^n)$, then we need to calculate:

$$E[\psi(\theta_1^n)|X_1^n, \alpha, \eta, \gamma] \propto \int \psi(\theta_1^n) \prod_{i=1}^{n} l(X_i|\theta_i, \gamma) m^{\eta,\alpha}(d\theta_1^n)$$

The simplest, straight forward way to do this is by a simple Monte Carlo integration by sampling a vector $\theta_{1,h}^n$ from $m^{\eta,\alpha}$, and then estimating $\hat{E}[\psi(\theta_1^n)|X_1^n, \alpha, \eta, \gamma]$ by

$$\hat{E}[\psi(\theta_1^n)|X_1^n, \alpha, \eta, \gamma] \propto \frac{1}{H} \sum_{h=1}^{H} \psi(\theta_{1,h}^n) \prod_{i=1}^{n} l(X_i|\theta_{i,h}, \gamma)$$

The problem with this method is that for many sampled vectors $\theta_{1,h}^n$, the term

$$\prod_{i=1}^{n} l(X_i|\theta_{i,h}, \gamma)$$

* Supported by Grant #R01-CA54852-01 from the National Cancer Institute and by Grant #MH15758 from the National Institute of Mental Health.

may be so small that, for computational purposes, it is equal to zero or makes the computation numerically unstable.

For a simple example to illustrate this, assume that $(X_i|\theta_i, \gamma)$ has a normal distribution with mean θ_i and variance one, and that m^η is a discrete distribution with 2 equally probable atoms at 5 and -5. Therefore, the x_i's are in two clusters around 5 and -5. Now if we generate a vector $\theta^n_{1,h}$ for the Monte Carlo integration, then $\prod l(X_i|\theta_{i,h}, \gamma)$ will be for all computational purposes zero, unless X_i and $\theta_{i,h}$ have the same sign for all i. Thus, we only get a "useful" sample vector $\theta^n_{1,h}$ with probability 2^{-n}. If n is 10, only one out of about every 1,000 samples is useful, the rest are ignored. If n is 50, then only 1 out of about every 10^{15} samples is useful. This simple Monte Carlo integration method, therefore, fails in this example.

Although this example may look extreme, similar results could be achieved in more routine problems. Suppose that n equals 2 and that there was a 25% probability of drawing a useful vector $\theta^n_{1,h}$ where $\prod^n_{i=1} l(X_i|\theta_{i,h}, \gamma)$ was not negligible. Then for the same model, when n is increased to 50, we would get a useful sample vector in only about 1 out of every 10^{15} samples. This situation will likely occur when the prior (m^η) for θ_i is more spread out than the likelihood function. The only case where there will be no trouble is when there is both a strong prior and a strong likelihood and they both agree with each other.

The authors suggested computational methods suffer from similar difficulties as the simple Monte Carlo integration procedure. For example, the authors suggest a vector θ^n_1 be sampled from m^η and then functions like equation 4.5(i) be calculated, which have the form

$$l^*(x_D|t, \gamma) = \prod_{j \in D} l(x_j|\theta_i, \gamma).$$

This product is not over all n terms, however, the number of terms is the number of elements in the cluster and this could still be quite large. So, for example, if we think that there are 2 clusters, with n of 50, then we expect the product to be of about 25 terms. Therefore, instead of only 1 in every 10^{15} samples being useful, we have only 1 in every 33 million samples being useful.

The solution to efficient sampling is to sample the vector θ^n_1 conditional on the data. This could be done with a Gibbs sampler (see Escobar 1988, 1991) or through importance sampling (see Escobar, 1991).

To sample from the posterior distribution, $(\theta|X^n_1, \eta, \alpha, \gamma)$, with Gibbs sampler, we must specify the conditional distributions. The conditional distributions are:

$$(\theta_i|\theta_j, j \neq i, X^n_i, \alpha, \eta, \gamma) \sim q_0 f(\theta|X_i, \eta, \gamma) + \sum_{j \neq i} q_j \varepsilon_{\theta_j},$$

where $f(\theta|X_i, \eta, \gamma)$ is the posterior distribution of θ given X_i when m^η is the prior for θ, and where ε_{θ_j} is a point mass at θ_j. The proportions q_0 and the q_j's are chosen so that X_i is "likely" from θ_i. They are defined as:

$$q_0 \propto \alpha f(X_i|\eta, \gamma),$$

$$q_j \propto l(X_i|\theta_j, \gamma),$$

and where the sum of q_0 and the q_j's is equal to one. The function $f(X_i|\eta, \gamma)$ is the marginal for X_i when m^η is the prior for θ_i, and the function $l(X_i|\theta_j, \gamma)$ is the likelihood of X_i given θ_j.

When this method was used to estimate a vector of normal means, the algorithm was fairly fast. For n of 50, it took 1.5 minutes to calculate the 50 means on a DEC 3100. In West (1990), this algorithm was used to calculate density and distribution functions with n of 136. This author reports that it took 1 hour to calculate these functions on a DEC 3100; however, he states that calculating the density and distribution functions dominated the calculation times.

3. The α parameter. The α parameter effects the number of expected mixtures; therefore, if we are going to use a Dirichlet process prior to calculate the posterior probability of the number of different clusters, we must realize that the value of α used in the prior will have an effect.

To see this, look at the value of K_n used in the paper, and let us calculate the expected number of clusters conditional on α and n. (Assuming a diffuse m^η,)

$$E[K_n|\alpha, n] = \sum_{i=1}^{n} \frac{\alpha}{\alpha + i - 1} \cong \alpha \ln \left[\frac{\alpha + n}{\alpha} \right]$$

When this formula is evaluated, if α is about n^{-1}, then we expect only one cluster. If α is about n^2, then we expect about $n - \frac{1}{2}$ clusters. So the size of α gives strong prior information on the expected number of clusters.

If α is unknown, then we may want to do a hierarchical analysis. That is, we can put a prior on α, calculate the likelihood, and then apply Bayes theorem.

To make calculating the prior easier, we could make the prior discrete. To help elicit the prior, we could use the relationship between the prior and the expected number of clusters from the above formula. Since there is a logarithmic relation between n and the number of clusters, we could choose to have the discrete atoms of the prior for α to be something like n^{-1}, n^0, n^1, and n^2, or some other sequence of n^q with q spread out between -1 and 2. For a flat prior we would just put equal weights on these atoms.

To calculate the likelihood, we need to evaluate the following expression:

$$l(X_1^n|\alpha(s), \eta, \gamma) = \int \prod_{i=1}^{n} l(X_i|\theta_i, \gamma) m^{\eta, \alpha(s)}(d\theta)$$

Again, the way to evaluate this expresion is by Monte Carlo integration where we draw our Monte Carlo samples conditional on the data. I do not know how to analyse the above expression with the Gibbs sampler, but it can be done with an importance sampling algorithm which is similar in form to the Gibbs sampler. Sample an H-vector of $\theta_{1,h}^n$ by the following rule:

$$(\theta_{i,h}|\theta_{j,h}, j < i, \alpha(s), \eta, \gamma, x_1^n) \sim q_0 f(\theta|X_i, \eta, \gamma) + \sum_{j=1}^{i-1} q_j \varepsilon_{\theta_{j,k}},$$

where q_0 and the q_j's are defined as before. Then we estimate the likelihood, $\hat{f}(x_1^n|\alpha(s), \eta, \gamma)$ by:

$$\hat{f}(X_1^n|\alpha(s), \eta, \gamma) = \frac{1}{H} \sum_{h=1}^{H} \prod_{i=1}^{n} \left[\frac{\alpha(s) f(X_i|\eta, \gamma) + \sum_{j=1}^{i-1} l(X_i|\theta_{j,h}, \gamma)}{\alpha(s) + i - 1} \right].$$

REPLY TO THE DISCUSSION

Michael Escobar's comment is very welcome; it draws attention to an aspect of our paper, the complexity of which was somewhat overlooked, but which, nevertheless, is capable of attracting major interest, as shown all along in this meeting.

Clearly our final remark was not deemed to report actual experience of the authors but rather intended to draw attention to the necessity of checking consistency of numerical approximations. On this account, we find Escobar's comment useful as far as it "dualizes" the problem of numerical stability with that of statistical consistency and his proposal definitely improves the approximation we suggested.

In his Section 3, note that the approximation

$$E(K_n \mid \alpha, \eta) \approx \alpha \ln(\frac{\alpha + n}{\alpha})$$

may be dangerous when α is a function of n (it gives 0 instead of 1 with $\alpha = \frac{1}{n}$). Better approximations may be obtained through the digamma function $\psi(\cdot)$ since

$$E[K_n \mid \alpha, \eta] = \alpha[\psi(\alpha + n) - \psi(\alpha)].$$

Last, but not least, we would like to point out that in a sense the objective of our paper is preliminary to any numerical computation: the practical use of results on Bayesian exact estimability is to display functions of the parameters for which numerical procedures may possibly be meaningful.

ADDITIONAL REFERENCES IN THE DISCUSSION

Escobar, M. D. (1988). *Estimating the Means of Several Normal Populations by Nonparametric Estimation of the Distribution of the Means.* Ph.D. Thesis, Yale University.

Escobar, M. D. (1991). Estimating normal means with a Dirichlet process prior. *Tech. Rep.* **512**, Carnegie Mellon University.

BAYESIAN STATISTICS 4, pp. 147–167
J. M. Bernardo, J. O. Berger, A. P. Dawid and A. F. M. Smith, (Eds.)

Model Determination using Predictive Distributions with Implementation via Sampling-Based Methods

A. E. GELFAND, D. K. DEY and H. CHANG
University of Connecticut, USA

SUMMARY

Model determination is divided into the issues of model adequacy and model selection. Predictive distributions are used to address both issues. This seems natural since, typically, prediction is a primary purpose for the chosen model. A cross-validation viewpoint is argued for. In particular, for a given model, it is proposed to validate conditional predictive distributions arising from single point deletion against observed responses. Sampling based methods are used to carry out required calculations. An example investigates the adequacy of and rather subtle choice between two sigmoidal growth models of the same dimension.

Keywords: MODEL ADEQUACY; MODEL CHOICE; PREDICTIVE DISTRIBUTIONS; CROSS-VALIDATION; SAMPLING BASED METHODS; SIGMOIDAL GROWTH MODEL; LOGISTIC GROWTH CURVE MODEL; GOMPERTZ MODEL.

1. INTRODUCTION

Responsible data analysis must address the issue of model determination which divides into two components: model assessment or checking and model choice or selection. That is, apart from rare situations, the model specification is never "correct". Rather the questions are (i) is a given model adequate? (ii) within a collection of models under consideration, which model is the best choice? The questions are distinct. We need not envision a collection of models to address (i). Conversely, amongst a collection of models several may be adequate (or perhaps all may be inadequate) but we still seek the "best" one. Nonetheless, in practice the questions are typically investigated concurrently. This paper adopts a predictive viewpoint for answering them with resultant overlapping methodology.

The literature on model assessment and model choice is, by now, enormous. Of course, our modelling framework here is Bayesian whence our approach to these problems is as well. We restrict ourselves to parametric models where the amount of data is fixed and which are expressible in the form Likelihood × Prior where the Likelihood is an explicit readily evaluable function of both data and the parameters with the prior a readily evaluable function of the parameters.

Model adequacy is considered with regard to the form of the product. Box (1980) is persuasive on behalf of the predictive stance in arguing that the adequacy of the model can not be assessed from the posterior distribution of the model parameters. Berger (1985) observes that "Bayesians have long used [predictive distributions] to check assumptions." A few additional references are Jeffreys (1961), Box and Tiao (1973), Dempster (1975), Rubin (1984) and Geisser (1985).

Turning to model selection, we need to clarify our objective. For a fixed likelihood the prior may be varied. This case is typically viewed not as one of model selection but rather one of model robustness (see Berger (1984) for a review). More recent work is summarized in Gelfand and Dey (1991). In this work the estimative side using the resultant posterior is usually investigated with regard to sensitivity to such variation.

Rather, the usual business of model selection is the specification of the likelihood or joint distribution of the data. For instance, in the case of response models one needs to specify the parametric form of the error distribution (equivalently, a transformation of the response variable), a parametric form for the mean and a parametric form for the variance. (See in this regard the recent work of Carlin and Polson (1991) and George and McCulloch (1991). There is an implicit trade-off here (Smith, 1986), e.g., complex form for the mean with pure Gaussian error versus simple specification of the mean with say a mixture of Gaussians as error. Hence judicious choice of the collection of models entertained is essential.

Our discussion is concerned with model choice in this broad sense emphasizing that predictive distributions enable arbitrary model comparisons. Moreover the predictive viewpoint is typically consonant with the intended use of the model. In the special case where models under consideration are nested so that a subset of components of the parameter vector may be viewed as a discrepancy parameter measuring departure from a baseline model (Box 1980) model choice reduces to posterior inference for this subset of parameters. In the absence of nesting, comparison of posteriors between models typically conveys little information. The models need not share the same dimensionality. Moreoever, even if they do, from one to another, the parameters will have different interpretations.

Model determination is closely related to other data analytic issues such as residual analysis, influence measures and outlier detection. We offer no elaboration here but note the recent papers of Pettit and Smith (1985), Geisser (1987), Chaloner and Brant (1988), Verdinelli and Wasserman (1991) and Guttman (1991).

The entire proposed enterprise rests upon our ability to obtain desired predictive distributions and to calculate expectations under these distributions. Analytic evaluation of the required integrals is generally hopeless with effective approximations only available for simpler cases. To accomplish needed integrations we propose the use of sampling-based methods as discussed in Rubin (1988) and in Gelfand and Smith (1990). Such calculation is very computer-intensive, particularly when undertaking more complex and realistic modeling. However, the routine availability of enormous computing power makes this no obstacle. In particular this implies that many contending models may eventually be considered. As Geisser (1988) notes, "this could create havoc ... for those who adhere to stringent versions of... Bayesian approaches" However we feel that the *art* of data analysis should not be bound by overly formal inferential frameworks.

This paper is essentially a synthesis and extension of earlier work in Bayesian model determination. Our main contributions are the fleshing out of the cross-validation approach and the presentation of straightforward computing procedures to implement such cross-validation. In Section 2 we detail the proposed predictive approaches for model determination. In Section 3 we describe the Monte Carlo techniques for performing the required calculations. An illustrative example is discussed in Section 4 and we offer brief conclusions in Section 5.

2. PREDICTIVE APPROACHES FOR MODEL DETERMINATION

2.1 *Predictive Densities*

Box (1980) discusses the complementary roles of the predictive and posterior distributions in Bayesian data analysis noting that the posterior distribution provides a basis for "*estimation* of parameters conditional on the adequacy of the entertained model" while the predictive distribution enables "*criticism* of the entertained model in the light of current data." We concur noting further that in comparing models predictive distributions are directly comparable while posteriors are not. The predictive distribution (or marginal likelihood) is the joint marginal of the data. This distribution may be used in various ways to examine model determination questions. Our approach, which is argued for below, is a cross-validation one and leads to the examination of a collection of conditional distributions arising from this joint distribution.

We use customary notation in letting upper case letters denote random variables and lower case letters denote the observed realizations of these variables in our sample. In particular, the observed value of the rth response, Y_r, in our sample is y_r. Let $\boldsymbol{Y}$ denote the $n \times 1$ data vector, and let $\boldsymbol{Y}_{(r)}$ denote the $n-1 \times 1$ data vector with Y_r deleted. Let $X_{n \times p}$ denote the matrix of explanatory variables whose r^{th} row, X_r is associated with y_r. Let $X_{(r)}$ denote the matrix which is X with the rth row deleted. All densities for the data will be denoted by f, all densities for the parameters will be denoted by π with arguments providing clarification. More precisely for a given model let $\boldsymbol{\theta}$ denote the vector of model parameters. Then $f(\boldsymbol{Y}|\boldsymbol{\theta}, X)$ is the joint density of the data given $\boldsymbol{\theta}$ and $\pi(\boldsymbol{\theta})$ is the prior density for $\boldsymbol{\theta}$ whence $f(\boldsymbol{Y}|\boldsymbol{\theta}, X) \cdot \pi(\boldsymbol{\theta})$ is the model specification. Both $\boldsymbol{Y}$ and $\boldsymbol{\theta}$ are presumed continuous.

Assuming that the integral exists the predictive density is $f(\boldsymbol{Y}) = \int f(\boldsymbol{Y}|\boldsymbol{\theta}, X)\pi(\boldsymbol{\theta})d\boldsymbol{\theta}$. In model selection, interest focuses on $f(\boldsymbol{Y}|\boldsymbol{\theta}, X)$ and π is often chosen to be vague. But if π is improper then $f(\boldsymbol{Y})$ necessarily is , making it awkward to use in model checking. Note however, that

$$f(Y_r|\boldsymbol{Y}_{(r)}) = \frac{f(\boldsymbol{Y})}{f(\boldsymbol{Y}_{(r)})} = \int f(Y_r|\boldsymbol{\theta}, \boldsymbol{Y}_{(r)}, X)\pi(\boldsymbol{\theta}|\boldsymbol{Y}_{(r)})d\theta \tag{1}$$

is proper since $\pi(\boldsymbol{\theta}|\boldsymbol{Y}_{(r)})$ is. The density $f(Y_r|\boldsymbol{\theta}, \boldsymbol{Y}_{(r)}, X)$ is immediate if the Y_r are conditionally independent given θ as well as, for example, if $f(\boldsymbol{Y}|\boldsymbol{\theta}, X)$ is multivariate normal.

Suppose that $f(\boldsymbol{Y})$ is proper and strictly positive over its domain. Then $f(\boldsymbol{Y})$ is equivalent to the set $f(Y_r|\boldsymbol{Y}_{(r)}) : r = 1, \ldots, n$ in the sense that each uniquely determines the other (Besag, 1974). Hence, in terms of model assessment, examining the observed $\boldsymbol{y}$ with respect to $f(\boldsymbol{Y})$ is the same as with respect to the set of $f(Y_r|\boldsymbol{Y}_{(r)})$. It may be easier to work with with the latter distributions since each is univariate.

We briefly argue that, in checking against the observed y_r, $f(Y_r|\boldsymbol{Y}_{(r)})$ is the preferred univariate predictive distribution for Y_r. If $f(\boldsymbol{Y})$ is proper we could consider the univariate marginal $f(Y_r)$. Of course the $f(Y_r)$ do not determine $f(\boldsymbol{Y})$ but more importantly $f(Y_r)$ ignores the remaining observations, $\boldsymbol{y}_{(r)}$. In practice, were we to attempt prediction of a new, not necessarily independent, $Y_{(r)}$ at $X_{(r)}$ we would use a posterior distribution for $\boldsymbol{\theta}$ in creating the desired predictive distribution. In assessing the model this seems appropiate as well; we should check how well the model predicts in the manner in which we would use it to predict. But then should we use $f(Y_r|\boldsymbol{y})$ or $f(Y_r|\boldsymbol{y}_{(r)})$? The former is used for prediction at a new vector say X_0 but for validation at an X_r for which we have already observed y_r the latter seems preferable. That is, if we propose to check the predictive distribution for Y_r against an observed y_r we should not use that y_r to determine this distribution. Note that $f(Y_r|\boldsymbol{y}_{(r)})$ may be quite different from $f(Y_r|\boldsymbol{y})$ even in the case that the Y_r are conditionally independent given $\boldsymbol{\theta}$.

Such cross validation is well established in the Bayesian literature dating at least to Stone (1974) and Geisser (1975). Frequentist model diagnostic approaches adopt a similar point of view (see e.g., Belsley, Kuh and Welsch, 1980 or Cook and Weisberg, 1982). Cross-validation schemes other than single point deletion may be helpful and will share the same advantages described above. However, in the sequel we use $f(Y_r|\boldsymbol{y}_{(r)})$ exclusively.

2.2. *Model Adequacy*

The predictive distributions, $f(Y_r|\boldsymbol{y}_{(r)})$, are to be checked against y_r for $r = 1, 2, \ldots n$ in the sense that, if the model holds, y_r may be viewed as a random observation from $f(Y_r|\boldsymbol{y}_{(r)})$. To do this we consider $g(Y_r; y_r)$, called a checking function by Box(1980), whose expectation under $f(Y_r|y_{(r)})$ we will calculate and denote by d_r. The set of d_r will be used for model assessment. Computation of the d_r is discussed in Section 3. Once obtained the approach is exploratory. In fact, since each d_r is a function of the entire data vector $\boldsymbol{Y}$ they will be strongly dependent making formal inference very difficult. Our strategy is a Bayesian analogue to the well accepted frequentist strategy of examining studentized residuals, DFFITS, DFBETAS etc., (again see Belsley, Kuh and Welsch, 180, or Cook and Weisberg, 1982).

We will look at several choices of g. For example

(i) $g_1(Y_r; y_r) = y_r - Y_r$ yielding $d_{1r} \equiv y_r - \mu_r$ where $\mu_r = E(Y_r|\boldsymbol{y}_{(r)})$. The d_{1r} are natural deviations or residuals mentioned in Geisser (1987, p. 138). With $\sigma_r^2 = \text{var}(Y_r|\boldsymbol{y}_{(r)})$, standardizing yields $d'_{1r} = d_{1r}/\sigma_r$. The quantity $\sum(d'_{1r})^2$ could be used as an index of model fit. Many large $|d'_{1r}|$ cast doubt upon the model but retaining the sign of d'_{1r} allows patterns of under or over fitting to be revealed. If the $f(Y_r|y_{(r)})$ are assumed approximately normal then a normal plot of the d'_{1r} may be informative as well.

(ii) $g_2(Y_r; y_r) = 1_{A_r}(Y_r)$ where $A_r = (-\infty, y_r)$ yielding $d_{2r} = P(Y_r \leq y_r|\boldsymbol{y}_{(r)})$. Viewing y_r as a random observation from $f(Y_r|\boldsymbol{y}_{(r)})$ implies $d_{2r} \sim U(0,1)$. Because of the dependence amongst the d_{2r} it would be wrong to expect them to exhibit the spread associated with independent uniform samples. Nonetheless an adequate model should manifest d_{2r} which are roughly centered about ·5 without many extreme values. Evidence to the contrary calls the model to question.

(iii) $g_3(Y_r; y_r) = 1_{B_r}(Y_R)$ where $B_r = \{Y_r : f(Y_r|\boldsymbol{y}_{(r)}) \leq f(y_r|\boldsymbol{y}_{(r)})\}$ yielding $d_{3r} = P(B_r|\boldsymbol{y}_{(r)})$. Again viewing y_r as a random observation from $f(Y_r|\boldsymbol{y}_{(r)})$ implies $f(y_r\ |\boldsymbol{y}_{(r)})$ is, itself, a random realization of $f(Y_r|\boldsymbol{y}_{(r)})$ whence $d_{3r} \sim U(0,1)$. Again the d_{3r} should be roughly centered about ·5 without many extreme values. The d_{3r}, adapted from Box (1980) also appear in Geisser (1987). Note that Y_r need not be univariate in this definition: g_3 may be used for other cross-validation schemes. In fact, Box proposed use of g_3 to assess the entire joint predictive distribution. Unfortunately, calculation of this multivariate probability will generally be difficult. This same measure is referred to as the surprise index in Aitchison and Dunsmore (1975). Assuming that Y_r is univariate and that $f(Y_r|\boldsymbol{y}_{(r)})$ is unimodal, the d_{3r} calculate a set of tail areas. If we further assume that $f(Y_r|\boldsymbol{y}_{(r)})$ is approximately a normal density, then the event B_r is approximately the event $\{(Y_r - \mu_r)^2/\sigma_r^2 \geq (d'_{1r})^2\}$. Thus $d_{3r} \approx P(\chi_1^2 \geq (d'_{1r})^2)$ relating d_{1r} and d_{3r}. In retaining the sign of the deviation, d_{1r} is preferable to the induced d_{3r}.

(iv) $g_\epsilon(Y_r; y_r) = (2\epsilon)^{-1} 1_{C_r(\epsilon)}(Y_r)$ where $C_r(\epsilon) = \{Y_r : y_r - \epsilon \leq Y_r \leq y_r + \epsilon\}$ yielding $d_{4r}(\epsilon) = = (2\epsilon)^{-1} P\left(C_r(\epsilon)|\boldsymbol{y}_{(r)}\right)$. To avoid specification of ϵ we take the limit as $\epsilon \to 0$ obtaining $d_{4r} = f(y_r|\boldsymbol{y}_{(r)})$. This quantity dates at least to Geisser and Eddy (1979) and is computed in the definition of B_r. In the case of conditionally independent Y_r given $\boldsymbol{\theta}$, they suggest the use of $\prod_{r=1}^n d_{4r}$ as a modification of $f(\boldsymbol{y})$ for assessing comparative validity of models. We might call $\prod_{r=1}^n d_{4r}$ a pseudo-predictive distribution or pseudo-marginal likelihood. Many small d_{4r} criticize the model but it may not be obvious what a small d_{4r} is. Following an idea of Berger (1985, p. 201) we might instead consider a relative likelihood leading to a modified d_{4r} such as $d'_{4r} = d_{4r}/\sup_y f(y|\boldsymbol{y}_{(r)})$ or $d''_{4r} = d_{4r}/E(f(Y_r|\boldsymbol{y}_{(r)})|\boldsymbol{y}_{(r)})$.

2.3. *Model Choice*

The standard Bayesian approach for model selection goes as follows. Suppose there are J proposed models with model M_j denoted as $f(\boldsymbol{Y}|\boldsymbol{\theta}_j; X, M_j) \cdot \pi(\boldsymbol{\theta}_j)$. If w_j denotes the prior probability of M_j then, by Bayes Theorem, the posterior probability of M_j is

$$p(M_j|\boldsymbol{Y}) = f(\boldsymbol{Y}|M_j) \cdot w_j / \sum_{j=1}^{J} f(\boldsymbol{Y}|M_j) \cdot w_j \tag{2}$$

where $f(\boldsymbol{Y}|M_j)$ is the predictive or joint marginal distribution of the data under model M_j. For observed $\boldsymbol{y}$ the model yielding the largest $p(M_j|\boldsymbol{y})$ is selected. Calculation of (2) is discussed in Section 3. Geisser and Eddy (1979) suggest a cross-validation version replacing $f(\boldsymbol{Y}|M_j)$ in (2) by the pseudo-predictive distribution.

There is a fundamental complication engendered in this formalism which was recognized as early as Bartlett (1957) and elegantly clarified by Pericchi (1984). Some models are implicitly disadvantaged relative to others using this approach even under a state of presumed indifference towards the models, i.e., $w_j = 1/J, = 1, \ldots, J$. Section 2.3.1 further investigates this complication including an extension of Pericchi's remedy. An alternative remedy is considered in Section 2.3.2 also using cross-validation ideas.

Another criticism is that, in most practical situations, we doubt that anyone including Bayesians would select models in this fashion. One doesn't really believe that any of the proposed models are correct whence attaching a prior probabilty to an individual model's correctness seems silly. The selection process is typically evolutionary with comparisons often made in pairs until a satisfactory choice (in terms of both parsimony and performance) is made. Such pairwise decisions would be made using the Bayes factor, $f(\boldsymbol{Y}|M_1)/f(\boldsymbol{Y}|M_2)$. But, if at least one of the $\pi(\boldsymbol{\theta}_j)$ is vague, interpretation of this factor is problematic. A possible remedy is suggested in Spiegelhalter and Smith (1982) using a reserved or imaginary training data set. In Section 2.3.3 we suggest simpler selection criteria based on the ideas in Section 2.2 and in the spirit of Box (1980, p. 427).

2.3.1. Neutralizing Differential Expected Increase in Information

A simple example due to Bartlett (1957) reveals a difficulty with the Bayes factor and the standard procedure. Suppose under Model 1, that $Y_1, \ldots, Y_n$ are i.i.d. $N(0,1)$ while, under Model 2, $Y_1, \ldots, Y_n$ are i.i.d. $N(\theta, 1)$ with $\theta \sim N(0, r^2)$. Then regardless of the data $\boldsymbol{Y}$, of n and of w_1 as $\tau^2 \to \infty$, $f(\boldsymbol{Y}|M_1)/f(\boldsymbol{Y}|M_2) \to \infty$ and $p(M_1|\boldsymbol{Y}) \to 1$. This example was extended to more general nested normal models in Smith and Spiegelhalter (1980). Pericchi (1984) identified the source of the complication: for a given experiment the expected increase in information about the model parameters varies with the specification of the model. His remedy is to weight the Bayes factor or to revise the prior probabilities w_j to achieve neutral discrimination with regard to what is expected to be learned about $\boldsymbol{\theta}_j$ under model M_j

In particular, using the usual information entropy measure, the information in the prior is $-\int \pi(\boldsymbol{\theta}) \log \pi(\boldsymbol{\theta}) d\boldsymbol{\theta}$ (making the rather strong assumption that this integral exists), the information in the posterior is $-\int \pi(\boldsymbol{\theta}|\boldsymbol{Y}) \log \pi(\boldsymbol{\theta}|\boldsymbol{Y}) d\boldsymbol{\theta}$ whence the expected increase in information about $\boldsymbol{\theta}$ from the experiment (Lindley, 1956) is

$$I(f, \pi) = \int \left(\int \pi(\boldsymbol{\theta}|\boldsymbol{Y}) \log \pi(\boldsymbol{\theta}|\boldsymbol{Y}) d\boldsymbol{\theta} \right) f(\boldsymbol{Y}) d\boldsymbol{Y} - \int \pi(\boldsymbol{\theta}) \log \pi(\boldsymbol{\theta}) d\boldsymbol{\theta}.$$

For two models with $I(f_1, \pi_1), I(f_2, \pi_2)$ Pericchi (1984) proposes revision of w_1 to $w_1' = \exp(I(f_1, \pi_1))/\{\exp(I(f_1, \pi_1)) + \exp(I(f_2, \pi_2))\}$. Equivalently the Bayes factor would be multiplied by $\rho = w_1'/(1 - w_1')$.

Under the linear model $Y = X\boldsymbol{\theta} + \epsilon$ with $X_{n\times p}, \epsilon \sim N(0, \sigma^2 I), \sigma^2$ known and $\theta \sim N(\mu_\theta, \sigma^2 V_\theta), \mu_\theta, V_\theta$ known, it follows from Stone (1959) that

$$I(f,\pi) = \frac{1}{2}\log(|I + V_\theta X^T X|). \tag{3}$$

For the example in Section 4 we propose a model choice between two nonlinear models with normal errors. But if we replace the mean of $Y_r, X_r\boldsymbol{\theta}$, by $\varphi(X_r;\boldsymbol{\theta}), I(f,\pi)$ no longer has the closed form (3). However a first order appproximation may be readily obtained. Assuming $\partial\varphi/\partial\theta_i$ exists for all $\theta_i, i = 1, 2, \ldots, p$ we may write

$$\varphi(X_r;\boldsymbol{\theta}) = \varphi(X_r;\boldsymbol{\theta}_0) + \sum_{i=1}^{p}(\theta_i - \theta_{0i}) \cdot \left.\frac{\partial\varphi}{\partial\theta_i}\right|_{\boldsymbol{\theta}_0}$$

so that $Y_r' \approx \sum a_{ri}(\boldsymbol{X}) \cdot \theta_i + \epsilon$ where $Y_r' = Y_r - \varphi(X_r;\boldsymbol{\theta}_0) - \sum a_{ri}(\boldsymbol{X}) \cdot \theta_{0i}$ and $a_{ri}(\boldsymbol{X}) = \left.\frac{\partial\varphi(X_r;\boldsymbol{\theta})}{\partial\theta_i}\right|_{\boldsymbol{\theta}_0}$. Hence using (3) $I(f,\pi) \approx \frac{1}{2}\log(|I + V_\theta A^T A|)$ where A is an $n \times p$ matrix such that $A_{ri} = a_{ri}(\boldsymbol{X})$. A practical choice for $\boldsymbol{\theta}_0$ might be the MLE. For two nonlinear models the resultant weight $\rho = \{|I + V_{\boldsymbol{\theta}_1} A_1^T A_1| / |I + V_{\boldsymbol{\theta}_2} A_2^T A_2|\}^{\frac{1}{2}}$. We may pass to noninformative prior specifications by setting $V_{\boldsymbol{\theta}_1} = V_{\boldsymbol{\theta}_2} = V$ and letting $V^{-1} \to \emptyset$ whence $\rho = \{|A_1^T A_1| / |A_2^T A_2|\}^{\frac{1}{2}}$.

2.3.2. A Maximum Expected Utility Approach

An alternative remedy modifies examination of (2) by formulating the problem of model choice as one of maximizing expected utility. Several authors (Box and Hill, 1967; San Martini and Spezzaferri, 1984; Poskitt 1987) discuss such an approach. The crucial concept is the introduction of a utility functional to capture "the utility of a model given the data". Utility structures incorporating posterior distributions as an argument will have limited applicability for model choice since the parameter vector may be interpreted differently from model to model. Use of predictive distributions avoids this problem. San Martini and Spezzaferri (1984) take the utility $U(f(Y_0|\boldsymbol{Y}), y_0)$, of the predictive distribution at a future unobserved Y_0, where Y_0 is the true unknown future value. From an argument in Bernardo (1979) they recognize that the unique proper local utility function has the form $b_0 f(y_0|\boldsymbol{Y}) + b_1(y_0)$.

In chosing between two models the expected utility solution is to select the model yielding the larger expected utility. It turns out that, regardless of b_0 and $b_1(y_0)$, we choose $M_1(M_2)$ if $w_1 K(f(Y_0|\boldsymbol{y}, M_1),\ f(Y_0|\boldsymbol{y}, M_2)) > (<) w_2 K(f(Y_0|\boldsymbol{y}, M_2),\ f(Y_0|\boldsymbol{y}, M_1))$ where $K(f_1, f_2) = \int f_1 \log(f_1/f_2)$ denotes the Kullback-Leibler divergence between densities f_1 against f_2. This criterion is appealing since $K(f_1, f_2)$ is interpreted as the expected or average information for discriminating in favor of f_1 against f_2. In the cross validation context we replace $K(f(Y_0|\boldsymbol{Y}, M_j), f(Y_0|\boldsymbol{Y}, M_j'))$ with $\sum_{r=1}^{n} K(f(Y_r|\boldsymbol{y}_{(r)}, M_j), Y_r|\boldsymbol{y}_{(r)}, M_j'))$. This substitution arises by replacing $f(Y_0|\boldsymbol{y}, M_j)$ with the pseudo-predictive distribution, $\prod_{r=1}^{n} f(Y_r|\boldsymbol{y}_{(r)}, M_j)$, as discussed in (iv) of Section 2.2. An alternative form is to choose $M_1(M_2)$ if

$$E_{f^*}[\log\frac{\prod f(Y_r|\boldsymbol{y}_{(r)}, M_1)}{\prod f(Y_r|\boldsymbol{y}_{(r)}, M_2)}] > (<) 0, \tag{4}$$

where $f^* = w_1 \prod f(Y_r|\boldsymbol{y}_{(r)}, M_1) + w_2 \prod f(Y_r|\boldsymbol{y}_{(r)}, M_2)$.

Calculation of the Kullback-Leibler divergences is discussed in Section 3. Other information measures (see e.g. Csiszár, 1977) could be investigated as well. The expected utility approach is readily extended to $J > 2$ models.

2.3.3. *Ad hoc* Procedures

Each of the criteria developed in Section 2.2. may be converted to an *ad hoc* model choice procedure. Given two models, for $k = 1, 2, 3, 4$ associate $d_{kr}(M_j)$ with model $M_j, j = 1, 2$.

For $k = 1$ choose M_j with the smaller value of $D_{1j} = \sum(d'_{1r}(M_j))^2$
For $k = 2, 3$ choose M_j with the smaller value of $D_{kj} = \sum(d_{kr}(M_j) - \cdot 5)^2$
For $k = 4$ choose M_j with the larger value of $\prod d_{4r}(M_j)$. Equivalently choose $M_1(M_2)$ according to

$$D_4 = \log\left[\frac{\prod d_{4r}(M_1)}{\prod d_{4r}(M_2)}\right] > (<) 0 \qquad (5)$$

Expression (5) may be directly compared with (4). For the former we calculate *expected* utilities ; for the latter we calculate *observed* utilities. In fact $\exp(D_4)$ is a pseudo-Bayes factor. Given that the $d_{kr}(M_j)$ will have already been calculated for use in model assessment the additional computation for their use in these ad hoc model choice procedure is negligible.

3. COMPUTATIONAL APPROACHES

We propose the use of sampling-based methodology to calculate the various objects of interest in Section 2. Monte Carlo techniques have significantly advanced our ability to carry out integrations required for Bayesian inference. The literature for noniterative methods is substantial. We mention here the recent papers of Rubin (1988), Geweke (1989) and West (1990). The paper of Geweke provides many further references. Iterative approaches are discussed in Tanner and Wong (1987) and in Gelfand and Smith (1990).

3.1. *Monte Carlo Estimates of the* d_r

For a given model, computational effort focuses on the calculation of the

$$d_r = E(g(Y_r; y_r)|\boldsymbol{y}_{(r)}) = \int\int g(Y_r; y_r) f(Y_r|\boldsymbol{\theta}, \boldsymbol{y}_{(r)}, X) \cdot \pi(\boldsymbol{\theta}|\boldsymbol{y}_{(r)}) d\boldsymbol{\theta} dY_r.$$

If $(\boldsymbol{\theta}_s, Y_{rs}), s = 1, \ldots, B$ are samples from the joint conditional distribution for $\boldsymbol{\theta}$ and $Y_r, f(Y_r|\boldsymbol{\theta}, \boldsymbol{y}_{(r)}, X) \cdot \pi(\boldsymbol{\theta}|\boldsymbol{y}_{(r)})$ then a Monte Carlo approximation for d_r is $\hat{d}_r = B^{-1}\sum_{s=1}^{B} g(Y_{rs}; y_r)$. Sampling from $f(Y_r|\boldsymbol{\theta}, \boldsymbol{y}_{(r)}, X)$ is usually no problem; sampling from $\pi(\boldsymbol{\theta}|\boldsymbol{y}_{(r)})$ is. We return to this matter shortly.

If $\epsilon_r(\boldsymbol{\theta}; \boldsymbol{y}) = \int g(Y_r; y_r) f(Y_r|\boldsymbol{\theta}, \boldsymbol{y}_{(r)}, X) dY_r$ then $d_r = E(\epsilon_r(\boldsymbol{\theta}; \boldsymbol{y})|\boldsymbol{y}_{(r)})$, an expectation with respect to the posterior $\pi(\boldsymbol{\theta}|\boldsymbol{y}_{(r)})$. In certain cases $\epsilon_r(\boldsymbol{\theta}; \boldsymbol{y})$ can be calculated explicitly whence $\hat{d}_r = B^{-1}\sum_{s=1}^{B} \epsilon_r(\boldsymbol{\theta}_s; \boldsymbol{y})$. We need not draw the sample of Y_{rs}. Such savings in random variate generation is referred to as streamlining in Rubin (1988) and in Gelfand and Smith (1990). In fact, the estimate of the predictive density itself, $f(Y_r|\boldsymbol{y}_{(r)})$ requires only the $\boldsymbol{\theta}_s$ i.e. $\hat{f}(Y_r|\boldsymbol{y}_{(r)}) = B^{-1}\sum_{s=1}^{B} f(Y_r|\boldsymbol{\theta}_s, \boldsymbol{y}_{(r)}, X)$.

If $h(\boldsymbol{\theta})$ is an importance sampling density for $\pi(\boldsymbol{\theta}|\boldsymbol{y}_{(r)})$ and $\boldsymbol{\theta}_s, s = 1, \ldots, B$ are drawn from h, the above Monte Carlo estimates are modified to $\hat{d}_r = \sum_{s=1}^{B} g(Y_{rs}, y_r) \cdot v_{rs}$, or $\hat{d}_r = \sum_{s=1}^{B} \epsilon_r(\boldsymbol{\theta}_s; \boldsymbol{y}) \cdot v_{rs}$ and $\hat{f}(Y_r|\boldsymbol{y}_{(r)}) = \sum_{s=1}^{B} f(Y_r|\boldsymbol{\theta}_s, \boldsymbol{y}_{(r)}, X) \cdot v_{rs}$ respectively where $v_{rs} = [\sum_{s=1}^{B} \pi(\boldsymbol{\theta}_s|\boldsymbol{y}_{(r)})/h(\boldsymbol{\theta}_s)]^{-1} \cdot \pi(\boldsymbol{\theta}_s|\boldsymbol{y}_{(r)})/h(\boldsymbol{\theta}_s)$. As a related remark, if, for example, $f(Y_r|\boldsymbol{\theta}, \boldsymbol{y}_{(r)}, X)$ is a normal distribution then $\hat{f}(Y_r|\boldsymbol{y}_{(r)})$ is a finite mixture of normals. Theory developed in Johnson and Geisser (1983) shows that in such situations $f(Y_r|\boldsymbol{y}_{(r)})$ is exactly or approximately a t-distribution. But t-distributions arise as scale mixtures of normal distributions which can be arbitrarily well approximated by a finite mixture of normals.

Note that $\pi(\boldsymbol{\theta}|\boldsymbol{y}_{(r)}) \propto \tau(\boldsymbol{\theta})/f(y_r|\boldsymbol{\theta},\boldsymbol{y}_{(r)}),X)$ where $\tau(\boldsymbol{\theta}) = f(\boldsymbol{y}|\boldsymbol{\theta},X)\cdot\pi(\boldsymbol{\theta})$ so that

$$v_{rs} = \frac{\tau(\boldsymbol{\theta}_s)/(h(\boldsymbol{\theta}_s)\cdot f(y_r|\boldsymbol{\theta},\boldsymbol{y}_{(r)},X))}{\left[\sum_{s=1}^{B}\tau(\boldsymbol{\theta}_s)/(h(\boldsymbol{\theta}_s)\cdot f(y_r|\boldsymbol{\theta},\boldsymbol{y}_{(r)},X))\right]^{-1}}.$$

Rather than develop an $h(\boldsymbol{\theta})$ for each $\pi(\boldsymbol{\theta}|\boldsymbol{y}_{(r)})$ it would be more efficient to find a simple choice which we could sample and then use for all r. The form of v_{rs} suggests $h(\boldsymbol{\theta}) \propto \tau(\boldsymbol{\theta})$ i.e. $h(\boldsymbol{\theta}) = \pi(\boldsymbol{\theta}|\boldsymbol{y})$ would be a natural choice. We recall that the Gibbs sampler, as described in Gelfand and Smith (1990) for application to hierarchical Bayes models, produces observations essentially from the joint posterior $\pi(\boldsymbol{\theta}|\boldsymbol{y})$. Hence, if the Gibbs sampler is used to carry out Bayesian inference under the given model, the outputted $\boldsymbol{\theta}_s$ can be used directly as input to carry out computations needed for studying model adequacy and model choice. Implementation of the Gibbs sampler for challenging models will require tailored versions of the rejection method. See Carlin and Gelfand (1991), Gilks and Wild (1992), Ritter and Tanner (1991). If a noniterative approach has been employed, resulting in an importance sampling density $h(\boldsymbol{\theta})$ for $\tau(\boldsymbol{\theta})$, then the samples from $h(\boldsymbol{\theta})$ can, as well, be used directly in the above formulas. There is considerable literature on the creation of a good importance sampling density. In particular we note the recent work of Geweke (1989) and West (1990).

For model choice, additional calculations we may wish to make are the Kullback -Leibler divergences $K(f(Y_r|\boldsymbol{y}_{(r)},M_j), f(Y_r|\boldsymbol{y}_{(r)},M_j'))$. Since these are expectations with respect to $f(Y_r|\boldsymbol{y}_{(r)},M_j)$, in principle they can be handled in the same way as calculation of the $d_r(M_j)$. However, in practice, the calculation requires enormous storage, even if n is small, since these estimated f's are created under different models but must be merged to calculate K.

The standard approach to model choice requires updating of w_j to $p(M_j|\boldsymbol{y})$ as in (2). Except in simple cases the marginalization to obtain $f(\boldsymbol{Y}|M_j)$ is not available in closed form. Noniterative Monte Carlo integration may be developed as follows. If $\pi(\boldsymbol{\theta})$ is proper and $\boldsymbol{\theta}_{js}, s = 1,\ldots,B$ are a sample, $\hat{f}(\boldsymbol{Y}|M_j) = B^{-1}\sum_{s=1}^{B} f(\boldsymbol{Y}|\boldsymbol{\theta}_{js};X)$ If $\pi(\boldsymbol{\theta}_j)$ is improper but $h(\boldsymbol{\theta}_j)$ is an importance sampling density for $\tau(\boldsymbol{\theta}_j),\tau$ defined above, then $\hat{f}(\boldsymbol{Y}|M_j) = B^{-1}\sum_{s=1}^{B}\tau(\boldsymbol{\theta}_{js})/h(\boldsymbol{\theta}_{js})$.

Interestingly, the Gibbs sampler is less attactive here. By itself, it does not readily produce an estimator of $f(\boldsymbol{Y}|M_j)$. For a collection of J models it need not uniquely provide the posterior probabilities $p(M_j|\boldsymbol{Y})$ since Markovian updating using the $\boldsymbol{\theta}_j, j = 1,\ldots,J$ and a variable M to label the model may violate conditions for convergence (see e.g., George and McCulloch, 1991).

3.2. *Simplified Sampling for Nonlinear Models*

Suppose the model $Y_r = \varphi(X_r;\beta) + \epsilon_r, r = 1,\ldots,n$ where the vector of errors, $\epsilon \sim N(0,\sigma^2\cdot W)$, W known positive definite. With $\boldsymbol{\theta} = (\beta,\sigma^2)$ suppose $\pi(\boldsymbol{\theta}) = \pi_1(\beta)\cdot\pi_2(\sigma^2)$ where $\pi_2(\sigma^2)$ is inverse gamma $IG(a,b)$ i.e. $\pi_2(\sigma^2) \propto \exp(-b/\sigma^2)/(\sigma^2)^{a+1}$. We allow the improper limiting case as $a \to 0$ and as $b \to 0$. Then $\pi(\boldsymbol{\theta}|\boldsymbol{Y}) = \pi_1(\beta|\boldsymbol{Y})\cdot\ \pi_2(\sigma^2|\beta,\boldsymbol{Y})$ where $\pi_2(\sigma^2|\beta,\boldsymbol{Y})$ is $IG(a',b')$ with $a' = a + n/2, b' = b + \frac{1}{2}(\boldsymbol{Y}-\varphi)^T W^{-1}(\boldsymbol{Y}-\varphi), \varphi$ the vector of $\varphi(X_r;\beta)$ and $\pi_1(\beta|\boldsymbol{Y}) \propto \pi_1(\beta)/(b')^{a'}$.

Suppose, after transformation of β to domain R^p, that a noninformative prior is taken for β, i.e., $\pi_1(\beta) = 1$ (as in, e.g., Johnson and Geisser, 1983, p.138). If $f(X_r;\beta) = X_r\beta$ then, as is well known (see, e.g., Box and Tiao 1983, p.117), $\pi_1(\beta|\boldsymbol{Y})$ is exactly a multivariate student t-distribution and sampling-based approaches are not needed. For the nonlinear case let $\phi(\beta) = (\boldsymbol{Y} - \boldsymbol{\varphi})^T W^{-1}(\boldsymbol{Y} - \boldsymbol{\varphi})$ and let $\hat{\beta}$ be the MLE for β whence $\phi(\hat{\beta})$ is the error or residual sum of squares for the model. Assuming derivatives exist, to a second order approximation, $\phi(\beta) \approx \phi(\hat{\beta}) + \frac{1}{2}(\beta - \hat{\beta})^T H(\beta - \hat{\beta})$ where H has entries $H_{ut} = \left.\frac{\partial^2 \phi}{\partial \beta_u \partial \beta_t}\right|_{\hat{\beta}}$. H is, of course, proportional to the inverse of the sample Fisher information matrix. At the least, standard nonlinear regression software handles the independence case ($W = I$) and routinely provides $\hat{\beta}, \phi(\hat{\beta}), \hat{\sigma}^2 = \frac{\phi(\hat{\boldsymbol{\beta}})}{n-p}$ and $(H^*)^{-1} = 2\hat{\sigma}^2 H^{-1}$, the estimated asymptotic covariance matrix of $\boldsymbol{\beta}$. In $\pi_1(\beta|\boldsymbol{Y})$ replacing $\phi(\beta)$ by this approximation again yields a multivariate student t-distribution, say $t(\beta)$.

For noniterative Monte Carlo we immediately have a promising importance sampling density for $\pi(\boldsymbol{\theta}|\boldsymbol{Y})$ namely $t(\beta) \cdot \pi_2(\sigma^2|\boldsymbol{\beta}, \boldsymbol{Y})$. The work of van Dijk and Kloek (1985), Geweke (1989) and West (1990) suggest refinements to $t(\beta)$. Simplification occurs for the Gibbs sampler as well since it may be applied directly to $\pi_1(\beta|\boldsymbol{Y})$ using $t(\beta)$ as described in Carlin and Gelfand (1991). The resulting Gibbs replicates, say β_s, would then be used to sample σ_s^2 from $\pi_2(\sigma^2\ |\beta, \boldsymbol{Y})$ to obtain $\boldsymbol{\theta}_s$. The illustrative example in Section 4 is handled using noniterative Monte Carlo. Other models may admit similar conjugacies which ameliorate the computing burden.

4. AN ILLUSTRATIVE EXAMPLE

Our example compares two sigmoidal growth curve models of the same dimension. Consider the following data (Ratkowsky, 1983, p.88) recording as Y the dry weight of onion bulbs tops versus growing time X.

X	1	2	3	4	5	6	7	8	9
Y	16.08	33.83	65.80	97.20	191.55	326.20	386.87	520.53	590.03

X	10	11	12	13	14	15
Y	651.92	724.93	699.56	689.86	637.56	717.41

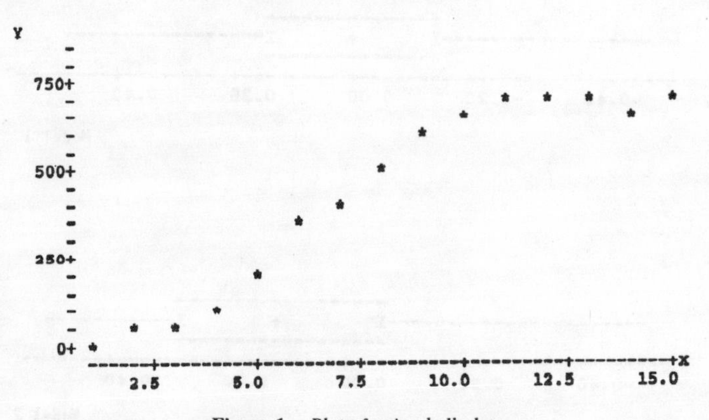

Figure 1. *Plot of onion bulb data.*

The data is plotted in Figure 1 suggesting sigmoidal behavior. We propose to investigate model adequacy for and choice between a logistic model,

$$Y_r = \beta_0(1 + \beta_1\beta_2^{X_r})^{-1} + \epsilon_r$$

and a Gompertz model,

$$Y_r = \beta_0 e^{-\beta_1 \beta_2^{X_r}} + \epsilon_r$$

where in either case we assume the ϵ_r iid $N(0,\sigma^2), r = 1,2,\ldots,15$. Under either model β_0 is interpreted as an asymptote while $\beta_2 \epsilon (0,1)$. We take $\beta_1 > 0$ to yield an increasing function of X. In both cases we reparametrize β to R^3 by setting $\beta_1' = \log \beta_1$, $\beta_2' = \log(\beta_2/(1-\beta_2))$ and then taking the prior $\pi(\beta_0, \beta_1', \beta_2', \sigma^2) = (\sigma^2)^{-1}$. In the notation of the previous section $\pi_1(\beta_0, \beta_1', \beta_2') = 1$ and $\pi_2(\sigma^2) = IG(0,0)$.

Logistic Model:

$$\hat{\beta}_0 = 702.876 \quad \hat{\beta}_1' = 4.454 \quad \hat{\beta}_2' = -0.008$$
$$\phi(\hat{\beta}) = 8913.991 \quad \hat{\sigma}^2 = 742.833$$

$$(H^*)^{-1} = \begin{bmatrix} 193.741 & -1.107 & -0.872 \\ -1.107 & 0.058 & 0.026 \\ -0.872 & 0.026 & 0.0133 \end{bmatrix}$$

Gompertz Model:

$$\hat{\beta}_0 = 723.053 \quad \hat{\beta}_1' = 2.502 \quad \hat{\beta}_2' = 0.564$$
$$\phi(\hat{\beta}) = 13616.000 \quad \hat{\sigma}^2 = 1134.667$$

$$(H^*)^{-1} = \begin{bmatrix} 486.053 & -3.842 & 2.311 \\ -3.842 & 0.0813 & 0.039 \\ 2.311 & 0.039 & 0.020 \end{bmatrix}$$

Table 1. *Maximum Likelihood Estimation for the Two Sigmoidal Growth Curve Models of Section 4.*

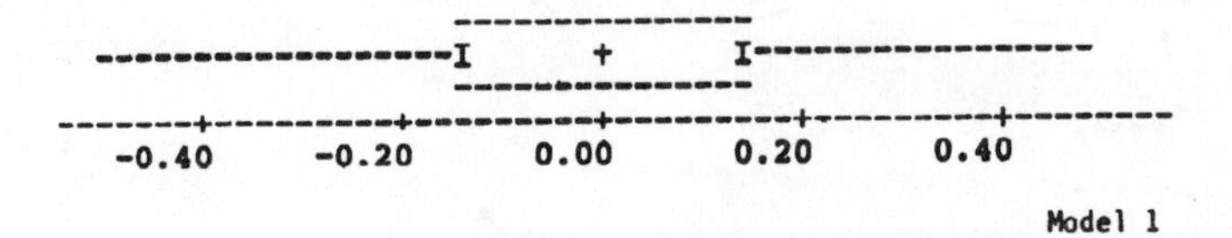

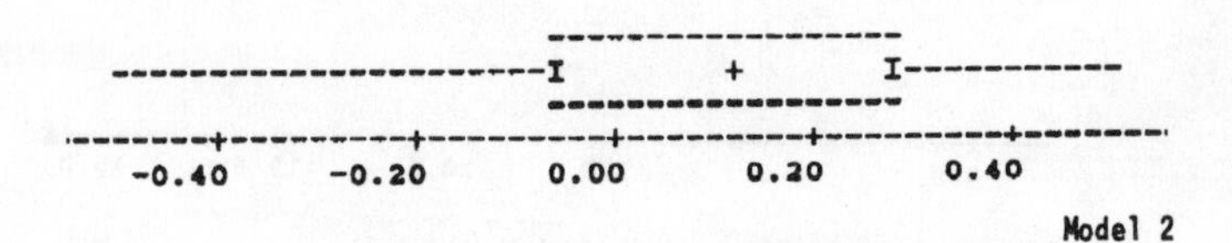

Figure 2. *Boxplots of* $d_{2r} - .5$ *for models 1 and 2.*

The results of a standard nonlinear regression fitting package (SAS PROC NLIN) for each model are given in Table 1. These estimates were used to obtain, for each model, a multivariate-t distribution which was then used as an importance sampling density for the noniterative Monte Carlo approach described in Section 3.2 with $B = 2000$. Table 2 provides the predictive means, $E(Y_r|y_{(r)})$, and the d_{ir} for each model.

Model 1

X_r	y_r	$E(Y_r\|y_{(r)})$	d'_{1r}	d_{2r}	d_{3r}	d_{4r}
1	16.08	15.88	0.0066	0.3846	0.9999	0.0364
2	33.83	31.00	0.0965	0.6125	0.7947	0.0361
3	65.80	59.19	0.2239	0.5091	0.8711	0.0349
4	97.20	110.26	−0.4347	0.3447	0.5418	0.0314
5	191.55	188.42	0.0967	0.5976	0.8754	0.0331
6	326.20	286.56	1.1858	0.8257	0.3467	0.0143
7	386.87	429.51	−1.3458	0.0963	0.2663	0.0110
8	520.53	524.68	−0.1244	0.4985	0.8740	0.0314
9	590.03	602.30	−0.3913	0.3181	0.6780	0.0304
10	651.92	647.43	0.1433	0.5822	0.8910	0.0336
11	724.93	665.16	2.2774	0.9832	0.0210	0.0028
12	699.56	687.16	0.3931	0.6559	0.6893	0.0304
13	689.96	697.74	−0.2425	0.3779	0.6734	0.0323
14	637.56	708.65	−2.9761	0.0012	0.0067	0.0006
15	717.41	698.73	0.5913	0.6784	0.5843	0.0266

Model 2

X_r	y_r	$E(Y_r\|y_{(r)})$	d'_{1r}	d_{2r}	d_{3r}	d_{4r}
1	16.08	0.38	0.4745	0.8105	0.4650	0.0280
2	33.83	5.40	0.8779	0.4938	0.6854	0.0205
3	65.80	30.25	1.1041	0.9247	0.1677	0.0158
4	97.20	97.22	−0.0007	0.6426	0.9840	0.0299
5	191.55	202.20	−0.2702	0.4809	0.9438	0.0275
6	326.20	314.97	0.2897	0.7406	0.8339	0.0258
7	386.87	436.20	−1.3458	0.0698	0.1539	0.0107
8	520.53	516.31	0.1157	0.4172	0.9202	0.0289
9	590.03	583.19	0.1916	0.7452	0.8931	0.0292
10	651.92	628.85	0.6543	0.8718	0.6351	0.0234
11	724.93	655.60	2.1501	0.9908	0.0207	0.0026
12	699.56	682.69	0.4533	0.5509	0.7938	0.0265
13	689.96	701.02	−0.2998	0.2456	0.5169	0.0276
14	637.56	717.28	−2.6886	0.0034	0.0038	0.0008
15	717.41	713.76	0.0947	0.6243	0.9778	0.0287

Table 2. *Predictive Means and d_{ir} for Models 1 and 2.*

Table 2 reveals that $X_r = 14$ and, to a lesser extent, $X_r = 11$ are troublesome points under both models. Plots of $d_{2r} vs X_r$ and $d_{4r} vs X_r$ for both models reveal no systematic patterns. For model 1 (logistic) $\overline{d}_2 = .4978, \overline{d}_3 = .6076$; for model 2 (Gompertz) $\overline{d}_2 = .5742, \overline{d}_3 = .5997$. For illustration, Figure 2 presents boxplots of $d_{2r} - .5$ for each model. Turning to the criteria of Section 2.3.3 we have $D_{11} = 18.27, D_{12} = 16.82; D_{21} =$

.9219, $D_{22} = 1.3160$; $D_{31} = 1.5657$, $D_{32} = 1.9487$ and $D_4 = 1.6863$. All told, both models seem to provide adequate fit with model 1 being preferable.

5. CONCLUSIONS

The predictive techniques proposed here for model checking and model choice are self-contained with respect to the experiment, accommodate both proper and improper priors, employ only univariate distributions and, using sampling-based methods are readily computed. The Monte Carlo tcchnology described here can be straightforwardly modified for use in other predictivist enterprises such as prediction of future observations, diagnostics for outlier/influential point detection (Johnson and Geisser, 1982, 1983) and optimal combination of models for prediction (Min ánd Zellner, 1990).

Methodology for effective model assessment and selection is available and implementable. As the art of Bayesian data analysis evolves and more challenging problems are tackled, judicious use of this methodology should become a standard component of the data analysis process.

ACKNOWLEDGEMENT

The first author's research was supported in part by NSF grant DMS 8918563.

REFERENCES

Aitchison, J. and Dunsmore, I. (1975). *Statistical Prediction Analysis,* Cambridge: University Press.

Bartlett, M. (1957). A comment on D. V. Lindley's statistical paradox, *Biometrika* **44**, 533–534.

Belsley, D. A., Kuh, E. and Welsch, R. E. (1980). *Regression Diagnostics*. New York: Wiley.

Berger, J. (1984). The robust Bayesian viewpoint. *Robustness of Bayesian Analysis* (J. Kadane, ed.), 63–124, Amsterdam: North-Holland.

Berger, J. (1985). *Statistical Decision Theory and Bayesian Analysis*. New York: Springer.

Bernardo, J. (1979). Expected information as expected utility. *Ann. Statist.* **7**, 686–690.

Besag, J. (1974). Spatial interaction and the statistical analysis of lattice systems. *J. Roy. Statist. Soc. B* **36**, 192–326, (with discussion).

Box, G. (1980). Sampling and Bayes' inference in scientific modeling and robustness. *J. Roy. Statist. Soc. A* **143**, 382–430, (with discussion).

Box, G. and Hill, W. (1967). Discrimination among mechanistic models. *Technometrics* **9**, 57–71.

Box, G. and Tiao, G. (1973). *Bayesian Inference in Statisical Analysis*. Reading, MA: Addison-Wesley.

Carlin, B. and Gelfand, A. (1991). An iterative Monte Carlo method for nonconjugate Bayesian analysis. *Statistics and Computing* **1**, 119–128.

Carlin, B. and Polson, N. (1991). Inference for nonconjugate Bayesian modeling using the Gibbs sampler. *Canadian Journal of Statistics* **1**, 119–128.

Cook, R. D. and Weisberg, S. (1982). *Residuals and Influence in Regression.* London: Chapman and Hall.

Csiszár, I. (1977). Information measures: a critical survey, *Transactions of the 7th Prague Conference on Information Theory, Statistical Decision Functions and Random Processes*. Boston, MA: Reidel.

Dempster, A. (1975). A subjective look at robustness. *Internat. Statist. Rev.* **46**, 349–374.

Geisser, S. (1975). The predictive sample reuse method with application. *J. Amer. Statist. Assoc.* **70**, 320–328.

Geisser, S. (1985). On the prediction of observables: a selective update. *Bayesian Statistics 2* (J. M. Bernardo, M. H. DeGroot, D. V. Lindley and A. F. M. Smith, eds.), Amsterdam: North-Holland, 203–230.

Geisser, S. (1987). Influential observations, diagnostics and discordancy test. *Appl. Statist.* **14**, 133–142.

Geisser, S. (1988). The future of statistics in retrospect. *Bayesian Statistics 3* (J. M. Bernardo, M. H. DeGroot, D. V. Lindley and A. F. M. Smith, eds.), Oxford: University Press, 147–158, (with discussion).

Geisser, S. and Eddy, W. (1979). A predictive approach to model selection. *J. Amer. Statist. Assoc.* **74**, 153–160.

Gelfand, A. E. and Dey, D. K. (1991). On measuring Bayesian robustness of contaminated classes of priors. *Statistics and Decisions* **9**, 63–80.

Gelfand, A. E. and Smith A. F. M. (1990). Sampling based approaches to calculating marginal densities. *J. Amer. Statist. Assoc.* **85**, 398–409.

George, E. and McCulloch, R. (1991). Variable selection via Gibbs sampling. *Tech. Rep.* University of Chicago.

Geweke, J. (1989). Bayesian inference in econometric models using Monte Carlo integration. *Econometrica* **57**, 1317–1339.

Gilks, W. and Wild, P. (1992). Adaptive rejection sampling for Gibbs sampling. *Appl. Statist.* (to appear).

Guttman, I. (1991). A Bayesian look at the question of diagnostics. *Tech. Rep.* **9104**, University of Toronto.

Jeffreys, H. (1961). *Theory of Probability.* Oxford: University Press.

Johnson, W. and Geisser, S. (1982). Assessing the predictive influence of observations. *Statistics and Probability in Honor of C. R. Rao.* (Kallianpur, Krishnaiah and Ghosh, eds.) Amsterdam: North-Holland, 343–358.

Johnson, W. and Geisser, S. (1983). A predictive view of the detection and characterization of influential observations in regression analysis. *J. Amer. Statist. Assoc.* **78**, 167–144.

Lindley, D. (1956). On a measure of information provided by an experiment. *Ann. Math. Statist.* **78**, 137–144.

Min C. and Zellner, A. (1990). Bayesian and non-Bayesian methods for combining models and forecasts with applications to forecasting international growth rates. *Tech. Rep.* University of Chicago.

Pericchi, L. (1984). An alternative to the standard Bayesian procedure for discrimination between normal linear models. *Biometrika* **71**, 576–586.

Pettit, L. I. and Smith, A. F. M. (1985). Outliers and influential observations in linear models. *Bayesian Statistics 2* (J. M. Bernardo, M. H. DeGroot, D. V. Lindley and A. F. M. Smith, eds.), Amsterdam: North-Holland, 473–494, (with discussion).

Poskitt, D. (1987). Precision, complexity and Bayesian model determination. *J. Roy. Statist. Soc. B* **49**, 199–208.

Ratkowsky, D. A. (1983). *Nonlinear Regression Modeling: A Unified Practical Approach.* New York: Marcel Dekker.

Ritter, C. and Tanner, M. (1991). Facilitating the Gibbs Sampler: the Gibbs stopper and the griddy Gibbs sampler. *Tech. Rep.* University of Wisconsin.

Rubin, D. (1984). Bayesianly justifiable and relevant frequency calculations for the applied statistician. *Ann. Statist.* **12**, 1151–1172.

Rubin, D. (1988). Using the SIR algorithm to simulate posterior distribution. *Bayesian Statistics 3* (J. M. Bernardo, M. H. DeGroot, D. V. Lindley and A. F. M. Smith, eds.), Oxford: University Press, 395–402, (with discussion).

San Martini, A. and Spezzaferri, F. (1984). A predictive model selective criterion. *The Statistician* **35**, 97–102.

Smith, A. F. M. (1986). Some Bayesian thoughts on modeling and model choice. *The Statistician* **35**, 97–102.

Smith, A. F. M. and Spiegelhalter, D. (1980). Bayes factors and choice criteria for linear models. *J. Roy. Statist. Soc. B* **42**, 213–220.

Spiegelhalter, D. and Smith, A. F. M. (1982). Bayes factors for linear and log-linear models with vague prior information. *J. Roy. Statist. Soc. B* **44**, 377–387.

Stone, M. (1959). Application of a measure of information to the design and comparison of regression experiments. *Ann. Math. Statist.* **39**, 55–72.

Stone, M. (1974). Cross-validatory choice and assessment of statistical predictions. *J. Roy. Statist. Soc. B* **36**, 111–147.

Tanner, M. and Wong, W. (1987). The calculation of posterior distributions by data augmentation. *J. Amer. Statist. Assoc.* **82**, 528–550, (with discussion).

Van Dijk, H. and Kloek, T. (1985). Experiments with some alternatives for simple importance sampling in Monte Carlo integration. *Bayesian Statistics 2* (J. M. Bernardo, M. H. DeGroot, D. V. Lindley and A. F. M. Smith, eds.), Amsterdam: North-Holland, 511–530, (with discussion).

Verdinelli, I. and Wasserman, L. (1991) Bayesian analysis of outlier problems using the Gibbs sampler. *Statistics and Computing* **1**, 135–139.

West, M. (1990). Bayesian computations: Monte Carlo density estimation. *Tech. Rep.* ISDS, Duke University.

DISCUSSION

A. E. RAFTERY (*University of Washington, USA*)

1. Introduction and Summary. It is a pleasure to congratulate the authors on an interesting and important paper that points out how sampling-based methods can make Bayesian diagnostics for model checking routinely available. Bayesian diagnostics are often similar to frequentist ones, but they have the great advantage of being systematically available through the predictive distribution, even for complex models. This is in contrast with frequentist diagnostics, which have to be developed from scratch for each new class of models, often requiring considerable ingenuity. The *interpretation* of Bayesian diagnostics is somewhat glossed over by the authors, however.

We part company to some extent on the issue of model choice. I am unconvinced by the arguments against the standard Bayesian procedure, namely the one based on posterior model probabilities. New results indicate that posterior model probabilities *can* be readily computed using sampling-based methods. Also, the standard Bayesian procedure *is* based on predictive distributions, in a prequential rather than a cross-validation sense.

2. Bayesian diagnostics for model checking. A real achievement of this paper is to show how sampling-based methods can be used to obtain Bayesian diagnostics systematically and routinely for a very wide class of models. When frequentist diagnostics are available they are often similar to Bayesian diagnostics. The great advantage of Bayesian diagnostics is that they are available quite generally from the predictive distribution, unlike their frequentist counterparts, which can require considerable ingenuity for each new class of models.

The authors have, however, rather glossed over the *interpretation* of their diagnostics. For example, in the nonlinear regression example, they conclude that points 11 and 14 are troublesome but that, all told, both models provide an adequate fit. What is the basis for this conclusion? Nothing is suggested beyond eyeballing the results, but there are certainly more precise criteria implicitly at work here, and they should be made explicit.

I would suggest that diagnostics not be used to reject the current model, but rather to guide the search for better models by indicating the direction of search, or the way in which the current model is inadequate. If this leads to the specification of an alternative model, then the current model can be compared with the alternative one using the posterior odds ratio (or posterior expected utilities if these can be specified); the current model will not be rejected unless the alternative one is decisively preferred. You don't abandon a model unless you have a better one in hand.

Even viewing diagnostics this way, as an exploratory tool rather than as a basis for inference, we still need some yardstick to calibrate our inspection of the results. Here it does seem that frequentist calculations are useful, and I suspect that such calculations implicitly underly the authors' interpretation of the results in their Table 2.

3. Model comparison: In support of the standard Bayesian procedure. The standard Bayesian procedure is given by the authors' equation (3), and amounts to basing inference on the posterior model probabilities. They raise two objections to this procedure which I will now briefly discuss.

3.1. "Bartlett's paradox". This is the observation due to Bartlett (1957) that if under M_1 the Y_i are i.i.d. $N(0,1)$, and under M_2 they are i.i.d. $N(\theta,1)$ with $\theta \sim N(0,\tau^2)$, then $p(M_1|Y) \to 1$ as $\tau^2 \to \infty$ regardless of the data; see the authors' Section 2.3.1.

This has been presented by the authors and by others that they cite as a major flaw of the standard Bayesian approach, but I do not find it too disquieting. Letting $\tau^2 \to \infty$ implies that $E[|\theta|]$ also becomes arbitrarily large, so it is not too surprising that, for any data set, $E[|\theta|]$

can be set large enough that the data prefer zero. Some prior information is almost always available that will limit the prior variance τ^2, and it is always important to investigate the sensitivity of $p(M_1|Y)$ to changes in τ^2. In practice, $p(M_1|Y)$ tends to be rather insensitive to changes in τ^2 over a wide range (see, e.g., Raftery, 1988). Thus, Bartlett's paradox seems to me to suggest that the use of highly diffuse points is not a good idea for model comparison.

It may be objected that it is desirable to have a "reference" procedure for model comparison. However, in my applied experience, reasonable proper priors are often readily accepted, especially when backed up with a serious sensitivity analysis; the likelihood is often the more controversial part of the analysis.

3.2. *The more serious criticism.* The authors write:

> "Another criticism is that, in most practical situations, we doubt that anyone including Bayesians would select models in this fashion. One doesn't really believe that any of the proposed models are correct whence attaching a prior probability to an individual model's correctness seems silly. The selection process is typically evolutionary with comparisons often made in pairs until a satisfactory choice (in terms of both parsimony and performance) is made."

Attaching a prior probability to a model is not any sillier than science as traditionally practised. Most of science is an attempt to find a model that predicts the observations to date well; it does not claim to have found the "truth" (if such a thing exists) or the "correct model". Science typically proceeds by adopting a *paradigm*, which means essentially *conditioning* on a collection of models, often with an explicit parametric form. Prior probabilities conditional on the adopted paradigm, or collection of models, do make sense.

Of course, if one does not so condition, the prior probability, and hence also the posterior probability of most models is zero. Since one does not believe the paradigm to be the "truth", this may make science as a whole seem silly, but its record of success argues in its favor. Note that the marginal likekihood, $f(Y|M_j)$, which is proportional to the posterior probability of M_j, is just the (predictive) probability of the data given the model M_j, and so is precisely the right quantity for evaluating the scientific theory defined by M_j.

Consider, for example, the question of whether smoking causes lung cancer, and suppose that the currently accepted way of addressing this issue is within the framework of the logistic regression model, logit(Pr[lung cancer]) = γl[smokes] + $\beta^T x$, where x is a vector of control variables. Conditionally on this framework (or "paradigm"), the issue becomes a comparison of the two models $M_1 : \gamma = 0$ and $M_2 : \gamma > 0$. Then a scientist's natural language statement "I am 90% sure that smoking causes lung cancer" is equivalent, given the framework, to the statement that $p(M_1) = 0.1$ and $p(M_2) = 0.9$ This does seem to make sense even if, unconditionally on the framework, $p(M_1) = p(M_2) = 0$. Of course, the natural language statement itself can be viewed as not being about "truth", but rather about future data and trends in scientific opinion. It might mean, for example, "I am 90% sure that future data will be better predicted by M_2 than by M_1", or "I am 90% sure than within T years the belief that smoking causes cancer will be generally accepted"; note that the latter two statements can be given standard betting interpretations. For an example where scientists might attach substantial prior probability to the smaller ("null") model, consider cold fusion.

The authors describe the standard Bayesian procedure as a model *selection* procedure, but it is considerably richer than that. When comparing two models that genuinely represent rival scientific hypotheses, the posterior odds ratio provides a summary of the evidence for one model against the other; unless the evidence is very strong, one model will not necessarily be selected.

Often, however, model form is not the object of primary scientific interest. The authors did not say what the main scientific question was in their growth curve example, but I suspect that it was not the choice between the two models that they considered. If interest focuses instead on some other quantity, Δ, such as the next observation, Y_{16}, or the asymptote, β_0, then *model selection is a false problem*, and it is important to take account of model uncertainty. The Bayesian approach provides an immediate way of doing this using the equation.

$$p(\Delta|Y) = \sum_{j=1}^{J} p(\Delta|Y, M_j)p(M_j|Y). \tag{1}$$

Hodges (1987) emphasized the importance of taking account of model uncertainty, pointing out that failure to do so leads to the overall uncertainty being underestimated, and hence, for example, to overly risky decisions.

If the posterior probability of one of the models is close to unity, or if the posterior distribution of Δ is almost the same for the models that account for most of the posterior probability, then, then $p(\Delta|Y)$ may be approximated by conditioning on a single model, namely by $p(\Delta|Y, M_i)$ for some i. This seems to be the main situation in which model selection, as such is a valid exercise. The "evolutionary" process to which the authors refer is in reality an informal search method for finding the main models that contribute to the sum in equation (1), and in this sense may be viewed as an approximation to the full (standard) Bayesian procedure. Clearer recognition of this might lead to more satisfactory model search strategies.

4. The standard Bayesian procedure and sampling-based methods. The key quantity for the implementation of the standard Bayesian procedure is the marginal likelihood, $f(Y|M_j) = \int f(Y|\boldsymbol{\theta}_j, X, M_j)\pi(\boldsymbol{\theta}_j)d\boldsymbol{\theta}_j$. The authors say that the Gibbs sampler does not readily produce an estimator of $f(Y|M_j)$. However, Newton and Raftery (1991) have recently pointed out the existence of a simple and general such estimator. They show that, given a sample from the posterior, *the marginal likelihood may be (simulation-consistently) estimated by the harmonic mean of the associated likelihood values.* This result applies no matter how the sample was obtained, whether directly using the analytic form of the posterior,by importance sampling, the Gibbs sampler, the SIR algorithm or the weighted likelihood bootstrap. There can be stability problems with this estimator, and slight modifications that avoid these are discussed in the cited reference.

The standard Bayesian procedure *is* a predictive approach since the marginal likelihood can be written

$$f(Y|M_j) = \prod_{r=1}^{n} f(Y_r|Y^{r-1}, M_j), \tag{2}$$

where $Y^{r-1} = (Y_1, \ldots, Y_{r-1})$. Note that the conditional densities on the right-hand side of equation (2) are conditional on the first $(r-1)$ observations, and *not* on all the other $(n-1)$ observations. Thus the standard Bayesian procedure is a "prequential" method in the sense of Dawid (1984), and not a cross-validation approach. Each conditional density on the right-hand side of equation (2) may be evaluated in a sampling-based way, using the same methods as the authors propose for their d_{4r}. It follows that this provides an alternative sampling-based way of calculating the marginal likelihood, and hence of implementing the standard Bayesian procedure.

Note also that equation (2) remains valid even if the observations are permuted. Thus, even if the model does not impose a natural ordering on the observations, "prequential diagnostics" may be obtained by sampling from the set of all permutations of the observations and averaging over diagnostics based on the conditional densities on the right-hand of equation (2).

If one replaces the conditional densities on the right-hand side of equation (2) by densities conditional on all the observations except the r-th one, one obtains the quantity that the authors denote by $D_4 = \prod_{r=1}^{n} d_{4r}$. This could be called a "pseudo-marginal likelihood", by analogy with the pseudo-likelihood concept introduced by Besag (1975). Using D_4 rather than $f(Y|M_j)$ is similar to using the pseudo-likelihood rather than the likelihood when the latter is available, which does not seem to be a very good choice. As an argument in favor of D_4, however, the authors point out that with improper priors D_4 is defined whereas $f(Y|M_j)$ is not. This strikes me as a disadvantage of improper priors rather than of the standard marginal likelihood.

I will attempt to summarize the various analogies and equivalences discussed in the following table.

Prequential analysis	Cross-validation
Likelihood	Pseudo-likelihood
Marginal likelihood ($f(Y\|M_j)$)	"Pseudo-marginal likelihood" "likelihood"(D_4)
Posterior model probability/ Bayes factor	Fixed-level significance test
BIC (Schwarz, 1978)	AIC, C_p

Entries in the same column are regarded as being related, either by being motivated by the same appproach or by being asymptotically equivalent. Entries in the same row are viewed as different approaches to the same task or concept. I prefer the entries in the left-hand column, headed "Prequential analysis", while the authors seem to incline to the entries in the right-hand column. Note that the difference can be important, especially with larger samples.

D. PEÑA (*Universidad Carlos III de Madrid, Spain*)

This paper contains three features that I really like: (1) it stresses the importance of model determination and model diagnosis in statistical analysis; (2) it advocates a cross-validatory assessment of the model using the predictive distribution; (3) it points out the difficulties of using a naive Bayesian analysis of model choice.

Model diagnosis has received an increasing interest in the Bayesian literature and I would like to add the works by Zellner (1975), Johnson and Geisser (1985) and Guttman and Peña (1988) to the references given in the paper.

The information about the model adequacy is contained in the joint predictive distribution $f(y)$ and a key problem is to devise simple procedures to reveal this information. A sensible first step is the one used in this paper. Using

$$f(y) = f(y_i|y_{(i)})f(y_{(i)})$$

we can look at the cross-validatory predictive distribution $f(y_i|y_{(i)})$ that is unidimensional. However, this procedure, although very useful, does not show some of the interesting multivariate features of the data, for instance, sets of similar points that are different from the rest of the data and which cannot be identified by univariate analysis because of masking. Also a set of outliers can be responsible for some other good points to appear as outlying, this situation has been called swamping in the statistical literature. To avoid these problems, we need to consider either the whole predictive density or the distributions $f(y_I|y_{(I)})$ and $f(y_{(I)})$, where y_I is a subset of observations. Of course, looking at all the possible decompositions of the data is an impossible task and we need to develop procedures to search for interesting combinations of points. Peña and Tiao (1992) may be a first step on this direction.

As far as the computation of $f(y_I|y_{(I)})$ is concerned it should be pointed out that the easiest way to understand its structure is to use:

$$f(y_I|y_{(I)}) = \int f(y_I|\theta, y_{(I)}) f(\theta|y_{(I)}) d\theta \tag{1}$$

instead of $f(y_I)/f(y_{(I)})$. The reason is that (1) is similar to the standard marginalization to compute the predictive, and therefore, standard techniques can be applied to obtain the distribution in a compact way. For instance, if $y \sim N(\mu, \sigma^2)$ with σ^2 known, and $\mu \sim N(\mu_0, \sigma_0^2)$, it is straightforward to show that

$$f(y_I|y_{(I)}) = \left(\frac{1}{\sqrt{2\pi}}\right)^I \sigma^{-(I-1)} (I\sigma_{(I)}^2 + \sigma^2)^{-\frac{1}{2}} \exp\left[\frac{1}{2}\left[\frac{S_I^2}{\sigma^2} + \frac{(\overline{y}_I - \hat{\theta}_{(I)})^2}{I\sigma_{(I)}^2 + \sigma^2}\right]\right]$$

where

$$\hat{\theta}_{(I)} = \frac{(n-I)\overline{y}_{n-I}\sigma^{-2} + \mu_0\sigma_0^{-2}}{(n-I)\sigma^{-2} + \sigma_0^{-2}}$$

and

$$\sigma_{(I)}^2 = ((n-I)\sigma^{-2} + \sigma_0^{-2})^{-1}$$

Therefore, the cross-validation predictive density for the subset I depends on the ratio $s_I^2 = \sum_{i \epsilon I}(y_i - \overline{y}_I)^2/(I\sigma^2)$ that is a key factor in the analysis of this subset.

L. R. PERICCHI (*Universidad Simón Bolívar, Venezuela*)

It would be a promising theoretical exercise to investigate the relationship between your interesting suggestions for selecting a model and the dimension of the model. As an extreme case, one may think of a model that encompasses both models 1 and 2 in your illustrative example, and then compare with models 1 and 2. Which model would your suggestions select?

L. I. PETTIT (*Goldsmiths' College London, UK*)

I would firstly like to comment on the measures of model adequacy discussed in Section 2.2 and fill in some of their history.

The possibility of using d_{1r} and other similar Bayesian residuals was discussed by Pettit (1986) and Geisser (1987). Chaloner and Brant (1988) suggest a different definition of Bayesian residuals using an idea going back to Zellner (1975). Geisser (1987) also discusses the use of d_{3r} which he describes as a discordancy ranking. The quantity d_{4r}, usually called the conditional predictive ordinate (CPO), was proposed by Geisser (1980) and used by Pettit and Smith (1983, 1985) and Pettit (1988) as a tool in outlier modelling. Pettit (1990) presents a number of results about the CPO for the normal distribution. The quantity $d_{4r'}$, is called the ratio ordinate measure by Pettit (1990). I think the idea of comparing a predictive distribution to its mode goes back to Roberts (1965).

As far as model choice goes, I have found the use of Bayes factors, which do not require a prior probability of an individual model's correctness (Section 2.3), to be very useful. The approach of Spiegelhalter and Smith (1982) to the problem of improper priors is important. Measuring the effect of individual observations on Bayes factors (Pettit and Young, 1990) leads to an expression which is the difference in logarithms of the CPO's for the two models and so ties in with the model adequacy ideas.

The computational methods discussed in this paper will be very useful in calculating all these quantities and it is therefore for me a very welcome paper.

F. SEILLIER-MOISEIWITSCH (*University of North Carolina, USA*)

The fundamental, and welcomed, stance of this paper is the shift of emphasis, in model determination, from parameters to observables: goodness-of-fit criteria are abandoned in favour of an assessment based on the model's ability to produce decent predictions. The selected model will indeed often be used on new observables.

Several questions arise regarding the checking functions the authors adopted. Which one did they find most useful? In particular, g_1 focuses on a single characteristic of the predictive distribution whereas g_3 and g_ϵ take into account the whole distribution function. Situations where the former is more informative are likely to be few. For model comparison, have they found the difference in logarithmic scores more revealing than looking at the difference in other scores?

Scoring rules can also be of use in checking the adequacy of a single model. The results of Seillier-Moiseiwitsch and Dawid (1992), developed for discrete outcomes and bounded scoring rules, can be adapted for continuous variables. These results assume a natural ordering in the realizations. The probability range can be partitioned into a fixed number of bins and the probabilities, under the predictive distribution, that the observable falls in each of these bins can be compared to the indicators of the realized Bernoulli process. The score constructed over a number of outcomes, once normalized, behaves asymptotically like a $N(0,1)$ random variable. The normalization is carried out with respect to the predictive distribution at each point.

The authors mention the difficulty in drawing formal inference from $\{d_r\}$ due to the strong dependency among these. Transformations that yield independent random variables could be considered. For instance, by conditioning on sufficient statistics in the probability integral transform, one is provided with a set of i.i.d. residuals (O'Reilly and Quesenberry, 1973). Let $\tilde{F}_n(y) = F(y|T_n)$ where T_n is a minimal sufficient statistic, $\{\tilde{F}_n(Y_i)\}$ are i.i.d. uniforms on $[0,1]$. Furthermore, $\tilde{F}_n(Y_1), \tilde{F}_n(Y_2|Y_1), \ldots, \tilde{F}_n(Y_\alpha|Y_1, \ldots, Y_{\alpha-1})$ also generate a set of α i.i.d. $U[0,1]$ random variables, where α is the number of components in the vector of minimally sufficient statistics. This conditional transform fits particularly well in a sequential sampling framework (Seillier-Moiseiwitsch, 1990). Indeed, if T_n is doubly transitive and adequate, then $\tilde{F}_{n-\alpha+1}(Y_{n-\alpha+1}), \ldots, \tilde{F}_n(Y_n)$ have the same distributional properties. If no natural ordering exists, the transform should be applied to the ordering sample (O'Reilly and Stephens, 1982).

REPLY TO THE DISCUSSION

We thank the discussants for their kind and generally positive remarks. We knew that our reference list for this active research area was very incomplete and appreciate the additional citations provided in the discussion. Pettit's historical perspective is a particularly welcome supplement.

Several discussants comment upon the close relationship between the model determination problem and the issues of diagnosing and modeling outliers. Also see Draper and Guttman (1987). We note that sampling-based approaches expedite calculations associated with these issues. See for instance, the recent paper of Verdinelli and Wasserman (1991).

Peña encourages us to investigate cross-validation schemes other than single point deletion in particular with regard to identifying masking and swamping. He suggests that $f(\boldsymbol{Y}_I|\boldsymbol{Y}_{(I)})$ be computed. We concur, noting that the methodology in Section 3 is pertinent to such computation. Our only reservation involves possible combinatoric problems as indicated in Peña and Tiao (1992).

Pericchi raises an interesting question which does not appear to have a simple answer. The difficulty is that, in general, it is not obvious what the model which "encompasses both models 1 and 2" is. In customary linear models it is clear; we merely augment the design matrix to do this. However, consider the two nonlinear models discussed in Section 4, i.e., model 1: $Y = \beta_0(1+\beta_1\beta_2^x)^{-1}+\epsilon$, model 2: $Y = \gamma_0 e^{-\gamma_1\gamma_2^x}+\epsilon$. The encompassing model which is additive in the mean structure will not be identifiable; the asymptote is $\beta_0+\gamma_0$. If we remedy this by setting $\beta_0=\gamma_0$ we can no longer retrieve model 1 or model 2 as a reduced model. Suppose we try a multiplicative form for the encompassing model $Y = \beta_0(1+\beta_1\beta_2^x)^{-1}e^{-\gamma_1\gamma_2^x}+\epsilon$. Now the reduced models are not identifiable; $\beta_1 = 0$ or $\beta_2 = 0$ produces model 2, $\gamma_1 = 0$ or $\gamma_2 = 0$ produces model 1.

Turning to the remarks of Seillier-Moiseiwitsch we agree that the checking function g_1 may be less informative than the others. Nonetheless examination of residuals is standard and familiar. Moreover, the resulting d_{1r} have an immediate connection with Bayesian residuals as discussed in Chaloner and Brant (1988). They consider the posterior distribution of the unobserved errors which, in our setting, leads to the distribution of $\epsilon_r|\boldsymbol{Y}_{(r)} = y_r - \varphi(X_r;\beta)|\boldsymbol{Y}_{(r)}$. The mean of this distribution is d_{1r}. Her suggestion to transform the $\{d_{ir}; r = 1,\ldots,n\}$ to a set of i.i.d. $U(0,1)$ variates is interesting but we suspect feasible only in certain simple cases. That is, preliminary reading of O'Reilly and Quesenberry (1973) yields several concerns. Their approach requires the joint predictive distribution, $f(\boldsymbol{Y})$, to be proper, requires an explicit expression for $f(\boldsymbol{Y})$ and in fact, appears to require that $f(\boldsymbol{Y})$ be an exponential family to effectively bring (minimal) sufficiency into play. A separate problem is that, even were we able to carry out the calculations, we worry about the inherent order dependence of the results since for general response model data no natural ordering need exist.

Finally, Raftery offers the lengthiest and most penetrating discussion. One of his main points concerns the computation of $f(\boldsymbol{Y})$. We agree that this can be done and, in fact, at the end of Section 3 mention the use of importance sampling densities to do so. Whether the posterior is a good choice is unclear since the resulting harmonic mean estimator may be unstable. Calculation of $f(\boldsymbol{Y})$ in a sequential fashion seems silly. In most cases the effort to compute $f(\boldsymbol{Y})$ directly would not be much greater than that required to compute an individual term in the factorization. We also note the aforementioned concern regarding the inherent order dependence which is not mitigated computationally by the suggestion to randomly sample permutations.

More importantly, we criticized the use of $f(\boldsymbol{Y})$ when it is not integrable and not because we couldn't compute it. We completely agree that the choice of likelihood is the critical problem and in fact say so in the introduction. We are less sanguine about the availability of proper priors. If they are developed through training data (imaginary or otherwise) is this not really similar in spirit to cross-validation?

Turning to our criticism of the standard Bayesian model choice procedure, there are no doubt situations where we may knowledgeably assign prior weights to models in which case we would certainly do so and obtain posterior odds. But "garden variety" specification of the likelihood with regard to features discussed in our introduction doesn't seem to readily lend itself to such weighting. However we thoroughly agree that Bayes factors (when interpretable) or pseudo-Bayes factors are vital objects to compute in comparing models. Still these factors may disadvantage some models relative to others. Hence we value the information obtained through other checking functions. A question requiring further analytic and empirical elaboration is, in the case of proper priors, how different will the Bayes factor and the pseudo-Bayes factor be particularly as n increases?

In conclusion we are invigorated by all of the discussion, critical or otherwise. Model determination is obviously a fundamental data analytic task. Illumination of its aspects in the Bayesian framework, particularly contentious ones, necessarily enhances our understanding of the task.

ADDITIONAL REFERENCES IN THE DISCUSSION

Besag, J. E. (1975). Statistical analysis of non-lattice data. *The Statistician* **24**, 179–195.

Chaloner, K. and Brant, R. (1988). A Bayesian approach to outlier detection and residual analysis. *Biometrika* **75**, 651–659.

Dawid, A. P. (1984). Present position and potential developments: some personal views. Statistical theory. The prequential approach. *J. Roy. Statist. Soc. A* **147**, 178–292, (with discussion).

Draper N. R. and Guttman, I. (1987). A common model selection criterion. *Probability and Bayesian Statistics* (R. Viertl, ed.), London: Plenum Press, 134–150.

Geisser, S. (1980). In discussion of G. E. P. Box. *J. Roy. Statist. Soc. A* **143**, 416–417.

Guttman, I. and Peña, D. (1988). Outliers and Influence: Evaluation by posteriors of parameters in the linear model. *Bayesian Statistics 3* (J. M. Bernardo, M. H. DeGroot, D. V. Lindley and A. F. M. Smith, eds.), Oxford: University Press, 631–640.

Hodges, J. S. (1987). Uncertainty, policy analysis and statistics. *Statist. Sci.* **2**, 259–291, (with discussion).

Johnson, W. and Geisser, S. (1985). Estimated influence measures of the multivariate general linear model. *J. Statist. Planning and Inference* **11**, 33–56.

Newton, M. A. and Raftery, A. E. (1991). Approximate Bayesian inference by the weighted likelihood bootstrap. *Tech. Rep.* **199**, University of Washington.

O'Reilly, F. J. and Quesenberry, C. P. (1973). The conditional probability integral transform and its applications to obtain composite chi-square goodness-of-fit tests. *Ann. Statist.* **1**, 74–83.

O'Reilly, F. J. and Stephens, M. A. (1982). Characterizations and goodness-of-fit tests. *J. Roy. Statist. Soc. B* **44**, 353–360.

Peña, D. and Tiao, G. C. (1992). Bayesian robustness functions for linear models. *Bayesian Statistics 4* (J. M. Bernardo, J. O. Berger, A. P. Dawid and A. F. M. Smith, eds.), Oxford: University Press, 365–389, (with discussion).

Pettit, L. I. (1986). Diagnostics in Bayesian model choice. *The Statistician* **35**, 183–190.

Pettit, L. I. (1988). Bayes methods for outliers in exponential samples. *J. Roy. Statist. Soc. B* **50**, 371–380.

Pettit, L. I. (1990). The conditional predictive ordinate for the normal distribution. *J. Roy. Statist. Soc. B* **52**, 175–184.

Pettit, L. I. and Smith, A. F. M. (1983). Bayesian model comparison in the presence of outliers. *Bull. Int. Statist. Inst.* **50**, 292–309.

Pettit, L. I. and Young, K. D. S. (1990). Measuring the effect of observations on Bayes factors. *Biometrika* **77** 455–466.

Raftery, A. E. (1988). Approximate Bayes factors for generalized linear models. *Tech. Rep.* **121**, University of Washington.

Roberts, H. V. (1965). Probabilistic prediction. *J. Amer. Statist. Assoc.* **60**, 50–62.

Seillier-Moiseiwitsch, F. (1990). Sequential probability forecasts and the probability integral transform. (Unpublished *Tech. Rep.*).

Seillier-Moiseiwitsch, F. and Dawid, A. P. (1992). On testing the validity of probability forecasts. *J. Amer. Statist. Assoc.* , (to appear).

Schwarz, G. (1978). Estimating the dimension of a model. *Ann. Statist.* **6**, 461–464.

Zellner, A. (1975). Bayesian analysis of regression error terms. *Amer. Statist.* **70**, 138–144.

BAYESIAN STATISTICS 4, pp. 169–193
J. M. Bernardo, J. O. Berger, A. P. Dawid and A. F. M. Smith, (Eds.)

Evaluating the Accuracy of Sampling-Based Approaches to the Calculation of Posterior Moments

JOHN GEWEKE
University of Minnesota, USA

SUMMARY

Data augmentation and Gibbs sampling are two closely related, sampling-based approaches to the calculation of posterior moments. The fact that each produces a sample whose constituents are neither independent nor identically distributed complicates the assessment of convergence and numerical accuracy of the approximations to the expected value of functions of interest under the posterior. In this paper methods from spectral analysis are used to evaluate numerical accuracy formally and construct diagnostics for convergence. These methods are illustrated in the normal linear model with informative priors, and in the Tobit censored regression model.

Keywords: DATA AUGMENTATION; GIBBS SAMPLING; MIXED ESTIMATION; MONTE CARLO INTEGRATION; TOBIT MODEL.

1. INTRODUCTION

Gibbs sampling (Geman and Geman, 1984) in conjunction with data augmentation (Tanner and Wong, 1987) constitutes a very useful tool for the solution of many important Bayesian multiple integration problems (Gelfand and Smith, 1990). The method is complementary to other numerical approaches, and is attractive and competitive in many statistical models. This is consistent with the current rapid growth in the application of the method to interesting problems by many Bayesian statisticians. It produces samples of functions of interest whose distribution converges to the posterior distribution, but whose constituents are in general never independently distributed and are identically distributed only in the limit. In this context treatment of the compelling questions of convergence and the accuracy of approximation has been informal.

The research reported here undertakes to place these matters on a more formal footing, and in so doing render the Gibbs sampler with data augmentation a more systematic, reliable, and replicable tool for solution of the Bayesian multiple integration problem. Its thesis is that the failure of the Gibbs sampling process to produce independently distributed realizations of the functions of interest is no inhibitor to learning about the means of these sequences. Indeed, the problem of inference for the mean of a stationary process is one of long–standing in time series analysis, with a number of solutions. We apply one of these solutions here, showing how it yields systematic assessments of the numerical accuracy of approximations to the expected values of functions of interest under the posterior. With a formal distribution theory for the approximations in hand, it is straightforward to construct diagnostics for convergence.

The paper begins with a brief exposition of the Gibbs sampler and data augmentation, in the next section. The main methodological results are presented in Section 3, together with a simple example constructed to provide some insight into the typical serial correlation structure of the Gibbs sampler. In Section 4 the proposed solution is applied to the multiple normal linear regression model with a proper normal prior on a subset of its coefficients. The Gibbs sampler proves to be very efficient in treating this problem, which is important in its own right in econometrics and is a key building block in applying the Gibbs sampler with data augmentation to Bayesian inference in many other statistical models. In Section 5 this application is extended to the Tobit censored regression model. The formal treatment of convergence and numerical accuracy appears essential to producing reliable results with the Gibbs sampler and data augmentation in this model. Conclusions, together with conjectures about future research, are presented in the final section.

2. THE GIBBS SAMPLER AND DATA AUGMENTATION

The generic task in applied Bayesian inference is to obtain the expected value of a function of interest under a posterior density. In standard notation, this object is expressed

$$E[g(\boldsymbol{\theta})] = \int_{\Theta} g(\boldsymbol{\theta})\pi(\boldsymbol{\theta})L(\boldsymbol{\theta};X)d\boldsymbol{\theta} \Big/ \int_{\Theta} \pi(\boldsymbol{\theta})L(\boldsymbol{\theta};X)d\boldsymbol{\theta},$$

where $\boldsymbol{\theta}$ is the finite–dimensional vector of parameters whose domain is a subset of Euclidean space Θ; X is the observed data; $g(\boldsymbol{\theta})$ is the function of interest; $\pi(\boldsymbol{\theta})$ is proportional to a proper or improper prior density; and $L(\boldsymbol{\theta};X)$ is proportional to the likelihood function. In what follows, we shall suppress the dependence of the likelihood function on the observed data, and denote the kernel of the posterior density $p(\boldsymbol{\theta}) = \pi(\boldsymbol{\theta})L(\boldsymbol{\theta})$. We shall refer to the calculation of $E[g(\boldsymbol{\theta})]$ as the Bayesian multiple integration problem.

2.1. *Other Approaches*

In some cases $p(\boldsymbol{\theta})$ and $g(\boldsymbol{\theta})p(\boldsymbol{\theta})$ are sufficiently simple form that it is possible to obtain an exact analytical evaluation of $E[g(\boldsymbol{\theta})] = \int_{\Theta} g(\boldsymbol{\theta})p(\boldsymbol{\theta})d\boldsymbol{\theta} / \int_{\Theta} p(\boldsymbol{\theta})d\boldsymbol{\theta}$, but the class of such cases is much smaller than the class of problems routinely studied in statistics and econometrics. Zellner (1971) provides a treatment of many of these cases, and the class has not grown much in the twenty years since that volume was written. However, a rich variety of methods of approximating $E[g(\boldsymbol{\theta})]$ has emerged, and continues to broaden. Series expansions of $p(\boldsymbol{\theta})$ and $g(\boldsymbol{\theta})p(\boldsymbol{\theta})$ provide one basis for these methods, including Laplace's method (Tierney and Kadane, 1986) and marginal inference (Leonard, Hsu and Tsu, 1989). Monte Carlo sampling from the parameter space Θ provides another line of attack, including importance sampling (Kloek and van Dijk, 1978; Geweke, 1989) and antithetic acceleration (Geweke, 1988). The various methods for solving the Bayesian multiple integration problem tend to be complementary. For example, series expansion methods lead to essentially instantaneous computations but provide approximations whose accuracy cannot easily be improved or systematically evaluated. Monte Carlo methods produce approximations whose accuracy is easily assessed and can be improved by increasing the number of iterations, but the computations may be quite time consuming. Both methods require preliminary analytical work, in the form of carrying out the series expansions or constructing the importance sampling density, that can be tedious.

The objective of this research is to facilitate the application of two other, closely related, Monte Carlo methods, which have become known as Gibbs sampling, and Gibbs sampling with data augmentation. The utility of the Gibbs sampler, as proposed by Geman and Geman (1984), for the generic task in applied Bayesian inference was recognized by Gelfand and Smith (1990). The use of data augmentation in the calculation of posterior densities was proposed by Tanner and Wong (1987). The potential of combining the two is immediately evident.

2.2. *The Gibbs Sampler*

The Gibbs sampler provides a method for sampling from a multivariate probability density, employing only the densities of subsets of vectors conditional on all the others. The method is easily described in the context of our generic Bayesian problem. Suppose the parameter vector is partitioned, $\theta' = (\theta'_{(1)}, \theta'_{(2)}, \ldots, \theta'_{(s)})$. Further suppose that the conditional distributions,

$$\theta_{(j)} | \{\theta_{(1)}, \ldots, \theta_{(j-1)}, \theta_{(j+1)}, \ldots, \theta_{(s)}\} \sim$$
$$p_{(j)}[\theta_{(1)}, \ldots, \theta_{(j-1)}, \theta_{(j+1)}, \ldots, \theta_{(s)}] \qquad (j = 1, \ldots, s)$$

are known, and are of a form that synthetic i.i.d. random variables can be generated readily and efficiently from each of the $p_{(j)}(\cdot)$. Now let $\theta^{(0)'} = (\theta^{(0)'}_{(1)}, \theta^{(0)'}_{(2)}, \ldots, \theta^{(0)'}_{(s)})$ be an arbitrary point in θ. Generate succesive synthetic random subvectors,

$$\theta^{(1)}_{(j)} | \{\theta^{(1)}_{(1)}, \ldots, \theta^{(1)}_{(j-1)}, \theta^{(0)}_{(j+1)}, \ldots, \theta^{(0)}_{(s)}\} \sim$$
$$p_{(j)}[\theta^{(1)}_{(1)}, \ldots, \theta^{(1)}_{(j-1)}, \theta^{(0)}_{(j+1)}, \ldots, \theta^{(0)}_{(s)}] \qquad (j = 1, \ldots, s)$$

For subsequent reference, denote the composition of the vector after step j of this conditional sampling process by $\theta^{(1,j)'} = (\theta^{(1)'}_{(1)}, \ldots, \theta^{(1)'}_{(j)}, \theta^{(0)'}_{(j+1)}, \ldots, \theta^{(0)'}_{(s)})$, and denote its composition after the last step by $\theta^{(1)} = \theta^{(1,s)}$. We shall refer to each of the conditional samplings as a *step*. We shall refer to the completion of the first s steps, resulting in the vector $\theta^{(1)}$, as the first *pass* through the vector θ.

The second and succesive passes are performed similarly. At the i'th step of the j'th pass,

$$\theta^{(j)}_{(i)} \Big| \left\{\theta^{(j)}_{(1)}, \ldots, \theta^{(j)}_{(i-1)}, \theta^{(j-1)}_{(i+1)}, \ldots, \theta^{(j-1)}_{(s)}\right\} \sim p_{(j)}\left[\theta^{(j)}_{(1)}, \ldots, \theta^{(j)}_{(i-1)}, \theta^{(j-1)}_{(i+1)}, \ldots, \theta^{(j-1)}_{(s)}\right]$$

and the composition of the vector is

$$\theta^{(j,i)'} = \left[\theta^{(j)'}_{(1)}, \ldots, \theta^{(j)'}_{(i)}, \theta^{(j-1)'}_{(i+1)}, \ldots, \theta^{(j-1)}_{(s)'}\right];$$

at the end of the j'th pass the composition of the vector is

$$\theta^{(j)'} = \left[\theta^{(j)'}_{(1)}, \ldots, \theta^{(j)'}_{(s)}\right]$$

Under weak conditions outlined in Gelfand and Smith (1990), which amount to the assumption that Θ is connected, $\theta^{(j)}$ converges in distribution to the limiting density $p(\theta)$. Moreover, the rate of convergence is geometric in the L_1 norm. To obtain these convergence results it is necessary only to assume that each subvector $\theta^{(j)}$ is visited infinitely often. Thus, many variants on the cyclical scheme outlined here are possible. For most applications the simplicity of the cyclical scheme seems to be compelling, but we shall return to this question in the last section of the paper.

Since $\theta^{(j)}$ converges in distribution to the posterior distribution $p(\cdot)$, the limiting distribution of $g(\theta^{(j)})$ is the same as the distribution of $g(\theta)$ under the posterior. Given independent realizations of $\theta^{(j)}$, the strong law of large numbers would at once motivate an approximation to $E[g(\theta)]$ using sample averages. Of course, succesive drawings are not independent, and in the applications reported in the literature the Gibbs sampling process is typically restarted many times, quasi-independently, in order to achieve a sufficiently good approximation to independence. However, creation of an approximation to independence is neither necessary nor desirable, and we shall return to this point in the next section.

The Gibbs sampler is an attractive solution of the Bayesian multiple integration problem when the conditional densities are simple and easy to obtain. In the special simple case $s = 1$, one is sampling directly from the posterior density and convergence trivially obtains in the first pass. This case is not inherently interesting, but it suggests that Gibbs sampling schemes with small s may have convergence and computational efficiency properties that are attractive relative to those with large s. At the other extreme, one can take s to be equal to the dimension of the parameter vector, and use one of several generic procedures for generating univariate random variables from an arbitrary distribution. Such schemes are likely to be impractical, since the integrand changes at each step of each pass.

The Gibbs sampler is a competitive solution of the Bayesian multiple integration problem when the form of the posterior density renders other methods awkward or inefficient. For example, the derivation of series expansions or importance sampling densities may be cumbersome, while at the same time the conditional densities $p^{(j)}(\cdot)$ are trivial. In the two examples taken up in this paper, series expansions and importance sampling densities can be constructed, but the Gibbs sampler is much simpler and computations with it are very fast.

2.3. *Data Augmentation*

In many instances the posterior density $p(\boldsymbol{\theta})$ does not immediately decompose into subvectors with convenient conditional densities. However, there always exists the formal possibility that one can reexpress the posterior density

$$p(\boldsymbol{\theta}) = \int_{Y^*} \tilde{r}(\theta|y^*)q(y^*)dy^*$$

$$\int_{Y^*} \tilde{q}(y^*|\theta)r(\theta)dy^*$$

and the conditional densities $\tilde{r}(\boldsymbol{\theta}|y^*)$ and $q(y^*|\boldsymbol{\theta})$ may be well suited to the Gibbs sampling scheme. (Of course, this may involve more than two-step passes: i.e., it may be necessary to further decompose $\boldsymbol{\theta}, y^*$, or both.) The introduction of y^* proposed by Tanner and Wong (1987), is known as data augmentation. The key to its utility is that the construction of y^* is frequently natural rather than artificial. Indeed, in many signal extraction problems and latent variable models, difficulties with the posterior density arise precisely because of the need to perform this integration. In these cases, it is often easy to draw $\boldsymbol{\theta}$ and y^* successively; it is not even necessary to write the posterior density explicitly. We shall return to such an example, previously studied using the Gibbs sampler by Chib (1990), in Section 5. In what follows, when we refer to the Gibbs sampler we shall implicitly include the possibility of data augmentation.

3. ASSESSING NUMERICAL ACCURACY AND CONVERGENCE

The Gibbs sampler, with or without data augmentation, suffers from the complications that the sequences produced are neither independent nor identically distributed. To date the literature has dealt with these problems in ways that are informal and computationally inefficient. Here, we suggest a careful and systematic treatment of the problem. This treatment has three attractions. (1) It is computationally efficient, using virtually all the sample evidence from the Gibbs sampling scheme. The example taken up in Section 5 yields drastic improvements in computational efficiency over that reported elsewhere, and there is sound reason to believe that is the case generally. (2) Using standard techniques in spectral analysis, the suggested treatment provides a standard error for the approximation of $E[g(\boldsymbol{\theta})]$ by corresponding sample averages of $g(\cdot)$ taken over the passes and steps of the Gibbs sampler. (3) Based on these distributional results, a diagnostic for nonconvergence of the Gibbs sampling scheme is constructed.

3.1. *Serial Correlation and the Efficient Use of Information*

We take up the dependence and convergence problems in succession. To begin, ignore the convergence problem and assume that the sequence $\boldsymbol{\theta}^{(j)}$ is identically but not independently distributed. In general, a fully efficient use of the realizations of the Gibbs sampling might entail the computation of the corresponding value of $g(\cdot)$ at each pass and step. To maintain this level of generality, consider the $s \times 1$ stochastic process

$$G(j) = \left(g(\boldsymbol{\theta}^{(j)}_{(1)}), g(\boldsymbol{\theta}^{(j)}_{(2)}), \ldots, g(\boldsymbol{\theta}^{(j)}_{(s)})\right)' \qquad (j = 1, 2, \ldots p).$$

The problem is to estimate the mean of $G(j)$, subject to the constraint that each mean of the $s \times 1$ vector is the same. Assume that the Gibbs sampling process, and the importance function $g(\cdot)$, jointly imply the existence of a spectrum for $\{G(j)\}$, and the existence of a spectral density $S_G(\omega)$ with no discontinuities at the frequency $\omega = 0$. The asymptotically efficient (in p) estimator of this mean is simply the grand sample average of all the $(g(\boldsymbol{\theta}^{(j)}_{(i)})$; in our notation, it is $(ps)^{(-1)} \sum_{j=1}^{p} \iota' G(j)$ where ι denotes an $s \times 1$ vector of 1's, and we shall refer to this estimator as $\overline{g}_p$. The asymptotic variance of this estimator is $(ps)^{(-1)} \iota' S_G(0) \iota$ (Hannan, 1970, 207–210). We may obtain a standard error of numerical approximation for the estimator by estimating $S_G(0)$ in conventional fashion. Moreover, estimation of the full spectral density may yield insights into the nature of the stochastic process implicit in the Gibbs sampling scheme, as will be suggested in some of the specific examples taken up subsequently.

While this method extracts the most information about $E[g(\boldsymbol{\theta})]$ given the realizations $\boldsymbol{\theta}^{(j)}_{(i)}$ the pertinent practical decision is how often to compute the function(s) of interest $g(\cdot)$ relative to the steps and passes of the Gibbs sampling process. The best decision would reflect the relative costs of drawing $\boldsymbol{\theta}^{(j)}_{(i)}$ and computing $g(\cdot)$, and the degree of serial correlation in the process $\{G(j)\}$. It is clear how this problem could be set up and the solution incorporated in sophisticated software, but we do not enter into these issues here. Instead, we conjecture that typically it will be satisfactory to compute $g(\boldsymbol{\theta}^{(j)}_{(i)})$ at the end of each pass. Since there are no computations within passes, $\{G(j)\}$ becomes a univariate stochastic process and the asymptotically efficient estimator of $E[g(\boldsymbol{\theta})]$ is $\overline{g}_{p\cdot} = p^{-1} \sum_{j=1}^{p} G(j)$, whose asymptotic variance is $p^{-1} S_G(0)$. The corresponding *numerical standard error* (NSE) of the estimate is $[p^{-1} \hat{S}_G(0)]^{1/2}$. In all results reported in this paper, $\hat{S}_G(0)$ is formed from the periodogram of $\{G(j)\}$ using a Daniell window of width $2\pi/M, M = (.3p^{1/2})$.

This method for assessing numerical accuracy can be applied to many variants on the basic sampling scheme for the $\boldsymbol{\theta}^{(j)}_{(i)}$. Many of the applications in which the sampling process is restarted many times can be analyzed in exactly the same way. For example, if every m'th pass is used in an effort to induce quasi–independence in the computed $G(j)$, the estimated spectral density may still be used to provide a measure of numerical accuracy. In fact, this computation would appear necessary in verifying a claim that $\{\boldsymbol{\theta}^{(j)}\}$ is a stochastic process which is essentially serially uncorrelated. Given the methods proposed here, of course, there is no analytical constraint requiring the construction of such a sequence in the first place.

3.2. *Assessing Convergence*

This formulation of the dependence problem also provides a practical perspective on the convergence problem. Given the sequence $\{G(j)\}$, a comparison of values early in the sequence with those late in the sequence is likely to reveal failure of convergence. Let

$$\overline{g}_p^A = p_A^{-1} \sum_{j=1}^{p_A} G(j), \qquad \overline{g}_p^B = p_B^{-1} \sum_{j=p^*}^{p} G(j) \quad (p^* = p - p_B + 1),$$

and let $\hat{S}_G^A(0)$ and $\hat{S}_G^B(0)$ denote consistent spectral sensity estimates for $\{G(j), = 1, \ldots, p_A\}$ and $\{G(j), j = p^*, \ldots, p)$ respectively. If the ratios p_A/p and p_B/p are fixed, with $(p_A + p_B)/p < 1$, then as $p \to \infty$,

$$(\overline{g}_p^A - \overline{g}_p^B)/[p_A^{-1}\hat{S}_G^A(0) + p_B^{-1}\hat{S}_G^B(0)]^{1/2} \Rightarrow N(0,1) \tag{3.1}$$

if the sequence $\{G(j)\}$ is stationary. We shall refer to the left side of this expression as the *convergence diagnostic* (CD). This application of a standard central limit theorem exploits not only the increasing number of elements of each sample average, but also the limiting independence of $\overline{g}_p^A$ and $\overline{g}_p^B$ owing to $(p_A+p_B)/p < 1$. These two conditions need to be kept in mind when using this diagnostic, as must considerations of power. In work to date we have taken $p_A = .1p$ and $p_B = .5p$. These choices meet the assumptions underlying (3.1), while attempting to provide diagnostic power against the possibility that $\{G(j)\}$ process was not fully converged early on.

3.3. *Preliminary Passes*

In practice, one is free to choose the start of the sampling process. From the initial and possibly arbitrary point $\boldsymbol{\theta}^{(0)}$, initial iterations may proceed before the sampling that enters into the computation of $\overline{g}_p$ begins. Indeed, a subsequent example will suggest that this process is critical to $\overline{g}_p$ whose numerical accuracy is reliably known. Computation of $g(\boldsymbol{\theta})$ is not required at this stage. In this paper the number of such presampling passes is treated as a subjectively chosen parameter of the experimental design. With some foundation of experience in this sort of exercise it should be possible to design algorithms for terminating the presample passes and initiating the computation of $\overline{g}_p$, based on (3.1) or similar computations.

3.4. *Relative Numerical Efficiency*

Variants of the Gibbs sampling procedure can be compared with each other and with other solutions of the Bayesian multiple integration problem by means of a convenient benchmark. Had the problem been solved by making p independent, identically distributed Monte Carlo drawings $\{\boldsymbol{\theta}_1, \ldots, \boldsymbol{\theta}_p\}$ directly from the posterior density, and $E[g(\boldsymbol{\theta})]$ estimated as the sample average of the $g(\boldsymbol{\theta}_i)$ over these drawings, then the variance of this estimate would be $\text{var}[g(\boldsymbol{\theta})]/p$, where $\text{var}[g(\boldsymbol{\theta})]$ is the posterior variance of $g(\boldsymbol{\theta})$. By contrast the variance of the Gibbs sampler is $S_G(0)/p$. Following Geweke (1989), we shall refer to the ratio of the former to the latter, $\text{var}[g(\boldsymbol{\theta})]/S_G(0)$, as the *relative numerical efficiency* (RNE) of the Gibbs sampling estimator of $E[g(\boldsymbol{\theta})]$. This quantity is of great practical interest for two reasons. First, it may be approximated by means of routine side computations in the Gibbs sampling process itself. While we cannot in fact construct i.i.d. drawings from the posterior density, the Gibbs sampling estimate $\hat{\text{var}}[g(\boldsymbol{\theta})]$ of $\text{var}[g(\boldsymbol{\theta})]$ can be formed in the same way that the Gibbs sampling estimate of $E[g(\boldsymbol{\theta})]$ is formed. The ratio $\hat{\text{var}}[g(\boldsymbol{\theta})]/\hat{S}_G(\boldsymbol{\theta})$ then approximates relative numerical efficiency. The other reason for considering RNE is its immediate relation to computational efficiency. The number of drawings required to achieve a given degree of numerical accuracy is inversely related to the relative numerical efficiency of the Gibbs sampling process for the function of interest: were RNE doubled then the number of drawings required would be halved, and so on.

It is worth noting that the RNE of the Gibbs sampling process is solely a function of the serial correlation characteristics of the process $\{G(j)\}$:

$$RNE = \text{var}[g(\boldsymbol{\theta})]/S_G(0) = \int_{-\pi}^{\pi} S_G(\omega)d\omega/S_G(0).$$

This formulation makes it clear that the relative numerical efficiency of the Gibbs sampling process depends on the power of the spectral density of $\{G(j)\}$ at $\omega = 0$, relative to the distribution of its spectral density across other frequencies. Thus, relative numerical efficiency may be quite different for different functions of interest. Furthermore, RNE is not bounded above by one: in principle, efficiency many times that achieved by i.i.d. sampling directly from the posterior density can be achieved by the Gibbs sampling estimator of $E[g(\boldsymbol{\theta})]$. Heuristically, positive serial correlation of $\{G(j)\}$ renders the Gibbs sampling estimator less efficient, and negative serial correlation in $\{G(j)\}$ renders it more efficient.

3.6. *A Constructed Example*

It may be helpful to illustrate these ideas in a constructed example simple enough that an analytical approach is possible. Consider the case of a bivariate normal posterior density for $\theta = (\theta_1, \theta_2)'$, with zero mean and $\text{var}(\theta_i) = \sigma_{ii}$, $\text{cov}(\theta_1, \theta_2) = \sigma_{12}$. Denote the stochastic process corresponding to the Gibbs sampler by $\{\tilde{\theta}\}$, and suppose that the Gibbs sampling design is

$$\tilde{\theta}_1^j = (\sigma_{12}/\sigma_{22})\tilde{\theta}_2^{j-1} + \varepsilon_{1j} \;\; \text{var}(\varepsilon_{1j}) = \sigma_{11} - \sigma_{12}^2/\sigma_{22}$$
$$\tilde{\theta}_2^j = (\sigma_{12}/\sigma_{11})\tilde{\theta}_1^j + \varepsilon_{2j} \;\; \text{var}(\varepsilon_{2j}) = \sigma_{22} - \sigma_{12}^2/\sigma_{11}$$

The spectral density of this bivariate process, at frequency $\omega = 0$, is

$$S_{\tilde{\theta}}(0) = (1-r^2)^{-1}\begin{bmatrix} \sigma_{11}(1+r^2) & 2\sigma_{12} \\ 2\sigma_{12} & \sigma_{22}(1+r^2) \end{bmatrix}, \quad r^2 \equiv \sigma_{12}^2/\sigma_{11}\sigma_{22}.$$

By comparison, the posterior variance of θ may be written,

$$\Sigma = (1-r^2)^{-1}\begin{bmatrix} \sigma_{11}(1-r^2) & \sigma_{12}(1-r^2) \\ \sigma_{12}(1-r^2) & \sigma_{22}(1-r^2) \end{bmatrix},$$

and their difference is

$$S_{\tilde{\theta}}(0) - \Sigma = (1-r^2)^{-1}\begin{bmatrix} 2\sigma_{11}r^2 & \sigma_{12}(1+r^2) \\ \Sigma_{12}(1+r^2) & \sigma_{22}r^2 \end{bmatrix}, \tag{3.3}$$

which is not positive semidefinite. Consequently, the relative numerical efficiency of some functions of interest of the parameters θ_1 and θ_2 would exceed one, while others would be less than one, if the functions of interest are to be evaluated at the end of each pass. If the functions of interest are evaluated after each step, then the pertinent difference is

$$2S_{\tilde{\theta}}(0) - \Sigma = (1-r^2)^{-1} \begin{bmatrix} \sigma_{11}r^2(1+3r^2) & \sigma_{12}(3+r^2) \\ \sigma_{12}(3+r^2) & \sigma_{22}(1+3r^2) \end{bmatrix}.$$

Since this matrix is positive definite, the relative numerical efficiency for any linear function of interest of the parameters must be less than one.

To examine these results in more detail, let $\theta_{11} = \theta_{22} = 1, \theta_{12} = .5^{1/2}$. Then the eigenvalues of $S_{\tilde{\theta}}(0) - \Sigma$ are 4.1213 and $-.1213$, with respective corresponding eigenvectors $(.5^{1/2}, .5^{1/2})$ and $(.5^{1/2}, -.5^{-1/2})$. The Gibbs sampling process induces positive correlation between $\tilde{\theta}_1^j$ and $\tilde{\theta}_2^j$. This raises the variance of $\tilde{\theta}_1^j + \tilde{\theta}_2^j$ and lowers the variance of $\tilde{\theta}_1^j - \tilde{\theta}_2^j$, relative to the case of independence. In the case of $\tilde{\theta}_1^j - \tilde{\theta}_2^j$, the reduction more than offsets the positive serial correlation in $\tilde{\theta}^j$.

Table 1 exhibits the numerical standard errors (NSE's) and relative numerical efficiencies (RNE's) of four alternative functions of interest. The population values are derived from $S_{\tilde{\theta}}(0)$ and Σ. The results presented for $p = 400$ and $p = 10,000$ are based on a starting value chosen from the posterior density (possible in this constructed example, but not possible generally) and no preliminary passes. The results for $p = 10,000$ agree quite well with the population values. Those for $p = 400$ agree well, except that the NSE for the function of interest $.5(\theta_1 - \theta_2)$ is somewhat too high and the corresponding RNE is therefore somewhat too low. This discrepancy can be traced to the smoothing of the periodogram in the formation of the estimate $\hat{S}_G(\theta)$: this function has a minimum at $\omega = 0$, and when $p = 400$ the estimate at $\omega = 0$ is an average of periodogram ordinates extending from $-\pi/3$ to $\pi/3$, which raises its value. When $p = 10,000$ the average extends from $-\pi/15$ to $\pi/15$ and this effect is negligible.

Table 2 provides some estimated spectral density ordinates from the case $p = 10,000$. The computations at frequencies other than $\omega = 0$ are not essential to the procedure, of course. However, they are easy to produce and are given here to further illustrate the differences in NSE's and RNE's for the different functions of interest. Note, in particular, that for the first three functions of interest power is greatest at $\omega = 0$. Whenever this is true, RNE must be less than one. For the fourth function of interest power is smallest at $\omega = 0$, implying that RNE must exceed one.

4. INFERENCE IN THE LINEAR MODEL WITH AN INFORMATIVE PRIOR

We turn now to a simple but important application of Gibbs sampling, the multiple normal linear regression model, with a proper normal prior on a subset of the coefficients. The example is simple because there is only a single parameter that prevents analytical solution of the whole problem, in the case of linear functions of interest of the coefficients, or integration by Monte Carlo sampling directly from the entire posterior density, in the case of nonlinear functions of interest of the coefficients. The example is important in itself, because the model is widely applied in many disciplines and informative normal priors are frequently a reasonable representation of prior knowledge. It is also important because the model and prior occur repeatedly as a key conditional distribution when more difficult problems are attacked using Gibbs sampling. (An example is provided in the next section.) The solution of this problem may then be applied in those cases. The strategy of constructing such "building blocks" seems well suited to the research program of constructing Gibbs samplers for many standard econometric models.

Function of interest, $g(\theta)$:	θ_1	θ_2	$.5(\theta_1+\theta_2)$	$.5(\theta_1-\theta_2)$
Population values:				
$E[g(\theta)]$	.000	.000	.000	.000
$sd[g(\theta)]$	1.000	1.000	.924	.383
NSE	$1.732/\sqrt{p}$	$1.732/\sqrt{p}$	$1.707/\sqrt{p}$	$.293/\sqrt{p}$
RNE	.333	.333	.293	1.707
Gibbs sampling, 400 passes (.09 seconds):				
$\hat{E}[g(\theta)]$	–.046	–.017	–.032	.016
$\hat{sd}[g(\theta)]$	.961	.998	.901	.384
NSE	.081	.088	.083	.016
RNE	.352	.318	.293	1.389
CD	–.137	.674	.291	–1.513
Gibbs sampling, 10,000 passes (1.89 seconds):				
$\hat{E}[g(\theta)]$	.014	.005	.009	.004
$\hat{sd}[g(\theta)]$	1.019	1.021	.945	.384
NSE	.0171	.0173	.0169	.0029
RNE	.355	.350	.311	1.810
CD	.147	.375	.267	–.588

Table 1. *Gibbs Sampling in the Constructed Bivariate Normal Example of Section 3.6.*

Functions of interest, $g(\theta)$: Frequency, ω:	θ_1	θ_2	$.5(\theta_1+\theta_2)$	$.5(\theta_1-\theta_2)$
$.0\pi$	2.916	2.969	2.861	.081
$.1\pi$	2.677	2.590	2.533	.100
$.2\pi$	1.890	2.000	1.826	.119
$.3\pi$	1.171	1.257	1.070	.144
$.4\pi$	.765	.742	.603	.150
$.5\pi$	.594	.602	.438	.159
$.6\pi$	.488	.484	.325	.160
$.7\pi$	.405	.414	.254	.155
$.8\pi$	.349	.390	.210	.159
$.9\pi$	.363	.353	.179	.178
π	.326	.342	.163	.171

Table 2. *Estimated Spectral Density Ordinates for the Constructed Bivariate Normal Example of Section 3.6.*

4.1. *The Model and the Prior*

To establish notation, write the multiple normal linear regression model,

$$y_i = x_i'\beta + \varepsilon_i, \quad \varepsilon_i \sim IIDN(0,\sigma^2) \quad (i = 1, \ldots, n),$$

where x_i is the i'th observation on a $k \times 1$ vector of explanatory variables. Alternatively the model can be expressed

$$y = X\beta + \varepsilon, \quad \varepsilon \sim N(0, I_n),$$

with each row of the $n \times 1$ vector y and $n \times k$ matrix X corresponding to a single observation. The likelihood function is

$$L(\beta, \sigma) \propto \sigma^{-n} \exp[-\sum_{i=1}^{n} (y - x_i'\beta)'(y - x_i'\beta)/2\sigma^2 = \sigma^{-n} \exp[(y - X\beta)'(y - X\beta)/2\sigma^2].$$

Since the likelihood function is the essential representation of the model, the developments reported in this section apply to any model that generates this likelihood function, including those with stochastic explanatory variables. The vector of unknown parameters is $\theta' = (\beta', \sigma)$.

The prior density is of the form $\pi(\beta, \sigma) = \pi_1(\beta)\pi_2(\sigma)$. The prior density for β can be expressed as $m(\leq k)$ linear combinations of β,

$$R\beta \sim N(r, T) \Leftrightarrow \pi_1(\beta) \propto \exp\left\{-\frac{1}{2}(R\beta - r)'T^{-1}(R\beta - r)\right\}$$

or

$$Q\beta \sim N(q, I_m) \Leftrightarrow \pi_1(\beta) \propto \exp\left\{-\frac{1}{2}(Q\beta - q)'(Q\beta - q)\right\},$$

where Q is a factorization of $T^{-1}, Q'Q = T^{-1}$, and $q = Qr$. When $m < k$ this prior is improper, but of course may be constructed as a limit of proper priors. The conjugate uninformative prior $\pi_2(\sigma) \propto \sigma^{-1}$ is assumed for σ, although what follows could easily be replicated for any of the family of proper, inverted gamma priors for σ, of which π_2 is a limit.

The posterior may be written

$$p(\beta, \sigma) \propto \sigma^{-(n+1)} \exp\left\{[\beta - \hat{\beta}(\sigma)]'[V(\sigma)]^{-1}[\beta - \hat{\beta}(\sigma)]\right\}, \tag{4.1}$$

with

$$\hat{\beta}(\sigma) = (X'X + \sigma^2 Q'Q)^{-1}(X'y + \sigma^2 Q'q) \tag{4.2}$$

and

$$V(\sigma) = \sigma^2(X'X + \sigma^2 Q'Q)^{-1}. \tag{4.3}$$

4.2. *Previous Approaches*

Analytical integration of the posterior density is not possible, even for linear functions of interest of the coefficients β. Most practical approaches, including the one taken here, rely on the observation that conditional on σ, the posterior density for β is multivariate normal. Theil and Goldberger (1961) suggested that σ^2 be fixed at $s^2 = (y - Xb)'(y - Xb)/(n - k)$, a procedure they termed "mixed estimation" because the mean and variance of the posterior density can then be computed using standard least squares regression software, appending the n entries in y with the m elements of sq, and the n rows of X with the m rows of sQ. Denote the point estimator

$$\hat{\beta}_{TG} = (X'X + s^2 Q'Q)^{-1}(X'y + s^2 Q'q)$$

and the corresponding variance

$$\hat{\text{var}}(\hat{\beta}_{TG}) = s^2(X'X + s^2Q'Q)^{-1}.$$

In part because of this convenient description, the Theil–Goldberger mixed estimator has proved popular and has been used in many applications.

Tiao and Zellner (1964) took up the problem of Bayesian inference for two normal linear regression models with the same coefficients but unequal variances in their disturbances. If one of the variances is known and the other is not, then the essentials of this problem are the same as the one posed here. They show that an asymptotic normal expansion of the posterior density yields the mean $\hat{\beta}_{TG}$ and the variance $\hat{\text{var}}(\hat{\beta}_{TG})$.

4.3. *The Gibbs Sampler*

We construct a two–step Gibbs sampler, based on the distribution of σ conditional on β, and the distribution of β conditional on σ. The posterior density of σ conditional on β is

$$p(\sigma|\beta) \propto \sigma^{-(n+1)} \exp[-(y - X\beta)'(y - X\beta)/2\sigma^2).$$

If we define $SSR(\beta) = (y - X\beta)'(y - X\beta)$, then

$$(SSR(\beta)/\sigma^2)|\beta \sim \chi^2(n).$$

The posterior density of β conditional on σ is normal as indicated by (4.1) - (4.3). However, it is computationally inefficient to invert the $k \times k$ matrix $X'X + \sigma^2Q'Q$ in each pass of the Gibbs sampling algorithm. Instead, let L be a factor of $(X'X)^{-1}, LL' = (X'X)^{-1}$. Let $L'Q'QL$ have diagonalization $P\Lambda P'$: i.e., Λ is a diagonal matrix of eigenvalues of $L'Q'QL$, and the columns of P are the corresponding, ordered eigenvectors normalized so that $P'P = PP' = I_k$. Finally, let $H = LP$. (These computations are only performed once, prior to the Gibbs sampling passes.) Then the variance of β conditional on σ may be expressed

$$(\sigma^{-2}X'X + Q'Q)^{-1} = H(\sigma^{-2}I_k + \Lambda)^{-1}H'.$$

This leads to a simple construction for β given σ. Construct $\varepsilon \sim N(0, I_k)$; scale ε_i by $(\sigma^{-2} + \lambda_i)^{-\frac{1}{2}}$ to form the vector ζ; form $\eta = H\varepsilon$: and then add the mean vector

$$(\sigma^{-2}X'X + Q'Q)^{-1}(\sigma^{-2}X'y + Q'q) = H(\sigma^{-2}I_k + \Lambda)^{-1}H'(X'y + \sigma^2Q'q),$$

using the right hand side of this expression to perform the computations. The number of multiplications required is proportional to k^3, the same as for matrix inversion, but the computations are nearly three times as fast for this method, suggesting that the factor of proportionality must be about one–third that for direct inversion.

4.4. *A Numerical Example*

This Gibbs sampling algorithm for Bayesian inference in the normal linear model with informative normal linear priors on the coefficients was coded in Fortran–77 using the IMSL Math/Library and IMSL Stat/Library. The results reported here were executed on a Sun Sparcstation 4/40 (IPC), in 64–bit arithmetic. We report results using 400 passes and 10,000 passes. For 400 passes, Gibbs sampling time was 0.32 seconds and the time required to form the periodograms and compute spectral density estimates at 21 ordinates was 4.2 seconds.

For 10,000 passes, Gibbs sampling time averaged 7.78 seconds and the spectral computations averaged 17.28 seconds.

The results reported here are based on artificial data from a model with $k = 3$ regressors. One regressor is an intercept term and the other two are orthogonal standard normal variates. The disturbance term is also standard normal. The coefficients are all 1.0. (Since the vector of coefficients β is generated jointly, conditional on σ, in our Gibbs sampling algorithm, results will not depend on the structure of the design matrix $X'X$; the orthogonal structure taken here is simply a convenient one.) The sample size is $n = 100$. Hence, with an uninformative prior the posterior distribution for β would be a multivariate Student t distribution with 97 degrees of freedom, mean approximately $(1, 1, 1)$, and variance approximately proportional to $(.01)I_3$. The informative prior employed is $\beta \sim N(r, (.01)I_3)$. Thus, the prior and the data are equally informative in the sense that the precision matrices associated with each are about the same.

	Mixed Estimate		Posterior, $p = 400$					Posterior, $p = 10,000$				
Parameter	Mean	s.d.	Mean	s.d.	NSE	RNE	CD	Mean	s.d.	NSE	RNE	CD
$r = (1, 1, 1)$												
β_1	.937	.073	.939	.065	.0032	1.028	−.631	.936	.066	.0007	.904	−.123
β_2	1.001	.073	.999	.069	.0033	1.075	−.152	1.001	.066	.0007	.970	−.555
β_3	1.045	.075	1.055	.064	.0034	.881	−.548	1.046	.068	.0007	.969	.251
σ^2			.823	.118	.0060	.968	.025	.818	.121	.0013	.851	−.457
$r = (1.5, 1.5, 1.5)$												
β_1	1.153	.073	1.185	.072	.0039	.830	1.017	1.185	.074	.0009	.720	−.041
β_2	1.222	.073	1.245	.070	.0036	.913	.521	1.249	.072	.0008	.736	−.202
β_3	1.288	.075	1.306	.072	.0034	1.111	.892	1.311	.072	.0007	1.069	−.913
σ^2			1.020	.155	.0095	.663	.857	1.019	.164	.0021	.590	−.498
$r = (2, 2, 2)$												
β_1	1.370	.073	1.638	.094	.0063	.557	−1.625	1.647	.095	.0012	.615	−.927
β_2	1.443	.073	1.685	.085	.0048	.775	−1.570	1.691	.092	.0011	.663	−.873
β_3	1.530	.075	1.746	.091	.0060	.579	−2.142	1.752	.089	.0010	.797	−1.754
σ^2			2.292	.409	3.146	.423	−2.991	2.333	.439	00620	.502	−.929
$r = (6, 6, 6)$												
β_1	3.101	.073	5.930	.110	.0059	.859	−.403	5.926	.102	.0011	.868	.208
β_2	3.210	.073	5.928	.098	.0051	.928	−.102	5.931	.100	.0010	1.067	−2.729
β_3	3.468	.075	5.949	.096	.0055	.759	1.216	5.943	.098	.0010	.982	−.459
σ^2			73.50	11.05	528.6	1.093	−1.451	74.33	10.74	10.65	1.017	−2.001
$r = (11, 11, 11)$												
β_1	5.266	.073	10.962	.105	.0050	1.086	−.242	10.963	.103	.0010	1.114	−3.05
β_2	5.420	.073	10.961	.095	.0045	1.119	.108	10.966	.099	.0010	.896	1.211
β_3	5.891	.075	10.975	.093	.0051	.847	−.773	10.972	.097	.0010	.960	−.757
σ^2			296.3	43.66	250.7	.758	−1.163	300.3	44.08	41.64	1.121	−.280

Table 3. *Linear Model with an Informative Prior. (See Section 4.4. for description.)*

The initial values for the algorithm are taken from the least squares estimates, one presample value was generated and discarded, and then the successive passes with computation of functions of interest at the end of each pass were initiated. Hence, $\beta^{(-1)} = b = (X'X)^{-1}X'y$ and $\sigma^{(-1)} = (s^2)^{1/2}$, where $s^2 = (y - Xb)'(y - Xb)/(n-k); \beta^{(0)}$ and $\sigma^{(0)}$ are discarded. This was done for the five alternative settings of the prior mean, r, reported in Table 3. When the prior mean is many standard deviations away from the sample mean, the initial values are violently unrepresentative of the posterior. Nevertheless, the diagnostics indicate no problems with convergence. Examination of the actual early values generated shows that the $\sigma^{(j)}$ sequence moves immediately from the neighbourhood of s^2, which is far too low when r is far from $(1, 1, 1)$, to values consistent with the mass of the posterior for σ. The contrast with the mixed estimates, in this regard is striking. The variance matrix (4.3) associated with this estimate implicitly takes s^2 as representative of σ^2, and does not reflect the larger plausible values which are implied when the sample and prior means are far apart. In the context of the asymptotic expansion of Tiao and Zellner (1964), that approximation is good to the extent that the data dominate the prior, a circumstance markedly uncharacteristic of situations in which the sample and prior means are far apart in the metric of sample precision.

The results in Table 3 strongly suggest that the Gibbs sampling algorithm provides an adequate solution to the problem of Bayesian inference in the normal linear model with an informative normal prior on the coefficients. Three aspects of the results support this conclusion. First, convergence beginning with the least square estimates is essentially instantaneous, even when these estimates provide parameter values at which the posterior density is quite low. Second, the relative numerical efficiency (RNE) for all parameters exceeds .5 and is often near 1. This might have been anticipated as an asymptotic result, since β and σ become independent as sample sizes increase without bound. In the examples studied here they are not, especially for values of r far from the sample mean. That the algorithm works so well in these circumstances is encouraging. Consistent with (but not implied by) the RNE values between .5 and 1.2, spectral densities of all parameters appear nearly flat, and are not reported here. Third, computation times are reasonable. The structure of the problem and the experience with this example indicate computation time of about $(8.5 \times 10^{-7})npk^3$ seconds. This implies reasonable desktop computing times for most of the econometric applications of this model.

5. INFERENCE IN THE TOBIT REGRESSION MODEL

Limited dependent variable models constitute one of the principal tools of applied econometrics. In some of these models the dependent variable is dichotomous, reflecting a decision to purchase or not purchase a durable good, whether or not to retire, etc. The probit and logit models are often used in these situations. In other cases decisions are of the form, "whether or not, and if so then how much?". This characterizes the form of many investment decisions, like the construction of a plant, and consumption decisions, like the purchase of an automobile. The Tobit model, introduced by Tobin (1958) is probably the most widely applied model in these situations. The monograph of Maddala (1983), and the three chapters of Grilliches and Intrilligator (1984) provide a thorough discussion of these and related limited dependent variable models.

5.1. *The Model and the Prior*

To establish notation, write the Tobit regression model,

$$y_i^* = x_i'\beta + \varepsilon_i, \quad \varepsilon_i \sim IIDN(0, \sigma^2) \quad (i = 1, \ldots, n), \tag{5.1}$$

$$y_i = \begin{cases} y_i^*, & \text{if } y_i^* \geq 0 \\ 0, & \text{if } y_i^* < 0. \end{cases} \quad (5.2)$$

We observe $\{x_i, y_i\}_{i=1}^n$; the $x_i's$ are $k \times 1$ vectors; the $y_i's$ are scalars; and the y_i^*'s are unobserved. For notational convenience, order the observations so that c censored observations (those for which $y_i = 0$) come first, followed by the $n - c$ uncensored observations. Let $y_2 = (y_{c+1}, \ldots, y_n)'$ and let $X_2' = [x_{c+1}, \ldots, x_n]$. Then the kernel of the likelihood function is

$$\Pi_{i=1}^c[1 - \Phi(x_i'\beta/\sigma)]\sigma^{(n-c)} \exp\{-(y_2 - X_2\beta)'(y_2 - X_2\beta)/2\sigma^2\}.$$

The prior density is exactly the same as that employed in the previous section: $\pi(\beta, \sigma) = \pi_1(\beta)\pi_2(\sigma)$ with the prior density for β expressed as $m(\leq k)$ linear combinations of β,

$$R\beta \sim N(r, T) \Leftrightarrow \pi_1(\beta) \propto \exp\{-\frac{1}{2}(R\beta - r)'T^{-1}(R\beta - r)\}$$

or

$$Q\beta \sim N(q, I_m) \Leftrightarrow \pi_1(\beta) \propto \exp\{-\frac{1}{2}(Q\beta - q)'(Q\beta - q)\},$$

where Q is a factorization of $T^{-1}, Q'Q = T^{-1}$, and $q = Qr; \pi_2(\sigma) \propto \sigma^{-1}$. The posterior density may be written

$$p(\beta, \sigma) \propto \Pi_{i=1}^c[1 - \Phi(x_i'\beta/\sigma)]\sigma^{(n-c)} \exp\{[\beta - \hat{\beta}(\sigma)]'[V(\sigma)]^{-1}[\beta - \hat{\beta}(\sigma)]\},$$

with

$$\hat{\beta}(\sigma) = (X_2'X_2 + \sigma^2 Q'Q)^{-1}(X_2'y_2 + \sigma^2 Q'q)$$

and

$$V(\sigma) = \sigma^2(X_2'X_2 + \sigma^2 Q'Q)^{-1}.$$

5.2. *Previous Approaches*

Maximum likelihood is well established as the principal method of frequentist inference in the Tobit censored regression model. Maddala (1983) and Amemiya (1984) provide thorough surveys. Bayesian inference in a closely related model that arises in biomedical applications has been discussed by Carriquiri *et al.* (1987) and Sweeting (1987). The most thorough Bayesian treatment of this model is provided by Chib (1990), who has implemented and compared Monte Carlo integration with importance sampling, Laplace approximations, and Gibbs sampling with data augmentation. The implementation of Gibbs sampling and data augmentation reported here extends Chib's treatment in three dimensions. First, the methodology of Section 3 is used to adduce evidence on convergence and the accuracy of the numerical approximations. Second, an informative prior is permitted, whereas Chib (1990) uses only uninformative priors. Third, the research reported here entails a controlled study of the effect of alternative starting values and convergence. The outcome suggests caution in interpreting informal diagnostics for convergence.

5.3. *The Gibbs Sampler with Data Augmentation*

Construction of a three–step Gibbs sampler with data augmentation is straightforward. Conditional of β and σ, y_i^* has a truncated normal distribution, constructed from $N(x_i'\beta, \sigma^2)$ truncated above at 0; i.e., the p.d.f. of y_i is

$$f(y_i^*|\beta,\sigma) = \begin{cases} [1-\phi(x_i'\beta/\sigma)]^{-1}\exp[-(y_i^* - x_i'\beta)^2/2\sigma^2)], y_i^* > 0; \\ 0, y_i^*; > 0 \end{cases}$$

for $i = 1, \ldots, c; y_i^* = y_i$ for $i = c+1, \ldots, n$. An algorithm for generating from the truncated univariate normal distribution, described in Geweke (1991) and considerably faster than either naive rejection methods or the conventional construction of Devroye (1986), is employed. This data augmentation constitutes the first step of the Gibbs sampler.

Conditional on $\{y_i^*\}_{i=1}^n$, the problem reduces to precisely the one set forth and solved in Section 4. This constitutes the second and third steps of the Gibbs sampler with data augmentation.

5.4. *A Numerical Example*

An artificial sample of size 200 was constructed, using a data generating process similar to Wales and Woodland (1980):

$$x_{i3}, z_{i2}, z_{i3} \text{ IID Uniform } (-1, 1);$$

$$x_{i2} = z_{i2} + z_{i3} + u_i, \text{ with } u_i \sim N(0, 1.312);$$

$$\varepsilon_i \sim N(0, .6428);$$

$$y_i^* = -1.1854 + 1.0x_{i2} + 10.0x_{i3} + \varepsilon_i;$$

$$y_i = \begin{cases} y_i^*, & \text{if } y_i^* \geq 0 \\ 0, & \text{if } y_i^* < 0. \end{cases}$$

The five random variables $x_{i3}, z_{i2}, z_{i3}, u_i, \varepsilon_i$ are mutually and serially independent. Of the generated sample, 114 obsevations are censored.

A small, full–factorial experiment was conducted. The total number of passes was taken to be either 400 or 10,000; no preliminary passes were taken, or the number of preliminary passes was set equal to the total number of passes; and four alternative initial values for $\theta = (\beta_1, \beta_2, \beta_3, \sigma^2)'$ were used. The four alternative starting values were:

(i) Uncensored ordinary least squares. $\beta^{(0)}$ is the least squares vector and $\sigma^{2(0)}$ is the corresponding value of s^2, from application of least squares to the full set of 200 observations with $y_i = 0$ for censored observations;

(ii) Censored ordinary least squares. $\beta^{(0)}$ is the least squares vector and $\sigma^{2(0)}$ is the corresponding value of s^2, from application of least squares to the 86 uncensored observations.

(iii) The augmented posterior mode. The augmented posterior density is a function of 118 variables. A computationally efficient method of finding this mode is to apply the Gibbs sampling algorithm with data augmentation, except that each vector is set to its conditional modal value rather than generating from the conditional posterior density.

(iv) The augmented posterior mode with censored $\sigma^{2(0)}$. This is a hybrid of (ii) and (iii), $\beta^{(0)}$ from (ii) and $\sigma^{2(0)}$ from (iii).

	400 passes				10,000 passes			
	0 preliminary		400 preliminary		0 preliminary		10,000 preliminary	
	Mean	CD	Mean	CD	Mean	CD	Mean	CD
Initial value: Uncensored OLS ($\beta_1 = 1.871, \beta_2 = .801, \beta_3 = 4.082, \sigma^2 = 2.357$)								
β_1	−1.272	3.098	−1.328	−3.719	−1.322	4.711	−1.318	3.663
β_2	.929	−1.242	.936	1.037	.936	−.682	.936	−1.517
β_3	9.941	−2.877	10.038	3.641	10.028	−4.487	10.025	−3.817
σ^2	.785	1.415	.714	1.358	.718	.901	.717	.617
Initial value: Censored OLS ($\beta_1 = -.898, \beta_2 = .928, \beta_3 = 9.423, \sigma^2 = 677$)								
β_1	−1.255	4.212	−1.281	−1.336	2.083	−1.311	1.151	
β_2	.935	−.468	.931	3.226	.936	−1.002	.935	1.600
β_3	9.912	−3.591	9.943	1.368	10.051	−1.754	10.010	−1.160
σ^2	757	1.455	.718	.390	.721	.672	.715	.125
Initial value: Augmented posterior mode ($\beta_1 = 1.043, \beta_2 = .928, \beta_3 = 9.624, \sigma^2 = .293$)								
β_1	−1.231	4.138	−1.304	−.650	−1.320	−1.176	−1.326	.817
β_2	.930	−2.532	.937	.380	.935	.966	.934	1.159
β_3	9.874	−4021	9.998	−.693	10.025	.947	10.038	−1.071
σ^2	.763	1.570	.711	.507	.717	1.032	.714	1.613
Initial value: Augmented posterior mode, censored σ^2 ($\beta_1 = 1.043, \beta_2 = .928, \beta_3 = 9.624, \sigma^2 = .677$)								
β_1	−1.298	5.920	−1.305	−3.194	−1.327	2.780	−1.314	−1.946
β_2	.927	−.616	.932	.431	.933	−1.363	.935	1.611
β_3	10.000	−4.781	10.005	3.891	10.038	−2.855	10.017	1.573
σ^2	.787	1.503	.694	−.543	.719	.602	.717	.615

Table 4. *Convergence Diagnostics, Tobit Censored Regression Model.*

Estimated posterior means and convergence diagnostics for all 16 cells are provided in Table 4, along with numerical values for the initial vectors. Table 5 provides greater detail, for some selected cells in which convergence diagnostics were satisfactory. Table 6 provides the estimated spectral densities of the sampled parameters from two of the cells. As a final check on the results reported here, the last cell in Table 5 was reexecuted, but using 1,000,000 preliminary passes rather than 10,000: the results were within the range anticipated from the NSE's for that cell. The computation times reported in Table 5 were realized on a Sun Sparcstation 4/40 (IPC), with software written in double precision Fortran–77 using the IMSL Math/Library and IMSL Stat/Library. These times correspond roughly to a 20-fold increase in speed over the similar computations of Chib (1990), who used a 16 Mhs 3MB personal computer with the Gauss programming language.

Censored OLS initial value; 10,000 passes; 10,000 preliminary passes
Execution times: Preliminary passes, 96.49; Gibbs sampling, 96.06; Spectral, 17.26

	Passes 1 – 1,000		Passes 5,001–10,000		All passes				
	Mean	St Dev	Mean	St Dev	Mean	St Dev	NSE	RNE	CD
β_1	–1295	.165	–1.316	.186	–1.311	.179	.0075	.057	1.151
β_2	.940	.062	.934	.059	.935	.059	.0011	.313	1.600
β_3	9.985	.275	10.018	.309	10.010	.300	.0124	.059	–1.160
σ^2	.715	.111	.714	.113	.715	.111	.0023	.226	.125

Augmented mode initial value; 10,000 passes; 0 preliminary passes
Execution times: Preliminary passes, 0.00; Gibbs sampling, 98.00; Spectral, 17.32

	Passes 1 – 1,000		Passes 5,001–10,000		All passes				
	Mean	St Dev	Mean	St Dev	Mean	St Dev	NSE	RNE	CD
β_1	–1.334	.235	–1.308	.202	–1.320	.201	.0088	.052	–1.176
β_2	.937	.063	.933	.060	.935	.060	.0011	.279	.966
β_3	10.040	.418	10.004	.340	10.025	.340	.0146	.054	:947
σ^2	.752	.630	.711	.111	.717	.225	.0047	.231	1.032

Augmented mode initial value; 10,000 passes;10,000 preliminary passes
Execution times: Preliminary passes,97.11; Gibbs sampling, 100.43; Spectral, 18.30

	Passes 1 – 1,000		Passes 5,001–10,000		All passes				
	Mean	St Dev	Mean	St Dev	Mean	St Dev	NSE	RNE	CD
β_1	–1.316	.190	–1.332	.195	–1.326	.195	.0086	052	817
β_2	.937	.061	.933	.061	.934	.060	.0011	.325	1.159
β_3	10.014	.313	10.049	.326	10.038	.327	.0143	.053	–1.071
σ^2	.728	.120	.715	.111	.714	.111	.0023	.235	1.613

Augmented mode with censored σ^2 initial value; 10,000 passes; 10,000 preliminary passes
Execution times: Preliminary passes,96.71; Gibbs sampling, 97.18; Spectral, 18.03

	Passes 1 – 1,000		Passes 5,001–10,000		All passes				
	Mean	St Dev	Mean	St Dev	Mean	St Dev	NSE	RNE	CD
β_1	–1.350	.189	–1.312	.187	–1.314	.188	.0083	.052	–1.946
β_2	.940	.062	.934	.061	.935	.060	.0011	.291	1.611
β_3	10.068	.319	10.017	.317	10.017	.317	.0138	053	1.575
σ^2	.722	.113	.718	.116	.717	.115	.0025	.215	.615

Table 5. *Tobit Censored Regression Model.*

The results of these experiments can be organized in several dimensions.

(1) 400 passes are generally insufficient for convergence. Only one of the eight cells, with augmented posterior mode initial values and 400 preliminary passes, performs satisfactorily. Poor convergence diagnostics correspond to estimated posterior means that are up to one-half posterior standard deviation from the values reported in Table 3. Thus, reasonable but unconverged values could be significantly misleading.
(2) Consistent with this finding, preliminary passes are important in producing reliable results. In many cells, preliminary passes are necessary to produce satisfactory convergence diagnostics.
(3) The augmented posterior mode exhibits strikingly better performance as an initial value than do the other initial values. From one perspective this is surprising. At the augmented posterior mode most of the y_i^* lie on the regression plane $x_i'\beta$– i.e., the corresponding ε_i^* are set to zero. Since 114 observations are censored this produces a very low value of $\sigma^{2(0)}$, and increasing the value of $\sigma^{2(0)}$ to a more reasonable value, which was done in the final cell of the experimental design in this diomension (see (iv), above) makes matters worse rather than better. The corresponding values of $\beta^{(0)}$ are no closer to the posterior mean than the other initial values in the experiment. From another perspective, this outcome is not surprising. At the augmented posterior mode the 118–dimensional density relevant for Gibbs sampling with data augmentation is high (by definition), and given the smoothness inherent in this problem movement in the various dimensions is more likely and hence convergence is more rapid starting from this point.

	400 passes and preliminary passes				10,000 passes and preliminary passes			
Frequency	β_1	β_2	β_3	σ^2	β_1	β_2	β_3	σ^2
$.00\pi$	.1315	.0090	.4047	.0367	.7360	.0110	2.0270	.0522
$.05\pi$	.1305	.0084	.4061	.0368	.2244	.0089	.6178	.0406
$.10\pi$	.1252	.0077	.3921	.0333	.0486	.0077	.1369	.0335
$.15\pi$	.0956	.0076	.2967	.0256	.0201	.0070	.0557	.0263
$.20\pi$	.0314	.0070	.1045	.0225	.0121	.0063	.0382	.0187
$.25\pi$	.0130	.0057	.0382	.0147	.0092	.0054	.0285	.0146
$.30\pi$	.0085	.0050	.0262	.0121	.0073	.0041	.0223	.0122
$.35\pi$	.0056	.0037	.0203	.0104	.0059	.0036	.0175	.0100
$.40\pi$	.0044	.0032	.0180	.0085	.0046	.0027	.0153	.0083
$.45\pi$	.0039	.0025	.0143	.0074	.0040	.0027	.0138	.0070
$.50\pi$	.0039	.0020	.0142	.0062	.0036	.0025	.0119	.0060
$.55\pi$	.0036	.0018	.0123	.0052	.0035	.0021	.0105	.0057
$.60\pi$	.0031	.0017	.0117	.0046	.0030	.0017	.0107	.0050
$.65\pi$	.0027	.0016	.0103	.0038	.0028	.0016	.0105	.0041
$.70\pi$	.0029	.0014	.0089	.0036	.0026	.0016	.0098	.0038
$.75\pi$	.0029	.0013	.0095	.0032	.0026	.0017	.0098	.0037
$.80\pi$	.0030	.0012	.0092	.0036	.0025	.0016	.0091	.0034
$.85\pi$	.0026	.0011	.0084	.0034	.0025	.0013	.0092	.0036
$.90\pi$	.0023	.0010	.0076	.0030	.0028	.0013	.0084	.0034
$.95\pi$	..0020	.0007	.0060	.0024	.0025	.0012	.0084	.0034
$.100\pi$	.0016	.0057	.0047	.0019	.0012	.0006	.0042	.0016

Table 6. *Spectral Densities of Sampled Parameters, Tobit Censored Regression Model.*

(4) There are quite substantial differences in the serial correlation properties of the sampled parameters, as indicated in Table 6. The parameters β_1 and β_3 exhibit very strong positive

serial correlation, to the extent that a poor picture of the pattern emerges with only 400 observations. Virtually without exception, these are the only parameters that exhibit poor convergence diagnostics. With a small number of observations it is impossible to distinguish between nonstationarity and high power at low frequencies. For our purposes the distinction is uninteresting, since either will lead to unreliable approximations of posterior moments if the number of passes is too small.

(5) As is necessarily the case, these parameters exhibit poor RNE's. The Gibbs sampler with data augmentation requires about 20 times as many passes as direct Monte Carlo sampling from the posterior would require independent samples (were that possible). From this perspective, the effective number of passes in the experiment is either 20 (when $p = 400$) or 500 (when $p = 10,000$) for β_1 and β_3. This is consistent with our inability to obtain satisfactory results for these parameters when $p = 400$.

6. CONCLUSIONS

Gibbs sampling with data augmentation is an attractive solution of the Bayesian multiple integration problem whenever the parameters of the (augmented) posterior density can be grouped in such a way that the distribution of any group conditional on the others is of a standard form from which synthetic sampling is straightforward. As our ingenuity in expressing posterior densities in this form increases, many standard models become amenable to Bayesian treatment using this method. It now appears that most of the standard applied econometric models can be cast in a form appropriate for Gibbs sampling. Even more promising, awareness of the Gibbs sampler and data augmentation are likely to suggest new models that are amenable to Bayesian inference, that are intractable using analytical or other numerical methods.

Yet one must be cautious. A great attraction of this approach, not to be minimized, is that it is in general straightforward to apply relative to other numerical methods. This means that less of the investigator's time is spent on arcane numerical issues, mistakes are less likely to be made, and incorporation into interactive software is more practical. But there is no small distance between having a method justified solely by a convergence result, and one which reliably produces approximations of integrals whose accuracy can be reliably assessed. Formal assessment of convergence and numerical accuracy are essential to rendering the Gibbs sampler a tool of replicable scientific studies, because of the pseudorandomness inherent in the method.

It is risky to speculate on productive avenues of future research based on the limited collective experience with the Gibbs sampler in Bayesian inference, but three seem clear. First, it is practical to write software that varies many aspects of the experimental design implicit in the Gibbs sampler, so as to produce satisfactory outcomes and economize on machine time. For example, preliminary passes can be used to compute convergence diagnostics, measure relative computation times for Gibbs sampling (on the one hand) and computation of functions of interest (on the other), and get at least a rough estimate of the spectral densities of the sampled processes at the zero frequency. This would determine the number of subsequent passes needed to attain desired numerical accuracy, and the appropriate points for computing functions of interest. Second, it would be very helpful to obtain additional insight on the relation of the internal structure of problems to the stochastic properties of the Gibbs sampler. The results presented here are suggestive, but serve mainly to raise questions. Can we use information about serial correlation to structure the pattern of steps within passes to achieve a greater computational efficiency? Is the use of the augmented posterior mode, which proved very helpful in the example in Section 5.4, an attractive starting point for most problems? Finally, the Gibbs sampler with data augmentation is inherently an asynchronously parallel algorithm, in which nodes compute inner products. This distinguishes it from other approaches, like Monte Carlo integration which is inherently distributed and series expansions which are inherently serial. Thus, there is at least the possibility that this method could prove practical and highly competitive in an enviroment of parallel architectures.

ACKNOWLEDGEMENTS

Financial support from National Science Foundation Grant SES–8908365 and research assistance from Zhenya Wang are gratefully acknowledged. Software and data may be requested by electronic mail to *geweke@atlas.socci.umn.edu*.

REFERENCES

Amemiya, T. (1984). Tobit models: a survey. *J. Econometrics* **24**, 3–61.

Carriquiri, A. L., Gianola, D. and Fernado, R. L. (1987). Mixed–model analysis of a censored normal distribution with reference to animal breeding. *Biometrics* **43**, 929–939.

Chib, S. (1990). Bayes inference in the Tobit censored regression model, (to appear).

Devroye, L. (1986). *Non–uniform Random Variate Generation*. New York: Springer.

Gelfand, A. E. and Smith, A. F. M. (1990). Sampling based approaches to calculating marginal densities. *J. Amer. Statist. Assoc.* **85**, 398–409.

Geman, S. and Geman, D. J. (1984). Stochastic relaxation, Gibbs distributions and the Bayesian restoration of images. *IEEE Transactions on Pattern analysis and Machine Intelligence* **6**, 721–741.

Geweke, J. (1988). Antithetic acceleration of Monte Carlo integration in Bayesian inference. *J. Econometrics* **38**, 73–90.

Geweke, J. (1989). Bayesian inference in econometric models using Monte Carlo integration. *Econometrica* **57**, 1317–1339.

Geweke, J. (1991). Efficient simulation from the multivariate normal and Student–*t* distributions subject to linear constraints. *Computing Science and Statistics: Proceedings of the Twenty–Third Symposium of the Interface*, (to appear).

Griliches, Z. and Intrilligator, M. (1984). *Handbook of Econometrics*. Amsterdam: North-Holland.

Hannan, E. J. (1970). *Multiple Time Series*. New York: Wiley.

Kloek, T. and Van Dijk, H. K. (1978). Bayesian estimates of equation systems parameters: an application of integration by Monte Carlo. *Econometrica* **46**, 1–20.

Leonard, T., Hsu, J. S. J. and Tsu, K.-W. (1989). Bayesian marginal inference. *J. Amer. Statist. Assoc.* **84**, 1051–1058.

Maddala, G. S. (1983). *Limited–Dependent and Qualitative Variables in Econometrics*. Cambridge: University Press.

Sweeting, T. J. (1987). Approximate Bayesian analysis of censored survival data. *Biometrika* **74**, 809–816.

Tanner, M. A. and Wong, W.–H. (1987). The calculation of posterior distributions by data augmentation. *J. Amer. Statist. Assoc.* **82**, 528–550.

Theil, H. and Goldberger, A. S. (1961). On pure and mixed statistical estimation in economics. *International Economic Review* **2**, 65–78.

Tiao, G. C. and Zellner, A. (1964). Bayes' Theorem and the use of prior knowledge in regression analysis. *Biometrika* **51**, 24–36.

Tierney, L. and Kadane, J. B. (1986). Accurate approximations for posterior moments and marginal densities. *J. Amer. Statist. Assoc.* **81**, 82–86.

Tobin J. (1958). Estimation of relationships for limited dependent variables. *Econometrica* **26**, 24–36.

Wales, T. J. and Woodland, A. D. (1980). Sample selectivity and the estimation of labor supply functions. *International Economic Review* **21**, 437–468.

Zellner, A. (1971). *An Introduction to Bayesian Inference in Econometrics*. New York: Wiley.

DISCUSSION

J. C. NAYLOR (*Nottingham Polytechnic, UK*)

Professor Geweke describes the application of methods in time series analysis to the approximation of posterior expectations using the Gibbs sampling algorithm. Such an approach is shown to deal with some of the perceived problems in using correlated and non–iid samples. A useful measure of relative numerical efficiency is also obtained which demonstrates that correlated samples may under certain circumstances be preferable. However, I think there are a number of difficulties in this application of these methods.

The tools available to describe the posterior density would normally include a variety of plots for marginal and predictive distributions for quantities of interest. Moments are of value if the posterior density is of appropriate form, for example multivariate normal, but plots would still be needed to verify this assumed form. The definition of the "generic task", Section 2, needs to be extended to the simultaneous computation of many expectations some of which may be over subspaces, see for example Naylor and Smith (1982).

Curiously the other approaches in Section 2.1 do not include methods of numerical integration, the only non–sampling approaches being asymptotic approximations. Professor Geweke kindly supplied data for the Tobit example, and an analysis using iterative Gauss–Hermite integration (Naylor and Smith 1982), easily produced posterior moments in agreement with those in Table 5. Additionally marginal posterior densities, which appeared normal, and predictive densities for selected values of the covariates, which appeared distinctly non–normal, were also obtained. Any of the methods in 2.1 might be expected to work for such a well behaved (i.e., normal) posterior. Can we assume that the time series approach will work just as well with non–normal posterior densities?

Alternative configurations of the Gibbs sampler include the generation of several sets of values on each pass or, equivalently, the combination of results from several applications of the algorithm. This has the disadvantage of discarding many of the points generated but does provide an uncorrelated sample from which posterior densities may be estimated.

If we are only interested in a selection of posterior moments this version of Gibbs has efficiency gains and the proposal to use time series methods is attractive. In the introduction a 'number of solutions' is mentioned; it is not clear why methods based on spectral analysis are in fact selected. These do not appear computationally efficient, requiring between 69% and 93% of the total computation time for the regression example.

The proposed convergence diagnostic (CD) is based on a comparison of, for example, the first 10% of a series with the last 50%. Superficially a decision for convergence in effect concludes that the first 10% had converged, the remaining 90% being just a check! More seriously, this CD appears to take the form of a frequentist Z–test in a situation that is not really appropriate:

(i) The test would, presumably, be applied on several 'passes' of the algorithm for each of several (many) functions of interest. How should results such as those in Table 3, in which 10% of the values would be declared 'significant at the 5% level', be interpreted? Perhaps a single 'key' function should be selected for the CD and a sequential test sought.

(ii) Often, in practice, data is collected in the hope of obtaining sufficient evidence to reject H_0. Further there is some sort of asymptotic guarantee of success in this endeavour if collection continues. The situation here is curiously reversed; we are looking for convergence in distribution to some asymptotic result and hence seek the collection of sufficient evidence to eventually accept H_0! A formal decision rule would involve a utility function in which the cost of continuing sampling would be balanced against the possible gains in information.

Finally, I should like to thank Professor Geweke for a thought provoking paper.

A. GELMAN (*University of California, USA*) and
D. B. RUBIN (*Harvard University, USA*)

Professor Geweke has presented a very clear exposition of the use of time series methods to monitor the convergence of the Gibbs sampler using one simulated sequence. Ripley (1987) also provides an excellent discussion combined with a brief historical overview of these techniques. Our use of basically the same technology has convinced us that such methods cannot be generally successful at monitoring convergence, because of long–range dependence that is undetectable from a single sequence, even when the parameter of interest has a unimodal distribution. Our article elsewhere in these proceedings (Gelman and Rubin, 1992) provides a striking example and shows how lack of convergence can be made apparent by examining multiple series and comparing within–series and between–series variabilities. Simple statistical procedures, even simpler than the time–series methods, can then be applied to multiple series to reliably monitor convergence in a larger class of problems than can be handled using single series.

Details of our methods appear in Gelman and Rubin (1991); our recommendations can be briefly summarized as follows: first independently simulate several, say m, series with starting points drawn from an approximately centered but overly dispersed distribution. (For example, a multivariate Cauchy, centered near the posterior mode and with scale determined by the second derivative of the log–posterior density, is often a good starting distribution. The required location and scale can often be obtained by ECM and SECM; see the article by Meng and Rubin, 1992, in these proceedings.) Second, discard the early values of each sequence, say the first half, leaving n iterates per sequence. Third, for each scalar parameter of interest, calculate the average of the m series variances, and the variance between the m series means. If the within–series variance is not appreciably larger than the between–series variance, then the series have not yet come close to converging to a common distribution. With a single sequence, the crucial between–series variance cannot be estimated, and as a result, methods such as in Geweke (1991) are generally unreliable.

N. G. POLSON (*University of Chicago, USA*) and
G. O. ROBERTS (*University of Nottingham, UK*)

This paper addresses the fundamental issue in the application of the Monte Carlo method, of how to estimate functionals of interest from observations of a stationary Markov process. Specifically, the Markov chain is defined explicitly in terms of the one–dimensional conditionals of the posterior, π. A convergence diagnostic is proposed to diagnose 'convergence' from one long run of the algorithm, and estimates are calculated by averaging along one long chain.

First, we discuss from a mathematical perspective a rigorous approach to forming Monte Carlo estimates, and secondly, we indicate why there is a need for diagnostics, together with empirical evidence of the efficiency of the Gibbs sampler in statistical applications.

To fix ideas, consider a general Markov chain defined on a finite space, V, and with stationary distribution π. In statistical applications, imagine V as a 'fine' discretisation of the parameter space $\Theta \in \mathcal{R}^k$, so that, typically, $|V| \sim O(P^k)$ for some large integer P. Let $H = E_\pi(h)$ be the functional of interest, and let K be the number of steps of the chain (possibly defined by Gibbs sampling). Consider the estimator $H_K = K^{-1}\sum_{n=1}^{K} h(X^{(n)})$ where $X^{(n)}$ denotes the nth iterate of the chain. Peskun (1973) demonstrates a central limit theorem for H_K namely

$$K^{1/2}(H_K - H) \xrightarrow{D} N(0, \sigma^2) \quad \text{as} \quad K \to \infty.$$

The author identifies σ^2 in terms of the spectral density function, and this provides the basis for his asymptotics. However Aldous (1987) criticises such asymptotics for not providing good estimates in $o(|V|)$steps of the chain. In the special case where π is uniform, Aldous proceeds to give a detailed analysis of the class of ergodic averages of the form

$$H_{T,N} = \frac{1}{N} \sum_{n=T+1}^{T+N} h(X^{(n)},$$

by deriving lower bounds on (T, N) in terms of the second eigenvalue of the transition matrix, and the density of the starting point. Intuitively, one selects T large enough to reduce bias, and N to obtain the desired accuracy.

In fact, in the operations research literature, many simulations issues have already been the focus of attention, for example the question of multiple versus one long simulation, see for example Kelton and Law (1984), Kelton (1986) and Whitt (1989). The general consensus among these authors is that one long run will lead to increase efficiency. However, the Gibbs sampler requires special attention, because the Markov Chain induced is typically highly complex, and knowledge of its convergence rate is rarely available. It is usually therefore sensible to run multiple replications of the algorithm. This involves a trade–off between a slight decrease in efficiency, and an increased knowledge of the Markov chain involved.

To provide polynomial time bounds (that is $o(|V|)$ steps) for the Metropolis algorithm and the Gibbs sampler, Applegate, Kannan and Polson (1990) bound the crucial second eigenvalue (the geometric rate of convergence for the Markov chain) in terms of the 'conductance' of the chain, see also Sinclair and Jerrum (1988). Conditions for the geometric convergence of the Gibbs sampler have been given (by Roberts and Polson 1991, and Schervish and Carlin 1990), and sometimes bounds on geometric rates are available. However, these general bounds require further attention in the case of the Gibbs sampler in particular examples, where theoretical estimates are invariably poor. Therefore stylised classes of statistical applications require extensive empirical work, and the use of carefully chosen diagnostics.

We note that the diagnostic proposed here confines attention to the functional of interest, and so conveys no information about convergence in distribution for iterates. This clearly restricts the use of the resulting sample to the estimating of the monitored functional. However, perhaps more seriously, it could also wrongly diagnose convergence, even of the functional of interest itself, especially if the starting value is close to the true value, but 'far' from the stationary distribution (measured in terms of time till 'convergence'). An approach that attempts to overcome these difficulties appears in Roberts (1992).

REPLY TO THE DISCUSSION

The theory that underlies the Gibbs sampler provides very general guidance for the design of procedures and the evaluation of accuracy. Convergence is known to be geometric, but known bounds on the rate are gross (as Polson and Roberts point out) and the key question of sensitivity to initial conditions remains. To this one might add that even if tight bounds could be obtained with substantial analytical effort on a case–by–case basis, the resulting methods would not be competitive in applied work. Thus, there exists analytical motivation for a broad class of procedures but the particulars remain to be worked out. This situation is very much like the one that arises in the serious application of asymptotic theory, either in a frequentist context or in the construction of approximate posterior densities.

This state of affairs leads naturally to a set of procedures familiar to all statisticians. The logic is straightforward. There is a set of circumstances which motivates the standard interpretation of the Gibbs sampling approximation to posterior moments, and which justifies related methods for evaluating the accuracy of this approximation. My paper presents one method; Gelman and Rubin provide another, and several others undoubtedly have been or will be developed. One hopes that the set of circumstances is pertinent to the problem at hand. If it is, then not only will various approximations and evaluations be valid, but there will also be other, observable relationships in the output of the Gibbs sampling experiment as well.

One of these relationships is the limiting standard normal distribution of the convergence diagnostic proposed in my paper. Dr. Naylor points out that it is unsatisfactory to have to work with a whole battery of such statistics when there are several functions of interest, as there almost always will be. I completely agree. There is a corresponding chi-square statistic that is a leading candidate for the "key" function he suggests, and this clearly merits further investigation.

A second set of relationships arises if the numerical experiments are organized as several replicates of the sequential sampling process described in my paper. Polson and Roberts propose multiple replications of the algorithm, presumably by using multiple independent initial conditions. It would appear that in practice these starting values must be drawn from some distribution other than the posterior, for it is the difficulty or impossibility of sampling directly from the posterior that is the most compelling practical justification for the Gibbs sampler. Gelman and Rubin provide a nice example of how such a scheme might be implemented, with starting values drawn from a distribution that would constitute a satisfactory importance sampling density for Monte Carlo integration in most cases. The effect of initial conditions is then embodied in the confounding of the importance sampling density with the Markov chain at each step in the sequence, but this is more tractable than the effects of an arbitrary starting value. In the output of the Gibbs sampling experiment, the variance of the Gibbs approximation in each replication, as assessed by the methods described in my paper, ought to be about the same for each replication and about equal, in turn, to the observed variance in the Gibbs approximation itself across the replications of the algorithm.

In fact there are many circumstances in which a simple modification of the methods proposed by Gelman and Rubin in their comment (and elsewhere) will lead to independent initial conditions drawn from the posterior density itself. If one can bound the ratio of the posterior to importance sampling density, this can be achieved through rejection sampling from the importance sampling density. One thereby avoids the occurrence of initial conditions far out in the tails of the posterior, from which convergence using the Gibbs sampler might be slow. In principle, one could implement i.i.d. sampling directly from the posterior density in this way. But in practice, the ratio of rejected draws to those accepted may be quite high– say, of the order of 10^5 or 10^6. This ratio might well be intolerable in simple Monte–Carlo sampling, but in Gibbs sampling it would be no more than a minor nuisance since it arises only at the initialization of the sequence. In general, the combination of Gibbs sampling with other methods of Monte Carlo integration, and in sufficiently low dimension the numerical integration methods cited by Naylor, may prove a fertile ground for providing practical and reliable evaluations of the accuracy of sampling–based approaches to the calculation of posterior moments.

ADDITIONAL REFERENCES IN THE DISCUSSION

Aldous, D. (1987). On the Markov chain simulation method for uniform combinatorial distributions and simulated annealing. *Probability in Engineering and Information Science* **1**, 33–46.

Applegate, D., Kannan, R. and Polson, N. G. (1990). Random polynomial time algorithms for sampling from joint distributions. *Tech. Rep.* **500**, Carnegie Mellon University.

Gelman, A. and Rubin, D. B. (1991). Honest inferences from iterative simulation. *Tech. Rep.* **307**, University of California.

Gelman, A. and Rubin, D. B. (1992). A single series from the Gibbs sampler provides a false sense of security. *Bayesian Statistics 4* (J. M. Bernardo, J. O. Berger, A. P. Dawid and A. F. M. Smith, eds.), Oxford: University Press, 627–633.

Kelton, W. D. and Law, A. M. (1984). An analytical evaluation of alternative strategies in steady state simulation. *Oper. Research* **32**, 169–184.

Kelton, W. D. (1986). Replication splitting and variance for simulating discrete–parameter stochastic processes. *Oper. Research Letters* **4**, 275–279.

Meng, X. L. and Rubin, D. B. (1992). Recent extensions to the EM algorithm. *Bayesian Statistics 4* (J. M. Bernardo, J. O. Berger, A. P. Dawid and A. F. M. Smith, eds.), Oxford: University Press, 307–320, (with discussion).

Naylor, J. C. and Smith, A. F. M. (1982) Applications of a method for the efficient computation of posterior distributions. *Appl. Statist.* **31**, 214–225.

Peskun, P. H. (1973). Optimum Monte–Carlo sampling using Markov chains. *Biometrika* **60**, 607–612.

Ripley, B. D. (1987). *Stochastic Simulation*. New York: Wiley.

Roberts, G. O. (1992). Convergence diagnostics of the Gibbs sampler. *Bayesian Statistics 4* (J. M. Bernardo, J. O. Berger, A. P. Dawid and A. F. M. Smith, eds.), Oxford: University Press, 777–784.

Roberts, G. O. and Polson, N. G. (1991). A note on the geometric convergence of the Gibbs sampler. (Unpublished *Tech. Rep.*).

Schervish, M. J. and Carlin, B. P. (1990). On the convergence of successive substitution sampling. *Tech. Rep.* **492**, Carnegie Mellon University.

Sinclair, A. J. and Jerrum, M. R. (1988). Conductance and the rapid mixing property of Markov chains: The approximation of the permanent resolved. *Proceedings of the Twentieth Annual Symposium on the Theory of Computing*, 235–244.

Whitt, W. (1989). The efficiency of one long run versus independent replications in steady state simulations. *Tech. Rep.* AT&T Bell Labs.

BAYESIAN STATISTICS 4, pp. 195–210
J. M. Bernardo, J. O. Berger, A. P. Dawid and A. F. M. Smith, (Eds.)

Non-informative Priors

JAYANTA K. GHOSH and RAHUL MUKERJEE
Purdue University, USA and Indian Statistical Institute, Calcutta, India
and *Indian Institute of Management, Calcutta, India*

SUMMARY

This paper reviews non-informative priors based on considerations of (i) entropy, (ii) matching what a frequentist might do and (iii) weak minimaxity. Some new results have also been presented.

Keywords: NON-INFORMATIVE PRIORS; ENTROPY; PROBABILITY MATCHING.

1. INTRODUCTION

The best way to think of a non-informative prior is, as Bernardo (1979) has argued, as a sort of origin or reference point against which any given prior reflecting some subjective opinion can be judged. In fact, it would be quite appropriate to call all candidates for non-informative priors reference priors. However, in what follows, the term reference prior will be used only for the priors proposed by Bernardo (1979) and Berger and Bernardo (1989) and the traditional terminology will be used for the other non-informative priors.

Non-informative priors have been proposed, often though not always, by appealing to one or more of the following notions:

(i) Entropy or information
(ii) Matching what a frequentist might do.
(iii) Weak minimaxity or bounded frequentist risk.

This paper is developed around these three themes. A related notion, which we will not consider, partly because of our lack of familiarity with the huge literature, is that of invariance. In a somewhat restricted set-up, invariance leads to the right-invariant Haar measure and in the more general but asymptotic set-up, where the Fisher information matrix is used to introduce a Riemannian metric, one ends up with the Jeffreys' prior. For the restricted set-up, an excellent reference is Berger (1985, Ch. 6), and a recent reference is Chang and Eaves (1990).

This paper is organized as follows. In Section 2 we begin with a rigorous treatment of the reference priors of Bernardo and propose a modification, based on a new entropy based principle, when there is a nuisance parameter. In this section we also offer an intuitive reason why the choice of a non-informative prior should depend on the choice of parameters of importance. In Section 2 we report results on priors, some of which are new, obtained by matching posterior and frequentist coverage probabilities and re-examine reference priors in this light. Berger and Bernardo (1989) have made such comparisons a basis for preferring their reference prior to several other candidates for non-informative prior. The calculation in Section 3 can also be used to examine theoretically to what extent matching fails for a given prior. In Section 4 we explore, briefly and heuristically, minimax concepts and how the priors arising from other considerations perform in this respect. Results of Sections 2 and 3 have been briefly reported in Ghosh (1991).

There is unlikely to be a consensus on what is an algorithm for producing a non-informative prior or even on the choice of such a prior in a given problem. In addition to the multitude of criteria referred to earlier, one must relate the choice, as DeGroot observed in his discussion of Bernardo (1979), to a particular decision problem being considered and the associated loss function. There are often several candidates, each fulfilling the different natural requirements, though not to the same extent. Moreover, these requirements are such that inevitably the likelihood principle and coherence will be violated. Perhaps none of these facts should be taken as too serious an objection to the use of non-informative priors if one thinks of them either as reference or starting points in a fully Bayesian analysis.

2. REFERENCE PRIORS

Reference priors were introduced by Bernardo (1979) and studied in a series of papers of which Berger and Bernardo (1989) is one of the most significant. It is still widely believed that Bernardo's original formulation may run into serious technical difficulties. In 1989, while lecturing at Purdue University, one of us suggested how this may be made rigorous and discovered a few days later that such a rigorous version already exists in the 1989 thesis of Clarke, a part of which is now available (with somewhat weaker assumptions) in Clarke and Barron (1990 a, b). Since this work is still not widely known, let us discuss it briefly.

Let

$$H(Z) = H(p) = -\int p(z)\log p(z)dz$$

be the Shannon entropy in Z or its density p. Then

$$H(Y,Z) = H(Y) + H_Y(Z),$$

where $H_Y(Z) = E(H(Z|Y))$ and $H(Z|y) = -\int p(z|y)\log p(z|y)dz$. The notations $p(\cdot)$ and $p(\cdot|\cdot)$ will often be used in a generic sense.

Let $X = (X_1,\dots,X_n), p(x|\theta)$ be the density of X under a parameter $\theta \in R^d$, and $p(\theta)$ be the prior. Whenever we consider asymptotics in this paper, we assume $X_1, X_2, \dots$ are i.i.d. and $p(x|\theta)$ satisfies various regularity conditions, but much of what we have to say may be expected to hold under more general conditions along the lines of Bickel and Ghosh (1990).

Lindley's measure of information in X is

$$\mathcal{I}(X, p(\theta)) = H(\theta) - H_X(\theta),$$

which may be written as a Kullback-Leibler divergence between the posterior and the prior as well as the expected divergence between $p(X|\theta)$ and $p(X)$, i.e.,

$$\mathcal{I}(X, p(\theta)) = E\left\{\log\frac{p(\theta|X)}{p(\theta)}\right\} = E\left\{\log\frac{p(X|\theta)}{p(X)}\right\}.$$

Bernardo (1979) suggested the larger the measure, i.e., the more informative the data, the less informative is the prior. He defined a reference prior as a p which maximizes $\mathcal{I}(X,p(\theta))$ in some asymptotic sense. We present now an algorithm for asymptotically maximizing $\mathcal{I}(X,p(\theta))$ which can be rigorously justified. Fix an increasing sequence of compact rectangles K, the union of which is the whole parameter space. In the following, K is fixed initially and n goes to infinity. Then, as shown in Clarke (1989), under suitable regularity conditions,

$$\mathcal{I}(X,p(\theta)) = \frac{d}{2}\log\frac{n}{2\pi e} + \int_K p(\theta)\log\{\det I(\theta)\}^{1/2}d\theta - \int_K p(\theta)\log p(\theta)d\theta + o(1), \quad (1.1)$$

where $I(\theta)$ is the per observation Fisher information matrix. Thus $\mathcal{I}(X, p(\theta))$ is the sum of a constant which does not depend on the prior and a term which converges to a functional $J(p, K)$. If we maximize $J(p, K)$ with respect to all priors over K, we get Jeffreys' prior concentrated on K. If we now make K tend to the whole parameter space we get improper Jeffreys' prior. It would be interesting to carry out a similar analysis also for the non-regular cases considered by Bernardo (1979).

Berger (personal communication) has pointed out the importance of checking how well the sequence of proper priors on the sequence of compacts K approximates the Jeffreys' prior.

Now suppose $\theta = (\theta_1, \theta_2)$, where θ_1 is a d_1-dimensional parameter of interest and θ_2 is a d_2-dimensional nuisance parameter. Let $p(\theta_2|\theta_1)$ be a given conditional prior density and $p(\theta_1)$ be the marginal, sought to be non-informative. The notation K is used in a generic sense. In this set-up Bernardo (1979) proposed the consideration of

$$\begin{aligned}\mathcal{I}(X, p(\theta_1)) &= E\left[\log\left\{\frac{p(\theta_1|X)}{p(\theta_1)}\right\}\right]\\ &= \mathcal{I}(X, p(\theta_1, \theta_2)) - \int_K p(\theta_1)\mathcal{I}(X, p(\theta_2|\theta_1))\, d\theta_1\end{aligned}$$

whose limiting behaviour is captured in the following formula, obtained by Clarke and one of us as an immediate consequence of (1.1). Assuming we can interchange the passage to limit and the integration with respect to θ_1, we get

$$\mathcal{I}(X, p(\theta_1)) = \frac{1}{2}d_1 \log\frac{n}{2\pi e} + \int_K p(\theta_1)\log\psi(\theta_1)d\theta_1 - \int_K p(\theta_1)\log p(\theta_1)d\theta_1 + o(1),$$

where $\psi(\theta_1) = \exp[\int p(\theta_2|\theta_1)\log\{(\det I)/(\det I_{22})\}^{1/2}\, d\theta_2]$, $I = I(\theta_1, \theta_2)$ is the per observation Fisher information matrix for θ, and $I_{22} = I_{22}(\theta_1, \theta_2)$ is the per observation Fisher information matrix for θ_2, given θ_1 is held fixed.

Asymptotic maximization with respect to $p(\theta_1)$ on compacts leads as before to

$$p(\theta_1) = \text{constant} \times \psi(\theta_1),$$

which is the reference prior for θ_1 for a given conditional $p(\theta_2|\theta_1)$. The reference prior for θ is now $p(\theta_1)p(\theta_2|\theta_1)$, *vide* Berger and Bernardo (1989).

In Berger and Bernardo (1989), $p(\theta_2|\theta_1)$ is not taken as given but chosen as a sequence of reference priors for θ_2 (given θ_1) normalized on a chosen sequence of compacts (actually θ_1-section of compact sets of (θ_1, θ_2)). In many examples this procedure has led to very satisfactory non-informative priors but a priori a natural thing to do would be to asymptotically maximize $\mathcal{I}(X, p(\theta_1))$ with respect to both $p(\theta_1)$ and $p(\theta_2|\theta_1)$. This can be done explicitly but leads to unacceptable priors. The reason why this is so is discussed below and, based on this discussion, a new procedure for getting non-informative priors is proposed. Some of the ideas occurring below also appear in the discussion of Bernardo (1979) by Copas and an observation of Jaynes quoted by Zellner (1990).

It has become clear from the asymptotic representation of the functionals $\mathcal{I}(X, p(\theta))$ and $\mathcal{I}(X, p(\theta_1))$ that we are essentially maximizing a functional J of the form $J_1 + J_2$, where

$$J_1 = \int p(\theta)\log\psi(\theta)d\theta, \qquad J_2 = -\int p(\theta)\log p(\theta)d\theta.$$

Maximizing the second functional would give one the uniform distribution, which, though a priori attractive, is unrelated to X and so often leads to unacceptable posteriors. Maximizing the functional J_1 would lead to a $p(\theta)$ which is degenerate and hence quite unacceptable as a non-informative prior. The reason why J works, when there is no nuisance parameter or the conditional density $p(\theta_2|\theta_1)$ of the nuisance parameter is specified, is the presence in J of both terms J_1 and J_2. The reason why maximizing J does not give satisfactory results for a nuisance parameter θ_2 and unspecified $p(\theta_2|\theta_1)$ is that J_2 now involves only $p(\theta_1)$. One may offer a similar explanation of why Bernardo's procedure, with a different measure of information loss, as suggested by DeGroot in his 1979 discussion, may not give satisfactory results unless one introduces a penalty term reflecting the deviation of the prior from the uniform. This means one would have to solve the problem of maximizing $J_1 + \lambda J_2$, without constraints, with respect to proper priors on a compact set K. The solution, using Holder's inequality, is

$$\begin{aligned} p(\theta) &= \text{constant } \times \psi^{1/\lambda}(\theta) \text{ on } K, \\ &= 0 \qquad\qquad\qquad\quad \text{outside } K. \end{aligned}$$

In the case where θ_2 is a nuisance parameter and one wants to determine both $p(\theta_1)$ and $p(\theta_2|\theta_1)$, the above approach suggests the maximization of

$$\int_K\int p(\theta_1,\theta_2)\log\left(\frac{\det I}{\det I_{22}}\right)^{1/2} d\theta_1 d\theta_2 - \lambda\int_K\int p(\theta_1,\theta_2)\log p(\theta_1,\theta_2)d\theta_1 d\theta_2.$$

With $\lambda = 1$, this would give

$$p(\theta_1,\theta_2) = \text{constant} \times \left(\frac{\det I}{\det I_{22}}\right)^{1/2}.$$

It would be interesting to examine the appropriateness of this choice in various special cases. Some other non-informative entropy based priors have been proposed by Zellner (1990).

We conclude this section with a brief intuitive discussion of why the choice of a non-informative prior should depend on the choice of the parameters of interest. We should like to draw an analogy with tests for randomness for a given finite sequence of zero's and ones, see for example Kolmogorov (1963). Of course, there is an upper bound to the number of tests a given finite sequence can satisfy. Similarly it would be too much to expect that the requirements of a non-informative prior can be satisfied by all functions of θ or even by all components of θ. In such a case, a prior which satisfies these requirements for one set of components will, in general, depend on which components have been chosen. This is true not only of the reference priors but also of the priors obtained in the next section.

3. MATCHING POSTERIOR AND FREQUENTIST PROBABILITIES

We begin by introducing a few notations from Bickel and Ghosh (1990). For simplicity we consider only a two-dimensional parameter vector $\theta = (\theta_1, \theta_2)$ with θ_2 the nuisance parameter. Throughout this section we assume θ_2 is orthogonal to θ_1 in the sense of Cox and Reid (1987). Let

$$\begin{aligned} &H_1\colon \theta_1 = \theta_{10}, \quad H_2 : \theta_1 = \theta_{10}, \theta_2 = \theta_{20}; \\ &\hat\theta = \text{unrestricted maximum likelihood estimate (mle)}, \\ &\hat\theta^{(1)} = \text{mle under } H_1, \theta_0 = (\theta_{10}, \theta_{20}), \\ &\lambda_1 = 2\{\log p(X,\hat\theta) - \log p(X,\hat\theta^{(1)})\} \\ &\quad = \text{(log) likelihood ratio statistic for testing } H_1, \end{aligned}$$

$$\lambda_2 = 2\{\log p(X,\hat{\theta}) - \log p(X,\theta_0)\}$$
$$= \text{(log) likelihood ratio statistic for testing } H_2,$$

Note that $\lambda_2 - \lambda_1 \geq 0$ and can be interpreted as the (log) likelihood ratio statistic for testing $H_3 : \theta_2 = \theta_{20}$, assuming H_1 is true. Under appropriate assumptions, λ_1 and $\lambda_2 - \lambda_1$ are χ^2's each with 1 d.f. in the usual frequentist sense. Introduce the signed square roots

$$T_1 = \lambda_1^{1/2}\text{sgn}(\hat{\theta}_1 - \theta_{10}), \qquad T_2 = (\lambda_2 - \lambda_1)^{1/2}\text{sgn}\left(\hat{\theta}_2^{(1)} - \theta_{20}\right).$$

We now write θ_1, θ_2 for θ_{10}, θ_{20}, and introduce a prior $p(\theta_1, \theta_2)$. It is shown in Bickel and Ghosh (1990, theorem 1) that the posterior density of T_1, T_2 given X can be written as

$$p_3(t,x) = \emptyset(t)\{1 + P_3 + Q_3\} + 0(n^{-2}),$$

where $\emptyset(t)$ is the two-dimensional standard normal density, P_3 is a polynomial in $n^{-1/2}$ of degree 3 and Q_3 is a polynomial in $n^{-1/2}t_1, n^{-1/2}t_2$ of degree 3. This has been used in Bickel and Ghosh (1990) to show that posterior density of λ_1 admits of a Bartlett correction and after such correction is approximable by a χ^2 density with 1 d.f., up to $0(n^{-2})$. The posterior density of T_1 is (*vide* Ghosh and Mukerjee (1991a))

$$\pi_3(t_1,x) = \emptyset(t_1)\left\{1 + n^{-1/2}G_1t_1 + n^{-1}G_2(t_1^2 - 1) + n^{-3/2}(\text{cubic in } t_1)\right\} + 0(n^{-2}),$$

where, with some abuse of notations, we use $\emptyset(t_1)$ to denote the one-dimensional standard normal density, and G_1, G_2 depend only on x (and p and its derivatives evaluated at $\hat{\theta}$). Let

$$E(\lambda_1|X) = 1 + n^{-1}B_1(X) + 0(n^{-2}), \quad A_{1-\alpha}(X) = \{\theta_1 : \lambda_1 \leq \chi^2_{1,\alpha}(1 + n^{-1}B_1(X))\},$$

where $\chi^2_{1,\alpha}$ is the upper α-point of a χ^2 with 1 d.f. Then the fact that the posterior distribution of λ_1 admits of a Bartlett correction means that the posterior probability

$$P(\theta_1 \varepsilon A_{1-\alpha}(X)|X) = 1 - \alpha + 0(n^{-2}).$$

Suppose we now wish to choose the prior p such that the frequentist probability

$$P_\theta(\theta_1 \varepsilon A_{1-\alpha}(X)) = 1 - \alpha + 0(n^{-2}) \quad (\forall \alpha)$$

uniformly on compact sets of θ. In general no prior is available if we require matching up to $0(n^{-2})$. On the other hand for all reasonable priors, matching holds up to $0(n^{-1})$.

A little reflection will show that this can be achieved by finding a constant $c_p(\theta)$, which depends on the choice of p, such that $B_1(X) - c_p(\theta)$ tends to zero in probability (for fixed θ) and matching it with the frequentist Bartlett correction term $F(\theta)$, where

$$E_\theta(\lambda_1) = 1 + n^{-1}F(\theta) + 0(n^{-2}).$$

This leads to a second-order partial differential equation for p,

$$\frac{\partial}{\partial \theta_1}\left[\frac{p_{10}(\theta)}{I_{11}} - \left\{\frac{K_{10\cdot 20}}{I_{11}^2} - \frac{K_{12}}{I_{11}I_{22}}\right\}p(\theta)\right] + \frac{\partial}{\partial \theta_2}\left\{\frac{K_{21}}{I_{11}I_{22}}p(\theta)\right\} = 0$$

where $p_{10}(\theta) = \partial p(\theta)/\partial\theta_1$,

$$K_{ij} = E_\theta\left\{\partial^{i+j}\log p(X_1|\theta)/\partial\theta_1^i\partial\theta_2^j\right\}, I_{11} = -K_{20}, I_{22} = -K_{02},$$

$$K_{ij \cdot i'j'} = E_\theta \left[\left\{ \partial^{i+j} \log p(X_1|\theta) / \partial\theta_1^i \partial\theta_2^j \right\} \left\{ \partial^{i'+j'} \log p(X_1|\theta) / \partial\theta_1^{i'} \partial\theta_2^{j'} \right\} \right].$$

The easiest way to get this equation is to calculate F in the following way, which is reminiscent of Stein (1985), Dawid (1991) and Bickel and Ghosh (1990). Having found $c_p(\theta)$ introduced earlier, one integrates with respect to p and then, after an integration by parts, makes p converge weakly to a distribution degenerate at θ.

Similar equations have been obtained using also the Bartlett corrected conditional likelihood ratio statistic of Cox and Reid (1987) as well as equal-tailed intervals and highest posterior density regions for the case where there is no nuisance parameter (Peers (1968), Ghosh and Mukerjee (1991 a, b, c)). The methods are similar to what has been sketched above.

As an example, consider i.i.d. observations from a normal N(mean θ_2, variance θ_1). Here use of λ_1 and the conditional likelihood ratio statistic leads to the solutions

$$p(\theta) = q_1(\theta_2)\theta_1^{-1} + q_2(\theta_2)\theta_1^{-3},$$

and

$$p(\theta) = q_3(\theta_2)\theta_1^{-1} + q_4(\theta_2)\theta_1^{-2},$$

respectively, where the q_i's are arbitrary functions of θ_2. A prior p satisfies both forms if it is of the form $q(\theta_2)\theta_1^{-1}$. The reference prior for θ_1 satisfies this.

Such matching with frequentist probability goes back to Welch and Peers (1963) and has been recently revived by Stein (1985) who was interested in frequentist confidence intervals. Lee (1989) considered a two-sided case where the solution has the unpleasant property of depending on $1-\alpha$. The work of Stein (1985) and Tibshirani (1989) (see also Peers (1965)) has important implications for the reference priors of the previous section. A brief review follows.

Suppose instead of choosing two-sided intervals for θ_1 we start with one-sided intervals, i.e., we choose $\theta_{1,\alpha}(X)$ (which depends on the prior p) such that

$$P(\theta_1 \leq \theta_{1,\alpha}(X)|X) = 1 - \alpha + 0(n^{-1}),$$

and require of our prior that

(a) $P_\theta(\theta_1 \leq \theta_{1,\alpha}(X)) = 1 - \alpha + 0(n^{-1}) \quad \forall\alpha,$

or

(b) (no nuisance parameter)
$P_{\theta_1}(\theta_1 \leq \theta_{1,\alpha}(X)) = 1 - \alpha + 0(n^{-1}) \quad \forall\alpha,$

or

(c) (integrated nuisance parameter given $p(\theta_2|\theta_1)$)
$\int P_\theta(\theta_1 \leq \theta_{1,\alpha}(X))p(\theta_2|\theta_1)d\theta_2 = 1 - \alpha + 0(n^{-1}) \quad \forall\alpha.$

For one-sided intervals, owing to lack of symmetry, matching is, in general, not possible beyond $0(n^{-1})$.

The solution for (b), due to Welch and Peers (1963) and Stein (1985), is the Jeffreys' prior. It can be shown easily that no such result is available if the dimension of θ_1 is more than one. The differential equation corresponding to (a), due to Tibshirani (1989) (see also Peers (1965)), is

$$-\partial(I^{11})^{1/2}/\partial\theta_1 = (I^{11})^{1/2} p_{10}(\theta)/p(\theta),$$

where $I^{11} = (I_{11})^{-1}$. The solution to this is

$$p(\theta) = \{I_{11}(\theta)\}^{1/2} \ell(\theta_2),$$

where $\ell(\theta_2)$ is an arbitrary function and, as before, $I_{11}(\theta)$ is the per observation Fisher information for θ_1 when θ_2 is held fixed. Using similar techniques one can solve (c). One gets the differential equation

$$\int \left\{ \frac{\partial}{\partial \theta_1} \left(\log \frac{p(\theta)}{I_{11}^{1/2}} \right) \right\} \frac{p(\theta_2|\theta_1)}{I_{11}^{1/2}} d\theta_2 = 0,$$

solving which one obtains

$$p(\theta_1) = \text{constant} \times \left\{ \int p(\theta_2|\theta_1) I_{11}^{-1/2} d\theta_2 \right\}^{-1}.$$

The implication of the solution to (a) is that, in general, the reference prior for θ_1 (in the sense of Berger and Bernardo (1989)) does not satisfy (a) but the reference prior for θ_2 does. It is easy to construct examples where the reference prior for θ_1 does not solve (a). It is of interest to note that the new non-informative prior for θ_1 proposed in Section 2 is also a solution to (a).

It is also of interest to note that the differential equation that one has to solve for (c) is very similar though not identical to an equation which is solved by the reference prior $p(\theta_1)$ of Berger and Bernardo (1989).

Suppose one had a point estimation problem with $\theta \in R^1$ and one wanted to match posterior and frequentist mean squares $nI(\theta)(\hat{\theta} - \theta)^2$, where $I(\theta)$ is the per observation Fisher information. It is clear that for all reasonable priors this holds up to $o(1)$ but, in general, for no prior can one match even up to $0(n^{-1/2})$. While the technical reason for this is readily apparent, namely, that one should consider $(-d^2 \log p(X|\theta)/d\theta^2)_{\hat{\theta}}(\hat{\theta} - \theta)^2$ instead of $nI(\theta)(\hat{\theta} - \theta)^2$, its intuitive implications are not clear.

For application of some of these ideas to sequential analysis, see Woodroofe (1986, 1989, 1990).

4. MINIMAX IDEAS

One of the weakest requirements of a point estimate T_n in the frequentist theory is that its risk $R(T_n, \theta)$ under θ should tend to zero as n tends to infinity. It is natural to expect that the Bayes estimate, say $\theta_n(p)$, corresponding to a non-informative prior should have this property. Clearly if $\theta_n(p)$ is minimax in following weak sense,

$$\sup_\theta R(\theta_n(p), \theta) \leq \Delta \inf_{T_n} \sup_\theta R(T_n, \theta), \tag{3.1}$$

where $\Delta (> 0)$ is a constant, and there is at least one uniformly consistent sequence of estimates T_n such that $\sup_\theta R(T_n, \theta) \to 0$, then $\theta_n(p)$ will also have this property. Under reasonable conditions $\theta_n(p)$ is minimax, i.e., satisfies (3.1) with $\Delta = 1$, if p is a least favourable distribution. Thus it is to be expected that often non-informative priors will be least favourable at least asymptotically.

That Bernardo's choice is in fact a least favourable choice follows from the fact noted first in Aitchison (1975) that $\mathcal{I}(X, p(\theta))$ is nothing but the Bayes risk when one predicts $p(X|\theta)$ using the entropy loss. Hence maximizing $\mathcal{I}(x, p(\theta))$ (asymptotically) would lead to an (asymptotically) least favourable p as noted in Clarke (1989).

Some of the priors obtained in Section 3, namely, the priors based on one-sided intervals and the prior based on highest posterior density, have (asymptotically) constant risk and are (asymptotically) Bayes among one-sided or two-sided intervals which are no larger. Hence one would expect them to be asymptotically minimax but this does not seem to imply consistency of any sort.

We have noted earlier that the method of matching fails to produce non-informative priors in a point estimation problem. However, the present method seems to work at least heuristically. Suppose for simplicity we have a one-dimensional parameter and the loss is $nI(\theta)(T-\theta)^2$. We choose a prior p such that for the Bayes estimate $\theta_n(p)$, the risk is

$$R(\theta_n(p),\theta) = \text{constant} + 0(n^{-1}), \tag{3.2}$$

which, being an equalizer, is expected to lead to the least favourable p if among solutions of (3.2) we minimize the constant. The equation (3.2) leads to a differential equation

$$2I^{-1}\frac{d^2}{d\theta^2}\log p(\theta) + I^{-1}\left(\frac{d}{d\theta}\log p(\theta)\right)^2 + 2I^{-2}\left(L_{001}+L_{11}\right)\left(\frac{d}{d\theta}\log p(\theta)\right)$$
$$+ I^{-2}L_{02} + I^{-3}\left(\frac{1}{2}L_{001}^2 + L_{11}^2 + 2L_{001}L_{11}\right) = \text{constant},$$

where $I = I(\theta)$,

$$L_{iju} = E_\theta\left\{(d\log p(X_1|\theta)/d\theta)^i(d^2\log p(X_1|\theta)/d\theta^2)^j(d^3\log p(X_1|\theta)/d\theta^3)^u\right\}, L_{ij} = L_{ij0}.$$

In several cases, including the binomial and exponential, one can check that the above method does lead to a least favourable prior which may be improper.

ACKNOWLEDGEMENT

We are indebted to Jim Berger and Bertrand S. Clarke for several helpful discussions and to Tom Sellke for letting us have a copy of his student's dissertation. The work of RM was supported by a grant from the Centre for Management and Development Studies, Indian Institute of Management Calcutta.

REFERENCES

Aitchison, J. (1975). Goodness of prediction fit. *Biometrica* **62**, 547–554

Berger, J. O. (1985). *Statistical Decision Theory and Bayesian Analysis*. New York: Springer.

Berger, J. O. and Bernardo, J. M. (1989). Estimating a product of means: Bayesian analysis with reference priors. *J. Amer. Statist. Assoc.* **84**, 200–207.

Bernardo, J. M. (1979). Reference posterior distributions for Bayesian inference. *J. Roy. Statist. Soc. B* **41**, 113–147, (with discussion).

Bickel, P. J. and Ghosh, J. K. (1990). A decomposition for the likelihood ratio statistic and the Bartlett correction, a Bayesian argument. *Ann. Statist.* **18**, 1070–1090.

Chang, T. and Eaves, D. (1990). Reference priors for the orbit in a group model. *Ann. Statist.* **18**, 1595–1614.

Clarke, B. S. (1989). *Asymptotic Cumulative Risk and Bayes' Risk Under Entropy Loss, with Applications*. Ph.D. Thesis, University of Illinois, Urbana-Champaign.

Clarke, B. S. and Barron, A. R. (1990a). Information theoretic asymptotics of Bayes methods. *IEEE Trans. Information Theory* **36**, 453–471.

Clarke, B. S. and Barron, A. R. (1990b). Entropy risk and the Bayesian central limit theorem (Unpublished *Tech. Rep.*).

Cox, D. R. and Reid, N. (1987). Parameter orthogonality and approximate conditional inference. *J. Roy. Statist. Soc. B* **49**, 1–39, (with discussion).

Dawid, A. P. (1991). Fisherian inference in likelihood and prequential frames of reference. *J. Roy. Statist. Soc. B* **53**, 79–109, (with discussion).

Ghosh, J. K. (1991). Paradigms of inference, paradoxes, reconciliation and application. *78th Indian Science Congress Proceedings* **2**.

Ghosh, J. K. and Mukerjee, R. (1991a). Bayesian and frequentist Bartlett correction for likelihood ratio and conditional likelihood ratio tests. (Unpublished *Tech. Rep.*).

Ghosh, J. K. and Mukerjee, R. (1991b). Characterization of priors under which Bayesian and frequentist Bartlett corrections are equivalent in the multiparameter case. *J. Multivariate Anal.* **38**, 385–393.

Ghosh, J. K. and Mukerjee, R. (1991c). Frequentist validity of highest posterior density regions in the multiparameter case. (Unpublished *Tech. Rep.*).

Kolmogorov, A. N. (1963). On tables of random numbers. *Sankya A* **25**, 369–376.

Lee, C. B. (1989). *Comparison of Frequentist Coverage Probability and Bayesian Posterior Coverage Probability, and Applications*. Ph.D. Thesis, Purdue University.

Peers, H. W. (1965). On confidence sets and Bayesian probability points in the case of several parameters. *J. Roy. Statist. Soc. B* **27**, 9–16.

Peers, H. W. (1968). Confidence properties of Bayesian interval estimates. *J. Roy. Statist. Soc. B* **30**, 535–544.

Stein, C. (1985). On the coverage probability of confidence sets based on a prior distribution. *Sequential Meth. Statist.* **16**, 485–514.

Tibshirani, R. (1989). Noninformative priors for one parameter of many. *Biometrika* **76**, 604–608.

Welch, B. L. and Peers, H. W. (1963). On formulae for confidence points based on integrals of weighted likelihoods. *J. Roy. Statist. Soc. B* **25**, 318–329.

Woodroofe, M. (1986). Very weak expansions for sequential confidence intervals. *Ann. Statist.* **14**, 1049–1067.

Woodroofe, M. (1989). Very weak expansions for sequentially designed experiments: Linear models. *Ann. Statist.* **17**, 1087–1102.

Woodroofe, M. (1990). Integrable expansions for posterior distributions for one parameter exponential families. *Tech. Rep.* **187**, University of Michigan.

Zellner, A. (1990). Bayesian methods and entropy in Economics and Econometrics. *10th International MaxEnt Workshop*, University of Wyoming.

DISCUSSION

N. G. POLSON (*University of Chicago, USA*)

The relationship between information theory and Bayesian statistics is an interesting one. The link arises from the decision problem of reporting a posterior density under a logarithmic utility function (Seidler, 1958; Bernardo, 1979b). Let $\theta \in \Re^k$ be the parameter of interest with *a priori* beliefs $p(\theta)$ and let x be a data sequence, then, in the light of the data, the experimenter with a logarithmic utility is honest and the optimal action is to report his true posterior beliefs $p(\theta|x)$. This leads to an expected utility given by the negative expected Shannon information $E_x[\int p(\theta|x)\log p(\theta|x)d\theta]$ where E_x denotes expectation with respect to the marginal beliefs $p(x)$. From a Bayesian perspective, information can be viewed as a change in expected utility (DeGroot, 1986), that is

$$I(\mathcal{E}) = E_x\left[\int p(\theta|x)\log p(\theta|x)d\theta\right] - \int p(\theta)\log p(\theta)d\theta = E_{x,\theta}\left[\log\left(\frac{p(\theta|x)}{p(\theta)}\right)\right]. \quad (1)$$

I make explicit here that $I(\cdot)$ is a functional on the space of experiments with elements $\mathcal{E}$, rather than a functional of the prior. The asymptotic version of (1) provides a useful approximation to $I(\mathcal{E})$, first used for optimal design of the linear model in Stone (1959). Polson (1988a) applies the asymptotic version of $I(\mathcal{E})$ to optimal design problems, sample size calculations and to characterise well-known families of likelihoods; for example, the normal, t family and the Huber family. Notice that, as a functional of $\mathcal{E}$, all of these problems can be viewed coherently with the Bayesian decision theoretic framework.

Let me now discuss a general framework for exploring the asymptotics of $I(\mathcal{E})$ based on a result of Ibragimov and H'asminsky (1973) who consider a location model in regular and non-regular situations. For extensions to non i.i.d. cases, for example, nonlinear models and stochastic differential equations, see Polson (1988b), Polson and Roberts (1990), respectively. The basic probabilistic tool for exploring the asymptotics of $I(\mathcal{E})$ is the LeCam-Ibragimov likelihood ratio process which is defined by

$$Z_{n,\theta}(\alpha) = \prod_{i=1}^{n} \frac{f(x_i|\theta + \phi(n,\theta)\alpha)}{f(x_i|\theta)} \tag{2}$$

for $\alpha \in \Re^k$ and a suitable normalising sequence of matrices $\phi(n, \theta)$. The key identity relating the likelihood ratio process to the Shannon information of the posterior is given by

$$p(\theta|\boldsymbol{x})^{-1} = \frac{p(\boldsymbol{x})}{f(\boldsymbol{x}|\theta)p(\theta)} = |\phi(n,\theta)| \int Z_{n,\theta}(\alpha) \frac{p(\theta + \phi(n,\theta)\alpha)}{p(\theta)} d\alpha.$$

Hence, the expected Shannon information of the posterior decomposes as

$$\begin{aligned} E_{\boldsymbol{x}} \left[\int p(\theta|\boldsymbol{x}) \log p(\theta|\boldsymbol{x}) d\theta \right] &= \\ = -E_\theta(\log|\phi(n,\theta)|) - E_{\boldsymbol{x},\theta} &\left(\log \int Z_{n,\theta}(\alpha) \frac{p(\theta+\phi(n,\theta)\alpha)}{p(\theta)} d\alpha \right) \end{aligned} \tag{3}$$

where E_θ denotes expectation with respect to the prior. Suppose now that $Z_{n,\theta}(\alpha) \Rightarrow Z_\theta(\alpha)$ where $\Rightarrow$ denotes weak convergence. Typically, $|\phi(n,\theta)|^{-2} = n^k |I(\theta)|$ where $|I(\theta)|$ is the determinant of Fisher information and $Z_\theta(\alpha) = \exp\left(\alpha Z - \frac{1}{2}\alpha^T\alpha\right)$, so that $E_{Z,\theta}[\log \int Z_\theta(\alpha)] = \frac{k}{2}\log(2\pi e)$. Moreover, suppose that the interchange of limit and integral is valid, then

$$\lim_{n\to\infty} \left(E_{\boldsymbol{x}} \left[\int p(\theta|\boldsymbol{x}) \log p(\theta|\boldsymbol{x}) d\theta \right] + E_\theta[\log|\phi(n,\theta)|] \right) = -E_{Z,\theta} \left[\log \int Z_\theta(\alpha) \right].$$

The regularity conditions required are tantamount to tightness conditions in order to govern tail behaviour and a uniform Lipschitz condition on the likelihood ratio of the form $E|Z_{n,\theta}(\alpha_1) - Z_{n,\theta}(\alpha_2)| \le C|\alpha_1 - \alpha_2|^\beta$ for some C and β (see lemma 3.1 of Ibragimov and H'asminsky, 1973). There is no explicit need for asymptotic normality.

In the regular i.i.d. case we obtain

$$\lim_{n\to\infty} \left(I(\mathcal{E}) - \frac{k}{2}\log\left(\frac{n}{2\pi e}\right) \right) = \int p(\theta) \log\left(\frac{|I(\theta)|^{1/2}}{p(\theta)} \right) d\theta.$$

If $|I(\theta)|^{1/2} \in L^1(\Theta)$ then an upper bound on the right hand side is given by $\log \int |I(\theta)|^{1/2} d\theta$ and equality is attained if and only if $p(\theta) \propto |I(\theta)|^{1/2}$, the Jeffreys' prior.

In the case of a nonlinear model, $y_i = \eta(x_i, \theta) + \varepsilon_i$, where x_i are design points and ε_i are independent and identically distributed, then (Polson, 1988a) we obtain

$$\lim_{n\to\infty} \left(I(\mathcal{E}) - E_\theta \left(\log \left| \sum_{i=1}^{n} I_i(\theta) \right|^{1/2} \right) \right) = -\int p(\theta)\log p(\theta) d\theta - \frac{k}{2}\log(2\pi e)$$

where $|\sum_{i=1}^n I_i(\theta)|$ is the determinant of Fisher's information and $I_i(\theta)$ is Fisher's information for the ith observation.

For a stochastic process X_t evolving according to $dX_t = S(\theta, t, X_t)dX_t + dB_t$, we obtain

$$\lim_{T\to\infty} \left(I(\mathcal{E}) - E_{XT}[\log |I_T(\theta)|^{1/2}] \right) = -\int p(\theta) \log p(\theta) d\theta - \frac{k}{2} \log(2\pi e)$$

where $I_T(\theta)$ is Fisher information, see Polson and Roberts (1990) for regularity conditions.

The paper contains useful results concerning the relationship between classical coverage properties and Bayesian credible regions. The use of the Bartlett correction, in a Bayesian framework, is clearly outlined and the ensuing approximations are typically valid to $0(n^{-2})$. In fact, in certain instances there is an exact match up between classical coverage intervals and Bayesian credible regions, see Bondar (1977). I also note that there is a relationship with the modified profile likelihood (Polson, 1987), which is a technique for eliminating nuisance parameters.

A non regular example sheds some light on the elimination of nuisance parameters using an information-theoretic approach (Polson, 1988b). Consider a location model with error density in the exponential power family, that is $f(x|\theta, \alpha) \propto \exp(-|x - \theta|^\alpha)$. Consider the information gain for θ given α, for $0 < \alpha < 1/2$, then the normalisation sequence depends on α and is $\phi = n^{-1/2\alpha+1}$ so that the rate of learning about θ depends on the nuisance parameter. An information-theoretic choice for the prior $p(\alpha|\theta)$ which is "noninformative" seems unclear. The product of normal means (Berger and Bernardo, 1989) also appears to be non-regular and it is not surprising that different priors result from using different parameterisations. A similar fate can befall improper priors when marginalising using alternative continuous reparameterisations (Dawid, Stone and Zidek, 1973, Seidenfeld, 1983). Stone (1979) provides a useful survey concerning problems that arise with the use of noninformative priors and the relationship with finite additivity. From a decision theoretic perspective, Parmigiani and Polson (1992) show that, in the case where Jeffreys prior depends on a design parameter, incoherence can arise. It should be noted, however, that Jeffreys did not explicitly endorse the rule in these design dependent cases.

Now, the elimination of nuisance parameters lies at the heart of any inferential framework. A Bayesian marginalises out the nuisance parameter with respect to the prior. Substantial effort has been directed at trying to avoid such a prior specification on the nuisance parameter. One such automatic approach is proposed by Bernardo (1979), Berger and Bernardo (1989) which, in my opinion, has an unBayesian flavour. It reminds me of Fisher's approach where, recognising difficulties with a classical approach, he proposed to factorise the likelihood function, then to argue fiducially in a stepwise fashion, and finally appeal to Bayes theorem, legitimately, to integrate out the nuisance parameters using the fiducial probabilities (see Fisher, 1956, p.114 & p.175).

Finally, concerning the notion of a "noninformative" or "ignorance" prior, let me note an insightful comment of Poincaré (1905): "If we were not ignorant there would be no probability, there could only be certainty. But our ignorance cannot be absolute for then there would be no longer any probability at all. Thus the problems of probability may be classed according to the greater or less depth of this ignorance".

J. O. BERGER (*Purdue University, USA*)

The paper presents a variety of exciting new ideas for the development of noninformative priors. I will focus on the most well established of the ideas, the use of asymptotic frequentist coverage of one–sided Bayesian credible sets, following Stein (1985) and Tibshirani (1989). The method is of particular interest to reference prior development because, as mentioned in the paper, it seems to suggest that the reference prior be determined by reversing the order of the parameters suggested in Berger and Bernardo (1992), when applying the reference prior algorithm.

To verify this "reversal" in the two parameter case, suppose $\theta = (\theta_1, \theta_2)$, where θ_1 is the parameter of interest and θ_2 is the nuisance parameter. Suppose that θ_1 and θ_2 are orthogonal, i.e. that the Fisher information matrix is $I(\theta)$ = diagonal $\{I_{11}(\theta),\ I_{22}(\theta)\}$. Following the notation of Section 2 of Berger and Bernardo (1992), define $\theta_{(1)} = \theta_2$ and $\theta_{(2)} = \theta_1$, so that the reference prior algorithm will be applied in the "reverse" order to that recommended in Berger and Bernardo (1992).

Lemma. *Suppose that the parameter space is* $\Theta = \Theta_1 \times \Theta_2$, *where* $\theta_1 \in \Theta_1$ *and* $\theta_2 \in \Theta_2$, *and that the compact sets chosen in application of the reference prior algorithm are, likewise, product sets. Then the reference prior, as defined by (2.2.6), is of the form*

$$\pi(\theta) = d(\theta_2)\sqrt{I_{11}(\theta)}$$

for some function d, providing the limit exists.

Proof. Still following Section 2 of Berger and Bernardo (1992), note that $h_1(\theta) = I_{22}(\theta)$ and $h_2(\theta) = I_{11}(\theta)$ (recalling that we have reordered θ as (θ_2, θ_1)). Thus (2.3.2) yields

$$\pi_2^\ell(\theta_1|\theta_2) = \frac{\sqrt{I_{11}(\theta)}1_{\Theta_1^\ell}(\theta_1)}{\int_{\Theta_1^\ell}\sqrt{I_{11}(\theta)}d\theta_1} \equiv \frac{\sqrt{I_{11}(\theta)}}{g_2^\ell(\theta_2)}\ 1_{\Theta_1^\ell}(\theta_1).$$

Next, observe that

$$E_1^\ell[(\log(I_{22}(\theta))|\theta_2] = \int_{\Theta_1^\ell}\log(I_{22}(\theta))\pi_2^\ell(\theta_1|\theta_2)d\theta_1 \equiv g_1^\ell(\theta_2).$$

Thus (2.3.3) yields

$$\begin{aligned}\pi_1^\ell(\theta) &= \frac{\pi_2^\ell(\theta_1|\theta_2)\exp\{\frac{1}{2}g_1^\ell(\theta_2)\}1_{\Theta_1^\ell}(\theta_1)1_{\Theta_2^\ell}(\theta_2)}{\int_{\Theta_2^\ell}\exp\{\frac{1}{2}g_1^\ell(\theta_2)\}d\theta_2}\\ &= \frac{\exp\{\frac{1}{2}g_1^\ell(\theta_2)\}1_{\Theta_2^\ell}(\theta_2)}{g_2^\ell(\theta_2)\int_{\Theta_2^\ell}\exp\{\frac{1}{2}g_1^\ell(\theta_2)\}d\theta_2}\cdot\sqrt{I_{11}(\theta)}\ 1_{\Theta_1^\ell}(\theta_1)\\ &\equiv g^\ell(\theta_2)\cdot\sqrt{I_{11}(\theta)}\ 1_{\Theta_1^\ell}(\theta_1).\end{aligned}$$

Passing to the limit in ℓ (which is assumed to exist) yields the desired result. ◁

This result, first brought to my attention by J. K. Ghosh, has caused me further uncertainty as to which parameter ordering to recommend in developing reference priors. Clearly, further study of examples in which the ordering matters will be necessary.

Certain limitations of the asymptotic frequentist coverage method should be mentioned. First, it is unclear how to extend the idea to higher dimensions. Second, finding orthogonal nuisance parameters can be extremely difficult. Finally, choice of $d(\theta_2)$ is left open; note that the "reverse" reference prior algorithm can be used to make this choice, providing the conditions of the above lemma are satisfied.

B. CLARKE (*Purdue University, USA*)

1. Introduction. Selecting the non-informativity principle most appropriate to a given problem should be based on a physical understanding which indicates the form of knowledge it is most important to seek. Since there are many non-informativity principles, it is important to know conditions under which each should be applied. Ghosh and Mukerjee review and contrast several principles which they group into three categories: information based, frequentist based, and risk based. We inquire as to methodological implications. To address this questions we consider categories that are physically based, conceptually based, and decision-theoretic. The principles described in their paper can be fit into these more general categories, and we try to elaborate some parallels implicit in their work so as to address methodological concerns.

First, we illustrate the three categories for an information based principle. It is seen that the optimizations coincide. That coincidence does not appear to hold in general: In two alternative formulations, based on the Hellinger and χ^2 distance, the quantities corresponding to the three categories can be written down and seen to be different. This also clarifies the connection between the first and third categories identified by Ghosh and Mukerjee. Their second category seems either conceptual or useful for evaluating a candidate prior. Here, we assume that the true distribution is a member of a known parametric family, satisfying various regularity assumptions which we have not formally investigated, and we follow the notation in the paper, noting only necessary departures from it.

2. Information based approach. The physical basis assumes that some agent is transmitting data to us. We compare the actual rate of transmission to its theoretical upper bound. So we want to know

$$p^* = \arg \sup_p I(X, p(\theta)), \tag{1}$$

the prior achieving the channel capacity. The closer the 'true' prior is to p^* the closer the transmission rate is to optimality.

Conceptually we consider the distance between the posterior and the prior which generated it, maximizing this over choices of prior. The result is the prior which is most changed by the data. One distance measure is

$$E \log \frac{p(\theta|X)}{p(\theta)}. \tag{2}$$

Alternatively, we could ask that the data and the prior be strongly dependent in the mutual information sense. Again, we maximize (2).

In the decision-theoretic basis physical assumptions enter, but only in the choice of loss function. The relative entropy is the right loss function for certain coding problems, in which case we seek the prior achieving.

$$\max_p \min_Q \int p(\theta) D(P_\theta^n || Q) d\theta, \tag{3}$$

where Q is a distribution.

When there are no nuisance parameters, the results of optimizing in (1), (2), and (3) are identical: Asymptotically, one obtains the Jeffreys' prior. In the presence of one set of nuisance parameters (1), (2), and (3) remain identical and the two-step prior of Berger and Bernardo is optimal.

3. Hellinger and χ^2. In the discrete case, the Hellinger distance H can be interpreted geometrically. It gives the distance between two probabilities, which are points on a simplex, as the Euclidean distance between two points on a sphere. The χ^2 distance can be interpreted in terms of the χ^2 test, so maximizing it represents a 'badness of fit' criterion.

Conceptually, we maximize the distance between prior and posterior:

$$E \int \left(\sqrt{p(\theta|X)} - \sqrt{p(\theta)}\right)^2 d\theta, \tag{4}$$

$$E \int \frac{(p(\theta) - p(\theta|X))^2}{p(\theta)} d\theta \quad \text{or} \quad E \int \frac{(p(\theta) - p(\theta|X))^2}{p(\theta|X)} d\theta. \tag{5a, b}$$

The decision-theoretic approach is to seek the least favorable priors

$$\arg \sup_p \min_Q \int p(\theta) H(P_\theta, Q) d\theta \tag{6a}$$

and

$$\arg \sup_p \min_Q \int p(\theta) \chi^2(P_\theta, Q) d\theta. \tag{6b}$$

4. Implications. We expect the optimizations in (4), (5) and (6) to yield different priors. Expressions (4) and (5) amount to asking that a density ratio be close to unity (although (5b) looks artificial). Modifying known results may give solutions or asymptotics.

Bayes estimators corresponding to the minimizations in (6) are normalized versions of $(\int p(\theta)\sqrt{p_\theta^n} d\theta)^2$ and $(\int p(\theta)[p_\theta^n]^2 d\theta)^{1/2}$. Expansions for the risk and Bayes risk of the Bayes estimators allow identification of the asymptotically least favorable priors. Preliminary work (Clarke and Sun) suggests that in the Hellinger case one gets $2(1 - O(1/n^{1/4}))$ and in the χ^2 case one gets $O(\sqrt{n})$, for the normal family.

These results suggest two theorems to seek: One to identify the reference prior as a function of the distance measure, the other to identify the form of the above risks, and Bayes risks, as a function of the loss. The latter would characterize the asymptotically least favorable prior. Consequently, specifying the physical context would give reference and least favorable priors immediately.

Although, the even harder work of understanding the physical problem and statistically modeling it remains, I found that this paper presented a framework which is valuable for understanding the issues in selecting a prior. The authors are to be thanked for their contribution.

REPLY TO THE DISCUSSION

Since we believe a non-informative prior is useful primarily as an approximate and tentative reference point, we do agree with the comment of Poincaré quoted by Professor Polson. The difficulty of defining "absolute ignorance" often comes out in the fact of the so-called non-informative priors being improper. However there is a deeper paradox or difficulty which seems to be inherent in the notion of information itself which makes it difficult to define "greater or less depth of this ignorance" in a completely satisfactory way. If one tries to define information in a prior in isolation from the observation X, one usually ends up with the uniform distribution, which, being non-invariant under both one to one and many to one transformations, can cause problems even when the parameter space consists of a finite number of points. On the other hand, if one defines information in a prior in the context of an experiment generating X, violations of coherence and likelihood principles become unavoidable. This is true both for the regular i.i.d. case and the non-linear model of Polson.

At the heart of the paradox lies the fact that information is generated by the interaction of the prior with the experiment or of the prior with the data as in a non-orthogonal design of treatments and blocks. Two logical but facile solutions are the following. In the first one treats only $H(p)$ as the information in the prior and Lindley's measure as a functional on the space of experiments. This will always lead to the uniform distribution as the least informative prior. In the second one accepts the fact that one must often introspect about θ when the data are already in hand and so what really matters is a posterior. Both the prior and the likelihood become irrelevant.

To sum up we think the real fundamental debate should be on how to measure the information in a prior rather than on which prior is non-informative. The alternative of trying to match what a frequentist would do needs careful examination also. A third thing to examine is the possible role of utility, especially in the context of nuisance parameter, as part of the general debate on the extent to which the elicitation of a subjective prior depends on the subjective utility function.

We may clarify once again the reason for our preference for the functional used by Bernardo. It has two good features. It has the right scale for θ and X unlike the measure $H(p) + H_\theta(X)$ in which the second term involving X dominates as $n \to \infty$ and so, on maximization, leads to unacceptable priors. Secondly the functional has a penalty term for deviation from the uniform. We have introduced a new measure in the case of a nuisance parameter which has these properties and whose maximization leads to a prior which is the same as the Berger-Bernardo prior in most cases and satisfies the matching property better.

The other important aspect of non-informative priors is that they are very often used. The reason for this is that eliciting a prior remains difficult. We can't see why the Berger-Bernardo priors are any more automatic and therefore more non-Bayesian than those of others including Jeffreys; in fact they need more judgement on the order of importance.

One of us has been asked why such a prior should be used anyway. If one has lots of data, the choice of prior doesn't matter. In the contrary case one should use one's subjective prior because it matters a lot which prior is to be used. The answer to this is that for a large data set one might as well use a non-informative prior since choice of a prior doesn't really matter. In the other case, one might want to report the results obtained by one's subjective prior as well as a non-informative prior, as a rough indication of how much the final inference depends on the prior.

We should point out in this context that other commonly used Bayesian tools, namely conjugate priors, Empirical Bayes and Hierarchical Bayes methodologies also suffer from dependence on the experiment. In the Neyman-Scott example, X_{ij} normal with mean μ_i and variance $\sigma^2, i = 1, \ldots, n, j = 1, 2$, a hierarchical prior gives a consistent estimate for σ^2 only if the prior is properly chosen. The details of this are available from Professor Malay Ghosh. A natural question to ask here is whether frequentist consistency of estimates should remain a desirable thing to a Bayesian.

Coming to more technical, rather than philosophical, matters, let us thank Professor Polson for drawing attention to the extensive literature on the asymptotic representation of Lindley's measure. We don't know if the earlier literature, barring the Russian work for a location parameter, is rigorous. Questions of rigour are important here because the discussants of Bernardo (1979) and many others have made this one of the critical issues.

Lastly let us express our pleasure in his asymptotic formula for a diffusion process. If $I_T(\theta)/T$ converges to a limit $I(\theta)$, as it does for a Brownian notion with drift, then the asymptotic maximization would again yield the prior $|I(\theta)|^{1/2}$. It would be nice to know when the limit $I(\theta)$ exists and its connections with the asymptotic normality of $X(T)$ as

$T \to \infty$.

Another technical question, which we have not been able to answer and so would like to put before others, is whether the priors obtained by matching can be obtained by maximizing a functional and whether such a functional has properties similar to those of an information functional.

Professor Berger asks whether we should use the reference prior for an orthogonal nuisance parameter instead of the reference prior recommended by Berger and Bernardo. Since the choice of an orthogonal nuisance parameter isn't unique, the reference prior for an orthogonal nuisance parameter isn't unique. Our favourite is the prior recommended in Section 2 which is invariant under different choices of an orthogonal nuisance parameter and has good matching properties as well as some information theoretic justification. In most of the examples we have checked it agrees with the Berger-Bernardo prior but this is not always so. We hope to carry out simulations in one of the latter examples and report elsewhere.

Professor Clarke has suggested some additional methods and a conceptual framework. An exploration of possible exact or approximate relations between different methods is of interest. A prior obtained by maximizing the expected divergence between posterior and prior will often be a minimax prior in a problem of predicting $p(X|\theta)$ with a suitable prediction loss. We checked this for one of Clarke's divergence measures.

ADDITIONAL REFERENCES IN THE DISCUSSION

Berger, J. O. and Bernardo, J. M. (1992). On the development of reference priors. *Bayesian Statistics 4* (J. M. Bernardo, J. O. Berger, A. P. Dawid and A. F. M. Smith, eds.), Oxford: University Press, 35–60, (with discussion).

Bernardo, J. M. (1979b). Expected information as expected utility. *Ann. Statist.* **7**, 686–690.

Bondar, J. V. (1977). A conditional confidence principle. *Ann. Statist.* **5**, 881–891.

Dawid, A. P., Stone, M. and Zidek, J. V. (1973). Marginalisation paradoxes in Bayesian and structural inference. *J. Roy. Statist. Soc. B* **35**, 189–233. (with discussion).

DeGroot, M. H. (1986). Changes in utility as information. *Recent Developments in the Foundations of Utility and Risk Theory* (L. Daboni *et al.* eds.), Dordrecht: Reidel.

Ibragimov, I. A. and H'asminsky, R. Z. (1973). On the information contained in a sample about a parameter. *2nd Intl. Symp. on Information Theory*, 295–309.

Parmigiani, G. and Polson, N.G. (1992). Bayesian design for random walk barriers. *Bayesian Statistics 4* (J. M. Bernardo, J. O. Berger, A. P. Dawid and A. F. M. Smith, eds.), Oxford: University Press, 715–721.

Poincaré, H. (1905). *Science and Hypothesis*. New York: Dover.

Polson, N. G. (1987). Discussion of Cox and Reid. *J. Roy. Statist. Soc. B* **49**, 24.

Polson, N. G. (1988a). Bayesian model choice: a decision theoretic criterion. *Tech. Rep.* University of Nottingham.

Polson, N. G. (1988b). On the expected amount of information from a nonlinear model. *J. Roy. Statist. Soc. B* , (to appear).

Seidenfeld, T. (1983). Probability, continuity, and transformations of continuous random variables. *Essays in epistemology and semantics* (H. Leblanc *et al.* eds.), 117–139.

Seidler, J. (1958). Relationships between information theory and decision functions theory. *Trans. 2nd. Prague conf. on information theory*, 579–592.

Stone, M. (1959). Application of a measure of information to the design and comparison of regression experiments. *J. Roy. Statist. Soc. B* **21**, 55–70.

Stone, M. (1979). A review and analysis of some inconsistencies related to improper priors and finite additivity. *Proc. of VIth International Congress of Logic, Methodology and Philosophy of Science*, 413–426.

BAYESIAN STATISTICS 4, pp. 211–226
J. M. Bernardo, J. O. Berger, A. P. Dawid and A. F. M. Smith, (Eds.)

Optimal Stopping for a Non-Communicating Team

PREM K. GOEL, CHANDRA M. GULATI and MORRIS H. DEGROOT*
The Ohio State University, USA, The University of Wollongong, Australia
and *Carnegie Mellon University, USA*

SUMMARY

We consider optimal solutions to a class of stopping problems in which the decision to stop sampling is made by a team. The model assumes that each member of the team observes a component of the observable random variable $\boldsymbol{X}$, with probability density function $f(\boldsymbol{x})$, and is not privy to the value obtained by other members of the team. After an observation is taken, the team members simultaneously cast votes either to stop or to continue sampling. A team decision to stop sampling is made on the basis of a vote counting algorithm. The cost of each observation to the team is $C(C > 0)$. Given a reward function $u(\boldsymbol{x})$, the reward to the team on stopping at the n-th observation is $u(\boldsymbol{x}_n)$. Let the random variable N denote the stopping time. The net gain to the team $\{u(\boldsymbol{X}_N) - NC\}$, is thus a random variable. We find the optimal stopping strategy, which maximizes the expected utility $E[\mathcal{U}\{u(\boldsymbol{X}_N) - NC\}]$ of the net gain, for a two member team with different vote counting algorithms. Examples are given to illustrate the results for various distributions of $\boldsymbol{X}$.

Keywords: OPTIMAL STOPPING; TEAM DECISION; AT-LEAST M-OUT-OF K; NON-COMMUNICATING TEAM; NET GAIN; UTILITY; DE-CENTRALIZED INFORMATION.

1. INTRODUCTION

Marschak and Radner (1972) ([MR] for short) define an *organization* as, "a group of persons whose actions agree with certain rules that further their common interests", and a *team* as an organization with "only common interests." The common interest is expressed mathematically by a single expected utility function $\mathcal{U}$ for the team. The [MR] monograph presents various formulations of the team organization problems in economic theory setting along with the underlying decision analyses for different information structures, corresponding to the level of information exchange between the team members, and various payoff functions in a comprehensive manner. A recent monograph by Kim and Roush (1987) presents the core material developed by Marschak in the 1950's and describes many aspects of teams, not previously studied, using modern techniques such as the theory of algorithms.

In this paper, it is assumed that each of the k members of a team can collect information about the state of the world by observing possibly different characteristics of same phenomena. The joint distribution of the observation vector $\boldsymbol{X} = (X_1, X_2, \ldots, X_k)$ is assumed to have a density function (p.d.f.) $f(x_1, x_2, \ldots, x_k)$ such that the mean $\boldsymbol{\mu} = (\mu_1, \mu_2, \ldots, \mu_k)$

* This work was initiated in collaboration with Morrie DeGroot. In the initial stages of the research, Morrie was a constant source of ideas during our mutual discussions. However, due to his untimely death, Morrie never saw the final product. We have taken the liberty of including him as a joint author, believing that he would have agreed to be a joint author had he seen the paper. Any errors or omissions are entirely our responsibility.

exists with $\mu_j < \infty, j = 1, 2, \ldots, k$. For $i = 1, 2, \ldots$ and $j = 1, 2, \ldots, k$, let X_{ij} denote the i-th observation obtained by the j-th member of the team. At stage i, the team member j only gets to look at the j-th component of the observation vector $\boldsymbol{X}_i$. The value of the observation obtained by a member is not revealed to any other member of the team.

It is also assumed that the team's cost of taking an observation vector at each stage of the sampling process is $C (C > 0)$. After an observation $\boldsymbol{x} = (x_1, x_2, \ldots, x_k)$ is taken, the team may choose to stop sampling and accept the gross payoff $u(\boldsymbol{x})$ or else reject the observation and buy another observation at a cost C. If sampling is stopped at the n-th observation vector $\boldsymbol{X}_n$, the team's utility of the net gain $G = u(\boldsymbol{x}_n) - nC$ to the team is given by $\mathcal{U}[G]$. When every team member votes to stop sampling or to continue sampling based only on the knowledge of the value of his component observation alone, we define it a *non-communicating team*, while [MR] call it as a case of *decentralized information*.

In the above formulation, the past observations cannot be recalled and the payoff depends only on the observation at which sampling is stopped. Here it is assumed that the voting to stop sampling or to continue sampling is done simultaneously by all members of the team after an observation is obtained. However, the team decision to stop sampling can be made according to an agreed upon vote counting algorithm, e.g., the sampling stops only if (i) all the members of the team vote to stop [consensus]; (ii) the team leader (say the first member) votes to stop [dictatorship] or (iii) at least m members of the team vote to stop ($m = 1$ means that any member can stop the sampling process) [democracy] etc. Given a vote counting algorithm, the objective is to determine a stopping rule N which maximizes the expected net gain $E[\mathcal{U}[u(\boldsymbol{X}_N) - NC]$. Problems of this type arise, for example, when a Board of Directors have to select a possible project, or when husband and wife are searching for jobs together and they hire an agent, who charges a fixed cost C for each round of job search. In these problems, the objective of the team is to arrive at a decision from the team's point of view.

This problem can be thought of as a generalization of the problems of sampling without recall [see, e.g., DeGroot (1970), Chapter 13]. Some other decision problems in which more than one decision maker is involved are considered in DeGroot and Kadane (1983) and Rose (1982). Kempthorne (1988) and others have considered problems so that a compromise is reached. In our proposed research, each member of the team takes an independent action to cast a vote and the stopping decision is made according to an agreed upon vote counting algorithm. Therefore, the need for a compromise after the information has been collected is avoided. Cyert and DeGroot (1987, page 26) point out that research into problems involving more than one decision maker is in an incipient state.

If the reward depends only on one player's (e.g., the first player's) observed value and the utility function $\mathcal{U}$ is linear, i.e.,

$$\mathcal{U}[u(\boldsymbol{X}_N) - NC] = u(X_{N1}) - NC, \tag{1.1}$$

the problem reduces to the well known problem of optimal stopping for one player [see, e.g., DeGroot (1968, 1970), and Leonardz (1973)]. This model also corresponds to a situation where the reward may depend on the observations of other members of the team, but is an additive function of the individual observation, and the team leader (say the first member of the team) makes a decision on the basis of his observation alone and decides to dictate his decision to the team by ignoring the recommendations of all other members. The one person problem in the economic context of a consumer searching for the lowest price, was addressed in a seminal paper by Stigler (1961). However, the problem addressed by Stigler is present in a variety of economic modelling contexts. Lippman and McCall (1976) provide a survey of literature on this problem within the paradigm of job search. Hall, Lippman and McCall (1979) discuss the one person optimal stopping problems for job search in which a non-linear expected utility must be maximized. Interesting essays, which extend the elementary search theory in various directions and test some of its empirical implications, are presented in Lippman and McCall (1979). A brief discussion of the one person problem is presented here, because it will set the stage for the later developments in the paper.

The player can take observations from a known continuous distribution with p.d.f. $f(x)$. If $u(x)$ is a monotone increasing function, we can assume, without loss of generality, that an appropriate transformation has been made so that f represents the density of $u(\cdot)$. Throughout this paper, it will be assumed that $E[u(X)] < \infty$. At stage $n, n = 1, 2, \ldots$, the player may either stop and receive the reward x_n or continue sampling by paying the cost C. The objective is to maximize the expected gain. Since no extra information can be gained at any stage about the p.d.f. f, it is well known that the player must follow the same strategy at all stages [see e.g., DeGroot (1968)].

At each stage, the optimal strategy is of the form: Continue sampling if $X < a$ and stop if $X \geq a$. Then the player picks the value of a, so that his expected net gain $G(a)$, where

$$G(a) = \frac{1}{1 - F(a)} \int_a^\infty x f(x) dx - \frac{C}{1 - F(a)}, \tag{1.2}$$

is maximized. Note that the first term in (1.2) equals the expected reward, $E[X|X \geq a]$, after one stops and the second term reflects the expected cost of sampling, since the stopping variable has a geometric distribution with mean $[1 - F(a)]^{-1}$. Now, the first derivative of $G(a)$ is given by

$$G'(a) = \frac{h(a)}{1 - F(a)} [T_f(a) - C] \tag{1.3}$$

where, $h(a)$ is the well known *hazard rate function* in survival analysis, i.e.,

$$h(a) = \frac{f(a)}{1 - F(a)}; \tag{1.4a}$$

and the functional $T_f(a)$ is equal to $E[\max(X - a, 0)], a \in \mathcal{R}$, i.e.,

$$T_f(a) = \int_a^\infty (x - a) f(x) dx = \int_a^\infty (1 - F(x)) dx, a \in R. \tag{1.4b}$$

Clearly the Mean Residual Life (MRL), $r(a)$, in reliability analysis, can be expressed in terms of $T_f(a)$ as follows:

$$r(a) = E[X - a|X \geq a] = \frac{T_f(a)}{1 - F(a)}. \tag{1.5}$$

The right hand side of (1.3) can also be expressed as $h(a)\{r(a) - C/[1 - F(a)]\}$. Note that the domain of $T_f(a)$ is the whole real line even though the support set of F may only be a finite interval or the positive part of the real line. The functional $T_f(a)$ has been studied extensively in Leonardz (1973), where it is established that as a function of a, $T_f(a)$ is: (a) non-negative, (b) continuous, (c) convex, (d) monotone decreasing and (e) differentiable for all a at which $F(a)$ is continuous. It follows from (1.3) that the derivative of G(a) is zero if and only if

$$T_f(a) = C. \tag{1.6}$$

Note that the equation (1.6) above is the same as the equation (2) in Section 13.5 of DeGroot (1970). The properties $(a) - (e)$ of $T_f(a)$ imply that (i) the equation (1.6) has exactly one solution a^* (say), which may sometimes be outside the support set of the distribution f, (ii) $G'(a) < 0$ for $a > a^*, G'(a) > 0$ for $a < a^*$, and $G''(a^*) = -h(a^*)$, which is negative if a^* is in the support set of F, and consequently $G(a)$ achieves its maximum at $a = a^*$, (iii) $G(a^*) = a^*$, and (iv) $a^* \geq 0$, if and only if $C \leq T_f(0)$. Hence, the optimal solution for the one player problem is given by:

Don't play the game if $C > E[\max(X,0)]$; otherwise, take at least one observation and stop sampling whenever $X \geq a^*$.

In this paper, we limit our discussion to the problem in which the team consists of only two members. It is also assumed that the characteristics observed by the two members are stochastically independent, i.e., $f(x_1, x_2) = f_1(x_1) * f_2(x_2)$ and that the reward function u and the utility function $\mathcal{U}$ are also both linear, i.e., $\mathcal{U}[u(x) - NC] = u_1(x_1) + u_2(x_2) - NC$. Without loss of generality, if u_1 and u_2 are monotone functions, then it can be assumed that f_i represents the distribution of $u_i(X_i), i = 1, 2$. Special attention will also be focussed on the case when $f_1 = f_2$. The problem can be considered under the following two different team decision algorithms for stopping; (P_2) Sampling is stopped at stage n only if both members vote to stop at that stage, or (P_1) Sampling is stopped at stage n if either of the two members votes to stop at that stage.

In Section 2, problem P_2 corresponding to the decision algorithm in which both members must vote to stop is considered. In Section 3, problem P_1 corresponding to the decision algorithm in which at least one member must vote to stop the process is considered. In each section, optimal strategies are found and conditions are derived under which strategies for both members are the same. Some examples are also presented to illustrate the results for the two problems. Section 4 contains some concluding remarks.

2. OPTIMAL STOPPING FOR DECISION ALGORITHM P_2

In this section, the problem of optimal stopping when both members must vote to stop the sampling process is considered. It is assumed that the characteristics observed by the members are stochastically independent each having continuous p.d.f.'s f_1 and f_2. Further, it is assumed that if the process is stopped after taking the N-th observation (x_{N1}, x_{N2}), the team receives a reward whose utility is $x_{N1} + x_{N2} - NC$, where C is the cost of taking observation at each stage $(C > 0)$. Since no information is gained about f_1 and f_2 from the observations, it is clear that under optimal strategy both team members must choose stopping sets S_1 and S_2 with the property that at each stage n, member 1 votes to stop if and only if $x_{n1} \in S_1$ and member 2 votes to stop if and only if $x_{n2} \in S_2$. It follows from standard dynamic programming considerations that $S1$ and S_2 are of the form $S_1 = (a, \infty)$ and $S_2 = (b, \infty)$, [see e.g., DeGroot (1968)] and the process is stopped if both players vote to stop. Let $G(a, b)$ denote the corresponding net expected utility. The team requires a strategy to maximize $G(a, b)$.

When $f_1 = f_2$, it is clear that both members are facing the same problem. It then seems intuitively reasonable that under optimal strategy both members should follow the same strategy. However, the following example proves that solutions other than a = b can be optimal. The only symmetry in the optimal solution is that members 1 and 2 play exchangeable roles, but it does not require the two of them to choose the same cut-off points.

Example 2.1. Suppose that the players are sampling from the same discrete distribution taking on the values –1, 0 and 2 with probabilities 0.5, 0.3 and 0.2 respectively. The Table 2.1 gives for each possible strategy (a, b), with both players voting to stop if $x_1 \geq a$ and $x_2 \geq b$, its expected net gain and the range of C over which it is optimal. In this example, the strategy (a, b) is equivalent to the strategy (b, a). Thus the roles of the two players are exchangeable.

Strategies (a, b)	Expected Net	Range of C for Gain Optimality
(2, 2)	4.0–25C	$0.00 \leq C \leq 0.08$
(2, 0)	2.8–10C	$0.08 \leq C \leq 0.18$
(2, –1)	1.9–5C	$0.18 \leq C \leq 0.30$
(0, 0)	1.6–4C	$0.30 \leq C \leq 0.40$
(0, –1)	0.7–2C	—
(–1, –1)	–0.2–C	—

Table 2.1. *Evaluation of Various Strategies.*

It is clear from Table 2.1 that for $0.08 \leq C \leq 0.18$, at each stage the players must decide in advance as to which of the two players will vote to stop when $x = 2$ and the other one will vote to stop when $x = 0$ or 2. The sampling is done until both the players vote to stop. The strategy $(0, -1)$ is not optimal for any C, and the strategy $(-1, -1)$ yields negative expected reward for all C. Finally, the maximum value of C for which this game may be played is $2E[X|X \geq O](P(X \geq 0))^2$, which is equal to 0.4. Note that if the reward X is changed to $(X + \nu), \nu > 0$, then the strategy $(a + \nu, b + \nu)$ is optimal for the same values of C for which (a, b) was optimal, and the strategies $(0 + \nu, -1 + \nu)$ and $(-1 + \nu, -1 + \nu)$ may now be feasible. In general, when sampling from a distribution with a positive support set, it is worthwhile to take at least one observation if $C < 2E[X]$. It must be emphasized that, even though in this example a negative cut-off point is not optimal, this is not true in general.

Now the problem where both $f_1(x_1)$ and $f_2(x_2)$ are continuous is considered. The net expected utility of the gain $G(a, b) = E[X_1 + X_2|X_1 \geq a, X_2 \geq b] - C/P[X_1 \geq a, X_2 \geq b]$. However, since, X_1 and X_2 are independent random variables, the gain G can be expressed as

$$G(a, b) = a + b + r_1(a) + r_2(b) - \frac{C}{(1 - F1(a))(1 - F2(b))}, \tag{2.1}$$

where $r_i(\cdot)$ denotes the MRL, defined in (1.5), corresponding to the distributions $f_i(\cdot), i = 1, 2$. The partial derivatives of $G(a, b)$ with respect to a and b are given by

$$G_a(a, b) = \frac{h_1(a)}{1 - F_1(a)} \left[T_{f_1}(a) - \frac{C}{1 - F_2(b)} \right], \tag{2.2a}$$

and

$$G_b(a, b) = \frac{h_2(b)}{1 - F_2(b)} \left[T_{f_2}(b) - \frac{C}{1 - F_1(a)} \right], \tag{2.2b}$$

where $h_i(\cdot)$, and $T_{f_i}(\cdot)$ denote the hazard rate function and the functional T_f, as defined in (1.4a) and (1.4b), for the distributions $f_i, i = 1, 2$. These derivatives are zero if

$$T_{f_1}(a) = \frac{C}{1 - F_2(b)} \tag{2.3a}$$

and

$$T_{f_2}(b) = \frac{C}{1 - F_1(a)}. \tag{2.3b}$$

In view of the properties of the functional T_f, listed in Section 1, and the properties (i) to (iv) of $G'(a)$, a function similar to $G_a(a,b)$ listed after (1.6), it is clear that for every fixed b, there is at most one solution $a^*(b)$ to (2.3a), and for every fixed a, there is at most one solution $b^*(a)$ to (2.3b). Furthermore, $a^*(b)$ and $b^*(a)$ are decreasing functions of b and a respectively. However, it is not necessary that there exist at least one point (a^*, b^*) which is a simultaneous solution to (2.3a) and (2.3b). On the other hand, there may be infinitely many simultaneous solutions to this system, all of which may or may not be global extrema.

For example, if $f_1(\cdot)$ and $f_2'(\cdot)$ are exponential distributions with parameters λ_1 and λ_2 with $\lambda_1 > \lambda_2$, then the hyper plane satisfying (2.3a) is always below that satisfying (2.3b). However, if $\lambda_1 = \lambda_2$, then the two equations are exactly the same and hence there are infinitely many solutions to this system of equations. For these distributions, the optimal solution to the problem requires analysis beyond solving for the critical points of (2.3a) and (2.3b) and examining the matrix of second derivatives at these points.

In fact, according to the main theory of [MR], "optimality with respect to each member separately holding the others fixed (person-by-person) is necessary and is sufficient if $G(a,b)$ is concave and differentiable in the a and b." Even though for individual distributions, one may have to examine the existence of a simultaneous solution to the above system, we examine the behavior of the Hessian of $G(a,b)$ in order to develop some understanding of the optimal solution for interesting classes of distributions. The pure second derivatives of $G(a,b)$ with respect to a and b at points satisfying either (2.3a) or (2.3b) are given as follows:

$$G_{aa}(a,b)|_{(a^*(b),b)} = -h_1(a^*(b)) \tag{2.4a}$$

$$G_{bb}(a,b)|_{(a,b^*(a))} = -h_2(b^*(a)). \tag{2.4b}$$

Furthermore, the mixed second derivative of $G(a,b)$ is given by

$$G_{ab}(a,b) = \frac{-Ch_1(a)h_2(b)}{[(1-F_1(a))(1-F_2(b))]}. \tag{2.4c}$$

Since the pure second derivatives are negative at the stationary points, it follows that for each fixed $b[a]$, $a^*(b)[b^*(a)]$ provide the maxima of $G(a,b)$. Furthermore, the determinant of the Hessian matrix of $G(a,b)$, evaluated at the stationary points (if any), is given by

$$\begin{aligned} H &= G_{aa}(a,b)G_{bb}(a,b) - G_{ab}^2(a,b)|_{(a^*,b^*)} \\ &= h_1(a^*)h_2(b^*)[1 - h_1(a^*)h_2(b^*)r_1(a^*)r_2(b^*)]. \end{aligned} \tag{2.5}$$

Again, since the pure second derivatives are negative at the stationary points, if $H > 0$, then (a^*, b^*) (if it exists) is a relative maxima, which is the unique global maxima; if $H = 0$, then the Hessian matrix is semi-definite and the stationary point may be a local maxima, or a local minima or neither and we need to examine the function further; and finally if $H < 0$, then the Hessian matrix is indefinite and (a^*, b^*) is a saddle point of the function $G(a,b)$.

We now further examine the properties of the stationary points of G. It follows from equations (1.5), that the equation (2.2a) evaluated at the solutions of (2.3b), and the equation (2.2b) evaluated at the solutions of (2.3a), can be written as

$$G_a(a, b^*(a)) = h_1(a)[r_1(a) - r_2(b^*(a))], \tag{2.6a}$$

$$G_b(a^*(b), b) = h_2(b)[r_1(a^*(b)) - r_2(b)]. \tag{2.6b}$$

It follows that if there exists a point (a^*, b^*) satisfying,

$$r_1(a^*) = r_2(b^*) = \frac{C}{(1 - F_1(a^*))(1 - F_2(b^*))}, \tag{2.7}$$

then there exists a stationary point of $G(a, b)$. If both the distributions have Decreasing Mean Residual Life, then $H > 0$ and therefore there is an interior point solution. If both distributions have Increasing Mean Residual Life, then the stationary point is a saddle point. One example of an IMRL distribution is a mixture of exponential distributions.

If the two MRL functions r_1, r_2 do not cross each other, as is the case with the exponential distributions with parameters λ_1 and $\lambda_2, \lambda_1 > \lambda_2$, then there is no stationary point of $G(a, b)$ and one needs to further examine the objective function G to determine the optimal solution, which will be a point on the boundary. It can be shown that if $r_1(\cdot) > r_2(\cdot)$ for all points, then we choose $b^* = -\infty$, and a^* satisfies $T_{f_1}(a^*) = C$, and a similar result holds if $r_1(\cdot) < r_2(\cdot)$. Thus, the solution is the same as that of the one member team problem. The above discussion can be summarized by the following result.

Theorem 2.1. *With linear utility function and an additive reward function, if there exists a point (a^*, b^*) satisfying (2.7), then it is the optimal solution to the two member team problem. Furthermore, if the two MRL functions do not intersect at all, then the optimal solution lies on the boundary in that the solution is the same as that of the one member team problem. In other cases, there may be infinitely many optimal solutions.*

Example 2.2. For the exponential distributions with parameters λ_1 and $\lambda_2, \lambda_1 > \lambda_2$, i.e., $E[X_1] < E[X_2]$, the optimal sequential procedure is as follows:

If $C > E[X_1]+E[X_2]$, then do not take any observation; if $E[X_2] < C < E[X_1]+E[X_2]$, then $a^* = b^* = 0$ and $G = E[X_1] + E[X_2] - C$; and finally if $0 < C < E[X_2]$, then $a^* = 0, b^* = (1/\lambda_2)$ ln $(1/C\lambda_2)$ and $G* = b^* + E[X_1]$.*

We now examine the optimal solution for the special case when $F_1 = F_2 = F$. The equations (2.7) and (2.5) respectively reduce to

$$r(a^*) = r(b^*) = C/[(1 - F(a^*))(1 - F(b^*)], \tag{2.8a}$$

$$H = h(a^*)h(b^*)[1 - h(a^*)r(a^*)h(b^*)r(b^*)]. \tag{2.8b}$$

Thus, if the MRL function of the distribution F is strictly monotone, then (2.8a) implies that $a^* = b^*$. It should be noted that when $h(a)$ is strictly increasing in a, i.e., $\log(1 - F(a))$ is a strictly concave function of a (or f is the density of an IFR distribution), then $r(a)$ is a decreasing function of a [see, Bryson and Siddiqui (1969)]. For such a random variable X, with support in $\mathcal{R}^+$, it is clear from (2.1) – (2.2b) that if $C > 2E[X]$ then the team does not take any observation and if $E[X] < C < 2E[X]$, then the optimal strategy is to take exactly one observation and stop, i.e., choose $a^* = b^* = 0$. This result can be generalized further when the support of the distribution is bounded below by a negative number a_0.

Furthermore, since, $r'(a) = h(a)r(a) - 1$, it is clear that the sign of H in (2.8b) is positive (negative) if the slope of $r(\cdot)$ at a^* and b^* is negative (positive) and $H = 0$ if the slope of $r(\cdot)$ at a^* and b^* is zero. Therefore, if the MRL function is strictly decreasing and the stationary point satisfies $H > 0$, then the optimal strategy for both the team members is the same and the maximum value of $G(a, b)$ is given by

$$G(a^*, a^*) = 2a^* + r(a^*). \tag{2.9}$$

The above discussion can be summarized in the following theorem.

Theorem 2.2. *With linear utility function and an additive reward function, if the distributions F_1 and F_2 are same, and possess a strictly monotone hazard rate, then the two members of the team have the same optimal cut-off points.*

It is clear from (2.9) that if $r(a^*)$ is greater than $-2a^*$, then a negative cut-off point is acceptable to the team. Note that in the one player problem, a negative cut-off point is not acceptable. Furthermore, when $F_1 = F_2$, the solution to the two-member team problem is similar to the one player problem with the cost C inflated to $C/(1-F(a^*))$. We now present examples of distributions in which (a^*, b^*) is a global maxima, and in which $H = 0$. We also present an example of a distribution in which a negative cut-off point is chosen in the optimal solution.

Example 2.3. Suppose that X_1 and X_2 have uniform distributions on the interval $[0, 1]$. Since $r(a) = 0.5(1-a), (0 \leq a \leq 1)$, is a strictly decreasing function, Furthermore, it follows that both players should follow the same strategy. Furthermore, it follows from (2.8a) and (2.9) that for $0 < C < 1/2, a^*$ satisfies

$$a^* = 1 - (2C)^{1/3} \tag{2.10}$$

and

$$G^* = G(a^*, a^*) = (1 + 3a^*)/2, \tag{2.11}$$

and if $1/2 \leq C \leq 1$, the optimal rule uses $a^* = 0$ and $G^* = 1 - C$. Thus under the optimal rule we take only one observation and stop. Note that as $C \to 0, a^* \to 1$ and $G^* \to 2$, i.e., the optimal strategy calls for stopping when a value close to 1 is observed by both the players. Table 2.2 gives the values of a^* and G^* for $C = .1(.1)1.0$.

C	0.01	0.1	0.2	0.3	0.4	≥ 0.5
a^*	0.729	0.415	0.263	0.157	0.072	0.0
G^*	1.593	1.123	0.895	0.735	0.608	$1 - C$

Table 2.2. *Optimal Stopping when Sampling from Uniform Distribution.*

Example 2.4. Suppose that the sampling distribution is exponential with parameter 1. Since $r(a) = 1$ for all a, it is not a strictly monotone function. It is easy to see that in this example, equations (2.3a) and (2.3b) are exactly the same and hence there is a degenerate situation in that $H = 0$. Furthermore, for $0 < C < 1$, there are an uncountable number of optimal strategies (a^*, b^*) satisfying $a^* + b^* = -\log C$ with $G^* = 1 - \log C$, and for $1 \leq C \leq 2$, only one observation should be taken with $G^* = 2 - C$. If $C > 2$, then no observation should be taken.

Example 2.5. Suppose that the two players are sampling from a standard Normal distribution. Let ϕ and Φ denote the p.d.f. and the c.d.f. of this distribution respectively. Since

$$r(a) = \frac{\phi(a)}{1 - \Phi(a)} - a, \tag{2.12}$$

and it is well known that $\log(1 - \Phi(a))$ is a strictly concave function of a, [see Karlin (1968)], it follows that the stationary point of G is the optimal solution under which the two members use same strategy a^* satisfying

$$r(a^*) = C/[(1 - \Phi(a^*))^2], \tag{2.13}$$

with G^* given by (2.9) and (2.13). It is easy to check that for $C \cong 0.4869, G^* = 0$ and $a^* \cong -0.515$.

C	0.05	0.10	0.15	0.20	0.25	0.30	0.35	0.40	0.45
a^*	.572	.308	.134	−.001	−.114	−.212	−.299	−.379	−.454
G^*	1.767	1.311	1.019	.796	.613	.456	.318	.193	.079

Table 2.3. *Optimal Stopping when Sampling from Normal Distribution.*

Table 2.3 gives values of a^* and G^* for $C = .1(.05)0.45$, showing that, in contrast to one member problem, a negative cut-off point may be optimal in the two member team problem.

3. OPTIMAL STOPPING FOR DECISION ALGORITHM P_1

In this section, the optimal stopping problem, in which the sampling process is stopped if either member votes to stop the process. It is assumed that that the characteristics observed by the members are stochastically independent each having continuous p.d.f.'s f_1 and f_2. Further, it is assumed that if the process is stopped after taking the N-th observation (x_{N1}, x_{N2}), the team receives a reward whose utility is $u_1(x_{N1}) + u_2(x_{N2}) - NC$, where C is the cost of taking observations at each stage ($C > 0$). If u_1 and u_2 are monotone increasing functions, one can assume that without loss of generality f_1 and f_2 represent the distributions of $u_1(X_1)$ and $u_2(X_2)$ respectively.

Since no information is gained about f_1 and f_2 from the observations, it is clear that each member should choose the same procedure throughout, except for switching labels in case the two are exchangeable. Under optimal strategy, both team members must choose stopping sets S_1 and S_2 with the property that at each stage n, member 1 votes to stop if and only if $x_{n1} \in S_1$ and member 2 must vote to stop if and only if $x_{n2} \in S_2$. It follows from standard dynamic programming considerations that S_1 and S_2 are of the form $S_1 = (a, \infty)$ and $S_2 = (b, \infty)$. Now the strategy (a, b) means that the sampling is stopped if either $x_{n1} \in S_1$ or $x_{n2} \in S_2$ or both.

Again, as before, it may be intuitively appealing to believe that when $f_1 = f_2$, both members should follow the same strategy under an optimal rule. However, this is not the case. The only symmetry in the optimal solution is that members 1 and 2 play exchangeable roles, and the two players are not required to choose the same cut-off points. For the Example 2.1 and the Problem P_1, in which either player may stop the process, the strategy (2, 2) dominates every other strategy for all $C(\leq 3/5)$ for which the game should be played. We, however, present another example to exhibit this phenomenon.

Example 3.1. Let us assume that the random variables X_1, X_2 are independently distributed each taking values $0, 1, 2$ with probabilities $.5, .3$ and $.2$ respectively. Table 3.1 gives for all possible strategies (a, b), the expected net gain and the range of C over which they are optimal. Note that the strategy $(1, 2)$ is equivalent to the Strategy 2 below because the roles of the two players are exchangeable.

Strategies I.D.	(a, b)	Expected Net Gain	Range of C for Optimality
1	$(2,2)$	$\frac{23}{9} - \frac{25}{9}C$	$0 \le C \le \frac{11}{40}$
2	$(2,1)$	$\frac{9}{4} - \frac{5}{3}C$	$\frac{11}{40} \le C \le \frac{3}{4}$
3	$(1,1)$	$2 - \frac{4}{3}C$	$\frac{3}{4} \le C \le \frac{3}{2}$
4	$(2,0)$	$1.5 - C$	—

Table 3.1. *Evaluation of Various Strategies.*

It is clear from Table 3.1 that the two players do not necessarily choose the same strategy. Strategy 4 is equivalent to stopping after the first observation. However, Strategy 3 strictly dominates the Strategy 4, and the maximum value of C for which the team's net expected payoff equals zero is equal to 1.5.

Now we consider the problem when the two distributions are continuous. Let $G_1(a,b)$ denote the corresponding net expected utility when either member can stop the process. The team follows a strategy to maximize

$$G_1(a,b) = E[X_1 + X_2 | X_1 \ge a \text{ or } X_2 \ge b] - C/P[X_1 \ge a \text{ or } X_2 \ge b]. \tag{3.1}$$

Let $K(a,b)$ denote

$$K(a,b) = E[X_1 + X_2] - C - F_2(b)E[\min(X_1,a)] - F_1(a)E[\min(X_2,a)], \tag{3.2}$$

and using the fact that

$$\begin{aligned} E[X_1 + X_2] =& E[X_1 + X_2 | X_1 \ge a \text{ or } X_2 \ge b] \times P[X_1 \ge a \text{ or } X_2 \ge b] \\ &+ E[X_1 + X_2 | X_1 < a, X_2 < b] \times P[X_1 < a, X_2 < b], \end{aligned} \tag{3.3}$$

we can write $G_1(a,b)$ as

$$G_1(a,b) = \frac{K(a,b)}{1 - F_1(a)F_2(b)}. \tag{3.4}$$

The partial derivatives of (3.4) with respect to a and b satisfy

$$G_{1a}(a,b) \le 0 \text{ iff } a + E[X_2 | X_2 < b] \ge \frac{K(a,b)}{1 - F_1(a)F_2(b)}, \tag{3.5}$$

$$G_{1b}(a,b) \le 0 \text{ iff } b + E[X_1 | X_1 < a] \ge \frac{K(a,b)}{1 - F_1(a)F_2(b)}. \tag{3.6}$$

It follows from (3.5) and (3.6) that the stationary points (a^*, b^*) satisfy

$$a^* + E[X_2 | X_2 < b^*] = \frac{K(a^*, b^*)}{1 - F_1(a^*)F_2(b^*)}, \tag{3.5a}$$

$$b^* + E[X_1 | X_1 < a^*] \ge \frac{K(a^*, b^*)}{1 - F_1(a^*)F_2(b^*)}. \tag{3.6a}$$

Just as in Section 2, it can be easily shown that the pure second derivatives at the stationary point of one equation, when the other variable is kept fixed, are negative. Hence properties similar to those discussed in Section 2 are obtained. Now, if $F_1 = F_2 = F$, then it follows that $a^* = b^*$ is a stationary point and if the function

$$\tau(a) = a + E[X|X < a] \tag{3.7}$$

is strictly monotone, then there is a unique stationary point. Furthermore, the determinant of the Hessian at the stationary point, if one exists, is positive if

$$s(a^*) = a^* - E[X_1|X_1 < a^*] < \frac{F(a^*)}{f(a^*)}. \tag{3.8}$$

It can be shown that (3.8) is equivalent to $\frac{d}{da}[s(a)]|a = a^* > 0$. Therefore (3.8) holds if s(a) is an increasing function of a. When both members use the same rule, the expected net gain is given by

$$G_1(a,a) = \frac{1}{1 - F^2_{(a)}}\{2E[X] - 2F(a)E[\min(X,a)] - C\}. \tag{3.9}$$

For $0 < C < 2E[X]$, the optimal a^* satisfies

$$(1 - F^2(a))a^* + (1 - 3F^2(a))E[X|X < a^*] = C - 2E[X] \tag{3.10}$$

and the maximum expected net gain is

$$G_1(a^*, a^*) = \tau(a^*). \tag{3.11}$$

We now present some examples to illustrate these ideas.

Example 3.2. Suppose that X_1 and X_2 have uniform distributions on the interval $[0, 1]$. Since $\tau(a)(= 3a/2, (0 \le a \le 1))$, is a strictly increasing function and (3.8) holds, it follows that both players should use the same strategy. From (3.10) and (3.11), it follows that for $0 < C < 1, a^*$ satisfies

$$a^{*3} - 3a^* + 2 = 2C \tag{3.12}$$

and

$$G_1^* = G_1(a^*, a^*) = 3a^*/2. \tag{3.13}$$

Note that as $C \to 0, a^* \to 1$ and $G_1^* \to 3/2$, i.e., the optimal strategy calls for stopping when a value close to 1 is observed by either player. Table 3.2 gives the values of a^* and G_1^*, for $C = 0.1(.1)0.9$.

C	0.1	0.2	0.3	0.4	0.5	0.6	0.7	0.8	0.9
a^*	.729	.628	.511	.426	.347	.273	.203	.134	.067
G_1^*	1.094	.913	.767	.639	.521	.410	.304	.201	.101

Table 3.2. *Optimal Stopping when Sampling from Uniform Distribution.*

Example 3.3. Suppose that the sampling distribution is exponential with parameter 1, as in Example 2.4. For this distribution, $\tau(a) = 1 + a - a\exp(-a)/[1 - \exp(-a)]$, which is a strictly increasing function of a, and (3.8) holds. Therefore a unique stationary solution exists. It follows from (3.10) and (3.11), that a^* satisfies

$$2e^{-a^*} - e^{-2a^*} + \frac{a^* e^{-2a^*}}{1 - e^{-a^*}} = C, \tag{3.14}$$

and $G_1^* = \tau(a^*)$. Table 3.3 gives values a^* and G_1^* for $C = 0.1(0.2)1.9$.

C	0.1	0.3	0.5	0.7	0.9	1.1	1.3	1.5	1.7	1.9
a^*	3.047	1.983	1.485	1.152	.898	.690	.511	.351	.205	.067
G_1^*	3.895	2.666	2.051	1.620	1.281	.995	.744	.516	.304	.100

Table 3.3 *Optimal Stopping when Sampling from Exponential Distribution.*

Example 3.4. As in Example 2.5, let us consider sampling from a standard normal distribution, for which $\tau(a) = a - \phi(a)/\Phi(a)$. From (3.10) and (3.11), it is clear that a^* satisfies

$$(1 - \Phi^2(a^*))\tau(a^*) + 2\Phi(a^*)\phi(a^*) = C \tag{3.15}$$

and $G_1^* = \tau(a^*)$.

C	.05	.10	.15	.20	.25	.30	.35	.40	.45
a^*	1.613	1.322	1.137	.998	.884	.787	.701	.625	.554
G_1^*	1.498	1.138	.897	.709	.551	.414	.290	.177	.073

Table 3.4. *Optimal Stopping when Sampling from Normal Distribution.*

Table 3.4 gives values of a^* and G_1^* for $C = 0.05(.05)0.45$. Note that, in contrast to Problem P_2, a negative cut-off point does not lead to an optimal strategy for the Problem P_1.

4. CONCLUDING REMARKS

It should be noted that in the case of complete knowledge of each other's observations, this problem is equivalent to one player taking observations from the distribution of the sum of the two random variables.

On evaluating the optimal values of the expected net gain in Example 2.4 for the exponential distribution, it is learnt that the optimal expected net gain to the team when both players must agree to stop sampling is less than when either player may stop the process. Intuitively, it seems reasonable that this relationship may hold in general, but this is not necessarily true here, as demonstrated in Tables 2.2 and 3.2 for the uniform distribution example. Note that the optimal value of the expected net gain when C is small is larger for the problem P_2 than for the Problem P_1, while the reverse is true when C is moderate. In addition, for the normal distribution example, the result is completely reversed from that for the exponential distribution example. Therefore it will be interesting to examine the classes of distributions in reliability theory for which the various directional relationships hold.

This class of problems will be further studied for two member teams when the reward or the utility function is nonlinear or when there are unknown parameters in the distributions of X_1 and X_2 or when the random variables are not stochastically independent. The generalization to a $k(>2)$ members team under the vote counting algorithm P_m in which at least $m(m = 1, 2, \ldots, k)$ members must agree to stop the process will also be studied. The optimal rule for $m \neq 1$, k is expected to exhibit interesting new features of the optimal solutions.

ACKNOWLEDGMENTS

This research was supported in part by the National Science Foundation under Grants # DMS–8906787, DMS–9008067 and INT–8913294 for travel to Australia under the U.S.-Australia Program. The U.S. Government is authorized to reproduce and distribute reprints for Governmental purposes notwithstanding any copyright notation thereon. We would like to acknowledge the help of Professor Murray Clayton, University of Wisconsin, whose critical reading caught some oversight on our part.

REFERENCES

Bryson, M. C. and Siddiqui, M. M. (1969). Some criteria for aging. *J. Amer. Statist. Assoc.* **64**, 1472–1483.

Cyert, R. M. and DeGroot, M. H. (1987). *Bayesian Analysis and Uncertainty in Economic Theory*. London: Chapman and Hall.

DeGroot, M. H. (1968). Some problems of optimal stopping, *J. Roy. Statist. Soc. B* **30**, 108–122.

DeGroot, M. H. (1970). *Optimal Statistical Decision*. New York: McGraw Hill.

DeGroot, M. H. and Kadane, J. B. (1983). Optimal sequential decisions in problems involving more than one decision maker. *Recent Advances in Statistics* (M. H. Rizvi, J. S. Rustagi, L. D. Siegmund, eds.), New York: Academic Press, 197–210.

Hall, J. R., Lippman, S. A. and McCall, J. J. (1979). Expected utility maximizing job search. *Studies in the Economics of Search* (Lippman, S. A. and McCall, J. J. eds.) Amsterdam: North Holland, 133–155.

Karlin, S. (1968). *Total Positivity 1*, Stanford, CA: Stanford University Press.

Kempthorne, P. J. (1988). Controlling risks under different loss functions: The compromise decision problem. *Ann. Statist.* **16**, 1594–1608.

Kim, K. H. and Roush, F. W. (1987). *Team Theory*. Chichester: Ellis Harwood Ltd.

Leonardz, B. (1973). *To Stop or Not To Stop*. New York: Halsted Press.

Lippman, S. A. and McCall, J. J. (1976). The economics of job search: a survey, Parts I and II. *Economic Inquiry* **14**, 155–189 and 347–368.

Lippman, S. A. and McCall, J. J. (1979). *Studies in the Economics of Search*. Amsterdam: North Holland.

Marschak, J. and Radner, R. (1972). *Economic Theory of Teams*. New Haven: Yale University Press.

Rose, J. (1982). A problem of optimal choice and assignment. *Operations Research* **30**, 172–181.

Stigler, G. J. (1961). The economics of information. *J. of Political Economy* **69**, 213–225.

DISCUSSION

M. K. CLAYTON (*University of Wisconsin, USA*)

It is a pleasure to comment on this paper. To me it is typical of many of the stopping problems in DeGroot (1970): while the problem is relatively straightforward to describe, the solution is nontrivial and has an elegant structure. My comments are directed primarily toward the structural aspects of this problem when the voting rule P_2 is used and observations are normally distributed.

The authors have promised to look at the problem where the members' observations are not independent. This seems most appropriate. For example, if the team consists of a husband and wife searching for jobs together, then their salaries might be positively correlated. To illustrate the effect of a correlation between the team members' observations, suppose X_1 and X_2 are bivariate normal with mean vector 0, unit variances, and correlation coefficient ρ. Table 1 gives optimal cutoff values and gains for two possible values of ρ. In these two examples it is possible to show, simply by finding the maximum of the appropriate gain function, that $a^* = b^*$. It would be reasonable to conjecture that this holds for arbitrary ρ.

	c	0.05	0.10	0.15	0.20	0.25	0.30	0.35	0.40	0.45
$\rho = 0.5$	a^*	.967	.674	.485	.341	.223	.121	.030	−.051	−.126
	G^*	2.475	1.931	1.581	1.315	1.097	.911	.746	.598	.463
$\rho = -0.5$	a^*	.169	−.061	−.221	−.352	−.467	−.572	−.673	−.770	−.866
	G^*	.944	.603	.384	.218	.082	−.034	−.137	−.229	−.313

Table 1. *Optimal strategies and gains for bivariate normal observations.*

The values in Table 1 can be compared with values in the authors' Table 2.3, which correspond to taking $\rho = 0$. For fixed cost, $\rho = 0.5$ provides larger optimal gains G^* than $\rho = 0$. This is not surprising: if the first member of the team is willing to stop with a positive value of X_1, then since observations are positively correlated the second member of the team is likely to be holding a positive value X_2 and jointly their gain will be larger than in the uncorrelated situation. In a different direction, if $\rho = -0.5$ it is not surprising that a^* can be negative, since if the first player stops with a negative value, then the second member's value is likely to be positive, and as a team the two players can end up happy. (Incidentally, if $\rho = -0.5$ and c is large then G^* is negative and in such cases it is optimal to take no observations.)

As mentioned, for the above examples $a^* = b^*$. It is perhaps surprising that this relationship could fail, but the authors show in their Example 2.1 that such can be the case. As they point out, "the need for a compromise after the information has been collected is avoided." However, the fact that sometimes $a^* \neq b^*$ suggests that a sort of compromise is enforced *before* the data are collected. Specifically, what is optimal for the team might not yield the best payoff for the individual members. A similar interpretation can be made when $a^* = b^* < 0$. Thus, while the members might not form a communicating team, it is at least a cooperating team. This leads to two questions: What is the cost of cooperating? What is the cost of failing to communicate?

In one form of a noncooperating team each player simply plays the one person game and ignores the other individual. Table 2 gives optimal cutoff values and gains for that problem assuming $N(0, 1)$ data. In comparing these values to those of the authors' Table 2.3, it is perhaps most relevant to compare a cost c for the team with a cost $c/2$ for a single player, under the rationale that in playing as a team the members split the costs. On the assumption that they also split the proceeds, we see that there is a considerable penalty for playing as a noncommunicating team. For example, if the team cost is 0.10 per observation, then from Table 2.3 the gain is 1.311 and the individual members .656. On the other hand, a player who goes it alone pays 0.05 and nets almost double, 1.256, from an optimal strategy.

c	0.025	0.05	0.075	0.10	0.125	0.15	0.175	0.20	0.225
$a^* = G^*$	1.569	1.256	1.055	.902	.778	.671	.577	.493	.416

Table 2. *Optimal strategies and gains for the one person problem with standard normal observations.*

Is it a bad idea, therefore, to be a team player? Not necessarily. The penalty in playing on a team arises in part when its members do not communicate. Suppose the team members observe X_1 and X_2 independent $N(0, 1)$. If the team members communicate, then as the authors point out, the problem reduces to that of the one person problem with data $Z = X_1 + X_2$. In Table 3 we tabulate the gains for a communicating team. The difference between these results and the authors' Table 2.3 is striking — there is a considerable cost in failing to communicate. Indeed, for sufficiently large costs, the gain for the communicating team is more than double that for the noncommunicating team. Since the team members must in any case communicate their votes, I wonder how realistic it is to assume that they do not communicate their observations.

c	0.05	0.10	0.15	0.20	0.25	0.30	0.35	0.40	0.45
$a^* = G^*$	3.138	2.511	2.109	1.805	1.555	1.342	1.154	.986	.832

Table 3. *Optimal strategies and gains for the two person communicating team with standard normal observations.*

This last example illustrates a curious relationship between the single player game and the two player game with stopping rule P_2. Specifically, if two players each play a one person game with cost 0.05 and and they pool their gains after they stop, then their expected gain is precisely equal to the gain in the communicating version of the two person game with cost 0.10. A moment's study of equation 11.9.3 of DeGroot (1970) shows that this will always hold if X_1 and X_2 are i.i.d. normal, although I believe that this fails in general. Is there a characterization of the normal distribution lurking here?

Finally, the authors have promised an investigation of this problem when the distributions are unknown. In this case the solution to the one person game can be elegant (DeGroot 1970), and challenging (e.g. Christensen 1986). For the two person game it is clear that, conditionally, the two players will follow different strategies, since as they take their observations their posterior distributions will differ. However, even if they start with the same prior, Example 2.1 suggests that it might be a challenge to characterize those situations where both players use identical stopping rules. I shall be interested in hearing of the authors' further research in this direction.

REPLY TO THE DISCUSSION

First for all, we would like to thank Murray Clayton for many insightful and interesting comments on our paper. We agree with most of his comments and believe that some of them will be very helpful in our future work in this area. We will mostly concentrate here on the comments related to the cost of not communicating and the cost of not cooperating, since we did not discuss these areas in our paper.

We also believe in Clayton's conjecture that $a^* = b^*$ holds for the bivariate normal distribution. .However, it is not clear that the stopping sets have to be of the form $\{X \geq a$ and $Y \geq b\}$ for all distributions, especially when X and Y are negatively dependent. We expect to examine these problems in our future work in which the observables are dependent random variables in general.

When one player can communicate the actual value of his/her observation to the other player, MR call it a case of *perfect information*. In practice, it may be possible to convey either the value of actual observation or the vote to stop sampling (i.e., a 0 or 1). However, the cost of transmitting these bits of information may not necessarily be the same. In the partial information case, when the vote needs to be conveyed, the players only have to transmit one bit. One can of course find the optimal solution to both these problems. Now,

Cost of failing to communicate
= Expected Net Gain from the ability to communicate full information
= Expected Net Gain (Sampling from $Z = X_1 + X_2$ and cost C^*)
–Expected Net Gain (2 player team with cost C).

If this cost is positive, the team will be prepared to pay the intermediary a higher cost of transmitting perfect information in order to gain more. Thus, one needs to consider trade offs between costs versus higher expected reward.

In comparing the team problem with the one person problem with half the costs, Clayton considers a non-cooperating team in which each player simply plays the one person game and ignores the other individual. It is correct to say that our team is cooperating in the sense that the team members can switch roles (or wear each other's hats). However, for the algorithm P_2, a more appropriate comparison corresponds to a one person problem with forced continuation, i.e., if Player 1 wanted to stop, but Player 2 doesn't vote to stop then he is 'forced to continue'. For the algorithm P_1, appropriate analogy is 'forced to stop'. For example, for a team of N players sampling from Uniform $(0,1)$ distribution with costs $C(0 < C < .5)$, the optimal cutoff point for the algorithm P_n is $a^* = 1 - (2C)^{1/(n+1)}$, and the expected net gain to the team is $G(a^*) = [(n-1) + (n+1)a^*]/2$. Thus the average gain per person $G(a^*)/n$, decreases as n increases. It stands to reason that if n is large, then under P_n, the team stops to vote after the first observation. The observation by Clayton on the curious relationship between the single player problem and the two player game with stopping algorithm P_2 for the normal distribution is certainly interesting. But we do not believe that this relationship will be true only for the iid normal variables.

Finally, we agree that this problem, when the distribution is unknown, is mighty hard. But, the problem is of sufficient interest to us that we will continue to explore it further. Once again, we thank Clayton for his insight into this problem and raising new questions for us to pursue.

ADDITIONAL REFERENCE IN THE DISCUSSION

Christensen, R. (1986). Finite stopping in sequential sampling without recall from a Dirichlet process. *Ann. Statist.* **14**, 275–282.

BAYESIAN STATISTICS 4, pp. 227–246
J. M. Bernardo, J. O. Berger, A. P. Dawid and A. F. M. Smith, (Eds.)

Parameterization Issues in Bayesian Inference

SUSAN E. HILLS and ADRIAN F. M. SMITH
University of Nottingham, UK and *Imperial College, UK*

SUMMARY

The adequacy and efficiency of numerical and analytic techniques for implementing parametric Bayesian methods can be highly dependent on the parameterization adopted in specifying the likelihood and prior forms. Issues relating to this problem will be reviewed and illustrated.

Keywords: BAYESIAN INFERENCE; PARAMETERIZATION; NUMERICAL INTEGRATION; LAPLACE APPROXIMATION; IMPORTANCE SAMPLING; GIBBS SAMPLING.

1. INTRODUCTION

Over the past decade, a great deal of effort has gone into developing accurate and efficient analytic and numerical procedures for implementing Bayesian methods. As is well known, such implementation presents difficulties because of the implicit need for (often high-dimensional) integrations, the latter being required for normalization, marginalization and interval or moment summaries. Implementation procedures are therefore typically, implicitly or explicitly, methods for approximating integrals.

But, in the case of parametric inference problems (our focus in this paper), the nature of such integrals will depend on the form of parameterization adopted in the likelihood and prior specification. This, in turn (perhaps in different senses and directions), will affect the accuracy, efficiency, and perhaps even the applicability of different implementation procedures, including normal likelihood approximations; the Laplace and related analytic approximations; adaptive quadrature based strategies; importance sampling and resampling Monte Carlo strategies; Markov chain based iterative sampling strategies, in particular, the Gibbs sampler.

In broad terms, many of these strategies are designed to be most efficient when a parameterization is used which, in some sense, produces posterior distributions "close to normality". When marginalization is a key issue, the required integration would clearly be side-stepped if a parameterization were adopted which led to parameters of interest being a posteriori independent of nuisance parameters (see, for example, Albert, 1989). Notions of approximate "normality" and "orthogonality" are therefore of basic concern in assessing proposed parameterizations and, more constructively, proposing better reparameterizations. But other concerns may also be important; particularly in the case of Monte Carlo methods, where choice of importance sampling functions and ease of random variate generation will also play a role.

Our objective in this paper is to review these parameterization issues in the context of the various implementation procedures we have cited. In Section 2, we focus briefly on each procedure and outline the issues we perceive to be of greatest concern in each case. In Section 3, we discuss some particular ideas for diagnostics and reparameterization.

2. PARAMETERIZATION ISSUES

2.1. *General Issues*

The parameterization of a statistical model is more or less arbitrary and the choice has potential relevance to both theoretical and practical issues.

From the perspective of Bayesian practice, ease of prior specification, efficiency of computation and focussed reporting of inferences of interest may each suggest a different desired parameterization. From the perspective of theory, forms of quadrature rule and error terms in analytic expansions will have different performances and magnitudes, respectively, under different parameterizations. Parameterization issues also arise in the context of sensitivity studies; as, for example, with the possibly changing forms of reference priors required for different marginal inferences of interest. We shall examine these various parameterization issues for a selection of Bayesian computation approaches.

2.2. *Normal Likelihood Approximation*

Perhaps the simplest of all Bayesian implementation procedures is to specify a constant (uniform) prior and assume (wish? pray?) that the likelihood surface is "close to normal". Notwithstanding the ill-defined nature of a "large sample", the latter is often appealed to as justification of the procedure, with little concern for the parameterization adopted.

Consider, for example, the likelihood contours in Figure 1.1, corresponding to samples from two independent exponential distributions with means λ and ϕ, with the likelihood parameterized in terms of ϕ and $\psi = \phi\lambda^{-1}$, the latter ratio of means being the parameter of interest. The same likelihood surface parameterized in terms of $\log(\phi)$ and $\log(\psi)$ is shown in Figure 1.2, and is seen to provide a picture much "closer to normality".

Similar sensitivity of likelihood contour shape to parameterization is common in non-linear regression models, where it is assumed that $y_i = f(\theta, x_i) + \epsilon_i, i = 1, \ldots, n$, for $f(\cdot, x)$ a non-linear function of $\theta = (\theta_1, \ldots, \theta_p)$ and ϵ_i independent $N(0, \sigma^2)$. If we first consider, for simplicity, the case $p = 1, \sigma^2 = 1$, and define

$$S(\theta) = \sum_{i=1}^{n} (y_i - f(\theta, x_i))^2,$$

with minimum at $\theta = \hat{\theta}$, it is easily verified that the log-likelihood is quadratic in $T(\theta)$, where

$$T(\theta) = \mathrm{sign}(\theta - \hat{\theta})(S(\theta) - S(\hat{\theta}))^{\frac{1}{2}}.$$

Thus, $T(\theta)$ could be regarded (if monotone) as a potential transformation to "likelihood normality", or, alternatively, used as a diagnostic for the "approximate likelihood normality" of the original parameter θ, by examining departures from a straight-line of the plot of $T(\theta)$ versus θ.

This latter use has been proposed (the so-called t-plot procedure) by Bates and Watts (1988), who, in the case of general p and σ^2 unknown, work, for $j = 1, \ldots, n$, with

$$T(\theta_j) = \mathrm{sign}(\theta_j - \hat{\theta}_j) \left[\frac{S(\theta_j) - S(\hat{\theta})}{S(\hat{\theta})/(n-p)} \right]^{\frac{1}{2}},$$

with $S(\theta_j), S(\hat{\theta})$ defined, respectively, as the minimum sum of squares over all $\theta_i, i \neq j$ for given θ_j, and over all θ_i.

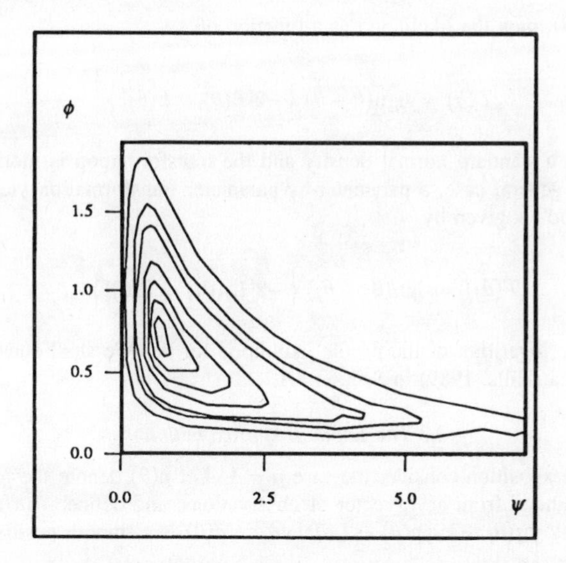

Figure 1.1. *Likelihood function for the parameters ψ and ϕ.*

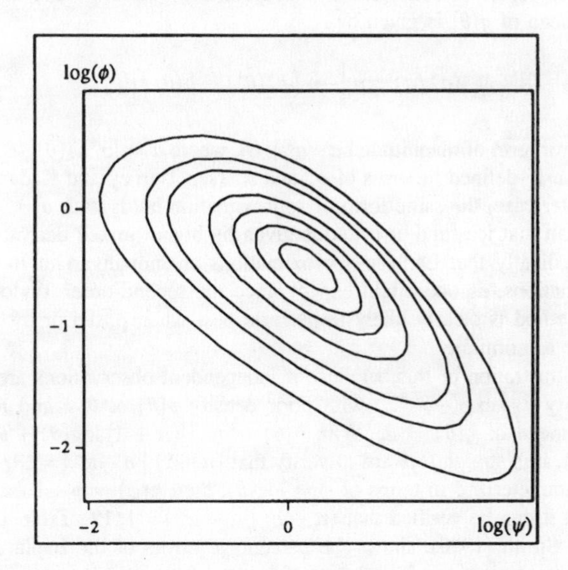

Figure 1.2. *Likelihood function for the parameters* $\log\psi$ *and* $\log\phi$.

This idea can be extended to general likelihoods (as pointed out by Tibshirani and Wasserman, 1989, but without reference to the earlier manifestation of the non-linear regression application in Bates and Watts). In the one-parameter case, if the log-likelihood is denoted by $L(\theta) = \log \ell(\theta)$, then the likelihood as a function of

$$T(\theta) = \text{sign}(\theta - \hat{\theta}) \left\{ -2[L(\theta) - L(\hat{\theta})] \right\}^{\frac{1}{2}}$$

is proportional to a standard normal density and the transformation is monotone if $\ell(\theta)$ is unimodal. In the general case, a parameter-by-parameter transformation yielding a "normal marginal likelihood" is given by

$$T(\theta_j) = \text{sign}(\theta_j - \hat{\theta}_j) \left\{ -2[L_j(\theta_j) - L(\hat{\theta})] \right\}^{\frac{1}{2}},$$

where $L_j(\cdot)$ is the logarithm of the profile likelihood for θ_j. We shall consider a Bayesian version of this idea (Hills, 1989) in Section 3.1.

2.3 *The Laplace Approximation*

For simplicity of exposition consider the case $p = 1$. Let $p(\theta)$ denote the prior density and $L(\theta)$ the log-likelihood from an n-vector of observations, and define $-nh(\theta) = \log p(\theta) + L(\theta), -nh^\star(\theta) = \log g(\theta) + \log p(\theta) + L(\theta)$, where $g(\theta)$ is a smooth positive function. Let $\hat{\theta}$ maximize $-h(\theta), \hat{\theta}^\star$ maximize $-h^\star(\theta)$ and define $\sigma = [h_2(\hat{\theta})]^{-\frac{1}{2}}, \sigma^\star = [h_2^\star(\hat{\theta}^\star)]^{-\frac{1}{2}}$, with subscripts denoting orders of derivative with respect to θ. Then the Laplace approximation to the posterior mean of $g(\theta)$ is given by

$$(\sigma^\star/\sigma) \exp\left[-n\left(h^\star(\hat{\theta}^\star) - h(\hat{\theta})\right)\right],$$

with first order error term of magnitude $| a - a^\star | /n$, where $a = [\sigma^4 h_4(\hat{\theta})]/8 + [5\sigma^6 h_3^2(\hat{\theta})]/24$ and $a^\star$ is analogously defined in terms of $\sigma^\star$ and $h^\star$ (see Tierney and Kadane, 1986). In the general p-parameter case, the same form of approximation holds with $\sigma, \sigma^\star$ suitably defined in terms of Hessian matrices and error terms given by higher mixed derivatives.

We see immediately that Laplace approximations are not invariant to (non-linear) parameter transformations, as one might expect since the second-order Taylor approximation underlying the method is clearly likely to perform best when posterior distributions are, in some sense, close to normality.

As a simple illustration of this, suppose n independent observations are taken from the exponential density $\theta^{-1}\exp(-\theta^{-1}x)$ with prior density $p(\theta) \propto \theta^{-1}$ and interest focussed on the posterior mean of $g(\theta) = \theta$. With $h(\theta) = n^{-1}[(n+1)\log(\theta) + \theta^{-1}n\bar{x}], h^\star(\theta) = h(\theta) - n^{-1}\log(\theta)$, it is straightforward to verify that $(n+1) \mid a - a^\star \mid = 13/12$. If, however, the model is reparameterized in terms of $\phi = \log(\theta)$, then $h(\phi) = \phi + \bar{x}\exp(-\phi), h^\star(\phi) = h(\phi) - n^{-1}\phi$ and it can be verified that $(n-1) \mid a - a^\star \mid = 1/12$. Table 1 below, adapted from Achcar and Smith (1990), shows the percentage errors of the Laplace approximation under the two parameterizations for a range of values of sample size n.

n	1	2	3	4	5	6	7	8	9
θ	26.0	11.7	6.6	4.3	3.0	2.2	1.7	1.3	1.1
$\log(\theta)$	3.9	1.4	0.7	0.4	0.3	0.2	0.2	0.1	0.1

Table 1. *Posterior mean for an exponential parameter. Percentage error of the Laplace approximation.*

Table 2 shows the percentage errors, for a range of values of sample size n, of the Laplace approximation to the posterior mean of the ratio, ψ, of two exponential parameters (as discussed in Section 2.2) under the parameterizations (ϕ, λ) and $(\log\phi, \log\lambda)$, with prior density specified by $p(\phi, \lambda) \propto \phi^{-1}\lambda^{-1}$.

n	2	3	4	5	6	7	8	9	10
(ϕ, λ)	16.2	5.4	2.5	1.3	0.8	0.5	0.4	0.3	0.2
$(\log\phi, \log\lambda)$	2.6	0.7	0.3	0.1	0.1		$\cdots$	$< 0 \cdot 1$	$\cdots$

Table 2. *Posterior mean for a ratio of exponential parameters. Percentage error of the Laplace approximation.*

Further illustrations of the behaviour of the Laplace approximation under reparameterization are provided by Achcar and Smith (1990), who show that reparameterization which leads to a locally uniform Jeffreys prior typically improves the accuracy of approximation. However, at the time of writing a clear theoretical underpinning for such reparameterization is still lacking. Given the tedium of detailed study of the error terms under reparameterization, this is possibly an area where symbolic computation could prove invaluable in stimulating theoretical insight.

2.4. *Adaptive Quadrature*

An adaptive quadrature approach to the integrations required in Bayesian calculation was presented in Naylor and Smith (1982), based on Gauss-Hermite product rules. In this approach, the Gauss-Hermite forms are clearly more efficient if marginal densities are "closer to normality", and the product form is clearly more efficient the closer we are to posterior "orthogonality" of parameters.

The reparameterization approach adopted is in two parts: first, individual parameters are transformed to the real line (for example, by using logarithms of positive parameters, logits of interval-constrained parameters); secondly, iterative orthogonalizing and centering and scaling transformation of the resulting real parameters is carried out, based on current estimates of the posterior mean and covariance matrix. Thus, if ϕ is the current working vector of real-valued parameters, for the next quadrature evaluations the working parameters vector ψ is defined by the constructive transformation

$$\psi_1 = \phi, \quad \psi_i = \phi_i + \sum_{j=1}^{i-1} \beta_{ij}\psi_j, \quad i = 2, \ldots, p,$$

with $\beta_{ij} = -\mathrm{cov}(\phi_i, \psi_j)/\mathrm{var}(\psi_j)$, followed by $\xi_i = (\psi_i - \mu_i)/\sqrt{2}\sigma_i$, where μ_i, σ_i are the mean and standard deviation of ψ_i, all moments being calculated with respect to the current estimate of the posterior density. Illustrations of the effectiveness of this strategy are provided in Smith *et al.* (1985, 1987); see, also, Marriott and Smith (1991) for applications in stationary time series analysis.

Although motivated by considerations of efficiency of product-rule quadrature, this orthogonalizing procedure is also highly relevant to other computational approaches, in particular, as we discuss later in Section 3.2, to the Gibbs sampler.

2.5. *Importance Sampling*

A variety of Monte Carlo importance sampling strategies for Bayesian calculations have been proposed. Clearly, in so far as these attempt to use "universal" forms of importance sampling density (such as normal or split-t forms), reparameterization to real-valued and, perhaps, orthogonal coordinates will also be of interest. In general, however, the primary concern is with the effective "shape" of the importance sampling density and the efficiency of random variate generation (see Section 3.3), and this may imply a totally different attitude to a given parameterization than that required by the Laplace or quadrature approaches.

As an example of this, consider the Gumbel bivariate exponential reliability model studied by Draper *et al.* (1988), which leads to a 3-parameter posterior density proportional to

$$\lambda_1^{-1}\lambda_2^{-1}\prod_{i=1}^{N}\lambda_1\lambda_2 e^{-(\lambda_1 x_{1i}+\lambda_2 x_{2i})}\left[1+\alpha(2e^{-\lambda_1 x_{1i}}-1)(2e^{-\lambda_2 x_{2i}}-1)\right],$$

for $\lambda_1, \lambda_2 > 0$, $-1 < \alpha < +1$. Quadrature strategies, or the diagnostic techniques to be discussed in Section 3.1, might suggest working with $\log(\lambda_1)$, $\log(\lambda_2)$ and $\log[(1+\alpha)/(1-\alpha)]$. However, inspection of the posterior reveals that from an importance sampling perspective it is much more straightforward to work with λ_1, λ_2 and α, perhaps designing an importance sampling density as a product of two gammas and a beta distribution.

2.6. *The Gibbs Sampler*

In the Gibbs sampling approach to Bayesian computations (see, for example, Gelfand and Smith, 1990), samples are generated by cycling, iteratively, through the conditional posterior densities $p(\theta_i|\theta_j, j \neq i), i = 1, \ldots, p$, where, for simplicity, notational dependence on the data is suppressed. Starting from some θ^0, after t cycles we have a realization θ^t, which, for large t, approximates a drawing from the joint posterior density $p(\theta)$.

It is straightforward to see that individual one-to-one reparameterization of the components θ_i does not affect the generated process. The inverse transformation of the output stream from such a reparameterization will be identical to that which would have obtained if the original parameterization were used (given the same starting values and random number stream). The only purpose of such component-wise reparameterization might therefore be to facilitate the required random variate generation. However, reparameterization to break high correlations among component parameters is of great importance here in determining the rate of convergence of the Gibbs sampler. We shall return to this in Section 3.2.

We also note in passing one of the great advantages of sampling-based approaches to Bayesian calculation. Namely, the ease with which, by suitable functions defined on the output stream of the Gibbs sampler, one can produce (marginal) summary inferences for any reparameterization $g(\theta)$ of interest.

3. DIAGNOSTICS AND REPARAMETERIZATION

3.1. *Diagnostics for Normality*

In Section 2.2, we discussed ideas relevant to diagnostics or reparameterization aimed at achieving approximate normal shape of the likelihood. An analogous diagnostic plot for assessing approximate marginal posterior normality can be developed as follows.

Suppressing notational dependence on the data, let $p(\theta)$ denote the joint posterior density, with a mode at $\hat{\theta} = (\hat{\theta}_1, \ldots, \hat{\theta}_p)$, and let $p_j(\theta_j)$ denote the profile posterior for θ_j, defined by maximizing $p(\theta)$ over $\theta_i, i \neq j$, for each θ_j. Our proposed diagnostic is then to plot $T(\theta_j)$ against θ_j, where

$$T(\theta_j) = \text{sign}(\theta_j - \hat{\theta}_j)\{-2[\log p_j(\theta_j) - \log p(\hat{\theta})]\}^{\frac{1}{2}}.$$

To motivate this procedure, consider the case of $p = 1$. If $p(\theta)$ is actually normal, we have

$$p(\theta) = p(\hat{\theta}) \exp\left\{-\frac{1}{2\sigma^2}(\theta - \hat{\theta})^2\right\}$$

for some σ^2, from which it follows immediately that $T(\theta)$ is linear in θ. In general, a Taylor expansion gives

$$\log p(\theta) = \log p(\hat{\theta}) + \frac{1}{2}H(\theta - \hat{\theta})^2 + O[(\theta - \hat{\theta})^3],$$

where H is the second derivative of $\log p(\theta)$ evaluated at $\hat{\theta}$. It follows that

$$T(\theta) = (\theta - \hat{\theta})\left[-H - 2\,O[(\theta - \hat{\theta})^3]/(\theta - \hat{\theta})^2\right]^{\frac{1}{2}}.$$

The first term in the square root expression is observed Fisher information (and does not depend on θ). If this term is large, it will dominate and $T(\theta)$ will be approximately linear in θ (the "large sample" case). Departure from linearity will then indicate the importance of the cubic term, which is related to Sprott's (1973) curvature measure for assessing normal approximation.

This $T(\theta)$ versus θ diagnostic plot (which we shall refer to as the Bayes t-plot) is discussed in detail in Hills and Smith (1991), where the behaviour of the diagnostic plot is studied for various stylized cases.

Suppose, for example, that in fact $\log(\theta - c)$ were distributed as $N(\mu, \sigma^2)$. In this case, it can be shown that

$$T(\theta) = \sigma^{-1}[\log(\theta - c) - (\mu - \sigma^2)],$$

which has an asymptote at $\theta = c$.

Based on this and other theoretical forms of $T(\theta)$ versus θ, "look-up" rules can be established (see Hills and Smith, 1992, for detailed derivations and illustration). In practice, accuracy of choices of constants (such as a, b, or c in the above) is not critical. In any case, a second diagnostic check can be carried out for the suggested reparameterization.

As an illustration, Figure 2.1 shows plots of $T(\theta)$ versus θ based on an actual posterior density for the joint parameterization $\phi, \psi = \phi\lambda^{-1}$ of a two sample exponential problem (cf. Section 2.2). These plots approximate the theoretical forms that would obtain were $\log\phi$ and $\log\psi$ to be normally distributed (in each case, the asymptotes look close to zero). With the joint posterior reexpressed in terms of $\log\phi, \log\psi$, the $T(\theta)$ versus θ plots can be repeated. The resulting forms, shown in Figure 2.2, are close to linearity, suggesting that the logarithmic transformations are satisfactory from a "closeness to normality" perspective.

In fact, this analysis was based on a selected data and prior specification for which the resulting joint posterior for $\log\phi, \log\psi$ actually has the contours shown in Figure 1.2. This is clearly reasonable from a "normality" perspective, but the marked negative correlation would obviously make a product quadrature rule somewhat inefficient. Figure 3.1 shows the posterior contours resulting from the reparameterization to $\log\eta_2, \log\psi$, where $\eta_2 = \phi\psi^{\frac{1}{2}}$, which is, of course, just a linear transformation of $\log\phi, \log\psi$.

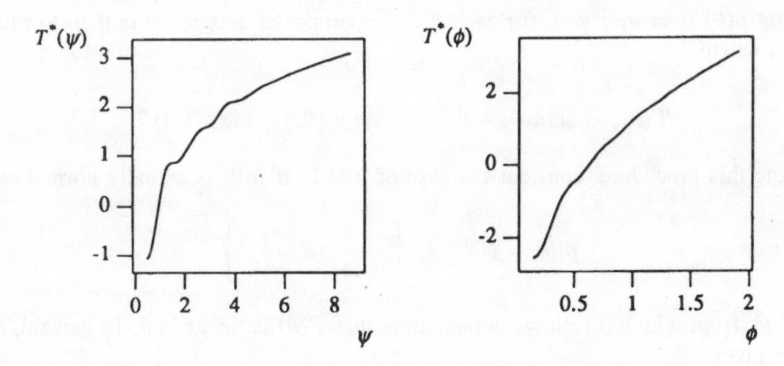

Figure 2.1. *Bayes t-plots for the ratio of exponential means problem, naive parameterization.*

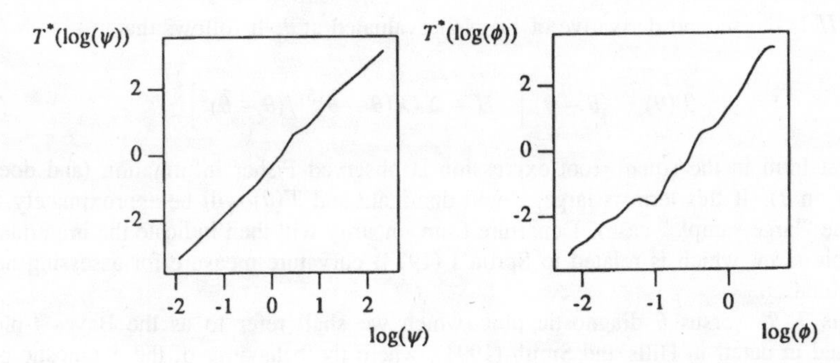

Figure 2.2. *Bayes t-plots for the ratio of exponential means problem,* log *parameterization.*

In addition to the obvious fact that this reparameterization produces nicely "circular" looking contours, it is interesting to note a link with the "orthogonal parameter" analysis of Cox and Reid (1987). These authors showed that, if ψ is taken as the parameter of interest, a particular orthogonal parameterization (the latter defined in the two parameter case as one which diagonalizes the expected Fisher information matrix) is given by $\eta_1 = \psi$ and $\eta_2 = \phi\psi^{1/2}$. That this version of "orthogonality" is not immediately suited to Bayesian calculations is obvious from Figure 3.2. The alternative orthogonal parameterization $\log \eta_2, \log \psi$ (essentially corresponding to the suggestion of the $T(\theta)$ versus θ plot, followed by a linear transformation to remove correlation) is clearly preferable.

3.2. *Correlation and the Gibbs Sampler*

If the Gibbs sampler is run in order to generate an "as if" independent sample from the joint posterior distribution (rather than simply in order to form estimates by ergodic averaging), this can be attempted either by replicate independent "short" runs of the process or by extracting multiple sample values from a single "long" run of the process. In the first case, output series need to be monitored by a stopping rule (formal or informal) which decides when "convergence" has been achieved: the required sample from the posterior is then formed by the final generated values from the stopped replicated series. In the second case, a monitoring process has to decide when to start extracting sample values from the output stream ("convergence") and what size "gaps" to leave between extracted values in order

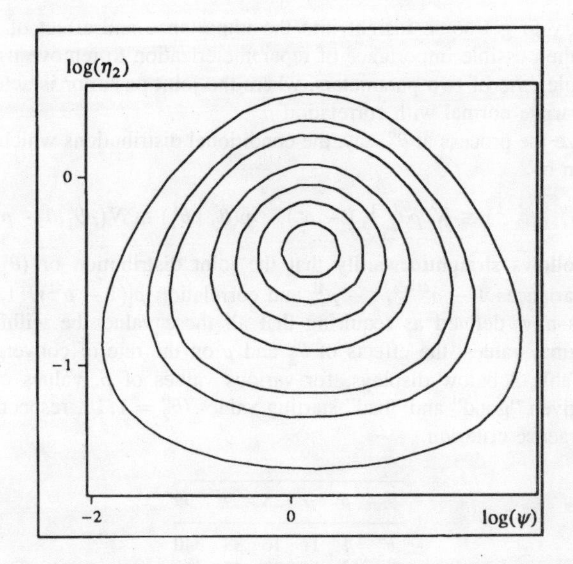

Figure 3.1. *Posterior density for the orthogonal parameterization* $\log(\psi)$ *and* $\log(\eta_2)$.

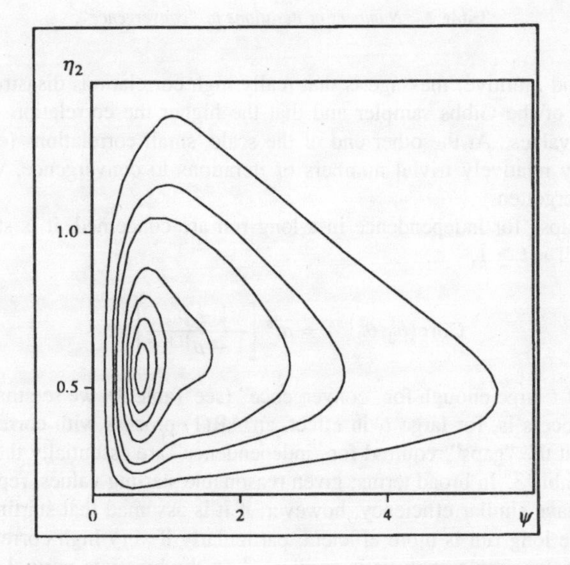

Figure 3.2. *Posterior density in the orthogonal parameterization* ψ *and* η_2.

to obtain an "independent" sample. The number of iterations to achieve "convergence" in either case is clearly a function of starting values and the correlation structure of the stochastic process generated by the Gibbs sampler. The size of "gaps" required for independence in the single "long" run also clearly depends on the correlation structure.

In order to try to get some insight into the importance and effect of such correlation (and hence into the possible importance of reparameterization to remove it), let us consider in detail the simple case of two parameters, where the joint posterior is actually zero-mean, unit-variance bivariate normal with correlation ρ.

If we initialize the process at θ_2^0, say, the conditional distributions which drive the Gibbs sampler are given by

$$p(\theta_1^t, \mid \theta_2^{t-1}) \equiv N(\rho\theta_2^{t-1}, 1-\rho^2), \quad p(\theta_2^t \mid \theta_1^t) \equiv N(\rho\theta_1^t, 1-\rho^2),$$

from which it follows straightforwardly that the joint distribution of (θ_1^t, θ_2^t) has means $\rho^{2t-1}\theta_2^0, \rho^{2t}\theta_2^0$, variances $1-\rho^{4t-2}, 1-\rho^{4t}$ and correlation $\rho[(1-\rho^{4t})/(1-\rho^{4t-2})]^{1/2}$. If "convergence" is now defined as requiring that all these values be within 10^{-k}, say, of their actual marginal values, the effects of θ_2^0 and ρ on the rate of convergence can easily be examined. Table 3 below displays, for various values of ρ, values of t required for "convergence", given "good" and "bad" starting values ($\theta_2^0 = 1, 10$, respectively) and using 10^{-3} as a convergence criterion.

ρ	.7	.8	.9	.99
$\theta_2^0 = 1$	11	16	33	340
$\theta_2^0 = 10$	14	22	45	460

Table 3. *Numbers of iterations to "convergence".*

The clear (and intuitive) message is that really high correlations disastrously slow down the convergence of the Gibbs sampler and that the higher the correlation the more serious are bad starting values. At the other end of the scale, small correlations (even of the order of $\rho = .8$) imply relatively trivial numbers of iterations to convergence, with bad starting values quickly forgotten.

So far as "gaps" for independence in a long run are concerned, it is straightforward to verify that, for all $s, t \geq 1$,

$$\text{Corr}\,(\theta_2^t, \theta_2^{t+s}) = \rho^{2s}\left[\frac{1-\rho^{4t}}{1-\rho^{4t+4s}}\right]^{\frac{1}{2}}.$$

For values of t large enough for "convergence" (see Table 3), we see that Corr $(\theta_2^t, \theta_2^{t+s})$ $\approx \rho^{2s}$; the θ_2 process is, for large t, in effect, an AR(1) process with correlation parameter ρ^2. It follows that the "gaps" required for "independence" are essentially the values given in the first row of Table 3. In broad terms; given reasonable starting values, replicate runs and a single long run have similar efficiency; however, if it is assumed that starting values will be chosen badly, one long run is more efficient, particularly if very high correlation is present. Of course, the above simple analysis is predicated on the bivariate normal assumption (and the specific goal of generating an "independent" sample). Perhaps speculatively, it would be interesting to investigate whether and how (perhaps by showing that, for a general Gibbs sampler, the highest pairwise correlation dominates and then linking the general process to the normal process by a coupling argument?) these quantitative insights might be used to guide pragmatic stopping and sampling rules.

In any case, perhaps the main message here is the importance of breaking high correlation, which can be achieved by means of an application after a few initial iterations (say, 10–15) of the orthogonalizing transformation defined in Section 2.4. Even if this only succeeds in reducing correlation, the kinds of simple calculations given above suggest that this will improve efficiency dramatically.

3.3. *Random Variate Generation*

In importance sampling or Gibbs sampling applications, required random variates (for example, parameter realizations from a conditional posterior in the Gibbs sampler) can obviously be obtained using any one-to-one transformation of an initial formulation and then subsequently be transformed back if required. Clearly, this does not affect the fundamental stochastic process involved, but it may well have pragmatic import in suggesting efficient random variate generation strategies. We shall briefly indicate two contexts in which this "reparameterization for simulation purposes" can be important. The first context involves the well-known ratio-of-uniforms technique for generating a realization from a (univariate) density proportional to $h(\theta)(> 0$, with $\int h(\theta)d\theta < \infty)$. The standard method involves generating a pair (u, v) uniformly from the region $C_h = \{(u, v) : 0 < u \leq [h(v/u)]^{1/2}\}$, and then setting $\theta = v/u$, which can be shown to be a realization from the required density. Typically, uniform generation from C_h is achieved by finding a bounding rectangle, generating uniform points in the latter and then testing for membership of C_h. Efficiency of the algorithm is then related to the probability that points generated in the rectangle are, in fact, in C_h.

Following specific suggestions of Kinderman and Monahan (1980), Wakefield *et al.* (1991) have shown that, for unimodal $h(\theta)$, the efficiency of the ratio-of-uniform procedure can be substantially improved by working with $\phi = \theta - \mu$, where μ is the mode of $h(\theta)$. This is proved by Wakefield *et al.* to be an optimal location transformation for symmetric $h(\theta)$ and numerical studies confirm its effectiveness also for a range of non-symmetric densities.

The second context involves a recently proposed adaptive rejection technique for generating random variates from a log-concave density: see Gilks and Wild (1992), with systematic Bayesian application in Dellaportas and Smith (1992). The basic idea is that an envelope function (and an internal "squeezing" function) can be easily constructed from just a few values on a log-concave function by joining tangents (and chords), and then exponentiating. The link with parameterization arises because log-concavity of a density may exist in one parameterization but not in another. For example, the normal is log-concave in both σ^{-1} and $\log \sigma$ (where σ is the scale parameter), but not in σ. Gilks and Wild (1992) provide a useful table summarizing the position for a range of commonly occuring density forms. Awareness of these reparameterization possibilities may assume increasing importance as the use of simulation techniques for Bayesian calculations increases.

ACKNOWLEDGEMENTS

This paper was prepared while the second author was University Distinguished Visiting Professor at the Department of Statistics, Ohio State University, Columbus.

REFERENCES

Achcar, J. A. and Smith, A. F. M. (1990). Aspects of reparametrization in approximate Bayesian inference. *Essays in honour of G. A. Barnard* (J. Hodges, ed.). Amsterdam: North-Holland.

Albert, J. H. (1989). Nuisance parameters and the use of exploratory graphical methods in a Bayesian analysis. *Ann. Statist.* **43**, 191–196.

Bates, D. M. and Watts, D. G. (1988). *Nonlinear Regression Analysis and its Applications.* New York: Wiley.

Cox, D. R. and Reid, N. (1987). Parameter orthogonality and approximate conditional inference. *J. Roy. Statist. Soc. B* **49**, 1–18.

Dellaportas, P. and Smith, A. F. M. (1992). Bayesian inference for generalized linear and proportional hazards models via Gibbs sampling. *Appl. Statist.* , (to appear).

Draper, N., Evans, M. and Guttman, I. (1988). A Bayesian approach to system reliability when two components are dependent. *Comp. Stat. and Data Anal.* **7**, 39–49.

Gelfand, A. E. and Smith, A. F. M. (1990). Sampling based approaches to calculating marginal densities. *J. Amer. Statist. Assoc.* **85**, 398–409.

Gilks, W. R. and Wild, P. (1992). Adaptive rejection sampling for Gibbs sampling. *Appl. Statist.* , (to appear).

Hills, S. E. (1989). *The Parameterisation of Statistical Models.* Ph.D. Thesis, University of Nottingham.

Hills, S. E. and Smith, A. F. M. (1992). Diagnostic plots for improved parameterization in Bayesian inference. *Biometrika* , (to appear).

Kass, R. E. and Slate, E. H. (1992). Reparameterization and diagnostics of posterior non-normality. *Bayesian Statistics 4* (J. M. Bernardo, J. O. Berger, A. P. Dawid and A. F. M. Smith, eds.), Oxford: University Press, 289–305, (with discussion).

Kinderman, A. J. and Monahan, J. F. (1980). New methods for generating student's t and gamma random variables. *Computing* **25**, 369–377.

Marriott, J. M. and Smith, A. F. M. (1991). Reparameterisation aspects of numerical Bayesian methodology for ARMA models. *Journal of Time Series Analysis*, (to appear).

Naylor, J. C. and Smith, A. F. M. (1982). Applications of a method for the efficient computation of posterior distributions. *Ann. Statist.* **31**, 214–225.

Smith, A. F. M., Skene, A. M., Shaw, J. E. H., Naylor, J. C. and Dransfield, M. (1985). The implementation of the Bayesian paradigm. *Comm. Statist. Theory and Methods* **14**, 1079–1102.

Smith, A.F.M., Skene, A. M., Shaw, J. E. H. and Naylor, J. C. (1987). Progress with numerical and graphical methods for practical Bayesian statistics. *The Statistician* **36**, 75–82.

Sprott, D. A. (1973). Normal likelihoods and their relation to large sample theory of estimation. *Biometrika* **60**, 457–465.

Tibshirani, R. and Wasserman, L. (1989). Some aspects of the reparameterization of statistical models. *Tech. Rep.* **449**, Carnegie-Mellon University.

Tierney, L. and Kadane, J. B. (1986). Accurate approximations for posterior moments and marginal densities. *J. Amer. Statist. Assoc.* **81**, 82–86.

Wakefield, J. C., Gelfand, A. E. and Smith, A. F. M. (1991). Efficient generation of random variates via the ratio-of-uniforms method. *Statistics and Computing* **1**, 129–134.

DISCUSSION

J. H. ALBERT (*Bowling Green State University, USA*)

This paper provides a good review of the parameterization issue in Bayesian inference. It seems that there are two compelling reasons for reparameterization. First, a number of good algorithms have been developed for summarizing a multivariate posterior distribution (for, example, the adaptive quadrature scheme of Naylor and Smith, 1982 and the Laplace algorithms in Tierney and Kadane, 1986) and all of these algorithms make some explicit assumptions about the shape of the posterior (such as multivariate normality). The efficiency and accuracy of these algorithms are typically very sensitive to the choice of parameterization. A second reason is in the situation where the parameter $\theta = (\theta_1, \theta_2)$, where θ_1 are the parameters of interest and θ_2 are nuisance parameters. Computation of the marginal posterior distribution of θ_1 is difficult if θ_1 and θ_2 are highly correlated and therefore it is desirable to find a reparameterization $(\theta_1, g(\theta_1, \theta_2))$ such that θ_1 and $g()$ are approximately independent.

For ease of communication, it seems that we should focus on simple summaries of the posterior. If the entire parameter set θ is of interest, then one simple summary is that θ is approximately normal with mean and precision given by the posterior mode and the negative hessian matrix evaluated at the mode. If $\theta = (\theta_1, \theta_2)$ has the posterior $\pi(\theta_1, \theta_2)$, where θ_1 is the parameter of interest, then the simplest approximation of the marginal posterior density of θ_1 is the "profile" posterior $\pi(\theta_1, \hat{\theta}_2(\theta_1))$, where $\hat{\theta}_2(\theta_1)$ is the posterior mode of θ_2

conditional on a fixed value of θ_1. These simple summaries may give poor approximations for "natural" parameterizations of the problem and one research goal should be to suggest parametrizations which improve the accuracy of these simple summaries. Hills and Smith describe some of the problems and one research goal should be to suggest reparameterizations diagnostics. Suppose that we focus on the posterior summary of multivariate normality. These diagnostics should ideally give the user some indication of the adequacy of the normal approximation and, if the approximation is inaccurate, suggest reparameterizations which improve the accuracy of the summary.

The paper reviews the use of Bayes t-plots which can be viewed as a Bayesian generalization of plots suggested by Bates and Watts for nonlinear regression. In Hills and Smith (1992), a number of these plots are displayed for a number of non-normal posterior distributions. I question the usefulness of these plots in practice. It seems difficult to assess the degree of nonlinearity in these plots and use the look-up tables to decide on a suitable reparameterization. In addition, these plots only suggest transformations of individual parameters and this class of transformations will be ineffective in certain problems. In practice, one typically applies naive transformations like changing positive parameters to logs and proportion parameters to logits. It is not clear that these plots will suggest non-naive transformations.

An alternative approach to reparameterization is to consider families of transformations and find the member of the class which gives the most accurate normality summary. This approach can be illustrated in the one parameter case where the posterior density is unimodal but significantly skewed. Let $l(\theta)$ denote the log-posterior of a positive parameter θ and let l'' and l''' denote the second and third derivatives of l. Then Albert (1989) shows that one can approximately remove the skewness of the posterior distributions of θ by a power transformation $\frac{\theta^{\nu}-1}{\nu}$, where the power $\nu = 1 + \frac{l'''(\hat{\theta})\hat{\theta}}{l''(\hat{\theta})3}$ and $\hat{\theta}$ is the posterior mode. (Applications of this method are given in Albert, 1990). One may question the usefulness of "normalizing" a one-dimensional posterior. This is useful in that it gives the simple summary $\frac{\theta^{\nu}-1}{\nu}$ distributed $N(\mu, \sigma)$, which can be used to compute probabilities and quantiles. In addition, in the application of the Gibbs sampler, it gives a simple algorithm for simulating from a unimodal distribution.

In the many-parameter case, it can be difficult to summarize the distribution accurately due to strong correlations between the parameters. In the two parameter case where $\theta = (\theta_1, \theta_2)$, it may be difficult to obtain the marginal posterior density of θ_1 due to strong non-linear correlations (see Racine-Poon, 1992, for an illustration of a highly correlated "banana" posterior). Following Tibshirani and Wasserman (1989), one can find an approximate independence transformation by assuming that θ_2, given θ_1, is approximately normally distributed with mean $\mu(\theta_1)$ and standard deviation $\sigma(\theta_1)$ and then transforming to $(\theta_1, (\theta_2 - \mu(\theta_1))/\sigma(\theta_1))$. In examples, Tibshirani and Wasserman find this transformation empirically and illustrate its success in simplifying the posterior. Althought this technique generalizes to a larger number of parameters, it is not clear that the algorithm is easy to implement in these higher-dimensional cases.

The correlation problem can also be present in sampling-based approaches to Bayesian computation. Hills and Smith show that the rate of convergence of the Gibbs sampler is significantly reduced in a bivariate normal posterior where the correlation is larger than .95. These correlation problems can occur naturally in multinomial regression problems discussed in Albert and Chib (1991). As a simple example of a multinomial probit model (Daganza, 1979), suppose that there exist independent unobserved latent variables $Z_1, \ldots, Z_N$, where $Z_i = (Z_{i1}, Z_{i2}, Z_{i3})$ is multivariate normal with mean $(z_{i1}, z_{i2}, z_{i3})\beta$ and identity variance-covariance matrix. One observes Y_i, the index of the maximum of $\{Z_{i1}, Z_{i2}, Z_{i3}\}$. To

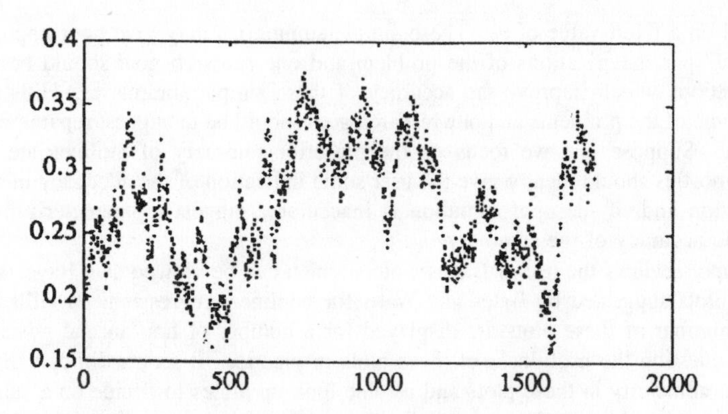

Figure 1. *Simulated values of β in one Gibbs sampling run for the multinomial probit example.*

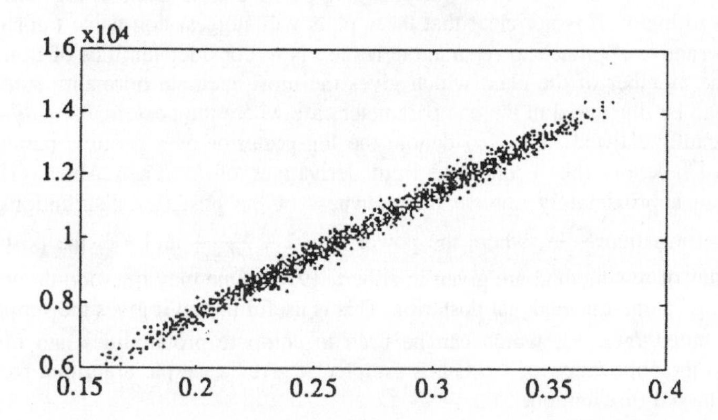

Figure 2. *Plot of simulated values of β against $\sum x_{ij} Z_{ik}$ for Gibbs run.*

implement the Gibbs sampler, one samples from the fully conditional posterior distributions of $(Z_1, \ldots, Z_N, \beta)$. Given β and $\{Y_i = k_i\}$, the $\{Z_i\}$ are independent with Z_i distributed $N((x_{i1}, x_{i2}, x_{i3})\beta, I)$ restricted such that $Z_{k_i} = \max\{Z_{i1}, Z_{i2}, Z_{i3}\}$. The distribution on β conditional on $\{Z_i\}$ has the usual normal error form. Given an initial estimate of β, one cycle of the Gibbs algorithm samples $\{Z_i\}$ and β from the above conditional distributions. The simulated values of β for a single chain of 1700 cycles is given in Figure 1. In this case it is difficult to get an accurate estimate of the marginal posterior density of β due to the strong correlations in the run. The cause for this problem is illustrated in Figure 2, which plots the simulated values of $\sum x_{ij} Z_{ij}$ against β. To remove this problem, one could reparameterize $(Z_1, \ldots, Z_N, \beta)$ by a linear transformation such that the new variables are approximately uncorrelated. However, with this transformation, the fully conditional distribution would no longer have the attractive form given above and it would be harder to perform the simulations. An alternative solution would be to deal with the correlation problem and devise some strategy to break up the strong correlations in the Gibbs run. In any event, this example illustrates the difficulties in applying general-use computational methods to badly behaved posteriors and the need to reparameterize to increase the efficiency of the method.

D. V. LINDLEY (*Somerset, UK*)

Bayesian ideas are often best discussed in the context of observables. Nevertheless there are circumstances in which parameters, though not observable, are useful. This is particularly true in scientific studies where the parameters have well-understood meanings for the scientist. At the last Valencia meeting I discussed the Hardy-Weinberg law where the two parameters have simple interpretations in the underlying genetics. Now a scientist is not going to take kindly to the idea of looking at $\theta_1^{1/3}\log\theta_2$ and $\theta_2^2 - \theta_1$, when his understanding is of θ_1 and θ_2. I therefore wonder if the authors have not concentrated too much on the mathematics, or the arithmetic, at the cost of the practical reality.

G. O. ROBERTS (*University of Nottingham, UK*)

I would like to add congratulations to the authors for an interesting summary of the parameterization methods in this area. I would also like to make some comments about correlation and the Gibbs sampler.

Speed of convergence for the Gibbs sampler is directly related to dependence, however it is only in the normal case that dependence is described directly by correlation. Under weak regularity conditions (see for example Roberts and Polson, 1991, and Schervish and Carlin, 1990), convergence of the Gibbs sampler occurs at a geometric rate, $\lambda(< 1)$. In the multivariate normal case, λ is a function of the off-diagonal entries in the correlation matrix, however in general, λ has no simple statistical interpretation.

Applications of the Gibbs sampler in Bayesian statistics are typically amenable to correlation reduction techniques, especially where parameters are location related.

Nevertheless, a reduction in correlation does not necessarily lead to increased efficiency of the algorithm. For instance, a high dependence between the location of x, and the variability of y in a bivariate problem where the correlation is zero, could frequently lead to slow convergence. An extreme example occurs when trying to sample from the uniform distribution on

$$\left[-\frac{1}{2},\frac{1}{2}\right]\times\left[-\frac{1}{2},\frac{1}{2}\right]\cup\left[\frac{1}{2},\frac{3}{2}\right]\times\left[\frac{1}{2},\frac{3}{2}\right]\cup\left[\frac{1}{2},\frac{3}{2}\right]\times\left[-\frac{3}{2},-\frac{1}{2}\right]$$

with parameters $\theta_1 = (1-\rho)x + \rho y$ and $\theta_2 = y$. It is easy to check that the iterates converge at a geometric rate $\forall\rho \in (0,1)$. However convergence is increasingly sluggish as $\rho \downarrow 0$, and for $\rho = 0$, the Markov chain becomes reducible, and convergence to the correct stationary distribution does not occur.

The above is somewhat pathological, but more realistic examples can occur, particularly when using mixture priors. The authors highlight the use of non-linear transformations, and it is to these methods that we must turn within the Gibbs sampler context, to avoid the possible anomalies mentioned above.

W. SEEWALD (*Ciba-Geigy, Basle, Switzerland*)

It appears to be crucial for the Gibbs sampler in order to run efficiently that the posterior distribution correlations are not too high. Even in the simple case of a one-way random effects model, this may be a problem if the between-group variance is small. Consider the model

$$\begin{aligned}
\alpha_i &= \mu + \delta_i, \quad \delta_i \sim N(0,\sigma_1^2), \quad i = 1,\ldots,I;\\
Y_{ij} &= \alpha_i + \varepsilon_{ij}, \quad \varepsilon_{ij} \sim N(0,\sigma_2^2), \quad j = 1,\ldots,J_i, \quad i = 1,\ldots,I;\\
\mu &\sim N(\mu_0,\sigma_0^2), \quad \sigma_1^2 \sim IG(a_1,b_1), \quad \sigma_2^2 \sim IG(a_2,b_2):
\end{aligned}$$

the last line defining priors for the independent hyperparameters. N denotes a Normal distribution, IG an inverse Gamma distribution; $\mu_0, \sigma_0^2, a_k, b_k$ are given constants. The set of conditional distributions used in the Gibbs sampler is

$$[\mu|Y, \alpha, \sigma_1^2, \sigma_2^2] = N\left(\frac{\sigma_1^2\mu_0 + \sigma_0^2\sum_i \alpha_i}{I\sigma_0^2 + \sigma_1^2}, \frac{\sigma_0^2\sigma_1^2}{I\sigma_0^2 + \sigma_1^2}\right) \tag{1}$$

$$[\alpha_i|Y, \mu, \sigma_1^2, \sigma_2^2] = N\left(\frac{J_i\sigma_1^2\overline{Y}_i + \sigma_2^2\mu}{J_i\sigma_1^2 + \sigma_2^2}, \frac{\sigma_1^2\sigma_2^2}{J_i\sigma_1^2 + \sigma_2^2}\right), \overline{Y}_i = \frac{1}{J_i}\Sigma_j Y_{ij} \tag{2}$$

$$[\sigma_1^2|Y, \mu, \alpha, \sigma_2^2] = IG(\ldots, \ldots), \quad [\sigma_2^2|Y, \mu, \alpha, \sigma_1^2] = IG(\ldots, \ldots) \tag{3}$$

If the between-group variance is large, everything is fine; however, if it is small (e.g., if the classical variance estimate is negative; see Table 1), one observes bad convergence due to high correlations (Figure 1).

gr.1	gr.2	gr.3	gr.4	gr.5	gr.6
508	507	498	505	506	499
520	517	519	521	518	515
512	516	510	506	519	514
503	500	498	506	500	501
497	497	500	499	500	495
488	491	495	485	491	491

Table 1. *Simulated data for six groups.*
Standard estimates of the variances: $\hat{\sigma}_1^2 = -17.63, \hat{\sigma}_2^2 = 113.5$

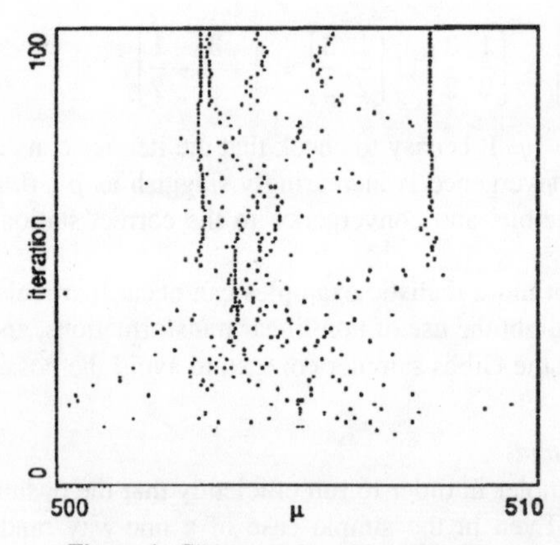

Figure 1. Gibbs sampler without blocking; μ, 5 replicates, 100 iterations

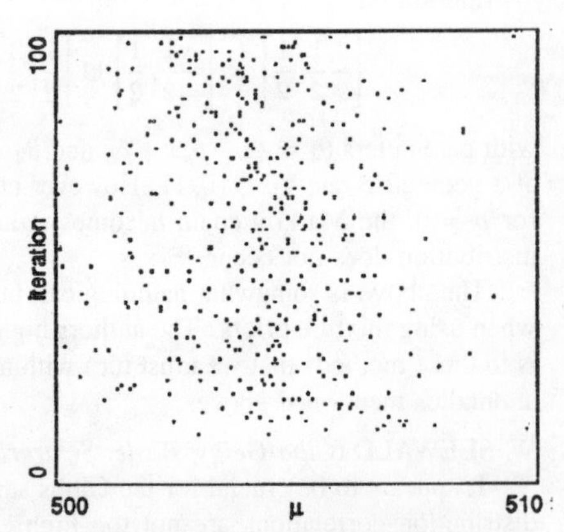

Figure 2. Gibbs sampler with blocking; μ, 5 replicates, 100 iterations

Among the various possibilities to get rid of the large correlations, I will mention only one, namely, "blocking"; in this case, this means treating the vector $\Phi = (\mu, \alpha_1, \ldots, \alpha_I)'$ as a single parameter in the Gibbs sampling scheme. Although this approach may not be easily generalized to more complicated models, it seems to work quite well in the actual example. Given σ_1^2 and σ_2^2, Φ is multivariate Normal with an inverse variance-covariance matrix which contains off-diagonal nonzero values only in its first row and column; so the easiest way to sample from it would be to sample its first component, μ, from its marginal distribution,

$$[\mu|\boldsymbol{Y}, \sigma_1^2, \sigma_2^2] = N(U, V), \frac{U}{V} = \frac{\mu_0}{\sigma_0^2} + \sum_i \frac{J_i \overline{Y}_i}{J_i \sigma_1^2 + \sigma_2^2}, \frac{1}{V} = \frac{1}{\sigma_0^2} + \frac{I}{\sigma_1^2} - \frac{\sigma_2^2}{\sigma_1^2} \sum_i \frac{1}{J_i \sigma_1^2 + \sigma_2^2} \quad (1')$$

and the remaining components, α_i, from its conditional distribution which is the same as (2). Thus, we obtain a sampling scheme which differs from the original one only in two points: first, the distribution of μ is not conditional on the α_i's; second, (1') and (2) are "linked together" in the sense that (μ, α) should always be sampled together and in this fixed order.

In a way, blocking is a technique which moves any high correlation between μ and α from the Gibbs sampler over to the random vector generator; while the Gibbs sampler cannot deal well with high correlations, the generation of multivariate normal random vectors with high correlations is not a problem (unless numeric instabilities begin to show up).

Figures 1 and 2 show the values of μ obtained by the Gibbs sampler without and with blocking, for five independent replicates and 100 iterations, using "bad" initial estimates.

N. D. SINGPURWALLA (*George Washington University, USA*)

I find this paper carefully crafted and masterfully exposited. The issue of parameterization —which appears to me to be a transformation of the parameters of a natural model (subjectively specified via some argument of indifference)— encompasses more than the effects of the performance of numerical techniques. It also has a bearing on the specification of a 'stopping rule'. This matter was flirted upon by Kass and Slate (1992) in Table 1 of their paper presented at the conference.

To elaborate on the above point, I will focus on the case of an exponential model with mean θ and failure rate $\lambda = \theta^{-1}$. Suppose that we place a gamma prior on θ. It can be shown [cf. Abel and Singpurwalla (1991)], using information theoretic arguments, that at any time a failure is more informative about θ than a survival, but that the exact opposite is true for λ. This is on initial reaction counterintuitive. As a consequence of the above, a question arises as to whether it is a transformation of the parameters that one must seek or should it be a transformation of the observables? Pursuing the former appears to me to be analogous to missing the boat but trying to catch up by taking a raft.

J. C. WAKEFIELD (*Imperial College, UK*)

I would like to report work, relevant to this paper, which I presented at the Workshop on Bayesian Computation via Stochastic Simulation, held at Ohio State University, February 15th-17th, 1991.

In particular I would like to discuss reparameterization for a population pharmacokinetics model, the anlysis being implemented via the Gibbs sampler. The data that I will analyze is a subset of that presented in Racine *et al.* (1986). Each of 10 individuals produces between five and eight blood samples, the concentration of drug within each of these samples being determined. Let t_{ij} denote the jth sampling time ($j = 1, \ldots, n_i$) for the ith individual ($i = 1, \ldots, I$) and y_{ij} the measured concentration at this time point. The assumed model is of hierarchical form, the first stage of which is given by

$$E[Y_{ij}] = f_{ij}(\phi_i, t_{ij}) = \frac{D}{V_i} \exp(-k_i t_{ij}) + \varepsilon_{ij}$$

where for illustration we naively assume $\varepsilon_{ij} \sim N(0, \tau_i^{-1})$, $\phi_i = (\phi_{i1}, \phi_{i2}) = (\log V_i, \log k_i)$ and D is the dose V_i and k_i represent respectively, the volume and elimination rate for the

ith individual. At the second stage an obvious assumption is that $\phi_i = (\log V_i, \log k_i)$ are realisations from a bivariate normal distribution with mean $\boldsymbol{\mu}$ and variance-covariance matrix Σ. At the final stage a normal prior is assumed for $\boldsymbol{\mu}$, and Σ is assigned an inverse Wishart distribution. Let $\boldsymbol{\tau} = (\tau_1, \ldots, \tau_I)$, with the τ_i's having gamma priors.

In terms of the Gibbs sampler the conditional distributions for the elements of ϕ_i are "awkward", the rest are straightforward. The conditional distribution for ϕ_i is given by

$$[\phi_i | \phi_j, j \neq i, \boldsymbol{\tau}, \boldsymbol{\mu}, \Sigma] \propto \exp\left[-\frac{\tau_i}{2}\sum_{j=1}^{n_i}(y_{ij} - D\exp(-\phi_{i1})\exp(-\exp(\phi_{i2})t_{ij}))^2\right]$$
$$\times \exp\left[-\frac{1}{2}(\phi_i - \boldsymbol{\mu})^T\Sigma^{-1}(\phi_i - \boldsymbol{\mu})\right].$$

For each element of ϕ_i the ratio-of-uniforms method was adopted (Wakefield, Gelfand and Smith, 1991). Using this approach the Gibbs sampler was successfully implemented. The form of the model indicates that we would expect large negative correlations between $\log V_i$ and $\log k_i$. Examination of the converged moments for the aforementioned dataset confirmed this, the correlations were approximately $-.9$. Consequently, orthogonalization, of the kind described in Section 2.4 of the paper, was utilized *within* each individual:

$$\psi_{i1} = \phi_{i1} \quad \text{and} \quad \psi_{i2} = \phi_{i2} - \frac{\text{cov}(\phi_{i2}, \psi_{i1})}{\text{var}(\psi_{i1})}\psi_{i1}.$$

Computationally the generation from the parameter set ψ_i is no more difficult than with the original, ratio-of-uniforms again being used. The elements of the population mean vector $\boldsymbol{\mu}$ give some overall indication of convergence. Figure 1 shows the sample mean estimate μ_{ϕ_1} plotted against iteration number. The estimate is based on 10 chains, orthogonalization being carried out at iteration 10. The solid line denotes the progress of the sampler under the original parameterization, the dotted line the orthogonalized version. Convergence is clearly being approached earlier in the latter case. With poorer initial estimates the difference would be even more marked. In this case it would probably be more sensible to orthogonalize several times since early iterations may not produce accurate estimates of the correlation structure.

Asymptotically, under the given model, τ_i is independent of ϕ_{i1} and ϕ_{i2}. If the more realistic error model

$$\text{var}(Y_{ij}) = \tau_i^{-1} f_{ij}(\phi_i, t_{ij})^\gamma$$

($\gamma \geq 0$) were used, however, then this would not be true. Consequently we would now wish to orthogonalize ϕ_{i1}, ϕ_{i2} and τ_i. In the original formulation the conditional distribution of τ_i has a gamma distribution and is consequently easy to generate from. A naive orthogonalization in the 'natural' order would not recover the simple form for τ_i. If, however, τ_i is placed first then its conditional distribution is unchanged.

A parameter which is of fundamental interest to the pharmacokineticist is the clearance. It is related to V_i and k_i via $Cl_i = V_i k_i$. It is often physiologically independent of V_i and so if we parameterize in terms of V_i and Cl_i we would, in general, obtain a more orthogonal set. This parameterization was adopted and, as expected, the correlations between $\log V_i$ and $\log Cl_i$ were close to zero. Unfortunately in this example, the conditional distributions for the $\log Cl_i$ are highly skewed and consequently the maximization of this distribution (which is required by the ratio-of-uniforms method) is more difficult. This shows that competing considerations must be taken into account when a 'good' parameterization is sought.

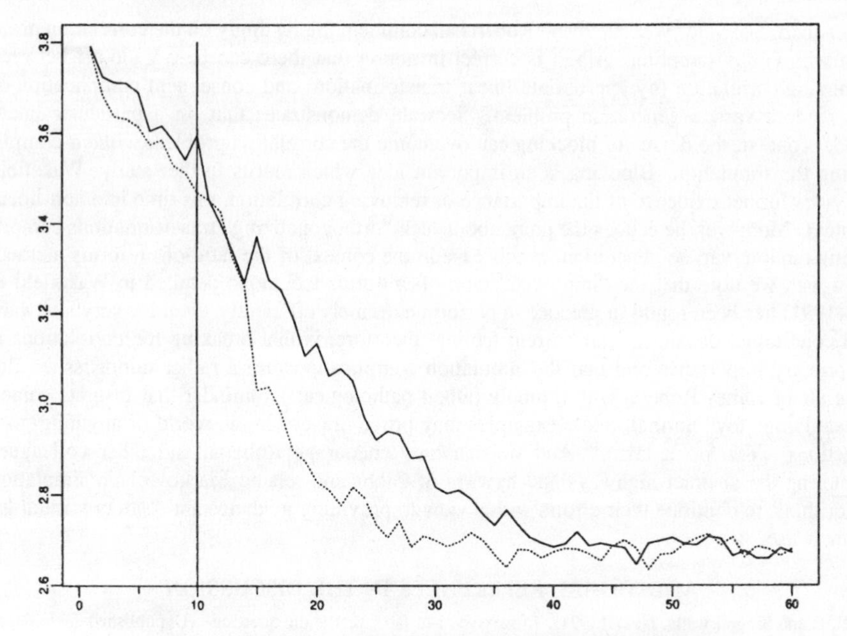

Figure 1. *Sample mean estimate for* μ_{ϕ_1} *versus iteration. Solid denotes original parameterization, dotted orthogonalized after the 10th iteration.*

REPLY TO THE DISCUSSION

We are grateful to the discussants for the interest they have shown in our paper and the issues they have raised.

Lindley and Singpurwalla query the whole enterprise of parameter transformation: the former from the perspective of scientific interpretability; the latter from the perspective of data transformation. In answer to Lindley we welcome the opportunity to underline that our main focus is on the question of which parameterization to use when performing certain kinds of numerical calculations or implementing certain kinds of analytic approximation procedures. The appropriate parameterization for these purposes may well not be the ideal one for eliciting prior information or the required one for reporting reference summaries. But there is no conflict! We know how to switch from one to the other, when required, using standard probability calculus techniques. The issue raised by Singpurwalla is intriguing —but somewhat baffling!— and we have to study the Abel and Singpurwalla paper before coming to a conclusion.

Albert is happy about the parameter transformation enterprise, but has doubts about the particular diagnostics idea that we present in Section 3.1. We shall simply have to differ on the issue of whether our diagnostics are useful in practice. We find them very useful! However, we readily accept that they are limited by virtue of working in one dimension at a time. Still working in single dimensions, Albert's Box-Cox type transformations are certainly interesting, but themselves have limitations because of the need for working through third derivatives (analytically tedious, numerically unstable). Perhaps more work is needed on the Tibshirani and Wasserman approach?.

Albert, Seewald, Wakefield and Roberts all comment interestingly on the correlation problem with Gibbs sampling. Albert is correct in noting that there can be a trade-off between removing correlation (by appropriate linear transformation) and consequent complication of the random variate generation problem. Seewald demonstrates that, in a particular linear model context, the device of blocking can overcome the correlation problem *without* complicating the simulation. Blocking is an important idea which merits further study. Wakefield provides further evidence of the importance of removing correlation, this time in a non-linear context. Moreover, he echoes the point about such "orthogonalizing" transformations complicating random variate generation, in this case in the context of the ratio-of-uniforms method. However, we note that the "improved" ratio-of-uniforms technique detailed in Wakefield *et al.* (1991) has been found in practice to perform extremely efficiently, even for very awkward full conditional densities. Our current feeling, therefore, is that breaking the correlations is of primary importance and that the simulation complications are a rather minor issue. But then along comes Roberts with a timely (albeit pathological!) reminder that insights gained by studying "toy" normal, linear examples may prove illusory in the world of anything-goes functions. Yes, point taken! And we can only encourage Roberts, and other colleagues exploring the abstract highways and byways of Gibbs and related Markov chain simulation algorithms, to continue their efforts with a view to providing guidance and "safety standards" in their use.

ADDITIONAL REFERENCES IN THE DISCUSSION

Abel, P. and Singpurwalla, N. D. (1991). To survive or to fail?; that is the question. (Unpublished *Tech. Rep.*).

Albert, J. H. (1989). On the use of power transformations to simplify posterior distributions. *Tech. Rep.* Department of Mathematics and Statistics, Bowling Green State University.

Albert, J. H. (1990). Algorithms for Bayesian computing and *Mathematica, Tech. Rep.* Department of Mathematics and Statistics, Bowling Green State University.

Albert, J. H. and Chib, S. (1991). Bayesian analysis of binary and polychotomus response data. *Tech. Rep.* Department of Mathematics and Statistics, Bowling Green State University.

Daganza, C. (1979). *Multinomial Probit*. New York: Academic Press.

Racine-Poon, A., Grieve, A. P., Flühler, H. and Smith A. F. M. (1986). Bayesian methods in practice: experiences in the pharmaceutical industry. *Applied Statistics* **35**, 93–120.

Racine-Poon, A. (1992). SAGA: Sample Assisted Graphical Analysis. *Bayesian Statistics 4* (J. M. Bernardo, J. O. Berger, A. P. Dawid and A. F. M. Smith, eds.), Oxford: University Press, 389–404, (with discussion).

Roberts, G. O. and Polson, N. G. (1991). A note on the geometric convergence of the Gibbs sampler. (Unpublished *Tech. Rep.*).

Schervish, M. J. and Carlin, B. P. (1990). On the convergence of successive substitution sampling. *Tech. Rep.* **492**, Carnegie-Mellon University.

BAYESIAN STATISTICS 4, pp. 247–266
J. M. Bernardo, J. O. Berger, A. P. Dawid and A. F. M. Smith, (Eds.)

Who Knows what Alternative Lurks in the Hearts of Significance Tests?

JIM HODGES
Rand Corporation, USA

SUMMARY

Hodges (1990) examined Bayesian and classical significance tests and found them to be more similar than is commonly believed. For a large class of significance tests, that paper constructed an alternative hypothesis and prior probability for the null hypothesis such that the posterior probability of the null is the P-value raised to the power $1-\delta$, for $0 < \delta < 1$; for small δ, the posterior probability of the null is uniformly arbitrarily close to the P-value. Many Bayesian writers have asserted that significance tests implicitly use an alternative, but have compared Bayesian and classical tests without knowing what that alternative is. We now know, and by examining it, this paper throws light on significance testing and on the likelihood principle. Two things are particularly relevant: first, many "pathologies" of P-values are shown to be properties of Bayesian hypothesis tests with proper alternative distributions; and second, the alternative that produces the P-value cannot be written as a parametric alternative but instead is necessarily a distribution specified directly on the observable used in the hypothesis test.

Keywords: SIGNIFICANCE TESTS; HYPOTHESIS TESTS; P-VALUES; COHERENCE; LIKELIHOOD PRINCIPLE.

1. INTRODUCTION

Bayesians have a long tradition of reviling significance tests as inherently unBayesian, going back at least to the first edition of Jeffreys' *Theory of Probability*. But Hodges (1990) showed that to an arbitrarily close approximation, the P-value is the posterior probability of the null hypothesis for a particular alternative, in a large class of cases to be detailed below. One could say that the alternative behind significance tests has been unmasked and can now be used as a club to beat on P-values. On the other hand, it now appears that P-values are not as unBayesian as we have been led to think. In view of the tradition of P-value bashing, it is of some interest to examine the new alternative hypothesis to see whether we must admit P-values to the Bayesian family after all.

It is also of interest to understand the implications this alternative has for the earlier Bayesian literature on the subject. A large portion of that literature could be summarized (with moderate injustice) as follows:

(1) We know significance testers have an alternative lurking somewhere.
(2) They don't tell us what it is, so we'll specify a convenient one for our purpose.
(3) The P-value differs from the posterior probability of the null that we compute from the convenient alternative.
  (3a) Therefore, P-values are defective.

As is now clear, step (2) is cheating, because the P-value does not test against step (2)'s strawman alternative. This violates the principle that you should beat up someone for what he claims to do and fails to deliver, not for what he doesn't claim to do. Also, step (3a) no longer works because it presumes that the Bayesian one is better; but it now appears that both solutions are equally Bayesian. Given the effort devoted to comparisons such as this, it is of interest to understand how the P-value alternative and the strawman differ, that the resulting posterior probabilities could be so different.

To answer these questions, Section 2 restates the construction in Hodges (1990) that yields the P-value as the posterior probability of the null, and explains why it works. Section 3 examines the P-value alternative by comparing it to the strawman alternative for a well-used special case, generalizing when possible. Section 4 admits that the new alternative has an odd property that might provide a new reason to hate significance tests, but argues that such a rejection would require the promulgation of a new principle internal to the Bayesian approach.

The bottom line is that you may continue to hate P-values, but not because they're unBayesian; and if you have been abstaining from exploratory uses of significance tests because you thought they are unBayesian, you need abstain no longer.

2. WHAT IS THE P-VALUE ALTERNATIVE AND WHY DOES IT WORK?

Let $X_1, \ldots, X_n$ be n observables, collectively denoted as $\boldsymbol{x}$. Suppose you have postulated a null hypothesis to explain them, and that you intend to make a significance test of that null hypothesis with the test statistic $T \equiv T(\boldsymbol{x})$. For cases in which T has an exact continuous distribution under the null, the P-value $P(T)$ — which can be viewed simply as a function of the observable $\boldsymbol{x}$ — has a uniform distribution under the null, conditional on any unknown parameters. Thus, for these cases, $P(T)$ has a uniform distribution with respect to $F(\boldsymbol{x})$, the marginal distribution function (with respect to the prior on the unknown parameters) of the observation $\boldsymbol{x}$ under the null. For the remainder of this paper, I will restrict consideration to models and test statistics with this property, although the result can be generalized broadly.

For significance tests in this class, construct the alternative density

$$p_{a,n}(\boldsymbol{x}) = K f(\boldsymbol{x}) \{P(t)^{\delta-1} - 1\}$$

where $f(\boldsymbol{x})$ is the density of $F(\boldsymbol{x})$, $0 < \delta < 1$, and $t = T(\boldsymbol{x})$. Then $K^{-1} = E(P(T)^{\delta-1}) - 1$, where the expectation is with respect to $F(\boldsymbol{x})$, i.e., $P(T)$ has a uniform distribution. Set the prior probability of the null hypothesis, π, so that it satisfies $(1 - \pi)K/\pi = 1$; this gives $\pi = K/(1 + K)$. Because we are only considering cases for which $P(T)$ has a uniform distribution, $K^{-1} = (1 - \delta)/\delta$ and $\pi = \delta$. For this alternative distribution and this prior probability on the null hypothesis, the posterior probability of the null is $P(t)^{1-\delta}$, which can be made uniformly arbitrarily close to the P-value $P(t)$ by making δ small. Small δ are of particular interest because of this approximation, but we will see that for any δ, the posterior probability of the null under $p_{a,n}(\boldsymbol{x})$ displays many of the "pathologies" associated with P-values, so that these pathologies can arise from proper alternative distributions. Note that this alternative can be constructed without cheating: if you supply the null model, a test statistic, and the prior distribution of parameters not specified in the null hypothesis, then the alternative can be constructed. The P-value alternative works for apparently any prior distribution for the parameters not specified in the null hypothesis, so that a P-value is consistent with any such prior.

Three aspects of the construction make it work. First is that the P-value has a uniform distribution under the null hypothesis, so it is possible to find the proportionality constant for $\delta > 0$ and $p_{a,n}(\boldsymbol{x})$ is a *bona fide* probability density in $\boldsymbol{x}$ for $\delta > 0$.

The second important feature of the construction is that $p_{a,n}(\mathbf{x})$ is a marginal distribution for $\boldsymbol{x}$, not an alternative specified in terms of the unknown parameters, which are integrated out to yield the marginal distribution for $\boldsymbol{x}$ used in a Bayesian hypothesis test. Section 3 shows that in considerable generality $p_{a,n}(\boldsymbol{x})$ *cannot* be such a parametric alternative. This is probably why earlier Bayesian writers did not find $p_{a,n}(\boldsymbol{x})$: to my knowledge, those writers worked exclusively with parametric alternatives.

The third important feature is that π, the prior probability of the null, is δ. Actually, π need not equal δ, but π/δ must be finite and bounded away from zero as δ approaches zero, or else the posterior probability of the null will go to 0 or 1. While this value of π may be distasteful to some, there is nothing illegal about it; you may prefer a different π, but $\pi = \delta$ is legitimate.

3. WHAT DOES THE *P*-VALUE ALTERNATIVE LOOK LIKE?

The exposition will use a familiar special case and generalize when possible. The special case is that $X_1, \ldots, X_n$ are independent and identically distributed normal random variables, with mean μ and variance 1. The null case is $\mu = 0$, the test statistic is $t = -|\bar{x}|\sqrt{n}$ (this expression simplifies later notation). Berger and Delampady (1987, pp. 317-318) (henceforth BD), to take an earlier instance of this special case, use the strawman alternative hypothesis $\mu \sim N(0, \tau)$, and produce a table in which the P-value — the (limiting) posterior probability of the null under the alternative $p_{a,n}(\boldsymbol{x})$ — differs from the posterior probability implied by their alternative. Let us compare these two alternatives.

For this setup, the (two-tailed) P-value alternative is

$$p_{a,n}(x) = \frac{\delta}{1-\delta}(2\pi)^{-n/2}\exp\left(-\sum_i x_i^2/2\right)\{2^{\delta-1}\Phi(t)^{\delta-1} - 1\}$$

for Φ the standard normal cumulative distribution function. The BD alternative yields an n-dimensional normal distribution for $\boldsymbol{x}$ with mean zero and covariance

$$\boldsymbol{I}_n + \tau \mathbf{1}\,\mathbf{1}^T$$

for $\mathbf{I}_n$ the n-dimensional identity matrix and $\mathbf{1}$ the n-vector of 1's. Both densities are invariant under orthogonal transformations of $\boldsymbol{x}$ that have $\mathbf{1}$ as an eigenvector, and neither is invariant under other transformations. Thus, it is convenient to represent $\boldsymbol{x}$ in $\mathbf{R}^n$ as a combination of a vector parallel to $\mathbf{1}$ and a vector parallel to $\sum x_i = 0$. I will use the representation $\boldsymbol{x} = \bar{x}\mathbf{1} + w\boldsymbol{v}$, where w is a scalar and $\mathbf{v} = -\sqrt{(n-1)/n}\,\mathbf{1} + \sqrt{n/(n-1)}\,\gamma$, γ being the n-vector $(0, 1, 1, \ldots, 1)^T$. Note that $\mathbf{v}$ and $\mathbf{1}$ are orthogonal and $\mathbf{v}$ has length 1. Thus w is the length of the component of $\boldsymbol{x}$ within the hyperplane $\sum x_i = 0$, and $\bar{x}$ is a convenient index of distance from $\sum x_i = 0$ parallel to $\mathbf{1}$. With $\boldsymbol{x}$ represented this way, the null, the P-value alternative, and the BD alternative are, respectively,

$$(2\pi)^{-n/2}\exp\left\{-\frac{1}{2}(w^2 + n\bar{x}^2)\right\}$$

$$\frac{\delta}{1-\delta}(2\pi)^{-n/2}\exp\left\{-\frac{1}{2}(w^2 + n\bar{x}^2)\right\}\left\{2^{\delta-1}\Phi(-\sqrt{n}|\bar{x}|)^{\delta-1} - 1\right\},$$

and

$$(2\pi)^{-n/2}(1 - n\tau)^{-1/2}\exp\left\{-\frac{1}{2}(w^2 + \frac{n}{1+n\tau}\bar{x}^2)\right\}.$$

(The symmetry exploited here is peculiar to the special case. In general, there is no symmetry, a counterexample being the test for the slope in simple linear regression, for which there is symmetry only if the regressors have an exploitable pattern.)

To examine the P-value alternative, note that it has three factors: the proportionality constant $\delta/(1-\delta)$, the null density, and a third factor that is roughly the reciprocal of the P-value. Thus, $p_{a,n}(\boldsymbol{x})$ modifies the null by pushing probability away from $\boldsymbol{x}$ in $\Sigma x_i = 0$ and toward values of $\boldsymbol{x}$ where the P-value is small. In particular, on the hyperplane $\bar{x} = 0$, the P-value is 1, so that $p_{a,n}(\boldsymbol{x}) = 0$ on that hyperplane. This feature of $p_{a,n}(\boldsymbol{x})$ figures prominently below. For fixed $\bar{x}$, the third factor of $p_{a,n}(\boldsymbol{x})$ is constant, so the P-value alternative is proportional to the null and the BD alternative for $\boldsymbol{x}$ varying with $\bar{x}$ fixed. Along lines parallel to **1**, both the second and third factors of $p_{a,n}(\boldsymbol{x})$ vary, and as $\bar{x}$ increases from 0, $p_{a,n}(\boldsymbol{x})$ rises to a mode and then falls off to zero. This can be seen in Figure 1, which shows $p_{a,n}(\boldsymbol{x})$ along the line $w = 0$, for three values of δ. By contrast, the BD alternative has a single mode at $\bar{x} = 0$ and declines monotonically to the right and left.

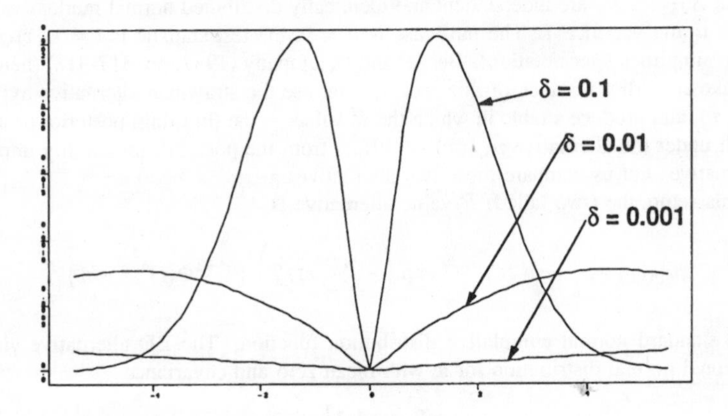

Figure 1. *The P-value alternative $p_{a,n}(\boldsymbol{x})$, viewed along the line $w = 0$, for the special case of the null model being iid $N(0,1)$, for $\delta = 0.1$, 0.01, and 0.001 and $n = 5$.*

(In general, $p_{a,n}(\boldsymbol{x}) = 0$ for values of $\boldsymbol{x}$ in the interior of $\mathbf{R}^n$ that give a P-value of 1. Modes are not necessarily as easily interpreted as in the special case. For example, in the simple linear regression case with a vector of regressors $\mathbf{z}$, $p_{a,n}(\boldsymbol{x})$ is infinite on the subspace $a\mathbf{1}$ and zero on the rest of the subspace $a\mathbf{1} + b\mathbf{z}$.)

Figure 1 suggests that as δ goes to zero for fixed n, the P-value alternative becomes an improper uniform distribution. This does not actually happen, for $p_{a,n}(\boldsymbol{x}) = 0$ on $\Sigma x_i = 0$ for all δ and n. However, as Figure 1 suggests, the mode in $\bar{x}$ does increase as δ approaches zero. To see this, consider non-negative $\bar{x}$, compute the derivative of $\log(p_{a,n}(\boldsymbol{x}))$ with respect to $\bar{x}$, and set it equal to zero to obtain this equation in $\bar{x}$ for the mode:

$$\bar{x} = \frac{1-\delta}{2^{1-\delta}\sqrt{n}} \quad \frac{\phi(-\sqrt{n}\bar{x})}{2^{\delta-1}\Phi(-\sqrt{n}\bar{x}) - \Phi^{2-\delta}(-\sqrt{n}\bar{x})}. \tag{1}$$

The right hand side is infinite for $\bar{x} = 0$, so the equation has a solution as long as the right hand side is less than $\bar{x}$ for some value of $\bar{x}$. Taking the limit of the right hand side as $\bar{x}$ approaches infinity and applying L'Hôpital's rule shows this to be the case for large $\bar{x}$ for $\delta > 0$, so that for $\delta > 0$, $p_{a,n}(\boldsymbol{x})$ has a finite non-zero mode in $\bar{x}$. However, the situation changes when $\delta = 0$: $p_{a,n}(\boldsymbol{x})$ becomes monotonically increasing in $\bar{x}$. This follows simply from noting that $-\exp(-n\bar{x}^2)$ is monotonically increasing, and that

$$\lim_{x\to\infty} \exp(-x^2/2)(\Phi(-x))^{-1}$$

is infinite.

Because the P-value alternative is zero on $\Sigma x_i = 0$ and particularly at the origin, it is easy to show that $p_{a,n}(\boldsymbol{x})$ cannot be derived by specifying a distribution for μ, i.e., as a parametric alternative. If it could, there would be a density $h(\mu)$ such that

$$\int h(\mu) \exp\{-\Sigma(x_i - \mu)^2\} d\mu$$

would be zero at $\boldsymbol{X} = 0$, which is impossible. This confirms what defenders of significance tests, such as Barnard, have been saying for decades: this significance test does not test against the parametric alternative $\mu \neq 0$.

This last property — that the P-value alternative cannot be constructed as a parametric alternative — appears to generalize widely. If $f(\boldsymbol{x}|\theta)$ is any parametric model for the vector $\boldsymbol{x}$ (within the class to which this paper is restricted), T is a test statistic with an exact continuous distribution for some null on θ, some $\boldsymbol{x}^*$ in the interior of $\mathbf{R}^n$ gives a P-value of 1, and $f(\boldsymbol{x}^*|\theta) \neq 0$ off a set of measure zero for any θ, then the only parametric alternatives that can yield $p_{a,n}(\boldsymbol{x})$ have their mass on the sets of measure zero on which $f(\boldsymbol{x}^*|\theta) = 0$; and if there are no such sets, $p_{a,n}(\boldsymbol{x})$ cannot be constructed as a parametric alternative. This is true, for example, of all the familiar two-tailed exact tests associated with regression and ANOVA.

The remaining feature to be considered is the behavior of $p_{a,n}(\boldsymbol{x})$ as n increases. Consider equation (1) above, and let n grow large with δ fixed. By an application of L'Hôpital's rule, the right hand side becomes approximately $(1-\delta)(\bar{x}+n^{-1})$, so that the mode is approximately $\bar{x} = (1 - \delta)/\delta n$, which goes to zero as n approaches infinity. Thus, as n grows large for fixed δ, the profile of $p_{a,n}(\boldsymbol{x})$ for any value of w comes to resemble a normal density with a thin pie slice taken out of the density right at its mode. For δ close to zero, the mode is approximately $1/\delta n$, which implies that as n gets large and δ gets small, $p_{a,n}(\boldsymbol{x})$ becomes rather flat.

To explore the limiting behavior of $p_{a,n}(\boldsymbol{x})$ further, consider Lindley's paradox (see, e.g., Berger and Delampady 1987). For our example, Lindley's paradox says: if $\bar{x} = k/\sqrt{n}$ (as n increases), then the P-value stays fixed at α corresponding to k, but the BD posterior probability of the null goes to 1. This has been considered a defect of P-values, but if a Bayesian test is constructed with the alternative $p_{a,n}(\boldsymbol{x})$ with $\delta > 0$, the posterior probability of the null stays fixed at $\alpha^{1-\delta}$ as n increases. Thus, this "paradox" depicts a difference between legitimate alternatives, not a difference between a Bayesian approach and another approach. The "paradox" can be worked in the other direction: if

$$\bar{x}^2 = \xi^2(n) = \{-2\log(B) + \log(1 + n\tau)\}(1 + n\tau)/n^2\tau,$$

so that $\bar{x}$ goes to zero at the rate $(\log(n)/n)^{1/2}$, then the BD Bayes factor stays fixed at B^{-1}, the BD posterior probability of the null stays fixed at $(1 + (1 - \pi)/B\pi)^{-1}$, and the P-value goes to zero. These two paths of $\bar{x}$ to the origin provide a means by which to characterize the different shapes of the BD and P-value alternatives as n grows. For non-negative $\bar{x}$, the ratio of $p_{a,n}(\boldsymbol{x})$ to the null and of the BD alternative to the null are, respectively:

$$\frac{\delta}{1-\delta}\left\{2^{\delta-1}\Phi^{\delta-1}(-\sqrt{n}\bar{x}) - 1\right\} \quad \text{and} \quad (1 + \tau n)^{-1/2} \exp\left\{\frac{1}{2}\ \frac{n^2\tau}{1 + n\tau}\bar{x}^2\right\}$$

Reinterpreting Lindley's paradox in these terms, the comparison of the two alternatives to the null can be summarized as follows:

- if $\bar{x} \to 0$ faster than $k/\sqrt{n}$, both alternatives become arbitrarily smaller than the null as n increases, but $p_{a,n}(\boldsymbol{x})$ goes to zero at the origin while the BD alternative reaches its mode,
- if $\bar{x} \to 0$ as $k/\sqrt{n}$ does, $p_{a,n}(\boldsymbol{x})$ maintains a ratio of $K(\alpha^{\delta-1} - 1)$ compared to the null, while the BD alternative becomes arbitrarily small with respect to the null,
- if $\bar{x} \to 0$ at the rate of, say, $(\log\log(n)/n)^{1/2}$, $p_{a,n}(\boldsymbol{x})$ gets arbitrarily large with respect to the null and the BD alternative arbitrarily small,
- if $\bar{x} \to 0$ as $\xi(n)$ does, $p_{a,n}(\boldsymbol{x})$ becomes arbitrarily large relative to the null while the BD alternative maintains the constant ratio of B^{-1},
- if $\bar{x} \to 0$ slower than $(\log(n)/n)^{1/2}$, both alternatives become arbitrarily larger than the null, but for every δ and τ there is an N such that for $n > N$, the BD alternative becomes arbitrarily larger than $p_{a,n}(\boldsymbol{x})$ (the proof being trivial).

Figure 2 depicts the behavior of the natural logs of the two ratios as functions of $\bar{x}$, for $n = 5$. The vertical lines marked "k/sqrt(n)" and "~sqrt(log(n)/n)" correspond to the two rates of convergence of $\bar{x}$ to zero, described above. At $\bar{x} = k/\sqrt{n}$, the top line maintains a constant height as n grows large, but the lower line becomes arbitrarily negative. At $\bar{x} = \xi(n)$ (roughly $(\log(n)/n)^{1/2}$), the lower line maintains a constant height as n increases, but the upper line becomes arbitrarily higher. To the left of $\bar{x} = k/\sqrt{n}$, the two curves become arbitrarily negative. Between $\bar{x} = k/\sqrt{n}$ and $\bar{x} = \xi(n)$, the top line becomes arbitrarily large and the bottom line arbitrarily small. To the right of $\bar{x} = \xi(n)$, both lines become arbitrarily large, and they eventually cross.

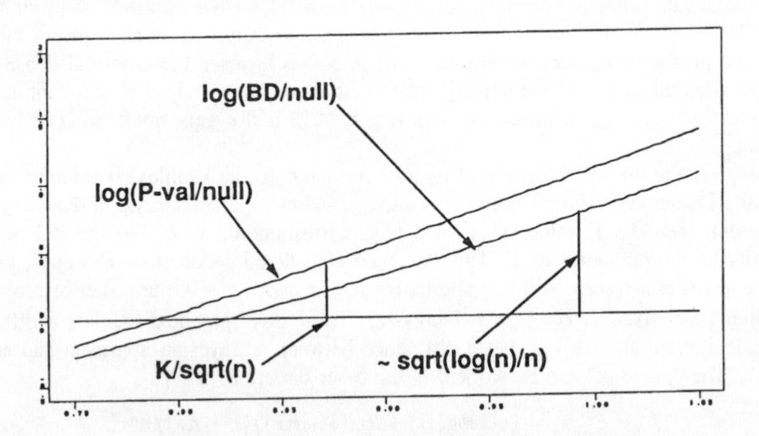

Figure 2. *Logs of the ratios of $p_{a,n}(\boldsymbol{x})$ to the null and the BD alternative to the null, as a function of $\bar{x}$ for any fixed w, in the special case of the null model being iid $N(0,1)$, for $\delta = 0.1$, $n = 5$.*

What does all this say about the alternative that produces the P-value? Generally, that it's just another alternative: apparent difficulties like Lindley's paradox are merely features of one Bayesian's alternative, which all must respect even if they prefer another. Many of the "pathologies" associated with P-values occur for $\delta > 0$, that is, with a proper alternative.

It is important that the P-value does not test against a parametric alternative, for this provides a way to examine $p_{a,n}(\boldsymbol{x})$'s implications for the likelihood principle (Berger and Wolpert, 1984). In one sense, $p_{a,n}(\boldsymbol{x})$ is trivially consistent with the likelihood principle, if the likelihood is the function $f(\boldsymbol{x}|\Psi)$, for Ψ taking the two values H_o and H_a. But this does not address the ways in which the principle has been used to argue against significance tests. Such arguments take the form, "you may not use the P-value to test $\mu = 0$ against $\mu \neq 0$ because the P-value uses aspects of the data not captured in the likelihood for μ." The development in this paper suggests a different argument: "you may not use the P-value to test $\mu = 0$ against $\mu \neq 0$ because the P-value does not test against any alternative specified in terms of μ." The likelihood principle adds nothing to this argument and would appear to be superfluous for this issue.

What *does* the P-value test against? It looks particularly for middling deviations from the values of $\bar{x}$ predicted by the model, and is less interested in the extremes of $\bar{x}$'s distribution than is the more familiar BD alternative. It must be admitted that when rendered as an alternative distribution, the P-value is unfamiliar and difficult to work with. With the backgrounds we have, it is unlikely that someone would, from scratch, directly specify $p_{a,n}(\boldsymbol{x})$, particularly for more complicated cases. But that does not imply that it should not be used.

4. A STRAW TO GRASP: AN ODD PROPERTY OF THE P-VALUE ALTERNATIVE

Suppose, to continue the special case, that we have $n+1$ observations. Construct $p_{a,n+1}(\boldsymbol{x})$ and the BD alternative and then, for each one, integrate out the last observation. The result for the BD alternative is the alternative that would have been obtained had we begun with n observations, but the result for $p_{a,n+1}(\boldsymbol{x})$ is not: in particular, the value at the n-dimensional origin is positive, while $p_{a,n}(0) = 0$. Unfortunately, the integrals involved are intractable and I have not been able to obtain more detailed results.

This result generalizes to cases of $n+1$ independent and identically distributed X_i. Suppose the X_i have common density $f(x_i|\theta)$ conditional on θ, and construct the P-value alternative $p_{a,n+1}(\boldsymbol{x})$. If there is some n-dimensional $\boldsymbol{x}^*$ in the interior of $\mathbf{R}^n$ such that the P-value is 1 for $\boldsymbol{x}^*$, then $p_{a,n}(\boldsymbol{x}^*) = 0$ and by a straightforward argument,

$$p_{a,n}(\boldsymbol{x}^*) \neq \int p_{a,n+1}(\boldsymbol{x}^*, x_{n+1})dx_{n+1}$$

unless the product of the individual densities at $\boldsymbol{x}^*$, $\Pi f(x_i|\theta)$, is zero for all θ. In the following discussion, I will refer to this property by saying that the BD alternative satisfies the integration property but that $p_{a,n}(\boldsymbol{x})$ does not.

Shall we grasp at this straw to maintain the age-old rejection of significance tests? Is it undesirable that $p_{a,n}(\boldsymbol{x})$ does not satisfy the integration property? There is apparently nothing unBayesian about it: the P-value alternative for $\delta > 0$ is a *bona fide* distribution, so we cannot object without inventing a new principle which would operate internal to the Bayesian approach. Still, the property does seem odd, for it seems to say that what we learn from the first n observations depends on how many other observations have been taken, even if we don't know their values. A sufficient condition for an alternative to satisfy the integration property is that it arises from a parametric model plus a probability distribution on the parameter values. I do not know if this condition is also necessary. If it is, then if we are to establish the integration property as a new principle, we are enshrining as principle the use of parametric alternatives. On the other hand, if some larger class of alternatives has the integration property, then *that* would be interesting, and defensible as a possible basis for a principle.

5. CONCLUSION

The motivation for Hodges (1990) was that Bayesians, like others, need tools for doing exploratory data analysis, but that Bayesians, unlike others, are mostly uncomfortable with standard EDA tools because they rely on the logic of significance tests (Box 1980). More candid Bayesians like Smith (1986) argue that the Bayesian approach cannot be used until a formal framework has been set up for a problem, but reject significance tests even in the informal work that precedes establishment of the formal framework. The point of the P-value alternative in Hodges (1990) was that Bayesians need not be so fastidious about making exploratory use of P-values, because they are a Bayesian construction to a close enough approximation that it isn't worth arguing over. I reiterate that point here: Bayesians, use P-values in exploration and feel good about it!

On a less polemical note, Bayesians have been saying for about a half-century that an alternative must lurk behind a significance test, so it is of some interest that the P-value alternative can be constructed at all. It is satisfying that $p_{a,n}(x)$ can not, in general, be expressed as an alternative in terms of the parameter specified in the null, just as defenders of significance tests have said all along. Instead, $p_{a,n}(x)$ tests the null model more generally, as Fisher's disjunction (Fisher 1973, p. 42) implies, although even after the examination in this paper, it is not obvious just what features of the model are being tested.

ACKNOWLEDGMENTS

In preparing this paper I had the helpful comments of John Adams, Bob Bell, and Mike Mattock of RAND, and of Paul von Batenburg of Touche Ross Nederlands. Sharon Koga prepared this paper in TEX.

REFERENCES

Berger, J. O. and Delampady, M. (1987). Testing precise hypotheses. *Statist. Sci.* **2**, 317–352, (with discussion).

Berger, J. O. and Wolpert, R. L. (1984). *The Likelihood Principle*. Hayward, CA: IMS.

Box, G. E. P. (1980). Sampling and Bayes inference in scientific modelling and robustness. *J. Roy. Statist. Soc. A* **143**, 383–430, (with discussion).

Fisher, R. A. (1973). *Statistical Methods and Scientific Inference*. New York: Hafner.

Hodges, J. S. (1990). Can/May Bayesians use pure tests of significance?. *Bayesian and Likelihood Methods in Statistics and Econometrics: Essays in Honor of George A. Barnard* (S. Geisser, J. S. Hodges, S. J. Press and A. Zellner, eds.), Amsterdam, North Holland, 75–90.

Smith, A. F. M. (1986). Some Bayesian thoughts on modelling and model choice. *The Statistician* **35**, 97–102.

DISCUSSION

D. J. POIRIER (*University of Toronto, Canada*)

Jim Hodges has tackled the daunting task of putting significance testing on a firm Bayesian foundation. In discussing this undertaking I am reminded of comments at the first Valencia meeting by Dawid (1980, p. 311):

> "I have learned to be wary of those who claim that they would like to reconcile the various opposing views on statistical inference. In my experience, the invariable consequence is, rather, a polarisation of attitudes and a great deal of fruitless apoplexy".

At the same meeting Kadane (1980, p. 317) pulled no punches regarding the desirability of Hodges' goal "... as a general matter, I believe that significance testing threatens the respectability of statistics more than any other single factor". Also, our Conference President has offered his assessment of significance tests: "They are widely used, yet are logically indefensible". (Lindley (1986, p 502)).

In an attempt to understand why Hodges tackles a topic in the face of skepticism of such Bayesian titans, I will discuss three essential ingredients of pure significance tests: (1) the use of sampling theory, (2) the use of unspecified alternatives, and (3) ambiguity over what is to be done if the null is rejected. Contrary to Hodges' claim, I will argue that his analysis is "unBayesian".

The issue regarding (1) was eloquently expressed by Jeffreys (1961, p. 385) in this assessment of the underlying logic of significance tests: "... a hypothesis that may be true may be rejected because it has not predicted observable results that have not occurred." The issue revolves around a litmus test of a statistician's pedigree, namely, the Likelihood Principle. I do not fully understand Hodges' remarks regarding the Likelihood Principle. There is an extensive literature on the Likelihood Principle versus significance test (e.g., Berger and Wolpert (1988, pp. 104–110)). In particular, Hodges fails to explain his implicit acceptance of a host of embarrassments (e.g., susceptibility to noninformative stopping rules) that go along with use of P-values. Honest Bayesians acknowledge their Achilles's heel (the prior) and address it through prior sensitivity analysis. It is time proponents of frequentist reasoning do the same, address their vulnerability to ambiguity over the sampling distribution, and acknowledge their willingness to consider data that could have been observed but were not.

As for (2), I have no trouble with the eloquent observation of Box (1980, p. 387) that: "... it seems a matter of ordinary human experience that an appreciation that a situation is unusual does not necessarily depend on the immediate availability of an alternative". Significance tests are intended to aid in that appreciation, i.e., to help assess, in the words of Barnard (1972, p. 129), "whether agreement between this hypothesis and the data is so bad that we must begin to exercise our imaginations to try to think of an alternative that fits the facts better". Indeed, Cromwell's Rule warns against the arrogance of dogmatically assuming all is well and cannot be made better, and I have adopted it as one of my Pragmatic Principles of Model Building (Poirier, (1988, p. 140)).

Such acknowledgements, however, are different than advocating statistical testing of null hypotheses without explicit alternatives. As Cornfield (1970, p. 28) noted long ago:

> "... the development of new hypotheses is ... no different from mathematics itself, which is concerned with methods of proving theorems, but has no advice on how to formulate new ones. Perhaps not the least of the advantages of the Bayesian outlook therefore is that it provides a clear-cut distinction between creative activity such as hypothesis formulation, which can be performed only by trained and imaginative people, and formal analysis which is in principle capable of reduction to routine performance by robots".

Hodges's attempt to provide a Bayesian foundation for testing without well-specified alternatives blurs Cornfield's distinction and obfuscates the difference between Bayesian techniques and the Bayesian outlook. The manner in which the alternative data density varies with x depends crucially on the choice of discrepancy function (tantamount to the choice of an alternative hypothesis) used to define the P-value, a matter about which the significance testing literature has little to say. Unless Hodges can argue that this density has a consistent and sensible interpretation across problems, however, these "foundations" seem to undermine, as much as support, significance tests.

In econometrics, where I usually hang my hat, pure significance tests have become popular in the rush by researchers to "out-test" their competitors. An example is the information test of White (1982), and the historical response to this test was typical: researchers sought alternative hypotheses for which the test had non-trivial power (e.g., Hall (1987)) in an attempt to understand how to use the test. Significance tests do not require alternatives for their derivation, but their conscientious use requires alternative hypotheses.

As for (3), proponents of significance tests often disarm their opponents by claiming such tests are only intended as "quick and dirty" methods for signalling the need to think more and reconsider the current window on the observable world. Who can argue with the advice to think more? Another sidestep is to argue that significance tests are intended for "assessing deficiencies" rather than testing. Such vagueness makes it impossible to define a "procedure" whose long-run performance is to be evaluated. This leaves frequentists without advice to give, but also conveniently "off the hook" since pretesting concerns cannot be directly raised. Hodges (1990) emphasizes the exploratory role of significance tests, but what are the rules of this game? Isn't this playing tennis without a net?

From the Bayesian perspective, I believe this vagueness is *less* devastating, although not without its annoyances. Once the researcher's mind has been jogged, by whatever manner, the troublesome issue is how to pick priors for a hypothesis suggested by the data. In memory of our dearly beloved and departed friend Morrie DeGroot, let me recall his characteristic wit and perception into the double-counting conundrum facing the Bayesian:

> "We open the newspaper in the morning and read some data on a topic we had not previously thought about. In order to process the data, we try to think about what our prior distribution would have been before we saw the data so we can calculate our posterior distribution. But we are too late. Who can say what our prior distribution would have been before we saw the data? We lost our virginity when we read the newspaper."
>
> DeGroot (1980, pp. 311–312)

How does one restore virginity lost? The importance of the question depends on the purity of the researcher. Once the researcher gives up the ideal state of the "single-prior Bayesian" and admits the need for sensitivity analysis in public research, one is left with the usual task of presenting a variety of mappings from "interesting" priors to posteriors and leaving it to the reader to decide whether the priors are sufficiently plausible to warrant serious consideration. Those who prefer "virgin priors" are likely "virgin data analysts".

Many Bayesian researchers have noted the possibility of assigning only $1 - \varepsilon$ prior probability to the hypotheses H_j $(j = 1, 2, \ldots, J)$ and reserving ε $(0 < \varepsilon < 1)$ probability for the hypothesis H_{J+1}: "something else". Provided the researcher specifies priors for the unknown parameters given H_j $(j = 1, 2, , \ldots, J)$ and the relative prior probabilities of these J hypotheses, standard posterior odds analyses permit comparison of the *relative* posterior probabilities of the hypotheses without specifying ε. If in the process the researcher's creative mind has a new insight leading to specification of "something else", then analysis can proceed straightforwardly given interesting sets of values for ε and priors for the parameters given H_{J+1}. The Bayesian moral is simple: only make *relative* probability statements about specifications explicitly entertained. Be suspicious of anyone promising more!

In summary, the thrust of my comments are echoes of other Bayesians who have come before. Criticism (like discovery) lies outside any current statistical paradigm. Perhaps that is the way it should be since a theory of criticism would amount to a theory of creativity. Exploratory data analysis may be best left as an art. I am willing to be convinced otherwise, but if anything, this paper has strengthened my conviction to leave art outside the protection of the Bayesian umbrella.

J. O. BERGER (*Purdue University, USA*)

I find the notion of specifying an 'alternative' distribution as a marginal distribution of x interesting, though from a practical perspective I agree with the author that the P-value alternative is "unfamiliar and difficult to work with." I presume others will address the sensibility of this alternative, and hence will confine my comments to three fundamental issues.

Issue 1. What does it mean to match P–values and posterior probabilities? The author has constructed an 'alternative' for which the posterior probability of the null is $P(t)^{1-\delta}$, which is approximately equal to the P–value, $P(t)$, for small δ. To be a bit more precise, suppose we agree that 'approximately equal' means $P(t)^{1-\delta} = P(t)(1+\varepsilon)$ for some specified small ε, which then implies that $\delta \cong -\varepsilon/\log P(t)$. Since this was to be set equal to the prior probability, π, of the null hypothesis, we have

$$\pi \cong -\varepsilon/\log P(t). \tag{1}$$

Now consider the usual Bayesian approach of specifying a parametric alternative to H_0 such as $H_1\colon \mu \neq \mu_0$, along with a conditional prior density, $g(\mu)$, on H_1. Supposing that $f(x|\mu)$ is the density (with $\mu = \mu_0$ specifying the null distribution and π again standing for its prior probability), the Bayes factor is

$$B_g(x) = f(x|\mu_0) \Big/ \int_{\{\mu \neq \mu_0\}} f(x|\mu) g(\mu) d\mu.$$

It is then easy to see that the posterior probability of H_0 and the P–value are equal if π is chosen to be

$$\pi = \left(1 + \frac{1}{B_g(x)}\left(\frac{1}{P(t)} - 1\right)^{-1}\right)^{-1}. \tag{2}$$

Frequently, $B_g(x)$ can be written as a function $\psi(P(t))$, in which case it seems that (1) says exactly the same thing as (2): *by appropriate choice of the prior probability π, one can always force agreement between the posterior probability and the P–value.* And note that (2) is usually a monotonically increasing function of $P(t)$, as is (1). It would be helpful for the author to clarify the distinction between his conclusion and the standard (2).

Issue 2. Do Bayesians denigrate a strawman? I don't think so. The author refers to the "strawman" $g_1(\mu) = \mathcal{N}(0, \tau)$ discussed in Berger and Delampady (1987). Although $g_1(\mu)$ was discussed in BD for certain purposes, the thrust of the article was to draw conclusions that are valid simultaneously for any 'reasonable' prior. The reason that Bayesians do not like P–values is that, *for any reasonable $g(\mu)$, the P–value will be very different from the Bayes factor $B_g(x)$.*

This can be said in many different ways. My current favorite is the "frequentist interpretation" in Example 6 of BD. It is very hard to read that example and afterwards take the familiar interpretation of P–values seriously, at least in situations with a parametric alternative. Of course, a feature of the reasoning here is that allowing freely varying π, when making comparisons between P–values and posterior probabilities, is senseless because of (2), so that comparisons require either fixing π (as in Example 6 of BD) or considering the Bayes factor. This also seems obvious from a pragmatic perspective: if one is trying to have the data say something about a null model, one must make sure to remove the influence of the (subjective) prior probability of the null model.

Issue 3. Are P–values useful in exploratory analysis? I do not think it is necessary to work hard to argue that P–values are useful in exploratory analysis. Virtually by definition, exploratory analysis cannot be done in a formal Bayesian way, meaning that adhoc indicators must be used. I am not convinced that P–values are particularly useful adhoc indicators and they may well do more harm than good in actual use, but 'adhoc' is not automatically bad if Bayes cannot be done.

Note that my argument here is not based on a feeling that parametric alternatives and nonparametric alternatives (typical in exploratory analysis) are fundamentally different; indeed, in BD several arguments are given to the effect that P–values are also misleading for nonparametric alternatives. Rather, the argument is that, for nonparametric alternatives, it may simply be impossible to do a believable Bayesian analysis.

L. M. BERLINER (*Ohio State University, USA*) and
C. ROBERT (*Université Pierre et Marie Curie, France*)

One might think that the evidence accumulated about the questionable value of P-values would outweigh the appeal of a formal similarity with Bayesian answers. However, since the debate is not quite closed, we take the opportunity of discussing this paper "to pound another nail in the coffin," to borrow an expression from Berger and Sellke (1987). First, note that the phenomenon exhibited by Hodges, namely the fact that $p(x)^{1-\delta}$ is the answer associated with the "marginal" distribution

$$m(x) = (\delta/(1-\delta))f(x)[p(x)^{\delta-1} - 1], \tag{1}$$

is in fact, a criticism in itself, since the "Bayesian" answer is $p(x)^{1-\delta}$, instead of $p(x)$. Hence, the P-value is not obtained as a Bayesian answer by Hodges, but only an approximation for small δ. To assess the approximation, consider the following values of $p(x)^{1-\delta}$ for various "interesting" values of $p(x)$ and various δ.

P-value	.01	.05	.10
δ			
.005	0.0102	0.0508	0.1012
.025	0.0112	0.0539	0.1059
.050	0.0126	0.0581	0.1122
.100	0.0158	0.0675	0.1259
.200	0.0251	0.0910	0.1585
.300	0.0398	0.1228	0.1995
.400	0.0631	0.1657	0.2512
.500	0.1000	0.2236	0.3162

Values of $p(x)^{1-\delta}$.

Note that for prior probabilities of the null as small as $\delta = .2$, the P-value is roughly 1/2 Hodges' answer of $p(x)^{1-\delta}$. On the other hand, for δ smaller than the P-value, yet in the range where Hodges' approximation is good, one notes that the P-value approximation actually can lend evidence *in favor* of the null. For intermediate values like $\delta = .05$ or $.1$ and when δ and the P-value are close, one must not only decide if the approximation is good, but also, whether or not one should pretend anything has actually been learned.

Our next point is that a procedure which corresponds "formally" to a Bayesian answer is not necessarily a good procedure. (Of course, Hodges only shows the P-value to be an approximation.) The literature is full of assessments of the behavior of the P-value from various points of view. Pertinent recent results of Hwang *et al.* (1991) are of interest. They demonstrate inadmissibility results concerning the P-value. In the case of continuous exponential families and two-sided hypotheses, one can view the P-value as an estimator of the indicator function at the null hypothesis, $I_{H_0}(\theta)$, and then evaluate an estimator of

$I_{H_0}(\theta)$ under quadratic loss. Hwang *et al.* (1991) show that the P-value is inadmissible in this case, and is thus suboptimal even from a frequentist perspective. Hwang and Pemantle (1990) have generalized this result to a large class of proper loss functions.

Next, we turn to some other operational and Bayesian criticisms of Hodges' suggestion:

(i) Note the following discrepancy in one-sided testing. Consider the normal distribution $X \sim N(\theta, 1)$ and the null hypothesis $H_0 : \theta \leq 0$. The P-value can then be written as the Bayesian posterior probability of the null corresponding to the uniform, improper prior. (See Casella and Berger, 1987, for generalizations.) Such results do not agree with Hodges' suggestion.

(ii) The P-value is not always uniquely defined in complicated settings. Different definitions of the P-value may lead to different forms for (1), leading to potential obvious contradictions to the likelihood principle.

(iii) (1) appears to be of no use at all in the presence of nuisance parameters.

(iv) Consider the independence of (1) with respect to the alternate hypothesis. For instance, hypothesis tests of $H_0 : X \sim N(0,1)$ against either $H_a : X \sim N(\theta, 1), \theta > 0$, or $H_{a'} : X \sim N(\theta, (1+\theta)^{-1}), \theta > 0$, produce the same P-value, $p(x) = P_0(X > x)$, since both alternatives are stochastically larger than H_0, and, hence, the same "marginal" (1). However, it seems to us that the marginal should be more skewed to the right in the second case than in the first. Problems also appear in nonparametric settings even when the P-value is naturally defined. For example, the following case points out a fundamental flaw in Hodges' suggestion. Consider $H_0 : X \sim f_0$ and $H_a : X \sim f_1$, where f_1 is stochastically larger than f_0. The P-value is $p(x) = P_0(X > x)$. Following Hodges literally, we should be testing H_0 against (1): We should replace one simple alternative with another! Such behavior can obviously be extended to composite alternatives in a fashion so that the marginal (1) does not even belong to the original collection of alternatives. This makes Hodges' suggestion questionable even in the context of exploratory work.

In general, "reconciliation" or approximation of P-values with Bayesian posterior probabilities of the null is hardly sufficient justification for suggesting the use of P-values. The analyst must ask whether or not the structure of priors leading, even approximately, to the P-value are sensible. Regarding Hodges' derivation, the fact that the marginal distribution (1) is only a "pseudo-marginal distribution" (there is no nondegenerate prior-likelihood pair leading to (1)), is extremely disturbing to the Bayesian who takes the Bayesian view of statistics seriously. Specifically, what sort of bona fide subjective reasoning could consistently lead to (1) as a meaningful distribution? Indeed, we take the view that Hodges' analysis actually suggests another reason *not* to use the P-value: It often approximates a Bayesian analysis corresponding to a ridiculous "prior" specification.

G. CASELLA (*Cornell University, USA*)

This article represents a valiant attempt to reconcile evidence in two-sided point-null hypothesis testing, an attempt for which Dr. Hodges is to be congratulated. Unfortunately, it appears to be impossible to reconcile frequentist and Bayesian evidence in this case. A cause of this is the fundamentally different approach to two-sided testing taken by frequentists and Bayesians. Much of the strange behavior displayed by the author's P-value alternative is directly attributable to the prior placing a point mass on the null hypothesis.

To obtain reconciliation between P-values and posterior probabilities means that a prior can be specified for which the P-value equals the posterior probability for all data values. That is, when testing $H_0 : \theta \in \Theta_0$ vs. $H_1 : \theta \in \Theta_0^c$ based on observing $X = x$, where

$X \sim f(x|\theta)$ and $\theta \sim \pi(\theta)$, evidence can be reconciled if

$$P(H_0|x) = \frac{\int_{\Theta_0} \pi(\theta|x)d\theta}{\int_{\Theta_0} \pi(\theta|x)d\theta + \int_{\Theta_0^c} \pi(\theta|x)d\theta} = P\text{-value} = P(x) \qquad \forall x,$$

where $\pi(\theta|x)$ is the posterior obtained from Bayes rule.

In a pair of papers, J. Berger and Sellke (1987) and Casella and R. Berger (1987a) examined reconciliation issues in both one-sided and two-sided testing. Berger and Sellke (1987) (and later Berger and Delampady, 1987) argued that reconciliation is impossible in two-sided testing, and saw this as a shortcoming of the P-values. Since, in two-sided testing, posterior probabilities are usually greater than P-values, it was concluded that P-values overstate evidence against H_0. However, Casella and Berger (1987a), in the spirit of the work of DeGroot (1973), showed reconciliation is possible in the one-sided case. They blamed the two-sided discrepancies on the Bayesian treatment of the point-null, and argued that point-mass priors place too much weight on H_0, causing the Bayesian posterior probability to be large (see also Casella and Berger, 1987b).

The different behavior in the one- and two-sided case was partially explained by Hwang *et al.* (1991). They showed that the P-value is generalized Bayes (and admissible) in the one-sided problem, but is not generalized Bayes in the two-sided problem. This latter result, which formalizes some of the author's ideas, shows that true reconciliation is impossible in point-null testing. A different approach to reconciliation was taken by Casella and Wells (1990). Using group structures, they established necessary and sufficient conditions for reconciliation, conditions that essentially eliminate the point-null case.

Since we know that, in point null testing, the P-value cannot be a posterior probability arising from a prior, even an improper prior, how do we interpret Hodges' result? Does the P-value alternative, which specifies a posterior unrelated to any prior, help us in understanding the discrepancies between P-values and posterior probabilities. Unfortunately, the answer seems to be no.

The solution proposed in the paper is a mathematical one: equate the P-value with the posterior probability and solve for the function $p_{a,n}(x)$ that must be the P-value alternative. This mathematical solution, however, is not a statistical solution, and arguing that $p_{a,n}(x)$ is an alternative density (or posterior probability) will anger frequentists as much as (I'm sure) it angers Bayesians. The function $p_{a,n}(x)$ cannot be updated (since it fails the "integration property") and doesn't allow a prior-based interpretation. Although Hodges says that "... the P-value alternative works for apparently any prior distribution ...", what is really the truth is that the P-value alternative works for no prior distribution.

To a frequentist, the P-value alternative is distasteful simply because of its strange behavior. In effect, to accept the P-value alternative means to defend it. Although I am not against using P-values for significance testing, I do not want to defend the P-value alternative.

The strange behavior of the P-value alternative is mainly a consequence of the fact that the P-value equals 1 at H_0. This behavior does not occur in one-sided testing, where the P-value only approaches 1 as the data go deeper in to H_0. That is, for $H_0 : \theta \leq 0$ vs. $H_1 : \theta > 0$, $\lim_{x \to -\infty} P(x) = 1$, but $P(x) < 1$, for $x > -\infty$. Since the P-value only equals 1 in the limit, reconciliation is possible. On the other hand, point-mass priors result in posterior probabilities that place too much weight on H_0. Until these particular features are addressed, true reconciliation can never obtain. I applaud the author's efforts at a mathematical reconciliation, and hope he will turn next to a more statistical reconciliation.

A. P. DAWID (*University College London, UK*)

The behaviour described in Section 4 of this paper is more than a mere curiosity — it displays a fundamental incoherence in the author's method of constructing a P-value alternative. The argument for the Bayesian position, as developed by de Finetti for example, starts from the idea that the decisions we take in different circumstances should, in a certain technical but intuitively forceful sense, *cohere*. When we consider the implications of this for a single experiment, we find that it is necessary and sufficient that the decision maker act as though he or she had a joint probability distribution for all the variables involved. Hodges claims that use of his P-value alternative is Bayesian because it has this property. However, coherence goes further than this, and also applies to the comparison of decisions made in different experiments — for example, with differing sample sizes. The importance of this aspect of coherence has been emphasised by Lindley (1978): some implications have been explored by Brown (1980) and Dawid (1988). As a simple example, consider the quadratic-loss estimation of a probability parameter θ on the basis of r successes in n trials. The minimax estimate is $\frac{r+\sqrt{n}/2}{n+\sqrt{n}}$, which is also the Bayes estimate if we use a $\beta(\sqrt{n}/2, \sqrt{n}/2)$ prior for θ. However, the dependence of this prior on n means that, whilst use of minimax for any one value of n could be defended as Bayesian, willingness to use it for all n is incoherent, and hence open to successful counterbetting. In exactly parallel fashion, the failure of the "integration property" means that willingness to use the P-value alternative for all n is incoherent and unBayesian.

Hodges speculates that the integration property requires putting a prior distribution on the model, but this, while sufficient, is clearly not necessary: the property is nothing other than Kolmogorov's consistency requirement for the existence of a joint distribution for an unlimited number of variables, and as such is satisfied for a collection $\{p_n(\cdot)\}$ if and only if there exist a fixed joint distribution for the infinite sequence of observations whose marginal density for the first n is $p_n(\cdot)$. Were this to be the case for $\{p_{a,n}(\cdot)\}$, the would-be coherent decision maker might have been able to take the P-value alternative more seriously.

E. I. GEORGE (*University of Chicago, USA*)

At this point in my career, I think of myself as both a Bayesian and a Frequentist. That is, I find both points of view enlightening, and neither without shortcomings. Having accepted ambiguity as a fact of life, I am willing to suffer any of the contradictions inherent in maintaining both perspectives. From my vantage point, I applaud this paper because it helps me understand significance testing in the light of Bayesian machinery. For example, I previously found it perplexing that Bayesian critics of P-value were able to find such large discrepancies between posterior probabilities and P-values. The paper resolves this dilemma for me by showing how the P-value alternatives can be quite far from the parametric alternatives which are so often chosen by Bayesians. I also find it enlightening to understand Lindley's paradox as a phenomenon resulting from a particular sequence of legitimate alternatives. It is curious that as opposed to a fence sitter like myself, an exclusive Bayesian and an exclusive Frequentist would each be inclined to reject this paper as irrelevant. The exclusive Bayesian would argue that statistical procedures should depend on the choice of prior and not vice-versa, whereas the exclusive Frequentist would argue that statistical procedures need no Bayesian justification.

I would like to comment on perhaps the most provocative conclusion of the paper, that the P-value alternatives does not satisfy the "integration property", namely that $p_{a,n}$ is not obtained from $p_{a,n+1}$ by integrating out the extra dimension. To me, this aspect of $p_{a,n}$ is not a defect but rather a property arising from a reasonable lack of invariance with respect to the defining construction. More precisely, $p_{a,n}$ is obtained by weighting those values of x which are "more extreme" according to the P-value. It is surprising that the more extreme values in R^ncannot be obtained by integrating or averaging over values in R^{N+1}. In fact, the argument could be made that alternative distributions which do satisfy the integration property are less extreme, and so make it more difficult to discriminate against the null. Thus, this feature provides further insight into the often cited discrepancies between posterior probabilities and P-values.

J. MORTERA (*University of Rome, Italy*)

The author states that P-values are a Bayesian construction and that we should use them and feel good about it. He reaches this conclusion by introducing a particular alternative density so that the P-value almost coincides with the posterior probability of the null hypothesis. By this same reasoning one can reach the completely opposite conclusion. For example, by substituting $P(t)$ with $1 - P(t)$ (which also has a uniform distribution) in the alternative density one has that the P-value, for small δ, coincides with the posterior probability of the *alternative* hypothesis! The point, of course, is that virtually anything is possible with an *ad hoc* alternative, and merely being Bayes with respect to some alternative should not be a source of much comfort.

DeGroot (1973) considers alternative densities constructed so as to "justify" P-values, but questions whether this class of alternative distributions can be derived from some "natural" assumptions. Are there any "natural" assumptions that justify the use of the bimodal alternative density $p_{a,n}(\boldsymbol{x})$ for the normal model in Section 2?

As Hodges states, taking a small δ (prior probability on the null) the P-value is close to the posterior probability of the null. If one is interested in testing a null hypothesis (versus a simple alternative) one surely has *some* prior belief in it being true. For $\delta \simeq 0$, there would hardly be any need to test!

There is a vast literature (see Berger's work) showing that the conclusions reached using P-values are often in disagreement with those reached even using a robust Bayesian approach. I don't think that P-values can be resuscitated for practical use.

G. PARMIGIANI (*Duke University, USA*)

"Who knows what alternative lurks in the heart of significance tests?" In a paper entitled "Doing what comes naturally: Interpreting a Tail Probability as a Posterior Probability or as a Likelihood Ratio" (1973), Morrie DeGroot constructed a family of alternatives that yields, as the title promises, the P-value as the posterior probablity of the null hypothesis. Dr. Hodges develops a different solution, and it is perhaps worthwhile to compare the two briefly.

Hodges proposes a simple alternative distribution on the sample space, given by:

$$p_{a,n}(\boldsymbol{x}) = \frac{\delta}{1-\delta}[(1 - G(t(\boldsymbol{x})))^{\delta-1} - 1]f(\boldsymbol{x}),$$

where f is the joint density of the data $\boldsymbol{x}$ and G is the cdf of the test statistic $t(\boldsymbol{x})$. The posterior probability of the null hypothesis can be made arbitrarily close to the P-value as δ becomes small. On the other hand, DeGroot assigned the alternative directly on the space of outcomes of the test statistic; he used the family of alternatives:

$$p^{\theta}_{a,n}(t) = (1 + \theta)G^{\theta}(t)g(t),$$

indexed by a parameter θ. If θ takes any nonnegative integer value with improper prior distribution $1/(1 + \theta)$, the posterior probability of the null hypothesis coincides with the P-value.

In both cases, the prior probability of the null hypothesis has to be, in some sense, very small. However, DeGroot's solution has the attractive feature that the alternative never depends on the prior probability of the null hypothesis. Also, DeGroot's family leaves the joint distribution unspecified, assuming that the statistic is all that is observed. So it is not possible to directly pose the question of the incoherence of probability assignments for varying n. A natural modification of DeGroot's family, that still leads to an exact P-value, is given by:

$$q^{\theta}_{a,n}(x) = (1+\theta)G^{\theta}(t(x))f(x).$$

This can be shown to suffer from the same marginalization problems that arise with Hodges' alternative. It would perhaps be interesting to investigate whether there exist a family of joint distributions that yields consistent marginalizations and generates DeGroot's alternative.

K. PÖTZELBERGER (*University of Economics, Vienna, Austria*)

In this talk Hodges presented a Bayesian interpretation of frequentist testing procedures based on the P-value. Formally, this can be done by defining a distribution on the alternative so that the corresponding Bayes factor is a function of the frequentist test statistic. However, this distribution on the alternative does not obey coherency. The distribution is, in many cases, not consistent in the sense that integrating out a part of the data will change the distribution, i.e., the distribution of the alternative depends on what might have been observed. Thus the Bayesian interpretation of the test turn out to exhibit the deficiencies of the test, rather than reconciling classical significance tests with Bayesian ideas.

We discuss the approach by showing that in certain situations using the marginal distribution of the observed data and computing the Bayes factor, can lead to an equivalent procedure. However, the level of the test may not be chosen freely any more. In the following example it has to be $\alpha = 0.32$.

Let $T(x)$ be a frequentist test statistic for a simple null hypothesis H_0 versus a composite alternative H_1. Let $f_n(x)$ denote the distribution of the observation $x = (x_1, \ldots, x_n)$ under the null hypothesis. Furthermore, we assume that under $H_0, ET = 0$ and $\sigma(T) = 1$ and that the null hypothesis is rejected, when $|T(x)| > c$. Define $h_n(x) = T^2(x) f_n(x)$ as the distribution of x under the alternative hypothesis. Then, formally, the corresponding Bayes factor is $B = 1/T^2(x)$, which leads hence to an acceptance of H_0 if $B \geq c_0 := 1/c^2$.

$h_n(x)$ is an exchangeable distribution. Denote by $h_{k,n}$ the marginal of $x^k = (x_1, \ldots, x_k)$ $(k \leq n)$, computed from h_n. Usually, $h_{k,n}$ differs from h_k, indicating that the interpretation of the test based on T is not coherent. One might, however, try to modify the model by replacing h_k by $h_{k,n}$, where $n > k$. This distribution depends on n. As an example we consider $x_i \sim^{iid} N(\theta, 1)$, with $H_0 = \{0\}$ and $H_1 = \{\theta \neq 0\}$. Let $T = \sqrt{n}\overline{x}_n$, so that $h_n \propto \overline{x}_n^2 f_n(x)$ and $h_{k,n}(x^k) = \epsilon h_k(x^k) + (1-\epsilon) f_k(x^k)$ with $\epsilon = k/n$. The corresponding Bayes factor is again a function of f_k/h_k. Precisely,

$$\frac{f_k}{h_{k,n}} = \frac{f_k}{\epsilon h_k + (1-\epsilon) f_k} = \frac{1}{\epsilon h_k / f_k + (1-\epsilon)},$$

so that $f_k/h_k \leq c_0$ if and only if

$$\frac{f_k}{h_{k,n}} \leq \frac{c_0}{\epsilon + (1-\epsilon)c_0}.$$

Again, $c_0/(\epsilon + (1-\epsilon)c_0)$ should be independent of n, which is the case only for $c_0 = 1$. We conclude that only for $c_0 = 1$, is a Bayesian interpretation of the frequentist test based on $\overline{x}_k$ possible. $c_0 = 1$ leads to $P(|\sqrt{k}\overline{x}_k| > c) \approx 0.32$.

REPLY TO THE DISCUSSION

My objective was to construct an alternative that yields the P-value as the posterior probability of the null, to examine it to see if it clarifies issues in hypothesis testing, and, in conjunction with Hodges (1990), to see whether it might make some Bayesians more comfortable about using P-values for exploratory purposes. This rejoinder clears up some areas of confusion and then considers new material raised in the discussion.

The point of the paper was not to defend P-values for testing a null hypothesis about a parameter θ against an alternative about θ (Berliner, Casella) or against an unstated alternative (Poirier). The point was to see what alternative the P-value tests against. As it turns out, in some generality that alternative cannot be expressed in terms of θ, which makes it plain why P-values appear to do poorly at testing against parametric alternatives: they're set up (implicitly) to test against something else (as George noted).

(In this connection, Mike Lavine of Duke University has pointed out that $p_{a,n}(\boldsymbol{x})$ is an exchangeable density for cases in which $\boldsymbol{x}$ is exchangeable under the null and T is symmetric in the x_i. This suggests that, in the limit as n approaches infinity at least, $p_{a,n}(\boldsymbol{x})$ can be expressed as some sort of mixture, although it is beyond me to derive it.)

The construction does not require cheating in the specification of δ (Berger): δ is fixed. Furthermore, $0 < P(\boldsymbol{x})^{1-\delta} - P(\boldsymbol{x}) < \delta(1-\delta)^{(1-\delta)/\delta}$, with the maximum occurring at $P(\boldsymbol{x}) = (1-\delta)^{1/\delta}$, so the difference between the P-value and the posterior probability is uniformly bounded and becomes arbitrarily small as δ does. Large values of δ are of no interest, although some P-value "pathologies" (such as Lindley's paradox) occur for all values of δ. For the two discrete cases I have worked through, one can set $\delta = 0$ without making the P-value alternative improper.

The final matter of confusion is whether nuisance parameters are a problem (Berliner, Casella). They are not: $f(\boldsymbol{x})$ is the marginal density of $\boldsymbol{x}$ under the null, with nuisance parameters integrated out against their prior. The construction works for any such prior because $f(\boldsymbol{x})$ is a factor of the alternative, so it does not appear in the Bayes factor.

Parmigiani pointed out DeGroot (1973), which I was remiss in not discussing. In DeGroot's construction, the null case obtains when his parameter $\theta = 0$, with the positive integral θ making up the alternative. If DeGroot's alternative is expressed as a single density on the test statistic t — by summing out θ from 1 to infinity — it is

$$Q^{-1} f(t)\{P(t)^{-1} - 1\} \qquad \text{for} \qquad Q = \sum_{k=0}^{\infty} (k+1)^{-1},$$

where the sum does not converge. If $p_{a,n}(\boldsymbol{x})$ is re-expressed in terms of the test statistic, then DeGroot's alternative is identical to $p_{a,n}(\boldsymbol{x})$ with $\delta = 0$. Note that the prior on DeGroot's parameter θ is improper, so that $\theta = 0$ — the null — has probability zero under his construction as it does in the limiting case of $p_{a,n}(\boldsymbol{x})$.

If the sample space of DeGroot's θ is restricted to the integers $0, 1, \ldots, K$, his distribution across θ is proper, and his alternative becomes

$$Q^{-1} f(t) \left\{ \frac{1 - F^{K+1}(t)}{1 - F(t)} - 1 \right\} \qquad \text{for} \qquad Q = \sum_{k=0}^{K} (k+1)^{-1},$$

where $F(t)$ is the cdf of $f(t)$. The prior probability of the null hypothesis is $Q^{-1} > 0$; thus if DeGroot's limiting case is approached by letting K become large, his construction, like mine, makes the prior probability of the null depend on the alternative. I believe it is impossible to re-express this proper-prior version of DeGroot's alternative to make it identical to $p_{a,n}(\boldsymbol{x})$, so it and $p_{a,n}(\boldsymbol{x})$ are distinct approaches to the P-value.

Given these similarities, DeGroot's motivation was intriguing:

> The purpose of this article is to present a few simple ideas which indicate how the calculation of tail areas can be made compatible with the principles of Bayesian statistics. These ideas, if successful, will serve the dual purpose of putting χ^2 tests, F tests, and other such procedures back into the repertory of the Bayesian statistician and of giving all statisticians the freedom that comes from being able to interpret the evidence exhibited in a tail area simply as a likelihood ratio or as a posterior probability. (p. 967)

If a Bayesian as stalwart as DeGroot could utter such a sentiment, one might wonder at the vehemence with which some discussants denounced it. Some had no problem with DeGroot's aim, and accept "*ad hoc* indicators" even without such a construction (Berger). Others accept the need for adhockery but reject P-values (Smith 1986). Still others balked at adhockery, inadmissibility, or incoherence (Berliner, Dawid, Mortera, or Poirier).

What shall we make of such objections? Dawid notes that $p_{a,n}(\boldsymbol{x})$ is coherent for any given n; if it weren't, then when I'm handed a sample of size n, my beliefs about it would be constrained by samples I could have gotten but didn't. (Sounds familiar?) For other cases, the response to $p_{a,n}(\boldsymbol{x})$'s incoherence makes an interesting contrast with the blasé response to Bernardo's incoherent reference priors. Berger addressed this issue in his talk here by noting that in any given analysis, he has only so much time to spend and might be willing to use an incoherent reference prior – risking a loss of utility – so he can concentrate his effort on aspects of the problem that could cost him more utility. (See also Bernardo, 1984.) The same reasoning applies to P-values used in exploration: You have only so much time, You may choose to spend less of it by using something simple, incoherent, *and available*, and thereby have more time for other aspects of the analysis.

Mortera and Poirier objected to adhockery without mentioning inadmissibility or incoherence. Mortera considered it absurd that replacing $P(\boldsymbol{x})$ by $1 - P(\boldsymbol{x})$ in $p_{a,n}(\boldsymbol{x})$ yields a proper density and a posterior probability of the null of $(1 - P(\boldsymbol{x}))^{1-\delta}$. To the contrary, she has constructed the usual test of whether the observations agree with the model "too well," as might be used to detect that Mendel's assistant fudged his data. Presumably Mortera is interested in such tests, but she could not make one in the iid normal example simply by changing τ in the BD alternative. Instead, she would need another alternative, one using iid normals with variance less than 1, or a t on 3 df scaled to have variance 1. Which is less *ad hoc*, the normal or the t? Plainly neither, and neither is less *ad hoc* than the alternative that Mortera constructed and scorned. Perhaps physicists derive models from first principles, but few if any others can; when a Bayesian casts about for a model and throws her line into the pool of handy specifications, how is this not *ad hoc*? Thus, I see no force to the objections of Mortera and Poirier regarding adhockery.

The conference President, Dennis Lindley, asked if our view of Bayesian statistics is too narrow. I would say yes, and the discussion illustrates why: our theory is written as if likelihoods are found in the cabbage patch, when in fact they are often created as part of the analysis. As the foregoing indicates, when we consider data analysis in its fullness, some comfortable old verities may not seem so comfortable anymore. At a minimum, it is clear that Bayesians are far from a consensus on how to think about real data analyses, and we should not hesitate to re-examine cherished beliefs in the search for a consensus.

ADDITIONAL REFERENCES IN THE DISCUSSION

Barnard, G. A. (1972). Review of *The Logic of Statistical Inference* by I. Hacking. *British J. Philosophy of Science* **23**, 123–190.

Barnard, G. A. (1980). Pivotal inference and the Bayesian controversy. *Bayesian Statistics* (J. M. Bernardo, M. H. DeGroot, D. V. Lindley and A. F. M. Smith, eds.), Valençia: University Press, 293–318, (with discussion).

Berger, J. O. and Sellke, T. (1987). Testing a point null hypothesis: The irreconcilability of P values and evidence. *J. Amer. Statist. Assoc.* **82**, 112–122.

Bernardo, J. M. (1984) Discussion of Geisser (1984). *Ann. Statist.* **38**, 247–248.

Brown, P. J. (1980). Coherence and complexity in classification problems. *Scandinavian J. Statist.* **7**, 95–98.

Casella, G. and Berger, R. L. (1987a). Reconciling Bayesian and frequentist evidence in the one-sided testing problem. *J. Amer. Statist. Assoc.* **82**, 106–111.

Casella, G. and Berger, R. L. (1987b). Comment on "Testing precise hypotheses" by J. O. Berger and M. Delampady. *Statist. Sci.* **2**, 344–417.

Casella, G. and Wells, M. T. (1990). Reconciliation, coherence, and P-values. *Tech. Rep.* **1100**, Cornell University.

Cornfield, J. (1970). The frequency theory of probability, Bayes theorem, and sequential clinical trials. *Bayesian Statistics* (D. L. Meyer and R. O. Collier, Jr., eds.). Itasca, IL: Peacock, 1–28.

Dawid, A. P. (1988). The infinite regress and its conjugate analysis. *Bayesian Statistics 3* (J. M. Bernardo, M. H. DeGroot, D. V. Lindley and A. F. M. Smith, eds.), Oxford: University Press, 96–110, (with discussion).

Dawid, A. P. (1980). Discussion of Barnard (1980). *Bayesian Statistics* (J. M. Bernardo, M. H. DeGroot, D. V. Lindley and A. F. M. Smith, eds.), Valencia: University Press, 311.

DeGroot, M. H. (1973). Doing what comes naturally: interpreting a tail area as a posterior probablity or as a likelihood ratio. *J. Amer. Statist. Assoc.* **68**, 966–969.

DeGroot, M. H. (1980). Discussion of Barnard (1980). *Bayesian Statistics* (J. M. Bernardo, M. H. DeGroot, D. V. Lindley and A. F. M. Smith, eds.), Valencia: University Press, 311–312.

Geisser, S. (1984). On prior distributions for binary trials. *Ann. Statist.* **38**, 244–247.

Hall, A. (1987). The information matrix test for the linear model. *Review of Economic Studies* **54**, 257–263.

Hwang, J. T., Casella, G., Robert, C., Wells, M. and Farrell, R. (1991). Estimation of accuracy in testing. *Ann. Statist.* **19**.

Hwang, J. T. and Pemantle, R. (1990). Evaluation of estimators of statistical significance under a class of proper loss functions. *Tech. Rep.* Cornell University.

Jeffreys, H. (1961). *Theory of Probability*. Oxford: Clarendon Press.

Kadane, J. B. (1980). Discussion of Barnard (1980). *Bayesian Statistics* (J. M. Bernardo, M. H. DeGroot, D. V. Lindley and A. F. M. Smith, eds.), Valencia: University Press 315–316.

Lindley, D. V. (1978). The Bayesian approach. *Scand. J. Statist.* **5**, 1–26, (with discussion).

Lindley, D. V. (1986). Discussion of "Test of significance in theory and practice". *The Statistician* **35**, 502–504.

Poirier, D. J. (1988). Frequentist and subjectivist perspectives on the problems of model building in economics. *Journal of Economic Perspectives* **2**, 121–170, (with discussion).

White, H. (1982). Maximum likelihood estimation of misspecified models. *Econometrica* **50**, 1–25.

BAYESIAN STATISTICS 4, pp. 267–287
J. M. Bernardo, J. O. Berger, A. P. Dawid and A. F. M. Smith, (Eds.)

Opinions in Dispute: the Sacco-Vanzetti Case*

JOSEPH B. KADANE and DAVID A. SCHUM
Carnegie Mellon University, USA and *George Mason University, USA*

SUMMARY

Bayesian methods are ideally suited for comparing the opinions of experts who disagree. In this paper we explore the historiography of the Sacco and Vanzetti case, a cause celèbre in American history of the 1920's. The evidence in this case, from the writings of prominent historians, is parsed using charting methods adapted from the work of Wigmore. Probability elicitation, using as experts two prominent historians of the case who disagree about the guilt of Sacco, is the next step to be undertaken to understand the nature and extent of their disagreement.

Keywords: ARGUMENT STRUCTURING; BAYESIAN MODELLING; ELICITATION; HISTORICAL APPLICATION; LEGAL APPLICATION.

1. INTRODUCTION

The subjective character of Bayesian inference becomes especially useful when there are differing opinions concerning a single subject. To illustrate the usefulness of Bayesian methods in modeling lack of agreement, it is important to eliminate uninteresting sources of disagreement. The first is apparent disagreement due to lack of careful consideration. One would not want to elicit casually considered probabilities and then to study how and why they differ. Consequently, expert opinion is a more fertile field for Bayesians interested in differing opinions than is non-expert opinion.

A second source of apparent disagreement is differing information. It might be argued that each instant of a person's life yields information that, in principle, could alter his/her view on any subject. If one takes such a broad view of information, it is impossible for two people to have the same information, and indeed the concept of information becomes so broad as to be meaningless. Taking a narrower view of information, then, it would be reasonable to expect experts to share specific facts that they find influential in shaping their views. By doing so, to a reasonable degree differences in opinion must be attributed to sources other than differing information. It is unnecessary and unuseful to get ensnared in speculation about whether infants at birth start with the same (reference ?) priors.

Thus, the criterion for an example of an interesting difference of opinion is that it should involve differences between experts who share a common information base. An additional, pragmatic, consideration is that the information base be reasonably static. Otherwise, opinions would have to be indexed by when they were assessed and by what information was available at the time. Recent, but not too recent, historical controversies represent a good field to find such problems. They should be recent enough that experts differ and care, but sufficiently

* This research was supported by NSF Grant SES–8900025 to Carnegie Mellon University and, in part, by NSF Grant SES–9007693 to George Mason University.

remote in time that new facts are no longer coming to light. This paper explores a Bayesian modeling of the Sacco and Vanzetti case. Experts continue to differ and to care about this case; new facts about this case have come to light but not at a very rapid rate. In Section 2 we briefly introduce and discuss this case. In Section 3 we present some methods for displaying and understanding the large mass of evidence that was produced during the trial of Sacco and Vanzetti; we also discuss various means by which experts may elicit probabilistic assessments about the strength of this evidence. In Section 4 we discuss some conclusions about how such elicitations may pinpoint and then resolve disagreement among these experts.

2. THE CASE OF SACCO AND VANZETTI

This case has certainly been a controversial episode in American history. On April 15, 1920 in South Braintree Massachusetts, five men killed two guards named Berardelli and Parmenter as they were delivering a payroll belonging to the Slater and Morrill Company. On May 5, 1920 two anarchists of Italian descent, Nicola Sacco and Bartolomeo Vanzetti were arrested. Tried together over the objections of their attorneys, they were convicted of the crime committed in South Braintree and were executed in 1927.

Whether they were given a fair trial and whether they were guilty (not the same thing), became questions that divided people along ideological lines. Anarchists quickly organized a defense committee. Communists and labor organizers used the case as a cause; but, ironically, the specific anarchist group to which Sacco and Vanzetti belonged had little ideological respect for the organizational control exerted by Communists and others. Equally passionately, conservatives in America argued that justice had been done and that the attacks on the case were, in fact, attacks on the legal system and on law and order. The controversy included Harvard Law Professor Felix Frankfurter, who wrote in defense of Sacco and Vanzetti, and Harvard President A. Lawrence Lowell, who headed a three-person committee advisory to Massachusetts Governor Fuller; this committee recommended to the governor that the trial was fair and that Sacco and Vanzetti should be executed.

While much of the debate has been ideological and, hence, independent of the evidence, there has been, from the very beginning, a small group of people interested in the factual evidence; this group has included both advocates and scholars. The eagerness with which clues were pursued means that entirely new evidence will probably not be found; there are, however, new analyses of existing evidence. Even though seventy years have passed, the divisions stirred by the case continue to inflame the passions of many persons.

In recent years, the defense of Sacco and Vanzetti has been championed by William Young and David Kaiser. Their book, *Postmortem: New Evidence in the Case of Sacco and Vanzetti* (1985), concludes that "the overwhelming probability is that... Sacco and Vanzetti were completely innocent of the South Braintree murders" (p. 164).

The other side of the argument has been carried in recent years by Francis Russell and James E. Starrs. Russell, an historical writer, wrote *Tragedy at Dedham* in 1962 and *Sacco and Vanzetti, the Case Resolved* in 1986. His conclusion is that Sacco was guilty and Vanzetti an accessory after the fact. James Starrs, Professor of Law and Forensic Sciences at George Washington University School of Law, has written a series of articles *Once More unto the Breech: The Firearms Evidence in the Sacco and Vanzetti Case Revisited* (1986). His conclusions are supportive of those of Russell.

It is important to focus the issue between these two sets of scholars. Since some of the evidence might be used in an argument about the fairness of the trial as well as in an argument about the guilt of the defendants, we have to exercise care in separating these two issues. It is not contradictory to believe that Vanzetti was not a participant in the crime and yet believe that he received a fair trial. The evidence was known at the time of trial to be weaker against Vanzetti than against Sacco; Vanzetti, however, declined the invitation of his attorney to plead Vanzetti's innocence and sacrifice Sacco (Russell, 1962, p.466). The issue in our studies concerns the guilt or innocence of Sacco and Vanzetti; fairness of the trial we take to be a separate issue.

By the time this research began, William Young had died. David Kaiser and Francis Russell had agreed to participate in our studies and to provide probability assessments. Before these assessments could be made, Francis Russell died; Professor James Starrs agreed to take Russell's place in our studies. However, before any probability elicitation is possible, we have to identify the events whose probabilities are subject to dispute. The Sacco and Vanzetti case, like many others, produced a true mass of details that have to be marshalled or organized in order to understand arguments made by both sides regarding the major facts in issue in this case; the next section describes how we attempted to make sense of this mass of details and how probability elicitation will take place in our studies.

3. STRUCTURING ARGUMENT FROM A MASS OF EVIDENCE

The entire transcript of the Sacco and Vanzetti trial together with subsequent proceedings has been published in several volumes (Henry Holt & Co. 1928). As this record shows, many defense and prosecution witnesses gave testimony; in addition, a variety of "real" or tangible evidence items were introduced at trial. If someone took the trouble to parse all of the testimony given in this trial as well as the accounts given of the tangible evidence, the individual details would surely number at least in the hundreds of thousands. So, how does any person begin to make sense out of masses of evidence such as that adduced in this trial ? Not all of these details are of equal importance as far as our present purposes are concerned; as in many trials, there is considerable irrelevancy in recorded evidence. Our time constraints did not permit us to parse all of the evidence given in this trial. Our first step in making this analysis tractable was to base our work on a summary of the trial record; this summary appears in Russell's work: *Tragedy in Dedham* (1962). Russell provides an account of the substance of all major items of real and testimonial evidence given during the trial and in the order in which this evidence was introduced. In short, Russell's work provides us with a "relevance filter". Now it is true, of course, that Russell's summary might not meet universal approval. With this in mind, we also included additional details of the trial evidence as they appear in the works of Young and Kaiser (1985) and Starrs (1986).

3.1. *Wigmore and the Structuring of Argument*

But, even upon examining these summaries of the trial evidence we are still left with a large mass of details, consideration of which surely influences any assessment of the probative value of the evidence for or against Sacco and Vanzetti. We still faced the question: how does anyone make sense out of all of this evidence? This question is very similar to the one asked years ago by the American jurist John H. Wigmore (1913, 1937). Wigmore considered the plight of an advocate preparing for trial and confronted with a mass of evidence on the basis of which persuasive arguments, bearing upon the major facts in issue in the case at hand, have to be constructed. What he did was to develop a method for the analysis and synthesis of a mass of evidence. The analytic part consists of a key list containing items of evidence, points or elements in the case to which the evidence is relevant, and intermediate reasoning stages that show possible sources of doubt in reasoning from the evidence to the major points or elements of the case being presented. The synthetic part consists of a chart showing the network of reasoning linkages involving items of evidence directly relevant on major points or elements of the case being presented; in addition, this chart shows the inferential bearing of ancillary evidence on various stages of reasoning from directly relevant evidence to major elements in the case.

We note, parenthetically, that Wigmore, himself, was a participant in the debates in the 1920s about whether or not Sacco and Vanzetti received a fair trial. As Russell records (1962, 371-372), Wigmore attempted to vindicate Massachusetts justice by attacking the position advocated by Felix Frankfurter that the trial was unfair. Wigmore and Frankfurter sparred about this matter in several printed articles.

Wigmore's method for charting complex arguments from a mass of evidence never achieved any popularity among practicing attorneys, the audience for whom they were originally intended; in fact, as William Twining records (1985, 164-166), these methods "went over like a lead balloon" in the legal profession. Part of the reason involves considerable tedium in applying the variety of quaint symbols Wigmore used to indicate various nodes and arcs on his evidence charts. In a recent paper, Tillers and Schum (1988) showed how Wigmore's evidence charts, first displayed in 1913, are directed acyclic graphs having fuzzy probabilities on their arcs. Within the field of evidence scholarship in law there are two contemporary advocates of Wigmore's methods for charting complex argument; they are Terry Anderson (Law School, University Of Miami, Florida) and William Twining (School Of Law, University College, London). In a recent work (Anderson & Twining, 1991) the many virtues of modified forms of Wigmorean analysis and synthesis are described; additional comments on Wigmore's methods are to be found in Twining (1985, 1990). In an effort to make Wigmore's methods more "user-friendly", a computer-based system for implementing Wigmore's methods has been constructed (Schum & Tillers, 1990). We have made use of this system in our Wigmorean analysis of the evidence in the Sacco and Vanzetti case.

There are several reasons why we have made use of Wigmore's methods as a basis for our present work. These methods simply allow us to lay bare for subsequent debate what we believe to be the essential stages in arguments from evidence to major hypotheses or facts in issue. Each stage in an argument we identify exposes a source of uncertainty we have to acknowledge in assessing the probative force of evidence. However, some persons reject any form of decomposition of arguments based upon evidence given at trial (see Twining, 1990 for a discussion of "atomism" vs "holism" in the evaluation of trial evidence). Wigmore's "atomistic" methods are entirely congenial to the decomposed Bayesian analyses we have in mind. Indeed, Wigmore's methods allow us to illustrate some very important structural and behavioral matters that arise in the analysis of complex probabilistic arguments.

As Wigmore himself noted (1937), any charting of an argument represents just the opinions of the person constructing the chart. There are no unique or normatively correct arguments; two equally-knowledgeable persons may justifiably perceive different reasoning routes from the same evidence to the same hypotheses (they may, of course, also discern different hypotheses that may be suggested by this evidence). Even earlier than Wigmore, John Venn (1907) noted that the number and the labeling of intermediate links in a chain of reasoning are always arbitrary to some extent. Thus, as we consider the arguments we have constructed (on behalf of the prosecution and defense) from the evidence in the Sacco and Vanzetti trial you may be able to see additional or different links in the chains of reasoning we have identified. Professor David Kaiser, whose knowledge of this case is very extensive, has been most helpful to us in checking the reasonableness of the arguments we have constructed so far.

3.2. *Example Key Lists and Charts*

In discussing our charting of the arguments the prosecution and defense seem to have been making, based upon the evidence they produced, we first have to identify the nature of the charge against Sacco and Vanzetti and then to show what the prosecution was obliged to prove, beyond reasonable doubt, in order to sustain this charge at trial. Sacco was charged with first-degree murder in the death of Alessandro Berardelli (one of the payroll guards) and Vanzetti was charged with being an accomplice to this murder; both charges, if sustained, carried a mandatory death penalty under Massachusetts law at the time. Ordinarily, a charge of first degree murder requires the prosecution to prove four essential points: (i) the victim was a human person, (ii) the defendant killed the victim, (iii) the killing was intentional, and (iv) this intention was fashioned beforehand (i.e., the killing was premeditated). But, under Massachusetts law in 1921 a first-degree murder charge could be sustained, without the premeditation element, if a murder was committed during the act of a felony which carried a life sentence as a penalty.

The prosecution alleged that Sacco and his accomplice Vanzetti, together with three other unnamed persons, murdered Berardelli during robbery of the payroll Berardelli and another guard were carrying at the time; the penalty for armed robbery under Massachusetts law at the time was life imprisonment. Thus, Sacco and Vanzetti were charged with first-degree murder. So, the major fact in issue in this case, or ultimate probandum (as Wigmore termed it) can be stated as:

H: Sacco, with the assistance of Vanzetti (and three other persons) committed first-degree murder in the death of Allesandro Berardelli on 15 April, 1920 in South Braintree, Massachusetts.

In keeping with Massachusetts law in 1921, this ultimate probandum may be parsed into three penultimate probanda as follows:

P(1): Allesandro Berardelli (a human person) died of gun shot wounds he received on 15 April, 1920 in South Braintree, Massachusetts

P(2): At the time he was shot, Berardelli (in company with another guard named Parmenter) was in possession of a payroll belonging to the Slater Morrill Co.

P(3): It was Nicola Sacco who, with the assistance of Bartolomeo Vanzetti (and others), intentionally took the life of Berardelli during a robbery of the payroll Parmenter and Berardelli were carrying.

At no time has there ever been any serious doubt that Berardelli was killed and that he was killed during a robbery of a payroll he was carrying at the time of his death. The major issue in this case has always been one of identity: were Sacco and Vanzetti involved in this crime? So, P(3) , as stated above, was the crucial element in the prosection's case.

Our present key-list contains hundreds of items and our present chart, containing 19 sectors or sections, measures 8 feet in length; this key-list and chart contains evidence on all three of the penultimate probanda listed above. Discussion of this entire chart would not serve our present purpose, which is to show how we intend to use Bayesian methods as a vehicle in attempts to settle disputes among experts who have different views about the verdict in this case. For this purpose we shall present just three sectors of our chart and their accompanying key-lists; each of these sectors and key-lists concerns penultimate probandum P(3).

The charts you see in Figures 1, 2, and 3 are much-simplified renderings of the charts Wigmore constructed; if you are curious to see what his charts looked like see Wigmore (1937) or Tillers and Schum (1988). Tables 1, 2, and 3 contain the key-lists that identify each symbol on the charts. To help you make sense of these charts we first need to distinguish between two important forms of evidence. The first form, which we label as directly relevant, is evidence that sets up a possible chain of reasoning to one of the penultimate probanda in this case. Directly relevant evidence can, of course, be given by either prosecution or defense. In the three figures the filled circles indicate directly relevant evidence given by the prosecution and the filled diamonds represent directly relevant evidence given by the defense.

The other form of evidence we describe as being indirectly relevant or ancillary evidence in the following sense; it is evidence that bears upon either the strength or the weakness of any link in a chain of reasoning set up by directly relevant evidence. As we shall discuss in Section 3.3, it is this ancillary evidence that allows us to make judgments of linkage probabilities; we have some interesting problems to consider when there is an absence of ancillary evidence at some link in a reasoning chain. In some situations, ancillary evidence may be evidence about other evidence. For example, one witness asserts that he saw defendant at the scene of a crime; this is directly relevant evidence since it sets up a chain of reasoning to a penultimate probandum. Then comes another witness who asserts that the first witness was not, at the time of his alleged observation, wearing the strongly corrective lenses he customarily wears. This evidence is indirectly relevant in the sense that it bears upon the credibility of the first witness. In our figures, ancillary evidence given by the prosecution is indicated by cross-hatched squares; ancillary evidence given by the defense is indicated by filled triangles. As we noted earlier, both testimonial and "real" or tangible evidence was introduced; we make no symbolic distinction between these two types of evidence.

In setting up a reasoning chain from an item of evidence to some penultimate probandum we specify certain propositions or events that are hypothetical in nature; these hypothetical events interposed between evidence and penultimate probanda Wigmore referred to as interim probanda. Such interim probanda may follow either directly relevant or ancillary evidence. For the prosecution, open circles represent interim probanda from directly relevant evidence and open boxes represent interim probanda from ancillary evidence. For the defense, open diamonds represent interim probanda for directly relevant evidence and open triangles represent interim probanda for ancillary evidence. It is by means of these assorted hypotheticals that we construct reasoning linkages based upon the evidence we have. In Section 3.3 below we shall have more to say about these linkages since some we have shown in these figures raise some interesting probabilistic issues in the analysis of inferential networks.

Here is one example of a reasoning chain we have constructed; it involves prosecution testimony from a police officer named Spears, who asserted at trial that Sacco had a 32-caliber Colt on him when he was arrested on 5 May, 1920 (his testimony is indicated by the filled circle numbered 61 in Figure 1). First, just because Spear testified to this event (or compound event, if you like) does not mean that this event actually happened; its actual occurrence (hypothesized) is indicated by open circle number 60. Now, even if Sacco had this weapon in his possession on 5 May, 1920, this would be just circumstantial evidence that one of the robbers had this same weapon during the crime (open circle number 62). In turn, this weapon being in the possession of one of the robbers during the crime would be just circumstantial evidence that this weapon was fired during the crime (open circle number 63). Now, one of the bullets extracted from the body of Berardelli was a 32-caliber Winchester (it was labeled Bullet III). Even if Sacco's weapon was fired during time of the crime, this is just circumstantial evidence that a Winchester bullet was fired from it (this event is numbered 66). Then, even if a Winchester bullet was fired from Sacco's weapon during the crime, this is just circumstantial evidence that the Winchester bullet extracted from Berardelli was the one that came from Sacco's 32 Colt (open circle 59). Finally, even if this Winchester bullet came from Sacco's colt, this would be just circumstantial evidence that Sacco was the

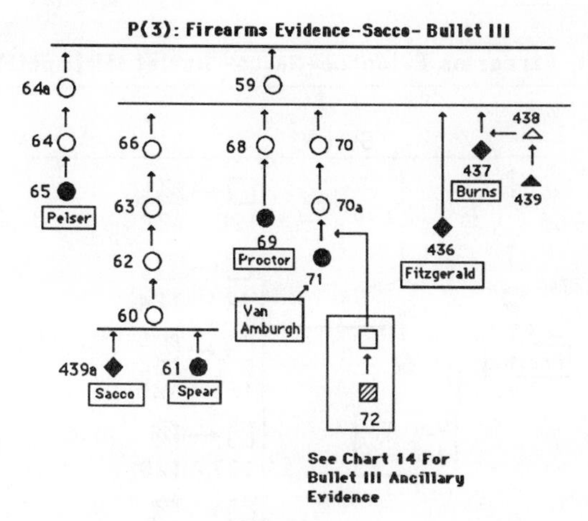

Figure 1.

P(3). Nicola Sacco, with the assistance of Bartolomeo Vanzetti [and others], intentionally took the life of Alessandro Berardelli during a robbery of the payroll Parmenter And Berardelli were carrying [i.e., Sacco was the one who pulled the trigger on the 32 Cal. Colt].

PROSECUTION:

59). Bullet III, a Winchester, came from Sacco's 32-Colt.
60). Sacco had a 32-Colt when arrested, 5 May 1920.
61). M. Spear testimony to 60).
62). Sacco had this Colt at the time of the crime.
63). Sacco's 32- Colt was fired during the crime.
64). Sacco fired a weapon at the scene of the crime.
65). Pelser testimony to 64).
66). A Winchester bullet was fired from Sacco's Colt during the crime.
67). This item deleted on correction.
68). Bullet III came from SOMEONE'S Colt 32.
69). Proctor testimony to 68).
70). Van Amburgh did believe that Bullet III came from Sacco's Colt.
70a). Van Amburgh was inclined to believe Bullet III came from Sacco's Colt.
71). Van Amburgh testimony to 70a).
72). Ancillary ballistics evidence Bullet III: see Chart 14.
73). Winchester Shell W, found at the scene, came from Sacco's Colt.

DEFENSE:

436). Fitzgerald testimony that Bullet III did not come from Sacco's Colt.
437). Burns' testimony that Bullet III did not come from Sacco's Colt.
438). Bullet II could have come from a Colt or a Bayard.
439). Burn's testimony on X-exam.
439a) Sacco's admission to 60 during the trial.

Table 1.

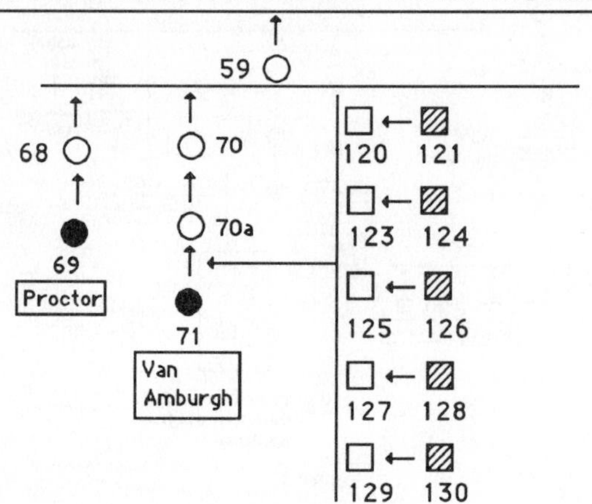

Figure 2.

PROSECUTION:

P(3). Nicola Sacco, with the assistance of Bartolomeo Vanzetti [and others], intentionally took the life of Alessandro Berardelli during a robbery of the payroll Parmenter And Berardelli were carrying.

59). Bullet III, a Winchester, came from Sacco's 32-Colt.

68). Bullet III came from SOMEONE'S Colt 32.

69). Proctor testimony to 68).

70). Van Amburgh did believe that Bullet III came from Sacco's Colt.

70a). Van Amburgh was inclined to believe that Bullet III came from Sacco's Colt

71). Van Amburgh testimony to 70a).

120) Bullet III was a Winchester.

121). Real evidence of a "W" stamped on Bullet III canalure.

122). Removed during revision

123). Bullet III had six lands and grooves.

124). All expert witness agreed to 123)

125). Bullet III had a left-twist.

126). Real evidence of 125) ?

127). Only Colts have rifling characteristics such as 123) and 125).

128). Evidence of 127 ?

129). Bullet III had similar lands/grooves to bullets test fired through Sacco's Colt.

130). Van A. testimony to 129).

DEFENSE:

The only defense counterevidence we find is shown on Chart 13.

Table 2.

one who pulled the trigger (this is essentially what P(3) says); someone else may have used Sacco's weapon during the crime.

Charts 13 and 14 shown in Figures 1 and 2 are especially important in the analysis of this case. The reason is that there is new and quite controversial evidence concerning Bullet III; indeed, this new evidence forms perhaps the major element in the current dispute between Young & Kaiser (1985) and Starrs (1986). Briefly, Young and Kaiser offer evidence suggesting that a bullet known to have come from Sacco's Colt during a test-fire was substituted by the prosecution for one of the bullets extracted from Berardelli's body the day of the crime; i.e., it is Young and Kaiser's belief that Sacco and Vanzetti were framed for the murder of Berardelli. Starrs, offering other evidence, disagrees. Our further charting will involve constructing revised as well as additional arguments based upon the new evidence these scholars have uncovered.

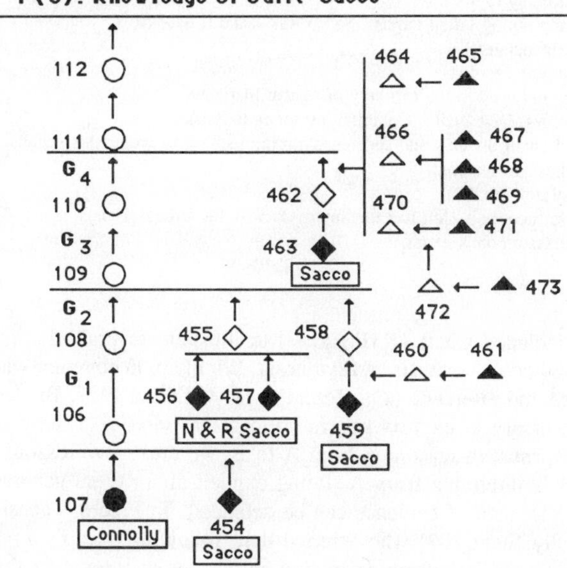

Figure 3.

P(3). Nicola Sacco, with the assistance of Bartolomeo Vanzetti [and others], intentionally took the life of Alessandro Berardelli during a robbery of the payroll Parmenter And Berardelli were carrying.

PROSECUTION:

106). Following his arrest on 5 May, 1920, Sacco made a threatening motion as if to draw a revolver that was hidden under his coat.
107). Connolly testimony to 106).
108). Sacco intended to draw his revolver.
G1: A person who makes a threat often intends to carry it out.
109). Sacco intended to use this weapon on the arresting officers.
G2: A person who draws a weapon often intends to use it
110). Sacco intended to escape from arrest.

G3: A person arrested will sometimes shoot arresting officers in order to escape.
111). Sacco knew he had committed a crime.
G4: Persons who escape from arresting officers often do so because of their knowledge of having committed a crime.
112). Sacco knew he had committed murder during the South Braintree robbery.

DEFENSE:
454). Sacco denial of 106).
455). Sacco said wife found revolver in a drawer and that he intended to shoot rabbits on the day of his arrest.
456). Sacco testimony to 455).
457). R. Sacco testimony to 455
458). Sacco carried a weapon because of his duties as a night-watchman.
459). Sacco testimony to 458).
460). Sacco was not a night-watchman in 1920.
461). Sacco admission on X-exam.
462). Sacco believed he was being arrested because of his political beliefs.
463). Sacco testimony to 462).
464). Immediately during initial inquiry, Sacco was asked if he were a radical or a socialist.
465). Sacco testimony to 464).
466). Sacco was involved in the recovery of radical literature.
467) - 469). Nelles, Delesandro, Columbo testimony to 466).
470). Sacco lied about his Colt and cartridges, during inquiry, to protect his friends in the radical movement.
471). Sacco testimony to 470).
472). Sacco's lies about his Colt had nothing to do with his radical friends.
473). Sacco admission on X-exam.

Table 3.

We have included Chart 18 (Figure 3) to illustrate several additional points about our Wigmorean methods of analysis and synthesis. Wigmore, in common with more recent scholars of evidence and inference (e.g. Toulmin, 1964; Cohen,1977; Binder & Bergman, 1984) believed it necessary to express generalizations that serve to license or warrant reasoning steps or stages; thus, in reasoning from A to B, we must give reasons why anyone should believe that B is inferrable from A. Being explicit about these generalizations is one way in which the relevance of evidence can be defended. In Figure 3 consider the testimony of Officer Connolly (item 107) who asserted that, following Sacco's arrest on the night of 5 May, 1920, he made a threatening motion as if to draw a revolver that was hidden under his coat. This evidence seems relevant to P(3), but precisely how ? The seven-stage chain of reasoning we have constructed on the basis of evidence 107 is just one way in which the relevance of Connolly's evidence might be defended. Consider item 106, that Sacco did, in fact, make this threatening motion as Connolly claims; what can be inferred from this interim probandum ? One possible generalization goes something like generalization G1 in the key-list in Table 3. This generalization asserts: "a person who makes a threat often carries it out". Anyone who agrees with this commonsense generalization is then entitled to infer, with some strength, that Sacco intended to draw his revolver (item 108). Other generalizations in the key list in Table 3 similarly license the further reasoning stages shown in Figure 3. In instances in which we believed there might be some disagreement about our chains of reasoning, we included specific generalizations such as the ones shown in the argument we constructed from Connolly's testimony.

Another reason for our compulsiveness in employing Wigmore's methods is that it helps us to find a specific place for every item of evidence we know about. One matter Wigmore repeatedly emphasized was that we have to account for all of the evidence we have. On our construal, Connolly's testimony (107) sets up a seven-stage reasoning chain or cascaded inference to P(3). In this cascaded inference or "inference-upon-inference", as it is called in the legal literature, we have at least six intermediate sources of uncertainty. One might think, as a result of so many sources of uncertainty, that Connolly's testimony might have little probative value. But, apparently, the defense took some pains to counter Connolly's testimony; they obviously thought it was damaging. Sacco took the stand in his own defense (so did Vanzetti). In rebuttal to Connolly's testimony, Sacco and his wife Rosina gave testimony that counters the interim stages 106, 109, and 111 in the argument we constructed, based upon Connolly's testimony. This is just one example of the means by which we have found a place for all of the evidence we know about that was given in this trial.

Finally, consider Sacco's testimony (item 463 in Figure 3) that he believed he was arrested because of his political beliefs. What Figure 3 shows is another inferential network involving just ancillary evidence introduced by the defense to support Sacco's assertion (items 465, 467, 468, 469, 471). But it also contains ancillary evidence given by Sacco on cross examination by the prosecution (473). We shall return to this interesting situation in 3.3 since it suggests that we may have to consider inferential networks that are embedded in other inferential networks.

3.3. *Probabilistic Issues*

We have not yet mentioned the directed arcs (indicated by arrows) in any of these charts. But even a casual observation of them shows that the direction of inference these charts describe is from the evidence to the penultimate probanda (the same is true in all other charts we have not shown). Thus, the reasoning depicted in these charts is "bottom up", inductive, or probablistic in nature. Construed in terms of odds and likelihood ratio, a Bayesian formalization for an inference about ultimate probandum H, based upon the entire mass of evidence E, can be expressed as:

$$P(H|E)/P(H^c|E) = [P(H)/P(H^c)][P(E|H)/P(E|H^c)],$$

where H^c is the event that ultimate probandum H is not true. The posterior odds $P(H|E)/P(H^c|E)$ is determined by the product of prior odds $[P(H)/P(H^c)]$ and the likelihood ratio for the mass of evidence $[P(E|H)/P(E|H^c)]$. This version of Bayes' rule applied to this case is an undecomposed or holistic representation since it involves the ultimate probandum P and the entire mass of evidence E. But we have decomposed H into three penultimate probanda (P(1), P(2), and P(3)), and we have decomposed the mass of evidence E into individual items of directly and indirectly relevant (ancillary) evidence and have, by means of our charting exercise, attempted to show the reasoning stages linking these individual items of evidence to the penultimate probanda and to each other. So, a Bayesian expression that takes account of all of this decomposition would be very complex indeed and would involve some very difficult assessments of possible conditional nonindependence among the evidence items. But, in our present work we shall not have to make this full-scale decomposition, at least not with respect to all of the evidence in this case. The reasons for this we shall explore in more detail in Section 3.4 below.

We do need to give further attention to prior odds $[P(H)/P(H^c)]$ and to the likelihood ratio $[P(E|H)/P(E|H^c)]$. First, our experts will make various probabilistic assessments that are required in the Bayesian analyses we contemplate. But there is one form of assessment that logic and intuition suggests experts cannot make in any believable way; such assessment involves the prior probability $P(H)$. Our experts have already thoroughly studied this case and its aftermath; they have intimate knowledge of the behavior and background of the defendants, and they also have knowledge of new evidence discovered since this case was tried. Thus, unless we were assured that they could become suddenly amnesic to all of this knowledge, it makes no sense to have them assess priors involving H or any of the penultimate probanda. What we are left with is their assessments of the probative force or value of the "old" and the "new" evidence they have considered; this involves their assessments of likelihoods and likelihood ratios.

Having just said, in effect, that our experts cannot assess "virgin" priors because of their knowledge of the case and the actors in it, we should also acknowledge that there was, at the time the case was tried, some basis for a prior judgement about the involvement of Sacco and Vanzetti in the crime that took place in South Braintree. Both Sacco and Vanzetti were involved in various anarchistic activities at the time including the distribution of literature. And, indeed, there were various acts of violence that were associated with anarchists in the Boston area. But, none of these activities ever involved the crime of robbery. Even some police officials at the time believed that the wrong men were being tried.

Further consideration of our evidence charts is informative about the ways in which our experts' probability assessments will be supported; there is some potential controversy here. Of course it is true that there are no possible relative frequencies to support any probability assessment made by any person having knowledge of this case. The events appearing throughout our evidence charts are singular or unique; they either happened on one occasion or they did not. We cannot play the world over and over again to chart their relative frequency of occurrence. So, upon what basis do our experts make the various probabilistic judgments we shall require of them ? One basis is provided by the various items of ancillary evidence given at trial and which we have charted; here is an example.

To set the stage for this example, consider Chart 18 in Figure 3 which contains stages in what we believe the prosecution must have been arguing based upon Connolly's testimony that Sacco went for his revolver following his arrest on 5 May 1920. This testimony refers to events which, if true, occurred considerably after the crime in South Braintree on April 15, 1920. Thus, to use Wigmore's term, this evidence is retrospectant in nature, one form of which concerns physical or mental traces left behind after a crime has been committed. The "trace" evidence of concern here refers to the "mental traces" involving Sacco's alleged knowledge of his guilt in the slaying of Berardelli; i.e., Sacco went for his revolver because he knew he shot Berardelli and wanted to avoid being arrested. As Figure 3 shows, the defense attempted to counter this argument in several ways, the first of which involves Sacco's direct denial (item 454) that he ever went for his revolver as Connolly alleged. Second, Sacco and his wife Rosina testfied that Sacco had the revolver on him on 5 May because he had intended to shoot rabbits that day. This seems to be an attempt to counter element 109 of the prosecution's argument; i.e., Sacco never had any intention of using this weapon on the arresting officers. In addition, Sacco further testified that he carried a revolver because of his duties as a night watchman (however, he admitted on cross-examination that he was not employed as a night watchman in 1920).

Now consider element 111 of the prosecution's argument; i.e., Sacco knew he had committed a crime. To counter this stage of the argument concerning Sacco's knowledge of guilt, Sacco testified (element 463) that he believed he was being arrested because of his political beliefs (and not because he had committed any crime). To support the counterargument from 462 against 111, the defense introduced five items of ancillary evidence (items 465, 467, 468, 469, and 471). This ancillary evidence is very interesting since it concerns

Sacco's involvement in the distribution of radical literature, questions immediately following his arrest about his involvement, and an explanation of why Sacco had lied about reasons for his having been armed at the time of his arrest. In addition, on cross-examination, Sacco admitted (element 473) that his lies about the weapon he was carrying had nothing to do with his fellow radicals. Here, finally, comes a probabilistic issue.

One way of looking at the defense counterargument involving prosecution element 111 is to say that they attempted to make element 462 appear improbable if element 111 were true and very probable, if element 111 were not true. In likelihood terms, suppose E = Sacco knew he had committed a crime and E^c = Sacco knew he did not commit a crime. In addition, let F = Sacco believed he was being arrested because of his political beliefs. Sacco never denied that he was associated with radical political activities. In probabilistic terms the defense argued that $P(F|E^c) >> P(F|E)$. To support such an assessment the ancillary evidence noted above was introduced to make element 462 disfavor element 111. The trouble, of course, is that not all of it favors the assessment: $P(F|E^c) >> P(F|E)$. On cross-examination, Sacco admitted that his lies about the Colt in his possession at the time of his arrest had nothing to do with his radical friends.

Now comes a point about which there may be controversy; how does one support non-frequentistic conditional probabilistic assessments, such as the ones mentioned above, in the absence of specific ancillary evidence ? For example, take the probabilistic linkage between Connolly's testimony (element 107) and the event to which he testifies (element 106); this linkage concerns Connolly's credibility. Upon what basis would any of our experts assess this probability in absence of specific ancillary evidence concerning Connolly's credibility ? None of our experts were at the trial itself and could not, for example, observe Connolly's demeanor and bearing while giving this testimony; this is just one conventional ground for testing the credibility of a witness. But we do have another ground for challenging Connolly's credibility; this comes in the form of Sacco's giving contradictory testimony (element 454). But, of course, we do not have specific credibility-related ancillary evidence about Sacco either as far as this item of testimony is concerned. In various instances which we shall discuss in Section 3.4, when specific conditional probability assessments are required, we shall have to rely upon our experts' stock of knowledge about this case and the actors in it that has come from their own study of this case. As we noted at the outset, our charting of the evidence in this case only involves what Russell, at least, took to be the major directly relevant and ancillary evidence.

A final probabilistic matter concerns taking our Wigmore charts to be "Bayesian inferential networks". In graph-theoretic analyses of inferential networks arcs connect one node to another. But an examination of any of our charts shows that, on occasion, we have one arc connected to another arc; this bears explanation as far as any probabilistic analysis is concerned. Again please consider Chart 18 in Figure 3 and note, for example, that there is an arc connecting the ancillary evidence to the arc connecting elements (nodes) 462 and 111. Such "arc-to-arc" linkages do have a very plausible interpretation. The role of ancillary evidence is to enable a person to assess the strength or weakness of an arc linking one stage of a reasoning chain to another; i.e., ancillary evidence allows us to assign probabilities on arcs set up by directly relevant evidence. So, in our present analysis what we have, in fact, are inferential networks embedded in other inferential networks. It could be argued that there is, in fact, just one "super" network and that arcs from ancillary evidence could be directly connected to arcs involving directly relevant evidence, if appropriate additional nodes could be identified. From our experience in such matters, these additional nodes are often quite difficult to identify in any plausible way.

3.4. *Elicitation Issues: Levels of Decomposition*

One matter is quite often overlooked when one contemplates applying a formal analysis to some actual problem; this matter involves the manner of elicitation of the subjective probabilities that such analyses (such as ours) often require. Let us first consider the three penultimate probanda in this case:

P(1): Allesandro Berardelli (a human person) died of gun shot wounds he received on 15 April, 1920 in South Braintree, Massachusetts

P(2): At the time he was shot, Berardelli (in company with another guard named Parmenter) was in possession of a payroll belonging to the Slater Morrill Co.

P(3): It was Nicola Sacco who, with the assistance of Bartolomeo Vanzetti (and others), intentionally took the life of Berardelli during a robbery of the payroll Parmenter and Berardelli were carrying.

As we noted earlier, there was just one issue in this case about which there was any dispute, namely, whether or not P(3) is true. No one disputed the fact that Berardelli was dead and that he was killed during a robbery of the payroll he and Parmenter were delivering at the time. Of the 18 charts in our Wigmorean analysis of this case, 16 concern the evidence bearing on P(3). Since nothing in the published works of our experts suggests that they disagree about the evidence bearing upon P(1) and P(2), we focus our attention mainly upon P(3). But there are some comparisons we could make that involve P(1) and P(2); these comparisons do require at least gross or undecomposed probability judgments (discussed later) on these two penultimate probanda.

Let us now consider possible levels of decomposition of our experts' probability assessment task. In keeping with our previous experience in obtaining probability assessments on masses of evidence in cascaded inference (Schum & Martin, 1982), we shall identify several basic levels of probability assessment task decomposition:

(i) Zero Task Decomposition (ZTD).
One thing we could do is simply to have our experts assess the likelihood ratio on H for the entire mass of evidence E; i.e., we have them assess the ratio: $[P(E|H)/P(E|H^c)]$. But, as we just noted, our major interest involves P(3) and not H. Suppose, instead, that we consider just that subset of evidence that bears upon P(3) and neither of the other two penultimate probanda; call this subset E_3. We could have the experts judge the likelihood ratio: $\Lambda E_3 = P[E_3|\ \text{P(3)}]/P[E_3|\ \text{P(3)}^c]$, where P(3)^c means that P(3) is not true. Such a judged ratio simply indicates how strongly a person believes the E_3 evidence, in the aggregate, favors P(3) over its complement. This is a very gross or holistic judgment since it focuses upon no specific item or subset of the evidence or upon no specific argument from any item or subset of the evidence. In addition, it does not consider the necessary distinction between directly relevant and ancillary evidence.

(ii) Partial Task Decomposition-Level 1 (PTD-1)
We could decompose the experts' assessment tasks in a number of ways, one of which involves sorting the evidence out in terms of various inferential issues relating to P(3). For example, again consider Chart 18 in Figure 3; this chart shows all of the evidence introduced about Sacco's knowledge of guilt (or lack thereof). Let E_{18} = the entire collection of evidence regarding Sacco's knowedge of guilt. We could have our experts assess values such as $\Lambda(E_{18}) = P[E_{18}|\text{P(3)}]/P[E_{18}|\text{P(3)}^c]$. We would require 16 such assessments, one for each collection of evidence we have charted on P(3). Such assessments are still holistic since they are made on the basis of an aggregate of items of directly relevant and ancillary evidence.

(iii) Partial Task Decomposition-Level 2 (PTD-2)
A further level of task decomposition considers individual items of directly relevant evidence. In Figure 3, Chart 18, consider Connolly's testimony and call it E_i^*. We could have experts assess likelihood ratios such as $\Lambda(E_i^*) = P[E_i^*|\text{P(3)}]/P[E_i^*|\text{P(3)}^c]$ for each item of directly relevant evidence. For our present charting of the evidence in this case, there are 77 items of directly relevant testimony and so we would have 77 assessments of $\Lambda(E_i^*)$ for each expert. This level of assessment is still holistic but obviously less so than the ZTD or PTD-1 assessments. These PTD-2 assessments are based upon aggregations of one item of directly relevant evidence together with all ancillary evidence given to support or weaken the chain of reasoning this item suggests. In addition, of course, this level of assessment does not even consider any steps or stages in an argument linking E_i^* with P(3). As we shall note again in Section 3.5, this level of decomposition, in common with ZTD, and PTD-1, does not focus upon any specific conditional nonindependencies that may be identified.

(iv) Partial Decomposition-Level 3 (PTD-3)
Yet another level of partial task decomposition arises in which we attempt to alert the experts to various conditional nonindependencies that can be identified. As an example, consider the testimony of Spear and of Pelser in Chart 13, Figure 1. Leaving aside the possibility that these two prosecution witnesses may have testified nonindependently, there is still some conditional nonindependence lurking in their joint testimony bearing (very indirectly) upon whether or not Bullet III, a Winchester, came from Sacco's Colt. Pelser testified (item 65) that Sacco fired a weapon at the scene of the crime; Spear testified (item 61) that Sacco had a 32-caliber Colt on him when he was arrested on 5 May 1920. Let E_{65}^* = Pelser's testimony and let E_{61}^* = Spear's testimony. Suppose someone asked you to assess $\Lambda(E_{61}^*) = P[E_{61}^*|\text{P(3)}]/P[E_{61}^*|\text{P(3)}^c]$. Very likely, your assessment might depend rather strongly upon whether or not you took into account Pelser's testimony; you might easily believe that Sacco's having a Colt on him when arrested is more probable, given P(3), if you also knew that Sacco fired a weapon at the scene of the crime. So, an assessment that takes this possibility into account is one that can be represented as $\Lambda(E_{61}^*|E_{65}^*) = P[E_{61}^*|\text{P(3)}\&E_{65}^*]/P[E_{61}^*|\text{P(3)}^c\&E_{65}^*]$. The conditional nonindependence here is reflected by the distinction between $\Lambda(E_{61}^*)$ and $\Lambda(E_{61}^*|E_{65}^*)$. So, at this level of decomposition, the probative force of individual items of directly relevant testimony are assessed where experts are instructed to make these assessments in light of other testimony involving possible nonindependencies that we could identify.

(v) Molecular Decomposition (MD)
At the molecular level, probabilistic assessments are made which incorporate individual links in chains of reasoning and ancillary evidence that supports or weakens these linkages. Another very important thing happens at this level as well; it is at this level of decomposition that specific conditional probabilities, and not simply their ratios, are required. This fact has recently provoked a good-natured argument about the likelihood principle in Bayesian analyses (Edwards, Schum, & Winkler, 1990). To illustrate MD in a simple case, consider Burns' testimony as shown in Chart 13, Figure 1. Burns', a defense expert witness, testified that Bullet III did not come from Sacco's Colt, as element 59 in the prosecution argument specifies. But, even if Bullet III did come from Sacco's Colt, this is just circumstantial evidence that Sacco himself pulled the trigger on this weapon in South Braintree on 15 April 1920; which is essentially what P(3) states. To assess the value of Burns' testimony on P(3) in Bayesian terms we need the

following conditional probabilities:

(1) The probability of Burns' testimony if 59 is true, possibly conditional upon P(3) or $P(3)^c$,

(2) The probability of Burns' testimony if 59 is not true, possibly conditional upon P(3) or $P(3)^c$,

(3) The probability of 59, if P(3) is true, and

(4) The probability of 59, if P(3) is not true.

Notice in Figure 1 that we have at least some ancillary evidence to support assessments (1) and (2).

It can easily be shown that, for this two-stage inference, we need these four conditional probabilities unless we believe that Burns' testimony depends only upon 59 being true or not, and not upon P(3) being true or not. Stated in other words, we believe that Burns' credibility is not conditional upon P(3) or $P(3)^c$ being true. In this case we need only obtain judgments of the ratio of (1) to (2). This fact about the way Bayes' rule operates has great behavioral significance as discussed in Edwards, Schum and Winkler (1990). People naturally find relative or ratio judgments easier to make than absolute judgments; unfortunately, in applications of Bayes' rule to cascaded inference we are allowed to make ratio judgments only at bottom stages of reasoning chains in which it is also the case that the event at the bottom stage (i.e., the evidence) is only conditional upon events at the next higher stage and upon none other. In all other cases we need assessments of individual conditional probabilities. Incidently, it is by means of these individual conditional probabilities that Bayes' rule accomodates to the rareness (or lack thereof) of events in reasoning chains. A major feature of Bayesian analyses of the probative force of evidence is that it responds both to ratios and to differences among its ingredients.

At this molecular level of decomposition the burden of assessment is enormous indeed, even for fairly simple inference problems. But, for a problem involving a mass of evidence such as we are now considering, this burden is unthinkable. We cannot easily account for all of the possible conditional probability assessments that would have to be made, based upon the arguments we have constructed in the 16 charts we have that concern P(3). But, to get some idea of what this number is, consider just the three charts we have included in this paper. Everywhere you see an open circle or an open diamond, there are at least two conditional probabilities required; an additional two more are required in linking the top node of each argument to P(3). This assumes that there are no conditional nonindependencies lurking among the events in a single reasoning chain or among stages in different reasoning chains. Another very heavy burden in MD is faced by the authors, namely, in developing appropriate formal expressions for the molecular decomposition of likelihood ratios. We have already written some formal expressions that will be appropriate; we will also obtain assistance from certain computer-assisted methods in which the writing of these formal expressions is unnecessary (e.g., Martin, 1980) So, MD is simply not possible for all of the evidence we have charted; nor is it necessary, as we shall now explain.

The kinds of probabilistic assessments we shall need from our experts are illustrated in Figure 4; determination of the specific number and form of these assessments must await our plotting of further arguments based upon the new evidence about which there is dispute. Figure 4 simply gives a general picture of the elicitation plan we have in mind. As shown, Sectors 13 and 14 contain the evidence regarding Bullet III and, as we mentioned, this evidence together with post-trial evidence forms a basic topic of disagreement among our experts. Thus, it may be necessary, in order to focus upon precise loci of disagreement among experts, to decompose the probability assessment here to the molecular level (MD).

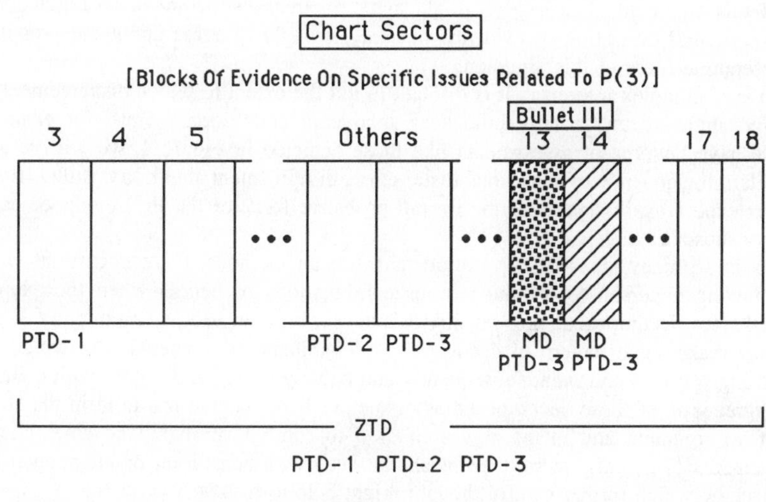

Figure 4.

For other blocks of evidence it may only be necessary for us to obtain a single likelihood ratio for the entire chart; e.g. see the PTD-1 judgment shown for Chart 3. In other cases, as shown, we may need likelihood ratios for individual items of testimony (PTD-2) or for items of testimony when there is recognizable conditional nonindependence (PTD-3). In some cases, Charts 13 and 14, for example, we might obtain more than one level of assessment at the same time; this provides very interesting information about judgmental consistency. Finally, we might ask our experts to make the entirely holistic ZDT assessments for the entire collection of evidence. This also provides for some very interesting comparisons.

3.5. *Bayesian Combinations and Dispute Resolution*

As we know, likelihood ratios for various items or bodies of evidence are multiplied together, provided that due consideration is made of possible conditional nonindependencies among items/bodies of evidence. In addition, various algorithms either exist or will be developed for determining Bayes-appropriate means for combining the molecular conditional probability assessments in the MD procedure. Various possible comparisons among individual and aggregate likelihood ratios are interesting to us; and there are some very interesting forms of sensitivity analysis we can undertake. Following is a brief account of the kinds of questions we can ask and, hopefully, answer during the remainder of our studies.

1. The MD assessments allow us to get to the nub of specific disagreements such as: whether or not the prosecution set up Sacco and Vanzetti by making a substitution involving Bullet III. We will chart Young & Kaiser's evidence and argument as well as Starrs' evidence and argument; then we will have them assess probabilities for links in their own arguments. In addition, we have entertained the possibility of having Starrs insert his own probabilities in the Young & Kaiser argument while having Kaiser insert his own probabilities in Starrs' argument; this provides the basis for some interesting comparisons. The reason is that there are two fundamental grounds for disagreement: (i) the structure of an argument, and (ii) the weight or force attached to various elements

of this argument. Our experts might agree about the structure of an argument from a particular collection of evidence but may strongly disagree about the probative or inferential force of this argument.

2. In very complex inference, it is difficult to tell the extent to which disagreement about any single matter actually influences an overall conclusion. Thus, for example, if we have patterns of assessments like those depicted in Figure 4, we will be able to determine how much, in a probabilistic sense, disagreement about, say, Bullet III affects our experts' assessments of the overall probative force of the P(3) evidence or about any subset of it.
3. As in so many areas of application of Bayesian methods, there is distinct virtue in allowing a person to try out various combinations of beliefs when their inference tasks are decomposed; i.e., sensitivity analysis is often useful. Frequently, a person may make some pattern of decomposed probabilistic assessments with which he/she strongly agrees and then, subsequently and equally-strongly, disagree with a Bayesian aggregation of these decomposed assessments. If the person has faith in the structure of an argument and in the algorithm used to combine probabilistic assessments for elements of this argument, then he/she is faced with behavioral or judgmental inconsistency which further careful thought might help to resolve.

4. CONCLUSIONS

In this paper we have described a study that is now in progress; our present conclusions concern only the methods we are employing in our work. The focus of our effort is upon the analysis and possible resolution of dispute about past events in which we employ a Bayesian view of probabilistic reasoning. There is unending dispute about the Sacco and Vanzetti murder case and we have identified persons who have expert knowledge about this case and who, in fact, disagree about certain quite crucial elements of it. A major ingredient in our studies involves the probabilistic assessments these experts will provide and which we shall combine and study using Bayesian methods. Our studies concern dispute about the evidence in this case and not about the fairness of the trial itself. The trouble is that this case, like many others, has produced a true mass of incomplete and inconclusive evidence that has come from sources having every gradation of credibility.

Our work so far has been essentially structural in nature. Any probabilistic analysis rests upon the structuring of argument. In our present work this structuring must be done with care so that we can identify the precise points at which there is disagreement among our experts. When there is a mass of evidence to be considered there are many arguments that need to be constructed; some of these arguments are complex and involve many reasoning steps and stages. Further, these arguments are often related in complex and interesting ways. We have employed a method for constructing arguments that was developed nearly eighty years ago by an American jurist named John H. Wigmore. Wigmore's methods have both analytic and synthetic components which we have illustrated by means of the tables and figures we discussed. Our view is that this method is very useful in constructing arguments based upon a mass of evidence, particularly when this method is employed with computer assistance. Our work thus far has involved structuring the arguments that both the prosecution and defense seem to have used during the trial. Further structural work will involve extending these arguments in order to accomodate new evidence and thoughts produced by our experts.

When the arguments we construct are acceptable to our experts we can then begin the process of having them make various forms of probabilistic judgments. We have discussed the various levels of "granularity" at which these judgments will be made. Our Bayesian methods will allow some interesting sensitivity analyses and comparisons involving the experts' probability elicitations. In the end, these experts may continue to disagree about the inferential force of collections of evidence we consider and they may even disagree about the inferential direction of certain evidence. But what our methods of analysis will provide is a very precise method for identifying the locus of their disagreement so that further study may be more efficient and productive.

Finally, though our work concerns historians of the Sacco and Vanzetti case and their disagreements, we believe our work will also be of interest to probabilists and to persons interested in study of the evidential foundations of legal disputes. Our study bears a relationship to current studies of inferential networks and the interesting evidential subtleties they reveal. Sacco and Vanzetti were found guilty by a twelve-person jury who were instructed that their standard of proof should be "beyond reasonable doubt". If our present Wigmorean analyses of the evidence and argument developed in this case are even reasonably correct, we have to conclude that there were sources of doubt that were never recognized during this trial. But, this is just one reason why historians and others will continue to disagree about whether or not the evidence in this case justified the verdict that was reached. We do not expect that our present Bayesian methods will lay to rest, once and for all, all dispute about this fascinating case.

REFERENCES

Anderson, T. and Twining, W. (1991). *Analysis of Evidence*. Boston: Little, Brown and Co.

Binder, D. and Bergman, P. (1984). *Fact Investigation: From Hypothesis to Proof*. St. Paul, Minn.: West Publishing Co.

Cohen, L. J. (1977). *The Probable and the Provable*. Oxford: University Press.

Edwards, W., Schum, D. and Winkler, R. (1990). Murder and [of ?] the likelihood principle: a trialogue. *J. Behavioral Decision Making* **3**, 75–89

Henry Holt and Co. (1928). *Sacco–Vanzetti Case Transcript*. New York.

Martin, A. (1980). A general algorithm for determining likelihood ratios. Cascaded Inference. *Tech. Rep.* **80–03**, Rice University.

Russell, F. (1962). *Tragedy at Dedham: The Story of the Sacco–Vanzetti Case*. New York: McGraw-Hill.

Russell, F. (1986). *Sacco and Vanzetti: The Case Resolved*. New York: Harper and Row.

Schum, D. and Martin, A. (1982). Formal and empirical research on cascaded inference in jurisprudence. *Law and Society Review* **17**, 105–151.

Schum, D. and Tillers, P. (1990). A technical note on computer–assisted Wigmorean analysis. *Tech. Rep.* **90–01**, George Mason University.

Starrs, J. E. (1986). Once more unto the breech: the firearms evidence in the Sacco and Vanzetti case revisited, Part I, *J. Forensic Sciences* **31**, 630–654; Part II, ibid, 1050–1078.

Tillers, P. and Schum, D. (1988). Charting new territory in judicial proof: beyond Wigmore. *Cardozo Law Review* **9**, 907–966

Toulmin, S. (1964). *The Uses of Argument*. Cambridge: University Press.

Twining, W. (1985). *Theories of Evidence: Bentham and Wigmore*. Stanford: University Press.

Twining, W. (1990). *Rethinking Evidence: Exploratory Essays*. Oxford: Basil Blackwell.

Venn, J. (1907). *The Principles of Inductive Logic*. New York: Chelsea Publishing.

Wigmore, J. H. (1913). The problem of proof. *Illinois Law Review* **8**, 77–103.

Wigmore, J. H. (1937). *The Science of Judicial Proof*. Boston: Little, Brown and Co.

Young, W. and Kaiser, D. E. (1985). *Postmortem: New Evidence in the Case of Sacco and Vanzetti*. Amherst, Mass.: University of Massachusetts Press.

DISCUSSION

M. GOLDSTEIN (*University of Durham, UK*)

I am very impressed by the care taken in this paper over the structuring of beliefs and evidence, and look forward to the finished account. Having never been on a jury, I can only speculate, but I would guess that, as well as focusing on the evidential structure, a juror will also try to form an overall impression of the probity of the various witnesses, by observing the way in which they give their testimony. I wonder where this fits onto the evidential charts.

I also wonder, on a technical level, whether it would make a difference to the analysis if the ultimate probandum that Sacco and Vanzetti were both guilty, were replaced by the probandum that Sacco and Vanzetti were both innocent. For example, suppose that the evidence shows that it is very unlikely that "Sacco was guilty but Vanzetti was innocent", but does not change the relative likelihoods of both the probandum of "both guilty" and also the probandum of "both innocent". Maybe, in this case, it does not matter, but in general perhaps the ultimate probandum might be better represented as a probability distribution over possibilities (guilty, guilty but insane, guilty by accident, etc.). This might increase the number of numerical specifications required, but each likelihood might then be easier to interpret, and so, hopefully, easier to specify.

REPLY TO THE DISCUSSION

1. Professor Goldstein's first comment concerns one of the forms of evidence commonly used in an assessment of witness credibility. This evidence conerns the *demeanor and bearing* of a witness while giving testimony and it allegedly bears upon the *veracity* of the witness; veracity is just one attribute of witness credibility. He asks whether or not there is a place for this form of evidence in our evidential charting methods; the answer is: definitely yes. To take account of the various attributes of witness credibility all we have to do is to further decompose the foundation or credibility-related stage of any reasoning chain based upon testimonial evidence. This further decomposition makes subsequent Bayesian analyses more complex but also more interesting in terms of what they reveal about the relationship between the credibility of a witness and the probative value of his/her testimony.

First, suppose the occurrence of event E is reported by Witness W_j. Let E_j^* represent the witness' assertion that event E occurred. We have to distinguish between E_j^* and E since W_j's telling us that E happened does not entail that it did. We might be inclined to believe that E_j^* is direct evidence on event E; i.e., there is a single step in reasoning from E_j^* to $\{E, \text{not } -E\}$. However, as we have shown elsewhere (Schum, 1989), this reasoning step can be decomposed by inserting two other event classes concerning what the witness believed and what the witness' senses recorded. When we do so, three attributes of a witness credibility are revealed: *veracity, objectivity, and observational sensitivity/accuracy*. Evidence about a witness' demeanor and bearing is commonly introduced as ancillary evidence bearing upon the veracity of a witness, namely, does this witness believe the event(s) he/she reported actually occurred? The trouble, of course, is that evidence about demeanor and bearing is quite inconclusive. Not every agitated witness is lying and not every composed witness is telling the truth.

What we have not done so far, but could easily do when it is required, is to make the decomposition noted above for witnesses whose testimony seems crucial. In the process we could find a place for demeanor and bearing evidence if, in fact, it were ever recorded for any witness in the Sacco-Vanzetti trial. We do, however, have anecdotal evidence about the behavior of certain witnesses that comes from Russell's book *Tragedy at Dedham*. The trouble is that is that we don't known how much of this "evidence" has been inserted by Russell in order to tell a good story.

2. Professor Goldstein's second comment concerns our definition and subsequent parsing of the ultimate hypotheses, possibilities, facts in issue, or "probanda" (to use Wigmore's terms) in the Sacco and Vanzetti case. He suggests that it would be quite informative to adopt a different scheme in which a variety of possibilities were entertained in addition to the major possibility asserted in our stated probandum: Sacco was guilty and Vanzetti was an accomplice. We agree that other sets of defined possibilities would be informative in studies of inferences based upon the evidence in the Sacco and Vanzetti case. However, our present analysis is constrained to follow the probanda, as they were originally asserted consistent with Massachusetts law at the time the trial was held. The reason is that the historians, whose disputes we are currently studying, have reacted to the possibilities entertained during the actual trial.

One reason for our agreement with Professor Goldstein on this matter concerns the issue of whether or not Sacco and Vanzetti should have been tried together. The attorneys for both Sacco and Vanzetti each filed severance motions for a separate trial. Judge Thayer denied these motions and so they were tried together. Trying these two persons together might have produced some awkwardness of the sort that Professor Goldstein's suggestions might eliminate. If the jury had found that Vanzetti was, indeed, innocent of the whole affair, then the stated ultimate probandum would have been disproven, even if Sacco were guilty.

ADDITIONAL REFERENCE IN THE DISCUSSION

Schum, D. (1989). Knowledge, probability and credibility. *Journal of Behavioral Decision Making* **2**, 39–62.

BAYESIAN STATISTICS 4, pp. 289–305
J. M. Bernardo, J. O. Berger, A. P. Dawid and A. F. M. Smith, (Eds.)

Reparameterization and Diagnostics of Posterior Non-Normality

ROBERT E. KASS and ELIZABETH H. SLATE
Carnegie Mellon University, USA

SUMMARY

We define and discuss diagnostics that indicate departure from marginal and joint posterior normality. We emphasize versions of these non-normality measures that are easily computed via second-order asymptotic expansions and illustrate their use. We discuss the goals of reparameterization and the benefits of approximate posterior Normality. We also mention "curvature" measures that are motivated by non-Bayesian considerations and indicate their relevance to Bayesian inference.

Keywords: ASYMPTOTICS; CURVATURE; LAPLACE'S METHOD; MODAL APPROXIMATION; POSTERIOR BIAS; POSTERIOR SKEWNESS; STANDARDIZED THIRD DERIVATIVES.

1. INTRODUCTION

It is widely appreciated that the parameterization of a statistical model affects the performance of numerical techniques used in making inferences. Working within the Bayesian paradigm, integration methods are of primary interest, and all those we know of can be improved by parameter transformation. This includes Gaussian quadrature, both Gauss-Hermite (Naylor and Smith, 1982) and subregion-adaptive Gauss-Legendre (Genz and Kass, 1991), Monte Carlo importance sampling (Geweke, 1989), Gibbs sampling (Gelfand and Smith, 1990), and approximation by Laplace's method (Tierney, Kass, and Kadane, 1989). Furthermore, Newton-like maximization routines, which are often used in conjunction with these methods, are also more effective in some parameterizations than in others.

The basic explanation for the improvement of numerical techniques following transformation of the parameter from, say, θ to ϕ, is that the posterior distribution of ϕ is "better behaved" than that of θ. In many cases this vaguely-defined concept may be made more specific by saying that the posterior distribution of ϕ is more nearly normal than is that of θ. In this paper we discuss several methods of assessing approximate posterior normality. When parameterizations are identified as improving the normal approximation to the posterior, they may be expected to improve performance of associated numerical techniques.

Identification of parameterizations that improve computation is not the only purpose of assessing approximate normality. More fundamentally, one would like to know when normality of a posterior distribution may be relied upon for inference. It is worth elaborating on this point just a bit.

A great many scientific inferences involve estimates. These are most often reported as numbers, but inductive reasoning based on Bayes' Theorem suggests that we think of such numbers as summaries of posterior distributions. This is the foundation for our teaching one of our most important elementary principles, that estimates without standard errors are meaningless. On the other hand, the appearance of full posterior distributions in results sections of scientific papers would seriously distract attention from the scientific work itself.

Furthermore, it should be recognized that each individual report of data soon becomes only one of many diverse sources of information about some set of phenomena. In the broader context of attempting to develop some explanation, or "theory", that would enable prediction of further observations, the ability to think easily about reported results becomes essential. With simplicity being paramount, the reporting of estimates together with standard errors becomes an eminently reasonable compromise. Our chief worry as statisticians is then to ensure that the range of probable values for an unknown quantity connoted by any given estimate and standard error is not grossly misleading. Since the connotation is provided by the normal distribution, our concern is to determine whether the normal approximation is grossly inadequate. We seek methodology that allows us to make this determination with minimal effort.

In Section 2 we list and briefly discuss several assessments of the normal approximation to the posterior for one-parameter families. Section 3 presents some applicable generalizations to multiparameter families. Since computation can be demanding and, in practice, only rough values of diagnostics are needed, we rely heavily on asymptotic approximations. Section 4 summarizes very concisely some work we have done on curvature measures that help diagnose poor performance of asymptotic normal approximations. Section 5 contains two examples, and Section 6 concludes with further discussion.

Notation We assume the parameter θ is m-dimensional and write the loglikelihood function as $\ell(\theta)$ and the log posterior as $\tilde{\ell}(\theta)$. The posterior mode and mean will be denoted by $\tilde{\theta}$ and $\bar{\theta}$, respectively, and $\hat{\theta}$ will be the MLE. Any second-order approximation to the mean (Tierney, Kass, and Kadane, 1989) will be denoted by $\bar{\theta}^*$. The usual first-order "modal" normal approximation to the posterior has mean $\tilde{\theta}$ and variance matrix $\tilde{\Sigma} = (-D^2\tilde{\ell}(\tilde{\theta}))^{-1}$. This is the approximation we will be considering throughout. In the one-dimensional case we write the modal standard deviation as $\tilde{\sigma}$.

2. SINGLE-PARAMETER FAMILIES

We begin with two criteria that involve special characteristics of the normal distribution. First, since the normal distribution has a log-quadratic density, we consider the third derivative of the log posterior, suitably standardized. Second, since the normal distribution is symmetric, we consider Pearson skewness. We then note that interval probabilities could be assessed, and the Kullback-Leibler number could be used as a global measure of adequacy of the normal approximation. We summarize some results for the exponential distribution and, in Section 2.2, comment on the theoretical relevance of the third derivative.

2.1. *Assessments*

Standardized third derivative. To be used as a diagnostic measure, the third derivative of the log posterior needs to be standardized so that the diagnostic gives the same value for both θ and $a + b\theta$, where a and b are constants (with $b \neq 0$). That is, the diagnostic should be *affine invariant*. Thus, we use $|\tilde{\ell}'''(\tilde{\theta}) \cdot \tilde{\sigma}^3|$. If we ignore the prior, or use a uniform prior, this measure becomes $|\ell'''(\hat{\theta}) \cdot (-\ell''(\hat{\theta}))^{-3/2}|$. The latter was originally proposed by Sprott (1973); the former was considered by Hills (1989) and Albert (1990).

Posterior Pearson skewness and bias. The Pearson skewness of a univariate distribution is (mean - mode)/(standard deviation). Thus, we consider the measure $(\bar{\theta} - \tilde{\theta})/\sigma$. There is another interpretation of this: analogously to the notion of bias of an estimator in non-Bayesian inference, we may consider $\tilde{\theta}$ to be an approximation to $\bar{\theta}$ and view $\tilde{\theta} - \bar{\theta}$ as the *posterior bias* of the mode. If we now standardize this quantity we again obtain (apart from sign) the Pearson skewness. In applications, especially when we move to multidimensional problems, we will substitute approximate versions of this quantity. Thus, in the one-parameter case we could use $(\bar{\theta}^* - \tilde{\theta})/\tilde{\sigma}$. We comment further on this in Section 3.

Interval probabilities. Another way to compare the normal approximation to the exact distribution is to take an interval that, according the normal approximation, is supposed to have probability $\tilde{p}$, and determine its exact posterior probability p. Then $\tilde{p}$ may be compared with p. We choose to examine posterior probabilities (calculated under the exact posterior distribution) $P\{\theta < \tilde{\theta} - 1.645\tilde{\sigma}\}$, $P\{\tilde{\theta} - 1.645\tilde{\sigma} \leq \theta \leq \tilde{\theta} + 1.645\tilde{\sigma}\}$, and $P\{\theta > \tilde{\theta} + 1.645\tilde{\sigma}\}$. Again, in multidimensional applications we will substitute a second-order asymptotic approximation p^* in place of p.

Kullback-Leibler number. The Kullback-Leibler discrepancy between the exact and approximate distributions is given by

$$\mathrm{KLD} = \int \log\left(\frac{p(\theta \mid y)}{\tilde{p}(\theta \mid y)}\right) p(\theta \mid y)\, d\theta,$$

where $\tilde{p}(\theta \mid y)$ is the approximate density for θ derived from the normal approximation and $p(\theta \mid y)$ is the exact posterior for θ. This furnishes a global assessment of the approximation. Unfortunately, when we move to nontrivial multidimensional problems, the Kullback-Leibler number becomes relatively difficult to compute.

Example: Exponential distribution. We consider a random sample from the exponential distribution with density $p(y_i \mid \theta) = \theta \exp(-\theta y_i)$ for $y_i > 0$, $i = 1, \ldots, n$. The prior on θ is taken to be gamma(α, β) so that the posterior is gamma(a,b), where $a = \alpha + n$ and $b = \beta + n\bar{y}$. Parameterizations of this model within the class of power transformations of θ have been investigated in detail by Slate (1991) using the foregoing criteria. We mention a few of the results here, focusing on the cube-root and log transformations for θ under the interval probability criterion (comparing $\tilde{p}$ to p). When $\phi = \theta^{1/3}$ we get $\tilde{\ell}'''(\tilde{\phi}) = 0$, and when $\phi = \log(\theta)$ it becomes variance-stabilizing, so that the prior of Jeffreys's general rule is uniform on ϕ. We note that related work on this example has been done by Achcar (1990).

The probability criterion was implemented by finding, for a given tail probability $\tilde{p}$, the effective sample size (the gamma posterior parameter a) needed to ensure that p would be adequately close to $\tilde{p}$. Here, the approximate .05 quantile is considered adequate if its true coverage is between one-half and twice what it should be, i.e., when $\tilde{p} = .05$, we require $.025 \leq p \leq .10$. Similarly, when $\tilde{p} = .95$, we require $.90 \leq p \leq .975$. (Other quantiles have been investigated by Slate, 1991.) Results for several values of the power ν are shown in Table 1. The cube root transformation ($\nu = 1/3$) outperforms the others tried in the sense that for each probability a smaller value of a is needed to achieve adequacy. For this parameterization, an effective sample size of 2 is adequate by our criterion. The log transformation ($\nu = 0$) is next best. If the effective sample size is at least 8 observations, then the log transformation also produces an adequate normal approximation. These observations will be used in discussing Example 1 of Section 5.

2.2. *Comments on the Role of the Third Derivative*

It is intuitive that when the third derivative of the log posterior vanishes an improved normal approximation should result, at least as the sample size increases. A formal statement of this may be found by examining an Edgeworth-type expansion for the distribution function (Johnson, 1970, equation (2.26)). Formally, when the third derivative at the mode vanishes,

Parameterization	a
original	38
mean-value	128
log	8
cube-root	2

Table 1. *Effective sample sizes needed under the tail probability criterion for the Exponential example. The value of the effective sample size a given here is the smallest such that the true coverage of both .05 tails is acceptable.*

the posterior satisfies

$$P\left\{\frac{\theta-\tilde{\theta}}{\tilde{\sigma}} \leq t\right\} = \Phi(t) + O(n^{-1})$$

which is an improvement over the usual $O(n^{-1/2})$ error. Similarly, it may be seen from an expansion of the posterior mean (Johnson, 1970; Tierney, Kass, and Kadane, 1989), that when the third derivative vanishes the order of the Pearson skewness (or the relative posterior bias of the mode) decreases from $O(n^{-1/2})$ to $O(n^{-3/2})$.

There is thus is fairly strong justification for using the standardized third derivative, but it must be remembered that this measure is local and has obvious limitations. These are emphasized when the Kullback-Leibler number is examined. The KLD may be expanded as

$$\text{KLD} = \log\left\{\frac{p(\tilde{\theta} \mid y)}{\hat{p}(\tilde{\theta} \mid y)}\right\} +$$

$$\frac{p(\tilde{\theta} \mid y)}{\hat{p}(\tilde{\theta} \mid y)}\left[\frac{1}{8}\tilde{\ell}^{(4)}(\tilde{\theta})\left(-\tilde{\ell}''(\tilde{\theta})\right)^{-2} + \frac{5}{12}\left(\tilde{\ell}'''(\tilde{\theta})\right)^{2}\left(-\tilde{\ell}''(\tilde{\theta})\right)^{-3} + O(n^{-2})\right],$$

where $p(\theta \mid y)$ is the true posterior for θ and $\tilde{p}(\theta \mid y)$ is the modal normal approximation. Since both the third and fourth derivatives of the log posterior appear, even asymptotically it will not in general be the case that a small third derivative will ensure a small Kullback-Leibler divergence.

3. MULTIDIMENSIONAL FAMILIES

It is not entirely trivial to obtain easily and rapidly-computed multidimensional generalizations of the criteria given in Section 2. For joint posterior normality assessment we consider standardized third derivatives, and for marginal assessment we consider Pearson skewness and interval probabilities.

3.1. *Third Derivative Summaries*

In multidimensional problems ($m > 1$), the third derivatives of the loglikelihood and log-posterior form three-way arrays. In order to simplify interpretation of such an array, it may be reduced to a single scalar summary. As in the one-parameter case, it will be important that the measure be affine invariant. There are two ways to generalize the standardized third derivative used in the one-parameter case. Letting the second and third partial derivatives be denoted by $\tilde{g}_{ab} = -\partial_{ab}\tilde{\ell}(\tilde{\theta})$ and $\partial_{abc}\tilde{\ell} = \partial_{abc}\tilde{\ell}(\tilde{\theta})$, and $(\tilde{g}^{ab})$ be the inverse of $(\tilde{g}_{ab})$, the two affine-invariant summaries are

$$\bar{B}^2 = m^{-2} \sum_{a,b,c,d,e,f} \tilde{g}^{ab}\tilde{g}^{de}\tilde{g}^{cf}\partial_{abc}\tilde{\ell}\,\partial_{def}\tilde{\ell}$$

and

$$B^2 = \sum_{a,b,c,d,e,f} \tilde{g}^{ad}\tilde{g}^{be}\tilde{g}^{cf}\partial_{abc}\tilde{\ell}\,\partial_{def}\tilde{\ell}.$$

These measures reduce to the standardized third derivative of Section 2.1 in the one-parameter case. When a uniform prior is used for θ, scalar summaries of the third derivative of the loglikelihood result.

The quantity $\bar{B}$ may be given an additional interpretation in terms of the posterior bias of the mode, again defined as $\tilde{\theta} - \bar{\theta}$. We standardize to produce the *relative posterior bias*, defined by

$$R = (\tilde{\theta} - \bar{\theta})^T\tilde{G}(\tilde{\theta} - \bar{\theta})$$

where $\tilde{G} = \tilde{\Sigma}^{-1}$ is the matrix having components $\tilde{g}_{ab}$. This is an affine-invariant scalar. Kass and Slate (1990) show that

$$R \doteq \frac{1}{4}m^2\bar{B}^2$$

with an error of order $O(n^{-2})$. This result not only provides additional interpretation of $\bar{B}$ as the leading term in the relative posterior bias of the mode, but it also gives a fast and convenient method of computing it, approximately. Letting $\bar{\theta}^*$ be a second-order approximation to the posterior mean $\bar{\theta}$, as discussed in Tierney and Kadane (1986) and Tierney, Kass, and Kadane (1989), we may define $R^* = (\tilde{\theta} - \bar{\theta}^*)^T\tilde{G}(\tilde{\theta} - \bar{\theta}^*)$ and obtain

$$m^2\bar{B}^2 \doteq 4R^*$$

which once again holds with error of order $O(n^{-2})$. The point is that existing computer code (Tierney, 1990a, b) is easy to apply to this calculation and, in particular, does not require the user to calculate loglikelihood derivatives analytically.

3.2. *Marginal Skewness and Bias*

In principle, Pearson skewness may be used as a measure of non-normality of any marginal posterior density. In practice, it may be difficult or time-consuming to compute. An alternative is to use approximations to the mean, mode, and standard deviation. Suppose the marginal Pearson skewness, which we denote by S_p, of a quantity $\gamma = g(\theta)$ is to be approximated. Kass, Tierney, and Kadane (1989) note that the mode $\tilde{\gamma}^*$ of the Laplace approximation to the marginal density of γ furnishes a second-order approximation to the mode of the marginal density itself. They then use

$$\hat{S}_p = (\bar{\gamma}^* - \tilde{\gamma}^*)/\tilde{\sigma}$$

where $\bar{\gamma}^*$ is a second-order approximation to the posterior mean of γ, and $\tilde{\sigma}$ is the first-order approximate standard deviation of γ. An alternative is

$$\hat{C} = (\bar{\gamma}^* - g(\tilde{\theta}))/\tilde{\sigma},$$

the point being that it is common to find $\tilde{\gamma}^* \neq g(\tilde{\theta})$. With existing code (Tierney, 1990a,b), both these measures are easily and rapidly computed, though $\hat{C}$ is perhaps slightly easier because it avoids the one additional step of obtaining $\tilde{\gamma}^*$.

3.3. *Marginal Probabilities*

DiCiccio, Field and Fraser (1990) (DFF) have derived an integrated form of the Laplace approximation which furnishes an accurate approximation to marginal tail probabilities. Let $\partial_i\tilde{\ell} = \frac{\partial\tilde{\ell}}{\partial\theta_i}$, $\partial_{ij}\tilde{\ell} = \frac{\partial^2\tilde{\ell}}{\partial\theta_i\partial\theta_j}$ and let $\hat{\theta}(t)$ maximize $\tilde{\ell}(\theta)$ subject to the constraint that $\theta_1 = t$. Continuing to let $\tilde{\theta}_1$ denote the first component of the joint mode for θ, if $t-\tilde{\theta}_1$ is $O(n^{-1/2})$, then

$$P(\theta_1 \leq t) = \Phi(r) + \phi(r)\left(\frac{1}{r} + \frac{\left|[-\partial_{ij}\tilde{\ell}(\tilde{\theta})]\right|^{1/2}}{\partial_1\tilde{\ell}\{\hat{\theta}(t)\}\left|[-\partial_{ab}\tilde{\ell}\{\hat{\theta}(t)\}]\right|^{1/2}}\right) + O(n^{-3/2}),$$

where the brackets indicate that a matrix has been formed of the contents,

$$r = r(t) = \operatorname{sgn}(t-\tilde{\theta}_1)(2[\tilde{\ell}(\tilde{\theta}) - \tilde{\ell}\{\hat{\theta}(t)\}])^{1/2},$$

and i and j vary over $1,\ldots,m$ and a and b range over $2,\ldots,m$.

4. CURVATURE DIAGNOSTICS

Recently we showed (Kass and Slate, 1990) how several curvature measures of nonlinearity, which have been used to diagnose non-normality of least-squares estimators in nonlinear regression (Beale, 1960; Bates and Watts, 1980), may be generalized (employing the α-connections of Amari, 1982) to produce diagnostics of non-normality of MLE's. We do not wish to give an extended discussion of that work here, but we do want to give a rough indication of some of the methodology because we believe it may be of use from a Bayesian point of view as well.

In nonlinear regression, the essential idea is to base nonlinearity measures on the vectors of second derivatives of the model function (which would be zero for a linear model) and to decompose them into components that are normal and tangential to the surface (at the least-squares estimate); suitably scaled as curvatures, the normal components are invariant with respect to reparameterization whereas the tangential components are not. The tangential components are thus used to define "parameter-effects curvatures" (Bates and Watts, 1980). It is generalizations of these we wish to consider here.

Let us confine our attention to the class of *exponential family nonlinear models*, which have product densities at $y = (y_1,\ldots,y_n)$ of the form $p(y \mid \eta) = \Pi_{i=1}^n p(y_i \mid \eta_i)$, with $\eta = (\eta_1,\ldots,\eta_n)$, where

$$p(y_i \mid \eta_i,\sigma) = \exp\{[y_i\eta_i(\theta) - \psi(\eta_i)]/\sigma^2 + b(y_i,\sigma)\}$$

with Θ being open in R^m and the mapping $\theta \to \eta(\theta)$ being an imbedding (i.e., one-to-one, infinitely differentiable with full-rank Jacobian) in the natural parameter space N of a regular exponential family of order n. Generalized linear models are exponential family nonlinear models, and when the dispersion parameter σ is known, exponential family nonlinear models become curved exponential families. This class of models includes, and may be considered a generalization of, normal nonlinear (and linear) regression models. As in the case of nonlinear regression, geometrical analysis of these models may be based on the imbedding $\theta \to \eta(\theta)$, ignoring the presence of the parameter σ and effectively treating it as if it were known. We therefore write $\mathcal{Q}_0 = \{p(y \mid \eta(\theta),\sigma) : \theta \in \Theta\}$, and take $\mathcal{Q} = \{p(y \mid \eta,\sigma) : \eta \in N\}$ to be the unrestricted exponential dispersion model. We then note that for each fixed σ, $\mathcal{Q}$ is a full exponential family of dimension n formed from the n-fold product of a one-dimensional full exponential family with itself. Let $\mathcal{Q}^{(1)}$ denote this one-dimensional exponential family.

Now, for each one-dimensional full exponential family there exists a parameterization such that the third derivative of the loglikelihood evaluated at the MLE vanishes. If β is such a parameterization for $\mathcal{Q}^{(1)}$, let us define ζ_i to be the parameter β as it occurs in the i-th member of the n-fold product in $\mathcal{Q}$ as defined above so that we obtain $\zeta = (\zeta_1, \ldots, \zeta_n)$. For example, if $\mathcal{Q}^{(1)}$ is the Exponential($1/\mu$) family, then $\beta = \mu^{-1/3}$ satisfies $\ell'''(\hat{\beta}) = 0$ and, according to the results in Section 2, when a conjugate prior is used, the posterior is acceptably normal for sample sizes greater than or equal to 2. We may parameterize $\mathcal{Q} = \mathcal{Q}^{(1)} \times \cdots \times \mathcal{Q}^{(1)}$ by $\mu = (\mu_1, \ldots, \mu_n)$, but if we instead use $\zeta = (\mu_1^{-1/3}, \ldots, \mu_n^{-1/3})$ then the product posterior on ζ will also be well-behaved for small sample sizes.

The intuitive idea behind the generalized curvature measures is simply that if the joint likelihood on ζ is approximately normal, then the likelihood restricted to $\phi = A + B\zeta$ for any vector A and $p \times m$ matrix B (with $p \leq m$) will also be approximately normal. The latter is the form of the parameter in a linear subspace (more generally, an affine subspace, since the constant vector A has been added so that the subspace may not include the origin). Such a subspace could be the representation in ζ-space of the subfamily $\mathcal{Q}_0$. Thus, we define "parameter-effects curvature" measures that assess departure from linearity of the parameterized model $\mathcal{Q}_0$, viewed as a subspace of $\mathcal{Q}$ when parameterized by ζ; these measures would be zero for a parameterization $\phi = A + B\zeta$. They are calculated for the m-dimensional surface $\zeta(\theta)$ at the point $\zeta(\hat{\theta})$. (Kass and Slate, 1990, show that when $\alpha = 1/3$ the α-connection coefficient satisfies $\overset{\alpha}{\Gamma}_{ijk}(\zeta) = 0$; this is the key to the geometrical foundation for the generalization; we concentrate here on the interesting case $\alpha = \frac{1}{3}$, but in that paper we consider other α-connections and more general models.)

We will not try to explain carefully the measures we wish to illustrate here, but do provide explicit formulae. The first two are formally analogous to two quantities that play a fundamental role in the theory of surfaces (Kass, 1989, notes that the first is analogous to what he considers to be a generalization of statistical curvature, and the second is analogous to the mean curvature). They are

$$\omega_Q^2 = \sum g^{ac} g^{bd} g_{ij} \cdot ((\partial_{ab}\zeta)_T)_i ((\partial_{cd}\zeta)_T)_j$$

$$m^2 \bar{\omega}_Q^2 = \sum g^{ab} g^{cd} g_{ij} \cdot ((\partial_{ab}\zeta)_T)_i ((\partial_{cd}\zeta)_T)_j$$

where (g_{ij}) is the $n \times n$ information matrix in terms of ζ at $\zeta(\hat{\theta})$, (g^{ab}) is the inverse of the $m \times m$ information matrix in terms of θ at $\hat{\theta}$, $(\cdot)_T$ signifies tangential components with respect to the information matrix (g_{ij}), and the summations are over all indices. The third curvature measure we will use is a generalization of the root-mean-squared curvature defined by Bates and Watts (1980). To define it we begin with a curve on the m-dimensional surface $\zeta(\theta)$ called a "lifted line" because it is the image of a line in θ-space. We let $c_v(t) = \zeta(\hat{\theta} + tv)$ at $c(0) = \zeta(\hat{\theta})$, where v is a vector in R^m. Thus, the curve is the image under $\theta \to \zeta(\theta)$ of a line through $\hat{\theta}$ in the direction of v in Θ; on the surface at $\zeta(\hat{\theta})$ its direction is $c_v'(0)$. We compute its curvature with respect to the inner product defined by Fisher information,

$$\kappa_{Q,T}(v) = \| c_v'(0) \|_{i(\zeta(\hat{\theta}))}^{-2} \cdot \| (c_v''(0))_T \|_{i(\zeta(\hat{\theta}))}$$

where the information inner product is used to define the norm $\| \cdot \|_{i(\zeta(\hat{\theta}))}$ (see Kass, 1989, and Kass and Slate, 1990) and we integrate this curvature over all directions for v, specifically,

$$\omega^2_{Q,RMS} = \frac{1}{S_m} \int_S (\kappa_{Q,T}(v))^2 \cdot dS$$

where the integral is over the sphere $\{v : \| c'_v(0) \|_{i(\zeta(\hat{\theta}))} = 1\}$ and $S_m = \pi^{m/2}/\Gamma(\frac{m}{2})$ is the surface area of the unit sphere in R^m.

Finally, we mention a result that connects the curvature measures of this section with the third-derivative measures of the previous one. We view third derivatives as second derivatives of first derivatives, and thereby derive a new third-derivative summary as a curvature-like second-derivative summary of the first-derivative vector. Specifically, for $h \in R^m$ we consider the "lifted line" $c_h(t) = (\partial_1 \tilde{\ell}(\tilde{\phi} + th), \ldots, \partial_m \tilde{\ell}(\tilde{\phi} + th))^T$, compute its curvature $\kappa(h) = \|c''_h(0)\|_{G^{-1}} / \|c'_h(0)\|^2_{G^{-1}}$, where $G^{-1} = \tilde{\Sigma}$ is the modal covariance matrix, and then take the spherical average to obtain

$$B^2_{RMS} = \frac{1}{S_m} \int_S \kappa(h)^2 \cdot dS$$

where the integral is over the sphere $\{h : \|c'_h(0)\|_{G^{-1}} = 1\}$. This quantity is analagous to $\omega^2_{Q,RMS}$ and may be considered a Bayesian version of a curvature measure similar to those of Bates and Watts. We find

$$B^2_{RMS} = \frac{m^2 \bar{B}^2 + 2B^2}{m(m+2)}.$$

We hope to include details in a revision of Kass and Slate (1990).

5. EXAMPLES

Example 1: Exponential survival model with covariate. We consider the 17 AG-positive patients from the leukemia data set and model of Feigl and Zelen (1965), which has been re-analyzed by many authors. The outcome (Y) is survival time in weeks and the predictor is white blood cell count (WBC), with $x = \log(WBC) - \text{mean}(\log(WBC))$. The model is

$$Y_i \sim \text{Exponential } (1/\mu_i)$$
$$\mu_i = \theta_1 \exp(-\theta_2 x_i)$$

independently, for $i = 1, \ldots, 17$, where $E(Y_i) = \mu_i$. Using a uniform prior on (θ_1, θ_2), the posterior mode is $\tilde{\theta} = (51, 1.11)$. From the Hessian of the log posterior at the mode we obtain approximate posterior standard deviations of 12 and .41 for θ_1 and θ_2, and their approximate correlation is .00.

In addition to θ, we also consider the parameterizations β and ϕ given by $\mu_i = (\beta_1 \cdot x_i^*)^{\beta_2}$, where $x_i^* = \exp(-x_i)$, and $\mu_i = \exp\{\phi_1 - \phi_2 x_i\}$. We note that Jeffreys's prior is uniform in the ϕ parameterization. Examination of the posterior contours (not displayed here, but given in Kass and Slate, 1990) shows that the posterior is well-behaved in the parameterization ϕ, while the parameterization θ is clearly worse and β is terrible. Our purpose here is to be able to reach these conclusions without the aid of the contours.

The curvature and third-derivative diagnostics are shown in Table 2. The measures ω_Q, $\bar{\omega}_Q$, and $\omega_{Q,RMS}$ indicate, appropriately, that ϕ improves on θ while β is disastrous. A similar conclusion is also obtained from $\bar{B}$ and B, which were calculated using a uniform prior in each parameterization. (When $\bar{B}$ and B are computed using the prior that is uniform

	Parameterization		
Diagnostic	ϕ	θ	β
$m^2\bar{\omega}_Q^2$	0.026	0.163	145.495
ω_Q^2	0.026	0.124	143.830
$\omega_{Q,RMS}^2$	0.010	0.052	54.145
$m^2\bar{B}^2$	0.237	1.472	198.158
B^2	0.237	1.119	523.282

Table 2. *Parameterization diagnostics for the Leukemia example.*

in ϕ and the appropriate Jacobian is used in the other parameterizations, similar results are obtained.)

We also examined the posterior probabilities of marginal approximate inference regions. Using the uniform prior on the ϕ parameterization as above, the probabilities of the marginal 95% regions were found using the DFF approximation. The results are shown in Table 3, in which $p2$ is the probability below the upper end point of the marginal 95% interval and $p1$ is the probability below the lower end point. The difference $p2 - p1$ is the probability for the interval. Again, the ϕ parameterization performs best in that both probabilities are nearer their nominal value. It may also be seen that θ_2 (= ϕ_2) is "more nearly normal" than either θ_1 or ϕ_1. Results were unavailable for β_1 due to numerical problems finding the constrained maximum required for the DFF formula. Kass, Tierney and Kadane (1989) found the Pearson skewness for θ_1 to be 0.48, and only 0.02 for θ_2 and 0.12 for ϕ_1. Overall, it is apparent from these diagnostics that the posterior on ϕ is adequately normal. In addition to verifying these conclusions by direct examination of the likelihood contours, we evaluated posterior probabilities for the approximate inference regions. We did so by taking the prior to be uniform on the parameterization ϕ and using Monte Carlo importance sampling for the computations. We found that the ellipsoids having putative 95% probability actually had probability .934(±.001), .866(±.001), and .440(±.002) for the parameterizations ϕ, θ, and β, respectively. Once again, these values are consistent with the indications of the diagnostics. We note that the approximate values for $m^2\bar{B}^2$ based on $4R^*$ are .237, 1.480, and 140.793 for the respective parameterizations, quite close to the values in Table 2 for the two reasonable parameterizations but understandably poor in the β parameterization.

	ϕ_1	ϕ_2	θ_1	θ_2	β_1	β_2
p2	0.945	0.976	0.852	0.973	0.512	0.992
p1	0.010	0.028	0.0004	0.032	*na*	0.008
p2 − p1	0.935	0.947	0.851	0.941	*na*	0.984

Table 3. *Probabilities for the marginal approximate 95% regions for the Leukemia example computed using the DFF approximation. p2 is the probability below the upper bound of the region (which should be .975) and p1 is the probability below the lower bound (which should be .025).*

Example 2: Two-component reliability model. As a second example we consider a reliability model for the lifetimes of two components, and data previously analyzed by Hills (1989, Section 3 of Chapter 3). The probability density function is

$$p(y_1, y_2|\lambda_1, \lambda_2, \psi) = \prod_{i=1}^{n} \lambda_1\lambda_2 \exp\{-(\lambda_1 y_{1i} + \lambda_2 y_{2i})\}$$
$$\left[1 + \frac{e^\psi - 1}{e^\psi + 1}(2\exp(-\lambda_1 y_{1i}) - 1)(2\exp(-\lambda_2 y_{2i}) - 1)\right].$$

The prior used by Hills and adopted by us is

$$p(\lambda_1, \lambda_2, \psi) = \begin{cases} \frac{\exp(\psi)}{\lambda_1\lambda_2(1+\exp(\psi))^2} & \lambda_1, \lambda_2 > 0 \\ 0 & \text{otherwise} \end{cases}$$

We also consider a parameterization studied by Hills, in which λ_1 and λ_2 are transformed to $\phi_1 = \log(\lambda_1)$ and $\phi_2 = \log(\lambda_2)$.

We mention just a few results. We find $m^2\bar{B}^2 = .691$ for the λ parameterization (while the second-order approximation gives $4R^* = .693$) and $m^2\bar{B}^2 = .455$ for ϕ ($4R^* = .456$), indicating a slight improvement when using ϕ in place of λ. We also note the good accuracy of the approximation $4R^*$ to $m^2\bar{B}^2$ in this case. The DFF approximate probabilities are shown in Table 4. Again, it appears that ϕ is slightly preferable. We would also conclude from these simple diagnostics that the normal approximation in the ϕ parameterization is tolerably accurate for most practical purposes, though the distribution of ψ is a bit skewed. These conclusions are consistent with two-dimensional marginal contours produced by Hills.

	λ_1	λ_2	ψ	ϕ_1	ϕ_2	ψ
$p2$	0.940	0.940	0.900	0.987	0.987	0.904
$p1$	0.005	0.005	0.021	0.046	0.045	0.023
$p2 - p1$	0.935	0.935	0.879	0.942	0.942	0.882

Table 4. *Probabilities for the marginal approximate 95% regions for the reliability example computed using the DFF approximation.* p1 *and* p2 *are as in Table 3.*

6. DISCUSSION

We began by linking the problems in our title. On the one hand, we noted the heuristic that transformation to approximate normality ought to improve many computations; on the other, we emphasized the desirability of normality, which can sometimes be achieved through reparameterization. We do not wish to blur the distinction between the two topics, however. For purely computational purposes quite arbitrary transformations might be considered while, for inference, interpretation is important and the class of transformations examined will necessarily be much more restricted. For computation, we might envision algorithmic data-dependent methods. There, non-normality assessments ought to remain helpful as a research tool, but ideas that start from a different heuristic, or bypass assessment, may well be successful. The line of investigation we have described, which has also been followed by several others we cited, has two goals. First, to develop and understand diagnostics that are easy to use and can indicate inadequacy of the normal approximation. Second, to determine parameterizations that are interpretable and are likely to lead to well-behaved posteriors.

The exponential survival model serves to illustrate the approach. Four types of diagnostics were used in that example: third-derivative summaries and curvature measures both indicated that, jointly, ϕ was an improvement over θ, while β was terrible; marginal Pearson skewness and interval probability approximations indicated that ϕ_1 was an improvement over θ_1. Overall, the posterior distribution of ϕ may be considered adequately normal for practical purposes. This is a case in which ϕ was the "obvious" parameterization. Yet, one may have doubts about improvement or magnitude of improvement and it is very helpful to have quick checks.

In that example there was also the structure of an exponential family nonlinear model. This allowed us to study in depth the one-parameter exponential family defining that structure, here based on an exponential distribution. We found that the cube-root transformation was very good according to all criteria and produced adequately normal approximations to the (conjugate) posterior for sample sizes greater than or equal to 2. Similarly, the log transformation gave adequate approximations for sample sizes greater than or equal to 8. (This happened not to depend on the value of the sufficient sample mean.) From this, when the two-parameter exponential model becomes linear in the log scale via reparameterization in terms of ϕ, roughly speaking, we would expect about $\dim(\theta) \cdot 8 = 16$ observations to be required for adequacy of the normal approximation. This is one way to look at the result that even with as few as 17 observations, the posterior on ϕ was adequately normal. We have not investigated such reasoning thoroughly, but we expect it to hold more generally (though with some qualification concerning the design, i.e., the x-variable's values).

The results involving the cube-root transformation of the exponential parameter are also of interest in conjunction with the curvature measures. Leaving aside what we found about the log transformation, we can be confident that, with quite small sample sizes, a model that is nearly linear in the cube-root ζ-space will have a nearly-normal posterior. This is what we observed for ϕ and it gives us another way of understanding the results. Although the curvature measures are not Bayesian in their motivation, they can be effective, as they were in this example. The reason is that, in the first place, the geometry is defined so that curvatures vanish when the observed third derivatives of the loglikelihood vanish. Secondly, although departure from this "optimal" situation is measured in a geometrical rather than an inferentially direct manner, the curvatures are based on expected third derivatives; and it has been found in many nonlinear regression examples that expected and observed second derivatives are generally quite close and the same may well be true for third derivatives in exponential family nonlinear models (and for much the same reason; see Kass and Slate, 1990, for a bit of further comment and references).

Like Kass, Tierney, and Kadane (1988), we have here emphasized speed and convenience of the methodology we are investigating. One way we like to think about these attributes is in terms of our ability to teach methods to students who have the background of a good course in linear regression. That is, we envision a system with at least the convenience of GLIM, that includes diagnostics that are as easily obtained as Cook's distance. Tierney's routines in NEW S and XLISP-STAT take a large step in this direction. Many of our calculations were done in these programming environments, and take little working time for analysis of a completely new example. Others, such as the DFF marginal probability approximation, were computed in FORTRAN, but this did not involve much programming effort, and it would be relatively easy to create general-purpose routines for use in NEW S or XLISP-STAT, as well. That work, together with additional research on the use of the diagnostics discussed here, will augment the very nice packages already available. We are rapidly approaching the time when inference based on widely applicable, simple but nontrivial models can become part of the elementary data analysis repertoire we expect so many of our students to master.

REFERENCES

Achcar, J. A. (1990). A Bayesian approach to reparameterization of the exponential distribution with type I censored data, *Tech. Rep.* **59**, ICMSC-Universidade De São Paulo, Brasil.

Achcar, J. A. and Smith, A. F. M. (1990). Aspects of reparameterization in approximate Bayesian inference, *Essays in Honor of George A. Barnard* (S. Geisser, J. S. Hodges, S. J. Press, A. Zellner, eds.), Amsterdam: North-Holland, 439–452.

Albert, J. H. (1989). On the use of power transformations to simplify posterior distributions. *Tech. Rep.* Bowling Green State University.

Amari, S. I. (1982). Differential geometry of curved exponential families—curvatures and information loss. *Ann. Statist.* **10**, 375–385.

Bates, D. M. and Watts, D. G. (1980). Relative curvature measures of nonlinearity. *J. Roy. Statist. Soc. B* **42**, 1–25.

Beale, E. M. L. (1960). Confidence regions in nonlinear estimation. *J. Roy. Statist. Soc. B* **22**, 41–76.

DiCiccio, T. J., Field, C. A. and Fraser, D. A. S. (1990). Approximations of marginal tail probabilities and inference for scalar parameters. *Biometrika* **77**, 77–95.

Feigl, P. and Zelen, M. (1965). Estimation of exponential survival probabilities with concomitant information. *Biometrics* **21**, 826–838.

Gelfand, A. E. and Smith, A. F. M. (1990). Sampling based approaches to calculating marginal densities. *J. Amer. Statist. Assoc.* **85**, 398–409.

Genz, A. and Kass, R. E. (1991). An application of subregion-adaptive quadrature to a problem in Bayesian inference. *Computer Science and Statistics: Proceedings of the 23rd Symposium on the Interface*, (to appear).

Geweke, J. (1989). Bayesian inference in econometric models using Monte Carlo integration. *Econometrica* **57**, 1317–1339.

Hills, S. (1989). *The Parameterisation of Statistical Models*. Ph.D. Thesis, University of Nottingham.

Johnson, R. A. (1970). Asymptotic expansions associated with posterior distributions. *Ann. Math. Statist.* **41**, 851–864.

Kass, R. E. (1989). The geometry of asymptotic inference. *Statist. Sci.* **4**, 188–234, (with discussion).

Kass, R. E. and Slate, E. H. (1990). Some diagnostics of maximum likelihood and posterior non-Normality. *Ann. Statist.* , (to appear).

Kass, R. E., Tierney, L. and Kadane, J. B. (1989). Approximate methods for assessing influence and sensitivity in Bayesian analysis. *Biometrika* **76**, 663–674.

Kass, R. E., Tierney, L. and Kadane, J. B. (1988). Asymptotics in Bayesian computation. *Bayesian Statistics 3* (J. M. Bernardo, M. H. DeGroot, D. V. Lindley and A. F. M. Smith, eds.), Oxford: University Press, 263–267, (with discussion).

Naylor, J. C. and Smith, A. F. M. (1982). Applications of a method for the efficient computation of posterior distributions. *Applied Statistics* **31**, 214–225.

Slate, E. H. (1991). *Reparameterization of Statistical Models*. Ph.D. Thesis, Department of Statistics, Carnegie Mellon University.

Sprott, D. A. (1973). Normal likelihoods and their relation to large sample theory of estimation. *Biometrika* **60**, 457–465.

Tierney, L. (1990a). Bayesian analysis in New S and Lisp. *Proceedings of the Statistical Computing Section of the A. S. A.*, (to appear).

Tierney, L. (1990b). *Lisp-Stat: An Object-Oriented Environment for Statistical Computing and Dynamic Graphics*. New York: Wiley.

Tierney, L. and Kadane, J. B. (1986). Accurate approximations for posterior moments and marginal densities. *J. Amer. Statist. Assoc.* **81**, 82–86.

Tierney, L., Kass, R. E. and Kadane, J. B. (1989). Fully exponential Laplace approximations to expectations and variances of nonpositive functions. *J. Amer. Statist. Assoc.* **84**, 710–716.

DISCUSSION

T. J. SWEETING (*University of Surrey, UK*)

I am rather fond of the signed root-log-likelihood ratio (SRLLR)

$$R = \text{sign}(\theta - \hat{\theta})[-2\{l(\theta) - l(\hat{\theta})\}]^{\frac{1}{2}}$$

In the case of a single parameter, this seems to be an ideal starting point for investigating normal parametrizations. To begin with, the SRLLR statistic is parametrization-invariant and the likelihood cast in terms of R is exactly normal. If computation is the only goal, then R is an obvious data-dependent parametrization to use. An alternative is the signed root-log-posterior ratio (SRLPR). Posterior distributions of these quantities are very often acceptably normal; in fact a relatively simple correction term to the latter yields posterior normality to $O(n^{-3/2})$ (DiCiccio, Field and Fraser, 1990 (DFF)). The SRLLR is also useful for deriving higher-order frequentist results. In particular, Bickel and Ghosh (1990) use Bayesian arguments to derive such results.

In the present paper it is argued that a major aim of reparametrization is to achieve simplicity and interpretability which, as the authors remark, demands a more restricted choice of transformations. In particular, one would hope to find a suitable data-free parametrization. The authors discuss four methods of assessing the adequacy of a given parametrization in the case of a single parameter. If this parametrization turns out to be poor, the first of these methods, based on third-order derivatives, can also guide us in our choice of reparametrization (cf. Sprott, 1973). The other three methods, on the other hand, do not appear to offer us any remedies when things are bad.

Let me return now to the SRLLR. The parameter $\phi = \phi(\theta)$ will have an acceptably normal likelihood if and only if it is approximately linearly related to R; that is, $R = a(y) + b(y)\phi$. Differentiation with respect to θ and evaluation at $\theta = \hat{\theta}$ gives $\phi'(\hat{\theta})/\phi(\hat{\theta}) = \frac{1}{3}l'''(\hat{\theta})/l''(\hat{\theta})$, which is precisely the expression one obtains from a direct third-order derivative analysis. If one now chooses ϕ so that this expression holds for all θ, then $\phi \propto \int[-l''(\theta)]^{1/3}d\theta$. In particular, if $l''(\theta)$ is proportional to a function of θ only, then this will yield a data-free parametrization. In the exponential example of Section 2, this gives the cube-root transformation. This way of looking at the (local) third derivative assessment suggests a simple global graphical assessment of a parametrization, since a plot of R against ϕ, over the range $|R| < 3$ say, should appear roughly linear. Some very recent work which uses such plots is reported in Hills and Smith (1992), where they are referred to as *t-plots* after Bates and Watts (1988).

In the exponential example we find that

$$R = \text{sign}(\theta\bar{y} - 1)[-2n\{\log(\theta\bar{y}) - (\theta\bar{y} - 1)\}]^{\frac{1}{2}}.$$

Suppose, for example, that a single observation y is available from the exponential (1) distribution. Figures 1a and 1b, which show the plots for θ and $\phi = \theta^{1/3}$ respectively, speak for themselves. The corresponding plot for ϕ when $n = 2$ is exceptionally good, and supports the authors' claim of an "effective sample size" of 2 for this parametrization.

Another general point to be made about the diagnostics in the paper is that they only indicate the suitability of a parametrization for the *data in hand*. In order to recommend a parametrization for general use, the methods need to be supplemented by theoretical and/or Monte Carlo investigation. This presents no problem in the exponential example, as R is a function of $\theta\bar{y}$. Thus when $n = 1$, for example, we just obtain the plot in Figure 1b with $\theta^{1/3}$ replaced by $(\theta y)^{1/3}$.

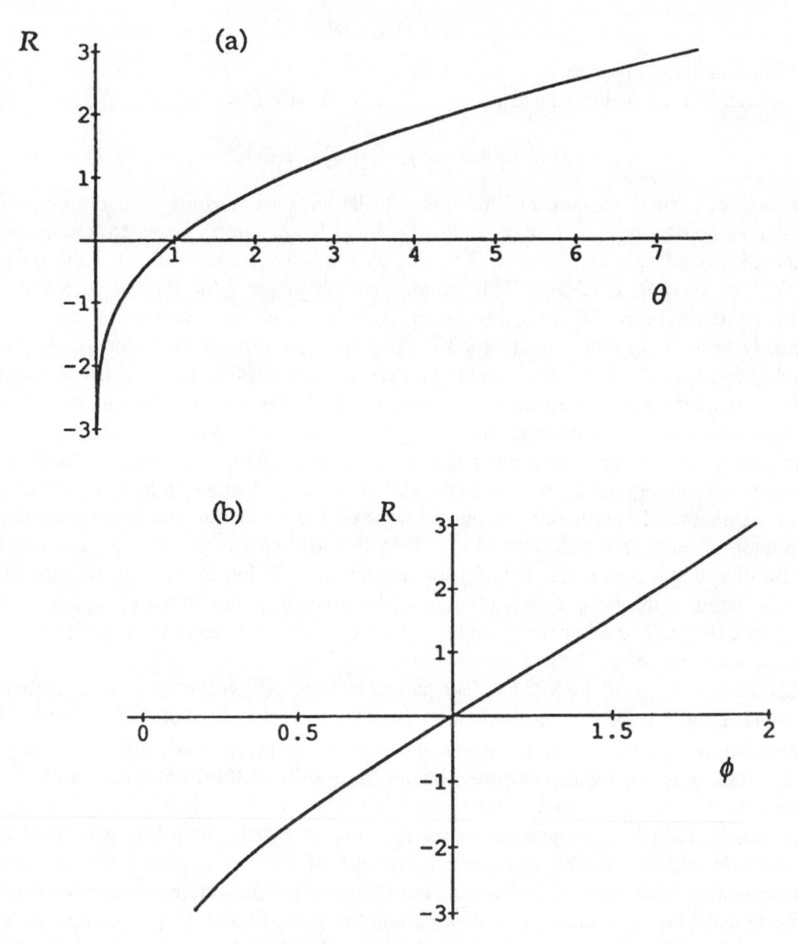

Figure 1. *Plot of R against (a) θ (b) $\phi = \theta^{1/3}$ for the example of Section 2; $n = 1$.*

Turning to the much more difficult multiparameter case, I found the ideas in the paper interesting, especially the curvature diagnostics in Section 4. I must confess that my initial reaction to a method based on assuming that infinity is adequately approximated by one was a little sceptical, but the method gains weight on seeing Figure 1b! Again, my main problem with these diagnostics is that it seems they cannot *guide* us in our choice of reparametrization. The authors mention the marginal posterior approximation for a single parameter based on the SRLPR statistic developed in DFF. Use of this formula when there are many parameters requires some care, however, as pointed out by DFF. The marginal posterior of θ_1 may be centred at a point quite distant from $\hat{\theta}_1$. DFF recommend first using the Tierney and Kadane Laplace approximation to the marginal posterior of θ_1, and then applying the one-dimensional version of the approximation given in Section 3.3 to this. In fact the examples given in DFF suggest that the distribution of this modified SRLPR is often acceptably normal without further adjustment.

In the two-parameter example 1, θ_1 and θ_2 are seen to be orthogonal since $\Sigma_i x_i = 0$ (generally, θ_2 and $\log(\theta_1) - \theta_2\bar{x}$ will be orthogonal). If independent priors are adopted for θ_1 and θ_2, then a likelihood-based Laplace approximation to the integrated likelihood of θ_1 (cf. Sweeting, 1987a) just introduces an extra factor $\theta_1^{1/2}$ into the profile likelihood of θ_1. Furthermore $\hat{\theta}_2(\theta_1) = \hat{\theta}_2$ exactly here, so the analysis of Section 2 gives $\theta_1^{-1/3}$ as the 'best' parametrization for the marginal posterior distribution of θ_1. It would now be of interest to examine t-plots for the conditional likelihood of θ_2 given θ_1. The form of R here seems to suggest that θ_2 would be a reasonable parametrization. Furthermore, the linearity of the t-plots is independent of θ_1 in this case.

One final comment: there are alternatives to using normal approximations for simple description. In Sweeting (1987b, 1988), for example, it is shown that in general regression models with nonnormal errors good χ^2 approximations to posterior distributions of scale parameters are available (so one could report a variance estimate along with its degrees of freedom). For numerical integration, one can then add Gauss-Laguerre quadrature to the list in Section 1.

Overall, I like the spirit of this paper. The techniques involve some quite advanced mathematics, but the *goal* is ultimate simplicity.

N. D. SINGPURWALLA (*George Washington University, USA*)

I found this paper clear and well written. I have two comments to make concerning the paper. The authors may also like to look at my comments on the paper by Hills and Smith (1992), also presented at the conference, which relates to the specification of a 'stopping rule', a topic touched upon by the authors in Table 1 of the paper.

Regarding the exponential distribution example, two issues come to mind. The first pertains to censoring, because $b = \beta+$(total time on test) and Table 1 focuses on a alone. Should b not also enter into the picture? The second pertains to the lack of symmetry in the values of a for $p = .05$ and $.95$.

Regarding the two-component reliability model of Example 2 given in the paper (taken from Hills), I would like to know if the prior that is chosen has grounding in the rationale behind the model, since the dependency parameter ψ should have a physical interpretation. I am presuming that λ_1 and λ_2 have the usual de Finettian interpretation of the reciprocal of the average life of an infinite collection of exchangeable lifelengths.

REPLY TO DISCUSSION

Reply to Sweeting. There are two main issues raised by Sweeting's discussion. The first and most fundamental question is how one might find some guidance in the choice of alternative parameterizations to compare with the original parameterization θ. As Sweeting pointed out, our work did not address this important part of the problem. In describing the goals of reparameterization we contrasted the quite different situations in which one allows arbitrary transformations, or one restricts to transformations resulting in *interpretable* parameterizations. We concentrated on the latter. Here, one needs two things: a class of possible transformations, and a criterion to use in selecting a member from that class. Our paper was mainly a discussion of possible criteria, with emphasis on the more practically important and much more demanding multidimensional case.

As far as classes of transformations are concerned, one big distinction is between those that operate on each parameter separately and those that do not. A simple example of the former applies power transformations to each parameter in turn. Choosing five standard powers to maintain interpretability, $\nu = 1, \frac{1}{2}, \frac{1}{3}, 0, -1$, we obtain a class having m^5 members. It is quite easy to apply our criteria to this class. We are currently turning our attention to defining genuinely multivariate transformations that are reasonably general but remain simple. This is a challenging problem, however.

The second issue involves the use of graphical methods, such as the one suggested by Sweeting and illustrated by Hills and Smith. Apparently, these are to be used analogously to the way normal scores are used in considering data transformations. A basic question here, though, is whether this is a good use for graphical techniques. We have found that, as far as transformations are concerned, normal scores plots provide only very crude information such as "heavy-tailed" or "slighly skewed" or "strongly skewed". The nice thing about normal scores plots is that they also allow one to look for outliers, as well. When we are interested solely in transformations it is not clear that graphics add anything to what could be obtained by trying a series of alternative transformations, such as the power transformations mentioned in the previous paragraph. (For instance, as both Sweeting and Hills and Smith point out, the signed-root-log-posterior-ratio will tend to provide the same information as the third derivative of the log posterior.) However, appropriately chosen graphics may be able to display other things as well (perhaps unsuspected irregularities such as multimodality would appear). Now the question of what to plot is raised. An alternative to plotting the signed-root-log-profile-posterior-ratio (SRLPPR) discussed by Hills and Smith is to plot $\Phi(\hat{F}^{-1}(\theta_j))$ (or $\hat{F}(\Phi^{-1}(\theta_j))$) vs. θ_j for each $j = 1, \ldots, m$, where $\hat{F}$ is the approximate marginal posterior c.d.f. found by integrating the Laplace approximation to the marginal density and Φ is the standard normal c.d.f. This is easy to compute and would seem more closely analogous to the normal scores plot and easier to interpret than the SRLPPR.

A much more minor reply concerns Sweeting's suggestion that we use DFF's modified version of their approximation (their equation (19)) instead of the one we did use (their equation (12)). This seems quite reasonable to us, though it involves perhaps a little more numerical effort. A further possibility is to apply the Laplace approximation to the marginal density, and then integrate it numerically.

Reply to Singpurwalla. It seems to us that transforming the data is quite different than transforming the parameters. It is, of course, true that there is often quite a bit of arbitrariness in the choice of model and that could be considered simultaneously, but we have assumed the model to be given in our work. We look forward to seeing the referenced work of Abel and Singpurwalla.

The censored exponential is, as Singpurwalla suggests, more involved than the simple exponential. Some work on that model has been done by Achcar (1990). As far as symmetry is concerned, though, whenever the posterior on a single parameter γ is skewed, the exact tail probabilities at $\tilde{\gamma} \pm c \cdot \tilde{\sigma}_\gamma$, where $\tilde{\gamma}$ and $\tilde{\sigma}_\gamma$ are the mode and modal standard deviation, will generally be different. That is all that is going on our example (where $c = 1.64$).

Concerning the interpretation of the reliability model in our Example 2, we must admit that we did not think much about the model and prior when we took it from the thesis of Hills. We will defer to her for justification of the prior.

A general comment on automatic transformations. At the meeting, some of the questioning from the floor involved the use of automatic data-dependent transformations for use in computation. Sweeting touched on this, too, suggesting that the SRLPR (presumably, the SRLPPR in the multiparameter case) could be used. We have thought a little about this, and intend to work on the problem. We mention here several possible approaches.

Beginning with Sweeting's suggestion, we note that SRLPPR is itself no closer to normality than the original parameterization until it is standardized using its mean and standard deviation. Thus, a numerical routine would have to incorporate calculation of mean and variance, which is somewhat involved, though certainly manageable (at least, to the required order of accuracy, which would be $O(n^{-3/2})$). A variation would replace the profile posterior with the Laplace approximation to the marginal posterior density, and then again would compute an appropriate standardization. An alternative is to use $\phi_j = \Phi^{-1}(\hat{F}(\theta_j))$ where $\hat{F}$ is defined as above. Thus, for example, if Laplace's method were applied in computing means, variances, and marginal densities of functions $g(\theta)$, but implemented using this ϕ parameterization, we would obtain an iterated version of the Laplace approximation, which ought to be quite accurate. Finally, one might try to automate the power transformations mentioned above, or one could fit to each margin some convenient family of densities such as Type IV and then transform using the appropriate c.d.f.'s. There are clearly many ideas here, and we hope to see some progress sometime soon.

ADDITIONAL REFERENCES IN THE DISCUSSION

Bates, D. M. and Watts, D. G. (1988). *Nonlinear Regression Analysis and its Applications.* New York: Wiley.

Bickel, P. J. and Ghosh, J. K. (1990). A decomposition for the likelihood ratio statistic and the Bartlett correction, a Bayesian argument. *Ann. Statist.* **18**, 1070–1090.

Hills, S. E. and Smith, A. F. M. (1992). Parameterization issues in Bayesian inference. *Bayesian Statistics 4* (J. M. Bernardo, J. O. Berger, A. P. Dawid and A. F. M. Smith, eds.), Oxford: University Press, 227–246, (with discussion).

Sweeting, T. J. (1987a). Discussion of Cox and Reid. *J. Roy. Statist. Soc. B* **49**, 20–21.

Sweeting, T. J. (1987b). Approximate Bayesian analysis of censored survival data. *Biometrika* **74**, 809–816.

Sweeting, T. J. (1988). Approximate posterior distributions in censored regression models. *Bayesian Statistics 3* (J. M. Bernardo, M. H. DeGroot, D. V. Lindley and A. F. M. Smith, eds.), Oxford: University Press, 791–799.

BAYESIAN STATISTICS 4, pp. 307–320
J. M. Bernardo, J. O. Berger, A. P. Dawid and A. F. M. Smith, (Eds.)

Recent Extensions to the EM Algorithm

XIAO-LI MENG and DONALD B. RUBIN
University of Chicago, USA and *Harvard University, USA*

SUMMARY

Two recent extensions to the EM algorithm have made it even more powerful and applicable in practice. Both extensions follow the theme of EM by using repeated complete-data computations to handle incomplete data. The Supplemented EM or SEM algorithm (Meng and Rubin, 1991a) enables users of EM to calculate the incomplete-data asymptotic variance-covariance matrix associated with the maximum likelihood estimate obtained by EM, using only the computer code for EM and for the complete-data asymptotic variance-covariance matrix. The Expectation/Conditional Maximization or ECM algorithm (Meng and Rubin, 1991b) replaces the M-step of EM with a set of conditional maximization (CM) steps, and thus can eliminate undesirable nested iterations with EM when its M-step is not in closed form. SEM and ECM can also be combined into a Supplemented ECM (SECM) algorithm for computing the incomplete-data asymptotic variance-covariance matrix for maximum likelihood estimates found by ECM. In this paper, we provide a general description of these new techniques, as well as brief discussions of their relevance in Bayesian computations and their relationships with the popular Gibbs sampler (Geman and Geman, 1984), and its extension, GIBS.

Keywords: BAYESIAN INFERENCE; ECM; GIBBS; GIBS; INCOMPLETE DATA; ITERATIVE SIMULATION; MAXIMUM LIKELIHOOD; MISSING DATA; SEM.

PROLOGUE

The content of our contribution is actually quite closely connected to several implicit themes of the Fourth Valencia International Meeting on Bayesian Statistics, in particular to the dominant theme of using the Gibbs sampler to iteratively simulate posterior densities. The Gibbs sampler is a potentially extremely powerful tool, but it can, in some cases, easily lead to incorrect answers, as illustrated in Gelman and Rubin (1992). The epilogue of this paper, entiled "Back to the Future", points out that returning to the consideration of past techniques, such as EM and extensions discussed in our article, actually helps to provide a more accurate view of the future use of iterative simulation techniques, such as the Gibbs sampler, than is conveyed by the current rush of enthusiastic but, at times, potentially naive contributions.

1. THE EM ALGORITHM

Since its general form was established by Dempster, Laird and Rubin (1977, hereafter DLR), EM has not only been applied to an enormous number of real data sets, but it also has helped to stimulate new research efforts in statistics, including data augmentation (Tanner and Wong, 1987) and other related methods summarized in Gelfand and Smith (1990). This notable success of EM is not that surprising, for it reflects the maturing of an old intuitive idea for handling missing data. As written in Little and Rubin (1987, p.129), "The EM algorithm formalizes a relatively old ad hoc idea for handling missing data: (1) replace missing values by estimated values, (2) estimate parameters, (3) reestimate the missing values assuming the new estimates are correct, (4) reestimate parameters, and so forth, iterating until convergence." The fundamental distinction between EM and its ad hoc predecessors is that at each iteration EM finds the conditional expectation of the missing sufficient statistics, or more generally the missing complete-data loglikelihood, rather than individual missing values. But the key idea of iterating between imputing "missing values" and estimating parameters has proven to be a very rich and useful one in incomplete-data problems.

More specifically, let $f(Y|\theta)$ be the density of the complete data $Y = (Y_{obs}, Y_{mis})$, where Y_{obs} is the observed data, Y_{mis} is the missing data (or latent data with data augmentation), and θ is a d-dimensional parameter with the parameter space Θ. Starting with an initial value $\theta^{(0)} \in \Theta$, EM performs one E-step and one M-step at each iteration. The E-step "imputes" the complete-data loglikelihood $L(\theta|Y) = \log f(Y|\theta)$ with its conditional expectation with respect to the conditional density $f(Y_{mis}|Y_{obs}, \theta^{(t)})$:

$$Q(\theta|\theta^{(t)}) = \int L(\theta|Y)\, f(Y_{mis}|Y_{obs}, \theta^{(t)})\, dY_{mis}. \tag{1.1}$$

The M-step then finds the next iterate $\theta^{(t+1)}$ by maximizing this expected complete-data loglikelihood, $Q(\theta|\theta^{(t)})$, that is, it finds $\theta^{(t+1)}$ such that

$$Q(\theta^{(t+1)}|\theta^{(t)}) \geq Q(\theta|\theta^{(t)}), \quad \text{for all} \quad \theta \in \Theta. \tag{1.2}$$

A key result of DLR shows that each iteration of EM increases the observed-data loglikelihood $L(\theta|Y_{obs})$, and the sequence $\{\theta^{(t)}, t \geq 0\}$ converges reliably to a (local) maximizer of $L(\theta|Y_{obs})$ (DLR, Wu, 1983). Replacing $L(\theta|Y)$ by $L(\theta|Y) + \log \pi(\theta)$ in (1.1), where $\pi(\theta)$ is the prior density for θ, one can apply EM to find the posterior mode of θ given the observed data, Y_{obs}.

Since EM converts the difficult problem of incomplete-data maximization into a sequence of simple complete-data maximizations, the implementation of EM is often straightforward, especially in modern computing environments. Just like any iterative method, however, there are some features of EM that are unattractive. Complaints about EM include: (i) its convergence will be slow in the presence of large amounts of missing information, (ii) it does not automatically provide the asymptotic variance-covariance matrix associated with the MLE, (iii) its E-step can be difficult to evaluate, especially when the complete-data density is not from an exponential family, and (iv) the M-step itself requires iteration when the complete-data maximization is not in closed form. We view (i) as a consequence of the simplicity and stability of EM, and although several attempts have been made in the literature (e.g., Louis, 1982, Lansky and Casella, 1990) to speed up EM, these methods more or less destroy the simplicity or stability of EM (e.g., with these methods $L(\theta^{(t+1)}|Y_{obs})$ can be less than $L(\theta^{(t)}|Y_{obs})$). In many modern computing environments, we believe that it is more important to have simplicity and stability than rapid convergence. The second complaint can also be viewed as arising from the simplicity of EM, which does not require the computation of any derivatives of $L(\theta|Y_{obs})$; nevertheless, as we describe in Section 2, by applying the SEM algorithm, one can compute the desired asymptotic variance-covariance matrix using only the code for EM and for the complete-data asymptotic variance-covariance matrix.

In principle, the third difficulty can be overcome by numerical evaluation of the E-step, such as by Monte Carlo (Wei and Tanner, 1990), but these methods typically only work well when the complete-data density is from an exponential family because only in these cases is the form of $Q(\theta|\theta^{(t)})$ as a function of θ preserved under numerical integration over Y_{mis}. For a nonexponential complete-data density, the difficulty with the E-step is generally unavoidable and is inherent in the statistical problem. In contrast, the difficulty with the M-step is computational, and can be overcome by methods such as the ECM algorithm described in Section 3, which replaces the M-step with a set of CM (conditional maximization) steps. Section 4 presents a combination of SEM and ECM – the SECM algorithm – for obtaining standard errors of maximum likelihood estimates obtained by ECM. Finally Section 5 provides a brief discussion on the use of these new techniques in Bayesian computation, as well as their relationships with the Gibbs sampler.

2. THE SEM ALGORITHM

Since it is usually desirable for purposes of statistical inference to have not only the estimate itself but also the variance associated with it, several efforts have been made in the literature (e.g., Louis, 1982, Meilijson, 1989) to supplement EM with an associated asymptotic variance-covariance matrix. But because these methods do not fully follow the theme of EM — repeated computations using complete-data tools, their use in practice has been limited. Using the fact that the rate of convergence of EM is determined by the fractions of missing information, Meng and Rubin (1991a) propose a Supplemented EM (SEM) algorithm for computing the asymptotic variance-covariance matrix, which can be applied to any problem where EM has been applied and the complete-data asymptotic variance-covariance matrix is available.

The SEM algorithm is built upon the following two identities

$$I_{obs} = I_{com} - I_{mis} \tag{2.1}$$

and

$$DM = I_{mis} I_{com}^{-1}, \tag{2.2}$$

where I_{obs} is the observed-data observed information matrix, I_{com} is the conditional expectation of the complete-data observed information matrix evaluated at the MLE $\hat{\theta}$, I_{mis} is the information matrix for the conditional density $f(Y_{mis}|Y_{obs}, \theta)$ evaluated at $\hat{\theta}$, and $DM = DM(\hat{\theta})$ is the Jacobian matrix of the mapping $M(\theta)$ defined by EM: $\theta^{(t+1)} = M(\theta^{(t)})$, evaluated at $\hat{\theta}$. Equality (2.1) has been called "the missing information principle" by Orchard and Woodbury (1972) because it has a very appealing interpretation: *observed information = complete information – missing information.* The matrix on the right hand side of (2.2) provides "the fractions of missing information", and thus identity (2.2) gives the well-known result established in DLR: *the rate of convergence of EM is governed by the fractions of missing information.* Combining (2.1) and (2.2), we can compute the observed information matrix I_{obs} as

$$I_{obs} = (I - DM)\, I_{com}, \tag{2.3}$$

and its inverse, which gives the desired asymptotic variance-covariance matrix,

$$V_{obs} \equiv I_{obs}^{-1} = V_{com} + \Delta V, \tag{2.4}$$

where

$$V_{com} = I_{com}^{-1}, \tag{2.5}$$

and

$$\Delta V = V_{com}\, DM\, (I - DM)^{-1} \tag{2.6}$$

can be viewed as the increase in variance due to missing information.

As described in Meng and Rubin (1991a), computing I_{com} or V_{com} is as easy as computing a complete-data observed information matrix or its inverse, especially for practically important cases with EM where the complete-data density is from an exponential family. The DM matrix is obtained by computing the rate of convergence of EM, that is, by numerically differentiating the function $M(\theta)$, which only requires the code for the E- and M-steps. Because both the rate of convergence of EM and the proportional increase in the variance are determined by the fractions of missing information, SEM has a nice "self adjustment" property that makes it numerically stable. When the increase in variance is large, the convergence of EM is slow and thus DM, and thereby ΔV, are computed accurately. In contrast, when the convergence of EM is fast, the accurately calculated V_{com} becomes the dominant term in (2.4) because the fractions of missing information, and thus the increase in variance, are small.

Another nice feature of SEM is that the accuracy of the resulting variance-covariance matrix, V_{obs}, can be deduced by inspecting its symmetry, because ΔV, although algebraically symmetric, is not necessarily numerically so unless all the computations involved are correct and accurate. This potential lack of numerical symmetry provides a highly desirable routine diagnostic for numerical accuracy and programming errors in practice, a feature that other commonly used numerical methods (e.g., quasi-Newton-Raphson) do not have. Since SEM can be applied to compute the observed information matrix I_{obs}, it is also useful in detecting whether EM has converged to a (local) maximum by checking the positiveness of the eigenvalues of I_{obs}. For Bayesian computations, one can apply SEM to compute the large-sample posterior variance by replacing I_{com} with

$$I_{com} - \frac{\partial^2 \log \pi(\hat{\theta})}{\partial\theta\, \partial\theta},$$

where $\pi(\theta)$ is the prior density for θ.

3. THE ECM ALGORITHM

An important reason for the popularity of EM is that its M-step uses the same computational methods as used for complete-data maximum likelihood estimation, which often are in closed form. There exist classes of practically important models, however, that do not possess closed-form MLEs even with complete data (e.g., some loglinear models for contingency tables). In such cases, other iterative numerical methods, such as Newton-Raphson, are necessary for implementing the M-step of EM. Although EM is still quite useful for formalizing problems in such cases, some simplicity of implementation is lost because the resultant EM algorithm involves nested iterations. DLR noticed this potential undesirable feature of EM, and suggested that one possible way to overcome this difficulty was to increase $Q(\theta|\theta^{(t)})$ of (1.1) rather than to maximize it at each M-step. DLR call such a generalization of EM a generalized EM (GEM), which, although still increasing the likelihood at each iteration, will not appropriately converge in general.

Motivated by practical applications, Meng and Rubin (1991b) propose a specific type of the GEM algorithm that is more broadly applicable than EM, but converges appropriately just as EM does. It is called the Expectation/Conditional Maximization (ECM) algorithm because it takes advantage of the simple structure of complete-data conditional maximizations, just as EM takes advantage of the simplicity of complete-data maximizations. More specifically, let $G = \{g_s(\theta),\ s = 1, \ldots, S\}$ be a set of S fixed (vector) functions. At the tth iteration, after performing the E-step, the ECM algorithm replaces the original M-step by a sequence of S CM steps: for $s = 1, \ldots, S$, find $\theta^{(t+\frac{s}{S})}$ that maximizes $Q(\theta|\theta^{(t)})$ over $\theta \in \Theta$ subject to the constraint $g_s(\theta) = g_s(\theta^{(t+\frac{s-1}{S})})$. In other words, the sth CM step within the tth iteration of ECM is to find $\theta^{(t+\frac{s}{S})}$ such that

$$Q\left(\theta^{(t+\frac{s}{S})}|\theta^{(t)}\right) \geq Q\left(\theta|\theta^{(t)}\right), \text{ for all } \theta \in \Theta_s^{(t)} \equiv \left\{\theta \in \Theta : g_s(\theta) = g_s(\theta^{(t+\frac{s-1}{S})})\right\}. \tag{3.1}$$

The next iterate of θ, $\theta^{(t+1)}$, is defined as $\theta^{(t+\frac{S}{S})}$, the output of the last CM-step within the tth iteration. It is easy to see from (3.1) that the resulting iterative algorithm is a GEM, and thus always increases the likelihood.

Since our goal is to maximize $L(\theta|Y_{obs})$ over the whole parameter space Θ, not over a subspace of Θ, the set of constraint functions, G, must satisfy certain conditions in order to guarantee the appropriate convergence of ECM. As shown in Meng and Rubin (1991b), a sufficient (and almost necessary) condition is that G be "space filling", that is,

$$closure\left\{\sum_{s=1}^{S} a_s\eta_s;\ a_s \geq 0,\ \eta_s \in T_s(\theta^{(t)})\right\} = R^d, \text{ for all } t \geq 0, \tag{3.2}$$

where $T_s(\theta)$ $(s = 1, \ldots, S)$ is the set of all feasible directions at θ determined by the constraint space $\{\theta' \in \Theta,\ g_s(\theta') = g_s(\theta)\}$. Intuitively, condition (3.2) simply guarantees that at any $\theta^{(t)}$, one can search in any direction in Θ for the maximum, and thus the resulting limit of ECM is a (local) maximizer of $L(\theta|Y_{obs})$ over Θ. In fact, under condition (3.2), almost all the convergence results established in DLR and Wu (1983) for EM, which itself can be viewed as a special case of ECM by setting $S = 1$ and $g_1(\theta) =$ constant, can be extended to ECM with little difficulty.

In practice, it is almost always true that for any s, $g_s(\theta)$ is differentiable, and the corresponding gradient of $g_s(\theta)$, $\nabla g_s(\theta)$, is of full rank at any $\theta \in \Theta_o$, the interior of Θ. Under this condition, (3.2) is equivalent to the following condition

$$\bigcap_{s=1}^{S} J_s(\theta^{(t)}) = \{\vec{0}\}, \text{ for all } t \geq 0, \tag{3.3}$$

where $J_s(\theta)$ is the row space of $\nabla g_s(\theta)$. The advantage of (3.3) is that its verification is typically straightforward. Because (3.2) and (3.3) do not involve the data, whenever the CM algorithm (that is, ECM in the absence of missing data) converges to a (local) complete-data MLE, the corresponding ECM converges to a (local) observed-data MLE, assuming it does not converge to a saddle point (a theoretical but not practical concern).

The key feature that makes ECM more applicable than EM in practice is that for many pragmatically important models where the complete-data MLEs need numerical iteration, there is a suitable choice of G that satisfies (3.3) and has all corresponding CM-steps in closed form. Examples include loglinear models for contingency tables with missing data (Meng and Rubin, 1991c) and multivariate-normal stochastic-censoring models with linear constraints on the means (Meng and Rubin, 1991d). A particularly useful choice of G, which always satisfies (3.3), is to set $g_1(\theta) = (\theta_2, \ldots, \theta_S)^t, \ldots, g_S(\theta) = (\theta_1, \ldots, \theta_{S-1})^t$, so that $(\theta_1, \ldots, \theta_S)^t$ partitions the vector parameter θ. The resulting algorithm defines a sub-class of ECM – the partitioned ECM (PECM), which finds $\theta_s^{(t+1)}$ $(s = 1, \ldots, S)$ as the maximizer of $Q(\theta|\theta^{(t)})$ given $\theta_j = \theta_j^{(t+1)}$ for $j \leq s-1$ and $\theta_j = \theta_j^{(t)}$ for $j \geq s+1$. PECM is particularly important and useful because it is commonly encountered in practice, and in the absence of missing data, is the same as ICM (Iterated Conditional Modes; Besag, 1986). It also has some conceptual and practical connections with the Gibbs sampler, as discussed in Section 5.

There is also an interesting variation of ECM that we call the Multi-Cycle ECM (MCECM) algorithm. A "cycle" starts with an E-step followed by one or more CM-steps. The ECM algorithm defined above only involves one cycle per iteration, and generally we can insert more E-steps within each iteration if the computation of an E-step is relatively cheap. For example, suppose we want to insert one E-step before each of s_k^{th} $(k = 1, \ldots, K \leq S)$ CM steps; then (3.1) is replaced by

$$Q\left(\theta^{(t+\frac{s}{S})} \,|\, \theta^{(t+\frac{s_k-1}{S})}\right) \geq Q\left(\theta \,|\, \theta^{(t+\frac{s_k-1}{S})}\right) \quad \text{for all } \theta \in \Theta_s^{(t)},\ s_k \leq s \leq s_{k+1}-1, \tag{3.4}$$

where $k = 1, \ldots, K$, $s_1 \equiv 1$ and $s_{K+1} \equiv S + 1$. The resulting iterative algorithm is a K-cycle ECM, and still increases the likelihood at each iteration because each cycle does so. All the convergence properties of an ECM also apply to an MCECM under condition (3.2). Obviously, MCECM involves more computation per iteration than ECM, since it performs more E-steps, but, surprisingly, this does not necessarily imply that MCECM converges more quickly than ECM in terms of number of iterations. Even more surprisingly, there are cases where ECM converges more quickly than EM! These examples and results on the rate of convergence of ECM and MCECM can be found in Meng (1990, 1991).

4. THE SECM ALGORITHM

Just as we have SEM to supplement EM with an asymptotic variance-covariance matrix, it is natural to want a supplemented ECM (SECM) algorithm to calculate the asymptotic variance-covariance matrix associated with the estimates obtained by the ECM algorithm. As we have seen in Section 2, equation (2.2), the SEM algorithm is built upon the fundamental relationship between the rate of convergence of EM and the fractions of missing information. The techniques used with SEM for computing the rate of convergence of EM can be applied to compute the rate of convergence of ECM, but the SEM formulae (2.3) or (2.6) cannot be used directly with ECM because these formulae use the rate of convergence of EM, which, in general, does not equal the rate of convergence of ECM. In fact, as shown in Meng (1990, 1991), under condition (3.3),

$$I - DM^{ECM} = (I - DM)(I - DM^{CM}) \tag{4.1}$$

where I is the $d \times d$ identity matrix, DM^{ECM} is the rate of convergence of ECM, and DM^{CM} is the rate of convergence of CM, that is, ECM without missing data. Equation (4.1) can be interpreted as *speed of ECM* = *speed of EM* × *speed of CM*, which is quite intuitive since ECM can be viewed as a composition of EM and CM.

From (4.1), we can easily find the SECM counterparts of (2.3) and (2.6),

$$I_{obs} = \left(I - DM^{ECM}\right)\left(I - DM^{CM}\right)^{-1} I_{com} \tag{4.2}$$

and

$$\Delta V = V_{com}\left(DM^{ECM} - DM^{CM}\right)\left(I - DM^{ECM}\right)^{-1}, \tag{4.3}$$

respectively. The matrix DM^{CM} depends only on V_{com} and $\nabla_s = \nabla g_s(\hat{\theta})$ $(s = 1, \ldots, S)$, the gradient of g_s at $\hat{\theta}$ (Meng, 1991):

$$DM^{CM} = \prod_{s=1}^{S} \left[\nabla_s (\nabla_s' V_{com} \nabla_s)^{-1} \nabla_s' V_{com}\right], \tag{4.4}$$

and thus can be computed once V_{com} and ∇_s $(s = 1, \ldots, S)$ are calculated. In other words, SECM uses the identical method as SEM to compute the rate of convergence of ECM, DM^{ECM}, and then applies (4.3) to calculate the increase in variance due to missing information. The extra work with SECM is the algebraic calculation of DM^{CM} given by (4.4), which does not involve any missing-data quantities once $\hat{\theta}$ is computed, and reduces to a zero matrix for EM. For multi-cycle ECM, the above SECM scheme does not work directly because the rate of convergence of MCECM has a more complicated expression than (4.1). Meng and Rubin (1991e) propose a modified SECM to handle this, which effectively converts any multi-cycle problem into a single-cycle one, and then applies (4.2) or (4.3).

5. DISCUSSION

As described in Meng (1990), the special case of ECM, PECM, has an interesting analogy with the Gibbs sampler (Geman and Geman, 1984), a currently popular device in Bayesian statistics. The Gibbs sampler is useful when it is difficult or impossible to draw from the joint density $\mathcal{L}(U_1, \ldots, U_S)$, but it is relatively easy to draw from all conditional densities $\mathcal{L}(U_s|U_{s'};\, s' \neq s)$ $(s = 1, \ldots, S)$. This situation is very similar to the situations where PECM is useful: the complete-data joint MLE for $(\theta_1, \ldots, \theta_S)$ is hard to calculate, but the complete-data conditional MLE of θ_s given $\{\theta_{s'}\,; s' \neq s\}$ is in closed form for all s. With the Gibbs sampler, given the current draws $(U_1^{(t)}, \ldots, U_S^{(t)})$, $U_1^{(t+1)}$ is drawn from $\mathcal{L}(U_1|U_2^{(t)}, \ldots, U_S^{(t)})$, and then $U_2^{(t+1)}$ is drawn from $\mathcal{L}(U_2|U_1^{(t+1)}, U_3^{(t)}, \ldots, U_S^{(t)})$, etc., until $U_S^{(t+1)}$ is drawn from $\mathcal{L}(U_S|U_1^{(t+1)}, \ldots, U_{S-1}^{(t+1)})$. Replacing the conditional draws by conditional maximizations, the CM-steps of PECM mimic exactly the same process. In this sense, PECM can be regarded as a deterministic version of the Gibbs sampler, just as data augmentation can be viewed as a stochastic generalization of EM (Wei and Tanner, 1990).

The difference between the Gibbs sampler and PECM in applications to missing-data problems, besides the obvious difference in operations, is that Gibbs uses the same operation (i.e., random draws) for both missing data and parameters, whereas PECM uses different operations for missing data (averaging) and for parameters (maximization). Besides these differences, however, there are theoretical and practical advantages in tracing analogies between these two iterative algorithms. Theoretically, because both methods share the same "alternating" scheme, some results, such as the effect of different orderings of alternations on the rate of convergence, obtained for one method may shed some light on the other. Also, the generalization from PECM to ECM indicates there is a possible analogous generalization for the Gibbs sampler, the generalized Gibbs sampler or "GIBS" — Generalized Iterative Bayesian Simulation, which has immediate applications (e.g., Bayesian analysis for loglinear models with incompletely observed contingency tables).

Practically, linking PECM with the Gibbs sampler provides an alternative approach to Gibbs for handling complicated missing data problems in which EM or data augmentation is hard to implement. Specifically, the simple structure of the conditional densities employed by the Gibbs sampler typically implies that the corresponding conditional modes are in closed-form or easy to obtain and thus implies that PECM can be used to calculate the posterior mode of the parameter of interest, and that the corresponding large-sample posterior variance can be obtained by SECM if the complete-data counterpart is available.

6. EPILOGUE — BACK TO THE FUTURE

In maximization problems, complications caused by the possible existence of local modes are well known. As a result, good practice typically uses several iterative sequences starting from "spread-out" or "over-dispersed" starting values. This approach is well-known and standard. With ECM, it is easy to run several sequences, especially with parallel processing machines, and see if they converge to the same value. Also, when using ECM for maximization, there are straightforward points to check for correct coding (e.g., the increase of likelihood at each iteration, the positive semi-definiteness of the variance-covariance matrix from SECM).

For iterative simulation problems, the complication of long-range dependence, especially with marginal summaries of an underlying multicomponent quantity, seems to be less well-known than it should be. In fact, the suggestion to use multiple starts for iterative simulation is often regarded as "wrong" by people who known the corresponding conclusion for maximization.

In fact, routine use of GIBS might begin with the use of ECM to search out modes and SECM to estimate the mass of any local modes, and might include the comparison of ECM and GIBS results to help detect errors (see Gelman and Rubin, 1991, for specific proposals). In any case, reliable future uses of the Gibbs sampler might look more like past reliable uses of maximization routines than the present uses of Gibbs.

ACKNOWLEDGEMENTS

This manuscript was prepared using computer facilities supported in part by the National Science Foundation Grant DMS 86-01732 and DMS 87-03942 awarded to the Department of Statistics at the University of Chicago, and by The University of Chicago Block Fund. This work was also partially supported by National Science Foundation grant SES 88-05433. Requests for relevant reprints should be made to Xiao-Li Meng, Department of Statistics, University of Chicago, 5734 University Ave., Chicago, IL 60637, USA.

REFERENCES

Besag, J. (1986). On the statistical analysis of dirty pictures. *J. Roy. Statist. Soc. B* **48**, 259–302, (with discussion).

Dempster, A. P., Laird, N. M. and Rubin, D. B. (1977). Maximum likelihood estimation from incomplete data via the EM algorithm (with discussion). *J. Roy. Statist. Soc. B* **39**, 1–38.

Gelfand, A.E. and Smith, A. F. M. (1990). Sampling–based approaches to calculating marginal densities. *J. Amer. Statist. Assoc.* **85**, 398–409.

Gelman, A. and Rubin, D. B. (1991). Honest inferences from iterative simulation. *Tech. Rep.* University of California, Berkeley.

Gelman, A. and Rubin, D. B. (1992). A single series from the Gibbs sampler provides a false sense of security. *Bayesian Statistics 4* (J. M. Bernardo, J. O. Berger, A. P. Dawid and A. F. M. Smith, eds.), Oxford: University Press, 625–631.

Geman, S. and Geman, D. (1984). Stochastic relaxation, Gibbs distribution, and the Bayesian restoration of images. *IEEE Transaction on Pattern Analysis and Machine Intelligence* **6**, 721–741.

Lansky, D. and Casella, G. (1990). Improving the EM algorithm. *Computing Science and Statistics: Proceedings of the Twenty–Second Symposium on the Interface.* Washington DC: ASA.

Little, R. J. A. and Rubin, D. B. (1987). *Statistical Analysis with Missing Data.* New York: Wiley.

Louis, T. A. (1982). Finding the observed information matrix when using the EM Algorithm. *J. Roy. Statist. Soc. B* **44**, 226–233.

Meilijson, I. (1989). A fast improvement to the EM algorithm on its own terms. *J. Roy. Statist. Soc. B* **51**, 127–138.

Meng, X. L. (1990). *Towards Complete Results for some Incomplete–Data Problems.* Ph.D. Thesis, Harvard University, Department of Statistics.

Meng, X. L. (1991). On the rate of convergence of the ECM algorithm. *Tech. Rep.* University of Chicago.

Meng, X. L. and Rubin, D. B. (1991a). Using EM to obtain asymptotic variance–covariance matrices: the SEM algorithm. *J. Amer. Statist. Assoc.* **86**, (to appear).

Meng, X. L. and Rubin, D. B. (1991b). Maximum likelihood estimation via the ECM algorithm: a general framework. *Biometrika* , (to appear).

Meng, X. L. and Rubin, D. B. (1991c). IPF for contingency tables with missing data via the ECM algorithm. *Proceedings of the Statistical Computing Section of the American Statistical Association*, (to appear).

Meng, X. L. and Rubin, D. B. (1991d). Fitting patterned normal models via the ECM algorithm. *Tech. Rep.* University of Chicago.

Meng, X. L. and Rubin, D. B. (1991e). Maximum likelihood estimation via the ECM algorithm: the asymptotic variance. *Tech. Rep.* University of Chicago.

Orchard, T. and Woodbury, M. A. (1972). A missing information principle: theory and application. *Proceedings of the 6th Berkeley Symposium on Mathematical Statistics and Probability* **1**, 697–715.

Tanner, M. A. and Wong, W. H.(1987). The calculation of posterior distributions by data augmentation. *J. Amer. Statist. Assoc.* **82**, 805–811.

Wei, G. C. G and Tanner, M. A. (1990). A Monte Carlo implementation of the EM algorithm and the poor man's data augmentation algorithms. *J. Amer. Statist. Assoc.* **85**, 699–704.

Wu, C. F. J (1983). On the convergence properties of the EM algorithm. *Ann. Statist.* **11**, 95–103.

DISCUSSION

C. P. ROBERT (*Université Paris VI, France*)

1. Introduction. Meng and Rubin (hereafter MR) develop in this paper several extensions of EM, along two main axes: (ii) computation of the associated information matrix and (iv) effective augmentation of the observed likelihood when the actual computation of the maximum of the completed-likelihood is not possible. These two points are considered by the authors as the main drawbacks of EM or, at least, as those which can be solved. The other mentioned problems are (i) slow convergence and (iii) E-step evaluation. I would like to point out two other major drawbacks of EM, namely that (v) the fixed points of EM contain local maxima, but also saddle-points of the likelihood function and that (vi) the algorithm is very unstable for small samples, i.e., very dependent on the starting point. In the special case of mixtures of distributions, an additional criticism is that (vii) the number of components has to be known beforehand; practically, this is an important restriction.

This why, instead of discussing the proposed extensions, I present here two stochastic extensions of EM, SEM and SAEM, because they deal rather efficiently with points (v) and (vii), bring partial solutions to (iii) and (vi), while being satisfactory for (i) and (iv). In these methods, introduced by Celeux and Diebolt (1985), 'S' stands for 'stochastic' and 'SA' for 'simulated annealing', as the missing data is simulated. This approach has been recently exploited in Wei and Tanner (1990).

In the last part of the paper, MR also discuss a formal connection between PECM and Gibbs sampling, since both approaches follow the same iterative scheme. If we focus again on the special case of *mixtures*, it appears that the connection is even stronger and takes place at a somehow more interesting level. In fact, Diebolt and Robert (1990) showed that Gibbs sampling can be used very efficiently in the computation of Bayes estimators for mixtures of exponential families. When the Bayes estimators are used in an iterative way called *prior feedback*, the resulting estimator is close to a local maximum of the likelihood. Therefore, the efficiency of Gibbs sampling is such that not only can a single Bayes estimator be computed in a reasonable time but even a sequence of these estimators can be produced for this alternative to EM and SEM. Obviously, the prior feedback algorithm has no real Bayesian justification and can be considered, at best, as an *empirical Bayes* approach. However, in practice, it seems to perform quite well, especially in delicate cases (see Robert and Soubiran, 1991), while being founded theoretically.

2. SEM algorithm. Instead of maximizing the expected complete-data likelihood, the SEM algorithm first simulates the missing data from the conditional density,

$$Y_{mis}^{(t+1)} \sim f\left(Y_{mis}|Y_{obs}, \theta^{(t)}\right)$$

(with MR notations) and then computes the maximum of the pseudo-completed loglikelihood, $L(Y_{obs}, Y_{mis}^{(t+1)}|\theta)$, thus producing the next estimator $\theta^{(t+1)}$. Obviously, the stochastic modification implies that one of the major features of EM disappears, namely that the observed likelihood is automatically increasing from one iteration to the next one. However, this feature is also responsible for (v), while SEM avoids saddle points.

In the special case when the missing data Y_{mis} belongs to a finite-state space, the Markov chain produced by SEM, $\theta^{(n)}$, is ergodic. Furthermore, the corresponding invariant distribution Ψ is such that $N^{-1/2}[\theta_\Psi - \hat{\theta}_N]$ converges to a $\mathcal{N}(0, \Sigma)$ distribution, where N is the sample size, θ_Ψ is the mean of Ψ and $\hat{\theta}_N$ the unique consistent solution of the likelihood equation (Celeux and Diebolt, 1986). After T_0 initializing iterations, the proposed estimator of θ is then approximating θ_Ψ by

$$\frac{1}{T}\sum_{t=1}^{T} \theta^{(t+T_0)}.$$

Celeux and Diebolt (1986) produce a large amount of experimental evidence which shows that, for 'reasonable' sample sizes, SEM is often preferable to EM. In some cases, it accelerates the convergence [(i)] and the stationary distribution Ψ is usually concentrated around a local maximum of the likelihood. Moreover, for mixtures, it allows mispecification about the number of components since an upper bound on the number of components is sufficient to insure convergence to the actual number if the sample size is large enough [(vii)].

At last, in some particular cases, SEM may even provide a good alternative when the E-step evaluation is too intricate. For instance, when considering *censored data*, it is usually much easier to simulate the censored observations than to work with the expected completed-likelihood (see Wei and Tanner, 1990 and Chauveau, 1990). A point still unsolved deals with the variance of the stationary distribution Ψ and its relation with the Fisher information [(ii)]. Note also that SEM only brings a slight improvement over EM for (vi), i.e. that it is almost as much unstable for small samples.

3. SAEM algorithm. Another method, called SAEM for Simulated Annealing EM, was proposed in Celeux and Diebolt (1990) in order to overcome some of the drawbacks of SEM. It depends on a *cooling temperature*, γ_n, such that the nth iteration step can be written as

$$\theta_{SAEM}^{(n)} = \gamma_n \theta_{SEM}^{(n)} + (1-\gamma_n)\theta_{EM}^{(n)}, \tag{3.1}$$

$\theta_{SEM}^{(n)}$ and $\theta_{EM}^{(n)}$ being determined through SEM and EM conditionally on the previous estimator, $\theta_{SAEM}^{(n-1)}$. The sequence γ_n is chosen to decrease slowly from 1 to 0. Therefore, the estimator in (3.1) is getting closer to an EM estimator as n increases.

The purpose behind this modification is threefold: (a) it allows a stochastic perturbation of EM at the beginning, in order to avoid saddle-points or close maxima attraction; (b) since γ_n is converging to 0, the resulting estimator can satisfy some *pointwise* convergence results and therefore gives more satisfactory answers than SEM (which only converges in distribution); (c) it should also increase the speed of the algorithm, as EM can be rather fast in the neighborhood of a main local maxima. In addition, the progressive diminution of the variance implies that SAEM is more stable than SEM for small samples. Under mild conditions, $\theta_{SAEM}^{(n)}$ is shown to converge a.s. to a local maxima of the likelihood if the cooling temperature is $cn^{-\mu}$ with $0 < \mu < 1$ (Celeux and Diebolt, 1990).

From a practical point of view, SAEM also enjoys the properties of SEM given in §2 and thus answers most of the criticisms addressed to EM, except for (ii). It has been observed through simulations that SAEM also works better for small sample sizes (Celeux and Diebolt, 1990). For instance, it also detects the proper number of components [(viii)]. Therefore, it appears preferable to both EM and SEM, even though it requires an additional computational burden. A remaining drawback is that SAM limit still depends on the starting conditions but experience has shown that convergence was actually more robust.

4. *Prior feedback.* We now turn to a new method called *prior feedback*, which can be considered as an alternative maximum likelihood algorithm relying on Bayesian tools. We only expose here the heuristics of this method, more detailed theoretic considerations being given in Robert and Soubiran (1991).

The method is based upon an iterative replacement (or *feedback*) of the hyperparameters of conjugate priors by the corresponding Bayes estimators ('posterior hyperparameters') until convergence. This approach has a typical empirical Bayes flavor, since we are actualizing the hyperparameters through the data. Prior feedback obviously calls for setups where conjugate priors are available, i.e. exponential families or mixtures of exponential families. Let us consider first the case of an exponential family,

$$f(x|\theta) = e^{\theta \cdot x - \psi(\theta)} h(x),$$

where conjugate priors are of the form

$$\pi(\theta|x_0, \lambda) = K(x_0, \lambda) e^{\theta \cdot x_0 - \lambda\psi(\theta)}.$$

It is well-known (see, e.g., Diaconis and Ylvisaker, 1979) that the prior and posterior expectations of the sample mean, $\nabla\psi(\theta)$, are related by

$$\mathrm{E}^{\pi}[\nabla\psi(\theta)] = \frac{x_0}{\lambda}, \qquad \mathrm{E}^{\pi}[\nabla\psi(\theta)|x] = \frac{x + x_0}{\lambda + 1}.$$

The prior feedback method then replaces the initial hyperparameter x_0 by x_1 such that

$$\frac{x_1}{\lambda} = \frac{x + x_0}{\lambda + 1}$$

and the resulting sequence of hyperparameters, (x_n), satisfies

$$\frac{x_n}{\lambda} = \frac{x + x_{n-1}}{\lambda + 1}.$$

It is straightforward to verify that the limit of the sequence is λx. Therefore, the limit of the associated sequence of Bayes estimators of $\nabla\psi(\theta)$ is x, maximum likelihood estimator, *for every choice of* λ. A similar result holds, when a is large enough, if we estimate a one-to-one function of θ, $h(\theta)$, and update x_n in terms of $\mathrm{E}^{\pi}[h(\theta)]$.

If we now consider the *mixture* model

$$x \sim \sum_{i=1}^{k} p_i e^{\theta_i \cdot x - \psi_i(\theta_i)} h_i(x),$$

we can still use conjugate priors,

$$\pi(\theta_i) = K_i(x_0^i, \lambda_i)e^{\theta_i \cdot x_0^i - \lambda_i \psi_i(\theta_i)}, \qquad p \sim \mathcal{D}(\alpha_1^0, \ldots, \alpha_n^0),$$

with $\sum_i \alpha_i^0 = m$. The analytical computation of the Bayes estimators gets too costly as the number of observations grows moderate. However, Gibbs sampling provides an efficient and fast approximation technique to compute the Bayes estimators. The hyperparameters are then updated in the following way:

$$\frac{x_t^i}{\lambda_i} = \mathrm{E}^\pi[\nabla\psi_i(\theta_i)|x_1, \ldots, x_n, x_{t-1}^i], \qquad \frac{\alpha_i^t}{m} = \mathrm{E}^\pi[p_i|x_1, \ldots, x_n, \alpha^{t-1}],$$

while m and λ_i remain constant $(1 \leq i \leq k)$. Note that, without Gibbs sampling, this algorithm could not be implemented at all (a detailed study of Gibbs sampling in the setting of mixtures, including convergence results, is given in Diebolt and Robert (1990)).

Analytic considerations in the case when only the weights are unknown, as well as simulations in various mixture setups, have shown that this algorithm was converging to a fixed point close to a local maximum of the likelihood (this can be checked by initializing EM or SEM at the resulting estimator). In addition, analytical results have suggested the following practical implementation: namely, to start the algorithm with large variances for the conjugate priors and then decrease the variance when stabilization occurs.

Further theoretical developments have indeed to be undertaken to establish more thoroughly the importance of prior feedback but the simulations we have observed show the potential of prior feedback, especially for small sample sizes. Actually, prior feedback appears to reach a stability missing in SAEM (and obviously SEM), while partially avoiding the artifacts of the likelihood.

S. E. FIENBERG (*York University, Canada*)

As with the EM algorithm, Meng and Rubin's ECM algorithm has been anticipated, in particular by Chen (1972) and Chen and Fienberg (1974, 1976). The problems considered in these earlier works relate to maximum likelihood estimation for loglinear models fitted to contingency tables. For the corresponding complete–data problems the minimal sufficient statistics are marginal totals and the well-known iterative proportional fitting algorithm can be viewed as a sequence of conditional maximization steps. The approach taken to the corresponding incomplete data problems by Chen and Fienberg intersperses component estimation steps with conditional maximization steps to achieve an algorithm that converges when MLEs exist and are unique.

In the discussion section, Meng and Rubin note that a possible generalization for the Gibbs sampler has immediate applications for Bayesian analysis for loglinear models with contingency tables. In fact, Epstein and Fienberg (1991) have developed such an approach which has the potential of linking the iterative proportional fitting algorithm and the Gibbs sampling in a sequence of interlocking steps analogous in some ways to those in the work of Chen and Fienberg.

REPLY TO THE DISCUSSION

We thank the discussants for their interest in our work and for their contributions.

With respect to the work of the scheduled discussant, Robert, we feel that his focus is rather different from ours. Our work is directed at finding extensions of EM that add practical utility while maintaining the stability and simplicity of EM. We strongly believe that the usefulness of any general statistical methodology for practice depends crucially on its simplicity and stability when applied to a variety of problems. The popularity of EM dramatically illustrates this theme.

The stochastic extensions of EM described in Robert's comment, however, may not maintain the stability of EM ("one of the major features of EM disappears"), may increase the complexity of implementation, and moreover may generate results that are difficult to interpret statistically (e.g., a distribution for $\theta - \hat{\theta}$ but neither its posterior nor sampling distribution). Of course, a sophisticated user may well be willing to apply a complicated method if it solves the problem, but the proposed stochastic extensions of EM cannot solve the statistical problems underlying Robert's criticisms of EM.

For example, Robert's "drawback" (vi) basically is concerned with the existence of many modes (possibly connected) of likelihoods or posterior densities, which is more likely to occur with small samples than large ones. It is well-known that with such functions, the convergence of any deterministic iterative algorithm, including EM, will depend on the starting point; similarly, his points (v) and (vii) can be taken as criticisms about all deterministic iterative algorithms. In fact, EM, with its monotone increase of the likelihood function, is far more stable than other commonly used iterative algorithms, such as Newton-Raphson, whose convergence depends heavily on the quadratic curvature of the loglikelihood.

Even if we had an algorithm with guaranteed convergence to the global mode, it would still not solve the statistical problem because summarizing a multi-modal posterior density by its global mode and the local curvature at it would not provide a complete statistical inference. In such cases, a stochastic iterative algorithm, such as GIBS, may be needed to map out the entire posterior density, and thereby to obtain a statistically adequate inference. The stochastic algorithm is used here, however, to simulate the entire density, not a mode. Although, at present, simulation techniques appear to be seductive to many statisticians, they are not appropriate for all problems. In analogy with an old Chinese saying: If you want to kill a fly, use a fly swatter; if you want to kill a turkey, use a gun. It is not helpful to complain that the fly swatter is an inappropriate tool because it cannot kill a turkey in a reasonable amount of time, and it is almost surely a waste of time to try to kill a fly using a gun!

With respect to Stephen Fienberg's comment, we feel the ECM situation in 1992 is likely to be very similar to the EM situation in 1977. Because the idea of ECM is so intuitive, it is natural to expect that many researchers have applied ECM to specific examples without realizing its generality. For example, in a paper entitled "Estimation of seemingly unrelated Tobit regressions via the EM algorithm", Hwang, Sloan and Adamache (1987) actually applied ECM but called it EM: their "M-step" (their equations (15) and (16)) is in fact a set of two CM-steps. If they had tried to apply EM, that is, if they had tried to implement the M-step (i.e., complete-data ML estimation), they would have found that it is not in closed form. Similarly, our reading of Chen (1972) and Chen and Fienberg (1974, 1976) did not reveal to us any recognition of the idea of efficiently combining two iterative maximization algorithms (i.e., EM and CM), or any suggestion that the techniques underlying IPF could be generalized to many other models.

Having not been able to obtain a copy of Epstein and Fienberg (1991), we are not able to comment knowledgeably on it. Judging from the title and the previous work mentioned above, however, we suspect its focus is again on a *special case* of a general technique, GIBS, but without the general formulation, which we believe is in fact quite subtle and not easy to define precisely.

The primary point of the SEM, ECM and GIBS formulations in our article is to foster the recognition of the generality of the ideas and thereby to stimulate new applications and recover old ones. Judging from the discussions, both written and oral at the conference, we are very encouraged. We hope our extensions of EM will be of general help to statisticians in practice, just as EM itself was.

ADDITIONAL REFERENCES IN THE DISCUSSION

Celeux, G. and Diebolt, J. (1985). The SEM algorithm: a probabilistic teacher derived from the EM algorithm for the mixture problem. *Comput. Stat. Quater.* **2**, 73–82.

Celeux, G. and Diebolt, J. (1986). L'algorithme SEM: un algorithme d'apprentissage probabiliste pour la reconnaissance des mélanges de densité. *Revue Statis. Appl.* **34**, 35–52.

Celeux, G. and Diebolt, J. (1990). Une version de type recuit-simulé de l'algorithme EM. *Notes aux Comptes Rendus Acad. Sciences de Paris* **310**, 119–124.

Chauveau, E. (1990). Algorithmes EM et SEM pour un mélange caché et censuré. *Rencontres Franco-Belges de Statistique*.

Chen, T. (1972). Mixed-up frequencies and missing data in contingency tables. Ph.D. Thesis, Department of Statistics, University of Chicago.

Chen, T. and Fienberg, S. E. (1974). Two-dimensional contingency tables with both completely and partially cross-classified data. *Biometrics* **30**, 629–642.

Chen, T. and Fienberg, S. E. (1976). The analysis of contingency tables with incompletely classified data. *Biometrics* **32**, 133–144.

Diaconis, P. and Ylvisaker, D. (1979). Conjugate priors for conjugate families. *Ann. Stat.* **7**, 269–281.

Diebolt, J. and Robert, C. (1990). Estimation of finite mixture distributions through Bayesian sampling. *Tech. Rep.* **121**, LSTA, Univ. Paris VI.

Epstein, L. D. and Fienberg, S. E. (1991). Using Gibbs sampling for Bayesian inference in multidimensional contingency tables. (Unpublished *Tech. Rep.*).

Hwang, C. J., Sloan, F. A. and Adamache, K. W. (1987). Estimation of seemingly unrelated Tobit regressions via the EM algorithm. *Journal of Business and Economic Statistics* **5**, 425–430.

Robert, C. and Soubiran, C. (1991). Estimation of a mixture model through Bayesian sampling and prior feedback. *Tech. Rep.* **138**, LSTA, Univ. Paris VI.

BAYESIAN STATISTICS 4, pp. 321–344
J. M. Bernardo, J. O. Berger, A. P. Dawid and A. F. M. Smith, (Eds.)

Hierarchical Models for Combining Information and for Meta-Analyses

C. N. MORRIS and S. L. NORMAND
Harvard University, USA*

SUMMARY

The science and art of combining results from similar and independent experiments sometimes is called "meta-analysis", especially in education, medicine and psychology. Hierarchical models provide an ideal perspective for classifying, understanding, and generalizing such analyses. We show how methods and perspectives already developed for the hierarchical Bayesian and the empirical Bayesian approaches may be used to enhance the science of combining information, especially for meta-analyses.

Keywords: EMPIRICAL BAYES; HIERARCHICAL BAYES; RANDOM EFFECTS; BORROWING STRENGTH; CARCINOGENIC; GIBBS SAMPLING; ADJUSTED LIKELIHOOD.

1. INTRODUCTION

The term "meta-analysis" generally refers to the practice of using statistical methods to combine data from related, but statistically independent studies for the purpose of summarizing information about certain treatment effects. Each of the k studies provides information about an effect parameter, which we denote θ_i here for the ith study, $i = 1, \cdots, k$, and which for convenience here will be one dimensional. In fact, the ideas presented here are equally applicable in much more general settings, but our main two concerns here are to point out that: (1) meta-analyists profitably could make greater use of existing hierarchical and empirical Bayes methods already developed, and (2) that these frameworks will lead naturally to appropriate extensions for further development of meta-analytic methods.

Significant applications of hierarchical models already have been made to meta-analysis. Variance component models have been proposed and developed extensively as random effects models, e.g. Bunker, Barnes and Mosteller (1977), DerSimonian and Laird (1986), Goodman (1989) and Hedges and Olkin (1985). These approaches are aimed principally at estimating the mean, μ, of the population of effects, with θ_i as a sample from that population. Less well-known is the work of DuMouchel (DuMouchel and Harris (1983), DuMouchel (1990)), Raudenbush and Bryk (1985), and Laird and Louis (1989) which focuses primarily on issues of borrowing strength from the other $k - 1$ related studies for the purpose of improving estimation of the remaining one.

This paper is aimed at showing that the hierarchical model, as pictured via the Hierarchical Diagram (Table 1), provides a natural way to do meta-analysis, from the perspectives of modeling and of theoretical development. (Equivalently, empirical Bayes modeling as

* Both authors are in the Department of Health Care Policy, Harvard Medical School, 25 Shattuck Street, Parcel B - 1st Floor, Boston, MA 02115. Professor Morris is also at the Department of Statistics, Harvard University, 1 Oxford Street, Cambridge, MA 02138.

in Morris (1983, 1988a) serves as a foundation for hierarchical modeling. Empirical Bayes terminology is not emphasized here because its meaning has been used by most statisticians in a more restricted way than in Morris (1983), and because the principal concerns here are of concepts, not terminology.) This paper treats: concepts and theory (Section 2); existing methods in meta-analysis, and the availability of off-the-shelf methods that are ready for meta-analysis, even if not developed for that purpose (Section 3); and an example (Section 4). In addition to off-the-shelf methods, Bayesian simulation methods, which are of intense interest currently because they simplify the problem of sampling from the posterior distribution, will be illustrated with data, for comparison with more traditional techniques.

	DESCRIPTIVE MODEL	INFERENTIAL MODEL
OBSERVED DATA $\mathbf{y}$	$y_i \mid \theta_i \overset{\text{indep.}}{\sim} f_i(y_i \mid \theta_i)$ $i = 1, \cdots, k$	$y_i \mid \alpha \overset{\text{indep.}}{\sim} f_i^*(y_i \mid \alpha)$ $i = 1, \cdots, k$ $\alpha \in \mathcal{A}$
UNKNOWN PARAMETERS θ	$\theta_i \mid \alpha \overset{\text{indep.}}{\sim} g(\theta_i \mid \alpha)$ $i = 1, \cdots, k$ $\alpha \in \mathcal{A}$ exchangeable	$\theta_i \mid y_i, \alpha \overset{\text{indep.}}{\sim} g_i^*(\theta_i \mid y_i, \alpha)$ $i = 1, \cdots, k$ $\alpha \in \mathcal{A}$

Table 1. *Hierarchical diagram for simultaneous inference of independent effects in k studies.*

We illustrate methods using a random sample of $k = 12$ studies drawn from data reported in Laird and Louis (1989). Each study investigates the carcinogenicity of a distinct chemical, yielding results for 12 chemicals corresponding to the 12 studies. In each study the National Cancer Institute subjected groups of male rats to a range of doses. Typically, groups consisted of 50 rats and the doses were at three levels. After two years, a binary variable indicating prescence or absence of malignant tumours was recorded for each rat in each study. Table 2 contains the 12 distinct chemical identification numbers along with the estimated slope, y_i, and its standard error, $V_i^{1/2}$, of the dose-response fit for each study. Interest lies in the chemical-specific assessment of carcinogenicity. Specifically, if the true slope, θ_i, of the dose-response fit is positive the corresponding chemical is carcinogenic, whereas a true negative slope indicates a protective chemical. In their analyses, Laird and Louis focus on constructing posterior intervals for the true slopes in order to classify the chemicals as carcinogenic or protective.

Study	1	2	3	4	5	6	7	8	9	10	11	12
Chemical	13	5	22	24	10	20	14	15	3	7	21	18
Slope, y_i	.291	1.120	1.620	−.200	.039	−.730	−1.431	-.437	.098	−.109	.637	.030
S.E., $V_i^{1/2}$	.205	.243	.253	.268	.279	.285	.352	.355	.362	.381	.409	.568

Table 2. *Tumour data.*
Studies have been sorted by the standard errors of their dose-response slopes.

2. THEORY AND OVERVIEW

Table 1 consists of two columns and four distributions. This hierarchical model is not the most general needed for meta-analysis, but it suffices to carry the main ideas, and it acknowledges the data limitations often encountered in practical meta-analysis. The effect θ_i is to be estimated for the i-th study, $i = 1, 2, \cdots, k$. The effect may be chosen to represent any quantity of interest, e.g. the true treatment-minus-control difference in the i-th study, or a ratio, or a proportion. Of course, the effects must be defined in the same manner for each of the studies being combined. Although one need not assume that all effects are equal, the methods provided here are advantageous only if the effects are similar to one another. Through most of this discussion we assume that the effects all follow the same distribution, based on information external to the sample - i.e. we assume that the effect distribution is exchangeable. This can be circumvented, but that is not discussed here. The lower left corner of Table 1 summarizes the distributional assumptions made thus far about the effects. The density g of θ_i is assumed known up to the value of an unobserved hyperparameter α, α being known to lie in a prespecified set $\mathcal{A}$. $\mathcal{A}$ may be chosen to be so large (infinite dimensional) that g is nonparametric, or $\mathcal{A}$ may have just one element, making g a fully specified prior distribution. Intermediate choices usually are of greater interest for meta-analyses, e.g. with α representing the mean and the variance of θ_i, and $\mathcal{A}$ the 2-dimensional space of such means and variances.

The left column of Table 1 is the Descriptive Model for this hierarchical problem. Both columns of Table 1 specify distributions for the observed data and the unobserved parameters jointly, but the Descriptive Model does so most usefully for applications modeling. The distribution of the data, $\{y_i\}$, conditional on the unobservables, $\{\theta_i\}$, is required and is specified in the upper left corner. In this Hierarchical Diagram for meta-analysis, we take the observations, y_i, as independent, assuming they come from statistically independent studies, and have densities f_i which are known fully when θ_i is known. (While this ignores the troublesome problem of accounting for unknown nuisance parameters, e.g. unknown variances, it nevertheless covers many cases of interest in meta-analyses.) For convenience, y_i is taken here to be an unbiased, or very nearly so, summary statistic for θ_i. It ordinarily would be a function of many observations, as would be a sample average, or the difference of averages, or a regression coefficient, a variance, a correlation, etc.

The right column of Table 1, the Inferential Model, is mathematically equivalent to the left, but with the data distribution marginalized, given the hyperparameter, and with the distribution of the effects θ_i now conditional on the data and the hyperparameters. Thus the right side of Table 1 is in more useful form for inference, the upper right corner providing a likelihood function for α and the lower right corner indicating how to proceed with α known. A fully Bayesian hierarchical model proceeds further by providing a completely specified distribution, $h(\alpha)$, for α, thereby adding a third level to the hierarchical diagram. This specification often is ignored in examples from the literature, which is justified by large k.

By using the model of Table 1 one can make inferences about the posterior distribution of α, given the data — called "random effects" analysis. One also can make inferences about a particular true value, say θ_1, given the data — called "borrowing strength" because data from all k experiments will be used to estimate θ_1.

The most common applications to random effects analyses involve estimating $\mu = \mathrm{E}\theta_i$, the mean of the population from which the studies were taken, usually for Normally distributed data. Using the perspective of the hierarchical model, we will show how the random effects approach extends to other situations, and how particular procedures can be evaluated, and if necessary, improved upon. Similar extensions and improvements are possible for borrowing strength, but so little has been done within the meta-analysis framework that our purpose here is to suggest that borrowing strength is an enormously underutilized approach to meta-analysis.

The information in Table 3 parallels that of Table 1, but it is specialized to the Normal distribution because it commonly is true that moderately large samples n_i for each individual study permit the assumptions: of approximate Normality (from the central limit theorem); of approximate unbiasedness (consistency); and of known variances. Conjugate Normal distributions are assumed for the exchangeable θ_i. The variances $\mathrm{Var}(y_i \mid \theta_i) = V_i$ might represent σ_i^2/n_i, or something more complicated, such as variances of regression coefficients. The key point is that V_i be permitted to depend on the study i. Dependence of the sampling variances V_i on the study occurs in almost all practical cases, and so it is indispensable even though it complicates methods substantially. Consequently, Stein's estimator (James and Stein, 1961), which can be used for borrowing strength purposes when "equal variances" (i.e. $V_1 = \cdots = V_k$) obtain, rarely is applicable to meta-analyses.

	DESCRIPTIVE MODEL	INFERENTIAL MODEL
OBSERVED DATA Y_i $i = 1, 2, \cdots, k$	$y_i \mid \theta_i \overset{\text{indep.}}{\sim} N(\theta_i, V_i \equiv \sigma^2/n_i)$ fixed effects $\theta_1 = \theta_2 = \cdots = \theta_k (= \mu)$	$y_i \mid \mu, \tau^2 \overset{\text{indep.}}{\sim} N(\mu, V_i + \tau^2)$ random effects for $\alpha = (\mu, \tau^2)$
UNKNOWN PARAMETERS θ_i $i = 1, 2, \cdots, k$	$\theta_i \mid \alpha \overset{\text{indep.}}{\sim} N(\mu, \tau^2)$ $\alpha = (\mu, \tau^2)$ unknown random effects if $\tau^2 > 0$ fixed effects if $\tau^2 = 0$	$\theta_i \mid y_i, \alpha \sim N(\theta_i^*, V_i(1 - B_i))$ $\theta_i^* \overset{\text{indep}}{\equiv} (1 - B_i) y_i + B_i \mu$ $B_i \equiv \frac{V_i}{V_i + \tau^2}$ borrowing strength

Table 3. *Hierarchical diagram for Normal data.*
The lower right corner provides an exchangeable conjugate distribution for the random effects distribution. The marginal distribution (upper right corner) provides information about the superpopulation in the lower left corner, i.e., $\alpha = (\mu, \tau^2)$; the random effects distribution. The conditional distribution for θ_i (lower left corner) shows how to make inferences about any particular effect θ_i when α is known.

In Table 3, $\alpha = (\mu, \tau^2)$ is an unknown two-dimensional hyperparameter and $\mathcal{A} = \{\alpha : -\infty < \mu < \infty \text{ and } \tau \geq 0\}$. Statistical questions that have been, or that could be, answered in meta-analytic studies, framed in terms of Table 3 and of these parameters, include:

1. What is μ, and how accurately do we know it? Is μ equal to some standard value (perhaps zero)?

2. What is τ^2? Is $\tau^2 > 0$? (If $\tau^2 = 0$, then the fixed effects model applies ($\theta_1 = \cdots = \theta_k = \mu$) and the analysis is maximally precise, and simplest. If $\tau^2 > 0$, the fixed effects analysis will overstate the accuracy.) And, even allowing for $\tau^2 > 0$, meta-analysis studies could address the following questions:

3. Is treatment 1 effective; e.g. is $\theta_1 > 0$? What is the probability that $\theta_1 > 0$? Even if $\tau^2 > 0$, can we estimate θ_1 more accurately, by borrowing strength from the other k-1 independent studies?

4. Is treatment 1 better than treatment 2; e.g. is $\theta_1 - \theta_2 > 0$? What is the probability that $\theta_1 > \theta_2$, given the data? How likely is it that treatment 1 is better than treatments 2, 3 and 4?

5. Does there exist a treatment i with $\theta_i > 0$? If so, which treatment possesses the maximal θ_i? What is an estimate of $\theta_{\max} \equiv \max(\theta_i)$, and what is an interval estimate? Can we identify a subset of treatments very likely to contain the most effective treatment? (These problems have been addressed in the ranking and selection literature without hierarchical models. Solutions based on hierarchical models permit a much wider range of applications and of questions to be answered. Efron and Morris (1975) and Berger and Deely (1988) do use the hierarchical modeling approach to do ranking and selection.)

Notice that questions 1 and 2 pertain to the random effects model, while 3, 4 and 5 require methods that borrow strength. Methods for borrowing strength, while well-developed, rarely have been proposed for meta-analysis. DuMouchel and Harris (1983), Raudenbush and Bryk (1985), Laird and Louis (1989), and DuMouchel (1990) provide notable exceptions to this.

Questions 3, 4 and 5 may be relevant especially to the conduct of meta-analyses, because one sometimes may be more interested in finding a particularly successful treatment, or a short list of potentially effective treatments, than in determining whether all k studies, on the average, involve effective treatments. For example, the random effects approach could verify a negative μ, while missing the point that study number 1 is very beneficial, i.e., $\theta_1 >> 0$. That is, several negative treatments may mask a particular positive one, in a meta-analysis. The effectiveness of the random effects approach has been challenged by those that favour the fixed effects approach, perhaps because it is hard to know the relevance of μ when the θ_i's are unequal. Shifting attention to estimation of a particular θ_i, or to some function of the ensemble of θ_i, such as $\theta_{\max}$, responds to such concerns. The approach discussed here also may be enlarged to include predictors of effectiveness, based on covariates observed for each study, by modeling the θ_i in the lower left corner of Tables 1 or 3 to have distributions depending on the covariates. That extension, which is available, e.g. (Morris, 1983) contains methods and references, would strengthen further the usefulness of the borrowing strength approach to meta-analysis, is more widely used.

3. SOME SIMPLIFIED METHODS

This section includes a few methods, mainly simple and easy-to-compute techniques (except possibly the Gibbs method), intended to illustrate the flavour, the range, the ease, and the value of the hierarchical perspective for meta-analysis. All methods described in this section refer to the Hierarchical Diagram and the Normal distribution of Table 3. A range of methods, too many to review here, for more complicated meta-analysis situations also is available from the literature, by interpreting empirical Bayes and hierarchical Bayes methods for meta-analysis.

3.1. *Fixed Effects*

In some situations prior information may be available indicating that $\theta_1 = \theta_2 = \cdots = \theta_k = \mu$, which in turn implies that the between-study variability, τ^2, is 0. This is a case of the fixed effects model, which assumes studies are homogeneous and are sampled from a single population. The population mean, μ is of central interest. It is estimated most efficiently, and without bias, if

$$\hat{\mu} = \sum_1^k W_i y_i / \sum_1^k W_i, \; \mathrm{Var}(\hat{\mu}) = 1/\sum_1^k W_i, \tag{1}$$

$W_i \equiv 1/V_i$ being the Fisher information for study i. Standard inferences about μ are available from the Normal distribution $\hat{\mu} \sim \mathrm{N}(\mu, 1/\sum W_i)$. Many fixed effects methods exist for more complicated models, but only (1) is needed for the model of Table 3, with all V_i known.

3.2. *Random Effects Methods: Estimating μ and τ^2*

Frequently the assumption of a homogeneous population is unreasonable. This may happen when the studies consist of collections of trials that range over calendar time. Case-mix, study design, and varying populations, are examples of study characteristics that might deviate from study to study. Because non-identical studies usually are combined in meta-analyses, the assumption that $\tau^2 = 0$ may not be valid. Methods for learning about μ and τ^2 from the Normal model, Table 3, follow.

3.2.1. Methods of DerSimonian and Laird for Random Effects Models

DerSimonian and Laird (1986) considered the following statistic to test for between-study variability. As before let $W_i = 1/V_i$ and define $\bar{y} = \sum_i^k W_i y_i / \sum_i^k W_i$, e.g., the weighted sample mean (optimal when $\tau = 0$). Since $Q_w = \sum_i^k W_i (y_i - \bar{y})^2$ is distributed as χ^2_{k-1} when $\theta_1 = \cdots = \theta_k$ ($\tau^2 = 0$), Q_w may be used to test $\mathrm{H}_O : \tau^2 = 0$ versus the alternative $\mathrm{H}_A : \tau^2 > 0$. If μ is unknown and τ^2 is known, then with $w_i(\tau) \equiv 1/(V_i + \tau^2)$,

$$\hat{\mu}(\tau) = \sum_1^k w_i(\tau) y_i / \sum_1^k w_i(\tau) \text{ and } \hat{\mu}(\tau) \sim N(\mu, 1/\sum_1^k w_i(\tau)) \tag{2}$$

is the optimal estimate. The DerSimonian and Laird estimator of μ when τ is unknown is

$$\hat{\mu}(\hat{\tau}) = \sum_1^k w_i(\hat{\tau}) y_i / \sum_1^k w_i(\hat{\tau}) \tag{3}$$

where $w_i(\hat{\tau}) = 1/(V_i + \hat{\tau}^2)$. The DerSimonian and Laird method uses a non-iterative estimate of τ^2 based on Q_w,

$$\hat{\tau}^2_{DL} = \max\left[0, \frac{Q_w - (k-1)}{\sum_1^k W_i - \frac{\sum_1^k W_i^2}{\sum_1^k W_i}}\right], \quad W_i \equiv 1/V_i \tag{4}$$

Perhaps guided by (2), DerSimonian and Laird suggest that $\mathrm{Var}(\hat{\mu}(\hat{\tau})) \equiv \hat{\sigma}^2_\mu = 1/\sum_1^k w_i(\hat{\tau})$ be used to approximate the variance of (3). They provide no formula for the variance of the estimate in (4).

3.2.2. Methods for Improper and Proper Bayesian Random Effects Inference

If we assign the prior distribution λ/χ^2_q to τ^2, then the prior density is $h(\tau^2) \propto \tau^{-(q+2)} \exp(-\lambda/2\tau^2)$, with respect to $d(\tau^2)$. For fixed μ this leads to the posterior density for τ^2, given $y = (y_1, \cdots, y_k)$,

$$h^*(\tau^2 \mid y) \propto \left\{\tau^{-(q+2)} \exp\left[-\frac{\lambda}{2\tau^2}\right]\right\} \times \left\{\prod_{i=1}^k \left(\frac{1}{V_i + \tau^2}\right)^{\frac{1}{2}} \exp\left[-\frac{(y_i - \mu)^2}{2(V_i + \tau^2)}\right]\right\} \tag{5}$$

The prior distribution $h(\tau^2)$ is proper only if $\lambda, q > 0$. The posterior distribution (5) is proper even for $\lambda = 0$ and even for negative values of q, provided $q > -k/2$. Therefore, we also will consider these formal cases with $\lambda = 0, q \geq -2$ (for example, if $\lambda = 0$ and if $k \geq 5$,with $q = -2$, τ^2 has a flat distribution that leads to minimax, admissible estimators when $V_1 = \cdots = V_k$, (Strawderman, 1971)). The posterior density of τ^2 in (5) when $\lambda = 0$ and $q > -2$ will have a spike, and therefore an isolated mode, near zero. This should be checked graphically; when checked for the example of the next section, it was seen to have a negligible effect. The maximum likelihood estimator for τ^2 can be obtained in the usual manner, using the second piece of the right side of equation (5) as the likelihood function.

Formula (5) can be converted into a full posterior distribution on (μ, τ^2), if (5) further is multiplied by a density function of μ (possibly conditional on τ^2). If, for convenience, we take that prior density function on μ to be flat, (5) is a posterior density for (μ, τ^2) jointly, besides being a posterior density on τ^2 for known μ. In this case, the conditional distribution of μ, given τ^2 and the data, is $N(\hat{\mu}(\tau), 1/\sum w_i(\tau))$, $\hat{\mu}(\tau)$ as in (2). The marginal density of τ^2, obtained by integrating μ out of (5), is proportional to

$$h^{**}(\tau^2 \mid y) \propto \left\{\tau^{-(q+2)} \exp\left[-\frac{\lambda}{2\tau^2}\right]\right\}\left\{\sum_{i=1}^{k} \frac{1}{V_i + \tau^2}\right\}^{-\frac{1}{2}} \prod_{i=1}^{k}\left\{\left(\frac{1}{V_i + \tau^2}\right)^{\frac{1}{2}} \exp\left[-\frac{(y_i - \hat{\mu}(\tau))^2}{2(V_i + \tau^2)}\right]\right\} \tag{6}$$

with $\hat{\mu}(\tau)$ given by (2). Exact inference for μ can be obtained by marginalizing the $N(\hat{\mu}(\tau), 1/\sum w_i(\tau))$ distribution (2) with respect to (6), but we will not study the exact full Bayesian issue of estimating μ here.

The behaviour of the density (5) for τ^2 when μ is known is simpler to study than (6), while (5) still provides useful and quite parallel insights. We thus consider μ to be known in the methods described in the remainder of this section. Exact inference of τ^2 with μ known can be obtained by evaluating (5) numerically and normalizing. In later sections, we refer to this method as the "exact" method.

Gibbs methods. We also consider a method for approximating the distribution of τ^2 in (5) by sampling from (5) with the Gibbs sampler (Gelfand and Smith, 1990) in which the density (5) is then estimated by a finite mixture density (the Rao-Blackwellized estimator). In the Gibbs approach, iteration j of the simulation consists of drawing a sample value for each potency, θ_i, $i = 1, \cdots, 12$, conditional on (τ^2, y) from a Normal distribution with mean $(1 - B_i^{(j)})y_i + B_i^{(j)}\mu$ and variance $V_i(1 - B_i^{(j)})$ with $B_i^{(j)} = V_i/(V_i + \tau^2_{(j-1)})$, $i = 1, 2, \cdots, 12$. This results in a vector of sampled effects, $\theta^{(j)} = (\theta_1^{(j)}, \theta_2^{(j)}, \cdots, \theta_{12}^{(j)})$. Then we make a random draw for τ^2 using $\tau^2 \mid \theta^{(j)} \sim (\lambda + \theta^{(j)}\theta^{(j)'})/\chi^2_{k+q}$ yielding a sampled value, $\tau^2_{(j)}$. To obtain the results of section 4, we replicated this process 100 times and cycled through 20 iterations. Convergence was monitored by plotting the quantiles of the sampled τ^2 between iterations. Starting values for the simulation were obtained following Morris' discussion of Tanner and Wong, (1987, pp. 542-543). (It should be noted that formula (2.6) on page 543 of the Tanner and Wong discussion is incorrect and should be replaced by

$$2l'(\tau^2) = -q\tau^{-2} + \lambda\tau^{-4} + \sum_{i=1}^{k} \frac{(S_i - V_i)}{(V_i + \tau^2)^2} - \tau^2 \sum_{i=1}^{k} \frac{1}{(V_i + \tau^2)} \tag{7}$$

with $S_i \equiv (y_i - \mu)^2$.) We then calculated approximations to the 100 order statistics from the posterior density for τ^2.

The adjusted Likelihood method. The adjusted likelihood method (Morris, 1988b) is as easy to implement as maximum likelihood and like the Gibbs method it offers accuracy improvements relative to maximum likelihood. The adjusted likelihood method is maximum likelihood if (i) the chosen approximating distribution is the Normal distribution and (ii) the likelihood function is used to approximate the posterior distribution. Unlike Gibbs, the accuracy of the adjusted likelihood method does not increase with computation time. For the hierarchical model of Table 3, it provides relatively simple formulae. Assuming that $\lambda/\tau^2 \sim \chi^2_q \equiv 2\text{Gam}(q/2, 1)$, then the Pearson approximation to the posterior density for τ^2 will be such that $\lambda_1/\tau^2 \sim \chi^2_{q_1} \equiv 2\text{Gam}(q_1/2, 1)$. Here $q_1 = 2(1 + \hat{\imath}\hat{\tau}^4)$ and $\lambda_1 = 2\hat{\imath}\hat{\tau}^6$ with $\hat{\imath}$ the observed information. The adjusted likelihood estimator of τ^2, which approximates $E\tau^2 \mid \text{Data}$, is obtained by solving (with $w_i(\hat{\tau})$ as in (3)).

$$\hat{\tau}^2 = \frac{\sum_{i=1}^k (S_i - V_i) w_i^2(\hat{\tau})}{\sum_{i=1}^k w_i^2(\hat{\tau})} + \frac{\lambda + (2-q)\hat{\tau}^2}{\hat{\tau}^4 \sum_{i=1}^k w_i^2(\hat{\tau})} \tag{8}$$

We denote the root in (8) by $\hat{\tau}^2_{AL(RE)}$. Equation (8) always converges in our experience, and to a non-negative value of τ^2, but convergence in general has not been proven.

Maximum Likelihood. For the maximum likelihood estimate, obtained from equation (8) with $\lambda = 0$ and $q = 2$, the estimator obtained by solving for the modal value of the second piece of (5) may be negative. In that case the maximum likelihood estimate is $\hat{\tau}^2 = 0$.

3.3. *Borrowing Strength*

We shift focus in the meta-analysis away from estimation of the random effects hyperparameters, μ and τ^2, and now consider estimation of the true effect θ_i of a particular study i, $1 \le i \le k$. As noted before, this is most relevant (when $\tau^2 > 0$) to subjects whose conditions (age, dosage, etc.) match up best with those for subjects of study i. When the hyperparameters (μ, τ^2) are known, we have from Table 3 that θ_i given $\mathbf{y}$ is Normally distributed with mean θ_i^* and variance $V_i(1 - B_i)$ where $B_i \equiv V_i/(V_i + \tau^2)$ is the shrinkage factor for the ith study. The larger the between-study variation τ^2, the smaller the shrinkage B_i of the observed study effects. The posterior mean $\theta_i^* = (1 - B_i)y_i + B_i\mu$ is seen as a compromise between two sources of information: the prior mean μ for the true study effect θ_i; and the observed effect y_i for the ith study.

When we do not know the parameters μ and τ^2, consistent estimates of them can be substituted into the formulae for θ_i^* and $V_i(1 - B_i)$, the posterior means and variances, to produce an approximate posterior distribution for θ_i, given the data (further approximating the posterior distribution by a Normal distribution). This will be sufficiently accurate for k large, but this approach can be improved upon significantly for small or moderate k. The Gibbs sampler applies equally to this situation. It may be used to approximate the posterior distribution of θ_i given the data, by marginalizing μ and τ^2 out of the conditional distribution, $g^*(\theta_i \mid y, \mu, \tau^2)$, and it has the advantage of not requiring the Normal approximation to the posterior distribution. The Gibbs sampler can be extended directly to more complicated models. Some disadvantages of the Gibbs approach are that: software is not widely available; it can be a computationally slow (and for some, expensive) procedure to provide accurate answers; and widely accepted guidelines for convergence and accuracy are not available.

It therefore is not a tool that can be used easily as one iterates through preliminary data analyses.

Alternatively, the adjusted likelihood method can be used easily. Needed for each study are the mean $\hat{B}_i \equiv \mathrm{E}(\mathrm{B_i} \mid \mathrm{y})$ and the variance $v_i \equiv \mathrm{Var}(\mathrm{B_i} \mid \mathrm{y})$, with $B_i = V_i/(V_i + \tau^2)$. Since $0 \leq B_i \leq 1$, and because the Beta distribution obeys this restriction with easily computed first two moments, we approximate B_i as a Beta(a_i, b_i) variable. The method simply chooses a_i and b_i so that the approximating Beta distribution has the same modal value and second derivative at the mode as the adjusted posterior density (or likelihood). Both densities (the Beta approximation and the exact posterior) are expressed relative to the measure $dB_i/B_i(1 - B_i)$ however. Then the modal value of a Beta(a_i, b_i) density, $B_i^{a_i}(1 - B_i)^{b_i}$, is $\hat{B}_i = a_i/(a_i + b_i)$, the mean of the Beta$(a_i, b_i)$ distribution. If the posterior density is exactly a Beta density, then $\hat{B}_i$ is exactly $\mathrm{E}(B_i \mid y)$, and $v_i \equiv \hat{B}_i(1-\hat{B}_i)/(a_i+b_i+1)$ is the exact variance. Otherwise, these are approximations, but better than ones arising from a Normal distribution approximation (equivalent to the maximum likelihood method) for the posterior distribution of B_i. The estimator for τ^2 resulting from the approximate posterior for B_i is $\hat{\tau}^2$ from (9).

$$\hat{\tau}^2 = \frac{\sum_{i=1}^{k}(S_i - V_i)w_i^2(\hat{\tau})}{\sum_{i=1}^{k} w_i^2(\hat{\tau})} + \frac{\lambda - q\hat{\tau}^2}{\hat{\tau}^4 \sum_{i=1}^{k} w_i^2(\hat{\tau})} \tag{9}$$

In principle, the solution to (9), denoted $\hat{\tau}^2_{AL(B_i)}$, could vary with the study, $i = 1, \cdots, k$. For this problem, there is just one root, $\hat{\tau}^2$ in (9), so $\hat{\tau}^2_{AL(B_i)}$ is the same for each study, greatly simplifying computation. (Note that $\hat{\tau}^2$ is not an estimate of τ^2. Rather, $\hat{B}_i = V_i/(V_i + \hat{\tau}^2)$ is to be regarded as an estimate of $B_i = V_i/(V_i + \tau^2)$. These notions of estimation differ because B_i is a non-linear function of τ^2. The notation $\hat{\tau}^2_{AL(B_i)}$ emphasizes this perspective; thus $\hat{\tau}^2_{AL(B_i)}$ differs from $\hat{\tau}^2_{AL(RE)}$ of (8). For the maximum likelihood method, the estimate of a non-linear function is the non-linear function of the estimate, but this small sample adjusted likelihood (ale) method does not permit that. It cannot, because it estimates an expectation).

4. APPLICATIONS AND EXAMPLES

We illustrate the concepts introduced in this paper using the tumour data from Laird and Louis (1989). We have no pharmacologic, tumour site, or dose information with which to supplement the descriptive model. Such information, if available, could be incorporated into the usual regression framework. Thus we assume that each dose-response slope, y_i, is independently Normally distributed for $i = 1, 2, \cdots, k = 12$ with unknown mean θ_i and known variance, V_i. We refer to θ_i as a "potency". V_i is the variance of the estimated dose-response coefficient reported for study i. We further assume that the $\{\theta_i\}$ are independently (exchangeably) Normally distributed with unknown mean, μ, and unknown variance, τ^2, for $i = 1, 2, \cdots, 12$.

Table 4 tabulates estimates of the hyperparameters, when available, using the methods discussed in Section 3. Methods with $(\lambda, q) = (0, -1)$ and $(0, -2)$ correspond respectively to τ flat and τ^2 flat. If we assume that the between-study variability is 0 (the fixed effects model), then the average potency effect is μ =.213 with an approximate 95% confidence interval given by (.044, .382), indicating a tendency for the chemicals to be carcinogenic. The estimate of the between-study variance is large compared with the within-study variance (V_i ranges from .04 to .32 in Table 2 and $\bar{V} \equiv \sum w_i(\hat{\tau})V_i/\sum w_i(\hat{\tau})$ is about .11 whereas

$\hat{\tau}^2$ ranges from .58 to .88 in Table 4, excluding the fixed effects model). Consequently $\hat{\sigma}^2_\mu$ in the fixed effects model is approximately one ninth of that for the random effects models. For simplicity, the adjusted likelihood estimates for τ^2, $\hat{\tau}^2_{AL(RE)}$, are calculated in Table 4 assuming μ is known and equal to its estimated value.

Method	$\hat{\mu}$	$\hat{\sigma}^2_\mu$	$\hat{\tau}^2$	Var(τ^2)
Fixed Effects	.213	$(.086)^2$	0	0
D & L	.092	$(.25)^2$	.6185	NA
MLE	.092	$(.24)^2$	.5806	$(.29)^2$
ALE ($\lambda = 0, q = -1$)	.091	$(.25)^2$	.7659	$(.47)^2$
ALE ($\lambda = 0, q = -2$)	.089	$(.27)^2$	.8852	$(.58)^2$

Table 4. *Parameter estimates for tumour data*

The fixed effects model assumes τ^2 is known and equals 0, eq. (1); D & L denotes the DerSimonian and Laird method, eq. (4); the MLE is calculated using (5); each ALE is calculated using equation (8), inserting $\hat{\mu}$ in place of each unknown μ. $\hat{\sigma}^2_\mu$ is calculated as $1/\sum w_i(\hat{\tau})$ although it should be noted that this estimator does not take into account the error associated with estimating τ. DerSimonian and Laird provide no estimate for the variance of τ^2. The MLE Var(τ^2) is calculated using expected Fisher information. The ALE Var(τ^2) is calculated as the variance of a Reciprocal Gamma random variable, with the two parameters (λ_1, q_1) of the Reciprocal Gamma distribution determined by matching the first two derivatives of the adjusted posterior density of τ^2 to the adjusted density of the Reciprocal Gamma distribution at the maximum. Then with $\tau^2 \sim \lambda_1/\chi^2_{q_1}$, $\hat{\tau}^2 = \mathrm{E}\tau^2 = \lambda_1/(q_1 - 2)$ and Var$(\tau^2) = 2\hat{\tau}^4/(q_1 - 4)$.

The average shrinkage factor for the ith study is 1.00, .15, .16, .15 and .13 using, respectively, the fixed effects model, the DerSimonian and Laird method, the maximum likelihood method, the adjusted likelihood method with $\lambda = 0, q = -1$ and the adjusted likelihood method with $\lambda = 0, q = -2$ (with respective ranges (1.00, 1.00), (.063, .34), (.068, .36), (.060, .33) and (.053, .30)). Note that as explained in Section 3.3, the estimates of the shrinkages for the adjusted likelihood method utilize $\hat{\tau}^2_{AL(B_i)}$ and hence, $\hat{\tau}^2$ values different from those that appear in Table 4. There is little shrinkage (maximum $\hat{B}_i = .36$ is for chemical 18 using the maximum likelihood estimate for τ^2). This small amount of shrinkage supports the hypothesis $\tau^2 > 0$, i.e., the fixed effects model for $\theta_1 = \cdots = \theta_k$ is inappropriate ($Q_w = 85.9$ on 11 degrees of freedom). Thus any optimistic claims about μ based on the small variance suggested by the fixed effects analysis, would be unwarranted.

The estimated mean posterior potencies, $\hat{\theta}^*_i = (1 - \hat{B}_i)y_i + \hat{B}_i\hat{\mu}(\hat{\tau})$, using a flat prior for τ ($\lambda = 0$, $q = -1$; $\hat{\tau}^2_{AL(B_i)} = (.81)^2$), are plotted in Figure (1), which displays approximate 95% posterior intervals for the potencies θ_i. The intervals are approximate in the sense that we use $V_i(1 - \hat{B}_i)$, the "naive" estimator of the posterior variance, for the potency associated with study i. More accurate measures of the posterior variance, which are inflated for uncertainty in the hyperparameters, are available and discussed later. The inappropriateness of a fixed effects model is visually apparent from this figure. The conventional parameter of interest, μ, is close to 0 in the methods that assume a random effects model, whereas $\mu = .213$ in the fixed effects model. Further, the effects of four chemicals are very different from 0 (chemicals 5, 14, 20, 22). They also are different from each other (14, 20 are less than 0 and 5, 22 are greater than 0). Partly because there is little shrinkage, only 2 chemicals (10 and 18) change rank, after shrinkage. Reranking can occur more often in other examples.

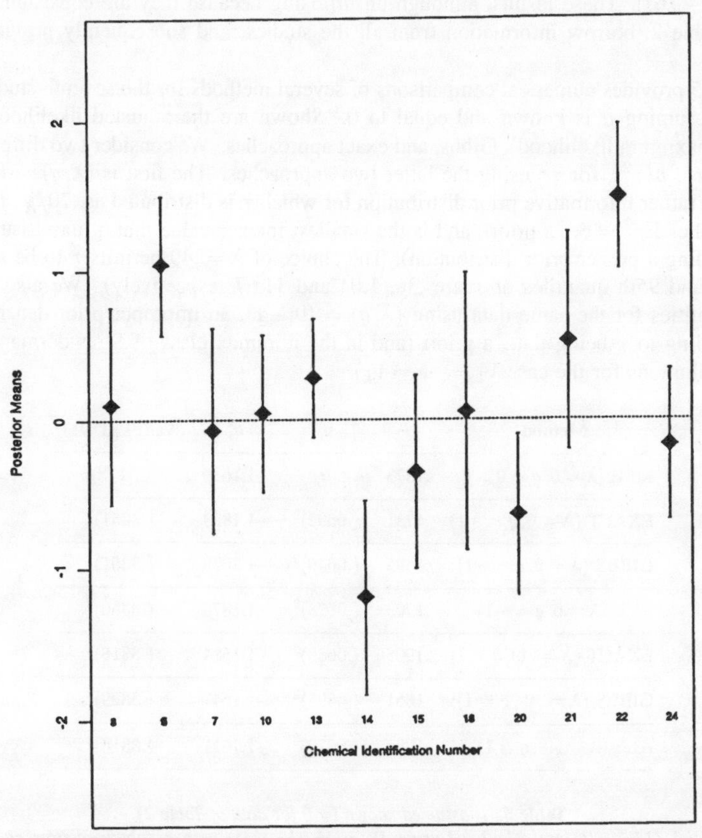

Figure 1. *Approximate 95% posterior intervals for potencies $\theta_1, \theta_2, \ldots \theta_{12}$. The intervals are calculated using the naive variance, $V_i(1-\hat{\beta}_i)$. The shrinkages are computed using the adjusted likelihood method with a flat prior for $\tau(\lambda = O, q = -1; \tau^2_{AL(B_i)} = (.81)^2)$ and assuming μ is unknown. Note that $i = 7$ corresponds to chemical identification number 14.*

If interest centers on classifying the chemicals, then the value $\theta_i = 0$ is of central interest. A chemical may be classified as carcinogenic if its average potency is positive, $\theta_i > 0$. We label chemical i as protective if $\theta_i < 0$. In particular, we have the approximation

$$P(\theta_i < m \mid y) \approx \Phi\left(\frac{m - \hat{\theta}_i^*}{s_i}\right) \tag{10}$$

to measure these carcinogenic/protective probabilities if the threshhold m is set to 0. In (10), $s_i^2 = V_i(1-\hat{B}_i) + (y_i - \mu)^2 v_i$ with $\hat{B}_i = \mathrm{E}(\mathrm{B_i} \mid \mathrm{y})$ and $v_i = \mathrm{Var}(\mathrm{B_i} \mid \mathrm{y})$. Other threshholds also could be assessed in this way. Using this notion of carcinogenicity, chemicals 5, 22, 13 and 21 are classified as potent ($P(\theta_i < 0 \mid y)$ = .000, .000, .0802 and .0748 respectively) whereas chemicals 20 and 14 clearly have a protective effect ($P(\theta_i < 0 \mid y)$ = .991 and .999) (this calculation uses $\hat{\tau}^2_{AL(B_i)} = (.81)^2$ corresponding to $\lambda = 0$ and q = -1 and replaces s_i^2 with $V_i(1 - \hat{B}_i)$). These results, although unsurprising because they are consistent with the data in Table 2, borrow information from all the studies, and subsequently provide tighter intervals.

Table 5 provides numerical comparisons of several methods for the seventh study (chemical 14) assuming μ is known and equal to 0. Shown are the adjusted likelihood (which includes maximum likelihood), Gibbs, and exact approaches. We consider two different prior distributions, $h(\tau^2)$, for τ^2 using the latter two approaches. The first is $(\lambda, q) = (.49, 1)$, a proper but rather informative prior distribution for which τ is distributed as $.70/\chi_1$ (the value $q = 1$ implies $E\tau^2 = \infty$, a priori, and is the smallest integer value that q may assume while still providing a proper prior distribution). The choice of $\lambda = .49$ permits τ to be small (the 5th, 50th and 95th quantiles of τ are .36, 1.04 and 11.67 respectively). We also calculate these quantities for the same data using $(\lambda, q) = (0, -1)$, an improper prior density on τ^2, corresponding to τ being flat , a priori (and in the minimax class of Strawderman's (1971) prior distributions for the case $V_1 = \cdots = V_k$).

Method	$\hat{B}_7$	v_7	θ_7^*	Var(θ_7 | Data)
MLE ($\lambda = 0, q = 0$)	.1873	0	−1.1630	$(.3173)^2$
EXACT ($\lambda = 0, q = -1$)	.1731	$(.0692)^2$	−1.1833	$(.3351)^2$
GIBBS ($\lambda = 0, q = -1$)	.1595	$(.0646)^2$	−1.2028	$(.3357)^2$
ALE ($\lambda = 0, q = -1$)	.1703	$(.0678)^2$	−1.1873	$(.3350)^2$
EXACT ($\lambda = .49, q = 1$)	.1905	$(.0688)^2$	−1.1584	$(.3316)^2$
GIBBS ($\lambda = .49, q = 1$)	.1861	$(.0698)^2$	−1.1647	$(.3329)^2$
ALE ($\lambda = .49, q = 1$)	.1886	$(.0679)^2$	−1.1611	$(.3316)^2$

Table 5. *μ assumed known (= 0 for data of Table 2)*
The MLE and ALE of B_7 are calculated using $\hat{B}_7 = V_7/(V_7 + \hat{\tau}^2)$ with $\hat{\tau}^2$ obtained from eq. (9). The estimates τ^2, $\hat{B}_7$ and v_7 for the exact and Gibbs methods are calculated numerically using the exact and estimated posterior density for τ^2. Var(B_7 | Data) $= v_7$ is calculated as $\frac{\hat{B}_7(1-\hat{B}_7)}{1+\hat{m}_7}$ where $\hat{m}_7 = \frac{\hat{\tau}^4(-l''(\hat{\tau}^2))}{\hat{B}_7(1-\hat{B}_7)}$. $\hat{\theta}_7^* = (1 - \hat{B}_7)y_7 + \hat{B}_7\mu$ and Var(θ_7 | Data) $= V_7(1 - \hat{B}_7) + (y_7 - \mu)^2 v_7$ with $V_7 = (.352)^2$ and $y_7 = -1.431$.

Included in Table 5 are estimates of $B_7 = V_7/(V_7 + \tau^2)$. B_7 is the true shrinkage factor associated with chemical 14 with a reported slope of $y_7 = -1.431$, far from $\mu = 0$. We have estimated B_7 in four ways: (i) and (ii) $\hat{B}_7 = V_7/(V_7 + \hat{\tau}^2)$ for the maximum likelihood and for the adjusted likelihood methods and (iii) and (iv) $\hat{B}_7 = E(B_7 \mid y)$, computed using the exact density and computed using the estimated posterior density obtained from the Gibbs sampler. Note that in this example, the adjusted likelihood estimate for B_7 is closer to the "exact" value of B_7 than is the estimate obtained from the Gibbs sampler. Also appearing in Table 5 is the variance, v_7, of B_7 conditional on the data, computed in several ways. The last two columns in Table 5 contain the approximate posterior mean and variance for θ_7. The maximum likelihood estimate for $\text{Var}(\theta_7 \mid \mathbf{y})$ is overly optimistic since v_7 is 0 for this case. Note that $s_7^2 = V_7(1-\hat{B}_7) + (y_7 - \mu)^2 v_7$ approximately (with μ assumed known) in Table 5, whereas the variance used in Figure 1 is the naive, and overly optimistic approximation to the variance, $V_7(1-\hat{B}_7)$.

The estimated posterior densities for τ^2 corresponding to the two prior distributions $(\lambda, q) = (.49, 1)$ and $(\lambda, q) = (0, -1)$ are displayed in Figure 2 together with the exact posterior densities obtained from (5). The estimated densities were obtained by evaluating the density estimates at 150 (unequally spaced) points over the range (.001, 4). From Figure 2, the exact and estimated posterior densities agree closely when the proper prior distribution is used, but they deviate for an improper prior distribution.

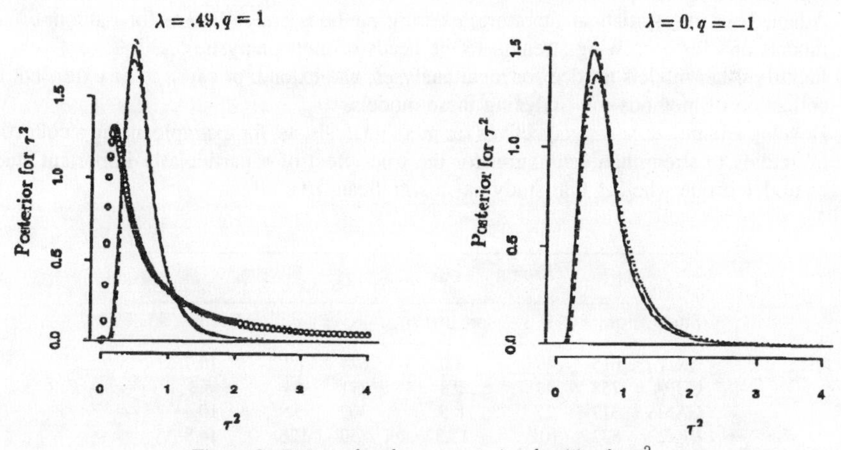

Figure 2. *Estimated and exact posterior densities for* τ^2.
Solid Line: Exact posterior density, eq. (5). Dotted Line: Estimated posterior using the Gibbs sampler. Dashed Line: Pearson approximation to the posterior density for τ^2. *The left plot uses a proper prior for* τ ($\tau \sim \frac{.70}{\chi_1}$: $\lambda = .49$ *and* $q = 1$, *denoted "ooo" in plot) and corresponding posterior values of* $\lambda_1 = 5.78, q_1 = 10.75$. *The right plot assumes a flat prior for* τ ($\lambda = 0$ *and* q = *-1) and posterior values of* $\lambda_1 = 5.67, q_1 = 9.28$. *Estimated posterior densities are based on 100 sampled values after 20 iterations evaluated at 150 (unequally spaced) points. Exact densities are evaluated at the same 150 points.* μ *is assumed known and equal to 0 in the calculations.*

In addition, Figure 2 shows the Reciprocal Gamma approximations (Pearson approximations) to the posterior densities of τ^2 in using the two prior distributions. As can be seen from the figure, the Pearson approximations coincide well with the exact densities, highlighting the accuracy of the adjusted likelihood method in this example.

5. SUMMARY AND CONCLUSIONS

When the fixed effects model is appropriate for meta-analysis, it provides the cleanest and simplest way to address the statistical problems, and to interpret the conclusions. It is not uncommon for that model to be inappropriate, however, as revealed by a statistical test, or possibly by non-sample information. In that case a fixed effects analysis will be misleading, being overly optimistic. Then the use of a random effects model, or more generally, of a hierarchical or empirical Bayes model, is in order. These methods provide a natural way to combine information from several different studies.

Random effects models focus on learning about the population of true effects for several possible studies. A host of techniques for random effects models are well-known and are available for making such inferences from data, including the DerSimonian-Laird methods for making inferences about τ^2. Despite the value of hierarchical and empirical Bayes methods already proposed for meta-analysis, we believe that these are underutilized, and that there needs to be much more awareness of the hierarchical modeling perspective.

Greater awareness of hierarchical models would permit meta-analysts and statisticians, in the presence of heterogeneous true effects, to do the following.

1. Calibrate and, when necessary, improve upon methods already used for meta-analysis, e.g. with small sample sizes.
2. Adapt, from the statistical literature, existing methods and software for random-effects models and for borrowing strength to the needs of meta-analysis.
3. Identify other models needed for meta-analyses, and extend, or cause to be extended, the collection of methods for analyzing these models.
4. Develop and utilize new perspectives for meta-analysis, as, for example, using a collection of studies to strengthen estimation for the true effect of a particularly important study, or to determine whether that study has a significant true effect.

	Aspirin			Placebo		
Study	n_i	d_i	Mortality(%)	n_i	d_i	Mortality(%)
UK-1	615	49	8.0	624	67	10.7
CDPA	758	44	5.8	771	64	8.3
GAMS	317	27	8.5	309	32	10.4
UK-2	832	102	12.3	850	126	14.8
PARIS	810	85	10.5	406	52	12.8
AMIS	2267	246	10.9	2257	219	9.7

Table 6. *Mortality in six clinical trials of aspirin in coronary heart disease.*
The table reports all-cause mortality for six major multicenter trials of aspirin and placebo during 1970-1979 in post-myocardial infarction patients (Canner, 1987; Goodman,1989). UK-1 = first United Kingdom trial, CDPA = Coronary Drug Project Aspirin trial, GAMS = German-Austrian Multicentre Study, UK-2 = second United Kingdom trial, PARIS = Persantine-Aspirin Reinfarction Study and AMIS = Aspirin Myocardial Infarction Study. n_i *and* d_i *denote, respectively, the number of patients and the number of deaths.*

To elaborate briefly on (2), the list of methods offered and discussed in this paper include a few of the many off-the-shelf methods that are available for hierarchical and empirical Bayes analysis. In particular, we have chosen not to discuss the range of models and approaches that would make use of covariates to explain the differences between the studies, and thereby reduce τ^2. With respect to (3), methods for checking the model, and particularly, for verifying exchangeability, will be required for the wider use of hierarchical models. These examples for extending (2) and (3) may both be used to improve a model that does not fit properly, as would be encountered with the aspirin data of Canner (1987) (see also Goodman, 1989), in which the last, and largest of six studies, gave a very different answer to the question of whether the rate of mortality is significantly reduced by the use of aspirin in post-myocardial infarction patients (Table 6). It would be worth investigating how the assessment of the effectiveness of aspirin is altered by the use of a long-tailed distribution for the effects θ_i, instead of the Normal distribution postulated for the hierarchical model of Table 3.

ACKNOWLEDGEMENS

Professor Morris gratefully acknowledges the help of Christine Waternaux who works with him on a closely related topic, as a part of a current National Research Council project on Combining Information. His work is supported partially by National Science Foundation Grant DMS 89-11562.

REFERENCES

Berger, J. O. and Deely, J. (1988). A Bayesian approach to ranking and selection of related means with alternatives to Analysis-of-Variance Methodology. *J. Amer. Statist. Assoc.* **83**, 364–373.

Bunker, J. P., Barnes, B. A. and Mosteller, F. (1977). *Costs, Risks, and Benefits of Surgery*. Oxford: University Press.

Canner, P. L. (1987). An overview of six clinical trials of aspirin in coronary heart disease. *Statistics in Medicine* **6**, 255–263.

DerSimonian, R. and Laird, N. (1986). Meta-analysis in clinical trials. *Controlled Clinical Trials* **7**, 177–188.

DuMouchel, W. H. (1990). Bayesian meta-analysis. *Statistical Methodology in the Pharmaceutical Sciences* (Donald A. Berry, ed.), New York: Marcel Dekker, 509–529.

DuMouchel, W. H. and Harris, J. E. (1983). Bayes methods for combining the results of cancer studies in humans and other species. *J. Amer. Statist. Assoc.* **78**, 293–315.

Efron, B. and Morris, C. N. (1975). Data analysis using Stein's estimator and its generalizations. *J. Amer. Statist. Assoc.* **70**, 311–319.

Gelfand, A. E. and Smith, A. F. M. (1990). Sampling-based approaches to calculating marginal densities. *J. Amer. Statist. Assoc.* **85**, 398–409.

Goodman, S. N. (1989). Meta-analysis and evidence. *Controlled Clinical Trials* **10**, 188–204.

Hodges, L. V. and Olkin, I. (1985). *Statistical Methods for Meta-Analysis*. New York: Academic Press.

James, W. and Stein, C. (1961). Estimation with quadratic loss. *Proceedings of the Fourth Berkeley Symposium on Mathematical Statistics and Probability* **1**, Berkeley, CA: University Press, 361–379.

Laird, N. and Louis, T. (1989). Empirical Bayes confidence intervals for a series of related experiments. *Biometrics* **45**, 481–495.

Morris, C. N. (1983). Parametric empirical Bayes inference: theory and applications. *J. Amer. Statist. Assoc.* **78**, 47–65.

Morris, C. N. (1988a). Determining the accuracy of Bayesian empirical Bayes estimators in the familiar exponential families. *Statistical Decision Theory and Related Topics IV* **1**, (S. S. Gupta and J. O. Berger, eds.) New York: Springer, 251–263.

Morris, C. N. (1988b). Approximating posterior distributions and posterior moments. *Bayesian Statistics 3* (J. M. Bernardo, M. H. DeGroot, D. V. Lindley and A. F. M. Smith, eds.), Oxford: University Press, 327–344, (with discussion).

Raudenbush, S. W. and Bryk, A. S. (1985). Empirical Bayes meta-analysis. *Journal of Educational Statistics* **10**, 75–98.

Strawderman, W. E. (1971). Proper Bayes minimax estimators of the multivariate Normal mean. *Ann. Math. Statist.* **42**, 385–388.

Tanner, M. and Wong, W. H. (1987). The calculation of posterior distributions by data augmentation. *J. Amer. Statist. Assoc.* **82**, 528–550.

DISCUSSION

W. E. STRAWDERMAN (*Rutgers University, USA*)

I congratulate the authors for a well written exposition on the utility of hierarchical models and especially of hierarchical and empirical Bayes methods in meta-analysis. The authors focus attention on a number of points which, while reasonably well known, are nevertherless worth highlighting.

1) Hierarchical Models are very general and are not confined to normal models exemplified in the paper. It is the structure of the similar, independent experiments which is hierarchical. Hence hierarchical models are natural and useful whatever the distribution of outcome variable(s) may be.
2) Meta-analysis via hierarchical models can usefully be employed for purposes of "borrowing strength" from several studies to improve the analyses of remaining studies as well as to "pool" the data from the several studies for purposes of making inferences about an overall summary parameter (e.g., the overall mean). Carl Morris of course has made fundamental contributions to the area of "borrowing strength" and it is natural that this aspect of Hierarchical Bayes procedures would be emphasized.
3) There is a wealth of interesting questions which can be naturally attacked by Bayes methods in these (and other) models. "Usual" methods would have a hard time dealing with some of these questions. Evaluating the probability that $\theta_1 - \theta_2 > 0$, and finding a small group of treatments containing an effective treatment are nice examples of such questions.
4) The authors emphasize the utility of Gibbs sampling and other similar procedures for finding "exact" results in hierarchical models. While Gibbs sampling ideas are widely represented in many papers at this conference, they are particularly appropriate in the context of this paper. These models can be extremely complex and difficult to analyze. Gibbs sampling shows great promise of allowing an extension of the class of models that can feasibly be analyzed.

The authors have focussed attention on a problem of currency and importance, have pointed out that there are existing techniques in the Bayes/Empirical Bayes literature that are underutilized in handling these problems and have provided an interesting and illuminating example demonstrating the practicality and utility of such hierarchical Bayes Methods. The paper is a useful addition to the meta-analysis literature and I thank the authors for a job well done.

B. CARLIN (*University of Minnesota, USA*)

First, congratulations to Morris and Normand on a fine paper. I certainly agree that the existing literature on Bayes and empirical Bayes methods offers tremendous "off-the-shelf" power which may be brought to bear in a wide variety of practically arising meta-analyses. With regard to specifics, I would like to make the following points:

1. The authors mention that the 95% credible intervals for the θ_i plotted in their Figure 1 are approximate since the naive estimate of posterior variance, $V_i(1 - \hat{B}_i)$, has been used. Since several papers (most notably Morris, 1983; Laird and Louis, 1987) have been written on the subject of "correcting" these naive intervals so that they do in fact have the advertised coverage probability, one might wonder whether such correction would change the results obtained, especially with regard to those chemicals having potencies naively judged as significantly different from zero. Using an approximate pivotal in conjunction with the bias correction method given by Carlin and Gelfand (1990), I obtained intervals only very slightly wider than those given in the paper. The four chemicals labelled "protective" or "potent" retained these designations; in particular, the most borderline of these chemicals, number 20, had its 95% interval widen from $(-1.17, -.12)$ to only $(-1.19, -.10)$. Thus the reporting of the naive EB intervals seems justified in this example.

2. In Figure 2 the authors offer a plot of $p(\tau^2|y)$, the marginal posterior distribution of the sampling variance for an arbitrary potency θ_i given $y = (y_1, \ldots, y_k)$. Their interest is apparently in ascertaining likely values of a future potency θ_{k+1}, but such uncertainty may

be quantified directly by considering $p(\theta_{k+1}|y)$, which is easily estimated by observing that

$$\begin{aligned} p(\theta_{k+1}|y) &= \int\int p(\theta_{k+1}|y,\mu,\tau^2)p(\mu,\tau^2|y)d\mu d\tau^2 \\ &= \int\int p(\theta_{k+1}|\mu,\tau^2)p(\mu,\tau^2|y)d\mu d\tau^2 \\ &\approx \frac{1}{m}\sum_{j=1}^{m} p(\theta_{k+1}|\mu_j,\tau_j^2), \end{aligned} \tag{1}$$

where $\{(\mu_j,\tau_j^2), j=1,\ldots,m\}$ are the currently available Gibbs samples obtained at convergence.

3. Table 1 gives the difference in raw mortality y_i from all causes for aspirin versus placebo in post-myocardial infarction patients in six major randomized multicenter trials conducted in the U. S. and Europe. Also shown are effective sample sizes n_i (average of number taking aspirin and number taking placebo) and estimated standard errors $\sqrt{\hat{V}_i} = \sqrt{\hat{\mathrm{Var}}(y_i)}$ for each study. In his talk, Morris referred to this dataset as a potential challenge for the meta-analysis tools discussed due to the presence of the "outlier" AMIS study, which has many more patients than the other studies and is also the only one not favouring aspirin. In the spirit of modelling with heavy tails as described by O'Hagan (1988), one could analyze this dataset using the model given in the authors' Table 3, but where the descriptive model for the unknown difference parameters is now $\theta_i|\mu,\tau^2 \overset{\text{ind}}{\sim} t_{v_i}(\mu,\tau^2), i=1,\ldots,6$, where the v_i are known degrees of freedom. The desired t errors for the θ_i are more tractably obtained using the reparametrization $\theta_i|\mu,\tau^2 \overset{\text{ind}}{\sim} N(\mu,\lambda_i\tau^2)$ where the latent variables λ_i are distributed independently as $v_i/\chi^2_{v_i} = \mathrm{IG}(v_i/2, 2/v_i)$, where IG denotes the inverse gamma distribution. Let us take $v_i = 5$ for all i, and also adopt the standard prior distributions $\sigma^2 \sim \mathrm{IG}(a,b), \tau^2 \sim \mathrm{IG}(c,d)$, and $\mu \sim N(\mu_0,\sigma_0^2)$. This prior specification along with our latent variable model enables simple conjugate forms for the complete conditional distributions needed for implementing the Gibbs sampler (see Carlin and Polson, 1992, for details). I chose to complete the prior specification by setting $a=3$ and $b=.00025$ (prior mean and standard deviation for σ^2 both equal to 2000, so that the $\sqrt{\sigma^2/n_i}$ will be roughly comparable to the estimated standard errors in Table 1), taking $c=3$ and $d=.25$ (prior mean and standard deviation for τ^2 both equal to 2), and setting $\mu_0=0$ and $\sigma_0^2=10$ (*a priori* indifference as to the benefit of aspirin).

Study Label	UK-1	CDPA	GAMS	UK-2	PARIS	AMIS
effective sample size (n_i)	619.5	764.5	313.0	841.0	608.0	2262.0
mortality difference in % (y_i)	2.77	2.50	1.84	2.56	2.31	–1.15
estimated standard error in % ($\sqrt{\hat{V}_i}$)	1.65	1.31	2.34	1.67	1.98	.90
lower 95% limit	–.26	–.17	–.96	–.11	–.47	–2.08
posterior mode	2.21	2.10	1.76	2.20	1.97	–.73
upper 95% limit	4.40	4.14	4.34	4.22	4.24	1.18

Table 1. *Aspirin data (Goodman, 1989).*

Running $G = 2500$ parallel replications of the Gibbs sampler for $i = 30$ iterations, I obtained the posterior modes and 95% equal tail credible sets given in Table 1. The point estimate for θ_6 has shrunk substantially back toward the positive realm, with 0 being well within the associated credible set. For the overall mean mortality difference μ I obtained an estimated posterior mode of 1.70 and 95% credible set $(-.21, 3.45)$, suggesting only a slight possibility of a negative overall mean of these data. However, using my equation (1) above to estimate the posterior of a potential future study θ_7, I obtained a posterior mode of 1.64 with associated credible set $(-1.75, 5.03)$, indicating a substantial probability that a future study will not support the use of aspirin in reducing mortality due to heart attack. As a side comment, simply treating the squares of the estimated standard errors as the (known) values of V_i in the analysis gives generally comparable results for θ_1 through θ_6, and quite comparable results for μ and θ_7. Of course a more complete analysis might employ beta distributions on the original survival proportions in each group; my goal here was simply to elaborate on the authors' point concerning the ready availability and applicability of standard Bayesian tools to meta-analyses.

W. DUMOUCHEL (*BBN Software Products, USA*) and
C. WATERNAUX (*Harvard University, USA*)

Morris and Normand are to be commended for their methodological contribution to meta-analysis. The usefulness of hierarchical models as a conceptual framework for meta-analysis must be brought more vigorously to the attention of applied statisticians. A few papers (quoted by Morris and Normand) have applied these models, yet their impact in the biomedical literature has been limited. The vast majority of meta-analyses in medicine estimate an overall parameter μ, e.g. the true effect for all studies. Analyses with hierarchical models are almost non-existent in clinical journals.

In our experience there is often considerable variation between studies. Controlled trials of psychopharmacological agents, for instance, are notorious for yielding a wide range of estimates of the efficacy of treatment even when the study protocols (dosage, length of treatment, control treatment) are reasonably similar. The differences in treatment response may be due to differences in patient characteristics. Hierarchical models are important tools because of their ability to model between-study variability either as a variance component here denoted by the parameter τ^2, or through the combination of a variance component and a regression model for the θ_i's when some of the between-study variability can be attributed to factors (covariates) known to have an effect on treatment.

Following the DerSimonian and Laird (1986) paper, simple variance components models have been sometimes used for meta-analysis (Berlin *et al.* 1989). The advantage of the DerSimonian and Laird approach is its relative simplicity. However, Morris and Normand point out that it is important to properly account for the variability in τ^2 and that the naive estimator of $\text{var}(\hat{\mu})$ obtained by substituting the estimate of τ^2 is still too small, although usually much larger than the estimate for the "fixed effect" models when the between-study variability is larger than the within-study variability. They provide improved estimates for τ^2 assuming that the prior for τ^2 and $\text{var}(\tau^2)$, assuming that the prior for τ^2 is of the form $g(\tau^2)d\tau^2 \propto \tau^{-(q+2)}\exp(-\lambda/2\tau^2)d\tau^2$, and computing the mode of an adjusted posterior density. The family of priors for τ^2 includes both proper (when $\lambda > 0$ and $q > 0$) and improper priors.

Morris and Normand emphasize the 'borrowing strength' component of the inferential model, that is, examining (and perhaps modeling) the individual $\hat{\theta}_i$'s rather than focussing on an overall estimate. This aspect of meta-analysis is often neglected. However, for researchers reluctant to give up the fixed effects models because, in the absence of a common μ, it is not clear what is being estimated, the ability to estimate the individual θ_i's may be an attractive feature of hierarchical models.

Prior distributions for the V_i's: Methods for combining information about μ from k experiments in the normal/normal case have been investigated for over fifty years (Cochran, 1937). Morris and Normand assume that the variances of the observed data (conditional on the θ_i's) , $\sigma_i^2 = V_i$ are known. DuMouchel (1990) assumes that the variances are known up to a proportionality constant $\sigma_i^2 = \sigma^2 V_i$, where the V_i's are known and the unknown parameter σ^2 has a scaled inverted chi-square distribution with q_σ degrees of freedom: $\sigma^{-2} \sim \chi^2(q_\sigma)/q_\sigma$. This assumption allows modeling some uncertainty in the V_i's while maintaining computational tractability since one only needs to perform one-dimensional numerical integrations with respect to τ^2/σ^2. The approach is a compromise between assuming that the variances are all known versus assuming that all the sampling variances are different and unknown. The latter assumptions, as in Box and Tiao (1973, ch.9) or Cochran (1954), lead to complicated but still approximate analyses.

Improper versus proper priors for τ^2: In Tables 4 and 5, Morris and Normand use three improper priors for τ^2:

(a) Flat prior for τ^2 $\quad (f_\tau(\tau)d\tau \propto \tau d\tau)$, $\quad [q = -2]$
(b) flat prior for τ $\quad (f_\tau(\tau)d\tau \propto d\tau)$, $\quad [q = -1]$
(c) flat prior for $\log \tau^2$ $\quad (f_\tau(\tau)d\tau \propto d\tau/\tau)$, $\quad [q = 0]$

Are these priors noninformative? Rules for when flat priors are harmless are given in the classic paper by Edwards, Lindman and Savage (1963). With a normal likelihood function for a mean μ, an improper uniform prior over the interval $-\infty < \mu < \infty$ gives virtually the same answer as the proper uniform prior over the interval $\overline{y} - 4\sigma/n^{1/2}, \overline{y} + 4\sigma/n^{1/2}$. When estimating the variance σ^2 of the normal distribution, a flat prior for $\log \sigma^2$ works well because the likelihood for σ^2 tends to zero fast enough to cancel out both improper tails of the prior. But with variance components estimation this is not true because $\tau^2 = 0$ has a non-zero likelihood value due to the presence of sampling variation when $\sigma > 0$. As a result, the posterior distribution for τ^2 when combined with the flat prior for log $\tau^2[q = 0]$ is improper. Therefore that prior is ruled out as an acceptable prior for Bayesian inference. Note that the inaccuracies of numerical integration or Gibbs sampling may cause one to miss this.

The first two priors used ($q = -2$ and $q = -1$ with $\lambda = 0$) yield proper posteriors but have the unattractive practical feature that they favor large τ's even though the possiblity of τ near 0 is usually highly likely a priori. Why not use a proper prior concentrated over a range of plausible values? It is relatively easy to choose q and λ to construct a prior that assigns 95% probability to an arbitrary range $0 < \tau_{low} < \tau < \tau_{high}$, where $(\tau_{low}, \tau_{high})$ are elicited from the meta-analyst.

More general models: The family of hierarchical models in DuMouchel (1990) can be applied to the data in Table 2 as follows [in this discussion we use notation agreeing with that of Morris and Normand- in particular the roles of σ and τ are interchanged from that of DuMouchel (1990)]:

$$(y|\theta, \sigma^2) \sim N(\theta, \sigma^2 V),$$

where $\mathbf{V}$ is a known $k \times k$ matrix and the scalar σ^2 has a scaled inverted chi-square distribution with q_σ degrees of freedom $(\sigma^2 \sim \chi^2(q_\sigma)/q_\sigma)$.

The distribution of θ is represented hierarchically by

$$(\theta|\mu, \tau^2) \sim N(X\mu, \tau^2 L),$$

where X is a known $k \times p$ matrix, μ is $p \times 1$, L is an elicited $k \times k$ covariance matrix and

$$\tau^2 \sim \lambda/\chi^2(q)$$

$$(\mu|\tau) \sim N(0, D \to \infty)$$

so μ has a diffuse prior. The model of Morris and Normand corresponds to the restrictions

$$\begin{aligned} q_\sigma &= \infty \\ V &= diag(V_1, \ldots, V_k) \\ X &= (11\ldots1)' \qquad [p=1] \\ L &= I \end{aligned}$$

but the more general formulation is often appropriate, and hardly more complex computationally.

In this model, both the likelihood function for θ based on y and the prior distribution of θ conditional on μ are multivariate Student t distributions. The posterior distribution of θ given y is a mixture of multivariate t distributions, each with degrees of freedom $q+q_\sigma+k-p$. Adopting the restrictions on V, X, and L given above, the posterior distribution of $\phi = q\tau^2/\lambda\sigma^2$ is

$$f(\phi|y) \propto \phi^{-(q+1)/2}\{\Pi w_i/\Sigma w_i\}^{1/2} c^{-(q+q_\sigma+k-1)}$$

where

$$w_i = W_i(\phi) = 1/(V_i + \lambda\phi/q)$$

$$c = c(\phi, y) = \frac{q/\phi + q_\sigma + S(\phi, y)}{q + q_\sigma + k - 3}$$

$$S(\phi, y) = \Sigma w_i (y_i - \overline{y})^2$$

$$\overline{y} = \overline{y}(\phi, y) = \Sigma w_i y_i / \Sigma w_i$$

Once the posterior distribution of ϕ is obtained, the moments μ, σ^2, τ^2 and θ are easily obtained as one-dimensional integrals with respect to $f(\phi|y)$, after normalizing so that $\int f(\phi|y)d\phi = 1$. See DuMouchel (1990) for details.

Using the data in Table 2 of Morris and Normand, we computed the posterior moments of μ, σ^2, τ^2, and θ for various proper prior distributions. We used the 18 priors corresponding to all combinations of

$$q_\sigma = 5, 5000$$

$$q = 0.001, 1, 5$$

$$q/\lambda = E[1/\tau^2] = 1/0.05, 1/.5, 1/5$$

To save space, we present only the lowest, median and highest values of the posterior means and standard deviations of τ^2, μ and θ_7 over the 18 prior distributions:

Parameter	Posterior Mean			Posterior St. Dev.		
	Lowest	Median	Highest	Lowest	Median	Highest
τ^2	0.40^2	0.75^2	1.61^2	0.16	0.52	1.10
μ	0.08	0.10	0.12	0.20	0.28	0.49
θ_7	–1.25	–0.72	–0.25	0.35	0.44	0.56

Larger, positive values of q correspond to greater prior certainty in the value of τ^2. Within the ranges given, the larger q is, the more sensitive are the posteriors to the values of λ/q. Thus all the Lowest values of the mean of τ^2 and of all posterior standard deviations in the above table were produced by the cases with $q = 5, \lambda/q = .05, P(0.14 < \tau < 0.56) = 95\%$ a priori, and all the corresponding Highest values by $q = 5, \lambda/q = 5, P(1.4 < \tau < 5.6) = 95\%$, while all the priors with the nearly noninformative $q = 0.001$ led to values very near the Median values in the table. The prior having $q = 1, \lambda/q = .5, P(0.32 < \tau < 22.6) = 95\%$ led to results very similar to the $q = 0.001$ case. For these data, the value of q_σ did not influence the posterior distributions much at all. When compared to the values obtained by Morris and Normand in Table 4, we find similar values for $\tau^2(.75^2$ when we set $q = 0.001$ vs. $.76^2$ when they set $q = 0$). But our posterior distribution for μ is a little more dispersed (our 0.10 ± 0.28 vs. their 0.09 ± 0.24) because the uncertainty in τ^2 is factored into our "exact" method but not in their MLE. The values in the columns of the table labeled "Lowest" and "Highest" show how more informative priors can affect the posterior distributions. The variation in the posterior mean for θ_7 shows that the amount of shrinkage can be quite affected by an informative prior on τ, with larger τ leading to less shrinkage and a lower estimate of θ_7.

In summary, we do indeed agree with Morris and Normand that Bayesian hierarchical models are great tools for meta-analyses, although we suggest avoiding an improper prior distribution for the heterogeneity variance τ^2 and extending the model to include covariate information whenever it is available.

S. E. FIENBERG (*York University, Canada*)

The authors give an excellent overview of Bayesian approaches to combining evidence in the form of "meta-analyses", but their example and the related discussion is to some extent misleading. In the notation of the paper, statisticians and scientists conducting a meta-analysis are interested in μ; they are not really interested in estimating the θ_i's. Of course, this does not mean that a Bayesian or a frequentist should ignore the θ_i's since estimation of μ depends on the model linking the θ_i's. Rather the estimation of between study variation is an integral part of meta-analysis (e.g., see Hedges, 1990).

Consider, for example, the recent meta-analyses attempt to assess the effects of the prophylactic use of aspirin to prevent heart attacks. There the quantity of interest is the mean of the predictive distribution of the "effect" for a new patient. The patient is only interested in whether or not to take aspirin and what the consequences will be. The only reason to be at all interested in a θ_i from an individual past study is that the conditions of that study are applicable directly to those under which the patient plans to take aspirin.

D. V. LINDLEY (*Somerset, UK*)

M. Jourdain said "Good Heavens! For more than forty years I have been speaking prose without knowing it". (J-B. Moliere: *Le Bourgeois Gentilhomme*.) I have the same feeling about meta-analysis as Jourdain did about prose. The key idea in the Bayesian paradigm, so admirably captured in Morris DeGroot's book, is that of coherence; of fitting ideas together satisfactorily so that they cohere. Consequently putting several analyses together is what Bayes is all about and the frequentists may be incoherent in their meta-analyses. The basic idea is a quantity of interest θ common to all the analyses. After the first analysis the Bayesian will have a posterior $p(\theta|x_1)$ where x_1 describes the data in the first experiment. This is the prior for the second analysis which will in turn produce its posterior $p(\theta|x_1, x_2) \propto p(x_2|\theta)p(\theta|x_1)$ given both analyses. (Here the experiments are supposed independent given θ.) And so on for any number of analyses. It will almost always happen that each experiment involves its own set of incidental parameters, ϕ_i for analysis i. In principle this causes no difficulty since they may be removed by integration with respect to their own probability distributions. Sometimes there is complexity in that θ is not common to all experiments, rather analysis i has θ_i. But these can be brought together with a hierarchical model and hyperparameters. Meta-analysis is a natural for the Bayesian and we should all be as grateful to the authors as Jourdain was to his philosophy professor.

M. WEST (*Duke University, USA*)

My comments relate to possible elaborations of the usual one-way classification model that is the focus of this paper.

Discussion at the presentation of this paper covered model elaboration involving forms of heavy-tailed alternatives to the normal data distributions $(y_i|\theta_i, V_i)$ and/or the distribution of the individual means $(\theta_i|\mu, \tau)$. Elaboration to scale mixtures of normals, such as Student t distributions, are natural, and, as a side issue, can accommodate the investigator's uncertainty about the V_i. Such models are discussed in West (1984) and, in a much more general framework, West (1985); computations of posterior and predictive distributions in these models are possible using iterative sampling techniques, exploiting the conditional conjugacy of scale mixture models. However, though useful in routine analysis of *large* datasets, extensive formalism along these lines may be rather heavy-handed in many instances of the type of problem the authors focus on due to small values of k typically encountered and the need to explore, rather more delicately the 'interesting' outlying cases.

Rather often, perhaps, the assumption of approximate (conditional) normality of the data will be more palatable (and less important in final inferences) than that of (conditional) normality (and certainly symmetry) of the θ_i. A very natural elaboration, and one which treats the issue of non-normality very much more delicately and appropriately than rather crude attempts at 'robustification' through the introduction of (arbitrary) heavy-tails, is a Dirichlet mixture model (West, 1991, and references therein) which includes the usual model presented here as a special limiting case. In the authors' 'descriptive' model, suppose the prior distribution function $G(.)$ of the density $g(.)$ is viewed as uncertain and assigned a Dirichlet process prior with base measure $N(\mu, \tau^2)$. Then, in expectation (over G), each θ_i has this normal prior, the basic structure of the usual model being retained *a priori*. However, the posterior of the θ_i is very non-standard, though computable using the extensions of techniques introduced by M. D. Escobar, discussed and illustrated in West (1992), see also Escobar and West (1991). This partially parametric model allows flexibly for non-normality of the θ_i distribution, and provides a natural setting in which to address, *a posteriori*, the very real practical issues of possible 'outlying' values and grouping or clustering of values amongst the θ_i. Further, posterior inferences about G are available under this model, addressing the basic meta-analytic issue.

REPLY TO THE DISCUSSION

We thank the discussants for their thoughtful comments. The main purpose of our paper is to demonstrate that hierarchical models provide a natural framework for meta-analysis. The elaborations and extensions provided by the discussants are most welcome.

Lindley's comments are right on target: meta-analysis is a good thing for Bayesians to be doing; and hierarchical modeling is a good thing for meta-analysts to be doing. Each group has been developing the tools and ideas needed by the other. It is time for the two groups to recognize that. Lindley gives a simple, general paradigm, that will be useful to all of us in thinking about meta-analysis applications.

Fienberg notes that statisticians and scientists really are interested in estimating μ. However, what does μ mean in the random effects setting when $\tau > 0$? The quantity of central interest in a meta-analysis varies with the person asking the question. Consider the example of aspirin use by post-myocardial infarction patients, to which Fienberg refers. A policy-maker could be interested in the average effect, μ, but a "new" patient would be more interested in the predictive distribution of his/her survival, given the past studies and his/her covariate vector. Carlin also notes that there are different needs. Regardless of the quantity of interest, the hierarchical model provides an appropriate and useful framework for modeling and inference.

We appreciate Strawderman's summary, with which we very much agree. His list outlines a nice research proposal. We regret that we did not say more in the original version of this text about the formal prior distributions used in part of the analysis, because Strawderman introduced them in his landmark 1971 paper (Strawderman, 1971) concerning minimax and admissible alternatives to Stein's estimator. We chose to work with those formal prior distributions because the resulting estimators possess those risk properties (for Normal distributions, when the variances V_i all take the same value). The concern expressed by DuMouchel and Waternaux about the potential harm that could be done when one uses improper priors is quite valid. Perhaps most statisticians, even Bayesian statisticians, do not realize that taking $\log(\tau)$ to be flat leads to a degenerate posterior mean that fully shrinks all components, regardless of what the data say. Formal priors that lead to admissible rules seem to be safe.

Carlin, DuMouchel and Waternaux, and West all provide elaborations to the Normal random effects distribution in the model of Table 3. Carlin and West are concerned particularly with the case of an outlying true effect. Carlin's analysis demonstrates the power available from using Gibbs methods for fitting hierarchical models in the aspirin data example of Table 6. His Gibbsian analysis handles easily the case of a long-tailed exchangeable prior distribution for the θ_i's, while simultaneously allowing for unknown V_i. We should point out that his first analysis of the potential error in using the "naive" posterior variance can be addressed by use of the ALE method for evaluating the posterior variance. Following the calculations of Section 4 and of Table 5, we note for the 7th case that v_7 is $(.068)^2$ (for the proper prior case), and so the corrected posterior standard deviation is only 4% larger than the naive value (larger amounts would be expected to occur with other data sets, if they give substantly larger shrinkages than does this one). Carlin's calculations agree exactly with this, for they are 4% wider.

Carlin's method of calculating the predictive distribution for a new experiment could be useful in more complicated settings. In the present case, and in others for which k is large, it would be close to the more simply-calculated value given by our formula (10).

DuMouchel and Waternaux show how fitting models that include regression effects and other nuisance parameters can be done exactly, subject only to a one-dimensional numerical integration involving the signal-to-noise ratio. Their approach doesn't need the Gibbs tool, so it requires less computation time and much less computational power. We expect that it could be packaged rather easily for wide use, and then would be very useful for meta-analyses.

Many prior distributions have been used by the discussants for this analysis. We wonder how they were assessed, and what their effect is on the tumour data analysis.

West has developed heavy-tailed and Dirichlet prior distribution (essentially, non-parametric random-effects distributions) methods in his earlier work. Dirichlet models may well be useful for meta-analyses, but West has not illustrated them here. It seems unlikely to us that they would give better answers for low-dimensional data sets like those of this paper ($k = 6$, and $k = 12$).

There seems to be wide agreement among these discussants, as well as among others at the conference, that hierarchical models and meta-analysis have a lot to do with each other. DuMouchel and Waternaux report that usually $\tau^2 > 0$ in practical meta-analyses. Others that have been involved in practical meta-analytic research have agreed. But the bulk of actual meta-analyses are conducted in the fixed-effects mode. Statisticians interested in hierarchical modeling, and in Bayesian and empirical Bayesian modeling, can provide a valuable scientific service by encouraging random effects models. This would involve development of user-friendly methods, concepts, and software, and participation in practical meta-analyses.

ADDITIONAL REFERENCES IN THE DISCUSSION

Berlin, J. A., Laird, N. M., Sacks, H. S. and Chalmers, T. C. (1989). A comparison of statistical methods for combining event rates from clinical trials. *Statistics in Medicine* **8**, 141–151.

Box, G. E. P. and Tiao, G. C. (1973). *Bayesian Inference in Statistical analysis*. Reading, MA: Addison-Wesley.

Carlin, B. P. and Gelfand, A. E. (1990). Approaches for empirical Bayes confidence intervals. *J. Amer. Statist. Assoc.* **85**, 105–114.

Carlin, B. P. and Polson, N. G. (1992). Inference for nonconjugate Bayesian models using the Gibbs Sampler. *Canadian Journal of Statistics* **20**. (to appear).

Cochran, W. G. (1937). Problems arising in the analysis of a series of similar experiments. *Journal of the Royal Statistical Society. Supp.* **4**, 102–118.

Cochran, W. G. (1954). The combination of estimates from different experiments. *Biometrics* **10**, 101–129 .

Edwards, W., Lindman H. and Savage L. J. (1963). Bayesian statistical inference for psychological research. *Psychological Review* **70**, 193–242.

Escobar, M. D. and West, M. (1991). Bayesian density estimation and inference using mixtures. *Tech. Rep.* **90-A16**, Duke University.

Hedges, L. V. (1990). Directions for future methodology. *The Future of Meta-Analysis* (K. W. Wachter and M. L. Straf, eds.), New York: Russel Sage Foundation, 11–26.

Laird, N. M. and Louis, T. A. (1987). Empirical Bayes confidence intervals based on bootstrap samples. *J. Amer. Statist. Assoc.* **82**, 739–750, (with discussion).

O'Hagan, A. (1988). Modelling with heavy tails. *Bayesian Statistics 3* (J. M. Bernardo, M. H. DeGroot, D. V. Lindley and A. F. M. Smith, eds.), Oxford: University Press, 345–359.

West, M. (1984). Outlier models and prior distributions in Bayesian linear regression. *J. Roy. Statist. Soc. B* **46**, 431–439.

West, M. (1985). Generalised linear models: scale parameters, outlier accommodation and prior distributions. *Bayesian Statistics 2* (J. M. Bernardo, M. H. DeGroot, D. V. Lindley and A. F. M. Smith, eds.), Amsterdam: North-Holland.

West, M. (1992). Modelling with mixtures. *Bayesian Statistics 4* (J. M. Bernardo, J. O. Berger, A. P. Dawid and A. F. M. Smith, eds.), Oxford: University Press, 503–524, (with discussion).

BAYESIAN STATISTICS 4, pp. 345–363
J. M. Bernardo, J. O. Berger, A. P. Dawid and A. F. M. Smith, (Eds.)

Some Bayesian Numerical Analysis

A. O'HAGAN
University of Nottingham, UK

SUMMARY

Bayesian approaches to interpolation, quadrature and optimisation are discussed, based on representing prior information about the function in question in terms of a Gaussian process. Emphasis is placed on how different methods are appropriate when the function is cheap or expensive to evaluate. A particular case of expensive functions is a regression function, where 'evaluation' consists of gaining observations (with the small added complication of measurement error).

Keywords: BAYES-HERMITE QUADRATURE; GAUSSIAN PROCESS; INTERPOLATION; NONPARAMETRIC REGRESSION; OPTIMISATION; QUADRATURE; SMOOTHING.

1. PRIORS FOR FUNCTIONS

Numerical techniques of interpolation, optimisation and quadrature have been studied for a very long time. They are all concerned with obtaining numerical approximations to specific properties of a given function — function values, locations of maxima and minima, integrals — that are not available exactly. In each case, the technique is based entirely on numerical evaluation of the function at a discrete set of points. The central theme of this paper is that all these problems are problems of inference. Inference is required about some unknown parameter — function value, integral, etc — based on data which are the function evaluations. Furthermore, of course, the mode of inference we shall use will be Bayesian. The formal structure is as follows.

The function in question is $f(\cdot)$, taking values $f(x)$ for $x \in \mathcal{X}$. We may or may not have an explicit expression for $f(\cdot)$, but in either case the numerical value $f(x)$ is unknown *a priori*. The prior distribution therefore consists of a distribution for the random function $f(\cdot)$. We then 'observe' the values $f = (f(x_1), \ldots, f(x_n))^T$ of the function at n discrete points in $\mathcal{X}$. The posterior distribution is simply the prior distribution conditioned on these values. Although in principle any prior distribution might be used, to reflect specific prior beliefs about $f(\cdot)$, as always it is the normal distribution that is the most tractable, and this paper will concentrate on techniques arising from representing prior information about $f(\cdot)$ in terms of a Gaussian process.

We shall assume the following hierarchical prior model. First,

$$f(\cdot)|\beta, \sigma^2 \sim N(h(\cdot)^T\beta, \sigma^2 v(\cdot, \cdot)). \tag{1}$$

That is, $f(\cdot)$ is a Gaussian process, conditional on hyperparameters β and σ^2, with mean $E(f(x)|\beta, \sigma^2) = h(x)^T\beta$ and covariance function $\mathrm{Cov}\,(f(x), f(x')|\beta, \sigma^2) = \sigma^2 v(x, x')$. The vector $h(\cdot)$ of q regressor functions $h(x) = (h_1(x), \ldots, h_q(x))^T$, and the covariance function $v(\cdot, \cdot)$ are known. The prior distribution of β and σ^2 is then given by

$$\beta|\sigma^2 \sim N(b_0, \sigma^2 B_0^{-1}), \tag{2}$$

$$\sigma^2 \sim s_0\chi^{-2}_{a_0}. \tag{3}$$

The special case of an uninformative prior distribution on the hyperparameters, represented by $B_0 = 0$, $s_0 = a_0 = 0$, so that $p(\beta, \sigma^2) \propto \sigma^{-2}$, will often be appropriate.

This is about the most general structure that yields tractable posterior analysis. Some further generalisation can be achieved by letting a function of $f(\cdot)$ have this distribution; for instance, we could let $f(x) = f_1(x) + f_2(x)f_3(x)$, where $f_1(\cdot)$ and $f_2(\cdot)$ are known functions and $f_3(\cdot)$ has the distribution (1)–(3). The mathematics would be essentially unchanged.

Gaussian process priors for functions have been proposed several times in a variety of contexts, as will be apparent from the references throughout this paper.

2. INTERPOLATION AND SMOOTHING

2.1. *Cheap and Expensive Functions*

Another theme of this paper is that the most appropriate methods for a given problem depend on how costly it is to obtain each evaluation $f(x_i)$ of the function. In interpolation, for instance, it is implicit that the function is expensive to evaluate. Otherwise, if we wanted $f(x)$ we would just evaluate it exactly rather than approximate it by interpolation. Interpolation was a very important technique before the ready availability of computers. Then for most complex functions the only recourse was interpolation in a book of tables wherein every figure was the result of somebody's very laborious computation. Although this is not true for the commonly used functions today, there is still no shortage of functions whose evaluation is impractical except by extensive use of the most powerful computers. See Sacks, Welch, Mitchell, and Wynn (1989).

Optimisation and quadrature (numerical integration) are important even for cheap functions. When maxima/minima and integrals cannot be found analytically, numerical methods are essential. Particularly in high dimensional spaces, solving these problems will typically require very large numbers of function evaluations. Many methods then become impractical, even with cheap function evaluations.

2.2. *Interpolation*

Given the prior distribution (1)–(3) and data f, the posterior distribution is easily obtained. See for instance O'Hagan (1978). We find

$$f(\cdot)|\beta, \sigma^2, f \sim N(m(\cdot), \sigma^2 w(\cdot, \cdot)), \tag{4}$$

where

$$m(x) = h(x)^T\beta + t(x)^T A^{-1}(f - H\beta), \tag{5}$$

$$t(x) = \begin{pmatrix} v(x, x_1) \\ \vdots \\ v(x, x_n) \end{pmatrix}, \qquad H = \begin{pmatrix} h(x_1)^T \\ \vdots \\ h(x_n)^T \end{pmatrix}, \tag{6}$$

$$A = \begin{pmatrix} v(x_1, x_1) & \dots & v(x_1, x_n) \\ \vdots & & \vdots \\ v(x_n, x_1) & \dots & v(x_n, x_n) \end{pmatrix}, \tag{7}$$

$$w(x, x') = v(x, x') - t(x)^T A^{-1} t(x'). \tag{8}$$

Also

$$\beta|\sigma^2, f \sim N(b, \sigma^2 B^{-1}), \tag{9}$$

where

$$b = B^{-1}(B_0 b_0 + H^T A^{-1} f), \tag{10}$$
$$B = B_0 + H^T A^{-1} H. \tag{11}$$

Finally,

$$\sigma^2 | f \sim s\chi_a^{-2},$$

where $a = a_0 + n - q$ and

$$s = s_0 + f^T\{A^{-1} - A^{-1}H(H^T A^{-1} H)^{-1} H^T A^{-1}\} f. \tag{12}$$

It follows that the posterior distribution of any particular $f(x)$ is t with degrees of freedom a, mean

$$m^*(x) = h(x)^T b + t(x)^T A^{-1}(f - Hb) \tag{13}$$

and variance $s\, w^*(x, x)/(a-2)$, where

$$w^*(x, x') = w(x, x') + \{h(x) - H^T A^{-1} t(x)\}^T B^{-1}\{h(x') - H^T A^{-1} t(x')\}. \tag{14}$$

The natural interpolant is (13). In particular of course, at the observation points x_i we have $m^*(x_i) = f(x_i)$ and $w^*(x_i, x_i) = 0$.

We have already said that interpolation is only relevant for expensive functions. A related point is that when the function is very expensive it can be worth spending a great deal of effort to locate good *design* points $(x_1, \ldots, x_n)$. Sacks, Welch, Mitchell and Wynn (1989) develop designs to minimise the integral of the posterior variance

$$\int_{\mathcal{X}} w^*(x, x)\, dx$$

over the space $\mathcal{X}$, and this is easily generalised as in O'Hagan (1978) to minimise a weighted average

$$\int_{\mathcal{X}} w^*(x, x)\, d\Omega(x).$$

Sacks and Schiller (1988) advocate minimising the maximum of $w^*(x, x)$ over a subset of $\mathcal{X}$.

2.3. *The Prior Covariance Function*

The interpolant (13) has a simple and natural form. The first term is the fitted regression $h(x)^T b$ corresponding to the regression model $h(x)^T \beta$ assumed for the prior mean. The estimate of the coefficient vector, b, is a familiar Bayesian, weighted average of the prior mean b_0 and a generalised Least Squares estimate $\hat\beta = (H^T A^{-1} H)^{-1} H^T A^{-1} f$. In the case of a non-informative prior, b reduces to $\hat\beta$.

The second term in (13) is the interpolant of the residuals $f - Hb$ from this fitted line, and ensures that $m^*(x)$ interpolates the actual observations. This term takes the form

$$t(x)^T A^{-1}(f - Hb) = \sum_{i=1}^{n} k_i v(x, x_i), \tag{15}$$

where the k_i's are the elements of the vector $A^{-1}(f - Hb)$. Indeed, these are the unique coefficients that make the right hand side of (13) interpolate the observations. The interpolant

(13) will have properties of smoothness, such as differentiability, if and only if the prior covariance function $v(\cdot, \cdot)$ has those properties. This is an important guide to the form of covariance function that is appropriate. In particular, the one-dimensional Markov form

$$v(x, x') = \exp\{-b|x - x'|\} \tag{16}$$

used by Blight and Ott (1975) produces an interpolant (13) that is not differentiable at the observed points. The stationary Gaussian form for $x \in R^p$,

$$v(x, x') = \exp\{-b(x - x')^T V^{-1}(x - x')\} \tag{17}$$

is used by O'Hagan (1991) and Sacks, Welch, Mitchell and Wynn (1989), and produces smooth interpolation, infinitely differentiable everywhere. O'Hagan (1978) also uses (17), but in a rather different model from (1)–(3). Wahba (1978, 1983) chooses covariance functions that make the interpolant a spline function.

2.4. *Example*

An interesting example of interpolation is drawing contours. Consider the contours of the bivariate density function proportional to $\pi(\theta, \phi) = (1+\theta^2)^{-2}(1+\phi^2)^{-2}$, where θ and ϕ are independent t random variables with 3 degrees of freedom. It is easy to solve $\pi(\theta, \phi) = c$ for $\theta = 0$, or $\phi = 0$ or $\theta = \pm\phi$. This gives eight fixed points on any contour line. Figure 1 shows three contours, at $c = 0.4, 0.03, 0.001$, drawn by interpolating these eight points in each case. In order to do this, I set up the unknown function as a vector-valued $f(x)$ taking values in the plane, with x taking values in $[0, 8)$. Then $f(0), f(1), \ldots, f(7)$ are the eight 'observed' points in sequence:

$$f(0) = (d_1, 0), f(1) = (d_2, d_2), f(2) = (0, d_1), f(3) = (-d_2, d_2), \ldots, f(7) = (d_2, -d_2),$$

where

$$d_1 = (c^{-1/2} - 1)^{1/2}, \quad d_2 = (c^{-1/4} - 1)^{1/2}.$$

The preceding theory is easily generalised to vector $f(\cdot)$ (as in O'Hagan (1978)). The covariance function was defined to be

$$\mathrm{Cov}\,(f(x), f(x')) = \exp\{-b||x - x'||^2\}I_2,$$

with

$$||x - x'|| = \min\,((|x - x'|, 8 - |x - x'|)$$

reflecting the fact that the contour is a closed loop. The case $q = 0$ was used, removing the regression term and giving $f(\cdot)$ a zero prior mean.

The smoothing parameter b in the covariance function was chosen to give good fit of the interpolated contour to the true value, in the following way. The value of the density $\pi(m^*(1/2))$ was found for the interpolated value $x = 1/2$, and b was adjusted to make this close to the required value c. The resulting interpolated contours are extremely good. $\pi(m^*(x))$ was found for a large selection of x values and found to be within 2% of the true c. (This is for the outermost, and most difficult to interpolate, contour. Accuracy to within 0.001% was achieved on the innermost contour).

An alternative approach, transforming to polar coordinates and representing the contour as a scalar function of the angle, proved to be less accurate (and could anyway fail for a non-convex contour).

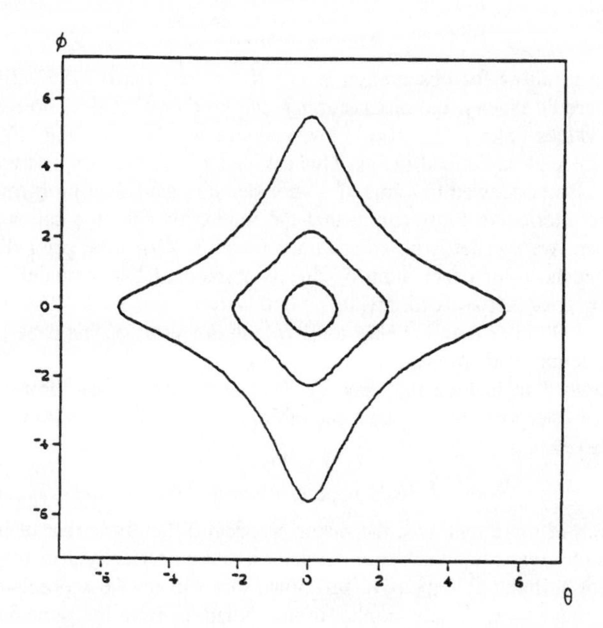

Figure 1. *Contours of product of independent t distributions.*

2.5. *Observing Derivatives*

There is a parallel theory for derivatives of $f(\cdot)$ itself. By considering the random variable $X_\delta = \delta^{-1}\{f(x+\delta) - f(x)\}$ and letting $\delta \to 0$ (for scalar x) it is easy to show that providing the random sequence X_δ converges, and providing that the indicated derivatives of $h(\cdot)$ and $v(\cdot, \cdot)$ exist, $f(\cdot)$ and all its derivatives are jointly normally distributed with

$$E(f^j(x)|\beta, \sigma^2) = h^j(x)^T\beta, \tag{18}$$

$$\text{Cov}\,(f^j(x),\, f^i(x')|\beta, \sigma^2) = \sigma^2 v^{ji}(x,\, x'), \tag{19}$$

where

$$f^j(x) = \mathrm{d}^j f(x)/\mathrm{d}x^j, \quad h^j(x) = \mathrm{d}^j h(x)/\mathrm{d}x^j \quad \text{and} \quad v^{ji}(x,\, x') = \mathrm{d}^{i+j} v(x,\, x')/\mathrm{d}x^j \mathrm{d}x'^i.$$

Since the covariance function (16) is not differentiable, nor is $f(\cdot)$ in that case. On the other hand, it turns out that (17) is sufficiently smooth for $f(\cdot)$ to be infinitely differentiable everywhere with probability one (if $h(\cdot)$ is). This is another argument in favour of (17) rather than (16) as a prior covariance function. We are generally dealing with functions that we know to be differentiable, so that (16) simply does not represent that prior belief.

We can now consider making observations not of $f(\cdot)$ but of its derivatives. The idea is potentially useful in all problems. Although derivatives are often unavailable when $f(\cdot)$ is expensive, for cheap functions we can generally 'observe' them. It may in fact be even cheaper to observe $f'(x)$ when we are also observing $f(x)$. In the theory of Section 2.2 we need only replace the functions $v(\cdot, \cdot)$ and $h(\cdot)$ in (6) and (7) by appropriate derivatives, using (18) and (19). The result (13) will then not only pass through any observed values $f(x_i)$ but agree with any observed derivatives.

2.6. *Smoothing*

It is also simple to allow for observation error. If the function (or derivatives) cannot be evaluated, 'observed', exactly, the observation f can be represented as normally distributed with mean true values $(f(x_1), \ldots, f(x_n))$ and variance matrix $\sigma^2 V_f$, e.g. $V_f = v_f I_n$. Then we simply add V_f to A as defined in (7). This can occur with very expensive functions, but it can of course also be viewed in terms of a very standard statistical problem. In regression modelling, as an alternative to assuming that the regression function follows some simple parametrised form, we can let it be an arbitrary function $f(\cdot)$. The prior distribution (1)–(3) then says that the prior expectation of $f(\cdot)$ is a standard linear model $h(\cdot)^T\beta$, but the Gaussian process prior allows it to deviate smoothly from that form. This is the approach used in Blight and Ott (1975) and O'Hagan (1978). In the terms of this paper, statistical data are expensive function evaluations.

With A redefined to include the error V_f, (13) no longer passes through the observed points. Instead of interpolation, we have smoothing. If V_f is large, $m^*(x)$ is just the fitted linear model $h(x)^T b$.

2.7. *Implementation*

In implementation of these methods, the major practical difficulty is that of inverting the A matrix. Given n observations, this is an $n \times n$ matrix, and numerically inverting it is an order-n^3 operation. For sufficiently expensive functions, this will not be a problem, and anyway only needs to be done once, for any number of interpolations from the same data. Otherwise, if the function is only moderately expensive to evaluate and n is large, interpolation at any given point can be computed using only a more feasible number of nearest observations to that point.

Although A may be inverted analytically if we use the covariance function (16), so reducing the computation to order-n^2, we have rejected this function because of its poor interpolation properties. Analytic inversion of A is not possible with (17) or other realistic covariance functions. In two or more dimensions, however, O'Hagan (1991) shows that dramatic reduction of effort may be achieved if the observations lie on a rectangular grid. See also the discussion of tetrahedral designs in Section 3.3 below.

3. QUADRATURE

3.1. *Inference about an Integral*

Let

$$k = \int_{\mathcal{X}} f(x)\,\mathrm{d}M(x), \tag{20}$$

the integral of $f(\cdot)$ with respect to some measure $M(\cdot)$ on $\mathcal{X}$. This formulation is very general. For instance $M(\cdot)$ might be Lebesgue measure on $\mathcal{X} = R^p$ for vector x, so that k is just the (Lebesgue) integral of $f(\cdot)$. Or if $M(\cdot)$ is Lebesgue measure on $C \subset R^p$, and $M(B) = 0$ if $B \cap C = \phi$, then $k = \int_C f(x)\,\mathrm{d}x$.

The posterior distribution of k is given by (9), (12) and

$$k|f,\beta,\sigma^2 \sim N(m_k, \sigma^2 w_k) \tag{21}$$

where

$$m_k = \int_{\mathcal{X}} m(x)\,\mathrm{d}M(x), \tag{22}$$

$$w_k = \int_{\mathcal{X}^2} w(x, x')\,\mathrm{d}M(x)\,\mathrm{d}M(x'). \tag{23}$$

Its marginal posterior distribution is therefore t with degrees of freedom a and mean

$$m_k^* = \int_{\mathcal{X}} m^*(x)\,\mathrm{d}M(x) = h^{*T}b + t^{*T}A^{-1}(f - Hb)\,, \tag{24}$$

where

$$h^* = \int_{\mathcal{X}} h(x)\,\mathrm{d}M(x), \qquad t^* = \int_{\mathcal{X}} t(x)\,\mathrm{d}M(x). \tag{25}$$

This technique is called Bayesian quadrature by O'Hagan (1991), where it is explored in some detail. Earlier treatments are described in the review of Diaconis (1988).

3.2. *Bayesian Application*

Integration of complex, intractable, and often high-dimensional, posterior densities is a major interest in Bayesian statistics. Consider, therefore, two separate Bayesian analyses. One gives rise to a posterior density which we only know as proportional to $f(\cdot)$ (obtained as the product of prior density and likelihood). The second Bayesian analysis is the one considered in this paper, providing Bayesian inference about $f(\cdot)$, and in particular about integrals of $f(\cdot)$. The first analysis is only of interest here to provide the context for $f(\cdot)$. Specifically, $f(\cdot)$ is proportional to a density function, and given a reasonable amount of information may be expected to be smooth, and probably unimodal.

Suppose that $f(\theta)$ is proportional to a density function for a random vector $\theta \in R^p$ (the parameter vector in the first Bayesian analysis). Then we wish to estimate

$$\int r(\theta) f(\theta)\,\mathrm{d}G(\theta), \tag{26}$$

identifying $\mathrm{d}M(\cdot)$ in (22) with $r(\cdot)\mathrm{d}G(\cdot)$ and $\mathcal{X}$ with R^p. The interpretation of (26) is that the distribution of θ is represented as a density $f(\theta)$ with respect to a measure $G(\cdot)$ on R^p and we wish to find the expectation of $r(\theta)$. Or strictly, since the density of θ is only proportional to $f(\theta)$, the expectation of $r(\theta)$ is (26) divided by $\int f(\theta)\,\mathrm{d}G(\theta)$, which is itself the case $r(\theta) = 1$ of (26). Inference about ratios of integrals is discussed in O'Hagan (1989).

Choice of the measure $G(\cdot)$ is an important part of the prior specification. It would *not* be reasonable for a typical density in the usual sense, of a density with respect to Lebegue measure, to be represented by the prior distribution (1)–(3). However, if $G(\cdot)$ is a suitable multivariate normal probability measure, then it becomes reasonable to model $f(\cdot)$ in this way. O'Hagan (1991) deals specifically with 'Bayes-Hermite' quadrature, in which $G(\cdot)$ is multivariate normal (and without loss of generality is assumed to be the standard normal, $N(0, I)$) and the covariance function also takes the 'Gaussian' form (17). These choices have the effect of allowing the integrals (25) to be performed analytically. It is essential to be able to do this when $f(\cdot)$ is cheap to evaluate, as is typically the case when it is the posterior density in some other Bayesian analysis. Otherwise, the estimate (24) depends on integrals which are no easier to evaluate than the original problem (20).

3.3. *Multidimensional Designs*

Inversion of A was mentioned as a practical problem in Section 2.6 and is even more important when $f(\cdot)$ is cheap. To integrate a high-dimensional function inevitably requires the number of function evaluations, n, to be large. Although the evaluations may themselves be cheap, inversion of A requires order-n^3 computation and so completely dominates the exercise. In contrast the Monte Carlo method, described in the context of Bayesian applications by van Dijk and Kloek (1980, 1984) and Geweke (1989), processes n function evaluations in order-n computations. (An important new variant on Monte Carlo methods is given by West (1991).) Although Bayesian quadrature makes vastly more efficient use of the information in each function evaluation, Monte Carlo can make so many more function evaluations that it is generally a superior technique in high dimensions. It should also be mentioned here that recent applications of the 'Gibbs sampling' technique are proving even more effective than Monte Carlo for such cases; see Gelfand and Smith (1990), Gelfand, Hills, Racine-Poon and Smith (1990).

Nevertheless, it is of interest to develop Bayesian quadrature of multiple integrals, for two reasons. First, for sufficiently expensive functions it is essential to make maximum use of function evaluations, for which purpose Bayesian quadrature is the best available technique. Second, by suitable choice of design points $x_1, \ldots, x_n$ it is possible either to make inversion of A analytically feasible or to reduce it to a much smaller problem. An example of the latter is the development of cartesian product designs in O'Hagan (1991), which exploits a Kronecker product form for A. We present here an even simpler design where A can be inverted analytically.

Suppose that the points $x_1, \ldots, x_{p+1}$ lie on a regular simplex in R^p. That is, the distance is the same between all pairs of points. Then with covariance function (17), A takes the intraclass form, where all diagonal elements are equal and all off-diagonal elements are equal. This matrix is trivial to invert and the following results are then easy to derive. We assume the Bayes-Hermite form, combining (17) with letting $G(\cdot)$ be the $N(0, I_p)$ measure. We also let $h(x) = 1$, so that the regression part of (1) consists just of an unknown mean β. Then let $(x_1, \ldots, x_{p+1})$ be a simplex centred on the origin, with all points at a distance d from the origin, so that the distance between pairs of points is $d\{2(p+1)/p\}^{1/2}$. Then the posterior variance of the simple integral $\int f(\theta)\,dG(\theta)$ is minimised by setting

$$d^2 = p(2+p)(1+2b)\ln(1+2b)\{2b(p+2+4b+4pb)\}^{-1}. \tag{27}$$

Setting $b = 1$ and evaluating (27) for various p we find the optimal tetrahedral designs shown in Table 1.

p	1	2	3	5	10	100
d	0.6704	0.9077	1.0849	1.3640	1.8792	5.7636

Table 1. *Distance d from origin of points in optimal tetrahedral design in p dimensions.*

As $p \to \infty$, $d^2/p \to (1+2b)\ln(1+2b)\{2b(1+4b)\}^{-1}$. Therefore in the case $b = 1$ shown in Table 1, d is asymptotically $0.5741\,p^{1/2}$.

Such designs are admittedly of little practical importance. They may provide a very cursory and quick exploration of a multidimensional function, but $p+1$ points in p dimensions is far too few for adequate integration. Conversely, the product designs require of the order of m^p observations, which is impractical when p is large. Good accuracy should be achievable with far fewer observations. Much work remains to be done on Bayesian quadrature in many dimensions, but there are some interesting points already to consider. For instance, the optimal tetrahedral designs expand in distance from the origin proportionally to $\sqrt{p}$ as p increases, and this is also observed in optimal product designs; see O'Hagan (1991).

Some idea of the likely patterns of good designs may be found from identifying small optimal designs. Optimal designs of $n = 3$, 4 and 5 points in $p = 2$ dimensions have been found for the model assumed in Table 1. The three point tetrahedral design in Table 1 proves to be the optimal three point design. The optimal four point design might be expected to be a square, but it is not. Nor is it a rectangle (a 2×2 product design), but a skew rhombus. The optimal five point design is a kind of flattened pentagon. These designs are shown in Figure 2. Note that the model is spherically symmetrical, so rotations of these designs around the origin are equally good.

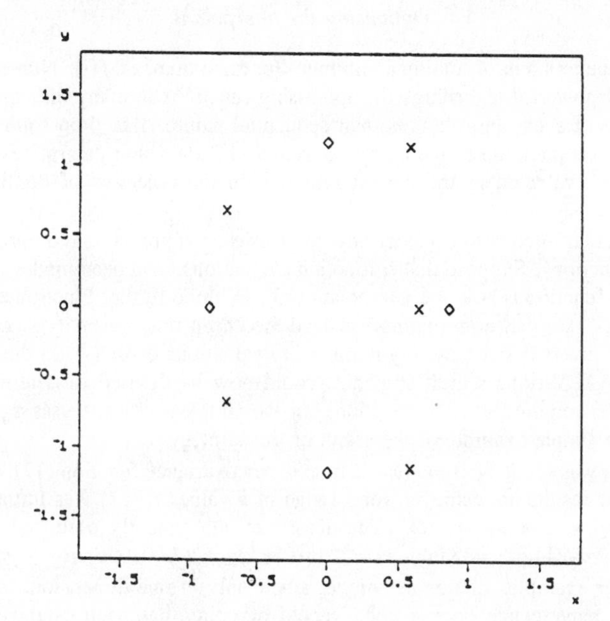

Figure 2. *Optimal* $4(=\diamond)$ *and* $5(=\times)$ *point designs in two dimensions.*

These findings suggest that optimal designs will not in general have any natural geometric form.

3.4. *The Value of Observing Derivatives*

The possibility of using observations of derivatives of the function in a quadrature problem does not seem to have been explored before. Yet the development in Section 2.5 above makes it clear that to do so adds no extra complexity to the analysis. I can report here only the results of some simple computations in one dimension, taking the Bayes-Hermite context again.

The optimal n point design, at least for small n, comprises only direct observations of the function, even when each observation is allowed to be either of $f(x_i)$ or its first derivative $f'(x_i)$. If this optimal n point design is used, observing $f'(x_i)$ *in addition* to $f(x_i)$ at all those points adds *no* information about the integral. (I would like to have a theoretical proof of this rather interesting empirical result.) All of this seems to suggest that nothing is to be gained from using derivatives (at least the first derivatives), but their use can be valuable in a sequential context. Suppose that the optimal two point design has been used, and we now decide to take an extra observation to improve the posterior variance of the integral. Then the best third observation is of a first derivative, rather than of the function itself. The same is true of adding one or two further points to the optimal three point design.

Further investigation is needed of the use of derivatives in quadrature.

4. OPTIMISATION

4.1. *Optimising the Interpolant*

Optimisation is the problem of finding a minimum (or maximum) of $f(\cdot)$. Numerical analysts have constructed powerful algorithms for optimising functions in many dimensions, and the important feature of such algorithms is their sequential nature. Based on some or all of the function evaluations made up to the current iteration, the algorithm determines one or more new points to be evaluated in the next iteration, with the objective of homing in on the minimum.

It is of particular interest to consider how the Bayesian approach may improve efficiency for expensive functions. Suppose that n function evaluations have been made, and the posterior mean of the function is now the interpolant (13). With no further function evaluations of the expensive $f(\cdot)$, we can now optimise instead the cheap function $m^*(\cdot)$. (Even for large n, we need only invert A once for any number of evaluations of $m^*(\cdot)$, so this is genuinely a cheap function.) Various search strategies could now be devised to determine the next evaluation of $f(\cdot)$ required at each iteration, but the simplest is just to set x_{n+1} to argmin $m^*(\cdot)$. Here is a simple example of the effect of this strategy.

The function was set at $f(x) = -x - 2 \ln x$. The covariance function (17) was used with $b = 0.1$. (Similar results are achieved for a range of b values.) $f(x)$ was initially evaluated at $x_1 = 0.1$, $x_2 = 2.5$, $x_3 = 4.9$. The algorithm subsequently produced $x_4 = 3.173$, $x_5 = 2.563$, $x_6 = 2.244$, $x_7 = 2.069$, $x_8 = 2.00744$, $x_9 = 2.000103$.

Other similar examples of this technique, albeit only in one dimension, have generally produced rapid convergence once a good spread of points has been established. Indeed, convergence is often as fast as the Newton–Raphson algorithm, which requires both first and second derivatives to be available (and so makes two evaluations at each step). In the case of expensive functions, derivatives will very often be unavailable, and standard optimisation methods using only function evaluations are very much slower. This Bayesian technique therefore promises to be highly effective in such cases.

4.2. *Response Surfaces*

Section 2.6 introduced the idea of a regression function as an expensive function. Observations can only be obtained by experimentation, which is typically costly compared with computation time, and are then subject also to experimental error. An important problem in industrial statistics is to optimise a regression function. The techniques of response surface methodology are also sequential, therefore. See Box and Draper (1987). Optimising a fitted non-parametric regression using the smoothing techniques of Section 2.6 could undoubtedly contribute to that methodology.

ACKNOWLEDGEMENTS

I would like to thank John Kent for assistance with the theory of derivatives of Gaussian processes, and Arieh Epstein for discussions on optimisation.

REFERENCES

Box, G. E. P. and Draper, N. R. (1987). *Empirical Model-Building and Response Surfaces*. New York: Wiley.

Diaconis, P. (1988). Bayesian numerical analysis. *Statistical Decision Theory and Related Topics IV* **1** (S. S. Gupta and J. Berger, eds.), New York: Springer, 163–175.

Gelfand, A. E., Hills, S. E., Racine-Poon, A. and Smith, A. F. M. (1990). Illustration of Bayesian inference in normal data models using Gibbs sampling. *J. Amer. Statist. Assoc.* **85**, 972–985.

Gelfand, A. E. and Smith, A. F. M. (1990). Sampling based approaches to calculating marginal densities. *J. Amer. Statist. Assoc.* **85**, 398–409.

Geweke, J. (1989). Bayesian inference in econometric models using Monte Carlo integration. *Econometrica* **57**, 1317–1340.

O'Hagan, A. (1978). Curve fitting and optimal design for prediction. *J. Roy. Statist. Soc. B* **40**, 1–42, (with discussion).

O'Hagan, A. (1989). Integrating posterior densities by Bayesian quadrature. *Tech. Rep.* **159**, University of Warwick, Coventry, UK.

O'Hagan, A. (1991). Bayes-Hermite quadrature. *J. Statist. Planning and Inference* , (to appear).

Sacks, J. and Schiller, S. (1988). Spatial designs. *Statistical Decision Theory and Related Topics IV* **2**. (S. S. Gupta and J. Berger, eds.), New York: Springer, 385–399.

Sacks, J., Welch, W. J., Mitchell, T. J. and Wynn, H. P. (1989). Design and analysis of computer experiments. *Statist. Sci* **4**, 409–435.

van Dijk, H. K. and Kloek, T. (1980). Further experience in Bayesian analysis using Monte Carlo integration. *J. Econometrics* **14**, 307–328.

van Dijk, H. K. and Kloek, T. (1984). Experiments with some alternatives for simple importance sampling in Monte Carlo integration. *Bayesian Statistics 2* (J. M. Bernardo, M. H. DeGroot, D. V. Lindley and A. F. M. Smith, eds.), Amsterdam: North-Holland, 511–530, (with discussion).

Wahba, G. (1978). Improper priors, spline smoothing and the problem of guarding against model errors in regression. *J. Roy. Statist. Soc. B* **40**, 364–372.

Wahba, G. (1983). Bayesian 'confidence intervals' for the cross-validated smoothing spline. *J. Roy. Statist. Soc. B* **45**, 133–150.

West, M. (1991). Bayesian computations: Monte Carlo density estimation. *J. Roy. Statist. Soc. B* , (to appear).

DISCUSSION

L. M. BERLINER (*Ohio State University, USA*)

Professor O'Hagan presents us a paper with two goals. The first goal is to review efforts in Bayesian numerical analysis based on Gaussian random function prior specifications. The second goal is to present some interesting applications of such specifications of current interest. Professor O'Hagan succeeds on both accounts; We should be grateful for this well-written and informative paper.

The first point for discussion I wish to raise involves the role of subjective Bayesian thinking in the use of Gaussian random field priors. In the second paragraph of the paper, O'Hagan writes "... this paper will concentrate on techniques arising from representing prior information about $f(\cdot)$ in terms of a Gaussian process". My question is "What prior information is actually represented?" Specifically, consider (1). One can readily ascribe meaning to the specification of the mean function of the process. However, I believe the specification of the covariance function $v(\cdot,\cdot)$ is a difficult problem. While we may find some comfort in being told that v controls the degree of smoothness of realizations of the process, I do not believe we have a clear picture of what typical realizations of a Gaussian process actually look like for various v. The motivation of my concern is my perception of the view users of these methods seem to take concerning results. Suppose we observe f at the design points without error. The posterior mean can be drawn as an interpolant of the data and, because of the advertised advantage of the Bayesian approach to function estimation, we can also associate posterior variances with our function estimates. The success of the approach may be well and good, but as a skeptical discussant, I suggest that the posterior mean may typically be smoother than typical realizations of the posterior or prior process. (This phenomenon may be amplified in the realistic, hierarchical model suggested by O'Hagan in that the posterior process involves t-distribution tails.) Has the posterior mean computation actually been averaged over realizations, which may or not seem reasonable, to produce a "nice" function? If "almost all" realizations of the process are unbelievable, what solace can we really take in the ability to find posterior variances?

To potentially help to alleviate the concerns raised above, consider the following:

(i) I think intuition can be obtained by thorough examination of the covariance function v. In one dimension, consider the common covariance function (see (17))

$$v(x, x') = \exp\{-b(x - x')^2\}.$$

This function is very well understood by statisticians, and, hence, we may have a reasonable "feeling" for the implied covariance on f. Perhaps we can learn a little more by investigating the implied (see (18) and (19)) functions (let f^1 denote the first derivate of f)

$$v^{0,1} = \text{Cov}\,(f(x), f^1(x')) = 2b(x - x')\exp\{-b(x - x')^2\}$$

and

$$v^{1,1} = \text{Cov}\,(f^1(x), f^1(x')) = 2b[1 - 2b(x - x')^2]\exp\{-b(x - x')^2\}$$

For $b = 2$, the three corresponding correlation functions are graphed as functions of $(x - x')$ below. Of course, both v and $v^{1,1}$ are even functions about zero, while $v^{0,1}$ is an odd function. Note that $f(x)$ and $f^1(x)$ are uncorrelated. The absolute value of the correlation between $f(x)$ and $f^1(x')$ grows as $|x - x'|$ does to a maximum of $e^{-1/2}$ at $|x - x'| = (2b)^{-1/2}$.

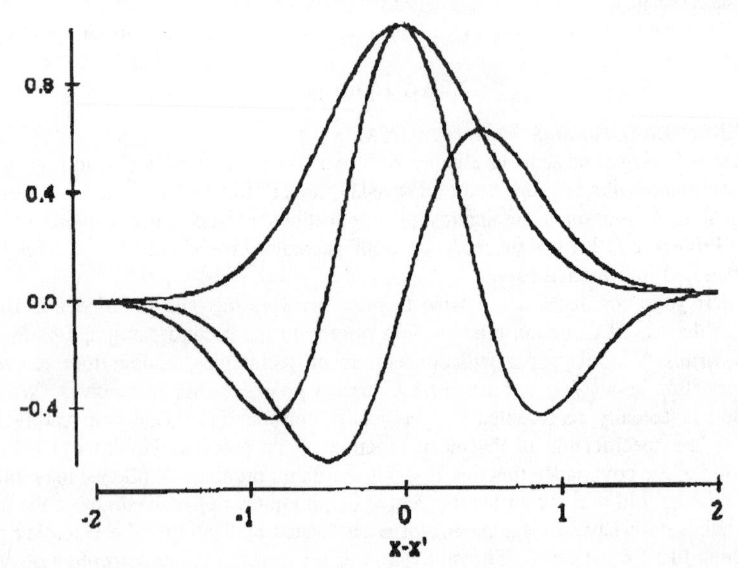

(ii) I think analysis beyond the reporting of the posterior means and variances should become a standard part of applied work in this area. For example, when possible, analysts might consider providing several "approximate" realizations from the appropriate posterior. If the realizations look ridiculous, the analysis may be questionable. Such realizations could be obtained as simulated random functions from the posterior. One possible method involves "brute force"; namely, we might approximate the process as an appropriate multivariate normal distribution. More sophisticated simulation techniques deserve investigation.

(iii) Depending on the application, the following suggestion may be useful. It may be desirable to try to model the real prior information available through the mean, or parametric hierarchical models for the mean, of the process. After exhausting prior information, the rest of the uncertainty is modelled as a Gaussian random process.

In the balance of this discussion, I will focus on the applications O'Hagan discusses. The main ongoing work described involves Bayesian interpolation, integration, and optimization based on Gaussian random fields. The primary contention of the author is that the analysis of such problems via Gaussian random process methods is appropriate when the function f of interest is extremely expensive to compute. The cost may involve time and/or money associated with experiments either in the "field" or on the computer. For such problems the notion of attempting to use data in an efficient manner as emphasized by O'Hagan is quite sensible.

In Section 3, O'Hagan presents discussion of Bayesian quadrature in which the integrand is modeled as a realization of a Gaussian process. The analysis seems reasonable, though some care should be given to rigor. For example, the estimate given in (22) need not exist if, as suggested as an example by the author, M is Lebesgue measure on R^p. Also, I doubt that the methods described by the author in the context of numerical Bayesian integration are, typically, serious competitors to sampling/simulation based methods. Forgive the repetition, but, I must ask when a Gaussian random process model is actually sensible for the likelihood-prior product? Also, sampling based methods such as Gibbs' Sampling enjoy the advantage of producing a sample from the posterior distribution, estimates of various marginal posteriors, as well as estimates of various posterior expectations. These comments apply to day-to-day Bayesian work. However, in defense of the current paper, it is difficult for me to imagine performing a Gibbs' Sampling analysis for a problem in which a *single* (efficiently programmed) computation of the function of interest requires several hours of supercomputer time.

O'Hagan presents optimization problems as another potential application of Gaussian random function analyses. While I agree that optimization problems may be a rich arena for the application of these methods, I think O'Hagan's discussion in this context falls short of what it should have been. First, the example analyzed is far too simple to be taken seriously as a test case. Though not exploited in the current paper, the extreme complexity functions often of interest in global optimization may lend credence to Gaussian random processes as priors. Also, there are various Bayesian and pseudo-Bayesian methods not mentioned by the author. Substantial, and directly pertinent, work based on Gaussian processes has been done in a series of contributions by J. Mockus (1989). Other work on optimization with a Bayesian flavor include Laarhoven (1988) and Laud, Berliner, and Goel (1989).

I was somewhat disappointed in the discussion of applications. I wish O'Hagan had told us more about the role of real Bayesian thinking for Gaussian processes in areas in which such methods are almost becoming routine. (See Wahba, 1989, for references.) For example, the computation of the posterior mean of a Gaussian random field is known as Kriging in the geostatistics literature; in some oceanography circles, a special case is known as "objective analysis". (What a great name!) Bayesian thinking has much to offer in these applications. For more evidence, see the exchange between Cressie (1990) and Wahba (1990).

In conclusion, I have emphasized that, in principle, in each application of the methods discussed, one should ask to what degree the Gaussian random field prior actually reflects prior beliefs. While I believe this question is fundamental, I am not a fool; I also believe in using procedures which seem to work even if a firm footing for the procedures is not available. Statisticians and other mathematical modelers must surely take such a view in order to actually do, rather than just talk about, something in practice. However, this operational view produces answers which must always be treated with a degree of suspicion commensurate with the level of belief in the model.

B. BETRÒ (*CNR-IAMI, Italy*)

In (continuous) optimization, we must distinguish between local problems, in which the function to be optimized (objective function) is assumed to be unimodal, and global problems, in which the global optimum is sought in absence of guaranteed unimodality. In the first case, a quadratic model of the function is adequate in a neighborhood of the extremum, and efficient methods exploit this local structure. It is hard to believe that the introduction of a stochastic model would improve the situation. Considering the example given in the paper, minimization of the function $x - 2\log x$ in the interval $[0,5]$ by subroutine E04ABF in the NAG library gives, after 9 function evaluations, 1.999999995 as an estimate of the minimum. Notice that E04ABF does not require evaluation of derivatives.

The fact with stochastic processes, in particular with Gaussian processes, is that their sample paths are far from being unimodal. Therefore they are better suited for global, rather than local, optimization problems. Unfortunately, as it is clear from (15), the optimization of the interpolant still requires the solution of a global optimization problem which, although dealing with a function cheaper than the original one, is not necessarily a simpler one. In one dimension the interpolant is possibly unimodal between the evaluation points, so that global optimization is affordable, but in the multidimensional case the situation obviously becomes dramatically worse.

There is a certain amount of literature on the use of Bayesian methods for global optimization. Other approaches, different from those modelling the objective function by a stochastic process, have been proposed. For a recent survey see Betrò (1991).

D. E. MYERS (*University of Arizona, USA*)

In examining a new derivation it is important to ask whether a new interpolator is obtained, or, if not, whether it provides new properties or whether it is more general than other developments. As given in the paper the interpolator is

$$m^*(x) = \Sigma k_i v(x, x_i) + \Sigma b_j h_j(x). \qquad (eq.13)$$

Using the uninformative prior on the hyperparameters wherein $B_0 = 0$ and using the notation in the paper, the coefficients in m^* are obtained as the solution of

$$\begin{bmatrix} A & H \\ H^T & 0 \end{bmatrix} \begin{bmatrix} k \\ b \end{bmatrix} = \begin{bmatrix} f \\ 0 \end{bmatrix} \qquad (M1)$$

k is the vector of k_i's. Under weaker conditions than used in the paper the interpolator can be rewritten as

$$m^*(x) = \Sigma\lambda_i(x) f(x_i) \qquad (M2)$$

where

$$\begin{bmatrix} A & H \\ H^T & 0 \end{bmatrix} \begin{bmatrix} \lambda_x \\ \mu_x \end{bmatrix} = \begin{bmatrix} t(x) \\ h(x) \end{bmatrix}. \qquad (M3)$$

The system (M3) can also be directly obtained by requiring that (M2) be unbiased and have minimum error variance, which is then given by $w^*(x, x)$. Eq. 13 is known in the numerical analysis literature as the radial basis function interpolator, Micchelli (1986). Both the thin plate and the smoothing spline are special cases. The form given by (M2) is known as the kriging estimator in the geostatistical literature. The equivalence of eq. 13 and M2 is well-known, Myers (1988). However either form is more general than indicated by the specific Bayesian analysis given in the paper. For example, conditional positive definiteness (with respect to the $h_{j,s}$) of the kernel function v is sufficient to ensure the invertability of the coefficient matrix in either (M1) or (M3) and hence to uniquely determine an interpolator. The assumption of normality precludes the use of conditional positive definite kernels. The prior should instead focus on specifying: i.) the type of generalized covariance, ii.) parameters in the generalized covariance, iii.) the functions $h_j, j = 0, \ldots, p$. Normality is neither needed nor particularly desirable, the prior on σ^2 is unimportant since the value of σ^2 has no effect on the interpolated values, only on the error variance. Note that the error variance does not depend on the data values nor do the coefficients in (M2). One of the advantanges of the use of (M2) is that it leads easily to the estimation of any linear functional of f, one simply applies that linear functional to the entries on the right hand side of (M3). This formulation also shows why one would minimize w^* in order to design a sampling plan. Sample plan design has received extensive consideration in the geostatistical literature, see Barnes (1989). Finally either form of m^* easily generalizes to the vector valued case as noted in Myers (1991).

J. PILZ (*Bergakademie Freiberg, Germany*)

I would like to comment on some issues raised in Section 2 on interpolation and smoothing. First, let me draw attention to geostatistical prediction techniques that have become known under the heading "kriging". From a statistical viewpoint, the kriging predictor is merely the best linear unbiased predictor of a random function (field) defined on some spatial domain. The natural interpolant (13) exactly coincides with the Bayesian kriging predictor introduced by Kitanidis (1986) and Omre (1987). The well-known (universal) kriging predictor comes out as a limiting case when $B_0 = 0$, i.e., when specifying a noninformative prior for (β, σ^2).

Pilz (1991b) developed a robust Bayesian version of (13) which only requires approximate knowledge of the first and second order moments of β: Assume

$$b_0 \in \mathcal{B} \quad \text{and} \quad B_0 \geq \bar{B}$$

where $\mathcal{B}$ is some subset of R^q supposed to be symmetric around some center point $b_1 \in R^q$ and $\bar{B}$ is some given positive definite matrix such that $B_0 - \bar{B}$ is positive semidefinite, i.e., the prior covariance matrix of $\beta|\sigma^2$ is bounded from above by $\sigma^2 \bar{B}^{-1}$. Then a robust Bayesian interpolant minimizing the maximum possible Bayes risk (posterior variance) is given by (13) with b replaced by

$$b^* = B^* \left((\bar{B}^{-1} + B_1)^{-1} b_1 + H^T A^{-1} f \right), B^* = \left((\bar{B}^{-1} + B_1)^{-1} + H^T A^{-1} H \right)^{-1}$$

where B_1 maximizes the trace functional

$$G(B_p) = \operatorname{tr} \left(H^T A^{-1} H + (\bar{B}^{-1} + B_p)^{-1} \right)^{-1}$$

defined on the set of all centered moment matrices $B_p = \int_{\mathcal{B}} (t - b_1) \times (t - b_1)^T P(dt)$ generated by the class of probability measures P over $\mathcal{B}$. Obviously, b^* is the three-stage hierarchical Bayes linear estimator of β, where, at the third stage, the hyperparameter $E\beta$ has mean b_1

and (least favourable) covariance matrix B_1. For further details and explicit form solutions for B_1 see Pilz (1991a), Sections 6.3, 15.5 and 17, where the case of an inadequate model for the mean function $h(\cdot)^T\beta$ is also treated.

In the geostatistics literature, the unknown covariance structure of the random function is inferred from the variogram $E(f(x) - f(x'))^2$. Under the assumption of covariance stationarity, the covariance function and the variogram are equivalent tools. For interpolation or smoothing it is very essential to capture the behaviour of the variogram near the origin. In this respect, a warning about the use of the Gaussian covariance function (17) is in order, even if one deals with smooth phenomena. Stein (1989) shows that slight misspecifications of this covariance function can have dramatic effects on prediction, he also proposes a smoooth alternative.

A full Bayesian approach to interpolation and smoothing would also require some prior modeling for the parameters of the covariance function, e.g., for the decay rate b in (16) or (17). Unfortunately, covariance parameters are not easy to handle in a Bayesian framework. For the case, however, that the covariance function admits a parametric linear model

$$v(x,x') = \theta_1 k_1(x-x') + \cdots + \theta_m k_m(x-x'), m \leq 3$$

where $\boldsymbol{\theta} = (\theta_1, \ldots, \theta_m)^T$ is an unknown vector of variance components and $k_1, \ldots, k_m$ are given correlation functions, Bayes invariant quadratic estimators for v (invariant w.r.t. translations $f + H\beta$ for all $\beta \in R^q$) have been derived by Stuchlý (1989), see also Pilz (1990).

A. H. SEHEULT (*University of Durham, UK*) and
J. A. SMITH (*University of Durham, UK*)

Professor O'Hagan discusses several problems in numerical analysis from a Bayesian perspective, concentrating on estimation and design issues in interpolation, integration and optimisation. A related problem is that of locating a zero, one which we considered in our poster presentation at the Meeting. Our model explicitly includes the location of the zero as random quantity, and the "slope" of the function is given a stationary covariance structure, leading to a nonstationary function process similar (but different) to that in O'Hagan (1978). A one-step-ahead sequential procedure selects the next design point to minimise the mean squared error from zero of the linear Bayes predictor of the function. The variance contribution to the mean squared error inhibits the next design point from being the zero of the predictor (the bias contribution) by not letting it wander too far from previous design points, and is in contrast to the optimisation method suggested by Professor O'Hagan where succesive points are chosen to optimise the predictor. However, in a series of papers, Schagen (1979, 1980a, 1980b, 1984) develops a sequential optimisation criterion which explores the function before gradually homing in on the optimum. This is achieved by incorporating (after each function evaluation) an estimate of the posterior probability of missing a "bump" in part of the function. Schagen also includes a method for estimating the "smoothing parameter" in the covariance function.

The distinction between expensive and cheap functions is important. Effective design will depend crucially on careful elicitation of prior information, especially for expensive functions. Moreover, fixed designs are questionable in this context; but if they are used experimenters should be free to modify them after contrasting observation with prediction. Also questionable in this context is slavish adherence to the widely-held, narrow view of Bayesian inference expounded by Lindley (1992) that "... the conditional distribution ... is a complete description of your uncertainty ... after the data have been seen". Expensive functions are often computer codes for complicated models of complex phenomena, and after each run of the code—each function evaluation—unanticipated relevant information often becomes available, and in such cases it is perhaps better to regard the conditional distribution as one input to posterior beliefs.

Finally, we believe that considerable care should be taken when attempting to compare the relative merits of Bayesian and classical methods of numerical analysis, especially when the former are developed in a context-free manner. The strength of Bayesian thinking is surely to regard and treat each application as unique: the danger of fixed methodology is that it encourages release from the burden of thinking!

REPLY TO THE DISCUSSION

The discussants have raised a number of issues, giving me helpful suggestions and a number of references, for which I am very grateful. As I tried to make clear, many if not all of the techniques I presented have been developed before, independently and often in quite different guises, by workers in several disparate fields. To those references I should add the work of Upsdell (1985), who derives a substantial amount of related theory.

Berliner, Myers and Pilz all draw attention to the work in geostatistics on methods known there as 'kriging'. Kriging developed as a classical 'best linear estimator' technique, as explained by Pilz, and as Myers shows, the interpolant (13) can be seen just as a solution of a set of linear equations. He thereby links it to work of numerical analysts on splines and radial basis functions. The Bayesian nature of these theories has also been demonstrated within those fields.

Myers points out that normality is not necessary to derive the interpolant. The robust formulation given by Pilz emphasises this fact, and (13) can be derived even more simply as a Bayes linear estimator as advocated by Goldstein (1988). The work of Seheult and Smith is also based on Bayes linear estimation. Myers further points out that the interpolant is independent of the variance parameter σ^2, and so the prior distribution I assume for σ^2 is irrelevant. Of course, posterior variances *will* depend on σ^2 and its prior distribution, and more work is needed on how to make use of the full posterior distribution rather than just its mean.

This brings in comments by Berliner and Betrò on the nature of sample paths. Sample paths of the posterior distribution will indeed be less well behaved than the posterior mean, and will indicate when it is reasonable to assume a Gaussian process prior distribution. Seheult and Smith stress the need to think carefully about prior information. They are right, of course, but to specify a prior distribution for a whole function in a realistic amount of time and effort necessitates compromise. Any prior distribution can only be an approximation to true prior beliefs, and sensitivity of posterior inference to that approximation should be looked for. It is clear that some posterior inferences will be sensitive to the Gaussian process prior assumption. To compute the posterior distribution of the number of local maxima (or modes), for instance, would not be sensible, since the Gaussian process is likely to produce 'bumpy' realisations.

Berliner makes a number of suggestions about assessing the realism of my prior modelling. I am not sure of the usefulness of his plots of covariance functions of derivatives. Their general form seems reasonable to me, and I cannot imagine having strong opinions about features like the separation $x - x^*$ at which the correlation between $f(x)$ and its first derivative $f'(x^*)$ is maximised. But I can see that some qualitative conclusions are important, such as that whenever $f'(x)$ exists it will be independent of $f(x)$.

Berliner makes useful comments on my quadrature ideas. In particular, he again reminds us of the need for care in modelling. The realism of the Gaussian process assumption depends on the underlying measure with respect to which I define the integral. For instance, the Bayes-Hermite formulation means that beliefs about the ratio of the density function being integrated to the approximating normal density, are represented as a stationary Gaussian process. If the tails of the density are heavier than the normal, this would not be at all realistic. More work is needed on prior formulations appropriate to a variety of contexts. It is worth saying that this is a strength of the Bayesian approach. Traditional numerical analysis takes account of context only in an anecdotal and unsystematic way. Careful attention to representing prior beliefs that truly reflect the real context, will guarantee posterior inferences that are appropriate to that context.

Betrò and Berliner comment on optimisation. Both rightly criticise my too-simple example, and both give references to several other Bayesian approaches to optimisation which I will need to study. The work of Seheult and Smith on finding a zero of a function is closely related. Their search strategy which does not initially home in too quickly, but spreads early points in a way that gives good information about the form of the function, sounds very interesting.

It seems that no discussion at Peñíscola was complete without a mention of Gibbs sampling, and Berliner obliges us here. It nicely reinforced my point about expensive and cheap functions. Sampling methods do seem to be the best we have for integrating high-dimensional *cheap* functions. Their inefficient use of function evaluations then matters less than their efficient coverage of the parameter space. We do not yet know how to deploy points deterministically in high dimensions in efficient, sparse ways and such that inference from them is practical for cheap functions. But Gibbs sampling is inappropriate for expensive functions, and I believe that research on methods for expensive functions will in the long run allow us to dispense with randomisation completely.

ADDITIONAL REFERENCES IN THE DISCUSSION

Barnes, R. (1989). A partial history of spatial sampling design. *The Geostatistics Newsletter* **3**, 10–12.

Betrò, B. (1991). Bayesian methods in global optimization. *Journal of Global Optimization* **1**, 1–14.

Cressie, N. (1990). Letters to the Editor: Reply to comment on Cressie. *Amer. Statist.* **44**, 256–258.

Goldstein, M. (1988). Adjusting belief structures. *J. Roy. Statist. Soc. B* **50**, 133–154.

Kitanidis, P. K. (1986). Parameter uncertainty in estimation of spatial functions: Bayesian analysis. *Water Resources Research* **22**, 499–507.

Laarhoven, P. J. M. van (1988). *Theoretical and Computational Aspects of Simulated Annealing*. Amsterdam: Centrum voor Wiskunde en Informatica.

Laud, P., Berliner, L. M. and Goel, P. K. (1989). A stochastic probing algorithm for global optimization. *Computing Science and Statistics. Proccedings of the 21st Symposium on the Interface* (K. Berk *et al.* , eds.). Alexandria, Virginia: ASA.

Lindley, D. V. (1992). Is our view of Bayesian Statistics too narrow?. *Bayesian Statistics 4* (J. M. Bernardo, J. O. Berger, A. P. Dawid and A. F. M. Smith, eds.), Oxford: University Press, 1–15, (with discussion).

Micchelli, C. (1986). Interpolation of scattered data: distance matrices and conditionally positive definite functions. *Constructive Approximation* **2**, 11–22.

Mockus, J. (1989). *Bayesian Approach to Global Optimization*. Dordrecht: Kluwer Academic.

Myers, D. E. (1988). Interpolation with positive definite functions. *Sciences de la Terre* **28**, 252–265.

Myers, D. E. (1991). An alternative Bayesian formulation for spatial interpolation. (Unpublished *Tech. Rep.*).

O'Hagan, A. (1978). Curve fitting and optimal design for prediction. *J. Roy. Statist. Soc. B* **40**, 1–42, (with discussion).

Omre, H. (1987). Bayesian kriging – merging observations and qualified guesses in kriging. *Math. Geology* **19**, 25–39.

Pilz, J. (1990). Bayes estimation of variograms and Bayesian collocation. *TUB Dokumentation Heft* **51**, Vol. II. 565–576.

Pilz, J. (1991a). *Bayesian Estimation and Experimental Design in Linear Regression Models*. New York: Wiley.

Pilz, J. (1991b). Robust Bayes linear prediction of regionalized variables. (Unpublished *Tech. Rep.*).

Schagen, I. P. (1979). Interpolation in two dimensions — a new technique. *J. Inst. Maths. Applics* **23**, 53–59.

Schagen, I. P. (1980a). Stochastic interpolating functions — applications in optimization. *J. Inst. Maths. Applics* **26**, 93–101.

Schagen, I. P. (1980b). The use of stochastic processes in interpolation and approximation. *Int. J. Comput. Math. B* **8**, 63–76.

Schagen, I. P. (1984). Sequential exploration of unknown multi-dimensional functions as an aid to optimization. *IMA J. Num. Anal.* **4**, 337–346.

Smith, J. A. and Seheult, A. H. (1991). Linear Bayes methods for locating zeros of deterministic functions. (Unpublished *Tech. Rep.*).

Stein, M. L. (1989). The loss of efficiency in kriging prediction caused by misspecifications of the covariance structure. *Geostatistics* **1** (M. Armstrong, ed.). Dordrecht: Kluwer Academic, 273–282.

Stuchlý, J. (1989). Bayes unbiased estimation in a model with three variance components. *Aplikace Mat.* **34**, 375–386.

Upsdell, M. P. (1985). *Bayesian Inference for Functions.* Ph.D. Thesis, University of Nottingham.

Wahba, G. (1990). Letters to the Editor: Comment on Cressie. *Amer. Statist.* **44**, 255–256.

BAYESIAN STATISTICS 4, pp. 365–388
J. M. Bernardo, J. O. Berger, A. P. Dawid and A. F. M. Smith, (Eds.)

Bayesian Robustness Functions for Linear Models

DANIEL PEÑA and GEORGE C. TIAO
Universidad Carlos III de Madrid, Spain and *University of Chicago, USA*

SUMMARY

This paper introduces two new diagnostic tools: the Bayesian Robustness curve (*BROC*) and the Sequential Bayesian Robustness curve (*SEBROC*). Both are built using the posterior odds for each possible number of outliers in a scale contaminated linear model. It is shown that these functions have a cross-validation interpretation, and can be useful to judge the robustness of the fitted model. The computation of these curves is carried out using ideas from stratified sampling.

Keywords: CROSS-VALIDATION; DIAGNOSIS; MIXTURE MODELS; MODEL SELECTION.

1. INTRODUCTION

The validation of a model (model diagnosis) is a key activity in statistical model building. Usually, this validation is made with the same sample data that has been used to build it. This is a serious limitation, because real data has normally many sources of heterogeneity: the parameters of the model can be changing over time, the sample may contains outliers or wrong measurement data, the model may be misspecified because of omitted variables, non-linear relations, and so on. Therefore, it would be better to use a sample for estimation and another for diagnostic checking, as it is often done in standard scientific practice.

In this paper we propose a portmanteau global checking of the model using a Bayesian approach that is based on the probabilities of 1, 2,... outliers in a heterogeneity model. The procedure can be interpreted as a cross-validation method, and leads to checking the model with out-of-sample data in a very complete way.

The paper is organized as follows. Section 2 reviews briefly some models for heterogeneity, and justifies choosing as the set-up of this paper the Box and Tiao (1968) formulation. Section 3 proposes two basic tools for model diagnosis: the Bayesian Robustness curve (*BROC*), and the sequential Bayesian Robustness curve (*SEBROC*). These curves are related to cross-validation methods and can be used for choosing between alternative regression models. Section 4 describes a procedure to carry out the required heavy computations using ideas from stratified sampling. Section 5 shows the behavior of the procedure in two examples. Finally, some concluding remarks are given in Section 6.

2. LINEAR MODELS FOR HETEROGENEITY

To allow for model heterogeneity, a useful model introduced by Tukey (1960) is the scale contaminated normal, that has been used extensively in the literature. For linear models, the representation is given by

$$y = X\beta + u \tag{2.1}$$

where y is an $n \times 1$ vector of responses, X a known $n \times p$ matrix of full rank p, β a $p \times 1$ vector of unknown parameters and u an $n \times 1$ vector of random errors that are distributed as

$$(1-\alpha)N(0,\sigma^2) + \alpha N(0,k^2\sigma^2) \tag{2.2}$$

where $N(0,\sigma^2)$ is the central distribution, $N(0,k^2\sigma^2)$ is the alternative distribution and $(1-\alpha)$ and α are, respectively, the probabilities of the errors coming from these distributions. Assuming k much larger than one means that the uncertainty of the alternative distribution is very large. This model has been used by Box and Tiao (1968) to investigate the effect of heterogeneity in linear models.

Abraham and Box (1978) approached this problem by introducing heterogeneity in the mean, instead of in the variance. Their model is

$$y = X\beta + \delta Z + u$$

where each element of the vector Z takes the value 1 with probability α, and 0 with $1-\alpha$. Guttman, Dutter and Freeman (1978) modified this additive model assuming that Z has exactly m non-zero elements, m being known. They determined the value of m by analyzing the model for $m = 0, 1, \ldots$ and so on.

The performance of these three models has been reviewed by Freeman (1980), and Eddy (1980) has shown that their estimation can be viewed as generalized least squares with the weights depending on the model. The latter author also suggested a mixed additive-multiplicative model combining the Box and Tiao and Abraham and Box formulations. Other relevant work on outliers in linear models using a Bayesian approach includes Pettit and Smith (1985), Guttman and Peña (1988), Chaloner and Brant (1988) and Guttman (1991).

The scale contaminated model used by Box and Tiao has two clear advantages: (1) the number of parameters is constant, and does not depend on the number of outliers as in the GDF model; (2) it can detect outliers that occur in both tails of the distribution, whereas the AB model cannot.

Box and Tiao (1968) assume that β and $\log\alpha$ are a priori locally independent and uniform,

$$P(\beta,\sigma) \propto \sigma^{-1}$$

and show that the marginal posterior distribution of β can be expressed as a weighted average of 2^n distributions obtained by considering all the possible cases of the n observations belonging to the alternative distribution. Calling $a_{(r)}$ the event that a particular set of r observations come from the alternative distribution and the remaining $n-r$ from the central one, the marginal posterior for β is

$$P(\beta|y) = \sum_{\forall r} P(a_{(r)}|y)P(\beta|a_{(r)},y) \tag{2.3}$$

where $P(\beta|a_{(r)},y)$ is a p-dimensional multivariate t distribution with mean

$$\hat{\beta}(r,\phi) = \hat{\beta} - \phi(X'X)^{-1}x_r'(I-\phi H_r)^{-1}(y_r - x_r\hat{\beta}) \tag{2.4}$$

where $\hat{\beta}$ is the usual least square estimator, $\phi = 1-k^{-2}$, y_r and x_r are, respectively, the $r \times 1$ vector and $r \times p$ matrix corresponding to the r observations that are assumed to come from the alternative distribution, and $H_r = x_r(X'X)^{-1}x_r'$ is the corresponding submatrix of the hat matrix H. The dispersion matrix is:

$$M_{(r)} = s_{(r)}^2(X'X - \phi x_r'x_r)^{-1} \tag{2.5}$$

where

$$s^2_{(r)} = \frac{1}{n-p}\left(\sum_{j\notin(r)}(y_j - \hat{y}_{j(r)})^2 + \frac{1}{k^2}\sum_{j\in(r)}(y_j - \hat{y}_{j(r)})^2\right) \tag{2.6}$$

$\hat{y}_{j(r)} = x'_{(j)}\hat{\beta}(r,\phi)$ and $x'_{(j)}$ is the j-th row of X. The weighting term, $p(a_{(r)}|y)$ in (2.3), is the posterior probability that the set of r observations comes from the alternative distribution and the remaining $n-r$ from the central distribution, and is given by

$$w_{(r)} = C\left(\frac{\alpha}{1-\alpha}\right)^r k^{-r}\left(\frac{|X'X|}{|X'X - \phi x'_r x_r|}\right)^{1/2}\left\{\frac{s^2}{s^2_{(r)}}\right\}^{\frac{(n-p)}{2}} \tag{2.7}$$

where C is the constant making the weights sum to unity, and s^2 the usual unbiased estimation of the residual variance.

3. THE BAYESIAN ROBUSTNESS CURVE

3.1. *Definitions*

Given two models M_1, M_0, and a set of data, y, the posterior odds for model M_1 against M_0 is defined by

$$P_{10} = \frac{P(M_1|y)}{P(M_0|y)} = \frac{P(y|M_1)}{P(y|M_0)}\frac{\pi(M_1)}{\pi(M_0)}$$

where $P(y|M_1)/P(y|M_0)$ is the Bayes factor, that is, the ratio of the predictive densities given the model, or the marginal likelihoods given the data, and $\pi(M_1)/\pi(M_0)$ the ratio of the a priori probabilities. Therefore, the posterior odds for the model M_1, exactly h observations are coming from the alternative distribution, against M_0, all the data comes from the central distribution, is given by

$$P_{h0} = \binom{n}{h}\left(\frac{\alpha}{1-\alpha}\right)^h F_{h,0}$$

where $F_{h,0}$ is the Bayes factor given by

$$F_{h,0} = \binom{n}{h}^{-1} k^{-h} \sum_{r\in S_h} \frac{|X'X|^{1/2}}{|X'X - \phi x'_r x_r|^{1/2}}\left\{\frac{s^2_{(r)}}{s^2}\right\}^{-(\frac{n-p}{2})} \tag{3.1}$$

and S_h is the set of all $\binom{n}{h}$ possibilities of having h outliers in n observations. Assuming that k is large, $\phi \doteq 1$, and this expression can be written

$$P_{h0} = \left(\frac{\alpha}{1-\alpha}\right)^h k^{-h} \sum_{r\in S_h} \frac{1}{|I - H_r|^{1/2}}\left(1 + \frac{h}{n-p}F_r\right)^{(\frac{n-p}{2})} \tag{3.2}$$

where

$$F_r = \frac{SS^2 - SS^2_{(r)}}{hs^2_{(r)}} \tag{3.3}$$

is the F value to test that this set contains outliers, that is, SS^2 is the least squares residual sum of squares for the whole sample, and $SS^2_{(r)}$ the residual sum of squares when the set

r is deleted. This test is usually called the Chow test (Chow, 1960) in the econometrics literature, and it is widely used for testing structural shifts in regression models.

Expression (3.2) is the usual posterior odds for choosing between nested linear models (Smith and Spiegelhalter, 1980). This relationship points out that choosing α and k, the prior probability ratio, can be done by the device of an imaginary training sample (Spiegelhalter and Smith, 1982). In this paper we will consider α and k as fixed, but will check the sensitivity of the results obtained to different values of k and α.

A global checking for outliers can be done by computing the posterior odds for $h = 1, 2, \ldots$, and plotting these ratios as a function of h. The resulting curve, $f(h)$, will be called the Bayesian Robustness curve (*BROC*) and will provide information about the possible number of outliers in the data set.

When there are several outliers the *BROC* curve can be unclear to identify the exact number, because of the masking effect. For instance, with r outliers it is possible, if the sample size is not very large and α small, that the probabilities for $r+1, r+2$ outliers are of similar size as that for r, which makes it very difficult to identify the exact number of atypical points in the sample.

This problem could be mitigated by looking at the sequential posterior odds

$$S_{h,h-1} = \frac{P_{h,0}}{P_{(h-1),0}} \tag{3.4}$$

The plot of $S_{h,h-1}$ against h will be called the sequential Bayesian Robustness Curve (*SE-BROC*).

3.2. *A Cross-Validation Interpretation of the Robustness Curve*

The ratio $s^2_{(r)}/s^2$ can be written

$$\frac{s^2_{(r)}}{s^2} = 1 - \frac{\phi}{(n-p)s^2}(e_r'(I - \phi H_r)^{-1} e_r) \tag{3.5}$$

where $e_r = y_r - x_r\hat{\beta}$ is the vector of the r least squares residuals corresponding to y_r that are assumed to come from the alternative distribution. This expression can be written as a function of the out-of-sample or predictive residuals as follows. Let

$$\hat{u}_r = y_r - x_r\hat{\beta}_{(r)} \tag{3.6}$$

a vector of r predictive residuals, where

$$\hat{\beta}_{(r)} = (X_{n-r}'X_{n-r})^{-1}X_{n-r}'y_{n-r} \tag{3.7}$$

is the coefficient β estimated by dropping the r observations, then as (Cook and Weisberg, 1982, p. 136),

$$\hat{\beta}_{(r)} = \hat{\beta} - (X'X)^{-1}x_r'(I - H_r)^{-1}e_r \tag{3.8}$$

we finally obtain

$$\hat{u}_r = (I - H_r)^{-1}e_r \tag{3.9}$$

so that, the posterior odds (3.2) can be written, assuming $\phi \doteq 1$

$$P_{h0} = \left(\frac{\alpha}{1-\alpha}\right)k^{-h}\sum_{r\in S_h}|I - H_r|^{-1/2}\left\{1 - \frac{1}{s^2(n-p)}(\hat{u}_r'(I - H_r)\hat{u}_r)\right\}^{-(\frac{n-p}{2})} \tag{3.10}$$

This expression points out the dependence of the Bayesian Robustness Curve for h outliers on the predicted residuals and the leverage of all the possible sets of h observations in the sample. Note that if all the data comes from the central distribution, $\hat{u}_r$ is a vector of multivariate normal variables with zero mean and covariance matrix $\sigma^2(I-H_r)^{-1}$. Therefore, the residual quadratic form has just the usual $\chi_r^2\sigma^2$ distribution and the weights of this quadratic form are proportional to their generalized standard deviations $|I - H_r|^{-1/2}$.

Also, using (3.1) and (3.5) we can write, with $\phi \doteq 1$,

$$P_{h,0} = \left(\frac{\alpha}{1-\alpha}\right)^h k^{-h} \sum_{r \in S_h} |I - H_r|^{-1/2} \left\{1 + \frac{\hat{u}_r'(I - H_r)\hat{u}_r}{s_{(r)}^2(n-p)}\right\}^{\frac{(n-p)}{2}} \tag{3.11}$$

that can be approximated for large n by

$$P_{h,0} \doteq \left(\frac{\alpha}{1-\alpha}\right)^h k^{-h} \sum_{r \in S_h} |I - H_r|^{-1/2} \exp\{Q_{(r)}/2\} \tag{3.12}$$

where,

$$Q_{(r)} = \hat{u}_r'(I - H_r)\hat{u}_r / s_{(r)}^2 \tag{3.13}$$

This result shows the relationship between the Bayesian Robustness curve and cross-validation ideas. The cross-validatory assessment of a statistical prediction function chooses among a set of predictors the one that minimizes a measure of out-of-sample error prediction. This discrepancy measure is built by dropping r observations, computing a predictor, $P(n-r)$, from the remaining $n-r$ observations, and making an out-of-sample error estimation using this predictor, for the r dropped observations. The procedure is repeated for all the sets of size r. This method was suggested for $r = 1$ by Stone (1974), that provides an interesting account of the historical roots of these ideas, and by Geisser (1975) that allowed for multiple dropping. See also Mosteller and Tukey (1968) and Snee (1977).

The Bayesian Robustness Curve can be seen as a discrepancy measure also built using cross-validation ideas, because by (3.11) the function depends on the out-of-sample or cross-validated residuals, $\hat{u}_t$, and their covariance matrix. One may indeed argue that the probabilistic set up in (2.1) and (2.2) provides an explicit theoretical justification for using cross-validation. There seem to be two key notions in cross-validation: one is the out-of-sample prediction, and the other is that for a given number of observations h to be treated as out-of-sample points, all subsets of observations of the same size h are considered and treated alike. Now in our set up, it is recognized that heterogeneity may exist in the sample and each observation has the same chance α of being heterogeneous (i.e., from the alternative model). This leads to expressions (3.10) and (3.11) which, for a given h, include all subsets of size h and use what seems to be the most natural measure for prediction discrepancy, the predictive density function itself. Further, in cross-validation one faces the problem of choosing h; but in the Bayesian set up here this is recognized through the prior probability of the number of heterogeneous observations.

3.3. *The BROC in Model Selection*

The *BROC* could be used as a complementary criterion to choose among regression models. Suppose that we have a set of linear models $\{M_i\}$ that differ in the explanatory variables included, but they have similar values of the residual variance and other secondary criteria as AIC, or Cp. Suppose that in this set of models there exists one, M_i, such that:

$$BROC(h; M_i) \leq BROC(h; M_j) \;\; \forall j, \forall h$$

where $BROC(h; M_j)$ is the Bayesian Robustness curve for M_j, then we will say that model M_i is the more robust of the set for out-of-sample forecasting, and will be the model selected. Note that for (3.10) if all the models have similar residual variances the most robust model is the one with minimum out-of-sample forecasting error.

4. COMPUTING THE BAYESIAN ROBUSTNESS CURVE

4.1. *Unmasking Potential Outliers*

The computation of the Bayesian curves introduced in this paper is a difficult problem when the sample size is large, even for moderate values of h, because of the huge number of sets $\binom{n}{h}$ to be considered. Also, when the number of outliers is small, the computation of the odds ratios on the right hand side of (3.10) for all sets of size h is very inefficient, because most of them will include points coming from the central distribution and, therefore, their values will be very close to zero.

A general procedure to compute predictive densities or Bayesian marginal posteriors is the Gibbs sampler (Gelfand *et al.* 1990) that has been applied to outliers problems (Verdinelli and Wasserman, 1990). However, we will explore here a different procedure that takes into account the basic structure of our problem.

Suppose we could classify the n sample points into two groups: the first would include interesting data points which have high potential of being outliers, the second the ones that should be considered "good" beyond any reasonable doubt. Let n_1 and n_2 be the number of observations in each group ($n_1 + n_2 = n$). Then, as

$$\binom{n}{h} = \sum_{j=0}^{h} \binom{n_1}{j}\binom{n_2}{h-j} = \binom{n_1}{h} + \sum_{j=0}^{h-1} \binom{n_1}{j}\binom{n_2}{h-j} \tag{4.1}$$

for $h \leq \min(n_1, n_2)$, we can compute all the combinations $\binom{n_1}{h}$ of interesting points, a sample of the mixed sets containing $j = 1, 2, \ldots, h-1$ interesting points, and a few of all the combinations $\binom{n_2}{h}$ of the good points. Also, n_1 will be the maximum value for h to be computed.

The key point in this approach is splitting the sample into these two groups. It is obvious that the groups cannot be chosen by just taking into account the individual probabilities, because of the masking and swamping effects. For instance, Table 1 presents a simulated sample of $N(0, 1)$ observations in which three outliers have been inserted at the last three positions. When computing the probabilities for each point being an outlier, $h = 1$, it is shown in the table that none of the three outliers has a large value, (masking effect) and that the most likely outlier is the 5th point (swamping effect).

i	1	2	3	4	5	6	7	8	9	10
y	.059	1.80	.26	.87	−1.45	−.70	1.25	3.5	4.0	4.0
$P_{10}(i)$	.029	.024	.028	.024	.084	.044	023	.047	.071	.071

Table 1. *Posterior odds, $P_{10}(i)$ for the data indicated, with $k = 5$; $\alpha = .10$.*

However, for $h = 2$ the joint probabilities give information about these effects. Let us call A_i the event: "point y_i is outlier". If (y_i, y_j) are any two good points, then the probability of both being outliers is

$$P(A_i A_j) = P(A_i|A_j)P(A_j) \simeq P(A_i)P(A_j)$$

because dropping A_j will change very slightly the fitted variance and, therefore, will not affect the probability of A_i. However, if y_j is an outlier

$$P(A_i A_j) = P(A_i|A_j)P(A_j) \neq P(A_i)P(A_j)$$

because dropping it will affect the variance of the remaining sample, and this will change the conditional probability of any other point —outlier or not— of being an outlier.

This fact suggests building the probabilistic matrix of "interactions"

$$d(i,j) = P(i,j) - P(i)P(j) \tag{4.2}$$

where $P(i,j)$ is the posterior odds for the pair (i,j) and $P(i)$ and $P(j)$ are those for the individual points. If the ith point is an outlier, all the column $d(i,.)$ is expected to have relatively large values. However, if there are other outliers that are somehow masked by the ith, these will show up as large values in the distribution of $d(i,.)$. Therefore, large values in this matrix will indicate outliers, and relatively large values in a column, possible masking between these points.

The visual inspection of this matrix is, in our experience, the most useful way to split the sample. However, an objective procedure can be organized as follows

(1) compute the probabilities (p_i) of each point being an outlier. Let $\bar{p}_1$ be the median of these probabilities and, $s_1 =$ median $|p_i - \bar{p}_1|/.6475$, a high breakdown point robust estimate of the variability. Then, include in the set A those points that satisfy

$$p_i \geq \bar{p}_i + c_1 s_1 \tag{4.3}$$

(2) compute the joint probabilities p_{ij}, and the differences

$$d_{ij} = p_{ij} - p_i p_j \tag{4.4}$$

and let $\bar{d}_i$ be the median of $\{d_{ij}, j = 1, \ldots, n\}$, and $s_i(d)$ the high breakdown point estimate of the standard deviation, median $|d_{ij} - \bar{d}_i|/.6745$. Then, if

$$d_{ij} \geq \bar{d}_i + c_2 s_i(d),$$

include the point j into the set A.

The values $\{c_i, i = 1, 2\}$ can be chosen using the Bonferroni method to take into account the size of the comparisons to be made. For moderate sample size c_1 could be equal to 3 and c_2 equal to 5.

When the number of outliers is small, $\bar{d}_i$ and $s_i(d)$ may be very close to zero, and there is a risk for this procedure to select too many points. In these cases, it is better to use directly the mean and the standard deviation instead of their robust estimates.

For instance, if we apply the above procedure to the data set in Table 1, none of the points are identified as outliers at the first stage based on the univariate probabilities (with $c_1 = c_2 = 3$). However, at the second stage, points $(5, 6)$ and $(8, 9, 10)$ are selected. Table 2 presents the differences d_{ij}, and it can be seen that there are clearly two sets of unrelated potential outliers $\{5, 6\}$ and $\{8, 9, 10\}$.

	1	2	3	4	5	6	7	8	9	10
1		·	·	·	·	·	·	·	·	·
2			·	·	·	·	·	·	·	·
3				·	·	·	·	·	·	·
4					·	·	·	·	·	·
5						.0028	·	·	·	·
6							·	·	·	·
7								·	·	·
8									.0023	.0023
9										.0061

Table 2. *Probabilistic matrix of interaction for the identification of masking effects. Only values greater than* .001 *are printed.*

4.2. *Computing Probabilities*

Let us assume that with the above procedure we classify the data into two groups: a group A of n_1 potential outliers and a group B of n_2 good points. Then, instead of analyzing the $\binom{n}{h}$ combinations needed for $BROC(h)$, we can compute the posterior odds for all the $\binom{n_1}{h}$ combinations of the potential outliers, and compute a random sample of the $\binom{n_1}{h-j}\binom{n_2}{j}$ terms, for each value of j.

	1	2	3	4	5	6	7	8	9	10
1		7^+	9^+	7^+	33	15	7^+	13	20	20
2			7^+	6^+	20	10	6^+	12	18	18
3				7^+	29	14	7^+	13	19	19
4					22	11	6^+	11	17	17
5						65^*	20	39^*	64^*	64^*
6							11	20^*	31^*	31^*
7								11	17	17
8									56^*	56^*
9										111^*

Table 3. 10^4 *posterior odds* $P_{20}(ij)$ *for data in Table 1. Points with * belong to the first group, and those with + to the third.*

To illustrate the advantage of this procedure, Table 3 presents the posterior odds $P_{20}(ij)$ for every pair of sample data of Table 1. It can be seen that the size and variability of the posterior odds depend on the number of potential outliers included in the pair. Thus, we could classify these values into three strata: The first, (marked by * in the table) contains pairs of 2 potential outliers from the set $\{5, 6, 8, 9, 10\}$. The distribution or values for the ten members in this group is skewed, with mean value 53.7×10^{-4} and large variability, with coefficient of variation .52. The second, (unmarked in the table) includes a good and a bad value, and the distribution of the 25 pairs has a mean of 16.9×10^{-4}, is less skewed and with smaller variability. Finally, the ten members of the third group of two good points, (marked by + in the table) have a small mean and very small variability, and the distribution is more symmetric. These properties are summarized in Table 4.

	Contribution to Global posterior odd	%	Mean	Std	*Number*
(*) 2 potential outliers	.0537	52.1	.00537	.00277	10
1 each group	.0424	41.2	.00169	.00054	25
(+) 2 good points	.0069	6.7	.00069	.00008	10
TOTAL	.1030	100%			

Table 4. *Statistics of the Stratification of the posterior odds for pairs.*

It is clear from this table that we can estimate the total posterior odds by computing all the values in the first strata, and by taking a medium sample in the second, and a small sample in the third.

This same pattern appears when considering sets of size 3, (Table 5): The mean and the variability decrease with the number of good points in the set.

Number of potential outliers	Contribution to posterior odd	%	Mean $\times$ 10^{-4}	Std $\times$ 10^{-4}	Number
3	.00095	49.2	9.52	14.56	10
2	.00082	42.5	1.63	0.90	50
1	.00014	7.2	0.27	0.44	50
0	.00002	0.1	0.02	0.00	10

Table 5. *Contribution to the Global posterior odds for $h = 3$ from the different strata.*

The previous analysis suggests the following procedure to compute the Bayesian Robustness curve:

(1) use the posterior odds for individual points and pairs and the procedure outlined in section 4.1, to split the sample into n_1 potential outliers and $n - n_1$ good points,

(2) compute the posterior odds for $h = 3, \ldots, n_1$ as follows: (i) all the $\binom{n_1}{h}$ terms corresponding to sets of potential outliers; (ii) a decreasing fraction f (usually a maximum of 10% is enough for moderate sample size) of the terms $\binom{n_1}{j}\binom{n_2}{h-j}$, for $j = h-1, \ldots, 0$ that include potential outliers and some good points. Then, estimate the contribution of each strata by $\binom{n_1}{j}\binom{n_2}{h-j}\bar{p}(j)$ where $\bar{p}(j)$ is the mean of the posterior odds obtained by sampling in this strata.

5. EXAMPLES

5.1. *Example 1*

As the first example we will use again the data from Table 1. Table 6 shows the posterior odds of 1 to 5 outliers for different values of α. It can be seen that, given the small sample size, the prior dominates the posterior, and the terms for h greater than one are always small.

h / α	1	2	3	4	5
.05	.211	.023	.002	.000	.000
.10	.445	.103	.020	.004	.001
.15	.707	.260	.080	.025	.008
.20	1.001	.521	.227	.100	.043

Table 6. *Posterior odds, P_{h0}, for data in Table 1 as a function of α for $k = 5$.*

Table 7 shows the prior ratio and the Bayes factor for several events in this data set. Although the Bayes factor is, as expected, the highest at the three true outliers, its size is not big enough to compensate for the small prior ratio.

$a_{(r)}$	prior ratio	Bayes factor	posterior odds
1	.053	.267	.0141
2	.053	.215	.0114
3	.053	.249	.0132
4	.053	.217	.0115
5	.053	.745	.0395
6	.053	.396	.0210
7	.053	.207	.0111
8	.053	.419	.0222
9	.053	.630	.0334
10	.053	.630	.0334
1,2	.0028	.071	.0002
8,10	.0028	.464	.0013
9,10	.0028	.892	.0025
1,2,3	.00015	.015	.000002
6,9,10	.00015	.345	.000052
7,8,9	.00015	.098	.000015
7,8,10	.00015	.201	.000032
8,9,10	.00015	3.506	.000526

Table 7. *Components of the posterior odds of the indicated observations being outliers for $\alpha = .05$, $k = 5$.*

Table 8 shows the sequential posterior odds for several values of α. It is clearly seen that, again, there is no evidence of outliers. This problem appears because both the size of the outliers and the sample size are very small.

α	h 1	2	3	4	5
.05	.211	.109	.087	—	—
.10	.445	.231	.194	.200	.200
.15	.707	.367	.307	.312	.320
.20	1.001	.520	.435	.440	.430

Table 8. *Sequential posterior odds $S_{h,h-1}$ for data in Table 1 as a function of α for $k = 5$.*

To show the effect of these two factors, we will begin by increasing the sample size and keeping constant the size of the outliers. To do so, ten new good points ($N(0,1)$) given in Table 9, have been added to make a sample of size 20. Table 10 shows the new posterior odds, and now there is a clear evidence of some outliers for $\alpha \geq .10$. For instance, for $\alpha = .15$ the most likely number of outliers is three or four.

n	11	12	13	14	15	16	17	18	19	20
y	.356	−.210	−.335	−.356	2.26	−.50	−1.19	−.88	.72	−.006

Table 9. *Ten additional points over the data in Table 1 to form a sample of 20 data points.*

α	1	2	3	4	5
.05	.514	.161	.060	.017	.003
.10	1.086	.718	.565	.347	.153
.15	1.726	1.881	2.266	2.210	1.547
.20	2.445	3.635	6.443	8.910	8.830

Table 10. *Posterior odds for a sample of 20 data with three outliers.*

The number of outliers is more clearly seen in the sequential posterior odds of Table 11. Apart from the first ratio, there is always a relative maximum for $h = 3$, that is greater than the one for $\alpha \geq .15$.

α	1	2	3	4	5
.05	.514	.313	.373	.219	.209
.10	1.086	.661	.788	.614	.441
.15	1.726	1.045	1.251	.975	.700
.20	2.445	1.486	1.772	1.381	.992

Table 11. *Sequential posterior odds for Data from Tables 1 and 9.*

To show the effect of doubling the relative size of the outliers, Table 12 shows the posterior odds for the sample data of Table 1 but with $y_8 = 7, y_9 = y_{10} = 8$, and Table 13 the breakdown into their components of the more relevant cases.

α	h 1	2	3	4	5
.05	.214	.0274	.0253	.0034	.0002
.10	.453	.1222	.2381	.0670	.0009
.15	.719	.3083	.9538	.4262	.0895
.20	1.018	.6188	2.7117	1.7167	.5109

Table 12. *Values of posterior odds P_{h0} for $k = 5$ and α, data from Table 1 with $y_8 = 7, y_9 = y_{10} = 8$.*

$a_{(r)}$	prior	likelihood	posterior
1	.053	.266	.0141
8	.053	.496	.0263
9	.053	.819	.0434
10	.053	.819	.0434
1,2	.0028	.071	.0002
8,10	.0028	.893	.0025
9,10	.0028	2.321	.0065
1,2,3	.00015	.000	.0000
7,8,10	.00015	.666	.0001
8,9,10	.00015	156.667	.0235

Table 13. *Selected posterior odds for Data from Table 1 with $y_8 = 7, y_9 = y_{10} = 8$.*

Although the masking effect has not changed and the posterior odds for $h = 1, 2$ in Tables 6 and 12 are similar, for $\alpha \geq .15$ there is now clear evidence of three outliers. This result is confirmed with the sequential posterior odds, shown in Table 14.

	1	2	3	4	5
.05	.214	.128	.923	.133	.063
.10	.453	.270	1.947	.281	.132
.15	.719	.429	3.093	.447	.210
.20	1.018	.607	4.382	.633	.298

Table 14. *Sequential posterior odds $S_{h,h-1}$ for $k = 5$ and α as indicated, same data as Table 13.*

In summary, the *BROC* can be a useful tool to identify the number of outliers given that the prior probability for the true number of outliers is not too small.

5.2. *Example 2*

As the second example, we will use the stack-loss data from Daniel and Wood (1980), that has been analyzed by a number of authors including Cook (1979), Gray and Ling (1984), Chaloner and Brant (1988) and Rousseeuw and Zomeren (1990). All of them agree that the set $\{1, 3, 4, 21\}$ includes outliers and that point 2 may also be an outlier. The posterior odds for $h = 1, 2, 3$ can be computed directly and their components are presented in Figures 1, 2 and 3. Figure 1 shows the posterior odds for each individual point with $\alpha = .05$ and $k = 5$. The global posterior odd for $h = 1$, that is the sum of the values shown in figure 1, is 1.175. Point $\{21\}$ stands out over all the others and contributes $67\%(.7898/1.175)$ to the global posterior odds. Figure 2 shows the posterior odds for pairs. The largest values is for set $\{4, 21\}$, and corresponds to .6332 that is 65% of the total $P_{20}(.9712)$. Finally, Figure 3 shows the posterior odds for groups of three. All the sets that stand out include the pairs $\{4, 21\}$, and correspond to the combination of this pair with points $\{1, 2, 3, 6, 13, 14, 15, 20\}$. These eight combinations account for 59.6% of the posterior odds $P_{30}(.3661)$ for size three.

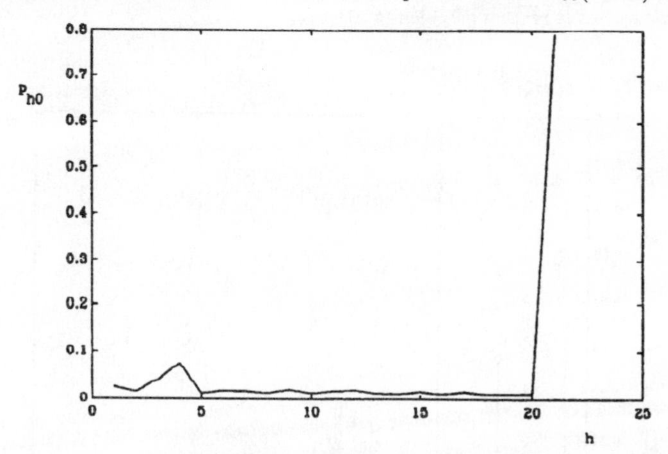

Figure 1. *Posterior odds for each point to be an outlier for the stack–loss data.*

To compute the *BROC* and *SEBROC* we will apply the procedure of Section 4. The probabilistic matrix of interactions, (4.4), $d(ij)$ has the following features: (i) most of the values are between .0000 and .0002; (ii) pairs $(1, 3)$, $(1, 4)$, $(3, 4)$ have values .0018, .0013, .0026; (iii) most of the pairs $(j, 21)$ are large, $(4, 21)$ being the largest with value .5747, and followed by $(2, 21)$, with .0157, and $(13, 21)$, with .0081. Therefore, the set of possible outliers is $\{1, 3, 4, 21\}$, and we may also include $\{2, 13\}$ to avoid a possible masking with point 21.

As the point 21 is dominating everything and is clearly an outlier, we decide to drop it and make the computation for the remaining sample of 20 points. Therefore, from now on the set of potential outliers is $\{1, 2, 3, 4, 13\}$.

Table 15 presents the posterior odds for this sample of 20 points. The approximate procedure has been applied to compute the posterior odds P_{40} and P_{50} using in both cases

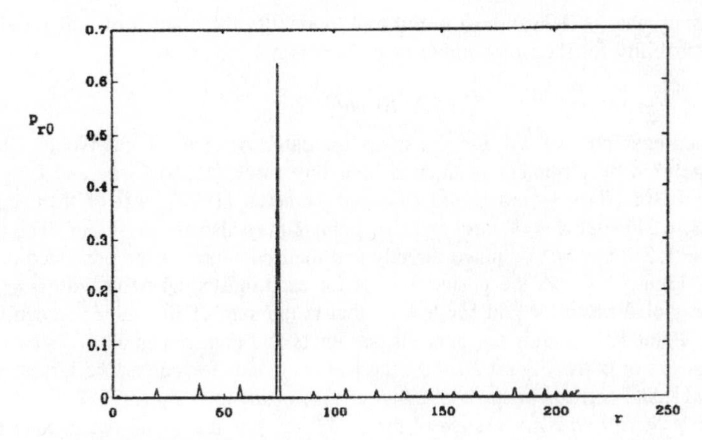

Figure 2. *Posterior odds for each possible pair of points to be outliers for the stack–loss data. The subsets* (21 *over* 2) = 210 *are ordered as follows:*$(1,2),(1,3),\ldots,(1,21),(2,3),\ldots,(20,21)$. *With this order the spike corresponds to the set 74 that contains the points* $(4,21)$.

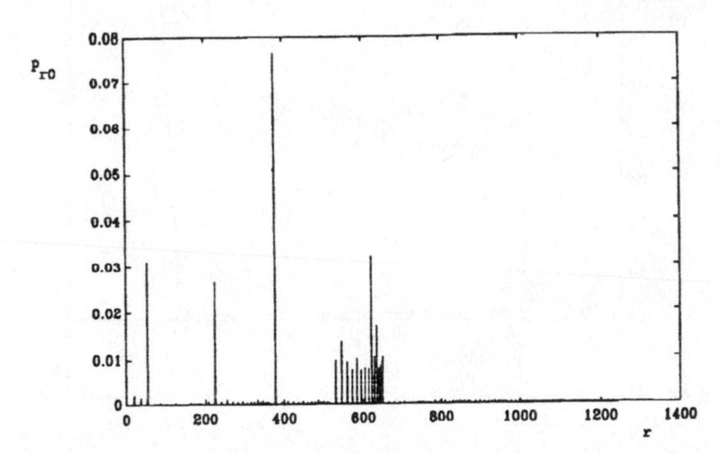

Figure 3. *Posterior odds for sets of size three. The sets* (21 *over* 3) = 1330. *are ordered as follows:* $(1,2,3)(1,2,4),\ldots,(1,2,21),(1,3,4),\ldots,(19,20,21)$. *The largest spike corresponds to* $(3,4,21)$ *and the next three to* $(1,4,21)$ *and* $(4,13,21)$.

a sample size in each strata of 20 points. Thus, for instance, for $h = 5$ the exact procedure requires $15,504$ evaluations of individual probabilities, and the approximate procedure only 101.

The analysis of the *BROC* for $\alpha = .10$ shows posterior odds greater than two for 1,2 and 3 outliers. The *SEBROC* factors have a maximum for three, and so it suggests that this is the number of outliers in the sample.

Daniel and Wood (1980) after a detailed analysis of this sample concluded (page 81) that

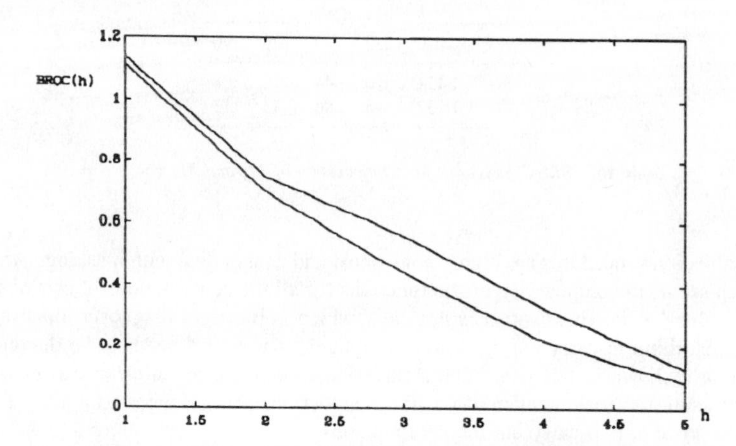

Figure 4. *BROC curves for models 1 and 2 for the stack–loss data.*

α	Method	1/0	2/0	3/0	4/0	5/0	2/1	3/2	4/3	5/4
	exact	1.141	.454	.223	.074	.015	.398	.492	.331	.202
.05	sampl.	–	–	–	.068	.019	.398	.492	.305	.279
	exact	2.408	2.024	2.097	1.469	.611	.840	1.036	.700	.416
.10	sampl.	–	–	–	1.432	.608	.840	1.036	.683	.425

Table 15. *Posterior odds for stack- loss data, point* 21th *deleted* ($k = 5$).

(i) points $\{1, 3, 4, 21\}$ represent transitional states, and (ii) the remaining seventeen points are well fitted by either

$$M_1 : \hat{y} = 14.1 + 0.71z_1 + 0.53z_2 + 0.0068z_1^2$$

or

$$M_2 : \hat{y} = 14.1 + 0.71z_1 + 0.51z_2 + 0.0254z_1z_2$$

where $z_1 = x_1 - 57.7$ and $z_2 = x_2 - 20.4$, where $\hat{s}^2(M_1) = 1.265$ and $\hat{s}^2(M_2) = 1.276$. Although the first model has smaller variance, the difference is very small, (the ratio is 1.008) and so an additional criterion could be to check the robustness of these models with out-of-sample forecasts. Figure 4 and Table 16 show the *BROC* curve for these 2 models. We see that model M_1 has smaller probability for any number of outliers and, therefore, will be the one to choose.

6. CONCLUDING REMARKS

Several authors have argued that using the same sample to fit and to assess the fit leads to underestimation of the error and small diagnostic power (Efron, 1983 and 1986; Gong, 1986; Mosteller and Tukey, 1977; Picard and Cook, 1984), but there are no general accepted guidelines about how to apply this idea in practical problems. The simplest one-at-a-time

	1/0	2/0	3/0	4/0	5/0
M_1	1.114	.70	.44	.22	.08
M_2	1.137	.75	.56	.33	.13

Table 16. *BROC Functions for Comparation of M_1 and M_2.$k = 5, \alpha = .05$.*

cross-validation procedures are highly non-robust and cannot deal with masking. An intuitive approach seems to compute diagnostic functions for all the combinations of sets of size r for several values of r. However, it is not clear which is the right diagnostic function to use, and the algorithm to carry out the heavy computations involved needs to be devoloped.

We have shown in this paper that a Bayesian approach offers a clear answer to the first question using the scale-contaminated linear model, and have suggested a simple stratified sampling procedure to carry out the computations.

The resulting Robustness curves, *BROC* and *SEBROC*, can be used as a global portmanteau check of the fitted model, and provide additional criteria for choosing among regression models.

REFERENCES

Abraham, B. and Box, G. E. P. (1978). Linear models and spurious observations. *Ann. Statist.* **27**, 131–138.

Box, G. E. P. and Tiao, C. G. (1968). A Bayesian approach to some outlier problems. *Biometrika* **55**, 119–129.

Chaloner, K. and Brant, R. (1988). A Bayesian approach to outlier detection and residual analysis. *Biometrika* **75**, 651–659.

Chow, G. C. (1960). A test for equality between sets of observations in two linear regressions. *Econometrica* **28**, 591–605.

Cook, R. D. (1979). Influential observations in linear regression. *J. Amer. Statist. Assoc.* **74**, 169–174.

Cook, R. D. and Weisberg, S. (1982). *Residuals and Influence in Regression.* New York: Wiley.

Daniel, C. and Wood, F. S. (1980). *Fitting Equations to Data.* New York: Wiley.

Eddy, W. F. (1980). Discussion of Freeman's paper. *Bayesian Statistics* (J. M. Bernardo, M. H. DeGroot, D. V. Lindley and A. F. M. Smith, eds.), Valencia: University Press, 370–373.

Efron, B. (1983). Estimating the error rate of a prediction rule: improvements on cross-validation. *J. Amer. Statist. Assoc.* **78**, 316–331.

Efron, B. (1986). How biased is the apparent error rate of prediction rule? *J. Amer. Statist. Assoc.* **81**, 461–470.

Freeman, P. R. (1980). On the number of outliers in data from a linear model. *Bayesian Statistics* (J. M. Bernardo, M. H. DeGroot, D. V. Lindley and A. F. M. Smith, eds.), Valencia: University Press, 349–365.

Geisser, S. (1975). The predictive sample reuse method with applications. *J. Amer. Statist. Assoc.* **70**, 320–328.

Gelfand, A. E., Hills, S. E., Racine-Poon, A. and Smith, A. F. M. (1990). Ilustration of Bayesian inference in normal data models using Gibbs sampling. *J. Amer. Statist. Assoc.* **85**, 412, 972–985.

Gong, G. (1986). Cross-validation, the jackknife and the bootstrap: excess error estimation in forward logistic regression. *J. Amer. Statist. Assoc.* **81**, 108–113.

Gray, J. B. and Ling, R. F. (1984). K-clustering as a detection tool for influential subsets in regression. *Technometrics* **26**, 305–330.

Guttman, I., Dutter, R. and Freeman, P. R. (1978). Care and handling of univariate outliers in the general linear model to detect spuriosity. A Bayesian approach. *Technometrics* **20**, 187–193.

Guttman, I. and Peña, D. (1988). Outliers and influence: evaluation by posteriors of parameters in the linear model. *Bayesian Statistics 3* (J. M. Bernardo, M. H. DeGroot, D. V. Lindley and A. F. M. Smith, eds.), Oxford: University Press, 631–640.

Guttman, I. (1991). A Bayesian look at the question of diagnostics. (Unpublished *Tech. Rep.*).

Mosteller, F. and Tukey, J. W. (1986). Data analysis including statistics. *Handbook of Social Psychology* **2** (G. Lindzey and E. Aronson, eds.), Reading, MA: Addison-Wesley.

Pettit, L. I. and Smith, A. F. M. (1985). Outliers and influential observations in linear models. *Bayesian Statistics 2* (J. M. Bernardo, M. H. DeGroot, D. V. Lindley and A. F. M. Smith, eds.), Amsterdam: North-Holland, 473–494, (with discussion).

Picard, R. P. and Cook, R. D. (1984). Cross-validation of regression models. *J. Amer. Statist. Assoc.* **79**, 575–582.

Rousseeuw, P. J. and van Zomeren, B. C. (1990). Unmasking multivariate outliers and leverage points. *J. Amer. Statist. Assoc.* **85**, 633–639.

Snee, R. D. (1977). Validation of regression models: methods and examples. *Technometrics* **19**, 415–428.

Smith, A. F. M. and Spiegelhalter, D. J. (1980). Bayes factors and choice criteria for linear models. *J. Roy. Statist. Soc. B* **42**, 213–220.

Spiegelhalter, D. J. and Smith, A. F. M. (1982). Bayes factors for linear and log-linear models with vague prior information. *J. Roy. Statist. Soc. B* **44**, 377–387.

Stone, M. (1974). Cross-validatory choice and assessment of statistical predictions. *J. Roy. Statist. Soc. B* **36**, 111–147.

Tukey, J. W. (1960). A survey of sampling from contaminated distributions. *Contributions to Probability and Statistics: Volume Dedicated to Harold Hotelling*. Stanford: University Press.

Verdinelli, I. and Wasserman, L. (1990). Bayesian analysis of outlier problems using the Gibbs sampler. *Tech. Rep.* **469**, Carnegie Mellon University.

DISCUSSION

W. POLASEK (*University of Basel, Switzerland*)

This is a trivial remark at a Bayesian conference, but the concept of an outlier, is certainly a subjective concept. Therefore, the easiest definition for an outlier is an (or set of) observation(s) which is (are) "non–normal". Since the normal distribution is often assumed for linear models, it might be better to refer to such points as "non–typical", i.e., points not following the majority of the others. Peña and Tiao's paper deals with outliers, a topic which has quite a long tradition in Bayesian statistics and especially in previous Valencia conferences (Freeman 1980, Pettit and Smith 1985). In order to evaluate their testing approach by Bayes–factors, we briefly review the Bayesian theory of influential points.

Detecting influential points. Bayesian models for outliers (i.e., extreme observations or contaminants) are concerned with two main goals: 1) identification: are there outliers? and 2) accommodation: what to do about them. Identification addresses the following questions: a) How many outliers are in my sample? and b) where are they located? While accommodation is concerned with questions of sensitivity like: a) How big is their local or global influence? b) How can we protect the inferences of our focus parameters against outliers? and c) What can we additionally learn for our inference problem (the data generating mechanism) by modelling the outliers. The last question could be asked for inferences or predictions.

Classical outlier concepts are designed for models which input minimal prior information. In their discussions of outliers, frequentists are always maliciously happy if they can find out about subjective assumptions in other models, or if they can disprove a claim that a particular method works "without any arbitrary inputs".

So Bayesians should be happy: They don't find themselves in a rush to minimize subjective inputs; on the contrary, they can find out what type of subjective input might be best for uncovering outliers or to protect against misleading inferences.

But despite a long tradition of outlier models (e.g., de Finetti 1961), there does not seem to exist an overall accepted methodology (either classical or in the Bayesian spirit) as to how to approach this very complex problem.

As with classical approaches, the Bayesian developments of the last two decades are full of technicalities. In general, estimation, testing, and sensitivity anlysis have been applied to the following important contamination models:

a) The location shift (LS) model: $(1-\alpha)f(x|\theta)+\alpha f(x+\delta|\theta)$;
b) the variance inflation (VI) model: $(1-\alpha)f(x|\theta)+\alpha f(x|k\theta)$.

Further modeling strategies involve the predictive or cross–validation approach (Geisser 1975), in particular the "conditional predictive ordinates" concept of Petit and Smith (1983), and the heavy tail (HT) approach of West (1984) and O' Hagan (1988).

Peña and Tiao propose a new tool for analysing outliers in the VI–model, the so–called Bayesian outlier curve (BROC) and the sequential Bayesian outlier curve (SEBROC). BROC and SEBROC rely on Bayes factors based on a fixed mixing factor α and the variance inflation factor k.

Classifying influential points. Classical statistics discriminates between outlying points in the factor space (leverage points), and outlying points in the observation space (e.g., studentized residuals). To extend this concept into many dimensions, Rousseeuw and Zomeren (1990) suggested a classification of all points based on: a) robust residuals r_i^* stemming from a high break–point estimation (e.g., least median of squares estimation), and b) robust distances based on a minimum volume ellipsoid (MVE) estimation.

Their analysis of the stack–loss data compares well with the analysis of Peña and Tiao, except for observation 2 which is not identified finally as an outlier: SEBROC and BROC indicate only 3 outliers not 4, but observation 2 appears as a candidate in the first round of computation. I suspect this might be a consequence of the normality assumption since the observation has a small residual, and the VI–based Bayesian analysis does not explicitly take leverage points into account.

The Bayesian counterpart of the robust residual and distance (R&D) scattergram diagnostics could be residuals from a fat–tailed posterior distribution as e.g., a mixture distribution as in West (1985) or the "extended power family" class of Albert *et al.* (1991), compared with Bayesian (posterior) leverage points, as they are derived in Smith (1983) or Pettit and Smith (1985). Bayesian leverage points modify the classical ones by incorporating the prior covariance matrix.

Sensitivity analysis. Surprisingly, the question of sensitivity analysis is not a major topic in the development of new tools. Outlier analysis is very much an exploratory concept now and we find an original Bayesian inferential method as a tool for exploration. This means that parameter inference for a tentative starting model is not always a major goal in data analysis, side aspects such as particular data patterns are quite important for the finally employed model and the interpretation of parameters.

Note that the splitting of the data set into a good and a bad group for the IV or LS model always generates a so–called information contact curve between the location parameter without the outliers and the location of the outliers. The scaling factor for this curve is the (inverse) variance inflation factor k for the outlier distribution. This means that, if we allow for large variances for the outliers, the weight for the location parameter shifts toward the location without the outliers. Large variances for the outlier part of the model is the protection against too much sensitivity for the posterior mean.

Conclusions. I found the (marked) probability diagnostic matrix for identifying the number of outliers quite interesting. The idea goes back to Box and Tiao (1968) but could be improved by ordering the observations and using a double display. One half of the matrix could contain the numerical values and the other graphical symbols for spotting the interesting patterns more easily. The inherent problem with model diagnostics is the large amount of information they produce. Making them more comprehensible would help to improve the Bayesian reporting process. Unfortunately, I missed this diagnostic matrix for Example 2, the stack–loss data. In particular, it would have been helpful to find out why observation 13 turns out to be an outlier in the first round, while it does not in the classical analysis (see Rousseeuw and Zomeren, 1990).

I found the discussion of the assessment of the prior distribution for the outliers somewhat lacking. For a large number of outliers we get a flat prior distribution and also flat contributions to the Bayes factor. This means that except for large outliers we will never have overwhelming data evidence and the prior–posterior sensitivity analysis will be high.

The efficiency of the sampling approach depends on the optimal selection of suspect points in the first round. Even then masking effects can occur and a conditional Bayes factor analysis (e.g., conditional on "observation 21 being an outlier") is necessary.

Summarizing, the authors have to be congratulated for proposing diagnostics based on Bayes factors in the variance inflation model for outliers with a promising suggestion for solving the numerical computation problems for groups of outliers.

K. CHALONER (*University of Minnesota, USA*)

I think that the mixture model used to define outliers in this paper has a fundamental flaw. "Outliers", generated from the alternative contaminating distribution, are not necessarily outlying. Indeed they are very likely to be in the centre of the data. This definition of an outlier seems, therefore, unrealistic.

An alternative approach to defining an outlier is to use the realized errors, as discussed in Zellner (1975) and Zellner and Moulton (1985). Chaloner and Brant (1988) use this concept to define outliers as outlying observations and calculate posterior probabilities of observations in a linear model being outliers. Outliers are by definition outlying. This simple approach detects the multiple outliers easily in the stack–loss data, including those that are masked.

A final comment on the stack–loss data is that it seems unneccesary, for this data set, to worry about whether there are three or four or more outliers in this collection of twenty one observations, because the data indicate that a linear model analysis is probably inappropriate.

G. CONSONNI (*Università di Pavia, Italy*) and
P. VERONESE (*Università di Milano, Italy*)

We would like to draw attention to the role played by the "variance–inflating" coefficient k, see (2.2). In the paper this parameter is considered to be fixed, and actually set equal to 5 in both examples used to illustrate the theory. No justification is offered for this numerical choice; on the other hand, we concur that choosing k (*before* looking at the data) might be a difficult enterprise. Clearly the right approach is to regard k as a random parameter in order to learn about it when data are observed. Have the authors suggestions concerning reasonable and manageable priors for k?

In this connection, we would like to emphasize that choosing a noninformative prior on k might lead to meaningless answers since B_{h_0} is zero, for any h and any data. To see why this is true, consider the Bayes factor, BF, say, $p(y|M_1)/p(y|M_0)$. It is well known that, with improper priors, the evaluation of BF is problematic, since the proportionality constants cannot be omitted. A reasonable way to avoid this difficulty is to: i) restrict the parameter space Θ to a compact subset, Θ_m, say so that the improper prior becomes proper on Θ_m; ii) evaluate the restricted BF, BF_m say, w.r.t. the above prior; iii)if $\Theta_m \uparrow \Theta (m \to \infty)$, it can be shown that BF can be computed as lim BF_m (see Consonni and Veronese, 1991, for further details).

Applying this procedure to the authors' setup, we have:

$\theta = (\beta, \sigma, k) \varepsilon \Theta = R^p \times R^+ \times (1, \infty)$. Choose $\Theta_m = \Phi_m \times \Omega_m$ with $\Phi_m \uparrow (R^p \times R^+)$ and $\Omega_m \uparrow (1, \infty)$. Assume k to be independent of (β, σ) so that $p(\beta, \sigma, R) = p(\beta, \sigma)p(p(R)$ with $p(\beta, \sigma)$ improper —as in the authors' paper— and $p(k)$ also improper. Let $C(\Phi_m)$ be the multiplicative constant that makes $p(\beta, \sigma)$ integrate to 1 on Φ_m and similarly for $C(\Omega_m)$ relative to $p(k)$. If $p(y|\beta, \sigma, k, M_1)$ is the likelihood under M_1 and $p(y|\beta, \sigma, M_0)$ that under M_0, then

$$BF = \lim_m BF_m = \lim_m \frac{C(\Phi_m)C(\Omega) \int_{\Phi_m} \int_{\Omega m} p(y|\beta, \sigma, k, M_1)p(\beta, \sigma)p(k)d\beta d\sigma dk}{C(\Phi_m) \int_{\Phi m} p(y|\beta, \sigma, M_0)p(\beta, \sigma)d\beta d\sigma}$$

which is zero, since $C(\Omega_m) \to \infty$ and the integrals are assumed finite over the whole parameter space, as is typical with standard improper priors.

Notice that BF need not be zero if prior over (β, σ), under M_0, is different from that under M_1, i.e., the priors depend on the assumed model. In this case the value of BF is not predetermined, provided $\lim C_1(\Phi_m)C_1(\Omega_m)/C_0(\Phi_m)$ is finite, where C_1 is the normalizing constant of the prior under M_1. Although model–dependent priors are not unusual, would they be appropriate, in the authors' opinion, for this problem?

E. I. GEORGE (*University of Chicago, USA*)

Ever since reading the seminal outlier paper of Box and Tiao (1968), I have been convinced that the Bayesian paradigm offered the most natural framework for getting a handle on the general problem of outlier identification. This paper nicely reinforces that point of view by showing how the Bayesian machinery has a cross–validation interpretation. Unfortunately, the calculation requirements for a complete Bayesian analysis are usually prohibitive, even for more moderate sample sizes. The problem is that one needs to obtain the data subsets which yield the largest posterior probabilities of being outliers. This then requires the calculation and comparison of close to 2^n quantities for n data points. What is usually done in this line of work, is to use some additional criterion which restricts attention to a manageable number of candidate subsets. Indeed, this paper proposes (in Section 4) several new criteria for this purpose. In this, the authors have succeeded admirably. I particularly like the idea of using the matrix of "interactions" to split the sample.

I would like to suggest that Gibbs sampling is another approach which may severely reduce the computational requirements of this problem. Although, as the authors point out, the Gibbs sampler could be applied to compute all of the marginal posterior probabilities, I have in mind a slightly different application. In George and McCulloch (1991), we used the Gibbs sampler as a search algorithm which efficiently identified high posterior probability models in the related problem of variable selection. In effect, I believe that the Gibbs sampler can be similarly used here to identify the high posterior probability subsets while bypassing the problem of computing all the subset probabilities. However, even the Gibbs sampler will be defeated by large sample sizes. Thus reduction methods such as those proposed by this paper will continue to be important. Perhaps using the Gibbs sampler in conjunction with these reduction methods will prove to be most efficient.

L. I. PETTIT (*Goldsmiths' College London, UK*)

We have to ask what we are trying to achieve with any model. In the case of outlier models are we trying to identify outliers or estimate parameters robustly? We also have to decide what we mean by an outlier model, as Pettit and Smith (1985) note:

> "There are a number of problems ... the first of these relates to the basic idea of an aberrant observation as one which is generated by a mechanism which differs from that generating the majority of the observations; without further elaboration, this concept does not embrace the intuitive notion of "surprisingness" or "outlyingness" which seem to underlie most people's understanding of the term "outlier".'

One question we might ask is for the probability that aberrant or contaminated observations do "outlie". In Pettit (1992a) this question is considered and, for example, it is shown that if we have a single observation from $N(0, 25)$ and nine observations from $N(0, 1)$ then the probability that the contaminant is the largest or smallest in the sample is 0.768.

Peña and Tiao calculate the overall Bayes factor that there is one 'outlier', (or contaminant). Although this may sometimes be of interest, I prefer to see how likely a particular extreme observation is to be a contaminant and thus identify the outliers. Looking at the expression for B_{h0} equation (3.2), we see that it is a monotonic increasing function of α and this helps to explain some of the results which are very sensitive to the choice of α. The more likely we think outliers are, the more likely we are to find them. The authors do not give any advice on choosing α. To my mind an outlier should be unusual or surprising and a value of α of about 0.05 is the sort of figure I would like to assume. Much larger values of α suggest a mixture model to me, not an outlier model. The stratified sampling method that they suggest will be very useful for this sort of mixture model but is it as helpful in calculating posterior estimates of parameters?

Peña and Tiao mention the possibility of determining the values of α and k by the device of an imaginary training sample (Spiegelhalter and Smith, 1982). I have recently (Pettit, 1992b) considered this approach for the Guttman, Dutter and Freeman (1978) model assuming vague priors for the unknown parameters and also for exponential models. The method seems to work well in these cases.

K. D. S. YOUNG (*University of Surrey, UK*)

Peña and Tiao use the Bayes factors comparing two models containing h outliers and no outliers. They adopt the Box and Tiao (1968) variance inflation model which contains observations with inflated variance with probability α. Observations which are generated with an inflated variance need not necessarily be extreme values.

Freeman (1980) considered two datasets, the second of which is given in Table 1.

–5.28	–1.10	–0.04	–0.02	0.12	0.71	1.35	1.46	1.74	3.89

Table 1. *Freeman's dataset 2.*

The first and last observations are outliers. Table 2 shows the posterior odds for 1 to 3 outliers for different values of α.

α	1	2	3
0.05	1.318	0.593	0.092
0.10	2.783	2.644	0.869
0.15	4.420	6.669	3.483
0.20	6.261	13.38	9.904

Table 2. *Posterior odds, B_{h0}, for data in Table 1 as a function of α for $k = 5$.*

This shows that even for a small sample size there is clearly one outlier and if α is made large enough 2. This is in contrast to the authors' Example 1, where they suggest that the small sample size is to blame for not finding the outliers. My question is why do we get satisfactory results for the Freeman dataset and not for Example 1?

The results are clearly strongly dependent on α. Is a value $\alpha = 0.2$ reasonable? This would imply that in a sample of size 10 there is only an 11% chance of obtaining uncontaminated data, but it is only for the larger values of α that the method seemed to work well.

REPLY TO THE DISCUSSION

We appreciate Polasek's comments and we welcome his suggestion of using graphical displays to present the information and to discover interesting patterns. As far as observation 2 in the stack loss data is concerned, Table 1 presents the conditional posterior probabilities that this observation comes from the alternative distribution using different conditioning sets, and it can be seen that this probability is always small.

A	$P(A)$	$P(2, A)$	$P(2/A)$
{21}	.7898	.0284	.04
{4,21}	.6332	.0267	.04
{3,4,21}	.0766	.0018	.02
{1,3,4,21}	.1099	.0038	.03

Table 1. *Probabilities of the indicated sets coming from the contaminating distribution for the stack-loss data.*

Also, following his request, we have included the probability diagnostic matrix of interactions for Example 2, the stack-loss data. These matrices are shown in Tables 2 and 3.

We agree with Consonni and Veronese that choosing a nonformative prior on k can lead to meaningless answers. Our recommendation would be to do a sensitivity analysis of the posterior odds with respect to this parameter in a range which is appropriate for the particular problem in hand.

Answering Pettit's question, we think that the stratified sampling method can also be useful in computing posterior estimates of the parameters, although the procedure should be adopted to identify the important components in the particular problem in hand. We are looking forward to seeing the application of the imaginary training sample to determine the values of α and k in this type of problems.

We cannot see the fundamental flaw indicated by Chaloner. Although it is true that data generated by the alternative contaminating distribution are not necessarily outlying, the probability that they are goes towards one when k increases. Besides, our objective here is to set up a model that allows for heterogeneity, and the model used does this. The stack-loss data is used in the paper to show the performance of the methods developed, and we agree that a linear model may not be adequate for this set of data.

The Freeman data set used by Young shows that the Box and Tiao model works very well when the situation is as clear as it is in her example. For instance, in her data set the size of the two extreme values measured by $(y - \bar{y}_{(I)})/s_{(I)}$, is -2.34 and 1.51 when $\bar{y}$ and s are computed with the 10 data points. These values increase to -4.29 and 2.08 when the most extreme value (-5.28) is deleted from the computation of $\bar{y}$ and s, and goes to -6.03 and 3.49 when the two outliers are deleted and the mean and variance are computed only by

	2	3	4	5	6	7	8	9	10	11	12	13	14	15	16	17	18	19	20	21
1		18	13																	−34
2		−1	−1																	157
3			26							1	1									−38
4					−1	−2	−1	−2				1	1						1	5745
5																				21
6																				37
7																				11
8																				.
9																				−5
10																				−8
11																				−30
12																				−44
13																				88
14																				19
15																				13
16																				1
17																				−14
18																				−1
19																				1
20																				5
21																				.

Table 2. *Probabilistic Diagnostic Matrix of Interaction for the stack-loss data. All values have been multiplied by* 10^3 *and those values smaller than 1 are omitted.*

	2	3	4	5	6	7	8	9	10	11	12	13	14	15	16	17	18	19	20
1		1	22																
2			7																
3			67																
4						−2	−1	−2				27	2	8	1	0	1	2	3
5																			

Table 3. *Probabilistic Diagnostic Matrix of Interactions for the stack-loss data. Observation 21 omitted. All values have been multiplied by* 10^3 *and those values smaller than 1 are omitted.*

using the homogeneous data. On the contrary, in our data set the initial standardized values for the three outliers are 1.10 and 1.36, they increase to 1.34 and 1.62 when the largest point (4) is deleted, to 1.8 and 2.14 when the two largest are deleted, and to 2.84 and 3.29 when the three heterogeneous observations are deleted. In summary, the two data sets are very different, because we have chosen the most difficult situation (with strong masking) to challenge our method.

Regarding the value of α, we should take into account that if k is not very large, values generated by the contaminating distribution are not necessarily outlying, as noted by Chaloner and Young. In our examples we are in the most difficult case in which k is small (the size of the outliers is small), and therefore $\alpha = .20$ does not necessarily mean a large probability of observing outliers. For instance, with $k = 4$ the probability of one point being apart more than 3σ from the mean is .46, and therefore with $\alpha = .20$ the joint probability of coming from the contaminated distribution and being further from the mean than 3σ is .09. Then, the probability of no outliers in a sample of size 10 is .39 instead of .11.

We thank George for his kind comments on our paper and are very pleased that this paper reinforces his Bayesian conversion. His suggestion of using Gibbs sampling in conjunction with the probabilistic interaction matrix is interesting and may stimulate further research on this topic.

ADDITIONAL REFERENCES IN THE DISCUSSION

Abraham, B. and Box, G. E. P. (1979). Bayesian analysis of some outliers problems in time series. *Biometrika* **66**, 229–236.

Albert, J., Delampady, M. and Polasek, W. (1991). A class of distributions for robustness studies. *Stat. Plan. and Inf.* (to appear).

Beckman, R. J. and Cook, R. D. (1983). Outlier. . .s. *Technometrics* **25**, 119–163, (with discussion).

Consonni, G. and Veronese, P. (1992). Bayes factors for linear models and improper priors. *In this volume.*

DeFinetti, B. (1961). The Bayesian approach to the rejection of outliers. *Proceedings of the Fourth Berkeley Symposium on Probability and Statistics* **1**, 199–210, Berkeley: University Press.

George, E. I. and McCulloch, R. E. (1991). Variable selection via Gibbs sampling. *Tech. Rep.* **99**, Graduate School of Business, University of Chicago.

Guttman, I. (1973). Care and handling of univariate or multivariate outliers in detecting spuriosity, a Bayesian approach. *Technometrics* **15**, 723–738.

Kitagawa, G. and Akaike, H. (1982). A quasi-Bayesian approach to outlier detection. *Ann. Math. Statist.* **34**, 389–398.

O'Hagan, A. (1979). On outlier rejection phenomena in Bayes inference. *J. Roy. Statist. Soc. B* **41**, 358–367.

O'Hagan, A. (1988). Modelling with heavy tails. *Bayesian Statistics 3* (J. M. Bernardo, M. H. DeGroot, D. V. Lindley and A. F. M. Smith, eds.), Oxford: University Press, 345–359, (with discussion).

Pettit, L. I. (1988). Bayes methods for outliers in exponential samples. *J. Roy. Statist. Soc. B* **50**, 292–309.

Pettit, L. I. and Smith, A. F. M. (1983). Bayesian model comparison in the presence of outliers. *Bull. Int. Statist. Inst.* **50**, 292–309.

Pettit, L. I. (1992a). Probabilities that contaminated normal observations are extreme. *Journal of Applied Statistics*, (to appear).

Pettit, L. I. (1992b). Bayes factors for outlier models using the device of imaginary observations. *J. Amer. Statist. Assoc.* , (to appear).

Polasek, W. (1984). Regression diagnostics for the general linear model. *J. Amer. Statist. Assoc.* **79**, 336–340.

Smith, A. F. M. (1983). Bayesian approaches to outliers and robustness. *Specifying Statistical Models* (J. P. Florens *et al.* eds.) New York: Springer, 13–35.

West, M. (1984). Outlier models and prior distributions in Bayesian linear regression. *J. Roy. Statist. Soc. B* **46**, 431–439.

West, M. (1985). Generalised linear models; scale parameters, outlier accommodation and prior distributions, *Bayesian Statistics 2* (J. M. Bernardo, M. H. DeGroot, D. V. Lindley and A. F. M. Smith, eds.), Amsterdam: North-Holland, 531–558, (with discussion).

Zellner, A. (1975). Bayesian analysis of regression error terms. *J. Amer. Statist. Assoc.* **70**, 138–144.

Zellner, A. and Moulton, B. R. (1985). Bayesian regression diagnostics with application to international consumption and income data. *J. Econometrics* **29**, 187–211.

BAYESIAN STATISTICS 4, pp. 389–404
J. M. Bernardo, J. O. Berger, A. P. Dawid and A. F. M. Smith, (Eds.)

SAGA: Sample Assisted Graphical Analysis

AMY RACINE-POON
CIBA-GEIGY, Switzerland

SUMMARY

Graphical diagnostic techniques are introduced via two illustrative examples. The first example involves reparameterization of a nonlinear estimation problem; the second involves identifying outliers in hierarchical models. The diagnostics are performed in the parameter space. Algorithms for generating samples from the posterior density are briefly discussed.

Keywords: SAMPLING RESAMPLING TECHNIQUES; GIBBS SAMPLER; DYNAMIC GRAPHICS; REPARAMETERIZATION; BOX-COX TRANSFORMATION; HIERARCHICAL MODELS; OUTLIERS.

1. INTRODUCTION

Recently considerable attention has been paid to the development of efficient algorithms for generating samples from high-dimensional posterior densities (see Gelfand and Smith, 1990, and Gelfand, *et al.* 1990). These sampling methodologies have been mostly used for the purpose of inference but not for diagnostics. Meanwhile, in the field of exploratory data analysis, due to the easy availability of highly efficient dynamic graphics, considerable progress has been made in multivariate exploratory data analysis (see Weihs and Schmidli, 1990). Most graphical techniques have so far been applied to the space of observations or the space of variables, like residual diagnostics and biplots. However, these graphical tools can also be applied to the parameter space when samples from the posterior density are available. We therefore have the means to visualize the forms and shapes of a high dimensional posterior density. This will give us insight into the impact on the inference of the choice of parameterization, transformation, model and prior, and also will provide us with a tool for diagnostics.

We shall illustrate sample assisted graphical analysis using two examples. The first involves the choice of reparameterization in nonlinear regression, whereas the second involves the identification of outlying individuals in hierarchical models. A brief review of sampling techniques will be given in Section 2; the examples will be given in Sections 3 and 4.

2. REVIEW OF SAMPLING TECHNIQUES

We briefly desribe two sampling algorithms; namely, a sample-resample technique and the Gibbs sampler. Details can be found in Rubin (1988), Gelfand and Smith (1990), Gelfand *et al.* (1990) and Smith and Gelfand (1992).

2.1. *Sample-Resample Techniques (SIR Algorithm)*

Suppose samples from the density g can be easily generated, but what is required is samples from another density f. We approximately resample from f as follows. Draw n samples $\{\theta_1, \theta_2, \dots \theta_n\}$ from g and resample from the discrete distribution $\{\theta_1, \theta_2, \dots \theta_n\}$ with probability w_i proportional to $f(\theta_i)/g(\theta_i)$. Rubin (1988) referred to this algorithm as the Sample Importance Resampling (SIR) algorithm.

How accurately these samples represent the density f, depends on n, the number of samples and the similarity between f and g. It is therefore essential to obtain a "good" g for efficient and reliable generation of random deviates from f. For the purpose of computing efficiency, the families of density functions to generate g are usually fairly standard, like the normal, beta or t density. To extend the scope of the f shapes and forms of density, f, which g would be used to approximate, one may consider weighted sum of densities from some standard families. We shall demonstrate in the context of Example 1, how, in combination with appropriate reparameterization, the normal and t families can be made to approximate some rather odd shapes of posterior density f.

We propose the following iterative algorithm to find a g, which may reasonably resemble f. We shall restrict the discussion to g here being in the multivariate normal family and shall later generalize to g being in the multivariate t family.

(a) *multivariate normal family*

step 1: generate m samples $\{\phi_{1,1}, \dots, \phi_{1,m}\}$ from $g^{(0)}$, a multivariate normal density with mean $\mu^{(0)}$ and covariance matrix $\Sigma^{(0)}$. Calculate the importance mean and covariance matrix $(\mu^{(1)}, \Sigma^{(1)})$ of the samples $\{\phi_{1,1}, \dots, \phi_{1,m}\}$ using w_{1j} as the weights. Here, w_{1j} is the importance weight assigned to $\phi_{1,j}$, which is proportional to $f(\phi_{1,j})/g^{(0)}(\phi_{1,j})$.

Step l: generate m samples $\{\phi_{l,1}, \dots, \phi_{l,m}\}$ from $g^{(l-1)}$, a multivariate normal density with mean $\mu^{(l-1)}$ and covariance matrix $\Sigma^{(l-1)}$. Calculate the importance mean and covariance matrix $(\mu^{(l)}, \Sigma^{(l)})$ of the samples $\{\phi_{l,1}, \dots, \phi_{l,m}\}$ using w_{lj} as the weight. Here, w_{lj} is the importance weight assigned to $\phi_{l,j}$, which is proportional to $f(\phi_{l,j})/g^{(l-1)}(\phi_{l,j})$.

One should monitor the convergence of $(\mu^{(l)}, \Sigma^{(l)})$ throughout the process. If the process converges then the final values for the importance mean and covariance can be used to generate g. However, if the process oscillates, this suggests that reparameterization or alternative choice of g appears to be necessary.

(b) *Multivariate t family*

Another alternative candidate for g is the multivariate t family, which may be more efficient when f is heavy tailed.

The iterative process proceeds similarly to that of the normal family, however an additional scale variable has to be generated, since the standard multivariate t_ν density can be considered as a scale mixture of multivariate normal densities in which the mixing density is a χ^2 density with ν degrees of freedom (see Johnson and Kotz, 1972).

step l: generate m samples $\{(\phi_{l,1}, \lambda_{l,1}), \dots, (\phi_{l,m}, \lambda_{l,m})\}$ where $\lambda_{l,j}$ is a χ^2_ν random deviate and $\phi_{l,j}$ is a multivariate normal deviate with mean $\mu^{(l-1)}$ and covariance matrix $\Sigma^{(l-1)}/\lambda_{l,j}$, respectively. Calculate the importance mean and covariance matrix $(\mu^{(1)}, \Sigma^{(1)})$ of the samples $\{\phi_{1,1}, \dots, \phi_{1,m}\}$ using $w_{lj} \cdot \lambda_{(l,j)}$ as the weight. Here, w_{lj} is the importance weight assigned to $\phi_{l,j}$, which is proportional to $f(\phi_{l,j})/g^{(l-1)}(\phi_{l,j})$.

Monitor the covergence of $(\mu^{(l)}, \Sigma^{(l)})$. Use the final value for the choice of g. If the process fails to converge, reparameterization may become necessary to obtain a better approximation of the samples from f. We shall later demonstrate in Section 3, how a reasonable reparameterization in the family of power transformation can be chosen.

2.2. *Gibbs Sampling Techniques*

The Gibbs sampler is a Markovian updating algorithm, which requires that samples from the full conditional densities are available, see Gelfand and Smith (1990) for technical detail.

Let $\{u_1, u_2, \ldots, u_k\}$ be the parameters in the model. At the l-th cycle, m replicates are generated, where if $\{u_{1,j}^{(l)}, u_{2,j}^{(l)}, \ldots, u_{k,j}^{(l)}\}$ denotes the jth replicate,

$$\begin{aligned} u_{1,j}^{(l)} &\sim p(u|u_{2,j}^{(l-1)}, \ldots, u_{k,j}^{(l-1)}, data) \\ u_{2,j}^{(l)} &\sim p(u|u_{1,j}^{(l)}, u_{3,j}^{(l-1)}, \ldots, u_{k,j}^{(l-1)}, data) \\ &\vdots \\ u_{k,j}^{(l)} &\sim p(u|u_{1,j}^{(l)}, \ldots, u_{k-1,j}^{(l)}, data) \end{aligned}$$

It is shown in Geman and Geman (1984) that under appropriate regularity conditions, $\{u_{1,j}^{(l)}, u_{2,j}^{(l)}, \ldots, u_{k,j}^{(l)}\}$ can be considered as a sample from the posterior density of $\{u_1, u_2, \ldots, u_k\}$, when l goes to ∞. However, issues of monitoring convergence are still very much a subject for research.

3. SAGA IN REPARAMETERIZATION

Reparameterization is an important issue in inference. A considerable number of classical inference procedures rely on approximate normality of an estimator. In Bayesian inference, although we do not rely on the normality approximation, numerical integration procedures can be made more efficient when the posterior density is similar to some well known density families. For example, the efficiency of Gaussian quadrature integration can be improved considerably when the parameterization of the parameters leads to a "Gaussian-type" of density (Smith *et al.* 1985). For sample-resample algorithms, one can gain efficiency and accuracy when the parameters of the model (after appropriate reparameterization) lead to approximate normality. However, it is quite difficult to imagine what parameterization would be adequate, and it is not obvious how to derive the "optimal" parameterization by algebraic or analytic means. If we could visualize, even crudely, what the posterior density before and after reparameterization resembles, it would be much easier to attempt various forms of parameterization and to make some reasonable choice within some standard reparameterization family.

We shall restrict our discussion to reparameterization which is applied to each parameter separately. Commenting on the version of this paper presented at the meeting, the invited discussant, Irwin Guttman, proposed exploring the family of Box-Cox power transformations of a parameter θ (Box and Cox, 1964). The Box-Cox transformation is defined to be,

$$\begin{aligned} \phi^{(\lambda)}(\theta) &= (\theta^{\lambda} - 1)/\lambda, \text{ when } \lambda > 0 \\ &= \log(\theta) \text{ , when } \lambda = 0. \end{aligned}$$

The λ will be chosen such that the empirical distribution of $\phi^{(\lambda)}(\theta)$ based on the generated samples resembles that of a normal distribution. Let $\hat{F}_n(\theta)$ denote the empirical cumulative distribution function (CDF) of θ based on n samples. Let $Z_n(\theta)$ be $\Phi^{-1}(\hat{F}_n(\theta))$, where Φ denotes the standard normal CDF. The $Z_n(\theta_i)$ should approximately resemble a sample from a standard normal distribution when n is sufficiently large. $Z_n(\theta)$ is a nondecreasing step function, which can be used to choose reparameterization. One way to judge which of the power transformations λ is most suitable is by using graphical display. The regressions of $Z_n(\theta_i)$ on $\theta_i^{(\lambda)}$ for various values of λ's can be carried out. The fitted curves for various λ

can be plotted against θ_i along with $Z_n(\theta_i)$ and these can be used as a criterion for judging the goodness of fit of the power of the transformations. Usually, the tail of the empirical CDF is rather problematic and it may be more reliable to use, say, the middle 95% of the samples for the diagnostic.

The following nonlinear regression example is used to demonstrate some aspects of SAGA in reparameterization.

Example 1. BOD (biochemical oxygen demand) levels are measured by Maske (1967), the data and details of the experiment are given in Bates and Watts (1988, p. 270 and p. 350). The BOD levels are believed to follow an exponential decay model,

$$y_i = \theta_1(1 - \exp(-\theta_2 \, \text{day}_i)) + \epsilon_i$$

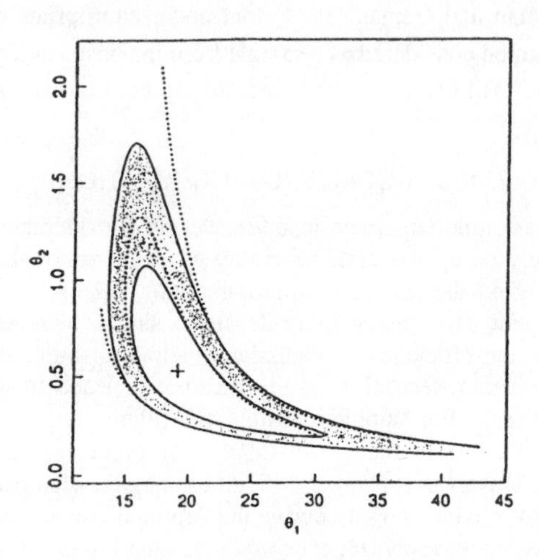

Figure 1. *Likelihood contours of* (θ_1, θ_2).

Such a problem may appear to be straightforward and indeed a maximum likelihood estimation procedure converged without any difficulty. However, as shown in Bates and Watts (p. 224), any inference based on asymptotic normality is extremely misleading. The likelihood contours of (θ_1, θ_2) display a "banana" shape with an extremely long tail (Figure 1). We therefore search for an alternative parameterization which will lead to more elliptical likelihood contours. This may result in more efficient approximation of the posterior density.

3.1. *Iterative Sample Algorithm*

We proceeded using an iterative sample-resample algorithm, as described in Section 2.1, in order to try to have a crude view of the shape of the likelihood. We started the sampling function using $g^{(0)}$ being bivariate normally distributed with the maximum likelihood estimate and its asymptotic covariance matrix as mean and covariance. The number of samples generated from the g's was 1000 per iteration. The importance weight here is proportional to the ratio of the likelihood $l(\theta_1, \theta_2)$ to $g^{(0)}(\theta_1, \theta_2)$.

The process failed to converge after 15 iterations; in fact, it oscillated. In order to have a closer inspection of the posterior density of (θ_1, θ_2) and to understand why the process failed to converge, we resampled 2000 samples from the 1000 samples of $g^{(15)}$, and graphically displayed these in a "Bubble scatter" plot (Figure 2). The size of the bubble is proportional to the frequency of the sample. There are several features of Figure 2 which become evident; namely, it displays a "banana" shape, and there are large bubbles at the two extremes of the figures. The large bubbles at the extreme indicate heavy tails of the likelihood (θ_1, θ_2). Now the reason why the process oscillated becomes clear, it is unlikely a normal density would capture the essence of such a shape of likelihood.

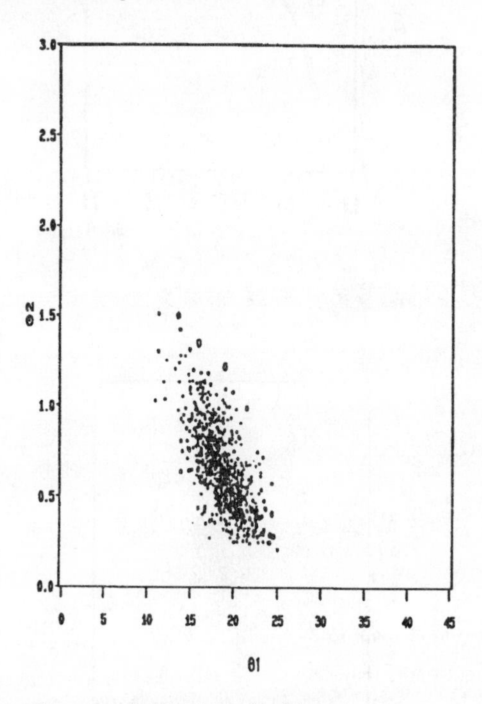

Figure 2. *Bubble plot of* (θ_1, θ_2) *after 15 iterations.*

We therefore proceed to select reasonable Box-Cox power transformations for the parameters (θ_1, θ_2), separately. We shall here illustrate only the reparameterization process of θ_2. Figure 3 illustrates for three different choices of λ's, namely, $-1, 0$ and 1, the least squares curve of $Z_n(\theta_2)$ with $\theta_2^{(\lambda)}$ plotted against the 2000 samples of θ_2. It is clear that lambda being 0, that is the logarithm, is probably a reasonably good reparameterization. Using similar techniques, we find that the reciprocal of θ_1 is an acceptable reparameterization of θ_1.

We then redisplay Figure 2 using the new reparameterization (ϕ_1, ϕ_2), where $\phi_1 = 1/\theta_1, \phi_2 = \log(\theta_2)$, in Figure 4. The banana shape of the samples is less evident, however the long tail (large bubble) phenomenon persisted.

We now have some reasons to believe that after the reparameterization, the approximation will be more accurate and the banana shape may disappear, hence the process may be stabilized. The iteration is therefore restarted using the new reparameterization and the

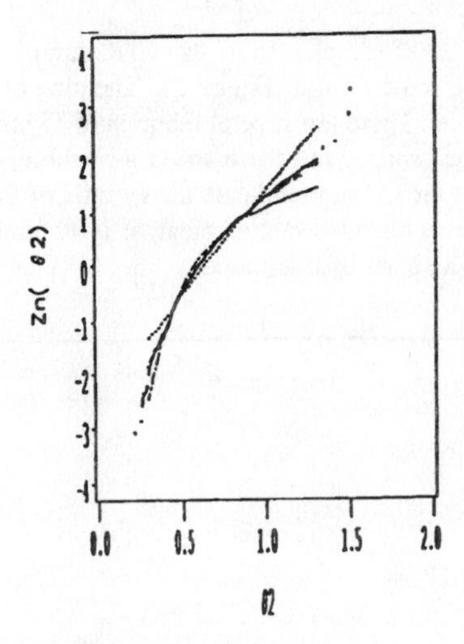

Figure 3. *Least curves of $Z_n(\theta_2)$ against θ_2 for* λ__.__.__ – 1, —0 *and* – – –1.

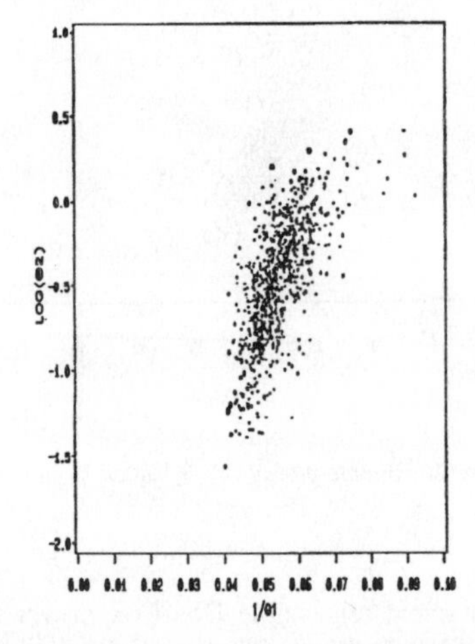

Figure 4. *Bubble plot of* $(\phi_1 = 1/\theta_1, \phi_2 = \log(\theta_2))$.

corresponding maximum likelihood estimate and its covariance structure as the initial values of the bivariate normal generator. We generated 1000 samples per iteration and the process appeared to be stable after 8 iterations. We then resample 2000 samples and examine the bubble scatter plot for (ϕ_1, ϕ_2), in Figure 5, which displays a rather promising elliptical shape with only a few large bubbles at the extreme. This indicates that the reparameterization is probably quite satisfactory for generating accurate samples from (ϕ_1, ϕ_2). We redisplay Figure 5 in the original parameterization (θ_1, θ_2) in Figure 6. Comparing Figures 2 and 6, one can easily see that Figure 6 captures the extreme long tail property much better than Figure 2. In addition, the banana shape of the likelihood is now much more obvious.

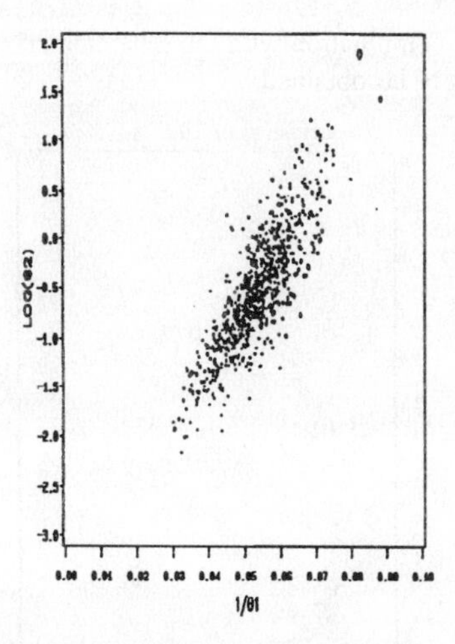

Figure 5. *Bubble plot of* $(\phi_1 = 1/\theta_1, \phi_2 = \log(\theta_2))$ *using samples after reparameterization.*

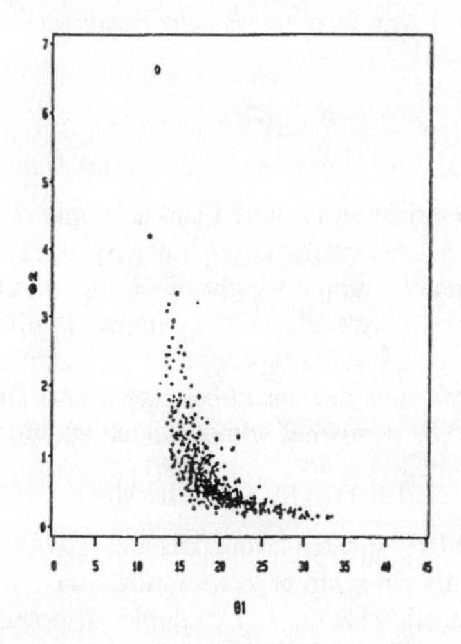

Figure 6. *Bubble plot of* θ_1, θ_2 *using samples after reparameterization.*

Since we noticed that, even when reparameterized in terms of (ϕ_1, ϕ_2), the posterior density is going to be heavy tailed, one can possibly obtain more accuracy and efficiency by choosing a family for g which is more heavy tailed than the normal family. We therefore redid the simulations with g from the bivariate t family. We have chosen to use a t_2 family, the initial $g^{(0)}$ having mean and covariance matrix derived from the maximum likelihood estimate using the (ϕ_1, ϕ_2) parameterization. The iterative process converges after a few iterations with 1000 samples. 2000 samples are then drawn, and the bubble scatter plot is displayed in Figure 7. The likelihood of (θ_1, θ_2) is very well approximated by Figure 7.

Hence using the t family in combination with the reparameterization, a representative sample from the likelihood can easily be obtained.

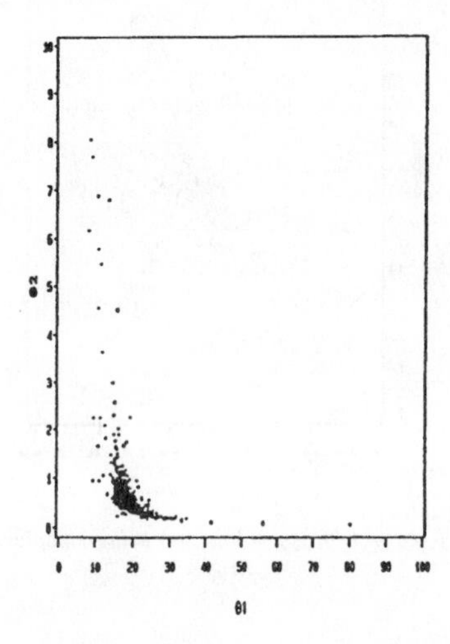

Figure 7. *Bubble plot of θ_1, θ_2 using samples after reparameterization based on a t family.*

3.2. *Sampling from the Posterior Density*

Once a reasonable reparameterization is chosen and a sample obtained from g, samples from the posterior density of (θ_1, θ_2) can easily be generated by using resample weight proportional to the product of the importance sample weight of (θ_1, θ_2) and the prior of (θ_1, θ_2). This is extremely convenient for sensitivity analysis of the impact of different choices of prior on the final inference for (θ_1, θ_2). Graphical comparison of the samples from the posterior density corresponding to the various priors can easily be carried out. Other summary statistics, e.g., sample percentiles, can also be compared without much additional effort.

4. SAGA IN OUTLIER DETECTION IN HIERARCHICAL MODEL

Hierarchical models (particularly population models and growth curve models) have attracted considerable attention recently. In some of these applications, one of the key requirements is to be able to perform diagnostic checks. For example, in population kinetics (Racine-Poon and Smith, 1989), the scientists would like not only to describe and understand the typical kinetics patterns and between individual (patients) variation of the population of interest, but also require a way of detecting high risk individuals (outliers) and important covariates (patient characteristics). However, it has been quite difficult to detect outlying individuals in a hierarchical model, even if the data from the individual are not sparse. To detect outlying individuals, one requires that diagnostic techniques be applied to the space of the individuals' parameters, which are unobservable. The following example shows that posterior samples from the individual parameters may be used to assist us to locate outliers and to identify some of the meaningful covariates.

Example 2. Dental measurements were taken on 11 (no. 1–11) girls and 16 (no. 12–27) boys recorded at the age of 8, 10, 12, 14 years, respectively. It is believed that this dental

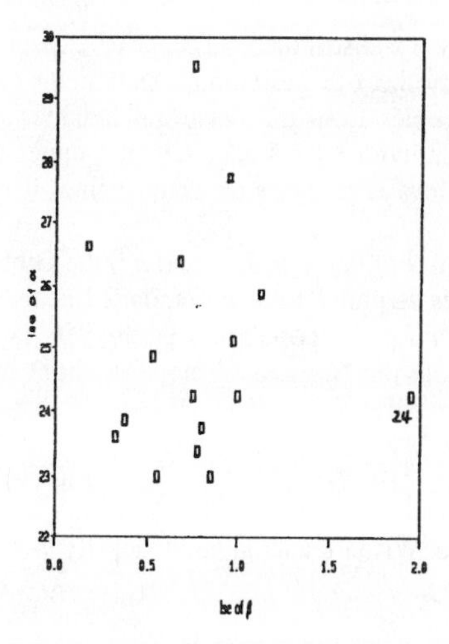

Figure 8. *Least squares estimates of slope and intercept at age 11 for the 15 boys.*

measurement and age relationship of children can be described by a straight line model (Fearn, 1976). To demonstrate the aspect of outlier detection in a hierarchical model, we shall use the data from the 16 boys. If one inspects the least squares estimates of the intercept at 11 years and the slope of the individuals (Figure 8), one may suspect that boy 24 is a good candidate for being an outlier, while his estimated slope appears to be considerably higher than the remaining 15 boys. However, this type of diagnostic is rather misleading, while there are two types of masking effects, namely the uncertainty in the estimate of the individual's parameter, and the between individual variation. It is rather difficult to judge whether the observed estimate of the individual is within reasonable variation of the population. One therefore requires methods to unmask these effects.

One possibility is to use a "normal t" population model to achieve the goal.

Normal t population model

The relationship of the individual measurements and the between individual variation are described by the following model.

Measurement within individual model

Let y_{ij} be the measurement of the i-th individual at x_{ij} for $i = 1, \ldots, I$ and $j = 1, \ldots, n_i$, that is,

$$y_{ij} = \alpha_i + \beta_i \cdot x_{ij} + \epsilon_{ij},$$

where ϵ_{ij} is independent $\mathrm{N}(0, \sigma^2)$. θ_i denotes the individual's parameter $(\alpha_i, \beta_i)^t$. Here the superscript t denotes the transpose of a vector.

Between individuals variation model.

The individual's parameter θ_i follows a bivariate t_ν distribution with mean and covariance θ, Σ, respectively. ν is the degrees of freedom of the t density. As described in Section 2, a t density can be considered as a scale mixture of normal density, with χ^2_ν as the mixing densities, so that

$$p(\theta_i|\lambda_i) = N(\mu, \lambda_i^{-1}\Sigma),$$

where $\nu\lambda_i$ are independently χ^2_ν distributed. The prior expectation of λ_i is 1. It is clear that if λ_i is small, it is likely the i-th individual's parameter θ_i will be far away from the population mean θ. The samples from the posterior densities of λ_i can therefore be used as an indicator of outlying individuals. When, say, the upper 95 sample percentile of the posterior density of λ_i is below 1, its prior expected value, it is a good indication that the i-th individual is an outlier.

In order to generate samples of $\theta_i, \lambda_i, \theta, \Sigma^{-1}$ and σ^2, the Gibbs sampler technique is used. The full conditional densities required for the standard linear "normal normal" population model with a conjugate prior for the population parameters is described in Gelfand *et al.* (1990). In order to incorporate the "normal t" model a slight modification is required. We take the following prior:

$$p(\mu, \Sigma^{-1}, \sigma^{-2}) = N(\eta, C)W((\rho R)^{-1}, \rho)G(\nu_0/2, \nu_0\sigma_0^2).$$

where W and G denote the Wishart and gamma density respectively. Defining $C_i^{-1} = \sigma^{-2}X_i^tX_i + \lambda_i\Sigma^{-1}, U^{-1} = \sum_i \lambda_i\Sigma^{-1} + C^{-1}, \lambda = (\lambda_1, \ldots, \lambda_I)$, the full conditionals defining the Gibbs sampler become

$$\begin{aligned}
p(\theta_i|\theta, \Sigma, \sigma^2, \lambda, y) &= N(C_i(\sigma^{-2}X_i^ty_i + \lambda_i\Sigma^{-1}\mu), C_i),\\
p(\mu|\theta, \Sigma^{-1}, \sigma^{-2}, \lambda, y) &= N(U(\Sigma^{-1}\sum_i \lambda_i\theta_i + C^{-1}\eta), U),\\
p(\Sigma^{-1}|\theta, \mu, \sigma^{-2}, \lambda, y) &= W((\Sigma_i\lambda_i(\theta_i - \mu)(\theta_i - \mu)^t + \rho R)^{-1}, I + \rho),\\
p(\nu_i\lambda_i|y, \theta, \mu, \Sigma^{-2}, \sigma^{-2}, \lambda_j, j = i) &= G((\nu + 2)/2, 1/2),
\end{aligned}$$

where $\nu_i = (\theta_i - \mu)^t\Sigma^{-1}(\theta_i - \mu) + \nu$.

$$p(\sigma^{-2}|\theta, \mu, \Sigma^{-1}) = G((\nu_0 + \sum_i n_i)/2, (\sum_i (y_i - X_i\theta_i)^t(y_i - X_i\theta_i) + \nu_0\sigma_0^2)/2).$$

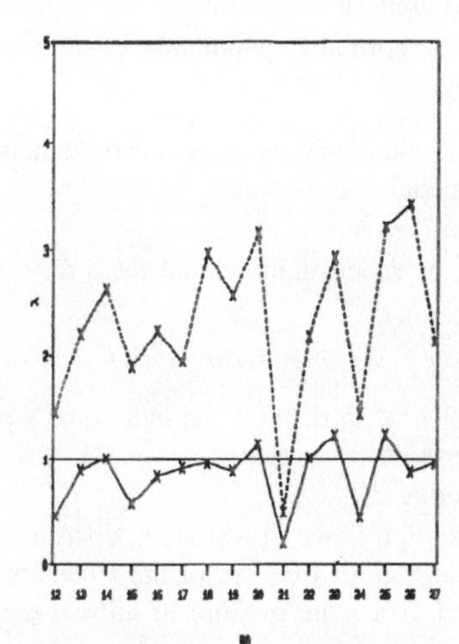

Figure 9. *Median and the 95 percentile of λ for the 15 boys.*

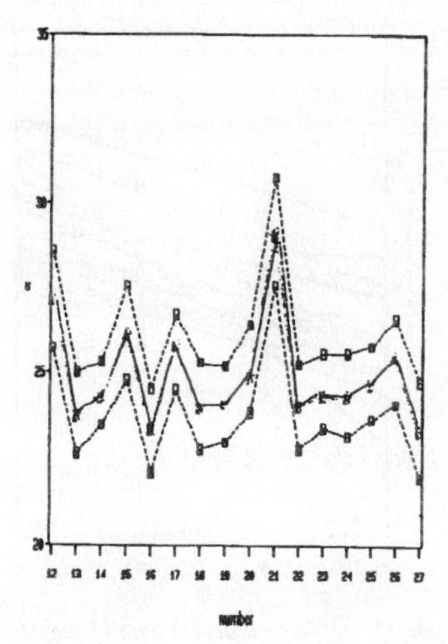

Figure 10. *Median 5 and the 95 percentile of the intercepts at age 11 for the 15 boys.*

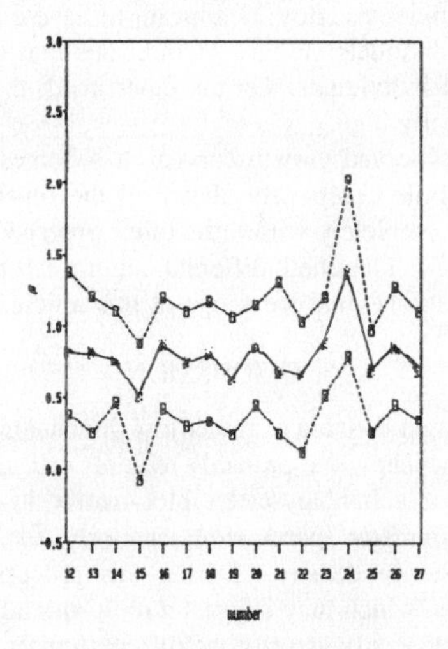

Figure 11. *Median 5 and the 95 percentile of the slopes for the 15 boys.*

For our examples, after 150 cycles of 100 replicates, we visually inspect the samples. Since the samples of the marginal posterior density of λ_i can be considered as a good indicator of an outlying individual, the 5 and 95 sample percentiles for each individual are calculated and graphically displayed in Figure 9. The sample percentiles of boy 24 do not appear to be exceptional, this indicates that he is probably not an outlier. However, the 95 sample percentile of boy 21 falls well below 1, this indicates that this probably is an outlying individual. This is quite contrary to our previous observation based on the plot of the least square estimates (Figure 8). We then proceed to graphical displays of the sample percentiles of α_i and β_i (Figure 10 and 11). In Figure 10, the 90% sample interval of α_{21} lies completely

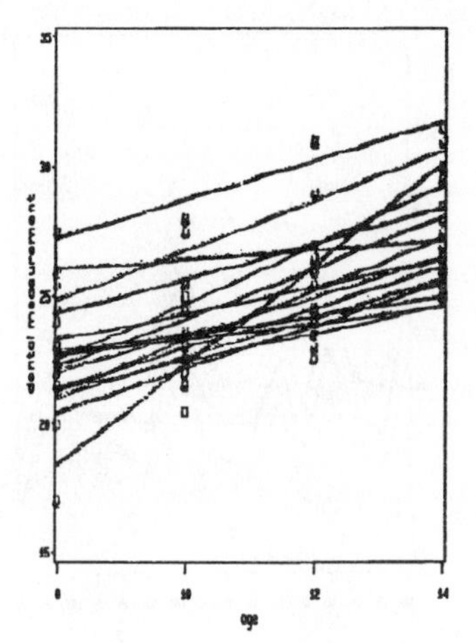

Figure 12. *Observed dental data of the 15 boys.*

above those of the other individuals, Boy 21 appears to have a much higher intercept at 11 years than the remaining individuals. Figure 11 indicates that the slope of boy 21 is quite similar to those of the other individuals. On the other hand, the slope of boy 24, although rather high, is not exceptionally high.

Figure 12 displays the observed growth curves. It becomes quite evident that the dental measurements of boy 21 lie completely above all the other observations, whereas the measurement of boy 24 lie completely within the other observed values.

Once the possible outlier is identified, different actions can be taken from the modelling point of view. However, this is beyond the scope of this article.

5. DISCUSSION

The two examples demonstrated here are of rather low dimension and are simple in nature. If the problem is of higher dimension, one probably requires efficient use of dynamic graphics. One possibility is the use of a bubble scatter plot matrix in combination with brushing to view several bivariate parameter spaces simultaneously, for the purpose of discovering unusual features of the posterior density. During this process, one may notice that the samples have several clusters, which may suggest a multi-modal posterior density. Dynamic graphic techniques allow us to easily identify the different mixtures and view their form and shape separately.

For the case of a population model, if the individual parameters have high dimension or the response function is nonlinear, one may need to view several of the individuals' parameters simultaneously to understand why some of the individuals are outliers. There can be individuals whose marginal posterior densities of the parameters are quite normal, however, the correlation between these parameters is different from the remaining individuals. In order to identify meaningful covariates, the samples (or summary of the samples) of the individual parameters should be viewed simultaneously with the various covariates. Again this requires dynamic graphics and the ability to identify clusters and structures in high dimensional space.

There are definitely a number of numerical and methodological difficulties to be overcome, and various graphical methods yet to be developed and implemented before SAGA becomes a useful tool. However, the possibility of applying sampling and graphical techniques does have some potential in the exploratory and iterative modelling part of the process of a statistical analysis.

REFERENCES

Bates, D. and Watts, D. (1988). *Nonlinear Regression Analysis and its Applications.* New York: Wiley.

Box, G. E. P. and Cox, D. R. (1964). An analysis of transformations. *J. Roy. Statist. Soc. B* **26**, 211–246.

Fearn, T. (1975). A Bayesian approach to growth curves. *Biometrika* **60**, 89–100.

Gelfand, A. E., Hills, S. E., Racine-Poon, A. and Smith, A. F. M. (1990). Illustration of Bayesian inference in normal data models using Gibbs Sampling. *J. Amer. Statist. Assoc.* **85**, 972–985.

Gelfand, A. E. and Smith, A. F. M. (1990). Sampling based approaches to calculating of marginal densities. *J. Amer. Statist. Assoc.* **85**, 390–409.

Geman, S. and Geman, D. (1984). Stochastic relaxation, Gibbs distributions and the Bayesian restoration of images. *IEEE Trans. of Pattern Anal. and Machine Intel.* **6**, 721–741.

Johnson, N. L. and Kotz, S. (1972). *Distributions in Statistics: Continuous Multivariate.* New York: Wiley.

Marske, D. (1967). *Biochemical Oxygen Demand Data Interpretation Using Sum of Squares Surfaces.* Ph.D. Thesis, University of Wisconsin, Madison.

Racine-Poon, A. and Smith, A. F. M. (1989). Population models. *Statistical Methods in the Pharmaceutical Sciences* (D. Berry, Ed.) New York: Marcel Dekker.

Rubin, D. B. (1988). Using the SIR algorithm to simulate posterior distributions. *Bayesian Statistics 3* (J. M. Bernardo, M. H. DeGroot, D. V. Lindley and A. F. M. Smith, eds.), Oxford: University Press, 395–402, (with discussion).

Smith, A. F. M. and Gelfand, A. E. (1992). Bayesian statistics without tears: a sampling-resampling perspective. *Amer. Statist.* , (to appear).

Smith, A. F. M., Skene, A. M., Shaw, J. E. H., Naylor, J. C. and Dransfield, M. (1985). The implementation of the Bayesian paradigm. *Commun. Statist., Theory and Methods* **14**, 1079–1102.

Weihs, C. and Schmidli, H. (1990). Omega: on-line multivariate graphical analysis. *Statist. Sci.* **5**. 175–226, (with discussion).

DISCUSSION

I. GUTTMAN (*University of Toronto, Canada*)

Amy Racine invites us to take a trip in wonderful Graphics-Land, giving us a short course on importance sampling and Gibbs sampling, and, during this trip, she gets us involved with Bubble baths, the SIR algorithm, served on ice, and/or gourmet dinners with nouvelle cuisine of Bayesian computing, served and cooked by Gibbs and sampling staff, with menu items that extend to outliers for desserts.

I congratulate her – it all just takes my breath away. In what follows, I express some minor reservations and quibbles, but the fact of the power of this paper remains impressive.

1. To begin at the beginning, let me first address the choice of title, for there is attached to this choice an attitude that, frankly, I worry about!.

S, A, G, A or SAGA, at the present time, stands for Sample Assisted Graphical Analysis. There is a kind of 'cart before the horse' impression that may be left here – for, after all, we do graphical analysis to aid in the inference process, and we are doing inference based on a given sample, the data. While Amy is no doubt referring to importance and/or Gibbs sampling methods, the reader or user will not discover this until well into the substance of her paper. My suggestion is the following: keep "SAGA" as acronym, but let it stand for "Statistical Analysis Graphical Assisted".

2. When talking about the SIR algorithm, using the multivariate t-family for the "easy to sample" density g, the weight, say $\bar{\omega}_{l,j}$, for the j-th observation $\phi_{l,j}$ found at the l-th

iteration, is taken to be such that

$$\bar{\omega}_{l,j} \propto \frac{f(\phi_{l,j})}{g^{l-1}(\phi_{l,j})}\lambda_{l,j}$$

where f is the true (but hard to sample) distribution of ϕ, with $\lambda_{l,j}$ an independent drawing on a χ^2_ν variable, independent of $\phi_{l,j}$. My concern is with the stability of $\bar{\omega}_{l,j}$. After all, g^{l-1}, the t-distribution used at the $(l-1)$st iteration, is a "kindred soul" to f, so that intuitively f is damped by g^{l-1} and not likely to cause much of the variation in $\bar{\omega}$'s, but this is then multiplied by a drawing from a χ^2_ν distribution, and we all know that this distribution is very variable – my suggestion is that instead of multiplying f/g^{l-1} by $\lambda_{l,j}$ we use $\lambda^{\varepsilon}_{l,j}$ with $\varepsilon \approx 1/2$. Finally, it would be nice to know what the author suggests for choices of v, the degrees of freedom of the χ^2_ν variable.

3. When illustrating the use of SAGA when the (exponential decay) model

$$E(y_i) = \theta_1(1 - \exp\{-\theta_2 x_i\})$$

is assumed, the question of how to produce the maximum likelihood estimate of (θ_1, θ_2) intervenes, and is not fully explained in the paper. My own favourite method is to regard the above model as a linear combination of the functions $[1, e^{-\theta x}]$ and use the 2-stage iterative procedure of Guttman, Pereyra and Scolnik (1973). There are further concerns: (i) It is not clear to me how exactly SAGA computed the Variance-Covariance Matrix that is needed as an essential end ingredient – is the observed information matrix used based on g, or f, or ...? (ii) The author states that bubble plots are "employed because of oscillations...? – what exactly is oscillating? (iii) Finally, as the author has acknowledged in her revised paper, I propose that SAGA include a Box-Cox type of analysis to ascertain what is the best re-parameterization from (θ_1, θ_2) – this would include the possibility of this analysis, based on the data, choosing the author's selection $\phi_1 = 1/\theta_1$ and $\phi_2 = \log\theta_2$.

4. The example that involves the hierarchical model of Lindley and Smith (1972) is beautiful. But it does invite comparisons – for example, to mention one aspect, outliers and influence detection methods abound in the literature, and it would have been nice to see some of these methods routinely applied as a part of the capability of SAGA. Further on this subject, with reference to the dental data:
   (i) It could be that Boy 24 is spurious and hence outlying, in that he is a member of a sub-population so that

$$E(y_{24}|\text{age}\, x) > E(y_j|\text{age}\, x), \quad j \neq 24,$$

   or, it could be that the l-th measurement on y for Boy 24 is such that

$$E(y_{24,l}|x) = E(y_{24,l'}|x) + a, \quad a \neq 0$$

   for all $l' \neq l$, which is to say, that $y_{24,l}$ is spurious because of a mean shift in the measurement process, causing an extreme observation, i.e., outlying, which is influential in the sense that the regression line for Boy 24 is pulled up. This means that we need diagnostics in SAGA which would distinguish between the above two types of spuriosity actually encountered.
   (ii) With reference to (i) above, again as part of SAGA, it would be nice to be able to make inference automatically about $\beta_{j,1} - \beta_{24,1}$ (β_1's are rates of change), and/or $E(y_j|x) - E(y_{24}|x)$ for $j \neq 24$.

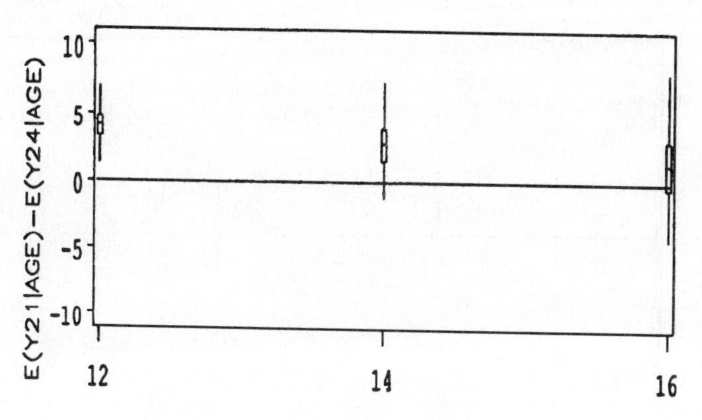

Figure A.

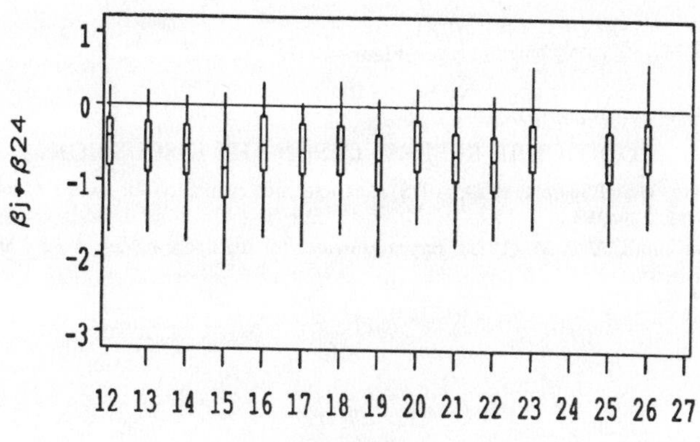

Figure B.

REPLY TO THE DISCUSSION

1. I do agree the choice of acronym of the discussant is much more appropriate.
2. Since the t density is used for generation only when one believes that the underlying density is going to be heavy tailed, it is therefore more natural to choose a low degree of freedom, say 1 or 2. Better choices for computing the weight are clearly possible. The weight that was chosen in the article down-weights extreme values, this turns out to be quite stable.
3. Since the "Box-Cox" proposal was so good, I have introduced it in the article. The oscillation that was observed is the oscillation of the major axis of the bubble plots.
4. and 5. Gibbs sampling provides samples of all the parameters in the model, that is all the individuals' slopes and intercepts and the population parameters and the error variance, which enable us to perform inference and diagnostics. Figure A displays the comparison of the profile of Boy 21 with Boy 24 at age 12, 14 and 16. Figure B, the inference of the differences in growth rates between each of the boys and Boy 24. Figure C is the residual diagnostics of Boy 24 for outlier detection. These were all carried out without much additional effort.

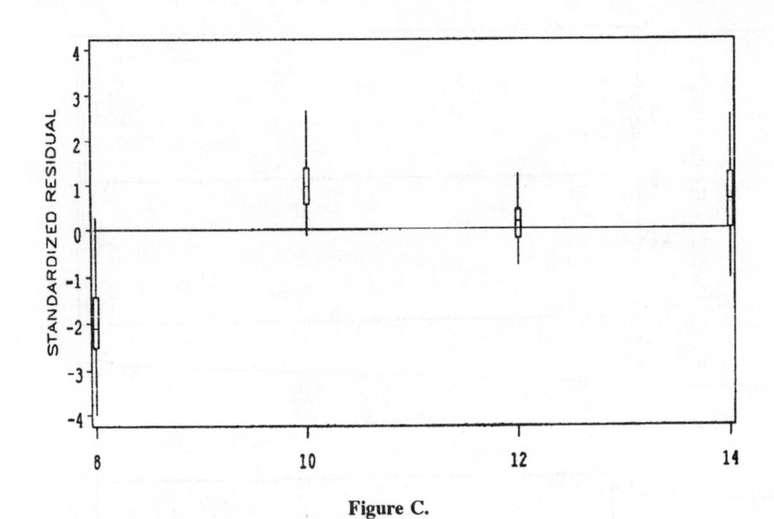

Figure C.

ADDITIONAL REFERENCES IN THE DISCUSSION

Guttman, I., Pereyra, V. and Scolnik, H. D. (1973). Least squares estimation for a class of non-linear models. *Technometrics* **15**, 209–218.

Lindley, D. V. and Smith, A. F. M. (1972). Bayes estimates for the linear model. *J. Roy. Statist. Soc. B* **34**, 1–41, (with discussion).

BAYESIAN STATISTICS 4, pp. 405–418
J. M. Bernardo, J. O. Berger, A. P. Dawid and A. F. M. Smith, (Eds.)

The Elusive Concept of Statistical Evidence

RICHARD M. ROYALL
Johns Hopkins University, USA

SUMMARY

We examine the fundamental rule for interpreting statistical data as evidence, the Law of Likelihood. The Law's rationale and implications are sketched, and its limitations are explored via an attempt to extend its scope to include composite hypotheses.

Keywords: LIKELIHOOD; STRENGTH OF EVIDENCE; SUPPORT.

INTRODUCTION

Our fundamental question is this: When is it correct to say that one statistical hypothesis is better-supported than another by a given body of observations?. Than is, when is the fact that a random variable X takes on a specific value x properly interpreted as evidence supporting hypothesis H_0 in favor of H_1? A second question concerns quantifying this evidence: Can we measure its strength? These questions lie at the heart of many applications of statistical methods in scientific inference. Questions of calibrating and revising degrees of belief and problems of choosing from a list of available actions are also important, but analyses aimed at objectively representing and interpreting the data as evidence are central to much of scientific research. A conspicuous example is the practice of using significance tests (p-values) to measure the strength of evidence.

Our questions are closely related to the work of Birnbaum (1962). He proved that, in the context of a parametric statistical model for a random variable X, the evidence in an observation, $X = x$, is represented by the likelihood function. Specifically, he proved that two observations that generate the same likelihood function are equivalent as evidence about the parameter; they have the same evidential meaning. This fact, that the likelihood function provides a mathematical *representation* of data as evidence, is called the Likelihood Principle. It implies that our questions about the *interpretation* of data as evidence might be restated as questions about interpretation of likelihoods. But it does not tell us how to interpret likelihoods, and for guidance in that endeavor, we must turn to a different principle, the Law of Likelihood. In the very limited context of comparing simple statistical hypotheses, our questions about evidence are answered explicitly by that law. Here we examine those answers and briefly explore some of the problems that arise when the questions pertain to composite hypotheses.

Time and space limitations have precluded efforts to give proper credit to those who first made or emphasized the points that will be presented. Nothing in the first three sections is original, and the entire work is intended simply to draw attention to unanswered questions about concepts that are surely fundamental to our discipline.

1. THE LAW OF LIKELIHOOD

Although the underlying concept has been traced by Edwards (1974) back to Venn (1876) and beyond, the Law of Likelihood appears to have been given a name only in 1965, by Hacking.

> **Law of Likelihood:** If hypothesis H_0 implies that the probability that a random variable X takes on the value x is $p_0(x)$, while hypothesis H_1 implies that the probability is $p_1(x)$, then the observation $X = x$ is evidence supporting H_0 over H_1 if and only if $p_0(x) > p_1(x)$, and the ratio $p_0(x)/p_1(x)$ measures the strength of that evidence.

This law asserts that evidence is measured by the likelihood ratio. Why should it be accepted? One favorable point is that it is the natural extension, to probabilistic phenomena, of scientists' established form of reasoning in deterministic situations: If H_0 implies that under specified conditions x will be observed, while H_1 implies that under the same conditions something else, not x, will be observed, and if those conditions are created and x is seen, then this observation is evidence supporting H_0 versus H_1. The Law of Likelihood simply extends this way of reasoning to say that if x is more *probable* under hypothesis H_0 than under H_1, then the occurrence of x is evidence supporting H_0 over H_1, and the strength of that evidence is determined by how *much* greater the probability is under H_0. The hypothesis that is better supported by an observation is the one that did the better job of predicting that observation.

Another important favorable point is that the Law of Likelihood passes the test represented by Bayes' theorem. That theorem says that in cases where two simple hypotheses have initial probabilities in an arbitrary ratio r, the effect of an observation with a likelihood ratio of k is to change the hypotheses' probability ratio to kr. The likelihood ratio is the precise factor by which the relative probabilities of the two hypotheses are changed by the observation. This implies that any viable concept of evidence must agree, in cases of simple hypotheses that have prior probabilities, with the Law of Likelihood.

A critical test of the Law of Likelihood is functional —does it work?. If we *use* the Law to evaluate evidence, will we be led to the truth?. Suppose H_0 is actually false and H_1 is true. Can we get observations that, according to the Law, are evidence for H_0 over H_1?. Certainly. Does this mean that the Law is invalid? Certainly not. Evidence, properly interpreted, can be misleading. This must be the case, for otherwise we would be able to determine the truth (with perfect certainty) from any scrap of evidence that is not utterly ambiguous. However, we might reasonably expect that strong evidence cannot be misleading very often. We might also expect that, as evidence accumulates, it will tend to favor a true hypothesis over a false one more and more strongly. Both of these expectations are met by the concept of evidence embodied in the Law of Likelihood.

A simple calculation shows that the probability of getting strong evidence in favor of the false hypothesis is small. Specifically, if H_1 is true and $k > 0$

$$\Pr(p_0(X)/p_1(X) \geq k) \leq \frac{1}{k}.$$

A much stronger result is also true: If an unscrupulous researcher sets out deliberately to find evidence supporting her favorite but erroneous hypothesis (H_0) over her rival's (H_1), which happens to be correct, by a factor of at least k, then the chances are good that she will be *eternally frustrated*. Specifically, suppose that the hypotheses pertain to a sequence of random variables $x_1, x_2, \ldots$ whose first n elements are the vector $\boldsymbol{X}_n$. Her hypothesis H_0 implies that the probability that $\boldsymbol{X}_n = \boldsymbol{x}_n$ is $p_{0n}(\boldsymbol{x}_n)$, while H_1 implies that the probability is $p_{1n}(\boldsymbol{x}_n)$. She checks after each observation to see whether her accumulated data are "satisfactory" (likelihood ratio favors H_0 by at least k), stopping and publishing only when this occurs. The probability that her scheme will succeed is no greater than $1/k$. That is, when H_1 is true.

$$\Pr\left(p_{0n}(\boldsymbol{X}_n)/p_{1n}(\boldsymbol{X}_n) \geq k, \text{ for some } n = 1, 2, \ldots\right) \leq 1/k$$

(Robbins, 1970).

In a more positive vein, the Law of Likelihood, together with the Law of Large Numbers, implies that the accumulating evidence represented by observations on a sequence $X_1, X_2, \ldots$ of independent random variables will eventually strongly favor the truth. Specifically, if p_1 is correct and p_0 identifies any other probability distribution, then the likelihood ratio $\prod_1^n p_0(X_i)/p_1(X_i)$ converges to zero with probability one. This means that we can specify any large number k with perfect certainty that our evidence will favor H_1 over H_0 by at least k if only we take enough observations.

Some statisticians agree that the likelihood ratio provides a proper ordering of sample points as evidence, i.e., if $p_0(x_1)/p_1(x_1) > p_0(x_2)/p_1(x_2)$ then $X = x_1$ is stronger evidence than $X = x_2$ for H_0 vs H_1; yet they doubt that a likelihood ratio of fixed magnitude, say, four, represents the same strength of evidence in all contexts. Doubts on this point come from failure to distinguish between the strength of the evidence, which is constant, and its implications, which vary according to the context (prior beliefs, available actions, etc.) of each application.

The key point here is that observations with a likelihood ratio of k are evidence strong enough to cause a k-fold increase in a prior probability ratio. The values of the prior probabilities do not matter, nor does their ratio. The effect is always the same. Thus when prior probabilities are present the numerical value of the likelihood ratio has a precise meaning. That meaning is retained quite generally: a likelihood ratio of k corresponds to evidence strong enough to cause a k-fold increase in a prior probability ratio, regardless of whether a prior ratio is actually available in a specific problem or not. The situation is analogous to that in physics where a unit of thermal energy, the BTU, is given a concrete meaning in terms of water — one BTU is that amount of energy required to raise the temperature of one pound of water at $39.2°F$ by one degree. But it is meaningful to measure thermal energy in BTU's in rating air conditioners and in other situations where there is no water at $39.2°F$ to be heated. Likewise the likelihood ratio, given a concrete meaning in terms of its effect on prior probabilities, retains that meaning in their absence.

2. A COUNTEREXAMPLE

The Law of Likelihood is intuitively attractive; in special situations where we *know* how to interpret evidence precisely (via its effect on the relative probabilities of hypotheses), the Law is clearly correct; and it works. But examples like the following have convinced some that the Law is false.

I shuffle an ordinary-looking deck of playing cards and turn over the top card. It is the ace of diamonds. According to the Law of Likelihood the hypothesis that the deck consists of fifty-two aces of diamonds (H_1) is better-supported that the hypothesis that the deck is normal (H_N) by a large factor, Pr(ace of diamonds$|H_1$)/Pr(ace of diamonds$|H_N$) $= 52$.

Some find this disturbing. Although the evidence is supposed to be strong, they would not be convinced that there are fifty-two aces of diamonds instead of a normal deck. Furthermore it seems unfair; no matter what card is drawn the Law implies that the corresponding trick-deck hypothesis (fifty-two cards just like the one drawn) is better-supported than the normal-deck hypothesis. Thus even if the deck *is* normal we will always claim to have found strong evidence that it is not.

As Edwards (1970) has observed, the first point rests on confusing evidence and belief. If drawing an ace of diamonds does not convince you that the deck has fifty-two aces of diamonds (H_1), this does not mean that the draw is not evidence in favor of that hypothesis over a normal deck (H_N). It means simply that the evidence is not strong enough to overcome your strong prior preference for H_N. Suppose I show you two decks, one normal and one actually composed of fifty-two aces of diamonds. I choose one of the decks, shuffle, and draw one card. It is an ace of diamonds. Now the belief that the deck is not the normal one looks quite reasonable. The evidence represented by the ace of diamonds is the same as before; it is the prior beliefs that have changed.

The second objection to the Law, that H_N is treated unfairly, rests on a misinterpretation: "evidence supporting H_1 over H_N" is *not* "evidence against H_N". Consider the fifty-two different trick deck hypotheses, $H_1, H_2, \ldots, H_{52}$, one stating that all fifty-two cards are aces of diamonds, one stating they are all fours of clubs, etc. Observing the ace of diamonds is evidence supporting H_1 over H_N. It is also evidence supporting H_N over H_2, H_3, etc., *decisively*. It is not evidence for or against H_N alone.

3. IRRELEVANCE OF THE SAMPLE SPACE

The Law of Likelihood says that the evidence in an observation, $X = x$, as it pertains to two probability distributions labelled θ_0 and θ_1, is represented by the likelihood ratio, $f(x;\theta_0)/f(x;\theta_1)$. In particular, the law implies that for interpreting the observation as evidence for $H_0 : \theta = \theta_0$ vis-a-vis $H_1 : \theta = \theta_1$, *only* the likelihood ratio is relevant. What other values of X might have been observed, and how the two distributions in question spread their remaining probability over the unobserved values is irrelevant — all that counts is what *was* observed and the relative probabilities of that observation under the two hypotheses.

For example (following Pratt, 1961), suppose the hypotheses concern the proportion θ of white balls in an urn, and twenty balls are drawn (with replacement). The result is reported in code, and you, knowing the code, will learn precisely how many draws produced white balls. I know only the code-word for "six", so that from the report I can determine only whether the outcome is "six" or "not six". Thus you will observe a binomial $(20, \theta)$ random variable X. I will observe Y, which equals six with probability

$$p = \binom{20}{6} \theta^6 (1-\theta)^{14}$$

and some other value, c, representing the outcome "not six", with probability $1 - p$. Your sample space consists of the twenty-one points $\{0, 1, \ldots, 20\}$ while mine consists of $\{6, c\}$.

Now suppose the experiment is done and six white balls are drawn. For any proportions θ_0 and θ_1, my evidence concerning $H_0 : \theta = \theta_0$ vis-a-vis $H_1 : \theta = \theta_1$ is the same as yours; the likelihood ratios for $Y = 6$ and for $X = 6$ are identical. Of course, if the experiment had a different outcome, your evidence and mine would have been different, but for the result that actually occurred, six white balls, my observation is equivalent, as evidence about θ, to yours. Any concept or technique for evaluating observations as evidence that denies this equivalence, attaching a different measure of "significance" to your report of this result than to mine, is invalid. Whatever can be properly inferred about θ from your report can be inferred from mine and vice-versa. The difference between our sample spaces is irrelevant.

Birnbaum (1962) called this conclusion the "Irrelevance of outcomes not actually observed". It calls into question analyses that use the most popular concepts and techniques in applied statistics (unbiased estimation, confidence intervals, significance tests, etc.) whenever these analyses purport to represent "what the data say about θ," i.e., to convey the meaning of the observations as evidence about θ. These conventional analyses are questionable because they all certify that there are important differences between observing $X = 6$ and observing $Y = 6$, whereas the two results are in fact evidentially equivalent.

4. COMPOSITE HYPOTHESES

When H_0 and H_1 are simple statistical hypotheses, the Law of Likelihood provides what seem to be entirely satisfactory answers to our questions about evidence in terms of the likelihood ratio. But when the hypotheses are composite, trouble appears; a composite hypothesis specifies a *set* of probability distributions and therefore a *set* of probabilities for the observation, so that for two composite hypotheses there is no definite likelihood ratio, and the Law of Likelihood is silent. Edwards (1972) argued that composite hypotheses are of little scientific importance and can therefore be ignored; few have accepted that position. Bayesians (e.g., Berger and Wolpert, 1988) suggest that we should examine ratios of average likelihoods, using as weight functions prior probability distributions conditioned on the respective hypotheses. This solution is rejected by many scientists and statisticians who, feeling that our questions about evidence for composite hypotheses *should* have answers that are independent of prior probabilities, turn to frequentist significance tests. However, the case of two simple hypotheses furnishes a clear counter-example to the validity of using tests to measure evidence — evidence is measured by the likelihood ratio, not by a tail area under one hypothesis.

When is it correct to say that one composite statistical hypothesis is better supported than another by a given body of observations? The Law of Likelihood may be silent for a very good reason: there may be no viable non-Bayesian concept of evidence that allows for comparison of composite hypotheses. Or there may be more than one.

4.1. *Finite Composites of Simple Hypotheses*

We first consider statistical hypotheses specifying that a discrete random variable X has one of a finite number of probability distributions indexed by a parameter θ. For an observation $X = x$ we let $\boldsymbol{p}_0(x)$ denote the vector of probabilities (i.e., the likelihood vector) under H_0; $\boldsymbol{p}_1(x)$ is the likelihood vector under H_1.

The intuitive concept that we explore is the one behind the Law of Likelihood, namely, that the support supplied by an observation is determined by how well the hypotheses in question fit that observation. Thus we seek to characterize when one vector of probabilities represents a better fit than another. If the degree to which evidence $X = x$ supports H_0 vis-a-vis H_1 can be represented by some function of the probability vectors $h(\boldsymbol{p}_0(x), \boldsymbol{p}_1(x))$, what can be said about h? We will denote generic vectors in the domain of h by $\boldsymbol{p}_0 = (p_{01}, p_{02}, \ldots, p_{0I})$ and $\boldsymbol{p}_1 = (p_{11}, p_{12}, \ldots, p_{iJ})$.

The Law of Likelihood provides a preliminary axiom for h:

Axiom 0. *If* $\boldsymbol{p}_0$ *and* $\boldsymbol{p}_1$ *are constant vectors* $(p_0, \ldots, p_0)$ *and* $(p_1, \ldots, p_1)$ *then*

$$h(\boldsymbol{p}_0, \boldsymbol{p}_1) = p_0/p_1.$$

Note that Axiom 0 establishes one condition under which we can measure statistical evidence as it relates to composite hypotheses, namely, when the likelihood vector under each hypothesis is constant. Note also that Axiom 0 is incompatible with Basu's (1975)

"Strong Law of Likelihood" which states that support for composite hypotheses can be measured by adding likelihoods over the hypotheses.

More generally, when the likelihood vectors are not constants it seems entirely compelling that the degree to which $X = x$ supports H_0 vs H_1 is at least $p_{0\,\min}/p_{1\,\max}$, but no greater than $p_{0\,\max}/p_{1\,\min}$, where $p_{0\,\min} = \min_i p_{0i}$, etc.; the support $h(\mathbf{p}_0, \mathbf{p}_1)$ is bounded by the two most extreme likelihood ratios. One reason this conclusion is compelling is that the extreme likelihood ratios are tight bounds on the factor by which the evidence $X = x$ can change the hypotheses' prior probability ratio. Can tighter bounds on the support ever be justified? In particular, can $X = x$ ever be properly characterized as evidence supporting H_0 over H_1 when some distribution under H_0 assigns less probability to x than some distribution under H_1? It is not clear that an affirmative answer is possible, but a negative one would certainly come as a surprise and disappointment to many who look to statistics for guidance in the evidential interpretation of the results of their experiments.

The following development leads to an affirmative answer. It also appears to lead to conclusions that are generally regarded as unacceptable. It is presented in hopes that it will stimulate further development and clarification.

We begin with two statements in the form of additional axioms. First, the greater the probabilities of $X = x$ under H, the better H fits:

Axiom 1: *If* $p_{0i} \geq p^*_{0i}, i = 1, \ldots, I$ *and* $p_{1j} \leq p^*_{1j}, j = 1, \ldots, J$ *then*

$$h(\mathbf{p}_0, \mathbf{p}_1) \geq h(\mathbf{p}^*_0, \mathbf{p}^*_1).$$

How well a composite hypothesis fits x is unaffected by the order in which the component hypotheses are listed:

Axiom 2: *If* $\mathbf{p}^*_0$ *is a permutation of* $\mathbf{p}_0$ *and* $\mathbf{p}^*_1$ *is a permutation of* $\mathbf{p}_1$ *then*

$$h(\mathbf{p}_0, \mathbf{p}_1) = h(\mathbf{p}^*_0, \mathbf{p}^*_1).$$

The next axiom generalizes Axiom 0. It says that if the vector of probabilities under H_0 equals some constant c times the vector under H'_0, then H_0 fits the observation c times better, and the degree of support for H_0 vs. any alternative H_1 is c times the support for H'_0 vs. H_1.

Axiom 3: *For any* $\mathbf{p}_0, \mathbf{p}_1$, *and scalar* $0 \geq c \geq 1/\max(p_{0i})$

$$h(c\mathbf{p}_0, \mathbf{p}_1) = ch(\mathbf{p}_0, \mathbf{p}_1).$$

A significant implication of Axiom 0 is that when all of their elements are equal, the *lengths* of the vectors $\mathbf{p}_0(x)$ and $\mathbf{p}_1(x)$ are irrelevant to the interpretation of $X = x$ as evidence for H_0 vs H_1. Examples like Pratt's (1961) show why this must be true: Consider an urn containing 100 balls and hypotheses specifying the proportion of white balls, θ, $H_0 : \theta = \frac{1}{10}$ and $H_1 : \theta = \frac{9}{10}$. The observation of a white ball is evidence supporting H_1 over H_0 by a factor of 9. Now consider two different hypotheses, H'_0 and H'_1, stating respectively that $\theta = \frac{1}{10}$ and $\theta = \frac{9}{10}$, but also stating that the non-white balls are red or green, in unspecified proportions. Now H'_0 consists of 91 simple hypotheses (specifying different proportions of red balls) and H'_1 consists of 11. It seems clear that a white ball is evidence supporting H'_1 over H'_0 by a factor of 9, the same as for H_1 over H_0. In fact the two pairs of hypotheses might refer to the same physical set-up, with the distinction between the two (H_0 vs H_1 or H_0' vs H_1') determined by whether the observer is red-green color blind or not. Clearly when white is drawn this evidence about the proportion of white balls does not depend on whether the observer might have detected the difference between red and green if a non-white ball had been drawn.

Variations on this example suggest:

Axiom 4: *If the sets of distinct elements in* $\boldsymbol{p}_0$ *and* $\boldsymbol{p}_1$ *are the same, then*

$$h(\boldsymbol{p}_0, \boldsymbol{p}_1) = 1$$

Let p_{0m}, p_{1m} denote the minima of the elements of $\boldsymbol{p}_0$, and $\boldsymbol{p}_1$ respectively, and let r_1 denote their ratio, p_{0m}/p_{1m}. Likewise define p_{0x} and p_{1x} as the maxima, with ratio $r_2 = p_{0x}/p_{1x}$. If $p_{0m} = p_{1m} = 0$ we have

$$\begin{aligned} h(\boldsymbol{p}_0, \boldsymbol{p}_1) &\geq h((0, p_{0x}, \ldots, p_{0x}), (0, \ldots, 0, p_{1x})) \\ &= (p_{0x}/p_{1x}) h((0, p_{1x}, \ldots, p_{1x}), (0, \ldots, 0, p_{1x})) \\ &= p_{0x}/p_{1x}. \end{aligned}$$

Here the inequality comes from Axioms 1 and 2, and the equalities from Axiom 3 and 4 respectively. But a similar argument establishes the reverse inequality so that $h(p_0, p_1) = p_{0x}/p_{1x}$. When p_{0m} and p_{1m} are not both zero, similar arguments shown that $h(\boldsymbol{p}_0, \boldsymbol{p}_1)$ lies between r_1 and r_2 and that $h(\boldsymbol{p}_1, \boldsymbol{p}_0)$ lies between $1/r_1$ and $1/r_2$. Thus we have

Theorem. *Under Axioms 1–4*

$$\begin{aligned} \min(r_1, r_2) &\leq h(\boldsymbol{p}_0, \boldsymbol{p}_1) \leq \max(r_1, r_2) \\ \min(1/r_1, 1/r_2) &\leq h(\boldsymbol{p}_1, \boldsymbol{p}_0) \leq \max(1/r_1, 1/r_2) \end{aligned}$$

and if $p_{0\,\mathrm{min}} = p_{1\,\mathrm{min}} = 0$ *then*

$$h(\boldsymbol{p}_0, \boldsymbol{p}_1) = r_2 = 1/h(\boldsymbol{p}_1 \boldsymbol{p}_0).$$

As an illustration, consider n independent Bernoulli trials obtained by rolling a cube and noting whether the upper face is green or not. The hypothesis H_0 that the proportion of green faces θ is no greater than $1/2$, consists of four simple hypotheses, $\theta = 0, 1/6, 1/3$, or $1/2$, while $H_1 : \theta > \frac{1}{2}$ consists of three, $\theta = 2/3, 5/6$, or 1. If green shows on each of three tosses, $h(\boldsymbol{p}_1, \boldsymbol{p}_0) \geq 8$ while 8 green on 10 tosses yields $h(\boldsymbol{p}_1, \boldsymbol{p}_0) = 6.62$.

If these axioms were accepted, this theorem would guide the interpretation of evidence regarding a parameter ϕ_1 in the presence of a nuisance parameter ϕ_2. If the value of ϕ_2 that maximizes the likelihood $p(x; \phi_1, \phi_2)$ is denoted by $\phi_{2U}(\phi_1)$ then $p(x; \phi_1, \phi_{2U}(\phi_1))$ is the upper profile likelihood function $\ell_U(\phi_1)$. Similarly we define the lower profile likelihood $\ell_L(\phi_1)$ in terms of the minimizing value of the nuisance parameter. The support for $\phi_1 = a$ vs $\phi_1 = b$ lies between $\ell_U(a)/\ell_U(b)$ and $\ell_L(a)/\ell_L(b)$.

4.2. *Composite Hypotheses with Interval Likelihoods*

So far we have confined attention to composite hypotheses which imply that possible values of the probability of the observation, $X = x$ comprise a (finite) vector. Now let us briefly consider broader hypotheses which imply that the possible values of this probability comprise an interval. For example, suppose X is a binomial (n, θ) random variable with n known and the hypotheses are $H_0 : \theta \leq \frac{1}{2}$ and $H_1 : \theta > \frac{1}{2}$. If $p(x;\theta)$ denotes the probability that $X = x$, then when $n/2 < x < n$ H_0 implies that the possible values of the probability comprise the interval $I_0(x) = [0, p(x; 1/2)]$; H_1 implies the interval $I_1(x) = [0, p(x; \hat{\theta})]$, where $\hat{\theta} = x/n > 1/2$. Since $I_1 = c(x)I_0$, where $c(x) = 2^n\hat{\theta}^x(1-\hat{\theta})^{n-x}$, we might extend the previous arguments to this case and conclude that $X = x$ supports H_1 over H_0 by $c(x)$. If $0 < \hat{\theta} < \frac{1}{2}$ the support is $1/c(x)$. For many of the composite hypotheses commonly considered in statistical analyses the results would take this same form: the evidence is measured by the ratio of the respective maxima of the likelihoods under the two hypotheses.

Many important problems involve continuous probability distributions with continuously varying parameters including nuisance parameters for which the lower profile likelihood $\ell_L(\phi_1) = 0$ for all ϕ_1. In this case the extensions of the above arguments would lead to interpreting the (upper) profile likelihood function as measuring relative support for values of ϕ. For example, if (ϕ_1, ϕ_2) are the mean and variance of a normal distribution, $\ell_U(\phi_1) = \left[\sum(x-\phi_1)^2/n\right]^{-\frac{n}{2}}$ would measure the relative support for values of the mean, and the degree to which an observed sample $\boldsymbol{X} = \boldsymbol{x}$ with positive mean supports $H_1 : \phi_1 > 0$ over $H_0 : \phi_1 \leq 0$ would be $\left[1 + t^2/(n-1)\right]^{n/2}$, where t is Student's t-statistic, $n^{1/2}\overline{x}/s$.

But this result is unsatisfactory in one important respect: the "profile likelihood function" is not a true likelihood function. In particular, it does not share a key property of likelihood functions: a true likelihood ratio cannot strongly favor the false hypothesis over the true one with high probability. That this is not true of profile likelihood in general is illustrated by the familiar example of Neyman and Scott (1948). If $X_{ij}/i = 1, 2$, and $j = 1, \ldots, n$ are independent normal random variables with $E(X_{ij}) = \phi_j, j = 1, \ldots, n$ as nuisance parameters and common variance ϕ_0 the profile likelihood ratio will, when n is large, favor the false value $\phi_0/2$ over the true ϕ_0 by a large factor with very high probability.

5. CONCLUSIONS

Three conclusions are suggested:

1. The essential concept in the interpretation of statistical data as evidence is the one (relative support) that is embodied in the Law of Likelihood. It is on this rock that statistical methods for evidential interpretation of data must stand.
2. Within a parametric probability model, statistical evidence is represented mathematically by the likelihood function, whose ratios measure relative support. This conclusion (the Likelihood Principle) derives much of its strength and significance directly from the Law of Likelihood, independently of other principles, such as Sufficiency and Conditionality.
3. For interpreting statistical evidence as it pertains to two simple hypotheses, Bayesian concepts and tools are not essential, but for composite hypotheses no satisfactory non-Bayesian methods have yet been found, and the position stated by Berger and Wolpert (1988) might well be the only one that is tenable: "... sensible use of the likelihood function seems possible only through Bayesian analysis". This does not imply that a personalistic Bayesian philosophy is necessary nor that unknown parameters must be perceived as random variables; it means simply that formal Bayesian methods might represent the best practical tools for objective interpretation of likelihood functions.

REFERENCES

Basu, D. (1975). Statistical information and likelihood. *Sankhya* **37**, 1–71.

Berger, J. O. and Wolpert, R. L. (1988). *The Likelihood Principle*. Hayward, CA: IMS.

Birnbaum, A. (1962). On the foundations of statistical inference. *J. Amer. Statist. Assoc.* **53**, 269–316.

Edwards, A. W. F. (1970). Likelihood (Letter to the editor). *Nature* **227**, 92.

Edwards, A. W. F. (1972). *Likelihood*. Cambridge: University Press.

Edwards, A. W. F. (1974). The history of likelihood. *International Statistical Review* **42**, 9–15.

Hacking, I. (1965). *Logic of Statistical Inference*. Cambridge: University Press.

Neyman, J. and Scott, E. L. (1948). Consistent estimates based on partially consistent observations. *Econometrika* **16**, 1–32.

Pratt, J. W. (1961). Review of *Testing Statistical Hypotheses*, by E. L. Lehmann. *J. Amer. Statist. Assoc.* **56**, 163–167.

Robbins, H. (1970). Statistical methods related to the law of the iterated logarithm. *Ann. Math. Statist.* **41**, 1397–1409.

DISCUSSION

S. FRENCH (*The University of Leeds, UK*)

Coming to the Valencia Conferences I am inevitably drawn to think of *small worlds*. Not those of David Lodge, however appropriate they may be to our regular gatherings, but those of Jimmie Savage: the small worlds that each of us construct when we build a statistical — or any other — model. Here, in reading the first words of this paper, I am dragged unwillingly into a small world not of my making and one with which I have little sympathy. It is a world populated by hypotheses —well, most statistical small worlds are— but in this world one of the hypotheses is *true*. Such a world is seldom a Bayesian one. Certainly it is not one I would construct. All models, all hypotheses are simply constructs which help me predict the future. None are true: some, however, may predict rather better than others. To be fair, in this paper there is no reference to the truth of an hypothesis immediately. We are to be concerned with whether "one statistical hypothesis is better-supported than another by a given body of observations". But within a few pages we find discussion of whether by using the Law of Likelihoods we will "be led to the truth".

I make these comments to indicate that my view of the world is so very different from that of Richard Royall's that I may easily misinterpret and misrepresent him. Add to that my almost complete ignorance of the literature on weight of evidence and likelihood ratios and you will appreciate that I am not the fairest choice of discussant. But that being said, let me begin.

First, a comment on the counter example in Section 2: while I cannot quarrel with the analysis that is presented, it did seem to miss something. The question as I understand it is: does the Law of Likelihood lead to sensible behaviour in weighing the evidence in favour of either a normal pack of cards or a pack of fifty two aces of diamonds. If the Law is to perform well in Royall's small world, it should lead us to the truth. The process of *leading to the truth* would not, I presume, be an instant one, but an incremental one. Rather than discuss the behaviour of the Law on just one piece of evidence —the first, what about the second piece of evidence? Consider that and, in the simplified small world with only two possibilities for the pack of cards, the Law of Likelihood will discover the 'truth'.

The nub of this paper centres on an axiomatic justification of a likelihood ratio based procedure. But the axiomatic system here is a very different kind of beast to that familiar to Bayesians. Look at, say, the axiomatic systems of Savage (1954) or DeGroot (1970) and you will not see a subjective probability or utility appearing in the axioms. Those axiom systems assert qualitative behaviour assumptions about belief and preference relations from which may be deduced the quantitative representation of the expected utility model. Thus, my predisposition if I were to investigate a quantitative measure of weight of evidence, would be to begin with a binary ordering of sets of evidence, discuss the properties that ordering should possess and seek to deduce the form of a quantitative measure consistent with these.

Consider, for instance, the Law of Likelihood in Section 1. I have no real problem with the assertion that "$X = x$ is evidence supporting H_0 over H_1 if and only if $p_0(x) > p_1(x)$". I should prefer to see the assertion in terms of an underlying qualitative probability ordering. Perhaps as: $X = x$ provides greater evidence in favour of H_0 over H_1 if and only if, before the observation, x given H_0 was more likely than x given H_1. Of course, "x given H_0" and "x given H_1" are two entities that need careful definition (see, e.g., Savage or DeGroot *op cit*); but that is to be encouraged. Careful definition invariably clarifies thought. The Law of Likelihood, as stated, then asserts that "$p_0(x)/p_1(x)$ measures the strength of evidence". Again I have no real quarrel with the result: but why does it hold? The difference $p_0(x) - p_1(x)$, might equally well be taken to represent strength of evidence. Well, it might if we didn't know Bayes' Theorem. But the use of Bayes' Theorem is surely justified for us by axioms such as CP introduced by DeGroot, following Villegas. This axiom, stated in behavioural terms, leads to the use of conditional probabilities, Bayes' Theorem and the likelihood ratios. For me the qualitative behavioural arguments discussing rational belief in Savage and DeGroot are the reason I have no quarrel with the Law of Likelihood as stated here.

Naturally Richard Royall may say that he assumed all those arguments in stating the Law of Likelihood. It was not his purpose to rehearse old ground. That I accept. My point is that the qualitative arguments are hidden beneath the quantitative statement of the Law of Likelihood. I happen to know and accept them, but I cannot see them explicitly in the Law's quantitative statement. What qualitative behavioural arguments are hidden beneath the quantitatively stated axioms of Section 4? I cannot see all of those argument nor do I know them.

Remember that I am in a small world neither of my choosing nor my making. The world of Section 4 contains several hypotheses, one of which is true. Our aim is to throw a lasso round a subset of these in such a way that I catch the true one amongst them. Being a Bayesian I cannot help feeling that the size of subset lassoed cannot be ignored — where 'size', of course, is measured in the metric of prior probability. But that feeling I shall try to ignore.

Axioms 0, 1 and 2 seem acceptable. They can be translated into behavioural terms quite easily. Things get interesting in Axiom 3 which essentially states that $h(\cdot,\cdot)$ will involve a ratio. It pretty much determines the form of $h(\cdot,\cdot)$. Yet I cannot express Axiom 3 to myself in persuasive behavioural terms. Note that I cannot appeal to Bayesian justifications of the use of the likelihood ratio between simple hypotheses. Because if I did, the 'size' of the two composite hypotheses, as measured some prior probability, would slip into the argument.

J. O. BERGER (*Purdue University, USA*)

It is recognized today, as virtually a foregone conclusion, that axiomatic developments of statistical inference or decision must either lead to a form of Bayesian analysis or to something that exhibits incoherence. (The argument for this statement is simply that hundreds of axiom systems have been tried, and all seem to end up in essentially the same place.) It is nevertheless useful to consider axiomatic developments from new perspectives, in the hope that discussion can be stimulated in a new community. Thus, in pursuing an axiomatic development of "statistical evidence" in the likelihood domain, nothing sensible that is non–Bayesian can be expected, but developing an axiomatic justification for Bayesianism in this domain would be valuable.

The discussions and examples in the paper are quite interesting, and the difficulty in setting up an axiom system here is highlighted. The source of the incoherency in the axiom system is Axiom 4. This is the axiom that is incompatible with Bayesian analysis. It is interesting to note that this is the axiom that is not explicitly justified in the development.

D. V. LINDLEY (*Somerset, UK*)

Whilst the likelihood principle is unassailable, the law of likelihood, at least under a reasonable interpretation of likelihood, is unsatisfactory for the following reason. It can happen that H_1 is *more* likely than H_2, and K_1 is *more* likely than K_2, yet the composite 'H_1 or H_2' is *less* likely than the composite 'K_1 or K_2'. This is surely ridiculous. Indeed it is one of the standard axioms in the Bayesian philosophy that no form of preference should have this property of reversal under unions. The paradox is essentially the same as Simpson's and depends on the reasonable definition of likelihood for a composite hypothesis as the integrated likelihood; here a weighted average of those for H_1 and H_2. The weights need not be the same for the two composites. For details see my discussion of the paper by Aitkin (1991) and the author's reply. The only satisfactory measures of support are probability-based. Likelihood will not do.

D. PEÑA (*Universidad Carlos III de Madrid, Spain*)

The set of axioms established by the author seem to be designed to lead to a measure of relative support which depends only on the most extreme probabilities. However, it is well known (Lindley, 1991) that the only way to obtain a coherent measure of the support is using the posterior probabilities of both hypothesis. In the finite composites of simple hypothesis case, if H_0 includes the set of values $\theta_{0i}, i = 1, \ldots, n_0$ and H_1 the set $\theta_{1j}, j = 1, \ldots, n_1$, the posterior odds are

$$\frac{P(H_0|\boldsymbol{x})}{P(H_1|\boldsymbol{x})} = \frac{\pi_0}{\pi_1} \cdot \frac{\sum \ell(\boldsymbol{x}|\theta_{0i})\pi_0 i}{\sum \ell(\boldsymbol{x}|\theta_{1j})\pi_1 j}$$

where $\ell(\boldsymbol{x}|\theta)$ is the likelihood, $\pi_0 = \sum \pi_{0i}$ the prior probability of H_0 and $\pi_1 = \sum \pi_{1j}$ those of H_1. An "objective" measure of the relative support for H_0 is the mean Bayes factor for each hypothesis:

$$B_{01} = \frac{n_0^{-1} \sum \ell(\boldsymbol{x}|\theta_{0i})}{n_1^{-1} \sum \ell(\boldsymbol{x}|\theta_{1j})}$$

that takes into account the dimension of each hypothesis, a crucial aspect that is not included in the system of axioms in the paper.

Let us move on to the continuous case, and assume that $H_0 : \theta \in \Omega; H_1 : \theta \in \overline{\Omega}$. Then, the relative support for H_0 against H_1, can be measured by the Bayes factor

$$B_{01} = \frac{\int_{\Omega} \ell(x|\theta) P_0(\theta) d\theta}{\int_{\overline{\Omega}} \ell(x|\theta) P_1(\theta) d\theta}$$

where $P_0(\theta)$ and $P_1(\theta)$ are the prior probabilities. The dependency from the priors can be removed assuming a reference prior. For instance, in the simplest case of the mean of a Normal population $(\theta, 1), H_0 : \theta \leq \theta_0; H_1 : \theta > \theta_0$ with the reference prior $P(\theta) \propto 1$, then it is straightforward to obtain that

$$B_{01} = \frac{\int_{-\infty}^{\theta_0} \exp\left[-\frac{n}{2}(\theta - \overline{x})^2\right] d\theta}{\int_{\theta_0}^{\infty} \exp\left[-\frac{n}{2}(\theta - \overline{x})^2\right] d\theta}$$

and, therefore, the Bayes factor is closely linked to the p-value in classical hypotheses testing. In my experience, scientists looking for an objective analysis in this problem will agree on a prior $N(\theta_0, \sigma_0^2)$ in which σ_0^2 can be roughly elicited asking for the range of possible values for θ. Then the relative support is

$$B_{01} = \frac{\int_{-\infty}^{\theta_0} P(\theta|\boldsymbol{x})d\theta}{\int_{\theta_0}^{\infty} P(\theta|\boldsymbol{x})d\theta}$$

where $P(\theta|\boldsymbol{x})$ is the posterior distribution, and, therefore, the Bayes factor is the ratio of the areas corresponding to the two hypotheses in posterior distributions for the parameters.

In summary, average likelihoods or areas under posterior distributions are a straightforward way to represent relative support for composite hypotheses, and its use avoids the problems associated with extreme likelihoods or posterior bayes factors (see Aitkin, 1991, for additional references on this problem).

REPLY TO THE DISCUSSION

Much of the discipline of statistics is concerned with three questions:
(i) What do these data say?
(ii) What is it reasonable to believe?
(iii) What should I do?

Each question represents an important component of most statistical inference problems. For example, a physician tests his patient for disease D. The test result might be modelled as a realization of a Bernoulli random variable, and if it is positive ($X = 1$) the physican might answer the three questions as follows:
(a) This result is evidence that this patient has D.
(b) This patient probably does not have D.
(c) This patient should be treated for D.

How can we determine which, if any, of these conclusions are valid? If the test has reasonable sensitivity and specificity ($\Pr(X = 1|D)$ and $\Pr(X = 0|\text{not}\, D)$, respectively) then it is clear that all three might be correct. Whether (b) is correct or not depends not only on the sensitivity and specificity but also on the probability of D a priori, which is determined by the prevalence of D in the population, the presence of other symptoms, etc. And even if (b) is correct, (c) might also be correct, depending on the consequences of treating when D is absent, not treating when D is present, etc. Thus to answer question (ii) it is clear that we must not only evaluate and interpret the new evidence ($X = 1$), but also combine that evidence with what we knew beforehand. And to answer (iii) we must specify as well what our options are and what costs are associated with the various possible actions when D is present and when it is not. These intuitive judgements are validated by the axiomatic developments of rules for coherent beliefs and for consistent behavior to which Simon French and Jim Berger refer.

But the present paper is concerned with conclusion (a) and question (i), which relate to a different area of statistics, namely what Birnbaum called "evidential interpretation" of data. It is often this question, not (ii) or (iii) that brings the scientist to the statistician's door. (It is the popular misconception that frequentists' significance tests address question (i) that accounts for those tests' continued prominence in statistical applications.) In contrast to the essentially Bayesian conclusions, (b) and (c), the validity of (a) depends on nothing more than the sensitivity and specificity of the diagnostic test; it is entirely independent of beliefs about D as well as of treatment options and costs. The assertion that (a) is valid, the Law of Likelihood, does not require a system of coherent opinions or preferences, although it is entirely consistent with such a system. Likelihoods do not measure degrees of belief, and degrees of belief do not measure the strength of statistical evidence; different axiom systems should be expected.

Of course I agree with Simon French that none of the probability models that we find mathematically tractable is true. Yet unlike him I find it entirely reasonable to test statistical procedures by asking "If H *were* true, then how would this procedure perform?" The literal impossibility of an infinite sequence of independent random variables does not preclude my interest in the law of large numbers or the central limit theorem.

I'm afraid that I have managed to mislead Simon French in Section 4, where the aim is *not* "to throw a lasso round a subset [of hypotheses] in such a way that I catch the true one amongst them". That is the aim of confidence interval procedures. Our aim is to interpret a given body as data as evidence relating to two specific hypotheses. Although the 'size' of a hypothesis obviously affects its plausibility or its probability of being true, it does not neccessarily affect the degree of support by data. For example in Axiom 0 the relative support, p_0/p_1, is the Bayes factor (the factor by which the data alter the relative probabilities of the two hypotheses), which is quite independent of the 'sizes', I and J, of the hypotheses.

Dennis Lindley and Daniel Peña suggest that Bayes factors are the (unique) right way to measure relative support, with Lindley further suggesting that Simpson's paradox reveals a fatal flaw in the Law of Likelihood. Let's take a closer look: The Law of Likelihood states that an observation $X = x$ is evidence supporting a simple hypothesis $H_1 : X \sim f_1$ over $H_2 : X \sim f_2$ if $f_1(x) > f_2(x)$. It supports $K_1 : X \sim g_1$ over $K_2 : X \sim g_2$ if $g_1(x) > g_2(x)$. But if $f_1(x) > f_2(x) > g_1(x) > g_2(x)$ then neither of the two composite hypotheses $C_1 : X \sim f_1$ or g_1 and $C_2 : X \sim f_2$ or g_2 implies a probability for the observation, so the Law of Likelihood makes no statement about the evidence for C_1 vis-à-vis C_2.

But suppose that as the discussants suggest we add to the problem prior probabilities $P(H_1|\, C_1) = p_1$ and $P(H_2|C_2) = p_2$. Now we have two new hypotheses, C_1^* and C_2^*. (C_1^* asserts that with probability $p_1\, X \sim f_1$ and with probability $(1 - p_1)\, X \sim g_1$, while C_2^* asserts that with probability $p_2\, X \sim f_2$, etc.) Although the Law of Likelihood makes no statement about the evidence for C_1 vis-à-vis C_2, the same is not true about the new hypotheses C_1^* and C_2^*; since

$$P(X = x|C_1^*) = p_i f_i(x) + (1 - p_i)g_i(x)$$

the relative support is given by the likelihood ratio, $P(X = x|C_1^*)/P(X = x|C_2^*)$, which can be less than unity if p_1 is small and p_2 large.

Thus within this (Bayesian) model it can be true that the effect of the observation $X = x$ is to increase the relative probability of H_1 (vs. H_2) and of K_1 (vs. K_2) while decreasing the relative probability of their weighted union C_1^* (vs. the differently weighted union C_2^*). This possibility no more justifies the conclusion that "Likelihood will not do" than the conclusion that "Probability will not do".

The weighted average of the likelihoods for H_1 and K_1, $p_1 f_1(x) + (1 - p_1)g_1(x)$, is the likelihood for the hypothesis C_1^* asserting that the weights, p_1 and $(1 - p_1)$, are the probabilities of H_1 and K_1 respectively. It is not the correct likelihood for a hypothesis C_1^{**} that specifies different probabilities for H_1 and K_1, and it is not a "reasonable definition of likelihood" for the hypothesis C_1 that says nothing about those probabilities.

ADDITIONAL REFERENCES IN THE DISCUSSION

Aitkin, M. (1991). Posterior Bayes factors. *J. Roy. Statist. Soc. B* **53**, 111–142, (with discussion).
DeGroot, M. H. (1970). *Optimal Statistical Decisions*, New York: McGraw-Hill.
Lindley, D. V. (1991). Discussion of M. Aitkin. *J. Roy. Statist. Soc. B* **53**, 130–131.
Savage, L. J. (1954). *The Foundations of Statistics*. New York: Wiley.

BAYESIAN STATISTICS 4, pp. 419–434
J. M. Bernardo, J. O. Berger, A. P. Dawid and A. F. M. Smith, (Eds.)

Bayesian Analysis of Linear Models*

MARK J. SCHERVISH
Carnegie Mellon University, USA

SUMMARY

We consider Bayesian methods for answering common questions in univariate and multivariate linear models. The types of questions we consider are all of the form "How far is some linear function of the parameters away from some specified value?" We construct random variables whose posterior distributions are useful for answering such questions. We compare the answers to those which classical statisticians try to derive from the usual F test. Of particular concern is the confounding of sample size with effect size, which the F test is unable to unravel.

Keywords: MULTIVARIATE LINEAR MODEL; F-TEST; ALTERNATE NONCENTRAL CHI-SQUARED; ALTERNATE NONCENTRAL BETA.

1. INTRODUCTION

Consider the well studied linear model with parameter $\Theta = (\boldsymbol{B}, \Sigma)$ with $\boldsymbol{B}$ being a vector of dimension p and Σ being a positive definite matrix. The conditional distribution of a vector $\boldsymbol{Y}$ given $\Theta = (\boldsymbol{\beta}, \sigma)$ is n-variate normal $N_n(\boldsymbol{x\beta}, \sigma^2 \boldsymbol{v})$ where $\boldsymbol{x}$ is some fixed, known design matrix and $\boldsymbol{v}$ is a known weight matrix. This model includes the usual analysis of variance and regression models as special cases. A common inferential procedure is to test a hypothesis of the form $H : \boldsymbol{a}_0 \boldsymbol{B} = \psi_0$ for some full rank matrix $\boldsymbol{a}_0$ and vector ψ_0. The purpose of this paper is to find posterior distributions for random variables which can be used to make inferences in a Bayesian framework that parallel and improve upon the usual hypothesis tests. Existing work in this area includes that of Dickey (1974) and Leonard (1974) as well as the pioneering efforts of Lindley and Smith (1972) and of Smith (1973). The focus of this paper is substantially different from that of Leonard (1974) who takes a primarily estimative approach to the problem. Dickey (1974), as well as considering the estimation problem, attacks the problem of testing hypotheses from the Bayesian point of view. The present work differs from Dickey (1974) by not giving any special status to the sharp null hypothesis H in the prior and loss function, and by considering instead the extent of the deviation from H.

We consider natural conjugate prior distributions for Θ of the form $\boldsymbol{B}$ given $\Sigma = \sigma$ has distribution $N_p(\boldsymbol{\beta}_0, \sigma^2 \boldsymbol{w}_0^{-1})$, where $\boldsymbol{\beta}_0$ and $\boldsymbol{w}_0$ are known. We give $1/\Sigma^2$ a gamma distribution, which is equivalent to Σ^2 having inverse gamma distribution, in particular $\Gamma^{-1}(a_0/2, b_0/2)$, where a_0 and b_0 are known positive constants. This prior is useful when the coordinates of $\boldsymbol{B}$ are a fixed selection of effects about which we wish to make inference. Because the effects $\boldsymbol{B}$ are always random in the Bayesian framework, the usual terms "random effects" and "fixed effects" models are inappropriate. For this reason we refer to the case of this paper as the *fixed selection* case. A different case was studied by Dickey (1974)

* This work was supported in part by National Science Foundation Grant DMS 88-05676 and by Office of Naval Research Contract N00014-91-J-1024.

in which the effects were considered as a sample from a large population of interest. We would call that case the *random selection* case.

In Section 2, we construct random variables which can be used to assess the degree of departure from a hypothesis. We also derive their posterior distributions. Particular attention is focused on the problem of separating the effect size from the sample size, which the usual F test is unable to do. In Section 4, we give an example of the use of the posterior distributions in the fixed selection model. Computer evaluation of the posterior probabilities described in Section 2 is discussed in Appendix A.

The purpose of this paper is to derive useful posterior distributions rather than to discuss a number of decision problems. The random variables, whose posterior distributions we find, resemble the test statistics used in the classical F tests. The advantage to the Bayesian approach is that it allows experimenters to make statements like "the probability is 0.03 that the means differ by more than 6", which are easier for them to understand than classical p values. The properties of the random variables whose posterior distributions we derive, are described in Section 3. In Section 5, we present some extentions to multivariate linear models.

Although the methods are fairly straightforward and the results are not surprising, it would appear that they help to facilitate understanding of how the Bayesian approach can improve upon Classical methods while maintaining a familiar setting.

2. THE FIXED SELECTION MODEL

The main results of this paper pertain to the fixed selection model which we consider next. In this model, the effects $\boldsymbol{B}$ are of interest in their own right and a natural conjugate prior might be appropriate. Such a prior says that conditional on $\Sigma = \sigma$, $\boldsymbol{B} \sim N_p(\boldsymbol{\beta}_0, \sigma^2 \boldsymbol{w}_0^{-1})$ and $1/\Sigma^2 \sim \Gamma(a_0/2, b_0/2)$. Here $\boldsymbol{w}_0^{-1}$ is a fixed positive definite matrix. This family of priors includes hierarchical priors like those of Lindley and Smith (1972) as long as the covariance matrices at each stage are known multiples of Σ^2 and the design matrices are known at each stage. The classical F test of $H : \boldsymbol{a}_0\boldsymbol{B} = \boldsymbol{\psi}_0$ involves the squared, estimated Mahalanobis distance between $\boldsymbol{a}_0\boldsymbol{B}$ and $\boldsymbol{\psi}_0$, namely

$$\frac{(\boldsymbol{a}_0\hat{\boldsymbol{\beta}} - \boldsymbol{\psi}_0)^T(\boldsymbol{a}_0(\boldsymbol{x}^T\boldsymbol{v}^{-1}\boldsymbol{x})^{-1}\boldsymbol{a}_0^T)^{-1}(\boldsymbol{a}_0\hat{\boldsymbol{\beta}} - \boldsymbol{\psi}_0)}{RSS/(n-p)},$$

where $\hat{\boldsymbol{\beta}}$ is the weighted least squares estimate of $\boldsymbol{B}$, and RSS is the residual sum of squares:

$$\hat{\boldsymbol{\beta}} = (\boldsymbol{x}^T\boldsymbol{v}^{-1}\boldsymbol{x})^{-1}\boldsymbol{x}^T\boldsymbol{v}^{-1}\boldsymbol{y}$$
$$RSS = (\boldsymbol{y} - \boldsymbol{x}\hat{\boldsymbol{\beta}})^T\boldsymbol{v}^{-1}(\boldsymbol{y} - \boldsymbol{x}\hat{\boldsymbol{\beta}}).$$

In the Bayesian framework, the posterior distribution of $\boldsymbol{B}$ given $\Sigma = \sigma$ and $\boldsymbol{Y} = \boldsymbol{y}$ is $N_P(\boldsymbol{\beta}_1, \sigma^2\boldsymbol{w}_1^{-1})$ where

$$\boldsymbol{w}_1 = \boldsymbol{w}_0 + \boldsymbol{x}^T\boldsymbol{v}^{-1}\boldsymbol{x}$$
$$\boldsymbol{\beta}_1 = \boldsymbol{w}_1^{-1}(\boldsymbol{w}_0\boldsymbol{\beta}_0 + \boldsymbol{x}^T\boldsymbol{v}^{-1}\boldsymbol{x}\hat{\boldsymbol{\beta}}).$$

Conditional on Σ, the squared Mahalanobis distance between $\boldsymbol{a}_0\boldsymbol{B}$ and $\boldsymbol{\psi}_0$ is

$$(\boldsymbol{a}_0\boldsymbol{B} - \boldsymbol{\psi}_0)^T(\boldsymbol{a}_0\boldsymbol{w}_1^{-1}\boldsymbol{a}_0^T)^{-1}(\boldsymbol{a}_0\boldsymbol{B} - \boldsymbol{\psi}_0)/\Sigma^2. \tag{1}$$

The expression in (1) as it stands is not a suitable measure of the deviation between $\boldsymbol{B}$ and the hypothesis for one important reason. Imagine that enough additional observations become available so that $\boldsymbol{w}_1$ changes to $2\boldsymbol{w}_1$. According to (1), the squared distance between $\boldsymbol{a}_0\boldsymbol{B}$ and $\boldsymbol{\psi}_0$ becomes twice as large even when $\boldsymbol{B}$ and $\boldsymbol{\psi}_0$ remain unchanged. It is desirable that the posterior distribution of the distance between $\boldsymbol{a}_0\boldsymbol{B}$ and $\boldsymbol{\psi}_0$ become more precise as $\boldsymbol{w}_1$ increases, but not that the actual distance become larger.

To overcome this difficulty, we consider a slightly modified measure of deviation of $\boldsymbol{B}$ from the hypothesis. Let $\Omega_H = \{\boldsymbol{\beta} : \boldsymbol{a}_0\boldsymbol{\beta} = \boldsymbol{\psi}_0\}$. Since the hypothesis is $H : \boldsymbol{a}_0\boldsymbol{B} = \boldsymbol{\psi}_0$, it stands to reason that there must exist p–dimensional vectors $\boldsymbol{\beta}$ such that $\boldsymbol{a}_0\boldsymbol{\beta} = \boldsymbol{\psi}_0$, otherwise the hypothesis makes no sense. Define

$$\rho(\boldsymbol{B}, \Omega_H) = \frac{1}{\ell}(\boldsymbol{a}_0\boldsymbol{B} - \boldsymbol{\psi}_0)^T(\boldsymbol{a}_0\boldsymbol{w}_1^{-1}\boldsymbol{a}_0^T)^{-1}(\boldsymbol{a}_0\boldsymbol{B} - \boldsymbol{\psi}_0), \tag{2}$$

where

$$\ell = \text{trace}[\boldsymbol{a}_0^T(\boldsymbol{a}_0\boldsymbol{w}_1^{-1}\boldsymbol{a}_0^T)^{-1}\boldsymbol{a}_0].$$

We will see in Section 3 that the random variable $\rho(\boldsymbol{B}, \Omega_H)$ can be interpreted as a coordinatewise weighted average squared distance between $\boldsymbol{B}$ and Ω_H.

The posterior distribution of $\rho(\boldsymbol{B}, \Omega_H)$ will be calculated in order to assess the magnitude of the deviation of $\boldsymbol{a}_0\boldsymbol{B}$ from $\boldsymbol{\psi}_0$. To parallel the classical F test, the posterior distribution of ρ/Σ^2 will also be calculated in order to assess how large the distance between $\boldsymbol{a}_0\boldsymbol{B}$ and $\boldsymbol{\psi}_0$ is relative to the error variance of the original observations.

Let the random variable in (1) equal H/Σ^2. Then $\rho(\boldsymbol{B}, \Omega_H) = \mathrm{H}/\ell$, and $\rho/\Sigma^2 = \mathrm{H}/(\ell\Sigma^2)$. Since ℓ is a constant, it suffices to find the posterior distributions of H and H/Σ^2. For ease of notation, let $\mathrm{T} = \Sigma^{-2}$ and consider $\Xi = \mathrm{H}/\Sigma^2 = \mathrm{HT}$ whose conditional distribution given $\mathrm{T} = \tau$ is non-central χ^2 with $q = \text{rank}[\boldsymbol{a}_0]$ degrees of freedom and squared non-centrality parameter $\tau\gamma$, where

$$\gamma = (\boldsymbol{a}_0\boldsymbol{\beta}_1 - \boldsymbol{\psi}_0)^T(\boldsymbol{a}_0\boldsymbol{w}_1^{-1}\boldsymbol{a}_0^T)^{-1}(\boldsymbol{a}_0\boldsymbol{\beta}_1 - \boldsymbol{\psi}_0).$$

(See Lehmann 1959, p. 312). The marginal posterior distribution of $\mathrm{T}\gamma$ is $\Gamma(a_1/2, b_1/[2\gamma])$ with $a_1 = a_0 + n$ and $b_1 = b_0 + RSS + (\boldsymbol{\beta}_0 - \boldsymbol{b})^T(\boldsymbol{w}_0^{-1} + (\boldsymbol{x}^T\boldsymbol{v}^{-1}\boldsymbol{x})^{-1})^{-1}(\boldsymbol{\beta}_0 - \hat{\boldsymbol{\beta}})$.

Theorem 1. *If the distribution of Ξ is non-central χ^2 with q degrees of freedom and (squared) non-centrality parameter λ given $\Lambda = \lambda$ and Λ has $\Gamma(a/2, b/2)$ distribution, then the density of Ξ is*

$$f_\Xi(\xi) = \sum_{i=0}^{\infty} \frac{\Gamma(.5a + i)}{i!\Gamma(.5a)}(1 - g)^{.5a}g^i \frac{(.5)^{.5q+i}}{\Gamma(.5q + i)}\xi^{.5q+i-1}\exp(-.5\xi),$$

where $g = 1/(1 + b)$. The mean of Ξ is

$$\mathrm{E}(\Xi) = q + \frac{a}{b} = q + \frac{ag}{1 - g}.$$

Proof. The conditional density of Ξ given $\Lambda = \lambda$ is

$$f_{\Xi|\Lambda}(\xi|\lambda) = \sum_{i=0}^{\infty} \exp(-.5\lambda)\frac{(.5\lambda)^i}{i!}\frac{(.5)^{.5q+i}}{\Gamma(.5q + i)}\xi^{.5q+i-1}\exp(-.5\xi).$$

The marginal density of Λ is

$$f_\Lambda(\lambda) = \frac{(.5b)^{.5a}}{\Gamma(.5a)}\lambda^{.5a-1}\exp(-.5b\lambda).$$

The joint density is

$$f_{\Xi,\Lambda}(\xi,\lambda)=\sum_{i=0}^{\infty}\frac{(.5)^{.5q+2i}(.5b)^{.5a}}{i!\Gamma(.5q+i)\Gamma(.5a)}\lambda^{i+.5a-1}\exp(-.5\lambda[b+1])\xi^{.5q+i-1}\exp(-.5\xi).$$

Integrating out λ gives the marginal density of Ξ

$$f_{\Xi}(\xi)=\sum_{i=0}^{\infty}\frac{b^{.5a}\Gamma(.5a+i)}{i!\Gamma(.5a)(1+b)^{.5a+i}}\frac{(.5)^{.5q+i}}{\Gamma(.5q+i)}\xi^{.5q+i-1}\exp(-.5\xi).$$

Plugging in the formula for g completes the proof of the density. The mean of Ξ given $\Lambda=\lambda$ is $q+\lambda$, and the mean of Λ is a/b. $\triangleleft$

The distribution in Theorem 1 was first derived by Geisser (1967). Schervish (1987) named it the *alternate non-central* χ^2 *distribution*, abbreviated ANC$\chi^2(q,a,g)$ due to the resemblance which it bears to the non-central χ^2 distribution. It was also derived by Rouanet and Lecoutre (1983) and called $L^2_{q,a}(a/b)$.

To apply Theorem 1, let

$$\begin{aligned} a &= a_1 \\ b &= b_1/\gamma \\ q &= q. \end{aligned}$$

It follows that the distribution of $\Xi=\mathrm{H}/\Sigma^2$ is $ANC\chi^2(q,a_1,g)$, where $g=\gamma/(\gamma+b_1)$. The mean of Ξ is $q+a_1\gamma/b_1$. In the special case of an improper prior with $a_0=-p$, $b_0=0$, and with $\boldsymbol{\beta}_0$ and $\boldsymbol{w}_0$ all 0s, we get $a_1=n-p$, $b_1=RSS$, $\boldsymbol{\beta}_1=\hat{\boldsymbol{\beta}}$, and $\boldsymbol{w}_1=\boldsymbol{x}^T\boldsymbol{v}^{-1}\boldsymbol{x}$. So the mean of Ξ/q is

$$1+\frac{a_1\gamma}{qb_1}=1+F,$$

where F is the usual F statistic for testing H. Just as the F statistic is an inflated measure of distance between $\boldsymbol{B}$ and $\boldsymbol{\psi}_0$, so too are Ξ and H. The factor $1/\ell$ adjusts H to separate the size of the deviation from the amount of information (sample size) available.

The distribution of $\mathrm{H}=\Xi/\mathrm{T}$ can be found by a similar method. The conditional distribution of HT given $\mathrm{T}=\tau$ is non-central χ^2 with q degrees of freedom and squared non-centrality parameter $\tau\gamma$.

Theorem 2. *If the distribution of* $\mathrm{H}\Lambda$ *is non-central* χ^2 *with* q *degrees of freedom and (squared) non-centrality parameter* $c\lambda$ *given* $\Lambda=\lambda$ *and* $\Lambda\sim\Gamma(a/2,b/2)$, *then the density of*

$$\Gamma=\frac{\mathrm{H}}{\mathrm{H}+b+c}$$

is

$$f_{\Gamma}(\gamma)=\sum_{i=0}^{\infty}\frac{\Gamma(.5a+i)}{i!\Gamma(.5a)}(1-g)^{.5a}g^{i}\frac{\Gamma(.5a+.5q+2i)}{\Gamma(.5q+i)\Gamma(.5a+i)}\gamma^{.5q+i-1}(1-\gamma)^{.5a+i-1},$$

where $g=c/(b+c)$.

Proof. The conditional density of H given $\Lambda=\lambda$ is

$$f_{\mathrm{H}|\Lambda}(\eta|\lambda)=\sum_{i=0}^{\infty}\exp(-.5c\lambda)\frac{(.5c\lambda)^{i}}{i!}\frac{\lambda^{.5q+i}}{2^{.5q+i}\Gamma(.5q+i)}\eta^{.5q+i-1}\exp(-.5\eta\lambda).$$

The marginal density of Λ is

$$f_{\Lambda}(\lambda) = \frac{(.5b)^{.5a}}{\Gamma(.5a)} \lambda^{.5a-1} \exp(-.5b\lambda).$$

The joint density is

$$f_{\mathrm{H},\Lambda}(\eta, \lambda) =$$

$$= \sum_{i=0}^{\infty} \frac{(.5c)^i (.5b)^{a/2}}{i!} \frac{1}{2^{2q+i}\Gamma(.5q+i)\Gamma(a/2)} \lambda^{.5a+.5q+2i-1} \eta^{.5q+i-1} \exp(-.5\lambda[b+c+\eta]).$$

Integrating λ out of this gives the density of H

$$f_{\mathrm{H}}(\eta) = \sum_{i=0}^{\infty} \frac{c^i b^{a/2}}{i!(b+c+\eta)^{.5a+.5q+2i}} \frac{\Gamma(.5a+.5q+2i)}{\Gamma(.5q+i)\Gamma(.5a)} \eta^{.5q+i-1}.$$

Now, make the change of variables from η to $\gamma = \eta/(\eta + b + c)$. The inverse is $\eta = (b+c)\gamma/(1-\gamma)$. The derivative is $(b+c)/(1-\gamma)^2$. The density of Γ is

$$f_{\Gamma}(\gamma) = \sum_{i=0}^{\infty} \frac{b^{.5a} c^i}{(b+c)^{.5a+i} i!} \frac{\Gamma(.5a+.5q+2i)}{\Gamma(.5q+i)\Gamma(.5a)} \gamma^{.5q+i-1} (1-\gamma)^{.5a+i-1}.$$

Setting $g = c/(b+c)$ and rearranging Γ function values produces the desired result. ◁

The distribution of Γ in Theorem 2 is called the *Alternate non-central Beta distribution* with parameters (q, a, g), and is denoted $ANCB(q, a, g)$. Notice that its density is a weighted average of beta densities with the same negative binomial weights as used in the $ANC\chi^2$ density. In contrast to the non-central Beta density of Lehmann (1959, p. 312) both the α and β parameters for the individual terms increase as q increases. In analogy to the relation between beta and F random variables, we can transform to $F = a\mathrm{H}/[q(b+c)]$, which would have *alternate non-central F distribution* denoted $ANCF(q, a, g)$ with density

$$f_F(x) =$$

$$\sum_{i=0}^{\infty} \frac{\Gamma(.5a+i)}{i!\Gamma(.5a)} (1-g)^{.5a} g^i \frac{\Gamma(.5a+.5q+2i)\left(\frac{q}{a}\right)^{.5q+i}}{\Gamma(.5a+i)\Gamma(.5q+i)} x^{.5q+i-1} \left(1+\frac{q}{a}x\right)^{-(.5a+.5q+2i)}.$$

The mean of the $ANCF(q, a, g)$ distribution is

$$\frac{a}{q} g + \frac{a}{a-2}(1-g).$$

The mean of H is $c + bq/(a-2)$. The $ANCF$ distribution, with a different scaling, was called ψ^2 by Rouanet and Lecoutre (1983). (See also Lecoutre, 1985.)

To apply Theorem 2, we set

$$\begin{aligned} a &= a_1 \\ b &= b_1 \\ q &= q \\ c &= \gamma. \end{aligned}$$

It follows that the distribution of $a_1\mathrm{H}/[q(b_1+\gamma)]$ is $ANCF(q, a_1, g)$ where $g = \gamma/(\gamma + b_1)$. Note that the three hyperparameters of the $ANCF$ distribution of a multiple of H are the same as the hyperparameters of the $ANC\chi^2$ distribution of Ξ. An example of the use of these distributions is given in Section 4. The rationale for the rescaling factor ℓ will be given in Section 4.

3. PROPERTIES OF THE DEVIATION MEASURE

It is often possible to write $\Omega_H = \{\beta : b_0\beta = \phi_0\}$ for some other matrix b_0 and vector ϕ_0. It is easy to show that these two sets are the same if and only if there exists a $q \times q$ non-singular matrix c such that $b_0 = ca_0$ and $\phi_0 = c\psi_0$. Let $g = a_0^T(a_0 w_1^{-1} a_0^T)^{-1} a_0$. Then

$$\rho(B, \Omega_H) = \frac{(B - v)^T g (B - v)}{\operatorname{trace} g},$$

for every $v \in \Omega_H$. In addition, if $b_0 = ca_0$ with c non-singular, then $g = b_0^T(b_0 w_1^{-1} b_0^T)^{-1} b_0$. It follows that ρ is identical for all equivalent representations of the hypothesis.

Next, consider a simple case in which w_1 is a diagonal matrix with λ_i in the ith diagonal entry. This would mean that the coordinates of B are conditionally independent given $\Sigma = \sigma$ with variances σ^2/λ_i. Now, suppose that the matrix a_0 is $p - 1 \times p$ with 1 in each (i, i) position, -1 in each (p, i) position and 0 everywhere else. Let ψ_0 be a vector of all 0s. It is easy to see that the ith diagonal entry of $(a_0 w_1^{-1} a_0^T)^{-1}$ is $\lambda_i - \lambda_i^2 / \sum_{j=1}^p \lambda_j$. It follows that

$$\begin{aligned}
(a_0 B - \psi_0)^T (a_0 w_1^{-1} a_0^T)^{-1} (a_0 B - \psi_0) = \\
&= \sum_{i=1}^{p-1} \lambda_i (\mathrm{B}_i - \mathrm{B}_p)^2 - \frac{1}{\sum_{j=1}^p \lambda_j} \left[\sum_{i-1}^{p-1} \lambda_i (\mathrm{B}_i - \mathrm{B}_p) \right]^2 \\
&= \sum_{i=1}^{p} \lambda_i \mathrm{B}_i^2 - \frac{1}{\sum_{j=1}^p \lambda_j} \left(\sum_{i=1}^{p} \mathrm{B}_i \right)^2 \\
&= \sum_{i=1}^{p} \lambda_i (\mathrm{B}_i - \overline{\mathrm{B}})^2 \\
&= \frac{1}{2 \sum_{j=1}^p \lambda_j} \sum_{i=1}^{p} \sum_{j=1}^{p} \lambda_i \lambda_j [(\mathrm{B}_i - \overline{\mathrm{B}}) - (\mathrm{B}_j - \overline{\mathrm{B}})]^2 \\
&= \frac{1}{2 \sum_{j=1}^p \lambda_j} \sum_{i=1}^{p} \sum_{j=1}^{p} \lambda_i \lambda_j (\mathrm{B}_i - \mathrm{B}_j)^2.
\end{aligned}$$

If we want to find the weighted average of all squared differences $(\mathrm{B}_i - \mathrm{B}_j)$ for $i \neq j$, we must divide the last sum above by the sum of $\lambda_i \lambda_j$ for all $i \neq j$, which equals $(\sum_{i=1}^p \lambda_i)^2 - \sum_{i=1}^p \lambda_i^2$. It is not difficult to show that

$$\operatorname{trace} a_0^T (a_0 w_1^{-1} a_0^T)^{-1} a_0 = \sum_{i=1}^{p} \lambda_i - \frac{\sum_{i=1}^p \lambda_i^2}{\sum_{i=1}^p \lambda_i}.$$

It follows that, in this case, $2\rho(B, \Omega_H)$ equals the weighted average of all squared differences between the coordinates of B.

In more general cases, let $\Omega'_H = \{\delta : a_0 w_1^{-1/2} \delta = \psi_0\}$. That is, $\delta \in \Omega'_H$ if and only if $w_1^{-1/2}\delta \in \Omega_H$. In other words, Ω'_H expresses the hypothesis in terms of the transform of B with independent (conditional on Σ) and identical variance coordinates. Then, note that $h = w_1^{-1/2} g w_1^{-1/2}$ is the projection matrix into the linear space parallel to Ω'_H. For each $v \in \Omega_H$

$$h w_1^{1/2}(B - v) = w_1^{-1/2} g (B - v).$$

It follows that the squared norm of the projection of $\boldsymbol{w}_1^{1/2}(\boldsymbol{B}-\boldsymbol{v})$ into the linear space parallel to Ω'_H is $(\boldsymbol{B}-\boldsymbol{v})^T\boldsymbol{g}(\boldsymbol{B}-\boldsymbol{v}) = \mathrm{H}$. The squared norm before projection is $(\boldsymbol{B}-\boldsymbol{v})^T\boldsymbol{w}_1(\boldsymbol{B}-\boldsymbol{v})$. If we divide this last by trace$\boldsymbol{w}_1$, we would obtain the weighted average of the squares of the principal components with the weights being the eigenvalues. Dividing H by trace$\boldsymbol{g}$ gives a corresponding weighted average for the projection.

We conclude this section with a note on the relationship between the non-central χ^2 distribution and the $ANC\chi^2$ distribution. Specifically consider what happens to the weights

$$h_i(a,g) = \frac{\Gamma(.5a+i)}{i!\Gamma(.5a)}(1-g)^{.5a}g^i$$

$$p_i(x) = \left(\frac{x}{2}\right)^i \frac{\exp(-.5x)}{i!},$$

where $x = \gamma\tau$ as $a \to \infty$ and $ag/(1-g) \to x$. It is straightforward to see that the weights $h_i(a,g)$ converge to $p_i(x)$. For the example to be discussed in Section 4, $a = 64$, and $g = .2270$. Table 1 gives the values of $h_i(64, .2270)$ and $p_i(18.79)$ for a few values of k in this example to show the extent of the difference.

i	$p_i(18.79)$	$h_i(64, .2270)$
0	0.0000	0.0003
1	0.0008	0.0019
2	0.0037	0.0072
3	0.0115	0.0185
4	0.0270	0.0367
5	0.0507	0.0600
6	0.0794	0.0840
7	0.1066	0.1035
8	0.1252	0.1145
9	0.1307	0.1156
10	0.1227	0.1076
11	0.1048	0.0932
12	0.0821	0.0758
13	0.0593	0.0583
14	0.0398	0.0425
15	0.0249	0.0296

Table 1. *Weights for non-central χ^2 and $ANC\chi^2$ for Example of Section 4.*

4. AN EXAMPLE

An experiment was conducted in which psychological tests were given to three groups of infants. The test was the Bailey scale of psychomotor development of Bailey (1965). One of the groups of infants lived under experimental conditions while the other two were control groups (for more details see Schervish and Elmer 1983). If we denote the mean test score in the experimental group as M_1 and the means of the control groups as M_2 and M_3, then the differences between the Ms were of interest to the experimenters. After consultation with the experimenters, the fixed selection model with the following parameters and hyperparameters was used:

$$\boldsymbol{B} = \begin{pmatrix} \mathrm{M}_1 \\ \mathrm{M}_2 \\ \mathrm{M}_3 \end{pmatrix}$$
$$\boldsymbol{v} = I$$
$$\boldsymbol{\beta}_0 = \begin{pmatrix} 88 \\ 97 \\ 100 \end{pmatrix}$$
$$\boldsymbol{w}_0^{-1} = 0.5I + 0.5J$$

where J is a matrix full of ones, $a_0 = 4$ and $b_0 = 1024$. Based on sample sizes of twenty from each group, the parameters of the posterior distribution were:

$$\boldsymbol{\beta}_1 = \begin{pmatrix} 95.40 \\ 98.95 \\ 110.92 \end{pmatrix}$$
$$\boldsymbol{w}_1 = 22I - 0.5J$$
$$a_1 = 64$$
$$b_1 = 9907.55.$$

The hypothesis is $H : \boldsymbol{a}_0\boldsymbol{B} = \boldsymbol{\psi}_0$ with

$$\boldsymbol{a}_0 = \begin{pmatrix} -1 & 1 & 0 \\ -1 & 0 & 1 \end{pmatrix}$$
$$\boldsymbol{\psi}_0 = \begin{pmatrix} 0 \\ 0 \end{pmatrix}$$

With this we calculate

$$\boldsymbol{a}_0^T(\boldsymbol{a}_0\boldsymbol{w}_1^{-1}\boldsymbol{a}_0^T)^{-1}\boldsymbol{a}_0 = \begin{pmatrix} 14.6667 & -7.3333 & -7.3333 \\ -7.3333 & 14.6667 & -7.3333 \\ -7.3333 & -7.3333 & 14.6667 \end{pmatrix}.$$

It follows that $\ell = 44$ and $\gamma = 2909.5$. The experimenters were interested in detecting differences of 10 points or more. In Section 3, we found that 2ρ could be interpreted as the weighted average squared distance between the coordinates of $\boldsymbol{B}$, at least when $\boldsymbol{w}_1$ is diagonal. Because $\boldsymbol{w}_1^{-1}$ is of the form $aI + bJ$, and $\boldsymbol{a}_0 J$ is all 0s, we can continue to use that interpretation here. Also, $\mathrm{H} = \ell\rho(\boldsymbol{B}, \Omega_H)$. (The hyperparameters of the $ANC\chi^2$ and $ANCF$ distributions are $(2, 64, .2270)$.) So we calculate

$$\Pr[2\rho(\boldsymbol{B}, \Omega_H) < 100] = \Pr\left(\frac{64\mathrm{H}}{2(9907.55 + 2909.5)} < 5.493\right)$$
$$= .2453$$
$$\Pr\left(\frac{2\rho(\boldsymbol{B}, \Omega_H)}{\Sigma^2} < 100\mathrm{E}\left[\frac{1}{\Sigma^2}\right]\right) = \Pr\left(\Xi < 50\frac{64}{9907.55}44\right)$$
$$= .2642.$$

Since 2ρ is an average squared deviation, $2\rho < 100$ means the average deviation (root mean square) is less than 10. It follows that there is about a 25% chance that the average deviation of the control group scores from the experimental group score is less than 10 points. Which probabilities one should calculate would be determined by the sort of inference one wished to make. The experimenters in this case were mainly interested in a description of the differences between the groups. Rouanet and Lecoutre (1983) suggest looking at the cumulative distribution function (CDF) of Ξ or H. We would prefer to use $\sqrt{2\rho}$, since it adjusts for the sample size and is in the original scale of measurement. Figure 1 has a plot of the CDF of $\sqrt{2\rho}$ for the problem just described. Note that one can extract all of the desired probabilistic information one desires concerning the distribution of ρ from this plot, including the fact that there is non-negligible probability that $\sqrt{\rho} < 10$.

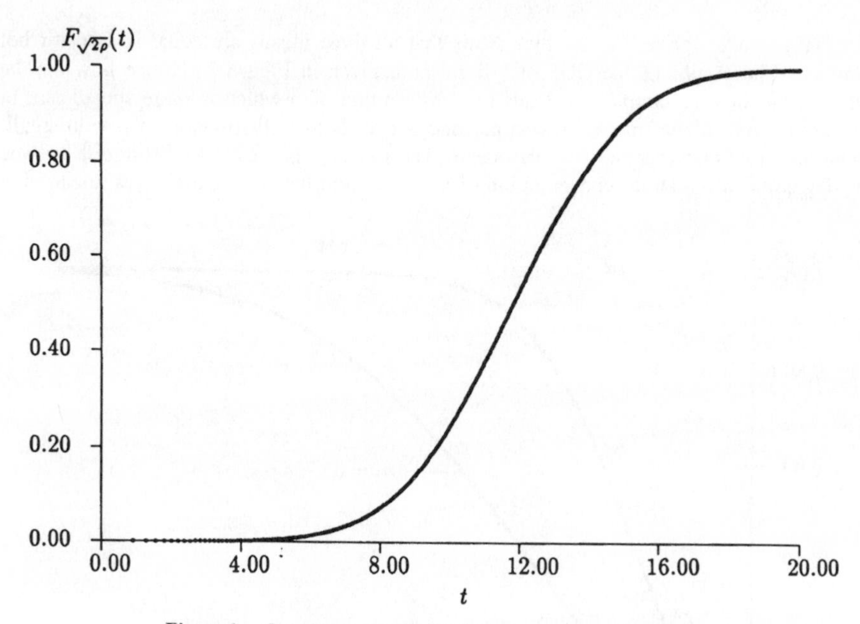

Figure 1. *Cumulative Distribution Function of Square Root of* 2ρ.

To compare to the usual F test, we use the data

$$\hat{\beta} = \begin{pmatrix} 95.6 \\ 98.6 \\ 111.5 \end{pmatrix}$$
$$RSS = 8723.49.$$

The F statistic is $(2843.4/2)/(8723.49/57) = 9.290$, with 2 and 57 degrees of freedom. This would be considered highly significant in the Classical sense. It is clear that the F test, in this case, fails to take into account the experimenter's idea of what is a large difference as well as the amount of precision due to the sample size.

As another example, consider the following data, generated to illustrate the extra information available from the measure ρ compared to the F test. Each data set consists of three samples of size m.

Statistic	Sample 1	Sample 2
m	5	20
$\overline{y}_1$	16.04	18.47
$\overline{y}_2$	19.89	19.78
$\overline{y}_3$	22.58	21.16
RSS	129.4	526.2

The $p-$value for testing the hypothesis that all three means are equal is .026 for both samples. The graphs of the CDF of $\sqrt{\rho}$ are both given in Figure 2. Notice how the data set with the smaller sample size leads to a distribution of ρ which is more spread out, but appears to have higher mean. The larger data set leads to a distribution of ρ with smaller mean but more concentrated near the mean. The $p-$value is unable to distinguish between the degree of information (concentration of the CDF) and the size of the effect (mean of ρ).

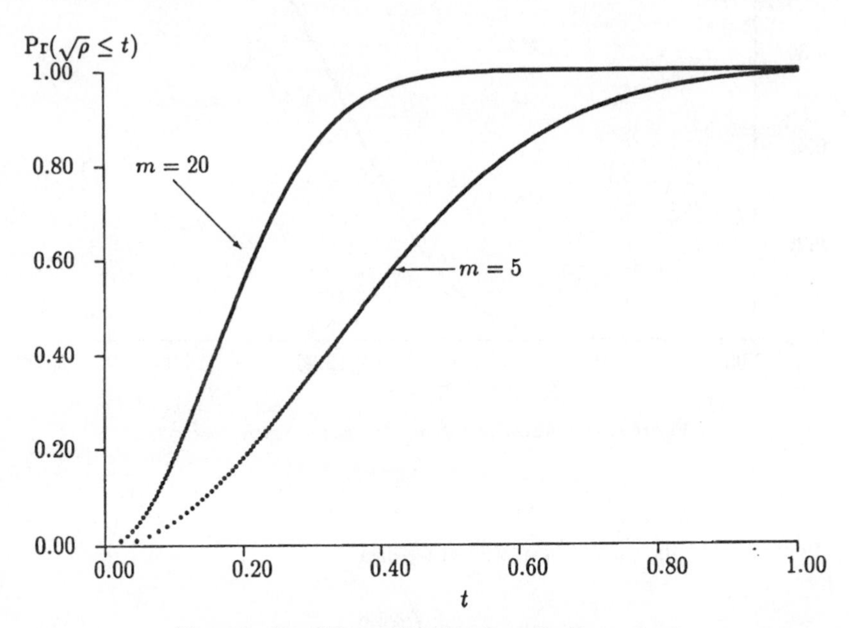

Figure 2. *Two Different CDFs With Same Significance Level.*

5. THE MULTIVARIATE CASE

We can generalize to multivariate cases the results given earlier. These correspond to matrix random quantities. Let the parameters be $\Theta = (\boldsymbol{B}, \Sigma)$, where $\boldsymbol{B}$ is a $p \times k$ matrix and Σ is a $k \times k$ positive definite matrix. We will suppose that the conditional distribution of $\boldsymbol{B}$ given $\Sigma = \boldsymbol{\sigma}$ is matrix normal $N_{p\times k}(\boldsymbol{\beta}_1, \boldsymbol{w}_1^{-1} \otimes \boldsymbol{\sigma})$ and that the marginal distribution of Σ is inverse Wishart $W_k^{-1}(a_1, a_1\boldsymbol{\sigma}_1)$. (Equivalently, the distribution of Σ^{-1} is Wishart $W_k(a_1 + k - 1, \boldsymbol{\sigma}_1^{-1}/a_1)$.) Recall that if $\boldsymbol{S} \sim W_k(a, \boldsymbol{b})$, then for every $k-$dimensional vector $\boldsymbol{\alpha}$,

$$\boldsymbol{\alpha}^T \boldsymbol{S} \boldsymbol{\alpha} \sim \Gamma\left(\frac{a}{2}, \frac{1}{2\boldsymbol{\alpha}^T \boldsymbol{b} \boldsymbol{\alpha}}\right).$$

Similarly, if $\boldsymbol{S} \sim W_k^{-1}(a, \boldsymbol{b})$ then

$$\boldsymbol{\alpha}^T \boldsymbol{S} \boldsymbol{\alpha} \sim \Gamma^{-1}\left(\frac{a}{2}, \frac{\boldsymbol{\alpha}^T \boldsymbol{b} \boldsymbol{\alpha}}{2}\right).$$

Let $\boldsymbol{\alpha}$ be a $k-$dimensional vector and let $\boldsymbol{\delta}$ be $p-$dimensional. Define the following quantities for later convenience:

$$\begin{aligned}
v &= \boldsymbol{\delta}^T \boldsymbol{w}_1^{-1} \boldsymbol{\delta} \\
u &= \boldsymbol{\delta}^T \boldsymbol{\beta}_1 \boldsymbol{\sigma}_1^{-1} \boldsymbol{\beta}_1^T \boldsymbol{\delta} \\
g &= \frac{u}{a_1 v + u} \\
c &= \boldsymbol{\alpha}^T \boldsymbol{\beta}_1^T w_1 \boldsymbol{\beta}_1 \boldsymbol{\alpha} \\
\mathrm{T} &= \boldsymbol{\alpha}^T \Sigma \boldsymbol{\alpha} \\
d &= \boldsymbol{\alpha}^T \boldsymbol{\sigma}_1 \boldsymbol{\alpha} \\
g' &= \frac{c}{a_1 d + c}.
\end{aligned}$$

We will consider four random variables.

$$\begin{aligned}
X_1 &= \boldsymbol{\delta}^T \boldsymbol{B} \Sigma^{-1} \boldsymbol{B}^T \boldsymbol{\delta} \\
X_2 &= \boldsymbol{\delta}^T \boldsymbol{B} \boldsymbol{\sigma}_1^{-1} \boldsymbol{B}^T \boldsymbol{\delta} \\
X_3 &= \boldsymbol{\alpha}^T \boldsymbol{B}^T w_1 \boldsymbol{B} \boldsymbol{\alpha} \\
X_4 &= \frac{X_3}{\mathrm{T}}.
\end{aligned}$$

First, consider X_1. The conditional distribution of $\Sigma^{-1/2} \boldsymbol{B}^T \boldsymbol{\delta}$ given Σ is $N_k(\Sigma^{-1/2} \boldsymbol{\beta}_1^T \boldsymbol{\delta}, vI)$. It follows that the conditional distribution of X_1/v given Σ is non-central χ^2 with k degrees of freedom and non-centrality parameter

$$\Lambda_1 = \frac{1}{v} \boldsymbol{\delta}^T \boldsymbol{\beta}_1 \Sigma^{-1} \boldsymbol{\beta}_1^T \boldsymbol{\delta}.$$

The distribution of Λ_1 is $\Gamma([a_1 + k - 1]/2, a_1 v/[2u])$. It follows from Theorem 1 that the marginal distribution of X_1/v is $ANC\chi^2(k, a_1 + k - 1, g)$.

Next, consider X_2. The conditional distribution of $\boldsymbol{\sigma}_1^{-1/2} \boldsymbol{B}^T \boldsymbol{\delta}$ given Σ is

$$N_k(\boldsymbol{\sigma}_1^{-1/2} \boldsymbol{\beta}_1^T \boldsymbol{\delta}, v \boldsymbol{\sigma}_1^{-1/2} \Sigma \boldsymbol{\sigma}_1^{-1/2}).$$

It follows that the marginal distribution of $\boldsymbol{\sigma}_1^{-1/2} \boldsymbol{B}^T \boldsymbol{\delta}$ is $k-$variate t with a_1 degrees of freedom, $t_{a_1;k}(\boldsymbol{\sigma}_1^{-1/2} \boldsymbol{\beta}_1^T \boldsymbol{\delta}, vI)$. This means that there exists a scalar random variable Y with $\Gamma^{-1}(a_1/2, a_1/2)$ distribution such that conditional on $Y = y$,

$$\boldsymbol{\sigma}_1^{-1/2} \boldsymbol{B}^T \boldsymbol{\delta} \sim N_k(\boldsymbol{\sigma}_1^{-1/2} \boldsymbol{\beta}_1^T \boldsymbol{\delta}, vyI).$$

Hence, the conditional distribution of $X_2/[vY]$ given Y is non-central χ^2 with k degrees of freedom and non-centrality parameter $u/[vY]$. The distribution of $1/[vY]$ is $\Gamma(a_1/2, a_1 v/2)$. It follows from Theorem 2 and the ensuing discussion that the marginal distribution of $a_1 X_2/[k(a_1 v + u)]$ is $ANCF(k, a_1, g)$.

Next, consider X_3. The conditional distribution of $\boldsymbol{w}_1^{1/2} \boldsymbol{B\alpha}$ given Σ is $N_p(\boldsymbol{w}_1^{1/2} \boldsymbol{\beta}_1\ \boldsymbol{\alpha}, \boldsymbol{\alpha}^T\ \Sigma\ \boldsymbol{\alpha})$. It follows that the conditional distribution of X_3/T given Σ is non-central χ^2 with p degrees of freedom and non-centrality parameter $\Lambda_3 = c/\mathrm{T}$. The distribution of $1/\mathrm{T}$ is $\Gamma(a_1/2, a_1 d/2)$. It follows from Theorem 2 and the ensuing discussion that the marginal distribution of $a_1 X_3/[p(a_1 d + c)]$ is $ANCF(p, a_1, g')$.

Finally, consider $X_4 = X_3/\mathrm{T}$. As we just saw, the conditional distribution of X_4 given Σ is non-central χ^2 with p degrees of freedom and non-centrality parameter Λ_3, where the distribution of Λ_3 is $\Gamma(a_1/2, a_1 d/[2c])$. It follows from Theorem 1 that the marginal distribution of X_4 is $ANC\chi^2(p, a_1, g')$.

6. CONCLUSION

A Bayesian approach is taken to the linear model. Posterior distributions are found for random variables which correspond to test statistics often used in the non-Bayesian framework. An important feature is that the confounding of effect size and sample size is relieved by the use of these random variables. In the fixed selection model, with natural conjugate prior, the distributions resemble the familiar non-central χ^2 and non-central F (or Beta). Computer evaluation of the posterior probabilities is described and an example is given to show how the distributions might be used. Some multivariate extensions are also given.

It is important when using such procedures to keep two things in mind. First, the choice of prior distribution is important. If one believes the random selection prior is better suited for a particular fixed selection model, then one should use it. Dickey (1974) considers this case. Secondly, the type of inference one wishes to make should determine the procedure one uses.

The procedures described in this paper for the fixed selection model are designed for an overall comparison of $\boldsymbol{a}_0\boldsymbol{B}$ and ψ_0. If specific functions of $\boldsymbol{B}$, such as $\boldsymbol{\lambda}^T \boldsymbol{a}_0 \boldsymbol{B}$ for a vector $\boldsymbol{\lambda}$, are of interest, procedures designed to deal with the specific functions of interest may be preferred.

Finally, it should be noted that the distributions and procedures described in this paper help to facilitate the presentation of Bayesian analysis of linear models in an elementary statistics course. The close similarity between the random variables and calculations performed here to those required for the F test make it possible to compare and contrast the Bayesian and Classical approaches without causing too much confusion on the part of the students. In particular, with improper priors, many of the calculations required to perform the F test are repeated exactly in the calculation of the distribution of H/Σ^2.

A. COMPUTER EVALUATION

Evaluation of the distribution functions of the $ANC\chi^2$ and $ANCF$ distributions require evaluation of the incomplete gamma and beta integrals. Continued fractions for these integrals are given by Abramowitz and Stegun (1965, p. 263 and p. 944). If Ξ has $ANC\chi^2(q, a, g)$ distribution, one can obtain the CDF of Ξ to within ϵ by

$$\sum_{i=0}^{N} h_i(a, g) I_G\left(\frac{t}{2}, i + \frac{q}{2}\right),$$

where $I_G(a, b)$ is the incomplete gamma ratio

$$I_G(x, b) = \frac{1}{\Gamma(b)} \int_0^x \exp(-z) z^{p-1} dz,$$

and N is the smallest integer such that

$$I_G\left(\frac{t}{2}, i + \frac{q}{2}\right) \left[1 - \sum_{i=0}^{N} h_i(a, g)\right] \leq \epsilon.$$

The reason we can use the extra factor of $I_G\left(\frac{t}{2}, i + \frac{q}{2}\right)$ before comparing 1 minus the sum of the weights to ϵ is that $I_G(x, b)$ is decreasing in b for fixed x.

Similarly, if B has $ANCB(q, a, g)$ distribution, one can calculate the CDF of B to within ϵ by

$$\sum_{i=0}^{N} h_i(a, g) I_B\left(t, i + \frac{q}{2}, i + \frac{a}{2}\right),$$

where $I_B(t, a, b)$ is the incomplete beta ratio

$$I_B(x, b, c) = \frac{\Gamma(b + c)}{\Gamma(b)\Gamma(c)} \int_0^x z^{b-1}(1 - z)^{c-1} dz,$$

and N is the smallest integer such that

$$1 - \sum_{i=0}^{N} h_i(a, g) \leq \epsilon.$$

$I_B(x, b, c)$ is not monotone as both b and c increase at the same rate, so we cannot make use of the extra factor as before.

REFERENCES

Abramowitz, M. and Stegun, I. A. (1965). *Handbook of Mathematical Functions.* New York: Dover.

Bailey, N. (1965). Comparisons of mental and motor test scores for ages 1-15 months by sex, birth order, race, geographical location, and education of parents. *Child Development* **36**, 378–411.

Dickey, J. (1974). Bayesian alternatives to the F-test and least-squares estimate in the normal linear model *Studies in Bayesian Econometrics and Statistics* (S. E. Fienberg and A. Zellner, eds.), Amsterdam: North-Holland, 515–554.

Geisser, S. (1967). Estimation associated with linear discriminants. *Ann. Math. Statist.* **38**, 807–817.

Lecoutre, B. (1985). Reconsideration of the F-test of the analysis of variance: the semi-Bayesian significance test, *Comm. Statist. Theory and Methods* **14**, 2437–2446.

Lehmann, E. L. (1959). *Testing Statistical Hypotheses.* New York: Wiley.

Leonard, T. (1974). A Bayesian investigation of the F-test procedure. *Tech. Rep.* **11**, University of Warwick.

Lindley, D. V. and Smith, A. F. M. (1972). Bayes estimates for the linear model. *J. Roy. Statist. Soc. B* **34**, 1–41.

Rouanet, H. and Lecoutre, B. (1983). Specific inference in ANOVA: from significance tests to Bayesian procedures, *British Journal of Mathematical and Statistical Psychology* **36**, 252–268.

Schervish, M. J. (1987). A review of multivariate analysis. *Statist. Sci.* **2**, 396–433.

Schervish, M. J. and Elmer, E. (1983). A residential program for treating victims of child abuse. *The Statistician* **32**, 131–137.

Smith, A. F. M. (1973). A general Bayesian linear model. *J. Roy. Statist. Soc. B* **35**, 67–75.

DISCUSSION

M. GHOSH (*University of Florida, USA*)

The ANOVA F-statistic is used rather blindly by most practising statisticians, often without realization of what this statistic actually signifies. I congratulate Professor Schervish for dealing a smart blow to this conventional approach, pointing out both theoretically, and with the aid of a nice example the ineffectiveness of the F-statistic as a measure of departure from a specified hypothesis. I agree with the author that this failure of the F-statistic is primarily due to its inability to "separate the effect size from the sample size".

However, while there are enough convincing arguments to discard the classical F-statistic, study of $\rho(\boldsymbol{B}, \Omega_{\mathrm{H}})$ seems slightly artificial on intuitive grounds, the main justification for its consideration being that this quantity is the direct Bayesian analog of the numerator of the usual F-statistic, (properly normalized). It would be more appealing intuitively to consider

$$\rho^*(\boldsymbol{B}) = (\boldsymbol{a}_0\boldsymbol{B} - \boldsymbol{\psi}_0)^T(\boldsymbol{a}_0\boldsymbol{a}_0^T)^{-1}(\boldsymbol{a}_0\boldsymbol{B} - \boldsymbol{\psi}_0)/\ell^*,$$

$$\ell^* = \mathrm{tr}[\boldsymbol{a}_0^T(\boldsymbol{a}_0\boldsymbol{a}_0^T)^{-1}\boldsymbol{a}_0] = p - 1.$$

I realize that Schervish's introduction of $\rho(\boldsymbol{B}, \Omega_{\mathrm{H}})$ is to create Bayesian awareness among students trained in the traditional frequentist mode. While I sympathize with this Bayesian salesmanship, I strongly feel that much of hypothesis testing is largely irrelevant for any Bayesian education.

Schervish's article brought reminiscences of my own work with Glen Meeden and Leone Low (Ghosh and Meeden, 1984, Ghosh, Low and Meeden, 1990). In order to emphasize the point made in the previous paragraph, I introduce briefly the problem considered in Ghosh and Meeden (1984).

Suppose in a large factory, there is a population of p workers. Each worker produces a large number of identical items in a workday. Let M_i denote the true average output of worker $i(i = 1, \ldots, p)$. Then $\boldsymbol{B} = (\mathrm{M}_1, \ldots, \mathrm{M}_p)^T$ is the vector of unknown parameters. The goal is to find out how much variation there is among the workers.

Ghosh and Meeden (1984) answered the above question by finding the posterior distribution of $\gamma = \sum_{i=1}^p (\mathrm{M}_i - \overline{\mathrm{M}})^2/(p-1)$, where $\overline{\mathrm{M}} = p^{-1}\sum_{i=1}^p \mathrm{M}_i$. Suppose observations are collected for a subset of workers of size $q(\leq p)$, and the selected workers are labeled $1, 2, \ldots, q$. Let X_{jk} denote the observed output of the jth worker on the kth day $(j = 1, \ldots, q; k = 1, \ldots, n)$. Then the minimal sufficient statistics is $(Y_1, \ldots, Y_q, S)$, where $Y_j = n^{-1}\sum_{k=1}^n X_{jk} (j = 1, \ldots, q)$, and $S = \sum_{j=1}^q \sum_{k=1}^n (X_{jk} - Y_j)^2$. Ghosh and Meeden (1984) found the posterior distribution of γ under the following hierarchical model:

(I). Conditional on $\boldsymbol{B}$, μ and $\sigma^2, Y_1, \ldots, Y_q$, and S are mutually independent with

$$Y_j \sim N(M_j, \sigma^2 n^{-1}), S \sim \sigma^2\chi^2_{(n-1)q}.$$

(II). Conditional on μ and σ^2, $\boldsymbol{B} \sim N(\mu 1_p, \sigma^2\lambda^{-1}\boldsymbol{I}_p)$, where $\lambda(> 0)$ is known.

(III). $\mu \sim$ uniform $(-\infty, \infty), (\sigma^2)^{-1} \sim$ Gamma $\left(\frac{1}{2}a, \frac{1}{2}b\right)$.

Suppose now we change the model slightly so that it fits into Schervish's framework. Let $q = p$, and consider the hierarchical model as given (I) – (III) with the only change that μ is *known*. Then, in Schervish's notations,

(i) $\boldsymbol{a}_0 = \begin{bmatrix} 1 & \cdots & 0-1 \\ \cdots & \cdots & \cdots \\ 0 & \cdots & 1-1 \end{bmatrix}, \Psi_0 = 0;$

(ii) $\boldsymbol{w}_1 = (n+\lambda)\boldsymbol{I}_p$;
(iii) $\rho(\boldsymbol{B}, \Omega_{\mathrm{H}}) = \ell^{-1}(n+\lambda)\boldsymbol{B}^T\boldsymbol{a}_0^T(\boldsymbol{a}_0\boldsymbol{a}_0^T)^{-1}\boldsymbol{a}_0\boldsymbol{B}$;
(iv) $\ell = (n+\lambda)\mathrm{tr}[\boldsymbol{a}_0^T(\boldsymbol{a}_0\boldsymbol{a}_0^T)^{-1}] = (n+\lambda)(p-1)$;
(v) $\boldsymbol{a}_0\boldsymbol{a}_0^T = \boldsymbol{I}_{p-1} - p^1\boldsymbol{J}_{p-1}$;
(vi) $(\boldsymbol{a}_0\boldsymbol{a}_0^T)^{-1} = \boldsymbol{I}_{p-1} + \boldsymbol{J}_{p-1}$;
(vii) $\rho(\boldsymbol{B}, \Omega_{\mathrm{H}}) = (p-1)^{-1}\sum_{i=1}^p(M_i - \overline{M})^2$

which matches perfectly the γ of Ghosh and Meeden (1984). However, if the model changes to

(I)' $\boldsymbol{Y} = (Y_1, \ldots, Y_p)^T|\boldsymbol{B}, \sigma^2 \sim N(\boldsymbol{\beta}, \sigma^2\boldsymbol{V})$;
(II)' $\boldsymbol{B} \sim N(\mu 1_p, \sigma^2\boldsymbol{w}_0^{-1})$;
(III)' $(\sigma^2)^{-1} \sim \mathrm{Gamma}(\frac{1}{2}a, \frac{1}{2}b)$,

then $\rho(\boldsymbol{B}, \Omega_{\mathrm{H}})$ changes, but $\rho^*(\boldsymbol{B})$ continues to equal γ. In my assessment, in the present context, it is more logical to consider the posterior distribution of γ rather than $\rho(\boldsymbol{B}, \Omega_{\mathrm{H}})$.

There are two minor comments which could possibly be topics of future research. First, when there is uncertainty about β_0, one should possibly go a step further, and use a hierarchical model using a flat prior for β_0. Second, to make the procedure robust in a different way, one could possibly invoke heavy-tailed priors, e.g., multivariate-t in such problems.

In summary, I like Schervish's paper and welcome any future research in this direction.

D. V. LINDLEY (*Somerset, UK*)

There is a rule that is central to the Bayesian paradigm that says: if in doubt about anything, express that doubt probabilistically. If the author is in doubt concerning $\boldsymbol{a}_0\boldsymbol{B}$, then calculate the probability distribution of $\boldsymbol{a}_0\boldsymbol{B}$ given all the information available, both from the prior and the data. In the case studied, this can be simple and leads to F-statistics. The argument of the paper seems directed towards the production of a single number even when $\boldsymbol{a}_0\boldsymbol{B}$ is a vector. This seems misplaced. Why try to force something into one dimension when it is naturally in more? Credible sets should not be replaced by intervals.

REPLY TO THE DISCUSSION

Let me respond first to Professor Lindley. I agree wholeheartedly that one must calculate the posterior distribution of $\boldsymbol{a}_0\boldsymbol{B}$. When this quantity is multidimensional, it may not be easy to determine how close it is to some hypothesised value. A single number measure of this distance can be a useful summary (albeit not a complete summary) of that one feature of the random vector. The distribution of this single number measure of distance may be easier to visualize than a collection of multidimensional credible regions. Although the latter contain more information, the information may be harder to extract. Something can still be learned from the single number measure.

I agree with Professor Ghosh that there are cases in which ρ^* is a better measure of deviation than the ρ in my paper. In the example which he presented, if the sample sizes are unequal (n_j instead of n) or in more general linear models, the distributions of ρ^* will not be available in closed form. (One needs $\boldsymbol{a}_0\boldsymbol{w}_1^{-1}\boldsymbol{a}_0^T$ to be a constant times the identity). The distribution can be expressed as a weighted average of familiar non-central distributions, and it would be worthwhile to be able to compute such distributions. It might even be feasible to cover such material in an advanced course after students have learned some simple analyses of linear models.

ADDITIONAL REFERENCES IN THE DISCUSSION

Ghosh, M., Low, L. Y. and Meeden, G. (1990). A hierarchical Bayes analysis of cross-classification models. *Bayesian and Likelihood Methods in Statistics and Econometrics* (S. Geisser, J. Hodges, S. J. Press and A. Zellner, eds.), Amsterdam: North-Holland, 277–301.

Ghosh, M. and Meeden, G. (1984). A new Bayesian analysis of a random effects model. *J. Roy. Statist. Soc. B* **46**, 474–482.

BAYESIAN STATISTICS 4, pp. 435–446
J. M. Bernardo, J. O. Berger, A. P. Dawid and A. F. M. Smith, (Eds.)

Warranties

N. D. SINGPURWALLA and S. P. WILSON
The George Washington University, USA

SUMMARY

Consumer products, such as automobiles, often carry warranties such as 5 years or 50,000 miles, whichever comes first. There are two issues of interest, the first being a specification of an optimum price-warranty combination, and the second the forecast of a reserve fund to meet warranty claims. The former involves a consideration of the item's reliability and the competitors' actions, and the latter involves the analysis of a time series in two dimensions for which the use of Bayesian methods is natural. In this expository paper we overview the warranty problem and suggest strategies for addressing the ensuing issues.

Keywords: RELIABILITY; INSURANCE; FAILURE MODELS WITH MULTIPLE SCALES; GAME THEORY; DYNAMIC LINEAR MODELS.

1. INTRODUCTION

Warranties have become a critical segment of the industrial environment and are a key means for communicating information about product quality. A warranty is a contractual agreement which requires the manufacturer to rectify a failed item, should failure occur during a period specified by the warranty. Whereas some warranties, such as say those for computers, do not restrict the amount of usage, other warranties, such as those for automobiles, impose restrictions on the amount of usage. Thus for example, it is common for automobiles to carry a warranty for 5 years or 50,000 miles whichever comes first. The cost of the warranty is built into the selling price of the item, and a problem faced by the manufacturer is a forecast of the amount of money that should be put into reserve to meet warranty claims. In the automobile industry, such amounts run into millions of dollars.

Since consumers cannot observe product quality at the time of purchase, only warranties can induce manufacturers to supply high quality goods. The longer the warranty the greater is the assurance of good quality, and the better is the protection to the consumer. However, consumers are risk averse so that high warranty levels lower the incentive to take care of the product. This *double sided moral hazard* (Emmons (1988)), together with competitive acts between manufacturers, leads to game theoretic considerations in setting optimum warranties.

It is clear from the above that the warranty problem is a multi-disciplinary one involving topics as diverse as economics, game theory, operations research and statistics. It appears that all attention has been restricted to the case of one dimensional warranties. Not much has been done with regard to two dimensional warranties, and also the issue of forecasting warranty claim reserves (cf. Robinson and McDonald (1991)). The determination of optimal two dimensional warranties requires the development of new bivariate failure models indexed by time and usage. In Section 2.1 we describe a family of bivariate failure models with the special feature that the index of failure is both time and usage. In Section 2.2 we outline some plausible utility functions, and in Section 2.3 we outline a game theoretic set-up for determining optimal warranties.

The question of forecasting warranty reserves, based on prior information and warranty claims data from previous versions of the product, involves the analyses of time series indexed by two dimensions, one dimension representing time and the other the model year. Our approach here involves a use of dynamic linear models of the type discussed by West and Harrison (1989), with caveats to account for the fact that the series pertaining to the previous model years serve as leading indicators; these are summarized in Section 3 of this paper. Section 3.1 deals with a description of the time series and Section 3.2 with the structure of dynamic models for forecasting under these series. Section 4 concludes with work in progress.

2. OPTIMAL TWO DIMENSIONAL WARRANTIES

The most common two dimensional warranty region is the rectangular one offered with automobiles, and is usually of the type, 5 years or 50,000 miles, whichever occurs first. A virtue of such warranties is that they are easy to implement. Their disadvantage is that they tend to encourage an above average use during the initial period of purchase, and therefore pose a risk to the manufacturer. On the other hand, such warranties protect the consumer from failures attributed to "infant mortality" which tend to occur during the initial period of use — see Section 3.1, and can therefore be discovered early on. Alternatives to the rectangular regions would be the circular, triangular and parabolic regions; details on these can be found in Singpurwalla and Wilson (1991).

2.1. *A Bivariate Failure Model Indexed by Time and Usage*

The current literature on reliability has considered probabilistic models of failure for a single item or for a system of items. A characteristic feature of such models is that they are indexed on a single variable such as the time to failure or number of miles to failure. Whereas such models are meaningful with one dimensional warranties, they are inadequate for two dimensional warranties, which require indexing by both time and usage. In what follows we introduce a generic bivariate model for failure which addresses the above need.

Suppose that the amount of usage that an item experiences at time t is $M(t)$; thus for example, for automobiles, $M(t)$ is the mileage at t. Suppose that the effect of $M(t)$ is to increase the failure rate of the item from α to $\alpha + \beta M(t)$, where β is a constant. This form of the failure rate is sensible; furthermore, it has also been entertained in the biostatistical literature (see for example, Kalbfleisch and Prentice (1980), p. 31). Since the amount of usage at time t is unknown, we shall assume that $M(t)$ is a Poisson random variable with parameter λt.

If T denotes the time to failure of the item and U denotes the usage at failure, then the joint distribution of T and U is of the form

$$P(T \leq t, U \leq u) = \sum_{n=0}^{u} \frac{\lambda^n(\alpha + \beta n)}{\beta^n n!} \left\{ \sum_{i=0}^{n} (-1)^i \binom{n}{i} \left(\frac{1 - \exp(-(\alpha + \lambda + \beta i)t)}{\alpha + \lambda + \beta i} \right) \right\} \tag{2.1}$$

and its joint density at t and u, for any $t \geq 0$ and any $u = 0, 1, 2, \ldots$, of the form:

$$f_{T,U}(t, u) = \frac{\alpha + \beta u}{u!} e^{-(\alpha+\lambda)t} \left[\frac{\lambda}{\beta}(1 - e^{-\beta t}) \right]^u \tag{2.2}$$

for details see Singpurwalla (1991).

Equations (2.1) and (2.2) are useful for obtaining the probability of failure in any two dimensional warranty region that is being contemplated. The conditionals and the marginals of T and U can also be found.

2.2. *Plausible Utility Functions*

Considerations which drive the choice of a utility function are p, the selling price of the item, v, the unit cost of manufacturing, c the cost of repair or replacement, and R the geometry of the warranty region. Let $E_M(R)$ be the expected number of renewals in the region R as assessed by the manufacturer. We are assuming that all repairs are such that the item is restored to perfection. Whereas the computation of $E_M(R)$ can, in principle, be analytically undertaken, see for example Hunter (1974), we have not been able to obtain closed form results for it using the model given by (2.2). However, since the marginal of U and the conditional of T given U, are known, the simulation of $E_M(R)$ is a natural alternative. If the number of items that are sold is n, where n is random, then in accordance with Glickman and Berger (1976), the manufacturer's total expected utility is

$$U_M(p,R) = a_1 E(n)\{p - v - cE_M(R)\}, \tag{2.3}$$

where $a_1 > 0$ is the utility of a unit of profit to the manufacturer, and $E(n)$ is the manufacturer's expecion of n. If N is the size of the market for the product, and if $\pi_M(p,R)$ denotes the manufacturer's probability that a consumer will purchase a product priced at p, and carrying a warranty described by a region R, then $E(n) = N\pi_M(p,R)$.

The desirability (to the consumer) of the region R, the effect of the unit price p, and the inconvenience (to the consumer) of failures during the warranty R will be reflected in $\pi_M(p,R)$, and given below is a plausible model for the same; note that $\pi_M(p,R)$ describes the manufacturer's views of the consumer's attitude towards p and R. Specifically, suppose that $P_C(R)$ is the probability of failure in the region R as perceived by the consumer; we assume that the manufacturer has knowledge of this probability (via marketing studies and the like). Let $E_C(R)$ be the expected number of failures in the region R, based on $P_C(R)$. Clearly the bigger R, the larger is $P_C(R)$. Thus attractive warranty regions (to the consumer) correspond to large values of $P_C(R)$. However larger probabilities of failure are also associated with the consumer's perception of poor quality, and thus impact negatively on the consumer's perception of inconvenience. The ability to capture the consumer's perception of a specified warranty, via the manufacturer's subjective assessment of $P_C(R)$ is a note– worthy aspect of our development. A utility function which captures the undesirability of R could be of the form $\exp\{-E_C(R)\} - b_1$, where b_1 lies between 0 and 1. The same is also true for the price p, for which a plausible form of the consumer's utility is $\exp\{-p\} - b_2$, where b_2 also lies between 0 and 1. The aforementioned three components of the consumer's utility can be combined, assuming additive utilities, as a convex combination involving the weights η_1, η_2 and η_3. The weights are chosen to reflect the degree of the consumer's preference over the three attributes, a specified warranty R, the price of the unit p, and the inconvenience of experiencing failures during the period of warranty. Thus to summarize, the consumer's expected utility (as perceived by the manufacturer) is of the form

$$U_C(p,R) = \eta_1(P_C(R)) + \eta_2[\exp\{-E_C(R)\} - b_1] + \eta_3[\exp\{-p\} - b_2]. \tag{2.4}$$

To obtain $\pi_M(p,R)$ from $U_C(p,R)$, we seek a function which maps $U_C(p,R)$ onto $[0,1]$, and is increasing in it. Several choices are possible, ranging from a simple linear translation and scaling, to logits and probits. However, the following specification has attractive features:

$$\pi_M(p,R) = .5[1 + \sin\{(((U_C(p,R) - u)/(U - u)) - .5)\pi\}], \tag{2.5}$$

where

$$u = \min U_C(p,R), \qquad \text{and}$$
$$U = \max U_C(p,R).$$

Once the above has been done, the manufacturer chooses that combination of p and R that maximizes $U_M(p, R)$. Issues pertaining to the maximization of $U_M(p, R)$ are discussed next.

2.2.1. Maximization of Expected Utility

Since any warranty region R is defined by two attributes, its shape and its dimensions, we find it convenient to identify each plausible region by $R_i(m, t), i = 1, 2, \ldots$, where the index i describes the shape, and the arguments m and t describe the dimensions.

In what follows, we shall assume that for any particular choice of $R_i(.,.), i = 1, 2, \ldots$, the manufacturer chooses those values of p, m and t, say p_i, m_i, and t_i, respectively, that maximize (2.3), the manufacturer's expected utility; denote this quantity by $U_M\{p_i, R_i(m_i, t_i)\}$. The manufacturer repeats this exercise for each choice of $R_i(.,.)$. The maximization has to be done with respect to at most 3 variables, and it is most likely that this will be done numerically. Assuming that the competitors' actions do not influence a manufacturer's utility, the manufacturer will choose that index i for which $U_M\{p_i, R_i(m_i, t_i)\}$ is a maximum. In reality of course, the competitiors' actions influence utilities and the manufacturer must now resort to game theoretic considerations. This is discussed next.

2.3. *Effect of Competitors' Actions*

Suppose that there are only two manufacturers, M_A and M_B, and suppose that each manufacturer has determined, for each warranty region under consideration, the optimum values of p_i, t_i and m_i; for convenience, we shall suppress the parenthetical arguments and denote these choices by $R_{i,A}$ and $R_{i,B}, i = 1, 2, \ldots, n$. Following game theoretic conventions, a tableaux (payoff matrix) decribing these choices and the associated expected utilities is shown in Table 1. The utilities here pertain to the two manufacturers and are denoted by $U_A(i, j)$ and $U_B(i, j)$, where the former denotes M_A's expected utility when M_A chooses $R_{i,A}$ and $R_{j,B}$; similarly the latter.

A plausible strategy for specifying the required expected utilities is based on the notion that the effect of a competitor's action is best captured via a change in the manufacturer's probability of the consumer purchasing the product. Specifically, suppose that $\mathcal{A}_i = \pi_M(p_{i,A}, R_{i,A})$ denotes A's probability that a consumer will purchase a product priced by M_A at $p_{i,A}$, and carrying a warranty of $R_{i,A}$, when there is no competition from M_B; similarly, define $\mathcal{B}_i = \pi_M(p_{i,B}, R_{i,B})$. Recall that these probabilities are based on the consumer's expected utilities, as perceived by the manufacturers.

Then, given knowledge of $\pi_M(p_{i,B}, R_{i,B})$, M_A will revise $\pi_M(p_{i,A}, R_{i,A})$ to $\pi_M(p_{i,A}, R_{i,A} | p_{i,B}, R_{i,B})$, via the relationship

$$\pi_M(p_{i,A}, R_{i,A} | p_{i,B}, R_{i,B}) = \mathcal{A}_i[(1 - \mathcal{A}_i)(1 - \mathcal{B}_i)]/(\mathcal{A}_i + \mathcal{B}_i); \tag{2.6}$$

similarly, M_B will revise $\mathcal{B}_i$. The revised probabilities will be used by M_A and M_B to compute their expected values for n. The development of (2.5) is based on the assumption that if the probability that a consumer does not buy from either manufacturer is $(1 - \mathcal{A}_i)(1 - \mathcal{B}_i)$, then $(\mathcal{A}_i + \mathcal{B}_i - \mathcal{A}_i\mathcal{B}_i)$, the probability that the consumer buys from either M_A or M_B, is to be apportioned in the ratio of $\mathcal{A}_i/(\mathcal{A}_i + \mathcal{B}_i)$ and $\mathcal{B}_i/(\mathcal{A}_i + \mathcal{B}_i)$ respectively.

Once the above is done, we have the necessary ingredients to proceed using the general approaches for solving two person nonzero sum games. A good introduction to nonzero sum games may be found in Luce and Raiffa (1957).

M_A's Choices → / M_B's Choices ↓	R_{1A}	...	R_{nA}
R_{1B}	$U_A(1,1)$ / $U_B(1,1)$	...	$U_A(n, 1)$ / $U_B(n,1)$
⋮	⋮	...	⋮
R_{nB}	$U_A(1,n)$ / $U_B(1,n)$	...	$U_A(n,n)$ / $U_B(n,n)$

Table 1. *Payoff Matrix for a Two Person Nonzero Sum Game Played by Two Manufacturers, M_A and M_B.*

3. FORECASTING WARRANTY RESERVES

In many organizations, the analysis of warranty claims data is undertaken to serve the needs of two departments, financial and engineering. Financial departments have an interest in tracking the amounts spent on warranties; their aim is to forecast warranty claims over the service life of the product, so that funds could be escrowed to meet such claims. Engineering departments are more concerned with frequency of repairs rather than costs; their aim is to improve on product design and to compare the observed data with engineering tests. In what follows, we give a brief description of the nature of the warranty claims data, and suggest some approaches for forecasting such claims.

3.1. *Description of Data*

Figure 3.1, taken from Robinson and McDonald (1991), is typical of the data considered by financial departments; it shows the cumulative warranty expense per unit, at various points in time, for different model years. Figure 3.2, also taken from McDonald and Robinson, shows the kind of information that is of interest to engineering groups; it is a plot of the cumulative frequency of repairs per 1000 vehicles, as a function of months in service, for several model years. On logarithmic scales the plots of Figures 3.1 and 3.2 (referred to as *log-log plots*) are usually nearly linear, with breaks occurring where warranty coverages change. To predict the cumulative number of repairs at ages greater than those that have been observed, it is therefore tempting to extrapolate the observed time series via such logarithmic plots, and to account for engineering judgment by making suitable adjustments to the forecasts. This practice is currently being followed by many organizations and appears to be the operating norm; however such extrapolations have proven to be misleading.

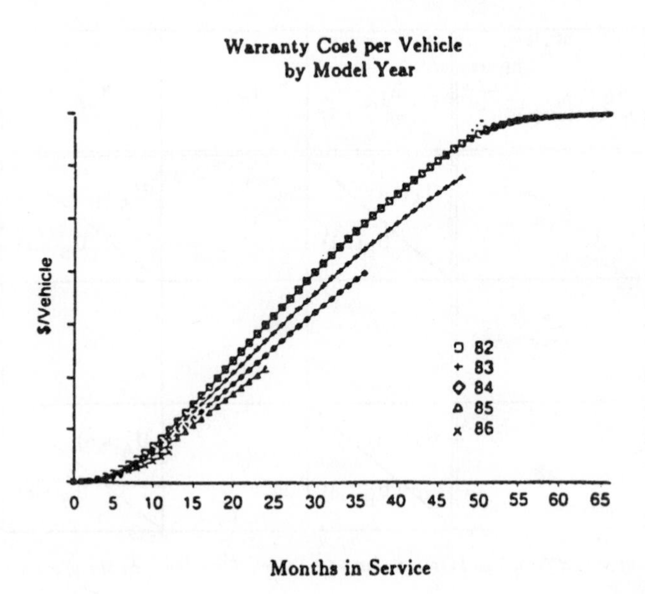

Figure 3.1. *Cumulative cost per car by months in service (from Robinson and McDonald).*

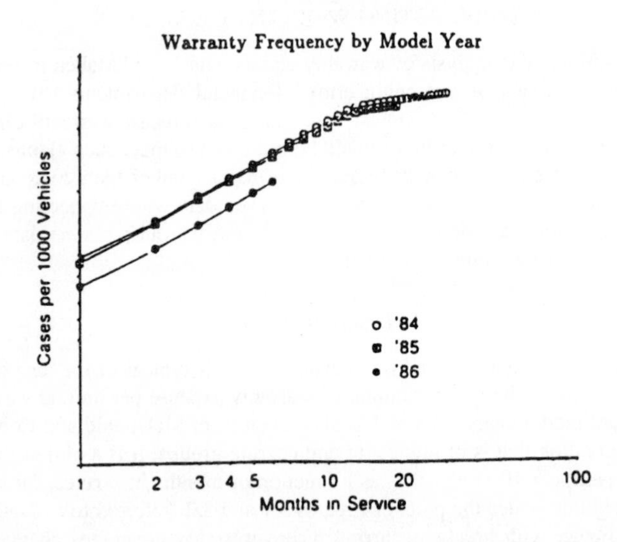

Figure 3.2. *Cumulative warranty frequency by months in service (from Robinson and McDonald).*

3.1.1. The Issue of Maturing Data

One of the nuances associated with warranty claims data is that a snapshot of the same series changes over time; this phenomenon is referred to as *maturing data.* Specifically, consider the plots of Figure 3.3., also taken from Robinson and McDonald. Here we show the repairs per 1000 vehicles for a 1986 model year automobile. The observed frequencies at say, 3 months in service, increase as the amount of data which goes into calculating the frequencies increases. The phenomenon of data maturation, which tends to diminish the forecasting capability of any method, suggests that graphical methods of extrapolation based on the log-log plot may be particularly ill suited. A consideration of dynamic linear models, with innovation terms which account for the added uncertainties due to a maturation of the data, appear to be more natural, and this is what we propose. The literature on dynamic linear models appears to exclude considerations of a *leading indicator series* (see Box and Jenkins (1976)) when analyzing the behaviour of a time series. Since warranty claims for previous model years serve as leading indicators for the series of current interest, what is needed is a formal mechanism for the incorporation of information from previous claims. An approach for undertaking the above is outlined in the following section

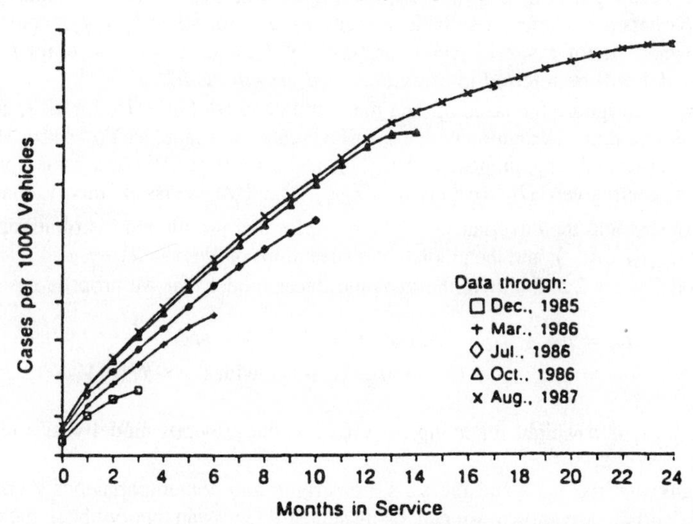

Figure 3.3. *Example of Data Maturation (from Robinson and McDonald).*

3.2. *Dynamic Linear Models with Leading Indicators*

Suppose that a snapshot of the available data is taken at some point in time, say τ. Let Y_{1t} denote the warranty claims for units manufactured in model year 1, and having an exposure to use of duration $t, t = 1, 2, \ldots, T_1$. Similarly, let $Y_{it}, t = 1, 2, \ldots, T_i$, denote the claims for the model year i, $i = 2, \ldots, p$. We assume that model year 1 is the furthest back in time from τ, and that model year p is the latest. It is therefore reasonable to expect that $T_1 > T_2 \ldots > T_p$. Assume that the item is warrantied for a maximum duration of exposure of say T; thus $T_i \leq T$. Recall, that due to a maturing of the claims data, all the values Y_{it} for which $T_i < T$, will change when a snapshot of the series is taken at any point other than τ. Given the Y_{it}'s our aim is to predict the $Y_{iT_{i+1}}, Y_{iT_{i+2}}, \ldots Y_{iT}$, for each i for which $T_i < T$.

Since our analysis of the data for model year 1 does not depend on any data from the previous model years we start off our forecasting scheme by looking at the time series for model year 1. Based on an examination of Figures 3.1 and 3.2, and also other similar series, we propose the *second order dynamic linear models* as suitable candidates to consider. Accordingly, we let

$$\begin{aligned} Y_{1t} &= \mu_{1t} + u_{1t}, && \text{with } u_{1t} \sim \mathcal{N}(0, U_{1t}), \\ \mu_{1t} &= \mu_{1t-1} + \beta_{1t-1} + v_{1t}, && \text{with } v_{1t} \sim \mathcal{N}(0, V_{1t}), \quad \text{and} \\ \beta_{1t} &= K_t\beta_{1t-1} + \omega_{1t}, && \text{with } \omega_{1t} \sim \mathcal{N}(0, W_{1t}), \end{aligned}$$

where K_t is chosen to describe various shapes of the underlying growth. When $K_t = 1$ for all t, the proposed model is an *ordinary growth model.* If for some $t = t_0, K_{t_0} = c_0 \neq 1$, and $K_t = 1$, otherwise, the model is known as a *growth model with a bend.* Such models are meaningful for describing claims resulting from warranty policies which involve a change in coverage. For example, with automobiles the nature of the coverage after 12 months in service changes, suggesting that $t_0 = 13$; see Figure 3.2. For the data of Figure 3.1 which shows an S-shaped tendency, a suitable strategy for K_t, for some $t_1 < t_2$, would be of the form $K_t = c_1 > 1$, for $t \leq t_1, K_t = 1$ for $t_1 < t \leq t_2$, and $K_t = c_2 < 1$, for $t \geq t_2$. The resulting model will be referred to as an *S-shaped growth model.*

If we were to ignore the issue of data maturation and set $U_{1t} = U_1$ for all t, and assume that U_1^{-1} has a gamma distribution with a specified scale and shape, let $V_{1t} = bU_1, W_{1t} = cU_2$, with b and c specified, and suppose that u_{1t}, v_{1t} and ω_{1t} are serially and contemporaneously uncorrelated, then given $D_{1t} = (Y_{11}, \ldots, Y_{1T_1})$, standard Gaussian theory enables us to obtain the means and the covariances of the posterior and the filtered distributions of $\theta^{(1)} = (\mu_{10}, \beta_{10}, \ldots, \mu_{1t_1}, \beta_{1t_1})$, and the predictive distributions of $Y_{1t}, t = T_1 + j, j+1, \ldots, (T - T_1)$.

For model years 2 and above, the dynamic linear model that we propose has the form

$$\begin{aligned} Y_{pt} &= \mu_{pt} + u_{pt}, \quad \text{with } u_{pt} \sim \mathcal{N}(0, U_{pt}), \quad \text{and} \\ \mu_{pt} &= \gamma\mu_{p(t-1)} + (1-\gamma)\mu_{(p-1)t} + v_{pt}, \text{with } v_{pt} \sim \mathcal{N}(0, V_{pt}) \end{aligned}$$

where $\gamma \in [0, 1]$ is a weight reflecting the effect of the previous model year's level on the model year $p, p \leq 2$.

Assuming that the u_{pt}'s and the v_{pt}'s are serially and contemporaneously uncorrelated, and that U_{pt} and V_{pt} are known, we can, using standard Gaussian theory obtain the means and the covariances of the posterior and the filtered distributions of $\theta^{(p)} = (\mu_{p0}, \mu_{p1}, \ldots, \mu_{pT_p})$, for every specified value of γ. Once the above is done, we can also find the needed predictive distributions. The details underlying the above are in Chen and Singpurwalla (1991). The key notion here is that the filtered means of the μ_{pt}'s determine the priors for the $\mu_{p+1,t}$'s, giving us an ability to correlate data from one year to the next and hence provide a degree of smoothing.

An alternative form for the system equation which is particularly germane when the underlying data displays a trend is to let

$$\begin{aligned} \mu_{pt} &= \mu_{p(t-1)} + \beta_{p(t-1)} + u_{pt}, && \text{with } u_{pt} \sim \mathcal{N}(0, U_{pt}), \quad \text{and} \\ \beta_{pt} &= \gamma K_t \beta_{p(t-1)} + (1-\gamma)\beta_{(p-1)t} + \omega_{pt}, && \text{with } \omega_{pt} \sim \mathcal{N}(0, W_{pt}), \end{aligned}$$

where the K_t are chosen to characterize the nature of the trend. Here γ is a weight reflecting the effect of the previous model year's trend on the model year p. Some details on the use of such models is in Chen and Singpurwalla (1991).

4. WORK IN PROGRESS AND FUTURE ACTIVITIES

The aim of this paper was to give an expository overview of the warranty problem and to indicate its multifaceted nature. The approaches that we have alluded to are at best preliminary and need further refinements, particularly with respect to the reasonableness and practicality of the underlying assumptions. Obviously, much more needs to be done with respect to the bivariate failure model of Section 2.1. The choice of meaningful and workable utility functions is another area that needs much attention. The importance of this topic overshadows the technical and computational issues posed by the modeling of failures problem. The efforts here would involve collaboration with specialists. The game theoretic set-up that we have formulated is restricted to only two manufacturers; extensions to more than two manufacturers would involve complexities with respect to the development of utility functions. Even so, the entries in the payoff matrix of Table 1 are at best tentative.

The material of Section 3 pertaining to the forecasting of warranty claims is more in the direction of mainline statistical activities but can stand improvements in several possible ways. It is here that the full force of modern Bayesian technology and experience with similar scenarios can be brought into play. For one, the specification of the shaping coefficient K_t could be made adaptive by incorporating prior distributions on both t_i and K_t. This would of course lead to nonlinear filtering problems calling for approximations and stochastic simulation techniques. The same is also true for the weighing constant γ. Finally, the question of how to meaningfully account for the maturing data needs to be addressed. One possibility is to make U_{pt} increase as a linear function of t. Another possibility is to use the previous model year estimate of $U_{(p-1)t}$. We are currently in the process of addressing such issues.

ACKNOWLEDGMENTS

Gary McDonald of General Motors introduced us to this problem and has been a continual source of encouragement and stimulation. Jim Lempke of Ford Motor Company has played a key role in supporting our understanding of the nuances of the warranty data and has persisted in supporting our efforts to develop a mechanism for dynamic linear models with leading indicators. Supported by the Office of Naval Research Contract N00014-85-K-0202 and Grant DAAL03-87-K-0056 the U.S. Army Research Office.

REFERENCES

Box, G. E. P. and Jenkins, G. M. (1976). *Time Series Analysis: Forecasting and Control*. San Francisco: Holden-Day.

Chen, J. and Singpurwalla, N. D. (1991). Dynamic linear models with leading indicators. *Tech. Rep.* **91/4**, The George Washington University.

Emmons, W. (1988). Warranties, moral hazard, and the lemons problem. *J. of Econ. Theory* **46**, 16–33.

Glickman, T. S. and Berger, P. D. (1976). Optimal price and protection period decisions for a product under warranty. *Mgt. Sci.* **22**, 1381–90.

Hunter, J. J. (1974). Renewal theory in two dimensions: basic results. *Adv. Appl. Prob.* **6**, 376–391.

Kalbfleisch, J. D. and Prentice, R. L. (1980). *The Statistical Analysis of Failure Time Data*. New York: Wiley.

Luce, R. D. and Raiffa, H. (1957). *Games and Decisions*. New York: Wiley.

Robinson, J. A. and McDonald, G. C. (1991). *Issues Related to Field Reliability and Warranty Data; Data Quality Control: Theory and Pragmatics*. New York: Marcel Dekker.

Singpurwalla, N. D. and Wilson, S. P. (1991). The warranty problem: its statistical and game theoretic aspects. *Tech. Rep.* **91/1**, The George Washington University.

Singpurwalla, N. D. (1991). Survival under multiple time scales in dynamic environments. *Proceedings of the Advanced Research Workshop in Survival Analysis and Related Topics* (P. K. Gelfand and J. Klein, eds.), (to appear).

West, M. and Harrison, J. (1989). *Bayesian Forecasting and Dynamic Models*. New York: Springer.

DISCUSSION

J. DE LA HORRA* (*Universidad Autónoma de Madrid, Spain*)

This is a stimulating and interesting paper about a very appealing subject: the problem of warranties for consumer products.

In the paper, there are two issues of interest. My comments will be focused on the first topic: how to obtain an optimum price-warranty combination. To fix ideas, let us consider the case of an automobile (this is the main example suggested by the authors).

Let p be the selling price of the car and let R be the warranty region. Of course, it is possible to consider any region R, but I am going to constrain my attention to rectangular warranties

$$R = \{(t, u) : t \leq t_W;\ u \leq u_W\}.$$

The reason for doing so is just a practical one: complicated warranties would be much more difficult to understand for ordinary people.

As a consequence, the utility for the manufacturer will be a function of p, t_W and u_W. I agree with the authors in considering as utility function:

$$U_M(p, R) = U_M(p, t_W, u_W) = a_1[N\pi(p, t_W, u_W)][p - v - cE[F]]$$

where N is the size of the market for that product (in other words, the number of possible buyers for a car of a fixed type), $\pi(p, t_W, u_W)$ is the probability that a possible consumer buys the car we are manufacturing (this probability depends on p, t_W and u_W), v is the unit cost of manufacturing, c is the cost of repair and, finally, F is the number of failures in the warranty region R. The main problem we have to solve is how to compute $E[F]$. If T denotes the time to failure and U denotes the usage at failure, the joint density for (U, T) given in the paper by (2.2) presents (in my opinion) some problems; for instance:

Is it reasonable to consider U as a discrete random variable? Is it reasonable to assume that $U|T = t$ is distributed according to a Poisson distribution?

There is a vast literature about families of bivariate distributions. The book by Mardia (1970) may be specially suitable: it is completely devoted to families of bivariate distributions.

As an example, let me propose a simple continuous probability model for U and T:

a) Suppose that the distribution for U is given by a Gamma density:

$$f(u|\theta) = \frac{\theta^\alpha}{\Gamma(\alpha)} u^{\alpha-1} \exp(-\theta u) \qquad \text{for } u > 0$$

where α is a suitable constant chosen by the manufacturer and the parameter θ measures the performance of the automobile: the smaller the θ, the larger is the number of miles to failure. The exponential distribution is widely used as the simplest model for lifetimes (see, for example, Mood *et al.* (1974; p. 113)) and is a particular case of Gamma distributions.

b) Suppose that the distribution for $T|U = u$ is given by a density $f(t|u, \theta) = f(t|u)$ (this density does not depend on θ). A Normal distribution is perhaps suitable (but this is not necessary for calculations below).

Therefore, with these assumptions, the distributions for U and T is given by the density $f(u, t|\theta) = f(u|\theta)f(t|u, \theta) = f(u|\theta)f(t|u)$. This model may be suitable for cars (in an approximate way). Next, we use $f(u, t|\theta)$ for obtaining the predictive distribution for (U, T), when we have got data $(u_1, t_1), \ldots, (u_n, t_n)$ from n previous cars. First of all, we have to

* This discussion has been partially supported by DGICYT under grant PB88–D178.

choose the prior distribution for θ. If we use the noninformative prior given by Jeffreys' rule:

$$\pi(\theta) \propto (I(\theta))^{1/2} \propto \frac{1}{\theta}$$

after some calculations, we have

$$f(u,t|(u_1,t_1),\ldots,(u_n,t_n)) = f(t|u)\frac{\Gamma((n+1)\alpha)(\Sigma u_i)^{n\alpha}u^{\alpha-1}}{\Gamma(\alpha)\Gamma(n\alpha)(u+\Sigma u_i)^{(n+1)\alpha}}.$$

Now, it is possible to compute $E(F)$ by simulation.

The model $f(u,t|\theta)$ is perhaps too simple. But the point is that, working with this continuous model, information coming from data (u_i,t_i) is used and this is obviously important. Of course, there are other alternatives:

a) We can use an informative prior.

b) We can consider more parameters in the model (but calculations become more complicated).

c) We can consider that $f(t|u,\theta)$ does depend on θ.

Finally, some words about $\pi(p,t_W,u_W)$, the probability that a possible consumer for a car of this type buys the automobile we are manufacturing. I agree with the authors in that $\pi(p,t_W,u_W)$ is a function of $U_C(p,t_W,u_W)$, the utility for the consumer when the price is p and the warranty is given by t_W and u_W. But, in my opinion, it is possible to give $U_C(p,t_W,u_W)$ in an easier way:

Let $\bar{p}$ be the mean price for cars of this type. $\bar{t}_W$ and $\bar{u}_W$ are analogously defined. Then we could define:

$$U_C(p,t_W,u_W) = \eta_1 g_1(p,\bar{p}) + \eta_2 g_2(t_W,\bar{t}_W) + \eta_3 g_3(u_W,\bar{u}_W)$$

where g_1 is a decreasing function of $p-\bar{p}$, g_2 is an increasing function of $t_W-\bar{t}_W$ and g_3 is an increasing function of $u_W-\bar{u}_W$ (for suitable functions g_1, g_2 and g_3). By using this utility, we are considering the effect of the competitors in an easy way, although the model proposed by the authors is perhaps more realistic. Now, we just have to maximize the utility $U_M(p,t_W,u_W)$. A further advantage of this approach is that it is not necessary to look for equilibrium points. The theory of equilibrium points in Game Theory is nice, mathematically speaking, but sometimes produces suspicious results for real life: remember the very well-known prisoner's dilemma (see, for instance, Owen (1968; p. 139)).

REPLY TO THE DISCUSSION

We would like to thank our discussant for his comments on our paper.

Firstly, we agree that rectangular warranty regions are the easiest to understand and, to the best of our knowledge, have been the only shape implemented so far. Other warranty regions are less intuitive (except perhaps the semi-infinite one we proposed), but we feel it is correct to investigate them in case they hold any advantages for either consumer or manufacturer.

The failure model presented in our paper was indexed by time and usage. We would like to emphasize here the novelty of our approach; we are not deriving an ordinary bivariate failure model in the sense of describing the failure of two components. Here we have only one component, and would like to describe its failure in terms of the two quantities time and usage. The model we presented was derived from a set of reasonable assumptions about the relationship between these two quantities.

There were two reasons for usage being a discrete quantity:

a) We felt it natural to model usage as a nondecreasing stochastic process. In many respects, the homogeneous Poisson process is the simplest such process. As this is our first attempt at such models, we chose the simplest case. We have shown that the resulting model is analytically tractable and can be used to obtain marginals, conditionals and moments.

b) In practice, a manufacturer will want to set his optimal warranty at only certain discrete values of usage and time i.e., in the automobile example time may be restricted to an integer number of years and usage to multiples of 1000 miles. So in practice we will maximize utility over a finite set of values for time and usage in the warranty. A continuous model is not essential.

We conclude our discussion of the model by stating that generalization of the model, including deriving models that are continuous in both time and usage, is the subject of present research by us. We are also looking at the placing of priors on the parameters in the model, so that we can allow for updating of our beliefs in them.

We think that Professor de la Horra's idea for the consumer's utility in the presence of competing manufacturers is a worthy one. Consumers compare a particular manufacturer's price and warranty with the average over all other manufacturers, and base their utility on this comparison. These utilities are used in the calculation of the manufacturers' utilities.

It is here that we feel game theory must enter. Each manufacturer has a set of strategies, and each strategy has a utility for each combination of its competitors' strategies. How does a manufacturer maximize his utility, given that his opponents have reasonable, if not complete, knowledge of his utilities and so can plan accordingly? We have in effect a n-person, non-zero sum game. We agree that equilibrium points may not provide a good solution to this game, but other types of 'solution' to such games exist (although again they may not be particularly sensible). This reflects more on the incompleteness of game theory in handling the psychology of such situations than on our model.

ADDITIONAL REFERENCES IN THE DISCUSSION

Mardia, K. V. (1970). *Families of Bivariate Distributions*. London: Charles W. Griffin.

Mood, A. M., Graybill, F. A. and Boes, D. C. (1974). *Introduction to the Theory of Statistics*. New York: McGraw-Hill.

Owen, G. (1968). *Game Theory*. Philadelphia: Saunders Co.

BAYESIAN STATISTICS 4, pp. 447–465
J. M. Bernardo, J. O. Berger, A. P. Dawid and A. F. M. Smith, (Eds.)

Learning in Probabilistic Expert Systems

DAVID J. SPIEGELHALTER and ROBERT G. COWELL
MRC Biostatistics Unit, Cambridge, UK and *University College London, UK*

SUMMARY

Probabilistic expert systems use a directed graphical structure to express conditional independence relationships, and conditional probability tables to summarise quantitative knowledge. We explore the consequences of assuming these probabilities to be parameters, where beliefs about those parameters are updated as data accumulate. A simple approximate Bayesian procedure is shown to be related to those used in 'unsupervised learning' and is investigated by simulations and applied to a difficult real example. The procedure has reasonable properties, although for certain missing data configurations the approximations used are clearly somewhat extreme, and further work is required to handle induced dependencies between parameters.

Keywords: BETA DISTRIBUTION; BAYESIAN INFERENCE; UNSUPERVISED LEARNING; DIRICHLET DISTRIBUTION; MOMENT MATCHING.

1. INTRODUCTION

Research into expert systems has provided a strong challenge for subjectivist Bayesian methodology. Early work claimed that formal probabilistic techniques were inadequate for handling uncertainty in expert systems (Shortliffe and Buchanan, 1975), and critics argued that too many numbers were required for large complex models, and that subjective estimates were either unavailable or unreliable. This has led to fifteen years of work on alternative representations, which tend to divide into two opposing camps: those, such as rule-based systems, that emphasise qualitative expert judgement and downplay the role of empirical data, and those that are entirely data-analytic and ignore background knowledge, such as neural networks and techniques in 'machine learning'.

In recent years, however, there has been a resurgence of interest in probabilistic reasoning, which has been intimately connected to a *graphical model* of the qualitative relationships between quantities of interest. It has been found that such graphs, embodying general conditional independence assumptions, provide a powerful tool for both visual representation of complex problems, and as a basis for computational algorithms. It is now recognised that the importance of such graphs extends well beyond probabilistic reasoning, and recent research has emphasised unifying schemes for handling belief functions, possibility theory, constraint satisfaction, decision analysis and so on (Oliver and Smith, 1990). Flexible software exists (see, for example, Andersen *et al.* (1989), and an increasing number of applications are being reported.

Up to now it has been generally assumed that the parameters of the model are given and fixed, and attention has focussed on issues concerning evidence propagation in such graphs; see, for example, Pearl (1988), Lauritzen and Spiegelhalter (1988) and Shenoy (1989). However, in this paper we shall consider procedures for learning about model parameters from accumulating data: this illustrates what we see is the great advantage of probabilistic reasoning over other proposals for representing beliefs in graphical models — the ability of Bayesian *probabilistic* reasoning to become Bayesian *statistical* reasoning. In the next section we shall show how by considering the initial subjective probability assignments as themselves being random quantities, we can adapt standard Bayesian techniques to handle case-to-case revision of beliefs concerning the parameters of the system being constructed. This allows the gradual adaptation from a judgement-based to a data-based model, instead of the either-or dichotomy described above.

Because data are almost inevitably incomplete, our sequential Bayesian updating procedure requires an extension of previous techniques in unsupervised learning (Titterington, Smith and Makov, 1985). A proposal based on moment-matching was made in Spiegelhalter and Lauritzen (1990) (henceforth denoted S&L (1990)) and this is described in Section 3, where some of its properties are also explored. We then use simulated data from an example given in Lauritzen and Spiegelhalter (1988) (denoted L&S (1988)) to investigate the effect on learning about a particular parameter when evidence becomes increasingly 'remote', in a sense related to the graphical structure of the model.

A real example concerned with the diagnosis of congenital heart disease provides a test of the learning procedure on very sparse data, and comments are made on its performance. In the final discussion we acknowledge that initial subjective judgements should not be kept in a model uncritically, but introduce the notion of sequential monitoring and criticism of prior judgements in the light of accumulating data. These ideas are embodied in software *BAIES*, which is the subject of a companion paper (Cowell, 1992).

2. THEORETICAL BASIS FOR LEARNING PROCEDURE

The basic model underlying probabilistic expert systems is to assume a set of random variables X_v, $v \in V$ are related to an acyclic directed graph in which each variable in V is represented by a node in the graph. The structure of the graph represents qualitative conditional independence assumptions among the variables in V. Specifically, each variable $v \in V$ has a set of 'parent' nodes $\mathrm{pa}(v)$, where for each $w \in \mathrm{pa}(v)$ there exists a directed link $w \to v$. The graph represents the assumption that if we know the values of $\mathrm{pa}(v)$, then v is independent of all other nodes in the graph that are not 'descendants' of v, where u is a descendant of v if there is a directed path from v to u. The parents of v are therefore intended to represent direct influences; this leads to such models being termed 'causal networks' or 'influence diagrams'. See Lauritzen *et al.* (1990) for a detailed discussion of the conditional independence properties expressed by such networks.

The quantitative component of the model comprises a conditional probability distribution for v for each configuration of parents $\mathrm{pa}(v)$; we shall use the abbreviated notation $p(v|\mathrm{pa}(v))$ to denote this table. The conditional independence assumptions lead to the joint distribution $p(V)$ having the modular form

$$p(V) = \prod_{v \in V} p(v|\mathrm{pa}(v)). \tag{1}$$

L&S(1988) show how by suitable adaptation of the graphical structure, one can exploit the decomposition (1) in order to provide straightforward calculation of the conditional probability $p(v|\mathcal{E})$, where v is any node and $\mathcal{E}$ is any evidence observed on the network, which may take the form of observed nodes, or possibly just likelihood terms on members of V.

Existing applications of probabilistic systems assume that $p(v|\mathrm{pa}(v))$ is fully specified for all v. This is clearly unrealistic, and S&L(1990) introduce the natural extension of considering these conditional probabilities as being generated by parameters θ_v, which are components of an overall parameterisation θ. Thus (1) becomes

$$p(V|\theta) = \prod_{v \in V} p(v|\mathrm{pa}(v), \theta_v). \tag{2}$$

Our major assumption is that of *global independence*, i.e., the parameters $\{\theta_v, v \in V\}$ are assumed *a priori* independent random variables and so $p(\theta) = \prod_v p(\theta_v)$. This assumption leads to the joint distribution of case-variables V and parameters θ being expressed as

$$p(V, \theta) = \prod_v p(v|\mathrm{pa}(v), \theta_v)p(\theta_v) \tag{3}$$

From (3) it is clear that θ_v may be considered, formally, as another parent of v in a general network, such as that shown in Figure 1 for the example introduced in L&S(1988). Thus, for example, θ_β is a random quantity whose realisation would provide the conditional probability distribution $p(\beta|\sigma)$, ie the incidence of bronchitis in smokers and non-smokers.

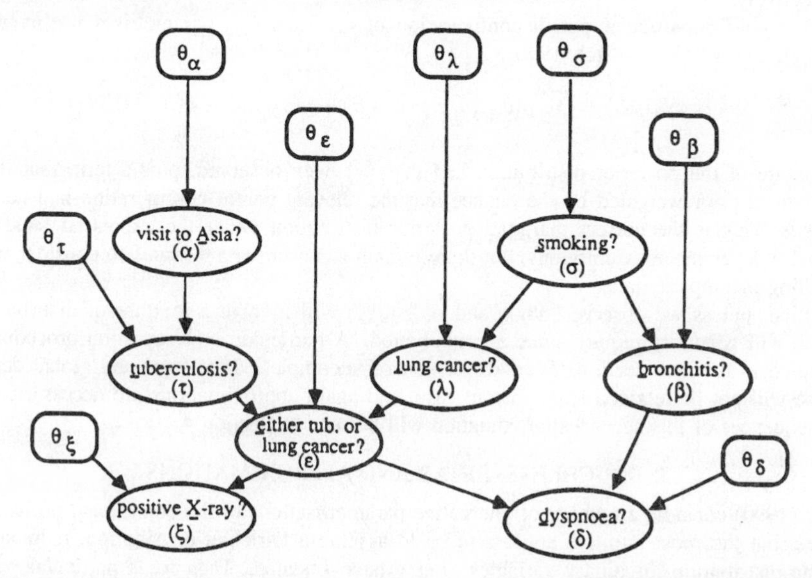

Figure 1. *Network from L&S(1988) with supplementary 'parameter' nodes, representing marginally independent random quantities $\theta_v, v \in V$ whose realisations specify the conditional probability tables for the network.*

Now

$$p(v) = \int p(v, \theta)\, d\theta = \int \prod_v p(v|\mathrm{pa}(v), \theta_v)p(\theta_v)\, d\theta_v = \prod_v p(v|\mathrm{pa}(v)) \tag{4}$$

where

$$p(v|\mathrm{pa}(v)) = \int p(v|\mathrm{pa}(v), \theta_v)p(\theta_v)\, d\theta_v$$

is the expectation of the conditional probability table for v. Hence when processing a new case we can simply use the current 'mean probabilities' within the standard evidence propagation techniques assuming known parameters.

After observing, say, evidence $\mathcal{E}$ on a current case, we wish to revise our beliefs concerning θ_v. It is straightforward to show that, given v and $\mathrm{pa}(v)$, θ_v is independent of all other nodes in the graph, and hence the appropriate posterior on θ_v is given by

$$p(\theta_v|\mathcal{E}) = \sum_{v,\text{pa}(v)} p(\theta_v|v, \text{pa}(v))\, p(v, \text{pa}(v)|\mathcal{E}) \tag{5}$$

where $p(v, \text{pa}(v)|\mathcal{E})$ is easily obtained from the evidence propagation procedure.

A further simplification is obtained if we are willing to assume *local independence*, by which we mean that θ_v breaks into components corresponding to the different configurations of $\text{pa}(v)$, which are then assumed mutually independent random quantities. For example, in Figure 1, we would break θ_β into $\theta_{\beta|\text{smoker}}$ and $\theta_{\beta|\text{non-smoker}}$, assumed to be marginally independent variables specifying the incidences of bronchitis in smokers and non-smokers respectively.

Let ρ be a particular parent configuration of v. Then $\theta_{v|\rho}$ parametrises $p(v|\text{pa}(v) = \rho, \theta_{v|\rho})$, and we obtain (S&L(1990))

$$p(\theta_{v|\rho}|\mathcal{E}) = \sum_v p(\theta_{v|\rho}|v, \rho)p(v, \rho|\mathcal{E}) + p(\theta_{v|\rho})(1 - p(\rho|\mathcal{E})), \tag{6}$$

a mixture of the posterior distribution had $v, \text{pa}(v)$ been observed, plus a term that is the unchanged prior weighted by the chance that the relevant parent configuration had not occurred. This is the correct marginal posterior distribution under the expressed local and global independence assumptions, but there remain some important issues to consider when handling incomplete data.

First, unless we observe both v and $\text{pa}(v)$, (6) will involve a mixture of distributions which will generally require some approximation. A particular approximation procedure is explored in the next section. Second, for general incomplete data, local and global dependence will not be retained from case to case, and again approximations are necessary. The consequences of this second approximation will be seen in Section 4.

3. DIRICHLETS, BETAS AND APPROXIMATIONS

S&L (1990) consider a number of alternative parametrisations of the conditional probability tables, but the most intuitive appears to be to assume a Dirichlet distribution, reducing to a beta distribution for binary variables. Let v have k states. Then for a particular parent configuration ρ, we assume a parametrisation $p(v|\text{pa}(v) = \rho, \theta_{v|\rho}) = \theta_{v|\rho}$, with $\theta_{v|\rho}$ having a Dirichlet distribution $\mathcal{D}[\alpha_1, \ldots, \alpha_k]$. We can think of the α_j as representing counts of past cases which are stored as a summary of our experience. For the next case, the conditional probability used for the jth category is from (4)

$$p(v_j|\text{pa}(v) = \rho) = E[\theta_{v|\rho}]_j = \alpha_j/\alpha$$

where $\alpha = \sum_{i=1}^k \alpha_i$ is the current 'precision' underlying our beliefs concerning $\theta_{v|\rho}$.

If we observe v to be in the jth state, we have by standard conjugate Bayesian updating that

$$\theta_{v|\rho}|v_j \sim \mathcal{D}[\alpha_1, \ldots, \alpha_j + 1, \ldots, \alpha_k],$$

denoted $\mathcal{D}_j$. Hence if both $\text{pa}(v)$ and v are observed, we have a simple accumulation of cases gradually revising our point estimates of the conditional probabilities underlying the system. However, assuming local and global independence, in general the revised distribution for $\theta_{v|\rho}$ is seen from (6) to be

$$p(\theta_{v|\rho}|\mathcal{E}) = \sum_j \mathcal{D}_j\, p(v_j, \rho|\mathcal{E}) + \mathcal{D}_0\,(1 - p(\rho|\mathcal{E})), \tag{7}$$

where $\mathcal{D}_0$ is the initial distribution $\mathcal{D}[\alpha_1, \ldots, \alpha_k]$. With general patterns of missing data (7) will need approximation to prevent an explosion of terms. A similar problem is faced in the area of 'unsupervised learning'; see, for example, Titterington, Smith and Makov (1985) and Bernardo and Girón (1988), in which (7) is generally approximated by a single Dirichlet distribution $\mathcal{D}[\alpha_1^*, \ldots, \alpha_k^*]$ with suitably chosen parameters. Techniques include the 'probabilistic teacher', in which the missing variables are randomly fixed according to their current posterior distribution, and the 'quasi-Bayes' (Smith and Makov, 1978) or 'fractional updating' (Titterington, 1976) procedure which would set

$$\alpha_i^* = \alpha_i + p(v_i, \rho|\mathcal{E}).$$

However, Bernardo and Girón (1988) criticise this procedure on the grounds that if ρ is observed, then $\alpha^* = \alpha + 1$, i.e., the precision is increased by one whatever the observed evidence on v. This will be discussed further below.

Our approach follows that of the 'probabilistic editor' (Titterington, Smith and Makov, 1985, p. 183), in which the approximating distribution attempts to match the moments of the correct mixture (7). There is a degree of arbitrariness in what aspect of the Dirichlet distribution is matched — we equate the 'average' variance of the two distributions. Let $\theta_{v|\rho}^i$ be the ith term in $\theta_{v|\rho}$, and let m_{ij} and ν_{ij} denote the mean and variance of $\theta_{v|\rho}^i$ in the jth term of (7). Then the true mean and variance of $\theta_{v|\rho}^i$ are given by

$$m_i = \sum_j m_{ij}\, p(v_j, \rho|\mathcal{E}) + m_{i0}\,(1 - p(\rho|\mathcal{E})),$$

and

$$\nu_i = \sum_j (\nu_{ij} + (m_{ij} - m_i)^2)\, p(v_j, \rho|\mathcal{E} \, + \, (\nu_{i0} + (m_{i0} - m_i)^2)\,(1 - p(\rho|\mathcal{E})).$$

The 'average variance' of the mixture distribution is defined to be

$$\overline{\nu} = \sum_i m_i \nu_i.$$

Suppose the approximating distribution obeys

$$\alpha_i^* = \alpha^* m_i$$

so that it has precision α^* and the correct mean. Then it will also have the correct average variance if

$$\alpha^* = \frac{\sum_i m_i^2 (1 - m_i)}{\sum_i m_i \nu_i} - 1.$$

We need to explore the properties of this technique, particularly with regard to the change in precision, and this is made easier by assuming for the moment that we observe ρ as part of $\mathcal{E}$, and hence the 0th term in (7) is absent. Denote $p(v_i|\mathcal{E})$ by p_i^*. Then it is straightforward to show that

$$m_i = \frac{\alpha_i + p_i^*}{\alpha + 1}$$

$$v_i = \frac{p_i^*(1 - p_i^*)}{(\alpha + 1)(\alpha + 2)} + \frac{m_i(1 - m_i)}{\alpha + 2}$$

which leads to

$$\alpha^* = \frac{(\alpha+1)^2 \sum_i m_i^2(1-m_i) - \sum_i m_i p_i^*(1-p_i^*)}{(\alpha+1)\sum_i m_i^2(1-m_i) + \sum_i m_i p_i^*(1-p_i^*)}.$$

This precision is maximised when $p_j^* = 1$ for some j, in which case $\alpha^* = \alpha + 1$. When $p_i^* = \alpha_i/\alpha$ for all i, then $\alpha^* = \alpha$, and thus there is no change in the parameters when the posterior distribution on v matches the prior distribution, i.e., no relevant evidence has been obtained. Thus our technique obeys the *desiderata* formulated by Bernardo and Girón (1988).

We note that it is quite feasible for α^* to be less than α, so that our precision *decreases* as we obtain evidence. This is easily seen by noting that

$$\alpha^* - \alpha = \frac{(\alpha+1)(\sum_i m_i^2(1-m_i) - \sum_i m_i p_i^*(1-p_i^*))}{(\alpha+1)\sum_i m_i^2(1-m_i) + \sum_i m_i p_i^*(1-p_i^*)},$$

and hence, for example, $\alpha^* - \alpha \leq 0$ if $p_i^* = 1/k$ for all i.

The binary case provides insight into the phenomenon of losing precision. We have exact matching of first and second moments, since $\nu_1 = \nu_2 = \overline{\nu}$. The change in parameters is given by

$$\alpha_1^* - \alpha_1 = \frac{(\alpha_1+1)\,\alpha\, p_1^*\,(p_1^* - \frac{\alpha_1}{\alpha})}{(\alpha_1+p_1^*)(\alpha_2+1-p_1^*) + (\alpha+1)p_1^*(1-p_1^*)}$$

$$\alpha_2^* - \alpha_2 = -\frac{(\alpha_2+1)\,\alpha\,(1-p_1^*)\,(p_1^* - \frac{\alpha_1}{\alpha})}{(\alpha_1+p_1^*)(\alpha_2+1-p_1^*) + (\alpha+1)p_1^*(1-p_1^*)}$$

It is clear that unless $p_1^* = 0$ or 1 then one of the parameters of the beta distribution must decrease — if we think of the parameters as representing implicit past cases, then a fraction of these are 'forgotten'. The change in total precision is given by

$$\alpha^* - \alpha = \frac{(\alpha+2)\,\alpha\,(p_1^* - \frac{\alpha_2+1}{\alpha+2})\,(p_1^* - \frac{\alpha_1}{\alpha})}{(\alpha_1+p_1^*)(\alpha_2+1-p_1^*) + (\alpha+1)p_1^*(1-p_1^*)}$$

which is therefore negative for p_1^* lying between $(\alpha_2+1)/(\alpha+2)$ and α_1/α. The maximum loss of precision can be found to occur when

$$p_1^*/(1-p_1^*) = \sqrt{\frac{1+\alpha_2^{-1}}{1+\alpha_1^{-1}}},$$

i.e., close to $p_1^* = .5$ for moderate α_1, α_2.

This is illustrated in Figure 2. The first example is that used by Bernardo and Girón (1988) to illustrate their procedure — their results appear very similar to our technique. The second example shows the effect of strong imbalance between α_1 and α_2, which will occur when the current estimated probability is extreme. Here we find that for a wide range of p_1^* there will be an overall drop in precision of up to the equivalent of four 'observations' — more extreme examples can be constructed in which up to a quarter of the precision can be lost from the evidence on a single case. This phenomenon occurs when an event which has a high estimated prior probability is contradicted by the available evidence, to the extent that the posterior probability approaches .5. This does not seem unreasonable behaviour, but its consequences are further examined in the next section.

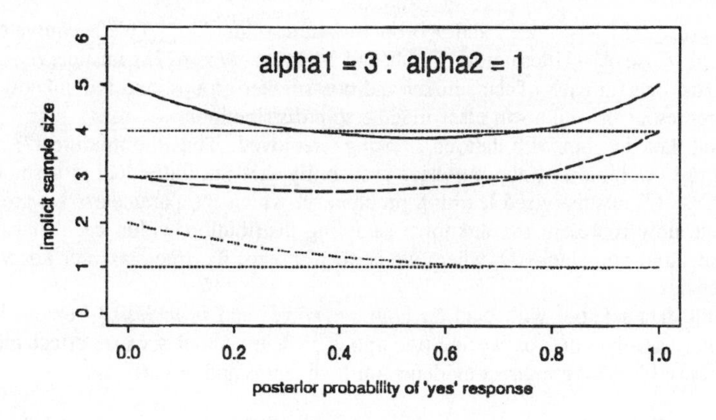

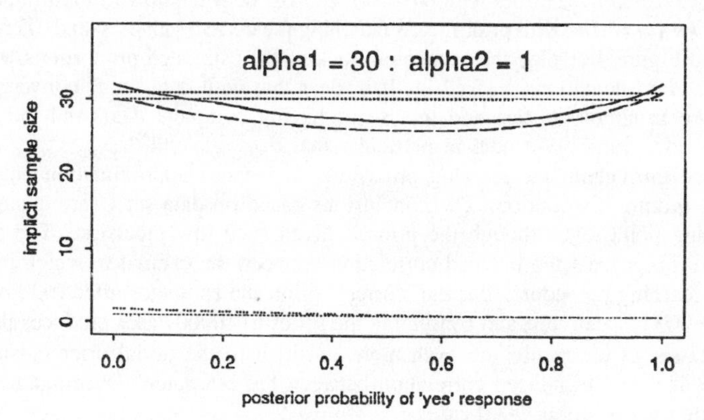

Figure 2. *Final implicit sample sizes α_1^* (dashes), α_2^* (grouped dots) and α^* (solid line) after observing evidence that leads to a posterior probability of a 'yes' response to be p_1^*, when the prior probability of such a response is α_1/α. The horizontal dotted lines are the original values of α_1, α_2 and α.*

4. SIMULATION EXPERIMENTS

The example illustrated in Figure 1 provides an opportunity to test the learning procedure with simulated data. 1000 cases were generated from the joint distribution derived from the conditional probability tables given in L&S (1988), with the slight adaptation that the proportion of non-smokers that have bronchitis was changed from .3 to .05. All the conditional probabilities are assumed to be correctly known, except those specified by θ_β i.e., the incidences of bronchitis in smokers and non-smokers. The purpose of the exercise was to investigate how well these proportions, which actually are .6 and .05 respectively, could be estimated in the light of extensive missing data.

Four data-sets were used, denoted A to D.

A The full data set, which provides a base-line for the other experiments.

B The full data-set, but with data on *bronchitis?* removed. Thus we observe the parent of β, but have only a sequence of posterior probabilities $p(\beta|\mathcal{E})$ from which to learn about $p(\text{bronchitis}|\text{smoker})$ and $p(\text{bronchitis}|\text{non}-\text{smoker})$. This is equivalent to two standard 'Case A' (Titterington, Smith and Makov, 1975, p178) unsupervised learning problems, one for each of the smoker and non-smoker groups, and the unknown proportions represent the unknown class membership distribution.

C The full data-set, but with data on *smoking?* removed. Thus the mixture (7) comprises two terms, weighted by the posterior probability that the individual is a smoker. This is a 'Case C' unsupervised learning problem, in which the parameters we are trying to estimate now represent the unknown sampling distribution within each of two classes (smokers and non-smokers), where we neither observe the true class nor know the class frequencies.

D The full data-set, but with data on both *smoking?* and *bronchitis?* removed. This is extremely challenging, as we are attempting to learn about a cause-effect relationship while only observing indirect evidence on both cause and effect.

For each of these data-sets, two prior distributions were used for each of $\theta_{\beta|\text{smoker}}$ and $\theta_{\beta|\text{non}-\text{smoker}}$: 'reference' priors with $\alpha_1 = \alpha_2 = .5$ for both unknown proportions, and 'correct' priors with α =1 but with prior means matching the correct values .6 and .05 respectively. Figure 3 and Figure 4 display the sequential track of the estimated proportions, with $+/-2$ posterior standard deviations marked in. It is clear that with data-set B convergence to the correct figure is almost as fast and precise as having complete data, and independent of the starting prior value. We note in particular that $\theta_{\beta|\text{smoker}}$ and $\theta_{\beta|\text{non}-\text{smoker}}$ retain local independence throughout the learning procedure, and hence approximation only comes in through the mixing distribution. The conclusions based on data-set C are clearly sensitive to the starting point, even though the prior is given such low precision. The plot for the 'reference' prior reveals the induced correlation between the estimates, which is ignored in the current learning procedure. For the 'correct' prior, the estimates after 1000 observations are .67 and .003 respectively, and continuing the series to 10000 cases produces little change.

For data-set D the results are even more sensitive to the initial prior mean, and there is evidence of extreme induced correlation between the estimates. Once again, the pattern changes little over a further 9000 cases.

The simulations, restricted as they are, suggest the learning procedure works well in the standard unsupervised learning problem in which the parent nodes are observed, may be inaccurate when trying to learn conditional probabilities with unobserved parent nodes, and could potentially be misleading if trying to learn about a link in a network on which no direct evidence is being observed.

5. DIAGNOSIS OF CONGENITAL HEART DISEASE

The precise diagnosis of congenital heart disease in the first days of life is a difficult and important task. As part of a project concerned with providing decision-support at Great Ormond Street Hospital for Sick Children data on 400 babies has been collected (Franklin *et al.* 1991) . A directed graphical model for part of the disease spectrum has been constructed in collaboration with the consultant paediatric cardiologists and is shown in Figure 5. Accompanying this model are extensive tables of subjectively estimated conditional probabilities, which are not shown here. Data are available on 168 babies with one of the six diseases covered by this network, but only nodes numbered 1, 2, 3 and 15 to 20 have been recorded. It would be feasible, if time consuming, to obtain data on the 'internal' nodes of the network from clinical records, but for the moment we are investigating sequential learning on the basis of such sparse data.

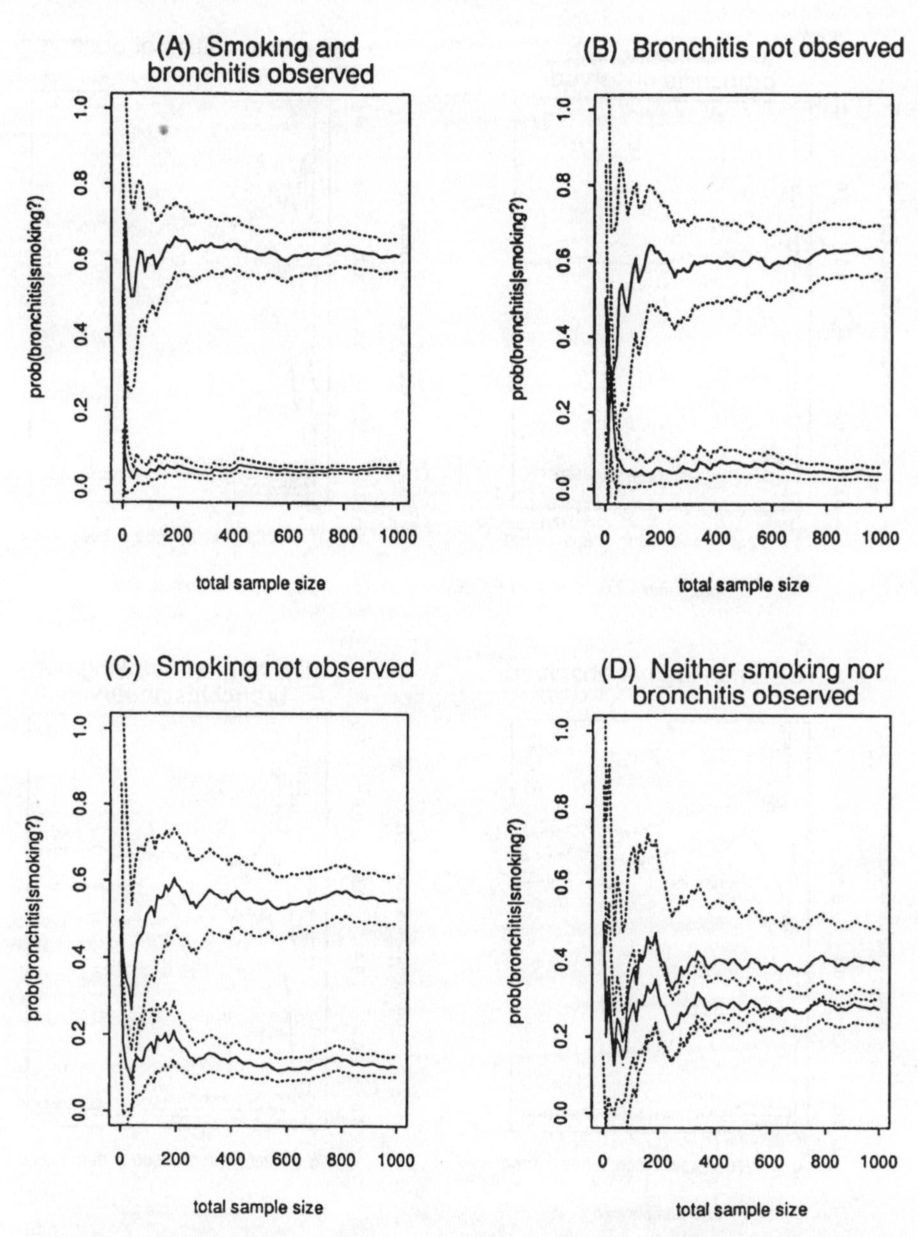

Figure 3. *Estimated incidence of bronchitis (+/ − 2 standard deviations as dotted lines) in smokers (upper solid line) and non-smokers (lower solid line), based on different types of missing data. The prior mean of each proportion was .5, with precision 1.*

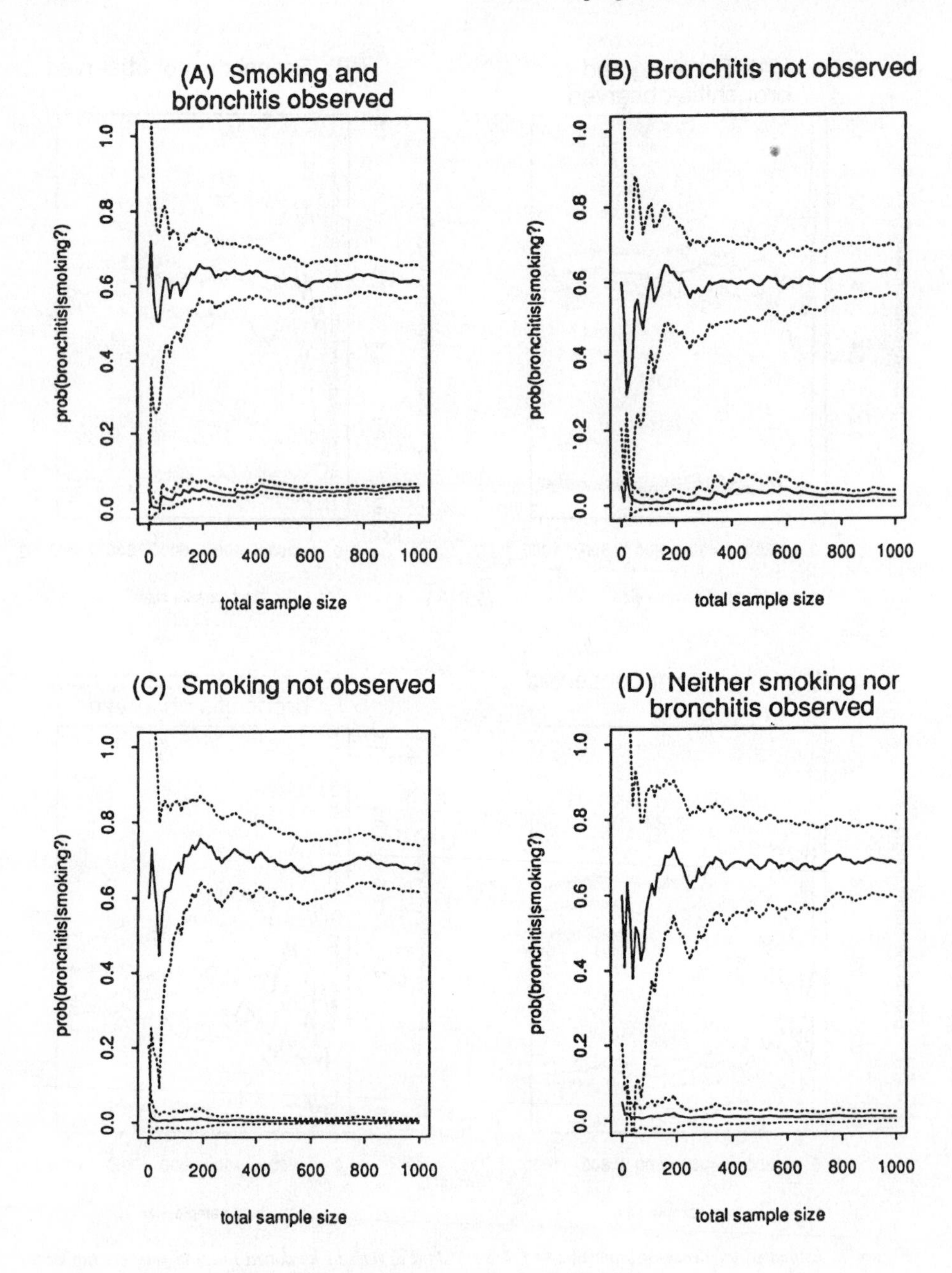

Figure 4. *Same data as before, but with the prior mean of each proportion being the correct values .6 and .05, with initial precision 1.*

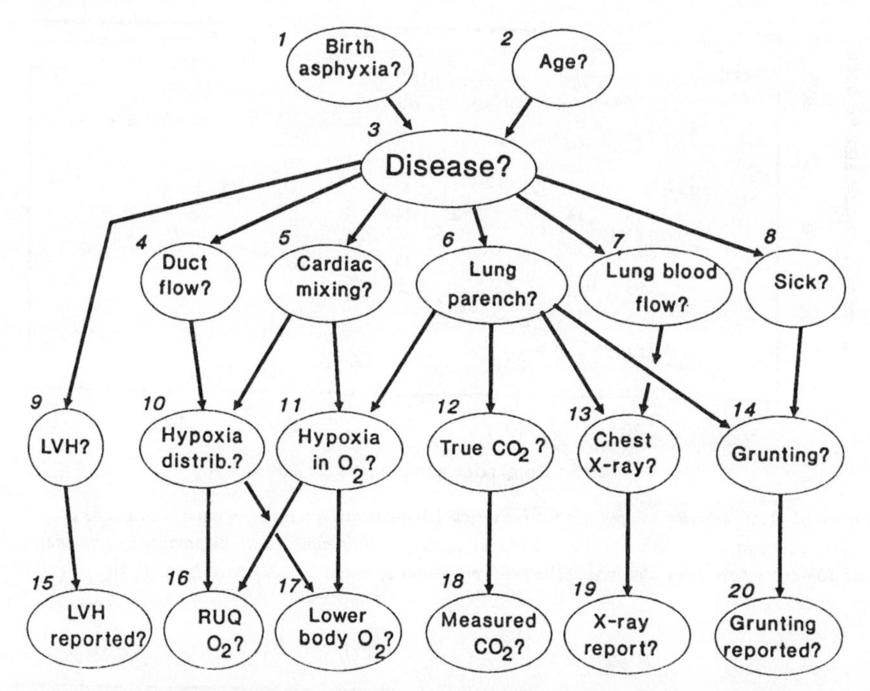

Figure 5. *Graphical model for the presentation of severely 'blue' babies. The structure represent the causal mechanism by which underlying disease produces a patho-physiological anomaly, which in turn leads to clinical manifestations. The final layer in the model reflects the possible observer errors in eliciting the clinical findings.*

Using reference priors produces no learning at all for nodes in which parents are not observed, since the initial uniform conditional probability estimates prevent the propagation of any evidence — essentially the graph remains disconnected apart from observed parent-child pairs. We have therefore used the expert prior means as a starting point, but have given each estimated conditional distribution a low precision $\alpha = 10$.

We first consider how much is learnt about each node from the accumulated data. We measure the knowledge associated with each node by the total implicit sample size, which is the sum of the precisions of its conditional probability tables. The total prior implicit sample size associated with each node therefore depends on the number of parent configurations; for example, node 13's parents have 3 states each, making 9 configurations, and hence node 13 has a prior sample size of 90. Figure 6 plots the prior sample size for each node against its posterior sample size after observing 168 cases. We see that nodes 2 and 3, which were always observed, have derived a full quota of experience from the data, while the other observed nodes have also benefited to some degree from the data (nodes 16 and 17 had considerable missing data). However, almost no evidence appears to have penetrated to the unobserved nodes, except for node 9, *LVH?*, which has benefitted from having both its parent and child observed, and hence being in the position of *bronchitis?* in data-set B of the last section.

It is also of interest to see if the estimated probabilities have been substantially revised

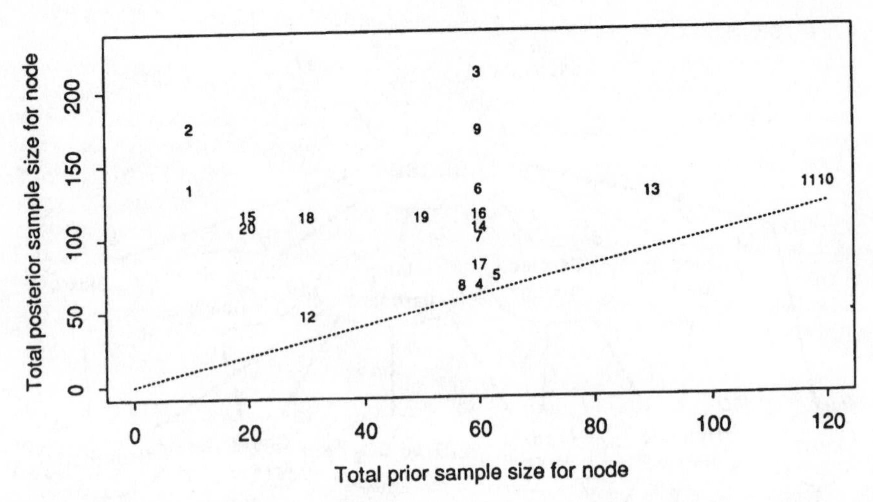

Figure 6. *Total posterior sample size for each node contrasted with total prior sample size. The dotted line shows equality, in which there has been no change in the precision attached to the probability estimates, even after 168 cases have been observed. (The prior precisions of nodes 5, 8, 11 have been slightly perturbed to improve readability.)*

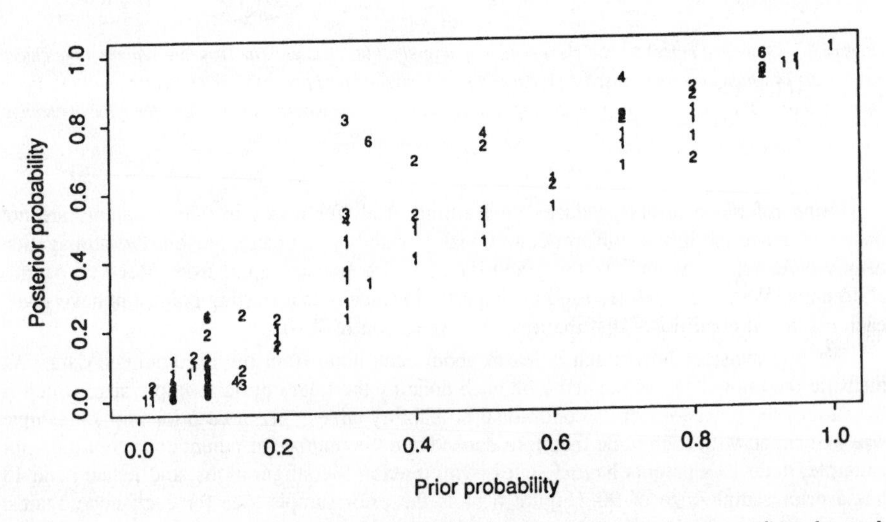

Figure 7. *Prior to posterior change in estimated probability of first category of each node, conditional on each parent configuration. The symbol shows the precision associated with the estimate,* $1 = 5 - 15$, $2 = 15 - 25$, *and so on.*

from their prior assessments. Figure 7 shows the prior against the posterior estimate of the probabilities of the first category of each node conditional on each parent configuration. Any change has been accompanied by an increase in precision, but on the whole the estimated conditional probabilities have changed little. However, this could be for a number of reasons: there might be little relevant evidence reaching the node, the initial estimates may be very good, or the initial estimates may be very influential and produce an inertia. Examining these alternative explanations is beyond the scope of this paper, but our general impression is that all these factors are operating simultaneously. Figure 6 shows that little evidence is reaching the inside of the graph, but Figure 7 shows that even when there is substantial evidence the estimates do not change unduly — this supports the idea of good prior estimates, which is reinforced by studies by Spiegelhalter *et al.* (1990, 1991). Finally, the simulation studies do suggest that if neither a node nor its parents are observed, then the prior estimates could exert a strong influence.

6. DISCUSSION

We have investigated a simple sequential learning procedure. It is similar to those adopted in unsupervised learning, and performs well in similar problems in expert systems, but problems arise in dealing with very sparse data on networks.

We first note that the issues of this paper are similar to those faced in learning on neural networks with unobserved intermediate nodes. It would be interesting to see if these techniques compare with those currently in use in three-level networks. Second, we should really be using non-sequential methods if we have the opportunity. Lauritzen (1991) has recently shown how the L&S (1988) algorithm permits an efficient implementation of the EM algorithm for graphical models with missing data, and this is likely to lead to superior results to those of our sequential techniques, although it is unclear how the initial prior estimates can be exploited.

Third, it is important to have formal techniques for monitoring the adequacy of the model being used. The *BAIES* software incorporates a variety of such monitors (Cowell, 1992), in particular a *global monitor* that sequentially accumulates the predictive probability, based on the current model, of all evidence observed. The comparison of global monitors provides a formal Bayes factor procedure for contrasting different model structures, or different prior opinions.

Finally, if we are going to use sequential learning it is clear from the simulation results that if there is substantial missing data there is a great need for techniques that can incorporate the induced dependencies between parameters. Dawid and Lauritzen (1989) provide a theoretical basis for such learning within the framework of *hyper-markov* models, but up to now there have been no implementable algorithms. Current work is focussing on computational techniques for these models.

ACKNOWLEDGEMENTS

We would like to thank Steffen Lauritzen and Phil Dawid for all their help.

REFERENCES

Andersen, S. K., Olesen, K. G., Jensen, F. V. and Jensen, F. (1989). *HUGIN* — a shell for building Bayesian belief universes for expert systems. *Proceedings of International Joint Conference on Artificial Intelligence*, San Mateo: Morgan Kaufman. 1080–1085.

Bernardo, J. M. and Girón, F. J. (1988). A Bayesian analysis of simple mixture problems. *Bayesian Statistics 3* (J. M. Bernardo, M. H. DeGroot, D. V. Lindley and A. F. M. Smith, eds.), Oxford: University Press. 67–78, (with discussion).

Cowell, R. G. (1992). *BAIES* – a probabilistic expert system shell with qualitative and quantitative learning. *Bayesian Statistics 4* (J. M. Bernardo, J. O. Berger, A. P. Dawid and A. F. M. Smith, eds.), Oxford: University Press, 595–600.

Dawid, A. P. and Lauritzen, S. L. (1989). Markov distributions, hyper-Markov laws and meta-Markov models on decomposable graphs, with applications to learning in expert systems. *Tech. Rep.* **89-31**, Aalborg University.

Franklin, R. C. G., Spiegelhalter, D. J., Macartney, F. J. and Bull, K. (1991). Evaluation of a diagnostic algorithm for heart disease in the neonate. *British Medical Journal* **302**, 935–939.

Lauritzen, S. L., Dawid, A. P., Larsen, B. N. and Leimer, H.-G. (1990). Independence properties of directed Markov fields. *Networks* **20**, 491–505.

Lauritzen, S. L. and Spiegelhalter, D. J. (1988). Local computations with probabilities on graphical structures and their application to expert systems. *J. Roy. Statist. Soc. B* **50**, 157–224, (with discussion).

Lauritzen, S. L. (1991). The EM algorithm for graphical association models with missing data. *Tech. Rep.* **91-05**, Aalborg University.

Oliver, R. M. and Smith, J. Q. (eds.) (1990). *Influence Diagrams, Belief Nets and Decision Analysis.* New York: Wiley

Pearl, J. (1988). *Probabilistic Inference in Intelligent Systems.* San Mateo: Morgan Kaufman.

Shenoy, P. P. (1989). A valuation-based language for expert systems. *Int. J. Approximate Reasoning* **3**, 383–411.

Shortliffe, E. H. and Buchanan, B. G. (1975). A model for inexact reasoning in medicine. *Mathematical Biosciences* **23**, 351–379.

Smith, A. F. M. and Makov, U. E. (1978). A quasi-Bayes sequential procedure for mixtures. *J. Roy. Statist. Soc. B* **40**, 106–111.

Spiegelhalter, D. J. and Lauritzen, S. L. (1990). Sequential updating of conditional probabilities on directed graphical structures. *Networks* **20**, 579–605.

Spiegelhalter, D. J., Franklin, R. C. G and Bull, K. (1990). Assessment, criticism and improvement of imprecise subjective probabilities for a medical expert system. *Uncertainty in Artificial Intelligence 5* (M. Henrion, R. Shachter, L. N. Kanal and J. F. Lemmer, eds.), Amsterdam: North-Holland, 285–294.

Spiegelhalter, D. J., Harris, N. L., Bull, K. and Franklin, R. C. G. (1991). Empirical evaluation of prior beliefs about frequencies: methodology and a case study in congenital heart disease. *Tech. Rep.* **91-4**, MRC Biostatistics Unit, Cambridge.

Titterington, D. M. (1976). Updating a diagnostic system using unconfirmed cases. *Applied Statist.* **25**, 238–247.

Titterington, D. M., Smith, A. F. M. and Makov, U. E. (1985). *Statistical Analysis of Finite Mixture Distributions.* New York: Wiley.

DISCUSSION

J. Q. SMITH (*University of Warwick, UK*)

For many years non-Bayesians have been stating that it is infeasible to build probabilistic systems for learning on complex networks. Spiegelhalter has been one of the regrettably few Bayesians to act to counter this criticism and to illustrate how well systems work when built properly. This paper represents an important staging post in the development of such systems. I hope the following comments might be useful in improving our current methodology.

High dimensional problems require a great deal of care. I shall therefore begin by listing four criteria against which a model construction should be continually appraised. This list is given in my own order of priority.

1. Objective Led Modelling Principle (O.L.M.P.) When constructing a learning process it is essential that we understand the object of the exercise. If a model is useful it should help us to act – the action space and possible objectives should therefore be *explicit.* Note in particular that these will only be influenced by values of *future* observables. No matter how complex we build a statistical model it can at best only approximate our belief structure. How good this approximation is can only be evaluated with regard to the context and decisions in which the system is being used. The Bayesian paradigm is a powerful engine but if we don't address the context of problems people want solved we never get out of first gear.

2. The Honesty Principle. We should try to include in our model plausible information rather than ignorance. Probability distributions are very good at representing information but kick against representing ignorance (a good illustration of this is given in the second example of this paper). It is far better to include plausible (if ill-judged) *consciously strong* prior information which we can re-evaluate at a later stage than to use ignorance priors which provide *implicit* strong prior information which we cannot disentangle from our predictions.

3. The Parsimony Principle (P.P.) The O.LM.P. and H.P. forces us towards building simple tailor-made systems. Complex heterogeneous systems have two very serious flaws which cannot be side-stepped. The first is that it can be virtually impossible to elicit (predictively) good prior information over many variables (H.P.). The second is that it is extremely difficult to understand how this prior information influences our inferences and to communicate this when talking about our *problem* (O.L.M.P.) rather than our model. I certainly do not advocate that we *never* use difficult high dimensional models but I do suggest that these should always be seen as a half-way house to constructing a simple model that both statistician and client are able to understand the implications of properly.

4. The Hypermarkov Principle (H.M.P.) In a parameterised model, the effect of our parameters on future observables will be both easier to interpret and also to calculate (P.P.) if our prior beliefs are related in their conditional independence (c.i.) structure to the algebraic form of the likelihood so that the corresponding posterior c.i. structure is the same as the prior. So, *if* such a model is plausible (H.P.) then it is often a good choice (see Smith (1990), Dawid and Lauritzen (1990)).

Now I will discuss the contents of the paper. My guess is that for non-systematic patterns of missing data the sequential algorithm they describe works well and so it was bold that they illustrated their methods with data that potentially might cause problems. Let us turn to the second (practical) example. First, it would have been very helpful for the authors to explain in what situation they envisage their model being used and the various costs involved (O.L.M.P.). Second, I have found that influence diagrams are very useful Bayesian elicitation tools but I am a little concerned about a blurring between causal structures and their implied conditional independences. In particular, I would like to ask the authors whether the conditional independences across variables in the same strata were *really* elicited (i.e., checked through questions like "given information A would B provide no more information about C") or whether they were just assumed (H.P.). My experience is that for realism such strata should be allowed to be dependent as in chain graph models (Wermuth and Lauritzen (1990)). Such dependency would have considerable influence on their ensuing inferences. Third, they state that strong prior information is necessary and this is introduced in a way to satisfy the H.M.P. However, this strong information may not be reasonably expected to give such c.i. if based on different structures of data from the given data set (H.P.). This makes me want to examine again the O.L.M.P. and P.P. and ask—exactly which probabilities are going to influence your future actions and can't you simplify this model in some way?.

Incidentally, these models are just big multinomials with cell probabilities as functions of the conditional probability parameters embedded in the graph. Since with full data on the cells mentioned the global Markov property ensures $\boldsymbol{\theta}_1, \boldsymbol{\theta}_2$ and $\boldsymbol{\theta}_3$ remain independent of everything else let us concentrate on the multinomial model on the top 3 layers. A quick calculation tells us this is a multinomial with 384 cells, N cases $= 168$, and 70 parameters. Before performing any numerical claculations we can therefore immediately deduce that the system is very sparse and start thinking about simplifying the model.

The first simulation example is instructive. I will discuss the first problematic situation called Case C. Here we observe β and λ but σ is not observed. We are interested in estimating the probability parameters $\boldsymbol{\theta}$ associated with β and we are fortunate here that there are no other observables in the ancestor set (λ, σ, β) so we can consider these variables separately from the rest of the graph. Marginalising out σ we note that we can write our likelihood $L(\boldsymbol{\theta}) = L^+(\boldsymbol{\theta})L^-(\boldsymbol{\theta})$

$$L^+(\boldsymbol{\theta}) = (\psi^+)^{\#bl}(1-\psi^+)^{\#\bar{b}l} \qquad \psi^+ = (10/11)\theta^+ + (1/11)\theta^-$$
$$L^-(\boldsymbol{\theta}) = (\psi^-)^{\#bl}(1-\psi^-)^{\#\bar{b}l} \qquad \psi^- = (1/189)\theta^+ + (188/189)\theta^-$$

where, for example, $\#\bar{b}l$ denotes the number of patients without bronchitis but with lung cancer. Clearly ψ^+ and ψ^- are easy to estimate directly and stably and hence (θ^+, θ^-) through the equation above. Furthermore, contrary to the conjecture in the paper strong (conjugate) prior information on (θ^+, θ^-) can be introduced for example by using independent Betas on ψ^+ and ψ^-. With systematic missing observations the H.M.P. suggest that we marginalise the problem onto the observed variables and their parameters, estimate and then project back onto the original problem. In some circumstances this will alleviate the necessity of sequential approximations altogether. Note that estimating *all* the parameters of (β, λ, σ) from observing data on (β, λ) is a multinomial experiment on 4 cells and 5 free parameters and so is clearly unidentified. By following the H.M.P. we address this issue explicitly.

Now I will turn to their sequential learning procedure. To help me discuss this I will lay down two criteria which I consider important requirements for approximate probabilistic learning.

(1) In a Bayesian expert system there should exist no (function of) *relevant* parameter θ for which

$$\hat{\theta}\{x_t\}_{t\leq n} \xrightarrow{dist} \theta \qquad \text{as } n \to \infty$$

when θ is unidentifiable from $\{x_t\}_{t\geq 1}$.

By a relevant θ I mean a parameter θ on which the predictive distribution of variables on which our objectives, losses, etc. may depend (O.L.M.P.). If we violate (1) then the system will claim almost certainty about statements we can *logically* deduce nothing from the data provided. This could clearly be catastrophic—much more so than confessing to more ignorance than we formally should have.

(2) If the data is not ordered then the results of the approximation should not be (very) sensitive to the *order* of data input.

I think the proposed system violates (2), especially in Case C. There are two ways to circumnavigate these problems—(i) input the data in some fixed order determined a priori as an algorithm for efficiency, e.g., by introducing fullest information first, or more sensibly (ii) model the order explicity as a time series. This is actually not significantly more difficult than the static model, Queen and Smith (1990, 1992), in this genre of system. It is more plausible than the static model (because operational and population characteristics change over time) and also can be used to ensure that the first criterion is never violated. We are currently developing simple Dirichlet time series for exactly this problem.

Finally, in the first example, when we reset probabilities $p(s|l) = p(s|l) = 0.5$ we obtain an unidentifiable system in case C with

$$L(\boldsymbol{\theta}) = \psi^{\#b}(1-\psi)^{\#\bar{b}} \qquad \psi = 1/2\theta^+ + 1/2\theta^-$$

where $\boldsymbol{\theta}$ are the bronchitis parameters. Their system approximates mixtures of 2 Betas with a single Beta and it is easy to check

$$\mathcal{E}(\theta^+) \to \mathcal{E}(\theta^-) \to \frac{\#b}{\#b + \#\bar{b}}$$

but some algebra gives that $\text{Var}(\theta^{+})$ and $\text{Var}(\theta^{-}) \rightarrow 0$. Criterion 1 is therefore not satisfied unless $\boldsymbol{\theta}$ is not relevant (O.L.M.P.)—in which case why has it been included? Notice that diagnostics based on the predictive performance of models are unlikely to identify the problem.

Real practical problems like the ones addressed by these authors provoke interesting theoretical issues which can feed back into the solutions proposed. Unlike classical statisticians we don't need to study solutions in search of an application. Please, please as Bayesians let us only address theoretical issues which derive from actual, rather than hypothetical contexts.

M. GOLDSTEIN (*University of Durham, UK*)

This paper is pretty much state of the art on probabilistic expert systems, and so provides an opportunity to make a general observation about the methodology. Learning, in this paper, is governed by an assumption of global independence between the parameter nodes. In most cases many of the parameter nodes in the influence diagram should be joined, as learning that prior judgements about one parameter were faulty may cast doubt on judgements for other parameters, the same background data will be input into judgements on subcollections of parameter nodes, nodes may be joined according as to which specialists made the relevant prior judgements, and so forth.

So why use global independence assumptions? Mainly, so that the network does not become too complicated to specify and analyse. But this raises the basic question as to whether we should try to express all of our learning and belief propagation within a single network. Coherence is a great tool for linking up our judgements, but in complex problems the requirements of full coherence may be so stringent that we are forced to lie, or suppress our knowledge, merely to enable our analysis to proceed. When this happens, strict coherence may become an obstacle, rather than an aid, and our knowledge may be better expressed as a collection of smaller, interrelated networks, capturing, in detail, limited aspects of our beliefs.

M. WEST (*Duke University, USA*)

The 'explosion' of mixture posterior distributions encountered here is familiar in sequential analyses of multi-process models for time series (West and Harrison, 1989, Chapter 12). In practical application of such models it has been found expedient to monitor the growth of such mixtures to identify and aggregate 'similar' components. The systematic 'collapsing' of a mixture of several Dirichlets to a single Dirichlet (however the collapsing is performed) must be questioned as a routine and automatic procedure; multi-modality, forms of skewness, and other important characteristics of the mixture can be lost in such a process. However, as discussed in my paper at this meeting (West, 1992), mixtures of larger numbers of standard forms may often be rather well approximated (certainly in terms of probabilities, if not densities) by mixtures of significantly fewer components. Thus, a mixture of, say one hundred Dirichlets might be very adequately collapsed to a mixture of ten with a suitable aggregation rule, whereas approximating a mixture of ten by a single component is typically much less reliable. The authors might therefore like to consider the multi-process technique of evolving mixtures; a prior mixture of ten components expands to a posterior mixture of one hundred through a ten component likelihood function, and this might then be reduced to a suitable approximating ten components before proceeding to further updating.

REPLY TO THE DISCUSSION

We are grateful for the fairly sympathetic comments concerning our approach to a difficult practical problem.

Jim Smith's principles lead him to a preference for fairly simple models that are purpose-built for specific objectives: both he and Michael Goldstein feel that such models can better represent reasonable beliefs than 'global' models. We basically agree that strong quantitative and qualitative assumptions are best made and tested in restricted, targeted models, but unfortunately expert systems research is characterised by the attempt to construct models than can be used for a variety of functions. For example, our model for congenital heart disease might be used for diagnosis, therapy selection, selecting tests, explanation, teaching, prediction, and possibly as a guide to structuring data-bases. We have elicited loss functions for errors in decision-making and intend incorporating these into a full influence diagram — in addition we need to see whether much simpler models are adequate for specific diagnostic tasks. Similarly we had already thought that the layered structure that arises naturally in this context points to the use of chain graphs, allowing undirected links between nodes on the same level.

Dr. Smith suggests our exploration of the learning algorithm with systematic missing data was 'bold' — perhaps it was positively foolhardy to immediately test the procedure on a difficult problem it was never really designed to solve. We have repeated the learning experiments on the 'Asia' example with data missing *at random*: Figure A shows the performance, starting with prior mean .5 and precision 1 for the two conditional probabilities under study, where each node for each case has independently been set to missing with probabilities .2, .4, .6 and .8 respectively. We can see that the procedure converges in a reasonable fashion. If learning was done just on cases for which both *smoking* and *bronchitis* are observed, the total final precision (equivalent sample size) of the two estimates would be around 640, 360, 160 and 40 respectively for the four situations examined. With the learning procedure, these precisions were 752, 491, 292 and 140 for 20%, 40%, 60% and 80% missing data. Thus in this context, learning with, say, 40% missing *observations* provided precisions approximately equivalent to those that would be obtained had there been about 50% missing *cases*, with complete data on the remainder. This seems the right sort of behaviour.

Jim Smith's careful analysis of systematic missing data is very welcome: it seems quite appropriate to collapse out the nodes that are not to be observed, learn on the collapsed graph, and then transform back if necessary. We agree that the type of over-confident convergence displayed in some of our experiments is highly undesirable, although by his first principle perhaps such behaviour should be judged in relation to its effect on the predictions of interest — if we are not really interested in the intermediate parameters of a system it may well not matter if the estimates are inconsistent.

Mike West suggests that collapsing a single Dirichlet after each case is too extreme, and a larger mixture could be carried along. We agree that even a small number of components might be an improvement: however, we feel the main problem with our technique is the continuing assumption of global and local independence of the parameters in spite of the dependencies that should really be accumulating. If we used discrete distributions for our parameters instead of Dirichlets, the updating could be done without any need for approximation, but we believe the problems with the convergence would still hold.

Michael Goldstein rightly comments that this assumption of global independence is often unlikely to be appropriate, even with complete data, and like Smith suggests simpler models. We are currently incorporating a variety of diagnostics for poor predictive performance in the system, and our intention is to compare the behaviour of simple and complex models.

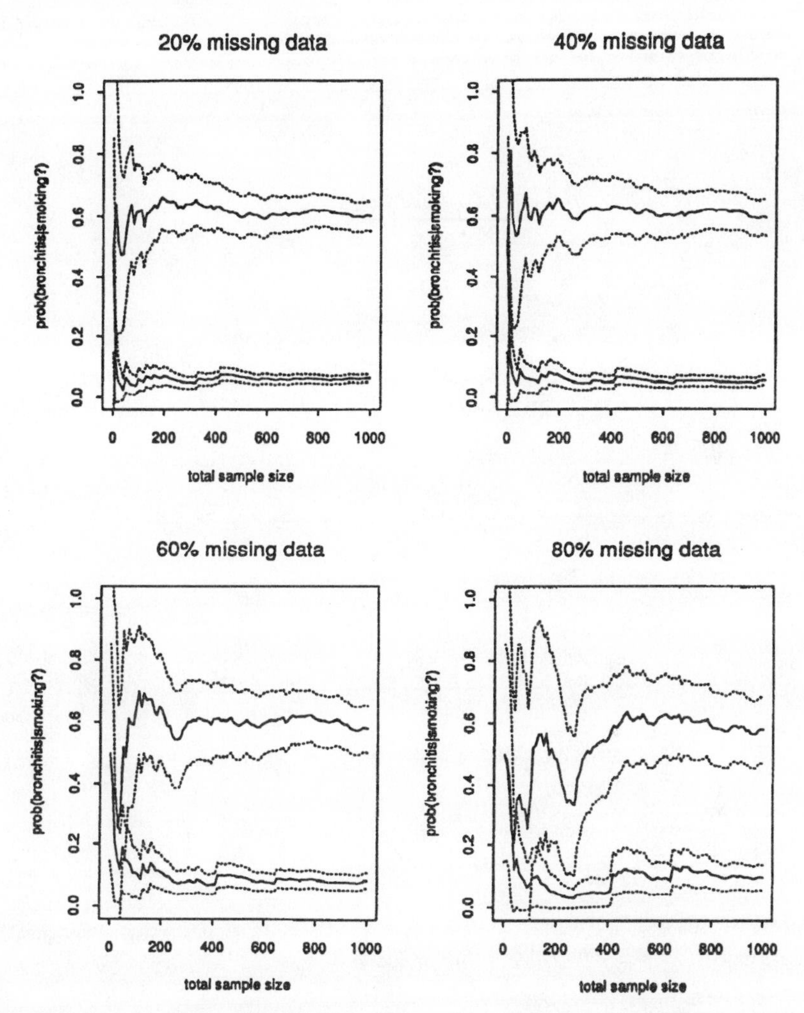

Figure A. *Estimated incidence of bronchitis in smokers and non-smokers, based on different proportions of missing data — each node on each case has independently been set to missing with the stated probability. The prior mean for each proportion was .5, with precision 1. (see Figure 3 in main paper).*

ADDITIONAL REFERENCES IN THE DISCUSSION

Queen, C. M. and Smith, J. Q. (1990). Multiregression dynamic models. *Tech. Rep.* **183**, University of Warwick.

Queen, C. M. and Smith, J. Q. (1992). Dynamic graphical models. *Bayesian Statistics 4* (J. M. Bernardo, J. O. Berger, A. P. Dawid and A. F. M. Smith, eds.), Oxford: University Press, 741–751.

Smith, J. Q. (1990). Statistical Principles on Graphs. *Influence Diagrams, Belief Nets and Decision Analysis* (R. M. Oliver and J. Q. Smith, eds.), New York: Wiley, 89–119.

Wermuth, N. and Lauritzen, S. L. (1990). On substantive research hypotheses, conditional independence graphs and graphical chain models. *J. Roy. Statist. Soc. B* **52**, 21–50.

West, M. and Harrison, P. J. (1989). *Bayesian Forecasting and Dynamic Models*. New York: Springer.

West, M. (1992). Modelling with mixtures. *Bayesian Statistics 4* (J. M. Bernardo, J. O. Berger, A. P. Dawid and A. F. M. Smith, eds.), Oxford: University Press, 503–524, (with discussion).

BAYESIAN STATISTICS 4, pp. 467–481
J. M. Bernardo, J. O. Berger, A. P. Dawid and A. F. M. Smith, (Eds.)

Advances in Bayesian Experimental Design*

ISABELLA VERDINELLI
University of Rome, Italy and *Carnegie Mellon University, USA*

SUMMARY

In this paper, current issues and recent advances in the Bayesian design of experiments are outlined. The unifying view of Bayesian decision theory is emphasized in order to include old and new optimality criteria derived from suitable utility functions. A special case is considered to illustrate the behavior of different designs and to show their sensitivity to the particular features of the experiment. Further research topics are briefly outlined and future directions to be taken are also described.

Keywords: BIASED COIN DESIGN; PREDICTION; SHANNON INFORMATION; UTILITY FUNCTION.

1. INTRODUCTION

The literature on Bayesian design, so far, has focused mainly on paralleling the results of non Bayesian design. Consequently, attention has been given to criteria based on the posterior covariance matrix of a multivariate normal model, as the natural extension of criteria based on the covariance matrix of the least squares estimator considered in the non Bayesian theory of design (Kiefer, 1959, 1961, 1974, 1975; Fedorov, 1972; Silvey, 1980). Some of the mathematical developments in classical experimental design have thus found a Bayesian parallel, as for example the general equivalence theorem of Kiefer and Wolfowitz (1959) that was cast in a Bayesian perspective by Chaloner (1984).

If inference is the main purpose of an experiment, within a linear multivariate normal model, reporting posterior densities or looking for Bayesian estimators leads us to consider functions of the posterior covariance matrix. Obviously, the purpose of the experiment influences the choice of the design, and many different issues can be taken into account. Recently, new criteria have been proposed to represent different goals that can arise in planning an experiment. These criteria will be reviewed here and some others will be presented, within the unifying framework of the Bayesian decision theory. The choice of a utility or loss function has often been emphasized as the fundamental issue in statistical inference (see, for example Savage, 1954; DeGroot, 1970; Lindley, 1972). In the theory of experimental design, the assessment of a utility function plays a crucial role. An important point is that the type of utility used depends strongly on the nature of the experiment. If the outcome of the observations is patient survival, we will surely use a different utility than if the outcome is the reaction of rats to an experimental drug. As I will show in this paper, the optimal design, in turn, depends heavily on this choice of utility. This is in contrast to Kiefer's goal (Kiefer, 1975) of searching for universally optimal designs. Thus, the aim of this paper is to show how a carefully chosen utility function leads to a design tailored to the specific experimental problem. The Bayesian approach to statistical problems is mainly concerned with reporting posteriors, rather than producing estimators. This point

* This research was partially supported by the Italian National Research Council (CNR). I would like to thank Rob Kass, Nick Polson and Larry Wasserman for useful discussions and suggestions.

is considered throughout the paper and special attention is given to expected utility functions based on Shannon information.

The general problem of looking for a Bayesian design is illustrated in Section 2, using the multivariate normal model with known variances. Section 3 deals with new utility functions proposed in the recent literature. In Section 4, the one way analysis of variance model is used to show the behavior of designs that maximize the different utility functions. Section 5 deals with the issue of randomized designs. Some discussion on computation and robustness of designs is given in Section 6.

2. OPTIMAL DESIGN CRITERIA IN THE BAYESIAN MULTIVARIATE NORMAL MODEL

The basic framework in the Bayesian design of experiments can be described within the multivariate normal regression model: $y \sim N(A_1\theta_1, \sigma^2 I_n)$ where σ^2 is known, and a design matrix A_1 has to be chosen when inference is required about the vector of parameters θ_1. Suppose that the prior information for the vector θ_1 is modeled as $\theta_1 \sim N(A_2\theta_2, C_2)$. The posterior distribution for θ_1 will then be $N(\theta_1^*, D(A_1))$, where θ_1^* is the posterior mean and $D(A_1)$ is the posterior covariance matrix, written as a function of the design matrix A_1. It is desirable that the posterior covariance matrix be small, in some sense. For this reason A_1 is chosen to maximize some function of its inverse $M(A_1) = D^{-1}(A_1)$. Following Kiefer (1975), Giovagnoli and Verdinelli (1983, 1985) considered functions ϕ on the class of inverse posterior covariance matrices as optimality criteria for the choice of a design. A_1^* is said to be Bayesian ϕ-optimal if $\phi\,(M(A_1^*)) = \max_{A_1 \in \mathcal{A}} \phi\,(M(A_1))$. Instances of such criteria are $\log|M(A_1)|$, $-trace(M(A_1))$ and $\lambda_{\min}(M(A_1))$, where $\lambda_{\min}$ indicates the minimum eigenvalue. These criteria, denoted respectively by Bayesian D-, A- and E-optimality, are the analogs of the same non-Bayesian criteria defined on the inverse covariance matrix of the least squares estimator $\hat{\theta}_1$.

The Bayes D-optimality criterion is especially appealing, as it can be derived from the maximization of the expected gain in Shannon information in the multivariate normal model with known variance. Several authors (Lindley, 1956; Stone, 1959a, b; DeGroot, 1962) suggested the use of the expected gain in Shannon information given by an experiment (Shannon, 1948), as a utility function. This was later seen by Bernardo (1979) and DeGroot (1986) within a decision-theoretic framework, as the gain in the expected utility given by an experiment. The experimenter would then choose a design A_1^* that maximizes the gain in expected utility or the expected Kullback-Leibler distance between the posterior and the prior densities:

$$U_1 = E_{\boldsymbol{y},\theta_1|A_1}\left[\log \frac{p(\theta_1|\boldsymbol{y}, A_1)}{p(\theta_1)}\right].$$

The use of the above in design theory has been discussed, first of all, by Stone (1959a, b) and later by Smith and Verdinelli (1980), Giovagnoli and Verdinelli (1983, 1985) and Verdinelli (1983), using the Lindley and Smith (1972) hierarchical model.

The Bayesian A- and E-optimality criteria can be derived within a decision theoretic framework using a quadratic loss function $L(\boldsymbol{d}, \theta_1) = (\boldsymbol{d} - \theta_1)^T W(\boldsymbol{d} - \theta_1)$, where $\boldsymbol{d}$ is an estimator for θ_1 and W is a symmetric positive definite matrix. The standard Bayes estimator $\boldsymbol{d}^* = \theta_1^*$, yields an expected loss given by *trace* $\{WD(A_1)\}$. A design that minimizes the expected posterior loss for $W = I$ is Bayes A-optimal. Bayes A-optimality was examined by Owen (1970) and by Brooks (1972, 1974). Chaloner (1984) considered some interesting modifications of it, using a different motivation. Bayes E-optimality criterion can be used if the aim of the experiment is to estimate linear combinations of the parameters, $\boldsymbol{c}^T\theta_1$, where

$\|c\| = w$ is fixed. The quadratic loss function with $W = I$ and a given c leads to the expected loss $c^T D(A_1)c$. Hence, for c fixed, the design A_1 should be chosen to minimize this expected loss. If, instead, only the norm w was fixed, a minimax argument would lead to a design that minimizes: $\sup_{\|c\|=w}\{c^T D(A_1)c\} = w^2\lambda_{\max}(D(A_1))$; this is the Bayes E-optimal design.

3. NEW UTILITY FUNCTIONS

The criteria of Section 2 concentrate on obtaining information about the parameters of the model. It is clear that, within a suitable decision theory framework, the choice of a utility or loss function that takes into account the goals of the experiment, produces various optimality criteria and designs. In some instances, it is necessary to account for other issues, besides information about the parameters when designing an experiment.

There are cases where prediction is considered more important than inference (see for example Geisser, 1966, 1971, 1984). In other cases, the value of the dependent variable might be of special concern. For example, in clinical trials, the value of the observations can represent a measure of how well patients fare. In such a case, the experimenter might be concerned not only with obtaining information about the parameters that represent the effects of, say, different drugs, but also with a large value of the output, which would indicate successful treatment of the patients. The clinician may allow experimentation only if the design proposed is expected to produce a good combination of outcome and information. In such experiments, it might be reasonable to consider a utility function that is a linear combination of two components. The total output is one component and the expected gain in Shannon information is the other component. The combined expected utility, as proposed in Verdinelli and Kadane (1992) is:

$$U_2 = E_{\boldsymbol{y},\theta_1|A_1}\left[\rho \boldsymbol{y}^T 1 + \beta \log \frac{p(\theta_1|\boldsymbol{y}, A_1)}{p(\theta_1)}\right].$$

The non negative weights ρ and β express the relative contribution that the experimenter is willing to attach to the two components of the combined utility and they affect the choice of the design maximizing U_2, only through the ratio $\gamma = \beta/2\rho$.

Let us now consider the problem of prediction, when designing an experiment. This concern can be important in settings like clinical trials or reliability. In reliability or quality control, for example, it is desirable to choose a design that allows us to predict the future life length of items in production or to keep the future level of output on target. For these types of problems, the Bayesian approach to experimental design can make use of predictive analysis. In particular, a logarithmic utility function on the class of predictive distributions can be used. Hence we can choose a design that maximizes the expected gain in Shannon information given by an experiment for the future observation y_{n+1} or, in other words, the expected utility provided by an experiment on the predictive density:

$$U_3 = E_{\boldsymbol{y}} E_{y_{n+1}|\boldsymbol{y},A_1}\left[\log \frac{p(y_{n+1}|\boldsymbol{y}, A_1)}{p(y_{n+1})}\right].$$

Note that $p(y_{n+1}|\boldsymbol{y}, A_1)$ is the usual predictive density of the future observation y_{n+1} given the vector $\boldsymbol{y}$ and A_1, while $p(y_{n+1})$ is the marginal density of the future observation y_{n+1} before the vector of data is observed and does not depend on A_1. This criterion has been discussed in a prediction context by San Martini and Spezzaferri (1984) when model selection was the aim of the analysis and by Verdinelli, Polson and Singpurwalla (1992), for an accelerated life test experiment.

A utility function that combines inference and prediction can be obtained by taking a linear combination of U_1 and U_3, namely:

$$U_4 = \rho\, E_{\boldsymbol{y}} E_{y_{n+1}|\boldsymbol{y},A_1}\left[\log\frac{p(y_{n+1}|\boldsymbol{y},A_1)}{p(y_{n+1})}\right] + \beta\, E_{\boldsymbol{y}|A_1} E_{\theta_1|\boldsymbol{y},A_1}\left[\log\frac{p(\theta_1|\boldsymbol{y},A_1)}{p(\theta_1)}\right].$$

Just as in U_2, the weights ρ and β express the relative contribution of the predictive and the inferential components of the utility. The weights affect the choice of the design maximizing U_4 only through their ratio ρ/β. This expected utility has the additional advantage, over U_2, that it combines two components that are measured in the same units, making it easier to assess the values of ρ and β. Values for these constants can be elicited using, for example, the method shown by Verdinelli and Kadane (1992). This might be simpler for U_4 than U_2.

4. AN EXAMPLE

This Section deals with a simple case of an experimental problem that shows how the designs produced by the utilities presented in Section 3 are sensitive to the goals of the experiment. Suppose that the experiment involves some new treatments and a control, described by a one way analysis of variance model. Suppose, further, that t new treatments are going to be examined and that they can be assumed to be exchangeable. This model can be used to describe a number of different experimental setups: a clinical trial, the reaction of laboratory animals to experimental drugs, an agricultural experiment or the comparison of new production processes with a standard one. Each of these situations can be represented by a careful choice of the utility function. The design problem for the one way analysis of variance model consists in choosing the optimal number n_i^*, $i = 0, 1, \ldots, t$ of experimental units to assign to each treatment, or the optimal proportions p_i^* of a given total of n subjects. An exchangeability assumption for the effects of the t new treatments α_i, $i = 1, \ldots, t$, say, can be modeled by assuming that the treatment effects are multivariate normally distributed and uncorrelated, with $E(\alpha_0) = \mu_0$, $Var(\alpha_0) = \tau_0^2$, where α_0 is the effect of the control, and for $i = 1, \ldots, t$, $E(\alpha_i) = \mu_1$, $Var(\alpha_i) = \tau_1^2$.

For this model, when dealing with the utilities U_1 and U_2, the proportions of the observations on the new treatments must be a constant (see Smith and Verdinelli, 1980 and Verdinelli and Kadane, 1992). Let us denote this common proportion by p, with $0 \leq p \leq 1/t$ while the proportion of observations on the control is $1 - tp$.

Suppose now, that the utility U_1 is to be maximized. This would be the case if the main goal of the analysis was to gather information on the effects of the control and of the new treatments, while little importance was given to the value of the output. The use of U_1 would be appropriate if the experiment was, for example, an agricultural trial performed on new fertilizers and the goal was to obtain information about their effects. In such a case, the interest would be in a Bayesian D-optimal design. Define $\delta_0 = \sigma^2 n^{-1}/\tau_0^2$ and $\delta_1 = \sigma^2 n^{-1}/\tau_1^2$. Maximizing U_1 with respect to p would lead to $p^*_{(1)} = \text{median}\{0, 1, \hat{p}\}$, where $\hat{p} = (t+1)^{-1}[1 - (\delta_1 - \delta_0)]$. In particular, large values of δ_1 imply $\hat{p} < 0$ and large values of δ_0 imply $\hat{p} > 1$, so that if τ_1^2 is small, no observations are taken on the new treatments and if τ_0^2 is small no observations are taken on the control. In other words, it is not worth doing the experiment if τ_1^2 is too small. Only values of the prior variances affect the D-optimal design. This is reasonable, since this criterion is concerned only with accurate parameter estimation. The proportion $p^*_{(1)}$ reduces to the non Bayesian D-optimal proportion $p_C^* = (t+1)^{-1}$ as the prior variances tend to infinity.

Now suppose that the utility U_2 is to be maximized to find the optimal proportion $p^*_{(2)}$ of observations on the new treatments. This would be the case if, for example, we were considering a clinical trial and there was concern about the immediate response of the patients in the trial. We are assuming here, that in every experiment, there is always interest in obtaining information about the parameters of the model, and both ρ and β in the expression of U_2 are nonzero. Following Verdinelli and Kadane (1992), let $A(\delta_0, \delta_1) = (1+\delta_0)^{-1} - (t\ \delta_1)^{-1}$, $B(\delta_0, \delta_1) = \delta_0^{-1} - t(1+t\delta_1)^{-1}$ and $\gamma = \beta/2p$.

If $\mu_1 - \mu_0 \leq \gamma n^{-1}\ A(\delta_0, \delta_1)$, then $p^*_{(2)} = 0$. This means that no observations are taken on any of the new treatments, unless the difference between the prior expectations of the new treatments and the control, $\mu_1 - \mu_0$ is large enough. If, instead, $\mu_1 - \mu_0 \geq \gamma n^{-1}\ B(\delta_0, \delta_1)$, then $p^*_{(2)} = 1/t$, that is no observations are taken on the control since it is expected the new treatments effect to be substantially superior to the control effect. Finally, if $\gamma n^{-1}\ A(\delta_0, \delta_1) < \mu_1 - \mu_0 < \gamma n^{-1}\ B(\delta_0, \delta_1)$ then $p^*_{(2)}$ is the positive root of the following second degree equation in p:

$$-t\ ndp^2 + [nd(1 - t\delta_1 + \delta_0) - \gamma(t+1)]p + nd\delta_1(\delta_0 + 1) + \gamma(1 + \delta_0 - \delta_1) = 0\ ,$$

where $d = \mu_1 - \mu_0$. Observations are made both on the new treatments and on the control only if the difference $\mu_1 - \mu_0$ has an intermediate value. We might say that the clinical trial is unethical if $\mu_1 - \mu_0$ is too extreme.

The sensitivity of the optimal design $p^*_{(2)}$ to the prior variances depends on the way in which $A(\delta_0, \delta_1)$ and $B(\delta_0, \delta_1)$ vary with the variance ratios $\delta_1 = \sigma^2 n^{-1}/\tau_1^2$ and $\delta_0 = \sigma^2 n^{-1}/\tau_0^2$. It can be shown that, if the prior knowledge about the effect of treatments and control is precise, only the prior expected value of the total output affects the choice of the optimal design. Specifically, when τ_0^2 and τ_1^2 are both small compared with σ^2/n, if $\mu_1 > \mu_0$ then $p^*_{(2)} = 1/t$ —no observations are taken on the control— and if $\mu_1 < \mu_0$ then $p^*_{(2)} = 0$. The case where τ_0^2 and τ_1^2 are both large compared with the variance of the observations σ^2/n, leads to the optimal $p^*_{(2)}$ given by the solution of the above equation, where δ_0 and δ_1 are replaced by zero. Finally, if a small value of τ_0^2 appears together with a large value of τ_1^2 it can be shown that it is optimal to take no observations on the control.

The case in which prediction is the concern of the experiment and the utility U_3 is to be maximized, is not especially meaningful in the one way analysis of variance model. It would imply that, for example, there was a specific reason for using, among t new production processes under consideration in an industrial trial, the h-th one for the future production. A more interesting use of U_3 to design an experiment for accelerated life testing has been presented in Verdinelli, Polson and Singpurwalla (1992). For the sake of comparison, it is easy to show that in the trial described, the use of utility U_3 leads to $p_h^* = 1$ and $p_i^* = 0$, for all $i \neq h$, an intuitively sensible result, given the type of problem.

On the other hand, the use of utility U_4 in a similar trial, appears more realistic. Special attention is given to the h-th production process, while information is also required on the other processes: the one in current use (control) and the remaining $t-1$. The optimal proportions of observations obtained in this last case are:

$$p_0^* = \frac{1 - \eta\delta_0 + t(\delta_1 - \delta_0)}{t + \eta + 1}, \quad p_h^* = \frac{(\eta+1)(1 - \delta_1 + \delta_0) + \eta t\delta_1}{t + \eta + 1}, \quad p_i^* = \frac{1 + \delta_0 - (\eta+1)\delta_1}{t + \eta + 1}$$

for $i \neq 0, h$, where $\eta = \rho/\beta$.

5. DESIGNS FOR RANDOMIZED CLINICAL TRIALS

An important issue in design, which has been recently raised within the Bayesian framework, is randomization. A discussion about randomization procedures can be found in Basu (1980) and the comments therein. Rubin (1978) presented some Bayesian motivations for randomization and, more recently, Kadane and Seidenfeld (1990) discussed Rubin's issues together with the general role of randomization from a Bayesian perspective. Another important contribution to the ethical basis of randomization was presented in the discussion of Ware (1989). It remains unclear if randomization has a role in the Bayesian framework. If randomization is required in a sequential trial we are faced with the problem of designing an experiment so that the final allocation satisfies a Bayesian optimality criterion. Randomization procedures for clinical trials have been recently considered in a Bayesian perspective by Ball, Smith and Verdinelli (1992) and Verdinelli (1992). The main goal of these papers was to obtain randomization probabilities that would, by the end of the trial, produce proportions of patients assigned to each treatment close to the optimal values required by a Bayesian D- or A-optimal design.

Consider a clinical trial described by the one way analysis of variance model in Section 4. Suppose we sequentially randomize the next experimental unit to one of the t treatment groups, the i-th of which has so far been allocated $n_i \geq 0$ units. If the next subject is allocated to group i, the resulting posterior precision matrix would be $M_i = \sigma^{-2}\, diag\{n_0, n_1, \ldots, n_i + 1, \ldots, n_t\} + C_2^{-1}$. Now suppose that the design criterion consists of maximizing some function ϕ of the posterior precision matrix. If $(\pi_o, \pi_1, \ldots, \pi_t)$ denote the randomization probabilities for allocating the next subject, we should choose these to maximize $\pi_0\phi(M_0) + \pi_1\phi(M_1) + \ldots + \pi_t\phi(M_t)$. However, it is immediately apparent from the linear form of the criterion that, in general, one of the π_i must be chosen to be 1 and the others 0. It follows that Bayesian decision theory applied with a design criterion on the inverse posterior covariance matrix, cannot provide non-degenerate randomization probabilities. A formal procedure, within the Bayesian framework, would consist of maximizing expected utility functions that take into account both elements of our design requirements, namely, the requirement of choosing a design to obtain efficient inference on the parameters and the requirement of deriving a non degenerate randomization rule. The first objective is accomplished by maximizing $\sum_{i=0}^{t} \pi_i\phi(M_i)$. One way to achieve the second goal is to use the entropy $-\sum_{i=0}^{t} \pi_i \log \pi_i$, that is maximized for $\pi_i = (t+1)^{-1}$, $i = 0, 1, \ldots, t$. To derive a method that allows randomization while obtaining an optimal Bayes design under a criterion ϕ, we choose $\pi_0, \pi_2, \ldots, \pi_t$ to maximize:

$$U_5(\pi_0, \pi_1, \ldots, \pi_t) = \sum_{i=0}^{t} \pi_i\phi(M_i) - c\{\sum_{i=o}^{t} \pi_i \log \pi_i\},$$

where c (≥ 0) is the trade-off coefficient between efficient inference ($c = 0$) and emphasis on randomization ($c \to \infty$). This utility function is entirely within the spirit of this paper, the main point being again to think carefully about the goals of the experiment at hand and then set an appropriate utility. Maximizing U_5 with respect to π_i leads to the general solution:

$$\pi_i^* = \frac{\exp\{c^{-1}\ \phi(M_i)\}}{\sum_{l=0}^{t} \exp\{c^{-1}\ \phi(M_l)\}} \qquad i = 0, 2, \ldots, t,$$

where the randomizing probabilities π_i^* depend on the optimality criterion ϕ and on the value chosen for c. As a special case, if $\phi(M_i) = log|M_i|$, we obtain: $\pi_i^* = |M_i|^{1/c}\{\sum_{l=0}^{t} |M_l|^{1/c}\}$.

6. DISCUSSION

So far, we have dealt with problems in the Bayesian design of experiments, connected with the framework of the multivariate normal linear model with known variance. The need for deriving tractable analytical results has overshadowed the need for more realistic assumptions, such as the relaxation of the hypothesis of known variance in the sampling distribution, the relaxation of the normality assumptions or considering problems related to non linear regression, to name just a few. A set of results for non-linear models are due to Chaloner (1987) and Chaloner and Larntz (1989, 1990), who examined computational solutions for choosing designs in non linear models. The implementation of software to help the practitioner to deal with these types of problems is going to be a very active area in the design of experiments. The recent techniques of substitution sampling methods will likely be usefully employed in this area as well. Even the case of unknown variance becomes algebraically complicated. For example, the utility U_1 leads to the computation of the Kullback-Leibler distance between two t-distributions. For this specific problem, a reasonable approximation to U_1 (see Polson, 1990) is the expected logarithm of the determinant of Fisher information that, incidentally, was used by Chaloner and Larntz in their work on designs for non linear models. Work is still in progress to examine the unknown variance case for the various utility structures considered in this paper.

Another area of Bayesian experimental designs that deserves attention is the assessment of the sensitivity of an optimal design to the choice of the prior and the model assumptions. The multivariate normal case described in the previous sections can be generalized by letting the prior vary in a class of conjugate priors Γ_C as considered in Berger (1990). In this way, instead of a single multivariate normal prior density for the parameters, a class of multivariate normal can be examined in order to find a design. In particular the class $\Gamma_C = \{N(A_2\theta_2, C_2) : \; A_2\theta_2 \in \mathcal{M}, \; C_2 \in \mathcal{S}\}$, where $\mathcal{M}$ and $\mathcal{S}$ are appropriate spaces, might lead to a closed form expression for design criteria. Optimal robust design should be analytically obtainable for some of the different utility functions examined here. Also, as pointed out earlier, numerical work can always be done if analytical expressions are not derivable. The problem of robustness of Bayes designs with respect to the prior distribution has been recently considered in two papers by Das Gupta and Studden (1991a, b), but more work still needs to be done in this direction. Another important point is the construction of designs that are robust to departures from the model assumptions. This is clearly another place where the use of computational techniques will be very helpful.

REFERENCES

Ball, F. G., Smith, A. F. M. and Verdinelli, I. (1992). Biased coin designs with a Bayesian bias. *J. Statist. Planning and Inference* , (to appear).

Basu, D. (1980). Randomization analysis of experimental data: the Fisher randomization test. *Ann. Statist.* **75**, 575–595.

Berger, J. O. (1990). Robust Bayesian analysis: sensitivity to the prior. *J. Statist. Plann. Infer.* **25**, 303–328.

Bernardo, J. M. (1979). Expected information as expected utility. *Ann. Statist.* **7**, 686–690.

Brooks, R. J. (1972). A decision theory approach to optimal regression designs. *Biometrika* **59**, 563–571.

Brooks, R. J. (1974). On the choice of an experiment for prediction in linear regression. *Biometrika* **61**, 303–311.

Chaloner, K. (1984). Optimal Bayesian experimental designs for linear models. *Ann. Statist.* **12**, 283–300.

Chaloner, K. (1987). An approach to design for generalized linear models. *Proceedings of the Workshop on Model–oriented Data Analysis*. Wartburg – Lecture Notes in Economics and Mathematical Systems, #297. Berlin: Springer-Verlag, 3–12.

Chaloner, K. and Larntz, K. (1989). Optimal Bayesian designs applied to logistic regression experiments. *J. Statist. Planning and Inference* **21**, 191–208.

Chaloner, K. and Larntz, K. (1991). Bayesian design for accelerated life testing. *J. Statist. Planning and Inference* , (to appear).

Das Gupta, A. and Studden, W. J. (1991a). Robust Bayes designs in normal linear models. *Ann. Statist.* **19**, 1244-1256.

Das Gupta, A. and Studden, W. J. (1991b). Towards a theory of compromise designs: frequentist, Bayes and robust Bayes. *J. Statist. Planning and Inference* , (to appear).

DeGroot, M. H. (1962). Uncertainty, information and sequential experiments. *Ann. Math. Statist.* **33**, 404–419.

DeGroot, M. H. (1986). Concepts of information based on utility. *Recent Developments in the Foundations of Utility and Risk Theory* (L. Daboni *et al.* eds.), Dordrecht: Reidel, 265–275.

DeGroot, M. H. (1970). *Optimal Statistical Decisions.* New York: McGraw–Hill.

Fedorov, V. V. (1972). *The Theory of Optimal Experiments* (W. J. Studden and E. M. Klimko, transl. and eds.) New York: Academic Press.

Geisser, S. (1966). Predictive discrimination. *Multivariate Analysis* (P. R. Krishnaiah, ed.), New York: Academic Press, 149–163.

Geisser, S. (1971). The inferential use of predictive distribution. *Foundations of Statistical Inference* (V. P. Godambe and D. A. Sprott, eds.) Toronto: Holt, Rinheart and Winston, 456–469.

Geisser, S. (1985). On the prediction of observables: a selective update. *Bayesian Statistics 2* (J. M. Bernardo, M. H. DeGroot, D. V. Lindley and A. F. M. Smith, eds.), Amsterdam: North-Holland, 203–229.

Giovagnoli, A. and Verdinelli, I. (1983). Bayes *D*–optimal and *E*–optimal block designs. *Biometrika* **70**, 695–706.

Giovagnoli, A. and Verdinelli, I. (1985). Optimal block designs under a hierarchical linear model. *Bayesian Statistics 2* (J. M. Bernardo, M. H. DeGroot, D. V. Lindley and A. F. M. Smith, eds.), Amsterdam: North-Holland, 655–661.

Kadane, J. B. and Seidenfeld, T. (1990) Randomization in a Bayesian perspective. *J. Statist. Planning and Inference* **25**, 329–345.

Kiefer, J. (1959). Optimal experimental designs. *J. Roy. Statist. Soc. B* **21**, 272–319, (with discussion).

Kiefer, J. (1961). Optimum designs in regression problems II. *Ann. Math. Statist.* **32**, 398–325.

Kiefer, J. (1974). General equivalence theory for optimum designs (approximate theory). *Ann. Statist.* **2**, 894–879.

Kiefer, J. and Wolfowitz, J. (1959). Optimum designs in regression problems. *Ann. Math. Statist.* **30**, 271–294.

Kiefer, J. (1975). Construction and optimality of generalized Youden designs. *A Survey of Statistical Design and Linear Models* (J. N. Srivastava, ed.), Amsterdam: North-Holland, 333–353.

Lindley, D. V. (1956). On the measure of information provided by an experiment. *Ann. Math. Statist.* **27**, 986–1005.

Lindley, D. V. (1972). *Bayesian Statistics, a Review*. Philadelphia, PA: SIAM.

Lindley, D. V. and Smith, A. F. M. (1972). Bayes estimates for the linear model. *J. Roy. Statist. Soc. B* **34**, 1–41, (with discussion).

Owen, R. J. (1970). The optimum design of a two–factor experiment using prior information. *Ann. Math. Statist.* **41**, 1917–34.

Polson, N. G. (1990). On the expected amount of information from a nonlinear model. *Tech. Rep.* Carnegie Mellon University.

Rubin, D. B. (1978). Bayesian inference for causal effects: the role of randomization. *Ann. Statist.* **6**, 34–58.

San Martini, A. and Spezzaferri, F. (1984). A predictive model selection criterion. *J. Roy. Statist. Soc. B* **46**, 296–303.

Savage, L. J. (1954). *The Foundations of Statistics*. New York: Wiley.

Shannon, C. E. (1948). A mathematical theory of communication. *Bell Systems Tech. J.* **27**, 379–423 and 623–656.

Silvey, S. D. (1980). *Optimal Design.* London: Chapman and Hall.

Smith, A. F. M. and Verdinelli, I. (1980). A note on Bayesian design for inference using a hierarchical linear model. *Biometrika* **67**, 613–619.

Stone, M. (1959a). Application of a measure of information to the design and comparison of regression experiment. *Ann. Math. Statist.* **30**, 55–70.

Stone, M. (1959b). Discussion of Kiefer. *J. Roy. Statist. Soc. B* **21**, 313–315.

Verdinelli, I. (1983). Computing Bayes D- and E-optimal designs for a two-way model. *The Statistician* **32**, 161–167.

Verdinelli, I. (1990). Procedure di randomizzazione nella statistica Bayesiana. *Atti della XXV Riunione Scientifica della SIS.* 35–41.

Verdinelli, I., Polson, N. and Singpurwalla, N. (1992). Shannon information and Bayesian design for prediction in accelerated life testing, (to appear).

Verdinelli, I. and Kadane, J. B. (1992). Bayesian designs for maximizing information and outcome. *J. Amer. Statist. Assoc.* , (to appear).

Ware, J. H. (1989). Investigating therapies of potentially great benefit: ECMO, *Statist. Sci.* **4**, 310–317.

DISCUSSION

J. DEELY (*University of Canterbury, New Zealand*)

Firstly let me express appreciation to Dr. Verdinelli for making sure that I had her paper with sufficient time to read it before the Conference. The miracles of e-mail never cease to amaze me. Secondly let us review what the main purposes of a discussant are:

1. reformulate what the paper is about in a way that is understandable to the audience,
2. as soon as possible refer to your own work,
3. pose a problem which hopefully is new and interesting and quite possibly solvable using different methods than those posed in the paper being discussed.

Experimental Design. The expression "Experimental Design" has come to mean many things to many people and hence I find the basic title so broad that it means very little. In its simplest form I think of experimental design along the following lines. There is a core of experimental units about which we want to know something because we have some purpose in mind. There is some information available about these experimental units and generally it has some influence upon what we want to know. In this context then the design of experiments can be stated as simply determining which experimental units to use and how many. The author in this paper deals with only the second question, namely how many in several different examples. This is done in the context of the normal linear model discussed fully by Lindley and Smith (1972). The author's stated purpose of "considering how the purpose of the experiment has an appropriate influence in the choice of design" is certainly admirable and in my view quite correct. It then follows that various utility functions should be used to accomplish specific purposes. In trying to discuss this topic the author has introduced new utility functions which she purports do accomplish this. All of these utility functions use in some way or another the Shannon information function.

In reading over this paper some specific technical questions are the following: i) How are a and b determined in the utility U_2? ii) Is $p(y_{n+1})$ not functionally dependent on A_1 in every case or is it just a result for the AOV example? iii) Again in U_1 how are the weights a, b in the linear combination determined? I would wonder just what type of elicitation is possible; iv) Example 2 is given without stating what the purpose of the experiment is in that particular case; v) How would the model handle unequal variances? vi) Can this approach be used in regression problems to determine what x values should be chosen in the experimental units?

My major misgiving about this paper centres around the utility functions and specifically I ask the question –"Can these utilities really make sense to an experimenter?" I think it is appropriate at this point to remind ourselves that there are two main reasons for research: i) to publish papers and ii) to solve practical problems.

The research outlined in this paper certainly qualifies it as a mechanism to accomplish (i) but I have doubts as to whether it can seriously accomplish (ii). To illustrate my misgiving more succinctly and at the same time to hark back to purpose 2 of a discussant, let me refer to the work done in Berger and Deely (1988).

Relevant Other Work. In this paper the usual AOV one way layout is studied and two functions are proposed for ranking the treatments. It is the case that in the AOV situation almost always the interest is in deciding which treatment is best after one has already rejected the hypothesis. Hence ranking the treatments in the analysis is an appropriate goal. Among other things they compute for each $i = 1, \ldots, k$

$$P_i = P(\theta_i \geq \theta_j, \forall j \neq i|\text{data}) \tag{1}$$

which says which treatment has the highest posterior probability. I would suggest that this criterion as a utility function is much easier to understand in practical experimental situations than those suggested by the author. In addition the quantity

$$\text{pred}(i) = P(Y_{i,n+1} > Y_{j,n+1} \forall j \neq i | P(Y_{i,n+1} > Y_{j,n+1} \forall j \neq i|\text{data}) \tag{2}$$

could be of interest in some experimental situations and again is more easily interpreted than those utilities given in this paper. That is, the posterior probability that a new observation on the i-th treatment exceeds all the observations from other treatments seems to me to be a very sensible and very easily understood utility. Now what does this have to do with design of experiments? Quite simply, one can use these two ideas of a utility function and try to design the experiment in such a way that these probabilities will be maximum for any given data set. In general, this may be a difficult task but for the normal linear model it has a fairly straightforward interpretation, namely assign the samples amongst the treatments in such a way that the terms

$$\frac{\bar{X}_i - \bar{X}_j}{\text{st.deviation term}}$$

are largest.

This follows from formula Eq.(3.2) in Berger and Deely (1988). To illustrate more specifically, consider the case of only two treatments. Then the optimal design, i.e., the allocation of samples to the treatments in such a way that the quantity (1) is maximized for any given data set, is to simply maximize

$$\left(\frac{\sigma_1^2}{n_1} + \frac{\sigma_2^2}{n_2}\right)^{1/2} \cdot (\bar{X}_1 - \bar{X}_2).$$

The reason for this will not be gone into here but it is simply that the quantities of interest are monotone functions of these various treatments. As usual these probability computations can be made for informative and non-informative cases.

How far these ideas can be extended is an open question at this time.

New Problem. This paper did stimulate me to ask how would de Finetti approach these kinds of questions. In that spirit, I would like to pose the following type of design of experiments problem. Suppose in a particular data-base that the records of 100 patients are available. To keep the matter simple, suppose we have the x_i values for each of 100 patients and we can measure with additional expense the y_i values. The values x_i for example could be sugar in the urine specimen whereas the y_i sample would be sugar in the blood. There are many experiments of this type where the x_i data is easily obtained and the y_i value can be obtained but often at additional expense or pain. Now discretise the problem and let the x_i values fall in one of four categories —very low, a little below normal, a little above normal and very high; and for each of the x_i values imagine the y_i value falls in one of three categories — very low, very high, or middle.

The purpose of our experiment is to be able to predict for the next patient, i.e., 101st patient, the y value after measuring that patient's x value. The problem is that we only have enough resources to measure the y_i values for 10 of the 100 patients in our data base. Which 10 of the 100 do we choose and why?

In conclusion, I would like to sincerely thank the author for providing a very stimulating paper and wish her continued success in devoting her energies to important problems. I would suggest in the spirit of my earlier comments, a new title with a new utility and then obtain a really wonderful new result. Apart from these minor suggestions for change, I really liked the paper.

S. FRENCH (*University of Leeds, UK*)

Doubtless I will be accused of riding an old hobby horse, but may I say how pleasing it is to see a paper emphasising the importance of considering carefully the utility function to use in an analysis. Qualitatively different utility functions lead to qualitatively different strategies. For that reason, among many others, I found this a particularly pleasing and stimulating paper.

I have one small technical suggestion. To maximise the expected utility U_2 of Section 3, it is necessary to determine the trade-off or ratio of weights: β/p. In the presentation, though not in the written paper, it was suggested that the decision maker might choose this trade-off by picking a point on a plot of the Pareto frontier in (total output) $\times$ (expected gain in Shannon Information) space. I would council strongly against that. Ask your decision maker for this trade-off directly —as in Keeney and Raiffa (1976). To ask for a point on a graph is to fall foul of all sorts of psychological biases arising from such things as arbitrary choices of scale for the two axes.

M. GOLDSTEIN (*University of Durham, UK*)

While preparing for our session on design for the Valencia Meeting, I looked at various conference proceedings on new work in experimental design. I was disappointed, but not altogether surprised, to find lots of new theory, but very few actual experiments being designed. I hope that the Bayesian literature on design does not develop the same way.

The hard experiments to design are those with a variety of objectives and a variety of costs (financial, ethical, etc.). We need theory which is relevant, and actually applied, to such experiments. It is not particularly a criticism of this paper, but it is perhaps a criticism of the field that the paper surveys, that we seem to be rather a long way from such considerations here.

B. TOMAN (*George Washington University, USA*)

It is clear from the results in this paper that the sensitivity of the optimal experimental design to the choice of the prior distribution is a very important issue. The author suggests considering a class of priors Γ_C which is a class of normal distributions with various mean vectors and covariance matrices. I would like to present one possible choice for such a class and the corresponding optimality criterion based on U_1. Similar criteria could be based on the other utility functions.

Consider the class $\Gamma_C = \{N(\theta_1, C); C = \text{diag}(c_0\tau_0^2, c_1\tau_1^2, \ldots, c_1\tau_1^2),\ c_{0L} \le c_0 \le c_{0H},\ c_{1L} \le c_1 \le c_{1H}\}$. This class of priors leads to a class of posterior distributions $\Gamma_{\text{post}C}$ for each of which the optimality criterion U_1 can be computed. We get (after some simplification):

$$\Phi_1 = (p + e_1\delta_1)^t(1 + e_0\delta_0 - pt),$$

$$\text{where } e_0 = 1/c_0, e_1 = 1/c_1, e_{0L} \leq e_0 \leq e_{0H}, e_{1L} \leq e_1 \leq e_{1H}.$$

The objective here is to find a value of p which achieves large values of Φ_1 for all posterior distributions in $\Gamma_{\text{post}C}$. If the experimenter wishes to treat all posterior distributions in $\Gamma_{\text{post}C}$ equally, a sensible optimality criterion for the selection of p is to maximize:

$$\Phi_2 = \int\int (p + e_1\delta_1)^t(1 + e_0\delta_0 - pt)f(e_0)f(e_1)de_0de_1,$$

where $f(e_0)$ and $f(e_1)$ are uniform distributions on the intervals $e_{0L} \leq e_0 \leq e_{0H}$ and $e_{1L} \leq e_1 \leq e_{1H}$. For the particular choice of $e_{0H} = e_{1H} = 1$ we get:

$$\Phi_2 = (1 - e_{0L})\left[(p+\delta_1)^{(t+1)} - (p + e_{1L}\delta_1)^{(t+1)}\right](2 + \delta_0 + e_{0L}\delta_0 - 2pt)/(2\delta_1(1+t)).$$

(This choice of e_{0H} and e_{1H} is sensible if the convariance matrix $C_0 = \text{diag}(\tau_0^2, \tau_1^2, \cdots, \tau_1^2)$ is elicited and the class Γ_C is supposed to contain distributions which are more "vague"). The optimal value of p is the solution to:

$$\left[(p+\delta_1)^{(t+1)} - (p + e_{1L}\delta_1)^{(t+1)}\right]t$$
$$-(t+1)(2 + \delta_0 + e_{0L}\delta_0 - 2pt)/2[(p+\delta_1)^t - (p + e_{1L}\delta_1)^t] = 0.$$

The following table gives the optimal values of p according to this criterion for $t = 2$ and several different values of $\delta_0, \delta_1, e_{0L}$ and e_{1L}. For comparison the optimal value of p corresponding to C_0 under $U_1(p^*)$ is also given. Note that the optimum proportion for the classical D-optimal design is $p = 1/3$.

$\delta_0 = 1.7E-2, \delta_1 = 5.6E-4, p^* = 0.34$			$\delta_0 = 1.7E-1, \delta_1 = 5.6E-4, p^* = 0.39$		
e_{0L}	e_{1L}	p	e_{0L}	e_{1L}	p
0.1	0.4	0.34	0.1	0.4	0.37
0.4	0.4	0.34	0.4	0.4	0.37
0.1	0.75	0.34	0.1	0.75	0.36
$\delta_0 = 3.3E-1, \delta_1 = 5.6E-4, p^* = 0.44$			$\delta_0 = 5.0E-1, \delta_1 = 5.6E-4, p^* = 0.5$		
e_{0L}	e_{1L}	p	e_{0L}	e_{1L}	p
0.1	0.4	0.39	0.1	0.4	0.42
0.4	0.4	0.40	0.4	0.4	0.40
0.1	0.75	0.38	0.1	0.75	0.41

Table 1.

Note that $e_{0L} = 0.1, e_{1L} = 0.4$ and $e_{0L} = 0.1, e_{1L} = 0.75$ are cases where Γ_C contains priors where the prior variance of the control is allowed to become proportionally much less informative than that of the new treatments.

It appears from this example that the optimal experimental design (p^*) can be quite robust unless the prior variance of the control is very small as compared to σ^2/n. However in cases where each observation is very expensive, i.e., n is small, p^* may not be robust. In such cases the use of the class Γ_C and some criterion such as Φ_2 would be useful.

YOU-GAN WANG (*University of Oxford, UK*)

When an experiment is designed it is helpful to set up a utility (loss) function which can incorporate both the output (cost) of experimentation and the importance of achieving accuracy. There is, of course, a conflict between the two. A good utility function should be easily understandable and give a clear indication of how to weight the different objectives. My comments are confined to the context of clinical trials.

Consider the example given in Section 4. The result on Bayesian D-optimality (maximizing U_1) is nice and the design may be suitable in phase I or II study of a clinical trial. However, the application of the U_2 function seems to be inappropriate although the mathematical results come out neatly. It is unclear to me how to explain the ratio γ, and there is no indication what kind of information we can use to determine the appropriate value γ. (It seems that the ratio only measures the relative importance of the two terms of the function). If we are clear which factors should be taken into consideration, we should go a little further by using a utility function which is expressed in terms of actual costs and benefits. In a recent paper (Wang, 1991), two criteria are considered, the total output of the trial and the expected mean of the "best" treatment determined after the trial. These two criteria reflect the interests of the patients within and outside the trial, respectively. (The response is assumed to be binary and the criteria considered may be applied to other cases as well.) It is unethical to focus the design on the second criterion only because one should also take accout of the interests of the patients involved in the trial. An obvious and reasonable way to overcome this difficulty is to use a linear combination of these two criteria. The ratio of the coefficients reflects the proportions of the patients within and outside the trial. In Dr. Verdinelli's paper, the use of linear combination seems to be inappropriate. It is unsatisfactory to use a linear combination of the total output and the expected gain in Shannon information just because the Bayesian D-optimality criterion can be achieved from the maximization of the expected gain in Shannon information.

REPLY TO THE DISCUSSION

I would like to thank all the discussants for their contributions.

One point that was raised by John Deely and You-Gan Wang is that it might be difficult to choose the ratio $\gamma = \beta/2\rho$ for the utility U_2. I agree with this and propose the use of the following method for assessing γ – a similar method can be used for determining the ratio of weights in the utility U_4. As shown in Section 4, in the one way analysis of variance model with treatments and control, a given γ leads to a value $p^*_{(2)}$ for the optimal proportion of observations on the new treatments. It is then possible to compute the corresponding expected total output and expected information.

We can now plot, for a set of values of γ, the expected total output against the expected information as in Figure 1. Each point on the curve represents a combination of the two components of the utility U_2 achieved by the optimal design that corresponds to the appropriate value of γ. In particular, Figure 1 was obtained for the following example. Assume that $n = 25$ patients are in a trial for testing a new drug to lower the level of a hormone. Suppose that the prior knowledge about the effects of an old treatment – the control – and of a new drug is such that a drop of 50 units of the hormone is expected under the control ($\mu_0 = 50$) with standard deviation $\tau_0 = 10$ and a drop of 70 units is expected under the new treatment, with a larger standard deviation $\tau_1 = 30$. Assuming $\sigma = 3$, we obtain the graph in figure 1. The upper left hand point of the curve maximizes information and represents the Bayesian D-optimal design that leads to $p^*_{(1)} = .502$. The lower right hand point maximizes expected output and corresponds to $p^*_{(2)} = 1$. The curve allows the experimenter to see the

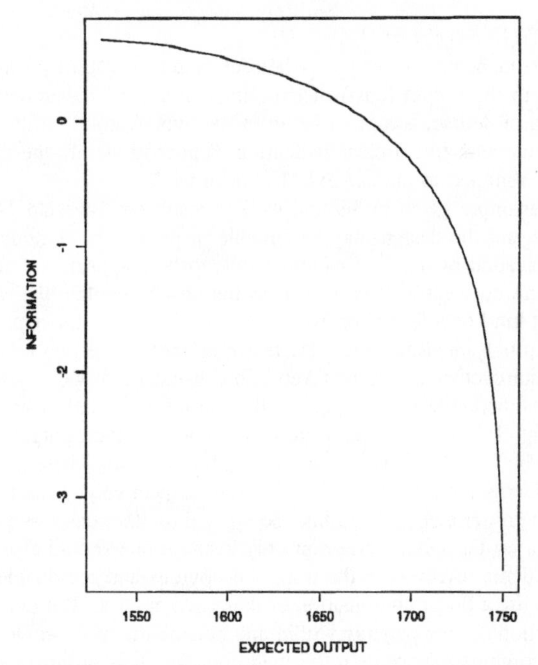

Figure 1. *Expected gain in information versus expected output for various values of* γ.

tradeoff between expected output and expected information. The ratio γ is determined by choosing a point on the curve. As Simon French points out, however, this method might have its drawbacks and he suggests asking for the tradeoff directly.

I now turn to John Deely's other questions. The density $p(y_{n+1})$ in U_3 and U_4 denotes the prior predictive distribution, or the marginal density of the future observation before the vector of data is observed. It does not depend on A_1, the design matrix for the first n observations only.

With this model, designs for the case of known, unequal variances can be found. The calculations will be more cumbersome. A more difficult problem to solve, as I pointed out in Section 6, is the case of unknown variances. The Bayesian design of some regression problems with these utilities has been considered in Verdinelli and Kadane (1992) and in Verdinelli, Polson and Singpurwalla (1992). Also, Pilz (1991) and Chaloner (1984) have dealt with Bayesian designs for regression.

Deely's main concern is how understandable the utility functions are to an experimenter. I agree that the concept of expected gain in Shannon information is difficult to grasp. The basic idea is that it gives a measure of how much information about the parameters is contained in the experiment. It expresses how much the prior has *changed* by observing the data. Even if the quantitative meaning of the utilities proposed is not easy to interpret, I believe that qualitatively they are understandable. On the other hand, Deely's expression (1) does not seem to measure any change caused by observing the data. Perhaps it might be better to consider some measure of the difference between expression (1) and $P(\theta_i \geq \theta_j, i \neq j)$. You-Gan Wang's suggestion of an alternative utility is to be welcomed; it is in the spirit of using utilities appropriate to the problem at hand. I would like to remark, once more, that the reason for using the utility U_2 is not based on the desire for reproducing Bayesian D-optimality as a special case, but rather (as in Lindley, 1956, Bernardo, 1979, DeGroot, 1986) on more fundamental decision theoretic motivations that regard the expected gain in Shannon information as a natural measure of the utility of an experiment.

Blaza Toman studies the important problem of the sensitivity of the design. She considers the model presented in Section 4 and allows for mispecification in the prior variances. The method she proposes is to average the determinant of the inverse posterior covariance matrix with respect to a distribution for the prior precision parameters, and use this as a new design criterion. Perhaps this is a way to approximate the fully Bayesian solution with unknown prior precision parameters. It is reassuring to see that the designs based on U_1 are robust in the examples she considers. It would be interesting to see further developments of this robustness procedure.

Both Michael Goldstein and John Deely are concerned about the practical use of the Bayesian theory of experimental design. I agree entirely with their concerns. My feeling is that this field is still relatively new and that practice will certainly follow, much as applications of Bayesian data analysis have become more common.

ADDITIONAL REFERENCES IN THE DISCUSSION

Berger, J. O. and Deely, J. (1988). A Bayesian approach to ranking and selection of related means with alternatives to analysis-of-variance methodology. *J. Amer. Statist. Assoc.* **402**, 364–373.

Keeney, R. L. and Raiffa, H. (1976). *Decisions with Multiple Objectives: Preference and Value Trade-offs.* New York: Wiley.

Pilz, J. (1991). *Bayesian Estimation and Experimental Design in Linear Regression Models.* New York: Wiley.

Wang, Y. G. (1991). Sequential allocation in clinical trials. *Comm. Statist. Theory and Methods* **20**, 791–805.

BAYESIAN STATISTICS 4, pp. 483–502
J. M. Bernardo, J. O. Berger, A. P. Dawid and A. F. M. Smith, (Eds.)

Recent Methodological Advances in Robust Bayesian Inference*

LARRY WASSERMAN
Carnegie Mellon University, USA

SUMMARY

Robust Bayesian inference is an approach to statistics that uses classes of models and priors. To use this approach, we must choose appropriate classes of priors and models, compute bounds on posterior expectations, succinctly summarize inferences and graphically display the results. I will discuss some recent attempts to address these issues.

Keywords: NEIGHBORHOODS OF PRIORS; ROBUST BAYESIAN INFERENCE; UPPER AND LOWER PROBABILITIES.

1. INTRODUCTION

I. J. Good (1959, 1961, 1962) started the robust Bayesian approach to inference. The purpose of this approach is to expose the imprecision of models and prior distributions. The landmark paper by Berger (1984) and the paper by DeRobertis and Hartigan (1981) started an explosion of research in this area. The theoretical underpinnings are now well understood. Nonetheless, robust methods are not a routine part of Bayesian data analysis. The reason is simple: Bayesian inference is computationally demanding – robust Bayesian inference inherits these computational demands and more.

In this paper, I will give a brief review of the robust Bayesian approach. Then I will consider methodological advances that make the robust Bayesian approach tractable. I will emphasize recent developments that I have been involved in; I will not attempt a comprehensive review. See Berger (1990, 1985, 1984) for excellent reviews and references.

2. THE ROBUST APPROACH

Consider a model consisting of a set of densities $\{f(y|\theta); \theta \in \Theta\}$ on the sample space $(Y, \mathcal{B}(Y))$ and let π be a prior probability measure on the parameter space $(\Theta, \mathcal{B}(\Theta))$. Let $\ell(\theta) = f(y|\theta)$ denote the observed likelihood function, π_y, the posterior distribution and $X \equiv X(\theta)$, some quantity of interest. Surely, the model and prior are only approximations. Therefore, it is important to investigate the imprecision in both $f(y|\theta)$ and π. In this paper, I will concentrate on imprecision in the prior distribution π. This is not to suggest that imprecision in the model is less important. My decision to focus on the prior is partly for simplicity and partly because the prior is usually the most contentious part of an analysis.

The analysis of imprecision can be divided into two schools of thought. The first, which I call the *sensitivity analysis approach*, imagines that the stated prior π is an approximation

* Some of this work is a review of recent research with several co-authors including Jay Kadane, Michael Lavine, Robert Wolpert and Fabrizio Ruggeri and was carried out with financial support from NSF grant DMS-9005858.

to a true, underlying prior π_0. This is consistent with Good's (1962) black box model and is the most commonly accepted approach. This school of thought may be subdivided into the nonparametric approach and the parametric approach. The nonparametric approach proceeds as follows. Since π can be regarded as being only approximately correct, we should embed π into a class of priors Γ. Then, the Bayesian paradigm may be carried out on the entire class of priors. But since the construction of π usually involves gross over-simplifications, we should use a class Γ that is not burdened with the same structural simplifications as π itself. Thus, if π were chosen from a conjugate class, we should not impose the restriction of conjugacy on the priors in Γ. In other words, we should choose Γ to be a large, nonparametric class. This is the approach suggested by Berger (1984, 1985, 1990) and it is the approach that I will explore in this paper.

The parametric approach to sensitivity analysis is to take Γ to be either some finite set of priors, or an infinite set of priors indexed by a finite dimensional parameter. This is a simple and useful thing to do. But such a Γ will not contain priors that differ from π in novel and unexpected ways. The conclusions may appear to be robust due to the restrictive nature of Γ and this might lull us into a false sense of security. A thorough analysis should include both types of sensitivity analysis.

The set of priors Γ defines a *lower probability* $\underline{P}$ and an *upper probability* $\overline{P}$ by $\underline{P}(A) = \inf_{\pi \in \Gamma} \pi(A)$ and $\overline{P}(A) = \sup_{\pi \in \Gamma} \pi(A)$. The second school of thought – an approach embraced by Smith (1961), Williams (1976), Fine (1988) and Walley (1981, 1991), among others – is to regard $\underline{P}$, rather than Γ, as fundamental. This approach is called the *theory of upper and lower probabilities*. The sensitivity analysis approach assumes that the axioms of probability are correct, but it admits that the particular probabilities used in an analysis may be incorrect. In contrast, the theory of upper and lower probabilities challenges the correctness of the axioms of probability. Despite the similarity between the approaches, there are some important differences. See Walley (1991) for a thorough discussion.

3. SETS OF PRIORS

The first few classes of priors that are discussed in this section are intended for the following type of analysis. Suppose we have selected a prior and we are interested in determining if the conclusions are highly sensitive to this choice of prior. This prior may be a default prior chosen for convenience – Jeffreys' prior, for example. Or, the prior might be an attempt to model someone's opinion. The question is whether more careful specification of the prior is required. The last class of priors that is presented in this section deals with the problem of modeling partial prior information.

The first nonparametric set of priors to be fully explored is the *ϵ-contaminated class*: $\Gamma_\epsilon^C = \{(1-\epsilon)\pi + \epsilon Q;\ Q \in \mathcal{S}\}$ where $\epsilon \in [0,1]$, $\mathcal{S} \subset \mathcal{P}$, $\mathcal{P}$ is the set of all priors and π is a given prior. The class Γ_ϵ^C and its many variants are tractable and have been thoroughly studied; see Huber (1973), Berger and Berliner (1986) and Sivaganesan and Berger (1989), for example.

Another nonparametric class is the *total variation class*: $\Gamma_\epsilon^T = \{Q;\ d(\pi, Q) \le \epsilon\}$ where d is the total variation metric defined by $d(P,Q) = \sup_{A \in \mathcal{B}(\Theta)} |P(A) - Q(A)|$. The class Γ_ϵ^T has not received much attention in Bayesian robustness despite the fact that it is intuitively appealing. But see Wasserman and Kadane (1990, 1991).

A third class, introduced by DeRobertis and Hartigan (1981), and called a *density ratio class* by Berger (1990), is defined by $\Gamma_{L,U}^R = \{\pi;\ \pi X/\pi Y \le UX/LY \text{ for all } X, Y \in \mathcal{L}^+(\Theta)\}$ where, for any measure μ we define $\mu X = \int X(\theta)\mu(d\theta)$, U and L are measures such that $L(A) \le U(A)$ for all measurable A, $\mathcal{L}(\Theta)$ is the set of bounded, measurable random variables and $\mathcal{L}^+(\Theta) = \{X \in \mathcal{L}(\Theta);\ X \geq 0\}$. An alternative definition is that $\Gamma^R_{L,U}$ is the set of probability measures with densities p, with respect to a dominating measure λ, that satisfy, $p(\theta)/p(\theta') \leq u(\theta)/l(\theta')$ for almost all θ, θ', where $l = dL/d\lambda$ and $u = dU/d\lambda$. Still another definition for this class is: the set of probability measures with densities p such that, for almost all θ, $l(\theta) \leq cp(\theta) \leq u(\theta)$ for some $c > 0$. Note that $\Gamma^R_{L,U} = \Gamma^R_{\alpha L, \alpha U}$ for every $\alpha > 0$. A special case of great interest is the class $\Gamma^R_k(\pi) \equiv \Gamma^R_{\pi, k\pi}$ where $k \geq 1$ is a constant and π is a fixed prior. This class may be thought of as a neighborhood around the target prior π. Every prior Q in the neighborhood has the property that the odds implied by Q differ from the odds implied by π by no more than k. This class is especially useful when π is a default prior chosen mainly for convenience.

A class that was investigated by Lavine (1988, 1991a, 1991b) is the *density bounded class* $\Gamma^B_{L,U}$ defined by $\Gamma^B_{L,U} = \{\pi; L(A) \leq \pi(A) \leq U(A)$ for all measurable $A\}$ where L and U are measures satisfying: $L(A) \leq U(A)$ for all A and $L(\Theta) < 1 < U(\Theta)$. Thus, π is in this class if its density p satisfies $l(\theta) \leq p(\theta) \leq u(\theta)$ for λ-almost all θ.

The classes $\Gamma^R_{L,U}$ and $\Gamma^B_{L,U}$ are very similar and they deserve further scrutiny. Their relationship is explicated in the following result, from Wasserman and Kadane (1991). The result says that a density ratio class may be decomposed into a set of density bounded classes.

Proposition. *Suppose that* l *and* u *are such that* $l(\theta) \leq u(\theta)$ *almost everywhere, and* $\int u(\theta)\lambda(d\theta) < \infty$. *Let* $L(A) = \int_A l(\theta)\lambda(d\theta)$ *and* $U(A) = \int_A u(\theta)\lambda(d\theta)$. *Then,*

$$\bigcup_{a \leq \alpha \leq b} \Gamma^B_{\alpha L, \alpha U} = \Gamma^R_{L,U} \subset \Gamma^B_{\hat{L},\hat{U}}$$

where $a = 1/U(\Theta)$, $b = 1/L(\Theta)$, $\hat{L}(A) = \int_A \hat{l}(\theta)\lambda(d\theta)$, $\hat{U}(A) = \int_A \hat{u}(\theta)\lambda(d\theta)$, $\hat{l}(\theta) = l(\theta)/U(\Theta)$ *and* $\hat{u}(\theta) = u(\theta)/L(\Theta)$. *Furthermore, if* $L(\Theta) < U(\Theta)$, *and if* l *and* u *are continuous functions, then the containment is proper.*

Finally, we turn to a different type of class. Suppose we have partial prior information that defines a class of priors Γ. Consider an example. Suppose n_1 patients are given treatment one and n_2 patients are given treatment two. Let X_i be the number of patients who survive on treatment i. Thus, X_i is binomial (n_i, p_i), $i = 1, 2$. The parameter is $\theta = (p_1, p_2)$. Suppose that the quantity of interest is $X(\theta) = p_1 - p_2$ and that we have a prior π_1 for p_1 and a prior π_2 for p_2 but we don't have a joint prior for θ. (This problem was studied in Lavine, Wasserman and Wolpert, 1991a). The given information defines a class of priors Γ^{π_1,π_2} which is the set of all joint priors with the given marginals. This class contains many unreasonable priors and it may be desirable to put further restrictions on the class. One way to do this is to add a density ratio constraint. For example, if we assume, a priori, that p_1 and p_2 are approximately independent, then we can use $\Gamma^{\pi_1,\pi_2}_k = \Gamma^{\pi_1,\pi_2} \cap \Gamma^R_k(\pi_0)$ where π_0 is the independence prior defined by π_1 and π_2. This class has the following interpretation: every prior in the class has the given marginals for p_1 and p_2 and is close to the independence prior π_0. Now, in our attempt to honestly represent partial prior information, we are led to a complicated set of priors. This raises the question of how we deal with such an unwieldy class. We address this in the next two sections.

4. PRIOR BOUNDS

Ultimately, we are interested in finding bounds on posterior quantities such as posterior expectations and posterior probabilities. Before doing that, we first consider the problem of bounding prior expectations. This is useful for two reasons. First, this will be a step towards the goal of computing posterior bounds. Second, it is reasonable to examine the implications of a prior class by finding bounds on prior quantities. I will describe four methods for finding prior bounds. I will emphasize finding upper bounds since finding lower bounds is similar.

Consider the problem of finding the upper expectation $\overline{E}X = \sup_{\pi \in \Gamma} \pi X$ where $\pi X = \int X(\theta)\pi(d\theta)$. Sometimes, $\overline{P}(A) = \sup_{\pi \in \Gamma} \pi(A)$ is easy to compute. This raises the following question: is it possible to get $\overline{E}X$ from $\overline{P}(A)$? In general the answer is no; see Huber and Strassen (1973), Walley (1991) and Wasserman and Kadane (1990). But it is possible to approximate the bound. To do so, define the upper Choquet integral $\overline{C}X$ (Choquet 1953, Huber 1973) by $\overline{C}X = \int_0^\infty \overline{P}(\{\theta; X(\theta) > t\})dt$ for $X \in \mathcal{L}^+(\Theta)$. Then $\overline{E}X \leq \overline{C}X$ so that $\overline{C}X$ gives a conservative bound. Two conditions are needed to make the bound exact. These are:

(i) $P(A) \leq \overline{P}(A)$ for every measurable A implies that $P \in \Gamma$ and
(ii) $\overline{P}(A \cup B) \leq \overline{P}(A) + \overline{P}(B) - \overline{P}(A \cap B)$ for every pair of measurable sets A and B.

The first condition is a closure property. The second condition says that $\overline{P}$ is "2-alternating". This condition arises in frequentist robustness (Huber 1973, Huber and Strassen 1973). Wasserman and Kadane (1990) exploited conditions (i) and (ii) to derive bounds in Bayesian robustness. The total variation class, the ϵ-contaminated class (with $\mathcal{S} = \mathcal{P}$) and the density bounded class all satisfy (i) and (ii). In section 5, I will point out that the Choquet integral can sometimes be used to get posterior bounds. But if (i) or (ii) are violated, the bound can be terrible. For example, the set of symmetric, unimodal priors (Sivaganesan and Berger 1989) fails both (i) and (ii). It is then possible to find a random variable X that makes $\overline{C}X - \overline{E}X$ arbitrarily large. The density ratio class satisfies (ii) but not (i). This approach has limitations, but it adds theoretical insight into the problem.

The second way to find $\overline{E}X$ is to identify the maximizing measures in Γ directly. For example, consider the class $\Gamma^B_{L,U}$ and let X be bounded and continuous. Let $X_t = \{\theta; X(\theta) > t\}$ and let M_t be the measure that is equal to U over X_t and equal to L otherwise. Lavine (1988, 1991b) shows that there is a t such that M_t integrates to one and that this probability measure is the maximizing measure, that is, $\int X(\theta)M_t(d\theta) = \overline{E}X$. Similar arguments are used in O'Hagan and Berger (1988). Obviously, some ingenuity is needed here, but the payoff is great if the maximizing measures can be found.

A third method is to discretize the parameter space and use linear programming techniques. This is discussed in Lavine, Wasserman and Wolpert (1991a,b). For example, consider the class Γ^{π_1,π_2}_k defined at the end of section 3. Using the proposition from that section, we have

$$\begin{aligned}\Gamma^{\pi_1,\pi_2}_k &= \Gamma^{\pi_1,\pi_2} \bigcap \Gamma^R_k(\pi_0) = \Gamma^{\pi_1,\pi_2} \bigcap \left(\bigcup_\alpha \Gamma^B_{\alpha\pi_0,\alpha k\pi_0} \right) \\ &= \bigcup_\alpha \left(\Gamma^{\pi_1,\pi_2} \bigcap \Gamma^B_{\alpha\pi_0,\alpha k\pi_0} \right) = \bigcup_\alpha \Lambda_\alpha, \text{ say}\end{aligned}$$

where $1/k \leq \alpha \leq 1$. Then, $\overline{E}X = \max_\alpha \overline{E}_\alpha X$ where $\overline{E}_\alpha X = \sup_{\pi \in \Lambda_\alpha} \pi X$. The latter maximization, if performed over a finite grid in the parameter space, involves maximizing a linear quantity subject to equality constraints (the given marginals) and inequality constraints (the density bounded class). This maximization is easily performed using the simplex method. Many problems can be converted into forms that are easily handled by linear programming techniques.

The fourth method is a Monte Carlo technique that was introduced in Wasserman and Kadane (1991). To illustrate the idea, consider the class $\Gamma^R_k(\pi)$. Let $(\theta_1, \ldots, \theta_N)$ be a random sample from the prior π. Let $x_i = X(\theta_i)$ and let $x_{(1)} \leq x_{(2)} \leq \ldots x_{(N)}$ be the ordered values. Now define,

$$\widehat{E}_k^R X = \max_{1 \leq i \leq N} \{(1 - i/N)\Delta + 1\}^{-1} \left\{ \Delta \sum_{j=i}^{N} x_{(j)}/N + \overline{x} \right\}$$

where $\Delta = k - 1$ and $\overline{x}$ is the sample average of the x_i's. Wasserman and Kadane (1991) show that $\widehat{E}_k^R X - \overline{E}_k^R X = O_P(1/\sqrt{N})$, where $\overline{E}_k^R X = \sup_{\pi \in \Gamma_k^R} E_\pi(X)$. Similar formulae exist for the classes Γ^T and Γ^B. This method is most useful for high dimensional problems. But the real payoff comes in computing posterior bounds, which we now turn to.

5. POSTERIOR BOUNDS

Lavine (1988) made a keen observation, namely, that any method for finding prior bounds can be turned into a method for finding posterior bounds. This is due to the following result (Lavine 1988, Lavine, Wasserman and Wolpert 1991a, 1991b, Walley 1991, Srinivasan and Truszczynska 1990).

Let $\overline{E}(X|y)$ be the upper bound on the posterior expectation of X having observed y. Then, under weak conditions, $\overline{E}(X|y)$ is the solution to the equation $s(\lambda) = 0$ where $s(\lambda) = \overline{E}(h_\lambda)$, $h_\lambda(\theta) = X(\theta)(\ell(\theta) - \lambda)$ and $\lambda \in \mathbf{R}$. Thus, the posterior problem has been "linearized". This generalizes the approach used by DeRobertis and Hartigan (1981). One way to implement this idea, is to choose a grid of values for λ, say, $(\lambda_1, \ldots, \lambda_j, \ldots, \lambda_M)$. For each λ_j we compute $s(\lambda_j)$. This can be done by any of the techniques in Section 4. Finally, we solve for λ numerically in the equation $s(\lambda) = 0$.

An important observation is that if $\pi_{\lambda_j} \in \Gamma$ maximizes the expectation of h_{λ_j}, then $\pi_{\lambda_{j+1}}$ is often close to π_{λ_j}. This means that in trying to find $\overline{E}(h_{\lambda_{j+1}})$, we can save time by using π_{λ_j} as a starting point. Unfortunately, π_λ is not continuous in λ; examples can be constructed where π_{λ_j} is far from $\pi_{\lambda_{j+1}}$. But experience suggests that the discontinuities are rare.

The maximizations that must be performed at each stage of the linearization algorithm can be demanding. Sometimes it is possible to approximate the bounds. Let $\mathcal{G} = \{\Gamma_1, \ldots, \Gamma_k\}$ where each $\Gamma_i \in \mathcal{G}$ satisfies two properties. Property 1 is that Γ_i should be more tractable than the original class of interest Γ and property 2 is that $\Gamma \subset \Gamma_i$ for $i = 1, \ldots, k$. Define $\widetilde{E}X = \min\{\overline{E}_1 X, \ldots, \overline{E}_k X\}$ where $\overline{E}_i X = \sup_{\pi \in \Gamma_i} \pi X$. Set $\widetilde{s}(\lambda) = \widetilde{E}(h_\lambda)$ and let $\widetilde{E}(X|y)$ be the solution of $\widetilde{s}(\lambda) = 0$. Then, $\overline{E}(X|y) \leq \widetilde{E}(X|y)$ so that a conservative approximation to the bound is produced. For example, in the two binomial problem with fixed marginals that was described in Section 3, we have that $\Gamma_k^{\pi_1,\pi_2} \subset \Gamma_i$ for $i = 1, 2$ where $\Gamma_i = \Gamma^{\pi_i} \bigcap \Gamma_k^R(\pi_0)$ and Γ^{π_i} is the set of all joint distributions with only the ith margin fixed. Applying the approximation leads to much simpler computations. Preliminary results suggest that the approximation is surprisingly accurate. Note that the approximation takes place at each step of the linearization algorithm – we are not approximating the posterior bound directly.

A second approach to posterior bounds is to exploit the 2-alternating condition. Specifically, suppose that the set of priors Γ satisfies conditions (i) and (ii) from Section 4. Walley (1981) and Wasserman and Kadane (1990) showed that the upper posterior probability of a set A is given by $\overline{r}(A)/(\overline{r}(A) + \underline{r}(A^c))$ where $\overline{r}(A) = \overline{C}(\ell I_A)$ and $\underline{r}(A^c) = \underline{C}(\ell I_{A^c})$. Here, ℓ is the likelihood, I_A is the indicator function and $\underline{C}$ is the lower Choquet integral.

Surprisingly, this leads to explicit formulae for certain classes of priors – the total variation class and the density bounded class, for example.

Another approach is to use the Monte Carlo method at each stage in the linearization algorithm. However, a more direct approach is possible. Consider the class $\Gamma_k^R(\pi)$. Think of Bayes' theorem as an operator that takes priors into posteriors. Specifically, let B_y be the operator that takes a prior π to the posterior π_y when the datum is y. Thus, $\pi_y = B_y(\pi)$. Let $B_y(\Gamma) = \{B_y(\pi); \pi \in \Gamma\}$. Then the density ratio class has the following Bayes' invariance property: $B_y(\Gamma_k^R(\pi)) = \Gamma_k^R(B_y(\pi))$. Instead of applying Bayes' theorem to the whole class Γ_k^R, we can update π first and then find the density ratio class around that posterior. In other words, the processes of constructing neighbourhoods and Bayesian updating can be interchanged. To bound the posterior expectation of X, we use the Bayes' invariance property and conclude that we only need bound the expectation of X over $\Gamma_k^R(\pi_y)$. Now we appeal to the Monte Carlo method. To do so, we will need to draw a sample from the posterior π_y. Here, we can rely on recent methods like those in Tanner and Wong (1987) and Gelfand and Smith (1990). Then we apply the formula $\widehat{E}_k^R X = \max_{1 \le i \le N}\{(1 - i/N)\Delta + 1\}^{-1}\{\Delta \sum_{j=i}^N x_{(j)}/N + \overline{x}\}$ directly to the sample from the posterior. This gives the posterior bound for each k. Since we often need a sample from the posterior, we might as well use that sample to do these Bayesian robustness computations. I do not know how to use a posterior sample to get the bounds directly for other classes. This is a pragmatic argument in favor of the density ratio class.

6. SUMMARIZATION

Having carried out a robust Bayesian analysis, a natural question is: how do we summarize the results? Displaying results in Bayesian inference is easy – plots of marginals and credible regions give useful graphical and numerical summaries. Little progress has been made in Bayesian robustness on these issues. Now that computations are becoming more tractable, I believe that this is an area that deserves more attention.

When a specific function $X = X(\theta)$ is of interest and a class like Γ_k^R is used, we can plot the upper and lower posterior expectation of X by k. That is, we can plot $\overline{E}_k^R(X|y)$ and $\underline{E}_k^R(X|y)$ by k, where $\overline{E}_k^R(X|y) = \sup_{\pi \in \Gamma_k^R} E_\pi(X|y)$ and $\underline{E}_k^R(X|y)$ is the corresponding lower bound. It is reasonable to let k vary from 1 to 20. Similarly, for ϵ-contaminated classes we need not pick a single value of ϵ. We just plot the bounds by ϵ.

Another approach to numerical summarization is to find robust Bayesian credible regions. This means finding the smallest region with lower posterior probability above a given threshold. An early version of Berger and Berliner (1986) considered this approach but the method was not in the published paper and I am not aware of much follow-up on this idea. I would like to suggest an alternative approach. The idea is to use likelihood regions calibrated with the set of posteriors. That is, let $R_c = \{\theta; L(\theta) \ge c\}$ and set $\overline{F}(c) = \overline{P}(R_c|y)$ and $\underline{F}(c) = \underline{P}(R_c|y)$. The region R_c for which $\underline{F}(c) = 0.95$, for example, is the likelihood region with lower posterior probability 0.95. A graph of the likelihood along with the curves $\overline{F}(c)$ and $\underline{F}(c)$ provides a succinct summary of the robust Bayesian analysis.

Why use the likelihood? There are several reasons. The first is that frequentist, likelihoodist and Bayesian all agree that the likelihood plays a central role in inference, and we are all comfortable dealing with the likelihood. The second reason is more formal. Consider an ϵ-contaminated class around a prior π. In Wasserman (1989) I showed that, of all regions with a fixed posterior probability content under π_y, the most robust region is the likelihood region. Furthermore, Ruggeri and Wasserman (1990) showed that a similar result can be obtained by using the Frechet derivative of the posterior. Admittedly, this is not a convincing argument that likelihood regions are optimal. But there is a sense in which they are robust and this supports their use. Also, Piccinato (1984) showed that likelihood regions have a natural interpretation in Bayesian inference.

7. DISCUSSION

A common criticism of Bayesian methodology is that the computations are intractable. Recent progress in analytic approximations (Tierney and Kadane 1986, Kass, Tierney and Kadane 1988) and in Monte Carlo methods (Tanner and Wong 1987, Gelfand and Smith 1990) make this criticism obsolete. A second, and more common criticism, is that the analysis depends on an indefensible prior. Addressing the second criticism implores us to consider robust Bayesian methods. Unfortunately, this brings us in a full circle back to the first criticism. Robust Bayesian methodology involves computations that are more difficult than standard Bayesian methods. But there has been real progress lately, as I have tried to show. For example, if we have already drawn a sample from the posterior, then we can use the methods in Section 5 to quickly compute bounds on posterior expectations. This should be included as a routine part of our analysis. We get this for free once the hard work of drawing a sample from the posterior is done.

Despite these methodological steps forward, there remain many practical and theoretical problems. There is still much room for computational developments. Every advance in Bayesian computing is potentially an advance in robust Bayesian inference. The challenge is to find ways to apply Bayesian computing methods to robust Bayesian problems. Similarly, we should try to apply recent work on analytic approximations. DeRobertis and Hartigan (1981) derived an asymptotic approximation to the bounds for the $\Gamma^R_{L,U}$ class. Perhaps the second order asymptotics of Tierney, Kass and Kadane can be used to improve this result.

Interpreting the inputs and outputs of a robust Bayesian analysis needs more attention. When dealing with classes like Γ^R_k and Γ^T_ϵ we should do the computations for many values of k and ϵ. But it may be possible to calibrate k and ϵ. For a normal density, the value $k = 10$ gives a range of the mean equal to about one standard deviation. Perhaps we can use canonical examples like the normal to help us interpret the "parameters" k and ϵ.

Finally, I know of only a handful of real examples that put this type of robust Bayesian inference to use. We need to develop a catalogue of applications before we can be sure we have found, let alone solved, the important problems.

REFERENCES

Berger, J. (1984). The robust Bayesian viewpoint. *Robustness in Bayesian Statistics* (J. Kadane ed.), Amsterdam: North-Holland, 63–124, (with discussion).

Berger, J. (1985). *Statistical Decision Theory.* New York: Springer.

Berger, J. (1990). Robust Bayesian analysis: sensitivity to the prior. *J. Statist. Planning and Inference* **25**, 303–328.

Berger, J. and Berliner, M. (1986). Robust Bayes and empirical Bayes analysis with ϵ-contaminated priors. *Ann. Statist.* **14**, 461–486.

Berger, J. O. and Bernardo, J. M. (1992). On the development of the reference prior method. *Bayesian Statistics 4* (J. M. Bernardo, J. O. Berger, A. P. Dawid and A. F. M. Smith, eds.), Oxford: University Press, 35–60, (with discussion).

Choquet, G. (1953). Theory of capacities. *Ann. Inst. Fourier.* **5**, 131–295.

DeRobertis, L. and Hartigan, J. A. (1981). Bayesian inference using intervals of measures. *Ann. Statist.* **9**, 235–244.

Fine, T. (1988). Lower probability models for uncertainty and nondeterministic processes. *J. Statist. Planning and Inference* **20**, 389–411.

Gelfand, A. E. and Smith, A. F. M. (1990). Sampling-based approaches to calculating marginal densities. *J. Amer. Statist. Assoc.* **85**, 398–409.

Good, I. J. (1959). Could a machine make probability judgments?. *Computers and Automation* **8**, 14–16 and 24–26.

Good, I. J. (1961). Discussion of C. A. B. Smith "Consistency in statistical inference and decision". *J. Roy. Statist. Soc. B* **23**, 28–29.

Good, I. J. (1962). Subjective probability as the measure of a non-measurable set. *Logic, Methodology and the Philosophy of Science.* Stanford: University Press, 319–329.

Huber, P. J. (1973). The use of Choquet capacities in statistics. *Bull. Internat. Statist. Institute.* **45**, 181–191.

Huber, P. J. and Strassen, V. (1973). Minimax tests and the Neyman-Pearson lemma for capacities. *Ann. Statist.* **1**, 251–263.

Kass, R., Tierney, L. and Kadane, J. (1988). Asymptotics in Bayesian computation. *Bayesian Statistics 3* (J. M. Bernardo, M. H. DeGroot, D. V. Lindley and A. F. M. Smith, eds.), Oxford: University Press, 261–278, (with discussion).

Lavine, M. (1988). Prior influence in Bayesian statistics. *Tech. Rep.* **88-06**, Institute of Statistics and Decision Sciences, Duke University, Durham, North Carolina.

Lavine, M. (1991a). An approach to robust Bayesian analysis for multidimensional spaces. *J. Amer. Statist. Assoc.* **86**, 396–399.

Lavine, M. (1991b). Sensitivity in Bayesian statistics: the prior and the likelihood. *J. Amer. Statist. Assoc.* **86**, 400–403.

Lavine, M., Wasserman, L. and Wolpert, R. (1991a). Bayesian inference with specified prior marginals. *J. Amer. Statist. Assoc.* , (to appear).

Lavine, M., Wasserman, L. and Wolpert, R. (1991b). Linearization of Bayesian robustness problems. *Tech. Rep.* **518**, Carnegie Mellon University.

O'Hagan, A. and Berger, J. (1988). Ranges of posterior probabilities for quasiunimodal priors with specified quantiles. *J. Amer. Statist. Assoc.* **83**, 503–508.

Piccinato, L. (1984). A Bayesian property of the likelihood sets. *Statistica* **44**, 197–204.

Ruggeri, F. and Wasserman, L. (1990). Infinitesimal sensitivity of posterior distributions. *Tech. Rep.* **501**, Department of Statistics, Carnegie Mellon University.

Sivaganesan, S. and Berger, J. (1989). Ranges of posterior measures for priors with unimodal contaminations. *Ann. Statist.* **17**, 868–889.

Smith, C. A. B. (1961). Consistency in statistical inference and decision. *J. Roy. Statist. Soc. B* **23**, 1–37, (with discussion).

Smith, A. F. M. and Gelfand, A. E. (1992). Bayesian statistics without tears: a sampling-resampling perspective. *Amer. Statist.*, (to appear).

Srinivasan, C. and Truszczynska, H. (1990). On the ranges of posterior quantities. *Tech. Rep.* **294**, University of Kentucky.

Tanner, M. and Wong, W. (1987). The calculation of posterior distributions by data augmentation. *J. Amer. Statist. Assoc.* **82**, 528–550, (with discussion).

Tierney, L. and Kadane, J. (1986). Accurate approximations for posterior moments and marginal densities. *J. Amer. Statist. Assoc.* **84**, 710–716.

Walley, P. (1981). Coherent lower (and upper) probabilities. *Tech. Rep.* **22**, University of Warwick, Coventry.

Walley, P. (1991). *Statistical reasoning with imprecise probabilities.* London: Chapman and Hall.

Wasserman, L. (1989). A robust Bayesian interpretation of likelihood regions. *Ann. Statist.* **17**, 1387–1393.

Wasserman, L. (1990). Prior envelopes based on belief functions. *Ann. Statist.* **18**, 454–464.

Wasserman, L. A. and Kadane, J. (1990). Bayes' theorem for Choquet capacities. *Ann. Statist.* **18**, 1328–1339.

Wasserman, L. A. and Kadane, J. (1991). Computing bounds on expectations. *J. Amer. Statist. Assoc.* , (to appear).

Williams, P. (1976). Indeterminate probabilities. *Formal Methods in the Methodology of Empirical Sciences.* (M. Przelecki, K. Szaniawski and R. Wojcicki, eds.), Dordrecht: Reidel, 229–246.

DISCUSSION

G. CASELLA (*Cornell University, USA*) and
M. T. WELLS (*Cornell University, USA*)

1. Introduction. We congratulate Professor Wasserman on presenting an extremely lucid account of the developments in robust Bayesian methods. There has truly been an explosion in this methodology in both theoretical and applied work, and the author has done an admirable job in explaining these advances. Because the amount of information in this paper is so great, it would be impossible to comment on all of the points. Instead, we will discuss some of the general ideas of robust Bayesianess. We will illustrate some of the difficulties and advantages of this approach, and will also present a simple method of implementing a robust Bayesian analysis.

2. Some Difficulties. An important feature of a robust Bayesian analysis is the ability to "tour" the prior and examine the performance of posterior quantity as this tour is taken. By doing so the robustness of a particular procedure can be examined. As a simple example, suppose we observe $y \sim N(\theta, 1)$ and consider the prior distribution of θ to be Student's t with ν degrees of freedom, $\theta \sim t_\nu$. As we vary the parameter ν, we can tour the class of priors. Moreover, we know that as ν increases we have priors with lighter tails.

In Figure 1 we show the posterior probabilities of the set

$$C_a = \{\theta : \theta \in ay \pm 1.645\sqrt{a}\}, \quad \frac{1}{2} \leq a \leq 1, \tag{1}$$

and see behavior that we might expect. For small values of a (which correspond to sets arising from light-tailed priors) the posterior probability of $C_a, P(C_a|y)$ changes greatly as ν varies. For large values of a these probabilities are flat (as functions of ν) displaying more robustness. The choice $a = 1$ corresponds to a likelihood region, which we expect to be robust from Wasserman (1989). (We note in passing that computations with this class are very easy since $\theta \sim t_\nu$ is equivalent to $\theta|\tau \sim \mathrm{N}(0, \tau^2), \tau^2 \sim$ Inverted Gamma $(\nu/2, \nu/2)$. In this latter formulation the Gibbs sampler is extremely easy to implement.)

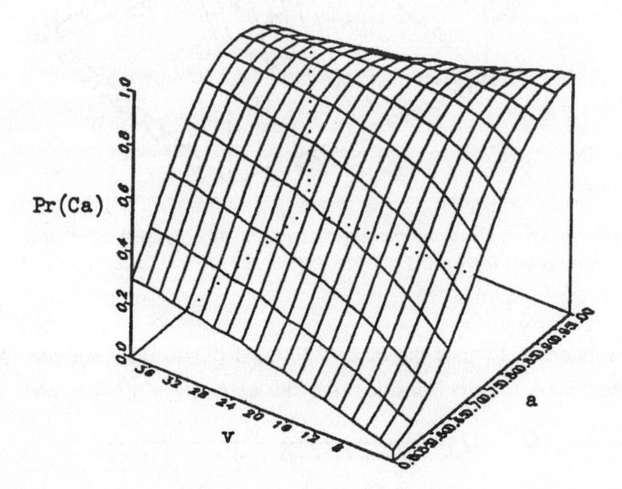

Figure 1. *Posterior probabilities of the set C_a given in (1), as a function of ν and a.*

Thus, the Student's t class displays properties that we think are important in a robust Bayes class of priors: it is easy to understand and easy to tour. Unfortunately neither of these properties is shared by the class of priors that seems to be the author's favorite, the density ratio class $\Gamma^R_{L,U}$. For any set C in the parameter space, the supremum (Q^S) and infimum (Q^I) of $P(C|y)$ are given by (Berger, 1990)

$$Q^S = \frac{\int_C \ell(\theta|y)U(\theta)d\theta}{\int_C \ell(\theta|y)U(\theta)d\theta + \int_{C^c} \ell(\theta|y)L(\theta)d\theta}$$
$$Q^I = \frac{\int_C \ell(\theta|y)L(\theta)d\theta}{\int_C \ell(\theta|y)L(\theta)d\theta + \int_{C^c} \ell(\theta|y)U(\theta)d\theta}, \tag{2}$$

where C^c is the complement of C. If we further define $\gamma^L(C) = \int_C \ell(\theta|y)\ L(\theta)d\theta \ / \int_\Theta \ell(\theta|y)\ L(\theta)d\theta$, and similarly $\gamma^U(C)$, then both Q^S and Q^I can be written as a function of $\gamma^L(C), \gamma^U(C)$ and $r = \int_\Theta \ell(\theta|y)U(\theta)d\theta / \int_\Theta \ell(\theta|y)L(\theta)d\theta$. For fixed r, a given set C will give rise to a pair $(\gamma^L(C), \gamma^U(C))$. We can now tour the prior class by varying γ^L and γ^U, and for each pair plot the quantity $Q^S - Q^I$, a measure of robustness. (As in Wasserman (1989) or Berger (1990), we interpret smaller values of $Q^S - Q^I$ as indicating robustness.) This is done in Figure 2 which, unfortunately, seems extremely difficult to interpret. Morever, it shows that sets C where γ^L and γ^U are close together are less robust than sets with γ^L and γ^U apart, with robustness increasing with the discrepancy.

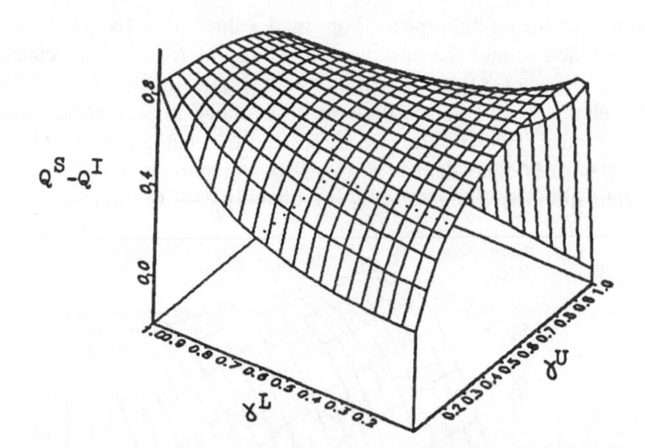

Figure 2. *The quantity $Q^S - Q^I$, using the formulas in (2), as a function of γ^U and γ^L, the posterior probablities based on the priors $U(\theta)$ and $L(\theta)$. The ratio of marginals, r, is set equal to 6.*

A simpler version of $\Gamma^R_{L,U}$ is the class Γ^R_K, but Γ^R_K still does not provide us with an easily interpretable class of prior densities. For this class $\gamma^U = \gamma^L = \gamma$, and

$$Q^S - Q^I = \frac{K\gamma}{K\gamma + (1-\gamma)} - \frac{\gamma}{\gamma + K(1-\gamma)}. \tag{3}$$

For fixed K, we obtain the curves of $Q^S - Q^I$ given in Figure 3. Again we conclude that within a given class Γ^R_K the most robust set has $\gamma = 0$. We are again unable to tour the class in a meaningful way.

It is possible to compare the performance of a procedure among different classes Γ_K^R. That is, K can vary in some range. For example, looking along a vertical line above $\gamma = .9$ in Figure 3 will give some idea of how the posterior probability of a set will vary with K. However, this does not describe the behavior of a set within a given class Γ_K^R.

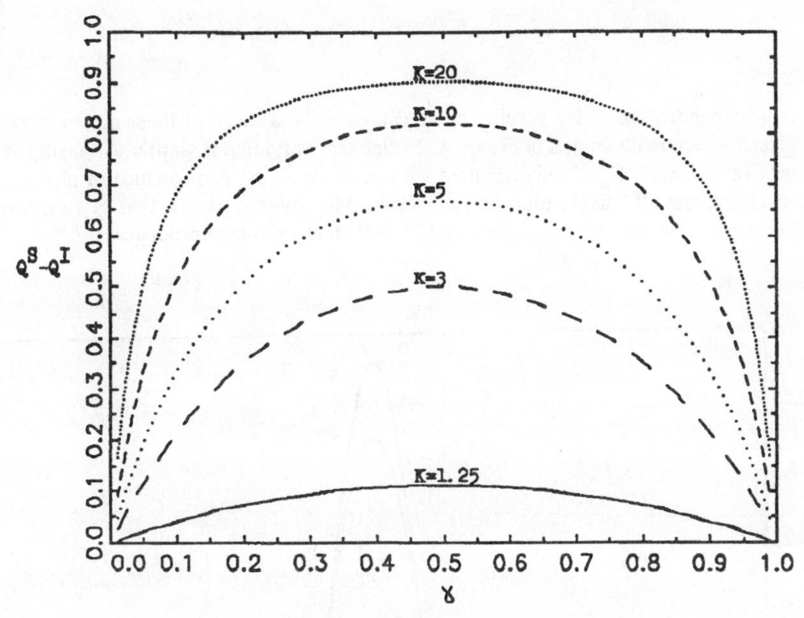

Figure 3. *The quantity $Q^S - Q^I$ for the class Γ_K^R. Here $\gamma^U = \gamma^L = \gamma$.*

3. Representing the Class of Priors. Since touring classes of priors such as $\Gamma_{L,U}^R$ is difficult, the following alternative is offered for consideration. Rather than evaluate a procedure using all priors in a class, perhaps we can choose a particular prior (in the class) which will result in a robust Bayes rule. Such an idea is in the spirit of the reference prior approach, whose history is marvellously summarized by Berger and Bernardo (1992).

We illustrate our idea for the case of ε-contamination classes $\Gamma_\varepsilon^C = \{(1-\varepsilon)\pi + \varepsilon Q; Q \in S\}$. Similar calculations can be performed for total variation classes such as Γ_ε^T, but only with difficulty for density ratio and density bounded classes. More complete details can be found in Casella and Wells (1992).

For the ε-contamination class, Γ_ε^C, consider the prior within the class that has minimum expected Fisher information. That is, we look for the prior π_ε^* that minimizes

$$i(\theta) = E_\theta[(\nabla \log p(\theta))(\nabla \log p(\theta))'], \tag{4}$$

where $p \in \Gamma_\varepsilon^C$ and $\nabla \log p(\theta)$ is the vector of first partial derivaties of $\nabla \log p(\theta)$. Thus, within the class Γ_ε^C we are operating in the spirit of Jeffreys (1937, 1961). Polson (1988) considered a similar approach.

Happily, the minimization problem described above has been solved by Huber (1981), in the context of frequentist robust statistical procedures, and that work applies here. We obtain as the noninformative prior

$$\pi^*_\varepsilon(\theta) = \begin{cases} \dfrac{1-\varepsilon}{\sqrt{2\pi}}\, e^{-\theta^2/2} & , \quad |\theta| \le m \\ \dfrac{1-\varepsilon}{\sqrt{2\pi}}\, e^{m^2/2-m|\theta|} & , \quad |\theta| > m \end{cases} \tag{5}$$

where m is dependent on ε. By varying ε, we can examine a range of these noninformative robust priors, which is illustrated in Figure 4. Notice that varying ε is similar to varying K in the density ratio classes Γ^R_K. However, here we can examine the performance of procedures against each π^*_ε, and effectively tour the prior class. Moreover, we know that as ε increases, our priors are getting heavier tails, and should result in more robust procedures.

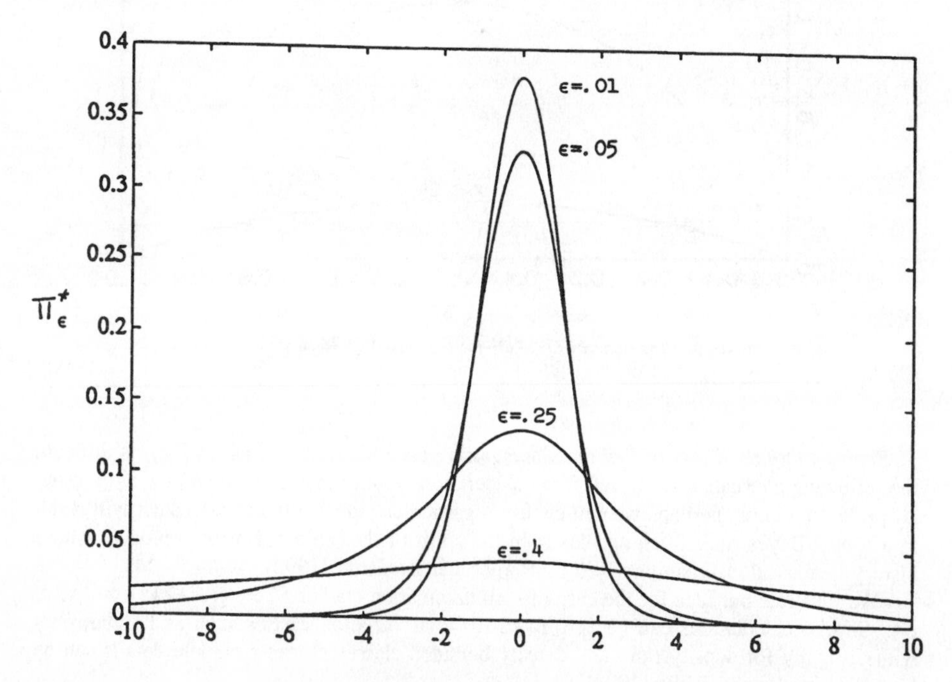

Figure 4. *The noninformative robust priors π^*_ε of (5).*

As an illustration of the performance of π^*_ε, we consider the example used by Wasserman (1989) to illustrate the robustness of the likelihood region. We can see that the noninformative robust Bayes region will be close to the HPD region when the data support the root prior ($Y = .5$), but will be closer to the likelihood region, and hence retain robustness, when the data disagree with the root prior ($Y = 4$).

	$Y = .5$		$Y = 4.0$	
	Set	δ	Set	δ
HPD	$(-1.27, 1.93)$	.148	$(1.07, 4.27)$	.857
Likelihood	$(-1.13, 2.13)$	.117	$(1.32, 6.68)$	.182
Noninformative	$(-1.30, 1.97)$	.126	$(1.43, 5.35)$	.213

Table 1. *For $Y \sim N(\theta, 1)$ and $\theta \in \Gamma_\varepsilon^C$ with $\pi = N(0,2)$ and $\varepsilon = .2$, credible regions and Wasserman's (1989) δ (equivalent to $Q^S - Q^I$) for three procedures.*

4. Conclusions. Robust Bayesian methodology offers the user of statistics a multitude of techniques to examine sensitivity of posterior answers to prior assumptions. Care must be taken, however, to ensure that the prior class is both useful and understandable. A number of prior classes promoted by the author, although containing interesting members and being mathematically tractable, are difficult to understand. The ability to tour through a class seems, to us, to be an essential property of any useful prior class. Moreover, this tour should take the user through meaningful prior densities.

In one sense, being a robust Bayesian can be reduced to a triviality. Simply use a likelihood region. The author will, of course, disagree with this since, as he stated in the paper, it is Bayesian calibration of the likelihood region that matters. However, since we know the likelihood region is robust using virtually any Bayesian calibration, the exact specification and consideration of the prior class is unnecessary. Thus, the most robust Bayesian isn't a Bayesian.

Two problems facing the robust Bayesian are the problem of examining the members of the prior class, and of obtaining Bayes (not likelihood) sets that can be considered robust. A strategy such as the noninformative robust Bayes strategy can accomplish both goals. It allows easy examination of all members of its class (as we vary ε) and results in Bayes credible sets that exhibit robust behavior. We hope that, in future research in robust Bayesian methods, these concerns are addressed.

J. O. BERGER (*Purdue University, USA*)

This is an excellent review and analysis of recent work in robust Bayesian methods of the author and collaborators. Overall, I find myself in complete agreement with the tone and conclusions of the paper, with one possible exception. This exception is the (tentative) conclusion that the density ratio class is superior to others.

From a computational perspective, the arguments of the author convince me that the density ratio class is wonderful, although the situation in higher dimensions is not quite clear (cf. Bose, 1990). However, the observation in Section 5 concerning the invariance property of this class — that the class of posteriors can be found as the density ratio class formed from the base posterior — has caused me to begin to worry about the class. To explain why, it is useful to detour into speculation concerning the real potential of robust Bayesian analysis.

In Bayesian analysis, one specifically elicits a finite number (possibly zero) of features of the prior. The real potential of the robust Bayesian approach lies in the possibility of assessing the adequacy of these limited beliefs in reaching a conclusion. If the range of answers obtained via the robust Bayesian methodology is too large, one knows that additional beliefs must be elicited (or more data obtained).

The key to realization of this potential is to embed the elicited beliefs in a class of priors that allows sufficient variability in unelicited features of the prior. The invariance property of the density ratio class seems to say that the variability in the posterior is fixed no matter what is encountered. This is intuitively troubling because other classes, such as the ε–contamination class, have the property that the class of posteriors can be large or small, depending on the data that is observed. This seems more consistent with the rather common view that specification of certain beliefs is sufficient for some data, and insufficient for other data.

To be a bit more specific, suppose one specifies a few quantiles of the prior, and fits them to a normal density. Compare, then, the use of the density ratio class and the ε–contamination class (perhaps with some constraints on the allowed contaminations), both using this normal prior as a base. In the density ratio class, the variation in the resulting class of posteriors will be essentially fixed for any observed data, while for the ε–contamination class the resulting class of posteriors can vary dramatically, being relatively small if the likelihood is compatible with the prior and very large if the likelihood is incompatible with the prior. The reason, of course, is that the "unelicited" tail of the prior is highly influential when the likelihood is incompatible with the prior, and the ε–contamination class has left the tail quite arbitrary while the density ratio class allows only limited tail deviation from the base prior. Of course, if one can honestly assess that the variation in density ratio should be no more than k from the base prior, then use of the density ratio class is fine, but it is not at all clear that such an assessment is routinely feasible.

This concern with density ratio classes is admittedly still rather vague, to a large extent because these issues are difficult to address in the abstract. The author's closing comment that we need to develop a catalogue of applications before drawing any firm conclusions is certainly also relevant here.

Another issue I found very interesting was the claimed robustness of the likelihood region as a Bayesian credible set. One reason I found this interesting is that "optimally Bayesian robust" procedures are often unappealing, compared with more *ad hoc* robust procedures. For instance, in estimation under squared error loss, optimal robust Bayesian estimators tend to be roughly equal to the midpoint of the range of posterior means. The midpoint can be a rather unsatisfactory estimator, however, in that it frequently does not correspond to sensible hierarchical Bayes estimators developed by putting a prior on the class of priors. The *ad hoc* solution of using the Bayes estimator for the Type–II maximum likelihood prior is often considerably more sensible. In the same vein, it is interesting to ask if the likelihood region corresponds, in some sense, to reasonable hierarchical Bayes credible sets.

D. V. LINDLEY (*Somerset, UK*)

One thing that the Bayesian argument teaches us when handling probability (the situation changes with decision-making) is that maxima and minima should be avoided in favour of averages. Thus likelihoods should be integrated, not maximized. This recipe seems appropriate when discussing robustness. It is often unsatisfactory to consider the worst that can happen over a class of priors because the prior that leads to that extreme behaviour is implausible. It is surely better to consider a hierarchy and handle the resulting hyperparameters. Thus, contemplating the robustness of $p(\theta)$ one might take a class $p(\theta, \alpha)$ of priors indexed by α with $p(\theta, 0)$ equal to $p(\theta)$. A probability $p(\alpha)$ for α enables the hierarchy to be studied. The central role is then played by the average 'prior' $\int p(\theta, \alpha)p(\alpha)d\alpha$. Doubts concerning anything should be expressed in terms of probability. Here this is done through the hyperparameter α.

E. MORENO (*Universidad de Granada, Spain*)

This paper presents a useful revision of most interesting classes of priors considered in the literature and Larry Wasserman is to be congratulated for both his clarity of exposition and his fine discussion of the methodology used to handle them.

I will focus my comment on the "linearization" method stated in Section 5 when it is applied to the class $\Gamma(L,U)$ of probability measures $\pi(\theta)$ defined by

$$\Gamma(L,U) = \{\pi(\theta) : L(C) \leq \pi(C) \leq U(C), \quad L(\Theta) < 1 < U(\Theta), \quad \pi(\Theta) = 1\}$$

where L and U are specified measures, Θ is the parameter space and C is any set in $B(\Theta)$. Due to "linearization", for a given quantity of interest $X(\theta)$, we essentially have to find $\bar{E}X = \sup_{\pi\in\Gamma} E^{\pi}X$ which involves two steps:

(i) To reduce the class Γ, whose dimension is infinite, to a class M with finite dimension where the problem may be solved, that is, the class M should satisfy $\sup_{\pi\in\Gamma} E^{\pi}X = \sup_{\pi\in M} E^{\pi}X$.

(ii) To explicitly find $\sup_{\pi\in M} E^{\pi}X$.

In Section 4 of the paper the class M is defined as $M = \cup_t M_t$, where

$$M_t = \left\{\pi(\theta) : \pi(\theta) = U(\theta)1_{X_t}(\theta) + L(\theta)1_{X_t^c}(\theta)\right\}$$

t being a real number, $X_t = \{\theta : X(\theta) > t\}$ and $X_t^c = \Theta - X_t$. This one-dimensional class M is claimed to be such that $\sup_{\pi\in\Gamma} E^{\pi}X = \sup_{\pi\in M} E^{\pi}X$. However, class M might contain no probability measure for simple quantities of interest $X(\theta)$. For example, let $X(\theta)$ be given by

$$X(\theta) = \frac{\theta - \theta_0 + \varepsilon}{\varepsilon} 1_{[\theta_0-\varepsilon,\theta_0]}(\theta) + 1_{[\theta_0,\theta_1]}(\theta) + \left(-\frac{\theta - \theta_1 - \varepsilon}{\varepsilon}\right) 1_{[\theta_1,\theta_1+\varepsilon]}(\theta),$$

where ε is some positive real number.

Denote $A = [\theta_0, \theta_1]$, and $A_t = \{\theta : (t-1)\varepsilon + \theta_0 \leq \theta \leq (1-t)\varepsilon + \theta_1\}$. Then

$$X_t = \begin{cases} \phi & \text{if } 1 \leq t \\ A_t & \text{if } 0 \leq t < 1 \\ \Theta & \text{If } 0 > t. \end{cases}$$

Therefore $M_t(\theta) = \mathrm{L}(\theta)$ for all $t \geq 1$, which is not a probability measure. $M_t(\theta) = U(\theta)1_{A_t}(\theta) + L(\theta)1_{A_t^c}(\theta)$ for each $0 \leq t < 1$, which are necessarily not probability measures, for instance if $U(A) + L(A^c) > 1$ it follows that $M_t(\Theta) = U(A_t) + L(A_t^c) = U(A) + U(A_t - A) + L(A_t^c) \geq U(A) + L(A^c) > 1$, for all $0 \leq t < 1$. Finally $M_t(\theta) = U(\theta)$ for all $t < 0$, which is not a probability measure.

In particular, the algorithm does not apply to finding ranges of prior (posterior) probabilities of a set A, except in case that $U(A) + L(A^c) = 1$, as π ranges over $\Gamma(L,U)$. In fact, for a given likelihood $f(x|\theta), \sup_{\pi\in\Gamma} P^{\pi}(A|x) = P^{\pi_0}(A|x)$ where $\pi_0(d\theta) = U(d\theta)$ $1_{A\cap\{\theta:f(x|\theta)\geq z_A\}}(\theta) + L(d\theta)1_{A\cap\{\theta:f(x|\theta)<z_A\}\cap A^c(\theta)}$, z_A being such that $\pi_0(\Theta) = 1$, provided that $U(A) + L(A^c) > 1$. In general, these ranges are obtained in Moreno and Pericchi (1992).

L. R. PERICCHI (*Universidad Simón Bolivar, Venezuela*)

We ought to be grateful to Dr. Wasserman for a clear and concise exposition of some recent methodological advances in Robust Bayesian Inference, useful for both the "sensitivity analysis" and the "lower probability" interpretations. However, as is suggested in the final paragraph, we should address real examples, as is our duty as statisticians, and make sure that this immense effort is not misinterpreted as just an excuse to develop beautiful mathematics.

I agree that there is plenty of room for further advances in methodology. But, in motivating these, effort should be devoted to the assessment of "natural classes". One such natural "neighbourhood" class is suggested in Pericchi and Walley (1991). In the situations where a conjugate prior exists, a "density ratio" class is formed with $l(\theta)$ as a conjugate elicited prior and $u(\theta)$ as the minimal height "reference" prior, i.e., u "touches" l at one point. Then updating, for example for posterior probability of sets, is quite simple because of conjugacy, although the proposal is not restricted to cases where conjugate priors exist. In that paper, we coincide with Dr. Wasserman's view that "likelihood sets" are the natural quantities of interest for robust analyses, but the class is used to calibrate their posterior probabilities. The neighbourhood described above, naturally encompasses the two most popular kinds of "precise" analyses. This proposal is readily applied for any dimension and it would be also interesting to try it with a "density bounded" class, although this is much more precise a priori than the "density ratio" class and the latter has some "pragmatic" advantages as the author points out. By the way, some advance in the prior assessment of a "density bounded" class is given in Moreno and Pericchi (1992) where explicit formulae for the prior and posterior extremes for sets are provided.

A problem very much overlooked in the Robust Bayes literature is that of representing "ignorance". If a precise approach is taken, then for unbounded parametric spaces any solution has to be improper. Merits of such an approach are those of being an "automatic" analysis and advances in the reference priors theory teach us about judicious parameterizations of the problem for the quantitities of interest at hand. However, there are some problems that such an approach is incapable of solving. One such problem is that of the existence of relevant subsets in the Student-t inference from reference priors, for example. The possibilities for representing "near ignorance" (an idea first suggested in the initial version of Peter Walley's book) if not ignorance, via classes of proper priors are wide. In the referenced paper by Pericchi and Walley a "near ignorant" class, based on a list of desiderata suited for likelihood sets is proposed and its properties explored for a normal location situation. In such a situation reference sets do not exist, however. Some progress is being made jointly with B. Sansó for the case of a normal location and scale likelihood aiming to propose an alternative to the Student-t analysis with minimum inputs from the user. Certainly, a lot has to be done in this area of near ignorance, and I expect that eventually it will become appealing to both Bayesians and non-Bayesians alike.

F. RUGGERI (*CNR-IAMI, Italy*)

Pragmatic arguments are given in favor of the density ratio class, because bounds might be obtained directly using a posterior sample and simple graphical summaries are possible. I accept the importance of such features but I question the fact that a class could be preferred to another one just because of its appealing properties. There are not better classes but just classes compatible with the prior knowledge. Should we give up some knowledge to get a nicer class, it could be a good initial attempt to explore robustness, but we should move to uglier classes, if our first analysis were not satisfactory.

As pointed out in a counterexample in Ruggeri (1991) (see also Ruggeri and Wasserman, 1990), caution should be paid in making statements about the robustness properties of the likelihood regions. Probably, such interpretation holds for the most common classes, but it has to be proved.

Another graphical summary in robust Bayesian analysis is provided by the concentration function (c.f.) of a measure Π w.r.t. another measure Π_0 (Cifarelli and Regazzini, 1987). Under very general conditions, the c.f. $\varphi(x)$ gives the minimum Π-measure achievable by the subsets sharing the same Π_0-measure x, for any $x \in [0,1]$. For the same subsets, the maximum Π-measure is given by $1-\varphi(1-x)$. Plotting the curves $\varphi(x)$ and $1-\varphi(1-x)$ gives an immediate perception of the discrepancy between the measures. Given a base measure Π_0 and a class Λ of measures Π, take the minimum of $\varphi(x)$ and the maximum of $1-\varphi(1-x)$ over all Π in Λ, for any x. Then the plot of such curves says something about the discrepancy between the measures in Λ and the base one. In the robust Bayesian framework, such an idea was developed by Fortini and Ruggeri (1990), who considered the ε-contaminated class Λ_ε^C.

REPLY TO THE DISCUSSION

I would like to thank all the discussants for their comments. I especially appreciate the thorough and thought-provoking discussion that was presented by George Casella and Marty Wells. Apparently, there is a point that I did not make clear. Classes of priors can be used for two purposes. First, they can be used to diagnose sensitivity to a specific prior. Second, they can be used to model partial prior information. With this distinction in mind, I will reply to each discussant separately.

Reply to Casella and Wells:

The first point raised by Casella and Wells is that classes of priors can be difficult to interpret. I agree. On the other hand, even probabilities are hard to understand. Who really grasps the implications of a probability in several dimensions? Priors on even two-dimensional spaces stretch our intuitive capabilities. Nonetheless, we get by using probability in complicated situations. The reason is that we have much experience with probability. Similarly, I think we can get by using a set of priors even if we don't fully grasp all its implications. What we need is experience with many examples.

Figure 1 gives us a representation of a t_ν prior. Figure 2 is an attempt to "tour" the $\Gamma_{L,U}^R$ class. The normalized versions of L and U are not extreme points in the class, so I'm not sure what we should expect to see in Figure 2. Large values of $|\gamma^L - \gamma^U|$ imply large values of $Q^S - Q^I$ but when $|\gamma^L - \gamma^U|$ is small, we can't say much. I agree that the figure is not revealing but I don't think I would have expected it to be. On the other hand, Figure 3 looks just like it should. Sets that are small have Q^S and Q^I both near zero so that $Q^S - Q^I$ is close to zero. Similarly for large sets. Hence the bow-shaped curves in Figure 3.

The important feature of a class like $\Gamma_k^R(\pi)$ is that it contains priors that are close to some target prior π. The structure of the class is not crucial. It is simply a tool that allows us to explore the effect of making small changes to the prior. Consider the following example.

A clinical trial was conducted to compare a new treatment called ECMO to a standard treatment. Of 9 infants treated with ECMO, 9 survived. On the standard treatment, 6 of 10 survived. The trial is discussed by Ware (1989). A Bayesian analysis is carried out by Kass and Greenhouse (1989) and is followed up by Lavine, Wasserman and Wolpert (1991a). For the purposes of illustration, let us consider a simple analysis of the data. Let X_i be the number of survivors on treatment i with ECMO being $i = 1$ and the standard treatment represented by $i = 2$. Then, X_i is binomial(n_i, p_i), $i = 1, 2$. We take the prior $\pi(p_1, p_2)$ to be uniform. Let $\Gamma = \Gamma_k^R(\pi)$ and consider all k from 1 to 20. The uniform prior makes every point equally likely. A prior Q is in Γ_{20}^R as long as it does not make any point more than 20 times more likely than any other point. The first graph in Figure 1 (of the rejoinder) shows the upper and lower bounds on the prior and posterior expectation of $p_1 - p_2$ as k varies from 1 to 20. The second graph shows the bounds on the prior and posterior probability that $p_1 > p_2$, that is, the probability that ECMO is superior.

The analysis is easy to interpret. Even if we think that a uniform prior is only approximately correct, we can conclude that ECMO is superior. I don't thoroughly understand the

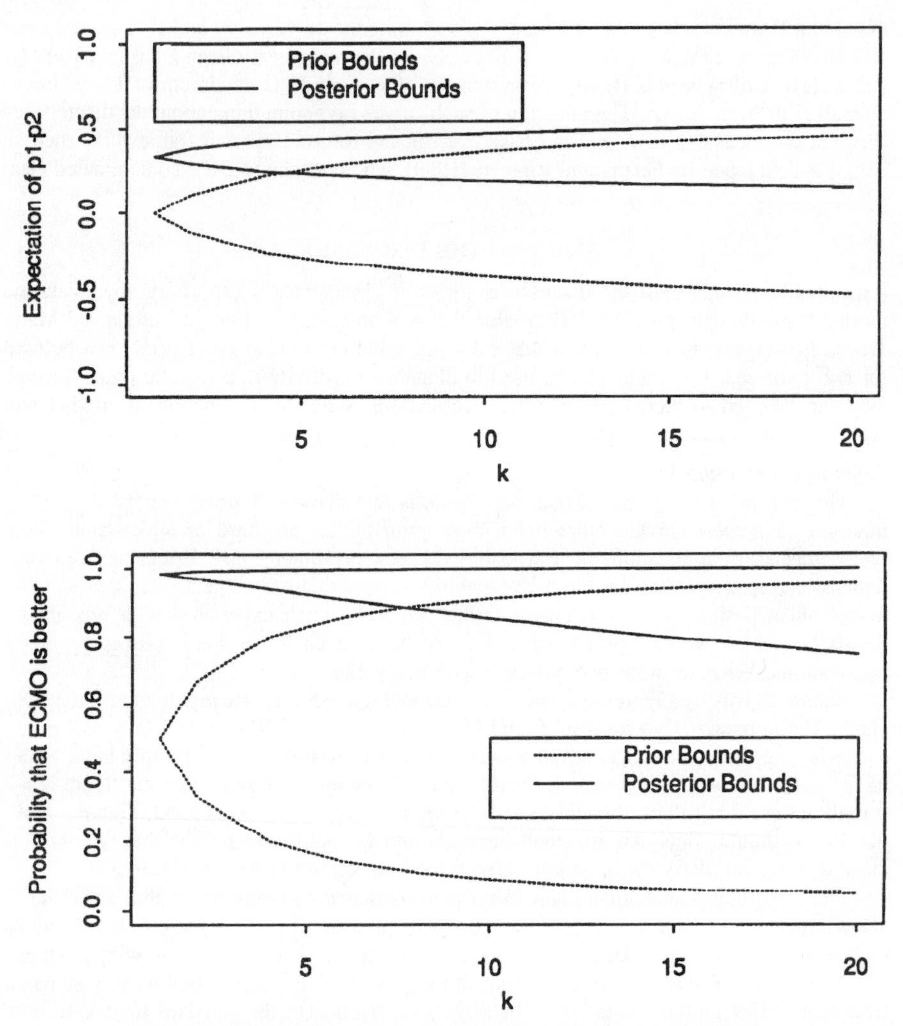

Figure 1. *Robust Bayesian analysis of the ECMO example using the $\Gamma_k^R(\pi)$ class where $\pi(p_1, p_2)$ is the uniform distribution.*

class Γ_k^R. But I think that we do understand the class well enough to use it in examples and get a feeling for what's going on.

The second point raised by the discussants is that it is possible to select a "robust prior" π^* from a set of priors Γ. As Casella and Wells point out, Polson (1988) has studied this problem. I like this idea. But there is no need to replace the robust Bayesian analysis with an analysis based only on π^*. We should do both. Switching the prior from π to π^* just because we have chosen to do a sensitivity analysis is questionable.

In summary, I agree with the main point that there is a need to deepen our understanding of classes of priors. But the methodology for computing answers in a robust Bayesian framework is readily available. We should put the technology to work and let experience be the teacher.

Reply to Berger:

Professor Berger is troubled by the Bayes' invariance property of the density ratio class $\Gamma_k^R(\pi)$. He says "the variability in the posterior is fixed no matter what is encountered." This is true if the parameter k is used as an omnibus measure of imprecision of the class. But k only measures one type of imprecision, namely, the error in quantities of the form $\pi X/\pi Y$, $X, Y \geq 0$. In other words, k measures the error in odds. To get a more complete measure of imprecision we need to examine $\overline{E}_k^R(X|y)$, $\underline{E}_k^R(X|y)$ and $\delta_k(X|y) = \overline{E}_k^R(X|y) - \underline{E}_k^R(X|y)$. Of course, these will depend on y so it does matter what data are encountered. It is true, however, that the error in odds remains constant. But we use this measure of error as a convenient device for defining the class; this doesn't imply that we must use that measure of error to define imprecision. However, I do agree that the particular density ratio class Γ_k^R does not allow a large range of tail behavior. Perhaps I put too much emphasis on this special case and not enough on the more general class $\Gamma_{L,U}^R$. I don't know whether likelihood regions correspond to reasonable hierarchical Bayes credible sets. I agree that this is an interesting question. Ruggeri raises some questions about these regions in his comment.

Reply to Lindley:

Professor Lindley reminds us that averaging is the Bayesian thing to do and that finding maxima and minima has a non-Bayesian flavour. In response to this, I should emphasize that I am not proposing a formal mode of inference based on the bounds. I contend that the bounds are useful as part of, not instead of, a fully Bayesian analysis. Here are two reasons why the bounds are useful: (1.) Suppose we have elicited partial prior information from an expert and this leads to a class Γ. To complete the analysis we need to do one of two things. First, we could try to elicit a prior over Γ. This is unreasonable given that we have been unable to get a complete prior over the parameter space. Second, we could try to get a completely specified prior. However, it would be useful to know how much we can say with just the partially specified information. For example, if the bounds on the posterior expectation for a quantity of interest are tight, then we don't need to bother getting a completely specified prior. (2.) Often, analyses are carried out based on some prior chosen mainly for convenience. Professor Lindley seems to accept this as reasonable when he writes (Lindley, 1990): "Berger puts is nicely when he says a subjective prior is better than a non-informative one, but the latter is 'very good' ". In such a case, we would like to check whether we would have obtained a much different answer if we had used a carefully elicited prior. The bounds from a robust Bayesian analysis will answer this. We could use a class of priors $p(\theta, \alpha)$ indexed by α as Professor Lindley suggests. But the robust Bayesian analysis may be easier. And surely we feel more secure if robustness obtains over a large class Γ, than if we use a more restrictive, parametric class of priors.

Every data analysis calls on scarce resources. The bounds from the robust Bayesian analysis are not meant as a formal substitute for precise probabilities, but as a way of diagnosing where our time and energy need to be applied.

Reply to Moreno:

Professor Moreno is correct – I was careless in my description of the maximizing measure for the density bounded class. If $U(\{\theta; X(\theta) = t\}) = 0$ for each t, then my statement is correct. Also, the Choquet integral gives the exact bound.

Reply to Pericchi:

Dr. Pericchi discusses some classes of priors that he and Peter Walley have studied. It is interesting to contrast their approach with the usual approach to sensitivity analysis. In their work, they seek "reasonable classes of priors" instead of "classes of reasonable priors" (Pericchi and Walley, 1991). The former are determined by examining the properties of the lower and upper probabilities that are induced by the class. If we take this approach, then our answer to Casella and Wells would be: "Don't tour around the class Γ at all; what's in them is not that important."

I have followed the work of Pericchi and Walley with great interest. Their attempts to model "near ignorance" are especially interesting. So far, their approach seems further from the routine implementation stage than more standard sensitivity analysis techniques. I look forward to future work in this area. I agree wholeheartedly that we should address real examples.

Reply to Ruggeri:

Ideally, classes of priors should be chosen to reflect prior knowledge. But as a matter of practice, we need reasonable defaults to fall back on. For models, we have defaults like normal distributions, exponential distributions, linear regression models and so on. We also have reasonable default priors: Jeffreys' prior, invariant priors and priors like those in Berger and Bernardo (1992). Similarly, we need default classes of priors. Of course, if we have reliable prior information, we should use it. But if we are trying to assess the sensitivity of our answer to a particular prior, then it is reasonable to use a convenient class like the density ratio class.

ADDITIONAL REFERENCES IN THE DISCUSSION

Berger, J. O. and Bernardo, J. M. (1992). On the development of reference priors. *Bayesian Statistics 4* (J. M. Bernardo, J. O. Berger, A. P. Dawid and A. F. M. Smith, eds.), Oxford: University Press, 35–60, (with discussion).

Bose, S. (1990). *Bayesian Robustness with Shape–constrained Priors and Mixture Priors*. Ph.D. Thesis, Purdue University.

Casella, G. and Wells, M. T. (1991). Noninformative priors for robust Bayesian inference. *Tech. Rep.* Cornell University Statistics Center.

Cifarelli, D. M. and Regazzini, E. (1987). On a general definition of concentration function. *Sankya B* **49**, 307–319.

Fortini, S. and Ruggeri, F. (1990). Concentration function in a robust Bayesian framework. *Tech. Rep.* **90.6**, CNR-IAMI, Milano.

Huber, P. J. (1981). *Robust Statistics*. New York: Wiley.

Jeffreys, H. (1937, 1961). *Theory of Probability*. Oxford: University Press.

Kass, R. E. and Greenhouse, J. B. (1989). Comment on Ware. *Statist. Sci.* **4**, 310–317.

Lindley, D. V. (1990). The present position in Bayesian statistics. *Statist. Sci.* **5**, 44–89.

Moreno, E. and Pericchi L. R. (1992). Bands of probability measures: a robust Bayesian analysis. *Bayesian Statistics 4* (J. M. Bernardo, J. O. Berger, A. P. Dawid and A. F. M. Smith, eds.), Oxford: University Press, 707–713.

Pericchi, L. R. and Walley P. (1991). Robust Bayesian credible intervals and prior ignorance. *Internat. Statist. Rev.* **58**, 1–23.

Polson, N. G. (1988). *Bayesian Perspectives on Statistical Modeling*. Ph.D. Thesis, University of Nottingham.

Ruggeri, F. (1991). Robust Bayesian analysis given a lower bound on the probability of a set. *Comm. Statist. Theory and Methods* **20**, 1881–1891.

Ware, J. H. (1989). Investigating therapies of potentially great benefits: ECMO. *Statist. Sci.* **4**, 298–340.

BAYESIAN STATISTICS 4, pp. 503–524
J. M. Bernardo, J. O. Berger, A. P. Dawid and A. F. M. Smith, (Eds.)

Modelling with Mixtures*

MIKE WEST
Duke University, USA

SUMMARY

Discrete mixtures of distributions of standard parametric forms commonly arise in statistical modelling and with methods of analysis that exploit mixture structure. This paper discusses general issues of modelling with mixtures that arise in fitting mixtures to data distributions, using mixtures to approximate functional forms, such as posterior distributions in parametric models, and development of mixture pruning methods useful for reducing the number of components of large mixtures. These issues arise in problems of density estimation using mixtures of Dirichlet processes, adaptive importance sampling function design in Monte Carlo integration, and Bayesian discrimination and cluster analysis.

Keywords: CLASSIFICATION AND DISCRIMINATION; DENSITY ESTIMATION; DIRICHLET MIXTURES OF NORMALS; POSTERIOR APPROXIMATIONS, SIMULATION.

1. INTRODUCTION

Several areas of application of discrete mixture distributions are reviewed here, involving recent methodological and computational developments in fitting mixtures to data and in approximating posterior distributions via mixtures of standard parametric forms.

The first area concerns the approximation of prior and posterior distributions by discrete mixtures of normal or, more generally, multivariate t distributions. This work stems from use of such mixtures as importance sampling functions in Monte Carlo integration, extending previous approaches to adaptive importance sampling. Mixtures have the flexibility to represent possibly very irregular forms of posterior densities, though there has to date been little application of mixtures, either as importance sampling distributions or as more direct approximations to posteriors, the main reason being the lack of analytic techniques for constructing suitable mixtures. A constructive approach, based on adaptive resampling and refinement of approximating mixtures (West, 1992), is reviewed and discussed, with illustration, in Section 2.

In data analysis using discrete mixtures, several themes arise according to application area. Direct fitting of mixtures of standard exponential family distributions, such as mixtures of exponentials or mixtures of normals, is technically possible now using iterative resampling schemes, such as Gibbs sampling. Some work in this area, with a focus on mixture deconvolution, discrimination and classification (Lavine and West, 1992), is discussed in Section 3.

Section 4 concerns discrete mixtures arising in nonparametric Bayesian density estimation using mixtures of Dirichlet processes. Some review of the basic structure of such models is given, and issues of posterior and predictive calculations in such models described. Particular methods of approximation of predictive distributions in these models lead to Bayesian

* Research reported here was partially financed by the National Science Foundation under grants DMS-8903842 and DMS- 9024793.

analogues of traditional kernel density techniques, and simpler forms based on mixtures of, typically, small numbers of components. Various practical issues in traditional kernel estimation, including smoothing parameter selection and modelling varying degrees of smoothness across the sample space, for instance, reduce to standard problems of modelling and inference in a model based framework. Other issues discussed include mixture deconvolution, inference about numbers of mixture components, and the modality characteristics of data distributions. More recent work on the computational side, involving the development of Gibbs sampling techniques for posterior analysis in mixtures of Dirichlet processes (Escobar, 1990; Escobar and West, 1991), is also discussed and illustrated.

2. MIXTURE MODELS FOR PRIORS AND POSTERIORS

2.1. *Sampling Posteriors and Kernel Reconstruction*

In simulation based analysis of posterior distributions, marginal posterior densities are typically approximated by some form of discrete mixture. With the availability of complete conditional distributions for the application of Gibbs sampling techniques, mixtures of various conditionals arise naturally. Elsewhere, kernel estimation techniques have been used by several authors to construct smooth approximations to posteriors when other, more efficient forms of approximation are unavailable (Gelfand and Smith, 1990; Wolpert, 1991). Suppose a simulation analysis provides an approximate sample from a posterior $p(\theta)$, with θ in p dimensions. Denote the sampled points $\Theta = \{\theta_j, j = 1, \ldots, n\}$. A kernel estimate of $p(\theta)$ has the form $n^{-1}\sum_{j=1}^{n} d(\theta|\theta_j, Vh^2)$ where $d(\theta|m, M)$ is a $p-$variate, elliptically symmetric density function with mode m and scale matrix M, V is an estimate of the variance matrix of $p(\theta)$, and h is a smoothing parameter. Important kernels are the multivariate t family, when the estimates are mixtures of equally weighted t distributions. More generally, in Monte Carlo with importance sampling, a posterior approximation $g_0(\theta)$ provides the random sample Θ, and importance weights are defined by $w_j \propto p(\theta_j)/g_0(\theta_j)$, subject to unit sum. Then posterior expectations are estimated by weighted sums such as $\mathrm{E}[a(\theta)] \approx \sum_{j=1}^{n} w_j a(\theta_j)$ for given functions $a(\theta)$. Now the kernel approach extends to give a posterior density estimate of the form

$$g(\theta) = \sum_{j=1}^{n} w_j d(\theta|\theta_j, Vh^2), \tag{1}$$

appropriately inserting the Monte Carlo weights (Wolpert, 1991; West, 1992).

Posterior reconstruction and graphical display using (1) require specification of h and V. Choice of smoothing parameter h may be guided by traditional kernel estimation theory, and also by the derivation of kernel densities as approximations to predictive distributions in Bayesian density estimation using mixtures of Dirichlet processes (Ferguson, 1983; West, 1990b) where the smoothing parameter h has interpretation as a function of model parameters and sample size. Routine recommendations (e.g., Silverman, 1976) are broadly applicable though tend to over-smooth more erratic, multimodal densities (West, 1990a, 1992). Choosing h as a slowly decreasing function of simulation sample size n, as traditionally recommended (Silverman, 1986), ensures large sample consistency of $g(\theta)$ as an estimate of the true posterior $p(\theta)$. For the scale matrix V, a natural choice is the Monte Carlo estimate of posterior variance $V = \sum_{j=1}^{n} w_j(\theta_j - \bar{\theta})(\theta_j - \bar{\theta})'$ where $\bar{\theta} = \sum_{j=1}^{n} w_j\theta_j$ is the Monte Carlo mean. However, as a direct approximation to $p(\theta)$ the mixture $g(\theta)$ will then be overdispersed — if the kernels are t on $a > 0$ degrees of freedom then the variance under $g(\theta)$ is $V(1 + h^2 a/(a-2))$, not V. Guided again by theory of Bayesian density estimation (West, 1990b), shrinkage of the kernel locations may be applied to reduce this overdispersion: with shrinkage factor x in $(0, 1)$, replacing the sampled points θ_j by $m_j = x\theta_j + (1-x)\bar{\theta}$ as the modes of the kernels, the resulting mixture has variance $V(x^2 + h^2a/(a-2))$ which reduces to V at $x^2 = 1 - h^2a/(a-2)$. In practice, h will be specified as a slowly decreasing function of sample size n (West, 1992) so that the shrinkage will be negligible with large sample sizes.

Incidentally, note that simulation and kernel reconstruction provide a technique by which any specified prior (that may easily be sampled) may be approximated by a mixture of distributions of standard forms – the kernels may be any standard form, not necessarily symmetric. At the second Valencia meeting, Diaconis and Ylvisaker (1985) discussed theoretical issues in approximating priors by mixtures of conjugate forms; see also Dalal and Hall (1983). More practically, Alspach and Sorenson (1972) developed approximating mixtures of normals in signal processing problems, although there are really no general, direct and automatic techniques for constructing such approximations. Sampling the prior, or an importance sampling function approximating the prior, and using a kernel technique with normal kernels is one possibility, which also allows neatly for approximate prior to posterior updating, as follows. Suppose the kernels in (1) are multivariate normal with modes m_j, and that n is large, say several thousands. Then each of the mixands will be closely concentrated about its mode m_j, since h will be small for large n. Multiplying by a likelihood function $l(\theta)$ we have posterior density $g(\theta)l(\theta) \propto \sum_{j=1}^{n} w_j d(\theta|m_j, Vh^2)l(\theta)$. With n large enough, hence h small enough, the likelihood function will be diffuse compared to each of the mixands, though not necessarily relative to the overall mixture. Hence a local log-quadratic approximation to the products $d(\theta|m_j, Vh^2)l(\theta)$ leads to approximate normal posteriors $d(\theta|m_j^*, V_j^*)$ multiplied by (data dependent) constants v_j, say; thus the posterior is approximately proportional to $\sum_{j=1}^{n} w_j^* d(\theta|m_j^*, V^*)$, with updated weights $w_j^* \propto w_j v_j$, and summing to unity. In some problems, and with informed priors, this can provide excellent approximations to posteriors of direct use; one area in which this has been employed is in providing initial guesses at posteriors in adaptive development and refinement of mixture approximations using importance sampling, as described below. This is of particular interest in application in sequential analyses, as in Section 2.4 below, where the priors to be approximated are quite typically fairly concentrated relative to likelihood functions, and so the mixands of kernel reconstructions are really very precise, and simple analytic updating as just described can be extremely effective.

2.2. *Adaptive Approximation of Posteriors*

In West (1992) mixtures of t distributions as just described are used as importance sampling distributions in a scheme to adaptively refine posterior approximations. The basic idea is to develop an initial guess at the posterior, say $g_0(\theta)$, to be used as a starting importance sampling function. Based on a sample of size n_0 from this distribution, a weighted kernel estimate is constructed as in (1), and used as a second guess; call this $g_1(\theta)$. Repeating this procedure, with a possibly different sample size n_1 from $g_1(\theta)$, we can revise the mixture importance sampling density to $g_2(\theta)$, and further if required. At each stage, the smoothing parameter may be chosen to increase or decrease the spread of the next importance function if that currently is thought to poorly represent the posterior, such as when the current distribution of the importance sampling weights varies dramatically away from uniformity; a more diffuse distribution will spread out the sampled points at the next stage, with a view to increasing support in regions of the parameter space previously undersupported. In cases when $g_0(\theta)$ is a rather crude initial guess, adaptive refinement is most appropriate, since then direct use of $g_0(\theta)$ will be highly inefficient, computationally and statistically. At the final stage, a relatively small sample may be sufficient for the desired accuracy, and often just one refinement may be sufficient to adjust a very crude approximation, say a single multivariate t density, to a mixture $g_1(\theta)$ of, say, several hundred t densities, much more closely representing the true $p(\theta)$. In approximating moments and probabilities, a Monte Carlo sample of a few thousand draws from $g_1(\theta)$ may do as well as, or better than, a sample of several times that from the original $g_0(\theta)$. At a final stage, the sampled points and weights provide for estimation of posterior expectations, as usual, and kernel reconstructions provide marginal posterior density and distribution function estimates.

An example, with $p = 5$ parameters, concerns a special case of a model introduced in Migon and Harrison (1985), and discussed in West and Harrison (1989, Section 14.4). The data comprise the first 56 weeks of observations on two series, Z_t, a binomial response variate, and X_t, an independent variable used to explain and predict Z_t. Z_t is a count, out of a total of 66 randomly sampled, of individuals who, when questioned, demonstrate their awareness of a current television advertisement; X_t is a summary measure of advertising expenditure in week t, referred to as the weekly TVR (standing for television rating). The model is a binomial regression of Z_t on current and past values of X_t, defined by $(Z_t|p_t) \sim \text{Bin}[66, p_t]$ with

$$\begin{aligned} p_t =& \alpha + E_t, \\ E_t =& \gamma - (\gamma - \rho E_{t-1})\exp(-\kappa X_t), \end{aligned} \tag{2}$$

for $t = 1, \ldots, 56$. The success probability p_t measures awareness, E_t represents the 'effect' of current and past TVR levels on current awareness, α is a lower threshold on the effect, the sum $\alpha + \gamma$ an upper threshold, ρ is a measure of rentention, or 'memory', of past advertising effects, and κ is a measure of the immediate, penetrative effect of current advertising. The second equation in (2) may be used to recursively evaluate each E_t, $(t = 1, \ldots, 56)$, for any specified values of the five uncertain quantities $(\alpha, \gamma, \rho, \kappa, E_0)$. For analysis using importance functions based on mixtures of multivariate T-9 distributions, transformation is made to real-valued parameters. Parameter restrictions are $1 > \rho >> 0$, $\kappa > 0$, $0.2 > \alpha > 0$, $0.8 > \gamma > 0$ and $E_0 < \gamma$, although the final restriction is ignored since E_0, and all subsequent values E_t, typically lie well below threshold. Parameter transforms used here are then $\theta_1 = \log(\alpha/(0.2 - \alpha))$, $\theta_2 = \log(\gamma/(0.8 - \gamma))$, $\theta_3 = \text{logit}(\rho)$, $\theta_4 = \log(\kappa)$, and $\theta_5 = \text{logit}(E_0)$. The initial prior for $\theta = (\alpha, \gamma, \rho, \kappa, E_0)'$ is a 5−variate t with 9 degrees of freedom whose mode and scale matrix are specified in line with the prior used in dynamic model analysis in Section 14.4 of West and Harrison (1989). The prior mode vector is $(0.0, 2.0, 2.0, -3.5, -0.75)'$ and the scale matrix is diagonal, the square root of diagonal elements given by $(0.15, 1.0, 0.2, 0.25, 0.25)$.

Illustrated analysis uses three stages of refinement, with sample sizes $n_0 = 1500$, $n_1 = 2000$ and $n_2 = 6500$, and with $g_0(\theta)$ simply the prior. Repeat analyses with varying, and larger, sample sizes verify the results of this particular analysis. Final approximations to the univariate posterior margins for ρ, κ, α and γ appear in Figure 1 (as full lines), and contours of some of the bivariate margins in Figure 2. The contours are graphed at heights of 0.001, 0.01, 0.1, 0.25, 0.5, 0.75 and 0.9 times the maximum. Superimposed on the posteriors in Figure 1 are, as dashed lines, the prior densities. Over the 56 weeks of data, variation in the regression variable X_t is such that there are periods when $X_t = 0$ for a while, leading to potential for learning about the 'memory decay' parameter ρ; the realisation of this potential is reflected in the figure for ρ. Similarly, variation in X_t when non-zero is necessary to inform about the 'penetration' parameter κ, and the graph reflects a reasonable degree of information for κ from this dataset. The posteriors for α and γ, however, differ only slightly from the priors; the response series Z_t never reaches very low or very high levels, so that

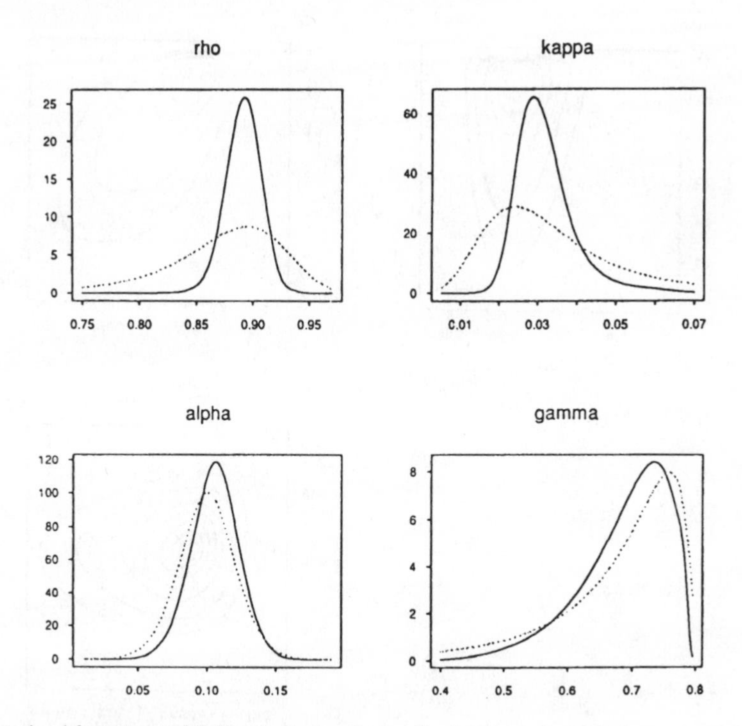

Figure 1. *Advertising model: Marginal priors (broken lines) and margins of final posterior (full lines).*

there is little opportunity to gain information about these two 'threshold' parameters.

2.3. *Reduction of Mixtures*

Sampling and, especially, evaluating the density functions of mixtures of multivariate t distributions is computationally expensive if the number of components is very large. Also, mixtures of very many components can rather often be very well matched by reduced mixtures of far fewer, as experiences in density estimation show (West, 1990b). West (1992) describes the reduction of weighted kernel estimates to mixtures of much smaller numbers of components, often lower than 10% of the original sample size, by simply dropping components of negligible weights and clustering others that are close. Hierarchical nearest neighbour clustering, 'averaging' the component of smallest weight at each stage with its nearest neighbour, has been found to be effective. This involves sequentially identifying the nearest neighbour of that θ_j with smallest weight at each stage, and so is computationally intensive in problems in several dimensions. It is both statistically and computationally effective, however, in the context of adaptive mixture refinement when the simulation sample size is large, say several thousands, and the mixture is later to be sampled and its density evaluated.

In the advertising model example, this mixture reduction was actually made between stages, reducing the mixtures of 1500, 2000 and 6500 components, respectively, to 500. The posteriors in the figures are based on the reduced 500 component mixtures. For comparison, Figure 3 redisplays the final univariate margins; the 500 component mixtures appear as full

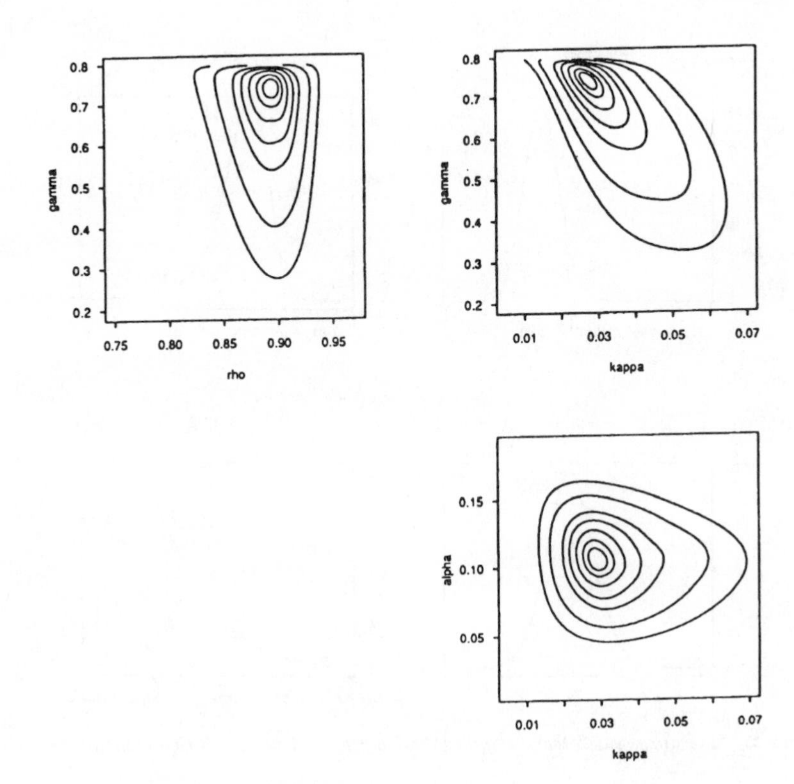

Figure 2. *Advertising model: Some bivariate posterior densities.*

lines, and the 6500 component mixtures from which they were constructed as broken lines. The adequacy of the reduced approximation is clear here, and similar aggreements are found between contour plots of bivariate margins.

2.4. *Sequential Analyses and Dynamic Models*

An original motivation for the development of the preceding sections was sequential analysis of non-linear dynamic models (West and Harrison, 1989). Simulation based analyses involve the propogation over time of discrete approximations to prior and posterior distributions based on approximate samples from those distributions, or from approximations to them. In West (1990a), a general approach is worked out in detail, with illustrations.

A very simple example illustrating the novel features of the computational problems arising in dynamic models is given by a non-normal, first-order polynomial model (Pole and West, 1988, 1990). Here the data y_t, $(t = 1, 2, \ldots)$, are modelled as $y_t = \theta_t + \nu_t$ and $\theta_t = \theta_{t-1} + \omega_t$ where t indexes time, θ_t is the level of the series at time t, ν_t and ω_t are independent, zero-mean noise terms with some specified, non-normal densities $f_\nu(.)$ and $f_\omega(.)$ respectively. Heavy-tailed noise distributions, such as Student T, permit the automatic modelling of outliers and jumps in level in the series, which may be a component of a larger system model. Let $D_t = \{y_t, y_{t-1}, \ldots, y_1\}$ for all t. Analysis involves sequential updating of distributions at each time point; assume at time $t-1$ that the current posterior for the level,

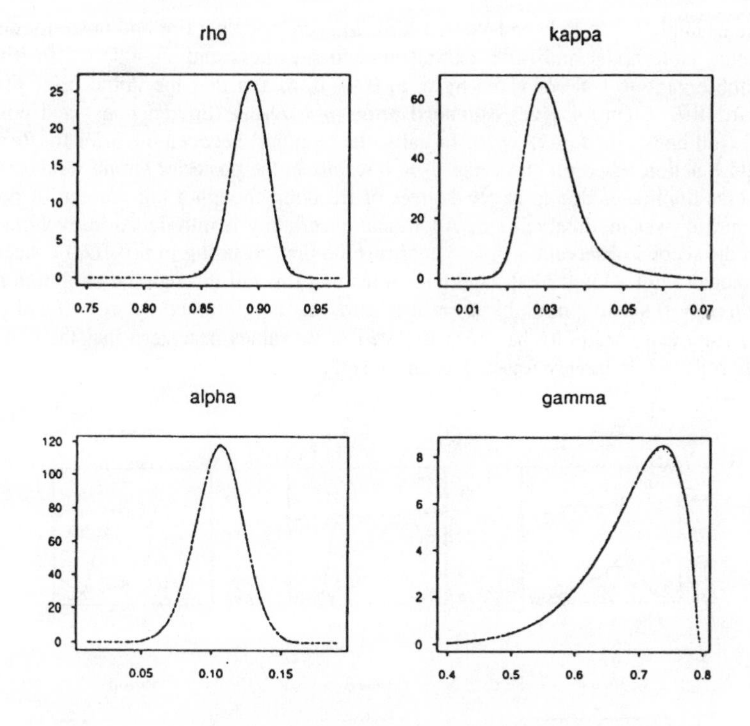

Figure 3. *Advertising model: Margins from final mixtures of 6500 components (broken lines), and 500 components (full lines).*

namely $p(\theta_{t-1}|D_{t-1})$, is available and summarised. The computations at time t involve the steps

$$\begin{array}{c} p(\theta_{t-1}|D_{t-1}) \rightarrow p(\theta_t|D_{t-1}) \rightarrow p(\theta_t|D_t) \\ \downarrow \\ p(y_t|D_{t-1}) \end{array} \tag{3}$$

The first step is the evolution, involving computing the implicitly defined convolution density $p(\theta_t|\, D_{t-1}) = \int f_\omega(\theta_t - \theta_{t-1})\; p(\theta_{t-1}|D_{t-1})\; d\theta_{t-1}$. Given an approximate sample from $p(\theta_{t-1}|\, D_{t-1})$, or from an appropriate importance function, this convolution may be evaluated directly by Monte Carlo. Similarly, the evolution may be directly simulated to generate points from $p(\theta_t|D_{t-1})$ given sampled values from $p(\theta_{t-1}|D_{t-1})$, or from a smooth reconstruction of the latter such as a weighted kernel estimate. So a Monte Carlo approximation may be transformed through the first step in (3). The forecasting step to simulate and evaluate $p(y_t|D_{t-1})$ may be similarly performed, since this involves just another convolution of distributions, $p(y_t|D_{t-1}) = \int f_\nu(y_t - \theta_t)p(\theta_t|D_{t-1})d\theta_t$. The final step involves updating the Monte Carlo approximation for $p(\theta_t|D_{t-1})$ to one for $p(\theta_t|D_t)$, processing the current observation y_t, and here the previous development of mixture approximations in Section 2.2 may be directly applied.

Take $f_\nu(.)$ to be standard Student T-5, and $f_\omega(.)$ to be T-5 with a scale factor of 0.1

(compare example 8.1 in Pole and West, 1988). In such models, prior and posterior densities can become multimodal and often exhibit marked skewness and shoulders. To illustrate, suppose observations 1–6 are given by 6, 6, 0, 0, 0, 6, and that the initial prior $p(\theta_1|D_0)$ is standard T-9. Figure 4 gives estimated priors $p(\theta_t|D_{t-1})$ (broken lines) and posteriors $p(\theta_t|D_t)$ (full lines), for $t = 1, \ldots, 6$. Initially, the 'conflict' between the prior for θ_1 and the likelihood function based on observing $y_1 = 6$ results in the posterior strongly favouring the prior (which dominates due to larger degrees of freedom) though a fair amount of posterior mass is spread over the interval 0–6. Additional uncertainty is introduced in evolving to θ_2, and then the second observation $y_2 = 6$ confirms the first, resulting in $p(\theta_2|D_2)$ concentrated near 6, though with a longer left-hand tail. Spreading this tail through the evolution to time 3 leads to $y_3 = 0$ shifting mass back towards zero, this is reinforced by $y_4 = 0$ and $y_5 = 0$, and then sufficient credibility has been invested in θ_6 values near zero that the conflicting observation $y_6 = 6$ is largely rejected as an outlier.

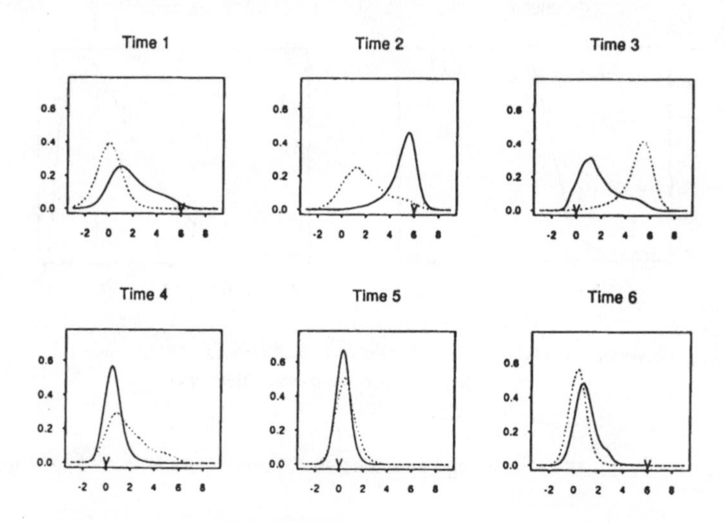

Figure 4. *Prior and posterior densities for level of non-normal process.*

The computations here use mixtures of T-9 distributions as approximating forms. Updating from $p(\theta_t|D_{t-1})$ to $p(\theta_t|D_t)$ at each t involves three successive refinements with sample sizes $n_0 = 250$, $n_1 = 350$ and, finally, $n_2 = 1500$. Following each refinement, these are clustered into mixtures of just 100 components before sampling for the next iteration. Thus the densities in the figure are mixtures of just 100 T-9 densities.

3. MIXTURES MODELS FOR DATA 1: DISCRIMINATION AND DECONVOLUTION

A common application of mixtures is in problems of discrimination, classification and clustering (Bernardo and Girón, 1985; Binder, 1978; Hartigan, 1975, Chapter 5; McClachlan and Basford, 1988, and Titterington, Smith and Makov, 1985). Simulation based analyses using Gibbs sampling (Gelfand and Smith, 1990), now provide for many relevant calculations to be performed in practical models, particularly models based on mixtures of standard exponential family distributions, such as mixtures of normals or exponentials. This is demonstrated and illustrated in Lavine and West (1992), in the context of mixtures of multivariate normals, and is currently being explored in exponential, and other, mixtures for application in survival and reliability analysis.

Suppose $p-$dimensional data y_j, $(j = 1, \ldots, n)$, are a random sample from a mixture of k normals, $y_j \sim \mathcal{N}[\mu_i; \Sigma_i]$ with probability θ_i, for $i = 1, \ldots, k$; k is known and we have mean vectors $\mu = \{\mu_i; i = 1, \ldots, k\}$, variance matrices $\Sigma = \{\Sigma_i; i = 1, \ldots, k\}$, and component probabilities $\theta = (\theta_1, \ldots, \theta_k)$. Introduce latent classification variables z_j, where $z_j = i$ implies that y_j is drawn from component i of the mixture, or classified into group i; then the z_j are conditionally independent with $P(z_j = i|\theta) = \theta_i$. Inference is typically required for μ, Σ, θ and the classifiers $z = \{z_j; j = 1, \ldots, n\}$. Prediction of future observations, and classification of future cases according to possible mixture component of origin, is also often at issue.

Write $y = \{y_j; j = 1, \ldots, n\}$. Then, with an appropriate prior structure, the joint posterior of all quantities $(z, \theta, \mu, \Sigma|y)$ factorises as follows.

(a) Fixing z means the data are classified as k independent normal samples, and a standard, conjugate analysis may be performed for each group if the component means and variances are assigned independent normal/inverse-Wishart priors (which may be based on previous analysis of completely classified training data). This leads to conditionally independent posteriors of the same form, with easily computed parameters.

(b) Assigning an initial Dirichlet prior to θ implies that $p(\theta|y, z, \mu, \Sigma) \equiv p(\theta|z)$ is also Dirichlet with simply updated parameters.

(c) For z we have, for all i and j, $P(z_j = i|y, \theta, \mu, \Sigma) \propto \theta_i p(y_j|\mu_i, \Sigma_i, z_j = i)$ and summing to unity over $i = 1, \ldots, k$. Here $p(y_j|\mu_i, \Sigma_i, z_j = i)$ is just the normal density function for group i, with mean vector μ_i and variance matrix Σ_i, evaluated at the point y_j.

This structure provides the basis for an iterative resampling scheme to draw approximate samples from the joint posterior, and such samples are also the basis of predictive inference for future cases. In particular, a further case y_f drawn from the mixture has predictive density $p(y_f|y)$ evaluated by Monte Carlo as a mixture over (many) conditional predictive t distributions. For such a case, classification is based on predictive classification probabilities $P(z_f = i|y)$, similarly approximated. Full details appear in Lavine and West (1992).

An example from that reference involves a $k = 3$ component mixture of bivariate normals. With priors based on a training sample of just 15 classified cases, a further 300 unclassified observations are processed using the Gibbs sampling analysis to produce an approximate posterior sample of size 500, and used to evaluate the predictive density for a future case y_f; the contours of this multimodal density appear in the first frame of Figure 5. The second frame provides contours of the predictive classification probabilities $P(z_f = 3|y)$ that may be used for discrimination – similar contours for groups 1 and 2 appear in Lavine and West (1991). Note that the usual linear or quadratic discrimination rules are clearly inappropriate – the contours of constant classification probability are neither linear nor quadratic. One particular feature of some interest is the fact that, while the chance of a future case coming from the third component naturally decreases away from the mass of the predictive density of that component, it *increases* again far from the mass, noticeably in the top right corner of the graph. This is a general phenomenon, and a feature that is completely missed by traditional discrimination techniques; on reflection, it is clear that, moving outwards in any direction from the mass of the predictive density, the classification probabilities will eventually favour the mixture components having largest variance resolved in that direction.

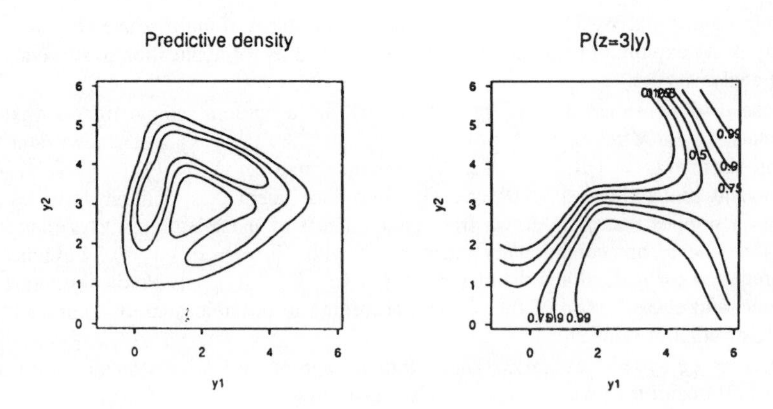

Figure 5. *Predictive contours and classification contours for normal mixture.*

4. MIXTURE MODELS FOR DATA 2: DENSITY ESTIMATION

4.1. *Mixtures of Dirichlet Processes*

Perhaps the most common use of mixture models is in non-parametric kernel density estimation (Silverman, 1976). Mixtures of Dirichlet processes (Ferguson, 1983) provide a natural model based foundation for Bayesian approaches to non-parametric density estimation, interpretation of kernel approaches as approximations to predictive densities, and insight into practical problems of smoothing parameter selection, local versus global smoothing, and multivariate and non-normal extensions (West, 1990b; Escobar and West, 1991). This class of models also provides a suitable setting for some problems of cluster analysis, although this remains to be explored in detail.

In the simplest normal model, scalar data y_j, $(j = 1, \ldots, n)$, are assumed conditionally independent with $y_j \sim \mathcal{N}[\mu_j; V]$, the means μ_j drawn independently from an uncertain prior $G(.)$, and $G(.)$ is modelled as a Dirichlet process with known base measure $\alpha G_0(.)$. Given V, and assuming $G_0(.)$ to be $\mathcal{N}[m_0; \tau V]$ for some scale factor τ, we can derive the following.
(a) k–configurations.
The n means $\{\mu_j\}$ are not necessarily distinct – there will be some number $k \leq n$ such that the μ_j are drawn from some k distinct quantities $\theta_1, \ldots, \theta_k$. The *configuration* of the means is defined in West (1990b) by an integer n–vector $c = (c_1, \ldots, c_n)$ whose elements take values between 1 and k, each such value appearing at least once, so that $c_j = i$ implies $\mu_j = \theta_i$. We label this arrangement of the means $C_k(c)$, and refer to it as a k–configuration. Further, let n_j be the number of the $\{\mu_j\}$ equal to θ_j, given by $n_j = \#\{c_i = j;\ i = 1, \ldots, n\}$, for $j = 1, \ldots, k$. Given $C_k(c)$, the setup is a one-way classification; the y_i are classified into k groups with group means θ_j. Further, the θ_j are independently drawn from the marginal prior $G_0(.)$ so that, given $C_k(c)$, posterior and predictive calculations can be peformed using standard theory.
(b) Predictions under $C_k(c)$.
Given the observed data $y = \{y_1, \ldots, y_n\}$, the following conditional predictive distributions for a future case y_f are simply derivable (West, 1990b). Firstly, given the μ_j, or just the distinct values θ_i, the conditional predictive distribution is easily obtained. Writing

$\mu = \{\mu_1, \ldots, \mu_n\}$, we have

$$\begin{aligned}(y_f|\mu, V, y) &\sim \alpha a_n \mathcal{N}[m_0; (1+\tau)V] + a_n \sum_{j=1}^{n} \mathcal{N}[\mu_j; V] \\ &\sim \alpha a_n \mathcal{N}[m_0; (1+\tau)V] + a_n \sum_{i=1}^{k} n_i \mathcal{N}[\theta_i; V]\end{aligned} \tag{4}$$

where $a_n = 1/(\alpha + n)$. This is a mixture of $k+1$ normals, one for each of the existing k groups and one corresponding to the possibility of a new group. Similarly, given the observed data y and knowledge of $\mathcal{C}_k(c)$ (though not of the actual values of the θ_i), it follows that

$$(y_f|V, \mathcal{C}_k(c), y) \sim \alpha a_n \mathcal{N}[m_0; (1+\tau)V] + a_n \sum_{j=1}^{k} n_j \mathcal{N}[m_j; (1 + w_j/n_j)V] \tag{5}$$

where, for each $j = 1, \ldots, k$, $m_j = w_j \bar{y}_j + (1 - w_j)m_0$ and $w_j = \tau n_j/(1 + \tau n_j)$, with $\bar{y}_j$ the sample mean of the n_j observations in group j.

Simple, but practically very important extensions provide for learning about V (West, 1990b). Suppose V^{-1} is assigned a gamma prior with specified shape $s_0/2 > 0$ and scale $S_0/2$, so that $v_0 = S_0/s_0$ is a prior estimate of V and s_0 the associated prior degrees of freedom. For each group $j = 1, \ldots, k$, define $R_j = (\bar{y}_j - m_0)^2 n_j/(1 + n_j\tau)$ and $S_j = \sum(x - \bar{y}_j)^2$, the sum being over all data points x in group j. Further, set $s = s_0 + n$ and $S = S_0 + \sum_{j=1}^{k}(S_j + R_j)$, and let $v_n = S/s$ be the corresponding posterior estimate of V. Then (5) is simply modified to a mixture of $(k+1)$ Student t distributions,

$$(y_f|\mathcal{C}_k(c), y) \sim \alpha a_n \mathcal{T}_s[m_0; (1+\tau)v_n] + a_n \sum_{j=1}^{k} n_j \mathcal{T}_s[m_j; (1 + w_j/n_j)v_n]. \tag{6}$$

Further generalisation to allow for different scale factors $V = V_j$ for group j is important in some applications – this induces different degrees of data-dependent smoothing across the sample space, a point of difficulty for non-Bayesian kernel approaches but trivially incorporated in these models. Escobar and West (1991) describe this extension, and show how (6) is modified; see also Ferguson (1983).

(c) Prior for k.

The number of groups k, $(1 \leq k \leq n)$, has a prior determined by α and n, given in Antoniak (1974), which favours larger values of k when α is large, and, for fixed α, decays rapidly as k increases. For $\alpha = 1$, corresponding to a reasonably imprecise prior, $\mathrm{E}(k) \approx \ln(n)$ for large n so that small numbers of mixture components are anticipated relative to n. For finite n, $P(k)$ is rather diffuse relative to a Poisson prior, but can be shown to be approximately Poisson for very large n.

(d) Multimodality and smoothing.

A major feature of mixture modelling is the capacity to model multimodal data distributions. For fixed α and n, (6) may be multimodal depending on the observed data configuration and the value of the prior variance τ; larger τ is consistent with more spread amongst the group means and hence greater chance of multimodality. West (1990b), and Escobar and West (1991), relate choice of τ to traditional smoothing parameter selection, and show how τ may be estimated (together with the observational variance V, possibly varying across groups).

It is clear, however, that the information about τ available from the data is often scant, and informed priors are needed in the same way that subjective choice of smoothing parameters is typically advocated by kernel estimators (Silverman, 1976). Some applications of density estimation focus on 'bump-hunting' to identify the number of modes, often using this as a proxy for inference about the number of components of a mixture (Silverman, 1981; Roeder, 1990). A key reason is that methods exist for inference about the modality of mixtures, but traditional approaches do not easily lead to inferences about numbers of components. The framework here, by contrast, provides priors, and hence posteriors, for the number of components k directly.

The number of modes h, say, will always be less than k, and the prior $P(h)$ will depend only on α, n and τ. By simulating from the model for given values of these parameters, the predictive densities (5) or (6) can be evaluated and the number of modes identified – this provides a Monte Carlo estimate of $P(h)$. Figure 6 provides summary information on $P(h)$ for the model with $V = 1$, $n = 50$ observations, the two cases $\alpha = 1$ and $\alpha = 5$, and as a function of the smoothing parameter τ over the range $0 < \tau \leq 350$. Here the model was simulated 10,000 times, the number of modes in (5) being counted by evaluation over a very fine grid each time. The figure indicates prior probabilities $P(h)$, at each value of τ, by circles with radii proportional to probability; also, the full line in each frame connects prior modes (to the nearest integer). Though larger h values are favoured for larger τ, the insensitivity of the prior for larger values, say $50 < \tau$, is apparent, and this is consistent with practical experiences that indicate robustness of posterior predictions across such ranges (West, 1990b).

4.2. *Simple Mixture Approximations to Predictive Distributions*

In practice, of course, the number of groups and configuration of the data into groups is unknown. Predictions should be based on the marginal predictive density defined by marginalisation over configurations, $p(y_f|y) = \sum p(y_f|C_k(c), y)P(C_k(c)|y)$, the sum being over all configurations and all k. The computations implied here are, for realistic sample sizes, impossible to perform, as is well known (Antoniak, 1974; Berry and Christensen, 1979; Ferguson, 1983; West, 1990b). In developing approximations, several authors identify the extreme configuration $C_n(1)$, where 1 is the unit n–vector, as leading to a Bayesian analogue of more traditional kernel procedures. Under $C_n(1)$, the μ_j are distinct and (6) becomes a mixture of $n+1$ terms; with n at all large relative to the initial precision α, this is close to a kernel form with n components located near the observations, since the component modes are $m_j \approx y_j$ for large n/α. This configuration, however, typically has very low prior probability $P(k = n)$, and in practice will usually have extremely low posterior probability too. West (1990b), develops an hierarchical search procedure to explore the space of configurations and to identify values of k and configurations $C_k(c)$ of high posterior probability. This search is trivially implemented, as described in that reference, and leads to simply computed relative posterior probabilities over a single chosen configuration for each value of k. These probabilities are typically maximum over smaller values of k, and decay rapidly otherwise. The favoured configurations lead to predictive densities that are mixtures of smaller numbers of components, and these may be used either directly as conditional predictions or approximately marginalised to estimate $p(y_f|y)$. The results in West (1990b) may be summarised as follows: (i) though the extreme n–configuration $C_n(1)$ typically will have little support under the posterior over configurations, the distribution function of the resulting 'Bayesian kernel density estimate' (equation (6) at $C_k(c) = C_n(1)$) will typically be close to that of other configurations with much greater support; (ii) the search algorithm identifies approximate

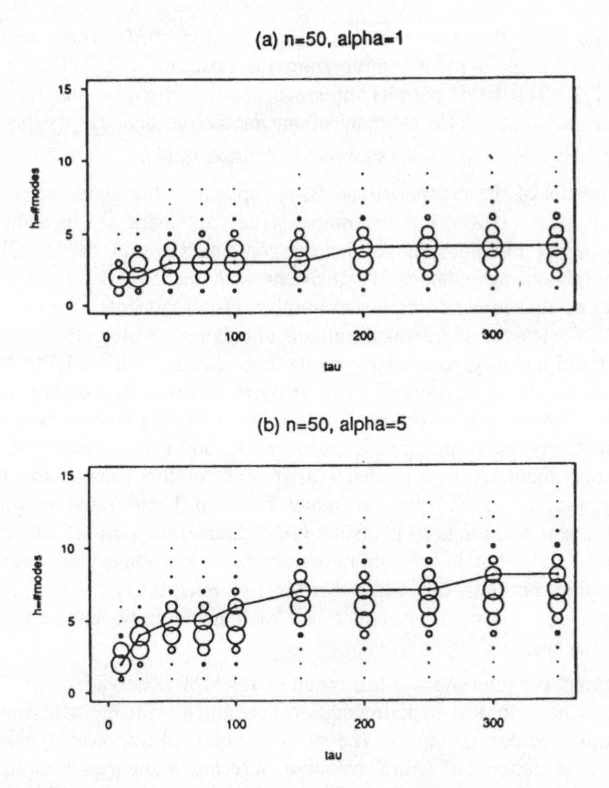

Figure 6. *Priors for number of modes h in predictions from mixture of Dirichlet processes.*

posterior modal configurations that are typically mixtures of small numbers of components, and the resulting predictive densities (6) are useful, and simple, approximations to marginal predictions; (iii) using the iterative sampling based analysis discussed below in Section 4.3, these simple mixture approximations can be compared with Monte Carlo approximations to the full marginal $p(y_f|D)$, and often turn out to be quite similar. The hierarchical search algorithm used to investigate configurations provides one means of identifying reduced mixtures, in point (ii), the computational benefits of reducing from n to $k << n$ being of interest, for example, in applications where n is very large and the predictive distributions is to be heavily evaluated, simulated, and so forth.

4.3. *Sampling Analyses*

Escobar (1990) introduced a Gibbs sampling technique that applies to produce approximate samples from the posterior distribution of the means $\mu = \{\mu_1, \ldots, \mu_n\}$ in (4). This is easily extended to analysis with V uncertain, as described and illustrated in West (1990b), and further to models with V varying across groups, and also to cover the case of uncertain τ; details appear in Escobar and West (1991). Restricting attention for theoretical discussion here to (4), suppose we have available a sample $\{\mu(r), V(r); r = 1, \ldots, N\}$ from the joint posterior $p(\mu, V|y)$. Thus, for each r, $\mu(r) = \{\mu_1(r), \ldots, \mu_n(r)\}$ is a single sample of the mean values, and $V(r)$ the corresponding sampled value of V. There will be some $k(r)$ distinct values in the set $\mu(r)$, a sample from the posterior distribution over the number of groups $P(k|y)$. The latter permits inference about k directly, and the former may be used to compute a Monte Carlo estimate of the predictive density $p(y_f|y)$ via $p(y_f|y) \approx N^{-1}\sum_{r=1}^{N} p(y_f|\mu(r), V(r), y)$, with summands defined in (4).

Complete details of the computational issues appear in the above references, where an example from Roeder (1990) is re-examined. The data there are measured velocities of a number of galaxies identified in six distinct conic sections of space. Clustering of the velocities of neighbouring galaxies is consistent with the Big Bang theory of the origin of the universe, so that interest lies in the number of components of a mixture of velocity distributions. While Roeder addresses the effects of uncertainty about density estimates on the assessment of multimodality, particularly on the hypothesis of unimodality, the scientifically interesting issue is that of clustering. The analysis reported in Escobar and West (1991) supports Roeder's broad conclusion that there are most likely four modes in the distribution, though the uncertainty about modality is considerable, and goes further to the more relevant issue of inference about k. In a model with $\alpha = 1$ for this sample size of $n = 82$, the prior $P(k)$ supports k heavily supports values between 3 and 7 but is quite diffuse over 1–11. Posteriors, across a range of plausible prior parameters, support rather large values of k between 5 and 7. The conclusions are rather typical of inferences with data consistent with heavily overlapping mixtures, but, while there is high uncertainty about k, the computations here at least provide a formal assessment of such uncertainty, unlike traditional density estimation approaches.

As an example, consider the raw histogram of $n = 159$ observations in Figure 7(a). This data comes from neurological experimentation concerned with the assessment of neuronal responses to simulated current inputs. The study of neuronal responses typically focusses on the identification of patterns of neural response in terms of the distributions of numbers of neural sites 'firing' and the distributions of magnitudes of response. This data comes from a collaborative study with Dr. D.A. Turner, of Duke Medical Center. The mixture models of this section are perfectly suited to the data structure in this area, since the underlying neurological science hypothesis is a mixture of normals for the sort of data displayed in the histogram. The uncertain number of components of the mixture and the distribution of the distinct means of the components relate to the numbers of neurons responding to external stimuli and the durations and magnitudes of responses. For illustration here, one analysis along the lines detailed in Escobar and West (1991) is summarised in Figure 7. The background to the data collection procedure provides a substantial amount of information on the observational variance V of the data. In this particular experiment, a similar dataset is available from a previous experiment designed to estimate V alone – based on similar neuronal measurements but without simulated stimuli. This routine calibration experiment provides $n = 159$ observations with roughly zero-mean and a sample variance of 0.047, which suggests a prior for V of the inverse gamma form with $s_0 = 158$ and $S_0 = 158 \times 0.047 = 7.43$. Other, more diffuse priors may be used, of course, but this is taken here – existing techniques directly guess a value for V based on such calibration data. Notice that this prior strongly supports values of V that are very small compared to the observed spread of the histogram in Figure 7(a), indicating that there is substantial variation in the stimulus response data above that contributed by background noise. The analysis summarised actually uses the model of Escobar and West (1991) that allows for different values of V across the distinct normal components, though the concentration of the prior (with 158 degrees of freedom) heavily restricts the values to be close, and a reanalysis with common V would

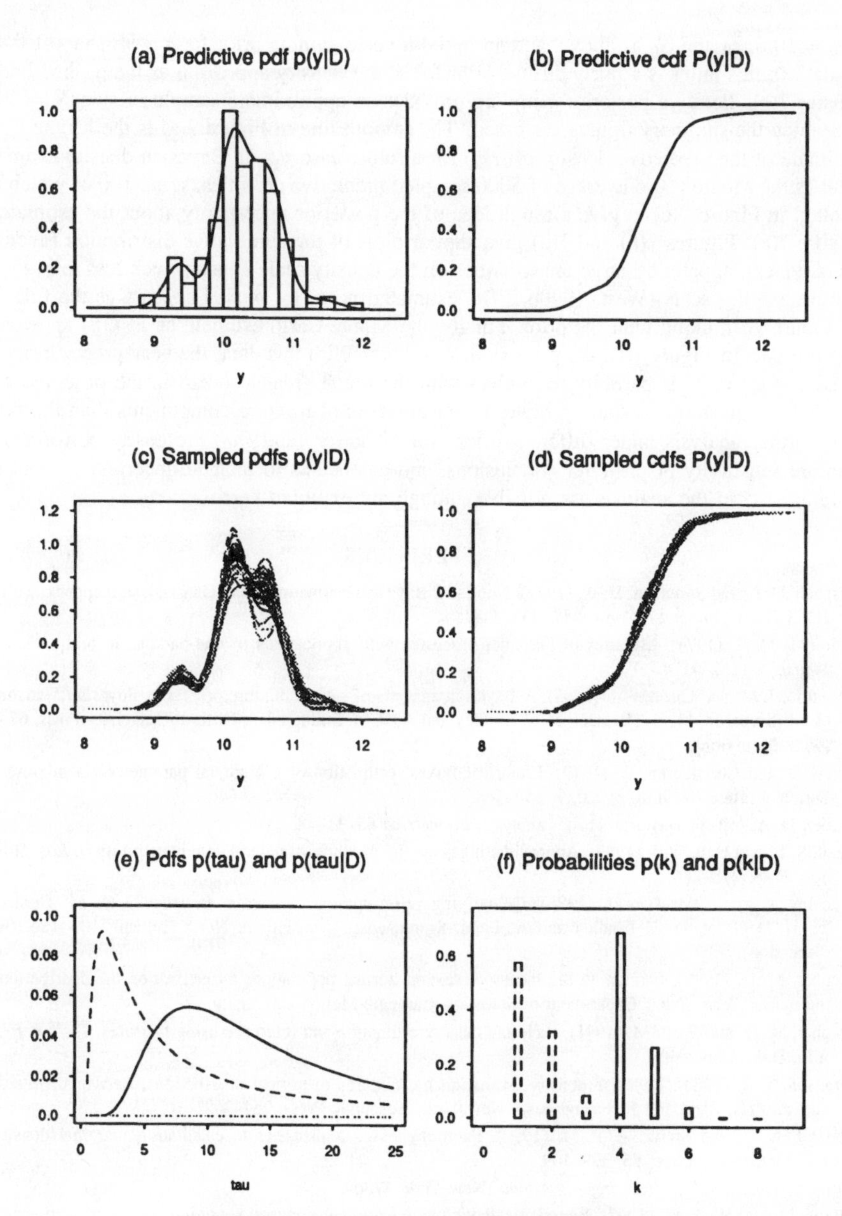

Figure 7. *Inferences for neurological response data.*

give substantially similar results. The analysis also includes the prior smoothing parameter τ, producing approximate posterior samples for the group means, hence the number of distinct components k, the observational variances, and τ. The initial priors are specified with

$m_0 = 10.0$, s_0 and S_0 as above, and an initial inverse gamma prior for τ with shape 1.0 and scale 5.0; this latter is a fairly diffuse prior for τ, the density appearing as the dashed line in Figure 7(e). Iterative posterior sampling provides an approximate sample of size $N = 5000$ on which the summary figures are based. The smooth line in Figure 7(a) is the Monte Carlo estimate of the predictive density $p(y_f|y)$ for a future case y_f – a Bayesian density estimate. This is the Monte Carlo average of 5000 sampled predictive densities, some 100 of which are plotted in Figure 7(c) to give a rough idea of the posterior uncertainty about the estimate in Figure 7(a). Figures 7(b) and 7(d) give similar plots of the cumulative distribution function; as is typical, apparently large uncertainties on the density scale appear much less relevant on the probability scale (West, 1990b). The estimated posterior $p(\tau|y)$ appears as the full line in Figure 7(e), along with the prior. Finally, the Monte Carlo estimate of $P(k|y)$ appears as the full bars in Figure 7(f), the prior as dashed bars. With this data, the very precise prior on small values of V is partially reconciled with the much greater spread in the observed data through the posterior favouring rather larger numbers of mixture components than the prior.

Further analysis under different priors for V and τ (and m_0) are easily performed to explore sensitivity of posterior conclusions, and extensions to include uncertainty about the parameter α in the analysis are possible, though not explored here.

REFERENCES

Alspach, D. L. and Sorenson, H. W. (1972). Nonlinear Bayesian estimation using Gaussian sum approximations. *IEEE Trans. Automatic Control* **17**, 439–448.

Antoniak, C. E. (1974). Mixtures of Dirichlet processes with applications to non-parametric problems. *Ann. Statist.* **2**, 1152–1174.

Bernardo, J. M. and Girón, F. J. (1985). A Bayesian analysis of simple mixture problems. *Bayesian Statistics 3* (J. M. Bernardo, M. H. DeGroot, D. V. Lindley and A. F. M. Smith, eds.), Oxford: University Press, 67–88, (with discussion).

Berry, D. and Christensen, R. (1979). Empirical Bayes' estimation of a binomial parameter via mixtures of Dirichlet processes. *Ann. Statist.* **7**, 558–568.

Binder, D. A. (1978). Bayesian cluster analysis. *Biometrika* **65**, 31–38.

Dalal, S. R. and Hall, W. J. (1983). Approximating priors by mixtures of natural conjugate priors. *J. Roy. Statist. Soc. B* **45**, 278–286.

Diaconis, P. and Ylvisaker, D. (1985). Quantifying prior opinion. *Bayesian Statistics 2* (J. M. Bernardo, M. H. DeGroot, D. V. Lindley and A. F. M. Smith, eds.), Amsterdam: North-Holland, 133–156, (with discussion).

Escobar, M. D. (1990), Estimating the means of several normal populations by estimating the distribution of the means. *Tech. Rep.* , Department of Statistics, Carnegie-Mellon University.

Escobar, M. D. and West, M. (1991). Bayesian density estimation and inference using mixtures. *Tech. Rep.* **90-A16**, Duke University..

Ferguson, T. S. (1983). Bayesian density estimation by mixtures of normal distributions. *Recent Advances in Statistics* (H. Rizvi and J. Rustagi eds.), New York: Academic Press, 287–302.

Gelfand, A. E. and Smith, A. F. M. (1990). Sampling based approaches to calculating marginal densities. *J. Amer. Statist. Assoc.* **85**, 398–409.

Hartigan, J. A. (1975). *Clustering Algorithms*. New York: Wiley.

Lavine, M. and West, M. (1992). Bayesian calculations for normal mixtures. *Canadian J. Statist.* , (to appear).

McClachan G. J. and Basford, K. E. (1988). *Mixture Models: Inference and Applications to Clustering*. New York: Marcel Dekker.

Migon, H. S. and Harrison, P. J. (1985). An application of non-linear Bayesian forecasting to television advertising. *Bayesian Statistics 2* (J. M. Bernardo, M. H. DeGroot, D. V. Lindley and A. F. M. Smith, eds.), Amsterdam: North-Holland, 681–696.

Pole, A. and West, M. (1988). Efficient numerical integration in dynamic models. *Tech. Rep.* **131**, University of Warwick.

Pole, A. and West, M. (1990). Efficient Bayesian learning in non-linear dynamic models. *J. Forecasting* **9**, 119–136.

Roeder, K. (1990). Density estimation with confidence sets exemplified by superclusters and voids in the galaxies. *J. Amer. Statist. Assoc.* **85**, 617–624.

Silverman, B. W. (1981). Using kernel density estimates to investigate multimodality. *J. Roy. Statist. Soc. B* **43**, 97–99.

Silverman, B. W. (1986). *Density Estimation for Statistics and Data Analysis*. London: Chapman and Hall.

Titterington, D. M., Smith, A. F. M. and Makov, U. E. (1985). *Statistical Analysis of Finite Mixture Distributions*. New York: Wiley.

West, M. and Harrison, P. J. (1989). *Bayesian Forecasting and Dynamic Models*. New York: Springer.

West, M. (1990a). Bayesian computations: sequential analyses and dynamic models. *Tech. Rep.* **90-A12**, Duke University.

West, M. (1990b). Bayesian kernel density estimation. *Tech. Rep.* **90-A02**, Duke University.

West, M. (1992). Bayesian computations: Monte-Carlo density estimation. *J. Roy. Statist. Soc. B* , (to appear).

Wolpert, R. L. (1991). Monte Carlo integration in Bayesian statistical analysis. *Contemporary Mathematics* **115**, 101–116.

DISCUSSION

F. J. GIRÓN* (*Universidad de Málaga, Spain*)

Actually, this paper is so full of good ideas, applications and suggestions that, due to space limitations, I will only comment on a couple of points that have attracted my attention. The first point refers to the important practical problem of reduction of mixtures, to be found in Section 2.3.

The problem of approximating a mixture of distributions by a smaller set of mixtures is a difficult one —I would rather say an ill-defined one— but an important one from a computational viewpoint. The theoretical results of Diaconis and Ylvisaker (1985) and Dalal and Hall (1983) are of little help in the *practical* problem of approximating mixtures.

One of the most frequently encountered problems in Bayesian statistical practice is that of approximating a finite, but large, mixture of multivariate Student t distributions. In most cases, the components in the mixture all have the same number of degrees of freedom. So let us consider the problem of approximating the mixture

$$\sum_{j=1}^{n} \omega_j t_p(\boldsymbol{\theta}|\boldsymbol{m}_j, \boldsymbol{V}_j; \nu); \tag{1}$$

by

$$\sum_{j=1}^{m} \omega'_j t_p(\boldsymbol{\theta}|\boldsymbol{m}'_j, \boldsymbol{V}'_j; \nu); \tag{2}$$

with m much smaller than n (say, e.g., 10%).

The problem of approximating mixtures of t distributions presents some difficulties. These difficulties, and the way to overcome them, become more apparent when one sees t distributions as scale mixture of normals. This representation of mixtures of t distributions as more general mixtures of normals also gives, in my opinion, a justification to the method for the reduction of mixtures West proposes.

For this problem the author suggests:

---

* Partially supported by the Dirección General de Investigación Científica y Técnica DGICYT as part of the project PB87-0607-C02-02.

i) Dropping components with "small" weights ω_j by "averaging" them with the nearest neighbour.
ii) Clustering components that are "close".
If one writes (1) as

$$(1) = \sum_{j=1}^{n} \omega_j \int N_p(\boldsymbol{\theta}|\boldsymbol{m}_j, \lambda \boldsymbol{V}_j) dInGa(\lambda|\nu/2, \nu/2)$$

$$= \int \left(\sum_{j=1}^{n} \omega_j N_p(\boldsymbol{\theta}|\boldsymbol{m}_j, \lambda \boldsymbol{V}_j) \right) dInGa(\lambda|\nu/2, \nu/2)$$

$$\approx \int \left(\sum_{j=1}^{m'} \omega'_j N_p(\boldsymbol{\theta}|\boldsymbol{m}'_j, \lambda \boldsymbol{V}_j) \right) dInGa(\lambda|\nu/2, \nu/2)$$

$$= \sum_{j=1}^{m'} \omega'_j t_p(\boldsymbol{\theta}|\boldsymbol{m}'_j, \boldsymbol{V}'_j; \nu),$$

then, this representation reduces the original problem to one of approximating mixtures of multivariate normal distributions. Suppose we want to approximate a small mixture of normals

$$\sum_{i=1}^{m} \alpha_i N_p(\boldsymbol{\theta}|\boldsymbol{m}_i, \lambda \boldsymbol{V}_i)$$

by a normal distribution of the form $N_p(\boldsymbol{\theta}|\boldsymbol{m}, \lambda \boldsymbol{V})$. It is well known that the best approximation to a single normal is given by,

$$\boldsymbol{m} = \sum_{i=1}^{m} \alpha_i \boldsymbol{m}_i \quad \text{and} \quad \lambda \boldsymbol{V} = \lambda \sum_{i=1}^{m} \alpha_i \boldsymbol{V}_i + \sum_{i=1}^{m} \alpha_i (\boldsymbol{m}_i - \boldsymbol{m})(\boldsymbol{m}_i - \boldsymbol{m})^T. \tag{3}$$

To make this approximation minimally dependent on the hyperparameter λ, it is obvious that either the components with large weights α_i must have very close means $\boldsymbol{m}_i$ or the components should have small weights α_i, so that the second term in (3) be small; and this is tantamount to the author's method. Also note that the approximation should only be good in a neighbourhood of $\lambda = 1$ if the number of degrees of freedom ν is large.

One additional inconvenience is the lack of identifiability of the preceding representation. An alternative way of writing (1) is

$$(1) = \sum_{j=1}^{n} \omega_j \int N_p(\boldsymbol{\theta}|\boldsymbol{m}_j, \lambda k \boldsymbol{V}_j) dInGa(\lambda|\nu/2K, \nu/2)$$

$$= \int \left(\sum_{j=1}^{n} \omega_j N_p(\boldsymbol{\theta}|\boldsymbol{m}_j, \lambda k \boldsymbol{V}_j) \right) dInGa(\lambda|\nu/2k, \nu/2)$$

$$\approx \int \left(\sum_{j=1}^{m''} \omega''_j N_p(\boldsymbol{\theta}|\boldsymbol{m}''_j, \lambda \boldsymbol{V}''_j) \right) dInGa(\lambda|\nu/2, \nu/2)$$

$$= \sum_{j=1}^{m''} \omega''_j t_p(\boldsymbol{\theta}|\boldsymbol{m}''_j, \boldsymbol{V}''_j; \nu),$$

and a different approximation results.

One possible way to avoid these difficulties is based on the following observation. Usually, mixtures of t distributions arise, in a natural way, as marginals of mixtures of normal-gamma or normal-inverted-gamma distributions of the form

$$\sum_{j=1}^{n} \omega_j N_p InGa(\boldsymbol{\theta}\sigma^2 | \boldsymbol{m}_j, \boldsymbol{S}_j, a_j, p_j) \tag{4}$$

with

$$\boldsymbol{V}_j = \frac{a_j}{p_j}\boldsymbol{S}_j, \quad \text{and} \quad p_j = \frac{\nu}{2}.$$

So a possibility is to approximate (4) by a mixture with a smaller number of terms, say,

$$\sum_{j=1}^{m} \omega_j' N_p InGa(\boldsymbol{\theta}\sigma^2 | \boldsymbol{m}_j', a_j, \nu/2)$$

and then marginalize to obtain the approximate mixture of t distributions.

My final comments refer to the sequential analyses of dynamic models in Section 2.4, where I follow the author's notation very closely. He considers the simplest dynamic model

$$\begin{array}{rll} \text{Observation equation} & y_t = \theta_t + \nu_t & f_\nu(\cdot) \\ \text{System equation} & \theta_t = \theta_{t-1} + \omega_t & f_\omega(\cdot); \end{array} \tag{5}$$

with $\theta_0 \perp\!\!\!\perp \nu_t \perp\!\!\!\perp \omega_t$ for all t.

Suppose ν_t and ω_t are distributed as scale mixtures of normals, say,

$$\nu_t \sim \int N(0, \nu V_t) dF_\nu(\nu) \quad \text{and} \quad \omega_t \sim \int N(0, \omega V_t) dF_\omega(\omega),$$

in an obvious notation. This assumption includes the important case of errors being distributed as t distributions with possibly different degrees of freedom.

From (5), it follows that

$$\left(\begin{matrix} y_t \\ \theta_t \end{matrix} \middle| \theta_{t-1}, D_{t-1} \right) \sim \int\int N\left[\begin{pmatrix} y_t \\ \theta_t \end{pmatrix}; \begin{pmatrix} \theta_{t-1} \\ \theta_{t-1} \end{pmatrix}, \begin{pmatrix} \nu V_t + \omega W_t & \omega W_t \\ \omega W_t & \omega W_t \end{pmatrix} \right] dF_\nu(\nu) dF_\omega(\omega).$$

Now, suppose θ_{t-1} given D_{t-1} is also distributed as a general mixture of normals

$$\theta_{t-1} | D_{t-1} \sim \int N(\theta_{t-1}; \mu_{t-1}(\lambda), \sigma_{t-1}^2(\lambda)) dF_\lambda(\lambda); \tag{6}$$

where λ may be a general hyperparameter.

From general properties of the multivariate mixtures of normals, it follows that the posterior distribution of $(y_t, \theta_t, \theta_{t-1} | D_{t-1})$ is the mixture

$$\int\int\int N\left[\begin{pmatrix} y_t \\ \theta_t \\ \theta_{t-1} \end{pmatrix}; \begin{pmatrix} \mu_{t-1}(\lambda) \\ \mu_{t-1}(\lambda) \\ \mu_{t-1}(\lambda) \end{pmatrix}, \begin{pmatrix} \nu V_t + \omega W_t + \sigma_{t-1}^2(\lambda) & \omega W_t + \sigma_{t-1}^2(\lambda) & \sigma_{t-1}^2(\lambda) \\ \omega W_t + \sigma_{t-1}^2(\lambda) & \omega W_t + \sigma_{t-1}^2(\lambda) & \sigma_{t-1}^2(\lambda) \\ \sigma_{t-1}^2(\lambda) & \sigma_{t-1}^2(\lambda) & \sigma_{t-1}^2(\lambda) \end{pmatrix} \right]$$

$$dF_\lambda(\lambda) dF_\nu(\nu) dF_\omega(\omega);$$

from which the posteriors of θ_t and y_t given D_{t-1} are easily derived. Finally, the posterior of θ_t given D_t, which is the final step in the sequential updating, turns out to be

$$\theta_t|D_t \sim \int N(\theta_t; \mu_t(\lambda,\nu,\omega), \sigma_t^2(\lambda,\nu,\omega))dF(\lambda,\nu,\omega|D_t); \qquad (7)$$

where

$$\mu_t(\lambda,\nu,\omega) = \frac{\nu V_t \mu_{t-1} + (\omega W_t + \sigma_{t-1}^2(\lambda))y_t}{\nu V_t + \omega W_t + \sigma_{t-1}^2(\lambda)};$$

$$\sigma_t^2(\lambda,\nu,\omega) = \frac{\nu V_t(\omega W_t + \sigma_{t-1}^2(\lambda))y_t}{\nu V_t + \omega W_t + \sigma_{t-1}^2(\lambda)};$$

$$dF(\lambda,\nu,\omega|D_t) \propto n(y_t; \mu_{t-1}(\lambda), \nu V_t + \omega W_t + \sigma_{t-1}^2(\lambda))dF_\lambda(\lambda)dF_\nu(\nu)dF_\omega(\omega);$$

and $n(.;.,;)$ denotes the density of a normal distribution.

This shows that if (6) is the true (not an approximation of the) distribution of $\theta_{t-1}|D_{t-1}$, then (7), which is the exact posterior of $\theta_t|D_t$, is also a (generally more complex) mixture of normals. This last result can, for example, be used to provide a Monte Carlo approximation to the true posterior. However, for these theoretical results to be of some practical value —thus, for example, to produce an approximate Kalman filter— this last infinite mixture should be approximated by a simpler mixture of normals.

I would like Professor West to comment on this issue of approximating —*reducing*, in his own words— a complex infinite mixture of normals by a simpler one, in a way similar (despite the infinite number of the terms in the mixture) to the one he employs in Section 2.3.

Finally, I just want to congratulate Mike West for this important piece of research on the facinating subject of mixtures.

REPLY TO THE DISCUSSION

Professor Girón comments initially on the reduction of mixtures of t distributions, and describes conditions under which the (very basic) weighted nearest-neighbour clustering technique in Section 2.3 will give rise to approximating mixtures 'close' to the original. These comments are wise, and echo discussion of West and Harrison (1989, Section 12.3.3). In the context of Section 2.3, these conditions will typically be approximately satisfied. There, reductions are made for mixtures of, say, several thousand components, to, say, several hundred, with dominating features (i) that components removed will indeed have negligible weight, and (ii) 'clustered' components will indeed be close so that the difference in means effect (equation (3) of discussion) will be small. With mixtures of fewer components, and in other applications, this may not be the case.

I appreciate the comments of Professor Girón on the use of the normal-scale mixture genesis of the t distribution to recast the problem of approximating mixtures of t distributions as that of approximating mixtures of normal/inverse gamma distributions. Though I am currently unclear as to the practical insights gained here, this structuring will certainly bear further study. Note that West and Harrison (Section 12.3.4) develop related though more general ideas for approximating mixtures of (general) multivariate normal/inverse Wishart distributions. The issue of identifiability is one I was not previously aware of, and may mitigate against fruitful practical implications of the scale mixing approach, though again it requires further study.

The scale mixing structure resurfaces in the second contribution of Professor Girón, concerning univariate dynamic models — first-order polynomial models (West and Harrison,

1989, chapter 2) — with Student t error distributions. The issue raised in discussion is that of approximating the resulting posteriors by finite discrete mixtures, in cases when the actual mixing distributions are continuous. The development in my paper of finite discrete mixtures as importance sampling distributions is geared to just this sort of problem. With the usual caveats applicable to importance sampling function design (and primarily the dominance of tails), the standard results (Geweke, 1989) apply to describe the expected behaviour of Monte Carlo approximations to posterior expectations based on any mixture approximation.

An issue not noted in the discussion is that of computation. The initial rationale for exploring mixture reduction is largely computational; if a mixture of several thousand multivariate distributions may be well approximated (as in often the case) by a mixture of a few hundred, the computations involved in evaluating the density and distribution function, and in sampling from the mixture, are enormously reduced. Simple clustering techniques can be statistically very effective in the sense of producing close approximation of vastly reduced mixtures to original mixtures of very many components, but the reduction process is computationally demanding, if not the dominant factor in computational time. So a computational trade-off arises between the costs of inference using the original mixture and those using the reduced mixture plus clustering computations. Careful exploration of this trade-off, and how it depends on the parameter dimension and the sizes of original and reduced mixture, are called for.

It is further desirable to develop, or import, efficient coding for reduction algorithms, and, more appropriately, look for alternative approximation algorithms. On the latter issue, we are interested in theory to guide the reduction process. Professor Girón states that *it is well-known that the best approximation* — of a mixture, or any other distribution, with finite variance — *(by) a single normal is given by* matching mean and variance matrix. Certainly this is true when 'best' is measured with respect to (one of many possible) density based distance measures, the Kullback-Leibler directed divergence being the archetype (e.g. West and Harrison, 1989, Section 12.3.4). Distance between densities (actually based on density ratios) clearly underlies the development of mixture importance sampling functions in the paper, and often this is natural. In the dynamic model setting and sequential analysis more generally, for example, a mixture approximation to a 'current' posterior is often to be used as the prior for further analysis, so the density function is to be used directly as input to Bayes theorem. Getting the approximation right in density terms is crucial. In other areas, however, a focus on other aspects of approximation may be more appropriate. There are many other distance measures between multivariate distributions that are not directly density based and that lead to different notions of goodness-of-fit of approximations. Some discussion of univariate cases, and the relationships between various measures, may be found in Titterington, Makov and Smith (1985, Section 4.5). I have made the point many times that *distribution* function estimation is typically, or should typically be, the objective of non-parametric data analysis usually couched in terms of *density* estimation, and that we should adopt loss functions that respect this.

On alternative clustering approaches, the real need is for theory to guide algorithmic development. Though mixture models, and various associated methods of collapsing and combining components, have, of course, been quite extensively studied in the context of cluster analysis *per se*, deeper theoretical underpinning for clustering algorithms is scant, and Bayesian approaches limited. The recent National Research Council Report on discriminant analysis and clustering (NRC, 1989) raises similar issues, noting in summary that cluster analysis '*is lacking in firm foundations and agreed upon methodology*'. Investigation of cluster analysis problems within multivariate generalisations of mixtures of Dirichlet processes (Section 4) may provide useful theoretical bases. As noted in Section 4, these models provide important ingredients for clustering issues in that they embody (restricted classes of) prior distributions over data configurations, and over the number of clusters (components), and likelihood functions for (at least approximately) updating to posterior distributions over clusters. Deeper exploration of these models may help us to develop more efficient (statistical and computational) algorithms for mixture approximation generally.

ADDITIONAL REFERENCES IN THE DISCUSSION

Geweke, J. F. (1989). Bayesian inference in econometric models using Monte Carlo integration. *Econometrica* **57**, 1317–1340.

National Research Council (1989). Board on Mathematical Sciences and Committee on Applied Theoretical and Applied Statistics: Report on discriminant analysis, classification, and clustering. *Statist. Sci.* **4**, 34–69.

BAYESIAN STATISTICS 4, pp. 525–546
J. M. Bernardo, J. O. Berger, A. P. Dawid and A. F. M. Smith, (Eds.)

Bayesian Hierarchical Logistic Models for Combining Field and Laboratory Survival Data

ROBERT L. WOLPERT and WILLIAM J. WARREN-HICKS
Duke University, USA

SUMMARY

Generalized linear regression models fit to multicollinear data sets can be unreliable for making predictions in data sets free from the multicollinearity. We overcome this problem in a study of brook trout response to acidification by constructing hierarchical Bayesian models to provide a coherent logical structure for synthesizing evidence from field observations and from related laboratory bioassay experiments, despite the great differences between field and laboratory settings and uncertainties about those differences.

Keywords: LOGISTIC REGRESSION; BAYESIAN HIERARCHICAL MODEL; MONTE CARLO INTEGRATION.

1. INTRODUCTION

Near multicollinearity in linear or generalized linear models can make it hard to identify model parameters or to make reliable predictions. Model-based predictions will depend on the uncertain values of model parameters, but collinear data give almost equal support for a range of parameter values along a curve in parameter space on which the likelihood function is nearly constant. The parameter choice may be irrelevant if new predictions are to be made from a data set sharing the same near multicollinearity, but model predictions in data sets free from multicollinearity or featuring a different multicollinearity structure will differ significantly.

Several methods have been proposed for selecting among parameter values supported equally well by the data. Popular *ad hoc* methods include ridge regression, in which parameter vectors close to the origin are favored, and model reduction, in which one or more regression coefficients are set to zero. The choice of a penalty function for ridge regression or of a set of variables to eliminate is problematic. A Bayesian analysis with an informative prior may be appropriate when expert opinion about parameter values is available, but we feel this situation is rare.

We propose a different approach, appropriate for applications where other sources of empirical information can be brought to bear on the problem of resolving the collinearity. The solution is to construct a Bayesian model with a *hierarchical* structure comprising submodels for both the primary problem of interest and a secondary problem in which supplementary empirical information is available. Additional parameters may be introduced to accommodate the differences (anticipated or discovered) between the primary and secondary submodels.

1.1. *The Application*

The models studied in this paper (motivated and described more fully elsewhere) are intended to predict the presence or absence of a specific species of fish (*Salvelinus fontinalis Mitchell*, commonly called brook trout) in Adirondack lakes, from measurements of water chemistry variables (pH and the concentrations of inorganic aluminum and calcium) often associated with *acidification*, or acid rain. Since calcium-rich soils buffer acidic rainfall, exchanging ionic calcium and other base cations for hydrogen ions in the runoff, the measured pH and concentrations of ionic aluminum and calcium are expected to exhibit multicollinearity in existing data sets. This expectation is borne out (Reckhow *et al.* 1987, Baker *et al.* 1988, and Reckhow, 1988), as illustrated in Figure 1.

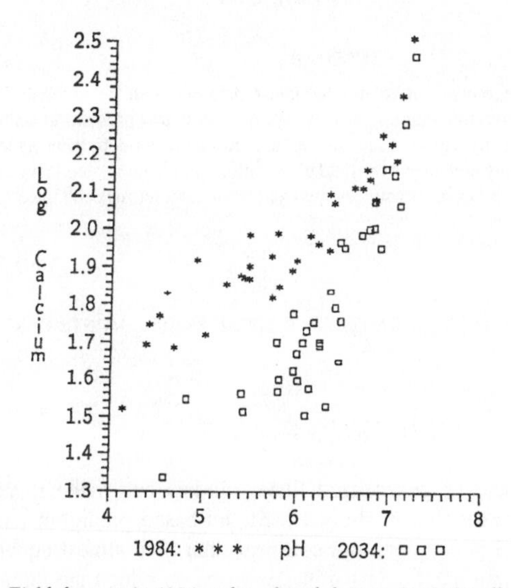

Figure 1. *Field data set in 1984 and predicted data set for 2034, illustrating changing collinearity.*

Despite this multicollinearity conventional binary regression models (logistic and probit, for example) can be quite effective in predicting fish response in lakes similar to those in the existing data sets (Warren-Hicks, 1990). The only effects of the collinearity are near-nonidentifiability of the individual regression coefficients, from a classical view, or large posterior variances for those parameters, from a Bayesian perspective— these play havoc with numerical optimization and integration methods, and lead to the very wide confidence and credible intervals commonly encountered in overparameterized models, but don't compromise the models' ability to fit the existing data.

Anticipated reductions in acid rain resulting from reductions in sulphur emissions mandated by the 1990 Clean Air Act are expected to reduce the rate of calcium leaching from soils into surface waters, raising the pH while lowering the concentration of calcium— causing future data sets to exhibit a different multicollinearity structure than that of current data sets, as illustrated in Figure 1. The models in this study were developed to assist the U.S. Environmental Protection Agency in predicting fish response in these future data sets, and so must overcome the obstacle of multicollinearity.

While field observations from data sets free from the multicollinearity are not yet available, there are supplementary data from laboratory bioassay experiments in which an early life stage of fish is exposed to controlled concentrations of selected chemical toxicants. These data may be assumed to reflect features of the actual relationship between fish survival and chemical concentration that are masked in the collinear field observations. However, they differ from the field survey data in fundamental ways: the field settings feature predators, a less reliable food supply, variations in chemical and physical features of the environment, and other conditions not present in the laboratory. Even more important, the indicator of fish response in the bioassay data is the survival *vs.* death of *individual fish* while the response variable in the field data is presence *vs.* absence for an entire *fish population*. These differences make it difficult to imagine how the laboratory data can be combined with the field data to improve model applicability and predictions.

1.2. *The Data*

The field data set was drawn from a survey of some 1469 Adirondack lakes conducted between 1984 and 1987 by the Adirondack Lake Survey Corporation (ALSC), as reported in Baker *et al.* (1990). It consists of data from 313 lakes: 177 from the 1984–1985 survey, and 136 from the 1986–1987 survey. Although the ALSC reported measurements of many variables from each lake, only four were considered in the present study: the presence or absence of brook trout, and the three water chemistry variables identified by Baker, *et al.* (1990) as having the greatest bearing on fish survival: pH and the concentrations of inorganic aluminum and calcium.

The bioassay data consist of survival observations of an early brook trout life stage (swimup fry). Similar studies from three different investigators were included (Baker, 1981; Ingersoll, 1986; and Holtze and Hutchinson, 1989). Each investigator manipulated pH and aluminum concentration (and, in one case, calcium concentration) in repeated trials, placing a number (usually 25) of fish in a flow-through exposure chamber and recording daily observations of the number of survivors. Observations were terminated and the experiment brought to a close after a predetermined period ranging from ten to twenty-one days. Variations in experimental protocol among investigators were minor and appear to be inconsequential (Warren-Hicks, 1990). The final bioassay data set consists of 163 trials including 4096 individual fish responses.

2. MODELS AND LIKELIHOOD FUNCTIONS

We consider four statistical models that specify the probabilities of recorded brook trout presence in lakes and of trout survival in bioassay experiments. To clarify their structures and relationships we introduce the mnemonic notation $\mathbf{M}_8^B$ for a model with an eight-dimensional parameter $\theta \in \mathbb{R}^8$, predicting *Both* Field and Laboratory data with separate, independent logistic submodels $\mathbf{M}_4^F$ and $\mathbf{M}_4^L$, respectively; $\mathbf{M}_6^H$ for an exchangeable six-dimensional *Hierarchical* model predicting both field and laboratory data; $\mathbf{M}_{10}^B$ for a more complex ten-dimensional model with independent generalized logistic submodels $\mathbf{M}_5^F$ and $\mathbf{M}_5^L$ reflecting features of the field and laboratory settings that are inconsistent with simple logistic models; and $\mathbf{M}_8^H$ for a non-exchangeable eight-dimensional Hierarchical model reflecting those same features.

Conditional on the observed vector x_i of explanatory variables and on an uncertain parameter vector θ, the indicator variables s_i of fish presence in the ith lake (for i in a set I^F indexing the lakes) and the survivor counts s_i of fish in the ith laboratory bioassay experiment (for i in a set I^L indexing the bioassays) are all taken to be independent binomial $\mathrm{Bi}(n_i, p_i)$

random variables, in each model we consider, with known n_i and with specified functional forms for the Field presence probabilities $p_i = p^F(\boldsymbol{x}_i, \theta)$, for $i \in I^F$, and Laboratory survival probabilities $p_i = p^L(\boldsymbol{x}_i, \theta)$, for $i \in I^L$. Thus in every case the log likelihood function can be written in the form

$$\ell(\theta) = c + \sum_{i \in I^F \cup I^L} [s_i \log(p_i) + (n_i - s_i) \log(1-p_i)], \qquad p_i = \begin{cases} p^F(\boldsymbol{x}_i, \theta) & \text{for } i \in I^F \\ p^L(\boldsymbol{x}_i, \theta) & \text{for } i \in I^L \end{cases};$$

only the dimension of θ and the form of the probability functions $p^F(\boldsymbol{x}_i, \theta)$ and $p^L(\boldsymbol{x}_i, \theta)$ change from one model to another. The explanatory variables in all cases consist of an intercept constant x_{i0}=1 and three recorded water chemistry indicators on a logarithmic scale: x_{i1}=pH, x_{i2}= $\log_{10}$ [Al], and x_{i3}= $\log_{10}$ [Ca].

2.1. $\mathbf{M}_4^F$: *A Logistic Regression Model for Field Observations*

The most commonly recommended statistical procedure for studying the relationship of binary data (in this case, indicators s_i of the presence of brook trout in a lake) to one or more explanatory variables is binary regression, especially multiple logistic regression (e.g., Cox, 1970; McCullagh and Nelder, 1989; Bishop, *et al.* 1975). In this procedure the success probabilities p_i depend on a *linear* functional of the vector $\boldsymbol{x}_i$ of explanatory variables $\{x_{ij}\}$, and so can be written in the form $p_i = G(\Sigma_j x_{ij} \beta_j^F)$ for some parameter vector β^F and specified cumulative distribution function (CDF) $G(x)$ (often the normal $\Phi(x)$ or the logistic $\Psi(x) = e^x/(1+e^x)$). We write the sum as the ith component $\boldsymbol{x}\beta_i^F$ of the matrix product $\boldsymbol{x}\beta^F$. Experimental evidence concerning β^F is reflected in the log likelihood function, which in the logistic case is simply $\ell_4^F(\beta^F) = \sum_{i \in I^F} \left[s_i \boldsymbol{x}\beta_i^F - n_i \log(1 + e^{\boldsymbol{x}\beta_i^F}) \right]$. Routine numerical methods suffice to find the point where $\ell_4^F(\beta^F)$ attains its maximum (the ***maximum likelihood estimator***, or MLE, $\hat{\beta}^F$). One manifestation of multicollinearity in model $\mathbf{M}_4^F$ is illustrated in the likelihood contour plot of Figure 2, showing the profile likelihood for the pH and aluminum logistic regression coefficients β_1^F and β_2^F in a neighborhood of $\hat{\beta}^F$. Note how poorly the logistic regression parameters are determined by the field observations.

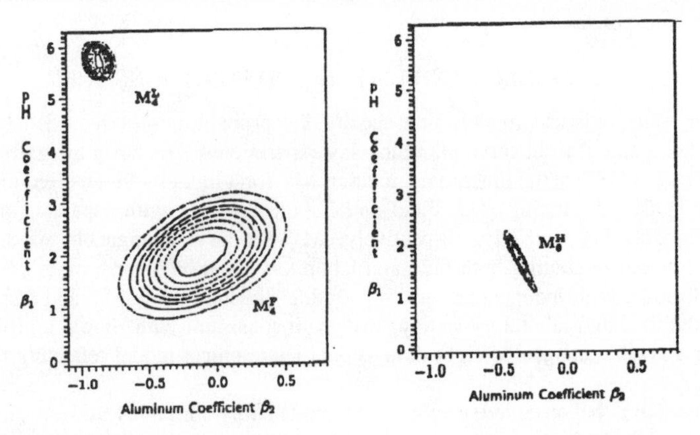

Figure 2. *Profile Likelihood Contour Plots for pH and [Al] coefficients in* $\mathbf{M}_8^B$ *and* $\mathbf{M}_6^H$.

Model:	M_8^B	M_6^H	M_{10}^B	M_8^H
Field:				
β_0^F	-10.04 ± 2.60	-8.53 ± 1.82	-17.34 ± 5.64	-20.63 ± 6.27
β_{pH}^F	1.72 ± 0.60	1.77 ± 0.34	2.02 ± 1.70	4.23 ± 1.29
β_{Al}^F	-0.10 ± 0.30	-0.28 ± 0.06	-0.88 ± 0.59	-0.67 ± 0.21
β_{Ca}^F	1.01 ± 1.51	0.22 ± 0.06	4.79 ± 3.16	$0.53 \pm 0\ 19$
$\log[r/(1-r)]$			3.15 ± 0.66	2.92 ± 0.49
Laboratory:				
β_0^L	-24.31 ± 0.93	-24.33	-24.31 ± 0.93	-24.32
β_{pH}^L	5.59 ± 0.20	5.59	5.59 ± 0.20	5.59
β_{Al}^L	-0.88 ± 0.06	$-0.88p$	-0.88 ± 0.06	-0.88
β_{Ca}^L	0.68 ± 0.13	0.69	0.68 ± 0.13	0.69
γ			10.10 ± 19.5	10.13 ± 20.3
Hierarchy:				
δ		-0.85 ± 0.42		-2.21 ± 0.85
$\log \sigma$		1.15 ± 0.19		0.28 ± 0.31

NOTE: Each entry is MLE±SE, the Maximum Likelihood Estimate plus-or-minus one Standard Error. The field and laboratory submodels of $\mathbf{M}_8^B$ are $\mathbf{M}_4^F$ and $\mathbf{M}_4^L$, respectively. In hierarchical models, $\beta^L = \sigma\beta^F + \delta$.

Table 1. *Maximum Likelihood Parameter Estimates.*

A more quantitative description of the multicollinearity in a $q+1$-dimensional model is given for the Frequentist by the Standard-Error covariance matrix SE for the sampling distribution of the MLE estimates $\hat{\beta}^F$, and for the Bayesian by the posterior covariance matrix $\mathfrak{T}$ for β^F under vague prior information. Under asymptotic normality SE $\approx [-\nabla^2 \ell_4^F(\hat{\beta}^F)]^{-1}$, the inverse observed information matrix at the MLE, while the *corrected* SE matrix (which is independent of the units in which the concentrations are measured) is the inverse of the lower-right $q \times q$ submatrix of $[-\nabla^2 \ell_4^F(\hat{\beta}^F)]$; this is proportional to the nonlinear analogue of what Weisberg (1985, p.43) calls the *corrected sum-of-squares* matrix $[\mathcal{X}^T \mathcal{X}]$ in linear models. In the field model $\mathbf{M}_4^F$ the corrected SE correlations derived from $[-\nabla^2 \ell_4^F(\hat{\beta}^F)]^{-1}$ are 0.274 for pH:[Al], -0.992 for pH:[Ca], and -0.367 for [Al]:[Ca] correlations. The high correlations among the parameters (especially pH:[Ca]) are one indication of multicollinearity in the model.

Another is the high condition number (39.9) of the corrected SE matrix. The calculation $\text{SE}(\hat{p}(\boldsymbol{x}))^2 \approx \boldsymbol{x}'[-\nabla^2 \ell_4^F(\hat{\beta}^F)]^{-1}\boldsymbol{x}$ shows that the condition number is approximately the ratio between the highest and lowest standard errors for predictions $\hat{p}(\boldsymbol{x}) = \Psi(\boldsymbol{x}^T \hat{\beta}^F)$ at all points $\boldsymbol{x}$ of fixed length $|\boldsymbol{x}|$; this ratio is about 40:1 in the present case.

The $\mathbf{M}_8^B$ column of Table 2 includes posterior means and (rather large) standard deviations for the parameters in model $\mathbf{M}_4^F$, while Table 3 and Figure 3 present the posterior predictive mean and standard deviation for $p(\boldsymbol{x}) = \Psi(\boldsymbol{x}^T \beta^F)$ at two points $\boldsymbol{x} = \boldsymbol{x}_a$ and $\boldsymbol{x} = \boldsymbol{x}_b$, and plots of the posterior predictive distributions, respectively, all under vague prior information. These illustrate that both the parameters and predictions at points like $\boldsymbol{x}_b$ far from the locus of multicollinearity are poorly determined by the data, while predictions at points like $\boldsymbol{x}_a$ close to the locus are well-determined— yet another illustration of multicollinearity.

Model:	$\mathbf{M}_8^B$	$\mathbf{M}_6^H$	$\mathbf{M}_{10}^B$	$\mathbf{M}_8^H$
Field:				
β_0^F	-10.54 ± 2.64	-9.23 ± 1.06	-21.61 ± 5.84	-22.69 ± 4.84
β_{pH}^F	1.76 ± 0.62	1.90 ± 0.21	2.75 ± 1.67	4.70 ± 0.99
β_{Al}^F	-0.12 ± 0.30	-0.30 ± 0.04	-1.34 ± 0.68	-0.74 ± 0.18
β_{Ca}^F	1.18 ± 1.56	0.23 ± 0.05	5.62 ± 3.62	0.57 ± 0.20
$\log[r/(1-r)]$			3.04 ± 0.57	2.70 ± 0.52
Laboratory:				
β_0^L	-24.38 ± 0.93	-24.28 ± 0.86	-24.38 ± 0.94	-24.24 ± 0.86
β_{pH}^L	5.61 ± 0.20	5.58 ± 0.19	5.61 ± 0.20	5.58 ± 0.19
β_{Al}^L	-0.89 ± 0.06	-0.88 ± 0.06	-0.88 ± 0.06	-0.87 ± 0.07
β_{Ca}^L	0.69 ± 0.13	0.68 ± 0.12	0.68 ± 0.14	0.66 ± 0.15
Hierarchy:				
δ		2.81 ± 0.61		2.70 ± 0.52
$\log \sigma$		1.08 ± 0.11		0.19 ± 0.22

NOTE: Each entry is Mean±SD, the posterior Mean plus-or-minus one posterior Standard Deviation, using a flat prior. In hierarchical models, $\beta^L = \delta + \sigma\beta^F$.

Table 2. *Posterior Parameter Means.*

Model:	$\mathbf{M}_8^B$	$\mathbf{M}_6^H$	$\mathbf{M}_{10}^B$	$\mathbf{M}_8^H$
Point x_a:	0.50 ± 0.08	0.52 ± 0.05	0.45 ± 0.12	0.58 ± 0.10
Point x_b:	0.49 ± 0.40	0.98 ± 0.01	0.62 ± 0.44	1.00 ± 0.01

NOTE: $x_a = [4.9, 1.3, 1.7]$ is near the locus of multicollinearity; $x_b = [5.5, -8.8, -0.3]$ is not.

Table 3. *Predictive Means & SD's of $p(x)$.*

2.2. $\mathbf{M}_4^L$: *A Logistic Regression Model for Laboratory Bioassay Experiments*

A logistic regression model similar to that above can be used to describe the number s_i of survivors from among n_i swimup fry in a controlled laboratory bioassay experiment (indexed by $i \in I^L$) with explanatory variable vector x_i. The likelihood function for the laboratory parameter vector β^L is simply $\ell_4^L(\beta^L) = \sum_{i \in I^L} \left[s_i x \beta_i^L - n_i \log(1 + e^{x\beta_i^L}) \right]$. The experimenter's ability to control the explanatory variables in x_i substantially reduces the multicollinearity which plagues the field observations, as reflected in the lower condition number (13.74) for the corrected SE matrix $[-\nabla^2 \ell_4^L(\hat{\beta}^L)]^{-1}$ and in the higher concentration of the profile likelihood function, as illustrated in Figure 2, indicating that the components of the regression vector β^L are much better-determined by the data (Warren-Hicks, 1990).

These two independent logistic regression models can be combined to form a single eight-dimensional model $\mathbf{M}_8^B$ with log likelihood function $\ell_8^B(\beta^F, \beta^L) = \ell_4^F(\beta^F) + \ell_4^L(\beta^L)$,

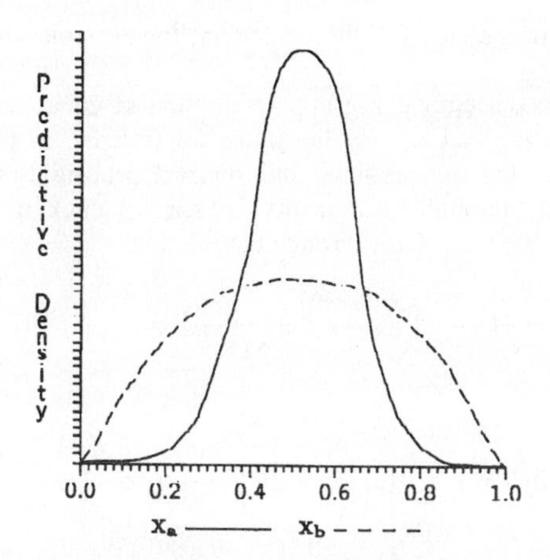

Figure 3. *Predictive distribution for $p(x_a)$ and $p(x_b)$ under vague prior information.*

or

$$\ell_8^B(\beta^F, \beta^L) = \sum_{i \in I^F} \left[s_i x \beta_i^F - n_i \log(1 + e^{x\beta_i^F}) \right] + \sum_{i \in I^L} \left[s_i x \beta_i^L - n_i \log(1 + e^{x\beta_i^L}) \right].$$

What is missing is some way to link the field and laboratory bioassay evidence together, i.e., to relate the logistic regression parameter vectors β^F governing the field presence and β^L for the bioassay survival. Without such a link the laboratory data are completely independent of the field data, and cannot help resolve the multicollinearity.

In the Bayesian approach the link is provided by a joint prior distribution for the two parameter vectors, $\pi^B(d\beta^F\, d\beta^L)$. Such a Bayesian approach seems especially appropriate for this application since it is only a *prior belief* that the laboratory data offer evidence about the field conditions that leads us to combine field and laboratory data.

At one extreme we might consider the case in which β^F and β^L are *independent* under the prior distribution, with conditional distribution $\pi^{L|F}(d\beta^L|\beta^F) = \pi^L(d\beta^L)$; in this case β^F and β^L would remain independent under the posterior distribution, and the laboratory data contribute nothing toward resolving the multicollinearity. At the other extreme we might consider a naïve pooled model with a prior distribution under which $\beta^F \equiv \beta^L$, with a unit point mass for the conditional distribution $\pi^{L|F}(d\beta^L|\beta^F) = \delta(\beta^L - \beta^F)d\beta^L$; in this case β^F and β^L are *identical* and laboratory bioassay data serves as a perfect substitute for field observations. We know of no biological argument that would justify an assumption that $\beta^L \equiv \beta^F$.

2.3. M_6^H: *An Exchangeable Hierarchical Bayesian Model*

A more appealing approach than either extreme is the intermediate course of a Bayesian analysis with a joint prior $\pi^B(d\beta^F\, d\beta^L)$ expressing the prior belief that, while the survival rates of swimup fry in laboratory bioassay experiments might be quite different from the fractions of lakes that support brook trout, still the "relative potency" of pH, inorganic aluminum, and calcium ought to be the same in the two settings: if a small increase of 0.01 in pH has the same effect as a 40% decrease in the inorganic aluminum concentration or a 3.9% increase in calcium concentration in the laboratory, for example, the same ought to hold in the field.

One way to describe a model embodying this assumption of equal relative potency is to introduce hyperparameters α_A and α_C for the *potencies* (relative to pH) of aluminum and calcium, respectively, so that the presence and survival probabilities depend on the three chemistry variables only through the quantity (which we think of as the "threat") $T_i = -\text{pH}_i + \alpha_A \log_{10}[\text{Al}_i] + \alpha_C \log_{10}[\text{Ca}_i]$, through the relations

$$\log\left(\frac{p_i^F}{1-p_i^F}\right) = \beta_0^F - \beta_1^F T_i \qquad \log\left(\frac{p_i^L}{1-p_i^L}\right) = \beta_0^L - \beta_1^L T_i$$

so the log likelihood is:

$$\begin{aligned}\ell(\alpha_A, \alpha_C, \beta_0^F, \beta_1^F, \beta_0^L, \beta_1^L) = &\sum_{i \in I^F} \left[s_i(\beta_0^F - \beta_1^F T_i) - n_i \log(1 + e^{(\beta_0^F - \beta_1^F T_i)}) \right] \\ &+ \sum_{i \in I^L} \left[s_i(\beta_0^L - \beta_1^L T_i) - n_i \log(1 + e^{(\beta_0^L - \beta_1^L T_i)}) \right].\end{aligned}$$

The Field and Laboratory submodels remain logistic, but they are no longer independent. A Bayesian analysis in which the laboratory data and field observations are related *only* through the relative potencies can be expressed in hierarchical form with a (hyper)prior distribution $\pi^\alpha(d\alpha_A\, d\alpha_C)$ and conditional distributions $\pi^{F|\alpha}(d\beta_0^F\, d\beta_1^F | \alpha_A, \alpha_C)$ and $\pi^{L|\alpha}\,(d\beta_0^L\, d\beta_1^L |\, \alpha_A\, \alpha_C)$. If these are identical the resulting model will be *exchangeable*, and will treat laboratory data and field observations in an entirely symmetric fashion.

Since our concern in this application is *not* symmetric, we use an equivalent but more convenient choice of parameters: we set $\theta = (\beta_0^F, \beta_1^F, \beta_2^F, \beta_3^F, \delta, \sigma) \in \mathbb{R}^6$, identify $\delta \in \mathbb{R}$ with $[\delta, 0, 0, 0]^T \in \mathbb{R}^4$, and treat $\beta^L = \sigma\beta^F + \delta$ as a derived quantity, leading to the log likelihood expression $\ell_6^H(\beta^F, \delta, \sigma) = \ell_4^F(\beta^F) + \ell_4^L(\sigma\beta^F + \delta)$, or

$$\begin{aligned}\ell_6^H(\beta_0^F, \beta_1^F, \beta_2^F, \beta_3^F, \delta, \sigma) = &\sum_{i \in I^F} \left[s_i \boldsymbol{x} \beta_i^F - n_i \log(1 + e^{\boldsymbol{x}\beta_i^F}) \right] \\ &+ \sum_{i \in I^L} \left[s_i(\sigma \boldsymbol{x} \beta_i^F + \delta) - n_i \log(1 + e^{(\sigma \boldsymbol{x} \beta_i^F + \delta)}) \right].\end{aligned}$$

Table 2 shows the posterior parameter means and standard deviations under the improper diffuse prior distribution $\pi^F(d\beta^F\, d\delta\, d\sigma) = d\beta_0^F\, d\beta_1^F\, d\beta_2^F\, d\beta_3^F\, d\delta\, d\sigma$, which some might regard as noninformative about β^F, σ, and δ.

It is possible to regard the role of the bioassay data as that of generating an *informative* prior distribution $\pi^\star(d\beta^F)$ for the field regression reflecting whatever limited information about the field setting is available from the laboratory bioassay experiments. The measure $\pi^\star(d\beta^F)$ is just the posterior distribution for β^F given $\boldsymbol{s}^L = \{s_i : i \in I^L\}$, with density:

$$\begin{aligned}\pi^\star(\beta^F) &\propto \iint \exp\left(\ell_4^L(\sigma\beta^F + \delta)\right) d\delta\, d\sigma \\ &= \iint \exp\left(\sum_{i \in I^L} [s_i(\sigma \boldsymbol{x}_i \beta^F + \delta) - n_i \log(1 + e^{(\sigma \boldsymbol{x}_i \beta^F + \delta)})] \right) d\delta\, d\sigma\end{aligned}$$

2.4. *Beyond Exchangeability*

A recorded presence of fish in the field data set would appear to be reliable, indicating that field investigators had caught at least one fish in the lake, but a false absence would be recorded if fish were in fact present in a lake but were not caught. In addition, fish may be absent from some lakes with quite suitable explanatory variables $\boldsymbol{x}_i$ for reasons unrelated to the explanatory variables. Thus the logistic regression model is in some ways unsuited to the field data set, since there is a substantial probability of a recorded absence even in a lake otherwise perfectly suited to brook trout. Under the logistic model the probability of presence $p^F(\boldsymbol{x}, \theta) = \Psi(\boldsymbol{x}^T \beta^F)$ is nearly one if $\boldsymbol{x}^T \beta^F$ is large, so "absence" ($s_i = 0$) lakes with otherwise suitable predictor variables $\boldsymbol{x}_i$ become highly influential and drive the likelihood to zero away from $\beta^F \approx 0$. The probit model, with its sharper tails, is even more susceptible to this phenomenon.

A simple remedy is to introduce a new parameter $r \in [0, 1]$, which might be interpreted as the conditional probability of a recorded presence (s_i=1) given an actual presence (or, perhaps, given water chemistry suitable for an actual presence), and to model the probability of recorded presence in the field as $p_i = r\Psi(\boldsymbol{x}\beta_i^F)$. Laboratory survival probability would remain $p_i = \Psi(\boldsymbol{x}\beta_i^L)$, since the adjustment is inappropriate under controlled conditions. This addition to the field-only logistic model yields a model $\mathbf{M}_5^F$ with parameter vector $\theta = [\beta^F, r] \in \mathbb{R}^5$.

In most of the bioassay trials the fish responded (i.e., died) within the first 3–5 days of the experiment, if they responded at all. For some individual bioassays, however, the empirical survival function still seemed to be falling at the end of the experiment. This suggests that some observations may be censored on the right, a possibility not reflected in the logistic model. One remedy is to introduce a new parameter $\gamma \geq 0$ and a simple survival model featuring a constant hazard h, equal to γ times the maximum of zero and an unobserved logistically-distributed random variable with location parameter $-\boldsymbol{x}\beta_i^L$ and unit scale. As $\gamma \to \infty$ this reverts to the logistic regression model, while for finite γ it leads to the following expression for the probability of survival for at least t_i days:

$$p_i = \int_{-\infty}^{\infty} e^{-\gamma t_i (h \vee 0)} \frac{e^{\boldsymbol{x}\beta_i^L + h}}{(1 + e^{\boldsymbol{x}\beta_i^L + h})^2}\, dh = z_i + \left(\frac{z_i}{1 - z_i}\right)^{\gamma t_i} B_{1-z_i}(1+\gamma t_i, 1-\gamma t_i)$$

where $z_i = e^{\boldsymbol{x}\beta_i^L}/(1+e^{\boldsymbol{x}\beta_i^L})$ is the probability of indefinite survival and the remaining term is the probability of a censored observation, i.e., an unobserved death subsequent to the t_i-day observation period. $B_z(a, b)$ denotes the unnormalized incomplete Beta function (Abramowitz and Stegun, 1964, *p.* 263), well-defined despite the unusual negative argument. With this addition to the laboratory logistic model we get the model $\mathbf{M}_5^L$ with parameter vector $\theta = [\beta^L, \gamma]$.

The models $\mathbf{M}_5^F$ and $\mathbf{M}_5^L$ can be combined to form a single ten-parameter generalized logistic model $\mathbf{M}_{10}^B$ with log likelihood function

$$\ell_{10}^B(\beta^F, \beta^L, r, \gamma) = \ell_5^F(\beta^F, r) + \ell_5^L(\beta^L, \gamma) = \sum_{i \in I^F \cup I^L} [s_i \log p_i + (n_i - s_i) \log(1 - p_i)]$$

where $p_i = r\, e^{\boldsymbol{x}\beta_i^F}/(1 + \boldsymbol{x}\beta_i^F)$ if $i \in I^F$, $p_i = z_i + \left(\frac{z_i}{1-z_i}\right)^{\gamma t_i} B_{1-z_i}(1+\gamma t_i, 1-\gamma t_i)$ if $i \in I^L$, and $z_i = e^{\boldsymbol{x}\beta_i^L}/(1 + e^{\boldsymbol{x}\beta_i^L})$.

2.5. $\mathbf{M}_8^H$: *A Non-exchangeable Hierarchical Model*

When the hierarchical model $\mathbf{M}_6^H$ is extended as above to overcome the inability of its simple logistic submodels to reflect unexplained absences in the Field data and censored deaths in the Laboratory data, the result is an eight-parameter model $\mathbf{M}_8^H$ whose Field and Laboratory submodels are no longer exchangeable. In the parameterization $\theta = (\beta^F, \delta, \sigma, r, \gamma) \in \mathbb{R}^8$, the likelihood is $\ell_8^H(\beta^F, \delta, \sigma, r, \gamma) = \ell_5^F(\beta^F, r) + \ell_5^L(\sigma\beta^F + \delta, \gamma)$, or

$$\begin{aligned}\ell_8^H(\beta_0^F, \beta_1^F, \beta_2^F, \beta_3^F, \delta, \sigma, r, \gamma) &= \sum_{i\in I^F}\left[s_i \log \frac{r\, e^{\boldsymbol{x}\beta_i^F}}{1+e^{\boldsymbol{x}\beta_i^F}} + (1-s_i)\log\left(1 - \frac{r\, e^{\boldsymbol{x}\beta_i^F}}{1+e^{\boldsymbol{x}\beta_i^F}}\right)\right] \\ &+ \sum_{i\in I^L} s_i \log\left(z_i + \left(\frac{z_i}{1-z_i}\right)^{\gamma t_i} B_{1-z_i}(1+\gamma t_i, 1-\gamma t_i)\right) \\ &+ \sum_{i\in I^L} (n_i - s_i)\log\left(1 - z_i - \left(\frac{z_i}{1-z_i}\right)^{\gamma t_i} B_{1-z_i}(1+\gamma t_i, 1-\gamma t_i)\right)\end{aligned}$$

where $z_i = \dfrac{\exp(\sigma\boldsymbol{x}\beta_i^F + \delta)}{1+\exp(\sigma\boldsymbol{x}\beta_i^F + \delta)}$.

3. NUMERICAL METHODS

Although variable-metric optimization methods may prove unstable in large logistic regression problems because of the near singularity of the log likelihood's Hessian away its maximum, we were able to find maximum likelihood estimates for the parameters of each model using an interactive modified Newton Raphson routine developed by E. Stallard (Manton and Stallard, 1988). In most cases we were also able to maximize the log likelihood using the Polak-Ribiere variant of the Fletcher-Reeves conjugate gradient method (Fletcher, 1987) as implemented by Press *et al.* (§10.6, 1986) or the conjugate gradient method (Powell, 1977), after making parameter transformations (e.g., replacing r with its logistic $\log\frac{r}{1-r}$) to render the optimization problems unconstrained. The results are summarized in Table 1.

Calculating the Bayesian posterior mean for $p^F(\boldsymbol{x})$ and for the parameter vector in model $\mathbf{M}_8^H$ requires evaluating the ratio of two eight-dimensional integrals. Tensor product quadrature rules are hopelessly inefficient for such a task, but Monte Carlo methods with appropriate variance-reduction techniques (described in Wolpert, 1991) permit us to calculate simultaneously $\bar{p}^F(\boldsymbol{x}_a)$, $\bar{p}^F(\boldsymbol{x}_b)$, the eight means in $\bar{\beta}^F$, and the eight variances and twenty-eight covariances in Σ to a precision of $\pm 1\%$ in a few minutes on a desk top Unix workstation. Similar techniques (also described in Wolpert, 1991) were used for calculating weighted kernel estimates of the univariate and bivariate marginal density functions for $p^F(\boldsymbol{x})$ and for β^F.

The most important calculations arising in the Bayesian analysis, those of the predictive mean $\bar{p}^F(\boldsymbol{x})$ and density $f_{\boldsymbol{x}}(p|\boldsymbol{x})$ of $p = p^F(\boldsymbol{x})$, are also the most inconvenient ones since they must be done anew for each value of $\boldsymbol{x}$ that proves to be interesting. In many cases the logistic $\eta = \log\frac{p}{1-p}$ is well approximated by a normal distribution with mean $\mu_{\boldsymbol{x}} = \boldsymbol{x}^T\bar{\beta}^F$ and variance $\sigma_{\boldsymbol{x}}^2 = \boldsymbol{x}^T\Sigma\boldsymbol{x}$ in terms of the posterior mean vector $\bar{\beta}^F$ and covariance matrix Σ for β^F, so the predictive density is approximately

$$f_{\boldsymbol{x}}(p|\boldsymbol{x}) \approx \frac{1}{p(1-p)\sigma_{\boldsymbol{x}}\sqrt{2\pi}} \exp\left(-\left(\log\frac{p}{1-p} - \mu_{\boldsymbol{x}}\right)^2/2\sigma_{\boldsymbol{x}}^2\right).$$

With enough data $p^F(\boldsymbol{x})$ too would have nearly a normal distribution, with approximate mean $\bar{p}^F(\boldsymbol{x}) \approx e^{\mu \boldsymbol{x}}/(1+e^{\mu \boldsymbol{x}})$ and variance $\sigma^2_{\boldsymbol{x}} \approx \bar{p}^F(\boldsymbol{x})^2(1-\bar{p}^F(\boldsymbol{x}))^2\boldsymbol{x}^T\Sigma\boldsymbol{x}$, but weak convergence to normality is much slower for $p^F(\boldsymbol{x})$ than for its logistic.

4. DISCUSSION

Four competing models have been introduced for the field and laboratory data in this study: the independent models $\mathbf{M}_8^B$ and $\mathbf{M}_{10}^B$, under which laboratory and field data are unrelated, and the hierarchical models $\mathbf{M}_6^H$ and $\mathbf{M}_8^H$, embodying an assumption of equal relative potency linking the laboratory and field submodels. As expected, the parameters appear better-determined by the data in the hierarchical models $\mathbf{M}_6^H$ and $\mathbf{M}_8^H$ than in the independent models, since they impose the two dimensional constraint of equal relative potency. Posterior means and standard deviations of the parameters in the models $\mathbf{M}_8^B$, $\mathbf{M}_{10}^B$, $\mathbf{M}_6^H$, and $\mathbf{M}_8^H$ are presented in Table 2. The posterior standard deviations display the expected reduction in parameter uncertainty for the models that combine laboratory and field data ($\mathbf{M}_6^H$, $\mathbf{M}_8^H$) compared to those that do not ($\mathbf{M}_8^B$, $\mathbf{M}_{10}^B$). In the combined models the laboratory information serves to help resolve the individual effects of pH, inorganic Al, and Ca on fish response to acidification. These individual effects are masked in the field-only models by the multicollinearity.

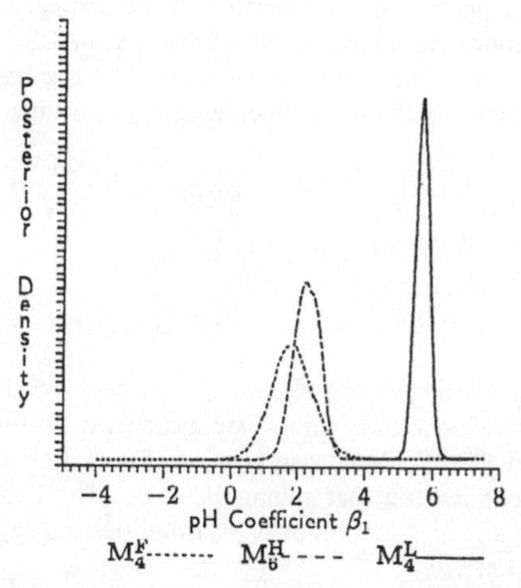

Figure 4. *Posterior Marginal Densities for Field pH Coefficient in* $\mathbf{M}_6^H$, *Field and Lab pH Coefficients in* $\mathbf{M}_8^B$.

Figure 4 presents overlaid plots of the posterior marginal densities for the field pH coefficient β_1^F from $\mathbf{M}_8^B$ and $\mathbf{M}_6^H$, along with the density for laboratory pH coefficient β_1^L from $\mathbf{M}_8^B$. The distribution of β_1^F under $\mathbf{M}_6^H$ is centered at about the same value as under $\mathbf{M}_8^B$, but the spread is much narrower (i.e., the uncertainty is much less) reflecting the influence of the laboratory data. Similar features were exhibited by plots (not shown) for the inorganic Al and Ca coefficients.

Opinions differ on what (if anything) would constitute a noninformative prior distribution, especially in multidimensional problems (see Bernardo, 1979, and Berger and Bernardo, 1989). We analyzed the data twice, first with a uniform prior and then with the Jeffreys prior proportional to the square-root of the determinant of the expected (Fisher) information matrix.

Since the results for the two priors were quite similar and the Jeffreys prior is expensive to compute we report only the results with a uniform prior.

4.1. *Prediction*

These models are intended to help investigators predict the effects on wildlife of various possible changes in the levels of environmental pollutants; as such, they should be judged on the basis of their simplicity and on their fit to the *field* data, and in particular on their ability to *predict* the presence or absence of trout populations for specified water chemistries that may differ from those in the present data set.

The field submodels of $\mathbf{M}_{10}^B$ and $\mathbf{M}_8^B$ (namely, $\mathbf{M}_4^F$ and $\mathbf{M}_5^F$) are simpler than the hierarchical models $\mathbf{M}_6^H$ and $\mathbf{M}_8^H$ in that they have fewer parameters. They also fit the existing field data set better, since the hierarchical models trade off some fit of the field data in order to fit the laboratory bioassay data better. Thus it would seem that the field-only models win on both grounds. But the hierarchical models were introduced in the hope that they would offer better predictions and less uncertainty in *future* data sets that differ from the present one by exhibiting less (or different) correlation among the predictor variables, so a definitive assessment of their performance in meeting this objective will have to await the availability of suitable data sets. While we cannot compare the models' predictions in a future data set, we *can* compare their predictions in a portion of the present data set reserved for that purpose: the 136 observations from the second (1986–1987) ALSC survey.

Each of the models introduced above can be viewed as a submodel of $\mathbf{M}_{10}^B$, in that its log likelihood can be written (up to an additive constant) as a function of ℓ_{10}^B:

$$\begin{array}{llllll}
\ell_8^B(\beta^F,\beta^L) & = & \ell_{10}^B(\beta^F,\beta^L,1,\infty) & \ell_{10}^B(\beta^F,\beta^L,r,\gamma) & = & \ell_{10}^B(\beta^F,\beta^L,r,\gamma) \\
\ell_4^F(\beta^F) & = & \ell_{10}^B(\beta^F,\hat\beta^L,1,\infty) & \ell_5^F(\beta^F,r) & = & \ell_{10}^B(\beta^F,\hat\beta^L,r,\infty) \\
\ell_4^L(\beta^L) & = & \ell_{10}^B(\hat\beta^F,\beta^L,1,\infty) & \ell_5^L(\beta^L,\gamma) & = & \ell_{10}^B(\hat\beta^F,\beta^L,1,\gamma) \\
\ell_6^H(\beta^F,\delta,\sigma) & = & \ell_{10}^B(\beta^F,\sigma\beta^F+\delta,1,\infty) & \ell_8^H(\beta^F,\delta,\sigma,r,\gamma) & = & \ell_{10}^B(\beta^F,\sigma\beta^F+\delta,r,\gamma)
\end{array}$$

Thus any model with log likelihood ℓ_6^H, ℓ_8^H, or ℓ_8^B can be viewed as model $\mathbf{M}_{10}^B$ with the parameter θ constrained to lie on a lower-dimensional submanifold $\Theta\subset\mathbb{R}^{10}$, and any Bayesian model with log likelihood ℓ_6^H, ℓ_8^H, or ℓ_8^B can be viewed as a Bayesian model for $\mathbf{M}_{10}^B$ with a prior distribution concentrated on that submanifold Θ. The choice among several such models can be viewed as a choice among prior distributions $\pi_j(d\theta)$ in a single uncontested model with log likelihood $\ell_{10}^B(\theta)$.

For Bayesians the task of comparing two or more possible prior distribution measures $\pi_j(d\theta)$ is straightforward, given a log likelihood function $\ell(\theta)$ and hyperprior probabilities p_j for the measures $\pi_j(d\theta)$. The posterior probability of the jth alternative is simply $p_j L(\pi_j)/\Sigma_i p_i L(\pi_i)$, proportional to the p_j-weighted integrated likelihood $L(\pi_j) = \int_\Theta \exp\left(\ell(\theta)-\ell(\hat\theta)\right)\pi_j(d\theta)/\int_\Theta \pi_j(d\theta)$ (the term $\ell(\hat\theta)$ is just a normalization factor). In particular, a comparison of any two such measures π_0 and π_1 depends on the data only through the *Bayes factor*

$$B = \frac{\int_\Theta \exp\left(\ell(\theta)-\ell(\hat\theta)\right)\pi_0(d\theta)}{\int_\Theta \exp\left(\ell(\theta)-\ell(\hat\theta)\right)\pi_1(d\theta)} \frac{\int_\Theta \pi_1(d\theta)}{\int_\Theta \pi_0(d\theta)} = \frac{L(\pi_0)}{L(\pi_1)}.$$

If all the prior distributions $\pi_j(d\theta)$ are proper (so numerator and denominator in this expression are finite), we can calculate the posterior probability of each measure and either choose the most probable one, *a posteriori*, or choose the one that minimizes some expression of loss that includes a cost for complexity.

Unfortunately the numerator or denominator or both above will be infinite if either prior measure π_j is improper, and here (as in most problems) noninformative prior distributions are improper. We circumvent this problem by calculating integrated likelihoods for a *reserved data set*, with respect to the (presumably proper) *posterior* of each measure π_j given a "training" data set.

Let I^R be a set indexing the lakes in the reserved data set, disjoint from $I^F \cup I^L$, and for $i \in I^R$ denote by $\boldsymbol{x}_i$ the recorded explanatory variables in the ith lake and by s_i an indicator of fish presence there. For any model (say, $\mathbf{M}_q^X$) we can compare $\{s_i\}$ with the posterior predictive means $\bar{p}_q^X(\boldsymbol{x}_i)$ given the "training" data sets I^F and I^L, under vague prior distributions. Let $\ell_5^R(\theta)$ be the same five-parameter log likelihood function as $\ell_5^F(\theta)$, but using the reserved data set I^R instead of the field data set I^F, and let $\hat{\theta}_5^R$ be the MLE. Some possible criteria for assessing the fit include:

1. Squared Residuals: MSR $= \Sigma_{i \in I^R} |s_i - \bar{p}_q^X(\boldsymbol{x}_i)|^2 / N$
2. Deviance: D $= 2(\ell_5^R(\hat{\theta}_5^R) - \ell_5^R(\hat{\theta}_q^X))$
3. Chi Square: χ^2 $= \Sigma_{i \in I^R} |s_i - \bar{p}_q^X(\boldsymbol{x}_i)|^2 / (\bar{p}_q^X(\boldsymbol{x}_i)(1 - \bar{p}_q^X(\boldsymbol{x}_i)))$
4. Predictive Integrated Likelihood: PIL $= \dfrac{\int \exp\left(\ell_5^R(\theta) - \ell_5^R(\hat{\theta}_5^R)\right) \exp\left(\ell_{10}^B(\theta)\right) \pi_q^X(d\theta)}{\int \exp\left(\ell_{10}^B(\theta)\right) \pi_q^X(d\theta)}$
5. Average Error: $\overline{\Delta}$ $= \Sigma_{i \in I^R} [s_i - \bar{p}_q^X(\boldsymbol{x}_i)] / N$

where $N = |I^R|$ is the number of observations in the reserved data set and where π_q^X is $\mathbf{M}_q^X$'s prior distribution, regarded as a measure on $\mathbb{R}^{10}$. The average error criterion assesses only model *calibration* while the others assess both calibration and *discrimination*. For models and data sets with moderate values of $p^F(\boldsymbol{x}_i)$ (bounded away from 0 and 1) criteria 1.–4. give similar results, but MSR (or, equivalently, the Brier score) is least sensitive and χ^2 most sensitive to prediction errors for extreme values of $p^F(\boldsymbol{x}_i)$.

Model: Criterion:	$\mathbf{M}_8^B$	$\mathbf{M}_6^H$	$\mathbf{M}_{10}^B$	$\mathbf{M}_8^H$
MSR:	0.0682	0.0729	0.0748	0.0748
D:	7.20	11.83	7.76	9.34
χ^2:	1.310	0.825	2.347	0.723
PIL:	0.0316	0.0020	0.0025	0.00003
$\overline{\Delta}$:	0.0199	0.0200	0.0165	0.0153

1. Criteria are Mean Squared Residuals, Deviance, Chi Square, Predictive Integrated Likelihood, and Average Error
2. Small values indicate good fit for MSR, D, and χ^2; large values for PIL; and values near zero for $\overline{\Delta}$.

Table 4. *Model Evaluation Criteria.*

By these criteria (summarized in Table 4) the hierarchical models are comparable to the field-only models, despite the influence of the laboratory bioassay data, and the modified logistic models $\mathbf{M}_5^F$ and $\mathbf{M}_8^H$ are outperformed by the simpler models $\mathbf{M}_4^F$ and $\mathbf{M}_6^H$ (apparently because the 1986–1987 ALSC survey data shows fewer anomalous absences than does the 1984–1985 survey; we hope to explore this further in the future). Since both surveys suffer from the same collinearity this does not offer an opportunity to confirm whether the hierarchical models fulfill their intended purpose, but it does show that these models fit existing data sets almost as well as field-only models.

5. CONCLUSIONS

The problems created by multicollinearity in a field data set were addressed using hierarchical Bayesian models for combining laboratory and field observations. Parameters in these models were shown to be well determined by the data, and model predictions of the presence and absence of brook trout in a reserved, independent data set were shown to be comparable to those of conventional models. Standard logistic regression models were modified to accommodate observed and anticipated features of the field and laboratory settings that were inconsistent with the naïve models, including false and coincidental absences in the field data and censored deaths in the laboratory data.

Data sets in the natural, social, and medical sciences often feature a collinearity structure that evolves over time due to changes in the environmental chemistry, changes in the social, economic, and demographic features of a population, and changes in medical technology, respectively. The methods presented here offer a way to resolve uncertainties arising in prediction problems with multicollinear data.

Bayesian methods provide a flexible and coherent framework for combining disparate information in the face of uncertainty about both the model and data. The predictive densities and marginal posterior distributions of model parameters that arise in the Bayesian analysis can help lend insight into the system under study. Model predictions and integrated likelihoods within a reserved portion of the data set provide a basis for model comparison and assessment.

6. ACKNOWLEDGEMENTS

We thank Joan Baker, Jim Berger, Michael Lavine, and Ken Reckhow for stimulating conversations and observations during the course of this work. We thank Neil Hutchinson, Eric Stallard, Joan Baker, the Electric Power Research Institute (EPRI), and the Adirondack Lake Survey Corporation (ALSC) for sharing data and computer programs. This work was partially supported by National Science Foundation grants DMS-8903842 and SES-8921227, the Environmental Protection Agency (contract 68-03-3439, as a part of its contribution to the National Acid Precipitation Assessment Program's 1990 *Assessment Report to Congress*, reporting on the regional biological effects of acidification), the Electric Power Research Institute (EPRI), the Sport Fishing Institute, and Kilkelly Associates, Inc.

We would like to thank the Duke University Institute of Statistics and Decision Sciences and the National Biomedical Simulation Resource for the use of their computing facilities. Machine readable FORTRAN 77 source code is available from the authors for the optimization, integration, and density estimation methods employed in this study, along with detailed descriptions of the data sets.

REFERENCES

Abramowitz, M. and Stegun, I. A. (1964). *Handbook of Mathematical Functions*. Washington, DC: National Bureau of Standards Press.

Baker, J. P. (1981). *Aluminum Toxicity to Fish as Related to Acid Precipitation and Adirondack Surface Water Quality*. Ph.D. Thesis, Cornell University, Ithaca.

Baker, J. P. and Schofield, C. L. (1982). Aluminum toxicity to fish in acidic waters. *Water, Air, Soil Pollution* **18**, 289–309.

Baker, J. P., Creager, C. S., Warren-Hicks, W., Christensen, S. W. and Godbout, L. (1988). *Identification of Critical Values for Effects of Acidification on Fish Populations*. Washington, DC: US Environmental Protection Agency.

Baker, J. P., Bernard, D. P., Christensen, S. W., Sale, M. J., Freda, J., Heltcher, K., Rowe, L., Scanlon, P., Stokes, P., Suter, G. and Warren-Hicks, W. (1990). *Biological Effects of Changes in Surface Water Acid-Base Chemistry*. Washington, DC: State of Science and Technology.

Baker, J. P., Christensen, S. W., Driscoll, C. D., Gallagher, J., Gherini, S., Holsapple, J., Kretser, W., Munson, R., Newton, R., Reckhow, K. and Schofield, C. (1990). *Final Report to the Adirondack Lakes Survey Corporation*. Ray Brook, NY: ALSC.

Berger, J. O. (1985). *Statistical Decision Theory and Bayesian Analysis*. New York: Springer.

Berger, J. O. and Wolpert, R. L. (1988). *The Likelihood Principle*. New York: IMS Press.

Bernardo, J. M. (1979). Reference posterior distributions for Bayesian inference. *J. Roy. Statist. Soc. B* **41**, 113–147, (with discussion).

Bishop, Y. M. M., Fienberg, S. E. and Holland, P. W. (1975). *Discrete Multivariate Analysis: Theory and Practice*. Cambridge MA: MIT Press.

Cox, D. R. (1970). *Analysis of Binary Data*. London: Chapman and Hall.

DuMouchel, W. H. and Harris, J. E. (1983). Bayes methods for combining the results of cancer studies in humans and other species. *J. Amer. Statistical Soc.* **78**, 293–315.

Efron, B. and Morris, C. (1972). Empirical Bayes on vector observations: an extension of Stein's method. *Biometrika* **59**, 335–347.

Efron, B. and Morris, C. (1973) Combining possibly related estimation problems. *J. Roy. Statist. Soc. B* **35**, 379–421, (with discussion).

Fletcher, R. (1987). *Practical Methods of Optimization*. New York: Wiley.

Good, I. J. (1980). Some history of the hierarchical Bayesian methodology. *Bayesian Statistics* (J. M. Bernardo, M. H. DeGroot, D. V. Lindley and A. F. M. Smith, eds.), Valencia: University Press, 489–519, (with discussion).

Hammersley, J. M. and Handscomb, D. C. (1964). *Monte Carlo Methods*. London: Methuen.

Holtze, K. E. and Hutchinson, N. J. (1989). Lethality of low pH and Al to early life stages of six fish species inhabiting precambrian shield waters in Ontario. *Can. J. Fish. Aquat. Sci.* **46**, 1188–1202.

IMSL (1987). *MATH/LIBRARY User's Manual, Version 1.0*. Houston, TX: IMSL.

Ingersoll, C. G. (1986). *The Effects of pH, Aluminum, and Calcium on Survival and Growth of Brook Trout (Salvelinus Fontinalis) Early Life Stages*. Ph.D. Thesis, University of Wyoming, Laramie.

Ingersoll, C. G., Galley, D., Mount, D., Mueller, M., Fernández, J. Hocker, J. and Bergman, H. (1990). Aluminum and acid toxicity to two strains of brook trout (*Salvelinus fontinalis*). *Can. J. Fish. Aquatic Sci.* (to appear).

Kalbfleisch, J. D. and Prentice, R. L. (1980). *The Statistical Analysis of Failure Time Data*. New York: Wiley.

Lindley, D. V. and Smith, A. F. M. (1972). Bayes estimates for the linear model. *J. Roy. Statist. Soc. B* **34**, 1–41, (with discussion).

Manton, K. G. and Stallard, E. (1988). *Chronic Disease Modeling*. London: Griffin.

McCullagh, P. and Nelder, J. A. (1989). *Generalized Linear Models*. London: Chapman and Hall.

Powell, M. J. D. (1977) Restart procedures for the conjugate gradient method. *Mathematical Programming* **12**, 241–254.

Press, W. H., Flannery, B. P., Teukolsky, S. A. and Vettering, W. T. (1986). *Numerical Recipes: The Art of Scientific Computing*. Cambridge: University Press.

Reckhow, K. H. (1988). A comparison of robust Bayes and classical estimators for regional lake models of fish response to acidification. *Water Resources Research* **24**, 1061–1068.

Reckhow, K., Black, R. W., Stockton, Jr. T. B., Vogt, J. D. and Wood, J. G. (1987). Empirical models of fish response to lake acidification. *Can. J. Fish. Aquat. Sci.* **44**, 1432–1442.

Smith, A. F. M., Skene, A. M., Shaw, J. E. H., Naylor, J. C. and Dransfield, M. (1985). The implementation of the Bayesian paradigm. *Commun. Statist. Theor. Meth.* **14**, 1079–1102.

Smith, A. F. M. and Spiegelhalter, D. J. (1980). Bayes factors and choice criteria for linear models. *J. Roy. Statist. Soc. B* **42**, 213–220.

San Martini, A. and Spezzaferri, F. (1984). A predictive model selection criterion. *J. Roy. Statist. Soc. B* **46**, 296–303.

Spiegelhalter, D. J. and Smith, A. F. M. (1982). Bayes factors for linear and log-linear models with vague prior information. *J. Roy. Statist. Soc. B* **44**, 377–387.

Warren-Hicks, W. J. (1990). *Predicting Brook Trout Population Response to Acidification: Using Bayesian Inference to Combine Laboratory and Field Data*. Ph.D. Thesis, Duke University.

Weisberg, S. (1985). *Applied Linear Regression*. New York: Wiley.

Wolpert, R. L. (1991). Monte Carlo integration in Bayesian statistical analysis. *Contemporary Mathematics* **116**, 101–115.

DISCUSSION

K. CHALONER (*University of Minnesota, USA*)

This paper has everything that a good statistical paper should have – it has real data, it addresses an important scientific issue of how acid rain affects brook trout, and it also contains some novel ideas for statistical analysis. The central idea can be thought of as using a model for relevant laboratory data to calculate a posterior distribution to be used as a prior distribution for the analysis of the field data.

The authors use a logistic regression model, with three explanatory variables, pH, Aluminium content and Calcium content. The purpose of the analysis is to predict in the field. It is argued that using the laboratory data for information on the logistic regression coefficients of the field data will reduce the problems of prediction. In the field, data observations are on lakes and the response is the presence or absence of any brook trout in an individual lake. In the laboratory, the observations are on individual fish and their survival or death under controlled conditions.

The authors primarily use a likelihood where the field data and the laboratory data are modelled as independent logistic regressions with related coefficients. The coefficients for the field data, β^F, are taken to be proportional to the coefficients for the laboratory data, β^L, by a factor σ, and there is a possible shift in location by δ. They also do some additional modelling to incorporate misclassification and possible differences between the three sets of laboratory data. They use a diffuse prior distribution with this combined likelihood. As they point out, this would be equivalent to first using the laboratory data and a diffuse prior distribution to derive a posterior distribution which can then be used as a prior distribution for the analysis of the field data. The derived prior distribution, however, would also be partly improper as there is a lack of identifiability in the first analysis between σ and β^L.

The question that I would like to raise for discussion is "why make everything so complicated?" Why not do a straightforward Bayesian analysis using an informative prior distribution? Why not systematically gather *all* relevant information, including the three laboratory experiments, other experiments, data from other parts of the world and anything else relevant? All this information can be reviewed and an informative prior distribution for β^F can be specified directly. The prior distribution does not necessarily have to be very precise as just a little bit of prior information may deal with the prediction problems induced by multicollinearity. A range of prior distributions could be specified.

The advantages of proceeding this way include:

1. The analysis, and the integration for marginal and predictive distributions, will be simpler. There will be only five parameters: three regression coefficients, an intercept and possibly a misclassification probability.
2. As we only have five parameters it will be much easier to think about adding other variables to the model. This is observational data and other variables might help explain the presence or absence of brook trout. Obvious candidates would be other chemicals such as phosphorous and sulphates. It would be also be easier to incorporate the other field data, referred to by the authors, on lakes which were limed or stocked. Other

variables are all kept constant in the laboratory data but can vary tremendously in the field.

3. Using the laboratory data to help specify an informative prior distribution makes it easier to understand the influence of the laboratory data on inferences about β^F. Using a range of informative distributions and examining the range of predictions should be straightforward. Wolpert and Warren-Hicks' approach makes the very strong assumption that relates the laboratory data to the field data in a very specific way. While these two responses are related it is hard to justify this specific relationship. The response in the field data is presence or absence of fish in the lake and in the lab data is survival or death of an individual fish.
4. We are not interested in the laboratory data directly, but, if we do model it, we must check and examine all the model assumptions.

So to summarize, while I completely agree that we should always try to use all the information available on any scientific problem I am not completely convinced that we should incorporate data which is not directly relevant into the likelihood rather than into the prior distribution. But this is, perhaps, debatable.

I do have some further very minor comments. First, the authors analyse 177 lakes of field data and keep 136 for model verification and assessment. They refer to another 1156 cases which were not analysed because the lakes were stocked or limed or had missing data or had no brook trout in the past. I would have thought that these cases would be at least as important, and even more important to include in the analysis than the laboratory data sets. Also, in the modelling of misclassification it perhaps might be helpful to take the probability of misclassification to be related to the explanatory variables, perhaps through the probability of presence, as if fish are scarce in a lake they will be harder to catch and so misclassification is more likely.

My final comment is that it is always easier to criticise data analysis than to do it and the authors should be congratulated on an excellent paper on a very important topic.

G. CONSONNI (*Università di Pavia, Italy*) and
P. VERONESE (*Università L. Bocconi, Milano, Italy*)

We would like to comment briefly on the authors' remarks concerning Bayesian model choice using noninformative priors (see sections 4.1.2 and 4.1.3). They correctly point out that, when π_0 and π_1 are improper, both the numerator and denominator of the Bayes factor are infinite. To "circumvent the problem" they compute, using a *reserve dataset*, proper posteriors for each π_i and take these as "priors" for the actual computation of the Bayes factor. Clearly, in so doing, they are no longer performing a "noninformative Bayesian analysis". We would like to point out, however, that the indeterminate from ∞/∞ of the Bayes factor can be solved for, by viewing improper priors as "limits" of restricted proper priors defined on subsets of the parameter space converging to the whole parameter space. Unfortunately, this procedure depends on the choice of the sequence of subsets which is typically an unnatural enterprise to do. An indirect default-solution to this problem is represented by the so-called imaginary training sample device of Spiegelhalter and Smith (1982).

For details on the general procedure for computing Bayes with improper priors see Consonni and Veronese (1992), in this volume.

A. E. GELFAND (*University of Connecticut, USA*)

As the authors note, in the case of improper priors choice amongst models cannot be handled by the use of Bayes factors. As an alternative, they propose computing predictive distributions for a "reserved" data set with respect to (proper) posteriors obtained from a "training" data set. The paper by Gelfand, Dey and Chang (1992), also in this volume, suggests another cross-validation approach which is self-contained and leads to comparison of pseudo predictive densities.

However, the model choice problem the authors face can, in certain cases, be handled in a different way. Suppose the situation involves either (i) direct comparison of nested models i.e., choice between full and reduced or (ii) comparison of a constrained vs. an unconstrained model which can typically be reparametrized to a nested case. Apart from precision parameters, suppose the full model involves parameters $\boldsymbol{\beta}_1$ and $\boldsymbol{\beta}_2$ while the reduced model sets $\boldsymbol{\beta}_2 = 0$. If $\boldsymbol{\beta}_2$ is univariate we could directly study where 0 falls relative to the posterior distribution of $\boldsymbol{\beta}_2$. In higher dimensions we could calculate P(A|data) where $\mathrm{A} = \{\boldsymbol{\beta}_2 : \mathrm{post}(\boldsymbol{\beta}_2|\mathrm{data}) \leq \mathrm{post}(0|\boldsymbol{Y})\}$ where post denotes the posterior density of $\boldsymbol{\beta}_2$. P(A|data) is a natural Bayesian probabilistic criterion.

Sampling based approaches for Bayesian computation such as Gibbs sampling or sampling/importance resampling routinely enable an estimate of the density post($\boldsymbol{\beta}_2$|data), as well as samples essentially from this posterior so that P(A|data) is readily approximated.

R. E. KASS (*Carnegie Mellon University, USA*)

Although it can be difficult to determine the degree of relevance of information from one source in analyzing data from another, there is surely much to be gained from trying. This paper gives a nice discussion of one such attempt. I wonder, though, about a particular technical detail. In section 4.1.3 the authors run up against a Bayes factor that takes the form ∞/∞. They propose a two-stage solution to this problem. I would like to make a remark about the indeterminacy, and then, more importantly, complain about their two-stage solution.

The authors have included the normalizing constants for the priors in their equation (4), which leads them to the statement that this indeterminate form occurs whenever the prior is improper. In fact, the usual way the Bayes factor is written does not include these normalizing constants, and one can often go ahead and compute the Bayes factor with improper priors; Jeffreys routinely did so. (In such cases the formal marginal density of the data, i.e., $\int \exp(\ell(\theta))\pi_j(d\theta)$ in their notation, is finite, but integrates to ∞.) However, the Bayes factor itself (i.e., the ratio of the formal marginal densities, without the prior normalizing constants) does take the indeterminate form the authors say they have encountered here when the *posterior* is improper, which it often is with hierarchical models (because the likelihood does not vanish at the boundary of the hyper-parameter space). Perhaps this is the problem. If so, they have gotten improper posteriors from the training data set and used them as priors for the reserve data set.

I should immediately add that improper posteriors sound much worse than they often are in practice, and it is possible that the authors didn't notice they had improper posteriors (if indeed they did have them, which I am only guessing about). The reason is that even when the likelihood function on the hyperparameters doesn't vanish at the boundary, it may be highly peaked. Thus, a flat prior that is appropriately truncated (to a compact set containing the maximum of the likelihood and not extending too far beyond the region where the peak of the likelihood is located) will give acceptable results. Furthermore, when using numerical methods, there is nothing to distinguish a flat prior from a flat prior that has been truncated; thus, when using numerical methods with a flat prior one may get acceptable results and not realize that, technically, the posterior is improper.

Thus, in a sense, all is well. My complaint is that the two-stage procedure the authors adopted combines the training and reserved data set in a manner that would be incoherent (i.e., would not follow the laws of probability) if a proper prior were used at the outset. For instance, the answer they get will depend on exactly how the data set is split. I would prefer to see an alternative solution to the problem.

In considering what might be done, I think it is important to keep in mind that the splitting of data is often (as here, in part) done in a non-parametric spirit of model assessment (see the paper by Gelfand, Dey, and Chang, 1992, in this volume, as well). By non-parametric I really mean alternative-free, that is, without specifying the type of departure from the model one is looking for. Various predictive methods can be useful for that rather loosely-defined general purpose. When it comes to selecting competing models, however, the appropriate tool is the Bayes factor. Unfortunately, a serious complication is that unless the data are very decisive, some sensitivity analysis will have to be performed. For this reason, which holds generally and is not a consequence of the indeterminacy in the particular case treated here, I do not see any way out of a full-fledged subjective approach. If public consumption requires simpler methodology (for either good or bad reasons), then I suggest reporting the Schwarz criterion (which is a crude approximation to the log of the Bayes factor) if it turns out to support conclusions reached with the subjective analysis. It is my own experience with a non-trivial model comparison problem that has led me to this point of view (Carlin, Kass, Lerch, and Huguenard, 1992).

U. MENZEFRICKE (*University of Toronto, Canada*)

These comments relate to the "training" set, mentioned in Section 4.1.3 of the written paper and also commented on in the talk. If $\pi_{0j}(j = 1, \ldots, k)$ is the (improper) prior distribution for the parameters of model M_j, then this prior can only be properly updated by using a training data set which is known to have been generated by model M_j. Denoting the training data X_{0j}, it is clear that one training data set is needed for each model. The resulting proper posterior distribution for model M_j, denoted π_{1j}, can then be used in the model choice problem, where it is to be decided which model generated a further data set, X_1. I think that, in this set up, it is justifiable to ignore the constants of proportionality for the improper prior distributions π_{0j}.

It seems to me, however, that the authors did not have a training set for each model considered. On the contrary, I think that the authors did not know with certainty which model generated their training data set, X_0. In this case the model choice problem arises not only for the further data set X_1, but also for the training data set X_0, and therefore the authors must not ignore the constants of proportionality for the improper prior distributions π_{0j}.

D. J. POIRIER (*University of Toronto, Canada*)

This is a most interesting paper blending together both theory and application. I have two brief unrelated comments.

Firstly, the use of laboratory data to augment the field experiments is most intriguing and potentially useful for predicting the presence of brook trout in lakes with water chemistry unlike those in the observed field experiments. It is this extrapolative nature, rather than multicollinearity, that is the primary culprit. If the authors had been able to design laboratory experiments, they could have chosen the levels for the various characteristics of water chemistry to best provide information on these predictions. Another option is to use a conjugate prior for the logistic model, which can be viewed as fictitious data from an experiment governed by the same parameters [see Koop and Poirier (1991)], and to optimally choose the levels for the various characteristics of water chemistry in the fictitious prior sample.

Secondly, the Jeffreys prior for the logistic model has many interesting properties. In particular, as in the i.i.d., binary case, the Jeffreys prior appears to be proper in the present setting. Hence it is a possible candidate for the Bayesian model choice comparisons in Section 4.

REPLY TO THE DISCUSSION

We would like to thank all the discussants for their thoughtful comments and helpful suggestions.

Professor Chaloner asks, "why not do a straightforward Bayesian analysis using an informative prior distribution?" Why not, indeed; it is well known that one cannot do better than to maximize subjective expected utility, and the straightforward Bayesian analysis she recommends can do that while our present, more baroque approach cannot. Moreover, she suggests that the simpler approach can free the modeler to develop the models in more productive ways, such as adding needed explanatory variables.

Of course the answer is that the present models were developed not only for the *scientific* objective of predicting accurately how fish populations will respond to changes in the level of environmental acidification, but also to the *political* and *sociological* objective of making those predictions in a manner that will be regarded as "objective" by dispassionate observers. Regulatory agencies may use predictive models to assist in setting allowed discharge limits and making similar decisions with great economic impact, and those who are most directly affected (and their consultants and lawyers) would have reason to contest any subjective prior information that had a material effect on the outcome. It can be argued that objectivity is really mythical (see Berger and Wolpert, 1988; Box, 1980; Berger and Berry, 1987), and that the choice of models is just as subjective and less subject to scrutiny than that of priors, but the fact remains that it is hard to persuade sceptics that a subjective Bayesian analysis is more objective than a data-driven noninformative Bayesian analysis. The models we study here grew out of a desire to use existing data to illuminate and possibly resolve uncertainties arising because of inadequacies of the present data set, without drawing heavily on our intuition or beliefs.

We share Professor Chaloner's regret that we only had access to a subset of 313 of the 1469 lakes in the ALSC survey, systematically selected in an effort to eliminate causes of fish absence other than acidification, and we share her worry about the effects of that selection process; we hope to study the entire data set in future work.

Although the probability of misclassification may well be related to other explanatory variables since, as Professor Chaloner notes, false "absence" misclassification is more likely when fish are scarce, that kind of misclassification is benign— it leads to a small systematic bias in the regression coefficients, but misclassifying borderline cases does little harm. The real problem comes from lakes with water chemistries that would lead to a *very* high predicted probability of presence, but in which no fish were reportedly found (possibly due to misclassification, or to inaccurate measurement or reporting of the predictor variables, or else due to other causes such as unsuitable physical or chemical conditions unrelated to the measured predictors). A logistic or probit regression will lose its sensitivity in such a case, driving to zero all the regression coefficients except the intercept term. This phenomenon is apparent in a comparison of the components of the MLE $\hat{\beta}^F$ for models $\mathbf{M}_4^F$ and $\mathbf{M}_5^F$, from Table 1; the predictions $p(x^T\hat{\beta}^F)$ are far more sensitive to changes in the chemistry variables under $\mathbf{M}_5^F$ than under $\mathbf{M}_4^F$. It is the problem of these highly influential data points that the misclassification coefficient r was intended to address.

Of course Professors Consonni and Veronese are correct that the ∞/∞ problem can be circumvented by taking sequences of proper priors converging weakly to the improper π_0 and π_1, for which the Bayes factors converge; they also observe that this approach would not be useful in general, since the limit is not unique (indeed it is easy to construct examples in which, for every number $0 \leq L \leq \infty$, there are sequences of proper priors such that the sequence of Bayes factors converges to L). Perhaps there are applications in which there is a "natural" sequence of priors to take, such as the trace of π_0 and π_1 on some distinguished increasing sequence of compact sets. We would welcome examples. We are not sure what they mean in suggesting that we "are no longer performing a noninformative Bayesian analysis," so we we plead *nolo contendere* and offer to pay the fine.

We welcome Professor Gelfand's suggestions and contributions addressing the timely and difficult problem of how Bayesians should select among competing models, and how we should test them for adequacy. The idea in Gelfand, Dey, and Chang (1992) seems similar to that here— fit a model to part of the data, and see how well it predicts the rest. We make an arbitrary division of the data into two parts of nearly equal size and use each but once for each measure; Gelfand, Dey, and Chang (if we understand correctly) remove a single observation from the data set and try to predict that one, and then repeat the process for each observation and average the results. Smith and Spiegelhalter (1980) use the "posterior" from a small imaginary sample as a proper surrogate for their improper prior. As Professor Gelfand observes, the problem can often be "reduced" to that of assessing a point null hypothesis against a diffuse alternative in as objective a way as possible.

We have misgivings about all these methods (our own included), and the others we've read about or thought about. Each of the methods seems to require either an arbitrary and not-irrelevant choice (how large a sample to reserve, how large an imaginary sample to take and what the imaginary observations are) or a reuse of the sample. As Professor Kass commented, "If you had a proper prior you wouldn't do that— so it can't be right for an improper prior, either."

We especially want to thank Professor Kass for his insightful comments, that helped deepen our understanding of the obstacles and choices we face in this problem. He suggests that we are too timid in normalizing our integrated likelihoods, ruling out Bayes factors for improper priors; and too daring in our two-stage solution, which might lead to improper posteriors. We make no apologies about our timidity, since our interest lies in comparing different-dimensional models (the four field models we consider have dimensions four, five, six, and eight, and are not all nested), and we feel that unnormalized Bayes factors are dangerous in such a setting— we see no way to avoid separate arbitrary scale factors, one each for the numerator and denominator, that will not simply cancel in the Bayes factor calculation.

The fear of an improper posterior, however, is alarming and seems well-founded for at least one of the improper prior distributions we considered ($\sigma^{-1} d\beta^F \, d\sigma \, d\delta$), since the likelihood is not integrable in a neighborhood of $\sigma=0$; we don't yet know about our other diffuse priors. As suggested, the log likelihood is in fact quite peaked, and falls off by several hundred away from its peak value as σ tends to zero, making the problem difficult to ascertain numerically.

We share Professor Kass's deeper concern about our two-stage process. It does indeed depend on the apparently arbitrary size of the reserved data set, in ways we do not yet understand; and since it does not reduce to an ordinary Bayes approach when the priors are in fact proper, then it cannot be coherent. Unfortunately it shares those two flaws with most (perhaps all?) other approaches to the problem, with the sole exception of a straightforward Bayesian approach with proper subjective prior distributions. Perhaps Nature is trying to tell us something.

We welcome Professor Menzefricke's comments and criticism of our use of a training set that does not, in fact, come from any of the priors under examination; the training set is a portion of the field and laboratory data sets, and is most unlikely to arise from a chance mechanism exactly like any of the simple models we consider. Unfortunately we don't really understand the implication of this criticism, or how in any realistic application one would encounter data "known to have been generated by the model." The integrated likelihoods $L(\pi_i)$ as defined in this paper *are* invariant under changes in scale, both for the possibly improper prior distributions (because of the factors $\int_\Theta \pi_i(d\theta)$) and for the likelihood functions (because of the factor $-\ell(\hat{\theta})$ in the argument of $\exp(\cdot)$).

Professor Poirier observes that the culprit in our problem is not multicollinearity *per se*, but rather it is our need to make *extrapolative* predictions, beyond the chemistries represented in our sample, rather than interpolative ones. That seems to us to be just a different perspective (more accurately, a different metric) on the same problem. The particular values of pH and the concentrations of aluminum and calcium in the future data sets do not range more widely than those in the present data set; it is just the particular *combination* of those variables that is new. The new points will be "extrapolative," in that they lie far from the center of our data *in the metric determined by the information matrix*, even though they are not remote in the Euclidean metric. If the present data set were more nearly orthogonal (so that all combinations of pH, [Al], and [Ca] in their respective ranges were represented) then the problem would be much less severe.

Professor Poirier also informs us that the Jeffreys prior may be proper in this case— an intriguing suggestion, since it might permit us to follow the usual straightforward Bayesian approach to the comparison of proper models.

Thank you again, to all the discussants. Perhaps if enough of us ponder the problem of identifying objective criteria for Bayesian model verification and model selection we can make some progress on this important problem and widen the acceptance of Bayesian methods in the empirical sciences.

ADDITIONAL REFERENCES IN THE DISCUSSION

Berger, J. O. and Berry, D. (1987). The relevance of stopping rules in statistical inference. *Statistical Decision Theory and Related Topics IV* **1**. (S. S. Gupta and J. O. Berger, eds.), New York: Springer, 29–72, (with discussion).

Box, G. E. P. (1980). Sampling and Bayes' inference in scientific modelling and robustness. *J. Roy. Statist. Soc. A* **143**, 383–430, (with discussion).

Carlin, B. P., Kass, R. E., Lerch, F. J. and Huguenard, B. R. (1992). Predicting working memory failure: a subjective Bayesian approach to model selection, *J. Amer. Statist. Assoc.* , (to appear).

Consonni, G. and Veronese, P. (1992). Bayes factors for linear models and improper priors. *Bayesian Statistics 4* (J. M. Bernardo, J. O. Berger, A. P. Dawid and A. F. M. Smith, eds.), Oxford: University Press, 587–594.

Gelfand, A. E., Dey, D. K. and Chang, H. (1992). Model determination using predictive distributions with implementation via sampling methods. *Bayesian Statistics 4* (J. M. Bernardo, J. O. Berger, A. P. Dawid and A. F. M. Smith, eds.), Oxford: University Press, 147–167, (with discussion).

Koop, G. and Poirier, D. J. (1991). Bayesian analysis of logit models. (Unpublished *Tech. Rep.*).

BAYESIAN STATISTICS 4, pp. 547–563
J. M. Bernardo, J. O. Berger, A. P. Dawid and A. F. M. Smith, (Eds.)

Bayesian Predictive Inference for Samples from Smooth Processes*

J. V. ZIDEK and S. WEERAHANDI
The University of British Columbia, Canada and *Bellcore, Canada*

SUMMARY

This paper extends a method of the authors to obtain a Bayes linear procedure for predicting the value at a specified time of a future sample path. It is supposed that data are available from a number of sample paths which are deemed to be related to the future sample path. The method is local and relies on a simple linear model which derives from the Taylor expansions of the processes at the point at which the inference is required. It is argued that the resulting Bayes linear predictor is approximately the same as that from any one of a family of complex Bayes linear predictors which obtain under the assumption that the processes possess several derivatives. An illustrative application to growth curve analysis is used to bring out some of the strenghts and weaknesses of the proposed method.

Keywords: LINEAR SMOOTHING; BAYES LINEAR PREDICTORS; NONPARAMETRIC REGRESSION; MULTIPLE TIME SERIES; SPLINES; KERNEL REGRESSION METHODS; LOCALLY WEIGHTED REGRESSION; MULTIPLE REGRESSION; LOESS; GROWTH CURVE ANALYSIS.

1. INTRODUCTION

Predictive inference for multiple time series is the subject of this paper. The proposed solution uses an extension of the method of Weerahandi and Zidek (1988, hereafter WZ), described by Cleveland and Devlin (1988) as a Bayesian version of locally weighted regression. A preliminary version of this paper is part of the work of Weerahandi and Zidek (1986).

Suppose data are obtained from each of a set of sample paths, like growth curves for example. The data vector is $Y = (Y_1^T, \ldots, Y_m^T)^T$ with $Y_i = (Y_i(t_{i1})), \ldots, Y_i(t_{in_i})^T$, an $n_i \times 1$ column vector for $i = 1, \ldots, m$. Assume $Y = S + N$, where S and N are partitioned in conformation with Y. The uncorrelated coordinates of N represent noise; they have mean zero and a common variance, σ_N^2. And $S_i = (S_i(t_{i1}), \ldots, S_i(t_{in_i}))^T$ where $S_i(t)$ is $P+1$ times differentiable in quadratic mean.

Of particular inferential interest is $\beta_{m+1}^T = (S_{m+1}(t_{m+1}), S_{m+1}^{(1)}(t_{m+1}), \ldots, S_{m+1}^{(p)}(t_{m+1}))$ where S_{m+1} represents a possibly as yet unsampled process and t_{m+1}, a possibly as yet unused sample point. Bracketed superscripts denote L_2 derivatives of the process and $p \leq P$. More generally, the object of inferential interest may be $\beta = (\beta_1^T, \ldots, \beta_m^T, \beta_{m+1}^T)^T$, where β_i is defined for all i as in the case $i = m+1$.

As is well known, the optimal linear, that is, Bayes linear procedure with respect to a quadratic loss function is

$$\beta_Y = E\beta + \alpha_{\beta Y}(Y - EY) \tag{1.1}$$

* Prepared in part under support to SIMS from the United States Environmental Protection Agency and in part from a grant provided by the Natural Science and Engineering Research Council of Canada.

where in general, for any two random vectors U and V, $\alpha_{UV} = \Gamma_{UV}\Gamma_{VV}^{-1}$, and $\Gamma_{UV} = E(U - EU)(V - EV)^T$. The reliability of β_Y is described by $\Gamma_{\beta \cdot Y}$ where, in general, for any two random vectors, U and V, $U \cdot V = U - [EU + \alpha_{UV}(V - EV)]$ and $\Gamma_{U \cdot V} = E(U \cdot V)(U \cdot V)^T$. A Bayes linear procedure is Bayes when Y and β have a joint Gaussian distribution. But even in general, linear procedures are commonly used as they are here because of their simplicity.

Nevertheless, the global modeling required to specify linear procedures can be very demanding. To simplify such modeling, and possibly enable a hierarchical construction, one may adopt the linear model,

$$Y = EY + A(\beta - E\beta) + E, \tag{1.2}$$

where $A = \alpha_{Y\beta}$ and E is independent of β. By construction, E has mean zero and a covariance matrix given by $C = \Gamma_{Y \cdot \beta}$. Model (1.2) yields

$$\Gamma_{YY} = C + A\Gamma_{\beta\beta}A^T \tag{1.3}$$

and

$$\Gamma_{\beta Y} = \Gamma_{\beta\beta}A^T. \tag{1.4}$$

From equations (1.3) and (1.4), an expression for $\alpha_{\beta Y}$ is readily found. A well known alternative expression is:

$$\alpha_{\beta Y} = \Gamma_{\beta \cdot Y}A^T C^{-1}, \tag{1.5}$$

where

$$\Gamma_{\beta \cdot Y} = (\Gamma_{\beta\beta}^{-1} + A^T C^{-1} A)^{-1}. \tag{1.6}$$

Even with the model (1.2) specifying the global prior distribution may be difficult-and possibly unnecessary. If sampling around the point of inference, $t = \tau$, is sufficiently intense and a semi-Markov-like property stated in Section 2 is believed to hold, then the data outside a window at $t = \tau$ would be relatively unimportant. Sometimes the data outside such a window are excluded (c.f. Muller, 1987), usually on heuristic grounds, we believe.

The implication of excluding the data outside a data window is that prior information is required only for certain local (not global) parameters, defined in Section 2.

Excluding data as above is desirable where permissible not only because this simplifies prior modeling, but also because this increases robustness. It avoids the possible consequences of misspecifying the global model, say by making a convenient choice like, say, that of an inappropriate weakly stationary model. And it avoids the potential impact of errors in those data which contribute little to the optimal procedure anyway.

To construct a "local" model, suppose remote data have been excluded so that the coordinates of Y are just the data for which $\Delta_{ij} = t_{ij} - \tau$ are small in magnitude. Taylor's theorem implies an approximate linear model given in the next section,

$$Y = A^0\beta + E^0. \tag{1.7}$$

E^0 includes the Taylor remainders, has expected value zero, is approximately uncorrelated with β, and has a covariance matrix, C^0; the component of the latter due to the Taylor remainder would be expected to be small. It will be argued heuristically that equations (1.2) and (1.7) are approximately equivalent under suitable regularity conditions. Zidek and Weerahandi (1990) go further in a special case and derive bounds on the errors from substituting (1.7) for (1.2) when the range of the coordinates of Y is an arbitrary finite dimensional inner product space, $p = 0$, and $P \geq 2$. The result is a Bayes linear procedure which is an approximation to the "true" Bayes linear procedure obtained by specifying completely the a priori model for the processes involved. Likewise, an approximation, $\Gamma^0_{\beta \cdot Y}$, is obtained for $\Gamma_{\beta \cdot Y}$. The latter is important since it indicates the uncertainty in the Bayes linear rule.

When Y and β have a joint Gaussian distribution, our approximations yield an approximate posterior distribution for β and hence β_{m+1}. This may be used to find attributes of interest for β such as credibility regions.

The problem of specifying the local prior distribution for β including the covariance hyperparameters is discussed in Section 2.

Our interest in the topic of this paper stems from discussions with Dr Ned Glick in 1978 when a crude version of the approximation given in WZ was conceived. The context of those discussions was a growth curve study where just one datum was obtained from each of a random sample of children. The situation is much different from that of classical time series analysis where there are a number of often equally spaced values from a single sample path, although the latter is a special case of the general situation considered here.

The example of Section 4 is also from a growth curve study but with several observations from each child's growth curve taken at the same equally spaced values. Our results are in close agreement with those Fearn (1975) obtained by a different Bayesian analysis.

2. APPROXIMATE BAYES LINEAR PROCEDURES

In this section, the model in equation (1.7) will be derived from (1.2). In this derivation, $U = O_p(\delta)$ will mean $|U/\delta|$ has bounded expectation for any $\delta > 0$ for any random vector, U; $u = O(\delta)$ has an analogous meaning when u is not a random object.

Let us adopt the following notation:

$$
\begin{aligned}
a_{ijr} &= \Delta^r_{ij}/r!, r = 0, 1, \ldots, P, \\
a_{ij} &= (a_{ij0}, \ldots, a_{ijp}), \\
\beta_{ir} &= S^{(r)}_i(\tau), r = 0, 1, \ldots, P, \\
\beta_i &= (\beta_{i0}, \ldots, \beta_{ip})^T, i = 1, 2 \ldots, m+1, \\
\beta &= (\beta^T_1, \ldots, \beta^T_{(m+1)})^T.
\end{aligned}
\tag{2.1}
$$

By the assumptions of Section 1 and Taylor's theorem

$$
S_{ij} = \sum_{r=0}^{P} a_{ijr}\beta_{ir} + O_p(|\Delta|^{(P+1)}_{ij}). \tag{2.2}
$$

Thus

$$
ES_{ij} = \sum_{r=0}^{P} a_{ijr}E\beta_{ir} + O_p(|\Delta|^{(P+1)}_{ij})
$$

and hence

$$
\begin{aligned}
\tilde{S}_{ij} &\triangleq S_{ij} - ES_{ij} \\
&= \sum_{r=0}^{P} a_{ijr}\tilde{\beta}_{ir} + O_p(|\Delta|^{(P+1)}_{ij})
\end{aligned}
$$

where $\tilde{\beta}_{ir} \overset{\Delta}{=} \beta_{ir} - E\beta_{ir}$ from which it follows that $\tilde{\beta}_{ir} = \tilde{S}_i^{(r)}(\tau)$.

A straight forward calculation now gives

$$\alpha_{S_{ij}\beta}\tilde{\beta} = \sum_{r=0}^{p} a_{ijr}\tilde{\beta}_{ir} + \sum_{r=p+1}^{P} a_{ijr}\alpha_{\beta_{ir}\beta}\tilde{\beta} + O_p(|\Delta|_{ij}^{(P+1)}) \tag{2.3}$$

and

$$E(S_{ij} \cdot \beta) = \sum_{r=p+1}^{P} a_{ijr}E(\beta_{ir} \cdot \beta) + O_p(|\Delta|_{ij}^{(P+1)}). \tag{2.4}$$

Combining equations (2.3) and (2.4) yields

$$S_{ij}(\beta) = \sum_{r=0}^{p} a_{ijr}\beta_{ir} + \sum_{r=p+1}^{P} a_{ijr}\beta_{ir}(\beta) + O_p(|\Delta|_{ij}^{(P+1)}). \tag{2.5}$$

where, in general, $U(V) = E(U) + \alpha_{UV}(V - E(V))$. Reinvoking equation (2.2) gives

$$S_{ij} \cdot \beta = \sum_{r=p+1}^{P} a_{ijr}\beta_{ir} \cdot \beta + O_p(|\Delta|_{ij}^{(P+1)}). \tag{2.6}$$

Since by definition, $S_{ij} = S_{ij}(\beta) + S_{ij} \cdot \beta$, a fundamental decomposition is obtained:

$$S_{ij} = \sum_{r=0}^{p} a_{ijr}\beta_{ir} + \sum_{r=p+1}^{P} a_{ijr}\beta_{ir}(\beta) + \sum_{r=p+1}^{P} a_{ijr}(\beta_{ir} \cdot \beta) + O_p(|\Delta|_{ij}^{(P+1)}). \tag{2.7}$$

By combining the second and third terms in equation (2.7), equation (2.2) is obtained, it being the basis on which Weerahandi and Zidek (1986) build their Bayes linear procedure.

We share with Zidek and Weerahandi (1990) the goal of finding a model which yields a local approximation to each member of the class of Bayes linear procedures implied by (1.2). We therefore use the decomposition in (2.7) to suggest an approximation to the model in equation (1.2). In particular, to order p, the approximation A_i^0 to A_i has ij th row defined by

$$A_{ij}^0\beta = a_{ij}\beta_i.$$

Equation (2.7) yields an estimate of the error in this approximation:

$$(A_{ij} - A_{ij}^0)\beta = \sum_{r=p+1}^{P} a_{ijr}\alpha_{(\beta_{ir}\beta)}\beta.$$

An analogous approximation, C^0 for C, is obtained below along with an estimate of the error in the approximation. Zidek and Weerahandi (1990) use these estimates of errors to bound the errors induced in β_Y and $\Gamma_{\beta \cdot Y}$ but for brevity we will not seek such bounds here.

From equation (2.6) we obtain,

$$\Gamma[(S_{ij} \cdot \beta)(S_{kl} \cdot \beta)] = \sum_{r=p+1}^{P} \sum_{s=p+1}^{P} a_{ijr}a_{kls}\Gamma_{ik}^{(rs \cdot \beta)} + O(|\Delta_{ij}|^{P+1} + |\Delta_{kl}|^{P+1}), \tag{2.8}$$

for all i, j, k, l, where $\Gamma_{ik}^{(rs\cdot\beta)} = \Gamma_{\beta_{ir}\cdot\beta,\beta_{ks}\cdot\beta}$ is the residual covariance between $S_i^{(r)}(\tau)$ and $S_k^{(s)}(\tau)$ when the linear effect Δ of β have been factored out. Assume like Weerahandi and Zidek (1988, 1990) in the special cases they treat, that $P \geq 2p+2$. Then to the order of the linear model for S based on β, the square root of the absolute value of the quantity in (2.8) is zero; locally the Taylor expansion has removed all variation and covariation in the S-processes. This was the heuristic basis for the model proposed by Weerahandi and Zidek(1986, 1988).

But a global approximant to the quantity in (2.8) is required which will yield a locally weighted predictive procedure and preserve its local character. Weerahandi and Zidek (1986, 1988), chose the approximant to be a diagonal matrix for simplicity. However in Zidek and Weerahandi (1990) where $p = 0$, the locally dominant term on the right of equation (2.4) is retained to capture the residual covariance structure (and enable bounds on approximation errors to be found). This term will also be retained here.

Let the approximation to C be given by

$$C^0_{(ij)(kl)} = \Sigma_{(ij)(kl)} + \eta_{(ij)(kl)} + \zeta_{(ij)(kl)}, \tag{2.9}$$

where $\Sigma_{(ij)(kl)} = 0$ unless $i = k$ and $j = l$ when $\Sigma_{(ij)(ij)}$ the variance of the noise in $Y_i(t_{ij})$ is positive. Furthermore

$$\eta_{(ij)(kl)} = \Delta_{(ij)(kl)}\Gamma_{ik}^{(p+1,p+1\cdot\beta)}$$

where

$$\Delta_{(ij)(kl)} = \frac{\Delta_{ij}^{(p+1)}\Delta_{kl}^{(p+1)}}{[(p+1)!]^2}e^{-(|\Delta_{ij}|+|\Delta_{kl}|)/\Delta_0}. \tag{2.10}$$

Finally

$$\zeta_{(ij)(kl)} = |\Delta_{ij}|^{(p+1)}|\Delta_{kl}|^{(p+1)}(|\Delta_{ij}| + |\Delta_{kl}|)\sigma_{(ij)(kl)}, \tag{2.11}$$

where the $\sigma_{(ij)(kl)}$ must be selected to assure positive definiteness of the resulting approximant; $(\sigma_{(ij)(kl)})$ could, in particular, be a diagonal matrix.

The approximation in equation (2.9) is meant to replace the higher order terms in (2.8) and insure, by making $\sigma_{(ij)(kl)}$ sufficiently large, that values of $Y_i(t_{ij})$ for which $|\Delta_{ij}|$ is large are "windowed out". This leads to the approximate model in equation (1.7), subject to the selection of the covariance hyperparameters in equation (2.9).

The more rigorous approach of Zidek and Weerahandi (1990) would entail partitioning Y as $(P^T, R^T)^T$ where P is the vector of data values in a window at $t = \tau$. The general results of Section 2 of Zidek and Weerahandi (1990) could now be applied. The error, $\beta_Y - \beta_P$, is (ibid, Theorem 1)

$$\Gamma_{\beta R\cdot P}\Gamma_{R\cdot P}^{-1}R \tag{2.12}$$

where $\Gamma_{\beta R\cdot P} = \Gamma_{\beta R} - \Gamma_{\beta R}\Gamma_{PP}^{-1}\Gamma_{RP}$ and $\Gamma_{R\cdot P}$ is defined in the Introduction.

Reduction to P from Y is only justified if it is believed that the relative error,

$$||\beta_Y - \beta_P||/||\beta_Y||, \tag{2.13}$$

is small. This is a generalized semi-Markov property. For an $AR(1)$ process (c.f. Zidek and Weerahandi (1990)), the error will be zero when P consists of just data points on either side of $t = \tau$. In general, the error in (2.13) need not be evaluated as long as it is believed to be small. This is the heuristic, presumably, which underlies all methods that use data windows. Of course, the approximation, β_P^0, to β_P derived from equation (1.7) is valid under the regularity conditions given here even if the relative error in equation (2.13) is not negligible, but then valuable information in the data at points remote from that at which inference is being made, $t = \tau$, may be lost.

To rigorously justify use of the model in equation (1.7) entails showing the error in (2.13) is small when β_Y and β_P are replaced by their corresponding approximants, β_Y^0 and β_P^0. Zidek and Weerahandi (1990) address this issue (when $p = 0$). It is plausible that this error will be small in the present context when $\sigma_{(ij)(kl)}$ in equation (2.11) is sufficiently large.

In summary, the analysis of this section has led us to the linear model in equation (1.7), where the ij th row of A^0 is given by

$$A_{ij}^0\beta = a_{ij}\beta_i$$

for all β, where E^0 is uncorrelated with β, and where $C^0 = \Gamma_{Y\cdot\beta}^0$ is determined by equation (2.9). This leads to an approximate Bayes linear procedure. It should be emphasized that we have been assuming the covariance hyperparameters are specified. We address the problem of specifying them in the next section where a particular implementation of our approximation is proposed.

3. EXCHANGEABLE GAUSSIAN PROCESS

A special case of the model in equation (1.7) will now be explored and a further approximation to the residual covariance introduced. For simplicity, the subscript, "0", imposed above on A and C to denote their approximations, will be suppressed.

The problem of specifying θ, the vector of covariance hyperparameters, will be addressed below, but suppose for now it has been specified. Assume the S_i 's and the noise processes are Gàussian, that the noise is homoscedastic and that the S_i 's are exchangeable. To be precise, suppose that in equation (2.9), $\Sigma_{(ij)(kl)} = \sigma^2$ or 0 according as $i = k$ and $j = l$ or not, and $\sigma_{(ij)(kl)} = \sigma_R^2$ or 0 according as $i = k$ and $j = l$ or not. Furthermore $\Gamma_{ik}^{(p+1,p+1\cdot\beta)}$ has a common value, say γ_B, for all pairs, $i \neq k$ and likewise $\Gamma_{ii}^{(p+1,p+1\cdot\beta)} = \gamma_W$ for all i. Assume $\gamma_B = \gamma_W = 0$, an assumption which is justified under reasonable conditions indicated in the Appendix of the preliminary version of this paper (Zidek and Weerahandi, 1991). Finally suppose that conditional on γ,

$$\beta_i|\gamma \overset{ind}{\sim} N_{(p+1)}(\gamma, \Lambda),$$

in the spirit of Lindley and Smith (1972, hereafter LS). Indeed, if γ is supposed to have a uniform (improper) prior distribution, the results of LS show that

$$\beta|Y, \theta \sim N_{(m+1)(p+1)}(D_0 d_0, D_0) \tag{3.1}$$

where in the notation of LS

$$D_0^{-1} = A_1^T C_1^{-1} A_1 + C_2^{-1} - C_2^{-1} A_2 (A_2^T C_2^{-1} A_2)^{-1} A_2^T C_2^{-1},$$

and

$$d_0 = A_1^T C_1^{-1} Y.$$

To translate these results in the present context, set $C_1 = C$, $A_1 = A$, $C_2 = Diag\{\Lambda, \ldots, \Lambda\}$, $A_2 = (I, \ldots, I)^T$, an $(m+1)(p+1) \times (p+1)$ matrix with I denoting the $(p+1) \times (p+1)$ identity matrix.

It is straightforward to show that

$$A_2^T C_2^{-1} A_2 = (m+1)\Lambda^{-1}.$$

Thus

$$C_2^{-1} A_2 (A_2^T C_2^{-1} A_2)^{-1} A_2^T C_2^{-1} = (m+1)^{-1} A_2 \Lambda^{-1} A_2^T.$$

The assumptions made above in this section entail

$$C = \sigma^2 I + \sigma_R^2 D$$

where $D = Diag\{D_1, \ldots, D_m\}$ and $D_i = Diag\{|\Delta_{i1}|^{2p+3}, \ldots, |\Delta_{in_i}|^{2p+3}\}$.

Now $D_0^{-1} = e - (m+1)^{-1} A_2 \Lambda^{-1} A_2^T$, where $e = Diag\{e_1, \ldots, e_{(m+1)}\}$, $e_i = v_i + \Lambda^{-1}$, $v_i = a_i^T c_i^{-1} a_i$ or 0 according as $i \leq m$ or $i = m+1$, $c_i = \sigma^2 I_{n_i} + \sigma_R^2 D_i, i = 1, \ldots, m$, and $a_i, i = 1, \ldots, m$ is the $n_i \times (p+1)$ matrix whose j-th row is a_{ij} , $j = 1, \ldots, n_i, i = 1, \ldots, m$, that is, $a_{ij} = (a_{ij0}, \ldots, a_{ijp})$, with $a_{ijr} = \Delta_{ij}^r / r!$ for $r = 0, \ldots, p$. In fact, $a_i, i = 1, \ldots, m$ is a submatrix of A (A^0 in the last section) defined just below equation (2.7) with $A = (Diag\{a_1, \ldots, a_m\}, 0)$.

Note that when, as in the next section, the n_i are identical, and the observations for each process, Y_i, are taken at the same time points, then the a_i are identical as are the c_i's. So the $v_i, i = 1, \ldots, m$ are identical in this case.

From the well known matrix equation (see LS), $(u + v^T t v)^{-1} = u^{-1} - u^{-1} v^T (t^{-1} + v u^{-1} v^T)^{-1} v u^{-1}$,

$$D_0 = e^{-1} + e^{-1} A_2 F^{-1} A_2^T e^{-1},$$

where $F = \Lambda \sum_{i=1}^m F_i^{-1} \Lambda$ and $F_i = v_i^{-1} + \Lambda$.

Of particular interest is the marginal posterior distribution of $\beta_{(m+1)}$ which is easily deduced from equation (3.1). It is

$$\beta_{(m+1)} | Y, \theta \sim N_{(p+1)}(\beta^*_{(m+1)}, \Sigma_{(m+1)}), \tag{3.2}$$

where $\Sigma_{(m+1)} = \Lambda + [\sum_{1=1}^m F_i^{-1}]^{-1}$, $\beta^*_{(m+1)} = [\sum F_i^{-1}]^{-1} \sum F_i^{-1} \hat{\beta}_i$, and $\hat{\beta}_i = (a_i^T c_i^{-1} a_i)^{-1} (a_i^T c_i^{-1} Y_j)$. In other words, $\beta^*_{(m+1)}$ is a weighted average of the local least squares estimators of the β_i. When the F_i's are equal as they are in the next section, the weights are identically equal to $1/m$. The distribution in (3.2) serves as a basis for inference about β_{m+1} if the covariance hyperparameters have been specified. However, in practice these parameters will not usually be specified at this first stage of hierarchical modeling and we now turn to this issue.

The variance of the noise process (in the homoscedastic case) is a global covariance hyperparameter unlike the other parameters in our approximate model which are local. Our theory is currently being extended to deal with global parameters. Inference about such parameters should be based on all the data, but for all approximately linear models (in the sense of Sacks and Ylvisaker, 1979), both repeated sampling and Bayesian inference will rely primarily on the data for which the model's local component is small, here the data in a window at τ. One approach to dealing with global parameters entails simultaneous inference across all τ by adopting models like those that produce splines (c.f. Wahba, 1982). It should be emphasized however, that the priors for splines are highly specialized and we believe they are insufficiently flexible as to adequately represent a reasonable spectrum of prior opinion. More work would seem to needed.

To deal with the remaining covariance hyperparameters (see equation (3.1)) the Bayesian paradigm entails putting a distribution on σ_R^2 and Λ. And provided that the conditional expectations of the S-processes given these parameters do not depend on them, it does not matter whether the approximation step in going from equation (1.2) to (1.7) is carried out before or after "marginalizing out" these covariance hyperparameters. However, it is not clear how well the marginal distribution from (1.2) would be approximated by that from its simpler relative, (1.7). Clearly the Gaussian model of Section 3 will be lost either way. For practical reasons, it is tempting to adopt the simple model in (1.7) as we do here. But this matter deserves further investigation.

The determination of the marginal posterior distribution of θ is straightforward. The result is

$$\pi(\theta|Y) = K|C|^{-1/2}|\Lambda|^{-m/2}|D_0|^{1/2}\exp[-1/2Y^TG^{-1}Y]\pi(\theta) \tag{3.3}$$

where K is the normalization constant whose exact value is unnecessary for our purposes, $G = C^{-1} - C^{-1}AD_0A^TC^{-1}$ and $\pi(\theta)$ is the prior density of θ .

By combining (3.2) and (3.3), the joint distribution of $\beta_{(m+1)}$ and θ is obtained and from this the marginal distribution of $\beta_{(m+1)}$. We will not discuss in general, the problem of determining the prior distribution of θ. But in the next section we will make a particular choice for the example considered there.

4. APPLICATION TO GROWTH CURVE ANALYSIS

The approximate model developed above is now applied to data used by Grizzle and Allen (1969) in their frequency theory analysis and Fearn (1975) in his Bayesian analysis of growth curves. These data consist of the Ramus heights (in mm), of 20 boys at 8, 8 1/2, 9 and 9 1/2 years of age. The data were collected to establish a normal growth curve for the use of orthodontists.

The example is a challenging one chosen to bring out strengths and weaknesses of our proposed method. As we will argue below, its use seems inappropriate here given the assumptions we have made in its development. And Fearn's analysis suggests the regression curve of Ramus height on age is very well approximated by a parametric (in fact, linear) model so our nonparametric approach seems unrealistic from the outset. Nevertheless, surprisingly good agreement with Fearn's results will be obtained.

Figure 1 depicts the data along with the results of our analysis. From a superficial examination, the inter-sampling points seem large; but our theory is designed for data which are clustered around the point of inference. However, sampling intensity is a relative concept and must be measured against the inherent variability of the process. So our immediate reaction to Figure 1 may be premature. Indeed, we do not yet know how to assess the adequacy of sampling intensity and there does not seem to be a generally satisfactory way of addressing this issue.

The number of derivatives to include in the model of equation (1.7) is somewhat arbitrary, although in some situations there may be physical models which dictate a useful upper bound. For example, if the underlying $S_i^{\prime}s$ were Brownian paths, then of necessity p would be zero. However, in this case the condition in the Section 2 that $P \geq 2p + 2$ would be violated; the justification given in Section 2 for using our method would be lost, although the method could still be applied and may even be justified on other grounds.

Derivatives should be included in equation (1.7) when inference about them is required. Point estimates of the derivatives might be obtained by numerical differentiation of the simultaneous estimate of the future sample path obtained by varying the point of inference, $t = \tau_{(m+1)}$. And adopting this simultaneous estimate is justified in that if the loss function

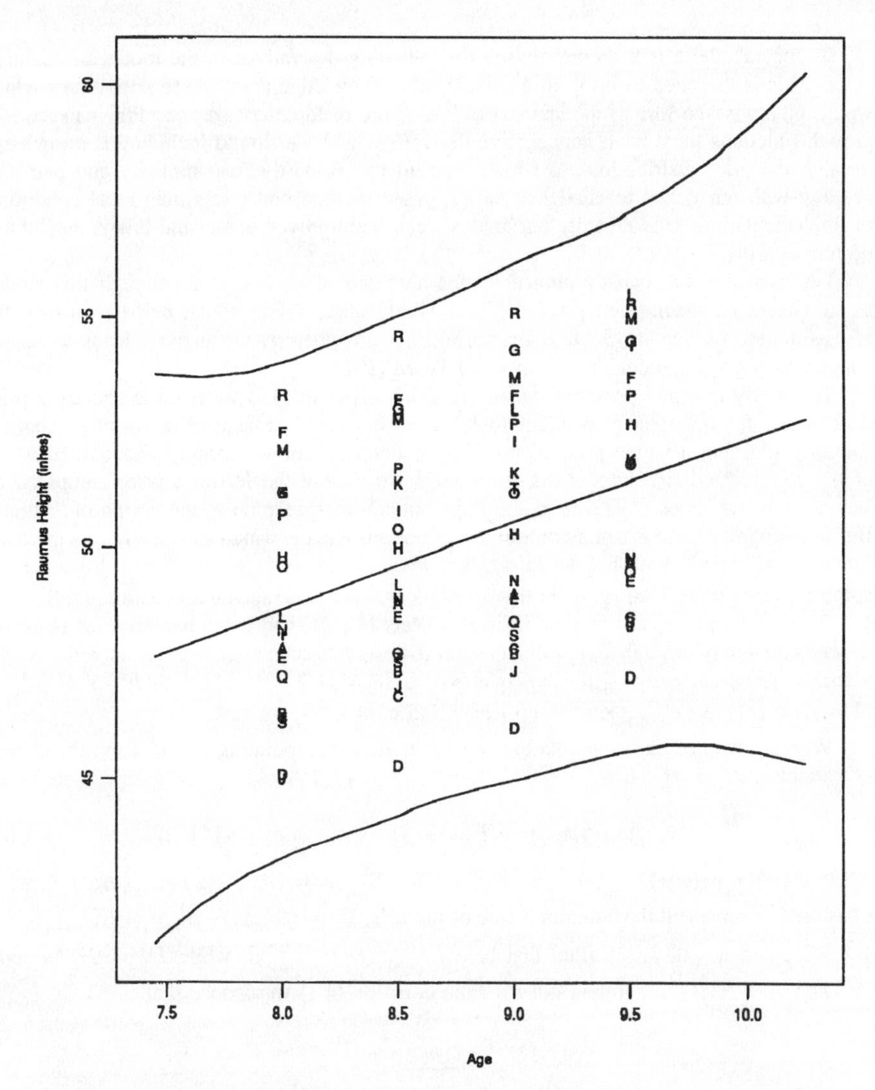

Figure 1. *Predicted Ramus Heights and 95% Pointwise Credibility Bounds Using The Locally Weighted Bayesian Smoother On the Grizzle and Allen (1969) Data.*

for estimating the curve were, say, integrated mean squared error over the domain of t, then the Bayes linear estimate at each time could be approximated as above. But even if simultaneous estimates were so obtained and the derivatives computed, the joint distribution of the derivatives and values of the process, would not be given by this approach.

As we have formulated the growth curve problem, estimates of the process derivatives are not required, and this leads us to choose $p = 0$. We believe the growth processes are quite smooth, at least thrice differential, so that we can view our inferential procedures as approximating the Bayes linear procedures which would obtain from imposing a global stochastic model on the underlying processes.

In general, there is a second reason for including derivatives in the model of equation (1.2). This is the need to bring in available prior knowledge about these derivatives which would otherwise be lost in the approximation of the residual covariance. For instance, the growth functions must have nonnegative derivatives and we should include this knowledge through the prior distribution. We have chosen $p = 0$, partly for simplicity and partly in keeping with our desire to challenge the proposed method under less than ideal conditions of implementation. Undoubtedly our analysis can be improved upon. and this is the subject of current work.

Let us make the realistic choice in this situation of $\sigma = 0$, even though this causes us to violate an assumption (in Zidek and Weerahandi, 1991) which helps to justify the approximation of Section 3. It is reassuring that nonetheless our analysis leads to results which are in good agreement with those of Fearn (1976).

To specify the joint posterior density function given in (3.3) we need to specify a prior distribution for θ. It is convenient to let $\omega = \sigma_R^{-2}\Lambda$. Our lack of knowledge about θ suggests adopting a vague prior distribution to describe our uncertainty about it. Because of the complicated structure of the covariance, we choose the Jeffrey's prior computed by Weerahandi and Zidek (1986) to give an operational interpretation of the notion of "vague". But we recognize there is some arbitrariness about this choice, which has the improper density function, $\sigma_R^{-4}\pi_1(\omega)$, with respect to $d\sigma_R^2 d\omega$, where $\pi_1(\omega) = (1 + \omega v)^{-1}$ and v denotes the common value of $v_i = a_i^T c_i^{-1} a_i$ in this special case (see Weerahandi and Zidek, ibid).

We may factor σ_R^2 out of a number of objects like D_0 and leave behind a factor which depends on θ only through ω; it will be useful to designate such resulting factors with a single asterisk. Thus, $D_0 = \sigma_R^2 D_0^*$ for example, where $D_0^* = (e^*)^{-1} + (e^*)^{-1}A_2(F^*)^{-1}A_2^T(e^*)^{-1}$, $e^* = Diag\{e_1^*, \ldots, e_{(m+1)}^*\}$, $e_i^* = v_i^* + \omega^{-1}$ and $v_i^* = a_i^T D_i^{-1} a_i$.

With the help of this new notation, we have, after incorporating the σ_R^2 from the change of variables, $d\Lambda = \sigma_R^2 d\omega$,

$$\pi(\beta_{(m+1)}, \omega, \sigma_R^2 | Y) = KF(\omega)(\sigma_R^2)^{-(n+2)/2} \exp[-(\sigma_R^{-2}T^*)/2], \tag{4.1}$$

where $F(\omega) = \pi_1(\omega)(\Sigma_{(m+1)}^*)^{-1/2}\omega^{-m/2}(D_0^*)^{1/2}$, $\Sigma_{(m+1)}^* = (m+1)\omega/m + (mv^*)^{-1}$, when v^* is used to represent the common value of the v_i^*'s, $T^* = (\beta_{(m+1)} - \beta_{(m+1)}^*)^2(\Sigma_{(m+1)}^*)^{-1} + Y^T D_2^* Y$, and it will be recalled that here $\beta_{(m+1)}^* = m^{-1}\sum \hat{\beta}_i$ while $D_2^* = D^{-1} - D^{-1} A^T(D_0^*)\, AD^{-1}$. By integrating out σ_R^2 from equation (4.1) we deduce that

$$[\beta_{(m+1)} - \beta_{(m+1)}^*]/\sigma_{(m+1)} | Y \sim t_{n-1} \tag{4.2}$$

and is distributed independently of ω with density

$$\pi(\omega | Y) = \tilde{K} F(\omega)(Y^T D_2^* Y)^{-n/2}, \tag{4.3}$$

where $\tilde{K}$ is the normalization constant, $n = \sum n_i$, t_{n-1} is the Student's t distribution with $n - 1$ degrees of freedom and $\sigma_{(m+1)} = (\Sigma_{(m+1)}^* Y^t D_2^* Y/(n-1))^{1/2}$. The inferences on $\beta_{(m+1)}$ can be based on (4.2) and (4.3) as illustrated below.

Numerical Illustration.

Returning to the application on Ramus heights of children described above, we now show how the foregoing theory can be used to make inferences on the Ramus height, say $H = \beta_{m+1}$, of a representative child who may or may not have been sampled yet. In

Figure 1, the Ramus heights of each boy are plotted against age at the times the heights were measured; the data points of the 20 boys are labeled A, B,..., and T.

Shown in the same figure are the estimated values of the expected growth curve, $\hat{H}(t) = \sum \hat{\beta}_i(t)/20$ and the 95 point-wise estimated Bayesian credibility band of $H(t)$. In estimating the mean growth curve, $\hat{\beta}_i(t_{m+1})$ was computed for a range of values of $t = t_{(m+1)}$ using the formula given in equation (3.2). The 95 point-wise credibility band for H was also obtained by varying $t = t_{(m+1)}$ using (4.2) in conjunction with the posterior distribution of ω given by (4.3) and the vague prior for ω described above. The analysis entails solving, for each t with the help of numerical integration, the appropriate equation to determine the 97.5 th percentile of the Student's t distribution with 79 degrees of freedom, but this is straightforward. The details may be found in Weerahandi and Zidek (1986).

Although we have not assumed a parametric model, interestingly the estimated normal growth curve of Ramus heights was found to be slightly concave but almost linear over the sampled range of ages. Hence the model assumed by Fearn (1975) seems reasonable; the inherent technical difficulties of his approach make the construction of Bayesian credibility intervals exceedingly difficult. An advantage of our nonparametric approach lies in its capability to handle even highly nonlinear growth curves such as those treated by Berkey (1982) without needing to identify appropriate parametric models.

5. CONCLUDING REMARKS

There has a been relative paucity of Bayesian research on nonparametric regression and smoothing, at least, if one ignores the theories of stationary time series and Kriging. It can be argued that these two theories have meaning only in a Bayesian framework, but neither is regarded as Bayesian. A paper more in the character of this one is that of O'Hagan (1979); but unlike O'Hagan, we are not concerned with adaptive modeling and focus on the implications of local smoothness for inference at a fixed point. The recent manuscript of O'Hagan (1989) on numerical quadrature as well as that of Angers and Delampady (1990) on smoothing use Bayesian approaches different from the one taken here. For a survey of other work on the analysis of continuous parameter processes, we refer the reader to the recent review of Sacks, Welch, Mitchell, and Wynn (1989).

There is an immense and rapidly growing list of repeated sampling school contributions to the theory of smoothing and nonparametric regression. Recently there have been some publications on the topic of this paper. References may be found in the forthcoming paper of Fraiman and Iribarren (1991).

Bayes linear procedures (see equation (1.1)) are often derived from linear models like that in equation (1.2) to enable hierarchical modeling. That is the case here; the model in equation (1.7) enables us to incorporate the knowledge that the processes are at least p times differentiable. We have justified this model as an approximation to that in (1.2). It is argued that the approximation will be good if these processes are actually $P \geq 2p+2$ times differentiable and the data are sampled with sufficient intensity near the point of interest. "Good" means that the resulting Bayes linear procedure derived from the model in (1.7) will be in good agreement with that derived from (1.2). It is also argued that the residual covariance matrices obtained from (1.2) and (1.7) will likewise be in close agreement under these circumstances. This justification is analogous to consistency in the frequency setting.

Our proposed method has a number of potential competitors including kernel, locally weighted regression (LWR) and spline methods. It has the simplicity of LWR methods. The model in equation (1.7) is just the conventional linear model of regression analysis and so susceptible to analysis by the host of methods and computational software which have been developed for such models.

But the potential domain of application of the method is extremely broad and so it may enjoy some advantages over its competitors. The theory in Section 2 addresses data from a fairly complex sampling plan in a seemingly natural way. As Zidek and Weerahandi (1990) show, vector valued processes can readily be accommodated. And in current work of the first author with Dr M. Delampady and Ms. Irene Yee, likelihood function estimators for time series are developed from the same starting point. The work of Joe, Ma and Zidek (1986) based on the proposed Bayesian approach suggests a computationally cheap alternative to cross validation. In fact, we view this paper as illustrating a general Bayesian framework for addressing problems of interpolating and extrapolating locally smooth functions, rather than as merely providing an additional method for tackling them.

When the processes which generate the data have P derivatives with $p \leq P \leq 2p+2$, the justification of this paper for the proposed method is lost. An appropriate approximation is unclear. Suppose (i) $m = 1$ and inference is about values of the single S-process; (ii) $P = p$ (iii) the process is weakly stationary; and (iv) the residuals to the left and right of $t = t_{m+1}$ from fitting the model in (1.7), are uncorrelated. Then it can be shown that the process is AR(p) and the resulting residual correlations may be taken to be approximately zero. This argument lead Weerahandi and Zidek (1988) to the approximation they chose, one which corresponds to that of Section 3. Clearly as $P \geq p$ increases and more of the process derivatives are buried in the model's residual term, so intuitively it seems that the residual correlation must increase. Therefore it would seem to be desirable to choose p as large as possible. But the price is increased modeling complexity and given the increased likelihood of prior model misspecification, it is not clear the tradeoff is worthwhile. Clearly this is a matter for further study.

The approach underlying the analysis of this paper can be applied without the justification described above provided the model expresses well the investigator's prior views. In particular it can be applied even when the data are not dense around the point of inference. In such situations values inferred from automatic procedures like spline fitting, become mathematical artifacts of their definitions. Indeed, such procedures can have other unforeseen properties. A simultaneous frequency property called "intriguing" (cf. Nychka (1988)) is observed by Wahba (1983) for point-wise intervals generated by a Bayesian approach. Nychka (1988) seeks to "remove some of the mystery" by interchanging the randomness in the function (here the S's) assumed by Wahba (1983) with the determinism of the observation times (here the t_{ij} 's) so that the latter are now endowed with stochastic uncertainty.

We believe the properties of the method proposed here may be more predictable. In particular, the investigator can ensure that values inferred from the proposed method are in accord with prior experience. However, accomplishing this would require that the investigator use an informative prior rather than that adopted in Section 3 and in the illustrative example of Section 4.

We will now summarize some of the concerns and open questions that remain. Given its fundamental importance, a better understanding of the extended semi-Markov property of Section 2 is required. This is needed to justify all inferential procedures, including those of this paper, which make local inferences either by local weighting or by relying only on data windows.

We need a better understanding of the role of the components of local curvature, like $\Gamma_{ik}^{(p+1)(p+1)\cdot\beta}$. These may be important for incorporating data at moderate distance from the point at which inference is being made, even when the extended semi Markov property holds. For simplicity this quantity was ignored in our analysis, albeit with some justification as provided by Zidek and Weerahandi (1991).

The hierarchical development of priors through linear models like that in equation (1.2) is important. But it is not clear how well the resulting marginal distribution is approximated through the adoption of the simpler model in (1.7) instead. Presumably the answer depends on the sampling intensity around the point of inference.

REFERENCES

Angers, J. F. and Delampady, M. (1990). Hierarchical Bayesian estimation of a function. *Tech. Rep.* **144**, Department of Statistics, University of British Columbia.

Berkey, C. S. (1982). Bayesian approach for a nonlinear growth model. *Biometrics* **38**, 953–961.

Cleveland, W. S. (1979). Robust locally weighted regression and smoothing scatterplots. *J. Amer. Statist. Assoc.* **74**, 829–836.

Cleveland, W. S. and Devlin, S. J. (1988). Locally weighted regression: an approach to regression analysis by local fitting. *J. Amer. Statist. Assoc.* **83**, 596–610.

Fearn, T. (1975). A Bayesian analysis of growth curves. *Biometrika* **62**, 89–100.

Fraiman, R. and Iribarren, G. P. (1991). Nonparametric regression estimation in models with weak error dependence. *Journal of Multivariate Analysis.* (to appear).

Grizzle, J. E. and Allen, D. M. (1969). Analysis of growth and dose response curves. *Biometrics* **25**, 357–381.

Heckman, N. E. (1988). Minimax estimates in a semiparametric model. *J. Amer. Statist. Assoc.* **83**, 1090–1096.

Joe, H., Ma, Wilson, and Zidek, J.V. (1986). A Bayesian nonparametric univariate smoothing method with applications to acid rain data analysis. *Tech. Rep.* **47**, Department of Statistics, University of British Columbia.

Lindley, D. V. and Smith, A. F. M. (1972). Bayes estimates for the linear model. *J. Roy. Statist. Soc. B* **34**, 1–41, (with discussion).

Muller, Hans–Georg (1987). Weighted local regression and kernel methods for nonparametric curve fitting. *J. Amer. Statist. Assoc.* **82**, 231–238.

Nychka, D. (1988). Bayesian confidence intervals for smoothing splines. *J. Amer. Statist. Assoc.* **83**, 1134–1143.

O'Hagan, A. (1978). Curve fitting and optimal design for prediction. *J. Roy. Statist. Soc. B* **40**, 1–42, (with discussion).

O'Hagan, A. (1989). Bayesian quadrature. *Tech. Rep.* Department of Statistics, University of Warwick.

Sacks, J. and Ylvisaker, D. (1978). Linear estimation for approximately linear models. *Ann. Statist.* **6**, 1122–1137.

Sacks, J., Welch, W. J., Mitchell, T. J. and Wynn, H. (1989). Design and analysis of computer experiments. *Statist. Sci.* **4**, 409–435.

Wahba, G. (1983). Bayesian "confidence intervals" for the cross validated smoothing spline. *J. Roy. Statist. Soc. B* **45**, 133–150.

Weerahandi, S. and Zidek, J. V. (1986). Analyses of multiple time series by Bayesian and empirical Bayesian nonparametric methods. *Tech. Rep.* **96**, Department of Statistics, University of British Columbia.

Weerahandi, S. and Zidek, J. V. (1988). Bayesian nonparametric smoothers for regular processes. *Canadian J. Statist.* **16**, 1, 61–73.

Zidek, J. V. and Weerahandi, S. (1990). Approximate Bayes linear smoothers for continuous processes. *Tech. Rep.* **105**, Department of Statistics, University of British Columbia.

Zidek, J. V. and Weerahandi, S. (1991). Bayesian predictive inference for samples from smooth processes. *Tech. Rep.* **106**, Department of Statistics, University of British Columbia.

DISCUSSION

J. PILZ (*Bergakademie Freiberg, Germany*)

I congratulate the authors for presenting a novel approach to Bayesian nonparametric smoothing, it's a real pleasure to have been invited to discuss their paper.

The authors give a sound and rigorous justification of using a local (approximately) linear model

$$Y = A\beta + \eta + N \tag{1}$$

where the i-th component (subvector) of β represents the i-fold application of the differential operator of the S-process at the point t at which inference (prediction) is required and the error term $E = \eta + N$ is made up of two random vectors, η is the remainder term of the Taylor series expansion and N represents noise. Accordingly, it is assumed that

$$\text{Cov } E = \sigma^2 I + \sigma_R^2 D \tag{2}$$

Moreover, it is argued that $\text{Cov}(\beta, \eta) \approx 0$ for small values of $|t_{ij} - t|$. The model assumptions are completed by assuming the β_i's to be exchangeable with $\beta_i \sim N(\gamma, \Lambda)$ and all the hyperparameters invoked to have a noninformative prior distribution.

The authors raise a number of important ideas and issues:

1. *Hierarchical linear modeling.* This seems to be a very convenient way for incorporating the prior information that the S-processes are at least p times differentiable. The hierarchy offers the possibility of a clear structuring and specification of prior knowledge, and the simplicity and comfortability of linear models allows the application of the bunches of methodology and software packages available by now for these models.
2. *The ideas of "windowing out" of data for which $|t_{ij} - t|$ is small.* This allows the prior specification to focus on local parameters, thus avoiding the risk of misspecification of global parameters. In this respect, the generalized semi-Markov property requiring that $||\beta(Y) - \beta(P)||/||\beta(Y)||$ is small may be a promising starting point for a more rigorous foundation of the intuitive appealing idea of using a data window. Alternatively, one could also define an equivalent measure for the approximate band structure of the precision matrix of the observations.
3. *Incorporation of prior knowledge and inference about derivatives.* This is a very important idea, which offers a great potential for improved prediction. In particular, prior knowledge about first order derivatives is an important source for revealing local trends and such knowledge will often be available, not only in growth curve analysis but also in other applications, e.g., in geophysical problems when analysing seismic records.
4. *Construction of credibility intervals.* This is often ignored in the literature on nonparametric smoothing, besides the cited papers by Wahba (1983) and Nychka (1988), two further notable exceptions are Wecker and Ansley (1983) and Silverman (1985).

The four ideas are not new, of course, but it is their combination which makes the proposed methodology a promising tool, although there is a long list of competitors in the literature on parametric and nonparametric smoothing. Parametric Bayes approaches were given e.g., in Fearn (1975), Blight and Ott (1975) and O'Hagan (1991), where, however, global linear models are considered. O'Hagan (1978) is concerned with local regression modeling and curve fitting, but he does not exploit the natural local structure implied by the regularity assumptions on the S-processes. Further, there is a vast literature on spline smoothing techniques, Bayesian approaches are well surveyed in Wahba (1990). However, the Bayesianity of spline smoothers is just an artifact of the definition of the underlying penalty function to be minimized, the derived prior is unnatural and inflexible (essentially unique). Moreover, the Bayesian models previously suggested in connection with nonparametric smoothing have mostly involved inference in infinite-dimensional spaces, except of Silverman (1985) and Eubank (1986), the latter, however, suffer again from the unnatural form in which specification of prior information is required. In my opinion, the inifinite-dimensional setting, apart from causing conceptual difficulties, simply yields an overspecification, in applications we should rather focus on finite-dimensional parameter spaces.

The authors mention kriging as a further competing methodology. I agree with them in stating that kriging has only a sound meaning within a Bayesian framework. The origins of kriging, both in a temporal setting and in a spatial setting, and its use in applications-oriented areas are discussed in Cressie (1989, 1990). The key idea is to use sliding neighborhoods of data locations chosen such that within these subregions (time intervals or spatial domains) the trend removed residuals satisfy the assumptions of covariance stationarity and then to find a best linear unbiased predictor (best linear weighted moving average of the observations), where the weights are based on the correlation structure of the observed residuals. This structure is inferred from the empirical variogram which measures the mean squared differences of the residuals at appropriate lags.

Bayesian approaches to kriging appear in Omre (1987) and in papers by de Waal and Groenewald (1991), Myers (1991) and Pilz (1991) given at this Valencia Meeting. The robust Bayesian kriging approach presented in Pilz (1991) seems to be ideally suited to the local linear model situation considered by the authors, see also my discussion on O'Hagan's paper in this volume.

Now, let me turn to a few minor quarrels:

(i) The problem of moderately sized differences $|t_{ij} - t|$ makes me feel somewhat uneasy. I would like to see a quantitative notion related to sampling intensity and process variability.

(ii) The choice of p is a further problem of concern. Some Akaike type information criterion taking account of the compromise of smooth residuals vs. modeling complexity would be helpful. However, I think that in applications it will be enough to choose $p \leq 2$.

(iii) The use of noninformative priors for the hyperparameters involved makes "backtracking" (i.e., checking whether the results are in good accordance with prior experience) complicated. The robust Bayesian kriging approach developed in Pilz (1991) would allow such backtracking, this approach could be modified to also include robustness against misspecification of the error covariance matrix when replacing $C = \text{Cov}E$ by some upper bound, $\bar{C}$ say. In the authors' formulation this would simply mean to specify some upper bounds for σ^2 and σ_R^2 in (2).

Finally, as it is always the case when being confronted with a new and promising methodology, it gives rise to a bunch of new problems, questions and hopes. Highly requested are:

- investigations of more complex settings, that is to see how the method works in unbalanced and nonequispaced situations along with $p > 0$ (the example on Ramus heights is too "smooth"!)
- extensions to vector-valued as well as space-time processes
- experimental design considerations, in this respect Goldstein's (1988) approach to the related design problem considered in O'Hagan (1978) could be promising
- developments of further strong Bayesian competitors, in particular I think of Bayesian kernel type predictors.

REPLY TO THE DISCUSSION

We are indebted to Dr. Pilz for clarifying a number of points in our presentation and for indicating possible improvements and promising extensions. We will confine our remarks to some of the "quarrels", "new problems, questions, and hopes" he has stated in his contribution.

We cannot address Dr. Pilz's well justified concern about the impact of "sampling intensity and process variability". And we are by no means alone for he raises a matter of fundamental importance. In discrete time series analysis, observations are taken from an underlying continuous time process. The resulting data will be of low quality in terms of its fitness for use if the process is highly variable and the sampling points not sufficiently close together as to adequately resolve it. Nevertheless this issue of sampling intensity does not seem to have been systematically addressed.

Qualitatively, the intensity of sampling must be sufficiently high as to enable sufficiently accurate predictive inference. It would seem possible in principle to address this as a design issue within the Bayesian framework, but we have not attempted to do so.

The sampling intensity issue is closely related to another fundamental issue and that is the selection of the bandwidth, in our paper, the implicit data window determined by the selection of the covariance hyperparameter in equation (2.11). This window's width is rigorously determined in Zidek and Weerahandi (1990) by analytical considerations. If the window is empty, then the proposed approximation would not necessarily be close to the "true" Bayes linear rule. It follows that sampling must be sufficiently intense as to put observations in this window. And there must be enough data in that window to insure under the generalized semi-Markov condition stated in the paper, that data outside the window will be of relatively little value. However, this answer, while mathematically precise, begs the question in that the "true" prior will not have been specified (otherwise it would be pointless to seek an approximation).

The issues of model complexity versus correlated residuals and of robustness raised by the discussant certainly deserve further attention. At this preliminary stage of our work, we are still mapping the main features of the terrain and have yet to explore these important details.

The extension to more complex settings has already been done to some extent. In Zidek and Weerahandi (1990) vector valued functions of a scalar parameter have been addressed. In unpublished material we have also looked at this approximation in a spatial setting. It is not surprising perhaps that the results are in qualitative agreement with the result obtained by kriging and thin plate splines, both of which are locally weighted averaging methods. Quantatively these methods do differ because in intepolating data from widely separated fixed monitoring sites, the values between stations are pure artifacts of the method and models adopted. The Bayesian methodologies proposed like that in our paper and that proposed, elsewhere by the discussant, are essential to ensure that the interpolated values are sensible.

In general, and in spatial analysis in particular, realistic indications of the relability of interpolants say, by credibility intervals, is extremely important. The results may well be used in regulation and policy making and exaggerated confidence in interpolants could have serious consequences. Nevertheless, there has been comparatively little attention given to this important subject in spatial analysis. The design question also becomes very important in this context since the costs of maintaining monitoring sites can be enormously expensive.

The idea of developing good Bayesian methodology seems uncontroversial, but it is not so clear that Bayesian versions of kernel estimators will have any value. Desirable methods will emerge from more fundamental considerations, we would think. And ultimately, satisfactory methods for smoothing may well have to be Bayesian in the light of the deficiencies of automatic data smoothing procedures. Hart (1991) shows that in kernel smoothing, a positive correlation in the data has a "disastrous effect on cross-validatory methods of choosing the bandwidth" and goes on to conclude that "under the best of circumstances it seems advisable to maintain a healthy scepticism about the answer obtained from automated procedures.

ADDITIONAL REFERENCES IN THE DISCUSSION

Blight, B. J. N. and Ott, L. (1975). A Bayesian approach to model inadequacy for polynomial regression. *Biometrika* **62**, 79–88.

Cressie, N. (1989). Geostatistics. *Amer. Statist.* **43**, 197–202.

Cressie, N. (1990). The origins of kriging. *Math. Geology* **22**, 239–252.

Eubank, R. L. (1986). A note on smoothness priors and nonlinear regression. *J. Amer. Statist. Assoc.* **81**, 514–517.

Fearn, T. (1975). A Bayesian approach to growth curves. *Biometrika* **62**, 98–100.

Goldstein, M. (1988). Adjusting belief structures. *J. Roy. Statist. Soc. B* **50**, 133–154.

Hart, D. P. (1991). Kernel regression estimation with time series data *J. Roy. Statist. Soc. B* **53**, 173–188.

O'Hagan, A. (1992). Some Bayesian numerical analysis. *Bayesian Statistics 4* (J. M. Bernardo, J. O. Berger, A. P. Dawid and A. F. M. Smith, eds.), Oxford: University Press, 349–363, (with discussion).

Omre, H. (1987). Bayesian kriging – merging observations and qualified guesses in kriging. *Math. Geology* **19**, 25–39.

Silverman, B. W. (1985). Some aspects of the spline smoothing approach to non-parametric regression curve fitting. *J. Roy. Statist. Soc. B* **47**, 1–52.

Wahba, G. (1990). *Spline Models for Observational Data*. Philadelphia, PA: SIAM.

Wecker, W. P. and Ansley, C. F. (1983). The signal extraction approach to nonlinear regression and spline smoothing. *J. Amer. Statist. Assoc.* **78**, 81–89.

CONTRIBUTED PAPERS

BAYESIAN STATISTICS 4, pp. 567–575
J. M. Bernardo, J. O. Berger, A. P. Dawid and A. F. M. Smith, (Eds.)

Use of the Student-t Prior for the Estimation of Normal Means: A Computational Approach*

JEAN-FRANÇOIS ANGERS
Université de Montréal, Canada

SUMMARY

Estimation of the mean of a multivariate normal distribution is considered. The components of the mean vector, θ, are assumed to be exchangeable which will be modeled, in a hierarchical fashion, with independent Student-t distributions as the first stage prior. The first stage Bayes estimator will be described in closed form when the Student-t priors have odd numbers of degrees of freedom.

Keywords: BAYES ESTIMATOR; HIERARCHICAL BAYES; NORMAL-STUDENT-T CONVOLUTION.

1. INTRODUCTION

Let $Y = (Y_1, Y_2, \ldots, Y_p)^t$ have a p-variate normal distribution with mean vector $\theta = (\theta_1, \theta_2, \ldots, \theta_p)^t$ and covariance matrix $\sigma^2 I_p$, where σ^2 is assumed to be known. The components of θ are believed to be exchangeable, and hence "shrinkage" estimation of them is desired.

The assumption of exchangeability of the components of θ can be modeled easily in a hierarchical Bayesian fashion, with a two stage prior. The first stage prior will be of the form $\pi_1(\theta \mid \mu, A) = \Pi_{j=1}^p \pi_{1,k_j}(\theta_j \mid \mu, A)$ where π_{1,k_j} denotes the Student-t distribution with k_j degrees of freedom and the hyperparameters μ and A represent the common location and scale parameters of π_{1,k_j}, for $j = 1, 2, \ldots, p$. The prior distribution on the hyperparameters μ and A will be denoted by $\pi_2(\mu, A) = \pi_{2,1}(\mu \mid A)\pi_{2,2}(A)$. If subjective information about the location of the θ_j's is available, it can be modeled through $\pi_{2,1}$; otherwise, $\pi_{2,1}$ will be chosen to be a noninformative prior. Typically, $\pi_{2,2}$ is chosen to be noninformative.

In Angers and Berger (1991), a generalized Bayes estimator was developed for the special case $k_j = 1$ for $j = 1, \ldots, p$. A numerical procedure to compute the estimator was also proposed. In this paper, the case where k_j is any odd integer will be investigated and a closed form expression will be given for the first stage Bayes estimator. In the last section, a numerical study will be performed to see how sensitive the proposed estimator is to the choice of the k_j's.

2. DEVELOPMENT OF THE HIERARCHICAL BAYES ESTIMATOR

If one uses the squared-error loss, it is well known (*cf.* Berger (1985) (Chapter 4)) that the hierarchical Bayes estimator will be given by

$$\widehat{\theta}_j = \frac{\int_0^\infty \int_{-\infty}^\infty \widehat{\theta}_{j|\mu,A} \left\{\Pi_{q=1}^p m(y_q \mid \mu, A)\right\} \pi_2(\mu, A)\, d\mu\, dA}{\int_0^\infty \int_{-\infty}^\infty \left\{\Pi_{q=1}^p m(y_q \mid \mu, A)\right\} \pi_2(\mu, A)\, d\mu\, dA}, \tag{1}$$

* This research was partially supported by NSERC.

where

$$m(y_q \mid \mu, A) = \int_{-\infty}^{\infty} f(y_q \mid \theta_q)\pi_{1,k_q}(\theta_q \mid \mu, A)\, d\theta_q,$$

$$f(y_q \mid \theta_q) = \frac{1}{\sqrt{2\pi\sigma^2}} \exp\left\{-\frac{1}{2\sigma^2}(y_q - \theta_q)^2\right\},$$

$$\pi_{1,k_q}(\theta_q \mid \mu, A) = \frac{\Gamma([k_q+1]/2)(k_q A^2)^{k_q/2}}{\sqrt{\pi}\Gamma(k_q/2)} \frac{1}{(k_q A^2 + (\theta-\mu)^2)^{[k_q+1]/2}},$$

$$\widehat{\theta}_{q|\mu,A} = \int_{-\infty}^{\infty} \theta_q \pi_{1,k_q}(\theta_q \mid \mu, A, y_q)\, d\theta_q,$$

$$\pi_{1,k_q}(\theta_q \mid \mu, A, y_q) = \frac{f(y_q \mid \theta_q)\pi_{1,k_q}(\theta_q \mid \mu, A)}{m(y_q \mid \mu, A)}.$$

The advantage of using equation 1 to compute $\widehat{\theta}_j$ is that only two 2-dimensional integrals have to be performed, even though the total number of parameters is $p+2$. However, since the integrands are a product of $p+1$ 1-dimensional integrals, an efficient way to compute these one dimensional integrals has to be used.

Beside the evaluation of equation 1, there exist other approaches one can use in order to compute $\widehat{\theta}_j$. The most well-known will be to use the empirical Bayes paradigm (*cf.* Morris (1983)). However, finding $\widehat{\mu}$ and $\widehat{A}$ can be as computationally intensive as the evaluation of equation 1. Using an approach similar to Spiegelhalter (1985), one can obtain an analytic expression for the unconditional prior of θ when $\pi_2(\mu, A) \propto A^{-a}$ ($a \geq 0$). This prior can be found by evaluating the following expression

$$\pi(\theta) = 2\pi i \sum_{q=1}^{p} \int_0^{\infty} \frac{\lambda^{a-2}}{(0.5[k_q-1])!} \frac{\partial^{(k_q-1)/2}}{\partial \mu^{(k_q-1)/2}} g_q\left(\theta_q + i\frac{\sqrt{k_q}}{\lambda}\right),$$

where k_q is an odd integer, and

$$g_q(\mu) = \frac{1}{\lambda^{p(k_q+1)}\left(\mu - \theta_q + i\frac{\sqrt{k_q}}{\lambda}\right)^{(k_q+1)/2}} \prod_{j \neq q} \frac{1}{\left[\frac{k_j}{\lambda} + (\mu - \theta_j)\right]^{(k_j+1)/2}}.$$

Doing so, $\widehat{\theta}_j$ will be given by

$$\widehat{\theta}_j = \frac{\int_{-\infty}^{\infty} \cdots \int_{-\infty}^{\infty} \theta_j \left\{\prod_{q=1}^{p} f(y_q \mid \theta_q)\right\} \pi(\theta)\, d\theta_1 \cdots d\theta_p}{\int_{-\infty}^{\infty} \cdots \int_{-\infty}^{\infty} \left\{\prod_{q=1}^{p} f(y_q \mid \theta_q)\right\} \pi(\theta)\, d\theta_1 \cdots d\theta_p}.$$

Unless p, $k_1, k_2, \ldots, k_p$ are small, equation 1 should be faster to evaluate than the previous equation. Futhermore, equation 1 allows the user to specify his own prior on μ and A while the previous equation is imposing a functional form for $\pi_2(\mu, A)$.

2.1. *Normal-Student-t Convolution*

In this subsection, μ and A will be assumed known.

Theorem 1. *Suppose that $Y_j \sim N(\theta_j, \sigma^2)$ independently for $j = 1, \ldots, p$ and that $\theta_j \sim \mathcal{T}_{k_j}(\mu, A)$ independently for $j = 1, \ldots, p$ where σ^2, μ and A are known. If $k_j = 2l + 1$, then*

$$\begin{aligned} m(y_j \mid \mu, A) &= \textit{marginal of } y_j \textit{ given } \mu \textit{ and } A \\ &= \frac{1}{2^{2l+1/2}\Gamma(l+1/2)\sigma} \\ &\quad \times \sum_{m=0}^{l} \frac{(2l-m)!}{(l-m)!} \left(2t_{(1)j}\right)^m \Re \mathrm{G}\left(\frac{m+1}{2}, t_j^2\right), \end{aligned} \tag{2}$$

$$\begin{aligned} \widehat{\theta}_{j|\mu,A} &= \textit{posterior mean of } \theta_j \textit{ given } \mu \textit{ and } A \\ &= y_j + \operatorname{sign}(\mu - y_j) \\ &\quad \times \frac{\sigma}{\sqrt{2}} \frac{\left[\sum_{m=0}^{l}(m+1)\frac{(2l-m)!}{(l-m)!}\left(2t_{(1)j}\right)^m \Im \mathrm{G}\left(\frac{m+2}{2}, t_j^2\right)\right]}{\left[\sum_{m=0}^{l}\frac{(2l-m)!}{(l-m)!}\left(2t_{(1)j}\right)^m \Re \mathrm{G}\left(\frac{m+1}{2}, t_j^2\right)\right]}, \end{aligned} \tag{3}$$

$$\begin{aligned} V_{j,j|\mu,A} &= \textit{posterior variance of } \theta_j \textit{ given } \mu \textit{ and } A \\ &= \sigma^2 - \left(y - \widehat{\theta}_{j|\mu,A}\right)^2 - \frac{\sigma^2}{2} \\ &\quad \times \frac{\left[\sum_{m=0}^{l}(m+2)(m+1)\frac{(2l-m)!}{(l-m)!}\left(2t_{(1)j}\right)^m \Re \mathrm{G}\left(\frac{m+3}{2}, t_j^2\right)\right]}{\left[\sum_{m=0}^{l}\frac{(2l-m)!}{(l-m)!}\left(2t_{(1)j}\right)^m \Re \mathrm{G}\left(\frac{m+1}{2}, t_j^2\right)\right]}, \end{aligned} \tag{4}$$

where $\Re \mathrm{G}\left(\frac{m+1}{2}, t_j^2\right)$ and $\Im \mathrm{G}\left(\frac{m+2}{2}, t_j^2\right)$ denote the real and the imaginary parts of the complex function

$$\mathrm{G}\left(a, t_j^2\right) = \frac{1}{\Gamma(a)\Gamma(a+1/2)} \sum_{m=0}^{\infty} \Gamma(a + m/2) \frac{(-2t_j)^m}{m!}, \tag{5}$$

and

$$t_j = t_{(1)j} - i t_{(2)j},$$
$$t_{(1)j} = \frac{\sqrt{2l+1}A}{\sqrt{2}\sigma},$$
$$t_{(2)j} = \frac{|\mu - y_j|}{\sqrt{2}\sigma}.$$

The proof of this theorem is achieved by proving the following lemmas.

Let $f_Y(\cdot)$ denote the density of $N(0, \sigma^2)$, $\pi_{1,k}(\cdot)$ the density of $\mathcal{T}_k(0, A)$ and let φ_Y and φ_μ be their characteristic functions.

Lemma 1. *If $k_j = 2l + 1$, then*

$$\varphi_\mu(z) = \frac{\sqrt{\pi}}{2^{2l}\Gamma(l+1/2)} e^{-\sqrt{2l+1}A|z|} \sum_{m=0}^{l} \frac{(2l-m)!}{m!(l-m)!} \left(2\sqrt{2l+1}A|z|\right)^m. \tag{6}$$

The proof of this lemma is obtained easily by using the change of variable $\eta = |\theta|/(\sqrt{k_j}A)$ in $\int_{-\infty}^{\infty} e^{iz\theta}\pi_{1,k}(\theta)\,d\theta$.

To compute the first stage marginal of y_j, one has to evaluate the integral

$$\int_{-\infty}^{\infty} |z|^m e^{\{-i|\mu-y_j|z\}} e^{(\sigma^2/2)z^2} e^{-\sqrt{2l+1}A|z|}\,dz,$$

for $m = 0, 1, \ldots, l$.

Lemma 2.

$$\int_{-\infty}^{\infty} |z|^m e^{\{-i|\mu-y_j|z\}} e^{(\sigma^2/2)z^2} e^{-\sqrt{2l+1}A|z|}\,dz = \frac{2m!}{(\sigma^2)^{m+1}}\Re\left(e^{t_j^2/2} D_{-(m+1)}(\sqrt{2}t_j)\right),$$

where $D_{-(m+1)}(\cdot)$ represents the parabolic cylinder function.

Proof.

$$\begin{aligned}
&\int_{-\infty}^{\infty} |z|^m e^{\{-i|\mu-y_j|z\}} e^{\sigma^2/2z^2} e^{-\sqrt{2l+1}A|z|}\,dz \\
&= \int_0^{\infty} z^m e^{\{i|\mu-y_j|z\}} \exp\left\{-\left(\frac{\sigma^2}{2}z^2 + \sqrt{2l+1}Az\right)\right\} dz \\
&\quad + \int_0^{\infty} z^m e^{\{-i|\mu-y_j|z\}} \exp\left\{-\left(\frac{\sigma^2}{2}z^2 + \sqrt{2l+1}Az\right)\right\} dz \\
&= 2\int_0^{\infty} z^m \cos(|\mu-y_j|z) \exp\left\{-\left(\frac{\sigma^2}{2}z^2 + \sqrt{2l+1}Az\right)\right\} dz \\
&= \frac{m!}{(\sigma^2)^{(m+1)/2}}\left\{e^{t_j^2/2} D_{-(m+1)}(\sqrt{2}t_j) + e^{\bar{t}_j^2/2} \mathrm{D}_{-(m+1)}(\sqrt{2}\bar{t}_j)\right\} \\
&= \frac{2m!}{(\sigma^2)^{(m+1)/2}}\Re\left(e^{t_j^2/2}\mathrm{D}_{-(m+1)}(\sqrt{2}t_j)\right).
\end{aligned}$$

◁

Consider the function $\mathrm{G}(a, t^2)$ defined by equation 5. Using the definition of the parabolic cylinder function (*cf.* Gradshteyn and Ryzhik (1980)), one has $\mathrm{D}_{-(m+1)}(\sqrt{2}t) = \sqrt{\pi}2^{-(m+1)/2}e^{-t^2/2}\mathrm{G}((m+1)/2, t^2)$, and substituting in the previous lemma, one has

$$\int_{-\infty}^{\infty} |z|^m e^{\{-i|\mu-y_j|z\}} e^{(\sigma^2/2)z^2} e^{-\sqrt{2l+1}A|z|}\,dz = \frac{2\sqrt{\pi}m!}{(2\sigma^2)^{(m+1)/2}}\Re(\mathrm{G}((m+1)/2, t_j^2).$$

Consequently, using the previous equation and equation 6, the first stage marginal of y_j is given by equation 2.

By Angers and Berger (1991), the first stage Bayes estimator of θ_j is given by

$$\widehat{\theta}_{j|\mu,A} = y_j + \frac{1}{m(y_j \mid \mu, A)}\int_{-\infty}^{\infty} \eta f_Y(\eta)\pi([\mu - y_j] - \eta)\,d\eta.$$

The last integral is given in the next lemma.

Lemma 3.

$$\int_{-\infty}^{\infty} \eta f_Y(\eta)\pi([\mu - y_j] - \eta)d\eta = \text{sign}(\mu - y_j)$$

$$\times \frac{1}{2^{2l+1}\Gamma(l+1/2)} \sum_{m=0}^{l} \frac{(2l-m)!}{(l-m)!}(m+1)(2t_{(1)j})^m \Im G(\frac{m+2}{2}, t_j^2).$$

Proof. Using the same techniques as in the proof of Lemma 2, we have

$$\int_{-\infty}^{\infty} \eta f_Y(\eta)\pi([\mu - y_j] - \eta)d\eta$$

$$= \frac{i\sigma^2}{2\pi} \int_{-\infty}^{\infty} ze^{-iz(\mu-y_j)}\varphi_Y(z)\varphi_\mu(z)dz$$

$$= \frac{i\sigma^2}{2\pi} \frac{\sqrt{\pi}}{2^{2l}\Gamma(l+1/2)} \sum_{m=0}^{l} \frac{(2l-m)!}{m!(l-m)!} \left(2\sqrt{2l+1}A\right)^m$$

$$\times \int_{-\infty}^{\infty} z|z|^m e^{-iz(\mu-y_j)} \exp\left\{-\left(\sigma^2/2z^2 + \sqrt{2l+1}Az\right)\right\} dz.$$

The last integral can be written

$$\int_{-\infty}^{\infty} z|z|^m e^{-iz(\mu-y_j)} \exp\left\{-\left(\frac{\sigma^2}{2}z^2 + \sqrt{2l+1}Az\right)\right\} dz$$

$$= -\int_{-\infty}^{0} |z|^{m+1} e^{-iz(\mu-y_j)} \exp\left\{-\left(\frac{\sigma^2}{2}z^2 + \sqrt{2l+1}Az\right)\right\} dz$$

$$+ \int_{0}^{\infty} z^{m+1} e^{-iz(\mu-y_j)} \exp\left\{-\left(\frac{\sigma^2}{2}z^2 + \sqrt{2l+1}Az\right)\right\} dz$$

$$= -\int_{0}^{\infty} z^{m+1} e^{iz(\mu-y_j)} \exp\left\{-\left(\frac{\sigma^2}{2}z^2 + \sqrt{2l+1}Az\right)\right\} dz$$

$$+ \int_{0}^{\infty} z^{m+1} e^{-iz(\mu-y_j)} \exp\left\{-\left(\frac{\sigma^2}{2}z^2 + \sqrt{2l+1}Az\right)\right\} dz$$

$$= -\text{sign}(\mu - y_j)2i \int_{0}^{\infty} z^{m+1} \sin(|\mu - y_j|z)$$

$$\times \exp\left\{-\left(\frac{\sigma^2}{2}z^2 + \sqrt{2l+1}Az\right)\right\} dz$$

$$= -\frac{\text{sign}(\mu - y_j)2i\sqrt{\pi}(m+1)!}{(\sigma^2)^{(m+2)/2}} \Im\left(e^{t_j^2/2} D_{(-m+2)}(\sqrt{2}t_j)\right)$$

$$= -\frac{\text{sign}(\mu - y_j)i\sqrt{\pi}(m+1)!}{2^{m/2}(\sigma^2)^{(m+2)/2}} \Im G(\frac{m+2}{2}, t_j^2).$$ ◁

The calculations needed to compute $V_{j,j|\mu,A}$ are similar to the ones required to evaluate $m(y_j \mid \mu, A)$ and $\widehat{\theta}_{j|\mu,A}$. Consequently, they will be omitted.

In order to obtain the hierarchical Bayes estimator, $\widehat{\theta}$, one will then have to integrate equation 3 with respect to the measure

$$\pi_2(\mu, A \mid y) = \frac{\prod_{q=1}^{p} m(y_q \mid \mu, A)\pi_2(\mu, A)}{\int_0^\infty \int_{-\infty}^\infty \left\{\prod_{q=1}^{p} m(y_q \mid \mu, A)\right\} \pi_2(\mu, A)\, d\mu\, dA},$$

where $m(y_q \mid \mu, A)$ is given by equation 3. The posterior expected loss of $\widehat{\theta}_j$ will then be given by

$$\rho(\widehat{\theta}_j) = \sigma^2 - (y_j - \widehat{\theta}_j)^2$$
$$- \frac{\sigma^2}{2} E^{\pi_2(\mu,A|y)} \left[\frac{\sum_{m=0}^{l} (m+2)(m+1) \frac{(2l-m)!}{(l-m)!} \left(2t_{(1)j}\right)^m \Re G\left(\frac{m+3}{2}, t_j^2\right)}{\sum_{m=0}^{l} \frac{(2l-m)!}{(l-m)!} \left(2t_{(1)j}\right)^m \Re G\left(\frac{m+1}{2}, t_j^2\right)} \right].$$

In the next section, the complex function $\mathrm{G}(a, t^2)$ will be studied.

3. THE FUNCTION $\mathrm{G}(a, t^2)$

In this section, the function $\mathrm{G}(a, t^2)$ will be studied and a scheme to compute it will be proposed. A quick way to compute $\mathrm{G}(a, t^2)$ will be to use a continued fraction. In order to do so, a recurrence relation for $\mathrm{G}(a, t^2)$ is needed.

Proposition 1.

$$\mathrm{G}(a+1, t^2) = \frac{1}{a+1/2} \left[\mathrm{G}(a, t^2) - t\mathrm{G}(a+1/2, t^2)\right].$$

Proof.
By Equation 5, we have that

$$\begin{aligned}
\mathrm{G}(a+1, t^2) &= \frac{1}{\Gamma(a+1)\Gamma(a+3/2)} \sum_{m=0}^{\infty} \Gamma(a+1+m/2) \frac{(-2t)^m}{m!} \\
&= \frac{1}{\Gamma(a+1)\Gamma(a+3/2)} \sum_{m=0}^{\infty} (a+m/2)\Gamma(a+m/2) \frac{(-2t)^m}{m!} \\
&= \frac{a}{\Gamma(a+1)\Gamma(a+3/2)} \sum_{m=0}^{\infty} \Gamma(a+m/2) \frac{(-2t)^m}{m!} \\
&\quad - \frac{t}{\Gamma(a+1)\Gamma(a+3/2)} \sum_{m=0}^{\infty} \Gamma(a+1/2+m/2) \frac{(-2t)^m}{m!} \\
&= \frac{1}{a+1/2} \frac{1}{\Gamma(a)\Gamma(a+1/2)} \sum_{m=0}^{\infty} \Gamma(a+m/2) \frac{(-2t)^m}{m!} \\
&\quad - \frac{t}{a+1/2} \frac{1}{\Gamma(a+1/2)\Gamma([a+1/2]+1/2)} \\
&\quad \times \sum_{m=0}^{\infty} \Gamma([a+1/2]+m/2) \frac{(-2t)^m}{m!} \\
&= \frac{1}{a+1/2} \left[\mathrm{G}(a, t^2) - t\mathrm{G}(a+1/2, t^2)\right].
\end{aligned}$$
◁

Using Proposition 1, one can compute the S-fraction expansion (*cf.* Wall (1948) (Chapter 18)) of the ratio $\mathrm{RG}(a, t^2) = \mathrm{G}(a+1/2, t^2)/\mathrm{G}(a, t^2)$.

Proposition 2.

$$\mathrm{RG}(a,t^2) = \frac{1}{t+}\,\frac{(a+1/2)}{t+}\,\frac{(a+1)}{t+\ldots}. \tag{7}$$

Proof.
Using Proposition 1, one has

$$(a+1/2)\mathrm{G}(a+1,t^2) = \mathrm{G}(a,t^2) - t\mathrm{G}(a+1/2,t^2).$$

Dividing both sides of the equation by $\mathrm{G}(a+1/2,t^2)$ and rearranging the different terms, one obtains

$$\mathrm{RG}(a,t^2) = \frac{1}{t-(a+1/2)\mathrm{RG}(a+1/2,t^2)}. \tag{8}$$

Replacing a by $a+1/2$ in equation 8 and $\mathrm{RG}(a+1/2,t^2)$ by the resulting expression leads to equation 7. $\triangleleft$

Proposition 3.
The continued fraction given by equation 7 converges $\forall t \in C$ such that $\Re(t) > 0$.

Proof. By means of equivalence transformations, equation 7 can be written as

$$\frac{1}{b_1+}\,\frac{1}{b_2+\ldots},$$

where

$$b_1 = t,$$

$$b_{2q} = 2^{2q-1}\left[\frac{\Gamma(a+q)}{\Gamma(a+1)}\right]^2 \frac{(2a)!}{(2a+2q-1)!}t,$$

$$b_{2q+1} = \frac{1}{2^{2q}}\left[\frac{\Gamma(a+1)}{\Gamma(a+q+1)}\right]^2 \frac{(2a+2q)!}{(2a)!}t.$$

Using Wall (1948) (Theorem 30.2), this continued fraction will converge if

C.1 $\Re(b_1) > 0$,
C.2 $\Re(b_{2q}) \geq 0$,
C.3 $|\Im(b_{2q-1})| \leq c\Re(b_{2q-1})$, for $c \in R^+$,
C.4 $\sum |b_{2q+1}|$ diverges.

Since $\Re(b_1) = \Re(b_{2q}) = t_1 = \sqrt{k}A/(\sqrt{2}\sigma)$, conditions C.1 and C.2 are obviously satisfied. By choosing $c = \Im(t)/\Re(t)$, as long as $\Re(t) > 0$, condition C.3 is also satisfied. Using Stirling's approximation, one can easily show that, if q is large enough,

$$\frac{(2a+2q)!}{\Gamma^2(a+q+1)} \simeq \frac{2^{2(a+q)}}{\sqrt{\pi}}\sqrt{a+q},$$

so $b_{2q+1} \simeq \frac{2^{2a}\Gamma^2(a+1)}{\sqrt{\pi}(2a)!}\sqrt{a+q}$. Consequently, condition C.4 is also satisfied.

Since $\mathrm{G}(a+1/2,t^2) = \left[\prod_{q=1}^{2a} \mathrm{RG}(q/2,t^2)\right]\mathrm{G}(1/2,t^2)$, a continued fraction to evaluate $\mathrm{G}(1/2,t^2)$ is required. Using Angers and Berger (1991), $\mathrm{G}(1/2,t^2)$ can be written as

$$\mathrm{G}(1/2,t^2) = \frac{2}{\sqrt{\pi}}e^{t^2}\int_t^\infty e^{-z^2}\,dz.$$

Using Wall (1948) (Chapter 18 and the equivalence transformations $r_{2q} = 1$, $r_{2q+1} = 1/2$), it can be easily seen that a continued fraction for $\mathrm{G}(1/2,t^2)$ is given by equation 7 with $a = 0$. It should be noted that for small values of $\Re(t)$, equation 7 (continued fraction) converges very slowly. In this case, the equation 5 (series) should be used instead. $\triangleleft$

4. NUMERICAL EXAMPLE

For the numerical example discussed in this section, the second stage prior was chosen to be $\pi_2(\mu, A) \equiv 1$. To integrate over the hyperparameters, we simply used Riemann sums over a bounded region. This region was chosen by looking at the values of $\prod_{q=1}^{p} m(y_q \mid \mu, A)$ for several values of A and μ (see Figure 1).

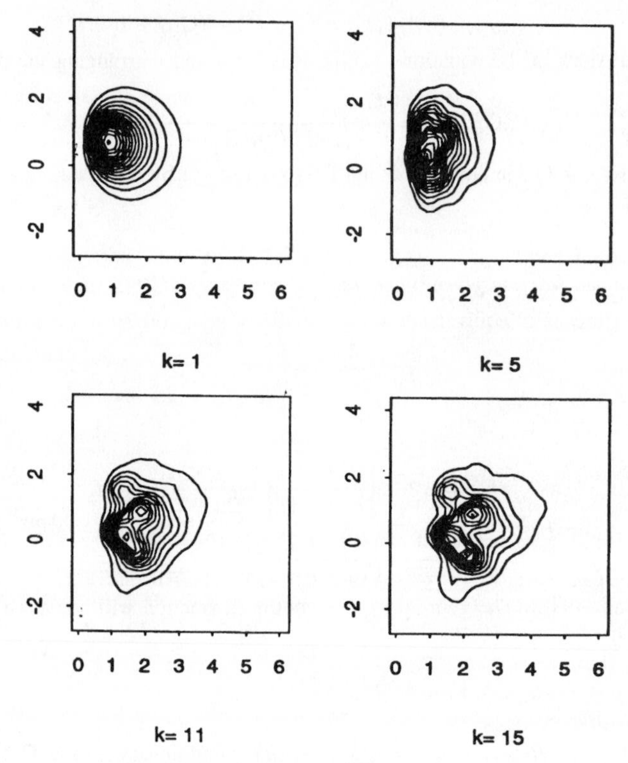

Figure 1. *Contour plot of* $\prod_{q=1}^{p} m(y_q|\mu, A)$.

This may not be the most efficient way to perform double numerical integrals and other techniques, such as Gibbs sampling and adaptative importance sampling, should eventually be considered.

The observation vector used in this section has been generated according to the following scheme:

1. generate $\theta_j \sim \mathcal{C}(0, 1)$, for $j = 1, \ldots, 10$,
2. generate $Y_j \sim N(\theta_j, 1)$, for $j = 1, \ldots, 10$.

To simplify the computer program, we chose $k_1 = \cdots = k_{10} = k$. In order to study the sensitivity of $\widehat{\theta}$ with respect to k, we computed $\widehat{\theta}$ for $k = 1, 3, \ldots, 15$.

In Figure 1, we give the contour plots of $\prod_{q=1}^{p} m(y_q \mid \mu, A)$ for k = 1, 5, 11 and 15. It should be noted that these curves are concentrated in the region $(0;\ 6) \times (-2,5;\ 4)$ which somehow justified the use of Riemann's sums.

In Table 1, the values of $\widehat{\theta}_j$ and its posterior expected loss for $j = 1, \ldots, 10$ and for $k = 1, 3, \ldots, 15$ are given. One can see that the effect of the number of degrees of freedom is very limited for the estimated values. The bigger change occurs at $j = 1$. In fact, the bigger $|y_j|$ is, the stronger the effect of k is. Furthermore, the change is almost monotonic. For $j = 4$ and 10, there is a reversal in the trend but this change is negligible (less than $0, 5\%$). It is also interesting to note that $\widehat{\theta}_1$ and $\widehat{\theta}_{10}$ seem to "diverge" while the others are shrunk toward $\overline{y} = 0, 418$. However, for those two cases, the posterior expected loss is very stable. For all the others, the posterior expected loss decreases when k increases.

		$\widehat{\theta}_j$ (posterior expected loss)							
$j\backslash k$	y_j	1	3	5	7	9	11	13	15
1	−4,55	−4,15 (1,08)	−4,45 (1,10)	−4,53 (1,09)	−4,57 (1,08)	−4,61 (1,08)	−4,64 (1,09)	−4,66 (1,09)	−4,69 (1,08)
2	−1,88	−1,26 (1,05)	−1,19 (1,08)	−1,18 (1,05)	−1,16 (0,98)	−1,10 (0,86)	−1,03 (0,71)	−0,96 (0,56)	−0,89 (0,41)
3	−0,54	−0,12 (0,81)	0,03 (0,71)	0,04 (0,66)	0,05 (0,62)	0,05 (0,58)	0,06 (0,54)	0,06 (0,50)	0,07 (0,46)
4	−0,44	−0,04 (0,79)	0,09 (0,69)	0,11 (0,66)	0,11 (0,62)	0,10 (0,59)	0,11 (0,56)	0,11 (0,53)	0,11 (0,50)
5	0,54	0,60 (0,70)	0,60 (0,65)	0,58 (0,65)	0,57 (0,66)	0,55 (0,66)	0,53 (0,67)	0,51 (0,66)	0,49 (0,66)
6	0,91	0,83 (0,70)	0,78 (0,65)	0,74 (0,66)	0,72 (0,66)	0,70 (0,65)	0,67 (0,64)	0,64 (0,63)	0,61 (0,62)
7	1,09	0,94 (0,71)	0,86 (0,66)	0,83 (0,66)	0,80 (0,65)	0,77 (0,64)	0,74 (0,62)	0,70 (0,60)	0,66 (0,57)
8	2,31	1,79 (0,87)	1,62 (0,80)	1,59 (0,78)	1,57 (0,74)	1,54 (0,68)	1,49 (0,60)	1,45 (0,52)	1,41 (0,45)
9	2,58	2,01 (0,92)	1,86 (0,88)	1,84 (0,86)	1,84 (0,82)	1,81 (0,76)	1,77 (0,69)	1,73 (0,61)	1,69 (0,54)
10	4,16	3,58 (1,14)	3,75 (1,24)	3,80 (1,25)	3,82 (1,25)	3,82 (1,25)	3,82 (1,25)	3,81 (1,24)	3,81 (1,24)

Table 1. *Hierarchical Bayes estimator of θ for several values of k.*

REFERENCES

Angers, J.-F., and Berger, J. O. (1991). Robust hiearchical Bayes estimation of exchangeable means. *Canadian J. Statist.* **19**, 39–56.

Berger, J. O. (1985). *Statistical Decision Theory and Bayesian Analysis.* New York: Springer.

Gradshteyn, I. S. and Ryzhik, I. M. (1980). *Table of Integrals, Series, and Products.* New York: Academic Press.

Morris, C. (1983). Parametric empirical Bayes inference: Theory and applications. *J. Amer. Statist. Assoc.* **78**, 47–65.

Spiegelhalter, D. J. (1985). Exact Bayesian inference on the parameters of a Cauchy distribution with vague prior information. *Bayesian Statistics 2* (J. M. Bernardo, M. H. DeGroot, D. V. Lindley and A. F. M. Smith, eds.), Amsterdam: North-Holland, 743–750.

Wall, H. S. (1948). *Analytic Theory of Continued Fraction.* New York: Von Nostrand.

BAYESIAN STATISTICS 4, pp. 577–586
J. M. Bernardo, J. O. Berger, A. P. Dawid and A. F. M. Smith, (Eds.)

Monte Carlo Bayesian Methods for Discrete Regression Models and Categorical Time Series

BRADLEY P. CARLIN and NICHOLAS G. POLSON
University of Minnesota, USA and *University of Chicago, USA*

SUMMARY

Discrete regression models and categorical time series are viewed as a constrained Bayesian hierarchical model. A Monte Carlo approach employing latent data variables is adopted, which leads to a conceptually simple and computationally feasible approach to this class of problems. We offer two illustrative examples. The first analyzes a binomial regression model and computes influence diagnostics based on Kullback-Leibler divergences between full and reduced dataset posteriors for a parameter of interest. The second example involves a state space model analysis of a binary time series produced by monitoring an infant's sleep patterns.

Keywords: BINOMIAL REGRESSION; GIBBS SAMPLER; INFLUENCE DIAGNOSTICS; KULLBACK-LEIBLER DIVERGENCE.

1. INTRODUCTION

In many fields of application the observed data are discrete valued and possibly correlated over time. Accordingly, there are many statistical techniques available for explaining and exploring possible mechanisms generating the data. The wide range of models includes: log-linear models, logistic regression models, categorical longitudinal models (Zeger *et al.* 1985), nonhomogeneous Markov chain models (Kaufmann, 1987) and binary time series models (Grether and Maddala, 1982; Heckmann, 1981). Liang, Zeger and Qaqish (1992) contains a recent survey of categorical regression models and their application.

The purpose of this paper is to develop a Monte Carlo framework for analyzing discrete regression models, and for prediction, filtering, and obtaining inferences in categorical time series. Our approach is to augment the likelihood for the data y with a vector of latent continuous random variables y^* in a way reminiscent of the work of Tanner and Wong (1987). In Section 2 we show how this approach leads to a simple algorithm for simulation from the joint posterior $p(\beta, y^*|y)$, and hence $p(\beta|y)$, where β denotes the vector of unknown model parameters. The algorithm is conceptually simple and provides Monte Carlo Bayesian inference where the frequentist approach is notoriously hard. Section 3 considers two examples: first, a Bayesian approach to estimating posterior distributions and computing influence diagnostics for binomial regression, and second, a state space model analysis of a binary dataset on an infant's sleep pattern.

2. LATENT VARIABLE MODELING

2.1. *Discrete Regression*

Suppose we wish to analyze data arising from the usual discrete regression setting, where we have continuous k-dimensional predictor variables x_i, and discrete univariate response variables y_i, $i = 1, \ldots, r$. Without loss of generality, we assume $y_i \in \{0, 1, 2, \ldots, n_i\}$ for all i. Consider the model structure

$$y_i = A(y_i^*), y_i^* = \boldsymbol{x}_i^T \beta + \epsilon_i, i = 1, \ldots, r. \tag{1}$$

Here, ϵ_i is an error innovation with mean 0 and cumulative distribution function F, β is a k-dimensional regression parameter, and $A(\cdot)$ is a many-to-one function which "discretizes" the unobserved continuous data y_i^*. Since A is not one-to-one, let A^{-1} denote its set theoretic inverse, and moreover suppose that $A^{-1}(y_i) = (a_i, b_i) \subset \Re$ for some a_i and b_i where $-\infty \leq a_i < b_i \leq \infty$. Hence $\mu(A^{-1}(y_i)) > 0$ for all possible y_i, where μ denotes Lebesgue measure. The desired discrete regression model may be viewed in the constrained regression framework given in (1) above as follows.

Suppose Y_i is discrete with cumulative distribution function $G(\cdot)$ (that is, $P(Y_i \leq y_i|\beta) = G(y_i|\beta)$). Then by taking $a_i = \boldsymbol{x}_i^T \beta + F^{-1}(G(y_i - 1|\beta))$ and $b_i = \boldsymbol{x}_i^T \beta + F^{-1}(G(y_i|\beta))$, the conditional density of y_i under model (1) becomes

$$\begin{aligned} P(Y_i = y_i|\beta) &= P(y_i^* \in A^{-1}(y_i)|\beta) = F\left(F^{-1}(G(y_i|\beta)\right) - F\left(F^{-1}(G(y_i - 1|\beta)\right) \\ &= G(y_i|\beta) - G(y_i - 1|\beta) \end{aligned} \tag{2}$$

as required. In applications, a flexible family arises by taking $G \sim$ Binomial(n_i, p_i) and choosing a link function in order to express p_i as a function of the covariates $\boldsymbol{x}_i$. It is easy to show that taking F to be a normal distribution in model (1) is equivalent to adopting the probit link function, $p_i = \Phi(x_i^T \beta)$ where Φ denotes the standard normal cdf. Albert and Chib (1991) observed that in the case of binary data ($n_i = 1$ for all i), the b_i constraint limits take a particularly simple form: if $y_i = 0$, $(a_i, b_i) = (-\infty, 0)$; if $y_i = 1$, $(a_i, b_i) = (0, \infty)$. A natural extension of the binary case is to consider a response with k categories having multinomial probabilities that depend on covariates through a link function, regression parameters, and possibly further noise (see McCullagh, 1980). This class of models, perhaps including *a priori* ordering constraints on the regression parameters, may also be analyzed in our sampling framework.

Returning to model (1), our interest focuses on estimating the k-dimensional parameter vector β; our model also involves the r latent data variables y_i^* as nuisance parameters. The addition of these parameters enables a simple expression for the complete conditional distribution of β; that is, the distribution of β given the observed data and all the remaining unknown parameters in the model (here, the y_i^*'s). But the complete conditionals for the y_i^*'s themselves are also readily available, and hence generation of a random sample from the marginal posterior distribution of β is easily accomplished using the Gibbs sampler (Gelfand and Smith, 1990). We do not attempt a review of the method here, merely pointing out that the ability to sample easily from each complete conditional distribution is all that is required. Albert and Chib (1991) show how to implement such a solution for binary response data. In the case of model (1), suppose we adopt a $N(\boldsymbol{\mu}, \Sigma)$ prior for β, and denote $(y_1^*, \ldots, y_r^*)$ by $\boldsymbol{y}^*$, $(y_1, \ldots, y_r)$ by $\boldsymbol{y}$, and $(\boldsymbol{x}_1, \ldots, \boldsymbol{x}_r)$ by X^T. Then the required complete conditionals are given by

$$\beta|\boldsymbol{y}^*, \boldsymbol{y} \sim \beta|\boldsymbol{y}^* \sim N(B\boldsymbol{b}, B), \text{ where } B^{-1} = \Sigma^{-1} + X^T X \text{ and } \boldsymbol{b} = \Sigma^{-1}\boldsymbol{\mu} + X^T \boldsymbol{y}^* \tag{3}$$

and

$$y_i^*|y_{j\neq i}^*, \boldsymbol{y}, \beta \sim y_i^*|y_i, \beta \propto N(\boldsymbol{x}_i^T\beta\,,\,1)\,I_{(a_i,b_i)} \tag{4}$$

where I_E denotes the indicator of the set E. For example, in our binary regression model where $y_i = A(y_i^*) = 1$ if $y_i^* > 0$ and $y_i = 0$ if $y_i^* < 0$, the complete conditional for y_i^* is normal truncated to $(-\infty, 0)$ if $y_i = 0$ and to $(0, \infty)$ if $y_i = 1$. Generation from truncated distributions like these is easily accomplished using a method given by Devroye (1986): If V_1 is a random variable having cumulative distribution function D, and V_2 is a truncated version of this random variable with support restricted to the interval $[a, b]$, then V_2 can be generated as $D^{-1}[D(a) + U(D(b) - D(a))]$, where $U \sim \text{Uniform}(0, 1)$.

Albert and Chib (1991) also introduce further parameters into the model which enable the normal (probit) link to be generalized to any t-link by writing the distribution of y_i^* as a scale mixture of normals. This idea also appears in a Gibbs sampling context in Carlin and Polson (1992).

2.2. *Discrete State Space Model*

Here we assume that the observed categorical time series, $\{y_t\}$, is a known function of an underlying continuous response process, $\{y_t^*\}$, which evolves according to a state space model. There are many examples of short binary time series in the applied sciences in the form of longitudinal data (see for example Zeger *et al.* 1985). We employ the hierarchical structure

$$y_t = A(y_t^*)\ ;\ y_t^* = H_t x_t + v_t\ ;\ x_t = C_t x_{t-1} + u_t\ ,\ t = 1, \ldots, T\ , \tag{5}$$

where x_t is an unknown $p \times 1$ state vector we wish to estimate, y_t^* is an unobservable $q \times 1$ latent data vector from a continuous valued process which gives rise to the observed discrete data $y_t \in \{0, 1, 2, \ldots, n_i\}$, C_t is a $p \times p$ matrix of constants, and H_t is a $q \times p$ matrix of constants. Again $y_t = A(y_t^*)$ is a known many-to-one function of y_t^*, and u_t and v_t are independent mean 0 error variables. Let $\boldsymbol{y} = (y_1, \ldots, y_T)$ denote the observed data, $\boldsymbol{y}^* = (y_1^*, \ldots, y_T^*)$ denote the latent data, $\boldsymbol{x} = (x_1, \ldots, x_T)$ the elements of the state, and x_0 the initial state, where we assume $x_0 \sim N_p(\mu_0, \Sigma_0)$. Typically one assumes $u_t \sim N_p(0, \Sigma)$ and $v_t \sim N_q(0, \Upsilon)$, where N_p denotes the p-dimensional normal distribution. Our goal is to determine estimates of the marginal densities of the model parameters, as well as the densities $p(x_{T+1}|\boldsymbol{y}, y_{T+1})$ (filtering) and $p(x_{T+1}|\boldsymbol{y})$ (one-step ahead prediction).

Similar to the previous subsection, let F be the cdf of v_t and G be the cdf of y_t given x_t. Then by choosing $a_t = H_t x_t + F^{-1}(G(y_t - 1|x_t))$ and $b_t = H_t x_t + F^{-1}(G(y_t|x_t))$, an argument very similar to that in equation (2) shows that our latent variable approach is equivalent to the desired discrete state space model. In the case where F is normal (as above), our model is equivalent to employing a generalized probit link function. Suppressing the conditioning on $(\mu_0, \Sigma_0, F_t, H_t)$, the likelihood specification for the latent variables $\boldsymbol{y}^*$ and the state parameters $\boldsymbol{x}$ is given by

$$p(y_1^*, \ldots, y_T^*, x_0, x_1, \ldots, x_T|\Sigma, \Upsilon) = g_1(x_0|\mu_0, \Sigma_0) \prod_{t=1}^{T} g_1(x_t|x_{t-1}, \Sigma) \prod_{t=1}^{T} g_2(y_t^*|x_t, \Upsilon) \tag{6}$$

where $g_1(\cdot)$ and $g_2(\cdot)$ are the appropriate normal densities. Assuming the matrices C_t and H_t to be known, in order to implement the Gibbs sampler we need to be able to generate from the following complete conditional distributions:

- $y_t^*|y_{j\neq t}^*, \Sigma, \Upsilon, \boldsymbol{y}, \boldsymbol{x}, x_0 \sim y_t^*|\Upsilon, y_t, x_t,\ \ t = 1, \ldots, T$

- $x_t|x_{j\neq t}, \boldsymbol{y}^*, \Sigma, \Upsilon, \boldsymbol{y} \sim x_t|x_{j\neq t}, y_t^*, \Sigma, \Upsilon, \quad t = 0, \ldots, T$
- $\Sigma|\boldsymbol{y}^*, \Upsilon, \boldsymbol{y}, \boldsymbol{x}, x_0 \sim \Sigma|\boldsymbol{x}, x_0$
- $\Upsilon|\boldsymbol{y}^*, \Sigma, \boldsymbol{y}, \boldsymbol{x}, x_0 \sim \Upsilon|\boldsymbol{y}^*, \boldsymbol{x}$

First, the complete conditional $y_t^*|\Upsilon, y_t, x_t$ can be obtained by noting that, by definition, $y_t^*|\Upsilon, x_t \sim N(H_t x_t, \Upsilon)$, and that given y_t this density is conditioned to the set $y_t^* \in A^{-1}(y_t) = (a_t, b_t)$. Therefore,

$$y_t^*|\Upsilon, y_t, x_t \propto N(H_t x_t, \Upsilon)\, I_{(a_t, b_t)}\,. \tag{7}$$

Secondly, by virtue of the conditioning on $\boldsymbol{y}^*$, the three remaining complete conditionals can all be determined from the usual continuous state space model: $y_t^* = H_t x_t + v_t$, $x_t = C_t x_{t-1} + u_t$. Carlin, Polson and Stoffer (1992) develop these complete conditionals, and also offer generalized versions to be used in nonlinear and nonnormal modeling scenarios. In the example of Section 3.2 below we provide more detail in the case of a univariate linear state space model.

2.3. *Categorical Time Series via a Markov Chain*

An interesting alternative approach to modeling categorical time series is suggested by Kaufmann (1987), where such a series is modeled as a nonhomogeneous Markov chain with an underlying regression structure. Specifically, let $\{y_t^*\}_{t=1}^T$ be a time series with c possible categories for each observation. Let $\boldsymbol{y}_t = (y_{t1}, \ldots, y_{tq})$ be an indicator of which category is observed at time t, where $q = c - 1$ and a vector of zeroes indicates category c. The Markov chain model is specified via the conditional probabilities $\{\pi_t\}$, where $\pi_t = P(y_{tj} = 1|\boldsymbol{y}_{t-1}, \ldots, \boldsymbol{y}_1)$ for $1 \le j \le q$. The regression structure appears by assuming that $\pi_t = h(Z_t^T\beta)$ where β is a $k \times 1$ vector of unknown parameters and $h(\cdot)$ is a link function. Here Z_t^T is a $k \times q$ matrix which is a known function of past observations *and* regressor variables $\{\boldsymbol{x}_t\}$. The likelihood function for the nonhomogeneous Markov chain is given by

$$L(\pi) = \prod_{t=1}^{T}\prod_{j=1}^{q} \pi_{tj}^{y_{tj}} \text{ where } \pi_t = h(Z_t^T\beta)$$

We can model this by introducing latent variables y_t^* that evolve according to the standard Bayesian regression model, $y_t^* = Z_t^T\beta + \epsilon_t$, $\beta = \beta_0 + \nu_t$, and assume that $y_t = A(y_t^*)$ for a suitable $A(\cdot)$. For example, to produce a binary time series, consider *clipping* the process y_t^*, that is $y_t = 1$ if $y_t^* \ge 0$ and $y_t = 0$ if $y_t^* < 0$, as in the case of a binary probit regression. In this example, the required complete conditional for y_t^* is given by $y_t^*|y_{j\neq t}^*, \beta, \boldsymbol{y} \sim N(Z_t^T\beta, 1) I_{A^{-1}(y_t)}$, $t = 1, \ldots, T$. Letting $\boldsymbol{Z}^T = (Z_1, Z_2, \ldots, Z_T)$, due to the conditioning on $\boldsymbol{y}^*$ the complete conditionals for $\beta|\boldsymbol{y}^*, \boldsymbol{y}$ can be obtained as $\beta|\boldsymbol{y}^*, \boldsymbol{y} \sim N(B\boldsymbol{b}, B)$ where $B^{-1} = \boldsymbol{Z}^T\boldsymbol{Z} + \Sigma_0^{-1}$ and $\boldsymbol{b} = \boldsymbol{Z}^T\boldsymbol{y}^* + \Sigma_0^{-1}\beta_0$.

As a typical example, consider the case where $\pi_t = h(Z_t^T\beta)$ with $Z_t^T\beta = \sum_{i=1}^{l} \alpha_i y_{t-i} + \sum_{i=1}^{k} \gamma_i x_{ti}$ for some explanatory variables x_t and parameters $\beta = (\alpha, \gamma)$ where we assume *a priori* that $\beta \sim N(\beta_0, \Sigma_0)$ with β_0, Σ_0 known. This then is a nonhomogeneous Markov chain of order l.

3. EXAMPLES

Example 3.1.: Binomial Regression Model. We apply the model introduced in Section 2.1 to the data presented in Table 1, which were discussed by Brown (1980) and recently reanalyzed by Bedrick and Hill (1990). The purpose of the experiment was to evaluate the ability of five binary preoperative prognostic variables to predict the presence or absence of cancer in the lymph nodes of 53 prostate cancer patients. The five prognostic variables are age (1 = under 60 years, 0 = not), stage (1 = biopsy indicates large or more irregular tumor, 0 = not), grade (1 = positive pathology reading, 0 = not), x-ray (1 = positive x-ray reading, 0 = not), and acid (1 = serum acid phosphatase level at least 60, 0 = not). Only 23 of the possible $2^5 = 32$ combinations of covariate values are present in the data. For the n_i patients having combination i, Table 1 gives the number of patients y_i who upon surgery proved positive for nodal involvement.

i	n_i	y_i	age_i	$stage_i$	$grade_i$	$xray_i$	$acid_i$	$KL(\{i\})$	$p(t_i^2)$	$p(\tilde{\Delta}_i)$
1	6	5	0	1	1	1	1	0.50	0.39	1.00
2	6	1	0	0	0	0	1	0.00	1.00	1.00
3	4	0	1	1	1	0	0	0.45	0.39	0.21
4	4	2	1	1	0	0	1	0.00	1.00	1.00
5	4	0	0	0	0	0	0	0.01	1.00	1.00
6	3	2	0	1	1	0	1	0.05	1.00	1.00
7	3	1	1	1	0	0	0	0.31	0.35	0.35
8	3	0	1	0	0	0	1	0.06	1.00	0.61
9	3	0	1	0	0	0	0	0.01	1.00	1.00
10	2	0	1	0	0	1	0	0.07	1.00	1.00
11	2	1	0	1	0	0	1	0.03	1.00	1.00
12	2	1	0	0	1	0	0	0.58	0.18	0.18
13	1	1	1	1	1	1	1	0.00	1.00	1.00
14	1	1	1	1	0	1	1	0.03	1.00	1.00
15	1	1	1	0	1	1	1	0.12	1.00	1.00
16	1	1	1	0	0	1	1	0.16	1.00	1.00
17	1	0	1	0	1	0	0	0.01	1.00	1.00
18	1	1	0	1	1	1	0	0.08	1.00	1.00
19	1	0	0	1	1	0	0	0.08	1.00	1.00
20	1	1	0	1	0	1	0	0.33	1.00	1.00
21	1	1	0	0	1	0	1	0.22	0.35	0.35
22	1	0	0	0	0	1	1	0.18	0.41	0.41
23	1	0	0	0	0	1	0	0.05	1.00	1.00

Table 1. *Nodal involvement data and influence measures.*

We apply our latent variable model assuming a vague prior specification on β by taking $\Sigma^{-1} = \mathbf{0}$. We successively generated from the complete conditional distributions given in (3) and (4) for $l = 30$ iterations, and replicated this entire process in parallel $G = 2500$ times, obtaining 2500 independent Gibbs samples from which to estimate the marginal posterior distributions of the components of β. (As all the generation is "one-for-one", this entire process takes only 2 minutes running FORTRAN on a DECStation 3100.) Using complete conditional density mixing as advocated by Gelfand and Smith (1990), we obtained smooth, unimodal univariate posteriors for the $\beta_j, j = 0, \ldots, 5$ (not shown). From these we found the β posterior mode vector $(-1.849, -.125, .805, .629, 1.126, 1.021)$, and the following 95% equal tail posterior credible intervals: for the intercept, $(-3.246, -.689)$; for age,

$(-1.201, .951)$; for stage, $(-.249, 1.898)$; for grade, $(-.472, 1.730)$; for x-ray, $(.040, 2.211)$; and for acid, $(-.009, 2.122)$. These posterior results are comparable to the standard probit model MLE vector of $\hat{\beta} = (-1.768, -.178, .796, .508, 1.005, .940)$ and associated estimated standard error vector $\hat{\sigma}(\hat{\beta}) = (.526, .434, .448, .467, .453, .441)$. The credible sets suggest that of the five prognostic variables, only the x-ray and serum acid test results are particularly good predictors of nodal involvement in this patient group.

Bedrick and Hill (1990) study the influence of each of the 23 covariate combination groups on β using exact conditional frequentist analyses. From a Bayesian point of view, Pettit and Smith (1985) define and compute a variety of influence measures within the framework of linear models. In particular, they recommend a symmetrised Kullback-Leibler divergence criterion. Carlin and Polson (1991) discuss how the Gibbs sampling approach may be used to evaluate various influence diagnostics based on Kullback-Leibler divergences. In particular, consider the influence diagnostic

$$\begin{aligned} KL(S) &= \int_{\Theta} p(\theta|\boldsymbol{y}) \log\left(\frac{p(\theta|\boldsymbol{y})}{p(\theta|\boldsymbol{y}(\bar{S}))}\right) d\theta = \\ &= E_{\theta|\boldsymbol{y}}[\log f(\boldsymbol{y}(S)|\theta)] - \log\left[E_{\theta|\boldsymbol{y}(\bar{S})} f(\boldsymbol{y}(S)|\theta)\right], \end{aligned}$$

where $\boldsymbol{y}(\bar{S})$ denotes the data with the set S deleted. Using the Gibbs sampler to generate samples $(\theta_1^{\boldsymbol{y}}, \ldots, \theta_G^{\boldsymbol{y}})$ from $p(\theta|\boldsymbol{y})$ and samples $(\theta_1^{\boldsymbol{y}(\bar{S})}, \ldots, \theta_G^{\boldsymbol{y}(\bar{S})})$ from $p(\theta|\boldsymbol{y}(\bar{S}))$, a Monte Carlo estimate of $KL(S)$ for suitably large G may be obtained as

$$KL(S) \approx \frac{1}{G}\sum_{j=1}^{G} \log f(\boldsymbol{y}(S)|\theta_j^{\boldsymbol{y}}) - \log\left(\frac{1}{G}\sum_{j=1}^{G} f(\boldsymbol{y}(S)|\theta_j^{\boldsymbol{y}(\bar{S})})\right). \tag{8}$$

In our case we let $\theta = \beta$ and $S = \{y_i\}$, so that $f(\boldsymbol{y}(S)|\theta_j^{\boldsymbol{y}}) = \binom{n_i}{y_i}[\Phi(\boldsymbol{x}_i^T\beta_j^{\boldsymbol{y}})]^{y_i}[1 - \Phi(\boldsymbol{x}_i^T\beta_j^{\boldsymbol{y}})]^{n_i - y_i}$, with a similar expression holding for $f(\boldsymbol{y}(S)|\theta_j^{\boldsymbol{y}(\bar{S})})$. Computing $KL(\{i\})$ for each of the 23 covariate combination groups entails running the Gibbs sampler 24 times, as the last term in (8) requires that the conditioning variables be from the appropriate "deleted" model. Carrying out these calculations, we arrive at the $KL(\{i\})$ column of Table 1. The exact conditional p-values computed by Bedrick and Hill (1990) for the standardized squared Pearson residuals, t_i^2, and the standardized squared deviance residuals, $\tilde{\Delta}_i$, are also shown for comparison. Notice that the orderings of points by the two frequentist approaches are comparable to that produced by our vague prior Bayes analysis. In particular, point #12 appears influential since it has negative x-ray and acid tests, yet an observed nodal involvement of 1/2. By contrast, the influence of point #1 seems due primarily to its large sample size. Yet no data point has a conspicuously large value of $KL(\{i\})$, comparable to the fact that none of the conditional p-values approaches the traditional 0.05 significance level.

Example 3.2.: Univariate Discrete State Space Model. Consider the infant sleep data plotted (as asterisks) versus time in Figure 1. This is binary data ($y_t = 1$ if the infant was judged to be in REM sleep during minute t, $y_t = 0$ otherwise) recorded during a 120 minute EEG study conducted and reported by Stoffer *et al.* (1988). As the existence of an underlying continuous "sleep state" variable x_t seems plausible, we shall fit a parsimonious specification of the model in subsection 2.2 above, where $p = q = 1$, $\Sigma_0 = \sigma_0^2$, $\Sigma = \sigma^2$, $\Upsilon = \tau^2$, $C_t = C$, and $H_t = 1$ (i.e., we assume the conditional expectation of the latent data y_t^* given the sleep state x_t is exactly the sleep state itself).

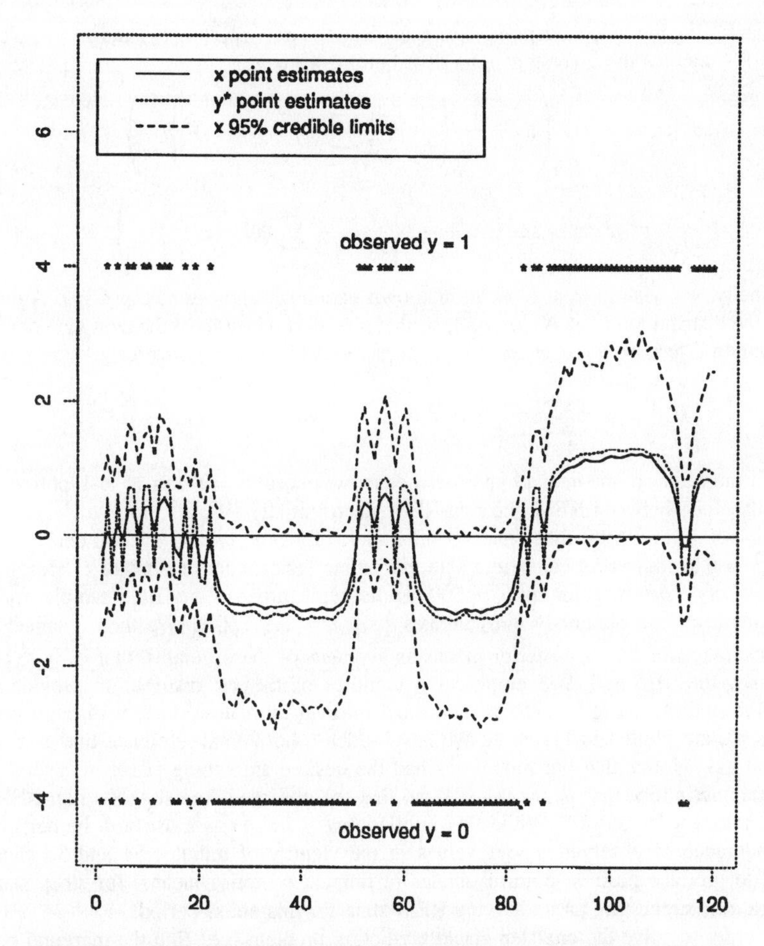

Figure 1. *Data and estimates, infant sleep data* $G = 500$, y_{121} *assumed unknown.*

We now list the complete conditional distributions necessary for implementation of the Gibbs sampler in this case. First for y_t^* we refer to equation (7), obtaining

$$y_t^*|\tau^2, y_t, x_t \propto \begin{cases} N(x_t, \tau^2)I_{(0,\infty)} & \text{if } y_t = 1 \\ N(x_t, \tau^2)I_{(-\infty,0)} & \text{if } y_t = 0 \end{cases}$$

Secondly for x_t, using Bayes Theorem and taking care with the endpoint cases we have that $x_t|x_{j\neq t}, y_t^*, \sigma^2, \tau^2 \sim N(B_t\boldsymbol{b}_t, B_t)$, where

$$\begin{array}{lll} B_t^{-1} = \sigma_0^{-2} + C^2\sigma^{-2}, & \boldsymbol{b}_t = \mu_0\sigma_0^{-2} + C\sigma^{-2}x_1, & t = 0 \\ B_t^{-1} = \sigma^{-2}(1 + C^2) + \tau^{-2}, & \boldsymbol{b}_t = C\sigma^{-2}(x_{t-1} + x_{t+1}) + y_t^*\tau^{-2}, & t = 1, \ldots, T-1 \\ B_t^{-1} = \sigma^{-2} + \tau^{-2} & \boldsymbol{b}_t = C\sigma^{-2}x_{T-1} + y_T^*\tau^{-2}, & t = T \end{array} \tag{9}$$

Next, assuming the independent *a priori* specifications $\sigma^2 \sim IG(a_0, b_0)$ and $\tau^2 \sim IG(c_0, d_0)$, where IG denotes the inverse gamma distribution, we have

$$\sigma^2|\boldsymbol{x}, x_0 \sim IG\left(a_0 + \frac{T}{2}, \left\{\frac{1}{b_0} + \frac{1}{2}\sum_{t=1}^{T}(x_t - Cx_{t-1})^2\right\}^{-1}\right), \text{ and}$$

$$\tau^2|\boldsymbol{y}^*, \boldsymbol{x} \sim IG\left(c_0 + \frac{T}{2}, \left\{\frac{1}{d_0} + \frac{1}{2}\sum_{t=1}^{T}(y_t^* - x_t)^2\right\}^{-1}\right).$$

Finally, we wish to treat C as an unknown parameter and estimate $p(C|\boldsymbol{y})$. *A priori* we adopt the distribution $C \sim N(\mu_C, \sigma_C^2)$, so that *a posteriori* we have the complete conditional distribution $C|\sigma^2, \tau^2, \boldsymbol{y}^*, \boldsymbol{y}, \boldsymbol{x}, x_0 \sim C|\sigma^2, \boldsymbol{x}, x_0 \sim N(B_C b_C, B_C)$, where

$$B_C^{-1} = \sigma^{-2}\sum_{t=1}^{T} x_{t-1}^2 + \sigma_C^{-2}, \text{ and } b_C = \sigma^{-2}\sum_{t=1}^{T} x_t x_{t-1} + \mu_C \sigma_C^{-2}. \tag{10}$$

For the values of the model hyperparameters, we chose $\mu_0 = 0$ and $\sigma_0^2 = 1$, corresponding to indifference between REM and non-REM sleep initially. We also took $a_0 = c_0 = 3$ and $b_0 = d_0 = 0.5$, or a prior mean and variance of 1.0 for both σ^2 and τ^2. We chose $\mu_C = 0.5$ and $\sigma_C^2 = 0.1$, indicating belief in a stationary time series where values of C close to 1 are very unlikely *a priori*. Running $G = 500$ parallel replications of the Gibbs sampler for $l = 25$ iterations each, we obtained Gibbs iterates $\{x_{tj}, j = 1, \ldots, G\}$. We then obtained simple point estimates of the x_t posterior means as averages of these iterates, that is, $\sum_{j=1}^{G} x_{tj}/G$. Similarly, the .025 and .975 empirical percentiles of the x_{tj} distribution provide a 95% posterior credible set for x_t. These point and interval estimates, along with point estimates for the y_t^*, are plotted in Figure 1. We have added a horizontal reference line at $x_t = 0$ as a visual aid. Notice that our model has had the desired smoothing effect in several places. For example, notice that $y_{54} = y_{58} = 0$, so that the observed data suggests non-REM sleep during minutes 54 and 58. While this implies that y_{54}^* and y_{58}^* must both be negative, the preponderance of observed $y = 1$ values in the vicinity of minutes 54 and 58 causes the model to produce positive point estimates (estimated posterior means) for sleep states x_{51} through x_{61}, suggesting an underlying REM state for this entire period.

In order to solve the one-step-ahead prediction problem (i.e., find the marginal posterior density of x_{121} given $\boldsymbol{y}$), notice that the nonappearance of y_{121}^* and x_{122} in the likelihood (6) means that a Monte Carlo mixture density estimate is available as

$$\hat{p}(x_{121}|\boldsymbol{y}) = \frac{1}{G}\sum_{j=1}^{G} N\left(C_j x_{(120,j)}\, ,\, \sigma_j^2\right).$$

This estimate is plotted as the solid line in Figure 2(b).

Now, for this infant y_{121} is actually available ($y_{121} = 1$), and so we may wish to solve the filtering problem (i.e., find the marginal posterior density of x_{121} given $\boldsymbol{y}$ *and* y_{121}). This is easily done in our context by including y_{121}^* and x_{121} as additional parameters in the sampling order and rerunning the algorithm. With Gibbs samples for these two parameters now available, we can obtain a mixture density estimate simply by mixing the appropriate complete conditional distributions from equation (9), namely

$$\hat{p}(x_{121}|\boldsymbol{y}, y_{121}) = \frac{1}{G}\sum_{j=1}^{G} N\left(B_{(121,j)} b_{(121,j)}\, ,\, B_{(121,j)}\right),$$

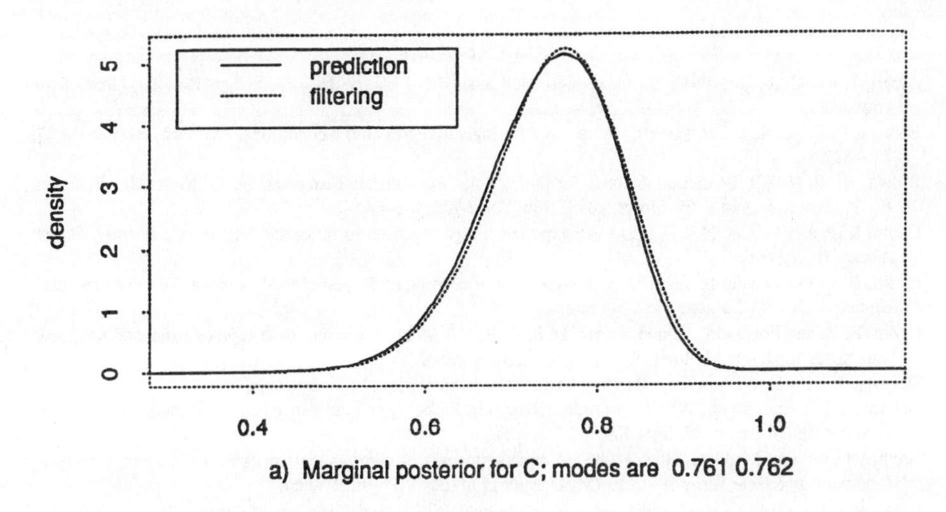

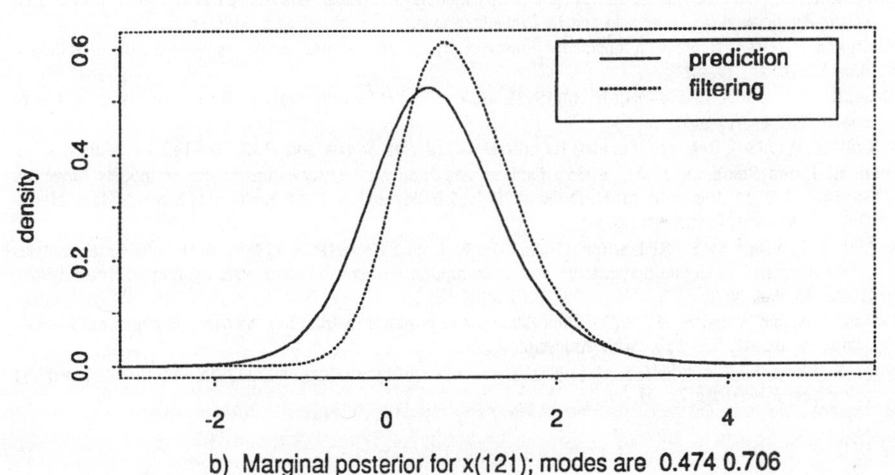

Figure 2. *Estimated posteriors, infant sleep data $G = 500$, prediction vs. filtering.*

where $B^{-1}_{(121,j)} = \sigma_j^{-2} + \tau_j^{-2}$ and $b_{(121,j)} = C_j\sigma_j^{-2}x_{(120,j)} + y^*_{(121,j)}\tau_j^{-2}$. This estimate is plotted as the dashed line in Figure 2(b). Notice that the model was "right" in its prediction that the infant would very likely continue in the REM state at minute 121, and as such the filtering posterior is even more strongly indicative of positive values for x_{121} (though negative values are still not completely ruled out).

In a similar vein, Figure 2(a) plots two estimated marginal posteriors for C, one conditional on y ("prediction") and the other conditional on y and y_{121} ("filtering"). Both of these curves are also mixture density estimates, obtained by mixing the complete conditionals given in (10) over the Gibbs replicates obtained in each case. Unlike the plots for x_{121}, in this case we would expect the filtering posterior to be only slightly less diffuse than that for prediction, and indeed this is what we find.

REFERENCES

Albert, J. and Chib, S. (1991). Bayesian regression analysis of binary data. *Tech. Rep.* Bowling Green State University.

Bedrick, E. J. and Hill, J. R. (1990). Outlier tests for logistic regression: a conditional approach. *Biometrika* **77**, 815–827.

Brown, B. W. (1980). Prediction analysis for binary data. *Biostatistics Casebook*, (R. G. Miller, Jr., B. Efron, B. W. Brown, Jr. and L. E. Moses, eds.), New York: Wiley, 3–18.

Carlin, B. P. and Polson, N. G. (1991). An expected utility approach to influence diagnostics. *J. Amer. Statist. Assoc.* (to appear).

Carlin, B. P. and Polson, N. G. (1992). Inference for nonconjugate Bayesian models using the Gibbs sampler. *Canad. J. Statist.*, (to appear).

Carlin, B. P., and Polson, N. G. and Stoffer, D. S. (1992). A Monte Carlo approach to nonnormal and nonlinear state space modeling. *J. Amer. Statist. Assoc.* , (to appear).

Devroye, L. (1986). *Non-uniform Random Variate Generation*. New York: Springer.

Gelfand, A. E. and Smith, A. F. M. (1990). Sampling based approaches to calculating marginal densities. *J. Amer. Statist. Assoc.* **85**, 398–409.

Grether, D. M. and Maddala, G. S. (1982). A time series model with qualitative variables. *Games, Economic Dynamics and Time Series Analysis* (M. Deistler *et al.* eds.), Physica: Wien.

Heckmann, J. J. (1981). Dynamic discrete probability models. *Structural Analysis of Discrete Data with Econometric Application* (C. F. Manski and D. McFadden, eds.), Cambridge, MA: MIT Press.

Kaufmann, H. (1987). Regression models for nonstationary categorical time series: asymptotic estimation theory. *Ann. Statist.* **15**, 79–98.

Liang, K.-Y., Zeger, S. L. and Qaqish, B. (1992). Multivariate regression analyses for categorical data. *J. Roy. Statist. Soc. B* , (to appear).

McCullagh, P. (1980). Regression models for ordinal data. *J. Roy. Statist. Soc. B* **42**, 109–142, (with discussion).

Pettit, L. I. and Smith, A. F. M. (1985). Outliers and influential observations in linear models. *Bayesian Statistics 2* (J. M. Bernardo, M. H. DeGroot, D. V. Lindley and A. F. M. Smith, eds.), Amsterdam: North-Holland, 473–494, (with discussion).

Stoffer, D. S., Scher, M. S., Richardson, G. A., Day, N. L. and Coble, P. A. (1988). A Walsh-Fourier analysis of the effects of moderate maternal alcohol consumption on neonatal sleep-state cycling. *J. Amer. Statist. Assoc.* **83**, 954–963.

Tanner, M. A. and Wong, W. H. (1987). The calculation of posterior distributions by data augmentation. *J. Amer. Statist. Assoc.* **82**, 528–550, (with discussion).

Zeger, S., Liang, K.-Y. and Self, S. G. (1985). The analysis of binary longitudinal data with time-independent covariates. *Biometrika* **72**, 31–39.

BAYESIAN STATISTICS 4, pp. 587–594
J. M. Bernardo, J. O. Berger, A. P. Dawid and A. F. M. Smith, (Eds.)

Bayes Factors for Linear Models and Improper Priors

GUIDO CONSONNI and PIERO VERONESE
Università di Pavia and *Università L. Bocconi, Milano, Italy*

SUMMARY

A new method is suggested to evaluate the Bayes factor for choosing between two nested models using improper priors for the model parameters. Within the above framework it is shown that commonly used vague priors always lead to an infinite Bayes factor. For normal linear models we identify a class of improper priors consistent with a finite Bayes factor. Furthermore we single out a subclass of such priors under which the Bayes factor is a function of the standard F-statistic. A numerical illustration is provided in the one-way analysis of variance setup.

Keywords: BAYES FACTOR; IMAGINARY TRAINING SAMPLE DEVICE; IMPROPER PRIOR; MODEL CHOICE; NUISANCE PARAMETER; TESTING PRECISE HYPOTHESIS.

1. INTRODUCTION AND SUMMARY

Consider the problem of choosing between two models, M_0 and M_1 with M_0 nested in M_1. It is easy to verify that this problem can be formulated as one in testing a precise hypothesis on the parameter of interest in the presence of a nuisance parameter. Setting aside the prior probabilities on M_0 and M_1, the problem reduces to the computation of the Bayes factor $B(x)$ in favour of the precise hypothesis, i.e. in favour of M_0, for given observation x.

The evaluation of $B(x)$ presents difficulties when the prior distribution of the parameters is improper, essentially because one cannot omit the normalizing constant as in standard prior-to-posterior analysis. This implies, in a routine approach, that $B(x)$ is defined only up to an *indeterminate* constant. Several attempts have been made to overcome this problem, see in particular Spiegelhalter and Smith (1982), and references therein. The main motivation behind these attempts is the desire to implement noninformative priors in Bayesian analysis. For a recent discussion of the motivations behind noninformative priors see Berger and Bernardo (1992).

In this paper a new method is adopted to choose between two linear models under vague (i.e. improper) priors. This method rests upon the idea that an improper prior may be regarded as a limit of proper priors defined on a sequence of compact subsets converging to the whole parameter space. We emphasize that this is not simply another adhockery added to the Bayesian toolkit, since it represents a coherent way of assessing probabilities according to de Finetti's approach. Further remarks on this issue may be found in Consonni and Veronese (1987).

Section 2 briefly examines the problem of testing a precise hypothesis $\theta = \theta^*$ with and without nuisance parameters using improper priors. While in the first case $B(x)$ is always infinite, this is no longer true when a nuisance parameter ϕ is present, provided θ and ϕ are not independent (unconditionally with respect to either hypothesis).

Section 3 explicitly deals with the problem of choosing between two normal nested linear models. Given a class of improper priors, we identify a subset of them leading to a finite

Bayes factor. Interestingly enough, the most commonly used priors for this problem do *not* belong to this subset.

For a fixed improper prior consistent with a finite Bayes factor, the actual evaluation of $B(x)$ depends upon the specific sequence of subsets of the parameter space associated with the given prior. A convenient method to assign indirectly such a sequence is based on the *imaginary training sample device* as formalized by Spiegelhalter and Smith (1982).

Finally, Section 4 identifies a class of improper priors leading to a Bayes factor function of the standard F-statistic. In this case, the Bayesian and frequentist approach share the same summarizing statistic for the data and so a comparison of the corresponding results becomes especially appropriate. To do so, we compute lower and upper bounds for $B(x)$ in the above class of priors. This is illustrated with a numerical example related to one-way analysis of variance models.

2. TESTING PRECISE HYPOTHESES WITH IMPROPER PRIORS

2.1. *Case A: No Nuisance Parameter*

Let X be a vector of observations having generalized density $f_\theta(x)$ for a given parameter $\theta \in \Theta$.

Consider the testing problem:

$$H_0 : \theta = \theta^* \quad \text{vs} \quad H_1 : \theta \neq \theta^*,$$

where θ^* is a fixed value in Θ.

Let $\pi_0 = Pr\{H_0\}$ and $\pi_1 = Pr\{H_1\} = 1 - \pi_0$. To start with, assume that the prior density on θ

$$p(\theta) = \pi_0 I_{\theta=\theta^*} + \pi_1 p_1(\theta) I_{\theta \neq \theta^*}$$

is *proper*, i.e. $p_1(\theta)$ is a proper density on Θ representing the prior on θ given H_1.

Then the posterior probability of H_0 can be written as

$$Pr\{H_0|X = x\} = \left\{\frac{1+\pi_1}{\pi_0} + \frac{1}{B(x)}\right\}^{-1}$$

where $B(x)$ is the Bayes factor in favour of H_0 having expression

$$B(x) = \frac{f_\theta^*(x)}{\int_{\theta \neq \theta^*} f_\theta(x) p_1(\theta) d\theta}$$

Assume now that $p(\theta)$ is *improper*. If we were to perform prior-to-posterior analysis on θ, we could typically consider the kernel, $g(\theta)$ say, corresponding to $p(\theta)$ since the "proportionality constant" is irrelevant. This, however, is no longer true in hypothesis testing since, as DeGroot (1982) argued, if $p_1(\theta) = cg_1(\theta)$, then $B(x)$ would depend upon the arbitrary choice of c. Hence the conclusion that "diffuse [i.e. improper] prior distributions ... are never appropriate for tests of significance".

While we agree with this conclusion, we would like to offer a different justification which will provide further insights especially when nuisance parameters are introduced.

The intuitive idea supporting our approach is that an improper prior is a "limiting version" of a proper distribution. Thus, for example, the analysis using the standard prior "constant over the real line" should be well approximated by a uniform prior over the interval $(-K, K)$ for a sufficiently large K; see also Consonni and Veronese (1987).

More formally, let $g_1(\theta)$ be the kernel of the improper prior under H_1 and let $\{\Theta_s\}$ be a sequence of compact sets converging to Θ such that

$$0 < C_{1s}^{-1} = \int_{\Theta_s} g_1(\theta) d\theta < \infty, \qquad \text{for all } s.$$

Define the "restricted Θ_s-prior under H_1" $p_{1s}(\theta)$, which is of course proper, as

$$p_{1s}(\theta) = C_{1s} g_1(\theta), \qquad \theta \in \Theta_s,$$

so that

$$Pr\{H_0|X = x\} = \lim_s Pr_s\{H_0|X = x\} \tag{2.1}$$

$$= \left\{ 1 + \frac{\pi_1}{\pi_0} \frac{1}{\lim_s B_s(x)} \right\}^{-1}, \tag{2.2}$$

where

$$B_s(x) = \frac{f_\theta^*(x)}{C_{1s} \int_{\Theta_s \backslash \theta^*} f_\theta(x) g_1(\theta) d\theta} \tag{2.3}$$

is the "Θ_s-restricted Bayes factor".

Equality (2.1) is meaningful, in a finitely-additive context, since "limits of coherent probabilities are coherent" see de Finetti (1974, Section 3.13).

Hence, by (2.2), it becomes sensible to compute $\lim B_s(x)$ and to set it equal to the Bayes factor $B(x)$ for this problem. If now, in accordance with the standard assumption behind the application of the "formal Bayes rule", the integrated likelihood

$$L_1(x) = \int_{\Theta \backslash \theta^*} f_\theta(x) g_1(\theta) d\theta$$

is positive and finite, then

$$B(x) = \lim_s B_s(x) = \frac{f_\theta^*(x)}{L_1(x) \lim_s C_{1s}} = \infty, \tag{2.4}$$

since $\lim_s C_{1s} = 0$.

As a consequence, using an improper prior to test a precise hypothesis without nuisance parameters always leads to $Pr\{H_0|X = x\} = 1$ for all x so that the analysis is useless. Notice, incidentally, that this result formalizes, in a general setting, a similar conclusion contained in Berger and Delampady (1987, p. 318) for normal models with conjugate priors having variance tending to infinity.

2.2. *Case B: Nuisance Parameter*

Let X have generalized density $f_{\theta\phi}(x)$ with $(\theta, \phi) \in \Theta \times \Phi$.

Given the same testing problem as in subsection 2.1, let $g_0(\phi)$ and $g_1(\theta, \phi)$ be the kernels of the improper priors on ϕ under H_0 and of (θ, ϕ) under H_1.

Let $\{\Theta_s\}$ and $\{\Phi_s\}$ be a sequence of compact sets converging, respectively, to Θ and Φ such that

$$0 < C_{0s}^{-1} = \int_{\Phi_s} g_0(\phi) d\phi < \infty$$

$$0 < C_{1s}^{-1} = \int_{\Phi_s} \int_{\Theta \backslash \theta^*} g_1(\theta, \phi) d\theta d\phi < \infty$$

for all s.

A straightforward extension of the analysis of subsection 2.1 leads to (2.2) where now

$$B_s(x) = \frac{C_{0s} \int\limits_{\Phi_s} f_{\theta^*\Theta}(x) g_0(\phi) d\phi}{C_{1s} \int\limits_{\Phi_s} \int\limits_{\Theta \backslash \theta^*} f_{\theta\phi}(x) g_1(\theta, \phi) d\theta d\phi}.$$

Under the usual assumption that the integrated likelihoods $L_i(x)$ are positive and finite we obtain

$$B(x) = \lim_s B_s(x) = \frac{L_0(x)}{L_1(x)} \lim_s \frac{C_{0s}}{C_{1s}} \tag{2.5}$$

Since both C_{0s} and C_{1s} tend to zero, the value of $B(x)$ in (2.5) is no longer necessarily ∞ as in (2.4). Hence, when a nuisance parameter is present, improper priors can lead to a finite Bayes factor and the corresponding analysis is no longer a priori meaningless.

Remark 1. For some choices of g_0 and g_1 it is still possible that $B(x) = \infty$ for all x. An important such case occurs when θ and ϕ are assumed unconditionally independent so that $g_1(\theta, \phi) = g_0(\phi) g^*(\theta)$, for some improper kernel g^*. Indeed, in this case

$$\lim_s = \frac{C_{0s}}{C_{1s}} = \int_\Theta g^*(\theta) d\theta = \infty.$$

3. CHOICE BETWEEN LINEAR MODELS

We apply the results of subsection 2.2 to the problem of choosing between two nested linear models

$$(Y | A_i, \theta_i, \sigma^2) \sim N(A_i \theta_i, \sigma^2 I), \qquad i = 0, 1,$$

where Y is an n-vector of observations ($n \geq 2$), A_i is known and of full rank p_i, θ_i is a p_i-vector of unknown parameters ($p_1 > p_0$) and σ^2 is unknown.

Letting $\theta_1 = (\theta_0 \vdots \theta)$, the choice between the above two models may be formulated as the following testing problem:

$$H_0 : \theta = 0 \quad \text{vs} \quad H_1 : \theta \neq 0,$$

with (θ_0, σ) representing the nuisance parameter.

Generalizing the analysis adopted by Spiegelhalter and Smith (1982), henceforth $S\&S$, we specify a *class* of *improper priors* on (θ_i, σ) having kernels

$$g_i(\theta_i, \sigma) = (2\pi)^{-p_i/2} \sigma^{h(p_i)}, \qquad i = 0, 1, \tag{3.1}$$

for some real function h (the factor $(2\pi)^{-p_i/2}$ is inserted merely for comparison purposes). Notice that $h(\cdot)$ should not be assumed constant for, otherwise, $B(x) = \infty$ according to Remark 1.

It may be checked that the condition for the integrated likelihoods to be finite leads to

$$h(p_i) < n - (p_i + 1), \tag{3.2}$$

so that

$$B(y) = \left(\frac{|A_1^T A_1|}{|A_0^T A_0|}\right)^{1/2} \left(\frac{1}{2}\right)^{(p_0 - p_1 + h(p_0) - h(p_1))/2} \times \frac{\Gamma((n - p_0 - h(p_0) - 1)/2)}{\Gamma((n - p_1 - h(p_1) - 1)/2)} \frac{Q_1^{(n - p_1 - h(p_1) - 1)/2}}{Q_0^{(n - p_0 - h(p_0) - 1)/2}} \times \lim_s \frac{C_{0s}}{C_{1s}} \tag{3.3.}$$

where $Q_i = y^T(I - A_i(A_i^T A_i)^{-1} A_i^T)y$ is the residual sum of squares under model i.

The evaluation of $\lim C_{0s}/C_{1s}$ depends, for fixed h, on the structure of the sequence of subsets of the parameter spaces relative to θ_0, θ_1 and σ. Symmetry considerations suggest taking the same sequence of intervals, $\{(a, b)\}$ say, $(a \to -\infty, b \to \infty)$, for each component of the vector θ_1. Letting $\{(c, d)\}(c \to 0^+, d \to \infty)$ be the corresponding sequence for σ, it is straightforward to check that $\lim C_{0s}/C_{1s}$ becomes:

$$C_{01} = \lim_{a,b,c,d} \frac{\int_c^d \sigma^{h(p_1)} d\sigma}{\int_c^d \sigma^{h(p_0)} d\sigma} (b - a)^{p_1 - p_0} (2\pi)^{-(p_1 - p_0)/2}. \tag{3.4}$$

Remark 2. An analysis of (3.4) leads to $B(y) = \infty$ for all y in *each* of the following three cases:

i) $h(p_0) = h(p_1)$;

ii) $h(p_i) > -1, \qquad h(p_0) < h(p_1)$;

iii) $h(p_i) < -1, \qquad h(p_0) > h(p_1)$.

As a consequence, the choice of $h(p_i) = -p_i - 1$ or $h(p_i) = -1$ adopted by $S\&S$ are inconsistent with a finite Bayes factor.

In all the remaining cases, it can be shown that, for each fixed h, there exist choices of a, b, c, d such that (3.4) is finite. For example, when $h(p_i) > -1$ (with $h(p_0) > h(p_1)$), one such choice is given by: $a = -b; c = 1/d; b = Kd^{(h(p_0) - h(p_1))/(p_1 - p_0)}$ where K is an arbitrary positive constant so that

$$C_{01} = (\pi/2)^{-(p_1 - p_0)/2} \left(\frac{h(p_0) + 1}{h(p_1) + 1}\right) K^{p_1 - p_0}.$$

The previous considerations have made explicit the dependence of the value of the Bayes factor on the choice of the sequence of subsets associated with the improper kernel.

This feature might seem disturbing at first sight, but it appears sensible since the kernel of an improper prior does not provide, *per se*, a complete description of the underlying probabilistic assessment. On the other hand, choosing a sequence of subsets does not seem a natural enterprise when assigning a distribution. This suggests searching for criteria leading to default procedures for the above choice or, equivalently, in our problem, for evaluating C_{01} in (3.4).

One such criterion, which tries to incorporate the concept of noninformativity usually associated with an improper prior, is the so-called *imaginary training sample device*. This was originally suggested by Good (1947) and later applied to the problem of model choice by $S\&S$.

In their words, such a "device" amounts to imagining "that a data set is available which: a) involves the smallest possible sample size permitting a comparison of M_0 and M_1 [i.e., the two competing models]; b) provides maximum possible support for M_0". They apply this method in order to fix the ratio c_0/c_1 appearing in *their* expression of the Bayes factor which is formally evaluated by setting the prior equal to $c_i(2\pi)^{-p_i/2}\sigma^{-p_i-1}$, where c_i is an "undefined constant". The major deficiency of $S\&S$'s approach is thus the *separation* of the evaluation of c_0/c_1 from the prior kernel.

On the other hand our approach is free from the above logical inconsistency, since the evaluation of C_{01} in (3.4) is automatically linked to the improper prior, which consists of *two* components, namely the kernel and the sequence of subsets converging to the whole parameter space. In summary we propose the following two-step procedure:

i) check that the prior kernels $g_i(\theta_i, \sigma)$ are compatible with a finite Bayes factor;

ii) use the imaginary training sample device to evaluate the constant C_{01}.

Notice that step ii) may be regarded as a way to bypass the "difficult" assignment of the sequence of subsets by indirectly fixing a default one, whose existence is guaranteed by step i).

4. BAYES FACTORS AS FUNCTIONS OF F-STATISTICS

The class of improper priors on (θ_i, σ) given in (3.1) leads to the Bayes factor (3.3) which depends on the data through the individual residual sum of squares Q_i. One may wonder at this stage which conditions on the above class of priors ensure that the data y enter (3.3) only through the ratio Q_0/Q_1 or, equivalently, through the standard F-statistic $\{(Q_0 - Q_1)/(p_1 - p_0)\}/\{Q_1/(n-p_1)\}$. Such a case is interesting for at least two reasons: i) it allows a comparison with frequentist and likelihood-based results (see e.g. Bertolino, Piccinato and Racugno, 1990); ii) it makes the implementation of the imaginary training sample device especially straightforward.

The answer to the above query is contained in the following

Proposition 1. Given the class of improper priors (3.1), a necessary and sufficient condition for the existence of a sequence of subsets converging to the parameter space such that the Bayes factor (3.3) is finite and a function of the F-statistic is $h(p_i) = K - p_i$ with $p_0 - 1 \leq K < n - 1$.

Furthermore the Bayes factor becomes

$$B(y) = B_K(y) = C_{01}\left(\frac{|A_1^T A_1|}{|A_0^T A_0|}\right)^{1/2}\left(1 + \frac{p_1 - p_0}{n - p_1}F\right)^{-(n-K-1)/2} \tag{4.1}$$

Proof. (3.3) is a function of Q_0/Q_1 if and only if $h(p_1) - h(p_0) = p_0 - p_1$ which leads to $h(p_i) = K - p_i$ with $K < n - 1$ because of (3.2). Furthermore, using Remark 2, it follows that $K \geq p_0 - 1$ to avoid an infinite value for $B(y)$. A direct substitution leads finally to (4.1).

Remark 3. If C_{01} in (4.1) is fixed according to the imaginary training sample device, then

$$C_{01} = \{|E_1^T E_1|/|E_0^T E_0|\}^{-1/2} \tag{4.2}$$

where E_i is the design matrix corresponding to the imaginary experiment under model i, see $S\&S$.

Remark 4. A robustness analysis on the class (3.1) suggests computing the lower and upper bounds of $B_K(y)$ in (4.1). It is straightforward to check that

$$\inf_K B_K(y) = B_{p_0-1}(y) = \overline{B}(y)$$

$$\sup_K B_K(y) = C_{01}\left(\frac{|A_1^T A_1|}{|A_0^T A_0|}\right)^{1/2} = \overline{B}.$$

Having set C_{01} as in (4.2), our expression (4.1) for $B(y)$ reduces, purely from a *formal* viewpoint, to $S\&S$'s expression (5) with $K = -1$. Such an equality is *formal* since the value $K = -1$ produces an infinite Bayes factor according to Proposition 1. Notice, however, that $B(y)$comes closer to the *numerical* value provided by $S\&S$ as p_0 gets smaller, i.e., as model H_0 becomes simpler.

4.1. *Bayes Factors for One-Way Analysis of Variance Models*

Given m groups with n_i observations in the i-th group, denote by M_0 the model prescribing the group means μ_i to be all equal and by M_1 the models which allows different group means. Fixing C_{01} in accordance with (4.2) gives $C_{01} = \{(m+1)/2\}^{1/2}$, see $S\&S$, so that the Bayes factor (4.1) becomes

$$B_K(y) = \left(\frac{m+1}{2n}\prod_1^m n_i\right)^{1/2}\left(1+\frac{m-1}{n-m}F\right)^{-(n-K-1)/2}$$

p-value	m	n^*	F	$\underline{B}(y)$	$\overline{B}$
0.05	4	10	2.87	0.38	25
		20	2.72	1.26	71
	8	10	2.14	2.07	2372
		20	2.07	19.27	26830
0.01	4	10	4.38	0.06	25
		20	4.05	0.20	71
	8	10	2.90	0.13	2372
		20	2.76	1.98	26830

Table 1. *One-way analysis of variance: values of* $\underline{B}(y)$, $\overline{B}$ *, for selected combinations of* m *and* $n_i = n^*(i = 1,\ldots,m)$*, corresponding to significant values of* F *at level* 5% *and* 1%.

A numerical comparison between p-values and $\underline{B}(y)$, $\overline{B}$ is contained in Table 1, which exhibits the well known discrepancy between p-values and Bayes factors, the latter being typically more favourable to M_0 than the former.

It is worth stressing that significant values of F represent, *at most*, i.e., in terms of $\underline{B}(y)$, only mild evidence against M_0, provided m and n^* are small (for $m = 4, n^* = 10$ and significance 5% $\underline{B}(y) = 0.38$) and point out mild or strong evidence *in favour* of M_0 as m or n^* increase.

ACKNOWLEDGEMENTS

This research was partially supported by C.N.R., contract N. 88.02970.10, and by M.U.R.S.T.

REFERENCES

Berger, J. O. and Bernardo, J. M. (1992). On the development of the reference prior method. *Bayesian Statistics 4* (J. M. Bernardo, J. O. Berger, A. P. Dawid and A. F. M. Smith, eds.), Oxford: University Press, 35–60, (with discussion).

Berger, J. O. and Delampady, M. (1987). Testing precise hypotheses. *Statist. Sci.* **2**, 317–352, (with discussion).

Bertolino, F., Piccinato, L. and Racugno, W. (1990). A marginal likelihood approach to analysis of variance. *The Statistician* **39**, 415–424.

Consonni, G. and Veronese, P. (1987). Coherent distributions and Lindley's paradox. *Probability and Bayesian Statistics* (R. Viertl, ed.), London: Plenum Press, 111–120.

de Finetti, B. (1974). *Theory of Probability*. New York: Wiley.

DeGroot, M. H. (1982). Discussion of "Lindley's paradox" by G. Shafer. *J. Amer. Statist. Assoc.* **77**, 336–339.

Good, I. J. (1947-50). *Probability and the Weighing of Evidence*. New York: Haffner.

Spiegelhalter, D. J. and Smith, A. F. M. (1982). Bayes factors for linear and log-linear models with vague prior information. *J. Roy. Statist. Soc. B* **44**, 377–387.

BAYESIAN STATISTICS 4, pp. 595–600
J. M. Bernardo, J. O. Berger, A. P. Dawid and A. F. M. Smith, (Eds.)

BAIES – A Probabilistic Expert System Shell with Qualitative and Quantitative Learning

ROBERT G. COWELL
University College London, UK

SUMMARY

BAIES (*B*ayesian *A*nalysis *I*n *E*xpert *S*ystems) is a probabilistic expert system shell program for analysing Bayesian networks. By treating the probability distributions of network models as arising from a set of Dirichlet distributions, *BAIES* can use updating algorithms in order to adapt the Dirichlet parameters, and hence revise the predictions made on future cases. Analysis of the predictions of the models against the data can be made both at the level of comparing models and also in criticising the local structure and prior distributions used within the models.

Keywords: BAYESIAN NETWORK; PROBABILISTIC EXPERT SYSTEM, EVIDENCE; LEARNING; MODEL CRITICISM; PREQUENTIAL ANALYSIS.

1. INTRODUCTION

Recent advances in the theory of Bayesian networks (Lauritzen and Spiegelhalter, 1988; Jensen *et al.* 1990) have made feasible the development of computer programs which can handle large network models. These ideas have been incorporated into a number of specialised and general purpose probabilistic expert systems, for example HUGIN (Andersen *et al.* 1989). *BAIES* is a probabilistic expert system shell program for analysing Bayesian networks which permits comparison and criticism of models and also qualitative and quantitative learning in models using data.

A Bayesian network model is specified by a *directed acyclic graph* (dag) $G = (V, E \subseteq V \times V)$, in which each vertex $v \in V$ represents a discrete random variable X_v and the directed edges E represent direct influences. The independence properties of a model (Dawid, 1979) are related to the topology of its network. Various authors have stressed the usefulness of this approach (e.g., Shachter, 1986; Pearl, 1988). By embedding the directed graph into a suitably chosen undirected graph, the total joint-probability distribution for the set of random variables may be expressed as a more manageable product of smaller distributions which depend upon the random variables associated with the vertices in the cliques in the undirected graph. A clique of a graph is a maximal (with respect to inclusion) complete subgraph (Golumbic, 1980). The cliques are organised into a structure called a *junction tree* (Jensen *et al.* 1990) for computational efficiency.

A deficiency of the Bayesian network approach is that both the qualitative structure of the associated network and the quantitative values of the conditional probabilities used to initialise the system need to be precisely known. (Exceptions arise for networks containing continuous nodes, (Shachter and Kenley, 1989)). In general this is unlikely; rather, (perhaps ideally), both the qualitative and quantitative nature of the Bayesian model should be derivable from a given set of data. *BAIES* has been designed to go some way to achieving this aim, which it does in the following manner.

BAIES takes for its input a set of candidate Bayesian network models $M = \{M_i\}$ over a set of discrete random variables $X = \{X_v\}$. Typically, the first model may represent the dependence and independence relationships between the random variables as agreed by one or more 'experts' in the domain being modelled, and the remaining models may be plausable alternatives of the first model – topological variations and/or different prior distributions. Alternatively, they may represent different expert assessments. Each model uses Dirichlet distributions to represent the conditional probability tables; the latter are taken as the means of the Dirichlet distributions, and are used to analyse individual data. The Dirichlet distributions are updated after each case has been processed by using the learning scheme described by (Spiegelhalter and Lauritzen, 1990) and summarized in Section 3 below.

Criticism of the local structure of each model and of the prior probability tables is achieved through the use of a prequential analysis (Dawid, 1984) of the predictions of the models against the data as the data are processed. Also, by calculating the Bayes factors overall comparison of the models can be achieved.

BAIES can be used in a batch mode to process a data file. In general this requires iterating the following steps after initialising the models. (1) Input evidence, (2) Propagate the evidence in the junction tree, (3) Calculate the marginal tables held in the dags, and update the conditional tables held in the dag-nodes, (4) Update the monitors, and finally (5) Re-initialise the potential tables held in the junction tree. Another option which can be included in this process is to select at the initialisation stage a particular random variable together with an integer $n \geq 1$. Then after every n-th data set of evidence has been processed, *BAIES* writes to files the values of the monitors in all of the models which contain the selected variable. This generates a sequence of data which can be analysed; for an application of this option see Spiegelhalter and Cowell (1991) in these proceedings.

The learning and model criticism capabilities of the *BAIES* expert system shell program are described in the next two sections. In addition to these, *BAIES* has the following useful facilities:

- An option to generate Monte-Carlo data files, both complete and with missing data – for use in numerical experiments.
- The ability to locate global maxima of a probability distribution – this can be used as an explanation facility.
- An interactive mode for the input and propagation of evidence – marginal, conditional, clique and separator tables can be viewed.
- Algorithms for using the Dirichlet distributions of one model to generate distributions of equivalent precision for structurally different models. In this way fair comparison of models to a common set of data is possible.

2. LEARNING ABOUT PROBABILITIES IN BAIES

2.1. *Sequential Learning*

BAIES implements the learning method of Spiegelhalter and Cowell (1991) (see also Bernardo and Girón, 1988), in which the elements of the conditional probability tables are taken to be random variables. Specifically, let v be a dag-node representing the random variable X_v having n states $X_v = (x_v^1, \ldots, x_v^n)$. Let $pa[v]$ denote the set of parents of v in the dag, and let $X_{pa[v]} = (X_u : u \in pa[v])$. Then within *BAIES*, for each configuration $\rho \in X_{pa[v]}$ of the state space of the parents $pa[v]$, the entries of the conditional probability table $P(X_v | X_{pa[v]} = \rho)$ are represented by the parameters of a Dirichlet distribution $\mathcal{D}[\alpha_1, \ldots, \alpha_n]$, such that

$$P(x_v^j | X_{pa[v]} = \rho) = \frac{\alpha_j}{\sum_i \alpha_i}.$$

The α's can can be regarded informally as a set of 'counts' (which are not necessarily integers). Now if both the state of X_v and its parent configuration $X_{pa[v]}$ is observed, then unity is added to the relevant slot in the relevant Dirichlet distribution. When incomplete evidence $\mathcal{E}$ is observed, a single Dirichlet distribution should be replaced by a mixture of Dirichlet distributions, as follows. Let $\mathcal{D}_j \equiv \mathcal{D}[\alpha_1, \ldots, \alpha_j + 1, \ldots, \alpha_n]$, with $\mathcal{D}_0 \equiv \mathcal{D}[\alpha_1, \ldots, \alpha_n]$. Then for the parent configuration $X_{pa[v]} = \rho$ the new mixture is given by:

$$(1 - P(\rho|\mathcal{E}))\mathcal{D}_0 + \sum_j P(x_v^j, \rho|\mathcal{E})\mathcal{D}_j.$$

The probabilities $P(\rho|\mathcal{E})$ and $P(x_v^j, \rho|\mathcal{E})$ may be obtained from the junction tree by the marginalisation of the potential of any clique containing both v and $pa[v]$, after the evidence $\mathcal{E}$ has been propagated. (Note that for complete data, the mixture reduces to the standard conjugate Bayesian updating, with only one term surviving in the expression above.) To avoid the explosion of terms this mixture would generate if it were itself used for analysing a second set of evidence (the theoretically exact approach, Spiegelhalter and Lauritzen, 1990), it is approximated by a single Dirichlet distribution whose mean and variance match the mean and average variance of the mixture.

Thus, following Spiegelhalter and Lauritzen (1990) let m_{ij} be the mean of the i-th slot of $\mathcal{D}_j$. Then the posterior mean is equal to:

$$m_i = (1 - P(\rho|\mathcal{E}))m_{i0} + \sum_{j=1}^{n} P(x_v^j, \rho|\mathcal{E})m_{ij}.$$

Approximating the Dirichlet mixture by a single Dirichlet distribution with parameters $(sm_1, \ldots, sm_k)$ for any $s > 0$ will give the correct mean. The scale factor s is chosen to give the same average variance, and is readily found (Spiegelhalter and Lauritzen, 1990). Note that it is possible for some entries to decrease using incomplete evidence (Bernardo and Girón, 1988; Spiegelhalter and Cowell, 1991). For a detailed example of results obtained with this facility, see Spiegelhalter and Cowell (1991). (An alternative scheme which keeps a limited number of Dirichlet distributions is currently being incorporated into *BAIES*; this is expected to yield a better approximation.)

2.2. *Non-sequential Learning*

The EM-algorithm (Dempster *et al.* 1977) has recently been formulated (Lauritzen, 1991) for discrete Bayesian networks so that the maximum likelihood estimate for the probability distribution of a model, given a set of incomplete data, can be obtained by local operations on the junction tree. This algorithm is implemented in *BAIES*, and it provides a non-sequential method for estimating the probability distribution of a model given data. It is an alternative to the sequential learning algorithm, and can be used as an independent method of assessing the performance of the latter when data are incomplete. Alternatively, in view of the generally slow convergence of the EM-algorithm, one could use the learning algorithm with the data to derive a reasonably good probability distribution, and then use the EM-algorithm to improve it. However because the EM-algorithm is a non-sequential method, the evaluation of monitors discussed in the next section is not possible.

3. MONITORS AND SCORING RULES

3.1. *Scoring Rules*

BAIES allows the comparison of model performance, individual conditional probability tables, and local model structure, through the use of *monitors* which use the following prequential scoring-rules (Dawid, 1984). Suppose that $P_m(Y)$ is a probability distribution for the random variable $Y = (y_i : i = 1, \ldots, n)$ after $m - 1$ data sets have been made (P_m will vary with m if learning about probabilities is occurring), and that $Y = y_j$ occurs on the m-th data set. Then we associate to the m-th data set a *logarithmic score*, S_m, together with an expectation $E(S_m)$ and variance $V(S_m)$, given by

$$
\begin{aligned}
S_m &= -\log(P_m(y_j)),\\
E(S_m) &= -\sum_k P_m(y_k)\log(P_m(y_k)),\\
V(S_m) &= \sum_k P_m(y_k)\log^2(P_m(y_k)) - E^2(S_m).
\end{aligned}
$$

From these the standardized test statistic, asymptotically standard normal when the model holds (Seillier-Moiseiwitsch and Dawid, 1992), is given by:

$$
Z_m \equiv \frac{\sum_{i=1}^m S_i - \sum_{i=1}^m E(S_i)}{\sqrt{\sum_{i=1}^m V(S_i)}}.
$$

3.2. *Global Monitors*

Each model has a global monitor, which is the accumulated sum of the logarithmic scores of the evidence observed, *ie.*

$$
-\sum_{i=1}^{m} \log(P(\mathcal{E}_i | \cup_{e=1}^{i-1} \mathcal{E}_e)),
$$

where m data-sets have been processed, and $P(\mathcal{E}_i | \cup_{e=1}^{i-1} \mathcal{E}_e)$ is the probability of the i-th data-set evidence $\mathcal{E}_i$ given all previous data-sets $\cup_{e=1}^{i-1} \mathcal{E}_e$. These probabilities are found by normalising the probability table of any clique after the evidence has been propagated. The rationale behind the use of the global monitor is that for data containing typical configurations 'good' models specified with 'good' conditional probability tables will accumulate a lower global monitor score than 'poor' models with 'poor' tables. The calculation of the expected score is not carried out – in general it is too laborious, except in the case of complete data on all the nodes.

3.3. *Parent-child Monitors*

Each dag-node has a parent-child monitor associated with its conditional probability table. Parent-child monitors do not test model structure directly, instead they test the quality of the (experts') assessed subjective prior probabilities. There is a separate monitor-entry for each parent configuration of the node v, and entries in the parent-child monitor of node v are only updated when both v and the configuration of all of its parents $pa[v]$ are observed, *ie.* complete data is observed on the node and its parents.

Let $P(X_v|X_{pa[v]}, \cup_{e=1}^{m-1}\mathcal{E}_e)$ denote the conditional tables for v given its parents $pa[v]$ after $m-1$ sets of data $\cup_{e=1}^{m-1}\mathcal{E}_e$ have been processed. (If the user sets the learning flag of X_v to OFF, then the conditional table will remain unchanged as the data is processed.) Suppose that the state $X_v = x_v^i$ is observed at the node v and that $X_{pa[v]} = \rho$ is the observed configuration of the parents, in the m-th set of data $\mathcal{E}_m$. Then the parent-child monitor for node v in the ρ-th slot is given the score:

$$S_m = -\log(P(X_v = x_v^i|X_{pa[v]}, \cup_{e=1}^{m-1}\mathcal{E}_e)).$$

Also calculated from the probability distribution are the expected score, its variance, and the reliability statistic. (S_m, $E(S_m)$ and $V(S_m)$ are all defined to be zero if the states of a node and its parents are not observed.) Although the monitors are evaluated only when complete data are available on a node and its parents, learning can take place using incomplete data. Thus kept in each dag-node is a set of counts of complete data together with the original conditional tables before any evidence has been propagated. These facilitate the calculation of two Bayes' standardisations which are also accumulated, *viz.* the score obtained from the original tables updated using only complete data, and the reference score obtained from a uniform prior table updated using only complete data.

3.4. *Node Monitors*

There are two types of node monitors, *conditional* and *unconditional*. Node monitors do not specify whether a local structure is correct or not. However they should identify problems of poor structure and/or poor probabilities associated with a node, although they cannot distinguish between these problems. Both types of node monitors for a node v are updated only if the state of v is observed.

Let $P(X_v|\cup_{e=1}^{m-1}\mathcal{E}_e)$ denote the current marginal table for v after $m-1$ sets of data $\cup_{e=1}^{m-1}\mathcal{E}_e$ have been processed, and suppose that in the m-th evidence set $X_v = x_v^i$ is observed. Then the unconditional monitor is given the score

$$S_m = -\log(P(X_v = x_v^i|\cup_{e=1}^{m-1}\mathcal{E}_e)).$$

The conditional node monitors are defined in a similar manner, but use marginals which are conditional upon the other evidence in the m-th evidence set. Again let $P(X_v|\cup_{e=1}^{m-1}\mathcal{E}_e)$ denote the current marginal table for v after $m-1$ sets of data $\cup_{e=1}^{m-1}\mathcal{E}_e$ have been processed, and suppose that in the m-th evidence set $\mathcal{E}_m$ the state $X_v = x_v^i$ is observed. Now suppose that all the evidence in $\mathcal{E}_m$ is propagated *except* the piece of evidence $X_v = x_v^i$; denote this evidence set by $\mathcal{E}_m\backslash X_v$. Then the new marginal for X_v given all the other evidence can be found. It is denoted by $P(X_v|\cup_{e=1}^{m-1}\mathcal{E}_e, \mathcal{E}_m\backslash X_v)$ and is used to calculate the score

$$S_m = -\log(P(X_v = x_v^i|\cup_{e=1}^{m-1}\mathcal{E}_e, \mathcal{E}_m\backslash X_v)).$$

For both types of node monitors, the expectations, variances and the reliability test statistics are also found. (If a state of a node is not observed, then S_m, $E(S_m)$ and $V(S_m)$ are all defined to be zero.) *BAIES* uses a propagation scheme called *fast retraction*, which enables the conditional probabilities to be found efficiently for all nodes using one propagation per evidence set; see (Cowell and Dawid, 1991) for theoretical details.

ACKNOWLEDGEMENTS

The development of *BAIES* was funded by the UK Science and Engineering Research Council. The author would like to thank A. P. Dawid and D. J. Spiegelhalter for their comments during the preparation of this paper.

REFERENCES

Andersen, S. K., Olesen, K. G., Jensen, F. V. and Jensen, F. (1989). HUGIN, a shell for building Bayesian belief universes for expert systems. *Proceedings of International Joint Conference on Artificial Intelligence*, Detroit: Morgan Kaufman, San Mateo, 1080–1085.

Bernardo, J. M. and Girón, F. J. (1988). A Bayesian analysis of simple mixture problems. *Bayesian Statistics 3* (J. M. Bernardo, M. H. DeGroot, D. V. Lindley and A. F. M. Smith, eds.), Oxford: University Press, 67–78, (with discussion).

Cowell, R. G. and Dawid, A. P. (1992). Fast retraction of evidence in a probabilistic expert system. *Statistics and Computing*, (to appear).

Dawid, A. P. (1979). Conditional independence in statistical theory. *J. Roy. Statist. Soc. B* **41**, 1–31, (with discussion).

Dawid, A. P. (1984). Present position and potential developments: some personal views. Statistical theory. The prequential approach. *J. Roy. Statist. Soc. A* **174**, 278–292, (with discussion).

Dempster, A. P., Laird, N. and Rubin, D. B. (1977). Maximum likelihood from incomplete data via the EM algorithm. *J. Roy. Statist. Soc. B* **39**, 1–38, (with discussion).

Golumbic, M. C. (1980). *Algorithmic Graph Theory and Perfect Graphs*. London: Academic Press.

Jensen, F. V. , Lauritzen, S. L. and Olesen, K. G. (1990). Bayesian updating in causal probabilistic networks by local computations. *Computational Statistics Quarterly* **4**, 269–282.

Lauritzen, S. L. (1991). The EM-algorithm for graphical association models with missing data. *Tech. Rep.* **91-05**, Aalborg University, Denmark.

Lauritzen, S. L. and Spiegelhalter, D. J. (1988). Local computations with probabilities on graphical structures and their application to expert systems. *J. Roy. Statist. Soc. B* **50**, 157–224, (with discussion).

Pearl, J. (1988). *Probabilistic Inference in Intelligent Systems*. San Mateo, CA: Morgan Kaufmann.

Seillier-Moiseiwitsch, F. and Dawid, A. P. (1992). On testing the validity of probability forecasts. *J. Amer. Statist. Assoc.* , (to appear).

Shachter, R. D. (1986). Evaluating influence diagrams. *Operations Research* **34**, 871–882.

Shachter, R. D. and Kenley, C. R. (1989). Gaussian Influence Diagrams. *Management Science* **35**, 527–549.

Spiegelhalter, D. J. and Cowell, R. G. (1992). Learning in probabilistic expert systems. *Bayesian Statistics 4* (J. M. Bernardo, J. O. Berger, A. P. Dawid and A. F. M. Smith, eds.), Oxford: University Press, 447–465, (with discussion).

Spiegelhalter, D. J. and Lauritzen, S. L. (1990). Sequential updating of conditional probabilities on directed graphical structures. *Networks* **20**, 579–605.

BAYESIAN STATISTICS 4, pp. 601–606
J. M. Bernardo, J. O. Berger, A. P. Dawid and A. F. M. Smith, (Eds.)

A Numerical Integration Strategy in Bayesian Analysis

PETROS DELLAPORTAS and DAVID E. WRIGHT
University of Nottingham, UK and *Plymouth Polytechnic South West, UK*

SUMMARY

This paper describes how embedded sequences of positive interpolatory integration rules, described in Dellaportas and Wright (1991), can be applied in Bayesian statistics. In particular, we discuss how a new numerical integration strategy can extend the applicability and increase the efficiency of currently available numerical methods. The proposed strategy is demonstrated using a 7-parameter proportional hazards model.

Keywords: BAYESIAN METHODS; INTEGRATION; EMBEDDED INTEGRATION RULES.

1. INTRODUCTION

A major obstacle to many practical applications of Bayesian methods is the difficulty in evaluating the various integrals required. It is often the case that special purpose numerical and analytic algorithms are designed to cope with such difficulties, but during the last decade, due to the great advances in computing hardware and software, various attempts to overcome this technical barrier via general implementation strategies have appeared in statistical literature. These include, for example, the numerical approximations of Smith *et al.* (1987), the analytic approximations of Tierney and Kadane (1986), the Monte Carlo integration strategies, see Geweke (1988), and more recently, the sampling based approaches described in Gelfand and Smith (1990). For a general review see Goel (1988), together with an attempt to classify the existing procedures given by van Dijk in the discussion thereafter.

In this paper we focus on numerical integration approximations for use in Bayesian statistics, and we present a strategy based on embedded sequences of positive interpolatory integration rules (PIIR's) derived from Gauss-Hermite product rules. We demonstrate how such a strategy can extend the applicability and increase the efficiency of currently available methods. The remainder of the paper proceeds as follows. Our proposed strategy is outlined in Section 2 and illustrated with an analysis of a proportional hazards model in Section 3. Concluding remarks are given in Section 4.

2. A NUMERICAL INTEGRATION STRATEGY

The first general purpose quadrature strategy for numerical integration in Bayesian analysis was introduced by Naylor and Smith (1982), and was subsequently refined and developed; see, for example, Smith *et al.* (1985, 1987), Naylor and Smith (1988). The strategy has been implemented via a computer package called 'BAYES FOUR'. See Naylor and Shaw (1985, 1986) for details. Aspects of its reliability, efficiency and accuracy have been examined by Shaw (1988) and Dellaportas (1990).

In Dellaportas and Wright (1991), we described how embedded sequences of positive interpolatory integration rules (PIIR's) based on Gauss-Hermite product rules can be derived by solving a series of systems of linear equations. Such sequences can also be derived using ideas from the theory of polynomial ideals; see, for example, Cools and Haegemans (1989).

If we represent the Gauss-Hermite product rule by $Q_m f$ where

$$Q_m f = \sum_{i=1}^{m} w_i f(x_i) \tag{1}$$

an embedded sequence of N rules of the form

$$\left.\begin{aligned} Q_m^{(1)} f &= \sum_{i \in S_1} w_i^{(1)} f(x_1) \\ Q_m^{(2)} f &= \sum_{i \in S_2} w_i^{(2)} f(x_i) \\ &\vdots \\ Q_m^{(N)} f &= Q_{mf} = \sum_{i=1}^{m} w_i^{(N)} f(x_i) \end{aligned}\right\} \tag{2}$$

with node sets $S_1 \subset S_2 \subset \ldots \subset S_{N-1} \subset (1, 2, \ldots, m)$ can be obtained. These correspond to N successive approximations finishing with the Gauss-Hermite product rule approximation $(w_i^{(N)} = w_i, i = 1, 2, \ldots, m)$.

As emphasized in Dellaportas and Wright (1991), the embedded sequences in (2) can be very useful because

(i) they provide a useful source of spatially distributed integration rules appropriate for use in Bayesian statistics and in particular in five dimensions and upwards.

(ii) they can improve the integration strategy of Naylor and Smith (1982), incorporating a facility to work through a sequence of embedded rules monitoring convergence after each stage and changing to a sequence with a different location and spread only when it is deemed necessary.

We propose a strategy similar to that described by Naylor and Smith (1982), based on the following steps. Of course, the initial parameter transformations and the orthogonalising linear transformations should also be incorporated within this strategy.

(i) Start the iterative strategy by selecting an appropriate base rule. In general, this will be larger than that used in the adaptive integration strategy of Smith *et al.* (1987) – typically 5 or more points in each dimension.

(ii) Apply an embedded sequence of PIIR's, centred using an initial estimate of the posterior mean φ and scaled using an initial estimate of the posterior covariance Σ, checking convergence after each approximation. If, in the early steps, there is evidence of considerable misspecification of (φ, Σ), stop and iterate again starting with the updated elements of (φ, Σ).

(iii) If the convergence in the sequence of PIIR's is rapid, then there is good indication that convergence is to the proper value of the integrand. This may well happen when the initial value of (φ, Σ) is not close to the true posterior (φ', Σ') but the integrand is 'well behaved'. In such cases we suggest the completion of the sequence and the derivation of marginal densities. Of course the option of verification using new approximations to (φ, Σ), possibly with a larger base rule, is available, especially when the initial rule is of low precision. Our experience, though, coincides with that of Rabinowitz *et al.*

(1987), who reported that their experiments with embedded sequences of PIIR's (in one dimension) show that rapid convergence is achieved with a PIIR embedded sequence, then in general this will be to the true value of the integrand. Convergence is detected using a measure Δ based on comparisons between current and previously estimated posterior moments, see Naylor and Shaw (1985). However, for given desired convergence criteria (for example $\Delta = 0.05$), we assume convergence if for three consecutive approximations in the embedded sequence we achieve $\Delta < 0.05$. Thus, the convergence may be detected automatically because the three approximations provide a better failsafe system.

(iv) If the convergence is slow, we suggest that there are two possible causes for this: misspecification of (φ, Σ) or a 'badly behaved' posterior kernel. Therefore, updating (φ, Σ) is recommended. If the iteration does not improve the convergence, then we may come to the conclusion that the assumptions are invalid. An increase in the size of base rule may overcome the problem, but a question arises —provided that we started with a large enough base rule— whether it is useful to proceed with a doubtful and expensive procedure or, maybe exploiting existing information from the posterior distribution (such as kurtosis or skewness), use an alternative Bayesian implementation strategy.

We note that the strategy proposed above is in essence an extension of the iterative strategy embodied in 'BAYES FOUR', at least as far as product rules are concerned. The important point is that we apply the product rule by proceeding through a sequence of embedded rules. The information in the sequence of approximations is used to enable us to stop early or diagnose convergence problems. In the standard 'BAYES FOUR' strategy the full product rule is used, effectively moving to the end of the sequence without exploiting any information about the development of the approximations.

3. AN ILLUSTRATIVE EXAMPLE

In this section we demonstrate the efficiency of our proposed strategy using a 7-parameter proportional hazards model. Other similar examples can be found in Dellaportas (1990) where, when based on the (objective) yardstick of function evaluations to convergence, the method proved efficient compared with existing numerical methods. However, use of an interactive adaptive algorithm is very much subjective and numerical illustrations could be misleading. In this paper, we feel that it suffices to illustrate the option of stopping relatively early without using all nodes of the full product rule, and illustrate how the convergence behaviour can provide information on whether an increase in size of the initial chosen product rule is desirable.

3.1. *A proportional Hazards Model*

Lawless (1982, p. 337) presented a set of a data which consists of survival times in months and corresponding regressor variables for 65 multiple myeloma patients (a subset from a more comprehensive set given by Krall *et al.* (1975)). The problem is to relate survival times for multiple myeloma patients to a number of prognostic variables. These prognostic variables are:

x_1: logarithm of a blood urea nitrogen measurement at diagnosis,
x_2: haemoglobin measurement at diagnosis
x_3: age at diagnosis,
x_4: sex: 0 male, 1 female,
x_5: serum calcium measurement at diagnosis.

To analyse these data, we used a Weibull regression model with proportional hazards, where the survival time has p.d.f

$$p(t \mid \beta, \rho) = \rho t^{\rho-1} e^{z\beta} \exp[-t^{\rho} e^{z\beta}], \qquad t > 0, \rho > 0,$$

where ρ is the shape parameter of the Weibull distribution, and the vector z of covariates defined in this case as follows:

$$z = \begin{bmatrix} 1 \\ x_1 - \bar{x}_1 \\ x_2 - \bar{x}_2 \\ x_3 - \bar{x}_3 \\ x_4 \\ x_5 - \bar{x}_5 \end{bmatrix}$$

Considering the $n + m = 65$ times as ordered in such a way that the first $n = 48$ times $t_1, t_2, \ldots, t_n$ are uncensored and the last $m = 17$ times, $t_{n+1}, t_{n+2}, \ldots, t_{n+m}$ are censored, the likelihood function can be written

$$l(\beta, \rho | \text{data}) = \left[\prod_{j=1}^{n} \rho t_j^{\rho-1} e^{z_j \beta} \right] \cdot \left[\prod_{j=1}^{n+m} \exp\left(t_j^{\rho} e^{z_j \beta} \right) \right],$$

where z_j denotes the vector of covariances for the j-th patient.

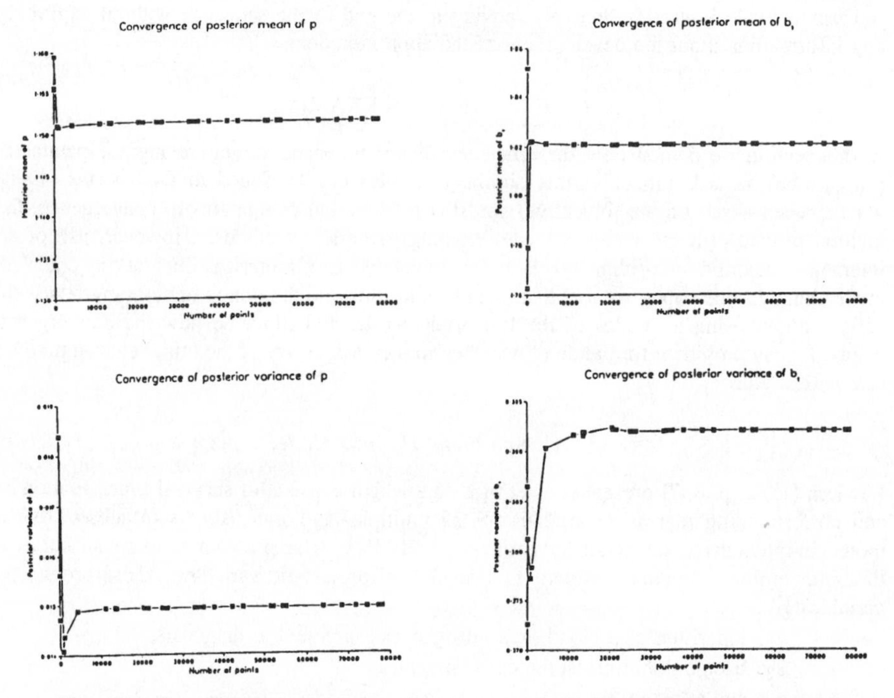

Figure 1. *Successive approximations obtained from an embedded sequence of integration rules.*

Using a locally uniform prior, $p(\beta, \rho)$ = constant, and the maximum likelihood estimates as our initial values for our numerical strategy, we applied an embedded sequence of PIIR's based on a 5^7 Gauss-Hermite product rule. The convergence of the posterior mean and variance vectors for two of the parameters are illustrated in Figure 1. The convergence of the other parameters is very similar, see Dellaportas (1990). Note that one iteration on an embedded sequence enables us to judge the behaviour of the kernel and, according to the remarks of Section 3, no further iterations are needed because the convergence is rapid. The results suggest that patients with higher blood urea nitrogen measurement at diagnosis (variable z_1) have longer survival times.

We emphasise here the importance of the embedded sequence of PIIR's in higher dimensions: the currently available integration rules for 7 dimensions are a 7-degree spherical rule with 452 nodes (Stroud (1971), p. 317–319, rule En: 7–2) and the 4^7 Gauss-Hermite product rule with 16384 nodes. A numerical integration strategy based on the mosaic of embedded integration rules provides more flexibility and efficiency.

4. CONCLUDING REMARKS

Our suggested numerical integration strategy can improve the currently available Bayesian quadrature strategy (Smith *et al.* (1985)) for the following reasons.

(1) It can increase the efficiency because working through embedded sequences of rules, the function evaluations are not discarded but re-used in subsequent iterations.
(2) It can improve the behaviour of convergence diagnostics because, being a generalisation of the existing algorithm, it is based on a much larger set of integration rules.
(3) It can improve the robustness of the algorithm to underlying assumptions, because indication of algorithm divergence is easier to detect via the series of integration rules.
(4) It can extend the applicability of 'BAYES FOUR' to up to 10 parameter problems.

REFERENCES

Cools, R. and Haegemans, A. (1989). On the construction of multi-dimensional embedded cubature formulae. *Number. Math.* **55**, 735–745.

Dellaportas, P. (1990). *Imbedded Integration Rules and Their Applications in Bayesian Analysis.* Ph.D. Thesis, Polytechnic South West, Plymouth, UK.

Dellaportas, P. and Wright, D. E. (1991). Positive imbedded integration in Bayesian analysis. *Statistics and Computing* **1**, 1–12.

Gelfand, A. E. and Smith, A. F. M. (1990). Sampling based approaches to calculating marginal densities. *J. Amer. Statist. Assoc.* **85**, 398–409.

Geweke, J. (1988). Antithetic acceleration of Monte Carlo integration in Bayesian inference. *Journal of Econometrics* **38**, 73–90.

Goel, P. K. (1988). Software for Bayesian analysis: current status and additional needs. *Bayesian Statistics 3* (J. M. Bernardo, M. H. DeGroot, D. V. Lindley and A. F. M. Smith, eds.), Oxford: University Press, 173–188, (with discussion).

Hills, S. E. (1989). *The Parametrisation of Statistical Models.* Ph.D. Thesis, University of Nottingham, Nottingham, UK.

Krall, J., Uthoff, V. and Harley, J. (1975). A step up procedure for selecting variables associated with survival. *Biometrics* **31**, 49–57.

Lawless, J. F. (1982). *Statistical Models and Methods for Lifetime Data.* New York: Wiley.

Naylor, J. C. and Shaw, J. E. H. (1985). *Bayes Four – User Guide. Tech. Rep.* **09–85**, University of Nottingham.

Naylor, J. C. and Shaw, J. E. H. (1986). *Bayes Four – Implementation Guide. Tech. Rep.* **01–86**, University of Nottingham.

Naylor, J. C. and Smith, A. F. M. (1982). Applications of a method for the efficient computation of posterior distributions. *Appl. Statist.* **31**, 214–225.

Naylor, J. C. and Smith, A. F. M. (1988). Econometric illustrations of novel numerical integration strategies for Bayesian inference. *Journal of Econometrics* **38**, 103–126.

Rabinowitz, P., Kautsky, J., Elhay, S. and Butcher, J. C. (1987). On sequences of imbedded integration rules. Numerical Integration: Recent Developments. *Software and Applications*, (P. Keast and G. Fairweather eds.) Dordrecht: Reidel, 113–139.

Shaw, J. E. H. (1988). *Numerical Integration and Display Methods for Bayesian Inference*. Ph.D. Thesis, University of Nottingham, UK.

Smith, A. F. M. (1988). What should be Bayesian about Bayesian software?. *Bayesian Statistics 3* (J. M. Bernardo, M. H. DeGroot, D. V. Lindley and A. F. M. Smith, eds.), Oxford: University Press, 429–435, (with discussion).

Smith, A. F. M., Skene, A. M., Shaw, J. E. H., Naylor, J. C. and Dransfield, M. (1985). The implementation of the Bayesian paradigm. *Commun. Statist. Theor. Meth.* **14**, 1079–1102.

Smith, A. F. M., Skene, A. M., Shaw, J. E. H., Naylor, J. C. (1987). Progress with numerical and graphical methods for practical Bayesian statistics. *The Statistician* **36**, 75–82.

Stroud, A. H. (1971). *Approximate Calculation of Multiple Integrals.* Englewood Cliffs, New Jersey: Prentice-Hall.

Tierney, L. and Kadane, J. B. (1986). Accurate approximations for posterior moments and marginal densities. *J. Amer. Statist. Assoc.* **81**, 82–86.

BAYESIAN STATISTICS 4, pp. 607–615
J. M. Bernardo, J. O. Berger, A. P. Dawid and A. F. M. Smith, (Eds.)

Reconciling Costs and Benefits in Experimental Design

M. FARROW and M. GOLDSTEIN
Sunderland Polytechnic, UK and *University of Durham, UK*

SUMMARY

We consider the design of experiments, the objective of which may be to learn about the values of several different quantities and which may involve costs of more than one type. In particular, medical experiments may give information on several different quantities and will generally involve ethical costs as well as financial costs. Experimenters may be unwilling to specify explicit trade-offs, for example between information gained and costs or between financial and ethical costs. We suggest an approach to exploring the set of possible designs and presenting the results of this exploration in a way that may help those who must choose the design to come to a decision. The degree of detail necessary in the prior belief specification depends upon the assumptions that can be made about the form of an underlying utility function. We illustrate our approach with a medical example.

Keywords: DESIGN OF EXPERIMENTS; LINEAR BAYES METHODS; UTILITY.

1. INTRODUCTION

1.1. *Learning about Many Quantities*

In some experiments there is a large number of quantities about which we wish to learn. In this paper we use as an example a repeated measurements experiment. Such experiments illustrate this point well. We may have several variables observed at each of several time points on individuals in a number of groups. In many applications, we will wish to learn about the means of all variables at all time points in all groups, albeit with some means or combinations of means more important than others. The question arises of how we can select an experimental design taking into account the amounts we will learn about quantities which may have different degrees of importance for us.

1.2. *Costs of More than One Type*

Counterbalancing the desire to learn more by observing more data are the costs involved in the experiment. In, for example, an industrial experiment it may be straightforward to calculate all of the costs in purely financial terms. In other areas of application there may also be costs of other types. In particular, in a medical experiment, there are likely to be both financial costs and ethical costs where the ethical costs involve such things as the discomfort experienced by a subject and possible damage to the subject's health as a result of taking part in the experiment.

In some medical experiments it may be reasonable to take the view that only ethical costs should be considered. Then we must, in some way, balance the cost to the experimental subjects of doing the experiment against the benefit to future patients.

There are, however, other medical experiments where the ethical benefit to future patients is less readily compared with the cost to the subjects. There are also experiments where the availability or cost of resources for carrying out the experimental work clearly plays a part, at least in deciding whether or not the experiment takes place at all. Thus the financial costs as well as the ethical costs are relevant to the design of some experiments. The question arises of how we can take into account costs of both types in selecting an experimental design.

1.3. *Prior Beliefs*

As well as assigning values or utilities to the costs and benefits of the experiment, we need to specify our prior beliefs about the quantities involved. In a multivariate repeated measurements experiment this involves a large number of unknowns. The conventional Bayesian approach would require specification of a prior probability density in many dimensions. To do this genuinely, without accepting convenient simplifying assumptions, would be a daunting task. However, the specifications which are actually necessary may be much simpler and may not require any imposed assumptions.

1.4. *Example*

Throughout this paper we refer to an experiment involving oral glucose tolerance tests (OGTT) of elderly subjects. This example is based on an actual experiment described by Wickramasinghe *et al.* (in submission). Analyses of the data are given by Farrow and Leyland (1991).

The OGTT is used to help measure the severity of disease suffered by a diabetic patient. After a period without food, patients are given glucose orally. Blood samples are taken and the concentration of glucose in each sample is determined. Roughly speaking, the quicker the glucose concentration returns to its fasting level the more healthy is the patient. In our example the concentration of C-peptide, which is believed to indicate insulin production, was also measured.

Elderly subjects in four groups were given the OGTT. Group 1 consisted of non-diabetic subjects. Groups 2, 3 and 4 consisted of diabetic subjects, each successive group representing a more severe degree of illness.

Blood samples were taken immediately before the glucose was given and then after 30, 60, 90 and 120 minutes. Thus we have four groups, two variables and five time points, $t = 0, \ldots, 4$.

Selecting a design involves choosing the numbers in each group, determining whether both variables should be measured in the case of each blood sample and determining whether the test should be curtailed before two hours for some patients. It should be noted that prolonging the test might have an adverse effect on the health of diabetic subjects.

2. BELIEF SPECIFICATION

To overcome the problem described in 1.3 above we adopt a partial belief specification sufficient for use in a linear Bayes analysis. Our approach is similar to that used in our earlier paper (Farrow and Goldstein, 1991).

Let the observation at time t on variable v for individual i in group g be

$$Y_{givt} = M_{gvt} + R_{givt},$$

where M_{gvt} represents the mean and R_{givt} represents the individual variation.

Our belief specification consists merely of prior means, variances and covariances for the set of means $\{M_{gvt}\}$ and variances and covariances for the individual variation $\{R_{givt}\}$. We leave the details for a future paper but we show the structure of our belief model by the influence diagrams in Figures 1, 2 and 3. The meaning of these diagrams is that any two nodes, neither of which is a direct predecessor of the other, are uncorrelated conditional on the values of their direct predecessors (if any), (see Goldstein, 1990).

Figure 1 shows how the observations, Y_{givt}, depend on the means, M_{gvt}, and the individual variation, R_{givt}, and how the uncertainty in the means for the four groups is modelled by uncertainty elements E_{gvt} and U_{jvt}. Each specific uncertainty element, E_{gvt} contributes to only one group while between-group correlations arise from the common uncertainty elements U_{jvt}.

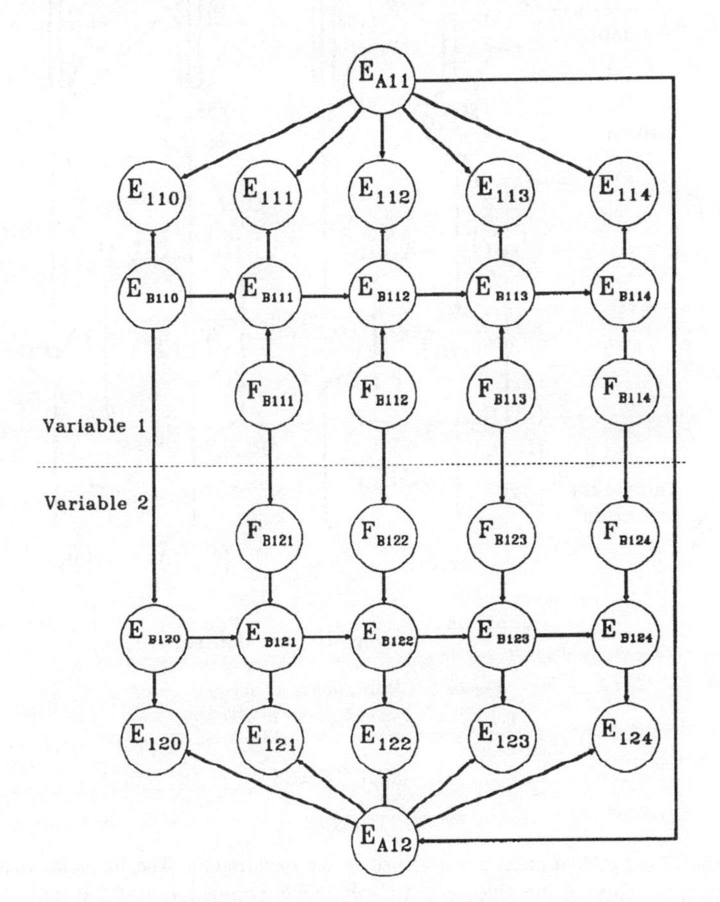

Figure 1. *Relations between groups.*
Only one variable and two observations per group are shown.

Figure 2 shows how the uncertainty elements for the two variables and five time points are related. Figure 3 shows how the individual variation elements of a typical individual are related. In each case the influence has been directed from Variable 1 to Variable 2 but this could equally well have been reversed.

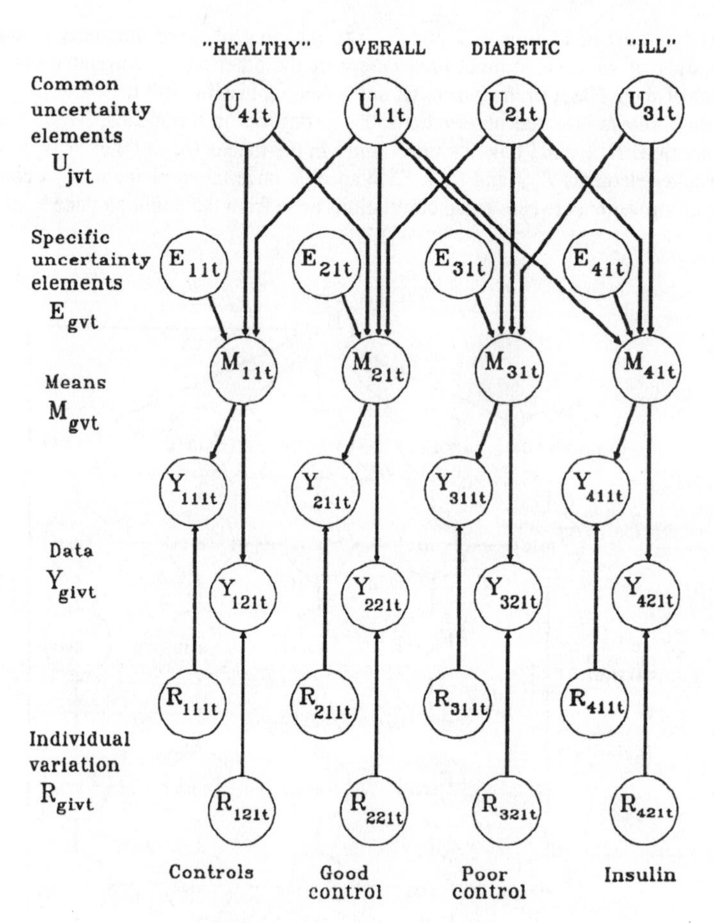

Figure 2. *Construction of* E_{g1t}, E_{g2t}
A corresponding model applies to U_{j1t}, U_{j2t}.

3. OBJECTIVES

3.1. *Costs*

Both financial and ethical costs are involved in the experiment. The financial costs consist of the laboratory costs of the glucose and C-peptide determinations and the staff, equipment and refreshment costs involved in taking the samples and looking after the subjects. The laboratory cost for C-peptide is much greater than for glucose. The sampling costs (staff etc.) per observation increase with the duration of the test and also with the severity of the subject's illness.

The ethical costs are more difficult to quantify although it is clearly undesirable, all other things being equal, to subject a diabetic person to a procedure which involves fasting followed by a dose of glucose then a further period with no food or medication. Furthermore the undesirability is greatly increased for those with more severe illness and increases with the duration of the test.

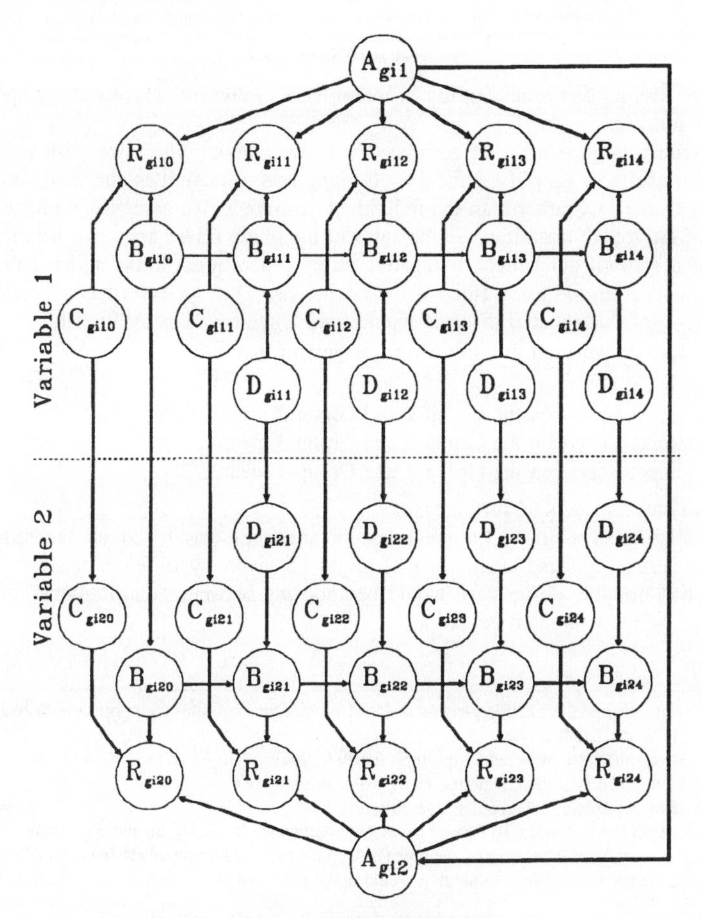

Figure 3. *Individual Variation.*

In principle we can elicit costs by asking questions based on comparisons of the undesirability of subjecting different numbers of subjects from the various groups to a two-hour OGTT and subjecting different numbers to different durations of OGTT. In practice, however, the consultant concerned preferred to work directly in terms of "units of ethical cost". One unit corresponded to taking only the initial blood sample from a non-diabetic subject. The costs were specified for members of each group and for durations of the test up to and some way beyond two hours. The ethical cost for a two-hour test on an insulin- dependent subject, for example, was given as 32 units.

By plotting the given costs against duration of the test and trying various functions of duration it was found that a reasonably good fit could be obtained by $q_{0g} + q_{1g}t + q_{2g}t^2$, where t is the time that the last sample is taken and the coefficients depend on the group g.

Although the real costs, in terms of subjects suffering undesirable consequences of taking part in the experiment, are unknown, we effectively obtained only expectations for these in this particular example. As a consequence we were unable to allow for a correlation between the costs and the observations obtained in a given design.

3.2. *Benefits*

We intend to discuss our benefit utility in more detail elsewhere. However our approach is basically as follows.

Faced with a large number of correlated unknowns about which we wish to learn, we have found it useful to simplify matters by forming sets of quantities and assigning relative importances to the sets rather than the individual quantities. We associate a utility function with each of the sets of quantities. In the spirit of the linear Bayes approach we may assign, to a particular state of our beliefs, a negative utility proportional to the squared distance of our expectations from the true values of the unknowns. Thus an experiment which reduced the expectation of this squared distance would have a positive expected utility.

In our example we used the following four sets of quantities:

1. The means for Group 1.
2. The differences between the Group 1 and Group 2 means.
3. The differences between the Group 2 and Group 3 means.
4. The differences between the Group 3 and Group 4 means.

Thus we had four benefit attributes.

Our elicitation of a utility for these benefit attributes was based on the reduction in standard deviation for these attributes that might be given by different designs. Thus in comparing two possible designs we would be choosing between "distributions" of rewards characterised by their standard deviations.

---

1. We need to know the relative weights attached to the four quantities. Suppose a 50% reduction in S_1 is worth 1 unit to you.
   (a) How much is a 50% reduction in S_2 worth to you?
   (b) How much is a 50% reduction in S_3 worth to you?
   (c) How much is a 50% reduction in S_4 worth to you?
2. For each of the following 10 lists of percentage reductions in S_1, S_2, S_3 and S_4, please imagine that you have the choice between the given reductions and a reduction of x% in each of S_1, S_2, S_3 and S_4. In each case, what value of x would make your two alternatives equally desirable?

(a)	90, 60, 60, 60	(f)	80, 60, 80, 60
(b)	60, 90, 60, 60	(g)	80, 60, 60, 80
(c)	60, 60, 90, 60	(h)	60, 80, 80, 60
(d)	60, 60, 60, 90	(i)	60, 80, 60, 80
(e)	80, 80, 60, 60	(j)	60, 60, 80, 80

3. Please indicate your order of preference if given a choice between the following 17 sets of percentage reductions in S_1, S_2, S_3 and S_4.

(a)	90, 60, 60, 60	(j)	60, 60, 80, 80
(b)	60, 90, 60, 60	(k)	60, 60, 60, 60
(c)	60, 60, 90, 60	(l)	65, 65, 65, 65
(d)	60, 60, 60, 90	(m)	70, 70, 70, 70
(e)	80, 80, 60, 60	(n)	75, 75, 75, 75
(f)	80, 60, 80, 60	(o)	80, 80, 80, 80
(g)	80, 60, 60, 80	(p)	85, 85, 85, 85
(h)	60, 80, 80, 60	(q)	90, 90, 90, 90
(i)	60, 80, 60, 80		

Table 1. *Benefit Utility Elicitation.*

The three questions tried out in the elicitation are given in Table 1. Here $S_1, \ldots, S_4$ are the standard deviations for the four attributes. Answers were obtained, in an interview, to Questions 1 and 3. Question 3 proved to be the most useful. Using the answers to Question 1 as a starting point, we were able to find a utility function, in which the four benefit attributes were mutually utility independent, such that ordering the corresponding expectations matched the preferences given in Question 3 almost exactly.

4. CONSIDERATIONS OF POSSIBLE DESIGNS

4.1. *Method*

In principle, we could determine an overall utility function and then search among possible designs for the design which maximises the expected utility. However, for two reasons, we might not attain this ideal.

Firstly there may be an excessively large number of possible designs from which to choose and no clear rules to guide our search. Thus we may choose a search algorithm which moves stepwise through the possibilities but which is not guaranteed to find the global optimum. The particular algorithm used for our example will be described elsewhere.

Secondly, there may be a reluctance on the part of our clients to specify trade-offs between the different costs and between costs and benefits. Clearly it is difficult to compare financial and ethical costs or to compare costs and benefits in a medical experiment. Therefore we present the results graphically to allow the client to make a final choice. We suggest a graph where the axes are, for example, ethical cost and financial cost. Each design considered produces a point on the graph which we can label with the expected benefit obtained. The designs considered are given by the steps of our stepwise search algorithm. In order to produce a spread of points and allow the client to make an informed choice we run the algorithm a number of times, each time using a different trade-off between costs and benefits or between ethical and financial costs. Thus, although we might not locate the "best" design we can display the characteristics of a large number of designs including those which are "good" under various different trade-off regimes.

4.2. *Results*

In Figure 4 we have plotted, for illustration, some points resulting from runs of our algorithm. In each case we have made the two costs additive and the cost and benefit utilities additive. In the first run we forced the addition of a subject with all observations made in each group at each cycle. In the others we varied the relative weights of the two costs. By treating the cost of one "complete" subject from each group as a standard we varied the proportion of this cost accounted for by the financial cost, setting this at 0%, 25%, 50%, 75% and 100%. In each case we stopped after 20 cycles if the algorithm had not already stopped itself. With the particular cost-benefit trade-off we used, the 0% and 25% runs gave identical results.

While we do not claim to be able to find an "optimal" design, it is clear from Figure 4 that useful comparisons can be made. In particular we can clearly improve on the simple design with equal group sizes and all observations made. For example, with a group size of 14 this design gives an expected benefit utility of 10.32 at a financial cost of 1383 and an ethical cost of 29568 while an expected benefit utility of 10.40 at a financial cost of 1290 and an ethical cost of 26752 can be obtained by Design A in Table 2. Here "complete" subjects have all ten observations made while "glucose-only" subjects have only the five glucose measurements. Similarly, Design B, in Table 2 gives a slightly reduced expected benefit utility of 10.21 at a financial cost of 1301 and an ethical cost of 23488. If financial costs were dominant we might consider Design C, giving an expected benefit utility of 10.25

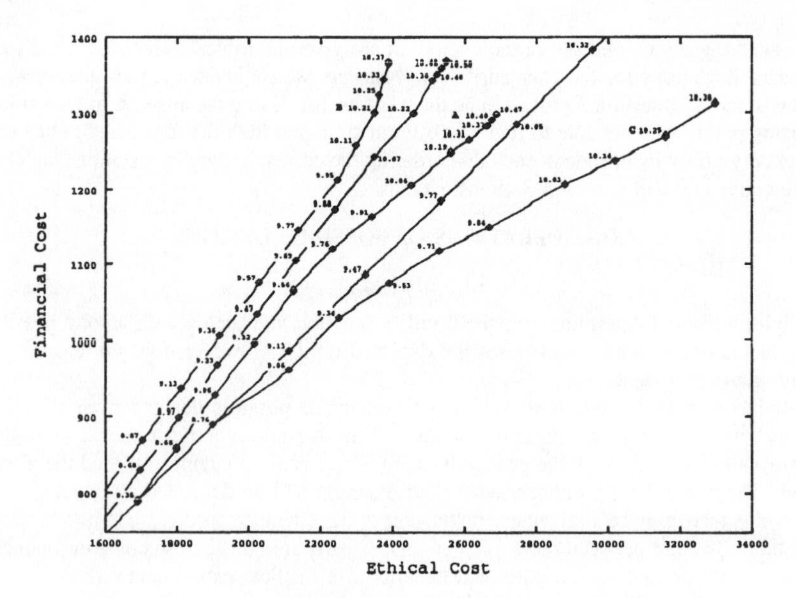

Figure 4. *Costs and benefits.*

at a financial cost of 1268 and an ethical cost of 31584, in which some subjects only provide the first two glucose measurements.

Design A

Group	1	2	3	4
"Complete" subjects	12	13	16	8
Glucose-only subjects	7	5	0	0

Design B

Group	1	2	3	4
"Complete" subjects	17	17	15	6
Glucose-only subjects	0	0	0	0

Design B

Group	1	2	3	4
"Complete" subjects	10	11	13	10
Glucose-only subjects	7	6	4	1
First two glucose only	0	0	0	6

Table 2. *Number of subjects of different types in three possible designs.*

ACKNOWLEDGEMENTS

The computation was done using the program $[B/D]$, developed, with the support of the UK Science and Engineering Research Council, by David Wooff and Michael Goldstein. A paper (Wooff, 1992) describing the operation of the program is included in this volume.

We are grateful to Dr. B. I. Chazan of Sunderland District General Hospital for his patient cooperation with our elicitations.

REFERENCES

Farrow, M. and Goldstein, M. (1991). Linear Bayes methods for crossover trials. (Unpublished *Tech. Rep.*)

Farrow, M. and Leyland, A. H. (1991). Interpretation of oral glucose tolerance test results. *Statistics in Medicine* (F. Dunstan, and J. Pickles, eds.), Oxford: University Press, 249–266.

Goldstein, M. (1990). Influence and belief adjustment. *Influence Diagrams, Belief Nets and Decision Analysis* (R. M. Oliver, and J. Q. Smith eds.), New York: Wiley, 143–174.

Wickramasinghe, L. S. P., Chazan, B. I., Farrow, M., Bansal, S. K. and Basu, S. K. (1991). C-peptide response to oral glucose and its clinical role in the elderly. (unpublished *Tech. Rep.*)

Wooff, D. A. (1992). $[B/D]$ works. *Bayesian Statistics 4* (J. M. Bernardo, J. O. Berger, A. P. Dawid and A. F. M. Smith, eds.), Oxford: University Press, 851–859.

BAYESIAN STATISTICS 4, pp. 617–624
J. M. Bernardo, J. O. Berger, A. P. Dawid and A. F. M. Smith, (Eds.)

The Use of Expert Judgement in the Context of a Postulated Mathematical Model

S. FRENCH*, R. M. COOKE** and F. VOGT**[1]
**University of Leeds, UK*, and ***Technical University of Delft, The Netherlands.*

SUMMARY

We consider the expert problem in which an analyst consults one or more experts for their views about uncertain quantities required in a decision or risk analysis. We extend the expert problem to the case in which the uncertain quantities are related in the analyst's mind by a physical model. The problem is first formulated as a Bayesian hierarchical model. Because of many computational difficulties, we explore a more pragmatic approach for circumstances in which the experts' judgements are brought together through a weighted linear opinion pool. Details are worked through for the case where the physical model is a simple normal regression model and the classically weighted linear opinion pool is used. We also note that the approach may be developed to allow the physical model to be drawn from the GLIM family. Unfortunately, applications of expert judgement seem to encounter physical models which are far from linear and these methods may still not be tractable in some important practical cases. For these we indicate a further approach, which may be more computationally feasible.

Keywords: CLASSICALLY WEIGHTED LINEAR OPINION POOL; EXPERT JUDGEMENT; GLIM MODELS; NORMAL LINEAR MODEL; THE EXPERT PROBLEM.

1. INTRODUCTION

Methods for incorporating expert judgements of the likelihood of uncertain events or quantities into decision, risk or other forms of analysis are not new. Cooke (1991), ESSRDA (1990), French (1985), Genest and Zidek (1986), and Wright and Ayton (1987) provide surveys of earlier work. Here we concentrate on the expert problem in which the person responsible for the analysis, hereafter referred to as the analyst, is distinct from the expert or group of experts from whom advice is sought. We confine ourselves to cases in which the experts give their judgements probabilistically and in which the analyst encodes her[2] uncertainty probabilistically.

Experience with this methodology has highlighted a problem. Namely, that the uncertain events and quantities are often related in the analyst's mind by a physical model. In risk analyses in particular, she would often use a parametric model to predict the uncertainties if she knew the parameter values to use. For instance, the following examples have arisen in applications.

Like other European countries implementing the *Post Seveso Directive*, the Netherlands uses dose response models in determining the zoning around industrial complexes handling toxic materials. The standard probit relation has been put forward for use within these:

$$\phi^{-1}(y(x)) = a_1 ln(x) + a_2 \tag{1.1}$$

---

[1] Frank Vogt is presently with Arthur Andersen, Den Haag, The Netherlands.

[2] We adopt a convention of taking the analyst to be female and the experts to be male. No connotation should be drawn from this.

where $\phi(.)$ is the cdf for the standard normal distribution, $y(x)$ the percentage of the exposed population which exhibits a biological effect (usually death) at a dose x of a particular toxic material. Fortunately, useful human data to which to fit such models is almost entirely lacking. Extrapolation from the data obtained on homogeneous animal populations in the laboratory to the heterogeneous human populations which might be exposed to the material in an accident is problematic. Thus the Dutch government has commissioned a group at the Technical University of Delft to obtain the 'most defensible' values, together with associated measures of uncertainty, of a_1 and a_2 based upon expert judgement.

Another example arises in risk studies of accidental releases at nuclear installations. Here the physical model may be a dispersion-deposition model. The dependent quantity y would be the time integrated surface concentration of contaminants at a given position. The independent variables x would include the position and time at which y is to be predicted, the composition and volume of the release, etc. The parameters would include meteorological variables such as wind speed, temperature, etc. Many forms of dispersion-deposition model have been proposed (CEC, 1990). Under the assumptions of constant meteorological conditions, several of these models reduce to a common form. Taking x_1 as the downwind distance, x_2 the crosswind distance and x_3 the vertical distance from the release point at time $x_4 = t, y$ is predicted by:

$$\begin{aligned} y(x_1, x_2, x_3, x_4) = & \frac{Q_0}{2\pi\sigma_{x_1}\sigma_{x_3}u} \exp\left(-\frac{x_2^2}{2\sigma_{x_2}^2}\right) \\ & \times \left\{\exp\left(-\frac{(x_3 - H)^2}{2\sigma_{x_3}^2}\right) + \exp\left(-\frac{(x_3 + H)^2}{2\sigma_{x_3}^2}\right)\right\} \end{aligned} \tag{1.2}$$

where Q_0 and the σ_x's are determined by the physical situation. We have given the simplified(!) form, which makes assumptions of constant wind field, flat terrain, etc, to emphasise the complexity that often occurs in practical applications.

The European Community is interested in the use of expert judgement to quantify the uncertainty in dispersion-deposition models and to predict the spread of contamination in the period immediately following an accidental release before physical data becomes available. Groups at Delft, Leeds and Warwick are working on these problems.

In these circumstances, one's first thought is that the analyst should ask the experts for advice on the values of the parameters in the models. However, to do so would be to ignore a central design principle for elicitation protocols. Experts should be asked about the values of (potentially) observable events and quantities (ESSRDA, 1990; Cooke, 1991). They should not be asked to provide distributions for abstract parameters.

Thus our problem may be summarised as: how should an analyst combine the information she gains from expert judgement with her belief in the appropriateness of a physical model to predict uncertain events and quantities in her analysis.

Section 2 provides a Bayesian formulation, one, however, that is subject to many computational difficulties. Section 3 explores a more pragmatic approach, which is developed for the case in which the experts' judgements are brought together through a weighted linear opinion pool. In Section 4 details are worked through for the case of a normal regression model and the generalization is noted to the case in which the physical model is drawn from the GLIM family. Unfortunately, applications of expert judgement encounter physical models which are far from linear. Thus the methods of this section may still not be tractable. Section 5 provides a further approach, which may be more computationally feasible.

2. A BAYESIAN APPROACH

Bayesian approaches to the expert problem are conceptually simple. The judgements provided by the experts are viewed as *data* (French, 1985). The problem, therefore, reduces to constructing a suitable prior distribution and a suitable likelihood function to describe the characteristics that the analyst expects in the experts' judgements, and then applying Bayes Theorem.

The former task is no harder than that of defining any prior distribution. Constructing the likelihood function is, however, far from easy. First, modelling potential cognitive biases in the experts is difficult. Second, the uncertain variables to be forecast are usually measured on a variety of scales. Some attempts to construct practically reasonable likelihood functions have been made: see Mendel and Sheridan (1987), Mosleh and Apostalakis (1986) and Wiper and French (1991). These efforts, however, perhaps serve more to show the inherent difficulty of the task than to provide solutions. Moreover, although these authors have found suitable approximations to work with conjugate distributions, in general it is clear that Bayes' Theorem will have to be applied numerically.

None the less, conceptually the Bayesian approach is straightforward. Here we formulate our problem along the lines of a hierarchical model (French, 1978; Lindley and Smith, 1972).

First, it is helpful to separate observational errors from modelling errors. Let b be a vector of parameters such that

$$y \sim \mathcal{P}_y(\cdot|b) \tag{2.1}$$

is the distribution describing the observational errors that the analyst expects in observing y. (The generic symbol $\mathcal{P}_u(.|h)$ is used for a probability distribution representing uncertainty about u given knowledge of $h; p_u(\cdot|h)$ is used for the corresponding density function.) In general, the location parameters within b will be determined by the physical model $M(a, x)$ to which the analyst subscribes. Here a is the vector of parameters within the model and x the vector of independent variables. Because of the possibility of modelling error, we assume that the analyst holds:

$$b \sim \mathcal{P}_b(\cdot|a, x). \tag{2.2}$$

For instance, assuming additive errors, (2.1) and (2.2) might be written:

$$y \sim b + \varepsilon \tag{2.3a}$$

$$b \sim M(a, x) + \eta \tag{2.3b}$$

(2.3a) corresponds to (2.1) with ε being the observational error whose distribution gives $\mathcal{P}_y(\cdot|b)$; while (2.3b) corresponds to (2.2) with η being the modelling error whose distribution gives $\mathcal{P}_b(\cdot|a, x)$. In practice, it is difficult to separate observational and modelling errors; but conceptually it may be useful to do so.

The analyst's prior knowledge about a is modelled by:

$$a \sim \mathcal{P}_a(\cdot). \tag{2.4}$$

Thus we have a three-stage hierarchical model:

$$a \sim \mathcal{P}_a(\cdot), \tag{2.5a}$$

$$b \sim \mathcal{P}(\cdot|a, x), \tag{2.5b}$$

$$y \sim \mathcal{P}_y(\cdot|b). \tag{2.5c}$$

To this model must be added the knowledge gained from the experts. Here we need to be clear about the judgements asked of them. They may be asked about the next observed

y or about the mean of a set of observations on y (in regression terms, the mean effects). Thus, if the expert judgements are given as a vector z, the analyst would form her likelihood function for the experts' judgements from distributions parameterised by either y or b:

$$z \sim \mathcal{P}_z(\cdot|y, c) \tag{2.6a}$$

$$z \sim \mathcal{P}_z(\cdot|b, c) \tag{2.6b}$$

where c is a vector of parameters which encode her view of the cognitive biases present in their judgements. NB: the parameter c and the distributions $\mathcal{P}_z(\cdot|\cdot,\cdot)$ differ in (2.6a) and (2.6b). The analyst may learn about values for c from the experts' performance on a calibration set of seed variables. The inclusion of such information may be achieved by augmenting several levels of the hierarchical model to reflect both the variables of real interest and the seed variables. Thus y and z would become (y, y_c) and (z, z_c), respectively, where y and z are the seed variables and the experts' predictions of them. Augmenting b similarly to (b, b_c) where b_c contains the parameters of the distribution for y_c and using, for the sake of example, the corresponding development of (2.6b):

$$\begin{aligned}
(a, c) &\sim \mathcal{P}_{a,c}(\cdot,\cdot), & (2.7a)\\
(b, b_c) &\sim \mathcal{P}_{b,b_c}(\cdot,\cdot|a, x), & (2.7b)\\
(y, y_c) &\sim \mathcal{P}_{y,y_c}(\cdot,\cdot|b, b_c), & (2.7c)\\
(z, z_c) &\sim \mathcal{P}_{z,z_c}(\cdot,\cdot|b, b_c, c). & (2.7d)
\end{aligned}$$

Note that (2.7a) models the analyst's prior knowledge of a and c. Using (2.7), the analyst's predictive distribution conditional on the elicited judgements (z, z_c) and the known values of the seed variables y_c is:

$$\begin{aligned}
p_y(y|x, z, z_c, y_c) \propto_y \int_a \int_{b,b_c} \int_c & p_{z,z_c}(z, z_c, |b, b_c, c) \\
& p_{y,y_c},(y, y_c|b, b_c) \cdot p_{b,b_c}(b, b_c|a, x) \\
& p_{b,b_c}(b, b_c|a, x) \cdot p_{a,c}(a, c) \cdot d(a, b, b_c, c)
\end{aligned} \tag{2.8}$$

A model of this form is explored in Wiper (1991).

3. A MORE PRAGMATIC APPROACH

The development which led to (2.8) was entirely conceptual. In practice, it would be far more straightforward. Given that the model $M(a, x)$ may have a more complicated form than (1.2) and that distributions in (2.7) may be far from conjugate, one must expect (2.8) to require numerical evaluation. The parameters a, b, b_c and c are multi-dimensional. Thus (2.8) is unlikely to lead to a practical method of assimilating expert judgement. One way forward is to retreat from a strict Bayesian formalism and to think more pragmatically.

Suppose that the analyst has little prior knowledge of the parameters a in the model $M(a, x)$. Thus the only information on a that she has is that implicit in the experts' judgements for the observations y. She could simply use a method for combining these expert judgements, but (i) this would lose her belief in the general form of the model $M(a, x)$ and (ii) she would be unable to predict y for values of x other than those at which the experts gave their judgements. None the less, let us suppose that she forms a predictive distribution for $y|x$ simply by combining the experts' judgements for $y|x$. By this we mean that she asks the experts, for example, for their distributions on y at the points specified by the independent variables x. From their replies she forms a distribution for $y|x$, perhaps by using a linear opinion pool with classical weights (Cooke, 1990, 1991). Let this distribution be $\mathcal{P}(\cdot|x, \mathcal{J})$, where $\mathcal{J}$ represents the information she gains from the experts' judgements.

Next suppose that (2.5b) and (2.5c) are combined to give:

$$p(y|a,x) = \int_b p_y(y|b) \cdot p_b(b|a,x) \cdot db, \tag{3.1}$$

the predictive distribution for y which she would use if she knew suitable values of a. We suggest that she approximates $P_y(\cdot|x, \mathcal{J})$, the distribution formed from the experts' judgements, by a distribution of the parametric form $\mathcal{P}_y(\cdot|a,x)$. Doing so has two effects. First,$\mathcal{P}_y(\cdot|a,x)$ introduces the form or shape of the relationship $M(a,x)$ between y and x in which she believes. Second, she may use the approximation $\mathcal{P}_y(\cdot|a,x)$ to predict y at values of x other than those at which the experts predicted y. It might be argued that the values of the parameters which provide the best fit to $P_y(\cdot|x, \mathcal{J})$ offer the 'most defensible' parameter values in the sense required in the probit example of Section 1.

Many have argued that to approximate a distribution by another, one should seek to minimise the relative entropy between the distributions. Bernardo (1987), for instance, argues this using the logarithmic scoring rule as a loss function. Thus we propose approximating – or fitting – $P_y(\cdot|x, \mathcal{J})$ by a distribution from the family $\mathcal{P}_y(\cdot|a,x)$ in such a way as to minimise the relative entropy between the two distributions:

$$\min_a \int_y ln\,(p_y(y)/p_y(y|a,x)) \cdot p_y(y) \cdot dy \tag{3.2}$$

In the next section we work through the details of (3.2) when $\mathcal{P}_y(\cdot|a,x)$ arises from a normal linear model.

4. NORMAL LINEAR MODELS

Suppose that $\mathcal{P}_y(\cdot|a,x)$ is a normal distribution, $N(Xa, V)$, where X is the design matrix arising from x and V is the *known* covariance matrix. Suppose further that the analyst has consulted E experts, who have provided their distribution $\mathcal{P}^e_y(\cdot|x), e = 1, 2, \ldots, E$. She has then formed $|\mathcal{P}_y(\cdot|x, \mathcal{J})$ by a linear opinion pool with Classical weights, w_e:

$$\mathcal{P}_y(\cdot|x, \mathcal{J}) = \sum_{e=1}^{E} w_e \mathcal{P}^e_y(\cdot|x). \tag{4.1}$$

Following (3.2), we wish to approximate - or fit - $\mathcal{P}_y(\cdot|x, \mathcal{J})$ by a normal distribution with density function $\phi_y(\cdot|Xa, V)$ by minimising with respect to a:

$$\begin{aligned} &\int_y ln\,(p_y(y)/\phi_y(y|Xa,V)) \cdot p_y(y|x,\mathcal{J}) \cdot dy \\ &= \text{constant term} - \int_y ln\,(\phi_y(y|Xa,V)) \cdot p_y(y|x,\mathcal{J}) \cdot dy \\ &= \text{constant term} + \sum_{e=1}^{E} \int_y \left(\frac{1}{2}(y - Xa)T_{V^{-1}}(y - Xa)\right) \cdot w_e \cdot p^e_y(y|x) \cdot dy \end{aligned} \tag{4.2}$$

This reduces to minimising with respect to a:

$$\sum_{e=1}^{E} w_e \cdot ((\mathcal{E}_{y|x} - Xa))\, V^{-1} \left(\mathcal{E}_{y|x} - Xa\right), \tag{4.3}$$

where $\mathcal{E}_{y|x}$ is expert e's expectation of $y|x$. Thus the analyst should fit the model parameters, a to the experts' expectations of $y|x$ using weighted least squares with the weighting deriving from the weights, w_e in the linear opinion pool and the assumed covariance matrix, V.

The above analysis can be carried through using a generalised linear model (GLIM) instead of a normal linear model. Minimising the relative entropy between two distributions corresponds to maximising the likelihood (see, e.g. Bernardo, 1987). In this case, replacing the normal density, $\phi_y(\cdot|Xa, V)$, by an exponential family density with canonical parameter linked to the linear predictor, Xa, leads to the need to maximise an expression which parallels the log likelihood of a GLIM model.

Suppose that $\mathcal{P}_y(\cdot|a, x)$ is a distribution from the exponential family with density:

$$p_y(\cdot|a, x) = exp\left\{[(\sum \theta_i y_i) - \sum \beta(\theta_i)]/\phi + \sum \gamma(y_i, \phi)\right\}, \tag{4.4}$$

where $\beta(\cdot)$ and $\gamma(\cdot, \cdot)$ are known functions, ϕ a known dispersion parameter and the components of y are assumed independent. Further suppose that the parameters θ_i are linked to the linear predictor, Xa via:

$$g\left(\frac{d\beta}{d\theta}(\theta_i)\right) = (Xa)_i, \tag{4.5}$$

where $g(\cdot)$ is a known link function. NB. $\cdot\frac{d\beta}{d\theta}(\theta_i)$ is the expectation of y_i given by (4.4): see McCullagh and Nelder (1983).

This leads us to minimise with respect to a:

$$\int_y ln\left(p_y(y)/p_y(y|a, x)\right) \cdot p_y(y|x, \mathcal{J}) \cdot dy.$$

$$= \text{constant term} - \sum_{e=1}^{E} \int_y \left([(\sum \theta_i y_i) - \sum \beta(\theta_i)]/\phi\right) \cdot w_e \cdot p_y^e(y|x) \cdot dy \tag{4.6}$$

$$= \text{constant term} - \sum_{e=1}^{E} w_e \left([(\sum \theta_i \mathcal{E}_{y_i|x}) - \sum \beta(\theta_i)]/\phi\right).$$

Minimizing (4.6) is the same as maximizing the log likelihood from a GLIM model with obsevations $\mathcal{E}_{y_i|x}$ and dispersion parameters (ϕ/w_e), which may be achieved by means of the EM algorithm.

Although analyses along the lines suggested above do provide a way forward in some cases, there are problems. First, the methods fit the probability models to the observations $\mathcal{E}_{y|x}$. These are not provided directly by the analyst. To take advantage of a calibration set the experts must provide probability statements not expectations . Generally, the experts are asked for quantiles from their distribution. A maximum entropy approximation to their full distribution is derived from these. Whilst it is, of course, trivially easy to calculate $\mathcal{E}_{y|x}$ from these approximations, the results are unlikely to be robust. Alternatively, one could ask the experts for their expectations, $\mathcal{E}_{y|x}$, as well as their quantiles. But the cognitive biases likely to be present in their articulation of quantiles may have different qualities to those present in their articulation of expectations. which would obviate the value of using a calibration set.

A second difficulty concerns the form of the underlying models $M(a, x)$. Since these may be considerably more complex than (1.2), it is unlikely that even the GLIM structure will enable the many non-linearities to be dealt with adequately.

5. APPROXIMATING BY DISTRIBUTIONS IN THE PARAMETER SPACE

Instead of using distributions of the form $\mathcal{P}_y(\cdot|a, x)$ to approximate $\mathcal{P}_y(\cdot|x, \mathcal{J})$, it may be more computationally straightforward to proceed as follows. Suppose that the analyst ignores the observational and modelling errors represented by distributions (2.5a) and (2.5b), and that she assumes that, if she knew the values of the parameters a, she could predict y perfectly. That is she accepts the deterministic model $M(a, x)$:

$$y = M(a, x). \tag{5.1}$$

However, she does not know a. Hence, she wishes to construct a distribution $\mathcal{P}_a(\cdot)$ such that this distribution induces a distribution on $y|x$ through the model (5.1) that approximates the distribution $\mathcal{P}_y(\cdot|x, \mathcal{J})$ Thus she wishes to solve:

Find a distribution $\mathcal{P}(\cdot)$ which minimizes the relative entropy between $\mathcal{P}_y(\cdot|x, \mathcal{J})$ and the distribution with density:

$$p_y(y|\mathcal{P}_a(\cdot)) = \int_{\{a|y=M(a,x)\}} p_a(a) \cdot da \tag{5.2}$$

An approximate solution has been implemented by (i) discretising the parameter space, $\{a\}$, and (ii) resolving the non-uniqueness of the solution (5.2) by choosing the distribution $\mathcal{P}_a(\cdot)$ which has maximum entropy. This gives the approximation:

$$p_a(a_j) \propto_a \frac{p_y(M(a_j, x)|x, \mathcal{J})}{\| \{k|M(a_k, x) = M(a_j, x)\} \|}, \tag{5.3}$$

where $\{a_k|k = 1, 2, \ldots, N\}$ is the set of points used to discretise the parameter space. This leads to a discrete approximation to the distribution over y which combines both the expert judgemental information and the analyst's belief in the model $M(a, x)$:

$$p_y(y_k|M(\cdot, \cdot), x, \mathcal{J}) = \sum_{\{j|y_k=M(a_j,x)\}} p_a(a_j), \tag{5.4}$$

where $\{y_k|k + 1, 2, \cdots, N'\}$ is the set of points in y-space corresponding to $\{a_k|k = 1, 2, \cdots, N\}$. This approach has been implemented in a software package PARFUM and is currently being evaluated: see Cooke and Vogt (1990).

The approaches of this and the previous sections can be combined in an obvious way. Essentially we are seeking to approximate $\mathcal{P}_y(\cdot|x, \mathcal{J})$ by a distribution arising from the three-stage model (2.5a, b, c). In section 4 and 5 we explored the result of assuming (2.5a) to give a degenerate distribution and approximating $\mathcal{P}_y(\cdot|x, \mathcal{J})$ by choosing a suitable value of a in (2.5b) and (2.5c). Immediately above we have explored the result of assuming (2.5b) and (2.5c) both to be degenerate distributions and approximating $\mathcal{P}_y(\cdot|x, \mathcal{J})$ by a suitable choice of $\mathcal{P}_a(\cdot)$. Combining all three stages of the model, if the analyst knows $\mathcal{P}_a(\cdot)$, gives a predictive distribution for $x|y$. Thus she may seek a distribution $\mathcal{P}_a(\cdot)$ such that this predictive distribution approximates $\mathcal{P}_y(\cdot|x, \mathcal{J})$. We are currently investigating ways that this approach might be made computationally feasible, but that task is far from trivial.

ACKNOWLEDGEMENTS

We are grateful to H. van der Weide and M. P. Wiper for helpful discussions. Financial support for the development of these ideas has been received from the Commission of the European Community.

REFERENCES

Bernardo, J. M. (1987). Approximations in statistics from a decision theoretical viewpoint. *Probability and Bayesian Statistics* (R. Viertl, ed.), London: Plenum Press, 53–60.

Commission of the European Communities (1990). *Radiation Protection: Proceedings of the Second International Workshop on Real-time Computing of the Enviromental Consequences of an Accidental Release to the Atmosphere from a Nuclear Installation.* Directorate-General Environment, Nuclear Safety and Civil Protection. Eur. 12320 EN/1, L-2920 Luxembourg.

Cooke, R. M. (1990). Statistics in expert resolution: a theory of weights for combining expert opinion. *Statistics in Science: the Foundations of Statistical Methods in Biology, Physics and Economics.* (R. M. Cooke and D. Constantini eds.), Dordrecht: Kluwer Academic, 41–72.

Cooke, R. M. (1991). *Experts in Uncertainty: Expert Opinion and Subjective Probability in Science*. Oxford: University Press.

Cooke, R. M. and Vogt, F. (1990). PARFUM: Parameter fitting for uncertain models. *Tech. Rep.* Technical University of Delft.

European Safety and Reliability Research and Development Association (1990). Expert judgement in risk and reliability analysis: experiences and pespectives. *Tech. Rep.* **2**, ESSRDA.

French, S. (1978). A Bayesian three-stage model in crystallography. *Acta Crystallographica* **A34**, 728–738.

French, S. (1985). Group consensus probability disributions: a critical survey. *Bayesian Statistics 2* (J. M. Bernardo, M. H. DeGroot, D. V. Lindley and A. F. M. Smith, eds.), Amsterdam: North-Holland, 183-201, (with discussion).

French, S., Cooke, R. M. and Wiper, M. P. (1991). The use of expert judgement in risk assessment. *Bull. Inst. Math. and Applications* **27**, 36–40.

Genest, C. and Zidek, J. (1986). Combining probability distributions: a critique and an annotated bibliography. *Statist. Sci.* **1**, 114-148.

Lewis, H. W. *et al.* (1978). Risk assessment review group report to the US Nuclear Regulatory Commision. *Tech. Rep.* **0400**, NUREG/CR. Nuclear Regulatory Commission, Washington, D. C.

Lindley, D. V. and Smith, A. F. M. (1972). Bayes estimates for the linear models. *J. Roy. Statist. Soc. B* **34**, 1–41.

McCullagh, P. M. and Nelder, J. A. (1983). *Generalised Linear Models.* London: Chapman and Hall.

Mendel. M. B. and Sheridan, T. B. (1987). Optimal estimation using human experts. *Tech. Rep.* Dept. of Mechanical Engineering, MIT.

Merkhofer, M. W. (1987). Quantifying judgemental uncertainty: methodology, experiences and insights. *Man and Cybernetics, IEEE Trans. Information Theory* **17**, 741–752.

Mosleh, A. and Apostolakis, G. (1986). The assessment of probability distributions from expert opinions with an application to seismic fragility curves. *Risk Analysis* **6**, 447–461.

US Nuclear Regulatory Commission (1989). Reactor Risk Reference Document. *Tech. Rep.* **1150**, NUREG, Washington, DC.

Wiper, M. P. (1990). *Calibration and the Use of Expert Probability Judgements.* Ph.D. Thesis, University of Leeds, UK.

Wiper, M. P. (1991). A normal model for combining expert judgement with the final form of a model. (Unpublished *Tech. Rep.*).

Wiper, M. P. and French, S. (1991). Combining expert opinions using a Normal–Wishart model. *Tech. Rep.* School of Computer Studies, University of Leeds.

Wright, G. and Ayton, P. (1987). *Judgemental Forecasting.* (G. Wright and P. Ayton, eds.) New York: Wiley.

BAYESIAN STATISTICS 4, pp. 625–631
J. M. Bernardo, J. O. Berger, A. P. Dawid and A. F. M. Smith, (Eds.)

A Single Series from the Gibbs Sampler Provides a False Sense of Security*

ANDREW GELMAN and DONALD B. RUBIN
University of California, USA and *Harvard University, USA*

SUMMARY

The Gibbs sampler can be very useful for simulating multivariate distributions, but naive use of it can give misleading—falsely precise—answers. An example with the Ising lattice model demonstrates that it is generally impossibile to assess convergence of a Gibbs sampler from a single sample series. This conclusion also applies to other iterative simulation methods such as the Metropolis algorithm.

Keywords: BAYESIAN INFERENCE; ITERATIVE SIMULATION; ISING MODEL; METROPOLIS ALGORITHM; RANDOM WALK.

1. INTRODUCTION

Bayesian inference is becoming more common in applied statistical work, partly to make use of the flexible modeling that occurs when treating all unknowns as random variables, but also because of the increasing availability of powerful computing environments. It is now often possible to obtain inferences using simulation–that is, to summarize a "target" posterior distribution by a sample of random draws from it, rather than by analytic calculations.

For many problems, direct simulation of a multivariate distribution is impossible, but one can simulate a random walk through the parameter space whose stationary distribution is the target distribution. Such a random walk may be considered an iterative Markov process that changes its distribution at every iteration, converging from the incorrect starting distribution to the target distribution as the number of iterations increases.

Such iterative simulation techniques have been used in physics since Metropolis *et al.* (1953) for studying Boltzmann distributions (also called Gibbs models) from statistical mechanics, with applications including solid state physics and the kinetic theory of gases. Geman and Geman (1984) applied the physical lattice models of statistical mechanics, and the associated computational techniques of iterative simulation, to image analysis, and coined the term "Gibbs sampler" for a particular technique. Since then, the Gibbs sampler has been applied with increasing frequency in a variety of statistical estimation problems, often to probability distributions with no connection to Gibbs models; Gelfand *et al.* (1990) provide a review.

Although the Gibbs sampler is becoming popular, it can be easily misused relative to direct simulation, because in practice, a finite number of iterations must be used to estimate the target distribution, and thus the simulated random variables are, in general, never from the desired target distribution. Various suggestions, some of which are quite sophisticated, have appeared in the statistical literature and elsewhere for judging convergence using one iteratively simulated sequence (e.g., Ripley, 1987, Chapter 6; Geweke, 1992; Raftery and Lewis, 1992). Here we present a simple but striking example of the problems that can arise

* This work was partially supported by the National Science Foundation and AT&T Bell Laboratories.

when trying to assess convergence from one observed series. (We use the terms "series" and "sequence" interchangeably.)

This example suggests that a far more generally successful approach is based on simulating multiple independent sequences. The suggestion to use multiple sequences with iterative simulation is not new (e.g., Fosdick, 1959), but no general method has appeared for drawing inferences about a target distribution from multiple series of finite length. Subsequent work (Gelman and Rubin, 1991) uses simple statistics to obtain valid conservative inferences for a large class of target distributions, including but not limited to those distributions that can be handled by one sequence.

2. ISING MODEL

The Ising model, described in detail in Kinderman and Snell (1980) and Pickard (1987), is a family of probability distributions defined on a lattice $Y = (Y_1, \ldots, Y_k)$ of binary variables: $Y_i = \pm 1$. It is a particularly appropriate model to illustrate potential problems of judging convergence with one iteratively simulated sequence since much of the understanding of it comes from computer simulation (e.g., Fosdick, 1959; Ehrman *et al.* 1960). The Ising model was originally used by physicists to idealize the magnetic behavior of solid iron: each component Y_i is the magnetic field ("up" or "down") of a dipole at a site of a crystal lattice. From the laws of statistical mechanics, Y is assigned the Boltzmann distribution:

$$P(Y) \propto \exp(-\beta U(Y)), \tag{1}$$

where the "inverse temperature" β is a scalar quantity, assumed known, and the "potential energy" U is a known scalar function that reflects the attraction of dipoles of like sign in the lattice—more likely states have lower energy. The Ising energy function is simple in that only attractions between nearest neighbors contribute to the total energy:

$$U(Y) = -\sum_{i,j} \delta_{ij} Y_i Y_j \tag{2}$$

where $\delta_{ij} = 1$ if i and j are "nearest neighbors" in the lattice and $\delta_{ij} = 0$ otherwise; in a two-dimensional lattice, each site has four neighbors, except for edge sites, which have three neighbors, and corner sites, with two neighbors each. The Ising model is commonly summarized by the "nearest-neighbor correlation" $\rho(Y)$:

$$\rho(Y) = \frac{\sum_{i,j} \delta_{ij} Y_i Y_j}{\sum_{i,j} \delta_{ij}}.$$

We take the distribution of $\rho(Y)$ as the target distribution, where Y is defined on a two-dimensional 100×100 lattice, and has distribution (1) with $k = 10{,}000$ and β fixed at 0.5.

3. ITERATIVE SIMULATION WITH THE GIBBS SAMPLER

For all but minuscule lattices, it is difficult analytically to calculate summaries, such as the mean or variance of $\rho(Y)$; integrating out k lattice parameters involves adding 2^k terms. Direct simulation of the model is essentially impossible except in the one-dimensional case, for which the lattice parameters can be simulated in order. However, it is easy to iteratively simulate the Ising distribution of Y, and thus of $\rho(Y)$, using the Gibbs sampler.

Given a multivariate target distribution $P(Y) = P(Y_1, \ldots, Y_k)$, the Gibbs sampler simulates a sequence of random vectors $(Y^{(1)}, Y^{(2)}, \ldots)$ whose distributions converge to the target distribution. The sequence $(Y^{(t)})$ may be considered a random walk whose stationary distribution is $P(Y)$. The Gibbs sampler proceeds as follows:

1. Choose a starting point $Y^{(0)} = (Y_1^{(0)}, \ldots, Y_k^{(0)})$ for which $P(Y^{(0)}) > 0$.
2. For $t = 1, 2, \ldots$:
   For $i = 1, \ldots, k$:
   Sample $Y_i^{(t)}$ from the conditional distribution:

$$P(Y_i \,|\, Y_j = Y_j^{(t-1)}, \text{ for all } j \neq i),$$

thereby altering one component of Y at a time; each *iteration* of the Gibbs sampler alters all k components of Y.

The proof that the Gibbs sampler converges to the target distribution has two steps: first, it is shown that the simulated sequence $(Y^{(t)})$ is a Markov chain with a unique stationary distribution, and second, it is shown that the stationary distribution equals the target distribution. The first step of the proof holds if the Markov chain is irreducible, aperiodic, and not transient (see, e.g., Feller, 1968, Section 15.7). The latter two conditions hold for a random walk on any proper distribution, and irreducibility holds as long as the random walk has a positive probability of eventually reaching any state from any other state, a condition satisfied by the Ising model.

To see that the target distribution is the stationary distribution of the Markov chain generated by the Gibbs sampler, consider starting the Gibbs sampler with a draw from $P(Y)$. Updating any component Y_i moves us to a new distribution with density,

$$P(Y_i|Y_j, \text{ all } j \neq i)\, P(Y_j, \text{ all } j \neq i),$$

which is the same as $P(Y)$.

4. RESULTS OF ITERATIVE SIMULATION

For the Ising model, each step of the Gibbs sampler can alter all k lattice sites in order. An obvious way to start the iterative simulation is by setting each site to ± 1 at random. For the Ising model on a 100×100 lattice with $\beta = 0.5$, theoretical calculations (Pickard, 1987) show that the marginal distribution of $\rho(Y)$ is approximately Gaussian with mean nearly 0.9 and standard deviation about 0.01.

Figure 1 shows the values of $\rho(Y^{(t)})$, for $t = 1$ to 2000, obtained by the Gibbs sampler with a random start, so that $\rho(Y^{(0)}) \approx 0$; the first few values are not displayed in order to improve resolution on the graph. The series seems to have "converged to stationarity" after the thousand or so steps required to free itself from the initial state. Now look at Figure 2, which zooms in on the first 500 steps of the series. Figure 2 looks to have converged after about 300 steps, but a glance at the next 1500 iterations, as displayed in Figure 1, shows that the apparent convergence is illusory. For comparison, the Gibbs sampler was run again for 2000 steps, but this time starting at a point $Y^{(0)}$ for which $\rho(Y^{(0)}) = 1$; Figure 3 displays the series $\rho(Y^{(t)})$, which again seems to have converged nicely. To destroy all illusions about convergence in any of these figures, compare Figures 1 and 3: the two iteratively simulated sequences appear to have "converged," but to different distributions! The series in Figure 3 has "stabilized" to a higher value of $\rho(Y)$ than that of Figure 1. We are of course still observing transient behavior, and an estimate of the distribution of $\rho(Y)$ based on Figure 1 alone, or on Figure 3 alone, would be misleadingly—falsely—precise. Furthermore, neither series alone carries with it the information that it has not stabilized after 2000 steps.

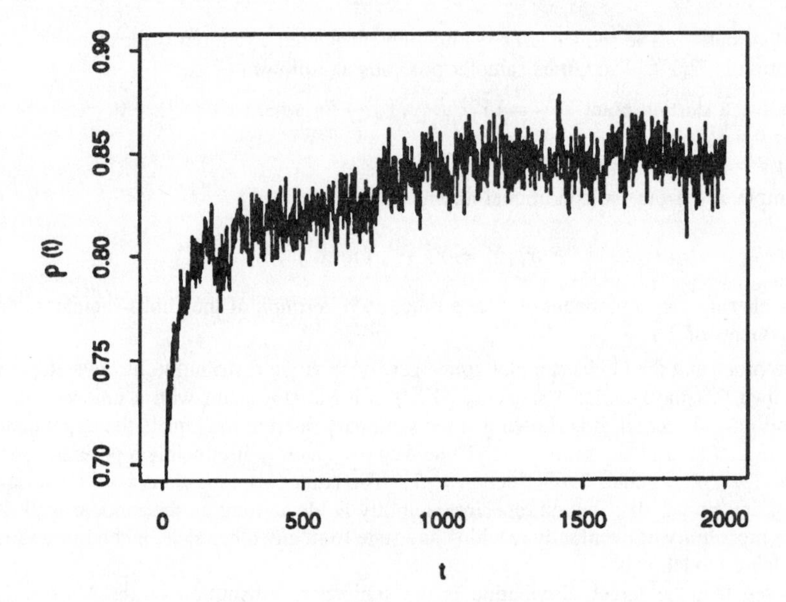

Figure 1. *2000 steps, starting at* $\rho = 0$

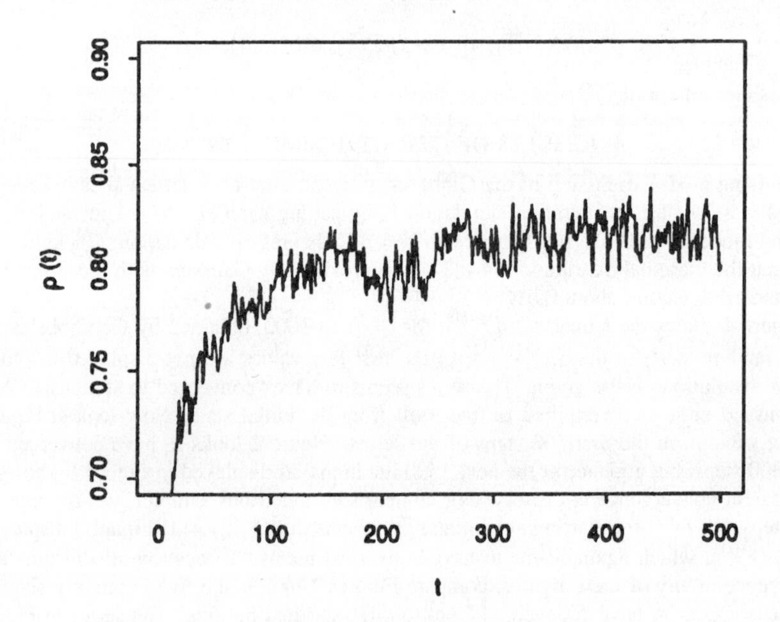

Figure 2. *First 500 steps*

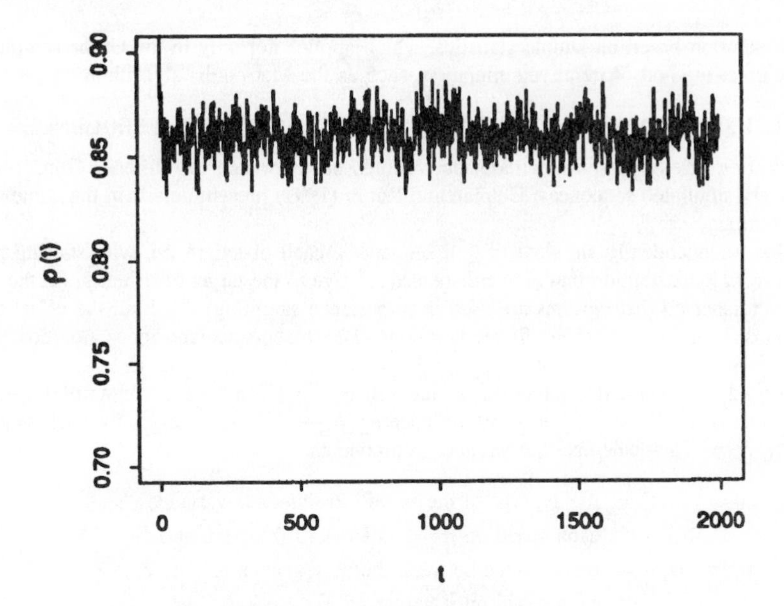

Figure 3. *2000 steps, starting at* $\rho = 1$

5. DISCUSSION

This example shows that the Gibbs sampler can stay in a small subset of its space for a long time, without any evidence of this problematic behavior being provided by one simulated series of finite length. The simplest way to run into trouble is with a two-chambered space, in which the probability of switching chambers is very low, but Figures 1–3 are especially disturbing because $\rho(Y)$ in the Ising model has a unimodal and approximately Gaussian marginal distribution, at least on the gross scale of interest. That is, the example is not pathological; the Gibbs sampler is just very slow. Rather than being a worst-case example, the Ising model is typical of the probability distributions for which iterative simulation methods were designed, and may be typical of many posterior distributions to which the Gibbs sampler is being applied.

A method designed for routine use must at the very least "work" for examples like the Ising model. By "work" we mean roughly that routine application should give valid conservative inferential summaries of the target distribution, that is, conservative relative to inferences that would be obtained from direct simulation of independent samples from the target distribution. Iterative simulation involves additional uncertainty due to the finite length of the simulated sequences, and so an appropriate inferential summary should reflect this in the same way that multiple imputation methods, which also involve the simulation of posterior distributions, reflect the uncertainty due to a finite number of imputations (Rubin, 1987).

In many cases, choosing a variety of dispersed starting points and running independent series may provide adequate diagnostic information, as in the example of Section 4. Nonetheless, for general practice, a more principled analysis of the between-series and within-series components of variability is far more convenient and useful. Gelman and Rubin (1991) offer such a solution based on simple statistics, which applies not only to the Gibbs sampler but to any other method of iterative simulation, such as the Metropolis algorithm.

6. USING COMPONENTS OF VARIANCE FROM MULTIPLE SEQUENCES

This section briefly presents, without derivations, our approach to inference from multiple iteratively simulated sequences; Gelman and Rubin (1991) present details in the context of a real example.

First, independently simulate $m \geq 2$ sequences, each of length $2n$, with starting points drawn from a distribution that is overdispersed relative to the target distribution (in the sense that overdispersed distributions are used in importance sampling). To limit the effect of the starting distribution, ignore the first n iterations of each sequence and focus attention on the last n.

Second, for each scalar quantity X of interest (e.g., $\rho(Y)$ in the above example), calculate the sample mean $\overline{x}_{i.} = \frac{1}{n}\sum_j x_{ij}$ and variance $s_i^2 = \frac{1}{n-1}\sum_j(x_{ij} - \overline{x}_{i.})^2$, for each sequence $i = 1, \ldots, m$. Then calculate the variance components,

$$\begin{aligned} W = {} & \text{the average of the } m \text{ within-sequence variances, } s_i^2, \\ & \text{each based on } n-1 \text{ degrees of freedom, and} \\ B/n = {} & \text{the variance between the } m \text{ sequence means, } \overline{x}_{i.}, \\ & \text{each based on } n \text{ values of } X. \end{aligned}$$

If the average within-sequence variance, W, is not substantially larger than the between-sequence variance, B/n, then the m sequences have not yet come close to converging to a common distribution. With only one sequence, between and within variabilities cannot be compared.

Third, estimate the target mean $\mu = \int XP(X)dX$ by $\hat{\mu} = \frac{1}{m}\sum_i \overline{x}_{i.}$, the sample mean of all mn simulated values of X.

Fourth, estimate the target variance, $\sigma^2 = \int (X-\mu)^2 P(X)dX$, by a weighted average of W and B, namely $\hat{\sigma}^2 = \frac{n-1}{n}W + \frac{1}{n}B$, which overestimates σ^2 assuming $P_0(X)$ is overdispersed. The estimate $\hat{\sigma}^2$ is unbiased under stationarity (i.e., if $P_0(X) = P(X)$), or in the limit $n \to \infty$.

Fifth, create a conservative Student-t distribution for X with center $\hat{\mu}$, scale $\sqrt{\widehat{V}} = \sqrt{\hat{\sigma}^2 + B/mn}$, and degrees of freedom $\nu = 2\widehat{V}^2/\widehat{\text{var}}(\widehat{V})$, where

$$\begin{aligned} \widehat{\text{var}}(\widehat{V}) = {} & \left(\frac{n-1}{n}\right)^2 \frac{1}{m}\text{var}(s_i^2) + \left(\frac{m+1}{mn}\right)^2 \frac{2}{m-1}B^2 + \\ & + 2\frac{(m-1)(n-1)}{mn^2} \cdot \frac{n}{m}\left[\text{cov}(s_i^2, \overline{x}_{i.}^2) - 2\overline{x}_{..}\text{cov}(s_i^2, \overline{x}_{i.})\right], \end{aligned}$$

and the variances and covariances are estimated from the m sample values of s_i^2, $\overline{x}_{i.}$, and $\overline{x}_{i.}^2$.

Sixth, monitor convergence of the iterative simulation by estimating the factor by which the scale of the Student-t distribution for X might be reduced if the simulations were continued in the limit $n \to \infty$. This potential scale reduction is estimated by $\sqrt{\widehat{R}} = \sqrt{\frac{\widehat{V}}{W}\frac{\nu}{\nu-2}}$, which declines to 1 as $n \to \infty$.

A computer program implementing the above steps appears in the appendix to Gelman and Rubin (1991) and is available from the authors.

7. PREVIOUS MULTIPLE-SEQUENCE METHODS

Our approach is related to previous uses of multiple sequences to monitor convergence of iterative simulation procedures. Fosdick (1959) simulated multiple sequences, stopping when the difference between sequence means was less than a prechosen error bound, thus basically using W but without using B as a comparison. Similarly, Ripley (1987) suggested examining at least three sequences as a check on relatively complicated single-sequence methods involving graphics and time-series analysis, thereby essentially estimating W quantitatively and B qualitatively. Tanner and Wong (1987) and Gelfand and Smith (1990) simulated multiple sequences, monitoring convergence by qualitatively comparing the set of m simulated values at time s to the corresponding set at a later time t; this general approach can be thought of as a qualitative comparison of values of B at two time points in the sequences, without using W as a comparison.

Our method differs from previous multiple-sequence methods by being fully quantitative, differs from single-sequence methods by relying on only a few assumptions, and differs from previous approaches of either kind by incorporating the uncertainty due to finite-length sequences into the distributional estimates. Current work focuses on the possibility of obtaining automatically overdispersed starting distributions and more efficient estimated target distributions for specific models.

REFERENCES

Ehrman, J. R., Fosdick, L. D. and Handscomb, D. C. (1960). Computation of order parameters in an Ising lattice by the Monte Carlo method. *Journal of Mathematical Physics* **1**, 547–558.

Feller, W. (1968). *An Introduction to Probability Theory and Its Applications* **1**. New York: Wiley.

Fosdick, L. D. (1959). Calculation of order parameters in a binary alloy by the Monte Carlo method. *Physical Review* **116**, 565–573.

Gelfand, A. E., Hills, S. E., Racine-Poon, A. and Smith, A. F. M. (1990). Illustration of Bayesian inference in normal data models using Gibbs sampling. *J. Amer. Statist. Assoc.* **85**, 398–409.

Gelfand, A. E. and Smith, A. F. M. (1990). Sampling-based approaches to calculating marginal densities. *J. Amer. Statist. Assoc.* **85**, 398–409.

Gelman, A. and Rubin, D. B. (1991). Honest inferences from iterative simulation. *Tech. Rep.* **307**, Department of Statistics, University of California, Berkeley.

Geman, S. and Geman, D. (1984). Stochastic relaxation, Gibbs distributions, and the Bayesian restoration of images. *IEEE Transactions on Pattern Analysis and Machine Intelligence* **6**, 721–741.

Geweke, J. (1992). Evaluating the accuracy of sampling-based approaches to the calculation of posterior moments. *Bayesian Statistics 4* (J. M. Bernardo, J. O. Berger, A. P. Dawid and A. F. M. Smith, eds.), Oxford: University Press, 169–193, (with discussion).

Kinderman, R. and Snell, J. L. (1980). *Markov Random Fields and Their Applications*. Providence, RI.: American Mathematical Society.

Metropolis, N., Rosenbluth, A. W., Rosenbluth, M. N., Teller, A. H. and Teller, E. (1953). Equation of state calculations by fast computing machines. *Journal of Chemical Physics* **21**, 1087–1092.

Pickard, D. K. (1987). Inference for discrete Markov fields: the simplest nontrivial case. *J. Amer. Statist. Assoc.* **82**, 90–96.

Raftery, A. E. and Lewis, S. (1992). How many iterations in the Gibbs sampler?. *Bayesian Statistics 4* (J. M. Bernardo, J. O. Berger, A. P. Dawid and A. F. M. Smith, eds.), Oxford: University Press, 763–774.

Ripley, B. D. (1987). *Stochastic Simulation*. New York: Wiley.

Rubin, D. B. (1987). *Multiple Imputation for Nonresponse in Surveys*. New York: Wiley.

Tanner, M. A. and Wong, W. H. (1987). The calculation of posterior distributions by data augmentation. *J. Amer. Statist. Assoc.* **82**, 528–550.

BAYESIAN STATISTICS 4, pp. 633–640
J. M. Bernardo, J. O. Berger, A. P. Dawid and A. F. M. Smith, (Eds.)

C_0–Coherence and Extensions of Conditional Probabilities

A. GILIO
Università "La Sapienza", Roma, Italy

SUMMARY

This paper gives a concise review of general results on the C_0-coherence concept, based on de Finetti's penalty criterion. Then, further results on coherent extensions of conditional probabilities are obtained, using the C_0 condition.

Keywords: C_0-COHERENCE; STRENGTHENING; PENALTY CRITERION; CONDITIONAL PROBABILITIES; CONVEX HULL; COHERENT EXTENSION; FUNDAMENTAL THEOREM OF PROBABILITIES; SIMPLEX METHOD.

1. INTRODUCTION

In the applications of statistical inference, we often need to assign probabilities on an arbitrary class $\mathcal{K}$ of conditional events, with no underlying structure. For example, this situation is familiar in the applications of expert systems and in such cases it is imperative that the coherence principle of de Finetti (1974) be used, in order to check consistency of numerical or qualitative probabilistic judgements (see Coletti, Gilio and Scozzafava (1990a), (1990b)).

Note that, from an axiomatic point of view, this principle can be looked on as the unique property that should be satisfied to prevent probability evaluations from being self-contradictory.

In the literature on foundations of subjective probability and Bayesian statistics the application of the coherence principle to the case of conditional events, on the basis of the betting criterion with strengthening, has been widely studied by many authors: see, e.g., Regazzini (1983), Holzer (1984), Scozzafava (1990).

Recently, modifying slightly the coherence condition based on the penalty criterion proposed by de Finetti, a general principle (C_0-coherence) has been studied and the concept of generalized atom for conditional events has been introduced by Gilio (1990).

When applied to conditional events, C_0-coherence does not require any strengthening and, with respect to the condition based on the betting criterion, this alternative approach to coherence constitutes an equivalent (but sometimes simpler) method to check admissibility of probabilistic evaluations and to obtain all the properties of conditional probabilities.

Moreover, the concept of generalized atom allows us to extend the pioneering ideas of de Finetti to conditional events: in fact, in this case too it is possible to characterize coherence geometrically in terms of the convex hull of points which represent the generalized atoms.

After a concise review on general aspects of C_0-coherence, in this paper we show some results on coherence of conditional probabilities. Given, for a family $\mathcal{F} = \{E_1|H_1, E_2|H_2, \dots, E_n|H_n\}$, a probabilistic assessment $\mathcal{P} = (p_1, p_2, \dots, p_n)$, we examine the extension of $\mathcal{P}$ considering a further event $E_{n+1}|H_{n+1}$.

We show in particular that, under a suitable condition of logical independence involving only the events E_j, H_j for which $H_{n+1} \wedge H_j \neq \emptyset$, the prevision point $\mathcal{P}' = (p_1, \dots, p_n, p_{n+1})$

is coherent if and only if $\mathcal{P}$ and p_{n+1} are coherent. An application to exchangeable events is considered too.

2. C_0-COHERENCE: A CONCISE REVIEW

Denote by $\mathcal{K}$ an arbitrary class of conditional events and by P a real function defined on $\mathcal{K}$. Moreover, given a finite family $\mathcal{F} = \{E_1|H_1, E_2|H_2, \ldots, E_n|H_n\} \subseteq \mathcal{K}$, consider the prevision point $\mathcal{P} = (p_1, p_2, \ldots, p_n)$, with $p_i = P(E_i|H_i), i = 1, 2, \ldots, n$.

As it is well known, on the basis of de Finetti's penalty criterion, to the point $\mathcal{P}$ can be associated a loss

$$\mathcal{L} = \sum_{i=1}^{n} H_i\,(E_i - p_i)^2 = \sum_{i=1}^{n} (E_i \,|\, H_i - P_i)^2 ,$$

where the symbol $E_i \,|\, H_i$ takes, respectively, the value 1 or 0 or p_i in the three cases $E_i = H_i = 1; E_i = 0, H_i = 1; H_i = 0$.

We denote the events and their indicators by the same symbol. Moreover, for the sake of simplicity and without loss of generality, we have assigned a unitary value to the arbitrary constants $k_1, k_2, \ldots, k_n$, which are present in the original expression of $\mathcal{L}$, given by de Finetti.

The geometrical characterization of coherence in terms of the *convex hull of constituents*, given by de Finetti, can be extended to the case of conditional events by using the concept of generalized atom (or *generalized constituent*) introduced by Gilio (1990). In this regard, given a *prevision* point $\mathcal{P} = (p_1, p_2, \ldots, p_n)$ for n conditional events $E_1 \,|\, H_1, E_2 \,|\, H_2, \ldots, E_n \,|\, H_n$ and denoting by $C_1, C_2, \ldots, C_s$ those atoms generated by the family $\{E_1, \ldots, E_n, H_1, \ldots, H_n\}$ such that $C_h \subseteq H_1 \vee H_2 \vee \ldots \vee H_n$, $h = 1, 2, \ldots, s$, the *generalized atoms* $Q_h = (\alpha_{h1}, \alpha_{h2}, \ldots, \alpha_{hn})$, are obtained in $\mathbb{R}^n$ from the C_h's putting

$$\alpha_{hi} = \begin{cases} 1, & \text{if } C_h \subseteq E_i \wedge H_i \\ 0, & \text{if } C_h \subseteq E_i^c \wedge H_i, \\ p_i, & \text{if } C_h \subseteq H_i^c \end{cases} \qquad \begin{matrix} i = 1, 2, \ldots, n \\ h = 1, 2, \ldots, s \end{matrix} . \tag{1}$$

It is easy to verify that to each constituent C_h there corresponds through (1) a value for each $E_i|H_i$ and then a value L_h of $\mathcal{L}$ given by

$$L_h = \sum_{i=1}^{n} (\alpha_{hi} - p_i)^2 = \overline{\mathcal{P}Q}_h^2. \tag{2}$$

We can now make the following definition

Definition 2.1. *The real function $P : \mathcal{K} \to \mathbb{R}$ is a C_0-coherent conditional probability on $\mathcal{K}$ if, for each n and for each finite family $\mathcal{F} \subseteq \mathcal{K}$, putting $\mathcal{P} = (p_1, p_2, \ldots, p_n)$, with $p_i = P(E_i|H_i)$, there does not exist a point $\mathcal{P}^* \neq \mathcal{P}$ such that $L_h^* \leq L_h$ for each h and with $L_h^* < L_h$ for at least one index h; i.e., such that $\mathcal{L}^* \leq \mathcal{L}, \mathcal{L}^* \neq \mathcal{L}$.*

Definition 2.1 differs from de Finetti's, which was applied to nonconditional events, because it requires, in place of $\mathcal{L}^* < \mathcal{L}$, the weaker condition $\mathcal{L}^* \leq \mathcal{L}, \mathcal{L}^* \neq \mathcal{L}$. This simple modification allows us to study coherence in a *unified* way: there is no need of *strengthening* when conditional events are considered. Moreover, it can be shown (see Gilio (1989) or Gilio (1990)) that our condition is equivalent to that one based on the *betting criterion with strengthening*.

Given a finite family $\mathcal{F}$ of n conditional events and a prevision point $\mathcal{P} = (p_1, p_2, \ldots, p_n)$, we denote, respectively, by $\mathcal{I}$ the convex hull of points $Q_h \in \mathbb{R}^n$, relative to the family $\mathcal{F}$ and prevision point $\mathcal{P}$, and, for each $J \subset \{1, 2, \ldots, n\}$, by $\mathcal{I}_J$ the convex hull of generalized atoms relative to the family $\mathcal{F} = \{E_j \,|\, H_j : j \in J\}$ and point $\mathcal{P}_J = (p_j : j \in J)$. The following results (see Gilio (1990)) can be proved.

Theorem 2.2. *If $\mathcal{P}$ is coherent then $\mathcal{P}_J$ is coherent, for each $J \subset \{1, 2, \ldots, n\}$.*

Theorem 2.3. *If $\mathcal{P}$ is coherent, then $\mathcal{P} \in \mathcal{I}$.*

Note that $\mathcal{P} \in \mathcal{I}$ amounts to say that there exist s nonnegative real numbers $\lambda_1, \lambda_2, \ldots, \lambda_s$, with $\sum_{h=1}^{s} \lambda_h = 1$, such that $\mathcal{P} = \sum_{h=1}^{s} \lambda_h Q_h$.

Theorem 2.4. *If there exist s positive numbers $\lambda_1, \lambda_2, \ldots, \lambda_s$, with $\sum_{h=1}^{s} \lambda_h = 1$, such that $\mathcal{P} = \sum_{h=1}^{s} \lambda_h Q_h$, then $\mathcal{P}$ is coherent.*

Theorem 2.5. *The prevision point $\mathcal{P}$ is coherent if and only if $\mathcal{P}_J \in \mathcal{I}_J$ for every $J \subseteq \{1, 2, \ldots, n\}$.*

For conditional events, Theorem 2.5 gives a geometrical characterization of coherence, in terms of convex hulls of generalized constituents. But this general theorem does not seem very useful for practical purposes: its application requires to check existence of (nonnegative) solutions of the $2^n - 1$ linear systems obtained from the conditions $\mathcal{P}_J \in \mathcal{I}_J$, $J \subseteq \{1, 2, \ldots, n\}$, $J \neq \emptyset$. Reductions in computational difficulties are achieved when some of the conditioning events $H_1, H_2, \ldots, H_n$ are equal. Specifically, we have

Theorem 2.6. *If $\mathcal{P} \in \mathcal{I}$ and $H_1 = H_2 = \ldots = H_n$, then $\mathcal{P}$ is coherent.*

In fact, if all H_i's coincide it can be verified from (1) that, for each pair of indices h, i, it is $\alpha_{hi} \in \{0, 1\}$. Then, if we arbitrarily choose another point $\mathcal{P}^* \neq \mathcal{P}$, for each h we have $Q_h^* = Q_h$. Moreover, as $\mathcal{P} \in \mathcal{I}$, it is $\mathcal{P} = \sum_{h=1}^{s} \lambda_h Q_h$. Then

$$\sum_{h=1}^{s} \lambda_h \overline{\mathcal{P}^* Q_h}^2 = \ldots = \sum_{h=1}^{s} \lambda_h \overline{\mathcal{P} Q_h}^2 + \overline{\mathcal{P}\mathcal{P}^*}^2 > \sum_{h=1}^{s} \lambda_h \overline{\mathcal{P} Q_h}^2,$$

that is $\sum_{h=1}^{s} \lambda_h (\overline{\mathcal{P}^* Q_h}^2 - \overline{\mathcal{P} Q_h}^2) > 0$. Hence, for at least one index h it must be $L_h^* > L_h$, which implies coherence of $\mathcal{P}$.

So, when $H_1 = H_2 = \ldots = H_n$, coherence of $\mathcal{P}$ amounts to existence of (nonnegative) solutions of the linear system obtained from the condition $\mathcal{P} \in \mathcal{I}$.

Corollary 2.7. *If $\mathcal{P}_J \in \mathcal{I}_J$ and, for all $i, j \in J$, $H_i = H_j$, then $\mathcal{P}_J$ is coherent.*

So, going back to the problem of verifying coherence through Theorem 2.5, if there are subsets $J_1, J_2, \ldots, J_r$ of indices, with cardinality $n_1, n_2, \ldots, n_r$, such that $H_i = H_j$ for $i, j \in J_h, h = 1, 2, \ldots, r$, we obtain a (not very large) reduction in computational task: in fact, there are $(\sum_{h=1}^{r} 2^{n_h} - 2r)$ systems for which the existence of solutions is ensured if conditions $\mathcal{P}_{J_i} \in \mathcal{I}_{J_i}, i = 1, 2, \ldots, r$, are satisfied.

In Gilio (1989) the following Theorem is proved.

Theorem 2.8. *Given a family $\mathcal{F} = \{E_1|H_1, E_2|H_2, \ldots, E_n|H_n\}$ such that $H_i \wedge H_j = \emptyset$ for $i \neq j$, a prevision point $\mathcal{P} = (p_1, p_2, \ldots, p_n)$, relative to the family $\mathcal{F}$, is coherent if and only if each value $p_i = P(E_i|H_i)$ is coherent, $i = 1, 2, \ldots, n$.*

In other words $\mathcal{P}$ is coherent provided $0 \leq p_i \leq 1$ for each i, with $p_i = 0$ if $E_i \wedge H_i = \emptyset$ and $p_i = 1$ if $H_i \subseteq E_i$.

In addition suppose that an event $E_{n+1}|H_{n+1}$ and the value $p_{n+1} = P(E_{n+1}|H_{n+1})$ are given. Then, the following Corollary is immediate.

Corollary 2.9. *If $H_{n+1} \wedge (H_1 \vee H_2 \vee \ldots \vee H_n) = \emptyset$, then under the hypotheses of Theorem 2.8, the prevision point $\mathcal{P}' = (p_1, p_2, \ldots, p_n, p_{n+1})$ is coherent if and only if $\mathcal{P} = (p_1, p_2, \ldots, p_n)$ and p_{n+1} are coherent.*

Our aim is to show that, under a suitable condition of *logical independence*, Corollary 2.9 can be extended to a more general setting.

3. NEW RESULTS AND APPLICATIONS

Consider a prevision point $\mathcal{P} = (p_1, p_2, \ldots, p_n)$, relative to an arbitrary family $\mathcal{F} = \{E_1|H_1, E_2|H_2, \ldots, E_n|H_n\}$. Given a conditional event $E_{n+1}|H_{n+1}$, with $P(E_{n+1}|H_{n+1}) = p_{n+1}$, we put $I = \{1, 2, \ldots, n\}$, $J = \{j \in I : H_{n+1} \wedge H_j \neq \emptyset\}$, $J_c = I \backslash J = \{j \in I : H_{n+1} \wedge H_j = \emptyset\}$, $\mathcal{P}_J = (p_j : j \in J), \mathcal{P}_{J_c} = (p_j : j \in J_c), \mathcal{P}' = (p_1, \ldots, p_n, p_{n+1})$.

Moreover, denote, by $\mathcal{L}, \mathcal{L}', \mathcal{L}_J, \mathcal{L}_{J_c}$ the losses associated respectively, to $\mathcal{P}, \mathcal{P}', \mathcal{P}_J, \mathcal{P}_{J_c}$.

Referring to the atoms $D_1, D_2, \ldots, D_t$ generated by the family $\{E_j, H_j : j \in J\} \cup \{H_{n+1}\}$, denote by $D_1, D_2, \ldots, D_r, r \leq t$, those atoms which are contained in H_{n+1}. Then, we have

Theorem 3.1. *If $E_{n+1} \wedge D_k \neq \emptyset$, $E^c_{n+1} \wedge D_k \neq \emptyset$ for every $k \leq r$, then the point $\mathcal{P}'$ is coherent if and only if $\mathcal{P}$ and p_{n+1} are coherent.*

Proof. By Theorem 2.2, coherence of $\mathcal{P}'$ implies coherence of $\mathcal{P}$ and p_{n+1}. ◁

Conversely, denote by $C_1, C_2, \ldots, C_s$ the atoms generated by $\{E_j, H_j : j \in I\}$, which are needed to determine the values of $\mathcal{L}$. Given a point $\mathcal{P}'^* = (p_1^*, p_2^*, \ldots, p_{n+1}^*) \neq \mathcal{P}'$, one has

$$\mathcal{L}' - \mathcal{L}'^* = \mathcal{L} - \mathcal{L}^* + H_{n+1}\left[(E_{n+1} - p_{n+1})^2 - (E_{n+1} - p^*_{n+1})^2\right] =$$

$$= (\mathcal{L}_J - \mathcal{L}^*_J) + (\mathcal{L}_{J_c} - \mathcal{L}^*_{J_c}) + H_{n+1}\left[(E_{n+1} - p_{n+1})^2 - (E_{n+1} - p^*_{n+1})^2\right].$$

Notice that for a constituent C_h compatible with a given D_k there corresponds, respectively

$$\mathcal{L}' - \mathcal{L}'^* = (\mathcal{L}_J - \mathcal{L}^*_J) + (\mathcal{L}_{J_c} - \mathcal{L}^*_{J_c}) = \mathcal{L} - \mathcal{L}^*,$$

or

$$\mathcal{L}' - \mathcal{L}'^* = \mathcal{L}_J - \mathcal{L}^*_J + (E_{n+1} - p_{n+1})^2 - (E_{n+1} - p^*_{n+1}),$$

according to whether $k > r$ or $k \leq r$. Since $\mathcal{P}$ is coherent there exist C_h, such that $\mathcal{L} - \mathcal{L}^* < 0$, and (at least) one atom D_k compatible with C_h. Then, if $k > r$, to C_h it corresponds

$$\mathcal{L}' - \mathcal{L}'^* = \mathcal{L} - \mathcal{L}^* < 0.$$

If, instead, $k \leq r$, it is $E_{n+1} \wedge D_k \neq \emptyset$, $E^c_{n+1} \wedge D_k \neq \emptyset$, and for $\mathcal{L}' - \mathcal{L}'^*$ both values

$$\mathcal{L}_J - \mathcal{L}^*_J + (1 - p_{n+1})^2 - (1 - p^*_{n+1})^2$$

or

$$\mathcal{L}_J - \mathcal{L}^*_J + (p_{n+1})^2 - (p^*_{n+1})^2$$

are possible.

Then, being $\mathcal{L} - \mathcal{L}^* < 0$, it must be $\mathcal{L}' - \mathcal{L}'^* < 0$ in at least one case. So, coherence of $\mathcal{P}$ and p_{n+1} implies coherence of $\mathcal{P}'$.

Note that if the set J is empty, i.e., $H_{n+1} \wedge H_j = \emptyset$ for every $j \in I$, then Theorem 3.1 reduces to the following Corollary.

Corollary 3.2. *Given a conditional event $E_{n+1} \mid H_{n+1}$ such that $H_{n+1} \wedge (H_1 \vee H_2 \vee \ldots \vee H_n) = \emptyset$ and putting $P(E_{n+1} \mid H_{n+1}) = p_{n+1}$, the prevision point $\mathcal{P}'$ is coherent if and only if $\mathcal{P}$ and p_{n+1} are coherent.*

Putting $I = J_1 \cup J_2 \cup \ldots \cup J_m$, with $J_h \cap J_k = \emptyset$ for $h \neq k$, introduce the subfamilies

$$\mathcal{F}_{J_i} = \{E_j | H_j \in \mathcal{F} : j \in J_i\}, \qquad i = 1, 2, \ldots, m.$$

Obviously, it is $\mathcal{F} = \mathcal{F}_{J_1} \cup \mathcal{F}_{J_2} \cup \ldots \cup \mathcal{F}_{J_m}$. Then, considering a prevision point $\mathcal{P} = (p_1, p_2, \ldots, p_n)$ and putting

$$\mathcal{P}_{J_i} = (p_j : j \in J_i), \qquad i = 1, 2, \ldots, m,$$

we have

Theorem 3.3. *If, for each $i \in J_h, j \in J_k$, with $h \neq k$, it is $H_i \wedge H_j = \emptyset$, then $\mathcal{P}$ is coherent if and only if, for each i, $\mathcal{P}_{J_i}$ is coherent.*

Proof. Coherence of $\mathcal{P}$ implies coherence of $\mathcal{P}_{J_i}$, for each i. ◁

Conversely, denoting by H_{J_i} the logical sum of the events H_j for which $j \in J_i$, $i = 1, 2, \ldots, m$, it is
(a) $H_{J_i} \wedge H_{J_k} = \emptyset$, for $i \neq k$;
(b) $\mathcal{L} = H_{J_1} \cdot \mathcal{L}_{J_1} + H_{J_2} \cdot \mathcal{L}_{J_2} + \ldots + H_{J_m} \cdot \mathcal{L}_{J_m}$.
So, recalling that $J_h \cap J_k = \emptyset$ for $h \neq k$, we have $\mathcal{L} = 0$ if $H_{J_1} = H_{J_2} = \ldots = H_{J_m} = 0$, otherwise it is $\mathcal{L} = \mathcal{L}_{J_i}$ if $H_{J_i} = 1$.
Then, for every point $\mathcal{P}^* = (p_1^*, p_2^*, \ldots, p_n^*) \neq \mathcal{P}$, coherence of points $\mathcal{P}_{J_i}$'s, implies that, for each i, in at least one case $\mathcal{L}_{J_i} - \mathcal{L}^*_{J_i} < 0$. Hence, from the relation

$$\mathcal{L} - \mathcal{L}^* = H_{J_1}\left(\mathcal{L}_{J_1} - \mathcal{L}^*_{J_1}\right) + H_{J_2}\left(\mathcal{L}_{J_2} - \mathcal{L}^*_{J_2}\right) + \cdots + H_{J_m}\left(\mathcal{L}_{J_m} - \mathcal{L}^*_{J_m}\right)$$

it follows $\mathcal{L} - \mathcal{L}^* < 0$, in at least one case. So, $\mathcal{P}$ is coherent.

Theorem 3.3 (which is mentioned without proof in Coletti, Gilio and Scozzafava (1990b) too) contains, as particular cases, Theorem 2.8 and Corollary 3.2.

In the general case, in which the condition of logical independence $E_{n+1} \wedge D_k \neq \emptyset$, $E^c_{n+1} \wedge D_k \neq \emptyset$, $k\lambda \leq r$, is not satisfied, the value p_{n+1} must belong to a suitable interval $[p', p''] \subseteq [0, 1]$ so that $\mathcal{P}'$ can be coherent. (See, e.g., Holzer (1985), Regazzini (1985), Gilio and Scozzafava (1988)).

Remark. If $H_1 = \ldots = H_n = H_{n+1}$ the interval $[p', p'']$ can be determined, on the basis of the de Finetti's fundamental theorem of probabilities, using the simplex method of linear programming (this method, referring to nonconditional events, was proposed by Bruno and Gilio (1980). See also Lad, Dickey and Rahman (1990)).

In fact, put $\mathcal{F}' = \mathcal{F} \cup \{E_{n+1} | H_{n+1}\}$ and denote, respectively, by Q and $\mathcal{Q}$ the generalized atoms relative to $\mathcal{F}', \mathcal{P}'$ and $\mathcal{F}, \mathcal{P}$. Notice that in this case $\alpha_{ij} \in \{0, 1\}$ for each i, j. Then, using Theorem (2.6), it can be seen that coherence of $\mathcal{P}$ and $\mathcal{P}'$ amounts to conditions

$$\mathcal{P} = \sum_{h=1}^{s} \lambda_h \mathcal{Q}_h, \qquad \mathcal{P}' = \sum_{k=1}^{r} w_k Q_k, \qquad r \geq s,$$

with:

$$\sum_{h=1}^{s} \lambda_h = \sum_{k=1}^{r} w_k = 1, \qquad \lambda_h \geq 0, \qquad w_k \geq 0.$$

Moreover, each generalized atom Q_k relative to $\mathcal{F}'$ and $\mathcal{P}'$ can be written as $Q_k = (\mathcal{Q}'_k, \alpha_{kn+1})$, where $\mathcal{Q}'_k$ is a generalized atom relative to $\mathcal{F}$ and $\mathcal{P}$. Obviously, if $r > s$ there are pairs of generalized atoms $Q_h = (\mathcal{Q}'_h, \alpha_{hn+1}) \neq Q_k = (\mathcal{Q}'_k, \alpha_{kn+1})$ such that $\mathcal{Q}'_h = \mathcal{Q}'_k$; i.e., Q_h and Q_k differ only because $\alpha_{hn+1} \neq \alpha_{kn+1}$.

Then, being $\mathcal{P}' = (\mathcal{P}, p_{n+1})$ we have

$$\mathcal{P}' = (\mathcal{P}, p_{n+1}) = \sum_{k=1}^{r} w_k(\mathcal{Q}'_k, \alpha_{kn+1}) = \left(\sum_{k=1}^{r} w_k \mathcal{Q}'_k, \sum_{k=1}^{r} w_k \alpha_{kn+1}\right),$$

with $\mathcal{P} = \sum_{k=1}^{r} w_k \mathcal{Q}'_k = \sum_{h=1}^{s} \lambda_h \mathcal{Q}_h$, $p_{n+1} = \sum_{k=1}^{r} w_k \alpha_{kn+1}$.

So, given the coherent point $\mathcal{P}$ and putting $\boldsymbol{w} = (w_1, w_2, \ldots, w_r)$, the interval $[p', p'']$ is determined by solving the two *linear programming* problems

$$\min_{w} \sum_{k=1}^{r} w_k \alpha_{kn+1} \quad , \quad \max_{w} \sum_{k=1}^{r} w_k \alpha_{kn+1},$$

subject to constraints

$$\sum_{k=1}^{r} w_k \mathcal{Q}'_k = \mathcal{P}, \quad \sum_{k=1}^{r} w_k = 1, \quad w_k \geq 0, \quad k = 1, 2, \ldots, r.$$

The values p', p'' are, respectively, the Min and Max of the objective function $\sum_{k=1}^{r} w_k\ \alpha_{kn+1}$.

An application. Referring to $N+1$ exchangeable events $E_1, E_2, \ldots, E_{N+1}$ and putting $S_N = E_1 + E_2 + \ldots + E_N$, in Lad and Deely (1991) (see their Theorem 2) it is proved that "... any vector $P_{N+1} \equiv (p_{0,N}, p_{1,N}, \ldots, p_{N,N})$ of conditional prevision assertions $p_{a,N} = P(E_{N+1}|S_N = a)$ for $a = 0, 1, \ldots, N$ is coherent as long as it lies within (or on the boundary of) the $(N+1)$-dimensional unit hypercube ...".

Without reference to exchangeability, this result would follow from Theorem 2.8 applied to the family $\mathcal{F} = \{E_{N+1}|H_0, E_{N+1}|H_1, \ldots, E_{N+1}|H_N\}$, with $H_a \equiv (S_N = a), a = 0, 1, \ldots, N$. What is surprising about their result (and not implied here) is that the additional assessments due to regarding the events exchangeably do not alter the conclusion of Theorem 2.8.

Other connections with Theorem 2.4 can be underlined. Firstly, it can be verified that the generalized constituents $Q_i = (\alpha_{i0}, \alpha_{i1}, \ldots, \alpha_{iN})$ relative to the family $\mathcal{F} = \{E_{N+1}|H_0, E_{N+1}|H_1, \ldots, E_{N+1}| H_N\}$ and the prevision point $\mathcal{P} = P_{N+1}$ are the following ones

$$\begin{array}{llll} Q_1 & = (0, p_{1,N}, \ldots, p_{N,N}), & Q_2 & = (1, p_{1,N}, \ldots, p_{N,N}), \\ Q_3 & = (p_{0,N}, 0, p_{2,N}, \ldots, p_{N,N}), & Q_4 & = (p_{0,N}, 1, p_{2,N}, \ldots, p_{N,N}), \\ \cdots & \cdots & \cdots & \cdots \\ Q_{2N+1} & = (p_{0,N}, \ldots, p_{N-1,N}, 0), & Q_{2N+2} & = (p_{0,N}, \ldots, p_{N-1,N}, 1). \end{array}$$

Then, recalling Theorem 2.4, if there exist $2N+2$ positive numbers $\lambda_1, \lambda_2, \ldots, \lambda_{2N+2}$, with $\sum_{h=1}^{2N+2} \lambda_h = 1$, such that $\mathcal{P} = \sum_{h=1}^{2N+2} \lambda_h Q_h$, the point $\mathcal{P}$ is coherent. Moreover,

$$p_{a,N} = \frac{\lambda_{2a+2}}{\lambda_{2a+1} + \lambda_{2a+2}}, \qquad a = 0, 1, \ldots, N. \tag{3}$$

Consider the *extended* family $\mathcal{F}' = \mathcal{F} \cup \{K_{a+1}\}$, where K_{a+1} is the *nonconditional* event $(S_{N+1} = a+1)$, and put $\mathcal{P}' = (\mathcal{P}, q_{a+1})$, with $q_{a+1} = P(K_{a+1})$. Since

$$(S_{N+1} = a+1) \equiv ((S_N = a) \wedge E_{N+1}) \vee ((S_N = a+1) \wedge E_{N+1}^c), \qquad (4)$$

it is easy to verify that each new generalized constituent Q_i' is simply obtained from Q_i by adjoining to it the component $\alpha_{i,N+1}$ given by

$$\alpha_{i,N+1} = \begin{cases} 1, & i = 2a+2 \text{ or } i = 2a+3 \\ 0, & \text{otherwise} \end{cases}. \qquad (5)$$

It follows that

$$q_{a+1} = \lambda_{2a+2} + \lambda_{2a+3}, \qquad a = 0, 1, \ldots, N-1. \qquad (6)$$

Moreover, we have $q_0 = \lambda_1$ and $q_{N+1} = \lambda_{2N+2}$.

Note that each vector of $2N+2$ positive numbers $(\lambda_1, \ldots, \lambda_{2N+2})$, with $\sum_{h=1}^{2N+2} \lambda_h = 1$, determines, for the events

$$E_{N+1}|H_0, E_{N+1}|H_1, \ldots, E_{N+1}|H_N, K_0, K_1, \ldots, K_{N+1},$$

a set of coherent probabilistic assertions, given by (3) and (6).

For example, the choice $\lambda_h = 1/(2N+2), h = 1, 2, \ldots, 2N+2$, gives $p_{a,N} = 1/2, a = 0, 1, \ldots, N; q_k = 1/(N+1), k = 1, \ldots, N q_0 = q_{N+1} = 1/2(N+1)$.

From the exchangeability hypothesis it follows that

$$q_{h+1} = q_h \cdot \frac{N-h+1}{h+1} \cdot \frac{p_{hN}}{1-p_{hN}}, \qquad h = 0, 1, \ldots, N. \qquad (7)$$

So, given λ_1, λ_2, we determine q_0, p_{0N} and, by means of (7), q_1. Then, using (6), we calculate λ_3; then, if we assign λ_4, using (3) we determine p_{1N}, and so on.

As we can see, Theorem (2.4) constitutes a useful and flexible tool to make probabilistic assessments.

In particular, if

$$\lambda_{2a+2} = \frac{a}{N-1} \cdot \lambda_{2a+1}, \qquad a = 1, 2, \ldots, N-1,$$

then

$$p_{aN} = P(E_{N+1}|S_N = a) = \frac{a}{N}, \qquad a = 1, 2, \ldots, N-1.$$

A detailed analysis of these "frequentist" assertions and other related aspects (with an outline of some undesirable implications) is developed in Lad and Deely (1991).

ACKNOWLEDGEMENTS

The author thanks two referees for their helpful suggestions.

REFERENCES

Bruno, G. and Gilio, A. (1980). Applicazione del metodo del simplesso al teorema fondamentale per le probabilità nella concezione soggettiva. *Statistica* **40**, 337–344.

Coletti, G., Gilio, A. and Scozzafava, R. (1990a). Coherent qualitative probability and uncertainty in artificial intelligence. *Proc. 8th Int. Congr. of Cybernetics and Systems* (C. N. Manikopoulos, ed.) New York: NJIT Press, (to appear).

Coletti, G., Gilio, A. and Scozzafava, R. (1990b). Conditional events with vague information in expert systems. *Lectures Notes in Computer Science* (B. Bouchon-Meunier, R. R. Yager and L. A. Zadeh., eds.), Paris: IPMV.

de Finetti, B. (1974, 1975). *Theory of Probability*. New York: Wiley.

Gilio, A. (1989). Probabilità condizionate C_0-coerenti. *Rendiconti di Matematica di Roma* **9**, 277–294.

Gilio, A. (1990). Criterio di penalizzazione e condizioni di coerenza nella valutazione soggettiva della probabilità. *Boll. Un. Mat. Ital.* **4-B**, 645–660.

Gilio, A. and Scozzafava, R. (1988). Le probabilità condizionate coerenti nei sistemi esperti. *Atti Giornate di Lavoro A.I.R.O. su "Ricerca Operativa e Intelligenza Artificiale"*, Pisa: Offset Grafica, 317–330.

Holzer, S. (1984). Sulla nozione di coerenza per le probabilità subordinate. *Rendiconti dell'Istituto di Matematica dell'Università di Trieste* **16**, 46–62.

Holzer, S. (1985). On coherence and conditional prevision. *Boll. Un. Mat. Ital.* **4**, 441–460.

Lad, F., Dickey, J. M. and Rahman M. A. (1990). The fundamental theorem of prevision. *Statistica* **50**, 19–38.

Lad, F. and Deely, J. (1991). Coherency conditions for finite exchangeable inference. *Tech. Rep.* University of Canterbury, Christchurch, New Zealand.

Regazzini, E. (1983). *Sulle probabilità coerenti nel senso di de Finetti*. Bologna: Editrice CLUEB.

Regazzini, E. (1985). Finitely additive conditional probabilities. *Rend. Sem. Mat. Fis. Milano* **55**, 69–89.

Scozzafava, R. (1990). Probabilità condizionate: de Finetti o Kolmogorov?. *Scritti in omaggio a L. Daboni*, Trieste: Lint, 223–237.

BAYESIAN STATISTICS 4, pp. 641–649
J. M. Bernardo, J. O. Berger, A. P. Dawid and A. F. M. Smith, (Eds.)

Derivative-free Adaptive Rejection Sampling for Gibbs Sampling

W. R. GILKS
MRC Biostatistics Unit, Cambridge, UK

SUMMARY

We propose a method for adaptive rejection sampling from log-concave densities. The density need only be known modulo a normalising constant. The method does not involve locating the mode of the density, and unlike the tangent-based adaptive rejection method of Gilks and Wild (1991), does not involve the evaluation of first derivatives of the log density. The method is well suited to sampling from complicated densities, such as those that arise in applications of Gibbs Sampling (Geman and Geman, 1984).

Keywords: BAYESIAN INFERENCE; GIBBS SAMPLING; LOG-CONCAVITY; REJECTION SAMPLING.

1. INTRODUCTION

Gibbs sampling (Geman and Geman, 1984) is a stochastic simulation technique for Bayesian analysis. Gelfand and Smith (1990) and Gelfand *et al.* (1990) demonstrate its applicability in a number of traditionally difficult areas, including variance components, errors-in-variables, missing data and growth curve problems. Gibbs sampling has also been applied in many other areas. However, in the main these applications have focussed on situations in which there is conjugacy between likelihoods and priors, for which the sampling involved in Gibbs sampling is straightforward.

To deal with non-conjugacy in applications of Gibbs sampling, (for example, in Generalised Linear Models), we propose a method of *derivative-free adaptive rejection sampling*. The method differs from the adaptive rejection technique of Gilks and Wild (1992) in the construction of the rejection envelope. Before describing the technique in detail, we briefly describe Gibbs sampling and the problem of non-conjugacy.

Gibbs sampling

To begin Gibbs sampling, initial values are assigned to each unknown model parameter β_i. In each of many iterations of the Gibbs sampler, for each parameter β_i in turn, the current value of the parameter is replaced by a value sampled from its full conditional distribution $[\beta_i|\{\beta_j, j \neq i\}$; data]. Geman and Geman (1984) show, under mild regularity conditions, that these values sampled for β_i represent a sample from the marginal posterior distribution of β_i; the pairs of values sampled for $\{\beta_i, \beta_j\}$ at each iteration represent a sample from the bivariate posterior margin of $\{\beta_i, \beta_j\}$; and so on. Sampled values are correlated between iterations, but this does not matter provided the Gibbs sampler is run sufficiently long, (until the samples generated at each iteration converge in distribution).

Conjugacy

Suppose we have a hierarchical model in which observations $\{Y_k\}$ are conditionally independent given parameter vector β. Then the full conditional distribution for β_i, the ith element of β, is proportional to

$$[\beta] \times \prod_k [Y_k|\beta] \tag{1}$$

considered as a function of β_i, where $[\beta]$ denotes the marginal (prior) distribution of β and $[Y_k|\beta]$ denotes the conditional distribution of Y_k given β. Sampling β_i from (1) may be straightforward when there is conjugacy between likelihood and prior, for then (1) reduces to a standard distribution.

Without conjugacy, sampling from (1) could be difficult and very expensive computationally, particularly when there are many independent obervations Y_k contributing to (1). Moreover, for each parameter, Gibbs sampling requires only one point to be sampled from the corresponding full conditional distribution: at the next iteration the full conditional will be different (through conditioning on different values of the remaining parameters). We therefore seek a sampling method which minimises the number of evaluations of (1).

In Section 2 we describe derivative-free adaptive rejection sampling. In Section 3 we evaluate the performance of the method.

2. DERIVATIVE-FREE ADAPTIVE REJECTION

To deal with the problem of non-conjugacy in the full conditionals in Gibbs sampling, we propose a method of adaptive rejection sampling. First we outline classical (non-adaptive) rejection sampling.

Rejection sampling

Rejection sampling requires the construction of a convenient upper bound to the density (the rejection envelope). A point is sampled from the rejection envelope, and is accepted with probability equal to the ratio of the density to its upper bound at that point. This is the rejection test. If the point is not accepted at the rejection step, a further point is sampled from the rejection envelope, rejection tested, and so on until a point is accepted. The efficiency of this method can be improved if a lower bound to the density can also be found (the squeezing function). Thus, in performing the rejection step, the sampled point can be accepted with probability equal to the ratio of the lower and upper bounds at that point, without needing to evaluate the density at that point (Ripley, 1987).

Adaptive rejection

Adaptive rejection (Gilks and Wild, 1991) improves rejection sampling by updating the envelope and squeezing functions whenever the density is evaluated. In Gilks and Wild (1991), the rejection envelope was constructed from tangents to the log density, necessitating evaluation of derivatives of the log density. In the version we describe here the envelope function is constructed in a way which does not require evaluation of derivatives. Both methods of adaptive rejection sampling require the density to be log-concave.

Log-concavity

Let $P = \{x, \ln f(x)\}$, a point on the logarithm of density $f(x)$. We say that $f(x)$ is log-concave with respect to x if

$$\textit{gradient of chord } P_2P_3 - \textit{gradient of chord } P_1P_2 \leq 0 \tag{2}$$

for any $x_1 < x_2 < x_3$ within its domain. An alternative definition of log-concavity would require that the second derivative of the log density is negative semi-definite everywhere within its support, but this would impose additional requirements of continuity and differentiability. We prefer (2) because the method for adaptive rejection sampling we describe below does not require the existence of continuous derivatives. Note also that the semi-definiteness in the inequality in (2) admits straight-line segments on the log density: i.e. exponential pieces on the density.

Virtually all densities in common use are concave on the logarithmic scale, with respect to (natural transformations of) both random variate and parameters (Devroye, 1986; Gilks and Wild, 1991). The only important exception is the t-distribution. Thus, usually all terms in full conditional distributions of the form of (1) are log-concave with respect to β_i; inducing log-concavity on the whole of (1). In particular, Generalised Linear models with all the usual link and error functions have log-concave full conditionals (Dellaportas and Smith, 1992). Therefore the method of adaptive rejection sampling we propose, requiring only that the density is log-concave, is applicable to a very large class of full conditionals.

Algorithm

We now describe derivative-free adaptive rejection sampling in detail.

Initialisation step:

To begin with the log density $\ln f(x)$ is evaluated (up to an additive normalising constant) at three or more points (for example, P1, P2 and P3 in Figure 1), such that at least one point lies to each side of the mode of the density. Note that this does not require calculation of the mode: it is sufficient to show that the chord joining the two leftmost points has a positive gradient, and the chord joining the two rightmost points has a negative gradient. A lower bound to the log density is constructed from chords joining the evaluated points of $f(x)$, together with vertical lines at the extreme points (P1 and P3 in Figure 1). An upper bound to the log density is constructed by extending these chords to their points of intersection. Note that the bounding property of these piece-wise linear functions is guaranteed by the log-concavity of $f(x)$.

Sampling step:

Exponentiating the piece-wise linear upper bound to $\ln f(x)$ yields a piece-wise exponential envelope function $e(x)$ which is easy to sample from (Figure 2). A point X is sampled from $e(x)$. A Uniform(0,1) random variate U is then independently sampled.

Squeezing step:

Exponentiating the piece-wise linear lower bound to $lnf(x)$ yields a piece-wise exponential squeezing function $s(x)$, (Figure 2). If

$$U \leq \frac{s(X)}{e(X)}$$

then X is accepted as the required sample from $f(x)$. Otherwise $f(X)$ is evaluated and the following rejection step is performed.

Rejection step:

If

$$U \leq \frac{f(X)}{e(X)}$$

then X is accepted as the required sample from $f(x)$. Otherwise X is rejected and the following updating step is performed.

Updating step:

The upper and lower bounds to $lnf(x)$ are updated to make use of the information gained in the rejection step (i.e. $f(X)$), by constructing chords passing through $f(X)$, (through P4 in Figure 3). Note that this does not merely improve the bounds in the P2-P3 segment: the upper bound in Figure 3 is improved from P1 through to $+\infty$. Thus new, improved, piece-wise exponential envelope and squeezing functions are constructed (Figure 4), which increase probabilities of acceptance at subsequent squeezing and rejection steps.

The sampling, squeezing, rejection and updating steps are repeated until a point X has been accepted (or until any required number of points X have been accepted). Through this process the envelope and squeezing functions rapidly adapt to the shape of the density.

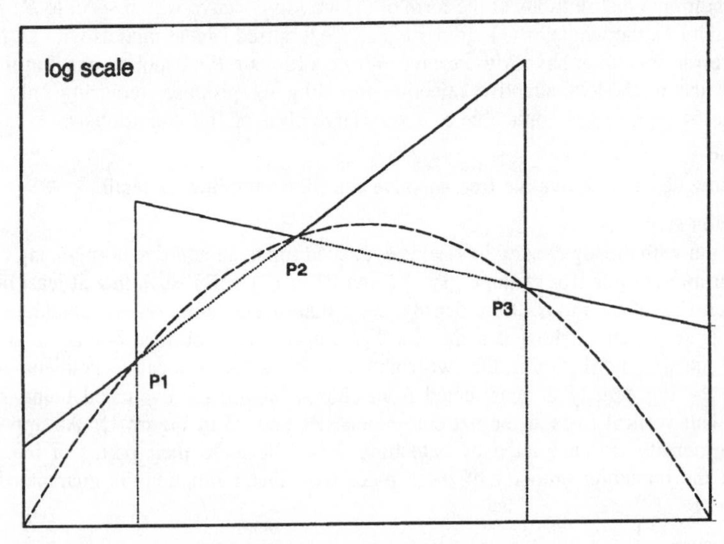

Figure 1. *Three initial points (P1, P2 and P3) on the log density* $\log f(x)$, *(broken line). Piece-wise linear upper bounds (solid line) and lower bounds (dotted line) are constructed by projecting chords through P1-P2 and P2-P3. At the extreme points (P1 and P3) vertical lines are constructed.*

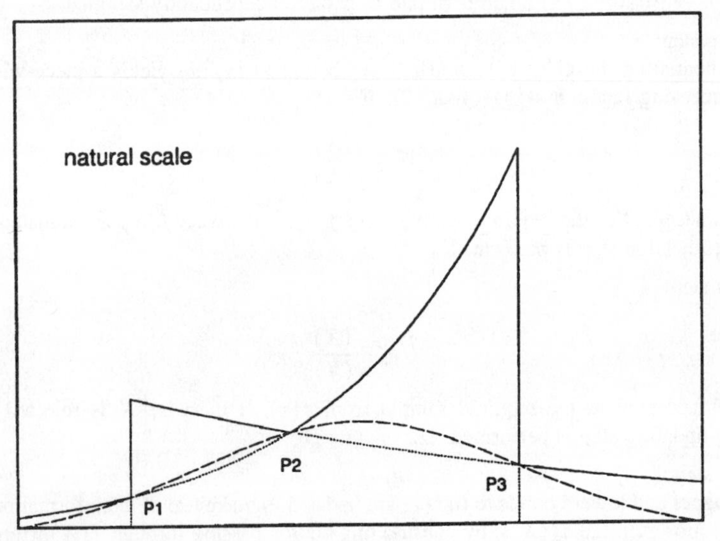

Figure 2. *The exponentiated form of Figure 1, showing the density* $f(x)$ *(broken line); the envelope function* $e(x)$ *(solid line); and the squeezing function* $s(x)$ *(dotted line).*

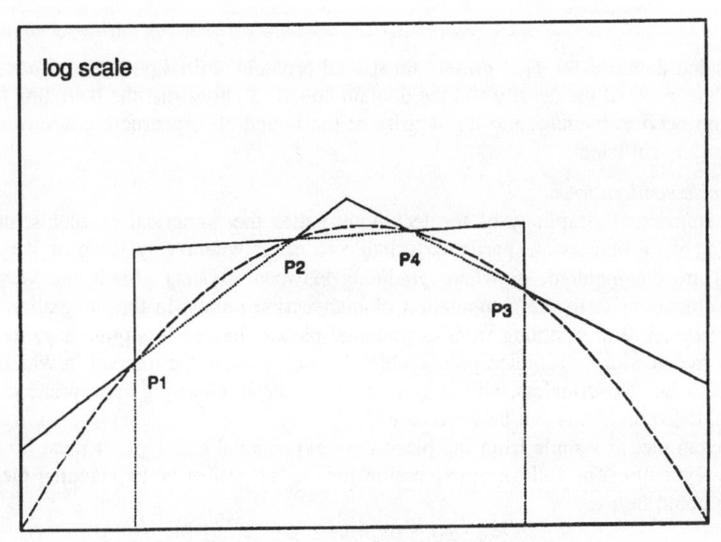

Figure 3. *A new point P4 is added to Figure 1. Chords through P2-P4 and P4-P3 are constructed, improving the upper bound from P1 to $+\infty$, and the lower bound from P2 to P3.*

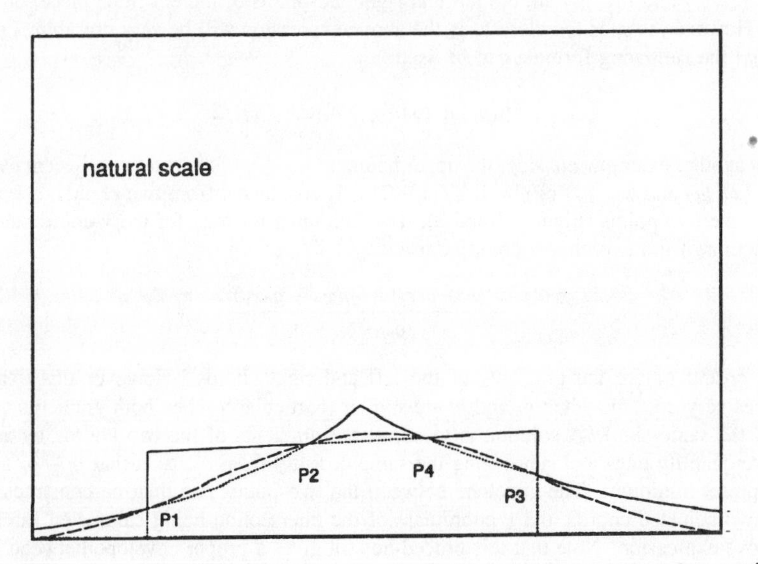

Figure 4. *The exponentiated form of Figure 3. The envelope and squeezing functions show a substantial improvement over those in Figure 2.*

Proof of derivative-free adaptive rejection sampling

The proof that the above method produces independent samples from $f(x)$ follows exactly as for the tangent-based version in Gilks and Wild (1992).

Bounds

Bounded domains for $f(x)$ present no special problem. Initial points need not intervene between the mode of the density and the domain bound. To construct the bounding functions, there is no need to evaluate the log density at the bound(s): constructing a vertical line at each bound is sufficient.

Numerical considerations

The conceptual simplicity of the technique belies the numerical problems met in implementing it. Problems are particularly liable to occur where very steep or very shallow gradients are encountered, or where gradients between adjacent chords are very similar. These problems arise in the computation of intersection points, in the integration of exponential pieces, and in sampling from exponential pieces. Indeed it is quite easy to calculate points of intersection or sampled points which lie well outside the interval in which they are supposed to lie. Nevertheless, with a great deal of care in choosing appropriate calculating formulae, these problems can be overcome.

For example, to sample from the piece-wise exponential envelope, it must first be integrated analytically. The following expression for the integral of an exponential piece of the envelope could be used:

$$\frac{\exp(y_r) - \exp(y_l)}{(y_r - y_l)} \times (x_r - x_l)$$

where (x_l, y_l) and (x_r, y_r) are the left and right end-points of the envelope piece, on the log scale. However, if y_r is too close to y_l the above expression will be very unstable. For small $|y_r - y_l|$ the following formula can be used:

$$[\exp(y_r) + \exp(y_l)] \times (x_r - x_l)/2.$$

As another example, consider the upper bound to $\log f(x)$ between two adjacent evaluated points (x_l, y_l) and (x_r, y_r) of $y = \log f(x)$. This is constucted from two chords, one through each of the two points (Figures 1 and 3). The following formula for the y-coordinate of the intersection of the two chords could be used:

$$\frac{g_r(y_l - g_l x_l) - g_l(y_r - g_r x_r)}{g_r - g_l}$$

where g_l and g_r are the gradients of the left and right chords. However this expression becomes very unstable when g_r and g_l are similar (particularly when both gradients are large and of the same sign). A solution is to adjust the gradients of the two chords by adding a positive quantity to g_l and subtracting the same quantity from g_r, such that $g_l - g_r$ achieves some preset minimum. The envelope between the two points can then be constructed from these two adjusted chords, the y-coordinate of the intersection being calculated safely from the above expression. Note that this procedure still gives a proper envelope between the two points.

3. EFFICIENCY

As noted in Section 1, in sampling from full conditionals of the form of (1), our aim is to minimise the number of evaluations of the log density. For the tangent version of adaptive rejection sampling, Gilks and Wild (1992) report that on average only three evaluations of the log density are required to accept one point, provided good initial points are chosen. We report a similar analysis here for the derivative-free version.

Table 1 shows that, for a standard normal density, with initial points at –1.0, 0.0 and 1.0, on average only about four density evaluations (including those at the three initial points) are required to accept one point using derivative-free adaptive rejection sampling. However, this gives a rather optimistic account of the performance of the method, since the position of the mode of the density (0.0 in this case) will not in general be known in advance. Table I shows that shifting the three initial points to one side can lead to substantially greater numbers of density evaluations, as can widening the intervals between those points, although even quite poor initial points produce only about eight density evaluations. With four initial points about five density evaluations (including those at the initial points) are required on average; shifting these initial points sideways now has a less detrimental effect. In general, moderate changes in the scale or location of the initial values do not have a substantial effect on the number of function evaluations required.

Initial x-values	Mean number of density evaluations	% with ≤ 4 density evaluations	% with ≥ 9 density evaluations
–0.5, 0.0, –0.5	5.0	43	2
–1.0, 0.0, 1.0	4.2	69	0
–2.0, 0.0, 2.0	4.7	43	0
–4.0, 0.0, 4.0	6.8	0	6
–8.0, 0.0, 8.0	8.7	0	55
–0.5, 0.5, 1.5	5.3	29	3
–0.1, 0.9, 1.9	7.5	0	17
–0.5, 1.5, 3.5	7.8	0	25
–0.1, 1.9, 3.9	8.6	0	50
–1.0, –0.5, 0.5, 1.0	4.8	44	0
–2.0, –0.5, 0.5, 2.0	4.7	48	0
–4.0, –0.5, 0.5, 4.0	5.6	20	2
–8.0, –0.5, 0.5, 8.0	6.5	8	9
–2.0, –1.0, 1.0, 2.0	4.8	43	0
–4.0, –1.0, 1.0, 4.0	5.5	19	1
–8.0, –1.0, 1.0, 8.0	6.0	4	3
–4.0, –2.0, 2.0, 4.0	6.0	2	1
–8.0, –2.0, 2.0, 8.0	6.2	0	2
–8.0, –4.0, 4.0, 8.0	7.5	0	17
–0.5, 0.5, 2.5, 3.5	6.3	8	6
–0.1, 0.9, 2.9, 3.9	8.5	0	46

Table 1. *Numbers of density evaluations (including those at initial x-values) required to sample one point from a standard normal density using derivative-free adaptive rejection sampling, for various initial x-values (1000 simulations for each).*

Table 2 reports a similar analysis for a highly skewed distribution: a Beta(11,2) density, which is defined on [0,1] and has its mode at 0.91. The skew seems hardly to have affected the number of density evaluations, and very bad initial points lead to on average fewer than 10 density evaluations.

Having sampled one point from the required density, further points can be sampled from the same density, with increasing efficiency, by continuing the adaptive rejection process. Table 3 shows, for the densities considered in Tables 1 and 2, the numbers of density evaluations required to sample 10 and 100 points. To sample 100 points only about 16 density evaluations are required.

Initial x-values	Mean number of density evaluations	% with ≤ 4 density evaluations	% with ≥ 9 density evaluations
.5, .9, .95, .99	5.3	18	0
.5, .6, .99, .999	5.9	4	0
.1, .2, .9999, .99999	8.1	0	34
.1, .2, .3, .99999	7.8	0	22
.1, .9999, .99999, .999999	9.6	0	85

Table 2. *Numbers of density evaluations (including those at initial x-values) required to sample one point from a Beta(11,2) density using derivative-free adaptive rejection sampling, for various initial x-values (1000 simulations for each).*

Density	Initial x-values	Sample size	Mean number of density evaluations	% with ≤ 4 density evaluations	% with ≥ 9 density evaluations
N(0,1)	−2.0, −1.0, 1.0, 2.0	10	7.7	0	28
		100	16.6	0	100
Beta(11,2)	0.5, 0.9, 0.95, 0.99	10	7.8	0	28
		100	15.4	0	100

Table 3. *Numbers of density evaluations (including those at initial x-values) required to sample 10 or 100 points from a standard normal or Beta(11,2) density using derivative-free adaptive rejection sampling (1000 simulations for each).*

4. CONCLUSION

Derivative-free adaptive rejection can be used as a black-box routine for efficiently sampling from a wide class of full conditional distributions arising in applications of Gibbs sampling (i.e., the class of log-concave densities). Provided that reasonably good initial points can be obtained, fewer than five density evaluations (including those at the initial points) will be needed on average to sample one point from the required density. Since iterations of the Gibbs sampler tend to be highly serially correlated, centiles of the rejection envelope for a given parameter at one iteration may be used to provide good initial points for the same parameter at the next iteration. On the basis of the results in Section 3, we would recommend using the 2.5%, 15%, 85% and 97.5% centiles from one iteration as initial points for the next iteration.

The tangent version of adaptive rejection (Gilks and Wild, 1992) requires fewer density evaluations than the derivative-free version (typically about three). However, the additional computational cost of calculating derivatives of the log density should also be taken into account. This additional cost will depend very much on the application, but if we were to assume that a derivative calculation is as expensive as a density evaluation, then the derivative-free version would be more efficient.

Moreover (and perhaps more importantly) the derivative-free version avoids the manual labour of deriving and programming the calculation of derivatives, which can be considerable in large and complex problems. One might consider using numerical derivatives with the tangent-based method. Again, this would typically result in more function evaluations than for the derivative-free method. Both methods of adaptive rejection tend to place evaluation points optimally, i.e., where the difference between the envelope and squeezing functions is greatest. A tangent-based method using numerical derivatives would be constrained to place its density evaluations points in tight pairs, thereby losing efficiency.

Gibbs sampling requires only one point to be sampled from each full conditional, whilst adaptive rejection performs most efficiently for second and subsequent samples from the same density. One way of exploiting this feature would be to sample by adaptive rejection many points from each full conditional, and use all of these sampled points in the construction of kernel density estimates: an empirical version of the 'Rao-Blackwellised' posterior density estimates proposed by Gelfand and Smith (1990). This would be straightforward and particularly advantageous where posterior density estimates of *functions* of the model parameters are required.

Finally, a referee has pointed out that there are wide classes of models involving standard distributions for which full conditional distributions of the parameters are not log-concave, for example: time-series models; dynamic models with dependent observations; and non-linear models. We are pleased to report that the above methodology can be extended to deal with non-log-concave densities (Gilks, Best and Tan, 1992).

A computer program for derivative-free adaptive rejection, written in C, is available from the author.

REFERENCES

Dellaportas, P. and Smith, A. F. M. (1992). Bayesian inference for generalised linear and proportional hazards models via Gibbs sampling. *Appl. Statist.* , (to appear).

Devroye, L. (1986). *Non-uniform Random Variate Generation*. New York: Springer.

Gelfand, A. E., Hills, S. E., Racine-Poon, A. and Smith, A. F. M. (1990). Illustration of Bayesian inference in normal data models using Gibbs sampling. *J. Amer. Statist. Assoc.* **85**, 972–985.

Gelfand, A. E. and Smith, A. F. M. (1990). Sampling based approaches to calculating marginal densities. *J. Amer. Statist. Assoc.* **85**, 398–409.

Geman, S. and Geman, D. (1984). Stochastic relaxation, Gibbs distributions, and the Bayesian restoration of images. *IEEE Trans. on Pattern Analysis and Machine Intelligence* **6**, 721–741.

Gilks, W. R., Best N. and Tan, K. (1992). Adaptive rejection sampling from densities which are not necessarily log-concave. *Tech. Rep.* MRC Cambridge, UK.

Gilks, W. R. and Wild, P. (1992). Adaptive rejection sampling for Gibbs sampling. *Appl. Statist.*, (to appear).

Ripley, B. (1987). *Stochastic Simulation.* New York: Wiley.

BAYESIAN STATISTICS 4, pp. 651–660
J. M. Bernardo, J. O. Berger, A. P. Dawid and A. F. M. Smith, (Eds.)

A Bayesian Justification for the Analysis of Residuals and Influence Measures*

F. J. GIRÓN, M. L. MARTÍNEZ and C. MORCILLO
Universidad de Málaga, Spain

SUMMARY

The usual approach to the analysis of residuals and influence measures in linear regression is based mostly on intuitive grounds. However, from a Bayesian perspective, the sequential updating of parameters via the Kalman filter applied to the regression model provides a sound basis for this analysis. Equivalence with the classical approach, and some possible generalizations are considered. First, we give some definitions of outlying and influential measures in a more general context.

Keywords: ANALYSIS OF RESIDUALS; BAYESIAN INFLUENCE MEASURES; PREDICTIVE DISTRIBUTIONS.

1. INTRODUCTION

Analysis of residuals, outlying observations and influence measures have been developed, from different points of view, mostly for the usual linear regression model. Many of these measures and procedures have been derived based on more or less intuitive and ad hoc grounds, depending on the particular model considered and the possible departures from the model. This explains the enormous quantity of discordance test found in the literature (see, e.g. Barnett and Lewis (1984)).

In this paper we consider these problems from a Bayesian perspective and try to give a unified approach to them. In this section we attempt to outline a general theory of outlying and influential observations with special emphasis on those models for which a Bayesian conjugate analysis (with or without prior information) is available. For these models, the sequential updating formulae of the hyperparameters of the conjugate family —from prior to posterior, in a Kalman filter-like fashion— give a sound basis for the definition and development of diagnostic tools. Section 2 develops the preceding theory for the usual linear regression model.

Let $\{f(x \mid \phi); \phi \in \Phi\}$ be a statistical model and $\boldsymbol{x} = (x_1, \ldots, x_n)$ a random sample from the model, where ϕ is a p–dimensional parameter vector, which may or may not include nuisance parameters. The case of more complex statistical models with dependent observations or changing parameters, as in autoregressive and dynamic models, can be treated in a similar way to that described here.

We concentrate here on two different but related problems of data analysis: one is the problem of deciding when one or more observations from the data D, where D denotes the vector $\boldsymbol{x}$ and possibly other concomitant variables that might be present in the model, can be regarded as *outliers*; the second one is the (possible) influence of these or other observations,

* This paper has been prepared with partial financial help from project number PB87–0607–C02–02 *Dirección General de Investigación Científica y Técnica* (DGICYT) granted by the *Ministerio de Educación y Ciencia*, Spain and by the *Consejería de Educación de la Junta de Andalucía*.

or some concomitant variables included in the model, on estimators of the parameter or some function of it.

Let us consider the case of a single observation first. From the viewpoint of the first problem one regards an observation as being inconsistent with the remaining ones when it cannot be predicted from these. From a Bayesian viewpoint, all the information about ϕ given by $D \backslash \{x_i\}$, usually represented by $D_{(i)}$, is contained in the posterior distribution $p(\phi \mid D_{(i)})$. From this we can compute the posterior predictive distribution of a future observation x, given by

$$p(x \mid D_{(i)}) = \int p(x \mid \phi)\, p(\phi \mid D_{(i)})\, d\phi. \tag{1.1}$$

Now, inconsistency of the observation x_i with the rest, $D_{(i)}$, is reflected in a very small posterior density for x_i. So, based on this intuitive idea, we propose the following definition which does not involve contemplating alternative hypotheses for the *outlying* observation as in other useful and interesting approaches to this problem, usually based on the Bayes factors, (see, e.g., Pettit and Smith (1985) and Pettit (1988). Our approach, based solely on this predictive distribution, is more closely related to the CPO (conditional predictive ordinate) criterion of Geisser (1980, 1985) and the related ROM (ratio ordinate measure) criterion proposed by Pettit (1990). Similar ideas of outlyingness when not contemplating alternatives are also found, with applications, in Geisser (1989, 1990)

Definition 1. *Observation x_i is an outlier or observation-influential, if x_i does not belong to a highest predictive density region of $p(y \mid D_{(i)})$ of probabilistic content $1 - \alpha$.*

We believe this definition to be more natural than the one based on the whole predictive density given all the data $p(y \mid D)$, as it reflects the surprise in predicting a new observation, say x_i, based on the knowledge collected up to this time when Bayes theorem is applied sequentially and x_i is the last observation.

This definition can be used to give a Bayesian analog of classic discordance tests, with no specific alternatives in mind, based on the probability of x_i not being in a H.P.D.

The extension of the definition to any subset of observations is straightforward.

Definition 2. *A subset $\mathcal{I} = \{i_1, \ldots, i_m\} \subset \{1, \ldots, n\}$ is outlying or observation-influential if the vector $x_{\mathcal{I}}$ does not belong to a high-dimesional H.P.D. of the posterior predictive density $p(y_1, \ldots, y_m \mid D_{(\mathcal{I})})$ of probabilistic content $1 - \alpha$.*

In general, the computation of this H.P.D. can be very complicated. Yet, in many instances, as for example, when there exists a predictive sufficient statistic, the computation of a H.P.D. for $y_1, \ldots, y_m$ is generally reduced to that of obtaining a H.P.D. for the predictive distribution of $p(t_m \mid D_{(\mathcal{I})})$, where $t_m = t_m(y_1, \ldots, y_m)$ is a sufficient statistic.

The next definition tries to capture the idea of influence of an observation (or a subset of observations) on the parameter ϕ of the model, or some subset or function of ϕ.

As all the relevant information regarding the parameter is contained in the posterior distribution, a measure of the influence of an observation, say x_i, should depend on any discrepancy measure between the posterior of ϕ given $D_{(i)}$ and D. A useful measure of discrepancy, which has been previously used by Johnson and Geisser (1982, 1983), Geisser (1985) and Guttman and Peña (1988), that takes into account the full distributions involved, i.e., $p(\phi \mid D_{(i)})$ and $p(\phi \mid D)$, is the directed Kullback-Leibler divergence.

Definition 3. *Observation x_i is more influential than observation x_j for parameter ϕ if*

$$\delta_{K-L}(p(\phi \mid \boldsymbol{D}_{(i)}), p(\phi \mid \boldsymbol{D})) \geq \delta_{K-L}(p(\phi \mid \boldsymbol{D}_{(j)}), p(\phi \mid \boldsymbol{D})),$$

where δ_{K-L} represents the directed Kullback-Leibler divergence.

This definition introduces a complete ordering on the individual observations according to their influence on the estimation of parameter ϕ. Obviously, this definition can be extended to any subset of observations, thus giving a complete order on the set of all possible partitions, according to their relative influence. If ϕ is not the parameter of interest, but instead any (measurable) function of it $\theta = f(\phi)$, definition 3 is obviously extended to this case.

One advantage of this definition is that it uses all the available information on the parameter excluding and including the suspected observations, and gives a measure of the relative influence of any subset of observations on (some function of) the parameter but, on the other hand, it does not provide a test for asserting the influence of a given subset of observations.

To establish a link with classical measures of influence, generally based on non-Bayesian ideas such as point estimators, etc., which, as Lindley (1990) remarks are extraneous to Bayesian inference and may cause difficulties, we suggest the next definition. To make it unambiguous let us recall that a Bayesian estimator can be regarded as a map from the space of (prior and/or posterior) distributions into the set of possible values of the parameter; thus, a Bayes estimator $\tilde{\phi}$ is a measurable mapping

$$\tilde{\phi}\colon p(\phi \mid \boldsymbol{D}) \mapsto \tilde{\phi}(p(\phi \mid \boldsymbol{D})) \in \Phi;$$

such as, e.g., the mean, mode or any other estimator obtained, for example, by minimizing a given loss function, conditional on their existence.

Now suppose $\tilde{\phi}$ is a well-defined Bayes estimator. The following definition makes precise the influence of an observation on a particular estimator.

Definition 4. *Observation x_i is estimation-influential (with respect to $\tilde{\phi}$) if $\tilde{\phi}(p(\phi \mid \boldsymbol{D}))$ does not belong to a H.P.D. of the posterior distribution $p(\phi \mid \boldsymbol{D}_{(i)})$ of probabilistic content* $1 - \alpha$.

Again, as in definition 1, the preceding definition is also based on the sequential aspects of Bayesian learning. Its generalization to a subset of observations is also straightforward.

When a conjugate analysis of the statistical problem is feasible, the preceding theory becomes more transparent. For illustration purposes, we consider the case of a single observation, the extension to a subset of observations being immediate.

Thus, suppose $\{C(\phi; \boldsymbol{\alpha}); \boldsymbol{\alpha} \in \boldsymbol{A}\}$ is a conjugate family for the statistical model contemplated, where $\boldsymbol{\alpha}$ is a vector of hyperparameters that fully characterizes the distribution (usually a simple extension or reparameterization of the sufficient statistic).

Suppose the random sample is ordered as $\boldsymbol{D} = \{x_1, \ldots, x_{i-1}, x_{i+1}, \ldots, x_n, x_i\}$ with observation x_i in the last place (the order of the remaining ones is unimportant for i.i.d. models). If the prior distribution is a member of the conjugate family, say, $C(\phi; \boldsymbol{\alpha}_0)$, then the posterior of $\phi \mid \boldsymbol{D}_{(i)} \sim C(\phi; \boldsymbol{\alpha}_{(i)})$ and also the full posterior $\phi \mid \boldsymbol{D} \sim C(\phi; \boldsymbol{\alpha}_n)$, where $\boldsymbol{\alpha}_n$, the final value of the hyperparameter given all the data, is updated by means of the recursive equation or filter

$$\boldsymbol{\alpha}_n = F(\boldsymbol{\alpha}_{(i)}; x_i).$$

In the conjugate case, from the posterior of $\phi \mid \boldsymbol{D}_{(i)}$ one can easily compute the predictive distribution of a future observation x given $\boldsymbol{D}_{(i)}$, which only depends on the hyperparameter

$\alpha_{(i)}$ and test if the observation x_i is consistent or not with this distribution to declare it outlying, which is generally equivalent to test the discrepancy between the x_i's and some function of $\alpha_{(i)}$, usually called *residuals*.

In this context too, the influence of an observation on the parameter ϕ is, according to definition 4, some measure of the discrepancy between $\alpha_{(i)}$ and α_n, which might be termed an *influence measure*.

As a simple illustration of these definitions, let us consider the case of a sample $x_1, \ldots, x_n$ from an exponential model with parameter λ,

$$f(x \mid \lambda) = \lambda e^{-\lambda x}.$$

Suppose we use a reference non-informative prior for λ, that is, $p_0(\lambda) \propto \lambda^{-1}$, so that the posterior $p(\lambda \mid \boldsymbol{D}_{(\mathcal{I})})$ is

$$p(\lambda \mid \boldsymbol{D}_{(\mathcal{I})}) \propto \lambda^{\nu-1} e^{-\lambda t_{(\mathcal{I})}}$$

where $\nu = n - m$ and $t_{(\mathcal{I})} = \sum_{i \notin \mathcal{I}}^{n} x_j$. Thus, given $\boldsymbol{D}_{(\mathcal{I})}$, $\lambda \sim Ga(t_{(\mathcal{I})}, \nu)$ and from (1.1) the predictive of $y_1, \ldots, y_m$ is

$$p(y_1, \ldots, y_m \mid \boldsymbol{D}_{(\mathcal{I})}) = \frac{\Gamma(n)}{\Gamma(\nu)} \frac{t_{(\mathcal{I})}^{\nu}}{(t_{(\mathcal{I})} + \sum_{j=1}^{m} y_j)^n}.$$

As this predictive density is a monotonic decreasing function of the sufficient statistic $s = \sum_{j=1}^{m} y_j$, the H.P.D. regions are of the form

$$\{(y_1, \ldots, y_m) \in \mathbb{R}^{+^m}; \quad s = \sum_{j=1}^{m} y_j \leq k\}$$

with k chosen in such a way that the predictive posterior probability equals $1 - \alpha$. As the region depends on the sufficient statistic s, this probability can be computed from the predictive density of s given $\boldsymbol{D}_{(\mathcal{I})}$, which turns out to be

$$p(s \mid \boldsymbol{D}_{(\mathcal{I})}) = \frac{1}{Be(m, \nu)} \frac{t_{(\mathcal{I})}^{\nu} s^{m-1}}{(t_{(\mathcal{I})} + s)^n};$$

that is, an inverted-beta-2 density. From well known relations with the beta or Snedecor's F distribution, percentiles of the predictive distributions are easily computed.

As a numerical example, let us consider the data shown in the following table, which reproduces Table 1 taken from Barnett and Lewis (1984), corresponding to a sample of 131 excess cycle times in steel manufacture.

Excess cycle time	1	2	3	4	5	6	7	8	9	10	11	12	13	14	15	21	32	35	92
Frequency	18	12	18	16	10	4	9	9	2	7	6	7	2	1	3	3	2	1	1

Table 1. *Data from Barnett and Lewis.*

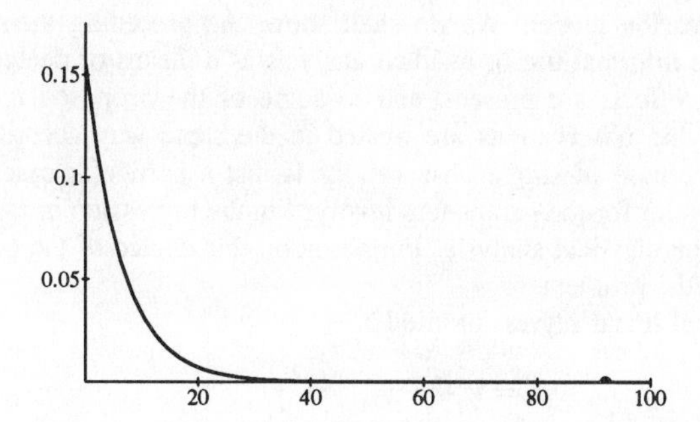

Figure 1. *Predictive distribution for 131st observation*

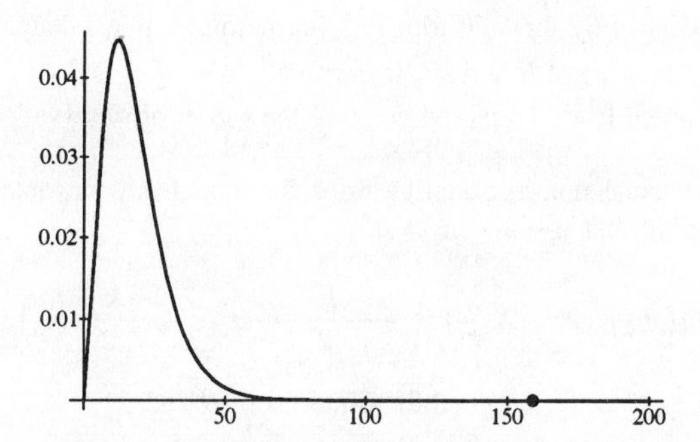

Figure 2. *Predictive distribution of the sum of the last four observations*

Figure 1 shows the predictive distribution of a single observation based on the first 130 observations, and the single observation 131, represented by a big point, which is far away from H.P.D.'s with large probabilistic content.

Figure 2 shows the predictive distribution of the sum of four observations based on the first 127 observations and the sum of the four upper observations which is also far away from H.P.D.'s with large probabilistic content.

However, in this example, no single observation (not even observation 131) is estimation-influential for the parameter λ, at least for the usual values, $1-\alpha$, of probabilistic content. In fact, if we take as estimator the mean of the posterior distribution, observation 131 becomes estimation-influential only for H.P.D.'s intervals of probabilistic content less than .669, that is, for $\alpha \geq .331$

In general, the property of being influential, according to definition 4, depends heavily — apart from the estimator being used— on the presence of explicative or concomitant variables in the model.

2. APPLICATION TO THE REGRESSION MODEL

In this section we shall study the particular case of detecting outliers and influential observations in the general regression model. As we shall show, the preceding theory gives a Bayesian justification to the informal use of residual analysis as a means of detecting single outliers, when no masking effects are present, and to some of the proposed measures of influence. Subsets of outlying observations are treated in the same way according to the preceding theory so that the case of single observations is just a particular case. We also assume the usual reference prior for the parameters involved in the regression model, in order to establish comparisons with classical analysis. Furthermore, this choice of the prior allows for a conjugate analysis of the problem.

Let us consider the usual linear regression model

$$y_i = \boldsymbol{x}_i'\boldsymbol{\theta} + u_i;$$

where $y_1, \ldots, y_n$ are random observable variables, $u_1, \ldots, u_n$ are independent errors normally distributed with mean 0 and unknown variance σ^2, $\boldsymbol{x}_i$ are $(k \times 1)$ vectors of independent variables, and $\boldsymbol{\theta}$ is a $(k \times 1)$ vector of unknown regression parameters.

As usual, for any subset $\mathcal{I}$ of m observations and for any vector or matrix involved in the model, we denote the corresponding subvector or submatrix by the subscript $_{\mathcal{I}}$, and the remaining subvector or submatrix by the subscript $_{(\mathcal{I})}$. Furthermore, if we denote by $\hat{\boldsymbol{\theta}}$ and $\hat{\sigma}^2$ the usual least squares estimators of $\boldsymbol{\theta}$ and σ^2, respectively, then $\hat{\boldsymbol{\theta}}_{(\mathcal{I})}$ and $\sigma^2_{(\mathcal{I})}$ denote the least squares estimates computed from the model deleting the subset of observations indexed by $\mathcal{I}$.

If we consider now the usual non-informative prior for $\boldsymbol{\theta}$ and σ^2, we obtain that the posterior given $\boldsymbol{D}_{(\mathcal{I})}$, assuming that $n - k - m > 0$, is

$$\boldsymbol{\theta}, \sigma^2 \mid \boldsymbol{D}_{(\mathcal{I})} \sim NIGa\Big(\hat{\boldsymbol{\theta}}_{(\mathcal{I})}, (\boldsymbol{X}'_{(\mathcal{I})}\boldsymbol{X}_{(\mathcal{I})})^{-1}, \frac{n-k-m}{2}\hat{\sigma}^2_{(\mathcal{I})}, \frac{n-k-m}{2}\Big). \tag{2.1}$$

From this, we obtain the posterior marginal distributions of $\boldsymbol{\theta}$ and σ^2

$$\boldsymbol{\theta} \mid \boldsymbol{D}_{(\mathcal{I})} \sim t\,(\hat{\boldsymbol{\theta}}_{(\mathcal{I})}, \hat{\sigma}^2_{(\mathcal{I})}(\boldsymbol{X}'_{(\mathcal{I})}\boldsymbol{X}_{(\mathcal{I})})^{-1}, n-k-m), \tag{2.2}$$

$$\sigma^2 \mid \boldsymbol{D}_{(\mathcal{I})} \sim IGa\Big(\frac{n-k-m}{2}\hat{\sigma}^2_{(\mathcal{I})}, \frac{n-k-m}{2}\Big). \tag{2.3}$$

As the vector of m remaining observations, say $\boldsymbol{z}$, follows the model

$$\boldsymbol{z} \mid \boldsymbol{\theta}, \sigma^2 \sim N(\boldsymbol{X}_{\mathcal{I}}\boldsymbol{\theta}, \sigma^2);$$

then, from properties of the normal-inverted-gamma distributions, the joint posterior distribution of $\boldsymbol{z}$ and σ^2 given $\boldsymbol{D}_{(\mathcal{I})}$ is

$$\boldsymbol{z}, \sigma^2 \mid \boldsymbol{D}_{(\mathcal{I})} \sim NIGa\Big(\boldsymbol{X}_{\mathcal{I}}\hat{\boldsymbol{\theta}}_{(\mathcal{I})}, \boldsymbol{V}_{\mathcal{I}}, \frac{n-k-m}{2}\hat{\sigma}^2_{(\mathcal{I})}, \frac{n-k-m}{2}\Big);$$

where $\boldsymbol{V}_{\mathcal{I}} = \boldsymbol{I}_m + \boldsymbol{X}_{\mathcal{I}}(\boldsymbol{X}'_{(\mathcal{I})}\boldsymbol{X}_{(\mathcal{I})})^{-1}\boldsymbol{X}'_{\mathcal{I}}$.

From this, the predictive distribution of $\boldsymbol{z}$ given $\boldsymbol{D}_{(\mathcal{I})}$ is

$$\boldsymbol{z} \mid \boldsymbol{D}_{(\mathcal{I})} \sim t_m(\boldsymbol{X}_{\mathcal{I}}\hat{\boldsymbol{\theta}}_{(\mathcal{I})}, \hat{\sigma}^2_{(\mathcal{I})}\boldsymbol{V}_{\mathcal{I}}, n-k-m);$$

and, from well known properties of the multivariate t distribution, it follows that the H.P.D. region of probabilistic content $1-\alpha$ is the ellipsoid in m-dimensional space given by

$$\left\{ z \in \mathbb{R}^m; \frac{(z - X_{\mathcal{I}}\hat{\theta}_{(\mathcal{I})})' V_{\mathcal{I}}^{-1} (z - X_{\mathcal{I}}\hat{\theta}_{(\mathcal{I})})}{m\hat{\sigma}^2_{(\mathcal{I})}} \leq F(m, n-k-m; 1-\alpha) \right\};$$

where $F(m, n-k-m; 1-\alpha)$ is the $1-\alpha$ percentile of the central F distribution with m and $n-k-m$ degrees of freedom.

Thus, according to definition 2, the subset of observations indexed by $\mathcal{I}$, i.e., $y_{\mathcal{I}}$ is outlying, if the quantity

$$\frac{(y_{\mathcal{I}} - X_{\mathcal{I}}\hat{\theta}_{(\mathcal{I})})' V_{\mathcal{I}}^{-1} (y_{\mathcal{I}} - X_{\mathcal{I}}\hat{\theta}_{(\mathcal{I})})}{m\hat{\sigma}^2_{(\mathcal{I})}} \geq F(m, n-k-m; 1-\alpha).$$

Note that the quantity $r_{\mathcal{I}} = y_{\mathcal{I}} - X_{\mathcal{I}}\hat{\theta}_{(\mathcal{I})}$ is the m-step ahead prediction error given by the Kalman filter when applied to the usual linear model or, as it is usually known, the jackknifed residual vector. Thus, the preceding test for detecting outliers in the subset of observations indexed by $\mathcal{I}$ is equivalent to

$$PD(\mathcal{I}) = \frac{r_{\mathcal{I}}' V_{\mathcal{I}}^{-1} r_{\mathcal{I}}}{m\hat{\sigma}^2_{(\mathcal{I})}} \geq F(m, n-k-m; 1-\alpha); \tag{2.4}$$

where $PD(\mathcal{I})$ denotes what we call the predictive distance between $y_{\mathcal{I}}$ and its prediction $X_{\mathcal{I}}\hat{\theta}_{(\mathcal{I})}$.

Moreover, by using the Kalman filter —without need to resort to well known matrix identities— a relation between the jackknifed residuals and the ordinary least squares residuals, on the one hand, and the "hat" matrix and the $V_{\mathcal{I}}$ matrix, on the other hand, is easily obtained; thus giving an alternative expression for the preceding test, based on more usual or classical measures of influence.

The recursive equations of the Kalman filter, for the linear model, are

$$\begin{aligned} \hat{\theta} &= \hat{\theta}_{(\mathcal{I})} + (X_{(\mathcal{I})}' X_{(\mathcal{I})})^{-1} X_{\mathcal{I}}' V_{\mathcal{I}}^{-1} r_{\mathcal{I}} \\ (X'X)^{-1} &= (X_{(\mathcal{I})}' X_{(\mathcal{I})})^{-1} - (X_{(\mathcal{I})}' X_{(\mathcal{I})})^{-1} X_{\mathcal{I}}' V_{\mathcal{I}}^{-1} X_{\mathcal{I}} (X_{(\mathcal{I})}' X_{(\mathcal{I})})^{-1} \\ (n-k)\hat{\sigma}^2 &= (n-k-m)\hat{\sigma}^2_{(\mathcal{I})} + r_{\mathcal{I}}' V_{\mathcal{I}}^{-1} r_{\mathcal{I}}. \end{aligned} \tag{2.5}$$

If we now define the matrix $H_{\mathcal{I}}$ as the submatrix of the hat matrix $X(X'X)^{-1}X'$ corresponding to the rows and columns indexed by $\mathcal{I}$, that is, $H_{\mathcal{I}} = X_{\mathcal{I}}(X'X)^{-1}X_{\mathcal{I}}'$, then, by left multiplying the second equation in (2.5) by $X_{\mathcal{I}}$ and right multiplying it by $X_{\mathcal{I}}'$, we get, after some matrix manipulation, that $H_{\mathcal{I}} = I_m - V_{\mathcal{I}}^{-1}$ or, equivalently, $V_{\mathcal{I}} = (I_m - H_{\mathcal{I}})^{-1}$.

The ordinary least squares residuals are defined by $e = y - X\hat{\theta}$, so the vector of residuals indexed by $\mathcal{I}$ is $e_{\mathcal{I}} = y_{\mathcal{I}} - X_{\mathcal{I}}\hat{\theta}$. Using the first equation in (2.5), after some algebra, one gets $e_{\mathcal{I}} = V_{\mathcal{I}}^{-1} r_{\mathcal{I}}$, or equivalently, $r_{\mathcal{I}} = V_{\mathcal{I}} e_{\mathcal{I}}$. Thus, the test for multiple outliers (2.4) can be written in terms of the usual least squares residuals and the hat matrix as follows

$$PD(\mathcal{I}) = \frac{e_{\mathcal{I}}' (I_m - H_{\mathcal{I}})^{-1} e_{\mathcal{I}}}{m\hat{\sigma}^2_{(\mathcal{I})}} \geq F(m, n-k-m, 1-\alpha). \tag{2.6}$$

For the case of testing a single observation, say $(y_i, \boldsymbol{x}_i)$, the predictive distribution of the i-th jackknifed residual is

$$r_i \mid \boldsymbol{D}_{(i)} \sim t\,(0, \hat{\sigma}^2_{(i)} v_i, n-k-1),$$

where $v_i = 1 + \boldsymbol{x}_i'(\boldsymbol{X}_{(i)}'\boldsymbol{X}_{(i)})^{-1}\boldsymbol{x}_i$.

Therefore, the i-th observation is regarded as a single outlier if

$$\left| \frac{r_i}{\hat{\sigma}_{(i)}\sqrt{v_i}} \right| \geq t(n-k-1; 1-\alpha).$$

This result is obviously equivalent to that obtained from equation (2.4) when $\mathcal{I} = \{i\}$, by recalling that the $F(1, n-k-1)$ distribution is the square of the $t(n-k-1)$ distribution.

Therefore, the quantity

$$t_i^* = \frac{r_i}{\hat{\sigma}_{(i)}\sqrt{v_i}},$$

which is usually called the i-th studentized jackknifed residual, is to be compared with the $1-\alpha$ fractile of the standard t distribution with $n-k-1$ degrees of freedom to declare the corresponding observation as outlying; thus showing the importance of residual analysis in the study of single outlying observations.

Let us now turn to the question of when and how a subset of observations can influence the estimation of the parameters in the regression model. We shall first study the influence on the regression coefficients $\boldsymbol{\theta}$, and then the influence on the estimation of the common variance σ^2. The joint influence on the pair $\boldsymbol{\theta}, \sigma^2$ could be treated in a similar way.

i) Influence measures for vector $\boldsymbol{\theta}$.

From the posterior distribution of $\boldsymbol{\theta} \mid \boldsymbol{D}_{(\mathcal{I})}$, given by (2.2), and the properties of the multivariate Student t, we get that the H.P.D. regions for $\boldsymbol{\theta}$, of probabilistic content $1-\alpha$, are of the form

$$\left\{ \boldsymbol{\theta} \in \mathbb{R}^k ; \frac{(\boldsymbol{\theta} - \hat{\boldsymbol{\theta}}_{(\mathcal{I})})'(\boldsymbol{X}_{(\mathcal{I})}'\boldsymbol{X}_{(\mathcal{I})})(\boldsymbol{\theta} - \hat{\boldsymbol{\theta}}_{(\mathcal{I})})}{k\hat{\sigma}^2_{(\mathcal{I})}} \leq F(k, n-k-m; 1-\alpha) \right\}.$$

Thus, according to definition 4, the subset of observations indexed by $\mathcal{I}$ will be influential on $\boldsymbol{\theta}$, with respect to the estimator $\hat{\boldsymbol{\theta}}$ (the mode of the posterior or the mean if $n-k-m > 1$) if

$$BD(\mathcal{I}) = \frac{(\hat{\boldsymbol{\theta}} - \hat{\boldsymbol{\theta}}_{(\mathcal{I})})'(\boldsymbol{X}_{(\mathcal{I})}'\boldsymbol{X}_{(\mathcal{I})})(\hat{\boldsymbol{\theta}} - \hat{\boldsymbol{\theta}}_{(\mathcal{I})})}{k\hat{\sigma}^2_{(\mathcal{I})}} \geq F(k, n-k-m; 1-\alpha);$$

where $BD(\mathcal{I})$ stands for *Bayesian distance*, which can be regarded as a sort distance between the usual least squares estimator of $\hat{\boldsymbol{\theta}}$ and the jackknifed estimator $\hat{\boldsymbol{\theta}}_{(\mathcal{I})}$.

Note that this expression is related to Welsch's (1982) distance for the detection of influential observations.

Simpler, but equivalent, expressions of this Bayesian distance can be given in terms of the two types of residuals, as follows

$$BD(\mathcal{I}) = \frac{\boldsymbol{r}_{\mathcal{I}}'\boldsymbol{V}_{\mathcal{I}}^{-1}(\boldsymbol{I}_m - \boldsymbol{V}_{\mathcal{I}}^{-1})\boldsymbol{r}_{\mathcal{I}}}{k\hat{\sigma}^2_{(\mathcal{I})}} = \frac{\boldsymbol{e}_{\mathcal{I}}'\boldsymbol{H}_{\mathcal{I}}(\boldsymbol{I}_m - \boldsymbol{H}_{\mathcal{I}})^{-1}\boldsymbol{e}_{\mathcal{I}}}{k\hat{\sigma}^2_{(\mathcal{I})}}.$$

For the case of a single observation i, the Bayesian distance $BD(i)$ can be written in terms of the studentized jackknifed residual t_i^* and the diagonal element h_{ii} of the hat matrix, as follows

$$BD(i) = \frac{t_i^{*2} h_{ii}}{k}.$$

Therefore, observation i is influential on $\boldsymbol{\theta}$ if the quantity

$$\frac{t_i^{*2} h_{ii}}{k} \geq F(k, n-k-1; 1-\alpha).$$

Measures of influence of subsets of observations on coordinates θ_i of the regression parameter $\boldsymbol{\theta}$, or linear combinations of the regression coefficients can be easily derived by obtaining the corresponding posterior distributions from equation (2.2) and then proceding in the usual way.

ii) Influence measures for σ^2.

Although, in general practice, the computation of influential measures for $\boldsymbol{\theta}$ is more important, sometimes it may be also of interest to derive influence measures for the scale parameter σ^2.

The posterior distribution of $\sigma^2 \mid \boldsymbol{D}_{(\mathcal{I})}$ is given by (2.3) or, equivalently, by

$$\frac{(n-k-m)\,\hat{\sigma}^2_{(\mathcal{I})}}{\sigma^2} \sim \chi^2(n-k-m).$$

Standardized H.P.D. intervals for σ^2 can be calculated from this distribution, as in Box and Tiao (p. 90, 1973):

$$\left(\frac{(n-k-m)\hat{\sigma}^2_{(\mathcal{I})}}{\bar{\chi}^2(n-k-m;\alpha)}, \frac{(n-k-m)\hat{\sigma}^2_{(\mathcal{I})}}{\underline{\chi}^2(n-k-m;\alpha)} \right);$$

where $\underline{\chi}^2(n-k-m;\alpha)$ and $\bar{\chi}^2(n-k-m;\alpha)$ represent the lower and upper limits, obtained by using a locally uniform prior on the metric $\log \sigma$.

Therefore, the subset of observations indexed by $\mathcal{I}$ will be influential on σ^2, with respect to the usual least squares estimator $\hat{\sigma}^2$, if either

$$\frac{(n-k-m)\,\hat{\sigma}^2_{(\mathcal{I})}}{\hat{\sigma}^2} \leq \underline{\chi}^2(n-k-m,\alpha) \quad \text{or} \quad \frac{(n-k-m)\,\hat{\sigma}^2_{(\mathcal{I})}}{\hat{\sigma}^2} \geq \bar{\chi}^2(n-k-m,\alpha).$$

In practice, when seeking for influential observations for the scale parameter σ^2, the second inequality never holds for the usual (small) values of α and for small m, so that we can regard the subset of observations scale or variance-influential only if the first inequality holds.

From the last equation in (2.5), the preceding test can be written, in terms of the least squares residuals and the hat matrix, as

$$MVR(\mathcal{I}) = n-k - \frac{\boldsymbol{e}'_{\mathcal{I}}(\boldsymbol{I}_m - \boldsymbol{H}_{\mathcal{I}})^{-1}\boldsymbol{e}_{\mathcal{I}}}{\hat{\sigma}^2} \leq \underline{\chi}^2(n-k-m,\alpha);$$

where $MVR(\mathcal{I})$ denotes what we called *modulated variance ratio*. It is easy to prove that the following relation between $MVR(\mathcal{I})$ and the predictive distance $PD(\mathcal{I})$ holds,

$$MVR(\mathcal{I}) = \frac{n-k}{1 + \left(\frac{m}{n-k-m}\right) PD(\mathcal{I})};$$

which establishes a link between the concepts of *observation-influence* (definition 2) and *estimation-influence* (definition 4) for the regression model.

For the case of single observations, the preceding formula simplifies to

$$t_i^2 \geq n - k - \underline{\chi}^2(n - k - 1, \alpha);$$

where t_i is the usual studentized residual, i.e.,

$$t_i = \frac{e_i}{\hat{\sigma}\sqrt{1 - h_{ii}}}.$$

If, instead of using the least squares estimator $\hat{\sigma}^2$ of σ^2, we had used the mean or mode of the posterior distribution of $\sigma^2 \mid D_{(\mathcal{I})}$ as a Bayes estimator, slightly different measures of influence would have been obtained.

REFERENCES

Barnett, V. and Lewis, T. (1984). *Outliers in Statistical Data.* New York: Wiley.

Box, G. E. P. and Tiao, G. C. (1973). *Bayesian Inference in Statistical Analysis.* Reading, MA: Addison-Wesley.

Geisser, S. (1980). Discussion on sampling and Bayes' inference in scientific modelling and robustness, by G. E. P. Box. *J. Roy. Statist. Soc. A* **143**, 416–417.

Geisser, S. (1985). On the prediction of observables: a selective update. *Bayesian Statistics 2* (J. M. Bernardo, M. H. DeGroot, D. V. Lindley and A. F. M. Smith, eds.), Amsterdam: North-Holland, 203–229, (with discussion).

Geisser, S. (1989). Predictive discordancy testing for exponential observations. *Canadian J. Statist.* **17**, 19–26.

Geisser, S. (1990). Predictive approaches to discordancy testing. *Bayesian and Likelihood Methods in Statistics and Econometrics* (S. Geisser *et al.* eds.), Amsterdam: North-Holland, 321–335.

Guttman, I. and Peña, D. (1988). Outliers and influence: evaluation by posteriors of parameters in the linear model. *Bayesian Statistics 3* (J. M. Bernardo, M. H. DeGroot, D. V. Lindley and A. F. M. Smith, eds.), Oxford: University Press, 631–640.

Johnson, W. and Geisser, S. (1982). Assessing the predective influence of observations. *Essays in Honor of C. R. Rao,* (Kallianpur, Krishnaiah and Ghosh, eds.), Amsterdam: North-Holland, 343–358.

Johnson, W. and Geisser, S. (1983). A predictive view of the detection and characterization of influential observations in regression analysis. *J. Amer. Statist. Assoc.* **78**, 137–144.

Lindley, D. V. (1990). The 1988 Wald memorial lectures: the present position in Bayesian statistics. *Statist. Sci.* **5**, 44–89, (with discussion).

Pettit, L. I. (1988). Bayes methods for outliers in exponential samples. *J. Roy. Statist. Soc. B* **50**, 371–380.

Pettit, L. I. (1990). The conditional predictive ordinate for the normal distribution. *J. Roy. Statist. Soc. B* **52**, 175–184.

Pettit, L. I. and Smith, A. F. M. (1985). Outliers and influential observations in linear models. *Bayesian Statistics 2* (J. M. Bernardo, M. H. DeGroot, D. V. Lindley and A. F. M. Smith, eds.), Amsterdam: North-Holland, 473–494, (with discussion).

Welsch, R. E. (1982). Influence functions and regression diagnostics. *Modern Data Analysis* (R. L. Launer and A. F. Siegel, eds.). New York: Academic Press, 149–169.

BAYESIAN STATISTICS 4, pp. 661–667
J. M. Bernardo, J. O. Berger, A. P. Dawid and A. F. M. Smith, (Eds.)

The Influence of Prior and Likelihood Tail Behaviour on the Posterior Distribution

M. A. GÓMEZ-VILLEGAS and P. MAÍN
Universidad Complutense de Madrid, Spain

SUMMARY

The tails of a continuous distribution may be classified as light, medium or heavy, depending on the limit of the hazard rate as x tends to infinity. We consider some likelihood and prior distributions with different right tails, and study the resultant posterior right tail. To this end we relate the concept and properties of regularly varying functions to the notion of tail weight.

Keywords: TAIL BEHAVIOUR; HAZARD RATE; BAYESIAN INFERENCE; REGULARLY VARYING FUNCTIONS.

1. INTRODUCTION

In an inference problem, it is very important to know the effect of thick tails. Many authors have studied the way that posterior distributions behave when the likelihood or the prior distribution are heavy-tailed. Theoretical results were given by Dawid (1973), complemented by Hill (1974), and later simplified by O'Hagan (1979).

Recently, O'Hagan (1990) has introduced the notion of *credence*, for inferences about a location parameter in the case of known variances. This may be used in certain situations to provide a measure of tail weight.

In this paper another way of assessing tail behaviour is considered and the effect on the posterior tail is studied. As pointed out in Maín (1988), the following classification of distributions according to their rigth tail behaviour may be introduced. Let X be a random variable with distribution function $F(\cdot)$ having infinite rigth endpoint and density function $f(\cdot)$, and suppose the limit

$$\lim_{x\to\infty} \frac{f(x)}{1-F(x)}$$

exists and is possibly infinite.

Definition 1.1.
(i) F is SR (outlier-resistant for the sum) if $\lim_{x\to\infty} \frac{f(x)}{1-F(x)} = \infty$.
(ii) F is SN (outlier-neutral for the sum) if $\lim_{x\to\infty} \frac{f(x)}{1-F(x)} = c$, *with* $0 < c < \infty$,
(iii) F is SP (outlier-prone for the sum) if $\lim_{x\to\infty} \frac{f(x)}{1-F(x)} = 0$.

Definition 1.2.
(i) F is RR (outlier-resistant for the ratio) if

$$\lim_{x\to\infty} \frac{xf(x)}{1-F(x)} = \infty.$$

(ii) F is RN (outlier-neutral for the ratio) if

$$\lim_{x\to\infty} \frac{xf(x)}{1-F(x)} = c, \qquad \text{with } 0 < c < \infty.$$

(iii) F is RP (outlier-prone for the ratio) if $\lim_{x\to\infty} \frac{xf(x)}{1-F(x)} = 0$.

We have the following Venn diagram:

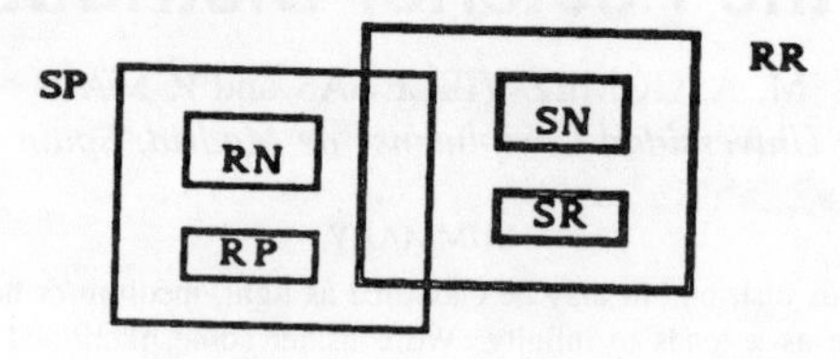

Figure 1. *Diagram of relations between the classes of distributions with different tail behaviour.*

These definitions are properly applied to location and scale families respectively, but for general distributions this classification can be summarized in the next definition.

Definition 1.3.
(i) F is a light-tailed distribution if

$$\lim_{x\to\infty} \frac{xf(x)}{1-F(x)} = \infty \quad \text{and} \quad \lim_{x\to\infty} \frac{f(x)}{1-F(x)} \neq 0;$$

that is, F is a RR distribution but not a SP distribution.
(ii) F is a medium-tailed distribution if

$$\lim_{x\to\infty} \frac{xf(x)}{1-F(x)} = \infty \quad \text{and} \quad \lim_{x\to\infty} \frac{f(x)}{1-F(x)} = 0;$$

that is, F is a RR and SP distribution.
(iii) F is a heavy-tailed distribution if

$$\lim_{x\to\infty} \frac{xf(x)}{1-F(x)} = c \ (0 < c < \infty) \quad \text{or} \quad \lim_{x\to\infty} \frac{xf(x)}{1-F(x)} = 0;$$

that is, F is a SP distribution but not a RR distribution.

Similarly, a classification of distributions depending on their left tail behaviour, for distributions with infinite left endpoint, can be established by considering

$$\lim_{x\to-\infty} \frac{f(x)}{F(x)} \quad \text{and} \quad \lim_{x\to-\infty} \frac{xf(x)}{F(x)}.$$

For distributions F with finite right endpoint x_F, we have to consider the tail behaviour of $G(y) = F(x_F - 1/y)$ at infinity. Then the definitions given for G are also applied to F. A similar analysis may be given for distributions with finite left endpoint.

Table 1 classifies some common distributions with regard to their tail behaviour.

Distribution	Left Tail	Right Tail
CAUCHY	Heavy	Heavy
PARETO	Heavy	Heavy
NORMAL	Light	Light
UNIFORM	Heavy	Heavy
EXPONENTIAL	Heavy	Light
LOGISTIC	Light	Light
LOGNORMAL	Medium	Medium
STUDENT-*t*	Heavy	Heavy

Table 1. *Tails of common distributions.*

2. RELATIONS BETWEEN PRIOR AND POSTERIOR TAIL BEHAVIOUR

Let X be a random variable with density function $f(x \mid \theta); \theta \in \Theta \subset \Re$, and let the prior distribution for Θ have density $\pi(\theta)$; then the posterior distribution given that $X = x$ has density

$$\pi(\theta \mid x) = \frac{\pi(\theta) f(x \mid \theta)}{p(x)}$$

where $p(x) = \int_\Theta (x \mid \theta)\pi(\theta) d\theta$ is the predictive density of X.

If right tail behaviour is considered, the prior and posterior limits are related by

$$\frac{\pi(\theta \mid x)}{1 - \Pi(\theta \mid x)} = \frac{\pi(\theta)}{1 - \Pi(\theta)} \frac{f(x \mid \theta)}{p_\theta^R(x)} \tag{2.1}$$

where $\Pi(\theta) = \int_{-\infty}^\theta \pi(u) du; \Pi(\theta \mid x) = \int_{-\infty}^\theta \pi(u \mid x) du$, are the corresponding distribution functions and $p_\theta^R(x) = \int_\Theta f(x \mid u) \pi_\theta^R(u) du$ is the predictive density corresponding to the right truncated prior

$$\pi_\theta^R(u) = \begin{cases} 0 & u < \theta \\ \frac{\pi(u)}{1 - \Pi(\theta)} & u \geq \theta \end{cases}$$

Proposition 2.1.

For a likelihood function with monotone nonincreasing right tail:

(i) if the posterior is heavy-tailed, then so is the prior.

(ii) if the prior is light-tailed, then so is the posterior.

Proof.

$$\frac{f(x \mid \theta)}{p_\theta^R(x)} = \frac{f(x \mid \theta)}{\int_\theta^\infty \frac{\pi(u)}{1-\Pi(\theta)} f(x \mid u) du}$$

but for θ large enough $f(x \mid \theta) \geq f(x \mid u) \quad u \geq \theta$ then

$$\frac{f(x \mid \theta)}{p_\theta^R(x)} \geq 1$$

and

$$\frac{\theta \pi(\theta \mid x)}{1 - \Pi(\theta \mid x)} \geq \frac{\theta \pi(\theta)}{1 - \Pi(\theta)}$$

Hence, both results follow on taking limits as θ tends to infinity. ◁

Proposition 2.2.

For a likelihood function with monotone nondecreasing right tail:
(i) if the posterior is light-tailed, then so is the prior.
(ii) if the prior is heavy-tailed, then so is the posterior.

Proof. The proof is similar to that of Proposition 2.1. ◁

Remark. These results confirm our intuitive considerations because a light-tailed distribution does not tend to throw up extreme observations, and this behaviour is enhanced when the likelihood decreases for larger values of the parameter. On the other hand, a heavy-tailed distribution is liable to show extreme observations, and with a likelihood increasing for larger values of the parameter this behaviour is strengthened. Left tail behaviour can be considered similarly.

3. INVARIANT TAIL BEHAVIOUR

In this section, some further results are needed. All of them are related to the notion of regularly varying functions. Karamata (1930), developed the theory of regular variation for positive functions of a positive argument. The fundamental role played by Karamata's theory in probability theory was already suggested by Gnedenko and Kolmogorov (1968), and later used by Feller (1971) in the theory of stable distributions and their domains of attraction.

Another important application is in extreme value theory, exhaustively developed by de Haan (1970); see also Seneta (1976). Roughly speaking, regularly varying functions are those functions which behave asymptotically like power functions multiplied by a factor which varies "more slowly" than a power function.

Definition 3.1. *A function R is said to be regularly varying at infinity if it is real-valued, positive and mesurable on (A, ∞), for some $A > 0$, and if for each $\lambda > 0$*

$$\lim_{x\to\infty} \frac{R(\lambda x)}{R(x)} = \lambda^{\rho}$$

for some real ρ which is called the exponent of regular variation.

Definition 3.2. *A function L which is regularly varying with exponent of regular variation $\rho = 0$ is called slowly varying. Moreover, a function R is regularly varying if and only if it can be written in the form*

$$R(x) = x^{\rho} L(x)$$

where $-\infty < \rho < \infty$ and L is slowly varying. Finally, if

$$\lim_{x\to\infty} \frac{R(\lambda x)}{R(x)} = \lambda^{-\infty} = \begin{cases} 0 & \lambda > 1 \\ 1 & \lambda = 1 \\ \infty & \lambda < \infty \end{cases}$$

then the function R is said to be rapidly varying with $\rho = -\infty$ (the case $\rho = \infty$ is handled similary).

Proposition 3.1.

(i) if F is a light-tailed distribution, then $1 - F$ is rapidly varying at infinity with $\rho = -\infty$.
(ii) if the density function f is nonincreasing, $1 - F$ is $-\infty$-varying and

$$\lim_{x\to\infty} \frac{f(x)}{1 - F(x)} \neq 0,$$

then F is light-tailed.

Proof. By Theorems 2.9.1. and 2.9.2. of de Haan (1970). ◁

Proposition 3.2

(i) If F is a medium-tailed distribution, then $1 - F$ is rapidly varying at infinity with $\rho = -\infty$.

(ii) If the density function f is nonincreasing, $1-F$ is $-\infty$-varying and $\lim_{x\to\infty} \frac{f(x)}{1-F(x)} = 0$, then F is medium-tailed.

Proof. The proof is similar to that of Proposition 3.1. ◁

Theorem 3.1.

If Z varies regularly with exponent γ and $Z_p^*(x) = \int_x^\infty y^p Z(y)dy$ exists, then

$$\frac{t^{p+1}Z(t)}{Z_p^*(t)} \to \lambda \qquad \text{as } t \to \infty \tag{4.1}$$

where $\lambda = -(p+\gamma+1) \geq 0$.

Conversely, if (4.1) holds with $\lambda > 0$, then Z and Z_p^* vary regularly with exponents $\gamma = -\lambda - p - 1$ and $-\lambda$, respectively. If (4.1) holds with $\lambda = 0$ then Z_p^* varies slowly (but nothing can be said about Z).

Proof. See Feller (1971). ◁

Proposition 3.3.

(i) If f is ρ-varying with $\rho \leq -1$, then F is a heavy-tailed distribution and $1 - F$ is $\rho + 1$-varying.

(ii) If $\lim_{x\to\infty} \frac{xf(x)}{1-F(x)} = c\,(0 < c < \infty)$, then f and $1 - F$ are regularly varying with exponents $\rho = -c - 1$ and $-c$ respectively.

(iii) If $\lim_{x\to\infty} \frac{xf(x)}{1-F(x)} = 0$, then $1 - F$ is slowly varying, but nothing can be said about f.

Proof. This is a consequence of Theorem 3.1. ◁

Now, let $L(\theta)$ be the standardized likelihood given an observation x: that is

$$L(\theta) = \frac{f(x\,|\,\theta)}{\int_\Theta f(x\,|\,\theta)d\theta} \qquad \theta \in \Theta;$$

and let the corresponding distribution function be $\mathcal{L}(\theta) = \int_{-\infty}^\theta L(u)du$.

In this context, tails are invariant in two general situations.

A. *$\mathcal{L}$ Heavy-Tailed, Π Medium or Light-Tailed*

Theorem 3.2. *Let*

$$\lim_{\theta\to\infty} \frac{\theta L(\theta)}{\int_\theta^\infty L(u)du} = c, \quad 0 < c < \infty;$$

then $1 - \mathcal{L}$ is $-c$-varying and L is $(-c - 1)$-varying. If, besides,

$$\lim_{\theta\to\infty} \frac{\theta\pi(\theta)}{1 - \Pi(\theta)} = \infty,$$

then $1 - \Pi$ is $-\infty$-varying and

$$\frac{\theta\pi(\theta\,|\,x)}{1 - \Pi(\theta\,|\,x)} = \frac{\theta\pi(\theta)}{1 - \Pi(\theta)} \frac{L(\theta)}{\int_\theta^\infty L(u)\frac{\pi(u)}{1-\Pi(\theta)}du};$$

the prior and posterior tail behaviour are the same.

Proof. See Appendix.

B. Π *Heavy-Tailed,* $\mathcal{L}$ *Medium or Light-Tailed*
In this case, similary to (2.1), we have

$$\frac{\theta\pi(\theta\,|\,x)}{1-\Pi(\theta\,|\,x)} = \frac{\theta L(\theta)}{1-\mathcal{L}(\theta)}\frac{\pi(\theta)}{\int_\theta^\infty \pi(u)\frac{L(u)}{1-\mathcal{L}(\theta)}du}$$

and similary to Theorem 3.2.

$$\lim_{\theta\to\infty}\frac{\pi(\theta)}{\int_\theta^\infty \pi(u)\frac{L(u)}{1-\mathcal{L}(\theta)}du} = 1,$$

so that the prior and likelihood tail behaviours are the same.

In conclusion, in a Bayesian problem with one component having much heavier tails, the posterior distribution has the same tail behaviour as the other component. As concluded by O'Hagan (1990), a heavy-tailed distribution is a less important information source than one with lighter tails.

APPENDIX

Proof of Theorem 3.2
Equation (2.1) can be written

$$\lim_{\theta\to\infty}\frac{\theta\pi(\theta\,|\,x)}{1-\Pi(\theta\,|\,x)} = \lim_{\theta\to\infty}\frac{\theta\pi(\theta)}{1-\Pi(\theta)}\lim_{\theta\to\infty}\frac{L(\theta)}{\int_\theta^\infty L(u)\pi_\theta^R(u)du}$$

and the comparisons of the prior and posterior tails depend on the last limit.

But L is $(-c-1)$-varying, then $\forall\varepsilon>0, \exists\theta_0$ such that $\forall\theta\geq\theta_0, \forall k>1$

$$(1-\varepsilon)k^{-c-1-\varepsilon} < \frac{L(\theta K)}{L(\theta)} < (1+\varepsilon)k^{-c-1+\varepsilon};$$

then

$$\int_1^\infty (1-\varepsilon)k^{-c-1-\varepsilon}\frac{\theta\pi(k\theta)}{1-\Pi(\theta)}dk < \int_1^\infty \frac{L(\theta k)}{L(\theta)}\frac{\theta\pi(k\theta)}{1-\Pi(\theta)}dk < \int_1^\infty (1+\varepsilon)k^{-c-1+\varepsilon}\frac{\theta\pi(k\theta)}{1-\Pi(\theta)}dk$$

and the last integral is to be calculated. But

$$I_1(\theta) = \int_1^\infty k^{-c-1+\varepsilon}\frac{\theta\pi(k\theta)}{1-\Pi(\theta)}dk = \int_\theta^\infty \left(\frac{u}{\theta}\right)^{-c-1+\varepsilon}\frac{\theta(u)}{1-\Pi(\theta)}du =$$

$$= 1 + (-c-1+\varepsilon)\int_\theta^\infty \frac{u^{(-c-1+\varepsilon)-1}}{\theta^{(-c-1+\varepsilon)-1}}\frac{1-\Pi(u)}{1-\Pi(\theta)}du,$$

and using Theorem 1.3.1. (a) from de Haan (1970), we get

$$\lim_{\theta\to\infty}\frac{\theta^{-c-1+\varepsilon}(1-\Pi(\theta))}{\int_\theta^\infty u^{(-c-1+\varepsilon)-1}(1-\Pi(u))du} = \infty,$$

because $1 - \Pi$ is $-\infty$-varying and nonincreasing. Therefore,

$$\lim_{\theta \to \infty} I_1(\theta) = 1.$$

Analogously, if we define

$$I_2(\theta) = \int_1^\infty k^{-c-1-\varepsilon} \frac{\theta \pi(k\theta)}{1 - \Pi(\theta)} dk$$

it can be proved that

$$\lim_{\theta \to \infty} I_2(\theta) = 1.$$

Summing up, $\forall \varepsilon > 0, \exists \theta_0$ such that $\forall k > 1$ and $\forall \theta \geq \theta_0$,

$$(1 - \varepsilon) I_2(\theta) < \int_1^\infty \frac{L(\theta k)}{L(\theta)} \frac{\theta \pi(k\theta)}{1 - \Pi(\theta)} dk < (1 + \varepsilon) I_1(\theta);$$

hence

$$\lim_{\theta \to \infty} \int_1^\infty \frac{L(\theta k)}{L(\theta)} \frac{\theta \pi(k\theta)}{1 - \Pi(\theta)} dk = \lim_{\theta \to \infty} \int_1^\infty \frac{L(u)}{L(\theta)} \frac{\pi(u)}{1 - \Pi(\theta)} du = 1$$

and

$$\lim_{\theta \to \infty} \frac{\theta \pi(\theta \mid x)}{1 - \Pi(\theta \mid x)} = \lim_{\theta \to \infty} \frac{\theta \pi(\theta)}{1 - \Pi(\theta)}$$

as asserted. ◁

REFERENCES

Dawid, A. P. (1973). Posterior expectations for large observations. *Biometrika* **60**, 664–667.

Feller, W. (1971). *An Introduction to Probability Theory and Its Applications* **2**, New York: Wiley.

de Haan, L. (1970). *On Regular Variation and Its Application to the Weak Convergence of Sample Extremes.* Amsterdam: Mathematical Centre Tracts.

Hill, B. M. (1974). On coherence, inadmissibility and inference about many parameters in the theory of least squares. *Studies in Bayesian Econometrics and Statistics* (S. E. Fienberg and A. Zellner eds.), Amsterdam: North-Holland, 245–274.

Gnedenko and Kolmogorov (1968). *Limit Distributions for Sums of Independent Random Variables.* Reading, MA: Addison-Wesley.

Karamata, J. (1930). Sur une mode de croissance régulière des fonctions. *Mathematica (Cluj)* **4**, 38–53.

Maín, P. (1988). Prior and posterior tail comparisons. *Bayesian Statistics 3* (J. M. Bernardo, M. H. DeGroot, D. V. Lindley and A. F. M. Smith, eds.), Oxford: University Press, 669–675.

O'Hagan, A. (1979). On outlier rejection phenomena in Bayes inference. *J. Roy. Statist. Soc. B* **41**, 358–367.

O'Hagan, A. (1990). Outliers and credence for location parameter inference. *J. Amer. Statist. Assoc.* **409**, 172–176.

Seneta, E. (1976). *Regularly Varying Functions.* New York: Springer.

BAYESIAN STATISTICS 4, pp. 669–674
J. M. Bernardo, J. O. Berger, A. P. Dawid and A. F. M. Smith, (Eds.)

Expected Logarithmic Divergence for Exponential Families

EDUARDO GUTIERREZ-PEÑA
*IIMAS, Universidad Nacional Autónoma de México, México
and Imperial College London, UK*

SUMMARY

Kullback-Leibler logarithmic divergence (Kullback-Leibler information) may be used within a Bayesian framework to produce a class of utility functions in statistical decision problems concerning parametric families of distributions. In such a case it is necessary to compute the expected value of the logarithmic divergence (ELD). We show that the ELD takes a simple form when the distribution of the sampling process belongs to a regular exponential family and the natural parameter has a distribution in the corresponding conjugate family. An application is outlined.

Keywords: KULLBACK-LEIBLER INFORMATION; EXPONENTIAL FAMILIES; ESTIMATION.

1. INTRODUCTION

Let X be a random variable with probability function $p(x|\theta)$ and suppose it is desired to discriminate between the hypotheses $H_0 : \theta = \theta_0$ and $H_1 : \theta = \theta_1$. The logarithmic divergence of $p(x|\theta_1)$ with respect to $p(x|\theta_0)$ is defined by

$$\delta(\theta_0 : \theta_1) = \delta\{p(x|\theta_0) : p(x|\theta_1)\} = \int p(x|\theta_0) \ \ln \frac{p(x|\theta_0)}{p(x|\theta_1)} dx,$$

and represents the mean information for discrimination in favour of H_0 against H_1 when H_0 is true (Kullback, 1959). The logarithmic divergence measures the discrepancy between two distributions: the greater $\delta(\theta_0 : \theta_1)$, the greater the discrepancy between $p(x|\theta_0)$ and $p(x|\theta_1)$.

Bernardo (1985) has proposed a Bayesian procedure for testing $H_0 : \theta = \theta_0$ vs. $H_1 : \theta \neq \theta_0$, within a decision theoretical framework, using this notion of discrepancy between distributions. Such a procedure is based on the following specification of the utility function

$$u(d_1, \theta) - u(d_0, \theta) = A\delta(\theta : \theta_0) + B \quad A \in \Re^+, B \in \Re$$

where d_i stands for "accept H_i" $(i = 0, 1)$. Maximizing the posterior expected utility leads to rejecting H_0 if and only if

$$E_{\theta|x}\{\delta(\theta : \theta_0)\} > -\frac{B}{A}.$$

The logarithmic divergence may also be used as a loss function in the point estimation problem, viewed as a statistical decision problem, yielding interesting results. In view of this, it would be useful to identify the form of the expected logarithmic divergence (ELD) under certain general conditions. It will be shown that the ELD takes a simple form when the distribution of the sampling process belongs to a regular exponential family and the natural parameter has a distribution in the corresponding conjugate family. In the next section the work of Diaconis and Ylvisaker (1979) on conjugate priors for exponential families is reviewed and the form of the expected logarithmic divergence is exhibited. In Section 3, the point estimation problem using the logarithmic divergence as loss function is treated in some detail. Finally, Section 4 contains some concluding remarks.

2. EXPECTED LOGARITHMIC DIVERGENCE

Let X be a random variable with probability function of the form

$$p(x|\theta) = a(\theta)b(x)\exp\{\theta' t(x)\} \tag{1}$$

where $\theta' = (\theta_1, \theta_2, \ldots, \theta_d)$, $t(x) = (t_1(x), t_2(x), \ldots, t_d(x))'$ and $\theta' t(x) = \sum_{i=1}^{d} \theta_i t_i(x)$. The probability function $p(x|\theta)$ is a density with respect to some σ-finite measure μ defined on the Borel sets of $\Re$ (typically μ is either a counting measure or Lebesgue measure). Let $\Theta = \{\theta \in \Re^d : M(\theta) < +\infty\}$ where

$$M(\theta) = \ln\left\{\int b(x)\ \exp\{\theta' t(x)\} d\mu(x)\right\} = -\ln\ a(\theta). \tag{2}$$

It can be shown that Θ is a convex subset of $\Re^d$. The family $\mathcal{F}$ of distributions with probability function of the form (1) and $\theta \in \Theta$ is called a *regular exponential family* if Θ is a nonempty open set in $\Re^d$ (*e.g.* Barndorff-Nielsen, 1978). Let $\mathcal{C}$ denote the family of the proper distributions on the Borel sets of Θ whose density with respect to the Lebesgue measure is of the form

$$p(\theta) = p(\theta; n_0, t_0) = H(n_0, t_0) a^{n_0}(\theta)\ \exp\{\theta' t_0\} \quad n_0 \in \Re^+, t_0 \in \Re^d \tag{3}$$

where

$$H(n_0, t_0) = \left\{\int a^{n_0}(\theta)\ \exp\{\theta' t_0\} d\theta\right\}^{-1},$$

provided that the integral exists. The family $\mathcal{C}$ is called a conjugate family of the exponential family $\mathcal{F}$. Clearly, $\mathcal{C}$ is also an exponential family and is closed under sampling.

Let

$$\nabla M(\theta) = \left[\frac{\partial M(\theta)}{\partial \theta_1}, \ldots, \frac{\partial M(\theta)}{\partial \theta_d}\right]' = \left[M_1(\theta), \ldots, M_d(\theta)\right]'.$$

It is readily seen that

$$E_{x|\theta}\{t(X)\} = \nabla M(\theta). \tag{4}$$

Theorem 1. *(Diaconis and Ylvisaker, 1979). Suppose* Θ *is open in* $\Re^d$*. If* θ *has a distribution* $p(\theta; n_0, t_0)$ *in* $\mathcal{C}$ *and* $n_0 > 0$*, then*

$$E_\theta\{\nabla M(\theta)\} = \frac{t_0}{n_0}. \tag{5}$$

Let $W(n,t) = -\ln\ H(n,t)$ and define

$$\nabla W(n,t) = \left[\frac{\partial W(n,t)}{\partial t_1}, \ldots, \frac{\partial W(n,t)}{\partial t_d}\right]' = \left[W_1(n,t), \ldots, W_d(n,t)\right]',$$

and

$$W_0(n,t) = \frac{\partial W(n,t)}{\partial n}.$$

We can now state our result.

Theorem 2. *Under the conditions of Theorem 1,*

$$E_\theta\{\delta(\theta:\theta_0)\} = W_0(n_0,t_0) + M(\theta_0) + \frac{1}{n_0}\Big\{d + \Big[\nabla W(n_0,t_0) - \theta_0\Big]' t_0\Big\} \tag{6}$$

Proof. For a distribution in $\mathcal{F}$, the logarithmic divergence is found to be

$$\begin{aligned}\delta(\theta:\theta_0) &= \int p(x|\theta)\ \ln\ \frac{p(x|\theta)}{p(x|\theta_0)}dx\\ &= M(\theta_0) - M(\theta) + (\theta-\theta_0)' E_{x|\theta}\{t(X)\}\\ &= M(\theta_0) - M(\theta) + (\theta-\theta_0)'\nabla M(\theta).\end{aligned}$$

Thus

$$E_\theta\{\delta(\theta:\theta_0)\} = M(\theta_0) - E_\theta\{M(\theta)\} + E_\theta\{\theta'\nabla M(\theta)\} - \theta_0' E_\theta\{\nabla M(\theta)\}.$$

Differentiation of the identity

$$\ln \int a^n(\theta)\ \exp\{\theta' t\} d\theta = W(n,t)$$

in n yields

$$\int \ln\ a(\theta) H(n,t) a^n(\theta)\ \exp\{\theta' t\} d\theta = W_0(n,t).$$

Hence

$$E_\theta\{M(\theta)\} = -W_0(n_0,t_0). \tag{7}$$

Now, from equation (5), we can write for each $i = 1,2,\ldots,d$

$$\ln \int M_i(\theta) a^n(\theta)\ \exp\{\theta' t\} d\theta = \ln\ t_i - \ln\ H(n,t) - \ln\ n.$$

If this identity is differentiated in t_i, and differentiation and integration are interchanged, it follows that

$$\int \theta_i M_i(\theta) H(n,t) a^n(\theta)\ \exp\{\theta' t\} d\theta = \frac{1}{n}\Big[1 + W_i(n,t) t_i\Big]$$

$(i = 1,2,\ldots,d)$. Therefore

$$E_\theta\{\theta'\nabla M(\theta)\} = \frac{1}{n_0}\Big\{d + \nabla W(n_0,t_0)' t_0\Big\}. \tag{8}$$

Finally, using (5)

$$E_\theta\{\delta(\theta:\theta_0)\} = W_0(n_0,t_0) + M(\theta_0) + \frac{1}{n_0}\Big\{d + \nabla W(n_0,t_0)' t_0\Big\} - \frac{1}{n_0}\theta_0' t_0$$

and the result follows. ◁

3. ESTIMATION

The problem of point estimation can be viewed as a statistical decision problem in which the decision space coincides with the parameter space.

Let $Z = (X_1, X_2, \ldots, X_n)$ denote a random sample of a random variable X with probability function (1) and suppose it is desired to estimate the value of θ. If the prior distribution of θ belongs to the conjugate family C, then the posterior distribution is seen to be

$$p(\theta|z) = p(\theta; n_1, t_1) = H(n_1, t_1)a^{n_1}(\theta)\ \exp\{\theta' t_1\}$$

where $n_1 = n_0 + n$ and $t_1 = t_0 + \sum_{i=1}^{n} t(x_i)$.

Consider the loss function

$$L(\hat{\theta}, \theta) = \delta(\theta : \hat{\theta}),$$

which reflects the discrepancy between $p(x|\theta)$ and $p(x \mid \hat{\theta})$. Denote by $\Delta_\theta(\hat{\theta})$ the posterior expected logarithmic divergence. Then by theorem 2

$$\Delta_\theta(\hat{\theta}) = E_{\theta|z}\{\delta(\theta : \hat{\theta})\} = W_0(n_1, t_1) + M(\hat{\theta}) + \frac{1}{n_1}\Big\{d + (\nabla W(n_1, t_1) - \hat{\theta})' t_1\Big\}.$$

It is readily seen that the value $\hat{\theta}^*$ of $\hat{\theta}$ minimizing $\Delta_\theta(\hat{\theta})$ is found by solving the equation

$$\nabla M(\hat{\theta}) = \frac{t_1}{n_1}. \tag{9}$$

Compare this with the maximum-likelihood estimate $\hat{\theta}_{ML}$, obtained by solving the equation

$$\nabla M(\hat{\theta}) = \frac{1}{n}\sum_{i=1}^{n} t(x_i).$$

It is straightforward to show that the mode of the posterior distribution of θ can also be computed by solving equation (9). Thus $\hat{\theta}^*$ and the mode of $p(\theta|z)$ coincide. The estimator $\hat{\theta}^*$ is also called the *generalized maximum-likelihood estimator* of θ (DeGroot, 1970; Section 11.4).

Now suppose we are not interested in estimating the value of θ, but rather we are concerned with the estimation of a transformation of θ. Furthermore, suppose that $\varphi = g(\theta)$ is a one-to-one differentiable transformation of θ such that g^{-1} is also differentiable. The posterior distribution of φ is found to be

$$p_1(\varphi|z) = p(g^{-1}(\varphi)|z)J(\varphi),$$

where $J(\varphi) = |\frac{\partial g^{-1}(\varphi)}{\partial \varphi}|$ denotes the Jacobian of the transformation.

Let $\delta_1(\varphi : \hat{\varphi})$ denote the logarithmic divergence between $p(x|\varphi)$ and $p(x|\hat{\varphi})$ (the distribution of X being parameterized in terms of φ). Then

$$\begin{aligned}
\Delta_\varphi(\hat{\varphi}) = E_{\varphi|z}\{\delta_1(\varphi : \hat{\varphi})\} &= \int \delta_1(\varphi : \hat{\varphi})\ p_1(\varphi|z)d\varphi \\
&= \int \delta\{g^{-1}(\varphi) : g^{-1}(\hat{\varphi})\}\ p(g^{-1}(\varphi)|z)J(\varphi)d\varphi \\
&= \int \delta(\theta : g^{-1}(\hat{\varphi}))\ p(\theta|z)d\theta \\
&= \Delta_\theta(g^{-1}(\hat{\varphi})).
\end{aligned}$$

It follows that $\Delta_\theta(\hat{\theta}) = \Delta_\varphi(g(\hat{\theta}))$. Actually $\hat{\theta}^*$ is such that $\Delta_\theta(\hat{\theta}^*) \leq \Delta_\theta(\hat{\theta})$ for all $\hat{\theta}$, so that

$$\Delta_\varphi(g(\hat{\theta}^*)) \leq \Delta_\varphi(g(\hat{\theta})) \text{ for all } \hat{\theta}$$

which in turn implies

$$\Delta_\varphi(g(\hat{\theta}^*)) \leq \Delta_\varphi(\hat{\varphi}) \text{ for all } \hat{\varphi}.$$

Hence $\hat{\theta}^*$ is invariant under one-to-one transformations, since $\hat{\varphi}^* = g(\hat{\theta}^*)$ minimizes $\Delta_\varphi(\hat{\varphi})$. The following theorem has been proved.

Theorem 3. *Let $p(x|\theta)$ belong to the regular exponential family $\mathcal{F}$ and suppose the prior distribution of θ belongs to the conjugate family C. If $\hat{\theta}^*$ denotes the value of $\hat{\theta}$ minimizing the posterior expected logarithmic divergence $E_{\theta|z}\{\delta(\theta : \hat{\theta})\}$ then*
(i) $\hat{\theta}^*$ is the mode of the posterior distribution of θ, and
(ii) $\hat{\theta}^*$, as an estimator of θ, is invariant under one-to-one transformations.

Note that minimization of $E_{\theta|z}\{\delta(\theta : \hat{\theta})\}$ in $\hat{\theta}$ is tantamount to maximization of

$$E_{\theta|z}\left[\int p(x|\theta) \ln p(x|\hat{\theta})dx\right].$$

On the other hand, if the logarithmic divergence of $p(x|\hat{\theta})$ with respect to $p(x|z)$ is considered, where $p(x|z)$ denotes the posterior predictive distribution of X, then

$$\begin{aligned}\delta\{p(x|z) : p(x|\hat{\theta})\} &= \int p(x|z) \ln \frac{p(x|z)}{p(x \mid \hat{\theta})}dx \\ &= \int p(x|z) \ln p(x \mid z)dx - \int p(x|z) \ln p(x|\hat{\theta})dx.\end{aligned}$$

Thus minimizing $\delta\{p(x|z) : p(x|\hat{\theta})\}$ in $\hat{\theta}$ is equivalent to maximizing

$$\int p(x|z) \ln p(x|\hat{\theta})dx,$$

but

$$\begin{aligned}E_{\theta|z}\left[\int p(x|\theta) \ln p(x|\hat{\theta})dx\right] &= \int p(\theta|z)\left\{\int p(x|\theta) \ln p(x|\hat{\theta})dx\right\}d\theta \\ &= \int \ln p(x|\hat{\theta})\left\{\int p(x|\theta)p(\theta|z)d\theta\right\}dx \\ &= \int p(x|z) \ln p(x \mid \hat{\theta})dx\end{aligned}$$

Therefore $p(x|\hat{\theta}^*)$ can be considered as the best approximation, among all models in $\mathcal{F}$, to the posterior predictive distribution $p(x|z)$ (note that, in general, $p(x|z)$ is not in $\mathcal{F}$).

Finally, consider the following approximation to the posterior distribution $p(\theta|z)$, for large n (*e.g.* Johnson, 1967)

$$p(\theta|z) \approx N(\hat{\theta}^*, \{n_1 I(\hat{\theta}^*)\}^{-1}),$$

where $I(\theta) = \frac{\partial^2 M(\theta)}{\partial\theta' \partial\theta}$ is the Fisher information for the parameter θ in the model (1). This approximation takes into account the prior distribution of θ, in contrast with the well-known normal approximation based on the maximum-likelihood estimate

$$p(\theta|z) \approx N(\hat{\theta}_{ML}, \{nI(\hat{\theta}_{ML})\}^{-1}).$$

4. CONCLUDING REMARKS

The ELD was shown to have a rather tractable form for exponential families, provided the distribution of θ belongs to the corresponding conjugate family. Minimization of the ELD resembles maximization of the likelihood function performed in the classical maximum-likelihood theory. As an instance, the estimate found by minimizing the ELD was shown to coincide with the corresponding *generalized maximum-likelihood estimate*.

ACKNOWLEDGEMENTS

I am indebted to Raúl Rueda, Manuel Mendoza and the referees for many helpful suggestions.

REFERENCES

Barndorff-Nielsen, O. (1978). *Information and Exponential Families in Statistical Theory.* New York: Wiley.

Bernardo, J. M. (1985). Análisis Bayesiano de los contrastes de hipótesis paramétricos. *Trab. Estadist.* **36**, 45–54.

DeGroot, M. H. (1970). *Optimal Statistical Decisions.* New York: McGraw-Hill.

Diaconis, P. and Ylvisaker, D. (1979). Conjugate priors for exponential families. *Ann. Statist.* **7**, 269–281.

Johnson, R. A. (1967). An asymptotic expansion for posterior distributions. *Ann. Math. Statist.* **38**, 1899–1906.

Kullback, S. (1959). *Information Theory and Statistics.* New York: Wiley.

BAYESIAN STATISTICS 4, pp. 675–688
J. M. Bernardo, J. O. Berger, A. P. Dawid and A. F. M. Smith, (Eds.)

Bayesian Models for Quality Assurance

TELBA Z. IRONY, CARLOS A. de B. PEREIRA and RICHARD E. BARLOW
The George Washington University, USA, Universidade de São Paulo, Brazil
and *University of California at Berkeley, USA*

SUMMARY

Two models for analyzing discrete time series of quality data are suggested. They enable analysts to extract information from data and from experts in a very effective way. The Additive model should be used for processes that degrade as time goes by (processes that age, for instance). The Multiplicative model fits processes that improve with time (processes that depend on learning). In the case that the process varies, sometimes getting better and sometimes getting worse, both models should be alternated according to the process behavior. Simulation is used to validate the proposed models and to evaluate their performances.

Keywords: QUALITY ASSURANCE; QUALITY AUDITS; PRODUCTION PROCESS; QMP – QUALITY MEASUREMENT PLAN; INFLUENCE DIAGRAM; STATISTICAL CONTROL; EXCHANGEABILITY; GAMMA DISTRIBUTION; BETA DISTRIBUTION; POISSON DISTRIBUTION.

1. INTRODUCTION

Quality audits are performed by inspectors in production processes in order to report product quality to management. A quality audit is a structured system of inspections done continuously on a sampling basis. Sampled product is inspected and defects are assessed whenever the product fails to meet engineering requirements. The results are combined into a rating period and compared to a quality standard which is a target value of defects per unit reflecting a trade-off between manufacturing cost, operating costs and customer need.

The models presented in this paper aim to replace the Quality Measurement Plan (QMP), which was implemented throughout AT&T Technologies in 1980 (see Hoadley (1981)). The QMP is a statistical method for analyzing discrete time series of quality audit data consisting of the expected number of defects given standard quality. The method is heuristic because the model's exact solution is mathematically intractable.

Two alternative models, namely the Additive model and the Multiplicative model, are presented in this paper. They are exact, tractable and much easier to use than the QMP.

The Additive model was formulated in order to deal with production processes that degrade as time goes by (processes that age, for instance). The Multiplicative model is appropriate for processes that improve with time (e.g. processes that depend on learning). In cases where the process varies, sometimes getting better and sometimes getting worse (behavior of an out-of-control process) both models should be alternated. This procedure is straightforward due to the fact that the posterior distributions in both models are of the same nature.

The models enable the analyst to extract information from the data and from the experts that are involved in the problem in a very effective way. The expert knowledge is intrinsically incorporated into the models. The information extracted from the available data updates the expert opinion via Bayes' theorem. Influence diagrams were very helpful to summarize the dynamics of the modeling process.

Simulation has been used to validate the proposed models and to evaluate their performance.

2. NOTATION AND ASSUMPTIONS

Suppose there are T rating periods: $t = 1, \ldots, T$ (current period). For period t, we have the following data from the audit:

$n_t =$ audit sample size; $x_t =$ number of defects in the audit sample;

$s =$ standard number of defects per unit.

$e_t = sn_t =$ expected number of defects in the sample when the quality standard is met.

The assumptions are the following: x_t has a Poisson distribution with mean $n_t\lambda_t$,where λ_t is the defect rate per unit. If λ_t is reparametrized on a quality index scale, the result is:

$\theta_t = \lambda_t/s =$ quality index at rating period t. In other words, $\theta_t = 1$ is the standard value. Therefore, we can write: $(x_t|\theta_t) \sim \text{Poi}(e_t\theta_t)$.

The parameter of interest is θ_T, the current quality index. The objective is to derive the posterior distribution of θ_T given the past data, d_{T-1}, and current data, x_T.

Here $d_{T-1} = (x_1, \ldots, x_{T-1})$ and d_0 is a constant.

The standard quality on the quality index scale is "one". "Two" means twice as many defects as expected under the standard. Hence, the larger the quality index, the worse the process.

3. THE ADDITIVE MODEL

3.1. *Assumptions*

This model is adequate for processes that degrade with time. It starts with a quality index θ, which may be thought of as the quality index for previous ratings. At each rating period t, an increment δ_t, also unknown, is added so that the quality index at rating period $t = 1$ will be $\theta_1 = \theta + \delta_1$, the quality index at rating period $t = 2$ will be $\theta_2 = \theta + \delta_1 + \delta_2$ and at $t = T$, the quality index will be $\theta_T = \theta + \delta_1 + \ldots + \delta_T$.

Here we are assuming that e_t is constant for all periods t (and we will call it e) but the model may be easily extended for the case in which e_t varies from period to period.

This model does not require the assumption of exchangeability between lots, allowing changes in the quality index from period to period. The following influence diagram represents the Additive model:

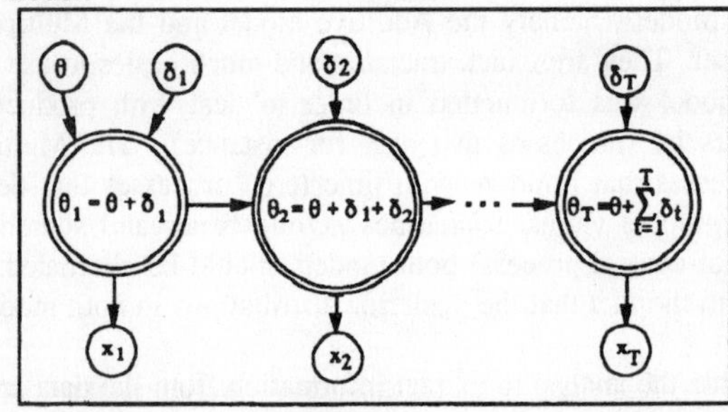

Figure 1.

Since usually many factors affect quality, there is a central limit theorem effect. Therefore unimodal prior distributions for θ and for the δ's are reasonable choices. A convenient assessment is a gamma distribution. Hence:

$\theta|\alpha_0, \beta_0 \sim \text{Gamma}(\alpha_0, \beta_0)$

$\delta_t|\alpha_t, \beta_t \sim \text{Gamma}(\alpha_t, \beta_t)$ where $\beta_t = \beta_0 + (t-1)e$ for $t = 1, \ldots, T$.

The choice of α_0 and β_0 is completely free, and will reflect the analyst's experience about the initial quality index θ. The assessment of prior distributions for the δ's is more delicate. The choice of α_t is also free but β_t is determined by β_0 and by the period in which the rating is being made. This feature makes the model work nicely. In order to fit a gamma distribution with any mean she pleases, the analyst has to pick the right α_t. Consequently, there will be a trade-off between the mean and the variance of the δ's. Usually the choice of the mean is more meaningful since the prior mean of δ_t expresses the average amount by which the analyst judges that the quality index has increased from period $t-1$ to period t. On the other hand, it is reasonable that the variance of δ_t becomes smaller and smaller as t increases and the analyst becomes more acquainted with the production process.

If it is judged that the means of the δ's are about the same for all rating periods, the variances will be decreasing. Whenever the variance of δ_t becomes too small, the assessment procedure must start all over again. In other words, t must be reset to 1 and new gamma distributions should be assessed for θ and δ_1 . These distributions should incorporate all the knowledge the analyst has gathered up to the current rating period. Then, α_0, β_0 and α_1 will be chosen freely and suitable means and variances for θ_1 will be assessed.

At any rating period T, the quality index is given by $\theta_T = \theta + \sum_{t=1}^{T} \delta_t$. It will be shown in the sequel that the posterior distribution for θ_T will be gamma with shape parameter $\sum_{t=0}^{T} \alpha_t + \sum_{t=1}^{T} x_t$ and scale parameter $(\beta_0 + Te)^{-1}$.

The influence diagram that represents this model has to be constructed as time goes by due to the dynamic nature of the model. The system may change at each rating period and three new nodes will be added to the influence diagram.

3.2. *The Solution*

Let us start with rating period $t = 1$. Based on previous knowledge about the production process, prior distributions for both, θ and δ_1 should be assessed. The corresponding influence diagram is:

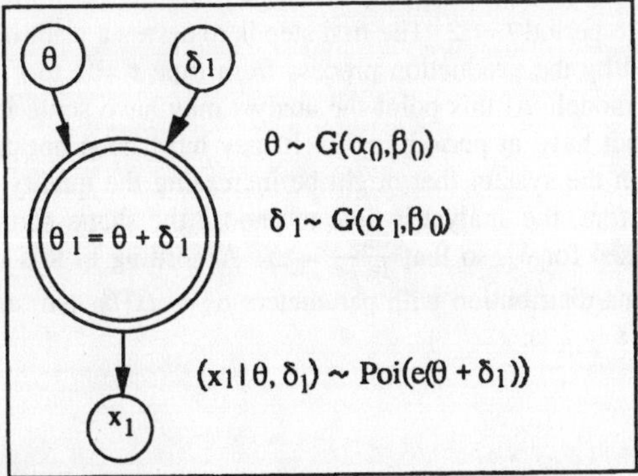

Figure 2.

Let $\theta \sim G(\alpha_0, \beta_0), \delta_1 \sim G(\alpha_1, \beta_0)$ and $\theta \amalg \delta_1 | \alpha_0, \alpha_1, \beta_0$[1].

Then $\theta_1 = (\theta + \delta_1) \sim G(\alpha_0 + \alpha_1, \beta_0)$.

Consequently:

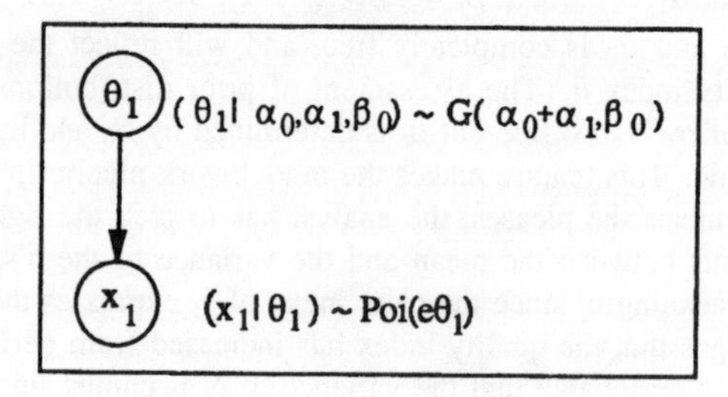

Figure 3.

Our objective is to compute the posterior distribution of θ_1, the current quality index at rating period $t = 1$, given the number of defects found in that period, x_1. This corresponds to an arc reversal operation in the influence diagram framework.

The updated influence diagram is:

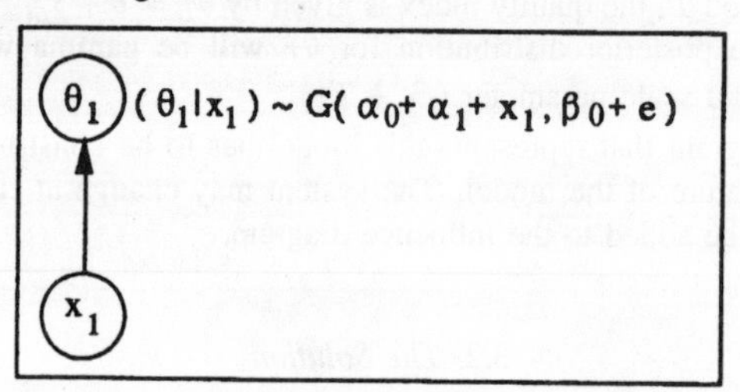

Figure 4.

Now we proceed to period $t = 2$. The first step is to assess a prior distribution for δ_2, the increment experienced by the production process from time $t = 1$ to $t = 2$, and here comes the advantage of this model. At this point the analyst may have some information about the process that she did not have at period $t = 1$. It may have been noticed, for instance, that there is a small flaw in the system that might be increasing the quality index by an amount Δ, on average. Therefore, the analyst is able to choose the shape parameter of the gamma distribution, α_2, assessed for δ_2, so that $\frac{\alpha_2}{\beta_0+e} = \Delta$. According to this reasoning, δ_2 will be assessed using a gamma distribution with parameters $\alpha_2 = \Delta(\beta_0 + e)$ and $\beta_2 = \beta_0 + e$. The influence diagram for $t = 2$ is:

(1) The symbol $\amalg$ means "is independent of". In this case, θ is conditionally independent of δ_1 given $\alpha_0, \alpha_1, \beta_0$.

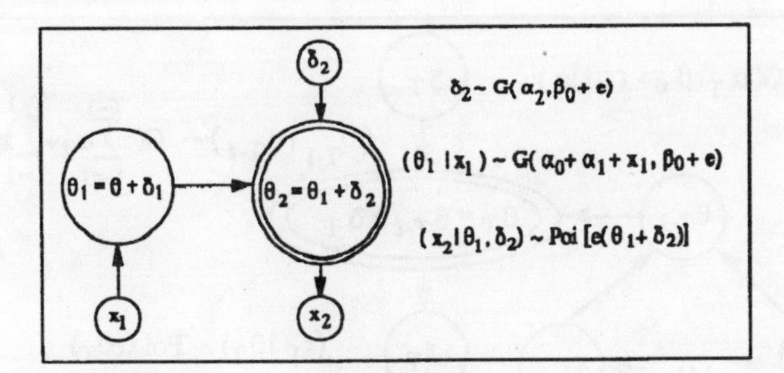

Figure 5.

By the same argument used previously, $\theta_2 = \theta_1 + \delta_2 \sim G\left(\sum_{t=0}^{2} \alpha_t + x_1, \beta_0 + e\right)$. Then:

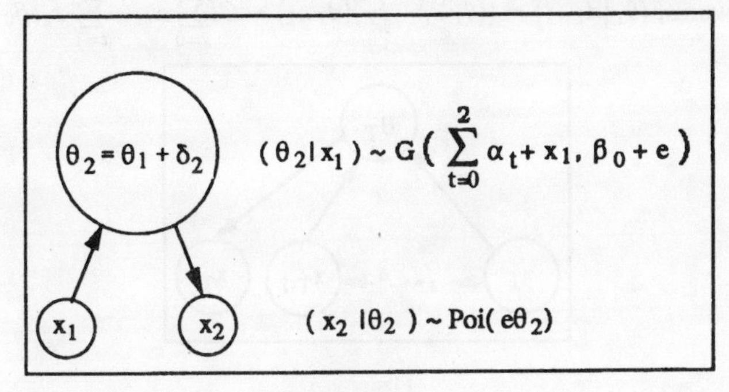

Figure 6.

The posterior distribution of θ_2 given x_1 and x_2 will be a gamma distribution with parameters $\left(\sum_{t=0}^{2} \alpha_t + \sum_{t=1}^{2} x_t\right)$ and $(\beta_0 + 2e)$. The resulting influence diagram is:

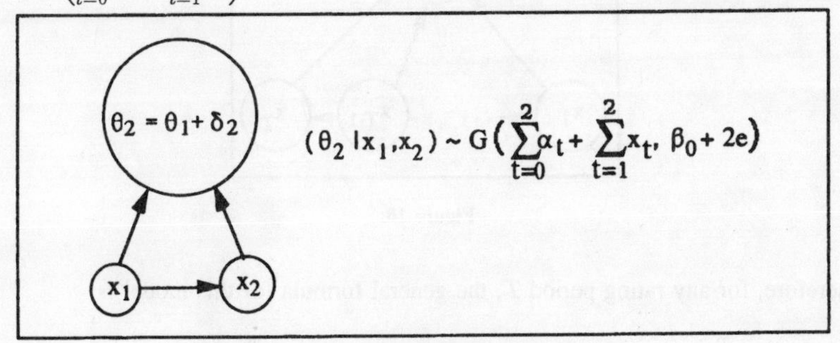

Figure 7.

By induction we are able to get the general influence diagram for any rating period $t = T$.

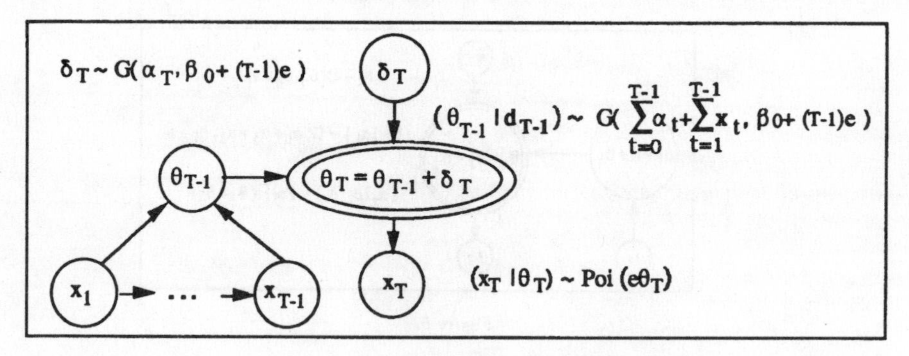

Figure 8.

In other words, $(\theta_T|\boldsymbol{d}_{T-1}) = ((\theta_{T-1} + \delta_T)|\boldsymbol{d}_{T-1}) \sim G(\sum_{t=0}^{T} \alpha_t + \sum_{t=1}^{T-1} x_t, \beta_0 + (T-1)e)$.

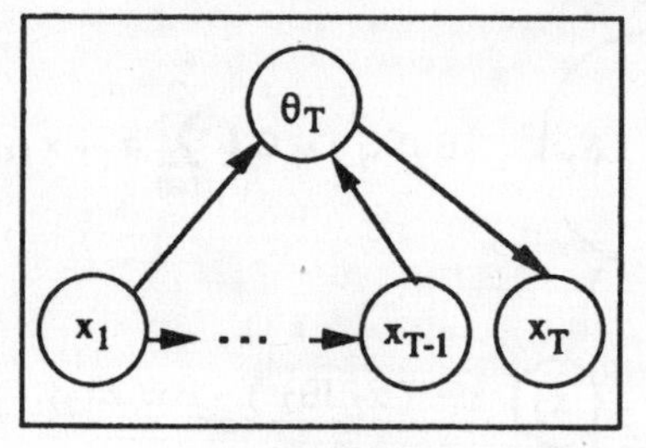

Figure 9.

The final influence diagram is obtained by performing the arc reversal operation.

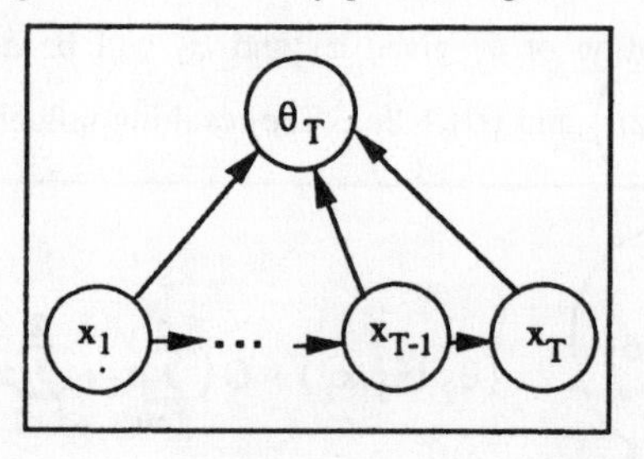

Figure 10.

Therefore, for any rating period T, the general formula for this model is:

$$(\theta_T|\boldsymbol{d}_T) \sim G(\sum_{t=0}^{T} \alpha_t + \sum_{t=1}^{T} x_t, \beta_0 + Te), \text{ where } \theta_T = \theta + \sum_{t=1}^{T} \delta_t \text{ and } \boldsymbol{d}_T = (x_1, \ldots, x_T).$$

4. THE MULTIPLICATIVE MODEL

4.1. *Assumptions*

This model should be used for processes that are judged to get better as time goes by. In this case, we also start with the initial quality index θ. At each rating period t, the quality index will be given by $\theta_t = \theta_{t-1}(1 - \delta_t)$ where $0 \leq \delta_t \leq 1, t = 1, \ldots, T$ and $\theta_0 = \theta$. δ_t is the proportion by which the analyst judges that the process got better at time t.

As before, $(x_t|\theta_t) \sim$ Poi $(e\theta_t)$ where e is the expected number of defects in the sample when the quality standard is met. We remain with the assumption that e is constant for all periods $t = 1, \ldots, T$. Again, the assumption of exchangeability is not needed and the following influence diagram will represent the Multiplicative model:

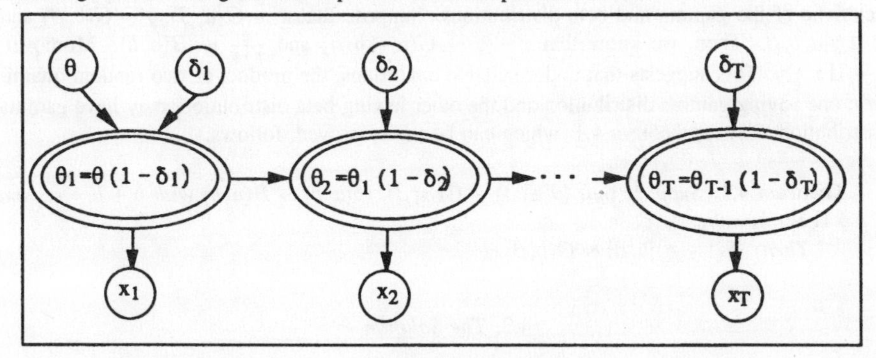

Figure 11.

The influence diagram representation requires an assessment of a joint distribution for the random quantities appearing in the model. The easiest thing to do is to assess prior distributions for θ and for the δ's, since it is been already assumed that $(x_t|\theta_t) \sim$ Poi $(e\theta_t)$.

The gamma distribution is a sensible assessment for θ.

Since $0 \leq \delta_t \leq 1$, beta distributions are reasonable assessments for the δ's.

After T rating periods, the analyst is interested in the posterior distribution for the quality index θ_T. The posterior distribution obtained for θ_T when the Multiplicative model is used is gamma. In symbols, $(\theta_T|\boldsymbol{d}_T) \sim G(\alpha_T, \beta_T)$. If the assessment for δ_T is a beta distribution with parameters a_T and $b_T(\delta_T \sim B(a_T, b_T))$, then $\alpha_T = b_T + x_T$. If the assessment for the initial quality index θ is a $G(\alpha_0, \beta_0)$, then $\beta_T = \beta_0 + Te$.

The Multiplicative model requires some constraints on the parameters of the distributions assessed to the δ's. These constraints will be understood as the model is explained and the solution is worked out. The first step is to assess a gamma distribution for $\theta : \theta \sim G(\alpha_0, \beta_0)$. The assessment for δ_1 should be of the form:

$\delta_1 \sim B(a_1, b_1)$ where $a_1 + b_1 = \alpha_0$ and $B(a_1, b_1)$ means a beta distribution with parameters a_1 and b_1.

At rating period t, we will have:

$(\theta_{t-1}|\boldsymbol{d}_{t-1}) \sim G(\alpha_{t-1}, \beta_{t-1})$ and $\delta_t \sim B(a_t, b_t)$ where $a_t + b_t = \alpha_{t-1}$ for $t = 1, \ldots, T$.

In other words, the assessments for the δ's must be such that the parameters of the beta distribution assessed for δ_t will depend upon the parameters of the posterior distribution of θ_{t-1} given $\boldsymbol{d}_{t-1}$.

The choice of α_0 and β_0, which will reflect the analyst's opinion about the process average, is free. The choice of a_t and b_t is constrained by the relation: $a_t + b_t = \alpha_{t-1}$. This means that there is a trade-off between the mean and the variance of δ_t and the analyst must keep it in mind. It is more intuitive to assess the mean of δ_t because it expresses the mean proportion by which the analyst judges the process is changing. Nevertheless, this choice is not completely free because it could lead to an unreasonable variance.

As in the Additive model, the influence diagram that represents this model has to be constructed as time goes by. At each rating period, new nodes referring to the current period are added to the influence diagram and the assessments of these nodes are made based upon the posterior distribution of the nodes referring to previous periods.

Before we proceed with the solution for this model, let us explore an interesting characteristic of the gamma and beta distributions. Suppose that $x \sim G(a, \beta), y \sim G(b, \beta)$ and $x \amalg y | a, b, \beta$. Then, we know that $x + y \sim G(a + b, \beta)$ and $\frac{x}{x+y} \sim B(a, b)$. Moreover, $\frac{x}{x+y} \amalg x + y$. This suggests that under suitable conditions, the product of two random quantities, one having gamma distribution and the other having beta distribution, may have gamma distribution. Hence, theorem 4.1, which can be easily proved, follows:

Theorem 4.1. *Suppose that* $(\theta | \alpha, \beta) \sim G(\alpha, \beta)$, $(\delta | a, b) \sim B(a, b)$ *with* $a + b = \alpha$, *and* $\theta \amalg \delta | a, b, \alpha, \beta$.

Then: $[\theta(1 - \delta) | b, \beta] \sim G(b, \beta)$.

4.2. *The Solution*

Starting with rating period $t = 1$, we have the following influence diagram.

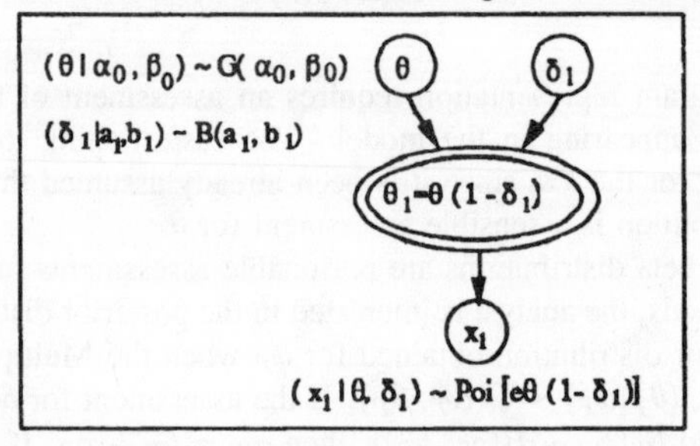

Figure 12.

By theorem 4.1, this diagram may be redrawn in the following way:

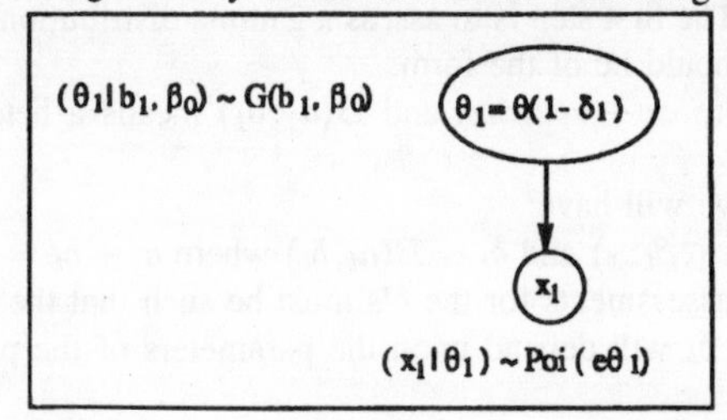

Figure 13.

The objective is to reverse the arc and to compute the posterior distribution of θ_1 given x_1.

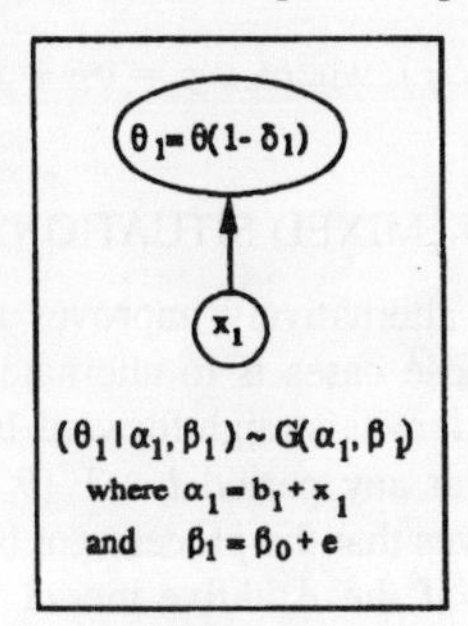

Figure 14.

The general solution, for any rating period $t = T$, may be obtained by induction. It is illustrated in the following sequence of influence diagrams:

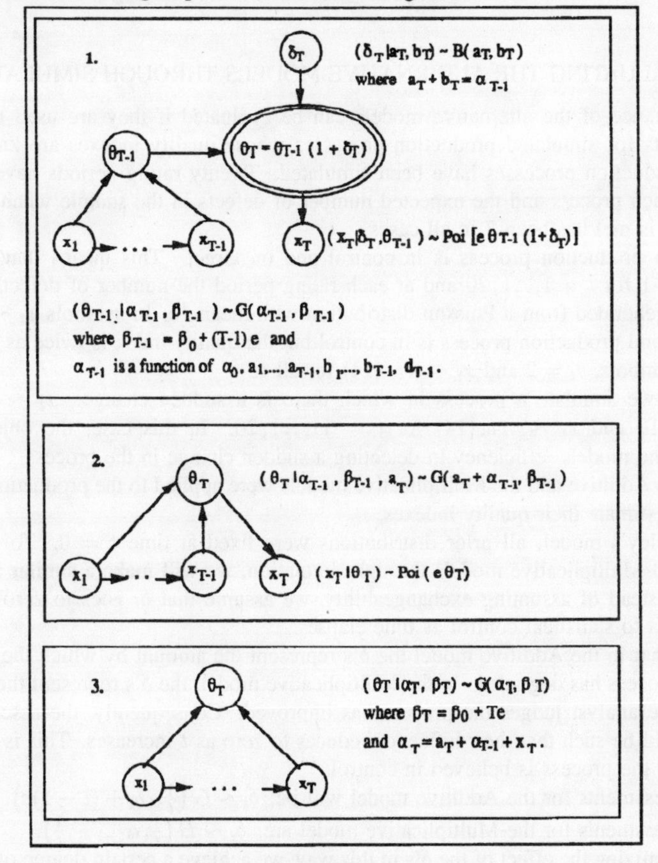

Figure 15.

In other words, the posterior distribution of the current quality index is given by:

$$(\theta_T|\alpha_T, \beta_T) \sim G(\alpha_T, \beta_T) \text{ where } \alpha_T = b_T + x_T \text{ and } \beta_T = \beta_0 + Te.$$

5. MIXED SITUATIONS

A situation in which the quality alternatively improves and deteriorates is surely of some interest. The best solution for those cases is to alternate the use of the Multiplicative and the Additive models. This procedure is straightforward because in both cases the posterior distribution of the quality index at any period $t-1$, $(\theta_{t-1}|\boldsymbol{d}_{t-1})$ is a gamma distribution. Consequently, if the analyst believes that the process got better at period t, she should switch to the Multiplicative model even if the Additive model was used up to that time. If it is believed that the process got worse at rating period t, the Additive model should be used.

Due to the mathematical tractability of both models, this is a much better idea than to use always the Multiplicative model, with the multipliers being allowed to span a range containing the value 1 or to use always the Additive model allowing the "increments" δ_t to be negative.

6. EVALUATING THE ALTERNATIVE MODELS THROUGH SIMULATION

The performance of the alternative models can be evaluated if they are used to construct control charts for simulated production processes whose quality indexes are known. Four different production processes have been simulated. Twenty rating periods have been considered in each process and the expected number of defects in the sample when the quality standard (e) is met has been 7 in all cases.

The first production process is in control and on target. This means that the quality index, θ_t, is 1 for $t = 1, \ldots, 20$ and at each rating period the number of defects in the lot, x_t, will be generated from a Poisson distribution with mean 1e. In symbols $x_t \sim \text{Poi}\,(7)$.

The second production process is in control but the quality index is twice as large as the target. In symbols, $\theta_t = 2$ and $x_t \sim \text{Poi}\,(14)$ for all t.

Finally, we simulate a process in which there is a sudden change. $x_t \sim \text{Poi}\,(7)$ for $t = 1, \ldots, 10$ and $x_t \sim \text{Poi}\,(14)$ for $t = 11, \ldots, 20$. In this case, the objective is to investigate the models' efficiency in detecting a sudden change in the process.

Both the Additive and the Multiplicative models were applied to the production processes in order to estimate their quality indexes.

In Hoadley's model, all prior distributions were fixed at time $t = 0$. To analyze the Additive and Multiplicative models through simulation, we will make a similar assumption. However, instead of assuming exchangeability, we assume that δ_t goes to zero that is, the process tends to statistical control as time elapses.

Recall that in the Additive model the δ's represent the amount by which the analyst believes the process has degraded. In the Multiplicative model, the δ's represent the proportion by which the analyst judges the process has improved. Consequently, the assessments for the δ's should be such that their influence reduces to zero as t increases. This is appropriate in cases that the process is believed in control.

The assessments for the Additive model will be: $\delta_t \sim G\left(\frac{1}{t}, \alpha_0 + (t-1)e\right)$.

The assessments for the Multiplicative model are: $\delta_t \sim B\left(\frac{1}{t}, \alpha_{t-1} - \frac{1}{t}\right)$.

By minimizing the effect of the δ's in this way, we achieve a certain degree of objectivity in the evaluation of the alternative models. It should be pointed out that the main difference between the Additive and the Multiplicative models is the role of the δ's. Since the δ's have no influence in our case, we do not expect to find a remarkable difference between the performances of the Additive and the Multiplicative models in the simulated processes.

Both models require a prior probability for the initial quality index θ. We begin with a gamma distribution having parameters $\alpha_0 = \beta_0 = 5$, which is equivalent to assessing mean 1 and variance 0.20 for θ. This is done for all simulated production processes.

The charts plotted in the sequel represent the analysis of the production processes via the alternative models. They are based upon the posterior probabilities for the quality index given the simulated data. The dots represent the posterior mean for the quality index at each rating period. A whisker represents one posterior standard deviation. Consequently, the intervals depicted in the charts are two standard deviations long and are centered on the posterior mean.

Processes in Statistical Control

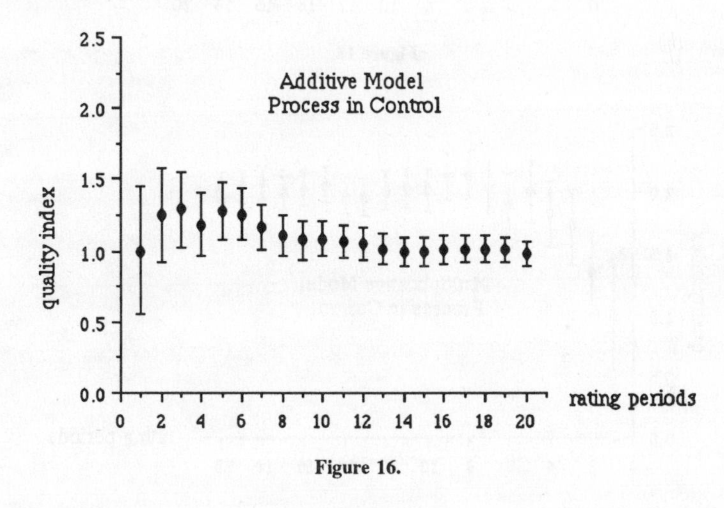

Figure 16.

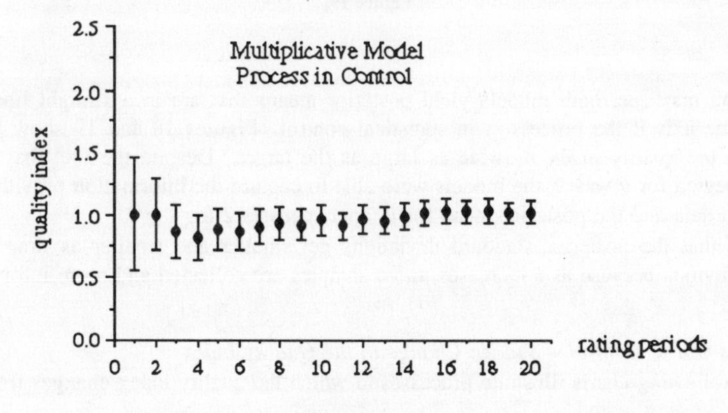

Figure 17.

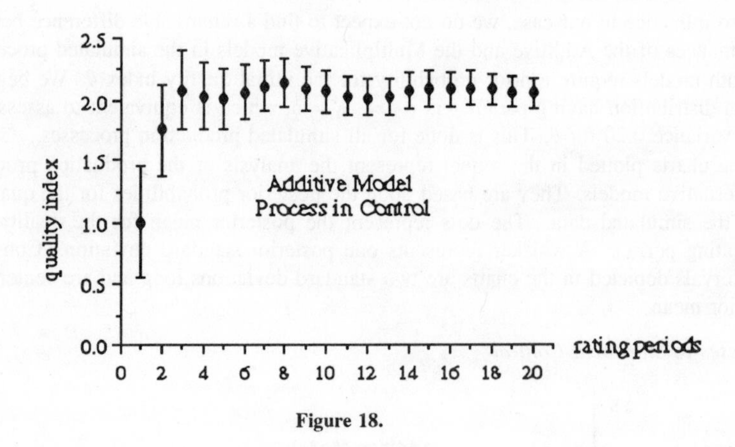

Figure 18.

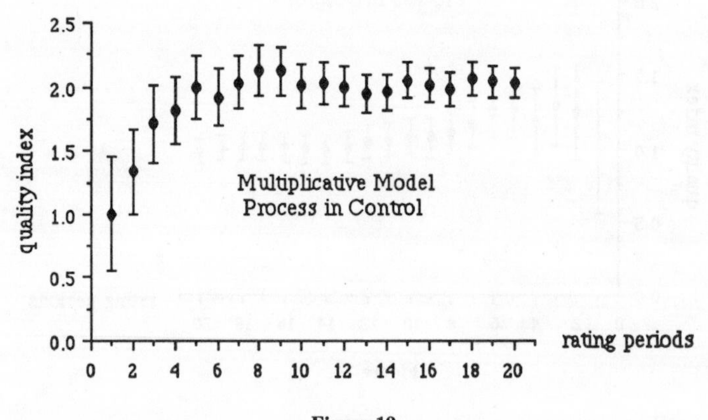

Figure 19.

As we may see, both models yield posterior means that are in a straight line parallel to the time axis if the process is in statistical control. Figures 18 and 19 show processes in which the quality index is twice as large as the target. Despite the fact that the prior mean assessed for θ was 1, the models were able to capture the information provided by the simulated data and the posterior mean for θ_t approximates 2 for $t \geq 2$.

Note that the posterior standard deviations get smaller and smaller as time goes by. This is obvious because as t increases, more samples are collected and more information is extracted.

Processes Out of Control – Sudden Change in the Quality Index

The following charts illustrate processes in which the quality index changes from 1 to 2 in the 11th rating period.

It is easy to observe in the charts that from rating period 11 on, the posterior means of the quality indexes start to increase systematically. Despite the fact that there is a sudden change in the quality index from 1 to 2 at rating period 11, the charts do not show a jump

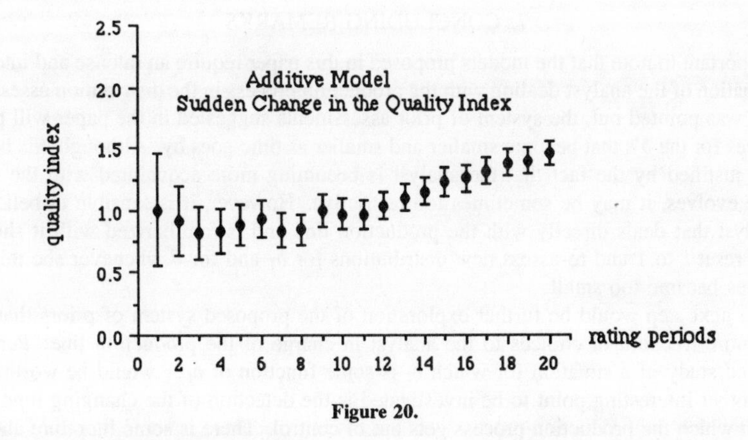

Figure 20.

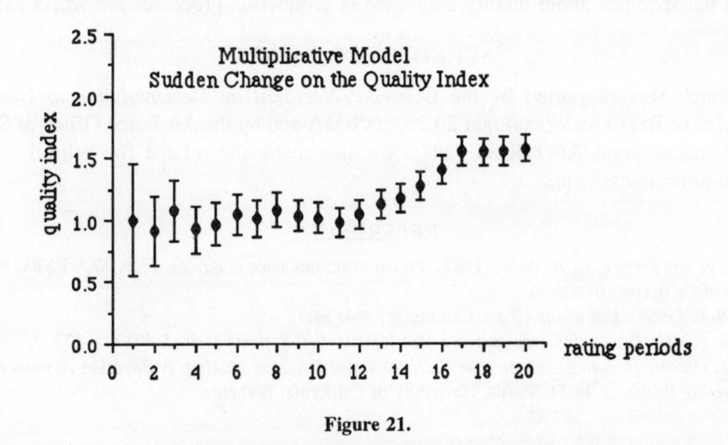

Figure 21.

from 1 to 2. Instead, they show a gradual change starting at period 11. This happens because by period 11, a great deal of information supporting the fact that the quality index should be around 1 have been assimilated. Good past history tempers an observed change. The more rating periods we have in the past, the greater is the inertia preventing the indication of a sudden change. In other words, the models are robust against statistical "jitter". They do not overreact to a few more defects.

Whenever a change of trend in the quality index is detected, one must suspect that there might be a sudden change in the process. In that case, the recommended procedure is to restrict the use of past data and to start recalculating the quality index from the point the change is detected. For example, in our case, the trend of change in the quality index started at period 11. Therefore, at period 11, t should be reset to 1, a wide prior distribution based on past data and expert judgement should be assessed to the quality index and the process must start all over.

7. CONCLUDING REMARKS

It is important to note that the models proposed in this paper require an intense and interactive participation of the analyst dealing with the production process in the distribution assessments.

As was pointed out, the system of prior assessments suggested in the paper will produce variances for the δ's that become smaller and smaller as time goes by. Although this behavior can be justified by the fact that the analyst is becoming more acquainted with the process as time evolves, it may be sometimes too restrictive. However, it is sensible to believe that an analyst that deals directly with the production line and is familiarized with it should be able to reset t to 1 and re-assess new distributions for δ_1 and for θ whenever she thinks the variances became too small.

The next step would be further exploration of the proposed system of priors that would allow more freedom of choices to the analyst in charge of the production line. Perhaps, a simulated study of a situation for which δ_t is some function of δ_{t-1} would be worthwhile.

Another interesting point to be investigated is the detection of the changing time for the cases in which the production process gets out of control. There is some literature about this topic but the specifics about quality assurance in production processes are worth exploring.

ACKNOWLEDGEMENTS

This research was supported by the *Conselho Nacional de Desenvolvimento Científico e Tecnológico* of Brazil under contract 20.2926/85-MA and by the Air Force Office of Scientific Research, under grant AFOSR-86-0208. We also thank the referee for helpful comments about the prior assessments.

REFERENCES

Barlow, R. E. and Pereira, C. A. de B. (1987). Probabilistic influence diagrams. *Tech. Rep.* **ESRC 87–7**, University of California, Berkeley.

Deming, W. E. (1986). *Out of the Crisis*. Cambridge, MA: MIT.

Hoadley, B. (1981). The quality measurement plan (QMP). *Bell Systems Tech. J.* **60**, 215–273.

Irony, T. Z. (1990). *Modeling, Information Extraction and Decision Making. A Bayesian Approach to Some Engineering Problems*. Ph.D. Thesis, University of California, Berkeley.

BAYESIAN STATISTICS 4, pp. 689–695
J. M. Bernardo, J. O. Berger, A. P. Dawid and A. F. M. Smith, (Eds.)

Multiobjective Bayesian Bandits

P. W. JONES
University of Keele, UK

SUMMARY

Modifications of the Bernoulli two armed bandit with finite horizon and independent beta priors are discussed. A terminal decision rule and a stopping rule are introduced. Optimal sequential designs and their characteristics are derived using a series of recurrence relations. The effects of changes in prior information are studied.

Keywords: BANDIT PROBLEM; BERNOULLI TRIALS; DYNAMIC PROGRAMMING; SEQUENTIAL OPTIMISATION; STOPPING RULES.

1. INTRODUCTION

The Bernoulli two armed bandit is a hypothetical machine with two arms, 1 and 2 which have unknown probabilities p_1, p_2 of winning one unit at each trial and $(1 - p_1), (1 - p_2)$ of obtaining nothing. The problem is to find a strategy, sequential design or sampling rule specifying which arm to pull at each trial so that the expected return, $E(R)$, to the experimenter is maximised in a fixed number of trials N or over a finite horizon. Several modifications to this basic bandit have been suggested, for example one probability known where the problem becomes one of optimal stopping, other reward distributions and infinite horizon with discounted rewards. In the last case the Gittins index may be used to determine the optimal strategy (Gittins (1989)). A review of bandit problems may be found in Berry and Fristedt (1985).

This paper considers the basic bandit problem described above where characteristics other than the expected return are studied. If one considers the application of bandits to the planning of sequential clinical trials where the two arms are treatments then the experimenter could be interested in making a terminal decision in which case maximising the probability of correctly selecting the superior treatment, $(P(CS))$, could be more important than maximising $E(R)$. Ethical considerations could dictate that minimising the number of patients on the poorer treatment, $E(N_{(1)})$, should be the objective and furthermore that early stopping of the trial should be possible. Wang (1991) considers the application of two armed bandits to sequential clinical trials and briefly discusses the problem of which criterion should be used in allocating treatments.

The modifications of the basic bandit are made in stages. Firstly the method of obtaining $E(R), E(N_{(1)})$ and $P(CS)$ is outlined for bandits where all observations are taken and optimisation is carried out with respect to one of them. Secondly a stopping rule is introduced and the expected sample size, $E(M)$, is added together with modifications of $E(R)$ and $E(N_{(1)})$ which facilitate easy comparison with the previous results. Finally a suboptimal scheme is discussed.

In all cases recurrence relations are given for the calculation of the characteristics and a numerical comparison with the fixed sample size or vector at a time (VT) sampling is made. The effects of changes in prior information are studied.

2. DISTRIBUTIONAL RESULTS

Suppose that p_1, p_2 are a priori independent and that p_i is assigned a beta prior with integer parameters (a_i, b_i) (beta (a_i, b_i)) with density proportional to

$$p_i^{a_i-1} q_i^{b_i-a_i-1}, 1 \leq a_i \leq b_i - 1, q_i = (1 - p_i)$$

then after v_i successes in t_i trials on arm i the posterior distribution of p_i is beta(r_i, n_i) where $r_i = a_i + v_i$ and $n_i = b_i + t_i$. The posterior expectation of p_i is $\hat{p}_i = \frac{r_i}{n_i}$ this is also the predictive probability that the next trial on arm i results in a 1.

In order to obtain $P(CS)$ and $E(N_{(1)})$, the posterior probability that $p_1 > p_2$ must be derived is

$$P(p_1 > p_2 | r_1, n_1, r_2, n_2) = \sum_{j=r_2}^{n_2-1} G(j)$$

where

$$G(j) = [(n_1 + n_2 - 1)\beta(r_1 + j + 1, n_1 + n_2 - r_1 - j - 1)] / [j(r_1 + j)\beta(r_1, n_1 - r_1)\beta(j, n_2 - j)]$$

and $\beta(\cdot, \cdot)$ is a beta function; monotonicity properties of this function are given in Jones and Madhi (1988).

Finally, the posterior predictive probability of obtaining r successes on a further n trials on arm 1 is given by

$$P_1(r, n | r_1, n_1) = \binom{n}{r} \frac{\beta(r_1 + r, n_1 + n - r_1 - r)}{\beta(r_1, n_1 - r_1)};$$

the result for arm 2 follows, note that this is $(\frac{1}{n+1})$ for a uniform prior for p_1 when $v_1 = t_1 = 0$. This probability is required when calculating $P(CS)$ for VT sampling when $v_1 = t_1 = 0$ and could be used to obtain the characteristics of designs using group sequential procedures although this is not discussed here.

3. TERMINAL DECISION

The results of sequential sampling may be expressed as a sample path starting at the origin (a_1, b_1, a_2, b_2) in four dimensional integer space and proceeding to (r_1, n_1, r_2, n_2) after v_i successes in t_i trials on arm i, $i = 1, 2$. The design gives which arm to pull at each point $(r_1, n_1, r_2, n_2), 0 \leq v_i \leq t_i \leq N - 1$. A design which is optimal with respect to one of the characteristics may be obtained by using dynamic programming, the other characteristic may then be calculated by using the dynamic programming recurrence relations but without using optimisation. The characteristics of any design are the values at the origin. The recurrence relations for each characteristic are defined below:

1. $E(R)$

Let $D(r_1, n_1, r_2, n_2)$ be the (optimal) expected return at (r_1, n_1, r_2, n_2) then

$$D(r_1, n_1, r_2, n_2) = (\max)[D_1(r_1, n_1, r_2, n_2), D_2(r_1, n_1, r_2, n_2)]$$

where

$$D_1(r_1, n_1, r_2, n_2) = \hat{p}_1[1 + D(r_1 + 1, n_1 + 1, r_2, n_2)] + (1 - \hat{p}_1)[D(r_1, n_1 + 1, r_2, n_2)]$$

is the (optimal) expected return when the next observation is on arm 1, $D_2(r_1, n_1, r_2, n_2)$ is defined similarly.

$D(r_1, n_1, r_2, n_2) = 0$ at termination (in this case where $t_1 + t_2 = N$).

To ease typographical problems in subsequent development, let

$$D_{1E} = D_{1E}(r_1, n_1, r_2, n_2) = \hat{p}_1[D(r_1 + 1, n_1 + 1, r_2, n_2)] + (1 - \hat{p}_1)[D(r_1, n_1 + 1, r_2, n_2)]$$

for arm 1, D_{2E} defined in a similar way for arm 2, the argument (r_1, n_1, r_2, n_2) will also be dropped.

2. $E(N_{(1)})$

$D = (\min)[D_1, D_2]$

where

$D_1 = P(p_1 < p_2 | r_1, n_1, r_2, n_2) + D_{1E}$

$D_2 = 1 - P(p_1 < p_2 | r_1, n_1, r_2, n_2) + D_{2E}$

$D = 0$ at termination.

3. $P(CS)$

$D = (\max)[D_{1E}, D_{2E}]$

$D = (\max)[1 - P(p_1 < p_2 | r_1, n_1, r_2, n_2), P(p_1 < p_2 | r_1, n_1, r_2, n_2)]$

at termination.

$P(CS)$ is obviously determined by the choice of terminal decision rule, if the objective is to maximise $P(CS)$ then the terminal decision should be based on the value of $P(p_1 < p_2 | r_1, n_1, r_2, n_2)$ on termination. However if optimising with respect to $E(R)$ or $E(N_{(1)})$ the terminal decision rule could be based on the relative values of r_1, r_2 or n_1, n_2 respectively. To enable direct comparisons to be made, the terminal decision for each design is based on $P(p_1 < p_2 | r_1, n_1, r_2, n_2)$.

Table 1 gives the values of the three characteristics when optimising one of them for $N = 25$ and two sets of priors (i) both uniform [beta(1, 2)] (ii) uniform v beta (1, 3) prior, other prior pairs were considered but the full results are not given here.

		(1,2) v (1,2)			(1,2) v (1,3)		
		$E(R)$	$E(N_{(1)})$	$P(CS)$	$E(R)$	$E(N_{(1)})$	$P(CS)$
	$E(R)$	15.6700	5.3399	.8542	13.7391	4.9880	.8577
	$E(N_{(1)})$	15.6625	5.3118	.8572	13.7372	4.9840	.8567
OPT	$P(CS)$	12.5	12.5	.8826	10.3315	12.8881	.8842
	VT	12.5	12.5	.8805	10.4167	12.5	.8826

Table 1. *Characteristics of optimal designs $N = 25$, no stopping.*

A further comparison is made with VT sampling where equal numbers of observations are made on each arm. In this case $E(N_{(1)}) = \frac{N}{2}$, $E(R) = \frac{N}{2}(\frac{a_1}{b_1} + \frac{a_2}{b_2})$ and

$$P(CS) = \sum_{i=0}^{N/2} \sum_{j=0}^{N/2} [P^*(i + a_1, N/2 + b_1, j + a_2, N/2 + b_2) P_1(i + a_1, N/2 + b_1 | a_1, b_1) P_2(j + a_2, N/2 + b_2 | a_2, b_2)]$$

$$P^*(i+a_1, N/2+b_1, j+a_2, N/2+b_2) =$$
$$= \max[P(p_1 < p_2|i+a_1, N/2+b_1, j+a_2, N/2+b_2),$$
$$1 - P(p_1 < p_2|i+a_1, N/2+b_1, j+a_2, N/2+b_2)].$$

The particular sample size of $N = 25$ was chosen to avoid excessive computation (there are 23,751 points in the space for $N = 25$ and 316,251 for $N = 50$ for example) and by noting the fact that as N increases $P(CS)$ increases slowly. It has been shown elsewhere that truncation produces little difference in the characteristics of optimal sequential designs (Lindley and Barnett (1965), Jones and Madhi (1986)).

From the Table it may be seen that optimising with respect to $E(R)$ and $E(N_{(1)})$ produces a design with similar characteristics. To check further that they are nearly coincident, the number of points at which the two designs give differing sampling decisions was checked for $N = 10$ and uniform priors, out of a possible 1001 points only 10 gave differing decisions. As the difference between the priors increase naturally $P(CS)$ increases, $E(N_{(1)})$ decreases and $E(R)$ approaches $N \max[a_1/b_1, a_2/b_2]$.

If the terminal decision is changed to comparing r_1, r_2 for $\max E(R)$ and n_1, n_2 for $\min E(N_{(1)})$ only a slight decrease (less than 1%) in $P(CS)$ is noted.

Maximising $P(CS)$ produces designs with rather poor characteristics with $P(CS)$ quite close to that obtained through VT sampling. Maximising $P(CS)$ obviously seeks those points at which there is maximum discrimination between the two arms. Assuming $P(CS)$ for VT sampling is close to the optimal value then a further justification of choosing a relatively small value for N is that $P(CS)$ for VT sampling at $N = 50$ is .9127 and for $N = 100$ is .9378 for uniform priors.

The choice of which characteristic to optimise must be dependent on the objectives of the experimenter.

4. STOPPING RULE

A number of adaptive stopping rules are available based on posterior (or current prior) values at the point (r_1, n_1, r_2, n_2) such as $|r_1 - r_2|, |n_1 - n_2|, P(p_1 < p_2|r_1, n_1, r_2, n_2)$ and $|\hat{p}_1 - \hat{p}_2|$, with sampling terminating once a preassigned difference or value is obtained. It is also possible to use the optimal D_1, D_2 in the dynamic programming equations although differences here could be difficult to interpret.

The particular rule considered here is the last and it is also used to define the terminal decision. The stopping rule becomes

stop if $|\hat{p}_1 - \hat{p}_2| \geq \Delta, \Delta$ *preassigned, continue otherwise.*

A further characteristic of interest under a stopping rule is the expected sample size $E(M)$, which may be calculated by using the following recurrence relation:

4. $E(M)$

$D = (\min)[D_1, D_2]$

$D_1 = 1 + D_{1E}$

$D_2 = 1 + D_{2E}$

$D = 0$ on termination.

The recurrence relations in Section 3 must be modified in order to make direct comparisons with the results in that section. The value of $P(CS)$ may be compared directly but it is difficult to interpret $E(R)$ and $E(N_{(1)})$ since they will depend on fewer observations. The modifications suggested below treat the sampling scheme as if it is in two stages, a sequential sample until termination with the remaining observation taken on the better arm. This gives:

5. $E(R^*)$

Similar to $E(R)$ except at termination

$$D = \begin{cases} (N - t_1 - t_2)\hat{p}_1 & \text{if } \hat{p}_1 > \hat{p}_2 \\ (N - t_1 - t_2)\hat{p}_2 & \text{otherwise} \end{cases}$$

6. $E(N^*_{(1)})$

Similar to $E(N_{(1)})$ except at termination

$$D = \begin{cases} (N - t_1 - t_2)P(p_1 < p_2 | r_1, n_1, r_2, n_2) & \text{if } \hat{p}_1 > \hat{p}_2 \\ (N - t_1 - t_2)[1 - P(p_1 < p_2 | r_1, n_1, r_2, n_2)] & \text{otherwise} \end{cases}$$

Table 2 gives the results for the two sets of priors and the sample size considered in Section 3. A small value of Δ causes early stopping and large Δ late stopping, the value of $\Delta = .4$ in Table 2 was chosen to give reasonable values for the characteristics especially $P(CS)$.

		$E(R^*)$	$E(N_{(1)^*})$	$E(M)$	$P(CS)$	$E(R)$	$E(N_{(1)})$
		$(1,2)$	v	$(1,2)$			
	$E(R^*)$	15.6595	5.3698	17.2444	.8507	9.6004	4.6861
	$E(N_{(1)^*})$	15.6531	5.3429	16.7424	.8527	9.2228	4.6191
OPT	$E(M)$	14.7796	7.9432	14.0172	.8519	7.0086	7.0086
	$P(CS)$	13.7344	10.1549	20.2330	.8821	10.1354	10.0484
		$(1,2)$	v	$(1,3)$			
	$E(R^*)$	13.7171	5.0537	14.5237	.8501	5.9935	4.0952
	$E(N_{(1)^*})$	13.7158	5.0506	14.5128	.8481	5.9853	4.0916
OPT	$E(M)$	13.2194	6.6003	13.4358	.8536	5.0063	5.6204
	$P(CS)$	11.0348	11.5806	19.8852	.8836	7.4926	11.4684

Table 2. *Characteristics of designs $N = 25$, stopping ($\Delta = \cdot 4$).*

The designs produced when optimising $E(R^*)$ and $E(N^*_{(1)})$ are little different from those when all observations are taken but with a considerable saving in observations. Early stopping has improved some of the characteristics when maximising $P(CS)$. Minimising $E(M)$ seems to be dominated by those designs which optimise $E(R^*)$ and $E(N^*_{(1)})$. The values of $E(R), E(N_{(1)})$ indicate that $\max P(CS)$ and $\min E(M)$ both search for points of maximum discrimination, unless these particular characteristics are of prime importance the above discussion suggests that either $\max E(R^*)$ or $\min E(N^*_{(1)})$ should be used.

It is obviously easier to truncate the sequential scheme rather than use an adaptive stopping rule. Some numerical results on the effects of truncation are given in Table 3. This shows that the designs for $\max E(R^*)$ or $\min E(N^*_{(1)})$ are fairly robust with respect to truncation and that the design for $\max P(CS)$ shows some improvements in $E(R^*)$ and $E(N^*_{(1)})$ over the corresponding values in Table 1.

		(1,2) v (1,2)			(1,2) v (1,3)		
		$E(R^*)$	$E(N_{(1)^*})$	$P(CS)$	$E(R^*)$	$E(N_{(1)^*})$	$P(CS)$
	$E(R^*)$	15.6527	5.4012	.8387	13.7251	5.0442	.8438
OPT	$E(N_{(1)^*})$	15.6467	5.3761	.8422	13.7232	5.0402	.8432
	$P(CS)$	13.9859	8.9709	.8529	11.8559	9.1506	.8562

Table 3. *Characteristics of designs, truncated at $M = 15, N = 25$.*

5. A SUBOPTIMAL DESIGN

Considerable computational effort is required to obtain the designs in the previous section due to the backward progress of the computation. In this section a suboptimal design is considered which proceeds forward from the origin and which need not be recomputed for each N.

Several sampling, stopping and termination rules are possible based on current prior distribution, for example the value of $P(p_1 < p_2|r_1, n_1, r_2, n_2) = P$ could be used where if $P < .5$ arm 1 is pulled, arm 2 otherwise, if $P > \Delta$, Δ preassigned then stop and make the terminal decision $p_1 < p_2$. Others could be based on the values of (r_1, r_2) or (n_1, n_2). However, to enable comparisons with the results of previous sections to be made, the stopping and termination rule used in the previous section based on $(\hat{p}_1 - \hat{p}_2)$ will be considered together with the sampling rule

if $\hat{p}_1 > \hat{p}_2$ pull 1
$\hat{p}_1 = \hat{p}_2$ choose at random
$\hat{p}_1 < \hat{p}_2$ pull 2

the characteristics of the design are given in Table 4 for $\Delta = .4$ and $N = 25$. A direct comparison with the values in Table 2 shows that this suboptimal design is very close to those which maximise $E(R^*)$ or minimise $E(N^*_{(1)})$.

	$(1,2)v(1,2)$	$(1,2)v(1,3)$
$E(R^*)$	15.5622	13.6695
$E(N_{(1)^*})$	5.7565	5.2021
$P(CS)$	.8228	.8351
$E(M)$	17.6868	14.6255
$E(R)$	9.8131	6.0095
$E(N_{(1)})$	5.1117	4.2479

Table 4. *Suboptimal design, $N = 25$, stopping ($\Delta = \cdot 4$).*

6. CONCLUSION

The designs involving $E(R)$ or $E(N_{(1)})$ are almost equivalent. Those which consider $P(CS)$ or $E(M)$ have poor properties. A stopping rule is worth considering whether adaptive or using truncation. Perhaps it is worthwhile using the suboptimal rule since it seems to have little effect on the characteristics.

This paper has not considered all combinations of sampling, stopping and termination rules but the ones considered give an illustration of the robustness of designs when quite considerable departures from the basic two armed bandit are allowed.

REFERENCES

Berry, D. A. and Fristedt, B. (1985). *Bandit Problems, Sequential Allocation of Experiments.* London: Chapman and Hall.

Gittins, J. C. (1989). *Multi-Armed Bandit Allocation Indices.* New York: Wiley.

Jones, P. W. and Madhi, S. A. (1986). Bayesian sequential methods for choosing the better of two binomial populations. *Sequential Analysis* **5**, 127–138.

Jones, P. W. and Madhi, S. A. (1988). Bayesian minimum sample size designs for the Bernoulli two armed bandit. *Sequential Analysis* **7**, 1–10.

Lindley, D. V. and Barnett, B. N. (1965). Sequential sampling: two decision problems with linear losses for binomial and normal random variables. *Biometrika* **52**, 191–199.

Wang, Y-G. (1991). Sequential allocation in clinical trials. *Comm. Statist. Theory and Methods* **20**, 791–805.

BAYESIAN STATISTICS 4, pp. 697–705
J. M. Bernardo, J. O. Berger, A. P. Dawid and A. F. M. Smith, (Eds.)

Bayesian Parametric Models for Lifetimes*

MAX B. MENDEL
University of California at Berkeley, USA

SUMMARY

The paper develops a systematic procedure for constructing probability models for unit lifetimes. It identifies the smallest class of such models that suffice for selecting a cost-optimal maintenance or operating policy. Lifetime-cost functions are introduced, replacing the classical failure rate functions for specifying model classes. It is shown how these classes can be represented through Bayesian parametric models. To illustrate, several common lifetime models and their finite population analogs are derived, including the exponential, the Weibull, Freund's multivariate exponential, and a new multivariate Weibull.

Keywords: MAINTENANCE & OPERATIONS COST; RELIABILITY THEORY; EXCHANGEABILITY; LIFETIME DISTRIBUTIONS.

1. INTRODUCTION

Reliability theory concerns the selection of a maintenance or operating policy from among a class of available policies. The outcome of such a policy depends on the lifetimes of the units, which are typically unknown. Barlow and Proschan (1975) represents a classical introduction to modelling unit lifetimes. They assume that unit lifetimes follow a frequentist parametric model of independent and identically distributed random variables. The "failure rate" is used to select a class of models such as the exponential and the Weibull. They define the policy whose expected cost is lowest to be the optimal policy.

Because of this focus on decision making, the appropriate lifetime models are instead Bayesian probability models (see also Singpurwalla (1988)). In fact, policies that minimize expected cost with respect to a frequentist distribution need not be preferred in practice; for instance, a frequentist may also consider the variance in the cost relevant. A popular method for obtaining Bayesian models is to adopt a frequentist model and simply provide the (frequentist) parameters with prior distributions. A Bayesian should justify the adoption of such models, for instance using information about the physical processes underlying the wear of the units. Moreover, the frequentist parameters are abstract quantities indexing possible "true models" (in the frequentist sense) without an operational meaning in terms of physically observables. It is unclear how a Bayesian provides such quantities with meaningful priors.

This paper exploits the symmetry in the decision problem to specify tractable classes of probability models. Since policies are judged only on the total cost they imply, we need only consider models that are invariant under transformations that preserve the uncertainty in the cost. Although such models may not take all available information about the units' wear into account, they are nevertheless sufficient for selecting optimal policies.

In Section 2, lifetime cost functions are introduced to describe the part of the total cost that is uncertain. Section 3 describes the natural symmetries implied by the empirical

* This work was supported by the Lawrence Livermore National Laboratories, Livermore, California, under contract No. 442446-26701. The author gratefully acknowledges the helpful comments and criticism of Richard Barlow and Yu Hayakawa.

structure of costs. Sections 4 and 5 present the resulting Bayesian parametric models for economically independent and dependent units, respectively. Section 6 concludes the paper.

2. COST AND LIFETIMES

The total expected cost of a policy is the sum of the fixed costs and the expected value of the costs that depend on the unknown lifetimes of the units. We describe the latter through "lifetime-cost functions." Consider a batch of N units with lifetimes

$$\boldsymbol{x} = (x_1, \ldots, x_N)$$

that are measured on appropriate, possibly different, scales, such as in "cycles" or "seconds."

Associated with each unit is also a lifetime cost. With $C_i = C_i(\boldsymbol{x})$ representing unit i's contribution to the total lifetime cost, we have N unit lifetime costs:

$$\boldsymbol{C} = (C_1, \ldots, C_N),$$

measured in dollars. Because of the additivity of money, we can sum unit lifetime costs to find the total lifetime cost and, after dividing by the number N of units, we find the average lifetime cost,

$$\theta = \frac{1}{N} \sum_{i=1}^{N} C_i. \tag{1}$$

Examples of lifetime cost functions for several engineering systems are given in Mendel (1991). Three properties defined there will be useful here. First when each unit's lifetime cost depends only on its own lifetime, the units are "economically independent." Second, when the lifetime cost functions are identically equal, the units are "economically identical." Finally, when units are both economically independent and economically identical, they are "economically exchangeable."

3. PARAMETRIC MODELS

A preference ordering among policies only partially determines a probability distribution on the space spanned by the lifetimes. If policies are only judged on their total cost, then they yield the same total cost on the level sets of the average cost. Therefore, by comparing policies we determine at most a distribution on the average cost function $\theta(\boldsymbol{x})$. On the other hand, this distribution suffices to distinguish among policies.

In practice one measures lifetimes rather than lifetime costs. To calculate conditional distributions, we have to extend the model to assign probabilities to the actual measurements. To accommodate for arbitrarily precise measurements, we assign a probability density value to each value of unit lifetimes. Such an extension does not affect the preference ordering among policies.

The following extension is natural; in particular, it makes the models independent of the money scale chosen to measure cost. First switch coordinates from lifetimes $\boldsymbol{x}$ to lifetime costs $\boldsymbol{C}$; see Figure 1. In these coordinates the lifetime space is a linear vector space and the level sets of the average cost function are simplices. Linear transformations preserve this vector-space structure and those transformations that also preserve simplices preserve the distribution of the average cost. These transformations are the group of linear isometries with respect to the ℓ^1 norm. Invariance under this group action leads to distributions that are

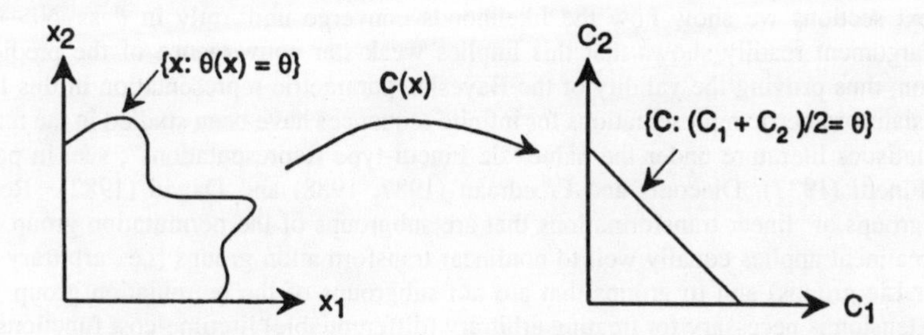

Figure 1. *Level set of the average cost in lifetime coordinates (left) and lifetime-cost coordinates (right) with* $N = 2$.

uniform (with respect to Hausdorf measure) on simplices. By switching back we find the distribution in the original lifetime coordinates.

We can represent the class of all invariant lifetime distributions through a one-dimensional Bayesian parametric model. We express the predictive density p of the first $n < N$ lifetimes as a mixture of fixed conditional distributions l as follows:

$$p(x_1, \ldots, x_n) = \int_0^\infty l(x_1, \ldots, x_n|\theta)\, \pi(\theta) d\theta. \tag{2}$$

Here the prior π is the density of the average lifetime cost. The average lifetime cost θ assumes the role of the statistical parameter; however, rather than its usual interpretation as an abstract quantity indexing possible "true" distributions, it is here defined *operationally* as a function of observable lifetimes and has a direct economic meaning. The likelihood l is a conditional density for a fixed value of the average lifetime cost and plays the role of the classical likelihood function when viewed as a function of the average lifetime cost. The likelihood is specified by the invariance and the prior by the preference ordering among policies.

The above analysis applies to batches of any pre-specified size N. When the maximum batch size cannot be specified beforehand, we require models that are valid for any batch size. Such models can be obtained by letting $N \to \infty$. We show that the Bayesian parametric representation converges in this limit: From Chebyshev's inequality we find for the probability that the average lifetime cost of the first N and the first $N' > N$ units deviate by more than ϵ:

$$P\left(\left|\frac{\sum_{i=1}^{N} C_i}{N} - \frac{\sum_{i=1}^{N'} C_i}{N'}\right| > \epsilon\right) \leq \frac{1}{\epsilon^2}\mathrm{E}\left(\frac{\sum_{i=1}^{N} C_i}{N} - \frac{\sum_{i=1}^{N'} C_i}{N'}\right)^2$$
$$\leq \frac{1}{\epsilon^2}\left(\frac{1}{N} + \frac{1}{N'}\right)\mathrm{E}[C^2],$$

The second inequality follows from the invariance which implies that the second moments of the lifetime costs are equal, say, to $\mathrm{E}[C^2]$. Since the last quantity can be made arbitrarily small by increasing N and N' sufficiently, we find that the average lifetime cost distribution converges, in the weak-star sense, to the distribution π of the limiting average lifetime cost,

$$\theta = \lim_{N\to\infty} \frac{1}{N} \sum_{i=1}^{N} C_i.$$

In the next sections we show how the likelihoods converge uniformly in θ as $N \to \infty$. A two-ϵ argument readily shows that this implies weak-star convergence of the predictive distribution, thus proving the validity of the Bayesian parametric representation in this limit.

Bayesian parametric representations for infinite sequences have been studied in the mathematical statistics literature under the name "de Finetti-type Representations"; see, in particular, de Finetti (1937), Diaconis and Freedman (1987, 1988) and Dawid (1982). Results exist for groups of linear transformations that are subgroups of the permutation group. The present treatment applies equally well to nonlinear transformation groups (i.e., arbitrary one-parameter Lie groups) and to groups that are not subgroups of the permutation group. The former extension is necessary for treating arbitrary (differentiable) lifetime-cost functions and the latter is necessary for modelling units that are not economically exchangeable. Finally, the existing arguments are typically based on existence proofs with the prior arising as an abstract mixing measure. This hides the economic interpretation of the parameter and its operational definition in terms of unit lifetimes, both of which are essential to the application of such models in reliability.

4. UNIVARIATE MODELS

In this section and the next, explicit expressions for the likelihood corresponding to particular lifetime-cost functions are derived. This section assumes that the units are economically independent. This will lead to univariate parametric models. That assumption is dropped in the next section where multivariate models are derived.

We start with the conditionally uniform distribution of the lifetime costs given a value θ of the average lifetime cost. Because of economic independence, we can change variables from lifetime costs to lifetimes for each unit separately. The local linear behavior of the coordinate transformation pertaining to unit i is given by the derivative,

$$c_i(x_i) = \frac{dC_i(x_i)}{dx_i},$$

of its lifetime cost. This derivative has an economic interpretation as the (local) tendency of the unit to dissipate cost when its lifetime is increased; following Mendel (1991), we refer to it as the unit's "lifetime cost force." After integrating out over x_{n+1} through x_N we find the likelihood:

$$l(x_1, \ldots, x_n|\theta) \propto \left(\prod_{i=1}^{n} \frac{c_i(x_i)}{\theta}\right) \left[1 - \frac{\sum_{i=1}^{n} C_i(x_i)}{N\theta}\right]^{N-n-1}. \tag{3}$$

Notice that lifetimes are statistically *dependent* conditional on the average cost. This expresses the fact that the lifetime of a unit partially determines the average cost and, hence, how much of the total lifetime cost can be dissipated by the remaining units. Unconditionally, lifetimes can still be statistically independent (choose the prior to be Erlang with N degrees of freedom). Notice also that we have not required units to be economically identical; however, if they are, their lifetimes become exchangeable random variables and vice versa.

For batches whose size is not bounded, we take the limit of the finite-batch likelihood in (3) as $N \to \infty$. From the very definition of the exponential function we find uniform convergence to the infinite-batch likelihood:

$$l(x_1, \ldots, x_n|\theta) \propto \prod_{i=1}^{n} \frac{c_i(x_i)}{\theta} e^{-C_i(x_i)/\theta}, \tag{4}$$

where θ is now a value of the limiting average cost. Notice that lifetimes are now statistically *independent* conditional on their average cost; when the batch size increases, the contribution of each unit to the average cost decreases and in the limit it disappears altogether. It is immediate that statistical independence arises only under these circumstances. When units are also economically identical (i.e., when they are economically exchangeable) we find that their lifetimes are independent and identically distributed conditional on the average cost and vice versa, consistent with the de Finetti theorem (see de Finetti (1937)).

Traditionally, unit lifetimes have been assumed to be independent conditional on the value of a parameter for both finite and infinite batches. If the parameter represents the average lifetime cost, then this assumption is incoherent when the batch is finite. If the parameter θ is abstract, it is not necessarily incoherent, but unnecessarily restricts the class of models (every mixture of independent variables can be represented as a mixture of finite-batch likelihoods, but not vice versa) by effectively requiring economic coherence with non-existing units. Furthermore, using models of independent variables as an approximation may have little practical value. The infinite-batch likelihood is merely notation for the limit as $N \to \infty$ of the finite-batch likelihood. Hence, in any real calculation of the former, the computing apparatus will truncate the series corresponding to this limit at some point and this point will effectively determine the maximum batch size.

Traditionally, unit lifetimes have also been assumed to be identically distributed. Some authors go so far as to introduce hypothetical units that are exchangeable with the existing ones to be able to pose parametric models and do inference. This is unnecessary. Because lifetime-cost functions may vary radically from unit to unit, the lifetimes can be very dissimilar statistically and still follow a one-dimensional parametric model.

Examples:

Exponential Models: The simplest non-trivial class of models corresponds to linear lifetime-cost functions, which implies a constant cost force. Specifically, with κ_i a constant for each unit, set:

$$C_i(x_i) = \kappa_i x_i \text{ and, hence, } c_i(x_i) = \kappa_i.$$

For instance, the operating cost of resistors on a constant potential source, such as heating elements or light bulbs on a voltage source or turbines on a reservoir of constant height, are linear; see Mendel (1991).

Substitution into the likelihoods of (4) and (3) gives a Bayesian exponential model and its finite population analog, respectively. For infinite batches of economically identical units, the ratio κ_i/θ does not depend on i and plays the role of the unknown classical failure rate.

Weibull Models: More generally, we can assume that a unit's lifetime cost is some power function of its lifetime. Specifically, with constants q and κ let

$$C_i(x_i) = \frac{\kappa}{q} x_i^q \text{ and, hence, } c_i(x_i) = \kappa x_i^{q-1}.$$

Such lifetime cost functions describe units that age in the sense of becoming more and more expensive to operate; see Mendel (1991). This example is easily generalized to units that are not economically identical by letting either κ or q, or both, depend on i.

Substitution gives the Bayesian Weibull model and its finite population analog. The usual "scale" and "shape" parameter correspond to $(\kappa/\theta)^{1/q}$ and q, respectively. The shape parameter, however, is *not* a random variable from the present perspective (it is not a function of the unknown lifetimes), but part of the specification of the lifetime cost. Assuming that q is unknown is incoherent with the assumption that only the lifetimes of the units are unknown. This is the Bayesian analog to the well-known classical result that no finite-dimensional sufficient statistic exists for estimating both the shape and scale parameter in a Weibull model.

Other examples of lifetime-cost functions are easily constructed and yield valid likelihood models as long as they are differentiable. Since polynomials can approximate any analytic lifetime-cost function arbitrarily closely, these form a versatile and mathematically tractable class. Some models —most notably the Gamma models— correspond to complicated lifetime-cost functions and are, hence, less practical.

5. MULTIVARIATE MODELS

To treat the general case it is convenient to divide the batch into N economically independent systems, each consisting of K economically dependent components[(1)]. This leads to a batch of NK units. With the index ij referring to the jth component of the ith system, we can arrange the lifetimes in a vector-valued sequence

$$\boldsymbol{x} = (\boldsymbol{x}_1, \ldots, \boldsymbol{x}_N) = \left(\begin{Bmatrix} x_{11} \\ \vdots \\ x_{1K} \end{Bmatrix}, \ldots, \begin{Bmatrix} x_{N1} \\ \vdots \\ x_{NK} \end{Bmatrix} \right),$$

and similarly for the lifetime costs $\boldsymbol{C}$. Summing the lifetime costs of the components in system i gives its total system lifetime cost:

$$||\boldsymbol{C}_i|| = \sum_{j=1}^{K} C_{ij}.$$

Summing over systems and dividing by NK gives the average lifetime costs:

$$\theta = \frac{\sum_{i=1}^{N} ||\boldsymbol{C}_i||/K}{N}.$$

Multivariate Bayesian parametric models are found by representing the predictive density of the components in the first n systems as a follows:

$$p(\boldsymbol{x}_1, \ldots, \boldsymbol{x}_n) = \int_0^\infty l(\boldsymbol{x}_1, \ldots \boldsymbol{x}_n|\theta)\, \pi(d\theta) \tag{5}$$

and this model representation remains valid when the number of systems is increased indefinitely.

To find the expression for the likelihood, we now transform vectors $\boldsymbol{C}_i$ to vectors $\boldsymbol{x}_i$. The local change of a K-dimensional volume element is give by the determinant,

$$\det[\boldsymbol{c}_i(\boldsymbol{x}_i)] = \begin{vmatrix} \frac{\partial C_{i1}(\boldsymbol{x}_i)}{\partial x_{i1}} & \cdots & \frac{\partial C_{i1}(\boldsymbol{x}_i)}{\partial x_{iK}} \\ \vdots & \ddots & \vdots \\ \frac{\partial C_{iK}(\boldsymbol{x}_i)}{\partial x_{i1}} & \cdots & \frac{\partial C_{iK}(\boldsymbol{x}_i)}{\partial x_{iK}} \end{vmatrix},$$

(1) Without too much loss of generality, we assume here that each system has the same number K of components

of the Jacobian matrix $[c_i(\boldsymbol{x}_i)]$ of the transformation. After integrating out over the $n+1$st through the Nth system, we find:

$$l(\boldsymbol{x}_1,\ldots,\boldsymbol{x}_n|\theta) \propto \left(\prod_{i=1}^{n} \frac{\det[c_i(\boldsymbol{x}_i)]}{\theta^K}\right)\left[1-\frac{\sum_{i=1}^{n}||C_i(\boldsymbol{x}_i)||/K}{N\theta}\right]^{K(N-n)-1}, \tag{6}$$

which converges uniformly to:

$$l(\boldsymbol{x}_1,\ldots,\boldsymbol{x}_n|\theta) \propto \prod_{i=1}^{n} \frac{\det[c_i(\boldsymbol{x}_i)]}{\theta^K} e^{-||C_i(\boldsymbol{x}_i)||/\theta}, \tag{7}$$

when $N \to \infty$. For infinite batches, the systems become statistically independent but the components within a system remain statistically dependent conditional on a value of the average cost.

Traditionally, the number of parameters in multivariate models have grown with the size of the vector. Here, the multivariate models still have a single statistical parameter θ about which the decision maker is uncertain. The quantity of interest remains the average cost, regardless of the economic dependence among units.

Examples:

Freund's Multivariate Exponential Model: Consider systems with costs:

$$||C_i(\boldsymbol{x}_i)|| = \kappa_{11}x_{i1} + \kappa_{22}x_{i2} + (\kappa_{12}+\kappa_{21})\min(x_{i1}, x_{i2}).$$

This could represent the operating cost of two resistors in series on a constant potential source; see Mendel (1991). We also have the Jacobian determinant:

$$\det[c_i(\boldsymbol{x}_i)] = (\kappa_{11} + \kappa_{12}\phi(x_{i1}; x_{i2}))(\kappa_{22} + \kappa_{21}\phi(x_{i2}; x_{i1})),$$

where $\phi(x_{i1}; x_{i2})$ is the structure function of a 2-component series system as a function of x_{i1} (i.e., $\phi(x_{1i}; x_{2i}) = 1$ if $x_{1i} < x_{2i}$ and 0 otherwise).

Substituting gives a Bayesian version of Freund's (1961) multivariate exponential and its finite population analog. Freund's original model has four (frequentist) parameters that correspond to linear combinations of κ_{11}/θ, κ_{12}/θ, κ_{21}/θ, and κ_{22}/θ. Thus, whereas Freund (ibid.) estimates four parameters, it is here only necessary to identify θ.

Multivariate Weibull Models: A multivariate Weibull model is obtained by considering the system lifetime-cost function whose qth derivative is constant. When $K = 1$ this reduces to the univariate case. Let $\dot{\boldsymbol{x}}_i = d\boldsymbol{x}_i/dt$ be the direction and the rate of a local movement away from the point $\boldsymbol{x}_i$, expressed per unit of clock time t. The qth derivative is a q-linear functional $D^q||C_i(\boldsymbol{x}_i)|| : (\Re^K)^q \to \Re$ such that

$$D^q||C_i(\boldsymbol{x}_i)||(\underbrace{\dot{\boldsymbol{x}}_i,\ldots,\dot{\boldsymbol{x}}_i}_{q \text{ entries}}) = \frac{d^q||C_i(\boldsymbol{x}_i)||}{dt^q},$$

where the derivatives on the right are all evaluated along $\dot{\boldsymbol{x}}_i$. If this functional is constant, we can represent it explicitly in lifetime coordinates by a symmetric tensor of order q. The system lifetime-cost function is recovered by integrating q times. For instance, when $q = 2$ and $K = 2$ we represent the functional by the 2-tensor:

$$\begin{bmatrix} \kappa_{11} & \kappa_{12} \\ \kappa_{12} & \kappa_{22} \end{bmatrix}.$$

Evaluating and integrating gives the familiar bilinear form of the Riemannian metric:

$$||\boldsymbol{C}_i(\boldsymbol{x}_i)|| = \frac{\kappa_{11}}{2}x_{i1}^2 + 2\kappa_{12}x_{i1}x_{i2} + \frac{\kappa_{22}}{2}x_{i2}^2.$$

The corresponding Jacobian is:

$$\det[\boldsymbol{c}_i(\boldsymbol{x}_i)] = \kappa_{11}\kappa_{12}x_{i1}^2 + \kappa_{11}\kappa_{22}x_{i1}x_{i2} + \kappa_{22}\kappa_{12}x_{i2}^2.$$

The multivariate "Weibull" models pertaining to these lifetime-cost functions appear to be new.

6. CONCLUSIONS

The paper presented the probability models that are sufficient for selecting cost-optimal maintenance and operating policies. Mathematically, a policy is a real-valued function δ that gives the total cost in dollars (the consequence) associated with each value of the units' lifetimes. This total cost is the sum of the fixed cost and the uncertain or random cost which, through the average lifetime cost, is a function of the unit lifetimes. Although in reliability utility is usually linear in money, we can introduce an arbitrary utility U for money and minimize its expectation,

$$\int_0^\infty \cdots \int_0^\infty U[\delta(\theta(\boldsymbol{x}))]\, p(\boldsymbol{x})\, dx_1 \cdots dx_N = \int_0^\infty U[\delta(\theta)]\, \pi(\theta) d\theta,$$

to find the optimal policy. Notice that this expectation only depends on the distribution of the average cost.

When the lifetimes of the first n units are measured, the preferred policy minimizes the posterior expected utility,

$$\int_0^\infty U[\delta(\theta)]\, \pi(\theta|x_1, \ldots x_n)\, d\theta.$$

Here, the posterior density of the average cost follows from the prior and the likelihood after applying Bayes's formula:

$$\pi(\theta|x_1, \ldots x_n) = \frac{l(x_1, \ldots, x_n|\theta)}{p(x_1, \cdots, x_n)}\, \pi(\theta).$$

The models developed in this paper apply *only* when a single lifetime-costs function applies to each unit. If any one of, say, m lifetime-cost functions are applicable, then the policy will depend on the m corresponding average lifetime costs and the parameter of the Bayesian model will be m-dimensional. The models become independent of the choice of lifetime-cost functions when they apply simultaneously to effectively *all* possible lifetime-cost functions, a limiting case. The parameter space then has the same dimensionality as the lifetime space and we have to consider the class of all possible distributions on the lifetime space. Such classes are too large to be practical, especially when the batch size is not bounded. The one-dimensional models presented here form the most tractable class. Only when these do not apply should one resort to higher dimensional classes.

REFERENCES

Barlow, R. E. and Mendel, M. B. (1991). De Finetti-type representations for life distributions. *J. Amer. Statist. Assoc.* **41**, (to appear).

Barlow, R. E. and Proschan, F. (1975). *Statistical Theory of Reliability and Life Testing*. Toronto: Holt, Rinehart, and Winston Inc.

Dawid, A. P. (1982). Inter-subjective statistical models. *Exchangeability in Probability and Statistics* (G. Koch and F. Spizzichino, eds.), Amsterdam: North-Holland, 217–232.

De Finetti, B. (1937). La prévision: ses lois logiques, ses sources subjectives. *Annales de l'Institut Henri Poincaré* (H. E. Kyburg, Jr. and H.E., Smokler, eds.), *Studies in Subjective Probability*. New York: Wiley, 53–118.

Diaconis, P. and Freedman, D. (1987). A dozen de Finetti-style results in search of a theory. *Ann. Inst. Henri Poincaré* **23**, 397–423.

Diaconis, P. and Freedman, D. (1988). Cauchy's equation and de Finetti's theorem. *Scandinavian J. Statist.* **17**, 235–250.

Freund, J. E. (1961). A bivariate extension of the exponential distribution. *J. Amer. Statist. Assoc.* **56**, 971–977.

Mendel, M. B. (1991). An economic approach to lifetime modelling. *Theory of Reliability* (C. A. Clarotti and F. Spizzichino eds.). *Tech. Rep.* **91-6**, University of California, Berkeley.

Mendel, M. B. (1989). *Development of Bayesian Parametric Theory with Applications to Control*. Ph.D. Thesis, Department of Mechanical Engineering, MIT, Cambridge, Mass.

Singpurwalla, N. D. (1988). *Foundational Issues in Reliability and Risk Analysis*. Philadelphia, PA: SIAM.

BAYESIAN STATISTICS 4, pp. 707–713
J. M. Bernardo, J. O. Berger, A. P. Dawid and A. F. M. Smith, (Eds.)

Bands of Probability Measures: a Robust Bayesian Analysis

ELÍAS MORENO and LUÍS R. PERICCHI
Universidad de Granada, Spain and *Universidad Simón Bolivar, Venezuela*

SUMMARY

In this paper the ranges of the posterior probabilities of an arbitrary set A, as the prior π ranges over a class of interval of quantiles $\Gamma_1, \Gamma_1 = \{\pi(\theta) : \alpha_k \leq \pi(C_k) \leq \beta_k, k = 1, 2, \ldots, n\}$, are calculated. Some illustrations are given, and the "quality" of the prior information given by Γ_1 is analyzed. It is concluded that for continuous parameters, a natural generalization of Γ_1, termed bands of probability measures $\Gamma(L, U) = \{\pi(\theta) : L(C) \leq \pi(C) \leq U(C)$, for any set $C, \pi(\Theta) = 1\}$ might be an appropriate model. This class allows considerable freedom on tail behaviour as long as the upper and lower measures L, U have different tails. This property is inherited from the class of interval of measures (DeRobertis and Hartigan, 1981), from which probability bands are obtained by restricting attention to those measures that integrate to one. Although the analysis of this restricted class is somewhat more involved, the restriction is quite natural and may lead to much smaller posterior ranges of the quantities of interest. General results for this model and some illustrations for two dimensional likelihoods are given. Comparison with interval of measures are also made.

Keywords: INTERVAL OF QUANTILES; INTERVAL OF MEASURES; PROBABILITY BANDS; ROBUST BAYES.

1. INTRODUCTION

One of the simplest classes of prior distributions that has been considered in Bayesian statistical inference is the quantile class Γ_γ defined by

$$\Gamma_\gamma = \{\pi(\theta) : \int_{C_k} \pi(d\theta) = \gamma_k, k = 1, 2, \ldots, n\} \tag{1}$$

where θ is the unknown parameter, the measurable sets $\{C_k, k = 1, 2, \ldots, n\}$ form a partition of the parameter space Θ, and γ_k are specified real numbers that satisfy $\Sigma\gamma_k = 1, \gamma_k \geq 0$. This class has been studied in a rather general setting by Moreno and Cano (1991).

A more flexible and realistic class that generalizes Γ_γ is

$$\Gamma_1 = \{\pi(\theta) : \alpha_k \leq \int_{C_k} \pi(d\theta) \leq \beta_k, k = 1, 2, \ldots, n\} \tag{2}$$

where α_k and β_k are the lower and upper probabilities of the set C_k respectively. To be coherent the conditions $\Sigma\alpha_k \leq \int_\Theta \pi(d\theta) = 1 \leq \Sigma\beta_k$ should be satisfied.

We find the range of the posterior probability of an arbitrary set A, as π ranges over the class Γ_1. However, Γ_1 does not give sensible posterior ranges for sets which are not generated by $\{C_k, k = 1, 2, \ldots, n\}$. This suggests that the information to be added to Γ_1 consists in the specification of lower and upper prior probabilities for all those measurable sets that make the likelihood a measurable function. For instance, if θ is a continuous real parameter, intervals of prior probabilities for any open set (θ_1, θ_2) must be elicited. In general this means that the prior information to be processed for an inference problem with likelihood $f(x \mid \theta)$, has to be elicited for all those subsets contained in the generating class of the σ-field $\mathcal{B}$ induced by the likelihood function $f(x \mid \theta)$, if ranges of sets other than the C_i's are of interest. This motivates the class,

$$\Gamma(L, U) = \{\pi(\theta) : L(C) \leq \pi(C) \leq U(C), \text{ for any } C \in \mathcal{B}, \int_{\Theta} \pi(d\theta) = 1\} \tag{3}$$

called bands of probability measures, where $U(.)$ and $L(.)$ are pre-determined fixed measures. Related to (3) is the class of interval of measures introduced by DeRobertis and Hartigan (1981),

$$\Gamma_{DH}(L, U) = \{\pi(\theta) : L(C) \leq \pi(C) \leq U(C), C \in \mathcal{B}\} \tag{4}$$

Note that $\pi(\theta) \in \Gamma_{DH}$ is *not* necessarily a probability measure. In their article, DeRobertis and Hartigan perform sensitivity analyses for different posterior quantities. This class is very interesting for various reasons. The elicitation process is simple and natural, and substantial prior imprecision can be modeled. Also the computation of posterior probabilities of sets is quite easy to handle and the behaviour with respect to the sample size of the posterior probabilities is quite realistic, see Pericchi and Walley (1991).

On the other hand it may be argued that this class is too large, in that it might contain measures which cannot be normed to one, that is, improper probability measures, thus motivating the restriction to those measures in the above class that integrate to one. Note that if robustness obtains for the larger, and easier to handle, interval of measures class, of course robustness will also obtain for the probability band class which is a sub class of the former. This article provides a comparison (prior and posterior) between Γ_{DH} and $\Gamma(L, U)$.

Previous analyses with classes (4) and (3) are in Bose (1990) and Lavine (1988) respectively. Bose imposes shape constraints on Γ_{DH} and Lavine developes numerical algorithms to calculate posterior ranges as the prior ranges over $\Gamma(L, U)$. Interesting relationships between classes (4) and (3) and related computing methods have been stated in Wasserman and Kadane (1991).

In this paper the emphasis has been placed on obtaining analytical results which give the general form of the solutions even for cases that cannot be derived from Lavine's algorithm. We also provide some guidelines for prior assessment.

2. RANGES OF POSTERIOR PROBABILITIES FOR THE CLASS Γ_1

Let $f(x \mid \theta)$ be the likelihood function and C an arbitrary set. The extrema of $P^{\pi}(C \mid x)$ as π ranges over Γ_1 are given in the following Theorem 1.

Theorem 1.

$$\inf_{\pi \in \Gamma_1} P^{\pi}(C|x) = \inf_{\gamma_i \in D} \frac{\sum_{i \in y} \underline{I}_i(x)\gamma_i}{\sum_{i \geq 1} \overline{I}_i(x)\gamma_i + \sum_{i \in y} \underline{I}_i(x)\gamma_i}$$

where y is a subset of indices of $\{1, \ldots, n\}$ defined by $i \in y$ if and only if $C_i \subset C$, $\underline{I}_i(x) = \inf_{\theta \in C_i} f(x|\theta), \overline{I}_i(x) = \sup_{\theta \in C_i \cap C^c} f(x|\theta), C^c = \Theta - C$, and $D = \{\gamma_i : \alpha_i \leq \gamma_i \leq \beta_i, i = 1, \ldots, n, \Sigma\gamma_i = 1\}$.

Proof. This follows from Corollary 1 in Moreno and Cano (1991). ◁

Remark. The $\inf_{\gamma_i \in D}$ that appears in Theorem 1 can be avoided if $\underline{I}_i(x)$ and $\bar{I}_i(x)$ are adequately re-ordered, so that no numerical computation is needed. For finding $\sup_{\pi \in \Gamma_1} P^{\pi}(C \mid x)$, use the general relationship $\sup_{\pi \in \Gamma_1} P^{\pi}(C \mid x) = 1 - \inf_{\pi \in \Gamma_1} P^{\pi}(C^c \mid x)$ and then Theorem 1.

Example 1.

Martz and Waller (1982) supposed that two engineers are concerned with the mean life θ of a proposed new industrial engine. The two engineers, A and B, quantify their prior beliefs about θ in terms of the probabilities given in Table 1.

i	Intervals C_i	Pr $[\theta \in C_i \vert A]$	Pr $[\theta \in c_i \vert B]$
1	$[0, 1000)$	.01	.15
2	$[1000, 2000)$	.04	.15
3	$[2000, 3000)$	.20	.20
4	$[3000, 4000)$	.50	.20
5	$[4000, 5000)$	.15	.15
6	$[5000, \infty)$	.10	.15

Table 1. *Specified prior probabilities of intervals.*

In Berger and O'Hagan (1989), O'Hagan and Berger (1988), and Berliner and Goel (1990) this example was analyzed for the prior beliefs of engineer A and B separately. Here we shall consider that for each C_i the interval $(\alpha_i, \beta_i) = (\min\{\Pr(C_i \mid A), \Pr(C_i \mid B)\}, \max\{\Pr(C_i \mid A), \Pr(C_i \mid B)\})$ contains all prior beliefs on the probability of C_i. That is, any opinion between those of engineers A and B is taken as reasonable. Therefore, from Table 1 we obtain the upper and lower probabilities α_i, β_i given in the third column of Table 2.

Observe that these upper and lower probabilities are coherent since $\Sigma\alpha_i = .7$ and $\Sigma\beta_i = 1.3$. Suppose as in Berger and O'Hagan (1988) that data becomes available in the form of two independent lifetimes which are exponentially distributed with mean θ. The observed life-times are 2000 and 2500 hours, so that the likelihood function is

$$f(2000, 2500|\theta) = \theta^{-2} \exp(-4500/\theta).$$

From theorem 1 it follows that the maximizing and minimizing probability measures are discrete. For instance, $\sup_{\pi \in \Gamma_1} P^{\pi}(C_3 \mid x)$ is obtained for the prior

$$\begin{aligned}\pi(\theta) = 0.15\; 1_{\{\theta=0\}}(\theta) + 0.15\; 1_{\{\theta=1000\}}(\theta) + 0.20\; 1_{\{\theta=2250\}}(\theta) \\ + 0.20\; 1_{\{\theta=4000\}}(\theta) + 0.15\; 1_{\{\theta=5000\}}(\theta) + 0.15\; 1_{\{\theta=\infty\}}(\theta)\end{aligned} .$$

The fourth column in Table 2 gives the ranges of the posterior probabilities, denoted by α_i^* and β_i^*, of the intervals C_i as the prior $\pi(\theta)$ ranges over Γ_1. Explanation of the last column in Table 2 is deferred to Section 3.

i	Intervals C_i	(α_i, β_i)	(α_i^*, β_i^*)	$(\alpha_i^{**}, \beta_i^{**})$
1	$[0, 1000)$	$(.01, .15)$	$(.00, .11)$	$(.001, .023)$
2	$[1000, 2000)$	$(.04, .15)$	$(.02, .25)$	$(.039, .194)$
3	$[2000, 3000)$	$(.20, .20)$	$(.22, .40)$	$(.235, .346)$
4	$[3000, 4000)$	$(.20, .50)$	$(.20, .60)$	$(.231, .561)$
5	$[4000, 5000)$	$(.15, .15)$	$(.10, .22)$	$(.122, .173)$
6	$[5000, \infty)$	$(.10, .15)$	$(.00, .15)$	$(.000, .012)$

Table 2. *Prior and posterior ranges for* $C_i, i = 1, \ldots, 6; x = (2000, 2500)$.

The fourth column shows that most of the intervals C_i have a very wide posterior range motivating a search for a more accurate prior information. On the other hand it is quite natural to enquire about the posterior sensitivity for sets other than C_i, for instance $C = [3000, 3500)$ for which the interval of prior probabilities is not known. It can be seen that the range of the posterior probabilities of C is $(\alpha_C^*, \beta_C^*) = (0, .86)$. This range is too wide to provide useful information and the infimum is zero which is not sensible. More generally, for all those measurable sets B such that $B \cap C_i \neq \phi, B \cap C_i \neq C_i$ for all i, it turns out that $(\alpha_B^*, \beta_B^*) = (0, 1)$ whatever observation x is considered.

Consequently, the class of priors to be considered could be the class of probability measures π such that $L(C) \leq \pi(C) \leq U(C)$ for any set C belonging to the σ-field $\mathcal{B}$ with respect to which the likelihood $f(x \mid \theta)$ is a measurable function for each x. Here to be coherent the lower L and the upper U should satisfy $L(\Theta) \leq 1 \leq U(\Theta)$.

Of course this is not the only way to obtain a reasonable Γ_1 maintaining the prior information. Shape constraints could be added. Related papers are Berger and O'Hagan (1988), O'Hagan and Berger (1989) and Moreno and Pericchi (1991).

3. RANGES OF POSTERIOR PROBABILITIES FOR THE CLASS Γ(L, U)

Consider the class $\Gamma(L, U)$ given in (3). To simplify notation we shall write $[f(x \mid \theta) \geq z] = \{\theta : f(x \mid \theta) \geq z\}$.

Lemma 1. *Let A be an arbitrary set in $\mathcal{B}$. Suppose that $f(x \mid \theta), L(d\theta), U(d\theta)$ satisfy $U([f(x \mid \theta) = z]) = 0$ for any $z \geq 0$. Then, we have*

(i) If $U(A) + L(A^c) \geq 1$ there exists a non-negative real number z_A such that $U(A \cap [f(x \mid \theta) \geq z_A]) + L(\{A \cap [f(x \mid \theta) < z_A]\} \cup A^c) = 1$.

(ii) If $U(A) + L(A^c) < 1$ there exists a non-negative real number z_A such that

$$U(A \cup \{A^c \cap [f(x|\theta) < z_A]\}) + L(A^c \cap [f(x|\theta) \geq z_A]) = 1.$$

Proof. This follows from the fact that L and U are measures and therefore continuous from above. ◁

Theorem 2. *Under the conditions of Lemma 1, for an arbitrary measurable set A,*

(i) If $U(A) + L(A^c) > 1$, then $\sup_{\pi \in \Gamma(L,U)} P^\pi(A|x) = P^{\pi_0}(A|x)$ where

$$\pi_0(d\theta) = U(d\theta)\, 1_{A \cap [f(x|\theta) \geq z_A]}(\theta) + L(d\theta)\, 1_{\{A \cap [f(x|\theta) < z_A]\} \cup A^c}(\theta),$$

z_A being such that $\pi_0(\Theta) = 1$.

(ii) *If* $U(A) + L(A^c) = 1$, *then* $\sup_{\pi \in \Gamma(L,U)} P^{\pi}(A|x) = P^{\pi_0}(A|x)$ *where*

$$\pi_0(d\theta) = U(d\theta)\, 1_A(\theta) + L(d\theta)\, 1_{A^c}(\theta).$$

(iii) *If* $U(A) + L(A^c) < 1$, *then* $\sup_{\pi \in \Gamma(L,U)} P^{\pi}(A|x) = P^{\pi_0}(A|x)$ *where*

$$\pi_0(d\theta) = U(d\theta)\, 1_{A \cup \{A^c \cap [f(x|\theta) < z_A]\}}(\theta) + L(d\theta)\, 1_{A^c \cap [f(x|\theta) \geq z_A]}(\theta),$$

z_A *being such that* $\pi_0(\Theta) = 1$.

Proof. (i) Note first that from Lemma 1 $\pi_0(d\theta) \in \Gamma(L,U)$. Suppose that assertion (i) is false. This means that there exists $\pi \in \Gamma(L,U)$ such that $P^{\pi}(A\,|\,x) > P^{\pi_0}(A\,|\,x)$, that is

$$\left\{1 + \frac{\int_{A^c} f(x|\theta)\pi(d\theta)}{\int_A f(x|\theta)\pi(d\theta)}\right\}^{-1} >$$

$$\left\{1 + \frac{\int_{A^c} f(x|\theta)L(d\theta)}{\int_{A\cap[f(x|\theta)\geq z_A} f(x|\theta)U(d\theta) + \int_{A\cap[f(x|\theta)<z_A} f(x|\theta)L(d\theta)}\right\}^{-1},$$

or equivalently it can written as

$$\frac{\int_{A^c} f(x|\theta)\pi(d\theta)}{\int_A f(x|\theta)\pi(d\theta)} < \frac{\int_{A^c} f(x|\theta)L(d\theta)}{\int_{A\cap[f(x|\theta)\geq z_A]} f(x|\theta)U(d\theta) + \int_{A\cap[f(x|\theta)<z_A]} f(x|\theta)L(d\theta)}.$$

Since $L(.) \leq \pi(.)$, we have

$$\int_{A\cap[f(x|\theta)\geq z_A]} f(x|\theta)U(d\theta) + \int_{A\cap[f(x|\theta)<z_A]} f(x|\theta)L(d\theta)$$
$$< \int_{A\cap[f(x|\theta)\geq z_A]} f(x|\theta)\pi(d\theta) + \int_{A\cap[f(x|\theta)<z_A]} f(x|\theta)\pi(d\theta)$$

and

$$\int_{A\cap[f(x|\theta)\geq z_A]} f(x|\theta)\{U(d\theta) - \pi(d\theta)\} < \int_{A\cap[f(x|\theta)<z_A]} f(x|\theta)\{\pi(d\theta) - L(d\theta)\}.$$

Since we are in case (i) and then $U(A\cap[f(x \mid \theta) \geq z_A]) + L(A\cap[f(x \mid \theta) < z_A]) + L(A^c) = 1$, it follows from the last inequality that

$$z_A\{1 - L(A^c)\} < z_A\{1 - \pi(A^c)\}$$

which is absurd since $L(A) \leq \pi(A)$ for any $A \in \mathcal{B}$. This proves assertion (i). Assertions (ii) and (iii) are similarly proved.

Similar analysis can be performed to find extrema of ratios of linear quantities of π, that is, quantities of the form $\{\int_\Theta f(\theta)\pi(d\theta)\}/\{\int_\Theta g(\theta)\pi(d\theta)\}$. ◁

Example 1. (continued)

As a first approximation, suppose that the engineers A and B assume that the prior probability they elicited for each C_i is uniformly spread out on the interval. Denote by L_0 and U_0 the obtained lower and upper measures. The last column in Table 2 gives posterior ranges of $P^\pi(C_i \mid x)$ as π ranges over $\Gamma(L_0, U_0)$. Comparing the last two columns in Table 2, we conclude that all the ranges have been reduced considerably by considering this particular $\Gamma(L_0, U_0)$. This reduction has been especially successful in the first and last intervals, i.e. C_1 and C_6.

Example 2.

Let $f(x_1, x_2 \mid \theta_1, \theta_2) = \frac{1}{2\pi} \exp\left\{\frac{-(x_1-\theta_1)^2-(x_2-\theta_2)^2}{2}\right\}$ and the class of priors $\Gamma(L, U)$ where $L(d\theta) = \frac{1}{2\pi(1.1)^2} \exp\left\{\frac{(\theta_1^2+\theta_2^2)}{2}\right\} d\theta$ and $U(d\theta) = \frac{1}{2\pi} d\theta$.

Let $H_0 : \theta_1 \leq \theta_2$ be the quantity of interest. For several observations $x = (x_1, x_2)$ the second column in table 3 gives ranges of $P^\pi(H_0 \mid x)$ as π ranges over $\Gamma(L, U)$ and in the third column the related ranges as π ranges over Γ_{DH}.

$\boldsymbol{x} = (x_1, x_2)$	$\underline{P}(H_0 \mid \boldsymbol{x}), \bar{P}(H_0 \mid \boldsymbol{x})$	$P_{DH}(H_0 \mid \boldsymbol{x}), P^{DH}(H_0 \mid \boldsymbol{x})$
$(0, 0)$	.39, .61	.29, .71
$(0.5, -0.5)$	.22, .41	.13, .49
$(1, -1)$	.10, .27	.04, .27
$(.5, 0)$	.30, .51	.20, .61
$(1.0, 0)$	.23, .43	.11, .52
$(1.5, 0)$	.14, .36	.06, .44

Table 3. *Ranges of the posterior probabilities of H_0.*

From second and third columns in table 3 it follows that the ranges of $P^\pi(H_0 \mid x)$ are substantially reduced when the class Γ_{DH} is restricted to the class $\Gamma(L, H)$. Note however that for the different observations considered no conclusive evidence is reached either in favour of nor against H_0. It should be also remarked that the prior imprecision implied in $\Gamma(L, U)$ is much smaller than that in Γ_{DH}, *a* point we explore next.

4. ELICITATION OF THE CLASS Γ(L,U)

Let $L(\Theta) = 1/\alpha, U(\Theta) = 1/\beta$, with $\alpha > 1, \beta < 1$. Denote the prior upper and lower probabilities of set C by $\bar{P}(C)$ and $P(C)$. We assume that the functional form of L and U, based for example on tail considerations, has been given. The following results help in the assessment of α and β, and give insight on the flexibility of the class to model partial prior knowledge.

Theorem 3. *Consider a set C. Then a priori, if:*
(i) $L(C) + U(C^c) \geq 1$, then $\underline{P}(C) = L(C)$,
(ii) $L(C) + U(C^c) < 1$, then $\underline{P}(C) = 1 - U(C^c)$,
(iii) $L(C^c) + U(C) \leq 1$, then $\overline{P}(C) = U(C)$,
(iv) $L(C^c) + U(C) > 1$, then $\overline{P}(C) = 1 - L(C^c)$.

Proof. It is immediate and we omit it. ◁

Denote by $\rho(C)$ the prior imprecision of set C, i.e. $\rho(C) = \bar{P}(C) - \underline{P}(C)$. We have then the following cases,

(a) If (i) and (iv) hold, then $\rho(C) = (\alpha - 1)/\alpha$,
(b) If (ii) and (iii) hold, then $\rho(C) = (1 - \beta)/\beta$,
(c) If (i) and (iii) hold, then $\rho(C) = U(C) - L(C)$,
(d) If (ii) and (iv) hold, then $\rho(C) = U(C^c) - L(C^c)$.

These expressions can be used for assessment. Note that when $U(d\theta)$ is improper a great variety of sets C will be in case (a), and then for all such sets $\rho(C) = (\alpha - 1)/\alpha$, that is a constant prior imprecision. This is a very inflexible prior assessment. In Example 2 $\rho(H_0) = 0.1735$. Thus, in working with $\Gamma(L, U)$ the assumption of an improper U leads to a very unreasonable prior imprecision.

If U is not improper a much more flexible and sensible prior assessment is obtained.

Regarding Γ_{DH} the inflexibility is the same as that of $\Gamma(L, U)$, and the imprecision for H_0 is now maximum.

ACKNOWLEDGEMENTS

We are indebted to James Berger and Peter Walley for useful discussions. We are also indebted to the referee for improving considerably the presentation of the paper.

REFERENCES

Berger, J. and O'Hagan, A. (1989). Ranges of posterior probabilities for unimodal priors with specified quantiles. *Bayesian Statistics 3* (J. M. Bernardo, M. H. DeGroot, D. V. Lindley and A. F. M. Smith, eds.), Oxford: University Press, 45–65, (with discussion).

Bose, S. (1990). *Bayesian Robustness with Shape-constrained Priors and Mixtures Priors.* Ph.D. Thesis, Purdue University.

Berliner, L. M. and Goel, P. K. (1990). Incorporating partial prior information: ranges of posterior probabilities. *Bayesian and Likelihood Methods in Statistics and Econometrics* (S. Geisser, J. S. Hodges and A. Zellner, eds.) Amsterdam: North-Holland, 397–406

DeRobertis, L. and Hartigan, J. A. (1981). Bayesian inference using intervals of measures. *Ann. Statist.* **9**, 235–244.

Lavine, M. (1988). Prior influence in Bayesian statistics. *Tech. Rep.* **88-06**, Duke University.

Martz, H. F. and Waller, R. A. (1982). *Bayesian Reliability Analysis.* New York: Wiley.

Moreno, E. and Cano, J. A. (1991). Robust Bayesian analysis with ϵ-contaminations partially known. *J. Roy. Statist. Soc. B* **53**, 143–155.

Moreno, E. and Pericchi, L. R. (1991). Robust Bayesian analysis for ϵ-contaminations with shape and quantile constraints. *Proc. Fifth Internat. Symp. on Applied Stochastic Model and Data Analysis* (R. Gutiérrez and M. J. Valderrama, eds.), Singapore: World Scientific Publishing, 454–470.

O'Hagan, A. and Berger, J. (1988). Ranges of posterior probabilities for quasi-unimodal priors with specified quantiles. *J. Amer. Statist. Assoc.* **83**, 503–508.

Pericchi, L. R. and Walley, P. (1991). Robust Bayesian credible intervals and prior ignorance. *Internat. Statist. Rev.* **59**, 1–23.

Wasserman, L. and Kadane J. B. (1991). Computing bounds on expectations. *Tech. Rep.* **504**, Carnegie Mellon University.

BAYESIAN STATISTICS 4, pp. 715–721
J. M. Bernardo, J. O. Berger, A. P. Dawid and A. F. M. Smith, (Eds.)

Bayesian Design for Random Walk Barriers

GIOVANNI PARMIGIANI and NICHOLAS G. POLSON
Duke University, USA and *University of Chicago, USA*

SUMMARY

We study the optimal design of absorption barriers and sample size for a simple random walk. High barriers yield more informative experiments, but may be more costly. We obtain optimal designs that are balancing the relative advantages of choosing a high barrier with few replications versus a lower barrier with more replications. We address the problem from a Bayesian, decision theoretic, viewpoint, by using a utility function based on a limiting form of Shannon information. After discussing properties of the optimal designs, we apply the results to a model describing the growth of cancer cells, and we illustrate how the selection of the prior distribution influences the solution.

Keywords: DESIGN; SHANNON INFORMATION; STOCHASTIC GROWTH MODEL; RANDOM WALK.

1. INTRODUCTION AND OVERVIEW

Consider a simple random walk N_t, $t = 0, 1, 2\ldots$ with unknown probability Θ of an upward step, and assume that the walk starts at zero and has absorbing barriers at m_1 and $-m_2$, where m_1 and m_2 are positive integers. An experimenter is interested in learning about Θ from the observation of the barrier hitting time and of which barrier the walk hits. He is allowed to replicate the walk n times and to select the values of m_1 and m_2, and wishes to do so optimally. In this paper we consider the optimal choice of both the absorption barriers and the number of replicates n.

This problem constitutes an ideal setting to illustrate how one can address simultaneously two of the fundamental trade-offs associated with experimental design. The first, and most often encountered, is that between cost of collecting information and cost of uncertainty. The second, also of practical relevance in a variety of experimental situations, is that between few replications of an informative and expensive design and more replications of a less informative but inexpensive design. Such trade-offs also arise, for example, in survival analysis and logistic regression.

In Section 2, we introduce a utility function accounting for the amount of information to be learnt about Θ and the cost of experimentation. The information measure used is based on a useful limiting form of Shannon information. In Section 3, we discuss properties of the optimal design. In particular, we show that if the cost is proportional to the number of steps of the walk, or if the 'exchange rate' between information and money is larger than $1/2$, then the simple Bernoulli trial ($m = 2$) with replication is the optimal design. When an additional penalty for replications is introduced, higher values of the barrier become attractive. Moreover, as long as the prior distribution is symmetric, it is optimal to select barriers equidistant from the starting point of the walk. Finally, in Section 4, we consider an application to the analysis of tumour growth, and illustrate it with various choices of prior distributions.

2. UTILITY AND INFORMATION

2.1. *Choice of the Utility Function*

Let $\mathcal{E}$ denote the experiment consisting of n exchangeable replications of the walk, with absorption barriers at m_1 and m_2. Suppose that each replication provides the observation of the time to absorbtion $X = \min\{x \mid N_x = m_1 \text{ or } N_x = -m_2\}$, and of the indicator $Y = I_{\{N_X = m_1\}}$ of which barrier the process hits.

For a specified prior distribution on Θ, a natural measure of the information contained in $\mathcal{E}$, as pointed out, for example, in Lindley (1956), is the gain in Shannon information

$$I_n = E\left\{\log\left(\frac{p(\Theta|\mathcal{E})}{p(\Theta)}\right)\right\},$$

where the expectation is taken with respect to the joint distribution of the parameter Θ and the experiment $\mathcal{E}$. In many design problems, however, I_n is a difficult quantity to study. Polson (1988) proposed to replace it with a quantity based on the asymptotic expansion of I_n derived in the i.i.d. case by Ibragimov and H'asminsky (1973). Let Ω be the support of p and consider the limit of I_n as the number of replications tends to infinity. Then, under suitable regularity conditions:

$$\lim_{n\to\infty}\left\{I_n - \frac{1}{2}\log\left(\frac{n}{2\pi e}\right)\right\} = \int_\Omega p(\theta)\log\left(\frac{i_{\mathcal{E}}(\theta)^{\frac{1}{2}}}{p(\theta)}\right)d\theta$$

where $i_{\mathcal{E}}(\theta)$ is Fisher information. Based on this limiting form we then adopt, as a criterion for the choice of the experiment $\mathcal{E}$, the expected normalized gain in information for large n, that is:

$$\frac{1}{2}\log\left(\frac{n}{2\pi e}\right) + \log\left(\frac{i_{\mathcal{E}}(\theta)^{\frac{1}{2}}}{p(\theta)}\right).$$

As with the experimentation cost, we assume that each replication entails a cost k, and that each of the Bernoulli trials underlying the random walk entails a cost c.

We can then write the utility associated with the experiment $\mathcal{E}$ for fixed θ as:

$$\mathcal{U}(\theta,\mathcal{E}) = \frac{1}{2}\log\left(\frac{n}{2\pi e}\right) + \log\left(\frac{i_{\mathcal{E}}(\theta)^{\frac{1}{2}}}{p(\theta)}\right) - \{cE(X|\theta) + k\}n. \tag{1}$$

This entails an expected utility function $U(m_1, m_2, n) = \int_\Omega \mathcal{U}(\theta,\mathcal{E})p(\theta)d\theta$ on the design parameters and the sample size. The choice of the optimal design and sample size consists then in the maximization of U with respect to m_1, m_2 and n.

The quantities k and c in the above expression are to be imagined as expressed in units of information. In most circumstances, the costs of replications and steps can be easily specified in terms of monetary amounts. If the information about Θ can also be translated into monetary terms, then k and c represent the ratio between the price of a replication (and a step) and the value of one unit of information. In general, however, the specification of the monetary value of one unit of information can be troublesome. In such cases, we find it attractive to think of the design problem as that of maximising the information under the constraint of a fixed expected cost for experimentation $[cE(X|\theta) + k]n$. Naturally, the first order conditions for the optimizations are the same, provided that one knows k/c and interprets the value of c in the unconstrained problem as the Lagrange multiplier of the constrained problem. Of further help in the choice of the appropriate cost parameters are trade-off plots where the expected cost of experimentation is graphed against the maximum information obtainable at that expected cost.

2.2. *Fisher Information*

An important feature of the experimental setting considered here is that each replication is more informative the longer it takes for the walk to hit one of the barriers. Intuitively, a longer walk corresponds to a larger number of Bernoulli trials. In particular, the Fisher information, on which the utility function adopted here is based, incorporates this feature in a natural way. An application of Wald's identity (see, for example, Cox and Miller, 1965), leads to the expression:

$$i_{\mathcal{E}}(\theta) = \frac{E(X|\theta)}{\theta(1-\theta)}. \tag{2}$$

The information in one replication is given by the information in a single Bernoulli trial, multiplied by the expected number of trials. In particular, in this case, it is useful to recall that:

$$E(X|\theta) = \frac{m_1\theta^{m_1}[\theta^{m_2} - (1-\theta)^{m_2}] - m_2(1-\theta)^{m_2}[\theta^{m_1} - (1-\theta)^{m_1}]}{(\theta^{m_1+m_2} - (1-\theta)^{m_1+m_2})(2\theta - 1)} \tag{3}$$

$$E(Y|\theta) = \frac{\theta^{m_1}[\theta^{m_2} - (1-\theta)^{m_2}]}{\theta^{m_1+m_2} - (1-\theta)^{m_1+m_2}}; \tag{4}$$

For ease of future reference, let us define $\varphi_\theta(m_1, m_2) = E(X|\theta)$ in (3). It can be verified that $\varphi_\theta(m_1, m_2)$ is increasing in both m_1 and m_2 for every θ. Figure 1 illustrates $i_{\mathcal{E}}(\theta)$ for various choices of m_1 and m_2. In passing, we note that the strong design dependence of Fisher information will affect the family of Jeffreys' priors for this experiment; as a consequence, the mechanical adoption of such priors will lead to choices of the design that conflict with standard expected utility theory and may even violate the dominance principle.

3. PROPERTIES OF THE OPTIMAL DESIGN

Based on Proposition 1, the problem of choosing the optimal design and sample size can be rewritten, neglecting irrelevant constants, as that of minimizing

$$\frac{1}{2}\log n + \frac{1}{2}\int_\Omega \log\varphi_\theta(m_1, m_2)p(\theta)d\theta - \left(c\int_\Omega \varphi_\theta(m_1, m_2)p(\theta)d\theta + k\right)n \tag{5}$$

with respect to (m_1, m_2, n). It can be verified that if the prior distribution is proper, the above integrals always exists, since φ is larger than one, and bounded above in θ for fixed design. The utility function depends on the design only through the function φ. For fixed θ and n, a higher φ increases the costs linearly and the information logarithmically. Note that if $c = 0$, increasing φ always increases expected utility so that the optimal design problem has no solution. Assume henceforth that $c > 0$.

To simplify the analysis we will from now on treat m_1, m_2 and n as real variables. For fixed m_1, m_2 it is possible to derive the optimal sample size.

Property 1: *The optimal sample size is:*

$$n^* = \frac{1}{2}\left(c\int_\Omega \varphi_\theta(m_1, m_2)p(\theta)d\theta + k\right)^{-1} \tag{6}$$

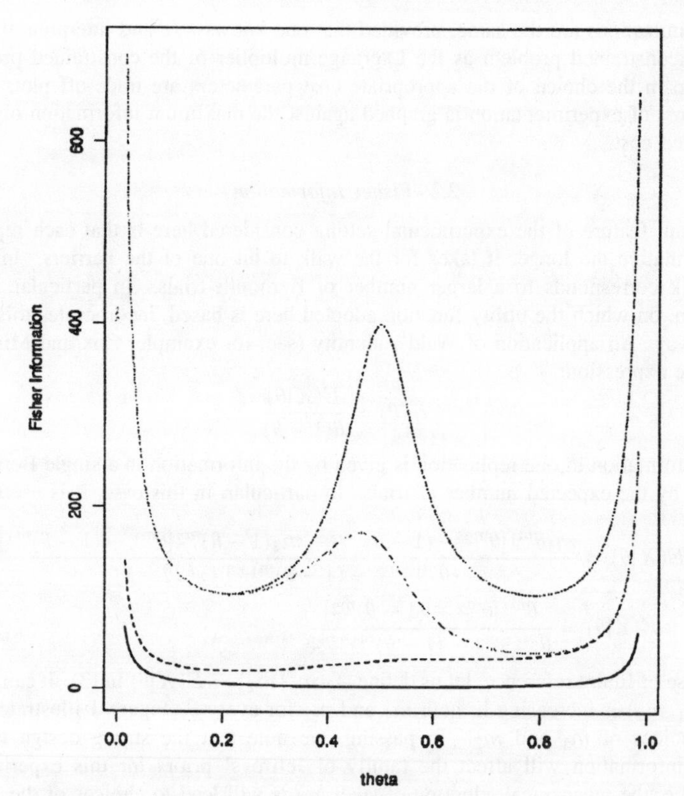

Figure 1. *Fisher Information as a function of Theta, for varying values of the design parameters; from the bottom,* $m_1 = m_2 = 1, m_1 = 4$ *and* $m_2 = 2, m_1 = 4$ *and* $m_2 = 10, m_1 = m_2 = 10$.

Proof. Differentiating (5) and setting the result to zero gives (6). Second order conditions are straightforward to verify. ◁

Substituting n^* in (5), and again neglecting irrelevant constants, reduces the optimal design problem to the maximization of:

$$U^* = \int_\Omega \log \varphi_\theta(m_1, m_2)p(\theta)d\theta - \log\left(\int_\Omega \varphi_\theta(m_1, m_2)p(\theta)d\theta + \frac{k}{c}\right). \quad (7)$$

A noticeable consequence of this on the analysis of the trade-offs is that the choice of the optimal design variables m_1 and m_2 only requires the specification of the ratio k/c, and therefore does not depend on the 'exchange rate' the experimenter decides to adopt between information and money. The exchange rate is only influencing the determination of n and the decision of whether or not to perform the experiment. This is a direct consequence of the fact that the cost component of the utility function is linear in n for fixed design parameters, and of the fact that a more expensive design affects the cost component of the utility function only through a term multiplying n. Finally, as expected, φ and n are inversely related. Heuristically, if c is increased, it becomes comparatively more convenient to replicate the experiment many times with small barriers, rather than doing a few replications with relatively higher barriers.

We now give sufficient conditions for optimality and use them to derive some useful properties of the optimal design. Let

$$\varphi_\theta^{(i)}(m_1, m_2) = \frac{\partial \varphi}{\partial m_i}, \qquad i = 1, 2. \tag{8}$$

A system of sufficient conditions for optimality is given by the following.

Property 2: *The optimum design variables m_1 and m_2 must satisfy:*

$$\int_\Omega \varphi_\theta(m_1, m_2)p(\theta)d\theta + \frac{k}{c} = \frac{\int_\Omega \varphi_\theta^{(1)}(m_1, m_2)p(\theta)d\theta}{\int_\Omega \frac{\varphi_\theta^{(1)}(m_1,m_2)}{\varphi_\theta(m_1,m_2)}p(\theta)d\theta} = \frac{\int_\Omega \varphi_\theta^{(2)}(m_1, m_2)p(\theta)d\theta}{\int_\Omega \frac{\varphi_\theta^{(2)}(m_1,m_2)}{\varphi_\theta(m_1,m_2)}p(\theta)d\theta}. \tag{9}$$

Proof. From (7):

$$\frac{\partial U^*}{\partial m_i} = \int_\Omega \frac{\varphi_\theta^{(i)}(m_1, m_2)}{\varphi_\theta(m_1, m_2)}p(\theta)d\theta - \frac{\int_\Omega \varphi_\theta^{(i)}(m_1, m_2)p(\theta)d\theta}{\int_\Omega \varphi_\theta(m_1, m_2)p(\theta)d\theta + \frac{k}{c}}, \tag{10}$$

Setting the above to zero yields the desired conditions. ◁

The following two properties give sufficient conditions for the simple Bernoulli trial to be the optimal design.

Property 3: *If $k = 0$ the optimal design is $m_1 = m_2 = 1$ for every choice of c and of the prior distribution. Consequently, the optimal sample size is $n^* = 1/(2c)$.*

Proof. From (3), we have: $\varphi_\theta(1, 1) = 1$ for every θ. Substituting into (9), all three terms are equal to unity. ◁

Property 4: *If $c > 1/2$ then $m_1 = m_2 = 1$ for every k and prior distribution such that $n^* \geq 1$.*

Proof. Rewrite (10) as:

$$\frac{\partial U^*}{\partial m_i} = \int_\Omega \frac{\varphi_\theta^{(i)}(m_1, m_2)}{\varphi_\theta(m_1, m_2)}p(\theta)d\theta - 2cn^* \int_\Omega \varphi_\theta^{(i)}(m_1, m_2)p(\theta)d\theta$$

We know $\varphi \geq 1$,therefore:

$$\int_\Omega \frac{\varphi_\theta^{(i)}(m_1, m_2)}{\varphi_\theta(m_1, m_2)}p(\theta)d\theta < \int_\Omega \varphi_\theta^{(i)}(m_1, m_2)p(\theta)d\theta$$

and both $\partial U^*/\partial m_i$ are negative if $n^* > 1/2c$. This is guaranteed if $c > 1/2$. ◁

Finally, we give a condition for the optimal design to be symmetric.

Property 5: *If the prior is symmetric about* $1/2$, *that is if* $p(\theta) = p(1-\theta)$, *then* $m_1 = m_2$ *satisfies the optimality conditions.*

Proof. From (3), we have that $\varphi_\theta(m_1, m_2) = \varphi_{1-\theta}(m_2, m_1)$, and, therefore, using (8), that $\varphi_\theta^1(m_1, m_2) = \varphi_{1-\theta}^2(m_2, m_1)$. Also, from the symmetry of the prior distribution,

$$\int_\Omega \varphi_\theta^{(1)}(m,m)p(\theta)d\theta = \int_\Omega \varphi_{1-\theta}^{(2)}(m,m)p(\theta)d\theta = \int_\Omega \varphi_\theta^{(2)}(m,m)p(\theta)d\theta$$

$$\int_\Omega \frac{\varphi_\theta^{(1)}(m,m)}{\varphi_\theta(m,m)}p(\theta)d\theta = \int_\Omega \frac{\varphi_{1-\theta}^{(2)}(m,m)}{\varphi_{1-\theta}(m,m)}p(\theta)d\theta = \int_\Omega \frac{\varphi_\theta^{(2)}(m,m)}{\varphi_\theta(m,m)}p(\theta)d\theta$$

so that the right hand equation of (9) is satisfied. ◁

4. AN APPLICATION TO CANCER GROWTH MODELS

A fruitful area of application for the problem considered in this paper is that of growth models. For example, Downham and Morgan (1973) and Downham and Green (1976) analyse the spread of cancer cells in a layer of competing normal cells. In this application, the position N_t of the walk represents the number of cancer cells after t cell divisions. The random walk is assumed to start at $N_0 = 1$, when the first abnormal cell appears, and to stop if no cancer cells are left, so that zero is the lower absorbtion barrier. Also, most empirical work is carried out based on experiments where there is an upper absorption barrier at a fixed number m of abnormal cells — typically related to the detectability threshold of the tumour. The parameter of interest is usually $\Gamma = \Theta/(1-\Theta)$, the relative division rate; Γ is believed to be larger than 1 and called carcinogenic advantage. Since the expected utility we adopt is invariant under one-to-one transformations of the parameter of interest, the conclusions reached hold for Γ as well. In this application the quantity k may be interpreted as the cost of one experimental unit, say a mouse, whereas c represents a cost for waiting.

This particular growth model obtains as a special case of the model discussed in the previous sections by setting $m_2 = 1$ and $m_1 = m - 1$, $m \geq 2$. The expected number of steps and the probability of hitting the upper barrier specialise to:

$$E(X|\theta) = \frac{m\theta^{m-1}}{\theta^m - (1-\theta)^m} - \frac{1}{2\theta - 1} \tag{11}$$

$$E(Y|\theta) = \frac{\theta - (1-\theta)}{\theta^m - (1-\theta)^m}\theta^{m-1}. \tag{12}$$

Since $E(X|\theta) = 1$ at $m = 2$, Equation (2) provides further intuition to the results of Hinkley (1979), who found that $m = 2$ is the choice that guarantees the smallest discrepancy between observed and expected information.

More generally however,

$$\frac{\partial \varphi_\theta(m)}{\partial m} = \frac{\theta^{m-1}\left[\theta^m - (1-\theta)^m(1 + m\log\theta(1-\theta))\right]}{(\theta^m - (1-\theta)^m)^2} > 0. \tag{13}$$

Therefore, as expected, the conditional information contained in a single replication is an increasing function of m, for every θ, so that Bernoulli trials are not necessarily optimal.

Let us now briefly illustrate the choice of the design variable m for various prior distributions. First, we take Θ uniform on $(0, 1)$; then $m = 5$ is optimal at $\frac{k}{c} = 1$. Moreover, if $\Theta \sim Beta(4, 1)$, $m = 18$ is optimal for $\frac{k}{c} = 1$. Note that as the mass is placed on higher values of θ, a higher value of m becomes more convenient. This dependence becomes extreme if the prior is uniform on $(\frac{1}{2}, 1)$: for example $\frac{k}{c} = 1$ implies $m = 773$. This is not an implausible specification for a stochastic growth, where the experimenter usually believes that $\Theta > \frac{1}{2}$. The reason for such a drastic change is that the latter prior assigns a much smaller probability to less informative replications —such as replications where the lower barrier is hit at the first step— and therefore makes the few replications with a higher barrier a more attractive choice.

ACKNOWLEDGEMENT

We thank Jay Kadane, Teddy Seidenfeld, Larry Wasserman and a referee for helpful comments.

REFERENCES

Cox, D. R. and Miller, H. D. (1965). *The Theory of Stochastic Processes.* London, Methuen.

Downham, D. Y. and Morgan, K. (1973). A stochastic model for a two-dimensional growth on a square lattice. *Bull. Internat. Statist. Institute.* **45**, 324–331.

Downham, D. Y. and Green, D. H. (1976). Inference for a two-dimensional stochastic growth model. *Biometrika* **63**, 551–554.

Hinkley, D. V. (1979). Likelihood inference for a simple growth model. *Biometrika* **66**, 659–662.

Ibragimov, I. and H'asminsky, R. (1973). On the information contained in a sample about a parameter. *Second Internat. Symp. on Information Theory,* (B. N. Petyrov and F. Csaki, eds.), Budapest: Aksdemiai Kiado, 295–309.

Kullback, S. and Leibler, R. A. (1951). On information and sufficiency. *Ann. Math. Statist.* **22**, 79–86.

Lindley, D. V. (1956). On a measure of the information provided by an experiment. *Ann. Math. Statist.* **27**, 986–1005.

Polson, N. G. (1988). *Bayesian Perspectives on Statistical Modelling.* Ph.D. Thesis, University of Nottingham.

BAYESIAN STATISTICS 4, pp. 723–730
J. M. Bernardo, J. O. Berger, A. P. Dawid and A. F. M. Smith, (Eds.)

Analysis of Multistage Survey as a Bayesian Hierarchical Model

M. E. PÉREZ and L. R. PERICCHI
Universidad Simón Bolívar, Caracas, Venezuela

SUMMARY

Using the hierarchical approach to the usual linear model we obtain a satisfactory Bayesian solution to multistage survey with unknown variances. This generalizes some results and corrects other results related to a unknown variances and regression relation, previously obtained by Scott and Smith (1969). Furthermore a reparameterization is employed which permits to reduce the dimension of the integration involved for HPD intervals. Finally, the results are applied in the context of a simulation study.

Keywords: HIERARCHICAL LINEAR MODEL; MULTISTAGE SURVEY; REPARAMETERIZATION.

1. INTRODUCTION

Let us suppose that the population under study has N clusters (units). The M_i elements in the ith cluster have values $y_{i1}, y_{i2}, \ldots, y_{iM_i}$. At first stage we select a sample, denoted by s, of n of the N clusters. At the second stage we select a sample, s_i, of m_i different elements from the M_i elements in the ith cluster. This procedure is called *two stage sampling*.

Scott and Smith (1969) specify a model for the population elements supposing that:

A1. The M_i elements in the i-th cluster are uncorrelated observations from a distribution with mean μ_i and variance σ_i^2.

A2. The means $\mu_1, \mu_2, \ldots, \mu_N$ are uncorrelated observations from a distribution with mean v and variance δ^2.

These assumptions are equivalent to specifying a superpopulation model in which we have exchangeability between elements in a given cluster and between the clusters means μ_i.

Supposing that the distributions in assumptions A1 and A2 are normal and that it is reasonable to approximate the prior density of v by $p(v) \propto$ constant, Scott and Smith derive the following posterior distribution when σ_i^2 and δ^2 are known:

$$p(\boldsymbol{y}, \boldsymbol{\mu}, v) \propto \exp\left\{-\frac{1}{2}\left[\sum_{i=1}^{N}\left(\frac{\mu_i - v}{\delta}\right)^2 + \sum_{i=1}^{N}\sum_{j=1}^{M_i}\left(\frac{y_{ij}-\mu_i}{\sigma_i}\right)^2\right]\right\} \tag{1.1}$$

The marginal posterior distribution of $\boldsymbol{\mu} = (\mu_1, \mu_2, \ldots, \mu_N)'$ is an N-variate normal with mean:

$$E(\mu_i) = \lambda_i y_{i.} + (1-\lambda_i)\overline{y} \tag{1.2}$$

where:

$$\lambda_i = \begin{cases} \frac{\delta^2}{\delta^2+(\sigma_i^2/m_i)} & i \in s \\ 0 & i \notin s \end{cases} \quad ; \quad y_{i.} = \sum_{j\in s_i}\frac{y_{ij}}{m_i} \quad ; \quad \overline{y} = \frac{\sum_{i=1}^N \lambda_i y_{i.}}{\sum_{i=1}^N \lambda_i}.$$

and covariance matrix C with elements:

$$c_{ij} = \begin{cases} (1-\lambda_i)^2 u^2 + (1-\lambda_i)\delta^2 & i = j \\ (1-\lambda_i)(1-\lambda_j)u^2 & i \neq j \end{cases}$$

where:

$$u^2 = \left[\sum_{i \in s} \left(\delta^2 + \frac{\sigma_i^2}{m_i} \right)^{-1} \right]^{-1}.$$

Expression (1.2) gives a neat solution. If the cluster i has been sampled ($i \in s$) the maximum likelihood estimator is pulled towards the overall mean. Otherwise, the mean is the overall sample mean.

When variances are unknown, Scott and Smith make the simplifying assumptions that $\sigma_i^2 = \sigma^2$, $m_i = m$. Choosing $p(v, \sigma^2, \delta^2) \propto \lambda \sigma^{-2} \delta^{-2}$, where

$$\lambda = \frac{\delta^2}{(\delta^2 + \sigma^2/m)},$$

the model is equivalent to the random effects model used by Box and Tiao (1968). Using some of their results, the posterior distribution of μ conditional on λ is a multivariate t-distribution. If f and F denote the density and c.d.f. of an F-variable with $(n-1, n(m-1))$ d.f., the marginal posterior density of λ is given by:

$$p(\lambda|s) = \frac{rf(\lambda r)}{F(r)} \tag{1.3}$$

where:

$$r = mn(m-1) \frac{\sum_{i \in s} (y_{i.} - \overline{y})^2}{(n-1) \sum_{i \in s} \sum_{j \in s_i} (y_{ij} - y_{i.})^2}.$$

They obtain the following expected values for the μ's:

$$E(\mu_i) = \begin{cases} \hat{\lambda} y_{i.} + (1-\hat{\lambda})\overline{y} & i \in s \\ \overline{y} & i \notin s \end{cases} \tag{1.4}$$

where:

$$1 - \hat{\lambda} = \frac{n(m-1)}{r(nm-m-2)} \frac{I_x(\frac{n-1}{2}, \frac{nm-m-2}{2})}{I_x(\frac{n-1}{2}, \frac{nm-m}{2})} \quad ; \quad x = \frac{m \sum_{i \in s} (y_{i.} - \overline{y})^2}{\sum_{i \in s} \sum_{j \in s_i} (y_{ij} - y_{i.})^2}$$

and $I_x(p, q)$ denotes the incomplete beta function.

It is important to observe that the prior distribution used by Scott and Smith depends on the size of the sample. This is an unfortunate feature, because there is no reason to suppose that our previous knowledge depends on the particular experiment we perform. Also, the result does not seem to relate to that obtained in the case of known variances. The assumption of equality in the sample sizes for each cluster appears to be very restrictive (see Lindley (1972)). Our main goal is to obtain an alternative approach to the case of unknown variances. In Section 2 we present that approach, and in Section 3 we make some extensions of the result. Finally, in Section 4 a simulation is analysed.

2. AN ALTERNATIVE APPROACH FOR THE NORMAL DISTRIBUTION, UNKNOWN EQUAL VARIANCES

The previous model can be analysed as a particular case of the General Bayesian Hierarchical Model developed in Lindley (1971) and Lindley and Smith (1972).

Following Lindley's ideas for the Hierarchical Bayesian model, and using them in the case of sampling, we suppose that $\sigma_i^2 = \sigma^2$ (we do not need to suppose that $m_i = m$), and assign prior distributions to σ^2 and δ^2 in the following way:

$$\frac{\nu_W \lambda_W}{\sigma^2} \propto \chi^2_{\nu_W} \quad ; \quad \frac{\nu_B \lambda_B}{\delta^2} \propto \chi^2_{\nu_B}$$

where $\nu_W, \lambda_W, \nu_B, \lambda_B$ are arbitrary parameters for the distribution. (For an explanation of the meaning of these parameters, see Lindley (1971).)

So, the prior distributions for σ^2 and δ^2 are inverse χ^2. If we choose

$$p(v) \propto \text{constant}$$

we obtain the following joint prior distribution:

$$p(v, \sigma^2, \delta^2) \propto \exp\left\{-\frac{\nu_W \lambda_W}{2\sigma^2} - \frac{\nu_B \lambda_B}{2\delta^2}\right\} \sigma^{\nu_W - 2} \delta^{\nu_B - 2}. \tag{2.1}$$

Using this prior we can see that the posterior distribution conditional on the sample is the product of two multivariate t kernels:

$$p(\mu|y) \propto \left\{ S_s^2 + \sum_{i \in s} m_i (y_{i.} - \mu_i) + \nu_W \lambda_W \right\}^{-\frac{1}{2}(M_s + \nu_W)}$$

$$\times \left\{ \sum_{i=1}^{N} (\mu_i - \mu_.)^2 + \nu_B \lambda_B \right\}^{-\frac{1}{2}(N + \nu_B - 1)} \tag{2.2}$$

where

$$S_s^2 = \sum_{i \in s} \sum_{j \in s_i} (y_{ij} - y_{i.})^2 \quad ; \quad M_s = \sum_{i \in s} m_i \quad ; \quad \mu_. = \frac{\sum_{i=1}^{N} \mu_i}{N}.$$

Now we can find the modal estimator for each μ_i:

$$\mu_i = \begin{cases} \lambda_i y_{i.} + (1 - \lambda_i)\mu_. & i \in s \\ \mu_. & i \notin s \end{cases} \tag{2.3}$$

where

$$\lambda_i = \begin{cases} \frac{S_B^2}{S_B^2 + (S_W^2/m_i)} & i \in s \\ 0 & i \notin s \end{cases}$$

$$S_W^2 = \frac{\{S_s^2 + \sum_{k \in s} m_k (\mu_k - y_{k.})^2 + \nu_W \lambda_W\}}{M_s + \nu_W} \quad ; \quad S_B^2 = \frac{\{\nu_B \lambda_B + \sum_{k=1}^{n} (\mu_k - \mu_.)^2\}}{N + \nu_B - 1}$$

S_W^2 and S_B^2 are the modal estimators for σ^2 and δ^2, respectively.

This result is analogous to that obtained by Scott and Smith when the variances are known, but in this case σ^2 and δ^2 have been replaced by their modal estimators, S_W^2 and

S_B^2. Additionally, we see that this estimator has the form of a shrinkage estimator. We can observe too that the inference obtained does not depend upon the sampling method employed.

These equations can be solved iteratively, selecting as initial iteration the (inadmisible for dimension bigger than 2) least squares estimators:

$$\mu_i^{(0)} = \begin{cases} y_{i.} & i \in s \\ y_{..} & i \notin s \end{cases}$$

$$S_W^{2\ (0)} = \frac{\sum_{i \in s} \sum_{j \in s_i} (y_{ij} - y_{i.})^2}{M_s - n} \quad ; \quad S_B^{2\ (0)} = \frac{\sum_{i=1}^{N} (\mu_i - \mu_.)^2}{N-1}$$

Summing over the i's, we can show that at the modal value:

$$\mu_. = \frac{\sum_{i=1}^{N} \lambda_i y_{i.}}{\sum_{i=1}^{N} \lambda_i} = \overline{y}$$

Although this result is intuitively appealing, it does not allow the calculation of HPD intervals. If we want to calculate the latter, we will need N-dimensional integrals, which are computationally expensive.

We can apply a useful reparametrization to the posterior obtained by Lindley for the usual linear model, making $\rho = \delta^2/\sigma^2$. Using this transformation, we obtain:

$$p(\boldsymbol{\mu}|\rho, \boldsymbol{y}) \propto \left[1 + (\boldsymbol{\mu} - \boldsymbol{\tau}(\rho))' \frac{C(\rho)}{\nu S^2(\rho)} (\boldsymbol{\mu} - \boldsymbol{\tau}(\rho))\right]^{-\frac{1}{2}(N+\nu)} \tag{2.4}$$

where

$$\nu = M_s + \nu_B + \nu_W - 1 \quad ; \quad \overline{y} = \frac{\sum_{i \in s} \frac{m_i y_{i.}}{m_i \rho + 1}}{\sum_{i \in s} \frac{m_i}{m_i \rho + 1}} \quad ; \quad \tau_i = \begin{cases} \frac{m_i \rho y_{i.} + \overline{y}}{m_i \rho + 1} & i \in s \\ \overline{y} & i \notin s \end{cases}$$

$$c_{ij}(\rho) = \begin{cases} a_i + b & i = j \\ b & i \neq j \end{cases} \quad ; \quad a_i = \begin{cases} m_i \rho + 1 & i \in s \\ 1 & i \notin s \end{cases} \quad ; \quad b = -\frac{1}{N}$$

$$S^2(\rho) = \frac{1}{\nu}\left(\rho S_s^2 + \rho \nu_W \lambda_W + \rho \sum_{i \in s} \frac{m_i}{m_i \rho + 1} (y_{i.} - \overline{y})^2 + \nu_B \lambda_B\right)$$

This conditional distribution is proportional to an N-variate Student-t with mean $\boldsymbol{\tau}(\rho)$ and variance $S^2(\rho) C^{-1}(\rho)$ with ν degrees of freedom.

To obtain the posterior distribution of $\boldsymbol{\mu}$, we have to know the conditional distribution of ρ given y. This conditional distribution, after tedious calculation (see Pérez and Pericchi (1991)), is:

$$p(\rho|\boldsymbol{y}) \propto \rho^{\frac{1}{2}(M_s + \nu_W - 3)} \prod_{i \in s} \frac{1}{(m_i \rho + 1)^{\frac{1}{2}}} \times S^2(\rho)^{-\frac{1}{2}\nu} \left[\sum_{i \in s} \frac{m_i}{m_i \rho + 1}\right]^{-\frac{1}{2}} \tag{2.5}$$

Finally, we can obtain the distribution of $\boldsymbol{\mu}$ conditional on the sample, by making

$$p(\boldsymbol{\mu}|\boldsymbol{y}) = \int_0^{\infty} p(\boldsymbol{\mu}|\rho, \boldsymbol{y}) p(\rho|\boldsymbol{y}) d\rho$$

which can be calculated using numerical methods, such as those implemented in the BAYES 4 package, and only requiring integrals of low orders.

Comparing (2.2) and (2.4), we see the advantages of the reparameterization. A cumbersome product of multivariate t's is replaced by a multivariate t, conditional on a simple parameter with known distribution.

We have the additional advantage that we can obtain posterior credible intervals for our estimates.

Using the properties of the multivariate t, we can also obtain the marginal distribution for each μ_i from (2.4). The conditional distribution of μ_i on ρ and on the sample is:

$$p(\mu_i|\rho, \boldsymbol{y}) \propto \left[1 + \frac{(\mu_i - \tau_i(\rho))^2}{\nu S^2(\rho) {c_{ii}}^{-1}(\rho)}\right]^{-\frac{1}{2}(\nu+1)} \tag{2.6}$$

with

$$c_{ii}{}^{-1}(\rho) = \begin{cases} \frac{1}{m_i\rho+1} + \frac{1}{(m_i\rho+1)^{\frac{1}{2}} \sum_{k\in s} \frac{m_k\rho}{m_k\rho+1}} & i \in s \\ 1 + \frac{1}{\sum_{k\in s} \frac{m_k\rho}{m_k\rho+1}} & i \notin s \end{cases}$$

Then, the marginal distribution of μ_i is:

$$p(\mu_i|\boldsymbol{y}) = \int_0^\infty p(\mu_i|\rho, \boldsymbol{y}) p(\rho|\boldsymbol{y}) d\rho$$

If we suppose ρ known, and we assign to σ^2 an inverse χ^2 prior and an uniform distribution for v:

$$\frac{\nu_W \lambda_W}{\sigma^2} \propto \chi^2_{\nu_W} \quad , \quad p(v) \propto \text{constant}$$

we obtain the following posterior distribution for $\boldsymbol{\mu}$:

$$p(\boldsymbol{\mu}|\boldsymbol{y}) \propto \left[1 + (\boldsymbol{\mu} - \boldsymbol{\tau})' \frac{C}{\nu S^2} (\boldsymbol{\mu} - \boldsymbol{\tau})\right]^{-\frac{1}{2}(N+\nu)}$$

which is proportional to an N-multivariate t with mean $\boldsymbol{\tau}$ and variance $S^2 C^{-1}$ with ν degrees of freedom. In this case, the mean and the modal value are the same:

$$E(\mu_i) = \tau_i = \begin{cases} \frac{m_i \rho y_{i.} + \bar{y}}{m_i\rho+1} & i \in s \\ \bar{y} & i \notin s \end{cases} \tag{2.8}$$

This result is similar to that obtained by Scott and Smith when variances are known. However, there is a change in the posterior distribution –we obtain a t distribution instead of a normal distribution. That is sensible, because we have less information about variances.

3. SOME EXTENSIONS OF THE RESULTS

Let us consider $\sigma_i^2 = k_i\sigma^2$, k_i known for each i. We can repeat the same procedure for this case, and we obtain analogous results, with m_i replaced by m_i/k_i in the weights of the estimators and in the posterior distributions.

On the other hand, there are situations in which the average value of the elements in a given cluster is thought to be related to some known variable, say X.

Since X provides information about the μ's, we must substitute our assumption A2 by:

A2'. The mean μ_i is normally distributed about $\alpha + \beta X_i$, with variance δ^2.

Let us suppose also that is reasonable to approximate the prior density for (α, β) by:

$$p(\alpha, \beta) \propto \text{constant}$$

Scott and Smith solve this problem for known variances. With variances unknown, we can again assign the inverse χ^2 priors to σ^2 and δ^2, and so we obtain the following posterior distribution:

$$p(\mu|y) \propto \left\{ S_s^2 + \sum_{i\in s} m_i(y_{i.} - \mu_i) + \nu_W\lambda_W \right\}^{-\frac{1}{2}(M_s+\nu_W)}$$

$$\times \left\{ \sum_{i=1}^{N} (\mu_i - \hat{\mu}_i)^2 + \nu_B\lambda_B \right\}^{-\frac{1}{2}(N+\nu_B-3)} \tag{3.1}$$

where $\hat{\mu}_i = \hat{\alpha} + \hat{\beta}(X_i - X_.)$, with $\hat{\alpha}$ and $\hat{\beta}$ the usual least squares estimators.

From this posterior distribution, we get the following modal estimators:

$$\mu_i = \begin{cases} \frac{m_i S_B^2 y_{i.} + S_W^2 \hat{\mu}_i}{m_i S_B^2 + S_W^2} & i \in s \\ \hat{\mu}_i & i \notin s \end{cases} \tag{3.2}$$

where:

$$S_W^2 = \frac{\{S_s^2 + \sum_{k\in s} m_k(\mu_k - y_{k.})^2 + \nu_W\lambda_W\}}{M_s + \nu_W} \quad ; \quad S_B^2 = \frac{\{\nu_B\lambda_B + \sum_{k=1}^{N}(\mu_k - \hat{\mu}_k)^2\}}{N + \nu_B - 3}.$$

Again, summing over i, it can be shown that:

$$\hat{\mu}_i = \overline{y} + \hat{\beta}(X_i - \overline{X})$$

$$\overline{y} = \frac{\sum_{i=1}^{N} \lambda_i y_{i.}}{\sum_{i=1}^{N} \lambda_i} \quad ; \quad \overline{X} = \frac{\sum_{i=1}^{N} \lambda_i X_i}{\sum_{i=1}^{N} \lambda_i}$$

$$\lambda_i = \begin{cases} \frac{S_B^2}{S_B^2 + (S_W^2/m_i)} & i \in s \\ 0 & i \notin s \end{cases} \quad ; \quad \hat{\beta} = \frac{\sum_{i=1}^{N} \lambda_i (X_i - \overline{X})(y_{i.} - \overline{y})}{\sum_{i=1}^{N} \lambda_i (X_i - \overline{X})^2}.$$

Except for a different parameterization for the X's, this result is analogous to that obtained by Scott and Smith in the case of known variances.

Cluster	μ_i	m_i	L.S.Est.	Modal Est.	Post. Mean	90% HPD Interval
1	20.2898	24	20.3545	20.4179	20.4083	(19.31 , 21.50)
2	26.6404	20	26.7337	26.6267	26.2906	(25.53 , 31.50)
3	19.4909	6	19.8054	20.0914	20.0393	(18.19 , 21.86)
4	22.1840	0	24.0557	23.0000	24.1949	(14.97 , 33.45)
5	25.4567	25	25.6007	25.5408	25.5815	(24.51 , 26.66)
6	22.2616	21	21.7833	21.8166	21.8218	(20.67 , 22.96)
7	25.9578	13	25.7973	25.6758	25.7556	(24.20 , 27.12)
8	27.1826	13	27.5175	27.3214	27.4316	(26.07 , 28.20)
9	23.7374	1	26.2812	25.0637	25.7897	(22.27 , 29.39)
10	26.7417	0	24.0557	23.0000	24.1949	(14.97 , 33.45)

Table 1. *Results of the simulation.*

4. NUMERICAL EXAMPLE

We simulate a sample from 10 clusters, with $v = 25$, $\sigma^2 = 4$ and $\delta^2 = 9$. In Table 1 appear the real value μ_i, the number of elements in the sample m_i, the least square estimator, the modal estimator, the posterior expectation and the 90% HPD. These two last features were calculated using the BAYES 4 program, and its graphic interface GR. We specify the prior distribution by $\nu_W = \nu_B = 1.01$ and $\lambda_W = \lambda_B = 10^{-4}$. This choice reflects weak knowledge about the variances (see Nazaret (1987)). We can see from this table that in all cases, the HPD interval contains the real value of the mean of the cluster. The smaller is m_i, the wider is the HPD interval.

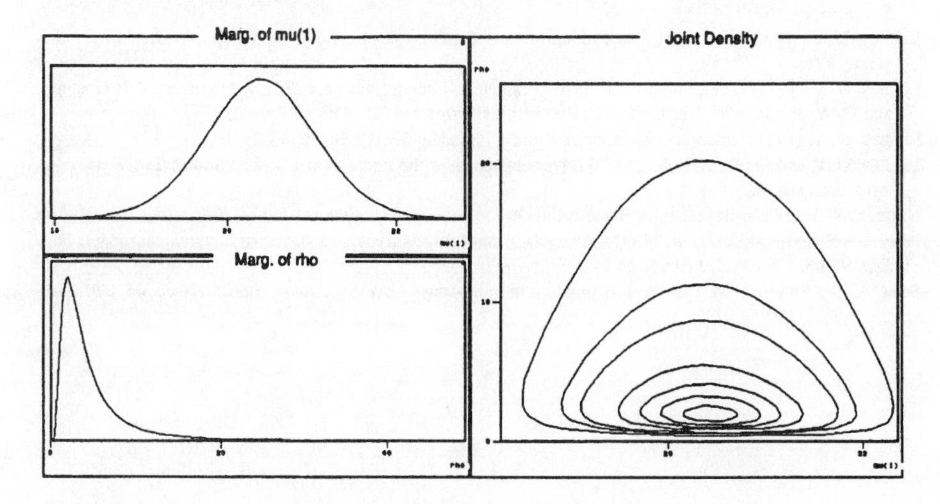

Figure 1. *Posterior Densities for μ_1 and $\rho(m = 24)$.*

In Figures 1 and 2 we show the joint posterior density of (μ_1, ρ) and (μ_4, ρ), respectively. We can see that the posterior distribution of the means is fairly symmetric, and its variance

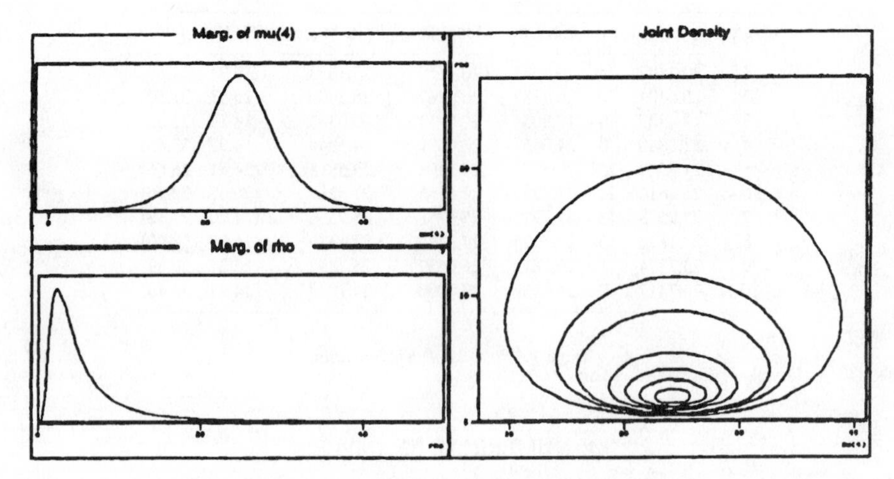

Figure 2. *Posterior Densities for* μ_1 *and* $\rho(m_4 = 0)$.

grows inversely with m_i. Group 4 has not been sampled. This is reflected in the wider range of the HPD interval.

REFERENCES

Box, G. E. P. and Tiao, G. C. (1968). Bayesian estimation of means for the random effect model. *J. Amer. Statist. Assoc.* **63**, 174–181.

Lindley, D. V. (1970). *Introduction to Probability and Statistics from a Bayesian Viewpoint.* Cambridge: University Press.

Lindley, D. V. (1971). The estimation of many parameters. *Foundations of Statistical Inference* (V. P. Godambe and D. A. Sprott, eds.), Toronto: Holt, Rinehart and Winston, 435–450.

Lindley, D. V. (1972). *Bayesian Statistics: A Review.* Philadelphia, PA: SIAM.

Lindley, D. V. and Smith, A. F. M. (1972) Bayes estimates for the linear model. *J. Roy. Statist. Soc. B* **34**, 1–42, (with discussion).

Nazaret, W. A. (1987). Bayesian log linear estimates for three-way contingency tables. *Biometrika* **74**, 401–410.

Pérez, M. E. and Pericchi, L. R. (1991). Analysis of multistage survey as a Bayesian hierarchical model. *Tech. Rep.* **91-05**, Universidad Simón Bolívar.

Scott, A. and Smith, T. M. F. (1969). Estimation in multi-stage surveys. *J. Amer. Statist. Assoc.* **64**, 830–840.

BAYESIAN STATISTICS 4, pp. 731–739
J. M. Bernardo, J. O. Berger, A. P. Dawid and A. F. M. Smith, (Eds.)

Bayes Factors and the Effect of Individual Observations on the Box-Cox Transformation

L. I. PETTIT
Goldsmiths' College, University of London, UK

SUMMARY

In this paper we consider the influence of individual observations on inferences about the Box and Cox (1964) power transformation parameter λ. Firstly we look at the effect of omitting observations on the posterior distribution of λ using the Kullback-Leibler divergence to measure the distance between the posterior conditional on all the data or on a reduced data set. Secondly we consider the effect of observations on the Bayes factors comparing particular choices of λ using the k_d approach of Pettit and Young (1990). We illustrate the methods using both regression data and designed experiments. We show that problems of masking that affect some classical diagnostics are not so serious. Finally we consider the value of the constant in a Bayes factor to test for additivity.

Keywords: ADDITIVITY; BAYES FACTOR; BOX-COX TRANSFORMATION; INFLUENTIAL OBSERVATION; KULLBACK-LEIBLER DISTANCE; MASKING.

1. INTRODUCTION

In this paper we shall consider the effects of observations on inferences when we are using the Box and Cox (1964) transformation. A number of authors have considered this problem from a classical viewpoint, see for example, Atkinson (1982, 1985, 1986, 1988), Cook and Wang (1983), Hinkley and Wang (1988) and Tsai and Wu (1990).

The basic ideas of the Box-Cox transformation are well known. We assume that for some λ to a sufficient approximation

$$y^{(\lambda)} = X\theta + \varepsilon\,,$$

where

$$y^{(\lambda)} = \begin{cases} \dfrac{y^{\lambda}-1}{\lambda} & \lambda \neq 0 \\ \log y & \lambda = 0 \end{cases}$$

and ε has distribution $N(0, \sigma^2 I)$.

The likelihood function relative to the original untransformed observations is

$$p(y|\theta,\sigma,\lambda) = \sigma^{-n} \exp\left\{-\frac{S_\lambda + (\theta-\hat{\theta}_\lambda)^T X^T X(\theta-\hat{\theta}_\lambda)}{2\sigma^2}\right\} J_\lambda\,,$$

where

$$J_\lambda = \prod_i y_i^{\lambda-1}\,,$$

$$\hat{\theta}_\lambda = (X^T X)^{-1} X^T y^{(\lambda)}\,,$$

$$S_\lambda = (y^{(\lambda)} - X\hat{\theta}_\lambda)^T(y^{(\lambda)} - X\hat{\theta}_\lambda).$$

Box and Cox (1964) suggested an outcome dependent prior

$$p(\theta, \sigma, \lambda) \propto p_0(\lambda)/(\sigma J_\lambda^{k/n})$$

where k is the dimension of θ.

Pericchi (1981) suggested an alternative non-informative prior

$$p(\theta, \sigma, \lambda) \propto p_0(\lambda)/\sigma^{k+1}$$

(see also Smith and Spiegelhalter, 1981) and we shall use this in our examples, although the methods could be used whatever prior distribution was chosen. It follows that the joint posterior is

$$p(\theta, \sigma, \lambda|y) \propto \sigma^{-n-k-1} \exp\left\{-\frac{S_\lambda + (\theta - \hat{\theta}_\lambda)^T X^T X(\theta - \hat{\theta}_\lambda)}{2\sigma^2}\right\} J_\lambda p_0(\lambda).$$

We shall assume $p_0(\lambda)$ uniform over the region of λ's under consideration.

Integrating with respect to θ and σ we find the posterior of λ is

$$p(\lambda|y) \propto J_\lambda (S_\lambda)^{-n/2}.$$

We can also find the Bayes factor comparing the choice of two particular values of λ,

$$B_{01} = \left(\frac{J_{\lambda_0}}{J_{\lambda_1}}\right)\left(\frac{S_{\lambda_0}}{S_{\lambda_1}}\right)^{-n/2}.$$

The question arises as to how to assess the influence of observations on the posterior distribution of λ or the Bayes factor. In the following sections we shall suggest a number of answers to this question and illustrate them using a simple linear regression example and a designed experiment.

2. INFLUENCE ON THE POSTERIOR DISTRIBUTION

Following the interest in detecting influential observations generated by Cook (1977), appropriate Bayesian diagnostics have been proposed by Johnson and Geisser (1983), Pettit and Smith (1985) and Guttman and Peña (1988). Suppose the full data are written y and we wish to measure the influence of a subset S of observations. Denoting the data omitting S by $y_{(S)}$, the influence measures may be written as (Pettit, 1986)

$$D[p(u|y); p(u|y_{(S)})]$$

where $D[.;.]$ is a distance measure and u is either a future observation or the parameter of interest. The distance measure used has been the Kullback-Leibler distance, both in its symmetric and asymmetric forms. For example here λ is the parameter of interest so we might look at

$$I(S) = \int \log \frac{p(\lambda|y)}{p(\lambda|y_{(S)})}[p(\lambda|y) - p(\lambda|y_{(S)})]\, d\lambda$$

the symmetric version of the Kullback-Leibler distance.

The Cook and Wang data.

Cook and Wang (1983) consider an artificial data set, 10 observations generated with a log transformation and the last observation with no transformation. The observations are given in Table 1. They were chosen so that the outlier has a high leverage value with which the methods in Atkinson (1982) have difficulty.

Case	x_i	y_i	Mode	Mean	Var	K–L	AK–L	k_d
			.705	.660	.101			
1	0.3	1.49	.848	.800	.128	0.20	0.20	–0.313
2	0.4	1.42	.626	.573	.147	0.06	0.13	0.378
3	0.5	1.77	.741	.702	.111	0.02	0.02	–0.105
4	0.6	2.18	.704	.662	.102	0.00	0.00	0.003
5	0.8	2.01	.768	.732	.091	0.06	0.06	–0.310
6	0.9	2.41	.736	.684	.118	0.01	0.02	–0.023
7	1.0	2.41	.832	.786	.099	0.15	0.08	–0.425
8	1.1	2.80	.744	.685	.130	0.02	0.04	0.009
9	1.2	3.25	.610	.554	.133	0.08	0.14	0.427
10	1.5	4.39	.644	.618	.050	0.31	0.28	0.050
11	2.6	6.20	–.182	–.208	.174	4.23	6.05	2.398

Table 1. *The Cook and Wang data with posterior modes, means and variances, symmetric Kullback-Leibler distances (K–L) and approximations (AK–L) and values of k_d comparing $\lambda_0 = 1$ versus $\lambda_1 = 0$.*

Case	k_d
1	–0.133
2	0.569
3	–0.018
4	–0.054
5	–0.333
6	0.037
7	–0.430
8	0.107
9	0.592
10	0.221
11	0.508
12	0.601
13	0.846

Table 2. *Values of k_d comparing $\lambda_0 = 1$ versus $\lambda_1 = 0$ for the augmented Cook and Wang data.*

We also show in Table 1 the symmetric Kullback-Leibler distances omitting one observation at a time, together with the posterior mode, mean and variance of λ. The asymmetric Kullback-Leibler distances give the same results. The posterior distribution of λ is clearly skew in this case but as an approximation to the symmetric Kullback-Leibler distance we calculated the distance between two normal distributions having the same means and variances. The results are shown in Table 1. Qualitatively they are similar to before, observation 11 is clearly the most influential but the approximation is not very good.

We illustrate the posterior distribution of λ conditional on all the data or omitting observation 11 in Figure 1. It is clear that observation 11 is highly influential on our inferences

Poison	Treatment A	B	C	D
1	0.31	0.82	0.43	0.45
	0.45	1.10	0.45	0.71
	0.46	0.88	0.63	0.66
	0.43	0.72	0.76	0.62
2	0.36	0.92	0.44	0.56
	0.29	0.61	0.35	1.02
	0.40	0.49	0.31	0.71
	0.23	1.24	0.40	0.38
3	0.22	0.30	0.23	0.30
	0.21	0.37	0.25	0.36
	0.18	0.38	0.24	0.31
	0.23	0.29	0.22	0.33

Table 3. *Box and Cox poisons data.*

about λ. Omitting observation 11 the mean is very different, consistent with a log transformation ($\lambda = 0$), but the uncertainty is also increased. The moments of λ and the plot were calculated using LISP-STAT (Tierney, 1990), see the appendix for details.

3. INFLUENCE ON BAYES FACTORS FOR SIMPLE LINEAR REGRESSION

As Box and Tiao (1973) point out when we use a Box-Cox transformation we usually look for a value of λ that is easily interpretable. Thus we often only wish to consider as small a number of λ values as plausible. In this case it may be of interest to consider the Bayes factor between a pair of values which the overall posterior indicate as being plausible. Pettit and Young (1990) have introduced a diagnostic to measure the effect of observations on a Bayes factor comparing models M_0 and M_1. This is defined as

$$k_d = \log_{10} B_{01} - \log_{10} B_{01}^{(d)},$$

where $B_{01}^{(d)}$ is the Bayes factor omitting the dth observation. Thus a positive value of k_d means that without observation d there is less evidence for model M_0. They also suggested the use of plots to illustrate these values.

We have in general for comparing λ_0 versus λ_1 that k_d is given by

$$k_d = (\lambda_0 - \lambda_1) \log_{10} y_d - \frac{n}{2} \log_{10} \left[\frac{RSS(\lambda_0)}{RSS(\lambda_1)} \right] + \frac{n-1}{2} \log_{10} \left[\frac{RSS_{(d)}(\lambda_0)}{RSS_{(d)}(\lambda_1)} \right].$$

The Cook and Wang data

For the Cook and Wang data we are interested in comparing $\lambda_0 = 1$ corresponding to no transformation and $\lambda_1 = 0$ corresponding to a log transformation. The overall log Bayes factor is $\log B_{01} = 0.655$ showing evidence for λ_0. The values of k_d are given in Table 1. It is clear that observation 11 is highly influential, its omission causes us to come to the opposite conclusion that a log transformation is needed.

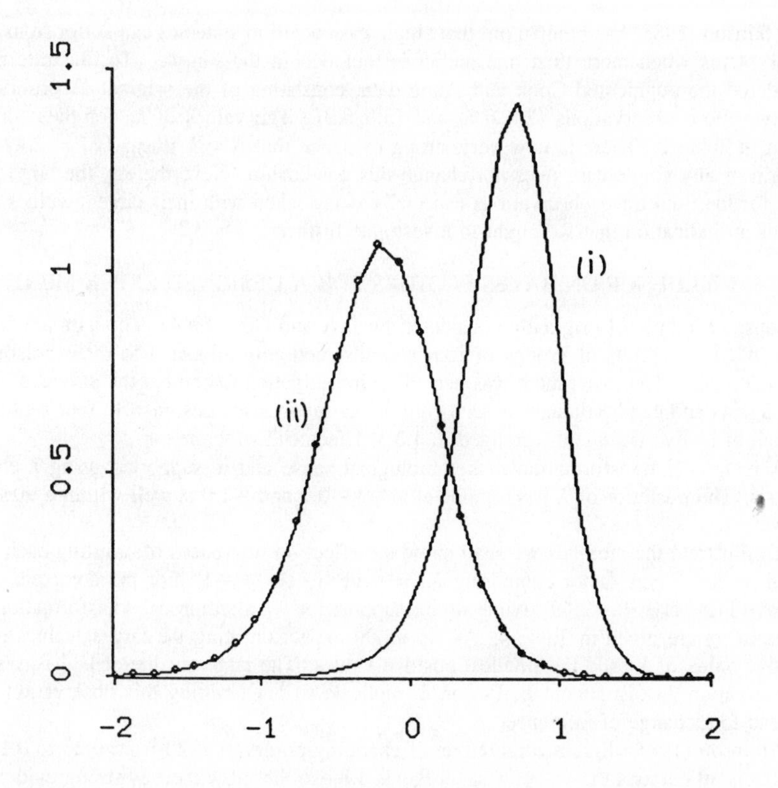

Figure 1. *Posterior distirbution of λ (i) conditional on all data (ii) omitting observation 11.*

	Treatment			
Poison	A	B	C	D
1	−0.27	0.27	−0.12	0.01
	0.19	−0.01	−0.07	0.17
	0.24	0.26	0.19	0.17
	0.10	0.21	0.08	0.15
2	0.04	0.07	0.07	0.07
	−0.26	0.04	−0.21	−0.73
	0.32	−0.26	−0.16	0.10
	0.60	−0.87	−0.08	−0.13
3	−0.42	−0.40	−0.26	−0.21
	−0.40	0.01	−0.30	−0.15
	0.45	−0.15	−0.31	−0.18
	−0.36	−0.42	−0.14	−0.07

Table 4. *Values of k_d for the Box and Cox poisons data.*

Atkinson (1988) has pointed out that single case deletion statistics can suffer from masking problems when more than one outlier is included in the sample. To illustrate this he considered the augmented Cook and Wang data, consisting of the original 11 observations plus two more observations $(2.8, 6.6)$ and $(3.0, 6.9)$. The values of k_d for these data are shown in Table 2. There is now very strong evidence that $\lambda = 1$ ($\log_{10} B_{01} = 2.421$) and deletion of any single case does not change this conclusion. Nevertheless, the large values of k_d for the final three observations especially when taken with their large x values would give us an indication that we ought to investigate further.

4. INFLUENCE ON BAYES FACTORS FOR A DESIGNED EXPERIMENT

We consider a data set originally considered by Box and Cox (1964). The data are survival times in 10 hour units of groups of four animals randomly allocated to three poisons and four treatments. The experiment was part of an investigation to combat the effects of certain toxic agents and can be thought of as a completely randomised design with four replications on each of twelve treatments arranged in a 3×4 factorial structure.

A reciprocal transformation makes biological sense and is supported using a classical analysis. The posterior of λ has a mode at about -0.7 and -1.0 is well within a 90% HPD interval.

To illustrate the methods we shall show the effect on inferences of omitting each observation on the Bayes factor comparing $\lambda_0 = 0$ versus $\lambda_1 = -1$. For the complete dataset we have $\log_{10} B_{01} = -2.49$ giving strong support for λ_1 a reciprocal transformation. The values of k_d are given in Table 4. As we might expect omitting the largest values gives a negative value of k_d and the smallest positive values. The largest value of $|k_d|$ corresponds to observation 24 (Treatment B, Poison 2, replicate 4) but omitting this observation would not lead to a change of inference.

Atkinson (1985) discusses the effect of changing observation 20 from 0.23 to 0.13. For this amended data set the value of $\log_{10} B_{01}$ is 4.86 so that now there is strong evidence for λ_0 a log transformation. Now $k_{20} = 7.97$ is very large and omission of observation 20 would lead us to change our inferences. The other values of $|k_d|$ are much smaller, the largest being $k_{24} = -0.66$.

5. A BAYES FACTOR FOR ADDITIVITY

Pericchi (1981) considers a Bayes factor for additivity, assuming that with the transformation λ normality and homoscedasticity are achieved. Let $\theta = (\theta_1, \theta_2)$, where θ_2 is the vector of interaction parameters with $\dim(\theta_1) = \nu_1, \dim(\theta_2) = \nu_2, \nu_1 + \nu_2 = k$. Assuming non-informative priors he finds the Bayes factor is proportional to

$$\left\{1 + \frac{\nu_2}{n-k} F(\lambda)\right\}^{-n/2}$$

where $F(\lambda)$ is the usual F statistic for additivity.

Spiegelhalter and Smith (1982) have discussed the form of Bayes factors when non-informative priors are used. Assuming two nested linear models

$$y \sim N(A_i\theta_i, \sigma^2 I_n) \quad i = 0, 1$$

and with prior

$$p(\theta_i, \sigma | A_i) = c_i (2\pi\sigma^2)^{-p_i/2} \sigma^{-1}$$

the Bayes factor is

$$B_{01} = \frac{c_0}{c_1}[|A_1^T A_1|/|A_0^T A_0|]^{1/2}\left[1+\frac{(p_1-p_0)}{(n-p_1)}F\right]^{-n/2}$$

The ratio of unspecified constants c_0/c_1 is determined using the device of imaginary observations. The result is that

$$\frac{c_0}{c_1} = [|E_1^T E_1|/|E_0^T E_0|]^{-1/2}$$

where E_0, E_1 are the design matrices for M_0, M_1 in a thought experiment which is the smallest possible to compare the two models.

The constant here can be obtained by extending the results in Spiegelhalter and Smith (1982) for a randomised complete block experiment to the case when we allow interaction. Thus we have an experiment with a treatments, b blocks, r replications and observations y_{ijk} assumed normally distributed with variances all equal to σ^2 and means $\mu+\beta_i+\alpha_j+(\beta\alpha)_{ij}$, $i=1,\ldots,b$, $j=1,\ldots,a$, $k=1,\ldots,r$ independent given the parameters.

We set $\beta_1=\alpha_1=(\beta\alpha)_{1j}=(\beta\alpha)_{i1}=0$ for identifiability. We can show that

$$A_1^T A_1 = r\begin{pmatrix} A_0^T A_0 & B \\ B^T & C \end{pmatrix},$$

where

$$A_0^T A_0 = \begin{pmatrix} ab & a1_{b-1}^T & b1_{a-1}^T \\ a1_{b-1} & aI_{b-1} & J_{b-1,a-1} \\ b1_{a-1} & J_{a-1,b-1} & bI_{a-1} \end{pmatrix}, \quad B = \begin{pmatrix} 1_{(a-1)(b-1)}^T \\ D \\ E \end{pmatrix}, \quad C = I_{(a-1)(b-1)}$$

and $J_{b-1,a-1}$ is a $b-1\times a-1$ matrix of ones,

$$D = \begin{pmatrix} 1_{a-1}^T & 0_{a-1}^T & \cdots & 0_{a-1}^T \\ 0_{a-1}^T & 1_{a-1}^T & \cdots & 0_{a-1}^T \\ \vdots & \vdots & \ddots & \vdots \\ 0_{a-1}^T & 0_{a-1}^T & \cdots & 1_{a-1}^T \end{pmatrix}, \quad E = \begin{pmatrix} I_{a-1} & I_{a-1} & \cdots & I_{a-1} \end{pmatrix}.$$

It follows that

$$(A_0^T A_0)^{-1} = \begin{pmatrix} \frac{a+b-1}{ab} & \frac{-1}{a}1_{b-1}^T & \frac{-1}{b}1_{a-1}^T \\ \frac{-1}{a}1_{b-1} & \frac{1}{a}(I_{b-1}+J_{b-1}) & 0_{b-1,a-1} \\ \frac{-1}{b}1_{a-1} & 0_{a-1,b-1} & \frac{1}{b}(I_{a-1}+J_{a-1}) \end{pmatrix}$$

so that

$$\begin{aligned} [|A_1^T A_1|/|A_0^T A_0|] &= r^{(a-1)(b-1)}|C - B^T(A_0^T A_0)^{-1}B| \\ &= \left(\frac{r}{b}\right)^{a-1}\left(\frac{r}{a}\right)^{b-1}. \end{aligned}$$

A minimal (imaginary) experiment in this case requires $(a-1)+(b-1)+(a-1)(b-1)+1+1$ observations. An example of such an experiment would be to include each treatment once in each block with a replication of the first treatment in the first block. With this layout we find that $A_1^T A_1$ is as before except that $r = 1$ and the first entry of $A_0^T A_0$ is $ab + 1$. Writing this as $E_0^T E_0$ we find that

$$(E_0^T E_0)^{-1} = \begin{pmatrix} \frac{a+b-1}{\Delta} & \frac{-b}{\Delta} 1_{b-1}^T & \frac{-a}{\Delta} 1_{a-1}^T \\ \frac{-b}{\Delta} 1_{b-1} & \frac{1}{a} I_{b-1} + g J_{b-1} & \frac{-1}{\Delta} J_{b-1,a-1} \\ \frac{-a}{\Delta} 1_{a-1} & \frac{-1}{\Delta} J_{a-1,b-1} & \frac{1}{b} I_{a-1} + h J_{a-1} \end{pmatrix},$$

where $\Delta = ab + a + b - 1$, $g = \frac{ab(b-1)+2-2a-b}{a(b-2)\Delta}$ and $h = \frac{ab(a-1)+2-2b-a}{b(a-2)\Delta}$.

It follows that

$$[|E_1^T E_1| / |E_0^T E_0|] = \frac{2}{\Delta} \left(\frac{1}{a}\right)^{b-2} \left(\frac{1}{b}\right)^{a-2}.$$

Hence the Bayes factor for additivity is

$$\left(\frac{ab + a + b - 1}{2ab}\right)^{1/2} r^{(a+b)/2-1} \left\{1 + \frac{\nu_2}{n-k} F(\lambda)\right\}^{-n/2}.$$

The effect of individual observations could be studied by calculating k_d for this Bayes factor.

For the Poisons data the value of the Bayes factor for additivity is 0.503 for a reciprocal transformation, giving very weak evidence that interaction is present, and 0.041 for no transformation, giving strong evidence for interaction.

6. DISCUSSION

In this paper we have considered the effect of observations on inferences about the parameter of a Box-Cox transformation. The results using k_d seem to be very useful, not suffering as much from masking as might be feared.

Much of the classical effort in this area has been concerned with approximating the maximum likelihood estimate of λ when the ith observation is omitted, Cook and Wang (1983), Hinkley and Wang (1988), Tsai and Wu (1990). Much of this work has been based on supposing an outlier model for the ith observation rather than omitting it so that the Jacobian is unchanged. However the results in Tsai and Wu (1990) indicate that this approach can be misleading and so we have not pursued this idea from a Bayesian perspective.

APPENDIX

The following functions can be defined in LISP-STAT to produce the posterior distribution of λ for a simple linear regression problem. An extension to multiple regression is straightforward. We assume the data has been read into variables `y-values` and `x-values`.

```
(defun bc (x p)
   (if (< (abs p) .0001) (log x) (/ (^x p) p)))

(defun f1 (theta)
   (def bcmodel (regression-model x-values (bc y-values (select theta 0))
:print nil))
     (+ (* (/ (* -1 (count-elements y-values)) 2)
(log (send bcmodel :residual-sum-of-squares)))
       (* (- (select theta 0) 1) (sum (log y-values)))))

(def lk (bayes-model #'f1 (list 1.0)))
```

Now `lk` can be used to obtain moments and plots in the usual way.

REFERENCES

Atkinson, A. C. (1982). Regression diagnostics, transformations and constructed variables. *J. Roy. Statist. Soc. B* **44**, 1–36, (with discussion).

Atkinson, A. C. (1985). *Plots, Transformations and Regression*. Oxford: University Press.

Atkinson, A. C. (1986). Diagnostic tests for transformations. *Technometrics* **28**, 29–37.

Atkinson, A. C. (1988). Transformations unmasked. *Technometrics* **30**, 311–318.

Box, G. E. P. and Cox, D. R. (1964). An analysis of transformations. *J. Roy. Statist. Soc. B* **26**, 211–246, (with discussion).

Box, G. E. P. and Tiao, G. C. (1973). *Bayesian Inference in Statistical Analysis*. Reading, MA: Addison-Wesley.

Cook, R. D. (1977). Detection of influential observations in linear regression. *Technometrics* **19**, 15–18.

Cook, R. D. and Wang, P. C. (1983). Transformations and influential cases in regression. *Technometrics* **25**, 337–343.

Guttman, I. and Peña, D. (1988). Outliers and influence: evaluation by posteriors of parameters in the linear model. *Bayesian Statistics 3* (J. M. Bernardo, M. H. DeGroot, D. V. Lindley and A. F. M. Smith, eds.), Oxford: University Press, 631–640.

Hinkley, D. V. and Wang, S. (1988). More about transformations and influential cases in regression. *Technometrics* **30**, 435–440.

Johnson, W. and Geisser, S. (1983). A predictive view of the detection and characterization of influential observations in regression analysis. *J. Amer. Statist. Assoc.* **78**, 137–144.

Pericchi, L. R. (1981). A Bayesian approach to transformations to normality. *Biometrika* **68**, 35–43.

Pettit, L. I. (1986). Diagnostics in Bayesian model choice. *The Statistician* **35**, 183–190.

Pettit, L. I. and Smith, A. F. M. (1985). Outliers and influential observations in linear models. *Bayesian Statistics 2* (J. M. Bernardo, M. H. DeGroot, D. V. Lindley and A. F. M. Smith, eds.), Amsterdam: North-Holland, 473–494, (with discussion).

Pettit, L. I. and Young, K. D. S. (1990). Measuring the effect of observations on Bayes factors. *Biometrika* **77**, 455–466.

Smith, A. F. M. and Spiegelhalter, D. J. (1981). Bayesian approaches to multivariate structure. *Interpreting Multivariate Data* (V. Barnett, ed.), New York: Wiley, 335–348.

Spiegelhalter, D. J. and Smith, A. F. M. (1982). Bayes factors for linear and log-linear models with vague prior information. *J. Roy. Statist. Soc. B* **44**, 377–387.

Tierney, L. (1990). *LISP-STAT*. New York: Wiley.

Tsai, C.-L. and Wu, X. (1990). Diagnostics in transformation and weighted regression. *Technometrics* **32**, 315–322.

REFERENCES

[illegible]

BAYESIAN STATISTICS 4, pp. 741–751
J. M. Bernardo, J. O. Berger, A. P. Dawid and A. F. M. Smith, (Eds.)

Dynamic Graphical Models

CATRIONA M. QUEEN and JIM Q. SMITH
University of Nottingham, UK and *University of Warwick, UK*

SUMMARY

Dynamic graphical models (DGM's) are defined to forecast multivariate time series for which there is believed to be symmetry between certain subsets of variables and a causal driving mechanism between these subsets. They are a specific type of graphical chain model (Wermuth and Lauritzen, 1990) which are both non-linear and non-normal. DGM's are a combination of dynamic multivariate regression models (Quintana, 1985, 1987, Quintana and West, 1987, 1988) and multiregression dynamic models (Queen and Smith, 1990) and as such they inherit the simplicity of both these models. If conditional normality is assumed across each subset, important quantities can be calculated and updating formulae can be found explicitly. The models are illustrated by a simple example.

Keywords: NON-LINEAR NON-SYMMETRIC MULTIVARIATE TIME SERIES; BAYESIAN FORECASTING; GRAPHICAL CHAIN MODELS; DYNAMIC MULTIVARIATE REGRESSION MODELS; MULTIREGRESSION DYNAMIC MODELS.

1. INTRODUCTION

In business multivariate time series there is often a research hypothesis (Wermuth and Lauritzen, 1990) concerning certain causal relationships between the processes. In this paper a model on a vector time series will be constructed by representing any given research hypothesis by a graph which holds for all time frames. This graph then defines a statistical model known as a *dynamic graphical model* (DGM) which is a subclass of graphical chain model (Lauritzen and Wermuth, 1984, 1989).

To illustrate how a DGM is derived for a multivariate time series, consider the problem of forecasting sales in a hypothetical small shampoo market. Consumers can choose between buying equally high priced conditioning shampoo and natural ingredients shampoo, or a cheaper standard shampoo. Because the standard shampoo is not very attractive, consumers would prefer to buy one of the more expensive shampoos if they can afford them. So it is expected that, at time t, the sales of the more expensive conditioning and natural ingredients shampoos, C_t and N_t respectively, will decrease as the number of affluent customers decreases, the rest having to buy standard shampoo. An index, I_t, of disposable income amongst consumers, at time t, may be a good indicator of consumer affluence. Let the sales of standard shampoo at time t be represented by S_t, then a research hypothesis of the causal links between the series at a single time period can be seen in the graphical representation of Figure 1.

If edge (C_t, N_t) of Figure 1 is made directed then this could be the graph of an influence diagram. An influence diagram takes an ordered set of random vectors $(\boldsymbol{X}_1, \ldots, \boldsymbol{X}_m)$, on which $\boldsymbol{X}_j \amalg \boldsymbol{X}_k \mid \boldsymbol{X}_l$ reads "$\boldsymbol{X}_j$ is independent of $\boldsymbol{X}_k$ given $\boldsymbol{X}_l$". An influence diagram consists of a directed graph G and the $m-1$ conditional independence statements:

$$\boldsymbol{X}_r \amalg Q(\boldsymbol{X}_r) \mid P(\boldsymbol{X}_r) \qquad 2 \leq r \leq m$$

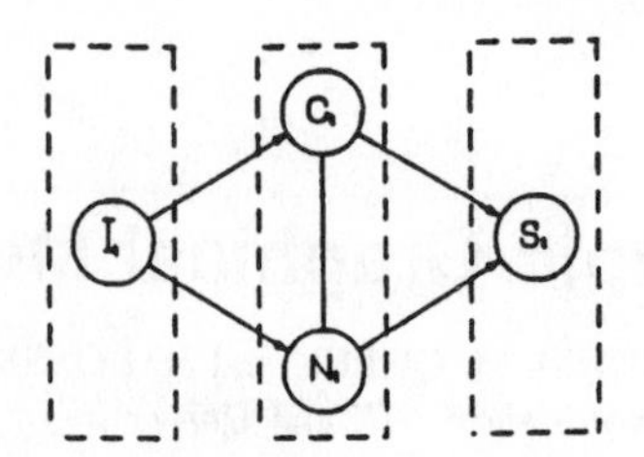

Figure 1. *Graph representing the research hypothesis of the shampoo market.*

where $P(\boldsymbol{X}_r) \subseteq \{\boldsymbol{X}_1, \ldots, \boldsymbol{X}_{r-1}\}$ and $Q(\boldsymbol{X}_r) = \{\boldsymbol{X}_1, \ldots, \boldsymbol{X}_{r-1}\} \backslash P(\boldsymbol{X}_r)$. $P(\boldsymbol{X}_r)$ is called the *parent set* of $\boldsymbol{X}_r$ and G has nodes labelled $(\boldsymbol{X}_1, \ldots, \boldsymbol{X}_m)$ and an edge from $\boldsymbol{X}_i$ to $\boldsymbol{X}_r$ iff $\boldsymbol{X}_i \in P(\boldsymbol{X}_r)$ (Howard and Matheson, 1981, Shachter, 1986 and Smith, 1989, 1990). G is called the *graph of the influence diagram.*

Formal causation between *processes* has been studied quite extensively in the economic literature. Granger (1969) defined causality in terms of properties of linear estimators. Florens and Mouchart (1985), modified this definition, relating non-causality explicitly to conditional independence. They state that U is a Granger *non-cause* of V if for all time t :

$$V_t \amalg \{U_k\}_{k<t} \mid \{V_k\}_{k<t}.$$

Their definition coincides with Granger's original one when the processes are jointly Gaussian. Queen and Smith (1990) give the obvious extension of this to a definition of *conditional causality* which states that U is a non-cause of V *given* W if for all time t :

$$V_t \amalg \{U_k\}_{k<t} \mid \{V_k\}_{k<t}, \{W_k\}_{k\leq t}.$$

In this earlier paper we introduce a new class of Bayesian forecasting models called multiregression dynamic models (MDM's) and prove the surprising result that MDM's with a particular influence diagram in a fixed time frame exhibit (conditional) Granger non-causality, as defined above, consistent with the heuristic ideas of non-causality used in research hypotheses. This result will be discussed later — it just corresponds to a stochastic generalisation of a result in Spiegelhalter and Lauritzen (1990) in a new setting. Thus, in the example above, if edge (C_t, N_t) had been directed in Figure 1, then the results of Queen and Smith (1990) would mean that $\{I_t\}_{t\leq 1}$ is a Granger non-cause of $\{S_t\}_{t\leq 1}$ given $\{C_t, N_t\}_{t\leq 1}$ when it is modelled by an MDM.

However, it is clear when considering the causal structure of the shampoo market problem that the implicit Granger non-cause is suspect and so to represent the series by an influence diagram is inappropriate. An MDM would require that N_t be conditioned on (C_t, I_t), but C_t would be conditioned on I_t alone, thus destroying the symmetry of the role between C_t and N_t in the given explanation of the dynamics of the market, as well as introducing spurious non-causes. A more satisfactory model would accommodate the symmetric relationship between $\{C_t\}$ and $\{N_t\}$ represented by the undirected edge (C_t, N_t) in Figure 1.

In this paper it will be shown that DGM's preserve the exchangeable form of the evolution over components C_t and N_t whilst $\{I_t\}_{t\leq 1}$ remains a conditional non-cause of $\{S_t\}_{t\leq 1}$ given $\{C_t, N_t\}_{t\leq 1}$. Thus both the causal and symmetric relationships of Figure 1 are accommodated. In terms of causal structures, these DGM's are represented by a subclass of chain graph (Wermuth and Lauritzen, 1990) at every time frame. Chain graphs are mixed and they partition variables into subsets, which will be called p-sets here, such that directed edges connect variables between p-sets consistent with the order of the partition (lower indexed p-sets to higher) and any pairs of variables within p-sets are joined by a non-causal undirected edge. This subclass is defined by imposing the following two conditions:

1. *all* variables in the same p-set of the chain graph are connected
2. if there is a directed edge from a variable in one p-set to a variable in another p-set, then there must be a directed edge from *every* variable in the first p-set to *every* variable in the second.

Figure 1 represents a chain graph of causality that lies in the subclass defined by 1 and 2 above. The partition of p-sets is $(\{I_t\}, \{C_t, N_t\}, \{S_t\})$. Figure 2 shows a research hypothesis of a real market containing 9 brands.

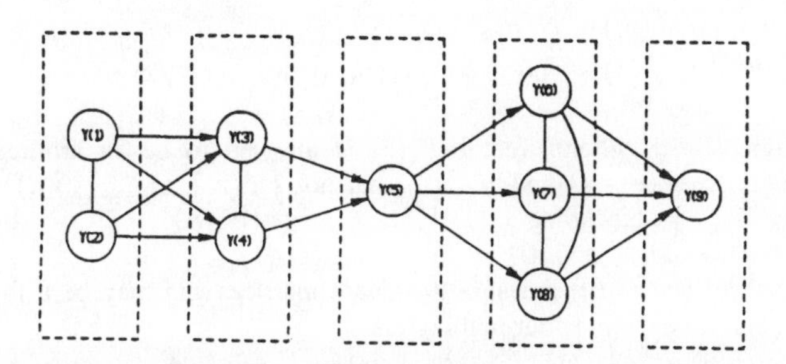

Figure 2. *Research hypothesis of a real market with p-sets* $\{Y(1), Y(2)\}$, $\{Y(3), Y(4)\}$, $\{Y(5)\}$, $\{Y(6), Y(7), Y(8)\}$, $\{Y(9)\}$.

DGM's are a combination of MDM's and *dynamic mulivariate regression* (DMR) models (Quintana, 1985, 1987, Quintana and West, 1987, 1988). As well as providing a class of time series models consistent with an assumed causal structure, these processes are extremely general and versatile. Furthermore, they allow closed-form updating relationships for processes which may be highly non-linear. Although the joint one-step ahead forecast distributions are not Gaussian, it is straightforward to calculate its moments. The models will be demonstrated by a working example of the simplified shampoo market.

2. MULTIREGRESSION DYNAMIC MODELS

MDM's model multivariate time series with a causal structure and are a stochastic generalisation of recursive simultaneous equation models (Harvey, 1989, Zellner, 1986) where $Y_t(r)$ is regressed against functions of a subset of contemporary variables which are listed before $Y_t(r)$ in a causal ordering.

Let $\boldsymbol{Y}_t{}^T = \{Y_t(1), \ldots, Y_t(n)\}$ represent the values of a vector time series at time t. If y_{tj} is the realisation of Y_{tj}, then use the conventional notation so that $\boldsymbol{y}^t_j = \{y_{1j}, \cdots, y_{tj}\}^T$. Let $\boldsymbol{\theta}^T_t = \{\boldsymbol{\theta}_t(1)^T, \boldsymbol{\theta}_t(2)^T, \ldots, \boldsymbol{\theta}_t(n)^T\}$ be the state vectors determining the distributions of $Y_t(1), Y_t(2), \ldots, Y_t(n)$ respectively, and let s_r be the dimension of the vector $\boldsymbol{\theta}_t(r)$, $1 \leq r \leq n$.

Suppose $\sigma(r)$ is the unknown observation variance for $Y_t(r)$. Call $\{\boldsymbol{Y}_t\}_{t\geq 1}$ a (conditionally normal) *Multiregression Dynamic Model* if the following system equation, n observation equations and the restrictions imposed on those equations given below, hold for all time.

Observation equations

$$Y_t(r) = \boldsymbol{F}_t(r)^T \boldsymbol{\theta}_t(r) + v_t(r), \qquad 1 \le r \le n \quad v_t(r) \sim N(0, \sigma(r))$$

System equation

$$\boldsymbol{\theta}_t = G_t \boldsymbol{\theta}_{t-1} + \boldsymbol{w}_t, \qquad \boldsymbol{w}_t \sim N(0, W_t^*)$$

Prior Information

$$(\boldsymbol{\theta}_t(r) \,|\, \sigma(r), \boldsymbol{y}^{t-1}(r)) \sim N(\boldsymbol{m}_{t-1}, C_{t-1}^*)$$
$$(\sigma(r)^{-1} \,|\, \boldsymbol{y}^{t-1}) \sim G\left(n_{t-1}(r)/2, \left(S_{t-1}(r) n_{t-1}(r)\right)/2\right) \qquad 1 \le r \le n$$

The s_r dimensional column vector $\boldsymbol{F}_t(r)$ is allowed to be an arbitrary but known function of $\boldsymbol{y}^t(1),\ldots,\boldsymbol{y}^t(r-1)$ and $\boldsymbol{y}^{t-1}(r)$, but not $\{\boldsymbol{y}^t(r+1), \ldots, \boldsymbol{y}^t(n)\}$ or $y_t(r)$; the $s \times s$ matrices $W_t^* = \text{blockdiag}\{\sigma(1)W_t(1), \ldots, \sigma(n)W_t(n)\}$, $C_{t-1}^* = \text{blockdiag}\{\sigma(1) C_{t-1}(1), \ldots, \sigma(n)C_{t-1}(n)\}$ and $G_t = \text{blockdiag}\{G_t(1), \ldots, G_t(n)\}$ are such that $\sigma(r) W_t(r)$, $\sigma(r)C_{t-1}(r)$ and $G_t(r)$ are $s_r \times s_r$ square matrices and may be a function of past vectors $\{\boldsymbol{y}^{t-1}(1), \ldots, \boldsymbol{y}^{t-1}(r)\}$, but nothing else.

The error vectors, $\boldsymbol{v}_t^T = \{v_t(1), \ldots, v_t(n)\}$ and $\boldsymbol{w}_t^T = \{\boldsymbol{w}_t(1)^T, \ldots, \boldsymbol{w}_t(n)^T\}$, where $v_t(r)$ is the observation error and $\boldsymbol{w}_t(r)$ is the s_r dimensional system error vector for $Y_t(r)$, $1 \le r \le n$, are such that variables within the sets $\{\boldsymbol{w}_t(r)\}_{1 \le r \le n}$, $\{v_t(r)\}_{1 \le r \le n}$ and $\{\boldsymbol{w}_t, \boldsymbol{v}_t\}_{t \ge 1}$ are mutually independent.

Using Lauritzen *et al.* (1990) and the Markov evolution of states given above, Queen and Smith (1990) generalise the non-stochastic results of Spiegelhalter and Lauritzen (1990) about the retention of the conditional independence under Bayesian sampling to prove that if

$$\amalg_{r=1}^n \boldsymbol{\theta}_{t-1}(r) | \boldsymbol{y}^{t-1}$$

then

$$\amalg_{r=1}^n \boldsymbol{\theta}_t(r) | \boldsymbol{y}^t \tag{1}$$

and

$$\boldsymbol{\theta}_t(r) \amalg \boldsymbol{y}^t(r+1), \boldsymbol{y}^t(r+2), \ldots, \boldsymbol{y}^t(n) | \boldsymbol{y}^t(1), \ldots, \boldsymbol{y}^t(r) \tag{2}$$

In the terminology of Spiegelhalter and Lauritzen (1990), for any time t, $\boldsymbol{\theta}_t(r)$ are globally Markov. In particular, equation 1 enables $Y_t(r) \,|\, \{Y_t(1), \ldots, Y_t(r-1)\}$, $1 \le r \le n$, to be forecast separately and updated in closed form. The joint one-step ahead forecast distribution of $\boldsymbol{Y}_t$ can be expressed as the product of the individual univariate forecast distributions, each of which (by equation 2) is the conditional one-step ahead forecast distribution of $Y_t(r) \,|\, \{y_t(1), \ldots, y_t(r-1)\}$, $1 \le r \le n$. Even when regression is linear, these models can yield highly non-Gaussian joint forecast distributions because of the appearance of unobserved components of the vector $\boldsymbol{Y}_t$ appearing in the individual conditional variances.

3. DYNAMIC MULTIVARIATE REGRESSION MODELS

Quintana and West developed DMR models as a multivariate version of the DLM (Harrison and Steven, 1976) where the evolution of its components exhibit a degree of symmetry. They allow conjugate normal-Wishart analysis of the mean and covariance matrix of the one-step ahead forecast distribution.

Let $\boldsymbol{Y}_t^T = \{Y_{t1}, \ldots, Y_{tq}\}$ be the q dimensional observation vector and $\Theta_t = \{\boldsymbol{\theta}_{t1}, \ldots, \boldsymbol{\theta}_{tq}\}$, where $\boldsymbol{\theta}_{tj}$ is the s dimensional state vector defining the distribution of Y_{tj} and is typically different for each series. Suppose that the covariance matrix for $\boldsymbol{Y}_t$ is unknown and is represented by the $q \times q$ matrix Σ.

Now, if $\boldsymbol{Y}_t$ is modelled by the DMR model, then the following observation, system equations and prior distribution for Θ_t and Σ hold for all time:

Observation equation

$$\boldsymbol{Y}_t^T = \boldsymbol{F}_t^T \Theta_t + \boldsymbol{v}_t^T, \qquad \boldsymbol{v}_t \sim N(0, V_t\Sigma)$$

System equation

$$\Theta_t = G_t\Theta_{t-1} + \Omega_t, \qquad \Omega_t \sim N(0, W_t, \Sigma)$$

Prior information

$$(\Theta_{t-1}, \Sigma \mid \boldsymbol{y}^{t-1}) \sim NW^{-1}_{n_{t-1}}(M_{t-1}, C_{t-1}, n_{t-1}S_{t-1})$$

where $\boldsymbol{F}_t$ is an s dimensional vector of independent variables; $\boldsymbol{v}_t^T = \{v_{t1}, \ldots, v_{tq}\}$ is the q dimensional observation error vector, where v_{tj} is the observation error for Y_{tj}; V_t is some known scalar; $\Omega_t = \{\boldsymbol{w}_{t1}, \ldots, \boldsymbol{w}_{tq}\}$ is the $s \times q$ system error matrix where $\boldsymbol{w}_{tj}$ is the system error vector for the DLM for Y_{tj}; G_t, W_t, C_{t-1} are $s \times s$ matrices, M_{t-1} is an $s \times q$ matrix and S_{t-1} is a $q \times q$ matrix. Note that $\boldsymbol{F}_t$, G_t, V_t, W_t and C_{t-1} are the same for each marginal univariate DLM of the q series in $\boldsymbol{Y}_t$. It is assumed that the observation errors, v_{tj}, and the system errors, $\boldsymbol{w}_{tj}$, are independent over time and the two series are mutually independent of one another. Ω_t is said to have a matrix-variate normal distribution, discussed by Dawid (1981), and Θ_{t-1} is said to have a normal/inverse Wishart distribution such that:

$$(\Theta_{t-1} \mid \Sigma) \sim N(M_{t-1}, C_{t-1}, \Sigma), \qquad \Sigma \sim W^{-1}_{n_{t-1}}(n_{t-1}S_{t-1})$$

and $E[\Sigma \mid \boldsymbol{y}^{t-1}] = S_{t-1}$.

4. DYNAMIC GRAPHICAL MODELS

Dynamic graphical models attempt to use the flexibility of MDM's whilst utilising any natural symmetries that might possibly exist between some of the components of the vector time series $\boldsymbol{Y}_t$. They are essentially MDM's where some of the components are vectors. Given a chain graph of N variables of the type described in the introduction, list the components as $\boldsymbol{Y}_t^T = \{\boldsymbol{Y}_t(1)^T, \ldots, \boldsymbol{Y}_t(n)^T\}$ where each vector $\boldsymbol{Y}_t(r)^T = \{Y_{t1}(r), \ldots, Y_{tq_r}(r)\}$ consists of the q_r variables in the rth p-set of the chain. Conditional on the values of their parents, $\boldsymbol{Y}_t(r)$ will follow a DLM if $q_r = 1$ and the symmetric DMR evolution if $q_r \geq 2$. Suppose that $\Theta_t(r) = \{\boldsymbol{\theta}_{t1}(r), \ldots, \boldsymbol{\theta}_{tq_r}(r)\}$ is the state parameter matrix defining the distribution of $\boldsymbol{Y}_t(r)$ and $\boldsymbol{\theta}_{tj}(r)$ is the s_r dimensional state vector defining the distribution of $Y_{tj}(r)$, $1 \leq r \leq n$. Let Σ denote the $N \times N$ unknown covariance matrix for $\boldsymbol{Y}_t$ whose r^{th}

diagonal entry is the $q_r \times q_r$ covariance matrix for $\boldsymbol{Y}_t(r)$, $\Sigma(r)$. Note that the DMR models for $\boldsymbol{Y}_t(r)$ $r = 1,\ldots,n$ are somewhat unusual as each variable within a p-set may have *functions* of unobserved contemporary variables as regressors. Because of the symmetry of the roles played by the variables within a partition p-set, it will be assumed here that $\boldsymbol{Y}_t(r)$ only has functions of *sums* of components over a parent p-set as regressors. For notational simplicity these sums will be labelled by:

$$\boldsymbol{X}_t(r)^T = \left\{ \sum_{j=1}^{q_1} Y_{tj}(1), \ldots, \sum_{j=1}^{q_{r-1}} Y_{tj}(r-1) \right\}$$

$$\boldsymbol{Z}_t(r)^T = \left\{ \sum_{j=1}^{q_{r+1}} Y_{tj}(r+1), \ldots, \sum_{j=1}^{q_n} Y_{tj}(n) \right\}$$

Explicitly, then, the DGM is given by:

Observation equations

$$\boldsymbol{Y}_t(r)^T = \boldsymbol{F}_t(r)^T \Theta_t(r) + \boldsymbol{v}_t(r)^T, \qquad r = 1,\ldots,n, \qquad \boldsymbol{v}_t(r) \sim N\left(0, V_t(r)\Sigma(r)\right),$$

System equation

$$\Theta_t = G_t \Theta_{t-1} + \Omega_t, \qquad \Omega_t \sim N(0, W_t, \Sigma),$$

Prior information

$$(\Theta_{t-1}, \Sigma \,|\, \boldsymbol{y}^{t-1}) \sim NW^{-1}_{n_{t-1}}(M_{t-1}, C_{t-1}, S_{t-1}{}')$$

$\boldsymbol{F}_t(r)$ is an s_r dimensional vector and is allowed to be a function of $\boldsymbol{x}^t(r)$ and $\boldsymbol{y}^{t-1}(r)$, but *not* $\boldsymbol{y}^t \backslash \{\boldsymbol{y}^t(r)\}$, $\boldsymbol{z}_t(r)$, $\sum_{j=1}^{q_r} \boldsymbol{y}_j^t(r)$ or $\boldsymbol{y}_t(r)$; $\boldsymbol{v}_t(r)$ is the q_r dimensional observation error vector; $V_t(r)$ $r = 1,\ldots,n$ is some known scalar; $\Theta_t = \text{blockdiag}\,\{\Theta_t(1),\ldots,\Theta_t(n)\}$ is the $s \times N$ parameter matrix; $\Omega_t = \text{blockdiag}\,\{\Omega_t(1),\ldots,\Omega_t(n)\}$ is the $s \times N$ system error matrix where $\Omega_t(r)$ is the $s_r{\times}q_r$ system error matrix for $\boldsymbol{Y}_t(r)$; $M_{t-1}{=}\text{blockdiag}\,\{M_{t-1}(1),\ldots,M_{t-1}(n)\}$ is an $s \times N$ matrix where $M_{t-1}(r)$ is an $s_r \times q_r$ matrix; $S_{t-1}{}'$ is an $N \times N$ matrix with the $q_r \times q_r$ matrices $S_{t-1}(1)n_{t-1}(1)$, $\ldots$, $S_{t-1}(n)n_{t-1}(n)$ on its diagonal; and $G_t = \text{blockdiag}\,\{G_t(1),\ldots,G_t(n)\}$, $W_t = \text{blockdiag}\,\{W_t(1),\ldots,W_t(n)\}$ and $C_{t-1} = \text{blockdiag}$ $\{C_{t-1}(1),\ldots,C_{t-1}(n)\}$ are all $s \times s$ matrices where $G_t(r)$, $W_t(r)$ and $C_{t-1}(r)$ are $s_r \times s_r$ square matrices which may be functions of past vectors $\boldsymbol{x}^{t-1}(r)$ and $\boldsymbol{y}^{t-1}(r)$, but nothing else. As for both the MDM and DMR models it is assumed that $\{v_{tj}(r)\}$ and $\{\boldsymbol{w}_{tj}(r)\}$ are both independent for $1 \le j \le q_r$, $1 \le r \le n$, and are mutually independent over time $t \ge 1$.

Note that $\{\sum_{j=1}^{q_1} \boldsymbol{Y}_t(1),\ldots,\sum_{j=1}^{q_n} \boldsymbol{Y}_t(n)\}$ are governed by an MDM whose conditional variances are estimated on-line through the estimation of Σ. Suppose that 1_{q_r} is a q_r dimensional vector such that $1_{q_r}{}^T = \{1,\ldots,1\}$. The (conditional normal) MDM across these regressors is then given by:

Observation equation

$$\sum_{j=1}^{q_r} Y_{tj}(r) = \boldsymbol{F}_t(r)^T \Theta_t(r) 1_{q_r} + \boldsymbol{v}_t(r)^T 1_{q_r}, \quad 1 \le r \le n, \quad \boldsymbol{v}_t(r)^T 1_{q_r} \sim N\left(0, V_t(r)\Sigma^*(r)\right)$$

System equation

$$\Theta_t(r)1_{q_r} = G_t(r)\Theta_{t-1}(r)1_{q_r} + \Omega_t(r)1_{q_r}, \quad 1 \le r \le n, \quad \Omega_t(r)1_{q_r} \sim N\left(0, \Sigma^*(r)W_t(r)\right)$$

Prior information

$$\left(\Theta_{t-1}(r)1_{q_r} \,|\, \boldsymbol{y}^{t-1}, \Sigma^*(r)\right) \sim N\left(M_{t-1}(r)1_{q_r}, C_{t-1}(r)\Sigma^*(r)\right)$$
$$\left(\Sigma^*(r)^{-1} \,|\, \boldsymbol{y}^{t-1}\right) \sim G\left(\frac{n_{t-1}(r)}{2}, \frac{S^*_{t-1}(r)n_{t-1}(r)}{2}\right)$$

where $\Sigma^*(r) = 1_{q_r}{}^T\Sigma(r)1_{q_r}$, $S^*_{t-1}(r) = 1_{q_r}{}^T S_{t-1}(r)1_{q_r}$ and $\boldsymbol{F}_t(r)$ and G_t are the same as in the DGM.

By substituting $\Theta_t(r)$, $\{\boldsymbol{x}^t(r), \boldsymbol{y}^t(1), \ldots, \boldsymbol{y}^t(r-1)\}$, $\{\sum_{j=1}^{q_r} \boldsymbol{y}^t_j(r), \boldsymbol{y}^t(r)\}$ and $\{\boldsymbol{z}^t(r), \boldsymbol{y}^t(r+1), \ldots, \boldsymbol{y}^t(n)\}$ for $\boldsymbol{\theta}_t(r)$, $\boldsymbol{x}^t(r)$, $\boldsymbol{y}^t(r)$ and $\boldsymbol{z}^t(r)$ respectively in Corollary 3.3 of Queen and Smith (1990), the analogous form of equations 1 and 2 for DGM's follow directly. That is, if $\amalg_{r=1}^n \Theta_0(r)$, then for all time t :

$$\amalg_{r=1}^n \Theta_t(r) \,|\, \boldsymbol{y}^t, \sum_{j=1}^{q_1} \boldsymbol{y}^t_j(1), \ldots, \sum_{j=1}^{q_n} \boldsymbol{y}^t_j(n) \tag{3}$$

and

$$\Theta_t(r) \amalg \boldsymbol{y}^t(r+1), \ldots, \boldsymbol{y}^t(n), \boldsymbol{z}^t(r) \,|\, \boldsymbol{y}^t(1), \ldots, \boldsymbol{y}^t(r), \boldsymbol{x}^t(r), \sum_{j=1}^{q_r} \boldsymbol{y}^t_j(r) \tag{4}$$

So, as with MDM's, the conditional distribution of $\boldsymbol{Y}_t(r) \,|\, \{\boldsymbol{Y}_t(1), \ldots, \boldsymbol{Y}_t(r-1)\}$, $1 \le r \le n$ for each p-set can be forecast separately and updated in closed form (by equation 3). The joint forecast distribution can once again be expressed as the product of these conditional distributions and is given by:

$$p\left\{\boldsymbol{y}_t \,|\, \boldsymbol{y}^{t-1}\right\} = \prod_r \int_{\Theta_t(r)} p\left\{\boldsymbol{y}_t(r) \,|\, \boldsymbol{x}_t(r), \boldsymbol{y}^{t-1}(r), \Theta_t(r)\right\} p\left\{\Theta_t(r) \,|\, \boldsymbol{y}^{t-1}\right\} d\Theta_t(r).$$

Now, in the definition of the DGM for $\boldsymbol{Y}_t$, it states that $G_t(r)$, $W_t(r)$ and $C_{t-1}(r)$ may be functions of $\boldsymbol{y}^{t-1}(r)$ and $\boldsymbol{x}^{t-1}(r)$ but nothing else. This, together with equation 4, allows the simplification of $p\left\{\Theta_t(r) \,|\, \boldsymbol{y}^{t-1}\right\}$ to:

$$p\left\{\Theta_t(r) \,|\, \boldsymbol{y}^{t-1}\right\} = p\left\{\Theta_t(r) \,|\, \boldsymbol{x}^{t-1}(r), \boldsymbol{y}^{t-1}(r)\right\}.$$

Therefore it is only necessary that the series of *sums* of the components in each of the first $r-1$ p-sets are known to find the conditional forecast distribution for $\boldsymbol{Y}_t(r)$.

5. A SIMPLE ILLUSTRATION OF A DYNAMIC GRAPHICAL MODEL

To illustrate the consequences of using a particular DGM, return to the original shampoo example. The simplest possible DGM is chosen, the linear DGM, where we regress on *linear* functions of parent p-set sums. Thus a *linear DGM* is defined so that

$$\boldsymbol{F}_t(r)^T = \{\boldsymbol{x}_t(r)^T, \tilde{\boldsymbol{x}}_t(r)^T\}$$

where $\tilde{\boldsymbol{x}}_t(r)$ is a set of possibly constant exogenous variables known at time t. Let $Y_t(1)$ and $Y_t(3)$ be the p-sets I_t and S_t respectively and $\boldsymbol{Y}_t(2)^T = \{Y_{t1}(2), Y_{t2}(2)\}$ represent the p-set $\{C_t, N_t\}$. Suppose that $\Sigma(1)$, $\Sigma(2)$ and $\Sigma(3)$ are the unknown variances/covariances of $Y(1)$, $\boldsymbol{Y}(2)$ and $Y(3)$ respectively. The linear DGM for this example is given by the following observation and system equations:

Observation equations

$$\begin{aligned} Y_t(1) &= \theta_t^{(0)}(1) + v_t(1), & v_t(1) &\sim N(0, \Sigma(1)) \\ \boldsymbol{Y}_t(2)^T &= \{1, y_t(1)\}\, \Theta_t(2) + \boldsymbol{v}_t(2)^T, & \boldsymbol{v}_t(2) &\sim N(0, \Sigma(2)) \\ Y_t(3) &= \{1, y_{t1}(2) + y_{t2}(2)\}\, \boldsymbol{\theta}_t(3) + v_t(3), & v_t(3) &\sim N(0, \Sigma(3)) \end{aligned}$$

System equations

$$\begin{aligned} \theta_t^{(0)}(1) &= \theta_{t-1}^{(0)}(1) + w_t, & w_t &\sim N(0, W_t(1)\Sigma(1)) \\ \Theta_t(2) &= \Theta_{t-1}(2) + \Omega_t(2), & \Omega_t(2) &\sim N(0, W(2), \Sigma(2)) \\ \boldsymbol{\theta}_t(3) &= \boldsymbol{\theta}_{t-1}(3) + \boldsymbol{w}_t(3), & \boldsymbol{w}_t(3) &\sim N(0, W(3)\Sigma(3)) \end{aligned}$$

Prior Information

$$\begin{aligned} \left(\theta_t^{(0)}(1) \,|\, D_0, \Sigma(1)\right) &\sim N(m_0(1), C_0(1)\Sigma(1)) \\ \left(\Sigma(1)^{-1} \,|\, D_0\right) &\sim G\left(n_0(1)/2, (S_0(1)n_0(1))/2\right), \\ \left(\Theta_0(2), \Sigma(2) \,|\, D_0\right) &\sim NW_{n_0(2)}^{-1}\left(M_0(2), C_0(2), S_0(2)n_0(2)\right), \\ \left(\boldsymbol{\theta}_0(3) \,|\, D_0, \Sigma(3)\right) &\sim N\left(\boldsymbol{m}_0(3), C_0(3)\Sigma(3)\right) \\ \left(\Sigma(3)^{-1} \,|\, D_0\right) &\sim G\left(n_0(3)/2, (S_0(3)n_0(3))/2\right), \end{aligned}$$

where

$$\Theta_t(2) = \begin{pmatrix} \theta_{t1}^{(0)}(2) & \theta_{t2}^{(0)}(2) \\ \theta_{t1}^{(1)}(2) & \theta_{t2}^{(1)}(2) \end{pmatrix},$$

$\boldsymbol{\theta}_t(3)^T = \left(\theta_t^{(0)}(3) \quad \theta_t^{(1)}(3)\right)$ and D_0 represents the knowledge of the system before any observations.

The monthly data set analysed is a gross modification of a non-seasonal market with 3 competitors when there was an increase in the interest rate at period 13 (all other sources of variation have been filtered). The DGM's retain nearly all the advantages of univariate DLM's and, in particular, the familiar technique of intervention analysis (West and Harrison, 1989) can be used at this time period on series I_t. The conditional forecast distributions for each series are found easily using DLM and DMR theory (West and Harrison, 1989) and the one-step ahead conditional forecasts for each distribution are given in Figure 3(a). Notice that some movement in $Y(3)$ has occurred after period 13, but less than in $\boldsymbol{Y}(2)$, reflecting the fact that secondary effects respond less strongly than primary effects. Notice from the non-elliptical contour plots in Figure 4, how the joint forecast density is non-Gaussian, even in this very simple linear DGM. This process, therefore, is very different from a multivariate normal time series.

It is straightforward, if somewhat tedious, to calculate the moments of the marginal forecast distribution explicitly (see Queen, 1991). In this particular example, let

$$\left(\Theta_t(2) \,|\, \boldsymbol{y}^{t-1}(2), \boldsymbol{y}^t(1), \Sigma(2)\right) \sim N(0, R_t(2))$$

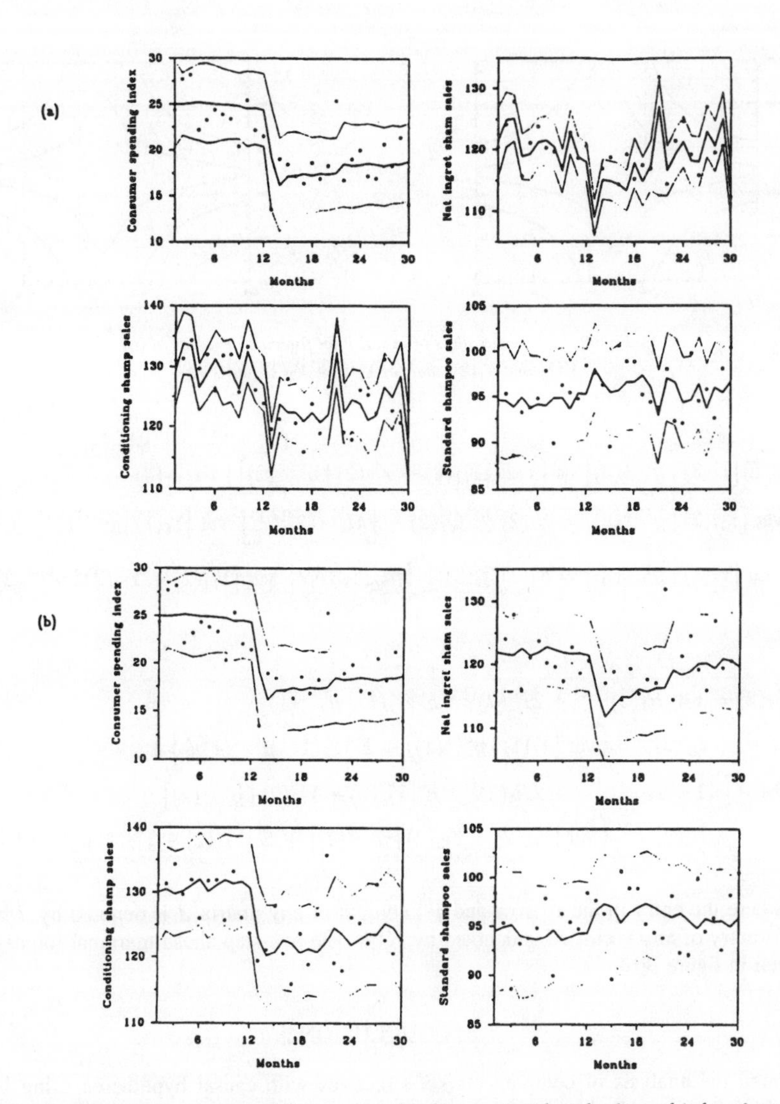

Figure 3. *Conditional (a) and marginal (b) one-step ahead forecasts of brand sales and index of consumer income in a shampoo market. The dots are the observations, the solid line gives the one-step ahead forecasts and the dotted lines represent the 90% confidence limits.*

and

$$\left(\boldsymbol{\theta}_t(3) \mid \boldsymbol{y}^{t-1}(3), \boldsymbol{y}^t(2), \Sigma(3)\right) \sim N(0, R_t(3)),$$

then by using the usual formulae relating conditional expectations and variances to their margins we obtain the marginal moments of components in p-sets:

$$E\left[\boldsymbol{Y}_t(2)^T \mid \boldsymbol{y}^{t-1}(2)\right] = \left\{1, E\left[Y_t(1) \mid \boldsymbol{y}^{t-1}(1)\right]\right\} M_{t-1}(2),$$

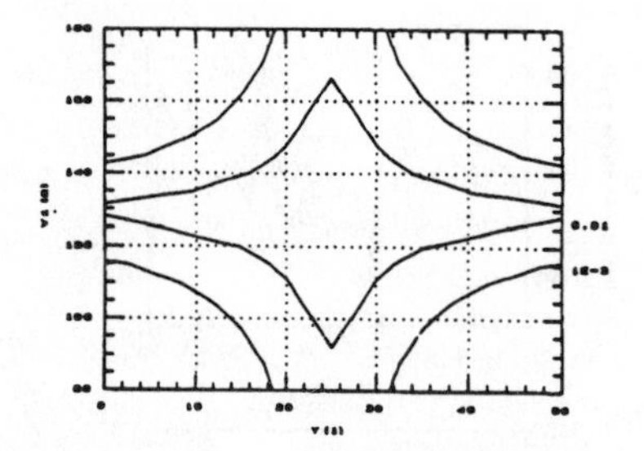

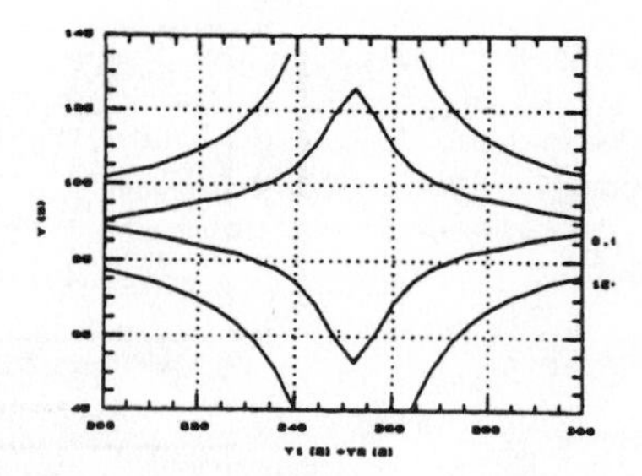

Figure 4. *Contour plots of initial joint forecast densities of (a) $Y(1)$ and $Y_1(2)$, and (b) $Y_1(2)+Y_2(2)$ and $Y(3)$.*

$$E\left[Y_t(3)\,|\,\boldsymbol{y}^{t-1}(3)\right] = \left\{1, E\left[Y_{t1}(2)+Y_{t2}(2)\,|\,\boldsymbol{y}^{t-1}(2)\right]\right\}\boldsymbol{m}_{t-1}(3),$$

$$\text{var}\left[Y_{tj}(2)\,|\,\boldsymbol{y}^{t-1}(2)\right] = S_t(2)^{(j,j)}Q_t(2) + \left[M_{t-1}(2)^{(2,j)}\right]^2 \text{var}\left[Y_t(1)\,|\,\boldsymbol{y}^{t-1}(1)\right], j=1,2,$$

$$\text{var}\left[Y_t(3)\,|\,\boldsymbol{y}^{t-1}(3)\right] = S_t(3)Q_t(3) + \left[\boldsymbol{m}_{t-1}(3)^{(1)}\right]^2 \text{var}\left[Y_{t1}(2)+Y_{t2}(2)\,|\,\boldsymbol{y}^{t-1}(2)\right],$$

where:

$$\begin{aligned}Q_t(2) &= 1 + R_t(2)^{(1,1)} + 2R_t(2)^{(1,2)}E\left[Y_t(1)\,|\,\boldsymbol{y}^{t-1}(1)\right] \\ &\quad + R_t(2)^{(2,2)}\left\{\text{var}\left[Y_t(1)\,|\,\boldsymbol{y}^{t-1}(1)\right] + E\left[Y_t(1)\,|\,\boldsymbol{y}^{t-1}(1)\right]^2\right\}, \\ Q_t(3) &= 1 + R_t(3)^{(1,1)} + 2R_t(3)^{(1,2)}E\left[Y_{t1}(2)+Y_{t2}(2)\,|\,\boldsymbol{y}^{t-1}(2)\right] \\ &\quad + R_t(3)^{(2,2)}\left\{\text{var}\left[Y_{t1}(2)+Y_{t2}(2)\,|\,\boldsymbol{y}^{t-1}(2)\right] + E\left[Y_{t1}(2)+Y_{t2}(2)\,|\,\boldsymbol{y}^{t-1}(2)\right]^2\right\}\end{aligned}$$

and where the entry in the ith row and j^{th} column of any matrix J is denoted by $J^{(i,j)}$ and the ith entry of any vector K is denoted by $K^{(i)}$. The one-step ahead marginal forecasts can be seen in figure 3(b).

6. CONCLUSION

Although the analysis of complex market structures with causal hypotheses using DGM's is in its infancy, preliminary results on real series are very encouraging. We believe that in the future they will extend the scope of the proven versatile DLM. In particular, they enable plausible strong and very non-symmetrical prior information to be incorporated in multivariate processes. This permits strong inferences about the development of the process to be made without requiring very large uncontaminated series, virtually unavailable in the business environment.

ACKNOWLEDGEMENTS

The UK Science and Engineering Research Council, together with Unilever Research, supported one of the authors of this research.

REFERENCES

Dawid, A. P. (1981). Some matrix-variate distribution theory: notational considerations and a Bayesian application. *Biometrika* **68**, 265–274.

Florens, J. P. and Mouchart, M. (1985). A linear theory of non-causality. *Econometrica* **53**, 157–175.

Granger, C. W. J. (1969) Investigating causal relations by econometric models and cross-spectral methods. *Econometrica* **37**, 424–438.

Harrison, P. J. and Stevens, C. F. (1976). Bayesian forecasting. *J. Roy. Statist. Soc. B* **38**, 205–247, (with discussion).

Harvey, A. C. (1989). *Forecasting Structural Time Series Models and the Kalman Filter.* Cambridge: University Press.

Howard, R. A. and Matheson, J. E. (1981). Influence diagrams. *Readings on the Principles and Applications of Decision Analysis* **2** (R. A. Howard and J. E. Matheson eds.), Menlo Park, CA: Strategic Decision Group, 719–762.

Lauritzen, S. L. and Wermuth, N. (1984). Mixed interaction models. *Tech. Rep.* **84-8**, Aarlborg University.

Lauritzen, S. L. and Wermuth, N. (1989). Graphical models for associations between variables, some of which are qualitative and some quantitative. *Ann. Statist.* **17**, 31–54.

Lauritzen, S. L., Dawid, A. P., Larson, B. N. and Leimer, H. G. (1990). Independence properties of directed Markov fields. *Networks* **20**, 491–505.

Queen, C. M. (1991). *Bayesian Graphical Forecasting Models for Business Time Series.* Ph.D. Thesis, University of Warwick.

Queen, C. M. and Smith, J. Q. (1990). Multiregression dynamic models. *Tech. Rep.* **183**, Statistics Department, University of Warwick.

Quintana, J. M. (1985). A dynamic linear matrix-variate regression model. *Tech. Rep.* **83**, Statistics Department, University of Warwick.

Quintana, J. M. (1987). *Multivariate Bayesian Forecasting Models.* Ph.D. Thesis, University of Warwick.

Quintana, J. M. and West, M. (1987). Multivariate time series analysis: new techniques applied to international exchange rate data. *The Statistician* **36**, 275–281.

Quintana, J. M. and West, M. (1988). Time series analysis of compositional data. *Bayesian Statistics 3* (J. M. Bernardo, M. H. DeGroot, D. V. Lindley and A. F. M. Smith, eds.), Oxford: University Press, 747–756

Shachter, R. D. (1986). Intelligent probabilistic inference. *Uncertainty and Artificial Intelligence.* (L. N. Kanal and J. Lemmer eds.), Amsterdam: North-Holland, 371–382.

Smith, J. Q. (1989). Influence diagrams for statistical modelling. *Ann. Statist.* **17**, 654–672.

Smith, J. Q. (1990). Statistical principles on graphs. *Influence Diagrams, Belief Nets and Decision Analysis* (R. M. Oliver and J. Q. Smith eds.), New York: Wiley, 89–120.

Spiegelhalter, D. J. and Lauritzen, S. L. (1990). Sequential updating of conditional probabilities on directed graphical structures. *Networks* **20**, 579–605.

Wermuth, N. and Lauritzen, S. L. (1990). On substantive research hypotheses, conditional independence graphs and graphical chain models. *J. Roy. Statist. Soc. B* **52-1**, 21–50.

West, M. and Harrison, P. J. (1989). *Bayesian Forecasting and Dynamic Models.* New York: Springer

Zellner, A. (1986). *Basic Issues in Econometrics.* Chicago: University Press.

REFERENCES

[illegible] (19[illegible]) [illegible] 48, [illegible]

[illegible] (1988). A [illegible]

[illegible]

[illegible]

[illegible] Academic Press.

[illegible] (1991) [illegible] 7, [illegible]

[illegible] (1992) [illegible]

[illegible] (1985) [illegible]

[illegible] (1990) [illegible]

[illegible]

[illegible] University of Warwick.

[illegible] (1993) [illegible]

[illegible]

[illegible] (1993) [illegible] 57, [illegible]

[illegible]

[illegible]

[illegible] (1988) [illegible]

[illegible]

[illegible]

[illegible] (1993) [illegible]

[illegible] Chicago: University Press.

BAYESIAN STATISTICS 4, pp. 753–762
J. M. Bernardo, J. O. Berger, A. P. Dawid and A. F. M. Smith, (Eds.)

Optimal Portfolios of Forward Currency Contracts

JOSÉ-MARIO QUINTANA
*The Chase Manhattan Bank, N. A., USA**

SUMMARY

The optimal management of portfolios of financial assets is an excellent challenge to the application of the Bayesian paradigm for the decision making process. The forward currency market, in particular, represents a free environment for constructing and rebalancing portfolios since there are virtually no constraints on the feasible asset allocations. First, the operational characteristics of the forward currency market are described. Second, both the Markowitz and the multi-period myopic utility optimizers are reviewed. In addition, their relationship with variable and fixed target optimizers with quadratic loss is explored. Third, a passive, an econometric, and a Bayesian forecasting model are cited. Finally, the impact of these statistical and optimization components on the overall performance is assessed by analyzing their results over the last eight years.

Keywords: BAYESIAN FORECASTING; EXCHANGE RATES; MYOPIC UTILITY FUNCTIONS; OPTIMAL PORTFOLIOS.

INTRODUCTION

Portfolio management stands on two legs: utility functions and probability beliefs. Thus, the formal Bayesian analysis should provide both the means for defining where to go, and a vision of how best to get there. The trading tools must be reliable because other market participants are ready to exploit any weaknesses, thereby making the journey treacherous.

The employment of quantitative methods for asset allocation is often regarded by practitioners as useful only if they are applied in conjunction with a protective shell. This shell takes the form of a set of constraints that reflect "financial wisdom." Yet, a sound methodology would allow for a freedom of choice since the more alternatives that are available the better must be the chances to succeed. The goal of this paper is twofold: to suggest that this strategy can be set to work and, at the same time, to stress that there is a rich field for further research.

1. THE FORWARD CURRENCY MARKET

Among financial markets the currency market enjoys a high degree of freedom for allocating assets via forward currency contracts. You can not only buy currencies with money that you do not have, but you can also sell currencies that you do not possess. The only requirement is to have credit acceptable to the dealers.

A forward currency contract is an agreement to exchange, in the future, one currency for another. The amount of each currency, and the delivery date are stated at the contract inception. At the maturity date, the contract is completed by either actual delivery, or

* The author's views are personal and not necessarily shared by The Chase Manhattan Bank.

settlement in cash (by reversing the trade just before the expiration). The latter form of completion is the most common.

The payoff x_{jt} for selling forward one home currency unit is:

$$x_{jt} = \frac{S_{jt}}{F_{jt}} - 1, \tag{1.1}$$

where t is the delivery date, S_{jt} is the future spot price, and F_{jt} is the forward price of the jth foreign currency unit. The quantities in (1.1) are in home (base) currency units.

There are forward currency contracts available in the foreign exchange market for various combinations of major currencies. In addition, synthetic forward exchange contracts between two arbitrary foreign currencies may be constructed by employing the most traded currency, the U. S. dollar, as a link currency (e.g., a forward exchange of yen for pounds can be accomplished by buying yen, selling U. S. dollars, buying back U. S. dollars, and selling pounds).

A casual spectator may conclude that the forward exchange rate (F_{jt}) is a consensus estimate of the future spot exchange rate (S_{jt}), according to the market participants. This would imply a null consensus expected payoff. However, according to The Interest Rate Parity Theorem (Sharpe, 1985, p. 548–550), the forward rates are a function of the spot rates and the prevailing interest rates:

$$F_{jt} = \frac{S_{j,t-1}(1+i_{0,t-1})}{(1+i_{j,t-1})}, \tag{1.2}$$

where $i_{0,t-1}$ and $i_{j,t-1}$ represent the home and foreign interest rates prevailing at the inception, and are applicable for the duration of the contract.

The reason for this relationship is that otherwise there would be an arbitrage opportunity (i.e., an occasion for making sure money with no explicit capital requirements) by borrowing one currency, lending the other, and simultaneously entering a partially offsetting forward contract. An arbitrage free condition can be interpreted as a financial parallel of the concept of coherence. Due to transaction costs and actual or potential exchange restrictions, the formula (1.2), in practice, only holds approximately. Nevertheless, The Interest Parity Theorem suggests the possibility of making a profit by using forward contracts; albeit undertaking an intrinsic risk.

2. PORTFOLIO OPTIMIZERS

A currency forward contract settled in cash represents, for all practical purposes, a bet (i.e., a fascinating object for a Bayesian observer). There are two main classes of participants in the forward market: the hedgers who seek to reduce their risk associated with the future exchange rates, and the speculators whose aim is to make profits while managing risk.

Henceforth, the focus is on the construction of optimal portfolios of U. S. dollar based, one month forward currency contracts, from the speculator perspective. Notice that there is no loss of generality, since, by allowing the buying or selling of any foreign currency, all the possible cross forward contracts are implicitly included.

Let x denote the payoff of a portfolio of forward currency contracts, then

$$x = \boldsymbol{x}'\boldsymbol{w}, \tag{2.1}$$

where $\boldsymbol{x} = (x_1, \ldots, x_n)'$ is the $n \times 1$ payoff vector, whose components are given by (1.1), and $\boldsymbol{w} = (w_1, \ldots, w_n)'$ is the $n \times 1$ vector of dollar amounts contracted.

2.1. *The Markowitz Optimizer*

The Markowitz approach to portfolio construction focuses on efficient portfolios: a portfolio is efficient if and only if any other portfolio with a higher expected payoff has a higher portfolio variance. The general form of the Markowitz mean-variance (MV) optimization is:

$$\max_{\boldsymbol{w}} \ (\mu - \frac{1}{2}\lambda\sigma^2), \tag{2.2}$$

where $\mu = \boldsymbol{\mu}'\boldsymbol{w}$ is the mean portfolio payoff, $\boldsymbol{\mu}$ is the $n \times 1$ mean payoff vector, $\sigma^2 = \boldsymbol{w}'\Sigma\boldsymbol{w}$ is the portfolio variance, Σ is the $n \times n$ variance payoff matrix, and λ is a non-negative risk-aversion parameter. The set of efficient portfolios is generated by varying the risk-aversion parameter.

The fact that the objective function in (2.2) is not observable presents a major problem in the performance evaluation of a speculative program. A favorite response to overcome this difficulty is to select a constant risk-aversion level for constructing the portfolios, and to replace μ and σ^2 by their sample counterparts observed during the assessment period. However, this is clearly grossly inaccurate since μ and σ^2 in (2.2) are the onestep ahead predictive mean and variance; not the several-steps ahead sample mean and variance. Incidentally, the sample optimization problem analogous to (2.2) is not only a difficult multi-period program, but it also has dubious financial goals. For example, in assessing risk, the focus should be on the *unexplained* variability as opposed to the *total* variability!

2.2. *The Multi-period Myopic Optimizers*

A program seldom ends at the forward maturity date. Therefore, in principle, the portfolio management should be a cumbersome, multi-period optimization problem. Nevertheless, the Bernoulli logarithmic utility function $U(1+x) = \log(1+x)$ is myopic (i.e., the multi-period problem can be solved sequentially onestep at a time).

Suppose that one unit of capital is used in the future to support the contract obligations (i.e., to cover a potential loss). Then the Bernoulli optimization is:

$$\max_{\boldsymbol{w}} \ \underset{\boldsymbol{x}}{E} \ \log(1+x). \tag{2.3}$$

It is easy to see that the myopic property of the Bernoulli utility function is due to its additivity over time.

Furthermore, if the payoffs are independent over time, admittedly a self defeating assumption from a time series forecaster viewpoint, then the power utility function

$$U(1+x) = \frac{(1+x)^{\gamma-1}}{\gamma}$$

is also myopic (Hakansson, 1971). The power (isoelastic marginal) utility optimization is:

$$\max_{\boldsymbol{w}} \ \underset{\boldsymbol{x}}{E} \ \frac{(1+x)^{\gamma}-1}{\gamma}, \tag{2.4}$$

where $1-\gamma$ represents the risk-aversion. Note that (2.3) is the continuous definition of (2.4) for $\gamma = 0$.

Unfortunately, even the single-period optimization of the power, including the logarithmic utility function, proves to be a formidable problem with no known closed solutions apart from simplified setups. Moreover, in the context of modern portfolio theory, forecasts are presented and used in the form of predictive mean and variances rather than full distributions. This leads us to consider tractable quadratic utility functions.

2.3. *The Fixed Target Optimizer*

The utility function of this optimizer is the negative of the square error between the future wealth $(1+x)$ and a predetermined target wealth $(1+p)$. In other words:

$$U(1+x) \;=\; -((1+x)-(1+p))^2 \;=\; -(x-p)^2. \tag{2.5}$$

Thus, the fixed target optimization is:

$$\max_{\boldsymbol{w}} \mathop{E}_{\boldsymbol{x}} -(x-p)^2 \;=\; -p^2+2p \max_{\boldsymbol{w}} \mathop{E}_{\boldsymbol{x}} \left(x-\frac{1}{2p}x^2\right), \tag{2.6}$$

where, of course, $p > 0$. It is revealing to note that $x - \frac{1}{2p}x^2$ in (2.6) is the second order Taylor's expansion of (2.4), for $1-\gamma = \frac{1}{p}$, at $x = 0$. Moreover,

$$\max_{\boldsymbol{w}} \mathop{E}_{\boldsymbol{x}} \left(x-\frac{1}{2p}x^2\right) \;=\; \max_{\boldsymbol{w}} \left(\mu - \frac{1}{2p}(\sigma^2+\mu^2)\right). \tag{2.7}$$

Note that the optimization process depends on the distribution of $\boldsymbol{x}$ only through the first two moments; a very convenient property. In addition, it is not difficult to show, by *reductio ad absurdum*, that a fixed target optimal portfolio is efficient.

2.4. *The Variable Target Optimizer*

The obvious drawback of the fixed target optimizer is that it does not represent rational behavior. In particular, in the event of an arbitrage opportunity, the optimizer settles for the target payoff, rather than for a riskless unlimited profit. This pitfall is avoided by allowing the target payoff to be a decision variable and by modifying (2.5) to become,

$$U(1+x) \;=\; 1+p-\frac{1}{2}\lambda((1+x)-(1+p))^2, \tag{2.8}$$

where λ represents the risk-aversion.

Thus, the variable target optimization is:

$$\max_{p,\boldsymbol{w}} \mathop{E}_{\boldsymbol{x}} \left(1+p-\frac{1}{2}\lambda(x-p)^2\right) \;=\; 1+ \max_{p,\boldsymbol{w}} \left(p-\frac{1}{2}\lambda(\sigma^2+(\mu-p)^2)\right). \tag{2.9}$$

By using standard methods, it is easy to verify that the optimal target payoff satisfies $p = \frac{1}{\lambda}+\mu$. Hence, the optimization (2.9) can be further reduced to:

$$1+\frac{1}{2\lambda}+\max_{\boldsymbol{w}} \left(\mu-\frac{1}{2}\lambda\sigma^2\right). \tag{2.10}$$

But this is equivalent to (2.2)! Thus, the variable target utility function (2.8) not only vindicates the Markowitz optimization from a Bayesian perspective, but also provides a means for assessing its performance.

Furthermore, a multi-period variable, or fixed target myopic analysis, can be induced by defining the overall utility to be weighted additive. In other words:

$$U(1+x_1,\ldots,1+x_n) \;=\; \delta_1 U(1+x_1) + \ldots + \delta_n U(1+x_n), \tag{2.11}$$

where $\delta_i > 0$ for each i. This is a fairly general utility function that represents several possible time attitudes including persistence, impatience, and no time preference (Fishburn, 1979, p. 93-97).

3. STATISTICAL MODELS

The Markowitz approach to portfolio construction is often regarded by practitioners as a fundamentally limited procedure that leads to deficient portfolios unless it is carefully controlled by imposing financially meaningful constraints; see, for example, Michaud (1989). Furthermore, any method relying on the first two moments is useless if they are nonexistent, as implied by Mandelbrot (1963). Nevertheless, it is obvious that any statement on the practical merits of an optimal portfolio construction method is vacuous unless a statistical model is included. It is frequently assumed that there is a true underlying model which generates the observations. However, the uniqueness and even the existence of a true model is always open to question (Dawid, 1986). Thus, a simple pragmatic approach is to consider the model as the way in which an observer looks at the observations and their context (West and Harrison, 1989). Brief descriptions of three diverging perspectives follow.

3.1. *A Passive Model*

The predictive mean of each forward contract payoff, according to this model, is the present value of the difference between the foreign and the domestic interest rates. This follows from (1.1), (1.2), and a martingale behavior assumption of the spot rate ($E\ S_t = S_{t-1}$). A justification for this martingale hypothesis is the central bank's intervention to "stabilize" the exchange spot rates due to economic and political goals (e.g., the exchange rate mechanism [ERM]). The predictive variance is simply the sample estimate of the observed payoff for, say, the 36 trailing months.

3.2. *An Econometric Model*

This model represents a typical approach: predictive mean payoffs are produced on a monthly basis using economic and financial input variables (short and long interest rates, inflation, trade balance, and industrial production). These payoff estimates are based on multiple regression that adapts slowly to structural change over time. The predictive variance is the same as it is for the passive model. This *cut and paste* forecasting procedure is a fairly common practice (or malpractice, depending on one's point of view).

3.3. *A Bayesian Forecasting Model*

The major premise of this time series model is that the effects of the relevant economic, financial, political, etc., factors are already reflected in the payoffs, but in an inefficient manner. The inputs are daily spot exchange rates and the prevailing one-month interest rates. The framework is dynamic and its statistical building blocks are found in Quintana (1987), Quintana and West (1987), and West and Harrison (1989).

4. PERFORMANCE

The two quadratic optimizers of Section 2, together with the three statistical models of Section 3 define six strategies for rebalancing portfolios on a monthly basis. To evaluate their performance a backtest was conducted using the following nine foreign currencies: Australian dollar, French franc, Deutschemark, Japanese yen, Dutch guilder, Swedish krona, Swiss franc, British pound, and Canadian dollar. First, a series of optimal portfolios were constructed according to each strategy. A common risk-aversion parameter set to the value 50 was employed (this parameter determines the leverage of the portfolios). Second, the performance of this series of portfolios was simulated using month-end rates (London close prices). The assessment period is from February 1983 through February 1991.

The performances of different strategies are illustrated in Figures 1 through 6. In each figure there are three time series, one for each statistical model. Figures 1, 3, and 5 correspond to results of the fixed target optimizer; whereas, Figures 2, 4, and 6 relate to the variable target optimizer.

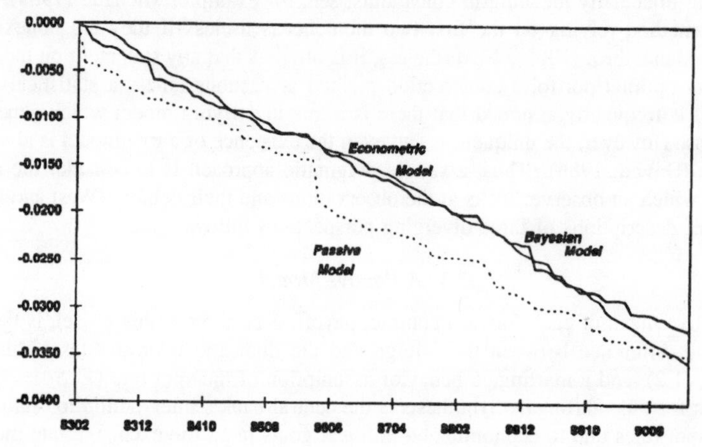

Figure 1. *Fixed Target Cumulative Utilities.*

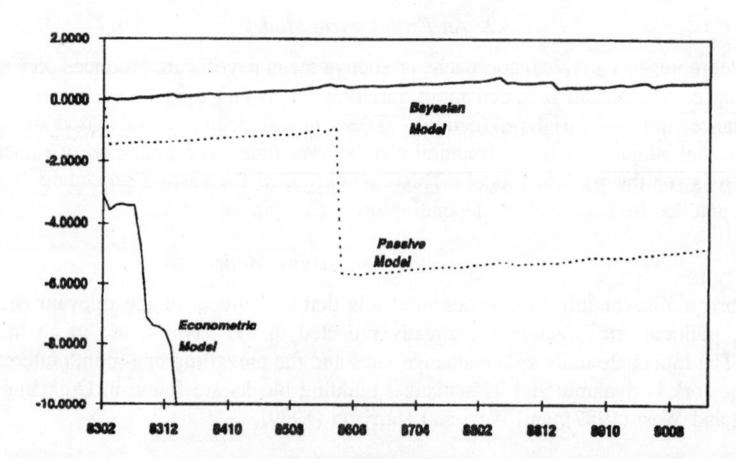

Figure 2. *Variable Target Cumulative Utilities.*

Figures 1 and 2 summarize the performance in terms of an equal weighted additive utility representing a no time preference attitude; see formula (2.10). For the fixed target cumulative utilities, shown in Figure 1, the Bayesian model persistently outperforms the passive model. The econometric model begins with a performance comparable to the Bayesian model, but at the end of 1989 it deteriorates and ends slightly worse than the passive model performance. Regarding the variable target cumulative utilities, appearing in Figure 2, there is a steady dominance of the Bayesian model over the other two. In addition, the passive model outper-

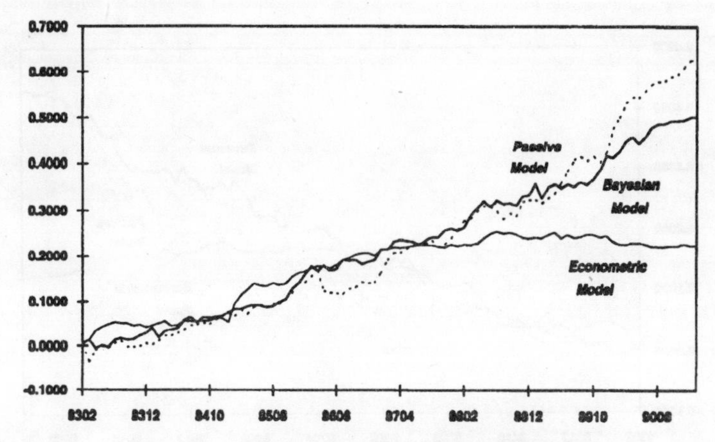

Figure 3. *Fixed Target Cumulative Payoffs Without Costs.*

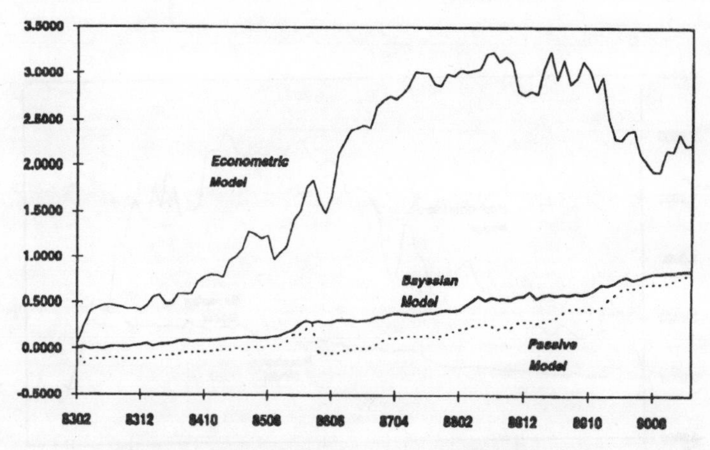

Figure 4. *Variable Target Cumulative Payoffs Without Costs.*

forms the econometric model. The latter performs so poorly that it literally disappears from the scene after a short period. Note the variable target optimization high sensitivity to the statistical models as opposed to the relative robustness of the fixed target optimization.

The fixed and variable target utility functions are, ultimately, a convenient artifice used to produce portfolios with desirable characteristics. One can not take these utility functions too seriously as the square error penalty seems to be too stringent; and therefore unrealistic. The final goal is to accumulate profits by taking an acceptable intrinsic risk. This process is depicted in Figures 3 and 4. The Bayesian and econometric fixed target optimizers have a steady accumulation of payoffs, but the performance of the econometric model decays in the second half of the period. The behavior of the passive model is relatively erratic although it ends with the highest overall payoff. The results for the variable target optimizers are

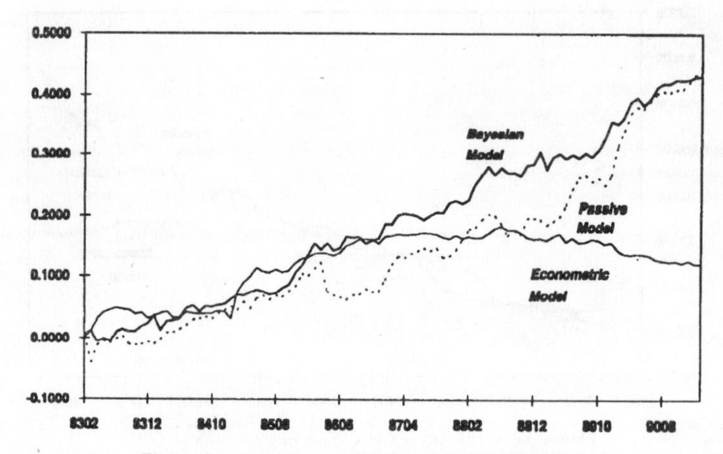

Figure 5. *Fixed Target Cumulative Payoffs With Costs.*

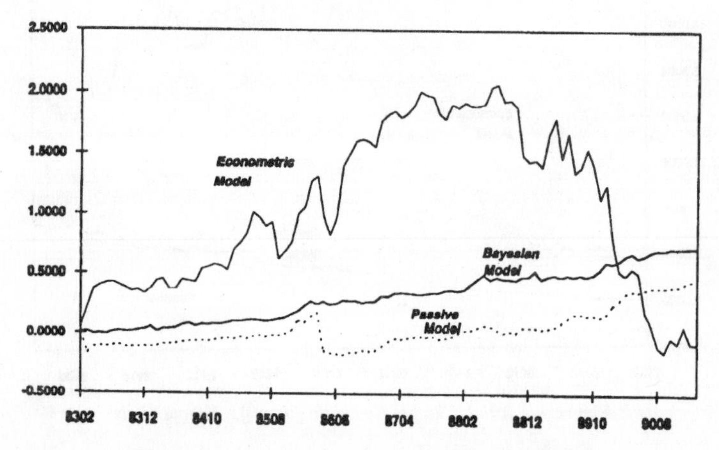

Figure 6. *Variable Target Cumulative Payoffs With Costs.*

strikingly different: the econometric model has the best overall payoff but this is accomplished at an unbearable risk level, whereas the passive and particularly the Bayesian model have a more reliable performance, albeit earning lower overall payoffs. It is also clear from Figures 3 and 4 that, as implied by the discussion in Section 2, the variable target optimizers produce higher payoffs than the corresponding fixed target optimizers.

The optimizations and the results shown in Figures 3 and 4 assume a world with no transaction costs. While the implementation of the costs in the optimization models is rather awkward, their inclusion in the simulated returns is not. These performances, net of transaction costs, are displayed in Figures 5 and 6 assuming that the portfolios are fully sold and bought each month (via outright forward contracts). The shift down of the cumulative payoffs due to transaction costs produces significantly different rankings. The fixed target optimization performance of the Bayesian model is now comparable to that of the passive model. The variable target optimizer performance of the econometric model is degraded from the top to the bottom. In addition, the Bayesian model now clearly outperforms the passive. This is a consequence of the very high leverage employed by the econometric and, to a lesser extent, by the passive model. The Bayesian model performance, in contrast, is affected only mildly; in this sense, it is rather robust.

Payoffs in percent net of transaction costs						
	Fixed target optimization			Variable target optimization		
Date	Passive	Econometric	Bayesian	Passive	Econometric	Bayesian
1983*	–0.7564	3.1253	2.3825	–12.5505	34.7412	3.3996
1984	5.0727	0.9987	2.9949	6.7933	24.1691	4.0305
1985	6.1168	9.1372	6.5042	15.5733	36.6866	11.5784
1986	–3.0212	2.4299	4.5536	–26.9571	42.1695	8.6695
1987	7.0612	1.4419	6.3514	12.4723	34.1755	9.4405
1988	5.3754	–0.6253	5.8543	8.5090	–44.8241	12.2612
1989	6.9900	–0.6261	7.4646	11.1067	–35.5094	11.6520
1990	16.4651	–2.8696	6.9951	28.7053	–77.5692	12.4641
1991**	1.3013	–0.5190	0.5952	2.3657	–14.9338	0.8186
Overall	5.3632	1.5125	5.3309	5.5602	–0.9935	9.0347

* (excluding January)
** (January and February only)

Table 4.1. *Net payoffs by statistical and optimization model.*

The yearly net payoffs are presented in Table 4.1. The positive and consistent performance of the Bayesian model is in contrast with the performances of the other two statistical models, for both the fixed and variable target optimizations. In particular, the average of the Bayesian model annualized payoff for the variable target optimization is a respectable 9.03%. This may seem rather modest relative to other investments such as international stocks and bonds; however, it is crucial to keep in mind that there are no explicit capital requirements for the forward contracts. For example, a long term investor, using this forward currency overlay to enhance the performance of any arbitrary portfolio, would have outperformed the underlying investment yearly by 9.03% on average! Of course, higher (lower) payoffs at a higher (lower) risk are attainable by decreasing (increasing) the risk-aversion parameter.

5. CONCLUSIONS

The advice coming from the performance results of Section 4 is obvious but often ignored: in a free mean-variance optimization context, the estimation of the variance should not be neglected. The negative impact of the statistical deficiencies is highly dependent on the type of selection rule that pinpoints a portfolio from the efficient set: as shown, the variable target optimization is very strict, whereas the fixed target optimization is fairly tolerant. Alternatively, the shortcomings of a poor variance estimation might be compensated by imposing limits on the asset allocations, thereby forcing diversification among the assets. However, the actual limits are often chosen by a trial-error backtesting process. This is an excellent recipe to produce overfitted models that typically yield disappointing real-time performances. In the end, all damage control procedures attack the symptoms rather than the origin of the problem.

It is reassuring that a Bayesian forecasting model performs relatively well in a free environment. However, this by no means implies that non-Bayesian models are useless. After all, from a predictivistic standpoint, any well defined procedure that produces one-step ahead distributions (or mean and variances), in terms of the previous endogenous and exogenous variables, is coherent. What is required is a forecasting model that captures the relevant aspects of exchange rate dynamics. A potential contender worth considering is the multivariate GARCH model (Bollerslev, Engle, and Wooldridge, 1988).

Finally, there are two critical features of portfolio management not yet included in this discussion. First, the cost of transacting must be an integral part of the optimization process. This is not only an optimization complication, but also a statistical intricacy because the future bid/ask spot spreads are unknown. Second, and most importantly, there is an implicit unrealistic assumption behind multi-period optimizers: information arrives in a discrete fashion. In contrast, portfolio management is a *continuous-time* undertaking in response to the flow of information. Although there has been a considerable effort to provide the formal means to attack this problem, the matter is not yet settled. For example, on the one hand, Merton (1971) and his followers have made significant progress in solving the optimization component using Itô's stochastic calculus. On the other hand, the nonlinear prediction counterpart based on the same complex technical platform is the subject of serious criticisms from the applications viewpoint. These latter difficulties are presented together with an interesting finitely additive alternative approach in Kallianpur and Karandikar (1985).

REFERENCES

Bollerslev, T., Engle, R. F. and Wooldridge, J. M. (1988). A capital asset pricing model with time-varying covariances. *J. of Political Economy* **96**, 116–131.

Dawid, A. P. (1986). A Bayesian view of statistical modelling. *Bayesian Inference and Decision Techniques: Essays in Honor of Bruno de Finetti* (P. K. Goel and A. Zeller, eds.), Amsterdam: North-Holland, 391–404.

Fishburn, P. C. (1979). *Utility Theory for Decision Making*. New York: Krieger.

Hakansson, N. H. (1971). On optimal myopic portfolio polices, with and without serial correlation of yields. *J. of Business* **44**, 324–334.

Kallianpur, G. and Karandikar, R. L. (1985). A finitely additive white noise approach to nonlinear filtering: a brief survey. *Multivariate Analysis* **6**. New York: Elsevier.

Mandelbrot, B. (1963). The variation of certain speculative prices. *J. Business* **36**, 394–419.

Merton, R. C. (1971). Optimum consumption and portfolio rules in a continuous-time model. *J. of Economic Theory* **3**, 373–413.

Michaud, R. O. (1989). The Markowitz optimization enigma: is 'optimized' optimal?. *Financial Analysis Journal* **45**, 31–42.

Quintana, J. M. (1987). *Multivariate Bayesian Forecasting Models*. Ph.D. Thesis, University of Warwick.

Quintana, J. M. and West, M. (1987). Multivariate time series analysis: new techniques applied to international exchange rate data. *The Statistician* **36**, 275–281.

Sharpe, W. F. (1985). *Investments*. Englewood Cliffs, NJ: Prentice-Hall.

West, M. and Harrison, P. J. (1989). *Bayesian Forecasting and Dynamic Models*. New York: Springer.

BAYESIAN STATISTICS 4, pp. 763–773
J. M. Bernardo, J. O. Berger, A. P. Dawid and A. F. M. Smith, (Eds.)

How many Iterations in the Gibbs Sampler?*

ADRIAN E. RAFTERY and STEVEN M. LEWIS
University of Washington, USA

SUMMARY

When the Gibbs sampler is used to estimate posterior distributions (Gelfand and Smith, 1990), the question of how many iterations are required is central to its implementation. When interest focuses on quantiles of functionals of the posterior distribution, we describe an easily-implemented method for determining the total number of iterations required, and also the number of initial iterations that should be discarded to allow for "burn-in". The method uses only the Gibbs iterates themselves, and does not, for example, require external specification of characteristics of the posterior density.

Keywords: HIERARCHICAL MODEL; MARKOV CHAIN; MONTE CARLO; QUANTILE; SAMPLING-BASED INFERENCE; SPATIAL STATISTICS.

1. INTRODUCTION

The Gibbs sampler was introduced by Geman and Geman (1984) as a way of simulating from high-dimensional complex distributions arising in image restoration. The method consists of iteratively simulating from the conditional distribution of one component of the random vector to be simulated given the current values of the other components. Each complete cycle through the components of the vector constitutes one step in a Markov chain whose stationary distribution is, under suitable conditions, the distribution to be simulated. Gelfand and Smith (1990) pointed out that the algorithm may also be used to simulate from posterior distributions, and hence may be used to solve standard statistical problems.

The Gibbs sampler can be extremely computationally demanding, even for relatively small-scale statistical problems, and hence it is important to know how many iterations are required to achieve the desired level of accuracy. Here we describe and investigate a simple method for doing this, first briefly mentioned in Raftery and Banfield (1991).

We focus on the situation where there is a single long run of the Gibbs sampler, as practiced by Geman and Geman (1984) and Besag, York and Mollié (1991), for example. Gelfand and Smith (1990) have instead adopted the following alogithm: (i) choose a starting point; (ii) run the Gibbs sampler for T iterations and store only the last iterate; (iii) return to (i). The choice between the two ways of implementing the algorithm has not been settled, and was the subject of considerable debate and controversy at the recent Workshop on Bayesian Computation via Stochastic Simulation in Columbus, Ohio in February, 1991.

Intuitive considerations suggest that one long run may well be more *efficient*. A heuristic argument for this might run as follows. Consider the following two ways of obtaining S

* This research was supported by ONR contract N-00014-88-K-0265 and by NIH grant no. 5R01HD26330-02. The authors are grateful to Jeremy York for providing the data for Examples 4 and 5, for helping with the analysis and for useful discussions and suggestions, and to Julian Besag and an anonymous referee for helpful comments. A Fortran program called "Gibbsit" that implements the methods described here may be obtained from StatLib by sending an e-mail message to *statlib@temper.stat.cmu.edu* containing the single line "send gibbsit from general". While the program is not maintained, questions about it may be addressed by e-mail to Adrian Raftery at *raftery@stat.washington.edu*.

values simulated from the posterior distribution. The first way consists of picking off every Tth value in a single long run of length $N = ST$. The second way is that of Gelfand and Smith (1990). In the first way, the starting point for every subsequence of length T is closer to a draw from the stationary distribution than the corresponding starting point in the second way, which is chosen by the user. Thus, the first way gives a result which is, at least, no worse than the second way. Sometimes, although not always, this may be exploited in the first way by reducing the value of T, to obtain the same result with less total iterations. A more formal argument along similar lines was presented by R.L. Smith in the concluding discussion at the Workshop on Bayesian Computation via Stochastic Simulation.

Gelman and Rubin (1992), on the other hand, have argued that, even if the one long run approach may be more efficient, it is still important to use several different starting points. The essence of their argument is that we cannot know, in the case of any individual problem, whether a single run has converged, and that combining the results of runs from several starting points gives an honest, if conservative, assessment of the underlying uncertainty. They illustrate their argument by showing that in the Ising model convergence can be quite slow. This example refers to the 10,000-dimensional binary state-space $\{-1, 1\}^{10,000}$, and is thus untypical of the parameter spaces that arise in typical statistical problems, but it should nevertheless be taken seriously. Here we suggest that combining internal information from a partial run with properties of Markov chains may provide an alternative way of solving the problem, without sacrificing the appealing simplicity of using a single long run. In particular, Markov chain theory r rovides results not just about ergodicity, but also about the (geometric) rate of convergence .o the stationary distribution, and the distribution of sample means. However, the method can easily be used when there are several runs from different starting points.

2. THE METHOD

2.1. *How Many Iterations to Estimate a Posterior Quantile?*

We consider the specific problem of calculating particular quantiles of the posterior distribution of a function U of the parameter θ. We formulate the problem as follows. Suppose that we want to estimate $P[U \leq u \mid y]$ to within $\pm r$ with probability s, where U is a function of θ. We will find the approximate number of iterations required to do this when the correct answer is q. For example, if $q = .025$, $r = .005$ and $s = .95$, this corresponds to requiring that the cumulative distribution function of the .025 quantile be estimated to within $\pm.005$ with probability .95. This might be a reasonable requirement if, roughly speaking, we wanted reported 95% intervals to have actual posterior probability between .94 and .96. We run the Gibbs sampler for an initial M iterations that we discard, and then for a further N iterations of which we store every kth. Typical choices in the literature are $M = 1,000$, $N = 10,000$ and $k = 10$ or 20 (Besag, York and Mollié 1991). Our problem is to determine M, N, and k. Note that when $k > 1$, we may still store and use all the N iterates, and the solution given here is then conservative.

We first calculate U_t for each iteration t, and then form $Z_t = \delta(U_t \leq u)$, where $\delta(\cdot)$ is the indicator function. $\{Z_t\}$ is a binary 0-1 process that is derived from a Markov chain by marginalization and truncation, but it is not itself a Markov chain. Nevertheless, it seems reasonable to suppose that the dependence in $\{Z_t\}$ falls off fairly rapidly with lag, and hence that if we form the new process $\{Z_t^{(k)}\}$, where $Z_t^{(k)} = Z_{1+(t-1)k}$, then $\{Z_t^{(k)}\}$ will be approximately a Markov chain for k sufficiently large.

No formal proof of this is presented here, but it does seem intuitively plausible. Here a data-based method, described below, is used to assess whether the assumption provides a reasonable approximation for the case at hand. A proof might go something as follows. The process $\{Z_t\}$ is ergodic and, if the underlying Markov chain is ϕ-mixing in the sense of Billingsley (1968), which will often be a direct consequence of the construction, then $\{Z_t\}$ is also ϕ-mixing with the same rate. Thus the maximum difference between $P[Z_t^{(k)} = i_0 \mid Z_{t-1}^{(k)} = i_1, Z_{t-2}^{(k)}]$ and $P[Z_t^{(k)} = i_0 \mid Z_{t-1}^{(k)} = i_1]$ eventually declines geometrically as a function of k, and so $\{Z_t^{(k)}\}$ is arbitrarily close to being a first-order Markov chain in that sense, for k sufficiently large.

In what follows, we draw on standard results for two-state Markov chains; see, for example, Cox and Miller (1965). Assuming that $\{Z_t^{(k)}\}$ is indeed a Markov chain, we now determine $M = mk$, the number of "burn-in" iterations, to be discarded. Let

$$P = \begin{pmatrix} 1-\alpha & \alpha \\ \beta & 1-\beta \end{pmatrix}$$

be the transition matrix for $\{Z_t^{(k)}\}$. The equilibrium distribution is then $\pi = (\pi_0, \pi_1) = (\alpha+\beta)^{-1}(\beta, \alpha)$, and the ℓ-step transition matrix is

$$P^\ell = \begin{pmatrix} \pi_0 & \pi_1 \\ \pi_0 & \pi_1 \end{pmatrix} + \frac{\lambda^\ell}{\alpha+\beta}\begin{pmatrix} \alpha & -\alpha \\ -\beta & \beta \end{pmatrix},$$

where $\lambda = (1-\alpha-\beta)$. Suppose that we require that $P[Z_m^{(k)} = i \mid Z_0^{(k)} = j]$ be within ε of π_i for $i, j = 0, 1$. If $e_0 = (1,0)$ and $e_1 = (0,1)$, then $P[Z_m^{(k)} = i \mid Z_0^{(k)} = j] = e_i P^m$, and so the requirement becomes

$$\lambda^m \leq \frac{\varepsilon(\alpha+\beta)}{\max(\alpha,\beta)},$$

which holds when

$$m = m^\star = \frac{\log\left(\frac{\varepsilon(\alpha+\beta)}{\max(\alpha,\beta)}\right)}{\log \lambda}.$$

Thus

$$M = m^\star k.$$

To determine N, we note that the estimate of $P[U \leq u \mid D]$ is $\overline{Z}_n^{(k)} = \frac{1}{n}\sum_{t=1}^n Z_t^{(k)}$. For n large, $\overline{Z}_n^{(k)}$ is approximately normally distributed with mean q and variance

$$\frac{1}{n}\frac{\alpha\beta(2-\alpha-\beta)}{(\alpha+\beta)^3}.$$

Thus the requirement that $P[q - r \leq \overline{Z}_n^{(k)} \leq q + r] = s$ will be satisfied if

$$n = n^\star = \frac{\alpha\beta(2-\alpha-\beta)}{(\alpha+\beta)^3}\left\{\frac{r}{\Phi(\frac{1}{2}(1+s))}\right\}^{-2}$$

where $\Phi(\cdot)$ is the standard normal cumulative distribution function. Thus we have $N = kn^\star$.
To determine k, we form the series $\{Z_t^{(k)}\}$ for $k = 1, 2, \ldots$. For each k, we compare the first-order Markov chain model with the second-order Markov chain model, and choose the smallest value of k for which the first-order model is preferred. We compare the models by

first recasting them as (closed-form) log-linear models for a 2^3 table (Bishop, Fienberg and Holland, 1975), and then using the BIC criterion, $G^2 - 2\log n$, where G^2 is the likelihood ratio test statistic. This was introduced by Schwarz (1978) in another context and generalized to log-linear models by Raftery (1986); it provides an approximation to twice the logarithm of the Bayes factor for the second-order model. One could also use a non-Bayesian test, but the choice of significance level is problematic in the presence of large samples of the size that arise routinely with the Gibbs sampler.

To implement the method, we run the sampler for an initial number of iterations, $N_{\min}$, and use this run to determine the number of additional runs required, as above. The procedure can be iterated, in that once the indicated number of iterations has been run, we may apply the method again to the entire run, reestimating α and β to determine more precisely if the number of iterations produced was in fact adequate. To determine $N_{\min}$, we note that the required N will be minimized if successive values of $\{Z_t\}$ are independent, in which case $M = 0$, $k = 1$ and

$$N = N_{\min} = \Phi^{-1}\left(\frac{1}{2}(1+s)\right)^2 q(1-q)/r^2.$$

For example, when $q = .025$, $r = .005$ and $s = .95$, we have $N_{\min} = 3,748$.

We also note that the user is not *required* to use only every kth iterate; if all the iterates are used the method proposed here will be conservative in the sense of possibly overestimating the number of iterations required. On the other hand, in the majority of cases that we have examined, the preferred value of k was, in fact, 1. Also, storage considerations often point to the desirability of storing only a portion of the iterates if this is reasonable.

The user needs to give only the required precision, as specified by the four quantities q, r, s and ε. Of these, the result is by far the most sensitive to r, since $N \propto r^{-2}$. It may often be more natural to specify the required precision in terms of the error in the estimate of a quantile rather than the error in the cumulative distribution function at the quantile, which is what r refers to. In order to see how r relates to accuracy on the former scale, we have shown in Table 1 the approximate maximum percentage error in the estimated quantile corresponding to a range of values of r, for $q = .025$. This is defined as

$$100\max\{\frac{F^{-1}(q \pm r)}{F^{-1}(q)} - 1\},$$

and is shown for three distributions: normal (light-tailed), t_4 (moderate tails), and Cauchy (heavy-tailed).

r	$N_{\min}$	Percent error		
	(s=.95)	N(0,1)	t_4	Cauchy
.0025	14982	2	4	11
.0050	3748	5	8	25
.0075	1665	8	13	43
.0100	936	11	19	67
.0125	600	14	26	101
.0150	416	19	37	150
.0200	234	31	65	402

Table 1. *Maximum percent error in the estimated .025 quantile.*

Suppose we regard a 14% error as acceptable, corresponding to an estimated .975 quantile of up to 2.24 in the normal distribution, compared with the true value of 1.96. Then, if we knew $p(U \mid y)$ to have light, normal-like, tails, Table 1 suggests that $r = .0125$ would be sufficiently small. However, with the heavier-tailed t_4 distribution, $r = .0075$ is required to achieve the same accuracy, while for the very heavy-tailed Cauchy, $r = .003$ is required, corresponding to $N_{\min} \approx 10,000$.

This suggests that if we are not sure in advance how heavy the posterior tail is, $r = .005$ is a reasonably safe choice (even for the Cauchy it is not catastrophic). It also suggests that the present method could be refined by using the initial set of Gibbs iterates to estimate the asymptotic rate of decay of the posterior tail nonparametrically with methods such as those of Hall (1982), and then choosing r in light of the estimate, perhaps by referring to a t-distribution with the appropriate degrees of freedom. At first sight it might appear that a component-wise reparametrization to lighten the tails would be a good remedy. However, we suspect that this would not be a real solution, and that the problem would reappear when the results were transformed back to the scale on which the quantity of actual interest is measured.

2.2. *Extensions*

Several quantiles: If there are Q quantiles to be estimated ($Q > 1$), it is possible to run the Gibbs sampler for $N_{\min}$ iterations, apply the method in Section 2.1 to each quantile individually and then use the maximum values of M, k and N. This will guarantee that for each quantile marginally, at least the specified accuracy will be achieved, or, equivalently, that the expected number of estimated quantiles that lie within r of the true value is at least sQ. A different and more demanding accuracy requirement is that with probability s, all Q quantiles lie within r of the true value. A conservative solution to this uses the Bonferroni bound idea and consists of replacing s by s/Q and proceeding as before.

Estimating probabilities: The Gibbs sampler is often used to estimate probabilities rather than distributions. Examples include the probability that the ground truth at a pixel is of a particular "color" in image reconstruction (Besag, 1986), the probability that an individual is of a given genotype in genetic pedigree analysis (Sheehan, 1990), and the probability that an individual has a specific disease based on an expert system diagnosis (Lauritzen and Spiegelhalter, 1988). The present method is directly applicable to this case, and indeed takes a slightly simpler form. The $\{Z_t\}$ process is given and does not have to be formed by truncation. One does not have to specify q, which is just the required probability. One simply specifies r and s and proceeds as before.

Independent iterates: When it is much more expensive to analyze a Gibbs iterate than to simulate it, as it may be in complex applications such as genetic pedigree analysis, it is desirable that the Gibbs iterates used be approximately independent. This can be achieved by making k big enough. To determine k, we can compare the independence model with the first-order Markov chain model for $\{Z_t^{(k)}\}$ and choose the smallest value of k for which the independence model is preferred. The models can be compared by recasting them as the independence and saturated models for a 2×2 table and using the BIC criterion. M is then calculated exactly as before, and we have $N = kN_{\min}$ by the approximate independence of the iterates.

3. EXAMPLES

We now apply the method to several examples, both simulated and real. In each case, we give the results only for $q = .025$, $r = .005$, $s = .95$ and $\varepsilon = .001$. The results are shown in Table 2 for all the examples. The value in the column headed $\hat{F}(F^{-1}(.025))$ should be between .02 and .03 for this specification. Results for other quantiles and other accuracy requirements, not shown here, were qualitatively similar.

Example	M	k	N	$\hat{F}(F^{-1}(.025))$
1. Indep. normal pars.	3	1	3,914	.023
2. Bimodal	4	1	4,256	.028
3. Cigar	36	3	26,916	.024
4. Spatial u_1	3	1	4,052	.024
5. Spatial smoothness	40	2	24,346	—

Table 2. *Results for the five examples.*

Example 1: Multivariate normal distribution with independent parameters. In this simulated example the method gave $k = 1$, a very small number of burn-in iterations ($M = 3$), and a value of N which is only slightly larger than the theoretical minimum (3,914 as against 3,748). Also, the result is within the specified bounds. While this is very much as one would expect, it is also a reassuring check on the performance of the method.

Example 2: A bimodal posterior distribution. Here we simulated, using the Gibbs sampler, from a mixture of two bivariate normal distributions, namely

$$\frac{1}{2}BVN(\mu_1, \Sigma) + \frac{1}{2}BVN(\mu_2, \Sigma),$$

where $\mu_1 = (-1, 1)^T$, $\mu_2 = (1, 0)^T$ and

$$\Sigma = \begin{pmatrix} 1 & .9 \\ .9 & 1 \end{pmatrix}.$$

The joint distribution is quite strongly bimodal, although the marginal distributions of the two components are not. The first 1,000 simulated values of the second component are shown in Figure 1. The result is surprisingly similar to that in Example 1. Again, $k = 1$, the amount of burn-in is negligible ($M = 4$), and $N = 4,256$ is not much larger than the theoretical minimum. The Gibbs iterates are slightly more highly correlated than in Example 1, and the value of N can be regarded as an index of this. Once again, the result is within the specified bounds.

Example 3: A cigar in ten dimensions. In order to investigate the effect of high posterior correlations between parameters, we used the Gibbs sampler to simulate from a 10-dimensional multivariate normal posterior distribution where each component had zero mean and unit variance, and all the pairwise correlations were equal to .9. This is a highly correlated distribution, where the first principal component (proportional to the mean of the parameters) accounts for 91% of the variance; the posterior distribution is concentrated about a thin "cigar" in 10-space. Note that this is a very poor parameterization for the Gibbs sampler.

The first 1,000 simulated values of the first parameter are shown in Figure 2. The results of applying the method are strikingly different from what we saw before. The amount of

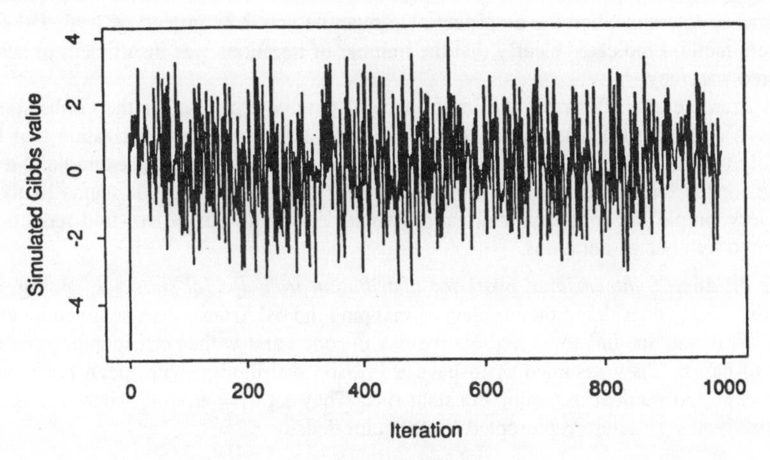

Figure 1. *Bimodal Example.*

burn-in is no longer negligible, although it is not huge ($M = 36$). The dependency structure of the binary sequence is more complicated than before, leading to $k = 3$, and the level of dependency is high, so that the required N is very large, at 26,916. After that number of iterations, the result was accurate. This phenomenon seems to be due to the high level of dependency in the sequence, and not primarily to the sampler being slow to converge to the desired distribution.

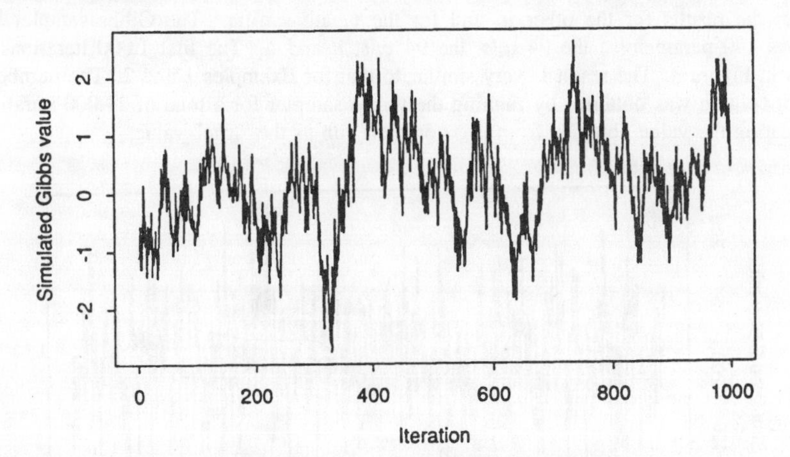

Figure 2. *Cigar in 10 dimensions example.*

It is of interest to consider the situation after 6,700 iterations; this is a large number, but substantially less than the prescribed 27,000. By that point, diagnostics based on changes in cumulative estimates suggest the Gibbs sampler to have converged. However, after 6,700 iterations, $1-\hat{F}(F^{-1}(.975)) = .045$, compared to the true value of .025, which is well outside the prescribed tolerance, and the empirical .975 quantile was 2.22 instead of 1.96. However, the present method indicated clearly that the number of iterations was insufficient to achieve the desired accuracy.

This example also illustrates the importance of parameterization for the Gibbs sampler (see also Wakefield, in discussion of Hills and Smith, 1992). A parameterization that leads to a highly correlated posterior distribution like the one considered in this example is a very poor one for the Gibbs sampler, and leads to considerable inefficiency. It seems likely that even a very simple linear reparameterization would lead to at least a five-fold reduction in the required number of iterations.

Example 4: An 190-dimensional posterior distribution from spatial statistics. Besag, York and Mollié (1991) considered the problem of mapping the risk from a disease given incidence data. Let x_i denote the unknown log relative risk in zone i and y_i the corresponding observed number of cases. They assumed y_i to have a Poisson distribution with mean $c_i e^{x_i}$, where c_i is the expected number assuming constant risk. They let $x_i = u_i + v_i$ where the u_i have substantial spatial structure represented by the joint density

$$p(u \mid \kappa) \propto \frac{1}{\kappa^{\frac{1}{2}n}} \exp\left\{ -\frac{1}{2\kappa} \sum_{i \sim j} (u_i - u_j)^2 \right\},$$

where $i \sim j$ denotes the fact that zones i and j are contiguous and κ is a spatial smoothness parameter. The v_i are assumed to be generated by Gaussian white noise with parameter λ. The prior distribution was $p(\kappa, \lambda) \propto \exp\{-.005(\kappa^{-1} + \lambda^{-1})\}$. Although this is improper, the resulting posterior distribution is proper. The main aim is to find the posterior distribution of x_i, but other features of the underlying mechanism may also be of interest.

Here we show only the result for u_1 for thyroid cancer deaths in 94 départements of France; the results for the other u_i and for the v_i are similar. The Gibbs sampler here involves 190 parameters: the 94 u_i's, the 94 v_i's, κ and λ. The first 1,000 iterations are shown in Figure 3. The result is very similar to that for Examples 1 and 2. The number in the last column was obtained by running the Gibbs sampler for a total of 11,000 iterations, and treating the value obtained from this complete run as the "true" value.

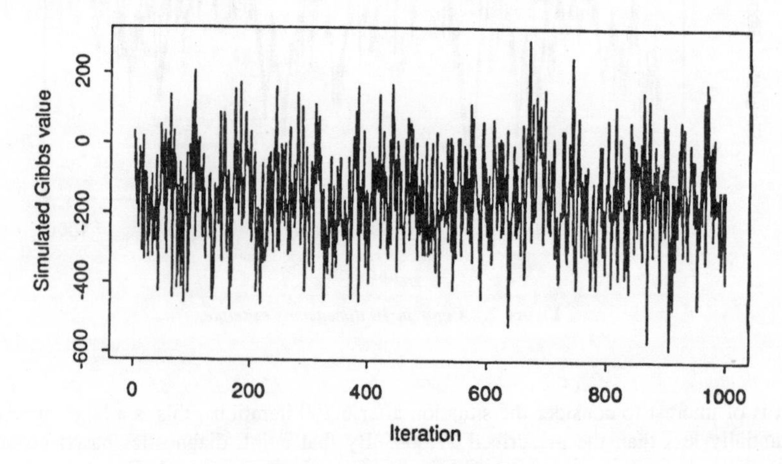

Figure 3. *Spatial example:* $u1$.

Example 5: The spatial smoothness parameter. We now consider separately the spatial smoothness parameter κ from Example 4. The first 1,000 Gibbs iterations are shown in Figure 4. The results are quite different from those for u_1, and are somewhat similar to those for Example 3. The dependency structure in the induced binary sequence is complex, leading to $k = 2$, and the dependency is high, leading to $N = 24,346$. The amount of burn-in, however, while not negligible, is fairly small ($M = 40$). It was not feasible to determine the correct answer in this case.

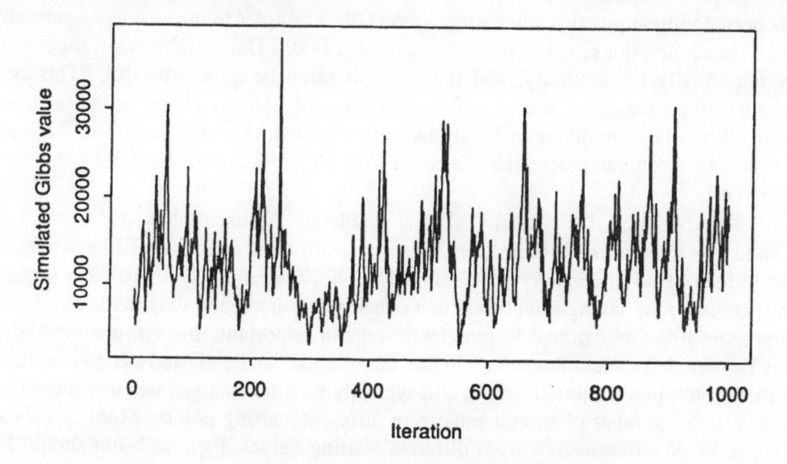

Figure 4. *Spatial example: kappa.*

While the difficulty with Example 3 could probably be resolved by appropriate reparameterization, the problem here seems more fundamental. Here the problem is due to the fact that κ sometimes gets "stuck" close to zero for several hundred iterations at a time. This is because having the u_i close together (i.e., high spatial smoothness) makes a small value of κ likely, while a small value of κ forces the u_i to be close together. Thus the Gibbs sampler gets caught periodically in a "vicious circle"; to escape it requires a rare event. The solution here may be the use of a different variation on Metropolis dynamics than the Gibbs sampler, perhaps involving simultaneous updating of some kind. This kind of problem seems likely to arise often in hierarchical models more generally. Note that the present method for determining the number of iterations would carry over to other forms of Metropolis dynamics.

4. DISCUSSION

We have proposed a method for determining how many iterations are necessary in the Gibbs sampler. This is easy to implement and does not require anything beyond an initial run from the sampler itself. It appears to give encouraging results in several examples. However, much more thorough investigation is required for various kinds of difficult posterior distributions.

For "nice" posterior distributions, the examples suggest that accuracy at the level specified for illustration in this paper can be achieved by running the sampler for 5,000 iterations and using all the iterates. However, when the posterior is not "nice", the required number can be very much greater. Example 3 suggests that poor parameterization can be one reason for massive inefficiency of the Gibbs sampler, and that even simple-minded reparameterization may have the potential to lead to substantial savings. Problems may also arise in hierarchical models where the Gibbs sampler sometimes has a tendency to get "stuck"; this is illustrated in Example 5.

Our experience suggests that the present method diagnoses such problems fairly well. When the prescribed number of iterations is much larger than N_{min}, there seem to be two ways to proceed. One is simply to run the sampler for the specified number of iterations; this seems the best course when iterates are computationally inexpensive. Otherwise it may well be worthwhile to reparameterize or to use a different Markov chain Monte Carlo scheme.

It has been common practice when running the Gibbs sampler to throw away a substantial number of initial iterations, often on the order of 1,000. Our results here suggest that this may not usually be necessary, and indeed, will often be quite wasteful. This is not too surprising given the geometric rate of convergence of Markov chains to the stationary distribution. When large numbers of iterations were required, this was due to the high level of dependence between successive iterates rather than to the failure of the Gibbs sampler to converge initially.

Thus, we suspect that, for typical statistical problems, the uncertainty due to the initial starting point that Gelman and Rubin (1992) capture with their methods will be a relatively small part of the overall uncertainty if the number of Gibbs iterations is realistically large. Of course, we are far from having established that conclusively here, and diagnostic checks such as those proposed by Gelman and Rubin (1992) remain important. Indeed, our method and theirs may be regarded as complementary in that our method can be viewed as determining the total number of iterations required, which will typically be little changed whether there is one long run or a small number of shorter runs from different starting points. More specifically, if there are to be R different runs from different starting values, then each run should have $NR^{-1} + M$ iterations, of which the first M are discarded. Thus the two methods could be synthesized by using our approach to determine the total number of required iterations, and using the method of Gelman and Rubin (1992), both as a further check for convergence, and also to incorporate uncertainty about the starting point.

It has also been common practice to use only every 10th or 20th iterate and to discard the rest. The results here also suggest that in many cases this is rather profligate. Indeed, in the "nice" cases, the dependency between successive iterates is weak and it makes sense to use them all, even when storage is an issue.

An alternative approach to determining the number of iterations starts by viewing the sequence of Gibbs iterates as a standard time series (e.g., Geyer, 1991; Geweke, 1992; Hills and Smith, 1992). If the quantity of interest is the mean of a function of the series, then the variance of such a mean is equal to the spectrum of the corresponding series at zero, which can be estimated using standard spectral methods. This requires the user to specify both a spectral window and a window width, and the estimate of the spectrum at zero can be quite sensitive to these choices.

Obtaining posterior quantiles defining Bayesian confidence intervals is often a key goal of an analysis. When this is the case, the present method exploits the natural simplification that arises from the implied dichotomization. Thus it avoids the need to specify quantities other than the required precision (such as spectral window widths), it yields a simple estimate of the number of "burn-in" iterations, and it provides a practical lower bound, N_{min}, on the number of iterations. This lower bound is known before the Gibbs sampler starts running.

It may be argued that often all that is required is a posterior mean and standard deviation, and that these are not quantiles. If this is indeed the case, and there is really no interest in the shape of the posterior distribution, then there may well be little point in running the Gibbs sampler at all, as cheaper methods are frequently available for posterior means and standard deviations. However, the posterior mean and standard deviation are often used to provide a *summary* of the posterior distribution. In that case, a robust measure location, such as the median, may be preferable to the posterior mean as a descriptive measure, and the median *is* a quantile. Also, the posterior standard deviation is often used as a way of obtaining an approximate confidence interval, say by taking the posterior mean plus or minus two posterior standard deviations. However, if a sample from the posterior is available, it seems worth calculating the required interval directly—again this will be defined by quantiles. Even if a single measure of posterior dispersion is required, it may well be better to use a more robust measure than the posterior standard deviation, such as a scaled version of the inter-quartile range; again this is defined by quantiles. Thus, appropriate summaries of the posterior distribution are often defined in terms of quantiles, even when at first sight it seems that a mean-like quantity is required.

One important message is that the required number of iterations can be dramatically different for different problems, and even for different quantities of interest within the same problem. Thus, it seems unwise to rely on a single "rule of thumb", and it would seem to be important to use some method, such as the one proposed here, to determine the number of iterations that are needed for the problem at hand.

REFERENCES

Besag, J. E. (1986). On the statistical analysis of dirty pictures. *J. Roy. Statist. Soc. B* **48**, 259–302, (with discussion).

Besag, J. E., York, J. and Mollié (1991). Bayesian image restoration, with two applications in spatial statistics. *Ann. Inst. Statist. Math.* **43**, 1–59, (with discussion).

Billingsley, P. (1968). *Convergence of Probability Measures*. New York: Wiley.

Bishop, Y. M. M., Fienberg, S. E. and Holland, P. W. (1975). *Discrete Multivariate Analysis*. Cambridge, MA: MIT Press.

Cox, D. R. and Miller, H. D. (1965). *The Theory of Stochastic Processes*. London: Chapman and Hall.

Gelfand, A. E. and Smith, A. F. M. (1990). Sampling-based approaches to calculating marginal densities. *J. Amer. Statist. Assoc.* **85**, 398–409.

Gelman, A. and Rubin, D. B. (1992). A single series from the Gibbs sampler provides a false sense of security. *Bayesian Statistics 4* (J. M. Bernardo, J. O. Berger, A. P. Dawid and A. F. M. Smith, eds.), Oxford: University Press, 625–631.

Geman, S. and Geman, D. (1984). Stochastic relaxation, Gibbs distributions and the Bayesian restoration of images. *IEEE Trans. Pattern Anal. Machine Intell.* **6**, 721–741.

Geweke, J. (1992). Evaluating the accuracy of sampling-based approaches to the calculation of posterior moments. *Bayesian Statistics 4* (J. M. Bernardo, J. O. Berger, A. P. Dawid and A. F. M. Smith, eds.), Oxford: University Press, 169–193, (with discussion).

Geyer, C. (1991). Monte Carlo maximum likelihood in exponential families. *Workshop on Bayesian Computation via Stochastic Simulation*, Columbus, Ohio.

Hall, P. (1982). On some simple estimates of an exponent of regular variation. *J. Roy. Statist. Soc. B* **44**, 37–42.

Hills, S. E. and Smith, A. F. M. (1992). Parametrization issues in Bayesian inference. *Bayesian Statistics 4* (J. M. Bernardo, J. O. Berger, A. P. Dawid and A. F. M. Smith, eds.), Oxford: University Press, 227–246, (with discussion).

Lauritzen, S. L. and Spiegelhalter, D. J. (1988). Local computations with probabilities on graphical structures and their application to expert systems. *J. Roy. Statist. Soc. B* **50**, 157–224, (with discussion).

Raftery, A. E. (1986). A note on Bayes factors for log-linear contingency tables with vague prior information. *J. Roy. Statist. Soc. B* **48**, 249–250.

Raftery, A. E. and Banfield, J. D. (1991). Stopping the Gibbs sampler, the use of morphology, and other issues in spatial statistics. *Ann. Inst. Statist. Math.* **43**, 32–43.

Schwarz, G. (1978). Estimating the dimension of a model. *Ann. Statist.* **6**, 461–464.

Sheehan, N. (1990). Image processing procedures applied to the estimation of genotypes on pedigrees. *Tech. Rep.* **176**, Department of Statistics, University of Washington.

BAYESIAN STATISTICS 4, pp. 775–782
J. M. Bernardo, J. O. Berger, A. P. Dawid and A. F. M. Smith, (Eds.)

Convergence Diagnostics of the Gibbs Sampler

G. O. ROBERTS
University of Cambridge, UK

SUMMARY

The emergence of the Gibbs sampler, and other associated stochastic simulation techniques, is clearly an important advance in the generation of samples from multi-dimensional distributions, which has particular application in Bayesian statistics. The advantages of the algorithm are its ease of implementation, its speed of convergence, and the lack of regularity conditions necessary to ensure convergence. Its major drawback, however, is that although convergence is well understood theoretically, diagnosis of convergence is often difficult. In this paper, we address the issue of convergence and its diagnosis. In particular, we focus our attention on the search for a one-dimensional summary statistic which attempts to describe the convergence mechanism.

Keywords: GIBBS SAMPLER; CONVERGENCE DIAGNOSTIC; REVERSIBLE MARKOV CHAIN.

1. INTRODUCTION

The Gibbs sampler, and other related stochastic simulation techniques, are notable for their speed of convergence under very general regularity conditions, which are responsible for their recent popularity in Bayesian statistics (see for example Geman and Geman, 1984, Tanner and Wong, 1987, and Gelfand and Smith, 1990). However although the algorithms are well understood theoretically (see for example Roberts and Polson, 1991, and Geman and Geman, 1984) and in fact convergence can be shown to happen at a geometric rate under a suitable distributional norm, accurate estimates of convergence rates are usually very difficult to obtain.

Therefore the major theoretical problem remaining is that of convergence diagnostics. Various ad-hoc methods involving monitoring of moments for the successive observations in the algorithm are often effective. However, it is easy to construct examples where such an approach wrongly diagnoses convergence.

We therefore search for a one-dimensional summary statistic, χ_n, which is calculated at each iteration, and has the following important properties. Letting $\mathbb{P}_n$ denote the distribution of the nth iterate of the Gibbs sampler, and $||\cdot||$ denote a distributional norm, we require that $||\mathbb{P}_n - \mathbb{P}|| \downarrow 0$, where $\mathbb{P}$ is the stationary distribution from which we wish to sample. Also let χ_n be an unbiased estimator of $||\mathbb{P}_n - \mathbb{P}||$, which is easy to calculate given the Gibbs sampler iterates themselves. We find $||\mathbb{P}_n - \mathbb{P}||$ to be a measure of dependence between initial conditions and the nth iterates of the algorithm. This conforms to the intuition that the higher the dependence between the k parameters, and therefore the higher the dependence between iterates, the slower the convergence of the algorithm.

We shall consider the case where $\mathbb{P}_n$ is a distribution with a continuous density f_n on $\mathbb{R}^k$, and therefore we shall consider integral norms and inner products over $\mathbb{R}^k$. Our search for a norm which converges monotonically to zero, leads us to consider inner products under which the operator generated by an iteration of the stochastic simulation technique, is self-adjoint. For this reason we are led to consider a symmetrized version of the Gibbs sampler introduced in Schervish and Carlin (1990).

2. CONVERGENCE

Suppose $f(\boldsymbol{\theta}) = f(\theta_1 \ldots \theta_n)$ is a continuous k-dimensional density function on $\mathbb{R}^k$ from which we wish to sample. The Gibbs sampler (Geman and Geman, 1984) assumes that the k one-dimensional conditionals $f(\theta_i|\theta_j,\ j \neq i)$ can be sampled, and is described as follows. Choose $\boldsymbol{\theta}^{(0)} \in \mathbb{R}^k$ and sample from $f(\theta_1|\theta_2^{(0)} \ldots \theta_k^{(0)})$ to produce $\theta_1^{(1)}$. Then, updating θ_1, we proceed to sample from $f(\theta_2|\theta_1^{(1)}, \theta_3^{(0)} \ldots \theta_k^{(0)})$ to produce $\theta_2^{(1)}$. This procedure is carried out for each of the parameters in turn, for example sampling from $f(\theta_i|\theta_j^{(1)},\ j < i,\ \theta_j^{(0)},\ j > i)$ to produce $\theta_i^{(1)}$. Eventually we obtain a new vector $\boldsymbol{\theta}^{(1)}$. The algorithm is repeated to produce a sequence of k-vectors $\{\boldsymbol{\theta}^{(n)},\ n \geq 0\}$ which form a Markov chain.

The stationary distribution for the Markov chain is $f(\boldsymbol{\theta})$ and, moreover, the process satisfies an ergodic-type result:

$$\frac{1}{n} \sum_{i=1}^{n} g(\boldsymbol{\theta}^{(i)}) \xrightarrow{n\to\infty} \int g(\boldsymbol{\theta}) f(\boldsymbol{\theta})\, \mathrm{d}\boldsymbol{\theta}$$

whenever the right hand side exists. Convergence, and the speed of convergence for this chain, is studied theoretically in a number of papers (for example, Geman and Geman, 1984, Tanner and Wong, 1987, Roberts and Polson, 1990, Schervish and Carlin, 1990, and Applegate *et al.* 1990). Unfortunately, although convergence can be shown to happen at a geometric rate under very general conditions, accurate bounds on geometric rates are not generally available, so that these results are of little use in the diagnostics of convergence.

The kernel of the Markov chain induced by the Gibbs sampler is given by

$$K(\boldsymbol{\theta}, \boldsymbol{\phi}) = \prod_{i=1}^{k} f(\phi_i|\phi_j,\ j < i,\ \theta_j,\ j > i).$$

It will be seen that the search for effective convergence diagnostics is equivalent to finding a suitable inner product on the function space that makes the transition operator P induced by K, and given by

$$Pf(\boldsymbol{\phi}) = \int_{\mathbb{R}^k} K(\boldsymbol{\theta}, \boldsymbol{\phi}) f(\boldsymbol{\theta})\, \mathrm{d}\boldsymbol{\theta},$$

self-adjoint.

Unfortunately, such an inner product is not readily available for the Gibbs kernel. Therefore we work with a related algorithm introduced in Schervish and Carlin, 1990.

Consider a Gibbs sampler iteration with the order of the parameters reversed so as to sample from $f(\theta_k|\theta_1^{(0)} \ldots \theta_{k-1}^{(0)})$ first. The kernel induced by this algorithm is given by

$$k(\boldsymbol{\theta}, \boldsymbol{\phi}) = \prod_{i=1}^{k} f(\phi_i|\theta_j,\ j < i,\ \phi_j,\ j > i).$$

Also let p be the operator induced by k, so that

$$pf(\boldsymbol{\phi}) = \int_{\mathbb{R}^k} k(\boldsymbol{\theta}, \boldsymbol{\phi}) f(\boldsymbol{\theta})\, \mathrm{d}\boldsymbol{\theta}.$$

Now consider the composite operator $T = pP$ corresponding to the application of a Gibbs sampler iteration, followed by a reversed order Gibbs sampler iteration. We shall call this operator an iteration of the reversible sampler. Let $L(\boldsymbol{\theta}, \boldsymbol{\phi})$ be the transition kernel associated with T, so that

$$L(\boldsymbol{\theta}, \boldsymbol{\phi}) = \int_{\mathbb{R}^k} K(\boldsymbol{\theta}, \boldsymbol{\eta}) k(\boldsymbol{\eta}, \boldsymbol{\phi}) \, \mathrm{d}\boldsymbol{\eta}.$$

We introduce the following inner product that makes T self-adjoint. Let

$$\mathcal{L}^2 = \left\{ \text{measurable functions } g \text{ on } \mathbb{R}^k \text{ such that } \int \frac{g^2(\boldsymbol{\theta})}{f(\boldsymbol{\theta})} \, \mathrm{d}\boldsymbol{\theta} < \infty \right\}.$$

Define

$$\langle g, h \rangle = \int \frac{g(\boldsymbol{\theta}) h(\boldsymbol{\theta})}{f(\boldsymbol{\theta})} \, \mathrm{d}\boldsymbol{\theta} \quad \text{for } g, h \in \mathcal{L}^2$$

making $\{\mathcal{L}^2, \langle \cdot, \cdot \rangle\}$ a pre-Hilbert space with norm $||g||2 = \langle g, g \rangle$. Moreover it is readily checked that T is self-adjoint under $\langle \cdot, \cdot \rangle$.

We remark that the reversible sampler is identical to the Gibbs sampler when $k = 2$, corresponding in particular to the data augmentation algorithm of Tanner and Wong (1987).

Note that although this inner product is desirable in many ways, it is unnatural from a statistical point of view, giving highest weight to those areas in $\mathbb{R}^k$ with the smallest density. A more natural inner product set up is available at the expense of further alternatives of the original Gibbs algorithm, and involving sampling from the square root of f. The approach will be considered in detail in a forthcoming paper by the author.

Lemma 1.

$$f(\theta) L(\theta, \phi) = f(\phi) L(\phi, \theta).$$

Proof.

$$\begin{aligned}
L(\boldsymbol{\theta}, \boldsymbol{\phi}) &= \int_{\mathbb{R}^k} K(\boldsymbol{\theta}, \boldsymbol{\eta}) k(\boldsymbol{\eta}, \boldsymbol{\phi}) \, \mathrm{d}\boldsymbol{\eta} \\
&= \int_{\mathbb{R}^{k-1}} f(\eta_1 | \theta_2 \ldots \theta_k) \ldots f(\eta_{k-1} | \eta_1 \ldots \eta_{k-2}, \theta_k) f(\phi_k | \eta_1 \ldots \eta_{k-1}) f(\phi_{k-1} | \eta_1 \ldots \eta_{k-2}, \phi_k) \\
&\qquad \ldots f(\phi_1 | \phi_2 \ldots \phi_k) \, \mathrm{d}\eta_1 \mathrm{d}\eta_2 \ldots \mathrm{d}\eta_{k-1} \\
&\qquad \text{(since the first sampling of the } k \text{ th coordinate is superfluous)} \\
&= \frac{f(\phi)}{f(\theta_2 \ldots \theta_k) f(\phi_2 \ldots \phi_k)} \int_{\mathbb{R}^{k-1}} f(\eta_1, \theta_2 \ldots \theta_k) \frac{f(\phi_2 \ldots \phi_k, \eta_1)}{f(\eta_1 \phi_3 \ldots \phi_k)} \frac{f(\eta_1, \eta_2, \theta_3 \ldots \theta_k)}{f(\eta_1 \theta_3 \ldots \theta_k)} \frac{f(\eta_1 \eta_2 \ldots \eta_k)}{f(\eta_1 \eta_2, \phi_4 \ldots \phi_k)} \\
&\qquad \frac{f(\eta_1 \ldots \eta_{k-1}, \theta_k)}{f(\eta_1 \ldots \eta_{k-2}, \theta_k)} \frac{f(\phi_1, \eta_1 \ldots \eta_{k-1})}{f(\eta_1 \ldots \eta_{k-1})} \, \mathrm{d}\eta_1 \ldots \mathrm{d}\eta_{k-1}.
\end{aligned}$$

The integral form is readily seen to be invariant to the interchange of $\boldsymbol{\theta}$ and $\boldsymbol{\phi}$ thus proving the result. ◁

Note that the relation in Lemma 1 ensures that the Markov chain induced by L is reversible (see for example Kelly, 1979).

Choose $\boldsymbol{\theta}^{(0)} \in \mathbb{R}^k$ and denote by $\{\boldsymbol{\theta}^{(0)}, \boldsymbol{\theta}^{(1)}, \ldots\}$ the Markov chain generated by successive iterations of the reversible sampler. We will write $f^{(i)}(\boldsymbol{\theta})$ for the density of $\boldsymbol{\theta}^{(i)}$, $i \geq 1$. Under mild regularity conditions, we show that $f(i) \in \mathcal{L}^2$. However first we describe the structure of $\mathcal{L}^2$ in terms of the eigenfunctions of L.

The spectral theorem (see for example Dunford and Schwartz, 1963) assures that since L is self-adjoint, then its eigenfunctions in $\mathcal{L}^2$ form an orthonormal set and, moreover, the space spanned by finite linear combinations of eigenfunctions is dense in $\mathcal{L}^2$. Thus any function $g \in \mathcal{L}^2$ can be written

$$g(\boldsymbol{\theta}) = \sum_{i=0}^{\infty} a_i e_i(\boldsymbol{\theta}) \text{ for almost all (with respect to } \langle \cdot, \cdot \rangle) \ \boldsymbol{\theta} \in \mathbb{R}^k,$$

where the e_i's ($i \geq 0$) are the eigenfunctions. Also since L is a stochastic transition operator, it is therefore a contraction, that is, the corresponding eigenvalues $\{r_i, \ i \geq 0\}$ each have absolute value less than or equal to unity. Since L is self-adjoint, the spectral theorem also tells us that the eigenvalues are all real.

By convention, we insist that $r_0 = 1$, so that $e_0(\boldsymbol{\theta}) = f(\boldsymbol{\theta})$. Much of the previous work on convergence has centred on showing that $\{r_i, \ i \geq 1\}$ is a set bounded in absolute value by some constant $r < 1$, so that convergence is therefore geometric in rate (for example, Roberts and Polson, 1990, and Geman and Geman, 1984, in the finite case). We shall not need geometric convergence here.

Lemma 2.
If

$$\mathbb{E}\left[\frac{L(\boldsymbol{\theta}^{(0)}, \boldsymbol{\theta}^{(1)})}{f(\boldsymbol{\theta}^{(1)})}\right] = \int_{\mathbb{R}^k} \frac{L^2(\boldsymbol{\theta}^{(0)}, \boldsymbol{\theta}^{(1)})}{f(\boldsymbol{\theta}^{(1)})} \, \mathrm{d}\boldsymbol{\theta}^{(1)} < \infty$$

then $f^{(i)}(\boldsymbol{\theta}) \in \mathcal{L}^2 \ \forall i \geq 1$.

Proof.
The above condition implies that $f^{(1)} \in \mathcal{L}^2$. Therefore,

$$f^{(1)}(\boldsymbol{\theta}) = \sum_{i=0}^{\infty} a_i e_i(\boldsymbol{\theta})$$

for almost all $\boldsymbol{\theta}$ (with respect to $\langle \cdot, \cdot \rangle$), and $||f^{(1)}||^2 = \sum_{i=0}^{\infty} a_i^2$. Now by induction $f^{(1)} \in \mathcal{L}^2$, and suppose $f^{(n)} \in \mathcal{L}^2$ then

$$f^{(n)}(\boldsymbol{\theta}) = \sum_{i=0}^{\infty} b_i e_i(\boldsymbol{\theta}) \ \text{ for almost all } \boldsymbol{\theta}$$

and so

$$f^{(n+1)}(\boldsymbol{\theta}) = L \sum_{i=0}^{\infty} b_i e_i(\boldsymbol{\theta}) = \sum_{i=0}^{\infty} b_i r_i e_i(\boldsymbol{\theta}) \ \text{ for almost all } \boldsymbol{\theta}.$$

Therefore,

$$||f^{(n+1)}(\boldsymbol{\theta})||^2 = \sum_{i=0}^{\infty} r_i^2 b_i^2 \leq \sum_{i=0}^{\infty} b_i^2 < \infty,$$

and therefore $f^{(n+1)} \in \mathcal{L}^2$. ◁

The norm of $f^{(n)}$ has the following highly desirable properties. Firstly it converges to $||f|| = 1$, and secondly this convergence is monotone, since it is clear that

$$||f_n^{(n)}||^2 = 1 + \sum_{i=1}^{\infty} a_i^2 r_i^2.$$

Let $g^{(n)} = f^{(n)} - f$, then clearly

$$||g^{(n)}||^2 = \sum_{i=1}^{\infty} a_i^2 r_i^2$$

and therefore the distance between $f^{(n)}$ and f under $|| \cdot ||$ converges monotonically to 0. This property is a direct result of using an inner product that makes L self-adjoint. In a non-self-adjoint set-up, there would exist a starting distribution such that convergence would not be monotone, making later diagnostic analysis potentially much more difficult.

It remains to find estimates for the norm, $||g^{(n)}||2$. Firstly we note that

$$||f^{(n)}||^2 = \mathbb{E}\left[\frac{f^{(n)}(\boldsymbol{\theta}^{(n)})}{f(\boldsymbol{\theta}^{(n)})}\right],$$

and therefore we search for a result that will enable us to manipulate terms such as

$$d(m, n) = \mathbb{E}\left[\frac{f^{(m)}(\boldsymbol{\theta}^{(n)})}{f(\boldsymbol{\theta}^{(n)})}\right].$$

Lemma 3.

$$d(m, n) = d(m+1, n-1) \qquad \forall m \geq 1,\ n \geq 2.$$

Proof.

$$d(m, n) = \mathbb{E}\left[\mathbb{E}\left[\frac{f^{(m)}(\boldsymbol{\theta}^{(n)})}{f(\boldsymbol{\theta}^{(n)})}|\mathcal{F}_{n-1}\right]\right]$$

where $\mathcal{F}_{n-1}$ denotes the σ algebra generated by$\{\boldsymbol{\theta}_0, \boldsymbol{\theta}_1, \ldots, \boldsymbol{\theta}^{n-1}\}$

$$= \mathbb{E}\left[\int \frac{f^{(m)}(\boldsymbol{\theta}) L(\boldsymbol{\theta}^{(n-1)}, \boldsymbol{\theta})}{f(\boldsymbol{\theta})}\, \mathrm{d}\boldsymbol{\theta}\right]$$

$$= \mathbb{E}\left[\int \frac{f^{(m)}(\boldsymbol{\theta}) L(\boldsymbol{\theta}, \boldsymbol{\theta}^{(n-1)})}{f(\boldsymbol{\theta}^{(n-1)})}\, \mathrm{d}\boldsymbol{\theta}\right]$$

$$= \mathbb{E}\left[\frac{f^{(m+1)}(\boldsymbol{\theta}^{(n-1)})}{f(\boldsymbol{\theta}^{(n-1)})}\right] = d(m+1, n-1) \quad \text{as required .} \quad \triangleleft$$

Corollary 1.

$$||f^{(n)}||^2 = \mathbb{E}\left[\frac{f^{(1)}(\boldsymbol{\theta}^{(2n-1)})}{f(\boldsymbol{\theta}^{(2n-1)})}\right] = \mathbb{E}\left[\frac{L(\boldsymbol{\theta}^{(0)}, \boldsymbol{\theta}^{(2n-1)})}{f(\boldsymbol{\theta}^{(2n-1)})}\right].$$

Therefore suppose we run m replications of the reversible sampler, giving rise to Markov chain realizations $\{\theta^{(i)}(j),\ i \geq 0\},\ j = 1, 2, \ldots, m$. We estimate $||g^{(n)}||^2$ by the unbiased estimator χ_n, where

$$\chi_n = \frac{1}{m} \sum_{j=1}^{m} \frac{L(\theta^{(0)}(j), \theta^{(2n-1)}(j))}{f(\theta^{(2n-1)}(j))} - 1.$$

Its properties are summarized by the following result, the proof of which is immediate.

Lemma 4.

$$\mathbb{E}[\chi_n] = ||g^{(n)}||^2 = \sum_{i=1}^{\infty} a_i^2 r_i^{2n} \downarrow 0 \text{ as } n \to \infty$$

and

$$\lim_{n\to\infty} \text{Var}\ (\chi_n) = \frac{1}{m}(||L(\theta^{(0)}, \cdot)||^2 - 1).$$

Remarks.

(i) If the full reversible kernel L is not available then the estimator can be modified, for example, in the following way. Let $\theta^{(1/2)}(j)$ be the k-tuple obtained by the first application of the forwards Gibbs sampler alone, for the jth replication. We assume here that each replication starts from the same point $\theta^{(0)}$.

$$\Lambda_n = \frac{1}{m(m-1)} \sum_{j \neq l} \frac{k(\theta^{(1/2)}(j), \theta^{(2n-1)}(l))}{f(\theta^{(2n-1)}(l))} - 1,$$

$\mathbf{E}[\Lambda_n] = ||g^{(n)}||^2$ as for χ_n and, moreover, it can be shown that

$$\lim_{n\to\infty} \text{Var}\ (\Lambda_n | \mathcal{F}_{1/2}) = \frac{1}{m}\left(\frac{m-2}{m(m-1)^2} \sum_{j,l} \langle k(\theta^{(1/2)}(j), \cdot), k(\theta^{(1/2)}(l), \cdot)\rangle + \right.$$
$$\left. + \frac{1}{m(m-1)^2} \sum_{j} ||k(\theta^{(1/2)}(j), \cdot)||^2 - 1\right)$$

and

$$\lim_{n\to\infty} \text{Var}\ (\Lambda_n) = \frac{1}{m}(||L(\theta^{(0)}, \cdot)||^2 - 1) + \frac{1}{m(m+1)} || \text{ Var }(k(\theta^{(1/2)}, \cdot))||^2.$$

Here $\mathcal{F}_{1/2}$ denotes the information obtained from the first operation of the forwards Gibbs sampler.

(ii) The important case of an unknown normalization factor for f can often be dealt with by importance sampling techniques which involve no further sampling. See Roberts and Polson (1991) for a description of these methods.

3. CONCLUSIONS

We have seen how a symmetrizing modification of the Gibbs sampler together with a carefully chosen norm, by which to assess convergence, enables us to describe theoretically the mechanism of convergence. Here we do not concern ourselves with the eigenvalues of the transition operator, or the related speed of convergence. Instead the essential feature for us is the monotonicity of convergence under the given norm. Moreover we find that the norm is easy to estimate at each iteration, with a statistic whose variance is conveniently bounded. These estimates involve no further sampling, even in the cases where either the reversible sampler kernel is not available, or the limiting distribution normalization constant is not known. Finally, convergence of the norm to zero cannot wrongly diagnose convergence.

This method therefore seems to have considerable advantages over existing methodology. However certain points are still somewhat unsatisfactory.

Firstly the integral norm used involving a multiplication factor of $(f(\theta))^{-1}$ is clearly not natural in any statistical sense. Nevertheless, it is the only way to produce a self-adjoint inner product without radically altering the algorithm itself.

Secondly it will not be possible in general to check the condition required for Lemma 2. However typically it will be easy to show that $L(\boldsymbol{\theta}^{(0)}, \boldsymbol{\theta})/f(\boldsymbol{\theta})$ is bounded for all $\boldsymbol{\theta}$. The need to check this condition might provide a motive for carefully choosing $\boldsymbol{\theta}^{(0)}$ such that the condition is easy to verify.

It would be useful to use the norm estimates described here directly for the Gibbs sampler. Since the norm is merely a measurement of dependence between $\boldsymbol{\theta}^{(0)}$ and $\boldsymbol{\theta}^{(n)}$, then certainly it would converge to zero for the Gibbs sampler. However monotonicity of convergence is not assured.

Finally, the necessity of using $\boldsymbol{\theta}^{(2n-1)}$ to estimate $||g^{(n)}||$ is clearly not a serious problem, although it might lead to a slight loss of efficiency.

We have not mentioned the issue of choosing a starting distribution for the algorithm. The reason for this is that there exists a dichotomy between the issues of speed, and diagnosis of convergence in this case. Starting far from stationarity is likely to lead to much more clear convergence diagnostics than if we start close to the stationary distribution, especially if the asymptotic variance of χ_n is not known. Furthermore our criterion for convergence requires that each replication achieves approximate independence from its starting point even if we start with an initial configuration drawn from the stationary distribution. The results used here pertaining to the case where L is unavailable, and involving estimating L by the first application of the Gibbs operator, can clearly be generalized to the case of different starting points. However we will still require at least two replications starting from each point.

It should be noted with caution that any attempt at convergence diagnostics based on a sequential decision rule violates the likelihood principle. Moreover, in spite of the fact that convergence may have been reached, dependence between the rule to stop after n iterations, and the observations in the nth iterate, means that we may be sampling from a subtly conditioned version of f. The worst effects of this can be removed by running the sampler for a fixed number of extra iterations after convergence has been diagnosed.

Of course, this paper contains merely the theoretical background to a practical problem, for which considerable computational difficulties might arise. Moreover, such a diagnostic can only be truly judged by its performance in practice. Therefore Roberts and Hills (1992) addresses these issues.

REFERENCES

Applegate, D., Kannan, R. and Polson, N. G. (1990). Random polynomial time algorithms for sampling from joint distributions. *Tech. Rep.* **500**, Carnegie Mellon University.

Dunford, N. and Schwartz, J. T. (1963). *Linear Operators, Part II: Spectral Theory*. New York: Interscience.

Gelfand, A. E. and Smith, A. F. M. (1990). Sampling based approaches to calculating marginal densities. *J. Amer. Statist. Assoc.* **85**, 398–409.

Geman, S. and Geman, D. (1984). Stochastic relaxation, Gibbs distributions and the Bayesian restoration of images. *IEEE Trans. on Pattern Anal. and Machine Intelligence* **6**, 721–741.

Kelly, F. (1979). *Reversibility and Stochastic Networks*. New York: Wiley.

Roberts, G. O. and Polson, N. G. (1990). A note on the geometric convergence of the Gibbs sampler. *Tech. Rep.* **91-01**, University of Nottingham.

Roberts, G. O. and Polson, N. G. (1991). A note on the stochastic simulation of joint densities. *Tech. Rep.* **91-06**, University Nottingham.

Roberts, G. O. and Hills, S. E. (1992). Assesing distributional convergence of the Gibbs sampler. *Tech. Rep.* **92-01**, University of Nottingham.

Schervish, M. J. and Carlin, B. P. (1990). On the convergence of successive substitution sampling. *Tech. Rep.* **4-12**, Carnegie Mellon University.

Tanner, M. A. and Wong, W. H. (1987). The calculation of posterior distributions by data augmentation. *J. Amer. Statist. Assoc.* **82**, 528–550, (with discussion).

BAYESIAN STATISTICS 4, pp. 783–789
J. M. Bernardo, J. O. Berger, A. P. Dawid and A. F. M. Smith, (Eds.)

An Evaluation of Robustness in Binomial Empirical Bayes Testing

S. SIVAGANESAN
University of Cincinnati, USA

SUMMARY

We consider the Binomial Empirical Bayes problem where (i) X_i, conditional on θ_i, have Binomial (n, θ_i) distribution and are independent, (ii) θ_i are i.i.d. from an unknown distribution G. As $i \to \infty$, one would learn the first n moments of G. We consider the problem of testing the null hypothesis $H_0 : \theta \leq \xi$ against the alternative $H_a : \theta > \xi$. Under the assumption that the first n moments have become known, we investigate the uncertainty in the posterior probability of H_0 by finding its ranges: (i) over all possible G having the known moments, (ii) over all unimodal G having mode at ξ and the known moments.

Keywords: BINOMIAL EMPIRICAL BAYES; TESTING; BAYESIAN ROBUSTNESS.

1. INTRODUCTION

We consider the binomial empirical Bayes problem where (i) one observes X_i, at stage $i = 1, 2, \ldots$ which, conditional on θ_i, have Binomial(n, θ_i) distributions and are independent; (ii) the θ_i's are i.i.d from an unknown distribution G. The binomial empirical Bayes problem is atypical in that, as $i \to \infty$, only the first n moments of G will become known. Thus, if $c_1, c_2, \ldots, c_n$ are the known values of the first n moments, all one can then assert about G is that it belongs to the class

$$\mathcal{G} = \left\{ \text{prob. dist. } G \text{ on } [0,1] : \int \theta^i dG(\theta) = c_i, \text{ for } i = 1, \ldots, n \right\}. \tag{1.1}$$

This inability to fully determine the distribution G (even after infinitely many stages have passed by) leads to a certain amount of uncertainty in the value(s) of the empirical Bayes procedure(s) to be used. It is therefore of interest from a robustness standpoint to determine how much uncertainty there remains in the value of an empirical Bayes procedure.

When estimation of θ_i is of concern and $X_i = x$, the posterior mean

$$\delta_G(x) = \frac{\int \theta^{x+1}(1-\theta)^{n-x} dG(\theta)}{\int \theta^x (1-\theta)^{n-x} dG(\theta)}$$

of θ_i would be of interest. The problem of evaluating the uncertainty in $\delta_G(x_i)$ was studied in Sivaganesan and Berger (1990), where the ranges of $\delta_G(x)$ were found under different settings. One interesting finding there was that, once the first n moments of G have become known, typically very little uncertainty would be left in the determination of $\delta_G(x)$. This was evident from the very tight ranges obtained for $\delta_G(x)$ as G varied over $\mathcal{G}$.

In this article we consider the posterior probability, given $X_i = x$ (at stage i), that $\theta_i \leq \xi$ defined by

$$P_G(\xi) = \frac{\int_0^\xi \theta^x (1-\theta)^{n-x} dG(\theta)}{\int_0^1 \theta^x (1-\theta)^{n-x} dG(\theta)}. \tag{1.2}$$

Here $\xi \in (0,1)$ is fixed. This quantity would be of interest, e.g., in testing the hypothesis $H_0 : \theta_i \leq \xi$ against $H_1 : \theta_i > \xi$ at stage i. Here we pay attention to evaluating the uncertainty or robustness, at stage i, of $P_G(\xi)$ when the first n moments of G have become known. We do this in Section 2 by calculating the range of $P_G(\xi)$ as G varies over the class $\mathcal{G}$. i.e., we find

$$\overline{P}(\xi) = \sup_{G \in \mathcal{G}} P_G(\xi) \quad \text{and} \quad \underline{P}(\xi) = \inf_{G \in \mathcal{G}} PG(\xi).$$

Indeed, the question of robustness, when a sufficient number of stages (or x_i's) have not accumulated to enable one to accurately know the first n moments of G, should be just as much of a concern. This would require a more extensive robustness study which also allows for the uncertainty in the estimation of the moments. (Such an approach is taken in Sivaganesan and Berger (1990) for the posterior mean.) We, however, limit our attention here to the case of known moments because: (i) this would give an indication of the extent of robustness that may be achieved when the moments themselves are estimated, (ii) it would be an integral part of any such more extensive study.

It turns out from the calculations in Section 2 that, unlike in the case of the posterior mean, $P_G(\xi)$ does not, in general, enjoy a sufficient degree of robustness w.r.t the class $\mathcal{G}$. This means that one is not likely to achieve robustness for $P_G(\xi)$ w.r.t the class $\mathcal{G}$. A common approach, in such instances, is to restrict consideration only to those G's which satisfy certain (other) features that one may (a priori) be confident about. In testing hypotheses of the form $H_0 : \theta_i \leq \xi$ against $H_1 : \theta_i > \xi$, one may often have prior knowledge that G is unimodal with mode ξ. In such cases, a more reasonable class for G is

$$\mathcal{G}_U = \left\{ G : G \text{ is unimodal with mode } \xi \text{ and } \int \theta^i dG(\theta) = c_i, \text{ for } i = 1, \ldots, n \right\}.$$

In Section 3, we consider this class and calculate the range of $P_G(\xi)$ by finding

$$\tilde{P}(\xi) = \sup_{G \in \mathcal{G}} P_G(\xi) \qquad \text{and} \qquad \underset{\sim}{P}(\xi) = \inf_{G \in \mathcal{G}} P_G(\xi).$$

The paper is organized as follows. In Section 1.2, we give some notation and, in Section 2, we review some results in moment theory and give the results for the class $\mathcal{G}$. In Section 3, we extend the results to the class $\mathcal{G}_U$ and give an example. Finally, we end with some comments and conclusions. Literature on empirical Bayes analysis is extensive. For general references on the topics of empirical and robust Bayes methods, the reader is referred to Berger (1985). For more references specifically related to this work, see Sivaganesan and Berger (1990).

1.2. *Notation*

Let c_i denote the ith moment of a probability distribution on the interval [0,1]. Thus, we define the n-moment space M by

$$\mathcal{M} = \left\{ (c_1, \ldots, c_n) : c_i = \int \theta^i dG(\theta), \text{ for some prob. dist. } G \text{ on}[0,1] \right\}.$$

Also, let square matrices $A_{k+1}, B_{k+1}, D_{k+1}$ and E_k (where the suffix indicates the order of the matrix) be defined by:

$$A_{k+1} = (c_{i+j-1}) \quad , \quad B_{k+1} = (c_{i+j-2} - c_{i+j-1})$$

$$D_{k+1} = (c_{i+j-2}) \qquad \text{and} \qquad E_k = (c_{i+j-1} - c_{i+j}).$$

Moreover, for $t \in [0,1]$, let $A_{k+1}(t)(B_{k+1}(t), D_{k+1}(t), E_{k+1}(t))$ be the matrices obtained by replacing the last column of $A_{k+1}(B_{k+1}, D_{k+1}, E_{k+1})$ by the (column) vector $(1, t, \ldots, t^{k+1})$. (Here, we will let $|A|$ denote the determinant of A, and $|A_0| = 1 = |A_{-1}|$, etc...)

For use in the next section, we also need to define the following quantities. For given $\xi \in (0,1)$, we let

$$a_k(\xi) = |A_{k+1}(\xi)|/\sqrt{|A_{k+1}||A_k|},$$

$$b_k(\xi) = |B_{k+1}(\xi)|/\sqrt{|B_{k+1}||B_k|},$$

$$d_k(\xi) = |D_{k+1}(\xi)|/\sqrt{|D_{k+1}||D_k|},$$

$$e_k(\xi) = |E_{k+1}(\xi)|/\sqrt{|E_{k+1}||E_k|},$$

$$\rho_1(\xi) = \begin{cases} \left[\xi \sum_{k=0}^{m} a_k^2(\xi)\right]^{-1} & \text{if } n = 2m+1 \\ \left[\sum_{k=0}^{m} d_k^2(\xi)\right]^{-1} & \text{if } n = 2m \end{cases} \tag{1.3}$$

$$\rho_2(\xi) = \begin{cases} \left[(1-\xi) \sum_{k=0}^{m} b_k^2(\xi)\right]^{-1} & \text{if } n = 2m+1 \\ \left[\xi(1-\xi) \sum_{k=0}^{m} e_k^2(\xi)\right]^{-1} & \text{if } n = 2m \end{cases} \tag{1.4}$$

2. CALCULATION OF $\underline{P}(\xi)$ AND $\overline{P}(\xi)$

First, we state some results from moment theory which we will use in the calculations. Proofs of these results can be found in Karlin and Studden (1966) and Krein and Nudelman (1977). The first result concerns the existence, for each $\xi \in (0,1)$, of a distribution $G_\xi \in \mathcal{G}$ having the point ξ in its support.

Result 1: *Let $c = (c_1, \ldots, c_n) \in \mathcal{M}$, and $\xi \in (0,1)$. Then,*

(i) If $n = 2m+1 (n = 2m)$, there exists a unique probability distribution $G_\xi \in \mathcal{G}$ whose support consists of $s = m+2 (s = m+1)$ points, one of which is ξ.

(ii) The mass that G_ξ assigns to the point ξ is given by

$$\rho(\xi) = \min\{\rho_1(\xi), \rho_2(\xi)\}. \tag{2.1}$$

(iii) When $n = 2m+1$, the support points of G_ξ are the solutions of the polynomial (of degree $s = m+2$) given by

$$Q_\xi(t) = \begin{cases} t(t-\xi) \sum_{k=0}^{m} \dfrac{|A_{k+1}(\xi)||A_{k+1}(t)|}{|A_{k+1}||A_k|} & \text{if } \rho(\xi) = \rho_1(\xi) \\ (1-t)(t-\xi) \sum_{k=0}^{m} \dfrac{|B_{k+1}(\xi)||B_{k+1}(t)|}{|B_{k+1}||B_k|} & \text{if } \rho(\xi) = \rho_2(\xi) \end{cases} \tag{2.2}$$

(iv) When $n = 2m$, the support points of G_ξ are the solutions of the polynomial (of degree $s = m + 1$) given by

$$Q_\xi(t) = \begin{cases} (t-\xi)\sum_{k=0}^{m} \frac{|D_{k+1}(\xi)||D_{k+1}(t)|}{|D_{k+1}||D_k|} & \text{if } \rho(\xi) = \rho_1(\xi) \\ t(1-t)(t-\xi)\sum_{k=0}^{m-1} \frac{|E_k(\xi)||E_k(t)|}{|E_k||E_{k+1}|} & \text{if } \rho(\xi) = \rho_2(\xi) \end{cases}$$

The next result concerns the extrema of the integral of a function $\Omega(t)$ over the interval $(0, \xi)$. This uses a notion known as a weak T-system for a set of functions. For definition and the associated properties of this notion, the reader may refer to Karlin and Studden (1966). For the purpose of this paper, it suffices to know that the functions $u_i(t) = t^i (i = 0, \ldots, n)$ and $u_{n+1}(t) = \Omega(t)$ form a weak T-system if $\Omega(t)$ is a polynomial of degree $\leq n$.

Result 2: *Let $\Omega(t)$ be a real-valued function defined on $[0,1]$. Suppose that the set of $n+2$ functions on $[0,1]$ given by $u_0(t) \equiv 1, u_i(t) = t^i (i = 1, \ldots, n)$, and $u_{n+1}(t) = \Omega(t)$ form a weak T-system. Then,*

$$\sup_{G\in\mathcal{G}} \int_0^\xi \Omega(t)dG(t) = \int_0^{\xi+} \Omega(t)dG_\xi(t),$$

and

$$\inf_{G\in\mathcal{G}} \int_0^\xi \Omega(t)dG(t) = \int_0^{\xi-} \Omega(t)dG_\xi(t),$$

where G_ξ is as in Result 1.

Now, we are ready to state the result concerning the calculation of $\underline{P}(\xi)$ and $\underline{P}(\xi)$. For use in the following, let $\theta_i (i = 1, \ldots, s)$ denote the support points of G_ξ. Note that the θ_i's can be found by solving the polynomial $Q_\xi(t)$ given in Result 1. Once the θ_i's are found, the mass, denoted α_i, that G_ξ assigns to θ_i can be easily obtained by solving the simultaneous equations

$$\sum_{i=1}^{s} \alpha_i \theta_i^k = c_k \quad \text{for} \quad k = 0, \ldots, s. \tag{2.3}$$

Theorem 1.

$$\overline{P}(\xi) = \sum_{\theta_i \leq \xi} \alpha_i \theta_i^x (1-\theta_i)^{n-x} / \; m(x), \tag{2.4}$$

and

$$\underline{P}(\xi) = \sum_{\theta < \xi} \alpha_i \theta_i^x (1-\theta_i)^{n-x} / \; m(x), \tag{2.5}$$

where

$$m(x) = \sum_{i=1}^{s} \alpha_i \theta_i^x (1-\theta_i)^{n-x}.$$

Proof. For $G \in \mathcal{G}$, the posterior probability $P_G(\xi)$ of $[0,\xi]$ is given by (1.2). Thus, observing that $m(x)$ is fully determined by the known moments of G, we have,

$$\overline{P}(\xi) = \sup_{G\in\mathcal{G}} P_G(\xi) = \sup_{G\in\mathcal{G}} \frac{\int_0^\xi \theta^x(1-\theta)^{n-x}dG(\theta)}{m(x)}$$
$$= \frac{\sup_{G\in\mathcal{G}} \int_0^\xi \theta^x(1-\theta)^{n-x}dG(\theta)}{m(x)}.$$

Now, let $\Omega(t) = t^x(1-t)^{n-x}$, a polynomial of degree n in t. Hence, the set of $n+2$ functions $u_0 \equiv 1, u_i(t) = t^i (i = 1, \ldots, n)$ and $u_{n+1}(t) = \Omega(t)$ form a weak T-system. Thus, using Result 2,

$$\overline{P}(\xi) = \frac{\int_0^{\xi+} \theta^x(1-\theta)^{n-x}dG_\xi(\theta)}{m(x)},$$

and

$$\underline{P}(\xi) = \frac{\int_0^{\xi-} \theta^x(1-\theta)^{n-x}dG_\xi(\theta)}{m(x)}.$$

Now, (2.4) and (2.5) follow by observing that G_ξ has support $\theta_i (i = 1, \ldots, s)$ with mass α_i at θ_i, and that

$$m(x) = \int \theta^x(1-\theta)^{n-x}dG(\theta) = \int \theta^x(1-\theta)^{n-x}dG_\xi(\theta). \quad \triangleleft$$

3. UNIMODAL DISTRIBUTIONS: CALCULATION OF $\underline{P}(\xi)$ AND $\overline{P}(\xi)$

In many instances of testing $H_0 : \theta_i \le \xi$ against $H_1 : \theta_i > \xi$, one may have prior knowledge that G is unimodal with mode ξ. Here, we restrict attention to

$$\mathcal{G}_U = \left\{ \begin{array}{c} \text{all unimodal distributions with mode at } \xi \text{ and first} \\ n \text{ moments } \mu_1, \ldots, \mu_n. \end{array} \right\} \tag{3.1}$$

Each $G \in \mathcal{G}_U$ can be written as a convex mixture of uniform distributions over intervals of the form (ξ, z) or (z, ξ) for $z \in [0,1]$. Letting $F(\cdot)$ be the mixing distribution, we would then have, for any measurable function $h(\theta)$,

$$\begin{aligned} \int_0^1 h(\theta)dG(\theta) &= \int_0^1 \left(\int_0^1 \frac{1}{x-\xi} I(\xi,z)(\theta)dF(z) \right) h(\theta)d\theta \\ &= \int_0^1 H(z)dF(z), \end{aligned} \tag{3.2}$$

where

$$H(z) = \frac{1}{x-\xi} \int_\xi^z h(\theta)d\theta \qquad \text{for} \qquad z \ne \xi. \tag{3.2}$$

Similarly, we have

$$\int_0^\xi h(\theta)dG(\theta) = \int_0^\xi H(z)dF(z).$$

Now, for $G \in \mathcal{G}_U$, we have using (3.2)

$$\mu_k = \int \theta^k dG(\theta) = \int H(z)dF(z) \qquad \text{for} \qquad k = 0, \ldots, n \tag{3.3}$$

where $H(z) = \sum_{i=0}^{k} z^i \xi^{k-1}/(k+1)$. Thus, letting c_i denote the ith moment of the distribution $F(\cdot)$, (3.3) becomes

$$(k+1)\mu_k = \sum_{i=0}^{k} \xi^{k-1} c_i \quad \text{for} \quad k = 0, \ldots, n.$$

Thus, c_k can be uniquely solved in terms of μ_k as follows:

$$c_0 = 1, \quad \text{and} \quad c_k = (k+1)\mu_k - \sum_{i=0}^{k-1} \xi^{k-i} c_i \quad \text{for} \quad k = 1, \ldots, n.$$

Hence, we have

Lemma: *$G \in \mathcal{G}_U$ if and only if the corresponding mixing distribution $F(\Omega) \in \mathcal{G}$.*

Now, using Results 1 and 2 of Section 2, let $F_\xi \in \mathcal{G}$ denote the unique distribution having the point ξ in its support. Also, let (as in Section 2) θ_i $(i = 1, \ldots, s)$ be the support points of F_ξ with mass α_i at θ_i. Then,

Theorem 2.

$$\tilde{P}(\xi) = \sum_{\theta_i \le \xi} \alpha_i H(\theta_i)/m(x),$$

and

$$\underset{\sim}{P}(\xi) = \sum_{\theta_i < \xi} \alpha_i H(\theta_i)/m(x),$$

where

$$H(z) = \begin{cases} \int_{\xi}^{z} \theta^x (1-\theta)^{n-x} d\theta/(z-\xi) & \text{if } z \ne \xi \\ \xi^x (1-\xi)^{n-x} & \text{if } x = \xi, \end{cases} \tag{3.4}$$

and $m(x) = \sum_{i=1}^{s} \alpha_i H(\theta_i)$.

Proof. As in the proof of Theorem 1,

$$\tilde{P}(\xi) = \frac{\sup_{G \in \mathcal{G}} \int_0^\xi \theta^x (1-\theta)^{n-x} dG(\theta)}{m(x)}.$$

Now, by using (3.2) and the Lemma, we have

$$\tilde{P}(\xi) = \frac{\sup_{F \in \mathcal{F}} \int_0^\xi H(z) dF(\theta)}{m(x)},$$

where $H(z)$ is as in (3.4). It is easy to verify that $H(z)$ is a polynomial of degree n in z. Thus, the system of functions $u_0 \equiv 1, u_i(t) = t^i (i = 1, \ldots, n)$, $u_{n+1}(t) = H(t)$ form a weak T-system. Hence, using Result 2 of (Section 2), we get

$$\tilde{P}(\xi) = \frac{\int_0^{\xi+} h(z) dF_\xi(z)}{m(x)}.$$

The rest of the proof follows the same lines as that of Theorem 1. ◁

Illustrative Example:

For simplicity, we consider examples with $n = 5$, although the method can be used with any larger n. Suppose that $\xi = 0.5$ and that, after a sufficient number of stages have passed by, the first $n = 5$ moments are known to be as follows:

$$c_1 = 0.50, c_2 = 0.30, c_3 = 0.20, c_4 = 0.14, c_5 = 0.11.$$

The ranges of $P_G(\xi)$ (over the class $\mathcal{G}$ and $\mathcal{G}_U$) can now be calculated using the methods described in sections 2 and 3, and are given below for various values of x.

x	0	1	2
$\underline{P}(\xi), \overline{P}(\xi))$	$(.84, .99)$	$(.53, .99)$	$(.18, .96)$
$\tilde{P}(\xi), \underset{\sim}{P}(\xi))$	$(.92, .97)$	$(.76, .90)$	$(.50, .74)$

DISCUSSION

We have shown how one may evaluate the sensitivity of the posterior probability that $\theta_i \leq \xi$ when the first n moments have become known. Besides its use in the case of known moments, we believe the treatment here can also be useful: (i) in giving an indication as to the extent of robustness that may exist when a sufficient number of stages have not passed to accurately know the moments, (ii) as a tool in carrying out a more extensive robustness study.

It is clear that the extent of robustness (indicated by the sizes of the ranges) depends strongly on the value of x. The fact that the ranges are very big for some value(s) of x clearly suggests that there can be a serious lack of robustness w.r.t. the class $\mathcal{G}$; i.e., knowing the first n moments of G cannot guarantee adequate robustness for $P_G(\xi)$. Such a phenomenon is likely to remain even for moderately large n. This points to a necessity, if one were to achieve reasonable robustness, to reduce the class $\mathcal{G}$ by the incorporation of prior knowledge.

The results from the examples for the unimodal class, $\mathcal{G}_U$, indicate a substantial improvement over those for the class $\mathcal{G}$. If testing of hypotheses is the only concern, one may (even) be satisfied with the degree of robustness evidenced in these examples. But, when the (posterior) probability in question is to be used for some other purposes, e.g., as a credible level, the robustness is likely to remain a concern. In such cases, robustness may only be achieved by the use of additional prior knowledge or assumptions. This is even more so if the moments themselves are not known accurately.

REFERENCES

Berger, J. O. (1985). *Statistical Decision Theory and Bayesian Analysis*. New York: Springer.

Karlin, S. and Studden, W. (1966). *Tchebycheff Systems: with Applications in Analysis and Statistics*. New York: Wiley.

Krein, M. and Nudelman, A. (1977). *The Markov Moment Problems and Extremal Problems*. Providence, RI: AMS.

Sivaganesan, S. and Berger, J. O. (1990). Robust Bayesian analysis of the binomial empirical Bayes problem. *Tech. Rep.* **90–49C**, Purdue University.

BAYESIAN STATISTICS 4, pp. 791–802
J. M. Bernardo, J. O. Berger, A. P. Dawid and A. F. M. Smith, (Eds.)

Incorporating Prior Knowledge in a Marked Point Process Model

KNUT SØLNA
The Norwegian Computing Centre, Norway

SUMMARY

A procedure for 3D stochastic modelling of geometric deformation of rock caused by tectonic activity will be presented. The rationale is to enable an analysis of the impact of faulting on fluid flow in petroleum reservoirs. The 3D stochastic model for the break patterns occurring in faulted zones is based on the theory of marked point processes. In order to generate realistic break patterns it is necessary to incorporate prior knowledge about the tectonic deformations, and a Bayesian framework making this possible will be specified. Each fault is associated with a prior marginal distribution and probabilistic modelling of spatial interaction between faults enables modelling of homogeneous release of the stress field. Using a Bayesian approach facilitates the incorporation of seismic information as a global constraint. The structural heterogeneities can be reproduced by stochastic simulation, which also makes possible a quantification of the uncertainty associated with the description.

Keywords: MARKED POINT PROCESS MODEL; PRIOR KNOWLEDGE; FAULT PATTERN.

1. INTRODUCTION

Many spatial phenomena are easily modelled by a spatial event variable. Such a variable can be considered as a number of points, each with a certain location in a two or three dimensional spatial domain. In the general case each event will also exhibit certain characteristics. These are often modelled as a set of values, a vector of marks, which describes the point. Such a spatial field model is called a marked point process. It appears that most of the work in Bayesian statistics concerning spatial phenomena utilizes a lattice based description. In this paper an application using a marked point process model in a Bayesian framework will be presented. This provides for a coherent incorporation of global constraining information. Global constraining information denotes direct or indirect observations of a function whose argument is the set of marked points in the whole region or in a subset of the region. Such information will often be available, and hence there is a need for models which articulate this type of information about spatial phenomena, and also for algorithms for efficient evaluation of these models.

Often the quantity of interest will be a non-linear function of the spatial field involved. Furthermore, the distribution for the spatial field will typically be of a very high dimension and have a complicated structure, consequently, one has no choice but to tackle the problem by simulation.

The outline of the paper is:
The problem in hand will be presented in the next section, thereafter the Bayesian framework used will be introduced. The model of the tectonic pattern will be presented in Section 4 along with some examples of realizations. The simulation procedure will be discussed in Section 5, followed by some closing remarks in Section 6.

2. THE SPATIAL PROBLEM

In many petroleum reservoirs considerable post-sedimentary tectonic activity has taken place. These reservoirs are therefore characterized by numerous fractures and faults. This is expected to have a significant impact on the production potential of oil. Faults denote structural discontinuities with a displacement of rock in the order of meters, whereas fractures designate structural discontinuities characterized by lack of cohesion but no displacement. The production consequences can be evaluated by numerical simulation of fluid flow. In order to do so, the tectonic pattern needs to be specified. In this paper a procedure for stochastic simulation of the pattern will be presented, such that the importance of tectonic deformation can be evaluated. In order to enable simulation of realistic break patterns, a Bayesian framework facilitating incorporation of prior knowledge, will also be presented. The resulting procedure allows integration of several sources of information of very different kinds.

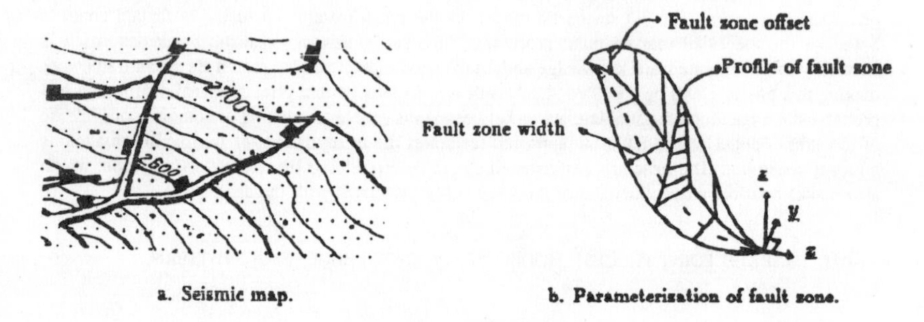

Figure 1. *The fault zones.*

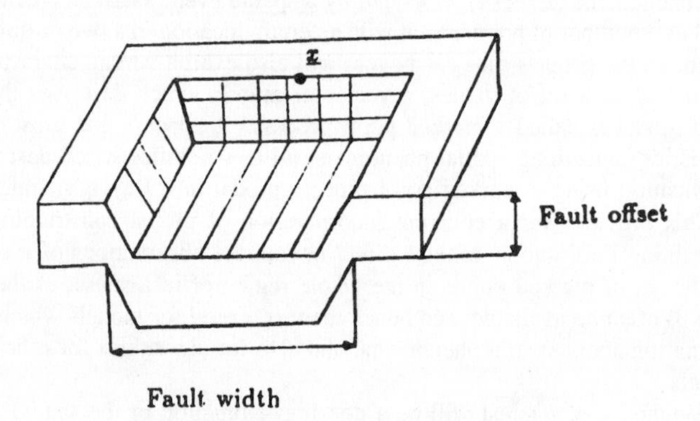

Figure 2. *Parameterization of the single fault.*

The tectonic pattern can be described as a fault zone made up of a large number of faults. The detailed break pattern induced by the faults will govern flow through the deformed rock structure. Associated with the fault zone there are a number of characteristics, most notably areal extent and total offset, which can be identified from seismic maps. A seismic map is presented in Figure 1a, the fault zones are the areas enveloped by the solid lines. However, due to the highly deformed rock in the interior of these zones, the break pattern of the fault zone is difficult to predict from seismic data. Consequently, information about the break pattern must be inferred from general geological knowledge, experience with similar reservoirs and information from wells. A fault zone is parameterized as shown in Figure 1b, and a single fault is parameterized as shown in Figure 2, see Walsh and Watterson (1988). A fault is defined by its width and its offset. A number of these are distributed in the fault zone such that their offset accumulates to the total offset of the fault zone, see Jamison (1989). That is, each fault induces a topological deformation of the rock formation and these transformations have to be coordinated such that the aggregated effect conforms to that found from seismic maps. Note that it is assumed that a fault cuts through the whole reservoir and deforms it accordingly.

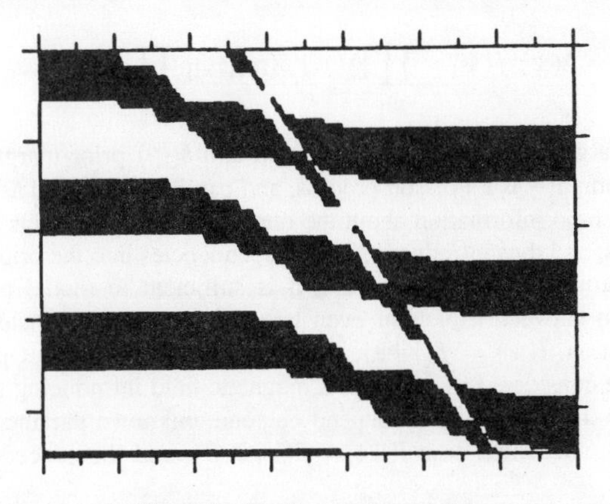

Figure 3. *Faulting of reservoir box.*

In Figure 3 a cross-section of a layered 3D reservoir box which has been transformed w.r.t. a realization from a fault zone model is shown. Note that a replication of the reservoir pattern has been assumed, such that new layers are shifted into the box from above when segments of the reservoir box are shifted vertically due to the fault zone offset. The shaded areas correspond to high permeable zones. In order to evaluate the impact of the break pattern on flow through the reservoir, average permeability across the box has been computed before and after transformation. The affiliated relative reduction in absolute permeability over the box is 25 %. If permeability is calculated based on a 2D cross-section rather than the complete box, the relative reduction is 60 %. Hence, it is essential to do this analysis in three dimensions. The model of the break pattern of a fault zone will be described shortly.

3. A BAYESIAN FRAMEWORK

Bayesian statistics offers a framework for incorporation of general and scene-specific prior knowledge into the analysis of a phenomenon by the formulation of an explicit probability model. Some principles relevant for the problem in hand will be briefly reviewed.

Consider a marked point process model of a spatial event variable. The field is defined by a set of marked points $\Psi = \{\psi_i = (x_i, m_i) \ : \ i \in S\}$. A single point is defined by its location x_i and its set of marks m_i. The number of points, n, may or may not be known. Often some prior knowledge is available about the process under study. In the present case the relationship between offset and width of faults for different rock types has been studied in detail. Furthermore, suppose that we have at our disposal a set of records $Y = \{y_k : k \in T\}$ which is related to the set of points Ψ. Y may only be indirectly related to Ψ, as is the case for seismic information. The present task is to predict the set of marked points Ψ based on prior knowledge about the underlying marked point process, the set of records Y, and a model for the statistical relationship between Y and Ψ.

Hence, a prior distribution for the points, $p(\Psi)$, must be specified. This distribution should reflect general and scene-specific knowledge about Ψ. Note that in general this information will concern only local characteristics of the phenomenon, such as marginal distribution for the marks of a point and interaction between neighbouring points. Global features cannot easily be included. For a marked point process, Ψ, the following parameterization will be assumed:

$$p(\Psi) = h(\Psi)\ f(\Psi) = \prod_i [\lambda(x_i)\ b(m_i|x_i)] \prod_{i<j} c(x_i, m_i; x_j, m_j).$$

$\lambda(\cdot)$ is the prior marginal distribution for location and $b(\cdot|\cdot)$ prior marginal distribution for marks given location. $h(\cdot)$ is a Poisson process, and can be interpreted as the distribution for a marked point given no information about the other points. $f(\cdot)$ models interaction between neighbouring points, and thus introduces spatial dependencies into the prior distribution. Note that the above parameterization implies that it is sufficient to model pairwise interaction, however, interaction between triplets or even larger sets could be considered. Furthermore, the likelihood function, $l(\Psi) = f(Y|\Psi)$, of a realization of the point process needs to be specified. Y can be considered as acting as a magnetic field introducing global ordering into the field. Also note that the field may depend on some unknown parameter set β, with prior distribution $q(\beta|Y)$. In which case $l(\Psi) = f(Y|\Psi, \beta)$, and the procedure is an empirical Bayes procedure.

Finally, Bayes theorem presents us with the posterior distribution:

$$\begin{aligned} P(\Psi, \beta) =& c_0\ l(\Psi)\ p(\Psi|\beta)\ q(\beta|Y) \\ =& c_0\ l(\Psi) \prod_i [\lambda(x_i|\beta)\ b(m_i|x_i, \beta)] \\ & \times \prod_{i<j} c(x_i, m_i; x_j, m_j|\beta)\ q(\beta|Y), \end{aligned}$$

with $c_0 = 1/f(Y)$ a normalizing constant. Hence, the result is a posterior distribution and not merely a point estimate. This distribution can be used for specification of a maximum a posteriori (MAP) estimate, calculation of credibility intervals or for drawing realizations of the marked point process.

Note that c_0 will be a function of the number of points, n, and will in general be hard to calculate. However, for the purpose of drawing realizations this difficulty can be circumvented. The above outline closely parallels Bayesian Image Analysis concepts, see Besag (1989) for a thorough discussion. In the next section the stated problem will be examined using this formalism. The simulation procedure for drawing realizations will be discussed in Section 5.

4. SPECIFICATION OF THE MODEL

The stochastic model is based on the theory of marked point processes, see Stoyan *et al.* (1987). Each fault is represented as a point with a number of characteristics or marks attached. A fault ψ_i is defined by its location x_i and its two marks; width, w_i, and offset,

o_i, see Figure 2. Note that the location of a fault is defined in a two dimensional domain; however, the offsets of the faults introduce a third dimension. Each fault is modelled as being parallel to the centreline of the fault zone and having the same orientation, that is each fault contributes to the total offset. The following parameterization of the posterior probability for the set of faults Ψ will be used:

$$\begin{aligned} P(\Psi) &= c_0\ l(\Psi)\ p(\Psi) \\ &= c_0\ l(\Psi)\ h(\Psi)\ f(\Psi) \\ &= c_0\ l(\Psi) \prod_i [\lambda(x_i)\ b(m_i|x_i)] \prod_{i<j} c(x_i; x_j). \end{aligned}$$

Note that it is assumed that bivariate interaction can be specified in terms of location only and that $\beta \equiv \beta_0$. The functions actually used will be specified below.

Poisson model

The Poisson model, $h(\Psi)$, is defined by the prior marginal distribution for fault intensity, $\lambda(x_i)$, and the prior marginal distribution for mark values $b(m_i|x_i)$. It will be assumed that the areal extent of the fault zone, D, is known. Initially the fault intensity will be chosen as being constant in this area. The number of faults, n, is chosen such that the expected deformation caused by the faults equals the total deformation which is assumed known. The distribution for fault width given location $b_w(w|x)$ is defined by:

$$W|x \equiv (d + R)g(x),$$

where d is a constant and R is an exponentially distributed random variable with mean λ. Note that if $g(x) = \textit{constant}$ the width distribution will be independent of location. The width distribution was truncated as shown above because it was decided not to model very short faults whose geometrical impact is minimal. The distribution for fault offset given fault width $b_o(o|w)$ is defined by:

$$O|w \equiv w^2 N,$$

where N is (truncated) normally distributed (μ, ω^2). Hence, the regression between offset and width is linear in the logarithmic domain. These distributions are chosen according to the results given in Walsh and Watterson (1988). The Poisson model is:

$$h(\Psi) = \lambda(x)\ b_w(w|x)\ b_o(o|w).$$

Prior distribution

In order to define the prior distribution $p(\cdot)$, the bivariate interaction term, $f(\cdot)$, has to be specified. In this model only pairwise interaction between location of faults has been included. The bivariate interaction term, $f(\cdot)$, models spatial repulsion between faults. Such an effect has been modelled because the stress is released homogeneously over the fault zone. The prior model used in this study is:

$$p(\Psi) = h(\Psi)\ f(\Psi) \propto h(\Psi)\ \exp\{-\sum_{i<j} u(||x_i - x_j||)\}$$

$$u(l) = \begin{cases} \delta_2/(\delta_3 + l) & \text{if } l \leq \delta_1 \\ \delta_2/(\delta_3 + \delta_1) & \text{otherwise,} \end{cases}$$

Introducing a finite range of influence, δ_1, is geologically sound, and was found to be important for efficient generation of realizations.

Posterior distribution

The final stepping-stone towards the complete model is the inclusion of the likelihood function $l(\Psi)$. As it is seen from Figure 1a the total offset over the fault zone can be found from the contour lines of the seismic map. In the study Y is the offset profile of the fault zone found from the seismic map. The seismic map is modelled as a gaussian field. Hence, given a certain realization with offset profile $o(\cdot)$, the accumulated difference between this profile and the measured profile, $y(\cdot)$, is a gaussian variable and defines the likelihood $l(\Psi)$:

$$l(\Psi) \propto \exp\{-[\int (y(x) - o(x))]^2/2\sigma^2 \, dx\}.$$

This term serves to reduce the likelihood of a large offset deviation. The posterior model is:

$$\begin{aligned} P(\Psi) =& c_0 \prod_i [\lambda(x_i)\; b_w(w_i|x_i)\; b_o(o_i|w_i)] \exp\{-\sum_{i<j} u(||x_i - x_j||)\} \\ & \times \exp\{-[\int (y(x) - o(x))]^2/2\sigma^2 \, dx\}. \end{aligned}$$

Note that $\beta \equiv \beta_0$. However, in a more elaborate model the parameters of the prior distribution could have been explicitly modelled by a prior probability model reflecting geological experience and updated with respect to scene specific information. There seems to be some invariance to scale w.r.t. fault patterns, hence observations of large features visible on seismic maps can be used for updating characteristics of small scale features. Furthermore, intense outcrop surveying makes specification of a prior model for fault intensity as a function of location in the fault zone possible. The parameter σ in $l(\Psi)$, reflecting uncertainty in faultzone offset, can be identified through the Bayesian interpolation procedure of the seismic surface.

Example

An example illustrating the structure of the stochastic model will be presented. The definintion of the faultzone geometry is synthetic and chosen for the purpose of illustrating the model. The parameters of the Poisson model, defining the geometry of a single fault, have been chosen according to geologists' expectations for a sand-stone reservoir. In the present case the parameters defining the repulsion effect in the Prior model have been chosen by geologists after visual inspection of simulated patterns.

In Figure 4 the fault zone is presented when the prior model has been used. The layout of the figure is as follows:

The upper display is a birds'-eye view of the fault zone. The zone is outlined by the dashed line and the location and width of the faults are indicated by the solid line segments. Note that the display is in relative coordinates with respect to the centreline of the fault zone.

The middle display is a horizontal view, facing the fault zone. The offset of the fault zone can be found as the vertical distance from the uppermost solid line to the dashed x-axis. The small areas outlined by the solid lines correspond to the hatched area in Figure 2, which is the part of the fault plane laid bare due to the vertical shift. The difference between the measured offset of the fault zone and the accumulated offset of the simulated set of faults can be seen as the bottom solid lines deviation from the axis.

The bottom display shows cross sections of the fault zone perpendicular to the centreline of the fault zone. The solid line is the average normalized fault zone profile of the realization. The dotted lines are three examples of normalized profiles.

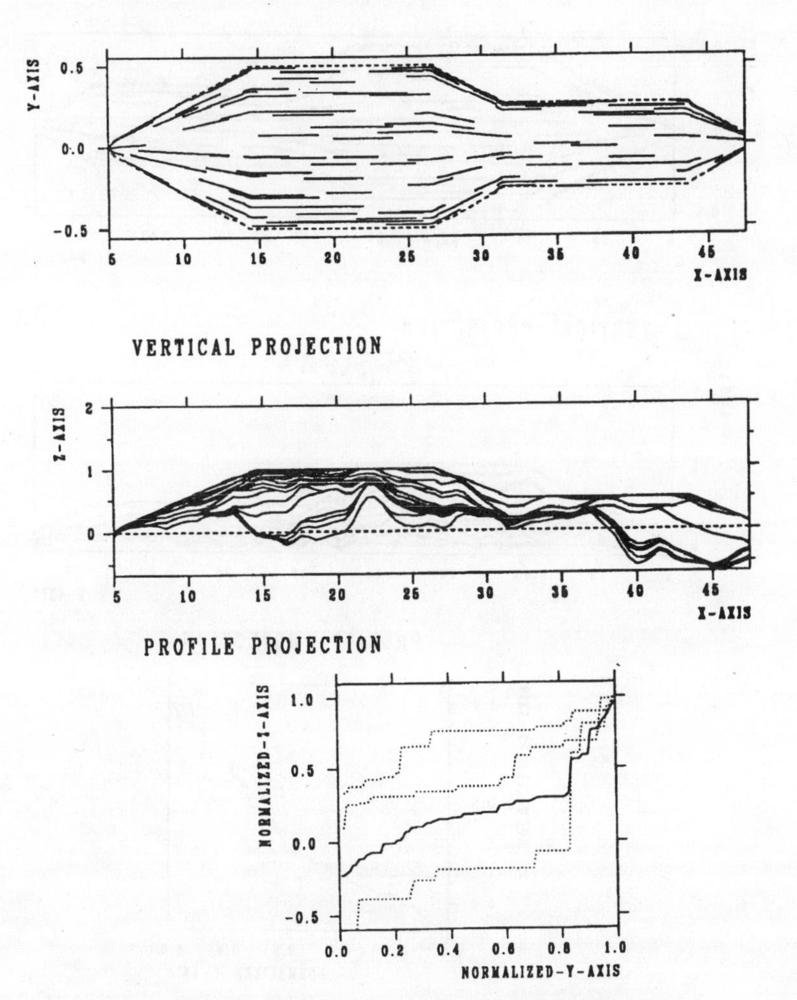

Figure 4. *Simulated fault zone with no offset constraint but with repulsion.*

From Figure 4 it is seen that the difference between the offset of the fault zone and the accumulated offset of the realization is rather large. Due to the term modelling bivariate spatial repulsion the faults are seen to be rather evenly distributed in the fault zone.

A realization from the posterior model is shown in Figure 5. It is seen that the degree of dispersion of faults is about the same as in Figure 4, and that the offset deviation in this case is very small.

The average fault zone profiles found in the bottom displays of Figures 4 and 5 are almost linear. It would be valuable to be able to control the expected profile through the specification of model parameters. Often the profile is assumed to be steeper close to the

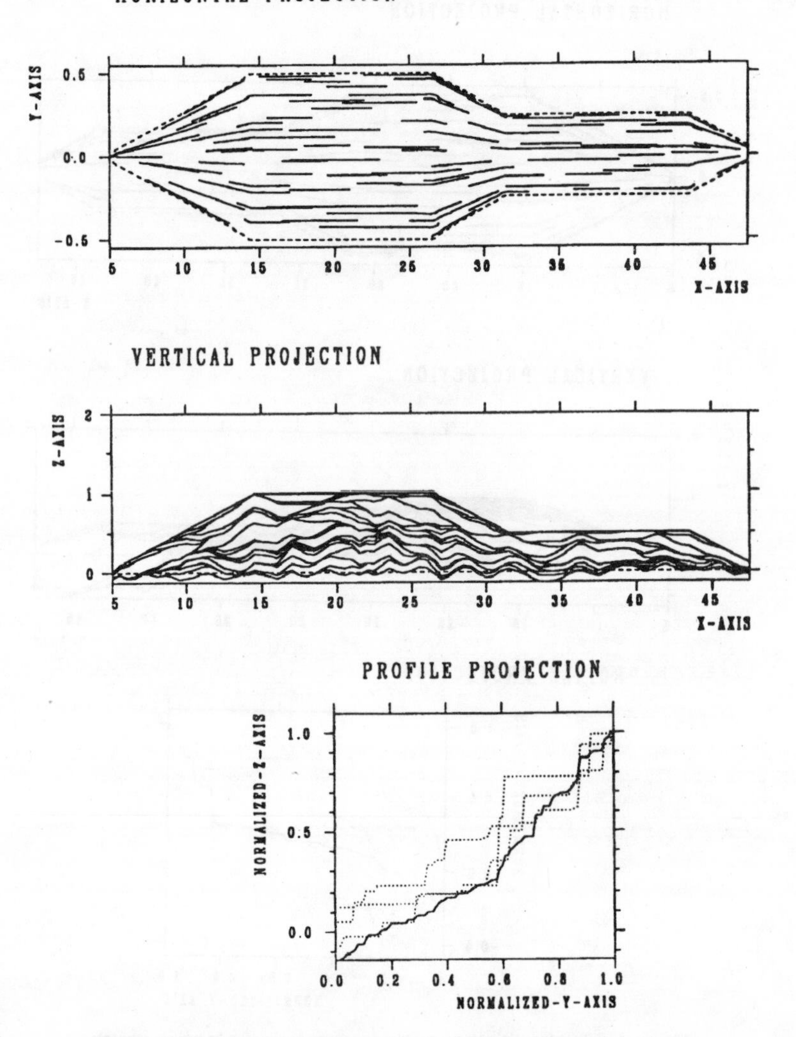

Figure 5. *Simulated fault zone with both offset constraint and repulsion.*

centreline, this will be the case for a less ductile reservoir rock. This can be obtained by specifying fault intensity and/or offset and width distributions as functions of the normalized distance from the centreline, and is illustrated in Figures 6 and 7. In Figure 6 such a profile has been realized by scaling intensity, whereas in Figure 7 the profile has been obtained by scaling offset and width. Also note that if fault zone offset changes relative to fault zone width, it might be desirable to let the parameters of the intensity and the mark distributions be a function of the fault zone centreline argument. That is, a number of parameters of the prior marginal distributions for intensity and mark of faults are updated w.r.t. scene specific information.

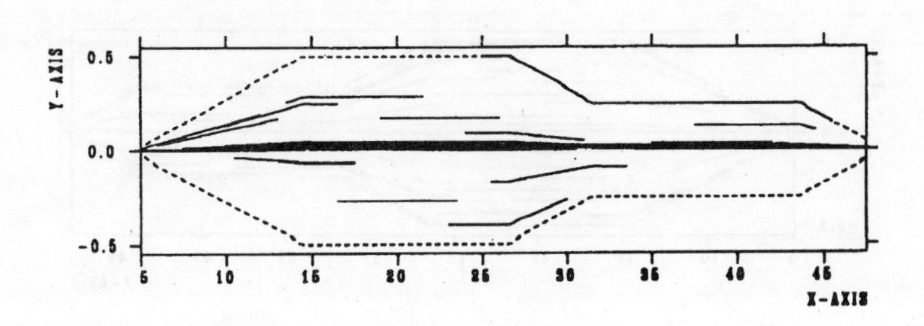

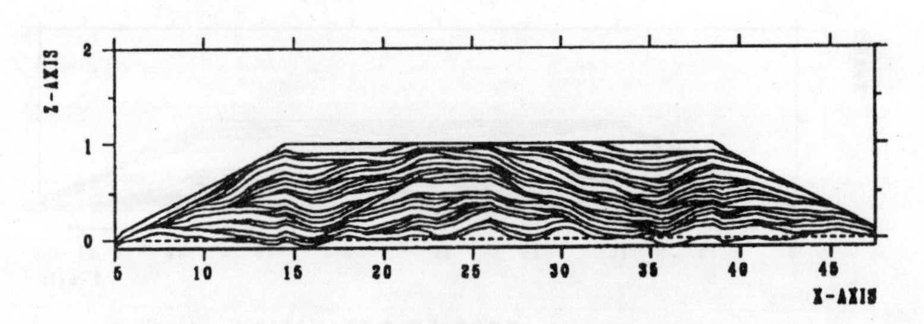

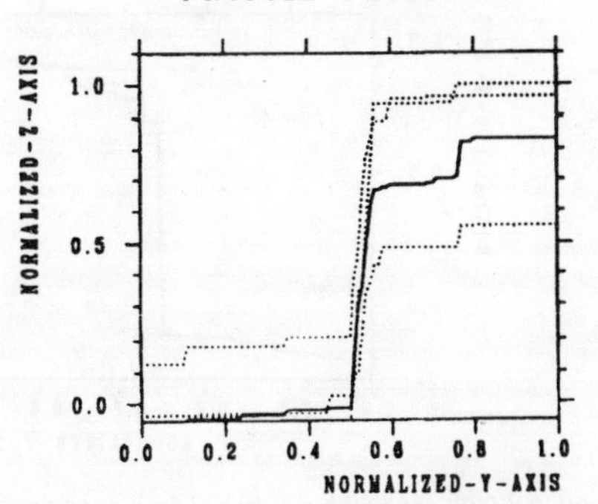

Figure 6. *Simulated fault zone with a required fault zone profile which is very steep close to the centre line. The profile is realized by varying fault intensity.*

5. SIMULATION PROCEDURE

Realizations from the Poisson model can be generated by assigning n, the number of faults, a value drawn from a Poisson distribution with parameter $\int_D d\lambda$, and thereafter simulating each location by rejection sampling. This corresponds to drawing locations according to a prior uniform distribution in D and accepting with probability $\propto \lambda(x)$. For each of the accepted locations a set of marks can be found from the prior marginal distribution for marks.

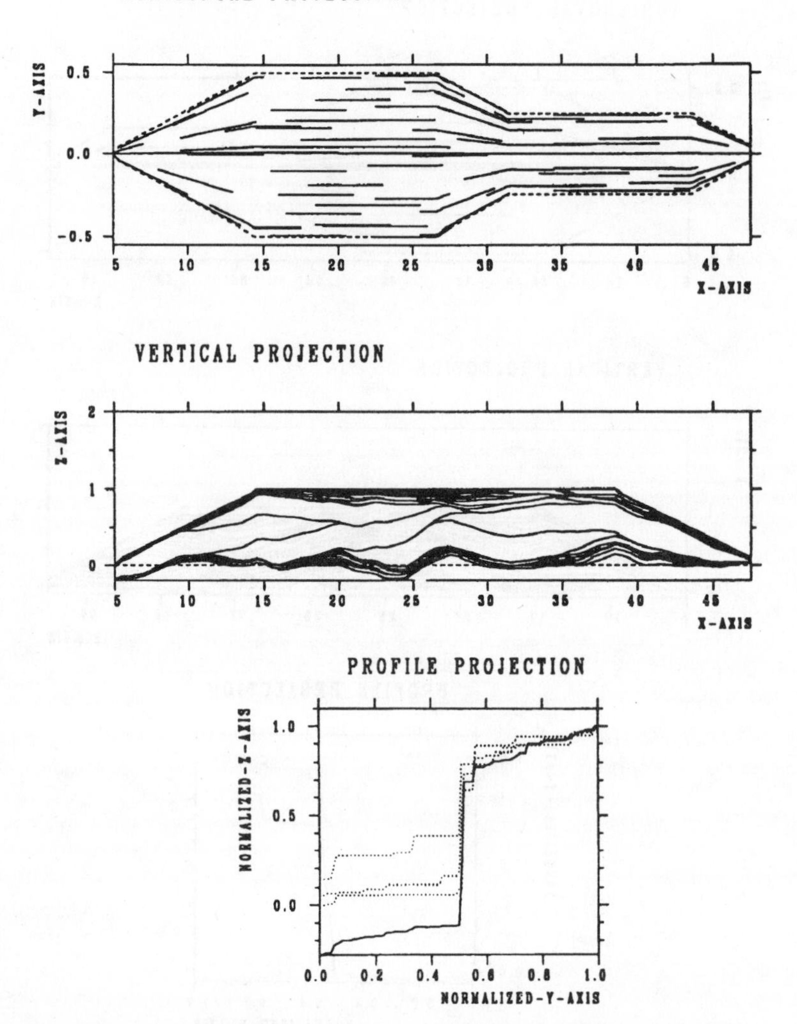

Figure 7. *Simulated fault zone with a required fault zone profile which is very steep close to the centre line. The profile is realized by varying offset/width.*

Drawing realizations from the posterior distribution is far more complicated, because in the general case all locations and mark values interact. Assume that the number of points, n, is fixed, see Preston (1977) for the case where the distribution of n is unknown. Realizations can be generated by drawing initial fields from the Poisson model and accepting with probability $\propto l(\Psi)f(\Psi|\beta)$. However, this might be terribly inefficient.

The key for efficient simulation of these type of phenomena is to consider one point at a time. The field is initialized by the Poisson model, and then one point is considered at a time in a birth-death simulation procedure. One point is removed, and a new point is found by sampling from the Poisson model and accepting with probability $\propto l(\Psi)f(\Psi|\beta)$ given the remaining $n-1$ points. This is the Ripley-Kelly algorithm, see Ripley and Kelly (1977). Selecting the points in sequence was found to be preferable for convergence.

It is easy to test the validity of the procedure. The process can be seen as a Markov chain with a finite statespace. Any state can be reached from any state after n steps, hence a unique limiting distribution exists. Furthermore, if $\widehat{\Psi}$ denotes Ψ after one replacement, then:

$$P(\Psi)\ h(\widehat{\Psi}|\beta)\ l(\widehat{\Psi})\ f(\widehat{\Psi}|\beta) = P(\widehat{\Psi})\ h(\Psi|\beta)\ l(\Psi)\ f(\Psi|\beta).$$

Hence:

$$P(\Psi) \propto l(\Psi)p(\Psi|\beta).$$

This argument is still valid if $\widehat{\Psi}$ is chosen as the next state with probability: MIN[1, $(l(\widehat{\Psi})$ $f(\widehat{\Psi}|\beta))/(l(\Psi)f(\Psi|\beta))$], and Ψ otherwise. This is the Metropolis algorithm, see Metropolis et al. (1953). Note that no point is rejected and that no constant of proportionality needs to be computed. This suggests that the Metropolis algorithm samples the spatial field more efficiently; moreover, given a computational limit, the influence on the result of the choice of initial state is smaller.

Simulation of β can be obtained by considering β as point $n+1$ with prior marginal distribution $q(\beta|Y)$. For computational efficiency it is important that the Poisson model is similar to the posterior marginal distribution. In this case the spatial model and the likelihood primarily serve the purpose of introducing order into the set of points rather than defining the posterior marginal distributions. That is, the Poisson distribution is rather sharp compared to the transition density $l(\Psi)$. Finally, note that it is common to use replications of the patterns in order to reduce border effects.

6. CLOSING REMARKS

A procedure for 3D stochastic modelling of geometric deformation of rock caused by tectonic activity has been presented. A marked point process model in a Bayesian framework has been used. Exploration of the stochastic fault zone model has demonstrated the representativity of fault zones occurring in outcrops and geologists' minds.

Furthermore, the framework has facilitated merging of different sources of information and the translation of geological insight into realistic 3D models of reservoir geometry. In the simple example shown above the parameters of the prior model have been considered as fixed and chosen based on geological experience only, or solely based on the set of records. In a more thorough study, a natural extension would have been to translate geological experience into a prior distribution of the parameters. This distribution could have been updated w.r.t. the set of records and included in the simulation procedure as indicated above.

The purpose of the fault-pattern model was to enable examination of the impact of faulting on fluid flow. Certainly, fluid flow also depends on sedimentary characteristics, and current work aims at identification of the relationship between typical fault-patterns, sedimentary characteristics and flow properties. In the general case the parameters of the prior model have to be chosen conditioned to observations of sedimentary characteristics.

Finally, note that the use of Bayesian methods in marked point processes is still in its infancy and more research is needed in order to formulate general frameworks for analysis of applications like the one discussed above.

ACKNOWLEDGEMENT

The study is performed in cooperation with Saga Petroleum a.s., and valuable geological expertise has been provided by Bjørn Tørudbakken, Nina Dahl, Johan Petter Nystuen and Ragnar Knarud. Valuable statistical expertise has been provided by Henning Omre, the Norwegian Computing Centre.

REFERENCES

Baddeley, A. and Møller, J. (1989). Nearest-neighbour Markov point processes and random sets. *Internat. Statist. Rev.* **57**, 89–121.

Besag, J. (1989). Towards Bayesian image analysis. *J. Amer. Statist. Assoc.* **6**, 395–407.

Jamison, W. R. (1989). Fault-fracture strain in Wingate Sandstone. *J. of Structural Geology* **11**, 959–974.

Metropolis, N., Rosenbluth, A. W., Rosenbluth, M. N., Teller, M. N. and Teller, E. (1953). Equations of state calculations by fast computing machines. *J. of Chemical Physics* **21**, 1087–1091.

Ripley, B. D. and Kelly, F. P. (1977). Markov point processes. *J. Lond. Math. Soc.* **15**, 188–192.

Stoyan, D., Kendall, W. S. and Mecke J. (1987). *Stochastic Geometry and its Applications*. Berlin: Akademie-Verlag.

Omre, H., Sølna, K. and Tørudbakken B. (1990). Stochastic modelling and simulation of fault zones. *Sciences de la Terre*, (to appear).

Preston, C. (1977). Spatial birth-and-death processes. *Bull. Internat. Statist. Institute.* **46**, 371–391.

Walsh, J. J. and Watterson, J. (1988). Analysis of Relationship between displacements and dimensions of faults. *J. of Structural Geology* **10**, 239–247.

BAYESIAN STATISTICS 4, pp. 803–811
J. M. Bernardo, J. O. Berger, A. P. Dawid and A. F. M. Smith, (Eds.)

Reliability Decision Problems under Conditions of Ageing

F. SPIZZICHINO
Università "La Sapienza", Rome, Italy

SUMMARY

Wear-out and infant mortality of items are fundamental concepts in the statistical analysis of failure and survival data. Our aim is the study of some general qualitative properties of Bayes decision procedures when statistical data are failure and survival times of units which undergo wear-out or infant mortality. In particular we obtain ordering properties of the acceptance regions in the "predictive" two-action decision problem, by using the notion of *majorization*. A prerequisite for our study is an appropriate formulation of the concept of wear-out from the Bayesian viewpoint: in the frequentist approach, situations of wear-out are modeled by the Increasing Failure Rate (IFR) property of the distributions of lifetimes of units; this characterization is no longer valid when probability is intended as a degree of belief. Indeed, even if wear-out is present, it may happen that, while progressively observing the survival of a unit, we become more and more optimistic about its residual lifetime, at least in an initial period. This happens when our initial state of information (about possible unknown parameters affecting the random mechanism of failure) is too pessimistic. In a recent paper, Barlow and Mendel gave a suitable formulation of wear-out (*Schur-concavity of the joint survival function*). We shall use this as a general condition for our results after showing that it is equivalent to a significant property of conditional survival probabilities for residual lifetimes, given the data about already observed survivals. An analysis of the particular case of "Schur-constant" survival functions will allow us to clarify the motivation for the present paper.

Keywords: SCHUR CONCAVE AND SCHUR CONVEX SURVIVAL FUNCTIONS; AGEING; MONOTONE ACCEPTANCE REGIONS.

1. INTRODUCTION

A problem of interest in Bayesian Life-testing is the "predictive" two-action decision problem which can be described as follows. Let $T_1, \dots, T_N$ be the lifetimes of (apparently) similar units $U_1, \dots, U_N$ in a lot; $T_1, \dots, T_N$ are exchangeable non-negative random variables; f_N denotes their joint density function and $\overline{F}_N$ denotes the corresponding joint survival function:

$$\overline{F}_N(t_1, \dots, t_N) \equiv P\{T_1 > t_1, \dots, T_N > t_N\}.$$

We fix $n < N$ and perform a life-testing experiment on $U_1, \dots, U_n$ in order to gain information concerning the behaviour of $U_{n+1}, \dots, U_N$, which, for the moment, are set apart for their future "operative life".

a_1 and a_2 are two actions; we must choose between a_1 and a_2 and our choice will yield a loss $l_i(t) = \sum_j \lambda_i(t_j)(i = 1, 2)$ depending on the lifetimes of $U_{n+1}, \dots, U_N$ through a suitable function $\lambda_i : \mathbb{R}_+ \to \mathbb{R}(i = 1, 2)$. It is rather natural to think of a_1 and a_2 as a "pessimistic" and an "optimistic" action, respectively; i.e. we shall assume $\lambda_1(t) - \lambda_2(t)$ to be non-decreasing; moreover, because of the special form assumed for $l_i(t)$, there is no loss of generality in considering the particular case $N = n + 1$. (Neverthless, the results we obtain may easily be extended to more general loss functions.)

The life-testing experiment on $U_1, \ldots, U_n$ can be of different kinds, and we shall denote the resulting data by the symbol $D_{(n)}$. Consider the *Bayes Decision Strategy*: accept a_2 if and only if

$$\mathbf{E}\left[l_1(T_{n+1}) - l_2(T_{n+1}) | D_{(n)} = \boldsymbol{d}\right] \geq 0. \tag{1.1}$$

In the present paper interest is focused on some ordering properties of the "acceptance" region Γ_n of those $\boldsymbol{d}$ for which the inequality in (1.1) is true. In this problem, we must take into consideration two issues which are typical of data from life-testing experiments:

(a) $D_{(n)}$ is, in general, an event of the form

$$T_1 = t_1, \ldots, T_h = t_h, T_{h+1} > s_1, \ldots, T_n > s_n, \text{ for some } h \leq n; \tag{1.2}$$

(b) In modelling the probabilistic behaviour of $T_1, \ldots, T_N$ (namely, in assessing the form of $\overline{F}_N$), we shall consider possible qualitative properties such as "wear out" or "infant mortality".

The event in (1.2) will be indicated by $\{D_{(n)} = (h, \boldsymbol{t}, \boldsymbol{s})\}$, where h is the *number of failures*, $\boldsymbol{t}$ is the vector of *failure times*, and $\boldsymbol{s}$ is the vector of *survival times*. Let the set $\mathcal{X}_n$ be defined by $\mathcal{X}_n \equiv \{\boldsymbol{d} = (h, \boldsymbol{t}, \boldsymbol{s}) | h = 0, \ldots, n; \boldsymbol{t} \in \mathbf{R}_+^{(h)}; \boldsymbol{s} \in \mathbf{R}_+^{(n-h)}\}$. Of course $\mathcal{X}_n$ is the space of possible statistical results in our life-testing experiment and the acceptance region Γ_n is a subset of $\mathcal{X}_n$. We assume that censoring is *non-informative*.

Before continuing, we recall the concepts of majorization and of Schur-functions Marshall & Olkin (1979). For $N \in \boldsymbol{N}$, the vector $\boldsymbol{t}' \equiv (t'_1, \ldots, t'_N)$ is said to be *majorized* by the vector $t'' \equiv (t''_1, \ldots, t''_N)$, denoted by $\boldsymbol{t}' \prec \boldsymbol{t}''$, if

$$\sum_{j=1}^{k} t'_{(j)} \geq \sum_{j=1}^{k} t''_{(j)} \quad k = 1, 2, \ldots, N-1, \quad \sum_{j=1}^{N} t'_{(j)} = \sum_{j=1}^{N} t''_{(j)}$$

where $t'_{(1)} \leq t'_{(2)} \leq \ldots \leq t'_{(N)}$ and $t''_{(1)} \leq t''_{(2)} \leq \ldots \leq t''_{(N)}$ are the *order statistics* of $(t'_1, \ldots, t'_N)$ and $(t''_1, \ldots, t''_N)$, respectively. Functions ψ for which $\boldsymbol{t}' < \boldsymbol{t}''$ implies $\psi(\boldsymbol{t}') \leq \psi(\boldsymbol{t}'')$ are said to be *Schur-convex*; ψ is *Schur-concave* if $\psi(\boldsymbol{t}') \geq \psi(\boldsymbol{t}'')$.

The central aim in this paper is to find suitable partial orderings $\hat{\prec}$ defined on $\mathcal{X}_n$ such that

$$\boldsymbol{d}' \hat{\prec} \boldsymbol{d}'', \quad \boldsymbol{d}' \in \Gamma_n \quad \Rightarrow \quad \boldsymbol{d}'' \in \Gamma_n. \tag{1.3}$$

In terms of the concept of majorization, results of this kind will be obtained in Section 3 by using the assumption of Schur-concavity for $\overline{F}_{n+1}$ or the stronger assumption of Schur-concavity for f_{n+1}. In particular we shall obtain results for conditionally i.i.d. lifetimes, with IFR (*Increasing Failure Rate*) unidimensional conditional distributions. Dual results can be obtained by replacing Schur-concavity by Schur-convexity and IFR by DFR (*Decreasing Failure Rate*).

In the next Section we shall discuss some preliminary topics:

(i) the necessity of a Bayesian formulation of wear-out and infant mortality
(ii) the related role of the concept of Schur survival functions
(iii) an analysis of the limiting case of Schur-constant survival functions which allows us to provide a motivation which originated the present study.

2. SCHUR SURVIVAL FUNCTIONS AND THE BAYESIAN FORMULATION OF AGEING

In the frequentist approach, wear-out is modeled by suitable properties (such as IFR, NBU, IFRA ...) of the failure rate of the lifetimes of units. In the subjectivist approach the use of such a theory is not appropriate when the "physical" behaviour of $U_1, \ldots, U_N$ is assessed, conditionally on some unknown parameter. Indeed, in such cases, our state of information on $T_1, \ldots, T_N$ is expressed by a predictive distribution, whose *ageing* properties are in general different from those of the conditional distributions. A heuristic analysis of the single-unit case will demonstrate this point.

Let us consider a unit U whose life-time is indicated by T. Our knowledge of the physical mechanism of failure suggests that the behaviour of U is influenced by a quantity Θ (which is unknown to us): our statistical model is expressed by the family of conditional densities $\{g(t|\theta)\}_{\theta \in L}$. We put

$$g(t|\theta) = r(t|\theta) \exp\left\{-\int_0^t r(\xi|\theta)\mathrm{d}\xi\right\} = r(t|\theta)\overline{G}(t|\theta),$$

where $r(t|\theta)$ is the *conditional failure rate.* The predictive density of T is

$$f(t) = \int_L r(t|\theta)\overline{G}(t|\theta)\mathrm{d}\Pi_0(\theta),$$

$\Pi_0(\theta)$ being the *initial* distribution of Θ. The predictive failure rate function of T is

$$r(t) = \frac{f(t)}{\overline{F}(t)} = \frac{\int_L r(t|\theta)\overline{G}(t|\theta)\mathrm{d}\Pi_0(\theta)}{\int_L \overline{G}(t|\theta)\mathrm{d}\Pi_0(\theta)} = \int_L r(t|\theta)\mathrm{d}\Pi_t(\theta),$$

where we put

$$\mathrm{d}\Pi_t(\theta) = \frac{\overline{G}(t|\theta)\mathrm{d}\Pi_0(\theta)}{\int_L \overline{G}(t|\theta)\mathrm{d}\Pi_0(\theta)};$$

$\Pi_t(\theta)$ is the *final* distribution of Θ, given the event $\{T > t\}$. It is clear that possible properties of the functions $r(t|\theta)$ which model the ageing process of U are in general not shared by $r(t)$. For instance, even if $r(t|\theta)$ is increasing in $t, \forall\theta \in L$, the resulting $r(t)$ may be decreasing, at least in a right neighborhood of 0. This happens when our initial state of information is too pessimistic and this pessimism is contradicted by the observation of a survival $T > \bar{t}$.

Consider now the case of $T_1, \ldots, T_N$ conditionally independent, identically distributed with an IFR conditional marginal univariate density; the unidimensional marginal of their joint predictive distribution is not IFR in general, but, in any case, we are in a situation of wear-out. It is of interest to determine which property of the joint distribution of independent IFR lifetimes is preserved under mixture. The joint survival function of independent IFR lifetimes is Schur-concave and this property is shared also by conditionally independent IFR lifetimes. On the other hand, Barlow and Mendel (1991) conjecture that, if $T_1, \ldots, T_N$ are conditionally i.i.d. and their joint survival function is Schur-concave, then they are conditionally IFR; so that, in the case of infinite populations, a Schur-concave joint survival function is the Bayesian equivalent of the frequentist concept of i.i.d. lifetimes with an IFR distribution depending on an unknown parameter. More generally, Barlow and Mendel argue that such a property is the appropriate Bayesian equivalent of the IFR property for exchangeable lifetimes (not necessarily conditionally i.i.d.). This is formalized in the following characterization of the property of Schur-concave joint survival functions which also provides a heuristic explanation.

Theorem 2.1. *$\overline{F}_N$ is Schur-concave if and only if, for any $s \in \mathbb{R}_+^N$ and any $\tau > 0$,*

$$s_i < s_j \Rightarrow P\{T_i > s_i + \tau | T_1 > s_1, \ldots, T_i > s_i, \ldots, T_j > s_j, \ldots, T_N > s_N\} \geq \\ \geq P\{T_j > s_j + \tau | T_1 > s_1, \ldots, T_i > s_i, \ldots, T_j > s_j, \ldots, T_N > s_N\}. \tag{2.1}$$

Proof.

$$P\{T_i > s_i + \tau | T_1 > s_1, \ldots, T_i > s_i, \ldots, T_j > s_j, \ldots, T_N > s_N\} = \frac{\overline{F}_N(s_1, \ldots, s_i + \tau, \ldots, s_j, \ldots, s_N)}{\overline{F}_N(s_1, \ldots, s_i, \ldots, s_j, \ldots, s_N)};$$

$$P\{T_i > s_i + \tau | T_1 > s_1, \ldots, T_i > s_i, \ldots, T_j > s_j, \ldots, T_N > s_N\} = \frac{\overline{F}_N(s_1, \ldots, s_i, \ldots, s_j + \tau, \ldots, s_N)}{\overline{F}_N(s_1, \ldots, s_i, \ldots, s_j, \ldots, s_N)}.$$

So we must show that $\overline{F}_N$ Schur-concave implies

$$\overline{F}_N(s_1, \ldots, s_i + \tau, \ldots, s_j, \ldots, s_N) \geq \overline{F}_N(s_1, \ldots, s_i, \ldots, s_j + \tau, \ldots, s_N). \tag{2.2}$$

This is immediate since $(s_1, \ldots, s_i + \tau, \ldots, s_j, \ldots, s_N) \prec (s_1, \ldots, s_i, \ldots, s_j + \tau, \ldots, s_N)$. Suppose now that (2.2) holds. We obtain, for $s_i < s_j$, $\frac{\partial}{\partial s_i}\overline{F}_N \geq \frac{\partial}{\partial s_j}\overline{F}_N$. This condition is well-known to be equivalent to Schur-concavity of $\overline{F}_N$, for a differentiable, permutation-invariant function $\overline{F}_N$ (see e.g. Marshall and Olkin (1979)).

Remark

By the result in Hardy, Littlewood and Polya (1952 p. 47), we can easily prove that (2.2) implies the Schur-concavity of $\overline{F}_N$ even without the assumption of differentiability.

Note that, in words, Theorem 2.1 says that, among two units which survived a life test, the "younger" is the "better" (if and only if $\overline{F}_N$ is Schur-concave): we do not compare the distribution of a residual life-time for a unit with the distribution of the residual life-time of the same unit when it was "younger" than it is now; for a fixed state of information, we compare the lifetimes of two units of different "ages". Note, moreover, that some of the values s in (2.1) can be equal to 0 (which says that the corresponding unit did not undergo the test at all). An aspect of wear-out is also indicated by the fact Marshall and Olkin (1974) that, for $\overline{F}_N$ Schur-concave,

$$P\{T_1 > s_1, \ldots, T_N > s_N\} \geq P\left\{T_1 > \sum_{j=1}^{N} s_i\right\}.$$

An analogous discussion can be carried out for the concept of infant mortality and its connection with Schur-convexity of $\overline{F}_N$. The interest in the results of the next Section will now be illustrated by studying, in some detail, the limiting case $\overline{F}_N$ *Schur-constant*:

$$\overline{F}_N(t) = \psi\left(\sum_{i=1}^{N} t_i\right), \tag{2.3}$$

for some given non-increasing function $\psi : \mathbb{R}_+ \to \mathbb{R}_+$. In the language introduced by Barlow and Mendel (1991), this is the case of "indifference with respect to ageing"; indeed, since $\overline{F}_N$ in (2.3) is both Schur-concave and Schur-convex, we can write, by applying Theorem 2.1,

$$P\{T_i > s_i + \tau | T_1 > s_1, \ldots, T_i > s_i, \ldots, T_j > s_j, \ldots, T_N > s_N\} \\ = P\{T_j > s_j + \tau | T_1 > s_1, \ldots, T_i > s_i, \ldots, T_j > s_j, \ldots, T_N > s_N\} \tag{2.4}$$

for $\tau > 0$. The identity (2.4) translates the frequentist idea that the age does not influence the distribution of the residual life-time into a proper subjectivist viewpoint.

For our discussion we need to emphasize that an absolutely continous $\overline{F}_N$ is Schur-constant if and only if f_N is Schur-constant.

Remark

It is interesting to note that, under (2.3), the marginal survival function of $T_1, \ldots, T_n (n = 1, \ldots, N-1)$ is again of the form $\overline{F}_n(t) = \psi(\sum_{i=1}^n s_i)$; indeed

$$\overline{F}_n(t) = P\{T_1 > t_1, \ldots, T_n > t_n, T_{n+1} > 0, \ldots, T_N > 0\} = \psi\left(\sum_{i=1}^n t_i\right).$$

In the limit as $N \to \infty$, (2.3) implies that $T_1, T_2, \ldots$ are conditionally independent, exponentially distributed (Diaconis and Ylvisaker, 1985, see Theorem 7). This is a de Finetti-type result which can be seen from a number of different viewpoints (Lauritzen (1988), Ressel (1985), Diaconis and Freedman (1987), Spizzichino (1988), Misiewicz and Cooke (1989), Barlow and Mendel (1991)).

The following result shows that, in a life testing experiment on $U_1, \ldots, U_N$ under indifference with respect to ageing, $(h, \sum_{i=1}^h t_i, \sum_{j=1}^{n-h} s_j)$ is a sufficient statistic. This is a well-known result (Barlow and Proschan (1988)) in the case of an "infinite population" (when $T_1, \ldots, T_N$ are conditionally independent, exponentially distributed).

For $\boldsymbol{d} \in \mathcal{X}_N$, let us denote by $f_{N-n}(\cdot|\boldsymbol{d})$ the conditional density of $T_{n+1}, \ldots, T_N$, given the result $\mathrm{D}_{(n)} = \boldsymbol{d}$; moreover for $\boldsymbol{d}' \equiv (h', \boldsymbol{t}', \boldsymbol{s}''), \boldsymbol{d}'' \equiv (h'', \boldsymbol{t}'', \boldsymbol{s}'') \in \mathcal{X}_N$ define

$$\boldsymbol{d}' \approx \boldsymbol{d}'' \quad \text{if and only if} \quad h' = h'', \sum_{i=1}^{h'} t_i' = \sum_{i=1}^{h''} t_i'', \sum_{j=1}^{n-h'} s_j' = \sum_{j=1}^{n-h''} s_j''.$$

Theorem 2.2. *Let $\overline{F}_N$ be of the form (2.3) and let $\boldsymbol{d}', \boldsymbol{d}'' \in \mathcal{X}_n$ be such that $\boldsymbol{d}' \approx \boldsymbol{d}''$. Then*

$$f_{N-n}(\boldsymbol{u}|\boldsymbol{d}') = f_{N-n}(\boldsymbol{u}|\boldsymbol{d}''). \tag{2.5}$$

Proof. $\overline{F}_N$ of the form (2.3) implies that the joint density function f_N is Schur-constant as well. It is

$$f_{N-n}(\boldsymbol{u}|\boldsymbol{d}') \propto \int_{s_1'}^{\infty} \cdots \int_{s_{n-h}'}^{\infty} f_N(\boldsymbol{t}', \boldsymbol{\xi}, \boldsymbol{u}) d\xi_1 \ldots d\xi_{n-h} \;;$$

$$f_{N-n}(\boldsymbol{u}|\boldsymbol{d}'') \propto \int_{s_1''}^{\infty} \cdots \int_{s_{n-h}''}^{\infty} f_N(\boldsymbol{t}'', \boldsymbol{\xi u}) d\xi_1 \ldots d\xi_{n-h}.$$

Define $\phi(\boldsymbol{\xi}) = f_N(\boldsymbol{t}', \boldsymbol{\xi}, \boldsymbol{u}) = f_N(\boldsymbol{t}'', \boldsymbol{\xi}, \boldsymbol{u})$; since $\phi(\boldsymbol{\xi})$ is Schur-constant, then

$$\int_{s_1}^{\infty} \cdots \int_{s_{n-h}}^{\infty} \phi(\boldsymbol{\xi}) d\boldsymbol{\xi}$$

is Schur-constant as well and then (2.5) holds. $\triangleleft$

The motivation which originated our study, can now be illustrated in a very simple way. Consider two possible results $d' \equiv (h, t', s''), d'' \equiv (h, t'', s'')$ for a life-testing experiment on $U_1, \ldots, U_n$ and suppose that $d' \approx d''$. By Theorem 2.2 we know that, under the assumption (2.3), d' and d'' contain the same information about $T_{n+1}, \ldots, T_N$. If such an assumption (expressing indifference with respect to ageing) is removed and is replaced, say, by the assumption that $U_1, \ldots, U_N$ undergo wear-out (infant mortality), it is natural to wonder: *which of the two results is more "optimistic"?* The discussion presented above shows that the condition (2.3) must be replaced by Schur-concavity (Schur-convexity) of $\overline{F}_{N\cdot}$, and the arguments in the next Section allow us to formalize the question and to give some answers to it.

3. MONOTONICITY PROPERTIES OF THE ACCEPTANCE REGION UNDER AGEING

Let us turn again to the problem described in the Introduction. Suppose that we observe an event of the form (1.2) for the lifetimes $T_1, \ldots, T_n$, and want to choose between the two actions a_1 and a_2; the choice of a_i yields a loss $\lambda_i(T_{n+1})$, where T_{n+1} is a lifetime such that $T_1, \ldots, T_n, T_{n+1}$ are exchangeable with a joint survival function $\overline{F}_{n+1}$. A result on the structure of the region Γ_n defined in (1.1) can easily be obtained under rather general conditions.

Theorem 3.1. *Suppose $\overline{F}_{n+1}(t_1, \ldots, t_{n+1})$ is such that, for $s' \prec s''$,*

$$\frac{\overline{F}_{n+1}(s'_1, \ldots, s'_n, u)}{\overline{F}_{n+1}(s''_1, \ldots, s''_n, u)} \le \frac{\overline{F}_{n+1}(s'_1, \ldots, s'_n, 0)}{\overline{F}_{n+1}(s''_1, \ldots, s''_n, 0)} \quad \forall u > 0. \tag{3.1}$$

Then we have

$$s' \equiv (s'_1, \ldots, s'_n) \prec s'' \equiv (s''_1, \ldots, s''_n), (0, s') \in \Gamma_n \Rightarrow (0, s'') \in \Gamma_n. \tag{3.2}$$

Proof. Since $\lambda_1 - \lambda_2$ is a non-decreasing function, using (1.1) we only need to show that

$$P\{T_{n+1} > u | T_1 > s'_1, \ldots, T_n > s'_n\} \le P\{T_{n+1} > u | T_1 > s''_1, \ldots, T_n > s''_n\}, \quad \forall u > 0. \tag{3.3}$$

Indeed, since

$$P\{T_{n+1} > u | T_1 > s_1, \ldots, T_n > s_N\} = \frac{\overline{F}_{n+1}(s_1, \ldots, s_n, u)}{\overline{F}_{n+1}(s_1, \ldots, s_n, 0)}$$

it follows that (3.3) is equivalent to (3.1). ◁

It can immediately be checked by differentiation that the condition (3.1) is true if $\overline{F}_{n+1}$ is Schur-concave and $\partial \overline{F}_{n+1}(t_1, \ldots, t_n, u)/\partial u$ is Schur-convex as a function of $(t_1, \ldots, t_n), \forall u \ge 0$; these conditions hold if $T_1, \ldots, T_n, T_{n+1}$ are conditionally i.i.d. with a unidimensional conditional IFR distribution. We have the implication $s' \prec s'', (0, s'') \in \Gamma_n \Rightarrow (0, s') \in \Gamma_n$ if we reverse the inequality (3.1); such a condition, in particular, holds if $T_1, \ldots, T_n, T_{n+1}$ are conditionally i.i.d. with a unidimensional conditional DFR distribution.

Under some stronger assumptions we can obtain a result for the case in which survival data are combined with failure data. The following notation will be used:

$$d' \equiv (h', t', s') \prec^* d'' \equiv (h'', t'', s'') \quad \text{if and only if} \quad h' = h'', t' \prec t'', s' \prec s''.$$

Let $f_n(\cdot|u)$ denote the conditional density of $T_1, \ldots, T_n$, given $T_{n+1} = u$, and $\phi_{\boldsymbol{d}}(u)$ be the "likelihood function" of T_{n+1} associated with a result $\boldsymbol{d} = (h, \boldsymbol{t}, \boldsymbol{s})$:

$$f_n(\cdot|u) = f_{n+1}(\cdot, u) \Big/ \int_0^\infty \cdots \int_0^\infty f_{n+1}(\xi_1, \ldots, \xi_n, u) d\xi_1 \ldots d\xi_n,$$

$$\phi_{\boldsymbol{d}}(u) = \int_{s_1}^\infty \cdots \int_{s_{n-h}}^\infty f_n(\boldsymbol{t}, \xi | u) d\xi_1 \ldots d\xi_{n-h}.$$

Theorem 3.2. *Assume that: a) f_{n+1} is Schur-convex; b) $\frac{\partial}{\partial u} f_{n+1}(t_1, \ldots, t_n, u)$ is a Schur-concave function of $(t_1, \ldots, t_n)$ for any fixed $u \geq 0$; c) $\frac{\partial}{\partial u} f_{n+1}(t_1, \ldots, t_n, u) \leq C(\boldsymbol{t}), \forall t \geq 0$ where $C(\boldsymbol{t})$ is integrable on $\mathbb{R}^n_+$. Let $\boldsymbol{d}' = (h, \boldsymbol{t}', \boldsymbol{s}'), \boldsymbol{d}'' = (h, \boldsymbol{t}'', \boldsymbol{s}'') \in \mathcal{X}_N$ be such that $\boldsymbol{d}' \prec^* \boldsymbol{d}''$. Then the likelihood ratio $\psi_{\boldsymbol{d}', \boldsymbol{d}''}(u) = \phi_{\boldsymbol{d}'}(u) / \phi_{\boldsymbol{d}''}(u)$ is a non-decreasing function of $u \geq 0$ and the following implication holds:*

$$\boldsymbol{d}' \prec^* \boldsymbol{d}'', \qquad \boldsymbol{d}'' \in \Gamma_n \Rightarrow \boldsymbol{d}' \in \Gamma_n. \tag{3.4}$$

Proof. $\psi_{\boldsymbol{d}', \boldsymbol{d}''}(u)$ is a differentiable function of u, and $\frac{\partial}{\partial u} \psi_{\boldsymbol{d}', \boldsymbol{d}''}(u) \geq 0$ if and only if

$$\left[\frac{\partial}{\partial u} \int_{s'_1}^\infty \cdots \int_{s'_{n-h}}^\infty f_{n+1}(\boldsymbol{t}', \boldsymbol{\xi}, u) d\xi_1 \ldots d\xi_{n-h}\right] \left[\int_{s''_1}^\infty \cdots \int_{s''_{n-h}}^\infty f_{n+1}(\boldsymbol{t}'', \boldsymbol{\xi}, u) d\xi_1 \ldots d\xi_{n-h}\right] -$$

$$\left[\frac{\partial}{\partial u} \int_{s''_1}^\infty \cdots \int_{s''_{n-h}}^\infty f_{n+1}(\boldsymbol{t}', \boldsymbol{\xi}, u) d\xi_1 \ldots d\xi_{n-h}\right] \left[\int_{s'_1}^\infty \cdots \int_{s'_{n-h}}^\infty f_{n+1}(\boldsymbol{t}', \boldsymbol{\xi}, u) d\xi_1 \ldots d\xi_{n-h}\right] \geq 0;$$

furthermore, because of c), we can write

$$\frac{\partial}{\partial u} \int_{s_1}^\infty \cdots \int_{s_{n-h}}^\infty f_{n+1}(\boldsymbol{t}, \boldsymbol{\xi}, u) d\xi_1 \ldots d\xi_{n-h} = \int_{s_1}^\infty \cdots \int_{s_{n-h}}^\infty \frac{\partial}{\partial u} f_{n+1}(\boldsymbol{t}, \boldsymbol{\xi}, u) d\xi_1 \ldots d\xi_{n-h}. \tag{3.5}$$

By using Theorem 2.1 in Marshall and Olkin (1974), it is easy to see that assumption a) implies the Schur-convexity of $\phi_{(h, \boldsymbol{t}, \boldsymbol{s})}(u)$ as a function of $\boldsymbol{t}$, and $\boldsymbol{s}$, for any fixed $u \geq 0$. Analogously, b) and (3.5) imply that $\int_{s_1}^\infty \cdots \int_{s_{n-h}}^\infty \frac{\partial}{\partial u} f_{n+1}(\boldsymbol{t}, \boldsymbol{\xi}, u) d\xi_1 \ldots d\xi_{n-h}$ is Schur-concave, as a function of $\boldsymbol{t}$, and $\boldsymbol{s}$; thus

$$\int_{s''_1}^\infty \cdots \int_{s''_{n-h}}^\infty f_{n+1}(\boldsymbol{t}'', \boldsymbol{\xi}, u) d\xi_1 \ldots d\xi_{n-h} \geq \int_{s'_1}^\infty \cdots \int_{s'_{n-h}}^\infty f_{n+1}(\boldsymbol{t}', \boldsymbol{\xi}, u) d\xi_1 \ldots d\xi_{n-h}$$

and

$$\frac{\partial}{\partial u} \int_{s'_1}^\infty \cdots \int_{s'_{n-h}}^\infty f_{n+1}(\boldsymbol{t}', \boldsymbol{\xi}, u) d\xi_1 \ldots d\xi_{n-h} \geq \frac{\partial}{\partial u} \int_{s''_1}^\infty \cdots \int_{s''_{n-h}}^\infty f_{n+1}(\boldsymbol{t}'', \boldsymbol{\xi}, u) d\xi_1 \ldots d\xi_{n-h}$$

from which we obtain $\frac{\partial}{\partial u}\psi_{\mathbf{d}',\mathbf{d}''}(u) \geq 0$. Recalling (1.1), we see that (3.4) is immediate if we can show that, for $\mathbf{d}' <^* \mathbf{d}''$,

$$\mathbf{E}\left[\lambda_1(T_{n+1}) - \lambda_2(T_{n+1})|\mathbf{d}_{(n)} = \mathbf{d}'\right] \geq \mathbf{E}\left[\lambda_1(T_{n+1}) - \lambda_2(T_{n+1})|\mathbf{d}_{(n)} = \mathbf{d}''\right]. \quad (3.6)$$

Equation (3.6) follows from standard arguments of stochastic comparison: for $k(\mathbf{d}) \equiv \{\int_0^\infty f_1(u)\phi_{\mathbf{d}}(u)du\}\gamma(u)^{-1} = \lambda_1(u) - \lambda_2(u)$ and $\eta(u) \equiv k(\mathbf{d}')\phi'_{\mathbf{d}}(u) - k(\mathbf{d}'')\phi''_{\mathbf{d}}(u)$, we must show that

$$\int_0^\infty \gamma(u)f_1(u)\eta(u) \geq 0.$$

Define $A \equiv \{u \geq 0|\eta(u) > 0\}$, $B \equiv \{u \geq 0|\eta(u) < 0\}, a = \inf_{u\in A}\gamma(u)$, and $b = \sup_{u\in B}\gamma(u)$. Since $\psi_{\mathbf{d}',\mathbf{d}''}(u)$ and $\gamma(u)$ are non-decreasing, it is the case that $a \geq b$ and

$$\begin{aligned}\int_0^\infty \gamma(u)f_1(u)\eta(u)du &\leq a\int_A f_1(u)\eta(u)du + b\int_B f_1(u)\eta(u)du \\ &= (b-a)\int_B f_1(u)\eta(u)du \geq 0. \quad \triangleleft\end{aligned}$$

Remark

Assumptions a) and b) are satisfied if $T_1,\ldots,T_n,T_{n+1}$ are conditionally i.i.d. with an unidimensional conditional density $g(t|\theta)$, given some parameter Θ, whose conditional failure rate $r(t|\theta)$ is a non-increasing, log-convex function for all values θ of Θ (this case includes Weibull distributions with unknown scale parameter and unknown shape parameter with values smaller than 1). If the assumptions a) and b) are replaced by a') f_{n+1} is Schur-concave and b') $\frac{\partial}{\partial u}f_{n+1}(t_1,\ldots,t_n,u)$ is a Schur-convex function of $(t_1,\ldots,t_n)$ for any fixed $u \geq 0$, then we have the implication $\mathbf{d}' \prec^* \mathbf{d}'', \mathbf{d}' \in \Gamma_n \Rightarrow \mathbf{d}'' \in \Gamma_n$. Assumptions a') and b'), in particular, hold if $T_1,\ldots,T_N$ are conditionally i.i.d. with a log-concave non-increasing conditional unidimensional density $g(t|\theta)$. Finally, we stress that Schur-concavity (convexity) of f_{n+1} implies Schur-concavity (convexity) for $\overline{F}_{n+1}$.

ACKNOWLEDGEMENTS

Partially supported by C.N.R., *Progetto Speciale per la Matematica Applicata and by M. U. R. S. T., Progetto Nazionale "Modelli Probabilistici e Statistica Matematica".*

REFERENCES

Barlow, R. E. and Proschan, F. (1975). *Statistical Theory of Reliability and Life Testing*. Toronto: Holt, Rinehart and Winston.

Barlow, R. E. (1985). A Bayesian explanation of an apparent failure rate paradox. *IEEE Trans. on Reliability* **2**, 107–108.

Barlow, R. E. and Proschan, F. (1988). *Life Distributions and Incomplete Data: Handbook of Statistics* **7**, (P. R. Krishnaiah and C. R. Rao, eds.) Amsterdam: Elsevier, 225–249.

Barlow, R. E. and Mendel, M. B. (1991). De Finetti-type representations for life distributions. *Reliability and Decision Making* (R. E. Barlow, C. A. Clarotti and F. Spizzichino, eds.), Amsterdam: Elsevier, (to appear).

Barlow, R. E., Irony, T. Z. and Shor, S. W. (1990). Informative sampling methods: the influence of experimental design on decision. *Influence Diagrams, Beliefs Nets and Decision Analysis*, (R. M. Oliver and J. Q. Smith eds.) New York: Wiley, 177–197.

Clarotti, C. A. and Spizzichino, F. (1989). The Bayes predictive approach in reliability. *IEEE Trans. on Reliability* **3**, 379–382.

Diaconis, P. and Ylvisaker, D. (1985). Quantifying prior opinion. *Bayesian Statistics 2* (J. M. Bernardo, M. H. DeGroot, D. V. Lindley and A. F. M. Smith, eds.), Amsterdam: North-Holland, 133–156, (with discussion).

Diaconis, P. and Freedman, D. (1987). A dozen de Finetti-style results in search of a theory. *Ann. Inst. H. Poincaré* **23**, 397–423.

Hardy, G. H., Littlewood, J. E. and Polya, G. (1952). *Inequalities*. Cambridge: University Press.

Karlin, S. and Rubin, H. (1956). The theory of decision procedures for distributions with monotone likelihood ratio. *Ann. Math. Statist.* **27**, 272–299.

Lauritzen, S. L. (1988). *Extremal Families and Systems of Sufficient Statistics*. New York: Springer.

Marshall, A. and Olkin, I. (1979). *Inequalities: Theory of Majorization and its Applications*. New York: Academic Press.

Marshall, A. and Olkin, I. (1974). Majorization in multivariate distributions. *Ann. Statist.* **2**, 1189–1200.

Misiewicz, J. and Cooke, R. (1989). l_p invariant probability measures with stochastic rescaling. *Tech. Rep.*, Delft University of Technology.

Ressel, P. (1985). De Finetti-type theorems: an analytical approach. *Ann. Prob.* **13**, 898–922.

Spizzichino, F. (1988). Symmetry conditions on opinion assessment leading to time-transformed exponential model. *Prove di Durata Accelerate e Opinioni degli Esperti in Affidabilità*. (C. A. Clarotti and D. V. Lindley eds.), Amsterdam: North-Holland, 83–87.

Spizzichino, F. (1990). A unifying model for the optimal design of life-testing and burn-in. *Reliability and Decision Making*, (R. E. Barlow, C. A. Clarotti, F. Spizzichino eds.), Amsterdam: Elsevier, (to appear).

BAYESIAN STATISTICS 4, pp. 813–820
J. M. Bernardo, J. O. Berger, A. P. Dawid and A. F. M. Smith, (Eds.)

Bayesian Analysis of Generalised Linear Models with Covariate Measurement Error

D. A. STEPHENS and P. DELLAPORTAS
Imperial College, UK and *University of Nottingham, UK*

SUMMARY

Use of generalised linear models when covariates are masked by measurement errors is appropriate in many practical epidemiological problems. However, inference based on such models is by no means straightforward. In previous analyses, simplifying assumptions were made. In this paper, we analyse such models in full generality under a Bayesian formulation. In order to compute the necessary posterior distributions, we utilize various numerical and sampling-based approximate and exact techniques. A specific example involving a logistic regression is considered.

Keywords: GENERALISED LINEAR MODELS; MEASUREMENT ERROR MODELS; NUMERICAL INTEGRATION; SAMPLING-BASED COMPUTATION; LOGISTIC REGRESSION.

1. INTRODUCTION

In the statistical analysis of epidemiological data, it is common to study regression models in which the explanatory variable is subject to measurement error. Possible models for data in these circumstances include extensions of conventional statistical models, such as the logistic regression model for success/failure data, the Poisson model in the analysis of contingency tables, and the proportional hazard model for survival data.

In the statistical literature, many specific types of measurement error regression models have been studied. For example, see Carroll *et al.* (1984), Stefanski and Carroll (1985), Tosteson *et al.* (1989) for analysis of binary data. Recently, however, the broader class of Generalised linear models (GLMs) has often been considered: a recent review has been given by Carroll (1987). A Bayesian approach to inference in this problem is largely absent from the literature, presumably due to the difficulty in evaluation of the various integrals required to compute the relevant posterior quantities. Recent developments in computation techniques, however, allow us to implement such models with remarkable ease.

2. NOTATION

Given a p-vector of covariates $\boldsymbol{X}$, response data $\boldsymbol{Y}$, and a probability density $p(\boldsymbol{Y}|\boldsymbol{X},\boldsymbol{\theta})$, suppose that interest lies in inference about the unknown and unobservable N-vector parameter $\boldsymbol{\theta}$. In addition, suppose that the actual values of the covariate, denoted $\boldsymbol{X}_a$, are not directly observed, but instead we can obtain a vector $\boldsymbol{X}_0$ approximating $\boldsymbol{X}_a$ with some error distribution, where $\boldsymbol{Y}$ is assumed to be conditionally independent of $\boldsymbol{X}_0$ given $\boldsymbol{X}_a$. Such models are termed *measurement error* or *error-in-variables* models. It is also common to distinguish between measurement error models and *Berkson* models, depending on whether $\boldsymbol{X}_0$ is modelled as a function of $\boldsymbol{X}_a$ or vice versa. For example, denoting the covariate error term ε_X, in the measurement error case, we might have for a single covariate value that

$$X_0 = X_a + \varepsilon_X,$$

with X_a and ε_X independent, and X_0 and ε_X correlated, whereas in the Berkson case, we would have

$$X_a = X_0 + \varepsilon_X,$$

with X_0 and ε_X independent, and X_a and ε_X correlated. In a classical statistical framework, analyses, estimation procedures etc. under the two alternative modelling assumptions may be fundamentally different; see, for example, Fuller (1987, chapter 1). In the Bayesian formulation, however, no distinction need be made, except purely for reasons of model specification. In a Bayesian analysis, the (unobservable) $\boldsymbol{X}_a$ are merely regarded as a further set of unknown parameters. In addition, measurement error models are commonly classified as either "structural", where $\boldsymbol{X}_a$ is assumed to be a fixed but random sample from a single unknown distribution, or "functional", where $\boldsymbol{X}_a$ is merely regarded as a vector of fixed unknowns. The difference in the analyses of the two types of model in a Bayesian framework lies solely in the prior used for the $\boldsymbol{X}_a$. In our example below, we shall assume a Berkson-type measurement error model.

In measurement error regression models, the response $\boldsymbol{Y}$ is related to $\boldsymbol{X}_a$ and $\boldsymbol{\theta}$ by the equation

$$\boldsymbol{Y} = \boldsymbol{f}(\boldsymbol{\theta}, \boldsymbol{X}_a) + \varepsilon_Y$$

for some function $\boldsymbol{f}$, where the ε_Y's are the observation error terms. Inferences about the unknown $\boldsymbol{\theta}$ and $\boldsymbol{X}_a$ conditional on the observed response $\boldsymbol{Y} = \boldsymbol{y}$ and observed covariate values $\boldsymbol{X}_0$ can be made via the posterior density given in the usual way by

$$p(\theta, \boldsymbol{X}_a|\boldsymbol{y}, \boldsymbol{X}_0) = \frac{l(\boldsymbol{y}|\boldsymbol{\theta}, \boldsymbol{X}_a, \boldsymbol{X}_0)p(\boldsymbol{\theta}, \boldsymbol{X}_a|\boldsymbol{X}_0)}{\int l(\boldsymbol{y}|\boldsymbol{\theta}, \boldsymbol{X}_a, \boldsymbol{X}_0)p(\boldsymbol{\theta}, \boldsymbol{X}_a|\boldsymbol{X}_0)d\boldsymbol{\theta}d\boldsymbol{X}_a}. \tag{1}$$

Note that, under our model specification, we may write

$$l(\boldsymbol{y}|\boldsymbol{\theta}, \boldsymbol{X}_a, \boldsymbol{X}_0) = l(\boldsymbol{y}|\boldsymbol{\theta}, \boldsymbol{X}_a).$$

It is evident from (1) that the Bayesian analysis proceeds in a similar fashion whether or not we assume a measurement error model or a Berkson model. In the measurement error case, $\boldsymbol{X}_0$ is truly observed, and thus we must have that in (1)

$$p(\boldsymbol{\theta}, \boldsymbol{X}_a|\boldsymbol{X}_0) \propto p(\boldsymbol{X}_0|\boldsymbol{X}_a, \boldsymbol{\theta})p(\boldsymbol{X}_a, \boldsymbol{\theta}).$$

In the Berkson case, $\boldsymbol{X}_0$ is presumed fixed by the experimenter and acts merely as a hyper-parameter in the "prior" for $\boldsymbol{X}_a$, and thus we have that $p(\boldsymbol{\theta}, \boldsymbol{X}_a|\boldsymbol{X}_0)$ can be immediately specified.

After specifying a suitable prior density, the evaluation of the density in (1) requires a numerical integration operation over a $p + n$ dimensional Euclidean space. We propose to use numerical approximation techniques to compute the required posterior quantities.

3. IMPLEMENTATION

We propose to solve the computation problem either directly, using the numerical integration strategy introduced by Naylor and Smith (1982), or indirectly, via sampling-based computation and the Gibbs sampler algorithm described by Gelfand and Smith (1990). The former will normally be adequate for a low dimensional problem, say $p+n < 6$ (and thus potentially suitable for the "structural" measurement error model problems mentioned above), but will require assumptions of approximate posterior normality, and a certain amount of expertise on the part of the user. The latter will arguably not be able to compete in terms of efficiency for low dimensional problems, but would actually be able to cope with any number of parameters (a necessary requirement for prospective use in "functional" measurement error modelling), and, theoretically, any form of realistic posterior surface. It should be noted that there are many other algorithms, based on, for example, analytic approximation and importance sampling techniques, designed for implementing Bayesian statistics; see Goel (1988) for a comprehensive list.

3.1. *Numerical Integration*

Bayesian numerical integration techniques rely on mixtures of Cartesian product, spherical, and importance sampling quadrature; see Naylor and Smith (1988), Smith *et al.* (1985). The general approach is based on an iterative scheme which exploits the approximate posterior normality of the required integrands, and gradually places integration points (nodes) over the posterior surface. Convergence is assessed via comparisons of consecutive estimates of posterior moments. In general, the class of regression models which we study here present smooth posterior surfaces, and the above strategy would be expected to perform adequately. We note, however, that it will rarely give the robustness and flexibility that the sampling-based algorithm offers.

3.2. *Sampling-Based Computation*

The mechanism by which the Gibbs sampler operates involves iterative sampling from the various full conditional (posterior) distributions of the unknown parameters; see Gelfand and Smith (1990), Gelfand *et al.* (1990) for full details of the algorithm. In addition, in Dellaportas and Smith (1992), specific numerical routines are used in GLM problems where the full conditional posterior densities are log-concave; see also Gilks and Wild (1992). The epidemiological models described in Section 1 are examples of the regression models studied by Dellaportas and Smith. We show below that the corresponding measurement error versions can be analysed using precisely the same tools under certain assumptions related to the error distribution of the explanatory variable, by exploiting the same property of log-concavity of the conditional posterior densities.

We shall not discuss here the important, but more routine, aspects of the Gibbs sampler, such as the merits or otherwise of using multiple independent runs, the assessment of convergence, parameterization and dependence etc.

4. EXAMPLE – AIR POLLUTION DATA

The data in Table 1 are given in Whittemore and Keller (1988), and relate to a study of the effect of exposure to nitrogen dioxide (NO_2) on reported respiratory illness. The response variable Y is binary, reflecting presence/absence of respiratory illness, and the explanatory variable $\boldsymbol{X}$ has three levels. The model assumed by Whittemore and Keller is a logistic regression with measurement error on the explanatory variable. The nature of this measurement error relationship is known precisely via a calibration study. For level $j = 1, 2, 3$, we have that

$$X_{aj} = \alpha + \beta X_{0j} + \varepsilon_{Xj}, \tag{2}$$

where $\alpha = 4.48, \beta = 0.76$, and ε_{Xj} is a random element having a normal distribution with zero mean and variance $\sigma_X^2 = 81.14$, with $\boldsymbol{X}_0 = (X_{01}, X_{02}, X_{03}) = (10, 30, 50)$. This is clearly a Berkson-type model. Here, X_{0j} and X_{aj} are interpreted as being some from of representative values for group j (X_{0j} as representing the "value" of the ordinal classification variable, X_{aj} as a "true average value" of that variable). We discuss the appropriateness of this interpretation in section six.

Respiratory illness	Bedroom NO_2 level (ppb) < 20	$20-39$	$40+$	Total
Yes ($Y=1$)	21	20	15	56
No ($Y=0$)	27	14	6	47
Total	48	34	21	103

Table 1.

Under the binomial GLM with logit link function and linear logistic predictor model used by Whittemore and Keller, the likelihood function $l(\boldsymbol{y}|\theta, \boldsymbol{X}_a)$ is given by

$$l(\boldsymbol{y}|\theta, \boldsymbol{X}_a) \propto \exp\left\{\sum_{j=1}^{3}\Big[y_{1j}\eta_j - (y_{1j}+y_{2j})\log(1+\exp(\eta_j))\Big]\right\} \tag{3}$$

where $\eta_j = \log(p_j(1-p_j)) = \theta_1 + \theta_2 X_{aj}$, and p_j and X_{aj} are the probability of presence of respiratory illness in an individual, and actual NO_2 exposure at level j respectively, for $j = 1, 2, 3$.

In our analysis, we shall assume that the prior density $p(\boldsymbol{\theta}, \boldsymbol{X}_a|\boldsymbol{X}_0)$ may be specified directly via (2) and the independence assumptions to be multivariate normal with expectation $\boldsymbol{\alpha} + \beta\boldsymbol{X}_0$ and covariance matrix $\sigma_X^2 \boldsymbol{I}_3$. Furthermore, we shall assume that $p(\boldsymbol{\theta})$ is bivariate normal with mean $\boldsymbol{\mu}$ and covariance matrix Σ, and thus from (2), we have that

$$p(\boldsymbol{\theta}, \boldsymbol{X}_a|\boldsymbol{X}_0) \propto \exp \left\{-\frac{1}{2}(\boldsymbol{\theta}-\boldsymbol{\mu})^T\Sigma^{-1}(\boldsymbol{\theta}-\boldsymbol{\mu}) - \frac{1}{2\sigma_X^2}(\boldsymbol{X}_a - \boldsymbol{\alpha} - \beta\boldsymbol{X}_0)^T(\boldsymbol{X}_a - \boldsymbol{\alpha} - \beta\boldsymbol{X}_0)\right\}. \tag{5}$$

Combining (3) and (5) through (1), we have an expression for the joint posterior density of $\boldsymbol{\theta}$ and $\boldsymbol{X}_a$. Clearly, however the integration operations necessary (to perform normalization and marginalization) are not analytically tractable, and thus we are forced to use computational techniques discussed in section three.

4.1. *Implementation of Numerical Integration*

To implement the numerical integration decribed in section (3.1), we utilize the BAYES FOUR software; see Naylor and Shaw (1991). Omitting technical details, we may use the results of Whittemore and Keller as initial values, and possibly after some exploratory iterations with spherical rules, we can obtain the desired posterior quantities with hopefully not more than 10^5 Cartesian grid points. The posterior marginal densities can be estimated pointwise over a set of nodes and reconstructed via spline interpolation.

4.2. *Implementation of the Gibbs Sampler*

To implement the Gibbs sampler in this problem, we require the functional forms (only up to proportionality) of the full conditional posterior densities of the $p+n$ parameters of $\boldsymbol{\theta}, \boldsymbol{X}_a$). From (3) and (5) it can be shown that the univariate full conditionals are given by

$$\begin{aligned} & p(\theta_1|\theta_2, \boldsymbol{X}_a, \boldsymbol{y}, \boldsymbol{X}_0) \propto \\ & \exp\left\{\sum_{j=1}^{3}\Big[y_{ij}\theta_1 - (y_{1j}+y_{2j}\log(1-\exp(\theta_1+\theta_2X_{aj}))\Big] - \frac{\tau_{11}}{2}(\theta_1-\mu_1')^2\right\} \\ & p(\theta_2|\theta_1, \boldsymbol{X}_a, \boldsymbol{y}, \boldsymbol{X}_0) \propto \\ & \exp\left\{\sum_{j=1}^{3}\Big[y_{ij}X_{aj}\theta_2 - (y_{1j}+y_{2j}\log(1-\exp(\theta_1+\theta_2X_{aj}))\Big] - \frac{\tau_{22}}{2}(\theta_2-\mu_2')^2\right\} \\ & p(X_{aj}|\theta_1,\theta_2, \boldsymbol{y}, \boldsymbol{X}_0) \propto \\ & \exp\left\{y_{ij}X_{aj}\theta_2 - (y_{1j}+y_{2j}\log(1-\exp(\theta_1+\theta_2X_{aj})) - \frac{1}{2\sigma_X^2}(X_{aj}-\alpha-\beta X_{0j})^2\right\} \end{aligned} \tag{6}$$

for $j = 1, 2, 3$, and where, if $[\Sigma^{-1}]_{ij} = \tau_{ij}$ for $i, j = 1, 2$,

$$\mu_i' = \mu_i - \frac{\tau_{ij}}{\tau_{ii}}(\theta_j - \mu_j)$$

for $i = 1, 2$ and $j \neq i$.

There are several techniques that allow us to sample from the densities in (6), even though they are not of a standard form. In particular, Dellaportas and Smith (1992) indicate that the adaptative rejection-sampling technique of Gilks and Wild (1992) can out-perform other maximization-based techniques in terms of function evaluations by "about 50%". The Gilks-Wild algorithm is only applicable to problems where the densities concerned are log-concave, that is, in the univariate case, where

$$\frac{\partial^2}{\partial\theta^2}\log(p(\theta)) < 0 \tag{7}$$

on the support of density $p(\theta)$ of parameter θ. It is easy to verify that the strict inequality in (7) is satisfied by each of the conditional densities in (6), and thus we propose to use the Gilks-Wild algorithm to implement the Gibbs sampler in this problem. In fact, provided that the error-term ε_Y is distributed with a log-concave density (see Devroye (1987) for a list of such densities), all common regression models will have log-concave full conditional posterior densities; see Dellaportas and Smith (1992).

5. RESULTS

We present the results in the form of marginal posterior density estimates for θ_1, θ_2, and the X_{aj}s, as depicted in figures 1 to 3.

Figures 1 and 2 depict the marginal posterior estimates for θ_1 and θ_2 under a vague prior specification for $\boldsymbol{\theta}(\Sigma^{-1} = 0$ in (5)) respectively. In each case, the bold curve represents the posterior density under a conventional GLM, computed via Gibbs sampling; the resulting sample moments of a sample of size 2000 for each parameter from the Gibbs sampler-based analysis correspond closely to those given by Whittemore and Keller. The dotted and solid

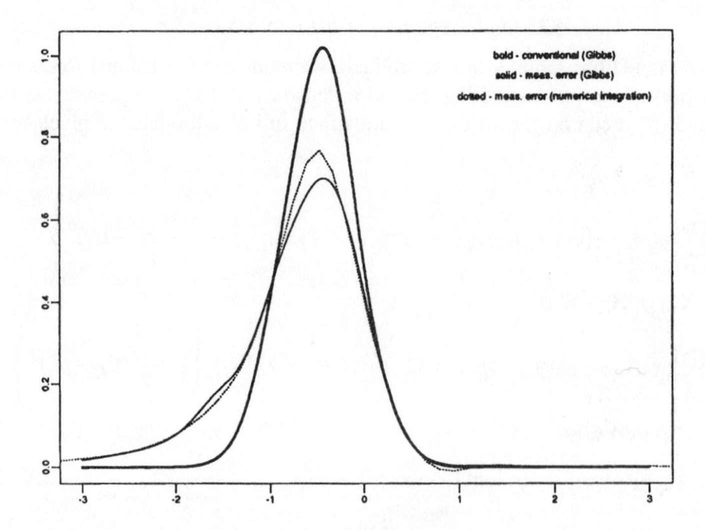

Figure 1. *Posterior density estimates for* θ_1.

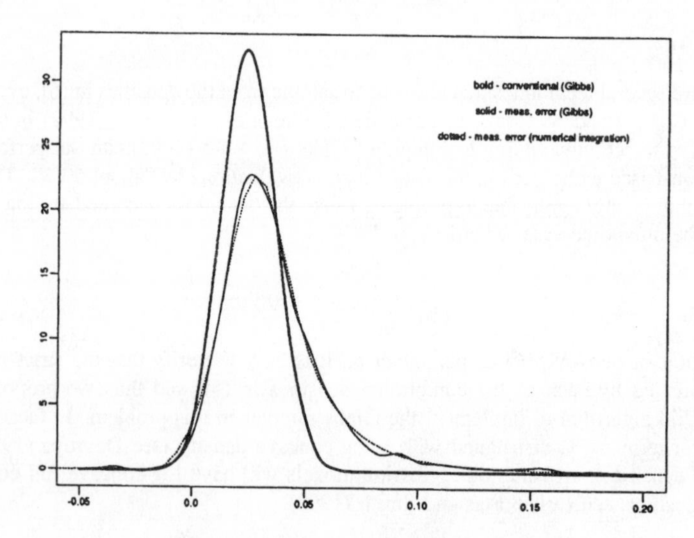

Figure 2. *Posterior density estimates for* θ_2.

curves represent the posterior density estimates computed via numerical integration (constructed using interpolation across a twenty point grid) and the Gibbs sampler (constructed using normal kernel density estimation) respectively, under the measurement error formulation. Clearly, these density estimates resemble each other reasonably closely, but have a larger variance and are more skewed than the posterior under the conventional model. Fur-

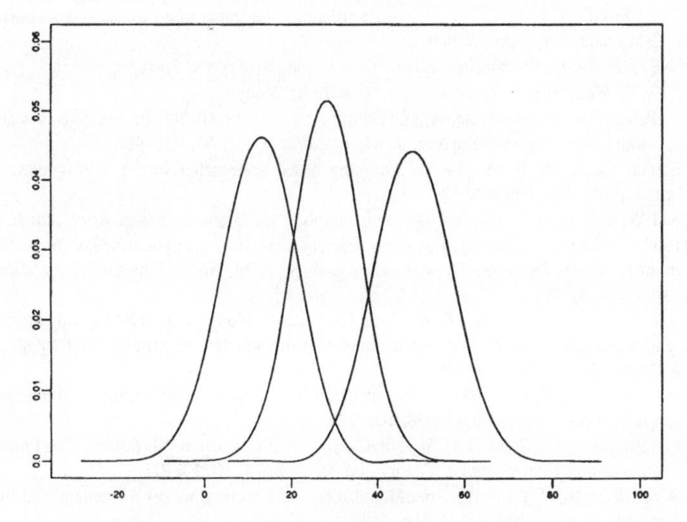

Figure 3. *Posterior densities for* X_{a1}, X_{a2}, X_{a3}.

thermore, in each case, the posterior standard deviation estimates are considerably larger than the standard errors given by the (maximum-likelihood) analysis of Whittemore and Keller. Figure 3 depicts the posterior density estimates for the elements of $\boldsymbol{X}_a$. Encouragingly, in each case, the posterior mode occurs at or near the "prior mode" target value. Note also that, although there is considerable overlap between the densities, the two "cut-off" values between adjacent densities (the ordinate values where the adjacent density functions are equal) occur close to the original data classification end-points of 20 and 40 as given by Whittemore and Keller.

6. DISCUSSION

The specific example studied was the air-pollution versus NO_2 exposure data of Whittemore and Keller (1988). We have assumed a binomial model for the binary response data, used a logit link, and a predictor for the transform of the canonical parameters which is linear in the covariate, which itself is subject to measurement error. This, however, is not completely satisfactory, as certain aspects of such a model are arguably inappropriate for these data. For example, the assumption of equal exposure within each group implicit in equations (2) and (3) is unrealistic, and should ideally be replaced by an assumption encapsulating some notion that each individual actually suffers a different level of exposure. The implications of such an assumption take the analysis outside of the range of conventional generalised linear regression models. Thus we are forced to retain our original interpretation of the X_as as being "true average" exposure level for each group.

REFERENCES

Carroll, R. J. (1987). Covariance analysis in generalised linear measurement error models. *Statistics in Medicine* **8**, 1075–1093.

Carroll, R. J., Spiegelman, C. H., Lan K. K., Bailey, K. T. and Abbott, R. D. (1984). On errors-in-variables for binary models. *Biometrika* **71**, 19–25.

Dellaportas, P. and Smith, A. F. M. (1992). Bayesian inference for generalised linear and proportional hazards models via Gibbs sampling. *Appl. Statist.* , (to appear).

Devroye, L. (1987). *Non-uniform Random Variate Generation.* New York: Springer.

Fuller, W. A. (1987). *Measurement Error Models.* New York: Wiley.

Gelfand, A. E., Hills, S. E., Racine-Poon, A. and Smith, A. F. M. (1990). Illustration of Bayesian inference in normal data models using Gibbs sampling. *J. Amer. Statist. Assoc.* **85**, 972–986.

Gelfand, A. E. and Smith, A. F. M. (1990). Sampling based approaches to calculating marginal densities. *J. Amer. Statist. Assoc.* **85**, 398–409.

Gilks, W. R. and Wild, P. (1992). Adaptive rejection sampling for Gibbs sampling. *Appl. Statist.* , (to appear).

Goel, P. K. (1988). Software for Bayesian analysis: current status and additional needs. *Bayesian Statistics 3* (J. M. Bernardo, M. H. DeGroot, D. V. Lindley and A. F. M. Smith, eds.), Oxford: University Press, 173–188, (with discussion).

Naylor, J. C. and Shaw, J. E. H. (1991). *Bayes Four User Guide.* Nottingham: Nottingham Polytechnic, UK.

Naylor, J. C. and Smith, A. F. M. (1983). Aplications of a method for the efficient computation of posterior distributions. *Appl. Statist.* **31**, 214–225.

Naylor, J. C. and Smith, A. F. M. (1988). Econometric illustrations of novel numerical integration strategies for Bayesian inference. *J. Econometrics* **38**, 103–126.

Smith, A. F. M., Skene, A. M., Shaw, J. E. H., Naylor, J. C. and Dransfield, M. (1985). The implementation of the Bayesian paradigm. *Comm. Statist. Theory and Methods* **14**, 1079–1102.

Stefanski, L. A. and Carroll, R. J. (1987). Conditional scores and optimal scores for generalised linear measurement-error models. *Biometrika* **74**, 703–716.

Tosteston, T. D. Stefanski, L. A. and Schafer, D. W. (1989). A measurement-error model for binary and ordinal regression. *Statistics in Medicine* **8**, 1139–1147.

Whittemore, A. S. and Keller, J. B. (1988). Approximations for errors in variables regression. *J. Amer. Statist. Assoc.* **83**, 1957–1966.

BAYESIAN STATISTICS 4, pp. 821–824
J. M. Bernardo, J. O. Berger, A. P. Dawid and A. F. M. Smith, (Eds.)

The James-Stein Estimator as an Empirical Bayes Estimator for an Arbitrary Location Family*

WILLIAM E. STRAWDERMAN
Rutgers University, USA

SUMMARY

A well known intuitive justification for the James-Stein estimator of the mean vector of a multivariate normal distribution is that it is an empirical Bayes estimator. The prior distribution is multivariate normal with the mean vector **0** and covariance matrix equal to $\sigma^2 I$ where σ^2 is unknown and is estimated from the data. Here, we derive a version of the James-Stein estimator of the location vector for an essentially arbitrary location family.

Keywords: EMPIRICAL BAYES; JAMES-STEIN; QUADRATIC LOSS.

1. INTRODUCTION

A well-known intuitive justification for the James-Stein estimator of the mean vector of a multivariate normal distribution is that it is an empirical Bayes estimator. The prior distribution is multivariate normal with mean vector **0** and covariance matrix equal to $\sigma^2 I$ where σ^2 is unknown and is estimated from the data. See, for example, Brandwein and Strawderman (1990) for details.

It is the purpose of this paper to show that a version of the James-Stein estimator may also be viewed as an empirical Bayes estimator of the location vector for an essentially arbitrary location family. It is already known that such estimators dominate the best invariant estimator for a wide variety of location models and loss functions (see Brandwein and Strawderman (1990) for some details and references). That such estimators are also, nearly universally, empirical Bayes procedures in location models may add to the attractiveness of the James-Stein estimator in practical settings.

2. A DERIVATION OF THE JAMES-STEIN ESTIMATOR AS AN EMPIRICAL BAYES ESTIMATOR

Let X be an observation from a location family

$$f(X - \theta), \tag{2.1}$$

in p dimensions. It is desired to estimate θ with loss equal to quadratic loss

$$L(\theta, \delta) = (\delta - \theta)' Q (\delta - \theta) \tag{2.2}$$

when Q is positive definite. For an arbitrary prior distribution $\pi(\theta)$, the Bayes estimator is $\delta_\pi(X) = E(\theta|X)$.

* Research supported by NSF Grant DMS-88-22622.

We take as a family of prior distributions for θ the family of conjugate priors of the form

$$\pi_n(\theta) = f^{*^n}(\theta) \tag{2.3}$$

where $f^{*^n}(\theta)$ denotes the n-fold convolution of $f(\theta)$ with itself, or alternatively θ is distributed as the sum

$$\sum_{i=1}^{n} Y_i$$

where the Y_i are iid random p-vectors with density $f(\cdot)$

It follows from Cohen, Rukhin and Strawderman (1987), that the Bayes estimator of θ with respect to $\pi_n(\theta)$ (for fixed n) is

$$\delta_n(X) = \frac{n}{n+1}X = (1 - \frac{1}{n+1})X. \tag{2.4}$$

By virtue of being unique, these Bayes estimators are admissible for each $n \geq 1$.

Suppose next that we know that the prior distribution is of the form $\pi_n(\theta) = f^{*^n}(\theta)$ but that n is unknown and must be estimated from the data – the classical Parametric Empirical Bayes setup.

Here is one way to estimate n that leads to a version of the James-Stein estimator.

Note that the marginal distribution of $X = (X - \theta) + \theta = \sum_{i=0}^{n} Y_i \sim f^{*n+1}(X)$ is that of an $(n+1)$–fold convolution of $f(\cdot)$ with itself (again note that $X - \theta = Y_0 \sim f(\cdot)$ independently of $\theta = \sum_{i=1}^{n} Y_i$). Hence, marginally,

$$\begin{aligned} EX'X &= E\left(\sum_{i=0}^{n} Y_i\right)'\left(\sum_{i=0}^{n} Y_i\right) \\ &= (n+1)E(Y_0'Y_0) \\ &= (n+1)\mathrm{tr}(\Sigma) \end{aligned}$$

where Σ, the covariance matrix of $X|\theta$, is known and tr stands for trace. Hence $X'X(\mathrm{tr}\Sigma)^{-1}$ is an unbiased estimator of $n+1$ and $\mathrm{tr}\,\Sigma(X'X)^{-1}$ is a "reasonable" estimator of $(n+1)^{-1}$

Therefore substituting $\mathrm{tr}\Sigma(X'X)^{-1}$ in (2.4)

$$\delta(X) = \left(1 - \frac{\mathrm{tr}\Sigma}{X'X}\right)X \tag{2.5}$$

may be viewed as an Empirical Bayes estimator of θ .

Note that if $\Sigma = \sigma^2 I_p$, $\mathrm{tr}\,\Sigma = p\sigma^2$ and (2.5) becomes

$$\delta_{JS} = \left(1 - \frac{p\sigma^2}{X'X}\right)X \tag{2.6}$$

which is essentially the usual James-Stein estimator with $(p-2)$ replaced by p.

3. OTHER EMPIRICAL BAYES ESTIMATORS

The estimators (2.4) and (2.6) are derived by substituting an estimate for the quantity $(n+1)^{-1}$ in (2.4). Any reasonable estimator of $(n+1)^{-1}$ might be considered and we suggest some alternatives in this section. Consider (marginally) unbiased estimators of $n+1$ of the form

$$cX'AX = c\left(\sum_{i=0}^{n} Y_i\right)' A \left(\sum_{i=0}^{n} Y_i\right).$$

By independence of $Y_0, \ldots, Y_n$, $EcX'AX = c(n+1)EY_0'AY_0 = c(n+1)\text{tr}A\Sigma$ implying $c = \text{tr}(A\Sigma)^{-1}$.

Hence estimators of the form

$$\delta(X) = \left(1 - \frac{\text{tr}A\Sigma}{X'AX}\right) X \tag{3.1}$$

may be considered Empirical Bayes estimators for the same class of priors as in section 2. One particularly attractive version results from the choice of $A = \Sigma^{-1}$ giving

$$\delta(X) = \left(1 - \frac{p}{X'\Sigma^{-1}X}\right) X \tag{3.2}$$

which (except for p in the place of $(p-2)$) corresponds to the James-Stein estimator for the Normal (θ, Σ) distribution.

The James-Stein estimator in the Normal case uses an unbiased estimator $(p-2)(X'X)^{-1}$ of $(n+1)^{-1}$. Hence a reasonable modification of our procedures might be to search for an unbiased estimator of $(n+1)^{-1}$ of the form $c(X'AX)^{-1}$. It seems difficult (if not impossible) to do this for general $f(\cdot)$.

However, at least for large values of n, $X = \sum_{i=0}^{n} Y_i$ is approximately normally distributed. This suggests that the value $p-2$ instead of p may correspond to more nearly unbiased estimators of $(n+1)^{-1}$ for the estimator (3.2).

Alternatively, maximum likelihood or minimaxity approaches may lead to reasonable procedures. We will not pursue these here as our main point is to demonstrate the rather surprising universality of the James-Stein procedure as an empirical Bayes estimator in location problems.

It does seem reasonable however to proceed one step further and study related hierarchical Bayes estimators. We do so briefly in the next section.

4. A HIERARCHICAL BAYES ESTIMATOR OF θ

Suppose the first stage prior is $f^{*n}(\theta)$ conditional on $N = n$ and that the (second stage) distribution of N is given by $p[N = n] = p_n$. Then the Bayes estimator is given by

$$\begin{aligned}\delta(X) &= E[\theta|X] = E[E(\theta|X, N)|X] \\ &= \left(\left(1 - \frac{1}{N+1}\right) X|X\right) \\ &= \left(1 - E\left(\frac{1}{N+1}|X\right)\right) X.\end{aligned} \tag{4.1}$$

The hierarchical Bayes estimator then estimates $(n+1)^{-1}$ via a Bayes estimate relative to quadratic loss instead of the method of moment estimators of the previous sections.

To get an explicit formula for the estimator (4.1) note that the distribution of $X|N = n$ is $f^{*(n+1)}(X)$. Hence the conditional distribution of $N|X$ is

$$p[N = n|X] = \frac{p_n f^{*(n+1)}(X)}{\sum_{i=1}^{\infty} p_n f^{*(n+1)}(X)}.$$

Therefore

$$E\left(\frac{1}{N+1}|X\right) = \sum_{n=1}^{\infty}\left(\frac{1}{n+1}\right) p_n f^{*(n+1)}(X) / \sum_{n=1}^{\infty} p_n f^{*(n+1)}(X). \tag{4.2}$$

Substitution of (4.2) into (4.1) gives the desired representation. The relative intractability of $f^{*n}(X)$ for distributions other than the multivariate normal suggests that (4.1) may be a difficult estimator to compute explicitly. Gibbs Sampling methods may be of aid in the calculation and evaluation of such procedures.

REFERENCES

Brandwein, A. C. and Strawderman, W. E. (1990). Stein estimation: the spherically symmetric case. *Statist. Sci.* **5**, 356–369.

Cohen, A., Rukhin, A. and Strawderman, W. E. (1987). A characterization of the multivariate normal distribution and some remarks on linear estimators. *J. Statist. Planning and Inference* **17**, 361–365.

BAYESIAN STATISTICS 4, pp. 825–835
J. M. Bernardo, J. O. Berger, A. P. Dawid and A. F. M. Smith, (Eds.)

On Asymptotic Posterior Normality in the Multiparameter Case

TREVOR J. SWEETING
University of Surrey, UK

SUMMARY

Current approaches to demonstrating asymptotic posterior normality in the multiparameter case fail to cover certain cases where the data arise from a stochastic process. In this paper we present a set of hypotheses for asymptotic posterior normality which are satisfied in such cases. Continuity-type conditions on observed information are imposed on suitable shrinking neighbourhoods and workable conditions relating to the tail behaviour of the posterior distribution are given. The theory is illustrated with an example of a two-parameter nonhomogeneous Poisson process.

Keywords: ASYMPTOTIC POSTERIOR NORMALITY; INFERENCE FOR STOCHASTIC PROCESSES; NONHOMOGENEOUS POISSON PROCESS.

1. INTRODUCTION

We are concerned in this paper with asymptotic posterior normality in the multiparameter case, with particular reference to data arising from stochastic processes. We present a technical improvement of the standard conditions for Laplace integration which extends the asymptotic results to certain examples not covered by the existing literature. A number of papers have appeared over the last decade or so which extend Walker's (1969) proof of asymptotic posterior normality in the i.i.d. case to cover general stochastic processes; we mention Heyde and Johnstone (1979), Basawa and Prakasa Rao (1980), Heyde (1982), Chen (1985), Sweeting and Adekola (1987) and Crowder (1988). In these papers a continuity-type condition is imposed on the observed information which involves either a *fixed* or a *shrinking* neighbourhood of the maximum likelihood (ML) estimate, $\hat{\theta}$, or the true underlying parameter value, θ_0. The simplest formulation involves fixed neighbourhoods and the results apply to a wide class of problems (see, for example, Heyde and Johnstone (1979)). It was shown in Sweeting and Adekola (1987), however, that it is necessary to use shrinking neighbourhoods for certain stochastic process applications; see also Dawid (1970) for problems when $\hat{\theta}$ is close to the boundary of the posterior support of θ. Chen (1985) introduces shrinking neighbourhoods, but the rate of shrinkage is not specified.

Although there is no major problem in adapting the fixed neighbourhood-type conditions to the multiparameter case, the formulation of shrinking neighbourhood conditions requires some care. In Adekola (1987) a multiparameter result is given which involves neighbourhoods of the form $|\theta - \hat{\theta}| \leq \delta_n$, but this result does not cover certain cases when the information on different parameters tends to infinity at different rates.

We begin by establishing the necessary notation. Let $(\Omega_t, \mathcal{A}_t)$ be a family of measurable spaces, where $t \in \mathcal{T}$ is a discrete or continuous time parameter. Let P_θ^t be probability measures defined on $(\Omega_t, \mathcal{A}_t)$, where the parameter $\theta \in \Theta$, an open subset of $\boldsymbol{R}^p$. Let x_t be the associated observation vector (that is, the data up to time t). Assume that, for each $t \in \mathcal{T}$ and $\theta \in \Theta$, P_θ^t is absolutely continuous with respect to a σ-finite measure μ_t and let

$p_t(x_t|\theta)$ be the associated density of P^t_θ. The log-likelihood function $l_t(\theta) = \log p_t(x_t|\theta)$ is assumed to exist a.e. (μ_t). Let $\mathcal{U}_t(\theta) = l'_t(\theta)$ be the vector of first-order partial derivatives of $l_t(\theta)$ and denote by $l''_t(\theta)$ the matrix of second-order derivatives, whenever they exist. Define $J_t(\theta) = -l''_t(\theta)$, the observed information at θ. Let M_p be the space of all real $p \times p$ matrices and M_p^+ be the space of all $p \times p$ positive definite matrices $A > 0$. Let $\lambda_{\max}(A)$, $\lambda_{\min}(A)$ denote the maximum and minimum eigenvalues of the symmetric matrix $A \in M_p$. The spectral norm $\|\cdot\|$ in M_p is $\|A\|^2 = \sup(|Ax|^2 : |x|^2 = 1) = \lambda_{\max}(A^T A)$. Finally $A^{1/2}$ will denote the left Cholesky square root (CSR) in M_p^+. We use properties of the CSR without comment; these properties, and the advantages of using the CSR in multiparameter asymptotic problems are fully discussed in Fahrmeir (1988).

Throughout the paper, we assume that the prior distribution on Θ satisfies the following condition.

C1. *(Prior distribution). The prior distribution of θ is absolutely continuous with respect to Lebesgue measure, with prior density $\pi(\theta)$ continuous and positive throughout Θ and zero on Θ^c.*

Under condition C1, from Bayes theorem θ has an absolutely continuous posterior distribution on Θ with density

$$\pi_t(\theta|x_t) = \pi(\theta)p_t(x_t|\theta)/p_t(x_t) \tag{1}$$

where $p_t(x_t) = \int_\Theta p_t(x_t|\theta)\pi(\theta)d\theta$. Suppose that, corresponding to the observed sequence (x_t), there exists a sequence $(\hat\theta_t)$ of local maxima of the likelihood function. Write $J_t = J_t(\hat\theta_t)$ and let $Z_t = J_t^{T/2}(\theta - \hat\theta_t)$. We wish to formulate conditions which will ensure that, with a suitable mode of convergence,

$$f_t(z|x_t) \to (2\pi)^{-p/2}e^{-|z|^2/2} \tag{2}$$

as $t \to \infty$ where

$$f_t(z|x_t) = |J_t|^{-1/2}\pi_t\left(\hat\theta_t + J_t^{-T/2}z|x_t\right) \tag{3}$$

is the posterior density of Z_t. The convergence (2) implies of course the convergence of the posterior distribution of Z_t to the standard p-dimensional normal distribution. In Section 2 we prove a basic result for the multiparameter case and investigate nonlocal behaviour of the posterior distribution. The main results are given in Section 3; the aim here has been to set the conditions at a level of generality most appropriate for practical use. The theory is illustrated in Section 4 with an example of a nonhomogeneous Poisson process.

2. PRELIMINARY RESULTS

We begin by giving a set of basic conditions which define 'well-behaved' sequences (x_t) for which (2) will hold. Lemma 2.1 below resembles Theorem 2.1 in Chen (1985), although our conditions and choice of neighbourhood are different. These have advantages when we consider practical criteria in the next section. In addition, conditions on the likelihood and prior are stated separately.

For $\phi \in \Theta, c > 0$ define the neighbourhoods $N_t(\phi, c) = \left\{\theta : \left|\{J_t(\phi)\}^{T/2}(\theta - \phi)\right| < c\right\}$ and write

$$\Delta_t(\phi, c) = \sup_{\theta \in N_t(\phi,c)} \|\{J_t(\phi)\}^{-1/2}J_t(\theta) - J_t(\phi))\{J_t(\phi)\}^{-T/2}\|.$$

We consider sequences (x_t) which satisfy the following conditions.

C2. *(Smoothness). The log-likelihood function $l_t(\theta)$ is twice differentiable with respect to θ throughout Θ.*

C3. *(Compactness). For each $t \in \mathcal{T}$ there exists a local maximum $\hat{\theta}_t$ of $l_t(\theta)$, and the set of limit points of $(\hat{\theta}_t)$ is compact.*

C4. *(Information growth). $J_t^{-1} \to 0$.*

C5. *(Information continuity). There exists a sequence (c_t) with $c_t \to \infty$ such that $\Delta_t\,(\hat{\theta}_t, c_t) \to 0$.*

C6. *(Nonlocal behaviour).*

$$\{p_t(x_t|\hat{\theta}_t)\}|J_t|^{1/2} \int_{\theta \in N_t^c(\hat{\theta}_t, c_t)} p_t(x_t|\theta)\pi(\theta)d\theta \to 0$$

where (c_t) is the sequence in condition C5.

Lemma 2.1. *Assume conditions C1 – C6 are satisfied by the sequence (x_t). Then (2) holds.*

Proof. Write $\theta_t = \hat{\theta}_t + J_t^{-T/2}z$. From (1) and (3) we have

$$f_t(z|x_t) = \left\{\pi(\theta_t)/\pi(\hat{\theta}_t)\right\}\left\{p_t(x_t|\theta_t)/p_t(x_t|\hat{\theta}_t)\right\}\left\{|J_t|^{-1/2}\pi(\hat{\theta}_t)p_t(x_t|\hat{\theta}_t)/p_t(x_t)\right\}$$

Conditions C1 and C3 imply that the first factor in curly braces tends to unity. For the second factor, Taylor expansion gives $l_t(\theta_t) - l_t(\hat{\theta}_t) = -(1/2)z^T[I + R_t(z)]z$ where $R_t(z) = J_t^{-1/2}(J_t(\theta_t^*) - J_t)J_t^{-T/2}$ and $\hat{\theta}_t^*$ lies on the line segment joining θ and $\hat{\theta}_t$. Thus $\theta_t^* \in N_t(\hat{\theta}_t, |z|)$ and hence $R_t(z) \to 0$ from C5. The second factor therefore tends to $e^{-|z|^2/2}$. The reciprocal of the final factor is

$$\{\pi(\hat{\theta}_t)p_t(x_t|\hat{\theta}_t)\}^{-1}|J_t|^{1/2} \int p_t(x_t|\theta)\pi(\theta)d\theta$$

Let M_1, M_2 denote the corresponding expressions with the integrals over $N_t(\hat{\theta}_t, c_t)$, $N_t^c(\hat{\theta}_t, c_t)$ respectively. It follows from C1, C3 and C6 that $M_2 \to 0$. Also

$$M_1 = \int_{|z| \le c_t} \left\{\pi(\hat{\theta}_t + J_t^{-T/2}z)/\pi(\hat{\theta}_t)\right\} \exp\left[-\frac{1}{2}z^T(I + R_t(z))z\right] dz \to (2\pi)^{p/2}$$

by Dominated Convergence, since for $|z| \le c_t$, $||R_t(z)|| \le \Delta_t(\hat{\theta}_t, c_t) \to 0$. ◁

We now obtain a more convenient form of the nonlocal condition C6, which can be difficult to check in its present form. The following condition is similar to, but more general than, hypotheses A5 and A6 given by Sweeting and Adekola (1987) for the single parameter case.

C6*. *For each $t \in \mathcal{T}$ there exists an open convex set C_t containing $\hat{\theta}_t$ which satisfies*
(i) $J_t(\theta) > 0$ *on* C_t
(ii) $\pi(\theta)$ *bounded on* C_t
(iii) $\{p_t(x_t|\hat{\theta}_t)\}^{-1}|J_t|^{1/2} \int_{\theta \notin C_t} p_t(x_t|\theta)\pi(\theta)d\theta \to 0$

Lemma 2.2. *Assume conditions C1 – C5. Then conditions C6 and C6* are equivalent.*

Proof. Write $N_t = N_t(\hat{\theta}_t, c_t)$. First note that condition C5 implies that $J_t(\theta) > 0$ eventually on N_t. Hence condition C6 implies C6* with the choice $C_t = N_t$. For the converse, in view of C6*(iii) it suffices to show that

$$H_t \equiv \{p_t(x_t|\hat{\theta}_t)\}^{-1}|J_1|^{1/2} \int_{\theta \in E_t} p_t(x_t|\theta)\pi(\theta)d\theta \to 0 \tag{4}$$

where $E_t = C_t - N_t$. Let $\theta \in E_t$ and consider the function $g_t(h) = l_t(\{1-h\}\hat{\theta}_t + h\theta), 0 < h < 1$. Since C_t is convex, $\hat{\theta}_t \in C_t$ and $J_t(\theta) > 0$ on C_t, it follows that g_t is concave, so that if $\phi = (1-h)\hat{\theta}_t + h\theta$ we have $l_t(\phi) - l_t(\hat{\theta}_t) > h[l_t(\theta) - l_t(\hat{\theta}_t)]$. Now take ϕ_t to be the point at which the line segment joining θ and $\hat{\theta}_t$ intersects the boundary ∂N_t of N_t. Then $\phi_t = (1-h_t)\hat{\theta}_t + h_t\theta$ with $h_t = |J_t^{T/2}(\theta - \hat{\theta}_t)|^{-1}c_t$ and it follows that

$$\begin{aligned} l_t(\theta) - l_t(\hat{\theta}_t) &< c_t^{-1}\left[l_t(\phi_t) - l_t(\hat{\theta}_t)\right]\left|J_t^{T/2}(\theta - \hat{\theta}_t)\right| \\ &= -\frac{1}{2}c_t^{-1}(\phi_t - \hat{\theta}_t)^T J_t(\phi_t^*)(\phi_t - \hat{\theta}_t))\left|J_t^{T/2}(\theta - \hat{\theta}_t)\right| \end{aligned}$$

where ϕ_t^* lies on the line segment joining ϕ_t and $\hat{\theta}_t$. But

$$\begin{aligned} (\phi_t - \hat{\theta}_t)^T J_t(\phi_t^*)(\phi_t - \hat{\theta}_t) &\geq c_t^2 \inf_{|x|=1} x^T \left[J_t^{-1/2} J_t(\phi_t^*) J_t^{-T/2}\right] x \\ &\geq c_t^2[1 - \Delta_t(\hat{\theta}_t, c_t)] \geq \frac{1}{2}c_t^2 \end{aligned}$$

for $t > t_0$ sufficiently large, by C5. Thus, if $t > t_0$, $l_t(\theta) - l_t(\hat{\theta}_t) \leq -\frac{1}{4}c_t|J_t^{T/2}(\theta - \hat{\theta}_t)|$ for all $\theta \in E_t$ and so

$$\begin{aligned} H_t &\leq |J_t|^{1/2} \int_{\theta \in R_t} \exp\left(-\frac{1}{4}c_t|J_t^{T/2}(\theta - \hat{\theta}_t)|\right)\pi(\theta)d\theta \\ &\leq c_t^{-p}\left[\sup_{\theta \in C_t} \pi(\theta)\right] \int_{|u|>c_t^2} e^{-|u|/4} du \end{aligned}$$

and (4) follows from C6*(ii). ◁

Thus use of condition C6* effectively reduces the region over which it is necessary to check nonlocal behaviour of the posterior distribution. Note in particular that if $C_t \subset K$ eventually, where K is a compact set in $\boldsymbol{R}^p$, then (ii) automatically holds under C1. The next result is often helpful when attempting to check C6(iii), provided that information on the different parameters does not vary too wildly.

Lemma 2.3. *Assume conditions C1 – C5 and C6*(i), (ii) hold. Suppose further that $C_t \supset C$, where C is a fixed neighbourhood of $\hat{\theta}_t$ and*

$$\sup_{\theta \in R_t - C_t} p_t(x_t|\theta) = \sup_{\theta \in \partial C_t} p_t(x_t|\theta) \tag{5}$$

where R_t is another neighbourhood of $\hat{\theta}_t$. Then if

$$\{\lambda \min(J_t(\hat{\theta}_t))\}^{-1} \log \lambda_{\max}(J_t(\hat{\theta}_t))$$

is bounded, we have

$$Q_t \equiv \{p_t(x_t|\hat{\theta}_t))\}^{-1} |J_t|^{1/2} \int_{\theta \in R_t - C_t} p_t(x_t|\theta)\pi(\theta)d(\theta) \to 0$$

Proof. From (5)

$$Q_t \le |J_t|^{1/2} \exp\left[\sup_{\theta \in \partial C_t} \left\{l_t(\theta) - l_t(\hat{\theta}_t)\right\}\right] \int_{\theta \in R_t - C_t} \pi(\theta)d\theta$$

Now from the proof of Lemma 2.2, if $\theta \in \partial C_t$ then for sufficiently large t we have

$$l_t(\theta) - l_t(\hat{\theta}_t) \le -\frac{1}{4} c_t |J_t^{T/2}(\theta - \hat{\theta}_t)| \le -\frac{1}{4}\{\lambda_{\min}(J_t)\}^{1/2} h$$

where $h = \inf_{\theta \in \partial C_t} |\theta - \hat{\theta}_t| > 0$ since $C_t \supset C$. Therefore

$$Q_t \le \{\lambda_{\max}(J_t)\}^{p/2} \exp\left(-\frac{1}{4} c_t \{\lambda_{\min}(J_t)\}^{1/2} h\right) \to 0$$

under the condition of the lemma. ◁

3. A GENERAL MULTIPARAMETER RESULT

In this section we obtain the main results of the paper. The conditions imposed are stochastic versions of C2 - C5, C6* with explicit reference to $\hat{\theta}_t$ removed, since for many applications it is easier to verify conditions cast in terms of a true underlying parameter value θ_0. All probability statements are to be understood as being with respect to θ_0. Write $J_{t0} = J_t(\theta_0)$ and $\mathcal{U}_{t0} = \mathcal{U}_t(\theta_0)$.

Let B_t be $\mathcal{A}_t$-measurable matrices in M_p^+ and write $W_t = B_t^{-1/2} J_{t0} B_t^{-T/2}$. We choose (B_t) to be any sequence satisfying

S. *(Information stability). (W_t) is stochastically bounded in M_p^+.*

The choice $B_t = J_{t0}$ trivially satisfies S, but very often it is more convenient to replace the J_{t0} by suitable matrices B_t,which may or may not be random. For example, a suitable choice is often $B_t = E\{J_{t0}\}$ when the expectation exists, or some asymptotic equivalent. Let $N_t^*(c) = \left\{\theta : |B_t^{T/2}(\theta - \theta_0)| < c\right\}$ and $\Delta_t^*(c) = \sup_{\theta \in N_t^*(c)} \|B_t^{-1/2}(J_t(\theta) - J_t(\theta_0))B_t^{-T/2}\|$.

D2. *(Smoothness). The log-likelihood function $l_t(\theta)$ is a.e. (μ_t) twice differentiable with respect to θ throughout θ.*

D3. *(Compactness). $(B_t^{-1/2}\mathcal{U}_{t0})$ is stochastically bounded.*

D4. *(Information growth). $B_t^{-1}\xrightarrow{p}0$.*

D5. *(Information continuity). $\Delta_t^*(c)\xrightarrow{p}0$ for every $c > 0$.*

D6. *(Nonlocal behaviour). For each $t \in \mathcal{T}$ there exists a nonrandom open convex set C_t containing θ_0 which satisfies*

(i) $P^t(J_t(\theta) > 0 \text{ on } C_t) \to 1$.

(ii) $\pi(\theta)$ eventually bounded on C_t

(iii) $\{p_t(x_t\,|\,\theta_o)\}^{-1}\,|\,B_t\,|^{1/2}\int_{\theta\notin C_t} p_t(x_t\,|\,\theta)\pi(\theta)d\theta\xrightarrow{p}0$.

Lemma 3.1 below is a version of Lemma A.1 (i) – (iii) in Sweeting and Adekola (1987), but the conditions imposed here are weaker.

Lemma 3.1. *Assume conditions D2 - D6(i) with $B_t = J_{t0}$. Then*

(i) With probability tending to one, there is a unique solution $\hat\theta_t$ of $l_t'(\theta) = 0$ in C_t at which point $l_t(\theta)$ assumes its maximum value over this region.

(ii) $J_{t0}^{T/2}(\hat\theta_t - \theta_0)$ is stochastically bounded.

(iii) $l_t(\hat\theta_t) - l_t(\theta_0)$ is stochastically bounded.

Proof. Write $X_t = J_{t0}^{-T/2}\mathcal{U}_{t0}$. The proof is an adaptation of Lemma 4 in Sweeting (1980). Instead of using the (uniform) asymptotic normality of the score statistic, we use only the stochastic boundedness in condition D3.

(i), (ii). Let $D_t = \partial N_t(\theta_0, c)$. If $(\phi - \theta_0)^T U_t(\phi) < 0$ for all $\phi \in D_t$ then there exists a local maximum $\hat\theta_t$ of $l_t(\theta)$ satisfying $|\,J_{t0}^{T/2}(\hat\theta_t - \theta_0)\,| \le c$. Let $\pi_t = P^t\Big(\sup_{\phi\in D_t}(\phi - \theta_0)^T U_t(\phi) \ge 0\Big)$. We show that $\overline{\lim}_{t\to\infty}\pi_t \to 0$ as $c \to \infty$.

If $\phi \in D_t$ then a.e. (μ_t),

$$(\phi - \theta_0)^T U_t(\phi) = (\phi - \theta_0)^T U_{t0} - (\phi - \theta_0)^T J_t(\theta^*)(\phi - \theta_0)$$

where $|\,J_{t0}^{T/2}(\theta^* - \theta_0)\,| \le c$. Take the sup and inf over the set $\{x \in \boldsymbol{R}^p : |x| = 1\}$. Then $\pi_t \le P^t(\sup x^T X_t \ge c \inf x^T V_t x)$ where $V_t = J_{t0}^{-1/2} J_t(\theta^*) J_{t0}^{-T/2}$. Now $\inf x^T V_t x = \mu_t$, the smallest eigenvalue of V_t. But from D5, $V_t\xrightarrow{p}I$ and so $\mu_t\xrightarrow{p}1$. Therefore $\pi_t \le P^t(|\,X_t\,| \ge \frac{1}{2}c) + P^t(\mu_t \le \frac{1}{2})$ so that

$$\overline{\lim_{t\to\infty}}\,\pi_t \le \overline{\lim_{t\to\infty}}\,P^t\left(|\,X_t\,| \ge \frac{1}{2}c\right) \to 0$$

as $c \to \infty$, from D3. The remaining assertion in (i) follows from D6(i).

(iii) We have $l_t(\hat\theta_t) - l_t(\theta_0) = -\frac{1}{2}(\hat\theta_t - \theta_0)^T J_t(\theta_t^*)(\hat\theta_t - \theta_0)$ and the result follows from (ii) and D5. ◁

Theorem 3.2. *Assume conditions C1, D2 - D6. Then the convergence (2) holds in probability.*

Proof. Initially take $B_t = J_{t0}$. We first show that the given conditions imply versions of D4 - D6 with θ_0 replaced by $\hat{\theta}_t$, which we denote by D4'– D6' . (All probability statements are still with respect to θ_0.) We have

$$G_t \equiv P^t\left(\|J_{t0}^{-1/2}(J_t - J_{t0})J_{t0}^{-T/2}\| > \varepsilon\right) \leq P^t(\Delta_t(\theta_0, c) > \varepsilon) + P^t(|\mathcal{Y}_t| > c)$$

where $\mathcal{Y}_t = J_{t0}^{T/2}(\hat{\theta}_t - \theta_o)$. Thus $\overline{\lim}_{t\to\infty} G_t \leq \overline{\lim}_{t\to\infty} P^t(|\mathcal{Y}_t| > c)$ and hence $G_t \to 0$ from Lemma 3.1(ii). That is,

$$J_{t0}^{-1/2} J_t^{1/2} \xrightarrow{p} I_p \tag{6}$$

It now follows from D4 that $J_t^{-1/2} \xrightarrow{p} 0$, which implies D4'. It also follows immediately from (6) and D5 that

$$\Delta_t'(\theta_0, c) \equiv \sup_{\phi \in N_t(\theta_0, c)} \|J_t^{-1/2}(J_t(\phi) - J_t)J_t^{-T/2}\| \xrightarrow{p} 0 \tag{7}$$

for every $c > 0$. Now choose t_0 sufficiently large so that $\|J_t^{-1/2} J_{t0}^{1/2} - I_p\| \leq 1/2$ for $t > t_0$. Then $|J_{t0}^{T/2}(\phi - \theta_0)| \leq (3/2)|J_t^{T/2}(\phi - \hat{\theta}_t)| + |\mathcal{Y}_t|$ and it follows that $N_t(\hat{\theta}_t, c) \subset N_t(\theta_0, (3/2)c + K)$ whenever $|\mathcal{Y}_t| > K$ and $t > t_0$. Thus for $t > t_0$

$$P^t(\Delta_t(\hat{\theta}_t, c) > \varepsilon) \leq P^t(\Delta_t'(\theta_0, \frac{3}{2}c + K) > \varepsilon) + P^t(|\mathcal{Y}_y| > K)$$

and D5' follows from (7) and Lemma 3.1 (ii). Finally, D6'(iii) follows again from (6) and Lemma 3.1(iii). Also note that, with probability tending to one, $\hat{\theta}_t \in C_t$ from Lemma 3.1(ii).

Let $E_t = \{\hat{\theta}_t \in C_t\} \cap \{J_t(\theta) > 0 \text{ on } C_t\}$ and condition on E_t for the remainder of the proof. Take an arbitrary subsequence and choose a further subsequence on which the convergence properties in D4' – D6' are almost sure. Then with respect to this subsequence conditions C2 - C6* hold almost surely, and hence (2) holds almost surely from Lemmas 2.1 and 2.2. We conclude that the convergence in (2) holds in probability; see, for example, Chow and Teicher (1978). The conditioning on E_t can now be removed, since $P^t(E_t) \to 1$. Finally, it is a relatively straightforward matter to show that if (B_t) satisfies condition S, then conditions D3 – D6 imply the same conditions with $B_t = J_{t0}$. ◁

In practice it is often the case that nonrandom B_t can be chosen so that $W_t \Rightarrow W > 0$ a.s. When the limit variable W is nondegenerate the model is *nonergodic*. Such models frequently occur in inference problems for stochastic processes: see, for example, Basawa and Scott (1983). Actually locally uniform versions of this condition and condition D5 imply the asymptotic normality of $(B_t^{-1/2}U_{t0})$ and hence condition D3 (Sweeting, 1992, Theorem 3.1). Alternatively, suppose that condition S holds and the B_t are nonrandom matrices satisfying

$$B_t^{-1/2} E(J_{t0}) B_t^{-T/2} \to I_p \tag{8}$$

and $E(J_{t0}) = E(U_{t0}U_{t0}^T)$. Then $E|B_t^{-1/2}U_{t0}|^2 \to p$ and condition D3 automatically holds.

The following result is a stochastic version of Lemma 2.3. We omit the proof, which proceeds in a similar way to that of Theorem 3.2; conditional on a suitable set E_t for which $P^t(E_t) \to 1$, every subsequence contains a further subsequence for which the hypotheses of Lemma 2.3 hold.

Lemma 3.3. *Assume conditions C1, S, D2 - D5 and D6(i), (ii). Suppose further that $C_t \supset C$ where C is a fixed neighbourhood of θ_0 and that, with probability tending to one,*

$$\sup_{\theta \in R_t - C_t} p_t(x_t|\theta) = \sup_{\theta \in \partial C_t} p_t(x_t|\theta) \tag{9}$$

where R_t is some nonrandom neighbourhood of θ_0. Then if

$$(\{\lambda \min(B_t)\}^{-1} \log \lambda_{\max}(B_t))$$

is stochastically bounded we have

$$Q_t \equiv \{p_t(x_t|\theta_0)\}^{-1}|B_t|^{1/2} \int_{\theta \in R_t - C_t} p_t(x_t|\theta)\pi(\theta)d\theta \xrightarrow{p} 0$$

4. AN EXAMPLE

We illustrate the theoretical results of Section 3 with a two-parameter nonhomogeneous Poisson process; such processes are used as models in a great variety of situations. Assume that the intensity function is $\lambda e^{\mu+\lambda t}$, where $\lambda > 0, \mu$, are both unknown. Suppose that the process is observed continuously over the time period $(0, t)$ yielding S_t events observed at times $x_1, \ldots, x_{S_t}$. Writing $\theta = (\mu, \lambda)$, the likelihood function here is proportional to $(\lambda e^{\mu})^{S_t} e^{\lambda \Sigma_i x_i} e^{-e^{\mu}(e^{\lambda t}-1)}$ and the observed information is found to be

$$J_t(\theta) = \begin{pmatrix} e^{\mu}(e^{\lambda t} - 1) & te^{\mu+\lambda t} \\ te^{\mu+\lambda t} & \lambda^{-2}S_t + t^2 e^{\mu+\lambda t} \end{pmatrix} \tag{10}$$

We verify conditions S and D3 – D6 with the choice

$$B_t = e^{\mu_0+\lambda_0 t} \begin{pmatrix} 1 & t \\ t & t^2 + \lambda_0^{-2} \end{pmatrix}$$

The CSR of B_t is

$$B_t^{1/2} = e^{(\mu_0+\lambda_0 t)/2} \begin{pmatrix} 1 & 0 \\ t & \lambda_0^{-1} \end{pmatrix}$$

and it is straightforward to show that $W_t = B_t^{-1/2} J_{t0} B_t^{-T/2} \xrightarrow{p} I_2$ on using the fact that $t^2(S_t e^{-(\mu_0+\lambda_0 t)} - 1) \xrightarrow{p} 0$. Condition S therefore holds. It is interesting to note in this example that it is not possible to use a *diagonal* normalizing matrix. Note also that (8) holds and hence condition D3 is satisfied. Condition D4 is elementary to check. Condition D5 is straightforward but tedious to check: first-order Taylor expansions give

$$(J_t(\theta) - J_{t0})_{ij} = (\mu - \mu_0)(J_{t\mu}(\theta_1))_{ij} + (\lambda - \lambda_0)(J_{t\lambda}(\theta_1))_{ij}$$

where θ_1 lies on the line segment joining θ and θ_0. Direct computation of the expression in D5 yields the required result, noting that $|B_t^{T/2}(\theta - \theta_0)| \le c$ implies $|\lambda - \lambda_0| \le c\lambda^{1/2} e^{-(\mu_0+\lambda_0 t)/2}$ and $|\mu - \mu_0| \le c(1 + \lambda^{1/2} t) e^{-(\mu_0+\lambda_0 t)/2}$.

Finally we investigate the nonlocal behaviour of the likelihood function using condition D6. First it is necessary to obtain a suitable convex region in which the log-likelihood function is concave. From (10) we see that $l_t(\theta)$ is concave if and only if

$$(1 - e^{-\lambda t})(S_t e^{-(\mu+\lambda t)} + (\lambda t)^2) > (\lambda t)^2 \tag{11}$$

Replace S_t by its asymptotic expectation $e^{\mu_0+\lambda_0 t}$; we return to the effect of this approximation below. With this replacement, on rearrangement (11) is found to be equivalent to $(\mu-\mu_0)+(\lambda-\lambda_0)t < \log[(\lambda t)^{-2}(e^{\lambda t}-1)]$ and we choose a suitable convex region contained in this region. Let $0 < \lambda_1 < \lambda_0$ and define C_t to be the region where

$$(\mu - \mu_0) + (\lambda - \lambda_0)t < ct$$

and $\lambda > \lambda_1$ (see Figure 1). For c sufficiently small, C_t will satisfy condition D6(i), using $S_t e^{-(\mu_0+\lambda_0 t)} \xrightarrow{p} 1$.

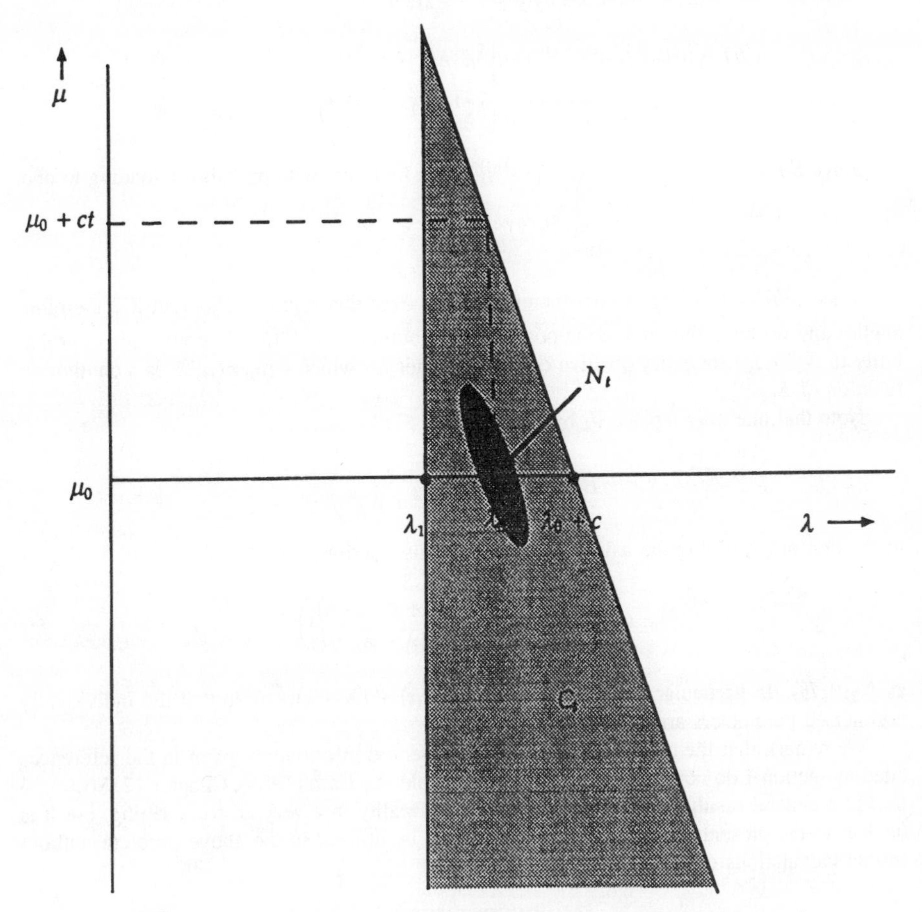

Figure 1. *The regions N_t and C_t for the nonhomogeneous Poisson process example*

Next consider condition D6(iii). Since $l_\mu = S_t - e^\mu(e^{\lambda t}-1)$, for fixed λ the log-likelihood $l_t(\theta)$ is maximized when

$$e^{(\mu-\mu_0)+(\lambda-\lambda_0)t} = S_t e^{-(\mu_0+\lambda_0 t)}(1 - e^{-\lambda t})^{-1} \tag{12}$$

Now when $\lambda > \lambda_1$, with probability tending to one this maximum occurs in C_t, and the hypotheses of Lemma 3.3 hold with R_t the region $\lambda > \lambda_1$. Finally consider the region $\lambda \le \lambda_1$. Again, temporarily replace S_t by $e^{\mu_0+\lambda_0 t}$. Then, with this approximation, for all θ we have

$$l_t(\theta) - l_t(\theta_0) \le e^{\mu_0+\lambda_0 t}\Big[(\mu-\mu_0) + (\lambda-\lambda_0)t + \log(\lambda/\lambda_0) \\ + 1 - e^{(\mu-\mu_0)+(\lambda-\lambda_0)t}(1-e^{-\lambda t})\Big]$$

Now use (12) with S_t replaced by $e^{\mu_0+\lambda_0 t}$. Then

$$\begin{aligned} l_t(\theta) - l_t(\theta_0) &\le e^{\mu_0+\lambda_0 t}\log\left[\lambda_0^{-1}\lambda(1-e^{-\lambda t})^{-1}\right] \\ &\le e^{\mu_0+\lambda_0 t}\log\left[\lambda_0^{-1}\lambda_1(1-e^{-\lambda_1 t})^{-1}\right] \le -ce^{\mu_0+\lambda_0 t} \end{aligned}$$

Since $S_t e^{-(\mu_0+\lambda_0 t)} - 1 = O_p(e^{-(\mu_0+\lambda_0 t)/2})$ we find that, with probability tending to one,

$$\sup_{\lambda \le \lambda_1}\,[l_t(\theta) - l_t(\theta_0)] < -c'e^{\lambda_0 t}$$

Since $|B_t| = O(e^{\lambda_0 t})$ D6(iii) therefore holds over this region. Theorem 3.2 therefore applies and we have shown that the posterior distribution of $J_t^{T/2}(\theta - \hat{\theta})$ converges in probability to $N_2(0, I_2)$ for every positive continuous prior for which $\sup_\mu \pi(\mu, \lambda)$ is a continuous function of λ.

Note that one may replace J_t by

$$\hat{B}_t = e^{\hat{\mu}+\hat{\lambda}t}\begin{pmatrix} 1 & t \\ t & t^2 + \hat{\lambda}^{-2} \end{pmatrix}$$

to the first order, giving the asymptotic posterior distribution of

$$e^{(1/2)(\hat{\mu}+\hat{\lambda}t)}\begin{pmatrix} (\mu-\hat{\mu}) + t(\lambda-\hat{\lambda}) \\ \hat{\lambda}^{-1}(\lambda-\hat{\lambda}) \end{pmatrix}$$

as $N_2(0, I_2)$. In particular, $e^{(1/2)(\hat{\mu}} + \hat{\lambda}t)[t^{-1}(\mu-\hat{\mu}) + (\lambda-\hat{\lambda})] \xrightarrow{p} 0$ so that the individually normalised parameters are asymptotically equal.

We remark that the conditions imposed on observed information given in the references cited in Section 1 do not apply in the present example. Le Cam (1986), Chapter 12, Section 4 proves a general result on asymptotic posterior normality in a very abstract setting, but it is unclear to the present author how this result could be utilized in the above problem without further calculations of the type given here.

REFERENCES

Adekola, A. O. (1987). *Asymptotic Posterior Normality for Stochastic Processes.* Ph.D. Thesis, University of Surrey.

Basawa, I. V. and Prakasa Rao, B. L. S. (1980). *Statistical Inference for Stochastic Processes.* London: Academic Press.

Basawa, I. V. and Scott, D. J. (1983). *Asymptotic Optimal Inference for Nonergodic Models.* New York: Springer.

Chen, C. F. (1985). On asymptotic normality of limiting density functions with Bayesian implications. *J. Roy. Statist. Soc. B* **47**, 540–546.

Chow, Y. S. and Teicher, H. (1978). *Probability Theory*. New York: Springer.

Crowder, M. J. (1988). Asymptotic expansions of posterior expectations, distributions and densities for stochastic processes. *Ann. Inst. Statist. Math.* **40**, 297–309.

Dawid, A. P. (1970). On the limiting normality of posterior distributions. *Proc. Camb. Phil. Soc.* **67**, 625–633.

Fahrmeir, L. (1988). A note on asymptotic testing theory for nonhomogeneous observations. *Stoch. Proc. Appl.* **28**, 267–273.

Heyde, C. C. (1982). Estimation in the presence of a threshold theorem: principles and their illustration for the traffic intensity. *Statistics and Probability: Essays in Honor of C. R. Rao.* (G. Kallanpur *et al.* eds.), 317–323

Heyde, C. C. and Johnstone, I. M. (1979). On asymptotic posterior normality for stochastic processes. *J. Roy. Statist. Soc. B* **41**, 184–189.

Le Cam, L. (1986). *Asymptotic Methods in Statistical Decision Theory*. New York: Springer.

Sweeting, T. J. (1980). Uniform asymptotic normality of the maximum likelihood estimator. *Ann. Statist.* **8**, 1375–1381.

Sweeting, T. J. and Adekola, A. O. (1987). Asymptotic posterior normality for stochastic processes revisited. *J. Roy. Statist. Soc. B* **49**, 215–222.

Sweeting, T. J. (1992). Asymptotic ancillarity and conditional inference for stochastic processes. *Ann. Statist.* **20**, (to appear).

Walker, A. M. (1969). On the asymptotic behaviour of posterior distributions. *J. Roy. Statist. Soc. B* **31**, 80–88.

BAYESIAN STATISTICS 4, pp. 837–842
J. M. Bernardo, J. O. Berger, A. P. Dawid and A. F. M. Smith, (Eds.)

BUGS: a Program to Perform Bayesian Inference using Gibbs Sampling

ANDREW THOMAS, DAVID J. SPIEGELHALTER and WALLY R. GILKS
MRC Biostatistics Unit, Cambridge, UK

SUMMARY

We describe a preliminary version of a general program for Bayesian inference using Gibbs sampling. A crucial feature of the program is the exploitation of conditional independence assumptions in the original statistical model to allow a smooth transition between model specification in a simple language, internal representation of the model in an object-oriented environment, and automatic derivation of the necessary sampling distributions. A useful starting point is the graphical representation of conditional independence assumptions. The eventual aim is to produce a program that can handle arbitrarily complex problems by decomposing them into modular components which then can be "intelligently" handled by the software.

Keywords: CONDITIONAL INDEPENDENCE; GRAPHICAL MODEL; SIMULATION.

1. INTRODUCTION

Gibbs sampling (Geman and Geman, 1984) has rapidly become acknowledged as a powerful computational tool for performing Bayesian inference in complex problems (Gelfand and Smith, 1990). Since the technique depends on sampling from the conditional distribution of each variable, given all others in the model, it is clearly advantageous to exploit conditional independence assumptions. An attractive, non-mathematical description of such assumptions is provided by a *graphical representation* of the model.

We describe a preliminary version of a general computer package for handling complex graphical models for which fragmentary data have been observed. The software architecture takes advantage of object-oriented programming techniques to provide a clean link between the language used to express a model, its internal representation in the program, and the computational procedures used for the analysis. Specifically, a variable (parameter or data) in the model is thought of as a node in a graph, whose local dependencies are specified by the user in a simple declarative form. The program then constructs for each node an object containing the node's value; links with other nodes; and procedures for acting on that node. The sampler can then operate directly on these objects using local computations.

We have provisionally called this program BUGS (Bayesian inference Using Gibbs Sampling). It is being developed in Top-Speed Modula-2 (Jensen and Partners, 1989) on an IBM-PC, although we hope to broaden the hardware platform.

The following sections describe the progress from graphical representation (Section 2) to model specification (Section 3) to computational algorithms (Section 4) and output analysis (Section 5). An example from Gelfand *et al.* (1990) is used for illustration. Throughout we make comments about proposed future development of various aspects of the software.

2. GRAPHICAL REPRESENTATION

Gibbs sampling is particularly appropriate for high-dimensional complex models: such models are typically characterised by extensive conditional independence assumptions. For example, in spatial problems such as image processing or cancer mapping, the true value of a parameter corresponding to a particular coordinate may be assumed to be conditionally independent of values at other coordinates given values for immediate neighbours, where neighbourhood may be defined in terms of pixel adjacency for images, or adjacent areas for maps. In hierarchical models in which individuals are nested within populations, repeated observations at the individual level will generally be assumed independent of population parameters given the parameters for a corresponding individual. In temporal models, observations at a specified time t may be assumed independent of observations at other times given the true underlying state β_t at time t. Further, various Markov assumptions may be made about the β_t's, providing additional conditional independence structure.

A very attractive device for representing conditional independence assumptions is a graphical model, in which nodes represent random variables and missing links represent conditional or marginal independence assumptions. BUGS currently only handles directed links, and so is not appropriate for the type of spatial problem outlined above. Hence we assume a model can be represented by a directed acyclic graph, in which it is natural to use terms such as *parents, ancestors, children* and *descendants* to describe the relationships between variables. The graph says that each node is conditionally independent, given its parents, of all other nodes that are not descendants. This implies that the joint density of all the variables in the graph is the product of the densities of each node conditional on its parents. For a full discussion of the independence properties expressed by directed acyclic graphs see Lauritzen *et al.* (1990). Such graphs are natural derivatives of early work in path analysis (Wright, 1934), and have been applied in expert systems (Pearl, 1988; Lauritzen and Spiegelhalter, 1988), epidemiology (Clayton, 1992), genetics (Thomas, 1991), and structural models in social science (Wermuth and Lauritzen, 1990, Spirtes and Glymour, 1990). In addition, there is increasing interest in using graphical models as a basis for multivariate analysis of data within a maximum-likelihood framework (Whittaker, 1990). We emphasise that BUGS currently does not contain a graphical interface, so explicit graphs have to be constructed manually.

We shall illustrate the principles using an example from Section 6 of Gelfand *et al.* (1990) concerning repeated measures of weight on rats which are assumed exchangeable. The mathematical expression of the model is as follows, where Y_{ij} denotes the weight of the ith rat ($i = 1, \ldots, 30$) on day $x_j (j = 1, \ldots, 5)$:

$$\begin{aligned} Y_{ij} &\sim N(\alpha_i + \beta_i\{x_j - \bar{x}\}, \tau_c) \\ \alpha_i &\sim N(\alpha_c, \tau_\alpha) \\ \beta_i &\sim N(\beta_c, \tau_\beta) \end{aligned}$$

where $\alpha_c, \tau_\alpha, \beta_c, \tau_\beta$, and τ_c are given 'noninformative' priors. $N(\mu, \tau)$ denotes a normal distribution with mean μ and precision τ. Note that the observation times $x_j (j = 1, \ldots, 5)$ were the same for each rat. Note also that Gelfand *et al.* (1990) did not normalise observation times through subtraction of the mean $\bar{x}$, and consequently they proposed a multivariate normal prior for (α_i, β_i).

Conditional independence assumptions in such models are fundamental to the computational algorithm, but are generally buried in the mathematical specification of the model. Our approach is to draw the qualitative structure of the model, as for the rat growth model in Figure 1, before considering further details of the model.

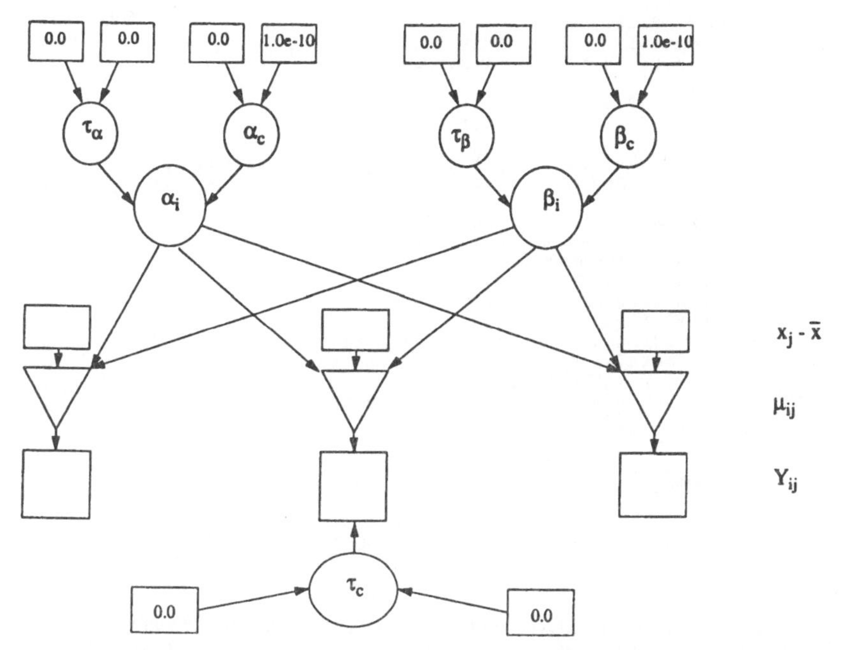

Figure 1. *The graph for the rat growth model (Gelfand et al. , 1990). Squares denote data nodes; circles denote unknown variables; rectangles denote constants; and triangles denote deterministic nodes.*

Four types of node are shown in Figure 1. Squares represent data, in our case the Y_{ij}'s. Circles represent unknown variables, with no distinction between unobserved data or parameters. Rectangles represent constants. These may be fixed in the experimental design, such as the nodes containing $\{x_j - \bar{x}\}$. Alternatively, they may be fixed parameters of prior distributions, such as the parameters for the prior distributions for the hyperparameters $\alpha_c, \tau_\alpha, \beta_c, \tau_\beta$, and the error precision τ_c. Triangles represent deterministic combinations of parent nodes. In our model, the linear predictors $\mu_{ij} = \alpha_i + \beta_i\{x_j - \bar{x}\}$ are explicitly represented as deterministic nodes in the graph. The introduction of deterministic nodes allows great freedom in specifying complex models in a highly modular form. In addition, by specifying deterministic nodes we can obtain marginal posterior distributions of functions of interest: for example $\alpha_0 = (\alpha_c - \beta_c\bar{x})$ in our model is the intercept at day 0 (see Section 5).

3. MODEL SPECIFICATION

We have developed a simple declarative language to present the statistical model to the Gibbs sampler. The language has two functions: firstly to describe the directed acyclic graph underlying the statistical model and secondly to tell the Gibbs sampler where to find the data sets for the model.

The specification file corresponding to the graph in Figure 1 is given in Figure 2. The file comprises three sections: a list of nodes found in the graph; the names of files containing the values of the data nodes and the initial values of the sampling nodes; and finally a list of conditional dependence relations.

BUGS reads the list of nodes and sets up a table of nodes in the computer's memory

```
model Rats;
   data in "c:\bugs\dat\rats.dat";
   inits in "c:\bugs\in\rats.in";
const
   N = 30,   # number of rats
   T = 5;    # number of time points
var
   x[T],mu[T,N],Y[T,N],alpha[N],beta[N],alpha_c,
   tau_alpha,beta_c,tau_beta,tau_c,alpha_0,x_bar;
{
  alpha_c ~ Normal(0.0,1.0E-10);
  beta_c  ~ Normal(0.0,1.0E-10);
  tau_c   ~ Gamma(0.0,0.0);
  tau_alpha ~ Gamma(0.0,0.0);
  tau_beta  ~ Gamma(0.0,0.0);
  for (i in 1:N) {
     alpha[i] ~ Normal(alpha_c,tau_alpha);
     beta[i] ~ Normal(beta_c,tau_beta);
     for (j in 1:T) {
        mu[j,i] <- alpha[i] + beta[i]*(x[j] - x_bar);
        Y[j,i] ~ Normal(mu[j,i],tau_c);
     }
  }
 alpha_0 <- alpha_c - x_bar * beta_c;
}
```

Figure 2. *The BUGS specification file for the rat growth model (Gelfand et al. 1990).*

in a format which easily interfaces with the S statistical package (Becker, Chambers and Wilks, 1988). The values of the nodes are then read in from a file where they have been previously entered as S objects. The conditional dependence relations are read and a directed acyclic graph of objects containing all the information needed to carry out Gibbs sampling is dynamically created.

4. COMPUTATIONAL ALGORITHMS

For directed graphical models, the full conditional distribution for any node β (i.e., its distribution given all other nodes in the model), is proportional to:

$$[\beta| \text{ parents of } \beta] \times \prod_i \left[i^{th} \text{ child of } \beta| \text{ parents of } i^{th} \text{ child of } \beta\right]$$

where $[a|b]$ denotes the conditional distribution of a given b. Each of the terms in the above expression is provided via the model specification. We refer loosely to $[\beta|$ parents of $\beta]$ as the *prior* for β, and $[i^{th}$ child of $\beta|\beta]$ as a *likelihood* contribution. Of course, in a directed graphical model, a prior for one node may be a likelihood contribution for another node, and likelihood contributions need not contain observed data.

All algorithms within BUGS for sampling from full conditional distributions utilise the above expression. These algorithms include adaptive rejection sampling (Gilks and Wild, 1992; Gilks, 1992); and the ratio-of-uniforms method (see for example Ripley, 1987). For each node, the sampling method of choice is determined via a small expert system, according to the following preference order:

Standard Density
Adaptive Rejection Sampling
Ratio-of-Uniforms
Inversion

The first choice is always a standard density if it is available. This possibility arises for continuous random variables where there is conjugacy between likelihood and prior components of the full conditional, or where there are no likelihood contributions. BUGS can recognise a number of conjugacy situations, including those where deterministic nodes intervene between likelihood and prior.

If conjugacy has not been recognised, the second choice is adaptive rejection sampling. When each of the likelihood and prior terms is log-concave with respect to the variable at the current node (as is usually the case), the full conditional will be log-concave (Gilks and Wild, 1992) and adaptive rejection sampling can be used to sample from the full conditional. BUGS can recognise a number of log-concavity situations, including all the usual generalised linear models (Dellaportas and Smith, 1992).

Otherwise, for continuous random variables, the ratio-of-uniforms method can usually be used. However, for discrete random variables, and sometimes for continuous random variables, inversion of the numerically evaluated cumulative distribution function may be the only feasible sampling method.

In the rat growth model introduced in Section 2, all nodes were updated using conjugacy.

At present, there is still much work to be done in fully implementing these sampling methods within BUGS. The number of recognised conjugacy and log-concavity situations needs to be extended, and as yet we have not implemented the ratio-of-uniforms method. We also need to devise rules for deciding when inversion would be preferable to the ratio-of-uniforms method.

5. OUTPUT ANALYSIS

Any node in the graph can be *inspected* (i.e. its current value displayed) at any time whilst the Gibbs sampler is running. Also, to summarise the history of any node, a *monitor* can be created, displayed or deleted at any time. At present BUGS has only one type of monitor, which reports the mean and standard deviation of the sampled values at a node. However, we plan to develop monitors which report centiles; produce kernel density plots; compare density plots; produce and compare time-series plots; calculate convergence diagnostics; and produce output files of sampled values for post-Gibbs analysis. Thus we view monitors as the primary vehicle for sampled data extraction and analysis.

As noted in Section 2, deterministic nodes can be created and monitored especially to record quantities of interest. For the rat growth model in Section 2 and Figures 1 and 2, the quantities of interest are the intercept α_0 and slope β_c of the population growth line (α_0 and β_c in Section 2). To estimate these BUGS was run for 1590 complete updates of the graph, in one long run. The first 59 iterations were discarded. Sample means for α_0 and β_c were found to be 106.5 and 6.19 with standard deviations 1.9 and 0.33 respectively. The calculations took 7 minutes on a 25 MHz IBM PC with a maths co-processor.

6. CONCLUSIONS

Gibbs sampling has provided a unifying framework for estimation for an enormous range of statistical problems where only approximate or problem-specific approaches existed previously. However, the utility of Gibbs sampling has been hampered by the lack of generic software for its implementation: generally each new problem has required a purpose built program, taking days or weeks to develop. We are aiming to remedy this situation with the BUGS program, which will provide a user-friendly environment for specifying, estimating and analysing a wide range of statistical models.

In this paper we have described how the elegance of the Gibbs sampling methodology leads naturally to graphical representation of the models; to model specification; to the internal object-oriented framework of BUGS, and to constructions for output analysis. Despite this, our task is considerable: we are still at an early stage.

REFERENCES

Becker, R. A., Chambers, J. M. and Wilks, A. R. (1988). *The New S Language: a Programming Environment for Data Analysis and Graphics*. Pacific Grove, CA: Wadsworth and Brooks.

Clayton D. (1992). A Monte Carlo method for Bayesian inference in frailty models. *Biometrics*, (to appear).

Dellaportas, P. and Smith, A. F. M. (1992). Bayesian inference for generalised linear models via Gibbs sampling. *Appl. Statist.* , (to appear).

Gelfand, A. E., Hills, S. E., Racine-Poon, A. and Smith A. F. M. (1990). Illustration of Bayesian inference in normal data models using Gibbs sampling. *J. Amer. Statist. Assoc.* **85**, 972–985.

Gelfand, A. E. and Smith, A. F. M. (1990). Sampling-based approaches to calculating marginal densities. *J. Amer. Statist. Assoc.* **85**, 398–409.

Geman, S. and Geman, D. (1984). Stochastic relaxation, Gibbs distributions, and the Bayesian restoration of images. *IEEE Transac. on Pattern Anal. and Mach. Intelligence* **6**, 721–741.

Gilks, W. R. (1992). Derivative-free adaptive rejection sampling for Gibbs sampling. *Bayesian Statistics 4* (J. M. Bernardo, J. O. Berger, A. P. Dawid and A. F. M. Smith, eds.), Oxford: University Press, 641–649.

Gilks, W. R. and Wild, P. (1992). Adaptive rejection sampling for Gibbs sampling. *Appl. Statist.* **41**, (to appear).

Jensen and Partners (1989). *Top-Speed Modula-2*. London: Jensen and Partners.

Lauritzen, S. L., Dawid, A. P., Larsen, B. N. and Leimer, H-G. (1990). Independence properties of directed Markov fields. *Networks* **20**, 491–505.

Lauritzen, S. L. and Spiegelhalter, D. J. (1988). Local computations with probabilities on graphical structures and their application to expert systems. *J. Roy. Statist. Soc. B* **50**, 157–224, (with discussion).

Pearl, J. (1988). *Probabilistic Reasoning in Intelligent Systems*. San Mateo, CA: Morgan Kaufmann.

Ripley, B. (1987). *Stochastic Simulation*. New York: Wiley.

Spirtes, P. and Glymour, C. (1990). Latent variables, causal models and overidentifying constraints. *J. Econometrics* **39**, 175–198.

Thomas, D. C. (1991). Fitting genetic data using Gibbs sampling: an application to nevus counts in 38 Utah kindreds. *Cytogenetics and Cell Genetics*, (to appear).

Wermuth, N. and Lauritzen, S. L. (1990). On substantive research hypotheses, conditional independence graphs, and graphical chain models. *J. Roy. Statist. Soc. B* **52**, 21–50, (with discussion).

Whittaker, J. (1990). *Graphical Models in Applied Multivariate Analysis*. New York: Wiley.

Wright, S. (1934). The method of path coefficients. *Ann. Math. Statist.* **5**, 161–215.

BAYESIAN STATISTICS 4, pp. 843–850
J. M. Bernardo, J. O. Berger, A. P. Dawid and A. F. M. Smith, (Eds.)

Empirical and Hierarchical Bayes Estimation in Multivariate Regression Models

A. J. VAN DER MERWE and C. A. VAN DER MERWE
University of the Orange Free State, South Africa

SUMMARY

Consider the linear multivariate regression model $Y = X_1\beta_1 + X_2\beta_2 + \varepsilon$, where $\varepsilon \sim N(0; I_n \otimes \Sigma)$. This paper is an extension of the work of Ghosh *et al.* (1989) and considers estimation of β_1 when it is suspected that $\beta_2 = 0$. Empirical Bayes and hierarchical Bayes estimators of β_1 are proposed. They serve as a compromise between $\tilde{\beta}_1$, the ordinary least squares estimator and $\hat{\beta}_1$, the restricted least squares estimator under the assumption $\beta_2 = 0$, and lean more towards $\hat{\beta}_1$ if the hypothesis $H_0 : \beta_2 = 0$ is true, and towards $\tilde{\beta}_1$ otherwise. The methods developed are applied to a real problem.

Keywords: MULTIVARIATE REGRESSION; HIERARCHICAL BAYES; EMPIRICAL BAYES; KRONECKER PRODUCT; SUBSET REGRESSION; MARGINAL DISTRIBUTION.

1. INTRODUCTION

Consider the general linear multivariate regression model $Y = X_1\beta_1 + X_2\beta_2 + \varepsilon$ where Y is a $n \times q$ matrix of observable random variables, X_ℓ a $n \times p_\ell$ matrix of known constants, β_ℓ a $p_\ell \times q$ matrix of unknown parameters ($\ell = 1, 2$) and ε a $n \times q$ matrix of unobservable random variables, referred to as errors. It is also assumed that ε is normally distributed with mean 0 and covariance matrix $I_n \otimes \Sigma$ where 0 is a $n \times q$ matrix of zeros, I_n the identity matrix of order n, Σ a $q \times q$ covariance matrix and $\otimes$ the Kronecker product sign. Also $(X'X)$ is nonsingular where $X = (X_1 : X_2)$ is a $n \times p$ matrix and $p = p_1 + p_2$.

The main purpose of this note is the estimation of β_1 when it is suspected that $\beta_2 = 0$. Let $\tilde{\beta}_1$ denote the usual least squares estimator (henceforth, referred to as OLSE) of β_1 and $\hat{\beta}_1$ the restricted least squares estimator (RLSE) of β_1 under the assumption that $\beta_2 = 0$. The problem is the development of certain shrinkage estimators (hierarchical Bayes estimators (HBE) and empirical Bayes estimators (EBE)), shrinking $\tilde{\beta}_1$ towards $\hat{\beta}_1$.

In order to develop shrinkage estimators of β_1 we first have to obtain the posterior distribution of β. Therefore, consider again the model $Y|\beta, \Sigma \sim N(X\beta, I_n \otimes \Sigma)$ where $I_n \otimes \Sigma$ denotes the covariance matrix between $\text{Vec}(Y)$ and $(\text{Vec}(Y))'$ where $\text{Vec}(Y)(nq \times 1) = (Y_{11} \ldots Y_{1q} \ldots Y_{n1} \ldots Y_{nq})'$. As in Ghosh *et al.* (1989) we are going to assume a normal prior on the unknown regression parameters viz $\beta \sim N(v, C \otimes A)$ where v is a $p \times q$ matrix, C a symmetric $p \times p$ p.d. matrix and A a $q \times q$ symmetric p.d. matrix. $C \otimes A$ denotes the covariance matrix between $\text{Vec}(\beta)$ and $(\text{Vec}(\beta))'$ where $\text{Vec}(\beta) = (\beta_{11} \ldots \beta_{1q} \ldots \beta_{p1} \ldots \beta_{pq})'$.

The matrix C will be taken as $(X'X)^{-1}$, the so called g-prior of Zellner (1986) (see also Arnold (1981) and Ghosh *et al.* (1989)).

Lemma 1.1 and Theorem 1.1, to be applied in Section 2, are now stated.

Lemma 1.1.

$$(I_n \otimes \Sigma + P_X \otimes A)^{-1} = (I_n \otimes I_q - P_X \otimes (I_q - B))(I_n \otimes \Sigma^{-1}) \tag{1.1}$$

where

$$P_X = X(X'X)^{-1}X' \quad \text{and} \quad B = (\Sigma + A)^{-1}\Sigma.$$

Proof. By making use of properties of Kronecker products and the binomial inverse theorem (Press (1982)) the result follows. ◁

Theorem 1.1.
(i) The marginal distribution of Y is normal with mean

$$E(Y|A, \Sigma, v) = Xv \tag{1.2}$$

and covariance matrix

$$\text{Var}(Y|A, \Sigma, v) = I_n \otimes \Sigma + P_X \otimes A. \tag{1.3}$$

(ii) The covariance matrix between $\text{Vec}(Y)$ *and* $(\text{Vec}(\beta))'$ *is given by*

$$\text{Cov}(Y, \beta|A, \Sigma, v) = X(X'X)^{-1} \otimes A. \tag{1.4}$$

(iii) The posterior distribution of β is normal with mean

$$E(\beta|Y, A, \Sigma, v) = v + (\tilde{\beta} - v)(I - B) \tag{1.5}$$

and covariance matrix

$$\text{Var}(\beta|Y, A, \Sigma, v) = (X'X)^{-1} \otimes \Sigma(I - B). \tag{1.6}$$

Proof. Using the theorem on conditional expectation (Anderson (1958) page 28 and Nel (1980)) as well as Lemma 1.1, equations (1.2)-(1.6) follow. ◁

2. EMPIRICAL BAYES ESTIMATION

Before going on, we need a few notations (Ghosh *et al.* (1989)):

$$F_{\ell j} = X_\ell' X_j (\ell, j = 1, 2), \tag{2.1}$$

$$F_{11\cdot 2} = F_{11} - F_{12}F_{12}^{-1}F_{21} \quad \text{and} \tag{2.2}$$

$$F_{22\cdot 1} = F_{22} - F_{21}F_{11}^{-1}F_{12}. \tag{2.3}$$

Also the OLSE of β is defined as $\tilde{\beta} = (X'X)^{-1}X'Y$ and the RLSE $\hat{\beta}_1 = (X_1'X_1)^{-1}$ $(X_1'Y)$. From the normal equations it follows that

$$X_1'Y = F_{11}\hat{\beta}_1 = F_{11}\tilde{\beta}_1 + F_{12}\tilde{\beta}_2 \tag{2.4}$$

and therefore

$$\tilde{\beta}_1 = \hat{\beta}_1 - F_{11}^{-1}F_{12}\tilde{\beta}_2. \tag{2.5}$$

As mentioned, our interest is in the estimation of β_1. Therefore, writing $v = \begin{pmatrix} v_1 \\ v_2 \end{pmatrix}$ where $v_i (i = 1, 2)$ is a $p_i \times q$ matrix, it follows (from (1.5) and (1.6)) that the posterior distribution of β_1 is normal with mean

$$E(\beta_1|Y, A, \Sigma, v_1) = v_1 + (\tilde{\beta}_1 - v_1)(I - B) \tag{2.6}$$

and covariance matrix

$$\text{Var}(\beta_1|Y, A, \Sigma, v_1) = F_{11\cdot 2}^{-1} \otimes \Sigma(I - B). \tag{2.7}$$

In an empirical Bayes setting, the parameters v_1, Σ and A are unknown and need to be estimated from the marginal distribution of Y (equations (1.2) and (1.3)). Since we want to shrink the OLSE $\tilde{\beta}_1$ of β_1 to the RLSE $\hat{\beta}_1 = (X_1'X_1)^{-1}X'Y$ of β_1 under $\beta_2 = 0$ we assume $v_2 = 0$.

The following theorem provides complete sufficient statistics for (v_1, Σ, A) based on the marginal distribution of Y.

Theorem 2.1.

Based on the marginal distribution of $Y, (\hat{\beta}_1, \tilde{\beta}_2' F_{22\cdot1}\tilde{\beta}_2, (Y - X\tilde{\beta})'(Y - X\tilde{\beta}))$ *are complete sufficient statistics for* (v_1, A, Σ).

Proof.

By using the arguments given in Van der Merwe and Van der Merwe (1991) and extending Theorem 2 (page 19) in Ghosh *et al.*, the result follows. ◁

The following theorem can also easily be proved:

Theorem 2.2.

The marginal distributions of $\hat{\beta}_1, \tilde{\beta}_2$ *and* $Y - X\tilde{\beta}$ *are normal and independent.*

Proof. The proof is given in Van der Merwe and Van der Merwe (1991). It is also shown that

$$E(\hat{\beta}_1 | A, \Sigma, v) = v_2, \tag{2.8}$$

$$\mathrm{Var}(\hat{\beta}_1 | A, \Sigma, v_1) = (X_1'X_1)^{-1} \otimes (\Sigma + A), \tag{2.9}$$

$$E(\tilde{\beta}_2 | A, \Sigma, v_2) = 0, \tag{2.10}$$

$$\mathrm{Var}(\tilde{\beta}_2 | A, \Sigma, v_2) = F_{22\cdot1}^{-1} \otimes (\Sigma + A). \tag{2.11}$$

From the normality of $\tilde{\beta}_2$, (2.10) and (2.11) it follows that

$$\tilde{\beta}_2' F_{22\cdot1} \tilde{\beta}_2 \sim W(\Sigma + A, p_2). \tag{2.12}$$

Accordingly the UMVUE of $(\Sigma + A)^{-1}$ is $(p_2 - q - 1)(\tilde{\beta}_2' F_{22\cdot1}\tilde{\beta}_2)^{-1}$.

Also from (2.8) and theorem 2.1, $\hat{\beta}_1$ is the UMVUE of v_1 (Note that $\hat{\beta}_1$ is also the MLE of v_1). Finally by writing

$$S = \frac{1}{n-p}(Y - X\tilde{\beta})'(Y - X\tilde{\beta}), \tag{2.13}$$

it follows that the best scale invariant estimator of Σ is

$$\frac{n-p}{n-p+q+1}S.$$

Plugging the above estimators for $v_1, (A + \Sigma)^{-1}$ and Σ in (2.6), it follows that an empirical Bayes estimator of β_1 is

$$EB(\beta_1) = \hat{\beta}_1 + (\tilde{\beta}_1 - \hat{\beta}_1)\left(I - \frac{(p_2 - q - 1)(n-p)}{n-p+q+1}\left(\tilde{\beta}_2' F_{22\cdot1}\tilde{\beta}_2\right)^{-1} S\right). \tag{2.14}$$

For the special case $q = 1$, equation (2.14) simplifies to equation (2.18) in Ghosh, *et al.* (1989). ◁

3. EXAMPLE

Example 3.1. This experiment is a 6×8 factorial experiment with unequal subclass numbers.

Factor $A \equiv$ Years : 1970; 1971; ...; 1975.
Factor $B \equiv$ Markets : Market 1; Market 2; ...; Market 8.
$Y \equiv$ Low-bid price for white bread.

The purpose of this experiment is to use regression analysis for the detection of collusive bidding. Collusive bidding may occur when the markets are not stable over time, i.e., if there is a market-time interaction — for example when the low-bid price in a particular market increases substantially in a given year in relation to the low-bid prices for the other markets. For further details see Mendenhall and Sincich (1989).

Two models can now be formulated:

Model 1 — Reduced Model

A stable market condition — Main effects only

$$\hat{Y} = X_1\hat{\beta}_1.$$

Model 2 — Complete Model (Full Model)

Unstable market condition — Interaction terms included

$$\hat{Y} = X_1\tilde{\beta}_1 + X_2\tilde{\beta}_2.$$

Testing the null hypothesis that "the market is stable" is equivalent to testing the null-hypothesis that the interaction parameters of Model 2 equal zero, i.e., $H_0 : \beta_2 = 0$. The calculated F-statistic is $F = 1.58$. The critical value of F for $\alpha = 0.05$; $f_1 = 35$ and $f_2 = 25.5$ df is approximately equal to 1.50. Since the calculated F-value exceeds the critical value, the null hypothesis is rejected and it is concluded that there is evidence of a year-market interaction.

As mentioned in the introductory paragraph, a third model (the empirical Bayes or adapted model) can also be formulated. This model is defined as

$$\begin{aligned}\hat{Y} = X_1 &\left[\hat{\beta}_1 + \left(\tilde{\beta}_1 - \hat{\beta}_1\right)\left(I - \frac{(p_2 - q - 1)(n-p)}{n - p + q + 1}\left(\tilde{\beta}_2' F_{22\cdot 1}\tilde{\beta}_2\right)^{-1} S\right)\right] \\ &+ X_2\tilde{\beta}_2\left[I - \frac{(p_2 - q - 1)(n-p)}{n - p + q + 1}\left(\tilde{\beta}_2' F_{22\cdot 1}\tilde{\beta}_2\right)^{-1} S\right].\end{aligned} \tag{3.1}$$

For the bread data $q = 1$; $n = 303$; $p = 48$; $p_1 = 13$; $p_2 = 35$. Only 40.97% of the values of the interaction terms are included in the adaptive model. This might seem somewhat surprising because there is evidence of a year-market interaction. The reason for this small percentage is that the sample size n as well as p_1 and p_2 are quite large while the calculated F-statistic is small.

In Figures 3.1, 3.2 and 3.3 the profile analyses for each of the three models are given. From Figure 3.3 it is clear that the empirical Bayes (adapted) model leans more towards Model 1 — the stable market condition — than towards Model 2 and although Figure 3.3 shows some signs of market instability, they are not nearly as severe as those portrayed by Model 2. It therefore seems that Figure 3.3 gives a better representation of the real market situation than either Figure 3.1 or 3.2. The reason for this is that the calculated F-statistic is only just significant, indicating the need for a compromised model. We therefore conclude that the instability of market conditions is not nearly as severe as originally thought and that collusive bidding did not really occur.

Reduced-Model

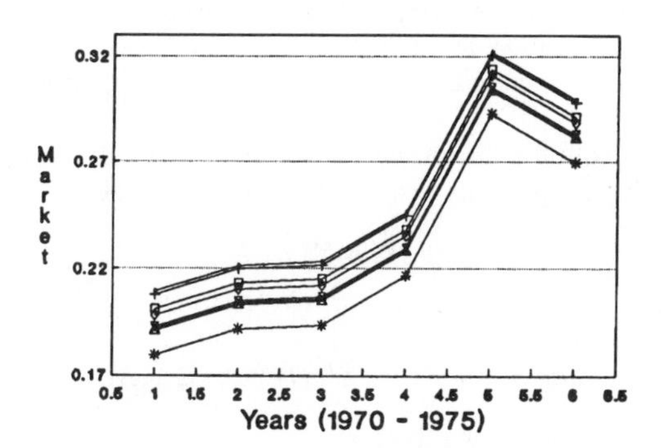

Figure 3.1.

Full-Model

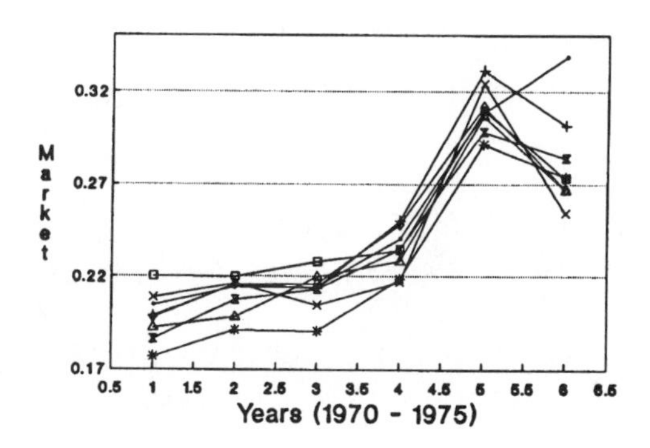

Figure 3.2.

Empirical Bayes-Model

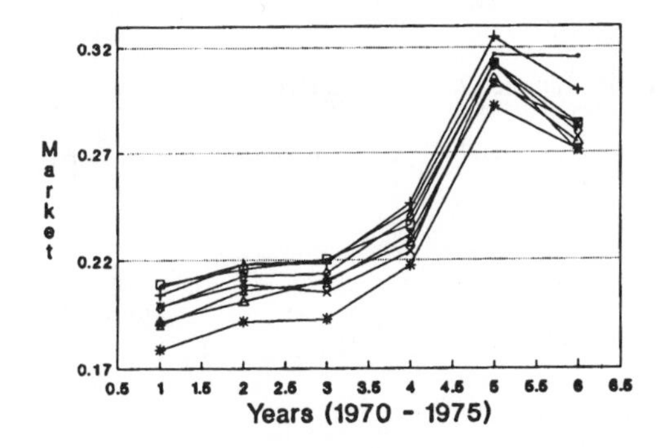

Figure 3.3.

4. HIERARCHICAL BAYES APPROACH

One problem with empirical Bayes estimators is that because there is no natural standard error, it is difficult to find confidence intervals or test hypotheses. (Press (1989)). There is however, a full Bayesian solution to this type of problem. The approach is based on the use of hierarchical priors.

Assume that v_1 is uniformly distributed on the R^{qp_1} space. By using (2.8) and (2.9) it can easily be proved that the posterior distribution of v_1 is normal with mean

$$E(v_1|Y, A, \Sigma) = \hat{\beta}_1 \tag{4.1}$$

and covariance matrix

$$\mathrm{Var}(v_1|Y, A, \Sigma) = (X_1'X_1)^{-1} \otimes (\Sigma + A). \tag{4.2}$$

This in turn leads to the following theorem:

Theorem 4.1. *The posterior distribution $\beta_1 \,|\, Y, A, \Sigma$ is multivariate normal with mean*

$$E(\beta_1|Y, A, \Sigma) = \hat{\beta}_1 + (\tilde{\beta}_1 - \hat{\beta}_1)(I - B) \tag{4.3}$$

and covariance matrix

$$\mathrm{Var}(\beta_1|Y, A, \Sigma) = F_{11\cdot 2}^{-1} \otimes \Sigma(I - B) + F_{11}^{-1} \otimes \Sigma B, \tag{4.4}$$

Proof.

$$\begin{aligned} E(\beta_1|Y, A, \Sigma) &= E\{E(\beta_1|Y, A, \Sigma, v_1)\} \\ &= \hat{\beta}_1 + (\tilde{\beta}_1 - \hat{\beta}_1)(I - B), \text{ from (2.6) and (4.1).} \end{aligned}$$

Also,

$$\begin{aligned} \mathrm{Var}(\beta_1|Y, A, \Sigma) &= E\{\mathrm{Var}(\beta_1|A, \Sigma, v_1)\} + \mathrm{Var}\{E(\beta_1|A, \Sigma, v_1)\} \\ &= F_{11\cdot 2}^{-1} \otimes \Sigma(I - B) + F_{11}^{-1} \otimes \Sigma B, \end{aligned}$$

from (2.7) and (4.2).

By assuming a uniform prior on A, it follows from (2.12) that the posterior distribution of A given Σ and the data is

$$f(A|Y, \Sigma) \propto \frac{1}{|\Sigma + A|^{p_2/2}} \exp\left\{-\frac{1}{2}(\Sigma + A)^{-1}(\tilde{\beta}_2' F_{22\cdot 1}\tilde{\beta}_2)\right\} \qquad A > 0 \tag{4.5}$$

which means that the density function of $H = (\Sigma + A)^{-1}$ is

$$f(H|Y, \Sigma) \propto |H|^{(p_2 - 2q - 2)/2} \exp\left\{-\frac{1}{2}H(\tilde{\beta}_2' F_{22\cdot 1}\tilde{\beta}_2)\right\} \qquad H > \Sigma^{-1},$$

a truncated $W((\tilde{\beta}_2' F_{22\cdot 1}\tilde{\beta}_2)^{-1}, p_2 - q - 1)$ distribution.

Furthermore, it is well-known that

$$U = (n - p)S \sim W(\Sigma, n - p).$$

Assuming that the prior on Σ is improper, i.e.,

$$p(\Sigma) \propto |\Sigma|^{(q+1)/2}, \tag{4.6}$$

the posterior distribution of Σ is given by

$$f(\Sigma|U) \propto |\Sigma|^{(n-p+q+1)/2} e^{-(\mathrm{tr}\Sigma^{-1}U)/2}. \quad \triangleleft \tag{4.7}$$

By using equations (4.3-6) and numerical integration, the unconditional posterior density function of $X_1\beta_1 + X_2\beta_2$ for year 1973 and Market 2 was obtained and is given in Figure 4.1.

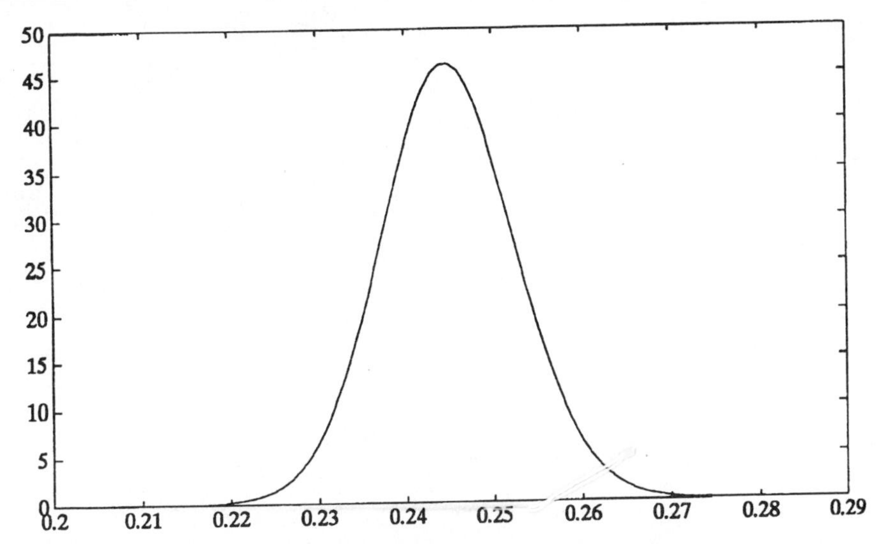

Figure 4.1. *Posterior Density Function — Year 1973 and Market 2.*

The modal value for this density function is 0.245 while the empirical Bayes-estimate is 0.247. The 95% Bayes credibility interval is 0.232 – 0.258. In the case of the full-model the maximum likelihood estimate is 0.250 and the 95% confidence interval: 0.23 – 0.27. For the reduced-model the maximum likelihood estimate is 0.245.

It therefore seems that for this example the hierarchical Bayes approach shrinks even more to the reduced-model than the empirical Bayes method.

ACKNOWLEDGEMENT

The authors are grateful to the referee and to Professor Berger for valuable comments and suggestions. We also wish to thank Mrs. L. Fourie for her typing of the original manuscript.

REFERENCES

Anderson, T. W. (1958). *An Introduction to Multivariate Statistical Analysis*. New York: Wiley.

Arnold, S. F. (1981). *The Theory of Linear Models in Multivariate Analysis*. New York: Wiley.

Berger, J. O. (1985). *Statistical Decision Theory and Bayesian Analysis*. New York: Springer.

Ghosh, M., Saleh, A. K. Md. E. and Sen, P. K. (1989). Empirical Bayes subset estimation in regression models. *Statistics and Decisions* **7**, 15–35.

Mendenhall, W. and Sincich, T. (1989). *A Second Course in Business Statistics. Regression Analysis*. London: Collier MacMillan.

Nel, D. G. (1980). On matrix differentiation in statistics. *South African Statist. J.* **14**, 137–193.

Press, S. J. (1982). *Applied Multivariate Analysis, using Bayesian and Frequentist Methods of Inference*. Melbourne, FL: Krieger.

Press, S. J. (1989). *Bayesian Statistics: Principles, Models and Applications*. New York: Wiley.

Van der Merwe, A. J. and Van der Merwe, C. A. (1991). Empirical and hierarchical Bayes estimation in multivariate regression models. *Tech. Rep.* **170**, University of the Orange Free State.

Zellner, A. (1986). On assessing prior distributions and Bayesian regression analysis with g-prior distributions. *Bayesian Inference and Decision Techniques*. (P. K. Goel and A. Zellner, eds.) Amsterdam: North-Holland, 233–243.

BAYESIAN STATISTICS 4, pp. 851–859
J. M. Bernardo, J. O. Berger, A. P. Dawid and A. F. M. Smith, (Eds.)

[B/D] Works

DAVID WOOFF
University of Durham, UK

SUMMARY

[B/D] is a computer language developed at the University of Hull, and more recently at the University of Durham, which implements a subjectivist statistical methodology based upon the analysis and revision of genuine beliefs about observables, using expectation rather than probability as a primitive. Within a traditional Bayesian framework, our environment assists in the analysis of partial belief specifications. The role of this paper is to survey very briefly our approach and the tool that we employ to solve problems within our approach, namely [B/D].

Keywords: BAYES LINEAR METHODS; BELIEF ANALYSIS; BAYESIAN SOFTWARE; PREVISION; PARTIAL BELIEF SPECIFICATION; EXCHANGEABLE AUTOCORRELATED REGRESSIONS.

1. INTRODUCTION

[B/D] (an acronym for *beliefs adjusted by data*) is a computer package which constitutes an implementation of the subjectivist theory advocated by Michael Goldstein following on from the ideas of Bruno de Finetti, in which *prevision* (expectation) is viewed as a primitive. [B/D] is the technical embodiment of this approach, which we feel to be both foundationally and practically satisfactory and justifiable, whereas we do not believe the same can always be said for the standard Bayesian methodology. In this paper we refrain from philosophical matters: these are discussed amply in the series of papers Goldstein (1981, etc.) However, note that within the context of a traditional Bayesian framework, [B/D] provides a standard Bayes treatment in the context of finite sample spaces and linear Bayes solutions to problems containing limited or partial belief specifications.

[B/D] has already been used to analyse a variety of problems within this (linear) subjectivist paradigm, principally to illustrate ongoing developments in theory and methodology. Other recent work, aimed at problem solving rather than describing [B/D], uses [B/D] as the central analytical package. See, for example, Farrow and Goldstein (1990, 1992).

In this paper we want to describe, very briefly, the essential features of the package to clarify the chief capabilities of [B/D]. We will do so generally and broadly by examining the sorts of questions that [B/D] answers. We will conclude with an example to highlight some of these features.

2. [B/D] AS A LANGUAGE

[B/D] is a computer language written to form the core of a general subjectivist statistical package. As a language it falls somewhere between a traditional high-level programming language (like Fortran and Pascal) and a statistical language (such as GLIM). Thus, [B/D] contains standard facilities such as if-then-else and for-step-do constructs, and also the special facilities for implementing the subjectivist methodology, for example the command which accomplishes the adjustment of one belief structure by another; and the command which constructs linear relationships.

The program can be used both interactively and in a batch mode. Whichever mode is used, the program can be given access to libraries of pre-existent subroutines, which are then used as required. These subroutines, or *macros*, are themselves simply collections of [B/D] commands; they will range from simple subroutines that you might write to perform a regular task, to large-scale applications such as a multiple linear regression module.

[B/D] is an *interpreted* language: when you are working interactively, any commands and macros that you enter are parsed and executed immediately. Most of the features of the package are written to allow the user the maximum flexibility of control, for example you specify the quantity and level of detail of output that you wish to see.

3. WHAT SHOULD THE PACKAGE DO?

Let us examine the essentials of a simple problem. Suppose that we begin with a collection of uncertain quantities $\mathcal{B} = [B_1, \ldots, B_k]$ of interest, over which we hold prior beliefs which we feel able to express in the form of previsions and covariances. (These form the minimal input into the program.) At this point we might check whether our inputs constitute coherent beliefs, and [B/D] provides us with commands to examine such joint coherence.

Next we need to consider what we are going to use to help us to reduce uncertainty in our collection of quantities. Typically we will consider another set of quantities, $\mathcal{D} = [D_1, \ldots, D_r]$ which we hope to be "informative", and for which we can specify not only previsions and covariances amongst the D_is but also previsions and covariances between the D_is and the B_js. The quantities in $\mathcal{D}$ are, in principle, all observable, although (if we are approaching the problem from a design angle) we might choose not to observe some, or indeed all, of them.

We have made various specifications (including, implicitly or explicitly in our specifications between $\mathcal{B}$ and $\mathcal{D}$, a model) which we want to explore. Typical questions that we might ask are as follows:

* Do our specifications seem sensible *a priori*? Does an analysis of relationships amongst these specifications yield any insights? How much latitude do we have in specifying coherent beliefs?
* How informative do we expect the data to be? Will we need to consider other sources of information if the data are insufficiently informative? Can we avoid the cost of observation on some quantities by choosing some optimal, or near-optimal, subset of the data quantities? Will the data quantities be equally informative for all B_i, or will they be very informative for some linear combinations of the B_is, and almost valueless for others?
* Having observed the data, are our revisions sensible? Do the data conflict with or support our prior expectations? With respect to the revision, do some subsets of the data conflict with other subsets?
* If our intention is to use the data for posterior design (for planning purposes), what design should we accept? Does the data, in combination with our specifications, now provide further insights?
* How should we generate models within the subjectivist approach? How do we choose between alternative models? When modelling, which features of the belief and data structures require our particular attention?
* How do we exploit exchangeability? How are models designed, and beliefs analysed within exchangeable situations?

Thus, a requirement of our package is that these questions be addressed, within the context of our subjectivist formulation. In addition, of course, the package must contain a variety of routines designed to make the approach as helpful and flexible as possible, so that it should contain tools for data exploration, graphical tools, aid to assist in the creation of complex structures, and so forth.

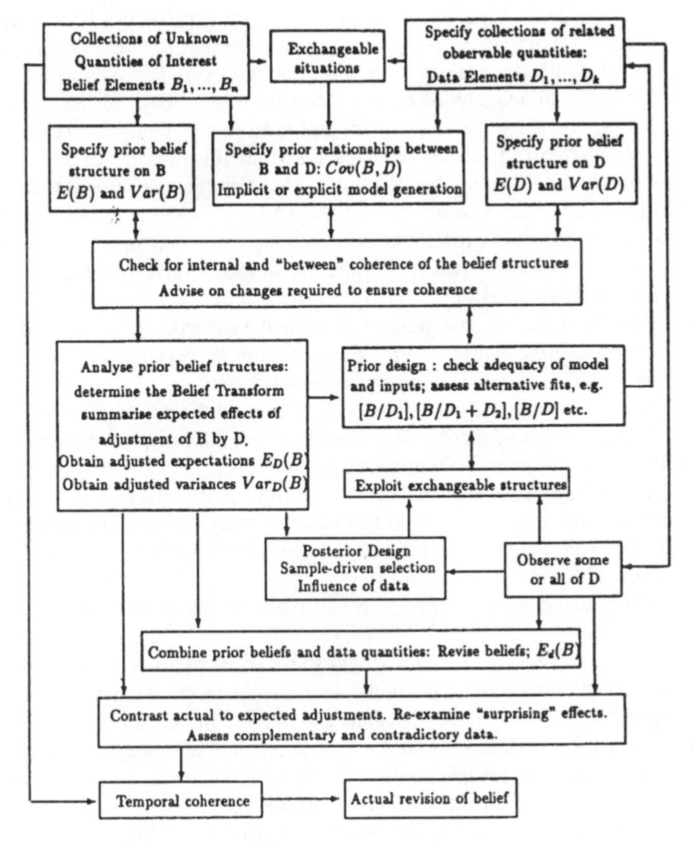

Figure 1. *[B/D] Mechanisms.*

4. WHAT FEATURES DOES [B/D] OFFER?

We use [B/D] to help us analyse our beliefs. The principal belief-analysis mechanisms of [B/D], and their inter-relationships, are shown in Figure 1. The main features of the package are:

* Powerful ways to help you explore and/or create large belief structures. Simple commands enabling the creation of highly complex models and error structures, for example time-dependent structures. Coherence checking and provision of coherence requirements. In general the structures that we create and analyse are the joint variance-covariance matrices specified over the quantities of interest, but note that some of the quantities over which beliefs are expressed might be functions of other quantities.

* Evaluation of the *belief transform* over $\mathcal{B}$ (quantities you wish to learn about) induced by $\mathcal{D}$ (observable quantities). The belief transform is essentially a representation of the linear operator transforming the covariance matrix over $\mathcal{B}$, via the entire covariance structure over $\mathcal{B}$ and $\mathcal{D}$, into a "posterior" covariance matrix over $\mathcal{B}$ which expresses expected reductions in variance. Amongst the properties of the belief transform are that the belief structure $\mathcal{B}$ is re-expressed as a collection of orthonormal eigenvectors whose corresponding eigenvalues summarise the expected change in beliefs induced by $\mathcal{D}$ for the linear combinations represented by the eigenvectors. We term this collection of eigenvector quantities the *belief grid*, and we refer to the eigenvectors themselves as *principal directions*. Each B_j (and any linear combination of the B_j's) can then be decomposed according to this belief grid, so that the potential effect of revision using $\mathcal{D}$ is quantified as additive variance reductions over these directions. The trace of the transform summarises the expected change in belief over the entire collection.
* Construction of the belief transform for alternative models and for alternative collections of data quantities: comparison of, and quantification of, the expected effects of revision for alternative models. For example, we compute the *adjusted variance ratio*, a scale-free measure expressing the relative worth of reduction in uncertainty, for our quantities. We can use such summary statistics for experimental design.
* Computation of the adjusted expectation, $E_D(B_i)$, and the adjusted variance, $\mathrm{Var}_D(B_i)$ for each B_i for alternative models D, and evaluation of the adjusted expectation, $E_d(B_i)$, for specific models whenever observations on the data elements, $D = d$ become available, yielding actual adjustments of belief. (You might prefer informally to use the word "posterior" for "adjusted" throughout this paper, although there is an important technical distinction between the two terms which we won't consider here: further discussion can be found in Goldstein (1985, 1986b).)
* Data-handling facilities: summaries, labelled plots, selection according to function, and so forth.
* Calculation of the *bearing*, a vector which summarises the *actual* changes in belief induced by $\mathcal{D}$. Just as the belief transform epitomises the changes in belief that are expected according to the prior belief specifications and a given model, so the bearing epitomises the changes that occur according to the data supplied. Each body of observations on $\mathcal{D}$ will have consequences for revision which can be summarised by the bearing in that (i) its length expresses overall actual change, and can be contrasted to its prior expectation as a diagnostic test for consistency of beliefs with data; and (ii) the change in expectation for any linear combination of quantities in $\mathcal{B}$ can be expressed as the prior covariance between that linear combination and the bearing. The bearing can be calculated for alternative or augmented models, and can thus be used for data-driven design.
* Diagnostic checks on the influence of portions of the data. We highlight contradictory effects via the *path correlation*, the correlation between two bearings, which shows whether, for example, the consequences for revision of two different bodies of data are contradictory or complementary. Determination of the *data trajectory* (successive related path correlations) for systematic augmentations or decompositions of a model.
* Full incorporation and analysis of exchangeable structures. Determination of optimal sample size. Computation of the limiting (very large sample) belief transform. Modelling which exploits exchangeable situations.
* Stepwise routines to allow efficient and parsimonious model building (or reduction), based upon the expected information content of a data structure or upon actual realisations of the structure.

Naturally, [B/D] contains a variety of other routines designed to make application of the core areas of the program, those relating to belief analysis, as easy as possible.

It has been said that Bayesian packages are worthless without some means of eliciting prior information, and it is hard to disagree with this view. In a sense [B/D] *is* an elicitation package: it helps to elicit genuine posterior beliefs, but we recognise that the provision of tools to aid prior elicitation is important. Hence we regard as highly desirable the implementation of an improved elicitation front-end to the package. Similarly, a complementary graphical approach to belief analysis can provide valuable insights. Consequently it is also our aim to extend the graphics capabilities of the program. Technically, both the theory and the implementation of the subjectivist approach await extension.

5. ILLUSTRATION

5.1. *Organisation*

To illustrate some of the aims and features of the package we turn to an example considered in a different form by Goldstein (1987a,1991) which relates to an industrial smelting process for aluminium. The main quantities of interest are (i) the concentration of alumina remaining in a solution at various time points; and (ii) the constants describing the underlying physical process. The situation is obfuscated by various operating features which we attempt to take into account through careful specification of error structures. A combination of experience and pragmatism leads us to an initial choice of the following model:

$$Y_{tr} = A_r + tB_r + E_{tr} + V_{tr} + H_{tr},$$

where

$$V_{tr} = V_{t-1r} + F_{tr},$$

and

$$H_{tr} = \phi H_{t-1r} + U_{tr}, \ \ t \geq 2$$

The components are as follows: (1) Y_{tr} is the alumina concentration at time t on run r; (2) A_r and B_r are intercept and slope constants which describe the underlying physical process (experience and physical "laws" strongly suggest that the process is linear) and which are considered *exchangeable between runs*; (3) E_{tr} represents an exchangeable "pure measurement" error structure, uncorrelated internally and with all other quantities; (4) V_{tr} expresses a random walk error structure, so that successive terms in the series contain uncorrelated jumps, F_{tr}, which are considered exchangeable and uncorrelated with all other quantities; (5) H_{tr} represents a stationary autoregressive error structure to account for the influence of alumina in suspension (considered similar in effect at neighbouring time points) where the U_{tr} are exchangeable and uncorrelated with each other and all other quantities. H_{1r} has a variance chosen to ensure that $\text{Var}(H_{tr})$ is constant over t, and there is no U_{1r} term. We could, if necessary, refine our model. At this stage, however, we consider that further detail would not afford significant changes in our inferences about the main quantities of interest: part of our analysis will provide evidence about the adequacy of our model.

Our initial input consists of these linear structures and specifications of first- and second-order moments over them. For this analysis, the non-zero expectation and covariance specifications were, for each r and $s \neq r$:

$E(A_r) = 1.4$ and $E(B_r) = 0.1$
$\text{Cov}(A_r, A_s) = 0.038$ and $\text{Var}(A_r) = 0.058$
$\text{Cov}(B_r, B_s) = 0.0016$ and $\text{Var}(B_r) = 0.0017$

$\text{Var}(U_{tr}) = 0.0204; \text{Var}(H_{1r}) = 0.04$ and $\phi = 0.7$
$\text{Var}(E_{tr}) = 0.01$ and $\text{Var}(F_{tr}) = 0.01$

In addition to these belief inputs, data became available in the form of $r = 3$ runs with $t = 13$ time points for each run (Table 1). We omit, for reasons of brevity and clarity, various technicalities, in particular a transformation of the time axis.

t	1	2	3	4	5	6	7	8	9	10	11	12	13
y_{t1}	1.79	2.14	2.13	2.07	2.08	1.88	1.94	2.01	2.35	2.23	2.58	2.48	2.82
y_{t2}	1.93	1.76	1.61	2.32	1.87	1.80	2.21	2.23	2.42	2.58	2.60	2.65	2.70
y_{t3}	1.54	1.48	1.57	1.28	1.50	1.79	1.88	2.11	2.48	2.28	3.39	3.44	2.80

Table 1. *Alumina concentration data.*

5.2. *Preliminaries*

In practice we would spend a great deal of effort in investigating and analysing our model, before and after observing data; we would fit alternative models and select the most appropriate, test for conflicts between the data and belief specifications in detail, and so forth. Instead, we shall limit ourselves to a very limited consideration of three aspects of the problem: overall features; design; and data assessment. For each such aspect we will show only a bare minimum of the results that we might have chosen to see.

The computer analysis was carried out by running a short (about 50 lines, including data declarations) [B/D] macro, briefly consisting of (a) the setting up of constants and explicit belief structures; (b) the specification of beliefs between several of the quantities functionally; (c) the assignment of linear relationships between the quantities; (d) the creation of the data-carrying quantities Y_t according to the specified model; (e) the input of data; (f) the statements needed to perform our three limited analyses.

5.3. *Analysis 1: Overall Features*

In our first analysis we examine the full adjustment of the collection of interest (the intercept and slope quantities) by our data quantities Y_{tr}. Exchangeability considerations permit us to view the $r = 3$ runs as replicated observations at each time point, and we write A and B as the intercept and slope quantities underlying this exchangeability representation.

Firstly we determine the *belief transform*, summarising expected changes in belief. We find that the belief transform has the following eigenvectors: $G_1 = 2.2A + 22.6B - 5.3$ and $G_2 = 4.6A - 10.6B - 5.4$, corresponding to eigenvalues 0.18 and 0.44 respectively. These eigenvectors constitute the belief grid of two directions; we expect to reduce our uncertainty in the first direction G_1 by about 82% and *orthogonally* for the second direction G_2 by about 56%. All other linear combinations have expected reductions in uncertainty according to the magnitude of their correlation with each direction. Information content, as expressed by the eigenvalues, ranges from zero (completely informative) to unity (non-informative), and information content over the complete belief structure is represented by the trace of the transform: the sum of the eigenvalues.

Related to these summaries are the *adjusted variances* for our belief elements (Table 2). For example, we expect to reduce uncertainty in the intercept quantity from an initial value of 0.0380 down to 0.0149, a reduction of some 60%. The reduction over the entire belief structure is expected to be about 69%.

Element	Initial	Adjusted	AV Ratio	%Change
A	0.0380	0.0149	0.3934	60.66%
B	0.0016	0.0004	0.2267	77.33%
Overall	2	0.6201	0.3101	68.99%

Table 2. *Adjusted variances.*

Next we examine the effect of incorporating actual observations by calculating *adjusted expectations* for our belief elements (Table 3). We find that B is revised downwards, fractionally; and that A is revised upwards, proportionately more, but not drastically so: by about 0.7 standard deviations relative to the reduction in uncertainty. Naively, this is our first indication that the data tend to support our prior beliefs.

Element	Initial	Adjusted	Std.
A	1.4000	1.5050	0.6913
B	0.1000	0.0982	–0.0523

Table 3. *Adjusted expectations.*

One way of exploring this issue is to determine the bearing, the vector which abstracts the effects of the actual adjustment. In this example, the bearing is $Z = 2.8A - 1.1B - 3.8$. All revisions are in the direction of Z in the sense that each change in expectation can be expressed as a covariance between Z and the particular belief element. For example, we can verify that the difference between the initial and adjusted expectation for the slope quantity (Table 3) is $\text{Cov}(A, Z) = 0.105$. Furthermore the length of the bearing expresses the (standardised) magnitude of overall actual adjustments. For example, the length of the bearing here is 0.29 and indicates that the largest change in prevision over our collection of belief elements is about 0.54 standard deviations. This is to be contrasted to the prior expectation for the bearing length, here equal to about 1.38. Large differences between these quantities serve as diagnostic warnings: an unexpectedly small value here might indicate that we were rather cautious in assessing prior variability.

5.4. *Analysis 2: Design*

Our second analysis touches on the issue of design. We look at the problem of finding a small, but informative, collection of data elements to use to predict a further observation, Y_{14} say, using the underlying Y_t quantities. One natural approach is stepwise selection of informative data elements, where selection can be based upon maximising the expected reduction in uncertainty (*a priori* design), or upon some aspects of the data, such as maximising actual changes in adjusted expectation (*a posteriori* design). If we are trying to learn about more than one quantity, we can order [B/D] to take into account, for example, the degree of variance reduction required for each quantity, as well as that required over the joint collection.

We carried out such a stepwise selection procedure to obtain a predictive collection for Y_{14}. Selection accorded to the minimisation of adjusted variance, and was constrained to stop after the second selection. On the first step Y_{13} was selected as the most influential quantity: it alone is expected to reduce uncertainty in Y_{14} by about 49%, whereas the least influential point, Y_1, would reduce uncertainty by some 10%. The second selection was Y_{11} (coincidentally, this quantity is also the most influential with respect to bearing length), although its contribution is minimal: a further relative reduction in variance of only 0.58%. Pragmatically, this suggests that the most appropriate predictor for Y_{14} is its immediate predecessor alone. However, we might be concerned at the failure of other observations to contribute further information: it may suggest that we re-examine our prior specifications.

5.5. *Analysis 3: Data Assessment*

In our third analysis we assess the relative effects of portions of the data diagnostically. We do this by calculating the *path correlation*, the correlation between two bearings. As we have already stated, a bearing summarises the direction and magnitude of actual changes. Hence bearings which point in different directions indicate contradictions within the data, whereas similar bearings are the result of complementary data.

As an example, we choose to look at the bearings which result from a full adjustment (as in the first analysis) of our belief elements by the data elements, where (i) we include data from the first run only; (ii) where we add in data from the second run; and (iii) where we include all the data. At stage (ii) we find the first two sets of data strongly complementary: their path correlation is close to unity and the bearings and their summary statistics are broadly similar. The output that we get at stage (iii), which contrasts the third set of the data to the first two sets, is shown in Table 4.

Our analysis shows that the third run is evidently "different": the path correlation of −0.9386 shows that, compared to the aggregation of the first two series, the third series is highly contradictory in the sense that changes in expectation are occurring in quite different directions (and our final bearing is thus the result of conflicting information tending to cancel each other). If we wished to, we could now probe more deeply into the data to try to pin-point the cause of the contradiction: in fact, we would find that the second half of the third series behaves differently and that this relates to some surprising aspects of the error structures for the third run.

Element	Current	Adjustment	Previous
a	2.7622	−1.4661	4.2283
b	−1.1496	5.6944	−6.8439
Constant	−3.7521	1.4831	−5.2352
Variance	0.2920	0.1336	0.7543
Expected	1.3799	0.1686	1.2112
Path Correlation		−0.9386	

Table 4. *Contradictory bearings.*

5.6. *Afterthoughts*

The example that we have given above is, in terms of [B/D]'s capabilities, essentially trivial. The linear structure is straightforward, and clean in the sense of imposing uncorrelatedness widely. Note, though, that this is a property of well-organised belief specification, rather than a programming requirement: [B/D] handles widespread interdependence very easily, provided that *you* are equal to the task of specification. Furthermore, the various tools available in [B/D] are designed to help you analyse these very complicated belief structures; the lack of space here allows only a superficial application of some of these tools within our example.

Whereas we can view this sort of example as trivial within the [B/D] approach (mathematically, at least: the role of the belief analyst is never trivial) it is not hard to see that a traditional Bayesian attempt at a genuine analysis becomes very complicated indeed, not least because the intercept and slope constants A_r and B_r are exchangeable over runs. Even were we able to perform such an analysis, it is not clear whether we would then be able to condense the information scattered therein over many dimensions into quantifications both meaningful and useful. This would be the case whether or not we felt that the traditional approach were foundationally sound.

Our restriction of attention to genuine limited belief specifications allows our analysis to be deeper, more thorough and more realistically informative: the information contained within our constructions is readily accessible, and not obscured by the essentially infinite number of pretence specifications that can frequently go into a traditional analysis. Hence many of the results available within the [B/D] approach are simply not available under other approaches, and we believe that these extra results, some of which we have considered above, are fundamentally useful.

6. PROGRAM IMPLEMENTATION

[B/D] is written in ISO-standard Pascal, level 0, for ease of portability. Current implementations are available for the IBM PC, running under DOS; and for SUN workstations, running under UNIX. The package is essentially complete, albeit subject to constant minor revision. Further revisions, to include additions to the general theory and methodology, are anticipated. Detailed documentation on both the theory underlying the approach and the use of the package, is under preparation by the author and Michael Goldstein. Documentation for an earlier incarnation of the package can be found in Goldstein (1987a) and Wooff (1987). We welcome any enquiries.

ACKNOWLEDGEMENTS

This research is financed with a grant from the UK Science and Engineering Research Council, to whom we are also grateful for earlier financial support. Thanks to Malcolm Farrow for providing the example, the data, and the belief specifications.

REFERENCES

Farrow, M. and Goldstein, M. (1990). Linear Bayes methods for crossover trials. (Unpublished *Tech. Rep.*).

Farrow, M. and Goldstein, M. (1992). Reconciling costs and benefits in experimental design. *Bayesian Statistics 4* (J. M. Bernardo, J. O. Berger, A. P. Dawid and A. F. M. Smith, eds.), Oxford: University Press, 607–615.

Goldstein, M. (1981). Revising previsions: a geometric interpretation. *J. Roy. Statist. Soc. B* **43**, 105–130.

Goldstein, M. (1985). Temporal Coherence. *Bayesian Statistics 2* (J. M. Bernardo, M. H. DeGroot, D. V. Lindley and A. F. M. Smith, eds.), Amsterdam: North-Holland, 231–248, (with discussion).

Goldstein, M. (1986a). Exchangeable belief structures. *J. Amer. Statist. Assoc.* **81**, 971–976.

Goldstein, M. (1986b). Separating beliefs. *Bayesian Inference and Decision Techniques* (P. K. Goel and A. Zellner, eds.) Amsterdam: North-Holland, 203–217.

Goldstein, M. (1987a). [B/D] Introduction and overview. *Tech. Rep.* **5**, University of Hull.

Goldstein, M. (1987b). Can we build a subjectivist statistical package?. *Probability and Bayesian Statistics* (R. Viertl, ed.), London: Plenum Press.

Goldstein, M. (1988a). Adjusting belief structures. *J. Roy. Statist. Soc. B* **50**, 133–154.

Goldstein, M. (1988b). The data trajectory. *Bayesian Statistics 3* (J. M. Bernardo, M. H. DeGroot, D. V. Lindley and A. F. M. Smith, eds.), Oxford: University Press, 189–209, (with discussion).

Goldstein, M. (1990). Influence and belief adjustment. *Influence Diagrams, Belief Nets and Decision Analysis* (J. Q. Smith and R. M. Oliver, eds.), New York: Wiley, 143–174.

Goldstein M. (1991). Belief transforms and the comparison of hypotheses. *Ann. Statist.*, (to appear).

Wooff, D. A. (1987). [B/D] Reference manual. Mathematics Research Reports, *Tech. Rep.* **6**, University of Hull.